Second International Conference on Sustainable Energy, Environment and Green Technologies (ICSEEGT 2022)

IOP Conference Series: Earth and Environmental Science
Volume 1084

Jaipur, India
24-25 June 2022

Editors:

Raj Kumar Satankar
Pooja Gupta
Mahesh Bundele

ISBN: 978-1-7138-6433-2
ISSN: 1755-1307

Printed from e-media with permission by:

Curran Associates, Inc.
57 Morehouse Lane
Red Hook, NY 12571

Some format issues inherent in the e-media version may also appear in this print version.

Institute of Physics
Dirac House, Temple Back
Bristol BS1 6BE UK

Phone: 44 1 17 929 7481
Fax: 44 1 17 920 0979

techtracking@iop.org

Additional copies of this publication are available from:

Curran Associates, Inc.
57 Morehouse Lane
Red Hook, NY 12571 USA
Phone: 845-758-0400
Fax: 845-758-2633
Email: curran@proceedings.com
Web: www.proceedings.com

TABLE OF CONTENTS

IOP Conf. Series: Earth and Environmental Science 1084 (2022) 011001 doi:10.1088/1755-1315/1084/1/011001

Second International Conference on Sustainable Energy, Environment and Green Technologies (ICSEEGT 2022)

(HYBRID MODE)

(June 24-25, 2022)

Editor(s):

Dr. Raj Kumar Satankar
Department of Mechanical Engineering, Poornima College of Engineering, Jaipur, India

Dr. Pooja Gupta
Department of Civil Engineering, Poornima College of Engineering, Jaipur, India

Dr. Mahesh Bundele
Principal & Director, Poornima College of Engineering, Jaipur, India

POORNIMA
COLLEGE OF ENGINEERING
Affiliated to RTU, Kota • Approved by AICTE & UGC under 2(f) • Accredited by NBA
ISI-6 RIICO Institutional Area, Sitapura, JAIPUR – 302022 (Rajasthan), INDIA
Phone: 9928555222 • Website: www.pce.poornima.org

IOP Conf. Series: Earth and Environmental Science 1084 (2022) 011001 doi:10.1088/1755-1315/1084/1/011001

PREFACE

The conference on Sustainable Energy, Environment, and Green Technologies (ICSEEGT 2022) is the second international conference organized at Poornima College of Engineering, Jaipur, Rajasthan, India. This event was organized in a hybrid mode. This conference aims to bring researchers from academia, government agencies, research laboratories, and the corporate sector to present their research on sustainable and renewable energy, the environment, transportation engineering and green technologies. Since green growth is pursuing economic development in an environmentally sustainable manner, therefore green growth must play a vital role in the development goal of improving living standards, accessing modern energy services, and using energy more efficiently. For this aim, the conference ICSEEGT 2022 provided an excellent platform to promote research in the subject area and exchange new ideas in these fields among academicians, engineers, scientists, and practitioners across the world.

The conference began with its inaugural session (in hybrid mode), seeking the blessings of Goddess Saraswati. The session witnessed addresses of dignified persons, Professor S.A Sherif, Department of Mechanical and Aerospace Engineering, University of Florida, USA, Professor L.M. Das, Former Professor, Centre for Energy Studies, Indian Institute of Technology, Delhi, India and Dr. L. N. Thakural, Senior Scientist, Surface Water Division, National Institute of Hydrology, Roorkee, India.

The inaugural session was followed by a keynote session of Professor P. Muthukumar, Professor, Indian Institute of Technology, Guwahati, India, on the *Energy Efficient and Environment Friendly Porous Radiant Burners for Cooking and Industrial Applications*. Later, four parallel technical sessions were organized on the first day, where authors and participants participated and presented their papers. Overall, 186 research papers were received for ICSEEGT 2022 from different organizations worldwide, out of which our expert team of reviewers accepted 102 papers. The research papers in this conference were invited considering these three tracks: sustainable/renewable energy, environment and transportation, green technologies, and urban development. The presenters and attendees participated with full enthusiasm in all the technical sessions. Out of 83 registered papers, 43 papers were presented in the technical sessions on the first day. The first Technical Session in offline mode was chaired by Dr. Narayan Lal Jain and Dr. Pran Nath Dadhich, and moderated by Mr. Rahul Sharma of Poornima College of Engineering, Jaipur, India. The second Technical Session was chaired by Dr Salifu Tahiru Azeko, Tamale Technical University, Tamale, Ghana: Mr. Sanjay Kumawat, Poornima College of Engineering, was the moderator for this session. Dr. Tanuj Chopra from Thapar Institute of Engineering and Technology, Patiala, India chaired the third Technical Session. The second and third session moderator were Mr. Divya Vishnoi and Mr. Sanjay Kumawat, Poornima College of Engineering respectively. Dr. Puneet Kumar Jain, National Institute of Technology, Raurkela, and Dr. Ram Niwash Mahia, National Institute of Technology, Hamirpur, India chaired the fourth Technical Session and moderated Dr. Monika Vardia from Poornima College of Engineering.

The day ended with a keynote session on *Recent Advances Materials on Energy Storage Technology and Sustainability* by Dr. M V Reddy, Senior Professional Researcher, Nouveau Monde Graphite (New graphite world) (NMG), Quebec, Montreal, Canada.

On the second day of this conference, ICSEEGT 2022, two more technical sessions were organized and about 40 more accepted papers were presented. The sessions were chaired by Dr. Pankaj Dumka, Jaypee University of Engineering and Technology, Guna, India and Dr. Pradeep Kumar Gautam, National

Institute of Technology Meghalaya, Dr. Surendra Kumar Saini, and Mr. Divya Vishnoi, PCE were the moderators for these sessions. There were three keynote sessions organized on this day. The first keynote talk, entitled *Noble Metal-Free Electro catalysts for Oxygen Reduction Reaction in Low-Temperature Fuel Cells*" was delivered by Dr. Wong Wai Yin, Research Fellow/ Senior Lecturer Fuel Cell Institute, Universiti Kebangsaan Malaysia. The second talk title was *Sustainable Livelihoods and Adaptation to Climate Change* and Dr. Ravindra S. Gavali, delivered it, representing National Institute of Rural Development & Panchayati Raj, Hyderabad. The third session was on *Hydrogen as a Future* Fuel by Dr. Dilip Sharma, Malviya National Institute of Technology, Jaipur, India.

At the end of the second day in the valedictory ceremony, Dr. Herb S. Rhee, Managing Director, Innovative Technology and Energy Centre (ITEC), Tanzania, joined as the chief guest. In contrast, Dr. Anand K. Plappally, Head, Center for Emerging Technologies for Sustainable Development (CETSD), IIT Jodhpur, India, Dr. Perumalla Janaki Ramulu, Director, Centre of Excellence for Advanced Manufacturing Engineering, Adama Science and Technology University Adama, Ethiopia participated as the guest of honor of the ceremony. On the contrary, the event was a great success with addresses and sessions of learned persons, wholehearted participation from all our authors, co-authors, attendees, and an expert team of reviewers, coordinators, staff members, and organizing team members.

ORGANIZING COMMITTEES

Chief Patrons:

- **Dr. S.M. Seth,** Chairman Emeritus Poornima Group, Former Director, National Institute of Hydrology, Roorkee, India
- **Sh. Shashikant Singhi,** Chairperson, Poornima Group, Jaipur, India

Patrons:

- **Sh. M.K.M, Shah,** Director General, Poornima Group, Jaipur, India
- **Sh. Hari Singh Shekhawat,** Director infrastructure, Poornima Group, Jaipur, India
- **Ar. Rahul Singhi,** Director, Poornima Group, Jaipur, India

General Chair:

- **Dr. Mahesh Bundele**, Principal & Director, Poornima College of Engineering, Jaipur, India

Organizing Chairs:

- **Pankaj Dhemla,** Vice Principal, Poornima College of Engineering, Jaipur, India
- **Dr. Narayan Lal Jain,** Head of the Department, Mechanical Engineering, Poornima College of Engineering, Jaipur, India
- **Dr. Pran Nath Dadhich,** Head of the Department, Civil Engineering, Poornima College of Engineering, Jaipur, India

Conveners/Organizing Secretary:

- **Dr. Raj Kumar Satankar,** Department of Mechanical Engineering, Poornima College of Engineering, Jaipur, India
- **Dr. Pooja Gupta,** Department of Civil Engineering, Poornima College of Engineering, Jaipur, India

Advisory Committee:

- **Dr. Omprakash Yadav,** Professor, North Dakota State University, USA
- **Prof. Abdul Ghani Albaa,** Visiting Professor, Environment Technology and Management, Princess Sumaya University for Technology, Jordan
- **Dr. Sandeep Choudhary,** Professor, Indian Institute of Technology, Indore, India
- **Dr. Rajat Rastogi,** Professor, Indian Institute of Technology, Roorkee, India
- **Dr. Rosalin Sahoo,** Professor, Indian Institute of Technology (BHU), Varanasi, India
- **Dr. Manoj Neergat,** Professor, Indian Institute of Technology, Bombay, India
- **Dr. K.S. Reddy,** Professor, Indian Institute of Technology, Madras, India
- **Dr. Avinash Kumar Agarwal,** Professor, Indian Institute of Technology, Kanpur, India
- **Dr. B. Ravindra,** Associate Professor, Indian Institute of Technology, Jodhpur, India
- **Dr. Rohit Goyal,** Professor, Malaviya National Institute of Technology, Jaipur, India
- **Dr. A. B. Gupta,** Professor, Malaviya National Institute of Technology, Jaipur, India
- **Dr. Dilip Sharma,** Professor, Malaviya National Institute of Technology, Jaipur, India
- **Dr. Jyotirmay Mathur,** Professor, Malaviya National Institute of Technology, Jaipur, India
- **Dr. G. D. Agrawal,** Professor, Malaviya National Institute of Technology, Jaipur, India
- **Dr. D. Dutta,** Director, NRSC, Hyderabad, India

Technical Program Committee:

- **Dr. Vitalis Anye**, Senior Lecturer, Department of Materials Science and Engineering, African University of Science and Technology, Abuja, Nigeria.
- **Dr. Kabiru Mustapha**, Senior Lecturer Department of Materials Science and Engineering, Kwara State University, Malete, Nigeria
- **Dr. Anand K Plappally**, Associate Professor, Indian Institute of Technology, Jodhpur, India
- **Dr. T. N. Verma**, Assistant Professor, Maulana Azad National Institute of Technology, Bhopal, India
- **Dr. Rahul Goyal**, Professor, Manipal University Jaipur
- **Dr. Yogesh Kumar**, Assistant Professor, National Institute of Technology, Patna
- **Dr. Dinesh Kumar Rajendran**, Assistant Professor, National Institute of Technology, Srinagar
- **Dr. P. Sudhakar Rao**, Assistant Professor, NITTTR, Chandigarh
- **Dr. K. B. Rana**, Assistant Professor, Rajasthan Technical University, Kota, India

IOP Conf. Series: Earth and Environmental Science 1084 (2022) 011002 doi:10.1088/1755-1315/1084/1/011002

Peer Review Statement

All papers published in this volume have been reviewed through processes administered by the Editors. Reviews were conducted by expert referees to the professional and scientific standards expected of a proceedings journal published by IOP Publishing Publishing.

- **Type of peer review:** Double Anonymous
- **Conference submission management system:** Morressier
- **Number of submissions received:** 186
- **Number of submissions sent for review:** 184
- **Number of submissions accepted:** 102
- **Acceptance Rate (Submissions Accepted / Submissions Received × 100):** 54.8
  Average number of reviews per paper: 2.38
- **Total number of reviewers involved:** 65
- **Contact person for queries:**
- **Name:** Dr. Raj Kumar Satankar
  Email: rajkumar.satankar@poornima.org
  Affiliation: Poornima College of Engineering, Jaipur, India

Sustainable Energy Business Model and Strategies for Marketing of Clean Energy Products in Rural Areas of India

Shiv Sankar Das[1], Bijaya Bijeta Nayak[2], Debashree Debadatta Behera[3], Rita Kumari Sahu[2], Sasmita Sahu[2] and Soumya Ranjan Giri[4]

[1] Zenith School of Management, Zenith Group of Institutions, Bhubaneswar–760002, Odisha, India

[2] School of Mechanical Engineering, KIIT University, Bhubaneswar–751024, Odisha, India

[3] Department of Mechanical Engineering, Centurion University of Technology and Management, Odisha–752050, India

[4] Zenith Institute of Science and Technology, Zenith Group of Institutions, Bhubaneswar, Odisha, India

Corresponding author E-mail: shivsankardas85@gmail.com

Abstract. Our world's problem can be resolved with the use of energy. Energy is the fundamental structure for economic development of any area, state, or a country. In today's world, energy access is a relevant issue. Availability of energy is required to improve health, agricultural productivity, and educational outcomes. The increased need for energy has led to a reliance on large-scale fossil fuel extraction and use. As a result, environmental degradation, pollution, and ecological imbalance have occurred. In order to find a solution, clean energy is vital. As a result, focusing on the shift from conventional to clean energy sources is critical. "Clean energy" is obtained mostly from non-polluting, zero-emission sources, as well as energy conserved via efficiency improvements. Access to clean energy can aid in the development of rural tribal communities. This paper focuses on a case analysis where sustainable energy business model was devised. The paper is divided into three segments. The first segment focuses on the introduction to clean energy, clean technology and clean energy products. Second segment focuses on the sustainable energy business model in rural areas of South Odisha, where a Digital Business Centre (DBC) was established. Finally the third segment focuses on the strategies for marketing of clean energy products and an Entrepreneurial-cum-SHG Model. Methodology/design/approach: In this study, a descriptive case technique was employed, and instances were purposefully chosen. Data was gathered to better comprehend the theoretical ideas of actors, factors, and their relationships, as well as to create the conceptual framework. The framework's applicability was tested using an action research approach (partial intervention in the ecosystem). Findings: Variety of players and variables were identified, and their relationships were established. Originality/value- An innovative strategy to clean energy intervention for income generation was established and a sense of entrepreneurial attitude was also developed.

1. Introduction

"Clean energy" is obtained mostly from non-polluting, zero-emission sources, as well as energy conserved via efficiency improvements [1]. Solar, wind, biomass, hydropower, and geothermal are the most popular clean energy sources. Any products, services, or process that decreases negative environmental consequences by implementing a large energy efficiency development, using sustainable resources, and pursuing environmental protection activities is referred to as "clean technology". Although there is no universal definition of "Clean Technology," it may be defined as a broad variety of products, services, and processes that employ clean materials and energy sources to drastically reduce the use of natural resources while also eliminating emissions and wastes [2].

It also contains a wide range of technologies relating to the use of Clean Energy Products (CEP), product recycling, the usage of e-vehicles, energy efficient motors, smokeless chulhas, and DC lighting systems, among other things. The items that utilise this technology are referred to as "clean energy products". Clean energy products assist to provide new sources of income, lower costs, decrease drudgery, and enhance health, all of which improve the quality of life in rural regions. In the current context, quality of life may be defined as improved living circumstances, such as access to contemporary amenities such as electricity and modern cooking fuel, among other things. "Clean energy services" refer to a range of services that include everything from site selection through construction project management, installation, and operation and maintenance. For example, installing a solar-powered house lighting system and providing training on how to use it, maintain it, and repair it when it breaks down.

2. Sustainable Energy Business Model

Presently, for any documents necessary carrying out any work one would require a photocopying or printing facility. People in the far flung rural areas of South Odisha, have to travel great distances to reach the business centre's, which are located outside of the community's reach. This results in a loss of half a day to a full day's pay. The Harsha Trust, as NGO based at South Odisha, did an average evaluation and revealed that the whole cost of photocopying one document results in an expense of Rs. 300 including travel, food, etc. Tribal people in remote areas of South Odisha endure a disproportionate share of the cost as a result of these expenditures. Recognising the importance of such basic business, one of the aforementioned business was required to be provided to the people.

Disruptive availability of power with changing voltage causes breakdown of costly equipment, resulting in further financial problem. One of the ideas for addressing this electrical problem was to adopt the use of clean energy products. As a result, the Harsha Trust acquired the concept of creating a solar-powered Digital Business Centre (DBC) in rural South Odisha. The DBC was founded in rural South Odisha by Harsha Trust. Solar-powered DBCs may provide the necessary electrical reliability while also decreasing voltage fluctuations, providing the residents in such areas with more economic stability. The solar-powered DBC can be easily replicated anytime there is a need for decentralised centre's to provide critical services to isolated and disadvantaged individuals.

2.1 Harsha Trust

Harsha Trust, is an NGO with more than a decade old grass-root organisation, has launched a livelihood security and quality of life improvement programme for approximately 100,000 people in 20,000 households from 500 villages in five districts of South Odisha, one of the country's most economically backward and left-wing extremist-affected regions, involving indigenous and other underprivileged communities. Harsha Trust has established incubation centre's in several parts of southern Odisha. The goal of the incubation centre is to help and encourage local entrepreneurs to create sustainable energy businesses that meet the needs of underprivileged areas. With the aid of other organisations such as NGOs working in the region, the incubation centre discovers youth groups and local businesses. They hold workshops for these incubatees to educate them on how to start a business, marketing of agro based products, farming techniques by using technological advanced products. Abha Innovations, Mukti Solar Energy, and Pioneer Power Solutions, all situated in South Odisha, are some of the enterprises nurtured by the incubation centre.

2.2 About the Study

The Digital Business Centre (DBC) was established in the Ganjam district of South Odisha. The former Planning Commission rated the districts of southern and western Odisha as the most backward. While farming is primarily for subsistence purposes, a small amount of their produce is sold or swapped in local markets when households lack other essentials.

According to Census 2011, just around 10 percent of the 20000 homes have access to electricity. It offered a wide range of solutions, from a single photocopier to a set of devices that serve a variety of digital services and are all powered by solar energy. This centre may help women become household entrepreneurs, enhancing their earning potential as well as their social standing. Furthermore, the establishment of the centres by differently abled individuals in rural areas may help them earn a new source of income. This is a modest case study of an entrepreneur from Markondi village in the Ganjan district of South Odisha.

Santanu Majhi of Ganjam's Markondi village intends to establish a business that would offer photography and printing services to the people of his Gram panchayat. People had to go 70 kilometers to acquire a Rs 1 photocopy. He saw that the region lacked consistent energy and decided to use a solar-powered photocopier, printer, and laptop. He bought a system worth of Rs 101500, with the Harsha Trust paying Rs 25000 in margin money and the remainder obtained with a loan of Rs 76500 from State Bank of India (SBI) at a rate of 13 percent over a four-year period. He got the system with the technical assistance of the Harsha Trust, and Abha Innovations installed them.Mr. Santanu was first assisted by the Harsha Trust in approaching the bank and comprehending and fulfilling the other procedures, conditions for loan acceptance. As a result, he secured a loan and began offering photography services, as well as assisting individuals with panchayat paperwork, applications, and other tasks. He currently earns between Rs 10000 and Rs 12000 per month from the service he does, and pays Rs 2100 per month towards loan repayments. In conjunction with the aforementioned implementation research, a model, as seen in figure 1, was constructed, along with the numerous components and actor connections stated in Table 1.

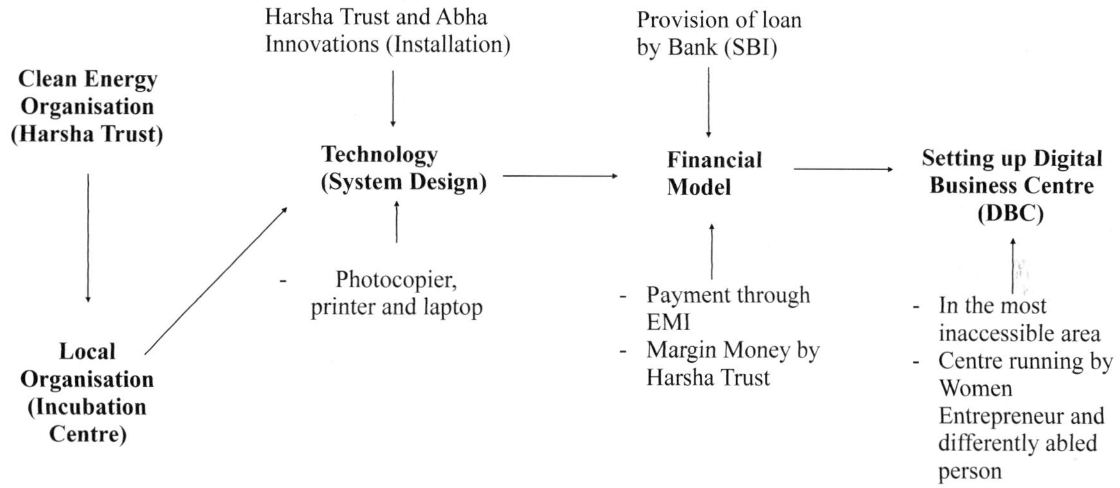

Figure 1. Sustainable Energy Business Model

The aforementioned study revealed several factors and participants involved in the introduction of a solar powered Digital Business Centre in a rural region of South Odisha. The pricing of the product, payment via an EMI plan, product features, Harsha Trust help in providing margin money, lack of stable electricity, inconvenience caused in acquiring services, and the loss of one day's income were all key factors in the development of solar- powered DBC. Finance-related variables, technology-related variables, capacity building variables, and infrastructure-related variables were all identified.

Table 1. Factors and Actors Linkages for Sustainable Business Model.

Major Factors	Actors	Linkages
Finance	Bank (SBI) Harsha Trust	Payment of loan amount in monthly instalments Margin money is provided.
Technology	System design, installation, and servicing were aided by the Harsha Trust and Abha Innovations.	Customized products are available. Product installation and servicing assistance Alternative energy source
Capacity Building	Harsha Trust and Incubation centre	The entrepreneur has been identified. Workshops were held to give technical and management training.
Infrastructure	Absence of photocopy centres	Remoteness of the installation of a solar-powered digital business centre

Incubation Centre and Abha Innovations played a significant part in the aforementioned concept, giving margin money, handholding support, design, and installation, respectively. The main funding choices were a bank loan and instalment payments, which made the intervention simple. For DBC engagement in rural regions, a partnership model was devised among all parties.

3. Marketing of Clean Energy Products in Rural Areas on India

The size of rural market for clean energy products is huge but the companies generally target the rural region closest to its area of operation – a convenient and obvious choice for them by which people living in distant are deprived of it. Hence, an effective distribution model should be in place for last mile consumer. The prices for all clean energy products should be kept low or the same unless it is manufactured in large quantity whether it is consumer electronic or FMCG. Direct marketing is an excellent option, by having a store in a rural location. For example, use of electrical stores, mobile sales and service outlets, cable television service provider, direct-to-home service shops etc. can act as a selling point for clean energy products. Financial incentives, government-led efforts, lower investment costs, increased dependability, information distribution, and environmental consciousness are critical variables for a larger marketing of clean energy products. An aware, convinced customer can make an adoption decision quicker than a un-aware customer. Depending on the location variables, different marketing and selling strategies can be used for distinct rural consumers. Documentation, quality, process, design and innovation, manufacturing capabilities, and, lastly, cost effectiveness are other variables that can aid in the promotion of the clean energy products in rural areas in India. Among the few which are highly required for marketing of clean energy products are door to door selling, selling in haats, melas, etc. There are different models of distribution, of which one such model is designed. Entrepreneurial-cum- SHG model is discussed in the later part of this paper.

Here, the consumers are sensitised for using a clean energy product, a solar operated light for lighting purpose and other income generation activities. Explaining and demonstrating the use of clean energy product to the SHG members at the village level during their weekly meetings, preferably in evenings is a low cost distribution model, which needs to be adopted for clean energy product distribution in the rural areas. Affordability, availability are the major concern for any rural consumer to think on getting a product. Having access to get a credit is a major concern for rural people to get hold to the product. They need to spend less but they deserve a good quality product. Many rural consumers neither have access to electricity, and nor do they have access to credit. When a reliable product say a solar light is introduced into the market its cost is naturally higher which in turn makes it not affordable to buy.

Affordable does not mean its cheap quality product but rather it should be able to generate repayments and help in income generation. Rural consumers should be provided with a low-interest credit which becomes a headache for both institutional and financing organisation. For example, if a household wants to purchase a solar operated home lighting system, it finds difficult to pay the full amount, now some of the organisations are giving these systems on credit and going for payments in a Equal Monthly Instalments (EMI) but it affects the organisation in terms of the running of the business the operational cost more precisely to say. Hence, a desirable system like jointly purchasing the product and giving it for rent, would be a good option for it. There are different branding and advertising strategy but for selling clean energy products. Strategies such as word of mouth where the company people can organise "tea party" at the home of satisfied customers, where people from around the village are invited to watch a demonstration of how the clean energy product is operating. They do often create noise so that people come outside to see what is happening which further acts like a free publicity. Other initiatives like banks themselves plays an important role in marketing the clean energy product at their own branch, when they identified a potential customer who could afford and benefit from the system they provide loan to purchase the product. Another approach is the demonstration model, where the seller puts a clean energy product in the home of a village's important or affluent resident as part of one-to-two month trial [3]. The show begins with a tea party, identical to the one described above. Technicians returned at the end of the trial period and proceed to unhook the product. The experience of living with the product, along with familiar and societal influences, finally can compel the majority of consumers to take out a loan and purchase the system.

Product sharing model is another approach where the rural consumers if cannot afford a high priced clean energy product even if they need it, they can cooperate with other consumers who wants to use the same product [4]. For example, if a solar operated drip irrigation system is purchased by a farmer at Rs 25000 than it can be shared with other farmers for irrigating the fields which could be nearby. The system can be given on rent, hourly basis or per day basis. This business model acts as an income generation activity for the farmer and subsequently helping him to lead a better quality of life. Use of clean energy products being an emerging industry, trade shows and exhibitions are another powerful avenue for marketing the products [5, 6]. Participating in an expo attract a good number of customers and subsequently helps in building a network of customers if the product is of good quality. It helps in scaling up the production and fulfilling the target. Clean energy products can be financed by giving rural consumers micro-credit. Micro-credit is the provision of small-scale loans. The name "micro" literally means "little," and according to the Reserve Bank of India's (RBI) micro credit special cell, borrowing amounts up to Rs 25000 might be classified as micro credit products, with this figure steadily increasing up to Rs 40000 over time.

Micro-credit providers through different schemes often comes with a relatively high interest rate because of the high costs of repayment collection, irregularity of loan payment and absence of securities. The loan extended to members of community-based savings, cooperatives and other finance institutions have high interest rate. The interest rates vary from 23 to 25 percent in India. Rural consumers are unable to reimburse the overall amount in which the interests represent a non-negligible amount. As a result a group based delivery model for financing of clean energy products is advised which can be used by different members of the Self Help Group (SHG) by involvement of various financial institutions. Rural consumers in the target community can join a cooperative to provide financial services (micro-savings, microcredit, micro-insurance, etc.) to themselves. Groups are made up of rural adolescents or women from the surrounding areas who may build support structures for micro-enterprises. This technique might be the fundamental unit of operation for Micro-Finance Institutions (MFIs). As MFIs are required to make collateral-free loans, group methods can assist in the creation of social collateral (peer pressure) that can successfully substitute physical collateral. In lending arrangements, collateral refers to a borrower's promise of certain property to a lender in order to ensure loan repayment. The collateral protects the lender against a borrower's default and may thus be used to balance the loan if the borrower fails to pay the principal amount and interest amount in accordance with the terms and conditions of the lending agreement. Using the group method, the whole financial process is delegated to the group rather than to the financial institutions. All financial operations, such as saving, borrowing, repaying debts, and keeping records, can be managed at the group level. A group of 10-20 people is formed using this strategy. These group members must save a set amount of money on a regular basis in a shared fund. The group members will decide on the amount and frequency of savings. Such a group is necessary to be affiliated to a financial institution in order to obtain a credit. Financial institutions will make loans in the name of the group, and the group as a whole will be accountable for repayment. The loan amount will be determined by the group's total collected savings. Before obtaining a loan, the group will select its members. Loans will be issued to a select member(s) of the group initially, followed by the remaining members. Most financial institutions want a percentage of the loan to be saved ahead of time, indicating the capacity to make regular payments and act as collateral. The criteria for distributing the loan among group members shall be determined by the group members themselves. With this financing, the entire group can launch a micro-enterprise together. Individuals may also utilise their loan for consumption or to address other pressing requirements. Members of the group will be jointly responsible for repaying each other's debts and will normally meet regularly or weekly to collect the repayments. Peer pressure and mutual obligation must function well to secure the payback process. Even if one member of the group defaults, the entire group is disqualified and will not be eligible for subsequent loans. As a result, the members will decide the borrower's creditworthiness. Because they will be involved in numerous group activities, this form of group-based delivery technique can assist to empower the group members. They will go to the bank, market, and organise the group meetings to boost their self-esteem. This concept will aid in the group's self-sufficiency and will also serve as an effective monitoring system for loan repayment and services they purchase.

3.1 Entrepreneurial-cum- SHG Model

In order to sensitise the spirit of entrepreneurship among the rural youth a model has been designed named Entrepreneurial-cum-SHG model for clean energy products intervention in the rural areas through the SHG or a cooperative.

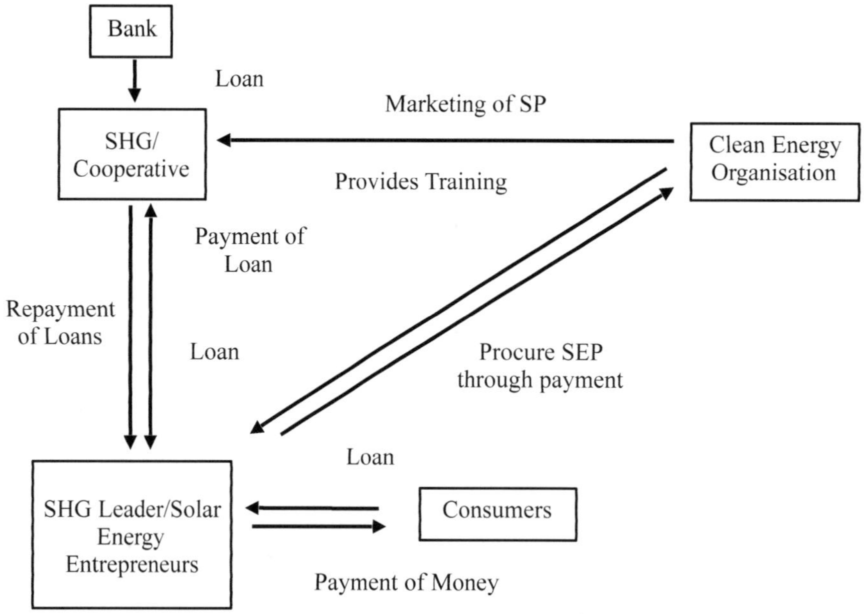

Figure 2. Entrepreneurial-cum- SHG Model

Figure 2 focuses on understanding how a SHG, which is a community based organisation can help in marketing of any clean energy product in the rural areas. Any SHG consist of 5-10 members that identify possible initiatives to be undertaken for the development of the livelihood in local context specific. The members in the SHGs are like minded group and consist of farmers, vendors, street hawkers, etc. who live in close vicinity to each other and can avail many schemes related to bank and others. Sometimes bank become reluctant to lend them loans for purchasing of any clean energy product, for instance the use of solar operated home lighting systems. Bank worries over the defaults and the high transaction costs associated with the small sized loans. SHG can play an important role by reducing the gap between the consumers and the bank.

In this model, the clean energy organisation approaches to a SHG to market their products, whether in the form of an CSR initiative or from marketing perspective. The SHG leader can act as a Clean Energy Entrepreneur (CEE).A CEE must be trained by the Clean Energy Organisation (CEO) to provide technical, managerial support to the consumers. Now, if the bank provide loans to SHG (here CEE needs to identify and submit loan proposals which can be finally sanctioned by the bank manager, before the loans are granted, more time and effort should be spend for training the SHGs members on banking habits, policies and how to maintain the books of accounts) then they would meet with the SHG members on a weekly basis who would pay the loan in instalments. The interest rates depends upon the bank which provides the loan to the SHG. Training sessions can be given to each SHG at the field level in a quarterly or semi-annually manner.

The important aspects of this model is that once the loans are provided to the SHGs, they can conduct meetings on weekly basis or every 15 days with the members, where instalments are collected. Conducting such weekly meetings can be a check against the default members and subsequently it creates a platform for a discussion. This model not only can offer an easy finance but also SHG members can deposit a nominal fee with SHG leader so as to promote banking habits. The socio-economic benefits is that by the use of clean energy products people can advance in their secondary sources of livelihood and maintain a good quality of life. From entrepreneurial perspective now if the loan is provided to the CEE directly through the bank, than the CEE can procure the Clean Energy Products (SP) from the clean energy organisations, companies and market it to the members, or generate possible leads, subsequently going for selling of the product. Consumers can take loans from the SHG through CEE (who is also a member of the SHG) and he or she can constantly monitor the actual utilisation of all types of loans to prevent non-payments or any diversion. For this purpose, a solar energy programme card can be prepared for keeping the track of payment of the instalments. Two cards can be prepared and one card will be maintained with the CEE and the other card will be kept by the consumer, bearing the same serial number. Duration for the loan of any clean energy product (like solar light, solar operated irrigation pump, solar dryer, etc.) can be kept for a short term period say 6-8 months depending on the interest and other terms and conditions of the bank. Consumers need to pay the amount within the particular time period.

3.2 Pay-as-You-Go (PAYG) Model

Another model for clean energy product marketing in the rural areas can be through Pay-as-You-Go (PAYG) business model. Here the consumers make a small deposit for the installation of the product and then pay regular instalments through mobile payment systems. It will have two main approaches that is energy as service approach, where the consumer pays for the electricity provided and does own the product and other approach is lease-to-own model, where the consumer becomes the owner of the product after a period of time. Out of these two models, lease-to-own is the most appropriate one. Two of the possible models has been recommended for clean energy product marketing. In this model the consumer pays a down payment for the product, which is attached to a smart meter. Regular payment has to be made by the consumer, if the consumer is unable to pay the amount than it will defer. The credit recharge, the payment point sends the consumer identification and recharge information to the service provider and it sends a recharge code. Then, the consumer uses this recharge code to unlock the system and gets the energy credits, further allowing the consumer to use the system till the energy credit gets exhausted. After the completion of 28 months the system permanently unlocks and the system is owned by the consumer. The meter locks the system after the consumer energy credit has run out. Another possible model where the consumer purchases the clean energy product is by paying some initial amount and takes the system and installs it. Consumer uses the system till the energy credits are exhausted, when finished the consumer brings the system attached to the charge controller to the service provider, where the provider attached the dongle to his/ her mobile and transfers the energy credits pertaining to the amount paid by the consumer. Dongle gets updated and the recharge is done. The consumer plugs the dongle into the charge controller and the system gets activated. This process goes on till full payment for the system is done and the system gets unlock for lifetime.

4. Conclusion

Access to and use of clean energy are essential prerequisites for a good quality of life of an individual. Clean energy products, process and associated technology play an important role for human development. It has the potential to improve health, livelihood and quality of life. For intervention of clean energy products access to finance is very much essential. Rural consumers have difficulty in access to loans. Provision of loans might not help in complete intervention of clean energy products, but it needs to be combined with easy repayment methods such as Equated Monthly Instalment (EMI) for easy affordability for rural consumers. For intervention of clean energy products, capacity building for individuals, organisations and other platforms is very much necessary. Capacity building plays an important role in awareness and training. Rural consumers need to be aware on the usefulness, advantage of using a clean energy product, availability and trained on its usage. They need to be aware on the availability of financing options offered by local banks and other financing institutions. Rural consumers prefer those products which are convenient, portable, easy to use and require less maintenance. Hence, the manufacturer needs to make optimum utilisation of technology for delivering these features those are relevant for the target audience. Rural consumers need to have the access to basic physical infrastructure and community engagement platforms. Although there are many models still prevailing in rural India for marketing of clean energy products, but there is a need for Co-creation, Co-existence, Collaboration, and Competition (4C) for successful intervention , marketing of clean energy products in rural areas of India. Figure 3 focuses on three critical aspects for an effective intervention of clean energy products in rural areas of India.

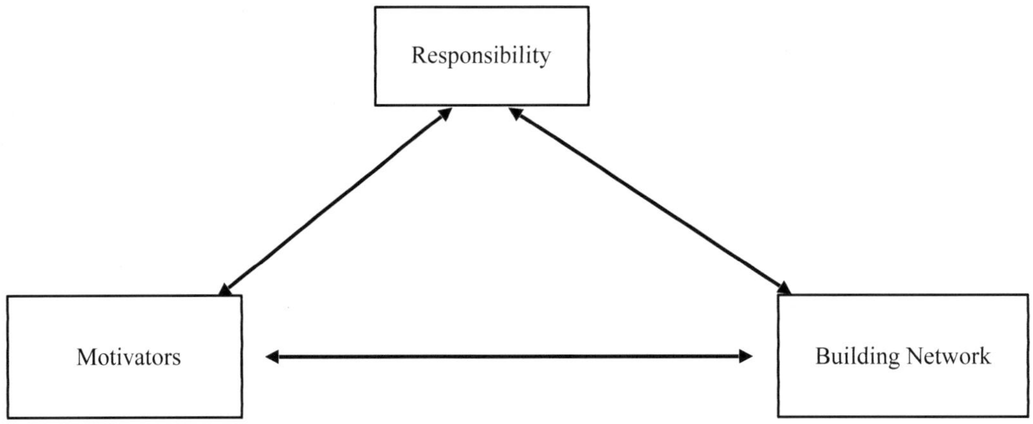

Figure 3. Triad Model for Successful Intervention of Clean Energy Products in Rural Areas in India

The first is a shared sense of responsibility among all stakeholders to adopt clean energy products; the second is a set of motivators, such as providing incentives, simple financial solutions such as EMI schemes, value addition, local job creation, and finally, network construction, as well as an effective monitoring mechanism and supporting infrastructure. These elements are required for creating an enabling environment for the marketing of clean energy products in rural areas of India.

References

1. Bairiganjan, S., and S. Sanyal. (2013). The last 50 mile: Using LVE Networks for increasing Clean Energy Access. New Ventures India, Hyderabad.
2. Jacobson, Arne, Bond, C., T., Lam, Hultman, & Nathan. (2013). Black Carbon and Kerosene Lighting: An Opportunity for Rapid Action on Climate Change and Clean Energy for Development. Retrieved from https:// www. osti. gov/ servlets /purl/22110332.
3. Kashyap, P. Rural Marketing. 2nd Edition, Dorling Kindersey (India) Pvt Ltd., Pearson, New Delhi, (2012).
4. Kotler, P. (1999). Marketing management: the Millennium Edition. Upper Saddle River, NJ: Prentice Hall.
5. Lay, J., Ondraczek, J., & Stoever, J. (2013). Renewables in the energy transition: Evidence on solar home systems and lighting fuel choice in Kenya. Energy Economics, 40, pp. 350–359.
6. Lin, N.H. (2007). The Effect of Brand Image and Product Knowledge on Purchase Intention Moderated by Price Discount. Journal of International Management Studies,8, pp. 121-132.

A Literature Review on Renewable Energy Resource and Optimization

Vijay Prakash Sharma[1], Devendra Kumar Somwanshi[2], Kalpit Jain[3], Raj Kumar Satankar[4]
[1]Assistant Professor, Manipal University Jaipur, Jaipur, India
[2,3]Assistant Professor, Poornima College of Engineering, Jaipur, India
[4]Associate Professor, Poornima College of Engineering, Jaipur, India

E-mail: [1]vijayprakashsharma@gmail.com, [2]imdev.som@gmail.com

Abstract. Step by step the energy request is broadening and in this way the necessities for a manageable source that won't hurt the climate are of prime significance. Several projections express that by 2050 the energy requesting will basically increase. In any case bigger part of the energy necessities is fulfilled by oil subordinates yet by the utilization of innocuous to the environment power designs could help in giving the energy requests. As we in general understand that feasible power is the energy which is either established by the environment or straightforwardly from the sun or from heat conveyed critical inside the earth. In this review paper the renewable energy s power and warmth, which is produced using sun based, wind, sea, hydropower, biomass, geothermal assets, bio engages and hydrogen and these resources are known as unlimited resources.

Keyword: Renewable Energy, Wind Energy, Hydropower, Biomass, Geothermal

1. Introduction

A most tremendous pursuit in our cutting-edge human progress is —energy of various plans. Different chief cut-off points can be at a stop without energy. In this way, energy is a piece of our life. There are different sort of energy utilization cycle such unpleasant energy in falling water, in stores of coal, oil and gas, and so on are most exceptional development [1].

Sustainability is become from typical cycles that are revitalized constantly. In its various plans, it gets directly from the sunlight, or from temperature conveyed tremendous exclusively from the earth [2]. Related with the explanation are control and warmth conveyed from sunshine based, wind, sea, hydropower, biomass, geothermal assets, bio fills and hydrogen grown from renewable resources [3].

Fuel sources which are recovered after a common time cycle are reliably known as unlimited wellsprings of energy like hydro power, sun arranged energy, wind energy, streaming energy, biomass fills, and so forth. In a specific district open feasible power sources are confined total, which relies on their particular qualities [4]. Precisely when economic power sources are disconnected at a higher rate than its regenerative rate is gets non-renewable. Energy conveyed from limitless common assets and advances.

The continuous energy request is expanding step by step an immediate consequence of different reasons like developing individuals, the yearning for managed supposition for conventional comforts and general money related and mechanical new development [5]. The power age framework is in a

general sense relied on imported oil and own vaporous petrol. Then again, as the data about the stores of oil-based products, on the off chance that they are eaten up at the continuous rate, the turned ignitable gas and coal will be depleted persistently 2020 to 2030.
To reduce the dependence on imported fuel and the pressure on oil gas, the ongoing power age system ought to be improved and all the while local energy resources should be examined and made. It very well may be referred to that concern for environment is a now a general issue and standard energy prompts ozone draining substances with disagreeable consequences for prosperity and climate [6]. In these points of view, furnishing of manageable power sources and movement of relative advancements is an astoundingly basic choice. Networks in natural locales and mostly in far away regions have near no potential outcomes to supply share on the public power.
Hardly examiners broke down limit of the breeze energy. A large portion of inspectors worked for developing the energy produced using wind. Split the difference of wind energy into existing power structure causes explicit difficulties like voltage rule, steadiness, and power quality. Various methods of reasoning like battery storing, diesel generator set were recommended for developing the security and managing the effectiveness of wind energy [7].

2. Literature Review

2.1. Building Integrated Photovoltaic Generation System

Kumar P., et al, 2009 highlights that significant amount of R & D work on PV / thermal technology was done in developed nations but very little work was done in India to build solar wall system which will help in saving natural resources like fossil fuels & issues like global warming & detoriation of environment & generate pollution free & clean renewable energy. Authors explain the advantages of early building design of ventilated PV systems into the building envelopes. In this innovation air was actively circulated behind the PV modules which maintain the temperature of cells to keep up with productivity during summer & a hot air was blown behind the PV modules which helps in melting ice formed on PV modules during winter. Building integrated PV (BIPV) was quickly creating idea of reconciliation of PV gadgets into building envelopes where PV can be coordinated into the façade of the structure as cladding or coating. Authors recommend various models like Metrological and Airflow model, Thermal model and Electrical model which work at the same time [8]. Haque, et al, 2013 presented an observational concentrate on assortment of sunlight based chargers in BIPV power age applications. Creators introduced four different sun powered charger types which included poly silicon twofold sided glass light through modules, single side tone polysilicon mistiness glass modules, formless silicon slight film modules and stack-type mu-translucent silicon layer modules. Authors implemented and designed BIPV experimental platform which specifies and defects different kind of solar panel structure power flows to explore the accuracy of PV simulating BIPV module replaces original building materials to convert sunlight into electricity for supplying household energy and to improve conversion efficiency of solar cell, module and system and reduce overall cost. In the experiment PV solar cells used where 4.2Kwp poly silicon, 1.89Kwp colour poly silicon, 0.75Kwp amorphous and 1-2Kwp stack type membrane. Each PV module was connected with appropriate inverter to record monitor indoor energy consumption, room temperature and humidity control, illumination, outdoor power equipment, sunlight intensity, environmental temperature, humidity and wind speed monitoring systems to provide long term testing on empirical research data. Data was recorded to the system before the experiment [9].
Majumder, et al, 2014 proposed PV/Wind/Biomass hybrid system for supplying electrical load in remote villages having 300 households through IRES. The village has good potential for generation of solar energy. Simulation of IRES was carried out in HOMER & 3744 simulations were done to compute the unit cost of energy. The results of simulation showed the lowest cost of energy Rs 12.462 Kwh with PV generator of 5 KW, wind turbine Rs 6 of 250 KW and hydro generator of 10 KW. HOMER modeled technical & economic factors involved. The values obtained using HOMER are not realistic & needs realistic & proper estimation & requires procedures to estimate values using statistical technique [10].

2.2. Issue Related to Building Integrated Photovoltaic Generation System

- The researchers analyzed BIPV by integrating PV devices into building envelopes in type of exteriors, rooftops or PV coating
- Authors determined different models like Meteorological & airflow Model, Electrical Model, Thermal Model [11].
- The solution approaches were found by Simulation models using MATLAB / Simulink.
- Authors concluded that the simulation studies verified different situation by using the functional load under the performance of system & indoor temperature was minimum during summer [12].
- Researchers concluded the solar divider system to be skilled to keep up with load interest during busy time [13].
- Authors analyzed the limitation of renewable energy resources and how to overcome by developing Integrated Renewable Energy System (IRES) [14].
- Researchers carried out Simulation in Hybrid Optimization Model for Electric Renewable (HOMER).
- Authors concluded that the model developed through HOMER was based on assumption of values and was not realistic and a statistical technique must be adopted using realistic estimations [15].

2.3. Wind Energy Generation

Lias Claire Dangelmaier, et al, 2011 portrayed the moderation measures to further develop the power framework recurrence execution and changing power framework age and activity and control. It additionally portrays changes to customary creates and prerequisites for existing and expected ages offices. Author had the chance to notice genuine working experience, issues connected with power framework recurrence control and reaction under different degrees of variable age of force framework. The Hawaii Electric Light Co. (HELCO) had experienced significant change in power generation in last decade wind power generation in last decade wind power generation had been increased from 15% to 11.8% in last 10 years. HELCO evaluated the system impacts of changes in generation mix and appointed a consultant. Consultant analyzed higher levels of variable generation increased load shedding events. HELCO implemented at least 2 out of 3 large steam units to be online operating to control [16].

Nicholas Miller, et al, 2011 introduced a contextual investigation on conceivable close term organization of wind-power on the Oahu power framework. The Oahu power framework was displayed in GE Multi Area Production Simulation and GE Positive Sequence Load Flow. New instruments were created to get to sub-hourly framework execution. These devices were utilized to distinguish framework execution and attributes during testing framework occasions, transient recurrence execution for important high wind possibilities. The above evaluation, various potential methodologies were considered to further develop framework activity, unwavering quality and financial matters. Fluctuation of force age from assets like wind and solar plant was difficult for framework tasks. Because of absence of accessible solar based information didn't allow definite examination of sun-oriented power in framework load. Simulation was done utilizing five essential devices. The outcomes gave the wind/sun oriented abbreviation start/come by unit, working hours by unit and absolute factor cost of activity. The framework execution during this specific time was broke down in powerful reproduction apparatus. The basic occasions like supported wind power drops in no less than 60 minutes, unpredictable wind power changes, HVDC link trip possibility occasion and burden dismissal possibility occasion were analyzed. The consequences of the study recommended attainable activity of the Oahu power framework with elevated degrees of wind and sun based power and proposed changes to introduce functional practices and other framework alterations to empower the activity of lattice with high inexhaustible infiltration. Framework change proposed like wind power, gauge and up guideline necessity expanded breeze energy by 7% by lessening wind energy shortening and diminished variable expense by 4% [17].

Mundackal J, et al, 2013 highlighted the issues related to renewable energy when connected to Grid. The power generated from this source vari es due to fluctuation in environmental conditions. Fluctuation in wind power when injected in to grid affects the power quality like active power reactive power variation of voltages harmonics of electrical behaviour of switching operation. Reconciliation of wind energy into existing power framework presents a specialized difficulties of requires thought of voltage guideline, dependability of force quality issue, which had extraordinary significance to the breeze turbine. In a steady speed wind turbine activity, change of wind speed prompts enormous variable of electrical power on the toil framework. A static compensator (STATCOM) was proposed for improving the power quality and provides energy saving and uninterruptable power. Mahmud et al, 2012 highlighted that, India was as yet reliant upon Coal and oil as significant energy sources. The problem of limited stock of fossil fuels & the adverse impact on global environment was a major concern. Although India is a major player in producing wind energy with support from Government of India and major costal stretch available. India stands on 3rd position in the world in producing wind energy but it was estimated that the energy demand will increase 3 times between 2005 to 2030 & India produce 70% of total energy from wind energy of total non-conventional energy sources still it faces lot of challenges & has great potential to generate wind energy in India. With a long shoreline of 7516 Km India has huge chance to tap the wind energy asset. States like Tamilnadu, Karnataka, Gujarat, Rajasthan, Kerala & Madhya Pradesh have wind energy installation. Due to lack of coordination with government authorities, non availability of land, huge initial investment are the major factors which lead to delay or failure of wind energy generation projects [18].

2.4 Issue Related to Wind Energy Generation

- Researchers analysed Wind Energy Potential in states like Andhra Pradesh, Gujarat, Karnataka, Kerala, Madhya Pradesh, Maharashtra, Rajasthan & Tamilnadu, Hawaii.
- featured the difficulties of the wind energy area like failure and serious dependability issues make negative difference [16].
- Researchers analysed dissected factors impacting the development of seaward wind energy in India like government support, monetary and share motivators, accessibility of nearby skill, capital for speculations and building and empowering R&D environments [17].
- Researchers analysed the wind power forecasting and reserve strategies.
- Researchers analysed the inclusion of Static compensator (STATCOM) and battery energy storage system (BESS) to rectify the power quality problems of wind power [18].
- Researchers analysed the wind power extraction with batteries and hysteresis based current controller.
- Researchers analysed incorporation of smart energy meters which are fit for estimating prompt voltage and flow of electrical circuit [19].

3. Conclusion

In current situation different metropolitan networks are logical going to benefit with supply of power conveyed by furnishing locally accessible harmless to the ecosystem power. Undeniable confirmation of specialty districts that would plainly benefit with such a method, evaluation and assessment of all helpful choices for power supply is crucial for outline and execute intervention drives in such manner. Other than a survey of energy supply circumstance is fundamental, it is imperative to lead energy frames in a couple of system and un-charged towns and evaluate innocuous to the environment power asset receptiveness in the equivalent.

Taking into account the disclosures of energy studies and other optional information plan, appropriate achievable power-based structures will be organized and their yearly power transport assessed. The piece of client moderateness and energy to pay the social classes for innocuous to the biological system power-based power supply while considering the designs organized in the assessment will in addition be dissected.

In this manner, by considering the above examination of economic power, one can in actuality handle the importance of creating electrical power from manageable power in a little while and additionally the obstacles like low reasonability and deficient disguising impact should be dealt with.
This study will give the data about the undeniable interest of energy and weight the board in future, this dissected information can be utilized to make the arrangement outline of power. Study will in addition be significant for those areas, which are unmindful of the manageable power creation by virtue of absence the sun-based energy.
Some consideration endeavours will correspondingly be done to cautious the nearby individuals and the more energetic students in regards to innocuous to the biological system power age and defending non-practical power sources and saving climate.

References

[1] Canada, A.H., “Solar voltaic generation power plants: 1000 MW to 10 kW photovoltaic plant design and application guide for the Pacific Northwest”, “Northcon 95. I EEE Technical Applications Conference and Workshops Northcon95”, 10-12 Oct. 1995, Portland, OR, USA.

[2] Suzuki, R. ; Kawamura, H. ; Yamanaka, S. ; Kawamura, H. ; Ohno, H. ; Naito, K., “Loss factors affecting power generation efficiency of a PV module”, “Photovoltaic Specialists Conference, 2002. Conference Record of the Twenty-Ninth IEEE” pp 1557-1560, 19-24 May 2002.

[3] Zhenhua Jiang ; “Power management of hybrid photovoltaic - fuel cell power systems”, “Power Engineering Society General Meeting, 2006. IEEE” Montreal, Que.

[4] Hassan Moghbelli, Robert Vartanian, “Implemenatation of the Movable Photovoltaic Array to Increase Output Power of the Solar Cells”, Proceedings of the International Conference on Renewable Energy for Developing Countries, 25-33, August 2006.

[5] Ladner-Garcia, H.P. ; O'Neill-Carrillo, E., “Determining realistic photovoltaic generation targets in an isolated power system”, “Power & Energy Society General Meeting, 2009. PES '09. IEEE” pp 1 - 5, 26-30 July 2009, Calgary, AB.

[6] Dai, Qinghui ; Chen, Jun, “Improving the efficiency of solar photovoltaic power generation in several important ways”, “Technology and Innovation Conference 2009 (ITIC 2009), International”, 1-3, 12-14 Oct. 2009, Xian, China.

[7] Hou, H.J.; Yang, Y.P.; Cui, Y.H.; Gao, S.; Pan, Y.X., "Assessment of concentrating solar power prospect in China," Sustainable Power Generation and Supply, 2009. SUPERGEN '09. International Conference on , vol., no., pp.1,5, 6-7 April 2009.

[8] Kumar, P. ; Biswas, S. ; Kumari, S., “Building integrated photovoltaic generation system”, pp 80-83, 16-17 Jan. 2014, Kalyani

[9] Haque, A.U. ; Nehrir, M.H. ; Mandal, P., “Solar PV power generation forecast using a hybrid intelligent approach”, “Power and Energy Society General Meeting (PES), 2013 IEEE” pp 1-5, 21-25 July 2013, Vancouver, BC.

[10] Majumder, R. ; Mukherjee, I. ; Tudu, B. ; Paul, D., “Review on feasibility of wind energy potential for India”, “Non Conventional Energy (ICONCE), 2014 1st International Conference”, pp 190-192, 16-17 Jan. 2014, Kalyani.

[11] Singh, S.N.; Singh, P.; Kumari, S.; Swati, "Optimal Design of Integration of Intelligent, Adaptive Solar (PV) Power Generator with Grid for Domestic Energy Management System," Recent Trends in Information, Telecommunication and Computing (ITC), 2010 International Conference on , vol., no., pp.345,347, 12-13 March 2010.

[12] Qing Ye; Taotao Ma; YuJun Gu; Tao Wang; Dacheng Wang; Yuhao Bai, "Research on dispatch scheduling model of micro-grid with distributed energy," Electricity Distribution (CICED), 2012 China International Conference on , vol., no., pp.1,5, 10-14 Sept. 2012.

[13] Nian Zhang; Behera, P.K., "Solar radiation prediction based on recurrent neural networks trained by Levenberg-Marquardt backpropagation learning algorithm," Innovative Smart Grid Technologies (ISGT), 2012 IEEE PES , vol., no., pp.1,7, 16-20 Jan. 2012

[14] Nouni MR, Mullick SC, Kandpal TC., “Photovoltaic projects for decentralized power supply in

India: a financial evaluation", Energy Policy 2006;34:3727– 38.
[15] Naksrisuk, C. ; Audomvongseree, K., "Dependable capacity evaluation of wind power and solar power generation systems", "Electrical Engineering/Electronics, Computer, Telecommunications and Information Technology (ECTI-CON), 2013 10th International Conference", pp 1-6, 15-17 May 2013, Krabi
[16] Dangelmaier, L.C. ; "System frequency performance of the Hawaii Electric Light System", "Power and Energy Society General Meeting, 2011 IEEE", pp 1-8, 24-29 July 2011, San Diego, CA
[17] Miller, N., Johal, H. ; Achilles, S. ; Roose, L. ; Griffin, J.P., "Integrating high levels of wind in island systems: Lessons from Hawaii", "Intelligent System Application to Power Systems (ISAP), 2011 16th International Conference" pp 1-8, 25-28 Sept. 2011, Hersonissos
[18] Mundackal , J. ; Varghese, A.C. ; Sreekala, P. ; Reshmi, V., "Grid power quality improvement and battery energy storage in wind energy systems", "Emerging Research Areas and 2013 International Conference on Microelectronics, Communications and Renewable Energy (AICERA/ICMiCR), 2013 Annual International Conference", pp 1-6, 4-6 June 2013, Kanjirapally
[19] Mahmud, M.A.; Pota, H.R.; Hossain, M.J., "Dynamic Stability of Three-Phase Grid-Connected Photovoltaic System Using Zero Dynamic Design Approach," Photovoltaics, IEEE Journal of , vol.2, no.4, pp.564,571, Oct. 2012.

Optimization of Renewable Energy Resource- A Review

Vijay Prakash Sharma[1], Devendra Kumar Somwanshi[2], Kalpit Jain[3], Dr. Raj Kumar Satankar[4]
[1]Assistant Professor, Manipal University Jaipur, Jaipur, India
[2,3]Assistant Professor, Poornima College of Engineering, Jaipur, India
[4]Associate Professor, Poornima College of Engineering, Jaipur, India

E-mail: [1]vijayprakashsharma@gmail.com, [2]imdev.som@gmail.com

Abstract. Bit by bit the energy demand is extending and thusly the necessities for a sustainable source that won't hurt the environment are of prime importance. A couple of projections express that by 2050 the energy solicitation will essentially increment. Anyway larger piece of the energy necessities is satisfied by oil subordinates yet by the use of harmless to the ecosystem power structures could help in giving the energy demands. As we as a whole realize that sustainable power is the energy which is either created by the climate or directly from the sun or from heat delivered significant inside the earth. In this review paper the renewable energy as power and warmth, which is made from sun based, wind, ocean, hydropower, biomass, geothermal resources, and bio empowers and hydrogen and these assets are known as inexhaustible assets.

Keyword: Renewable Energy, Wind Energy, Hydropower, Biomass, Geothermal

1. Introduction

A most gigantic search in our state-of-the-art human headway is —energy of different designs. Various principal limits can be at a stop without energy. So, energy is a piece of our life. There are different kind of energy usage cycle such rough energy in falling water, in stores of coal, oil and gas, etc are most outstanding construction [1].

Renewable is gotten from standard cycles that are recharged constantly. In its various designs, it gets directly from the sun, or from heat delivered significant inside the earth [2]. Associated with the definition are power and warmth delivered from daylight based, wind, ocean, hydropower, biomass, geothermal resources, bio fills and hydrogen got from renewable resources [3].

Fuel sources which are recuperated after a typical time cycle are consistently known as inexhaustible wellsprings of energy like hydro power, sun situated energy, wind energy, streaming energy, biomass fills, etc. In a particular region open sustainable power sources are restricted sum, which depends upon their singular characteristics feature [4]. Exactly when sustainable power sources are isolated at a higher rate than its regenerative rate is gets nonrenewable. Energy conveyed from inexhaustible ordinary resources and advances.

The ongoing energy demand is extending bit by bit a direct result of various reasons like growing people, the longing for dealt with assumption for ordinary solaces and general monetary and mechanical turn of events [5]. The power age system is fundamentally depended upon imported oil and own vaporous petroleum. On the other hand, as the information about the stores of oil based goods, if they are gobbled up at the ongoing rate, the pivoted combustible gas and coal will be drained continually 2020 to 2030.

To lessen the reliance on imported fuel and the tension on petroleum gas, the current power age framework should be enhanced and simultaneously native energy assets must be investigated and created. It could be referenced that worry for climate is a now a general issue and ordinary energy leads to ozone depleting substances with unfriendly ramifications for wellbeing and environment [6]. In these perspectives, outfitting of sustainable power sources and progression of relative developments is an astoundingly critical decision. Networks in rustic regions and chiefly in far off districts have close to no possible results to supply partake on the public power.

Scarcely analysts analyzed capacity of the breeze energy. Most of examiners worked for growing the energy made from wind. Compromise of wind energy into existing power structure causes specific hardships like voltage rule, relentlessness, and power quality. Different philosophies like battery stockpiling, diesel generator set were suggested for growing the security and dealing with the efficiency of wind energy [7].

2. Literature Review

2.1. Long Term Solar Power Generation

Alfred H. Canada, 1995 proposed the arrangement and application for 100MW to 10KW photovoltaic plant for Pacific Northwest environment based daylight based fuel openness assumptions and changing assessments for neighborhoods the ordinary dispatch able power from a 100MW gms SVC power plant and associated with the utility plant. Creators wrap up the electrical utilities can plans, present and work negligible cost SVG power plants and decide the point of interaction with any independent power producers PV module [1]. Rsuzuki, et al, 2002 considered the definite system for a PV structure to conclude the setback factors impacting impact age viability: Authors investigated the effect of daylight based irradiance, the event point and azimuth point of east and west by separating prompt and diffuse portion of sun based irradiance. Creators assumed that the distinction in power age viability with time can be surveyed by disengaging the effect of daylight based irradiance and episode point of quick and diffuse sun arranged irradiance [2]. Zhenhua Jiang, 2006 proposed an arrangement of cross breed framework for power the chiefs made from energy units and batteries composed with PV power systems to achieve nonstop first rate power. Author highlighted the disadvantage of energy parts was its low components which might drop the yield voltage during load pick. To vanquish these energy parts ought to be bigger than expected to meet apex power necessities [3].

Hassan Moghbelli, et al, 2006 proposed completing of portable energy in photovoltaic converter and considered its effect on extending the yield power credits diverged from the proper energy. Because of the development of the sun the light communicated of the surface with different point was not sensible for a decent PV energy. The proposed moveable PV energy follows the sun development and gets most outrageous radiation and produces ideal energy as the sunlight sends outwardly of sun situated cells [4]. Lander H. P., et al, 2009 tended to the particular necessities of harmless to the ecosystem power progressions in the island of UPERM and as far as possible attainable in the island. Maker highlighted the dependence on oil based goods for time of energy which has a negative viable and regular impact in the island. Creators highlighted the cut-off points to the accomplishment of potential energy age resources, for instance, huge cost of PV modules& the incredible idea of PV cells could restrict the execution of new advancement [5]. DAI Qinghui, et al, 2009 proposed a couple of systems to deal with the capability of sun oriented based force and coordinated study and examination of the strategies. Sun based energy has boundless holds yet the stream efficiency to change sun based energy over to change over sun situated energy into power was under 15% and with unreasonable expense of PV sheets it ends up being difficult to progress sun oriented energy for a colossal extension [6].

H J Hou, et al, 2009 presented evaluation of Concentrating Solar Power in Renewable Energies Generation. Using Concentrating Solar Collectors has possibly the most diminished cost and no impact on the environment in making economic power from daylight based atomic energy. Key parts of sun oriented based zenith are heliostats, the beneficiary, steam age system and limit structure. Daylight based zeniths have extraordinary long stretch opportunities for high change efficiencies. Informative dish systems are dish shaped reflector and are moderately little units. Each advancement

partakes in its own advantages, applications and inconveniences. The condition is no doubt sensible for creating CSP plants anyway the high capital cost and development of PSP systems are the rule limits for executing it. In any case, in since a really long time prior run with leveled out energy cost diminished and created development CSP will get ferocious with normal power plants [7]. K.C McCintyre et al, 2010 done examination of consolidating scaled down network by imitating PV cells and energy units Modeling and Optimization of Integrated. Customary resources like water sewer combustible gas, sun arranged energy wind turbine were standard and the openness of these resources was similarly pondered. Recreation results using HOMER showed usage of PV cells connected with the leaving network was a prevalent game plan. Re-enactment results showed energy parts and Micro turbine progression ought to have been worked out in view of high capital theory and replacement cost [8].
XiaoliXu, et al, 2010 done appraisal of web noticing data and characteristics limits of motor system all environment following versatile auto-following technique. Sunlight based power-based force age used PV cells which are inventive yet the adequacy was capricious. These systems vanquish the issues like low sun arranged azimuth following precision, low power age viability, weak versatility and huge cost. The system unites all environment auto following for sun powered based azimuth light sensation following and time continuing in versatile manner. Analysed results show all environment versatile auto following strategy further fosters the accompanying movement precision and constancy of daylight based Power age structure, further creates photo electric change adequacy and threatening to wind limit of sun oriented based cell plate and reduces power dispersal of power age framework [9]. Yahyavi, et al, 2010 researched the suitability of assessing power and energy per unit volume and the impact of stature in system efficiency. Creators researched the assortments in daylight-based radiation impact capability of a sun based energy power delivering system. Daylight based energy was at this point remembered to be a huge choice for a safeguarded and clean environment. Considering the examination Authors wrap up use of trackers increase the yield of energy by 57% and it was more capable to equip the structure with a tracker and raise the development of daylight-based force for a higher collection of sun energy per unit volume [10].

2.2. Issue Related to Long Term Sustainable Solar Power Generation

- Specialists separated the impact of hourly light event on a natural photovoltaic surface from quick and diffuse sky brightening [11].
- Examiners explored the impact of daylight-based irradiance and episode point of prompt and diffuse sun-based irradiance which change the power age capability with time [12].
- Examiners separated cream photovoltaic energy part power system for fruitful power the chiefs [13].
- Experts researched the improvement of viability of daylight-based force by using modified worldwide situating structure, MPPT, consistent battery storing charging advancement to reduce cost of sun arranged force [14].
- Examiners presented the usage of BESS Battery energy accumulating structure on demand side for ideal constraint of PV system to appraise stack and reduce CO2 spread [15].

3. Conclusion

In current circumstance various urban communities are likely going to benefit with supply of force conveyed by outfitting locally available environmentally friendly power. Unmistakable verification of specialty regions that would clearly benefit with such a procedure, assessment and examination of all useful decisions for power supply is essential to frame and execute mediation drives in such way. Other than a review of energy supply situation is essential, it is vital to lead energy outlines in a few framework and un-charged towns and assess harmless to the ecosystem power resource openness in the same.
Considering the revelations of energy studies and other discretionary data arrangement, suitable feasible power-based frameworks will be arranged and their yearly power transport evaluated. The piece of client moderateness and enthusiasm to pay the people groups for harmless to the ecosystem

power-based power supply while considering the structures arranged in the examination will moreover be analysed.

Thusly, by taking into account the above investigation of sustainable power, one can without a doubt grasp the meaning of producing electrical power from sustainable power before long and moreover the impediments like low viability and incomplete concealing influence ought to be taken care of.

This study will give the information about the unmistakable interest of energy and burden the board in future, this analysed data can be used to make the appointment illustration of force. Study will moreover be important for those areas, which are oblivious to the sustainable power creation on account of obliviousness the sun-based energy.

Some care undertakings will similarly be done to careful the close by people and the more youthful understudies regarding harmless to the ecosystem power age and safeguarding non-sustainable power sources and saving environment.

References

[1] Canada, A.H., “Solar voltaic generation power plants: 1000 MW to 10 kW photovoltaic plant design and application guide for the Pacific Northwest”, “Northcon 95. I EEE Technical Applications Conference and Workshops Northcon95”, 10-12 Oct. 1995, Portland, OR, USA.

[2] Suzuki, R. ; Kawamura, H. ; Yamanaka, S. ; Kawamura, H. ; Ohno, H. ; Naito, K., “Loss factors affecting power generation efficiency of a PV module”, “Photovoltaic Specialists Conference, 2002. Conference Record of the Twenty-Ninth IEEE” pp 1557-1560, 19-24 May 2002.

[3] Zhenhua Jiang ; “Power management of hybrid photovoltaic - fuel cell power systems”, “Power Engineering Society General Meeting, 2006. IEEE” Montreal, Que.

[4] Hassan Moghbelli, Robert Vartanian, “Implemenatation of the Movable Photovoltaic Array to Increase Output Power of the Solar Cells”, Proceedings of the International Conference on Renewable Energy for Developing Countries, 25-33, August 2006.

[5] Ladner-Garcia, H.P. ; O'Neill-Carrillo, E., “Determining realistic photovoltaic generation targets in an isolated power system”, “Power & Energy Society General Meeting, 2009. PES '09. IEEE” pp 1 - 5, 26-30 July 2009, Calgary, AB.

[6] Dai, Qinghui ; Chen, Jun, “Improving the efficiency of solar photovoltaic power generation in several important ways”, “Technology and Innovation Conference 2009 (ITIC 2009), International”, 1-3, 12-14 Oct. 2009, Xian, China.

[7] Hou, H.J.; Yang, Y.P.; Cui, Y.H.; Gao, S.; Pan, Y.X., "Assessment of concentrating solar power prospect in China," Sustainable Power Generation and Supply, 2009. SUPERGEN '09. International Conference on , vol., no., pp.1,5, 6-7 April 2009

[8] McIntyre, K.C. ; Clancey-Rivera, C. ; Tobin, M.C. ; Erickson, M.D. ; Zhang, X., “The feasibility of an environmentally friendly microgrid”, “North American Power Symposium (NAPS), 2010”, pp 1-6, 26-28 Sept. 2010, Arlington, TX

[9] XiaoliXu ; Qiushuang Liu ; YunboZuo, “A Study on All-Weather Flexible Auto-tracking Control Strategy of High-Efficiency Solar Concentrating Photovoltaic Power Generation System”, “Intelligent Systems (GCIS), 2010 Second WRI Global Congress” Volume:2, pp 375-378,16-17 Dec. 2010, Wuhan.

[10] Yahyavi, M. ; Vaziri, M. ; Vadhva, S., “Solar energy in a volume and efficiency in solar power generation”, “Information Reuse and Integration (IRI), 2010 IEEE International Conference”pp 394-399, 4-6 Aug. 2010, Las Vegas, NV.

[11] Aggarwal, A. ; Gupta, A. ; Kotwal, S., “Sustainable powering: Configuration of renewable sources exclusively for powering a house”, “Computer and Communication Technology (ICCCT), 2011 2nd International Conference”, pp 286-291, 15-17 Sept. 2011, Allahabad.

[12] Juuso, E.K. ; Yebra, L.J., “Optimisation of solar energy collection with smart adaptive control”, “Industrial Electronics Society, IECON 2013 - 39th Annual Conference of the IEEE” pp 7938-7943, 10-13 Nov. 2013, Vienna

[13] Kalkhambkar, V. ; Kumar, R. ; Bhakar, R., “Optimal sizing of PV-battery for loss reduction and

IOP Conf. Series: Earth and Environmental Science 1084 (2022) 012003 doi:10.1088/1755-1315/1084/1/012003

intermittency mitigation", "Recent Advances and Innovations in Engineering (ICRAIE), 2014", pp 1-6, 9-11 May 2014, Jaipur.

[14] Jinn-Chang Wu ; Chia-Wei Chou, "A Solar Power Generation System With a Seven-Level Inverter", "Power Electronics, IEEE Transactions" Volume:29 , Issue: 7, pp 3454 – 3462, 17 February 2014.

[15] Viehweger, C. ; Hartmann, B. ; Keutel, T. ; Kanoun, O., "Simulation of Shading Effects on the power output of solar modules for enhanced efficiency in photovoltaic energy generation", "Instrumentation and Measurement Technology Conference (I2MTC) Proceedings, 2014 IEEE International", pp 610-613, 12-15 May 2014, Montevideo.

Analysis of different buffer layers on CuO based solar cells using SCAPS-1D

Varun Mishra[1], Chandni Tiwari[1], Rahul Chauhan[2]

[1]Department of Electronics & Communication Engineering, Graphic Era (Deemed to be University), Dehradun, India, 248001
[2]Department of Computer Science & Engineering, Graphic Era Hill University, Dehradun, India, 248001
Email: *chandnitiwari.ece@geu.ac.in*

Abstract. We have used SCAPS-1D to do simulations on CuO based solar cells using ZnO, TiO_2, CdS and ZnSe as buffer layers. The solar cell characteristics are analysed by varying different material properties. The CuO/ZnO based solar cell showed highest efficiency ~20%. The interface defect however impacts the efficiency of this solar cell significantly. The interface defect and buffer layer thickness has shown little impact on the TiO_2 and ZnSe buffer layer based solar cell. Further, we have optimized the impact of absorber layer thickness, carrier concentration and absorber defect as well. The effect of buffer layer thickness and electron affinity has also been investigated. Further, we have tried to understand the effect of temperature on the performance of these solar cells. The increase in temperature has resulted in decreased efficiency attributing to increased recombination rate.

1. Introduction

Renewable energy sources have the potential to overcome the energy problems of the globe. Solar energy is one such renewable energy which is available in abundant. Solar cells are devices which are used to convert the solar energy in the form of electricity[1]. The main focus of photovoltaic market is centered on silicon based solar cells owing to its abundance and feasibility[2]. However, this imposes a limitation on efficiency and cost. Several alternative materials have been investigated to be used as absorber layer in solar cell in recent years. These materials are to offer high efficiency, low cost and clean synthesis technique.

Metal Oxide Semiconductors (MOS) are promising candidate for solar cells which offers easy synthesis techniques, abundancy, non-toxicity and chemical stability[3]. MOS based heterojunction solar cells which are entirely made up of oxides have attracted the attention of research community owing to its less complexity and easy production. Cupric Oxide (CuO) is one such metal oxide semiconductor which has been studied widely for solar cell applications[4][5]. Apart from that, it has been used in different applications[6][7][8]. It has intrinsic p-type characteristics and have band gap reported in the range of 1.0-2.1 eV which may be direct or indirect[9][10]. It offers low cost, non-toxicity and ease of fabrication.

The buffer layer plays an important role in heterojunction based solar cells. The incident light reaches the absorber layer by going through the buffer layer. Hence, absorption of photons in buffer layer should be minimum[2]. This puts a requirement of electrical resistance and low surface recombination. The large band gap buffer layers allow majority of light energy to pass to absorption layer however, band gaps difference between the layers should be compatible for depletion layer to overlap and to produce maximum efficiency. Hence, a comparative study of various buffer layers with CuO is required in order

to understand its compatibility which leads to maximum efficiency. Numerical simulations are one such technique which offers the flexibility to investigate the impact of various parameters on solar cell performance. Results of such simulations can be used to improve the working of solar cells. SCAPS-1D is one such open-source software developed by University of Ghent to understand the solar cells performance[11].
This is used for electrical and optical modelling of heterojunctions. This was primarily developed for investigation of Cadmium Telluride (CdTe) and Copper Indium Gallium Arsenide (CIGS) solar cells. This simulation software utilizes the available data of the materials to understand the output such as current and voltage under illumination and dark. The goal is targeted to examine the performance of various solar cells before its physical fabrication.
In this paper, we have investigated CuO based heterojunction solar cell with ZnO, TiO_2, CdS and ZnSe as buffer layer in SCAPS-1D. The parameters such as absorber thickness, Absorber carrier concentration, absorber defect concentration, interface defect, buffer layer thickness and buffer layer electron affinity are varied to understand its impact on solar cell performance.

2. Methodology

The device structure of the proposed solar cell is shown in Figure 1 (a). The glass substrate is coated with fluorine doped tin oxide (FTO) on which buffer layer is established. The absorber layer is situated on the buffer layer. The light energy is passed through glass substrate. Here, CuO acts as absorber layer and ZnO, TiO_2, CdS and ZnSe acts as buffer layers. The various device parameters required for simulation of these materials are shown in Table 1. The thickness of absorber layer and buffer layer is kept at 2 μm and 0.1 μm, respectively for all the solar cells. The solar cell devices are simulated using SCAPS, where continuity and coupled Poisson equations at interfaces and contacts for both the charge carriers with the suitable boundary conditions, are solved numerically.

Table 1. Parameters of different materials

Parameters	CuO[3]	ZnO[SACPS]	TiO_2[3]	CdS[SCAPS]	ZnSe[12]
Thickness (μm)	2	0.1	0.1	0.1	0.1
Band gap (eV)	1.51	3.37	3.2	2.4	2.9
Electron affinity (eV)	4.3	4.3	4.2	4.2	4
Dielectric permittivity (relative)	18.1	9	10	10	10
CB effective density of states (cm^{-3})	$2.2x10^{19}$	$2.2x10^{18}$	$2.2x10^{17}$	$2.2x10^{18}$	$2.2x10^{18}$
VB effective density of states (cm^{-3})	$5.5x10^{20}$	$1.8x10^{19}$	$6x10^{17}$	$1.8x10^{19}$	$1.8x10^{19}$
Electron thermal velocity (cm/s)	$1x10^{7}$	$1x10^{7}$	$1x10^{7}$	$1x10^{7}$	$1x10^{7}$
Hole thermal velocity (cm/s)	$1x10^{7}$	$1x10^{7}$	$1x10^{7}$	$1x10^{7}$	$1x10^{7}$
Electron mobility (cm^2/V s)	100	140	100	100	50
Hole mobility (cm^2/V s)	0.1	25	250	25	20
Shallow uniform donor density ND (cm^{-3})	0	10^{15}	10^{17}	10^{16}	10^{18}
Shallow uniform acceptor density NA (cm^{-3})	10^{16}	0	0	0	0
Absorber Defect (cm^{-3})	10^{11}	10^{11}	10^{11}	10^{11}	10^{11}
Interface Defect (cm^{-3})	10^{11}	10^{11}	10^{11}	10^{11}	10^{11}
Defect type at bulk/interface	Donor/Neutral				

3. Results & Discussions.

3.1. Basic Cell Performance

The I-V characteristics of different solar cells under illumination of AM 1.5 spectrum with a light power of 1000 W/m^2 at room temperature is shown in Figure 1(b). It is evident from the graph that all the solar cells have shown significant current density. Additionally, we can observe that current densities of TiO_2 and ZnSe based solar cells are almost equal. The obtained efficiencies of ZnO, TiO_2, CdS and ZnSe based solar cells are 20.13%, 19.42%, 16.37% and 19.63% respectively. The results are summarized in Table 2. The external quantum efficiencies (QE) of the different solar cells are shown in Figure 1(c). The observed difference of current densities of devices can be verified from QE curve. Since the absorber layer has prime effect on the absorbed photon energy, hence the QE curves of all the devices are relatively same. The QE covers large spectrum of solar energy owing to small bandgap of 1.6 eV. However, at lower wavelength the impact of buffer layer is seen in the QE curve.

Table 2. Efficiencies of different solar cells

	CuO/ZnO	CuO/TiO_2	CuO/CdS	CuO/ZnSe
η (%)	20.13	19.42	16.37	19.63

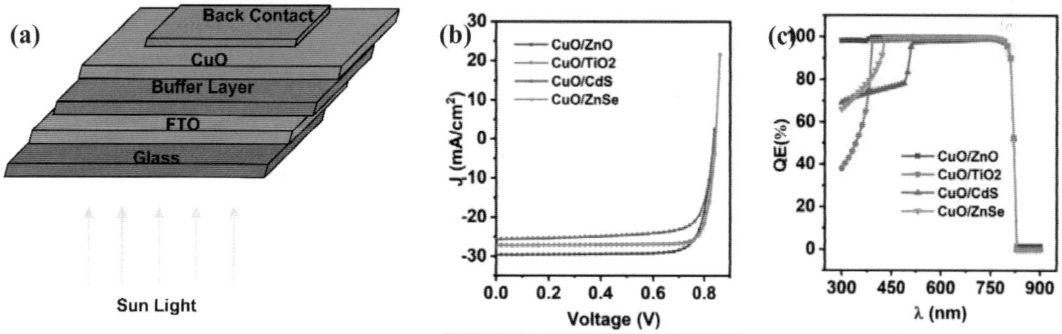

Figure 1. (a) Schematic of the proposed device, (b) I-V characteristics and (c) QE curve of different solar cells

3.2 Impact of absorber layer thickness

The influence of absorber layer thickness on different solar cell parameter is investigated for all the proposed devices and is summarized in Figure 2.

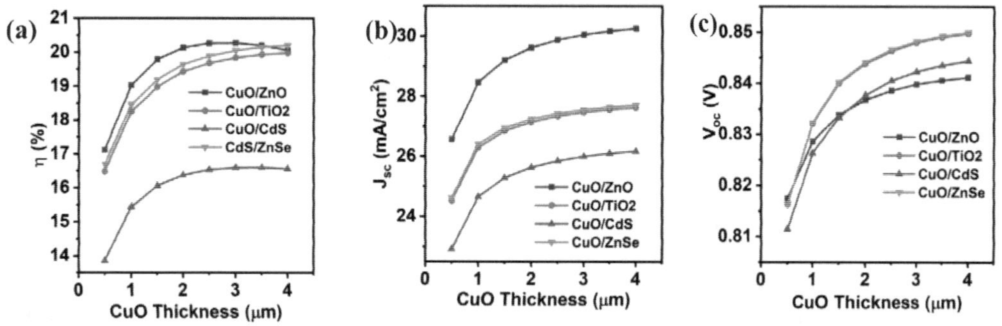

Figure 2. Impact of absorber layer thickness on (a) efficiency, (b) current density and (c) open circuit voltage

It is evident from Figure 2(a) that as the thickness of CuO is increased, the efficiencies of the cells increases and later on, it reaches to saturation after achieving a critical thickness. The thickness around

2 µm is significant for all the devices. Similar changes are observed for current density and open circuit voltage of all the devices in Figure 2(b) & 2(c). The observer layer thickness has constant effect on device parameters of all the solar cells. The impact of buffer layer can be seen on the initial and final values of all the parameters. We can also infer that open circuit voltage of TiO_2 and ZnSe based are largest.

3.3 Carrier Concentration

CuO is intrinsic p-type material where the carrier concentration depends on the defects and non-stoichiometry of the synthesized material. The carrier concentration of the absorber layer is varied in the range of 10^{12} to 10^{18} cm^{-3}. It can be observed from the Figure 3(a) that the efficiency starts increasing with increase in the concentration and then starts decreasing after reaching a particular value of carrier concentration. The minority carrier life time is reduced by increasing carrier concentration which in turn reduces the effective charge carriers accumulated at the contacts. Additionally, the increase in carrier concentration reduces the depletion width in absorber layer which has impact on separation of photogenerated charges which leads to reduction in short circuit current.

CuO has various intrinsic defects such as vacancies, antisites, interstitials, and substitution. The dominant defect present is acceptor which makes the system intrinsic p-type. Different donor and acceptor defects can be introduced with the help of SCAPS. We have considered single donor defect at the absorber layer and neutral defect at the interface of Absorber/buffer layer for ease of investigation. Here, we can see in Figure 3(b) that the moderate defect concentration in absorber layer has not affecting the efficiency significantly. This may be due to low scattering of defect photogenerated carriers. However, when the defect concentration is increased beyond 10^{14} cm^{-3}, the efficiency drops immediately. On the other hand, the interface defects have much more impact on efficiency of the solar cell at moderate values as well. The interface defect reduces the efficiency drastically for CuO/ZnO and CuO/CdS solar cell which is shown in Figure 3(c). Whereas, the CuO/ZnSe and CuO/TiO_2 solar cells does not get a significant change in efficiency by the interface defects.

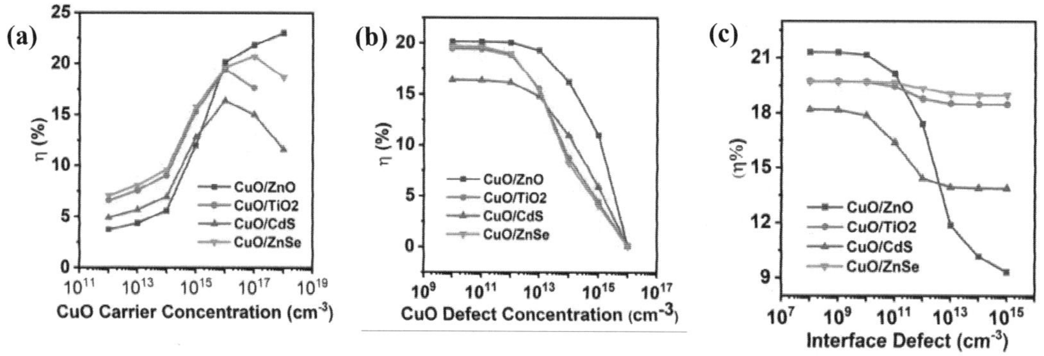

Figure 3. (a) Impact of absorber layer carrier concentration on efficiency, (b) Impact of absorber layer defect concentration on efficiency and (c) Impact of interface defect concentration on efficiency

3.4 Impact of buffer layer

Buffer Layer thickness plays an important role in solar cell energy conversion. The maximum absorption of the light should take place in absorber layer and hence thickness of the buffer layer needs to be optimized. It is evident from the Figure 4(a) that the efficiency of the solar cell is increasing by increased thickness of ZnO. Whereas, efficiency decreases for increased thickness of CdS and eventually becomes constant. The thickness of ZnSe and TiO_2 does not have much impact on the efficiency of the solar cell.

The electron affinity of the buffer layer affects the solar cell efficiency. The Figure 4(b) shows that the graph has an inverted "u" shape structure. The optimum range of electron affinity of buffer layer is from 4.1-4.4 eV where the solar cells show highest efficiency.

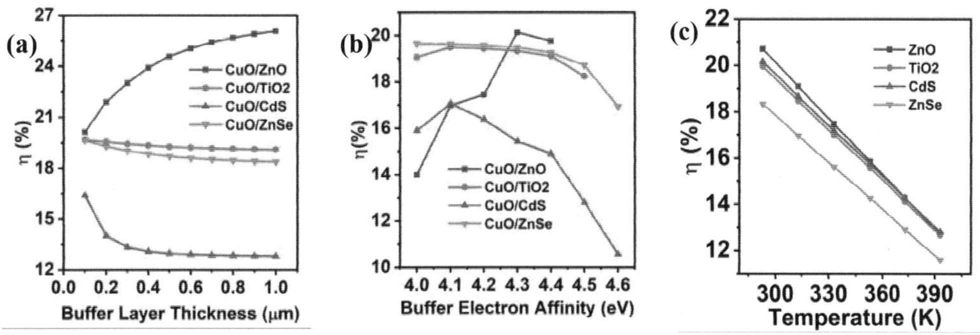

Figure 4. (a) Curve of buffer layer thickness on efficiency, (b) Impact of buffer layer electron affinity on efficiency and (c) Impact of temperature on efficiency

The temperature plays a significant role in achieving the highest efficiencies of the solar cell. The temperature range is varied from 20°C-120°C. It can be inferred from the Figure 4(c) that with increase in temperature, the efficiencies of the solar cells decrease. This is due to increased recombination rates due to increased carrier concentration.

4. Conclusion

In summary, we investigated the impact of buffer layer on the CuO based solar cell using SCAPS-1D software. The CuO/ZnO based solar cell resulted in highest efficiency however the efficiency drastically decreased with increased interface defect. The results obtained from CuO/TiO_2 and CuO/ZnSe solar cells are almost identical. These solar cells also do not get impacted by interfacial defects and the thickness of their buffer layers. The highest achieved efficiencies are 20.13%, 19.42%, 16.37% and 19.63% for ZnO, TiO_2, CdS and ZnSe buffer layes respectively. Hence, we can say that TiO_2 or ZnSe may be a better choice to be used as buffer layer for low cost CuO based solar cells.

Acknowledgement

The authors are grateful Dr. Marc Burgelman from University of Gent, Belgium who has provided the access to SCAPS-1D simulator.

References

1. S. Hussain, C. Cao, Z. Usman, Z. Chen, G. Nabi, W. S. Khan, Z. Ali, F. K. Butt, and T. Mahmood, Thin Solid Films **522**, 430 (2012).
2. S. H. Zyoud, A. H. Zyoud, N. M. Ahmed, and A. F. I. Abdelkader, Crystals **11**, 1454 (2021).
3. P. Sawicka-Chudy, Z. Starowicz, G. Wisz, R. Yavorskyi, Z. Zapukhlyak, M. Bester, Ł. Głowa, M. Sibiński, and M. Cholewa, Mater. Res. Express **6**, 085918 (2019).
4. A. H. Jayatissa, K. Guo, and A. C. Jayasuriya, Appl. Surf. Sci. **255**, 9474 (2009).
5. M. Rokhmat, E. Wibowo, Sutisna, Khairurrijal, and M. Abdullah, Procedia Eng. **170**, 72 (2017).
6. A. Joshi, A. Sharma, R. K. Bachheti, A. Husen, and V. K. Mishra, in *Nanomater. Plant Potential* (Springer International Publishing, Cham, 2019), pp. 221–237.
7. V. Rathi, B. Prasad, V. Mishra, and R. Kanojia, Mater. Today Proc. **60**, 2071 (2022).
8. V. Mishra, Y. K. Verma, S. K. Gupta, and V. Rathi, Silicon **14**, 2275 (2022).
9. T. Viet Pham, M. Rao, P. Andreasson, Y. Peng, J. Wang, and K. B. Jinesh, Appl. Phys. Lett. **102**, 032101 (2013).
10. M. Heinemann, B. Eifert, and C. Heiliger, Phys. Rev. B **87**, 115111 (2013).
11. G. K. Gupta and A. Dixit, Opt. Mater. (Amst). **82**, 11 (2018).
12. L. Et-taya, A. Benami, and T. Ouslimane, Sustainability **14**, 1916 (2022).

IOP Conf. Series: Earth and Environmental Science 1084 (2022) 012005 doi:10.1088/1755-1315/1084/1/012005

Comparison of multi linear regression and artificial neural network to predict the energy consumption of residential buildings

Rajat Nainwal[1], **Aniket Sharma**[2]

[1]Research Scholar, Department of Architecture, National Institute of Technology, Hamirpur, Himachal Pradesh, India
[2]Assistant Professor, Department of Architecture, National Institute of Technology, Hamirpur, Himachal Pradesh, India
Corresponding author's e-mail: rajatn@nith.ac.in

Abstract. The building sector along with its sub-sectors is the world's leading consumer of all forms of energy and consumes about 30% of the world's final energy consumption. With the increase in population, industrialization and urbanization trend, the energy sector has grown rapidly in the past decades to cater to the increase in energy demand. To cater this, many regulatory efforts in the form of energy efficiency and conservation codes offering guidelines and measures during and post design have been proposed in different countries of the world. In order to optimize the consumption of energy and achieving energy efficiency in buildings, an important role is played by the energy consumption prediction. In the due course of time, various energy prediction models have been developed by researchers, which are either based on Physical, Hybrid, or Data-driven methods. In this paper, the energy prediction results using two mostly used data-driven methods i.e. multi linear regression and artificial neural network methods, are compared. For this energy consumption data was collected and monitored from six dwelling units when the outside temperature range is 4°C to 15 °C. It has been observed that the monthly energy consumption predicted using ANN method is more accurate than the prediction using MLR. The variation in energy prediction values ranges between -0.302% to +1.752% in ANN, whereas in MLR it ranges between -2.112% to +2.448%.

1. Introduction

The building sector along with its sub-sectors is the world's leading consumer of all forms of energy and consumes about 30% of the world's final energy consumption [1]. With the increase in population, industrialization and urbanization trend, the energy sector has grown rapidly in the past decades to cater to the increase in energy demand. To cater this, many regulatory efforts in the form of energy efficiency and conservation codes offering guidelines and measures during and post design have been proposed in different countries of the world. Chandel *et. al.* [2] has reviewed various energy regulations of different countries to improve the energy performance of building sector.

Apart from these regulations, other way to achieve energy efficiency in buildings is through implementation of advanced strategies such as predictive and user-specific optimization of electrical systems. In order to optimize the consumption of energy and achieving energy efficiency in buildings, an important role is played by the energy consumption prediction, as achieving energy efficiency at building level is also important for achieving the sustainable growth of the country. It not only acts as an evaluation tool for determining the energy saving potential of a building but also an important

IOP Conf. Series: Earth and Environmental Science 1084 (2022) 012005 doi:10.1088/1755-1315/1084/1/012005

component of smart building approach. It enables the sequences for controlling and allows the occupants to optimize the building systems and reduces the energy demand in building. Not only the overall performance, but the prediction at the individual system level is also important. A building consumes energy in all the phases of its lifecycle i.e. the design and development phase, operation and controlling phase, and the demolition phase [3]. Out of these three stages, the energy consumed is maximum in operation and controlling stage followed by designing stage and demolition stage. Also, operation and controlling stage is long-lasting in building life cycle, and therefore accurate energy prediction of this stage will lead to significant improvement for energy consumption.

In the due course of time, various energy prediction models have been developed by researchers, which are either based on Physical, Hybrid, or Data-driven methods. The physical methods use thermodynamic equations for establishing a relationship between building's physical behavior and its interaction with the surroundings to predict its energy consumption. As the thermodynamic properties of the materials are known, these methods are also known as 'white-box' methods. For reducing the complications of thermal mechanisms arising in the building, certain assumptions are made which affect the model's performance and accuracy. In the data-driven methods, the prediction is done without knowing the internal relationship of parameters of the building, thus, known as 'black box' methods. These methods have gained much popularity in the recent years as they take less time, user-friendly and obtain more accurate results. The hybrid method combines physical and data-driven methods and hence also known as 'grey-box' methods. In this, a detailed accurate specifications about the building envelope and complete information of the building schedule is required which is very problematic, leading to thwarting the implementation of these approaches for the existing buildings [4].

Hence, the objective of this paper is to identify the accurate method for building energy consumption prediction among the two mostly used data-driven methods i.e. Multi-Linear Regression (MLR) and Artificial Neural Network (ANN). A methodology has been proposed to collect and analyze the data which helps in predicting the categorical energy consumption. The paper is divided into 6 sections. After the introductory section, section 2 discusses the methodology, section 3 discusses MLR method for energy prediction, section 4 discusses ANN method for energy prediction, results are discussed in section 5 followed by conclusion in section 6.

2. Methodology

The energy in any building is consumed by the various appliances used inside it, which are categorized as lighting, space heating, cooling, water heating, ventilation, and refrigeration and miscellaneous-use appliances. A residential building of three-storeys in composite climate having six dwelling units in total is selected for data collection. After reviewing the various available literature [5]–[7], it has been observed that the energy consumption pattern of a building is necessary for energy prediction. So, a questionnaire has been prepared and data from these six dwelling units was collected. Figure 1 shows the overall methodology for determining the break-up of energy consumption through different appliances of different categories resulting in determining the energy consumption pattern of each dwelling.

The break-up of per day energy consumption is determined and the usage profile of the appliances is developed. The total monthly consumption is then validated with the energy bills monitored by the state electricity board. Hence, the exact energy consumption value by the above categorized appliances was obtained. The same data was then entered into IBM SPSS software and used to predict the categorical as well as monthly energy consumption using MLR and ANN and then the values are compared and variations in the predicted energy consumption values are determined.

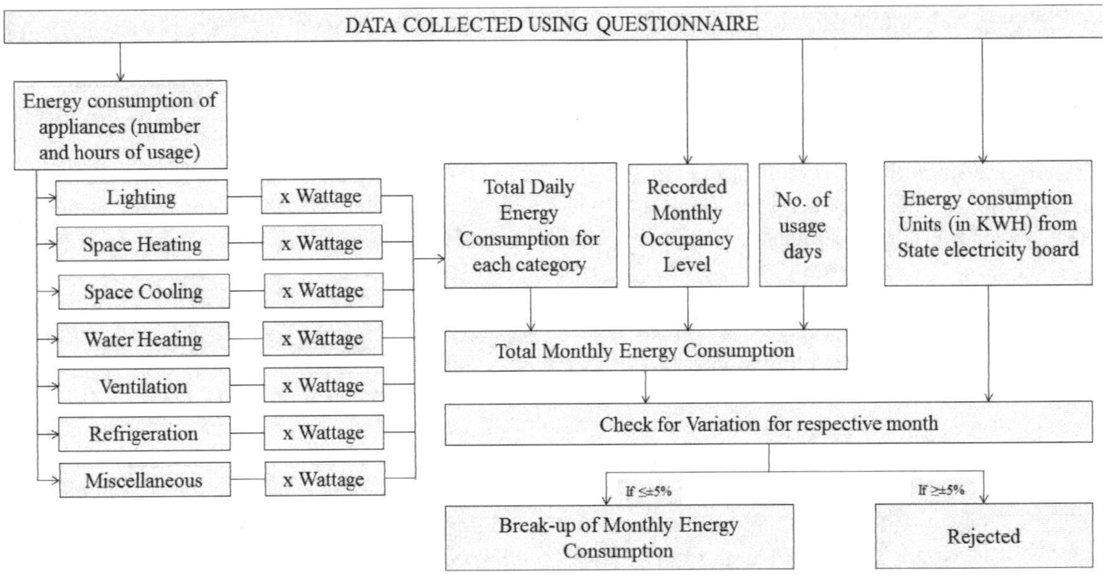

Figure 1. Methodology to be adopted for determining the break-up of energy consumption

3. Multi Linear Regression

Regression is a tool widely used for describing the change in dependent variables to independent variables which are used as inputs in the tool [8]. According to Zhao *et. al.* [9], regression models became popular in the domain of research, in order to solve the 3 problems. Firstly, for predicting the energy consumption based on climatic parameters. Secondly, for predicting energy indexes. Thirdly, for determining the important attributes of energy consumption, which helps in understanding the thermal behavior of the building. According to Ciulla *et. al.* [10], it correlates the building's energy consumption with the variables influencing its consumption, in a very simple way and allows the estimation of the requirement of building's energy, immediately. Different researchers had performed building energy consumption prediction using MLR in various studies [11]–[13]. For predicting the monthly energy consumption of the studied buildings, the relationship between dependent and independent variables proposed by Deng *et.al* [14] was used (shown as equation 1). Here, the monthly energy consumption is the dependent variable, whereas, occupancy, outdoor temperature, per day energy consumption are the independent variables.

$$Y = \beta_0 + \beta_1 X_1 + \beta_2 X_2 + e \tag{1}$$

Where, Y = Dependent variables; X_1 *and* X_2 *= Independent variables;* β_1 *and* β_2 *= Coefficients; and, e = Intercept of coefficients.*

4. Artificial Neural Networks

In the recent years, Artificial Neural Network (ANN) is the most widely used method for predicting the building's energy consumption as it is very good at dealing with the non-linear problems [9], [15]. Arditi *et. al.* [16], defined ANN as a system for handling information whose architecture is inspired by the structure of biological systems. Just like the working of the human brain which continuously learns and predicts things based on what eyes see, it also replicates the same as it performs learning and predicting things which are dependent on the input given by the user. The different nodes are operated in parallel and the communication is done through the connections known as Synapses. It can solve difficult problems just by using the combination functions (which are linear) and activation functions (which are non-linear). The schematic diagram of single layer of neuron having an activation function has been presented in Figure 2.

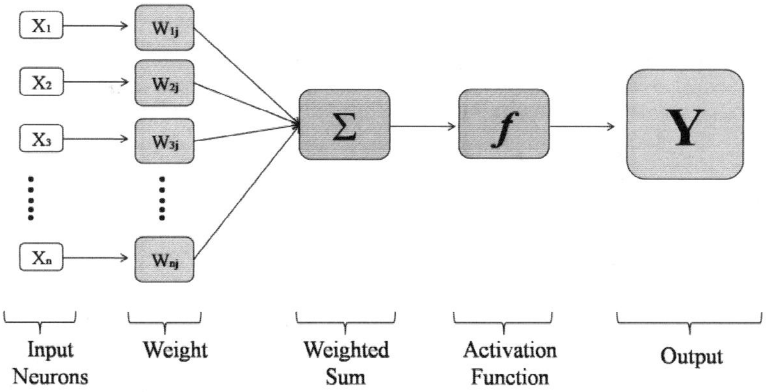

Figure 2. The schematic diagram of single layer of neuron having an activation function

Various ANN architectures have been used for predicting the building's energy consumption. These architectures are categorized into Feed Forward Neural Network (FFNN), Recurrent Neural Networks (RNN) and Convolutional Neural Network (CNN). In FFNN, connections from input to output layer transmitted the signals, passing through hidden layers. In RNN, a cycle is formed by the connections between layers that exhibits a dynamic temporal behavior. In CNN, operations like convolution and pooling are used to extract features automatically. In the study done by Bojic *et. al.* [17], the thermal behavior and energy consumption of the building had been predicted by proposing an ANN model having 5 input neurons with a hidden layer having 46 neurons. Another study done by Aydinalp *et. al.* [18], used neural networks for predicting the energy demand in residential sector required for lighting, space heating, space cooling and hot water. In studies [19], [20] the annual energy consumption has been predicted using ANN. In this study, the Multi-Layer Perceptron (MLP), a most widely used type of FFNN is used to predict the energy consumption of the selected building. The standardized formula of MLP is shown below:

$$O = AF(WX + b) \quad (2)$$

where, O is Output of the neuron, AF is Activation Function, W is weight, b is bias and X is input value. The most widely used AF is sigmoid (σ) and hyperbolic tangent (tan-h). Equation 3 and 4 shows the general representation of σ and tan-h AF.

$$\sigma(x) = 1 / (1+e^{-x}) \quad (3)$$

$$\tanh(x) = (1-e^{-2x}) / (1+e^{-2x}) \quad (4)$$

The input and output used for predicting the monthly energy consumption of the studied buildings are shown in Table 1.

Table 1. Input and output used for predicting the monthly energy consumption of the studied buildings

Category	Input used	No. of hidden layers	Activation Function used	Output
Lighting	Year of construction, flooring material, occupancy, outside temperature and per day energy consumption	1	Hyperbolic tangent	Total monthly energy consumption
Space Heating		1	Hyperbolic tangent	
Water Heating		1	Hyperbolic tangent	
Ventilation		1	Sigmoid	
Refrigeration		1	Sigmoid	
Miscellaneous		1	Hyperbolic tangent	

The wall material is similar in all the cases, so exempted to be used as an input.

5. Results and Discussion

After data collection, using methodology presented in Figure 1, energy consumption of different category appliances was determined for the month of January where the temperature ranges from 4°C to 15°C, so eliminating the cooling energy from the analysis. The physical condition of the dwelling units is presented in Table 2.

Table 2: Physical Condition of the studied dwelling units

Floor and No.	Year of construction	Built-up area (in Ft.^{2})	Flooring Material	Wall Material	Floor-to-floor height
G 1	2008	530.00	R.C.C.	Cement plastered 9" thick brick wall	3 m.
G 2	2008	530.00	R.C.C.	Cement plastered 9" thick brick wall	3 m.
F 1	2015	530.00	Tiles	Cement plastered 9" thick brick wall	3 m.
F 2	2015	530.00	Tiles	Cement plastered 9" thick brick wall	3 m.
S 1	2015	530.00	Tiles	Cement plastered 9" thick brick wall	3 m.
S 2	2015	530.00	Tiles	Cement plastered 9" thick brick wall	3 m.

Where, G, F and S stands for ground, first and second floor

The accurate duration of using different electrical appliances in all the 6 dwelling units of the building was observed and monitored for a day and the occupancy in these dwelling units was recorded for a month. The calculated total energy consumption per day (TCPD) by the different appliances is presented in Table 3 along with the recorded occupancy for the month.

Table 3: TCPD by different appliances with recorded monthly occupancy

Floor and No.	No. of Person	Occupancy	TCPD L	TCPD SH	TCPD WH	TCPD V	TCPD R	TCPD M	TMEC
G 1	5	80%	0.890	16.0	3.0	0.4	6.0	1.63	27.92
G 2	4	75%	1.320	4.8	2.0	0.4	2.88	0.98	12.38
F 1	1	75%	0.696	8.0	2.0	0.09	5.28	0.58	16.646
F 2	4	75%	0.728	4.80	4.0	0.4	5.28	1.28	16.488
S 1	3	75%	0.652	12.00	4.0	0.3	6.0	1.0	23.952
S 2	4	50%	0.678	5.20	6.0	0.12	2.64	4.53	19.198

Where, L is lighting, SH is space heating, WH is water heating, V is ventilation, R is refrigeration, M is miscellaneous, TCPD is total consumption per day (in KWH), TMEC is total monthly energy consumption (in KWH)

It is seen in Figure 3 that the maximum energy is used by the space heating appliances and minimum energy is used by the appliances for ventilation. Figure 4 presents the average monthly energy consumption by the appliances of different categories. It is seen in Figure 4 that the 42% of monthly energy consumption is used for space heating purpose and around 19% is used for water heating purpose. 23.5% of monthly energy consumption is done by refrigeration appliances, which is due to the fact that in 5 out of 6 studied cases, the refrigeration process independent of its occupancy.

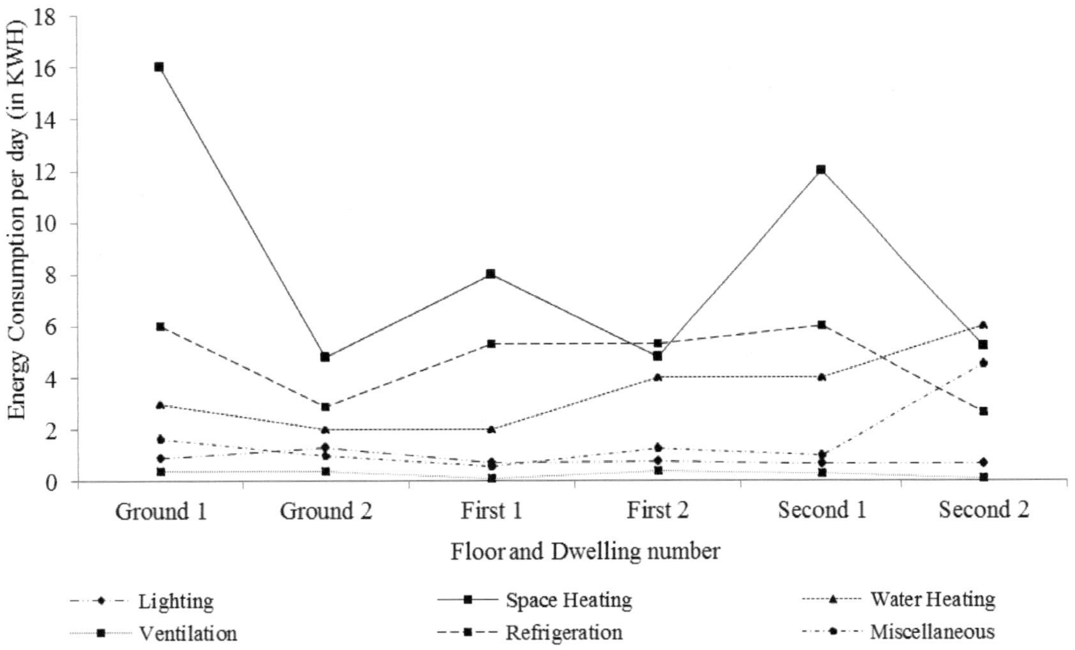

Figure 3. Energy consumption profile by different categories of appliances

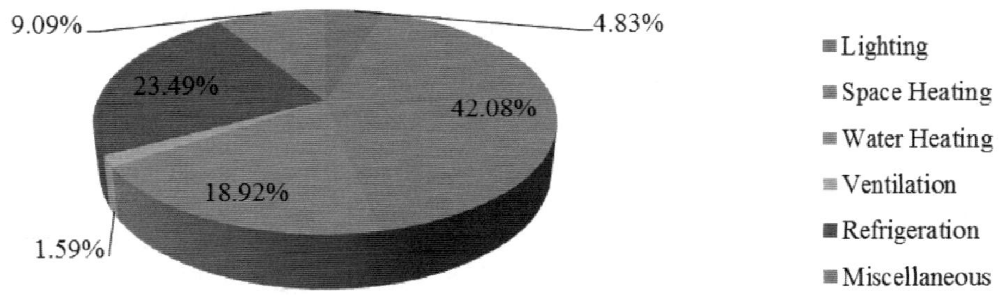

Figure 4. Average monthly energy consumption by different appliances

Using the same building conditions, appliances usage duration and occupancy schedule, the energy consumption for the month has been predicted by both MLR and ANN using equations 1 to 4. In ANN, the training and testing data is kept at the ratio of 80:20. The activation functions used for training and testing are hyperbolic tangent and sigmoid, respectively. There is 1 hidden layer having 50 neurons.

Table 4 presents the predicted energy consumption of each flat using the proposed methodology, MLR method and ANN method. The predicted energy consumption values are then compared with the actual energy consumption values monitored by the state electricity board using electricity bills. It is seen in Table 3 and Figure 5 that among the predicted values, the least variation is in the one predicted using ANN method with variations ranging from -0.302% to +1.752%. Hence, ANN method based prediction is most accurate among all.

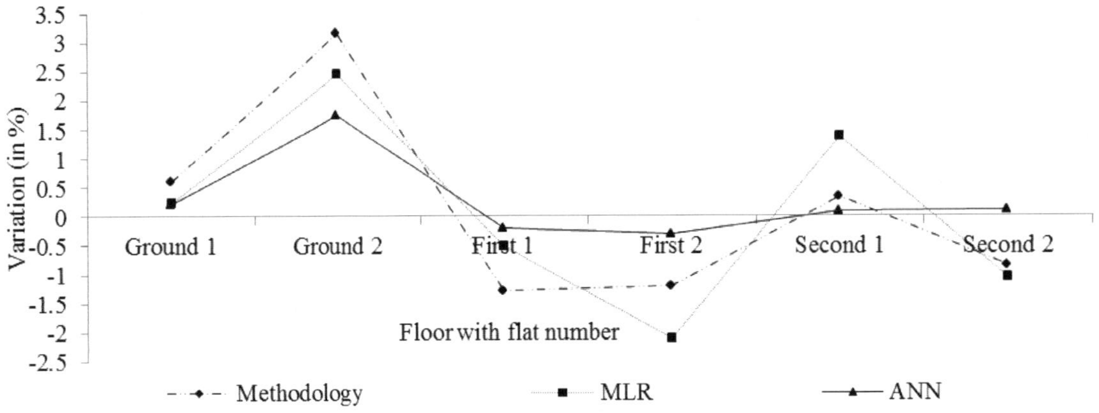

Figure 5. Variation in predicted energy consumption values by different methods

Table 4. Predicted monthly energy consumption of each dwelling unit using the proposed methodology, MLR and ANN method

Floor and No.		L	SH	WH	V	R	M	Total Monthly	Total Monthly bill	Variation (in %)
G1	Proposed Methodology	22.072	396.80	74.40	9.92	148.80	40.24	692.232	688	+0.615
	MLR	21.837	391.72	77.52	9.62	148.04	40.91	689.647		+0.239
	ANN	22.029	395.19	75.33	9.86	146.92	40.11	689.439		+0.209
G2	Proposed Methodology	30.690	111.6	46.50	9.30	66.96	22.78	287.83	279	+3.164
	MLR	30.762	110.05	46.03	9.42	66.99	22.58	285.832		+2.448
	ANN	30.759	109.25	46.65	9.11	64.84	23.28	283.889		+1.752
F1	Proposed Methodology	16.182	186.00	46.50	2.09	122.76	13.48	387.012	392	-1.272
	MLR	16.251	188.23	46.03	2.11	123.02	14.38	390.021		-0.504
	ANN	16.199	189.84	46.19	2.08	122.57	14.35	391.229		-0.196
F2	Proposed Methodology	16.926	111.6	93.00	9.30	122.76	29.76	383.346	388	-1.199
	MLR	16.995	110.05	91.59	9.42	123.02	28.73	379.805		-2.112
	ANN	16.927	115.09	92.99	9.85	122.57	29.40	386.827		-0.302
S1	Proposed Methodology	15.159	279.00	93.00	6.97	139.50	23.25	556.879	555	+0.338
	MLR	15.228	285.95	91.59	7.06	139.83	22.99	562.648		+1.378
	ANN	15.294	277.05	92.99	6.98	139.80	23.39	555.504		+0.090
S2	Proposed Methodology	10.509	80.60	93.00	1.86	20.46	70.21	276.639	279	-0.846
	MLR	10.462	79.58	93.62	1.80	20.30	70.31	276.072		-1.049
	ANN	10.512	82.27	93.49	1.87	20.60	70.57	279.312		+0.111

Where, G, F and S stands for ground, first and second floor. L is lighting, SH is space heating, WH is water heating, V is ventilation, R is refrigeration, M is miscellaneous All data is in KWH units.

6. Conclusion

The energy consumption pattern of the residential building located in composite climate is determined using the proposed methodology. The monthly energy consumption of all the six dwellings units of the building is then predicted using MLR and ANN method. The following conclusions are drawn:

1. Maximum energy is used for space heating in all the dwelling units i.e. 42% while minimum for ventilation purpose.

2. Energy used for space heating and water heating together contributes around 61% in the monthly energy consumption.
3. The monthly energy consumption predicted using ANN method has lower variation than prediction using MLR method. The variation in energy prediction values ranges between -0.302% to +1.752% in ANN, whereas in MLR it ranges between -2.112% to +2.448%.
4. ANN method has higher accuracy than the MLR method in the domain of energy consumption prediction.

References

[1] International Energy Agency, "IEA Database," 2021. [Online]. Available: https://www.iea.org/data-and-statistics/data-browser?country=WORLD&fuel=Energy consumption&indicator=TFCShareBySector. [Accessed: 26-Aug-2021].
[2] S. S. Chandel, A. Sharma, and B. M. Marwaha, "Review of energy efficiency initiatives and regulations for residential buildings in India," *Renew. Sustain. Energy Rev.*, vol. 54, pp. 1443–1458, 2016.
[3] L. Ngwepe and C. Aigbavboa, "A theoretical review of building life cycle stages and their related environmental impacts," 2015.
[4] A. Foucquier, S. Robert, F. Suard, L. Stéphan, and A. Jay, "State of the art in building modelling and energy performances prediction: A review," *Renew. Sustain. Energy Rev.*, vol. 23, pp. 272–288, 2013.
[5] Z. Wang and R. S. Srinivasan, "A review of artificial intelligence based building energy prediction with a focus on ensemble prediction models," *Proc. - Winter Simul. Conf.*, no. February, pp. 3438–3448, 2016.
[6] S. Dukkipati, R. K. Iyer, and A. Sreenivas, "An Assessment of Energy Data Management in India," no. October. Prayas Energy Group, Pune, 2014.
[7] "Commercial Buildings Energy Consumption Survey (CBECS) Questionnaire." United States Energy Information Administration, Washington, 2018.
[8] Z. Wang and R. S. Srinivasan, "A review of artificial intelligence based building energy use prediction: Contrasting the capabilities of single and ensemble prediction models," *Renew. Sustain. Energy Rev.*, vol. 75, no. October, pp. 796–808, 2017.
[9] H. Zhao and F. Magoulès, "A review on the prediction of building energy consumption," *Renew. Sustain. Energy Rev.*, vol. 16, no. 6, pp. 3586–3592, 2012.
[10] G. Ciulla and A. D'Amico, "Building energy performance forecasting: A multiple linear regression approach," *Appl. Energy*, vol. 253, no. June, p. 113500, 2019.
[11] N. Amjady, "Short-term hourly load forecasting using time-series modeling with peak load estimation capability," *IEEE Trans. Power Syst.*, vol. 16, no. 4, pp. 798–805, 2001.
[12] V. Bianco, O. Manca, and S. Nardini, "Electricity consumption forecasting in Italy using linear regression models," *Energy*, vol. 34, no. 9, pp. 1413–1421, 2009.
[13] A. D. Papalexopoulos and T. C. Hesterberg, "A regression-based approach to short-term system load forecasting," *IEEE Trans. Power Syst.*, vol. 5, no. 4, pp. 1535–1550, 1990.
[14] H. Deng, D. Fannon, and M. J. Eckelman, "Predictive modeling for US commercial building energy use: A comparison of existing statistical and machine learning algorithms using CBECS microdata," *Energy Build.*, vol. 163, pp. 34–43, 2018.
[15] M. A. Mat Daut, M. Y. Hassan, H. Abdullah, H. A. Rahman, M. P. Abdullah, and F. Hussin, "Building electrical energy consumption forecasting analysis using conventional and artificial intelligence methods: A review," *Renew. Sustain. Energy Rev.*, vol. 70, no. December, pp. 1108–1118, 2017.
[16] D. Arditi and O. B. Tokdemir, "Comparison of Case Based Reasoning and Artificial Neural Networks," *J. Comput. Civ. Eng.*, vol. 13, no. 3, pp. 162–169, 1999.
[17] S. A. Kalogirou and M. Bojic, "Artificial neural networks for the prediction of the energy consumption of a passive solar building," *Energy*, vol. 25, no. 5, pp. 479–491, 2000.

[18] M. Aydinalp-Koksal and V. I. Ugursal, "Comparison of neural network, conditional demand analysis, and engineering approaches for modeling end-use energy consumption in the residential sector," *Appl. Energy*, vol. 85, no. 4, pp. 271–296, 2008.
[19] S. L. Wong, K. K. W. Wan, and T. N. T. Lam, "Artificial neural networks for energy analysis of office buildings with daylighting," *Appl. Energy*, vol. 87, no. 2, pp. 551–557, 2010.
[20] S. Fiorelli and A. H. Neto, "Comparison between detailed model simulation and artificial neural network for forecasting building energy consumption," *Energy Build.*, vol. 40, pp. 2169–2176, 2008.

Effect of Pozzolanic Material on Drying Shrinkage Characteristics of Blended Cementitious Mortar

Nitesh Thikare[1], Shalini Yadav[2], Ranjit Patil[3]

[1.] Research Scholar, Department of Civil Engineering, Rabindranath Tagore University, Bhopal India.

[2.] Professor, Department of Civil Engineering, Rabindranath Tagore University, Bhopal India.

[3.] Assistant Professor, Department of Civil Engineering, Priyadarshini Bhagwati College of Engg., Nagpur India.

E-mail: nthikare7391@gmail.com

Abstract: To maintain sustainability, the production of Portland cement is a major contribution to greenhouse gas emissions. Various methodologies are used for the purpose of reduction of Portland cement consumption in mortar are desirable to minimize the greenhouse gases. Based on past studies, the authors have studied fly ash behavior with the cement and its effect on mortar and concrete i.e. usage of fly ash (FA) contents in mortar as one of the sources of cement replacement, which makes mortar less permeable due to its fineness and pozzolanic response. This paper briefly discusses the cement replacement with a different fineness of FA from various hoppers at different percentages and does experimental examinations related to the compressive strength (CS) and drying shrinkage (DS) mortar. The FA was used for the research work from the Khaperkheda power plant, Nagpur. FA is used as a replacement with cement in a 12.5 % proportion up to 37.5% with a mix proportion for the mortar is 1:3. Casting of mortar for compressive strength and drying shrinkage is done as per the Indian Standard (IS) code. The result elucidated that, up to optimum level, the CS of mortar was increased and further observed increase in FA content decreased the CS. For drying shrinkage, the value of shrinkage is found lesser than the controlled cement mortar due to its fineness.

Keywords: Fly Ash, Mortar, Compressive Strength (CS), Drying Shrinkage (DS)

1. Introduction

The industrial waste, fly ash is obtained from thermal power plants which have pozzolanic characteristics after burning of pulverized coal. At present, India produces around 120 Million tons of ash per year. Due to the rapid increase in industrialization and population more consumption of power is required for the country. In India around, 80 percent of electricity (power) generation depends on thermal power plants. The source of generation of power in plants is due to the burning of pulverized coal, which generates 35 % of ash. As the requirement for power increases, day-by-day also increases ash production. The utilization of ash is major factor in India, about 70 percent of ash is used in cement industries [1]

The utilization of fly ash in recent times is more important and is used in mortar, concrete, landfilling, etc. Fly ash has become a popular element in the production of high-strength mortar. The cementitious supplementary material FA has already been implemented in accordance with IS 4031-1988. Fly ash has some amount of cementitious characteristics like cement hence it is used as an SCM in concrete or mortar [2]. Due to loss of moisture, shrinkage will occur at every cementitious product. As the multidimensional contraction of the mortar due to loss of moisture, the DS is evaluated on the largest side of the body. The total DS in mortar may occur due to various forms of shrinkage (plastic, drying, autogenous, etc.). DS is formed by the evaporation of internal water in the mortar sample. During the hydration process, the volume change of unstrained cement paste is found nearly equal to the loss of moisture [3].

2. Characterization of Material

2.1. Study of Fly Ash

FA is generated by burning pulverizing coal in a thermal power plant. It contains major cementitious components such as silica (Si), calcium (Ca), alumina (Al), and iron (Fe) in the large amount. Si and al are found about 75 to 95 percent in fly ash. When coal is burned, it can be classified into two types - crystalline and non-crystalline, which includes, SiO_2, $3Al_2O_3$ $2H_2O$, Fe_2O_3, Fe_3O_4, FeO, K_2O $Al_2O_3.6SiO_2$.

Classification of FA is depending on the reactive calcium oxide present in ash i.e. <10% and >10%. If it's less than 10%, calcium oxide (CaO), is classified as class-F FA and if it contains more than 10%, CaO, it is classified as class-C FA. Class-F FA is pozzolanic in nature and is formed by burning older anthracite and bituminous coal. For the pozzolanic property it requires Portland cement, quick lime, and water, after reacting with each other it will form cementitious compounds. Alternatively, the formation of geo-polymer is done by the addition of sodium silicate into class-F FA. Class-C FA is generated by the ignition of lignite or sub-bituminous coal. To get self-cementing properties, Coal is added. Class-C FA has more than 20% lime and it gets hardened with the addition of water into it and gets more strength as time passes without any addition of activator. Alkali, as well as sulfate contents, are greater in Class C FA. Chemical requirement as per IS 3812-1981 and Physical Characteristics requirement as per IS 3512-1981 is given below in Tables 1 and 2.

Table 1. Chemical requirement as per IS 3812-1981

Characteristic (% by mass)	Requirement
$SiO_2 + Al_2O_3 + Fe_2O_3$	70.0
SiO_2	35.0
MgO	5.0
SO_3	2.5
Na_2O	1.5
LOI	12.0

Table 2. Physical Characterization requirement as per IS 3512-1981

Characteristic	Requirement
Fineness	320 in m^2/kg
Lime reactivity	4.0 in N/mm^2
Drying shrinkage (maximum)	0.15%
Soundness	0.8%

2.2. Study of Materials

Fly Ash Utilization: The pozzolanic character, spherical shape, and relative regularity of FA all contribute to its uses. In order of decreasing frequency, fly ash recycling includes Class F type FA. The physical characteristics of FA were obtained from the Khaparkheda power plant, Nagpur as in mentioned in Table 3.

Table 3. Specific Gravity (SG) and fineness of FA

Hoppers	SG (gm/cc)	Fineness
Hopper 1	2.13	309
Hopper 2	2.13	448

Ordinary Portland Cement (OPC): 43 grade OPC is formulated by thoroughly mixing calcareous and argillaceous materials in proper proportion then the mix is burned at high temperature in the kiln, clinkering and grinding resulting the formation of cement clinker. To manufacture cement that meets IS requirements, the appropriate percentage of gypsum is added to the clinker. Ordinary Portland Cement grades are determined by the average compressive strength requirement after 28 days. Table 4 shows, the chemical compositions and physical properties of cement used in the study.

Table 4. Observed chemical compositions and physical properties of cement

Oxide	Cement	Oxide	Cement
SiO_2	19.71 percent	Na_2O_3	0.25 percent
Al_2O_3	5.20 percent	LOI	0.96 percent
Fe_2O_3	3.73 percent	Consistency	33 percent
CaO	62.91 percent	Setting Time (Initial)	56 min
MgO	2.54 percent	setting time (Final)	180 min.
SO_3	2.72 percent	SG	3.14

3. Experimental Program

Fly ash assessment for use of additional cement materials begins with mortar testing. Mortar contains cement, water, and fine aggregate. With the proper proportion of mortar i.e. cement sample 37.5%, 25%, 12.5%, 0% replaced with the FA. The mixing proportion of the material is given in table 6 which contains a 1:3 proportion of cement to sand for two hopper fly ash.

3.1. Study of compressive strength of mortar

The standard IS 4031: 1988-Part 7 is used for the determining strength of masonry cement. The cube used for compressive strength tests is 50 mm cubes. The quantity of mixing of material is given in table 6 which contains a 1:3 proportion of cement to sand. The amount of water required is calculated by flow tests are determined by a flow of 110±5 % with 25 drops in 15 seconds as shown below in table 5:

Table 5. water requirement of fly ash of hoppers by using flow-table test

Physical test	Water Requirement
Hopper 1	103 % @ 14 % water
Hopper 2	106 % @ 14 % water

Table 6. Compressive strength quantity for cubes of 50*50*50mm size

Compressive Strength	% replacement	sand(gm)	Cement(gm)	FA (gm)	Water(ml)
Plane sample	-	2250	750	-	450
Hopper 1	12.5 %	2250	656.25	93.75	432.6
Hopper 1	25 %	2250	562.5	187.55	432.6
Hopper 1	37.5 %	2250	468.75	281.25	432.6
Hopper 2	12.5 %	2250	656.25	93.75	445.2
Hopper 2	25 %	2250	562.5	187.55	445.2
Hopper 2	37.5 %	2250	468.75	281.25	445.2

3.2. Study of Drying Shrinkage of mortar

The materials used for drying shrinkage are cement and sand with a 1:3 proportion used for the testing of mortar. The mould used for drying shrinkage is mould of 25 mm x 25 mm x 280 mm. The mortar is placed in the beam in two layers by using thumbs and forefingers for pressing the mortar into the corners. On the next day remove the mould and immerse it in water for six days for curing. After curing the length of the specimen, the specimen was computed by the length comparator apparatus. The specimen was protected from humidity loss by maintaining the relative humidity at 50±5 % and a temperature of 27±2°C. Repeat measurement of sample length 28 days after initial measurement. The quantity of mixing of material for DS is shown in table 7.

Table 7. Drying Shrinkage quantity for beam of 25 mm x 25 mm x 280 mm size

Compressive Strength	% replacement	sand(gm)	Cement(gm)	FA(gm)	Water(ml)
Plane sample		900	300	-	180
Hopper 1	12.5 %	900	262.5	37.5	173.04
Hopper 1	25 %	900	225	75	173.04
Hopper 1	37.5 %	900	187.5	112.5	173.04
Hopper 2	12.5 %	900	262.5	37.5	178.08
Hopper 2	25 %	900	225	75	178.08
Hopper 2	37.5 %	900	187.5	112.5	178.08

4. Testing Results and Discussion

4.1. Compressive Strength (CS)

The specimen of each mould is cast with varying percentages of fly ash with 12.5%, 25%, and 37.5%, of different fineness (hopper 1 and 2), and the test was done for 7 and 28 days. CS in the produced mortars ranged between 12.31 and 18.23 N/mm^2 on day 7, and between 21.36 and 33.52 N/mm^2 on day 28. Based on the results, the mortar with the highest degree of fineness has the highest CS.

Figure 1 shows, CS for hoppers 1 and 2 for different replacements of FA of Khaparkheda power plant. The foregoing findings show that when the percentage of fly ash replacement increases, the CS for mortar in hoppers 1 and 2 increases slightly.

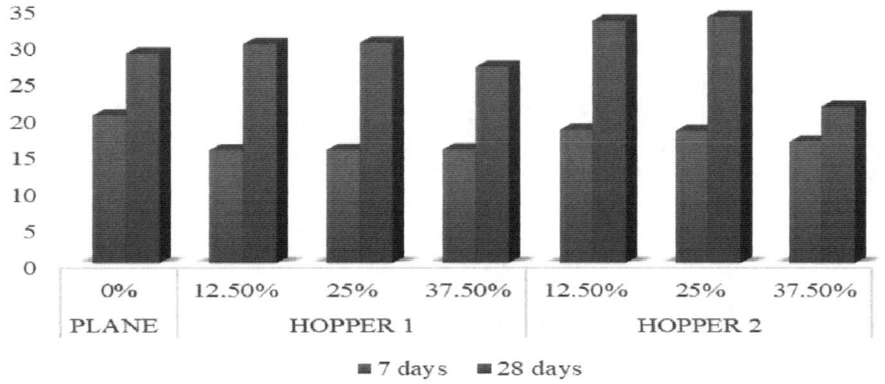

Figure 1. Compressive strength of fly ash mortar for 7- days and 28-days strength for various replacements.

For the 28 days, strength in hopper 1 for a various percentage of replacement gives somewhat equal in compressive strength, while for another hopper compressive strength is greater as compared to hopper 1 but for 37.5% replacement shows sudden fall of strength means that as a percentage of fly ash increases above optimum value the strength will get decreases.

4.2. Drying Shrinkage Result

Below graph shows the Drying Shrinkage of 12.5%, 25%, and 37.5% replacement of fly ash for two hoppers of the Khaparkheda thermal power plant.

Drying shrinkage in various percentages increases from hopper 1 to 5. For 12.5%, in hopper 2 drying shrinkage is more as compared to others. Drying shrinkage in various percentages increases from hopper 1 and 2. For 12.5%, in hopper 2 drying shrinkage is lesser. As the percentage of replacement of FA increases up to 25% shrinkage will be increased and from 25% to 37.5% shrinkage will decrease.

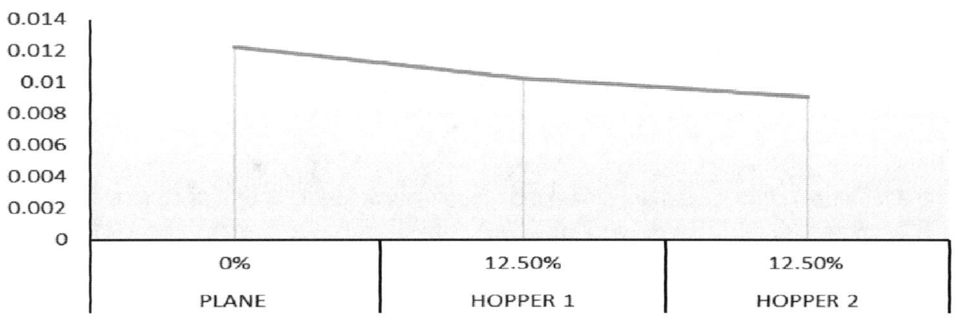

Figure 2. DS with respect to the fineness of fly ash with 12.5% replacement.

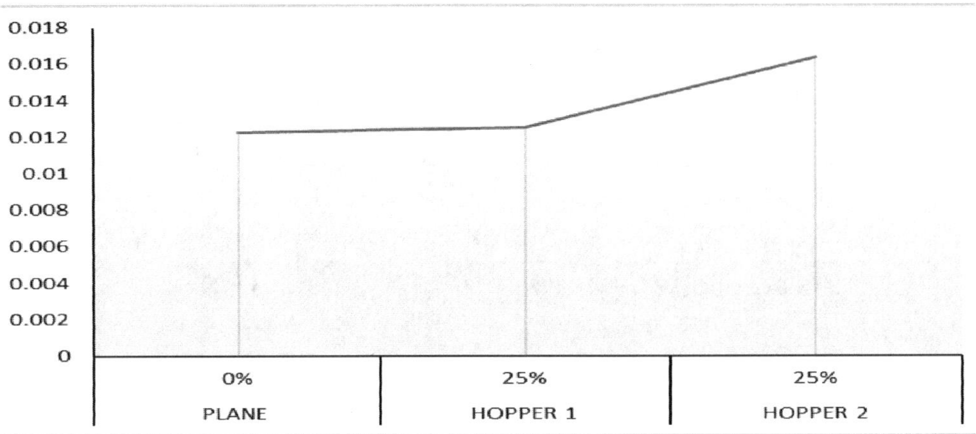

Figure 3. DS with respect to the fineness of fly ash with 25% replacement

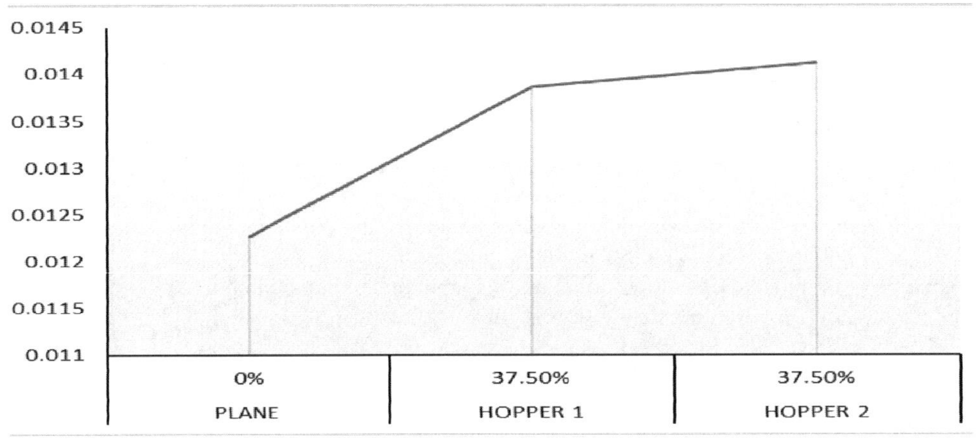

Figure 4. DS with respect to the fineness of fly ash with 37.5% replacement

5. Conclusion

From the detailed analysis of the study found various results in hopper-1and hopper-2 of different fineness respectively which conclude that:

(i) Shrinkage is a volumetric reduction in concrete that is often caused by water loss.
(ii) In the majority of cases, shrinkage is determined by measuring longitudinal strain.
(iii) The fineness ratio of different types of fly ash tends to vary somewhat yet consistently.
(iv) Test results also indicated that the different fineness of fly ash had effects on the CS as well as DS.
(v) The fine fly ash with a large surface area was more reactive, resulting in a strength enhancement.
(vi) The finer fly ash requires less water as it has a spherical form and smooth surface
(vii) Due to its fineness, FA helped in packing or filling the voids to increase the CS of the mortar.

References

[1] Malathy R, Appa Rao G 2002, "Drying shrinkage of cementitious composites incorporated with silica fume", *27th Conference on Our World in Concrete Structures*. Singapore, Article Online Id: 100027043

[2] Federico Lopez-Flores 1982, "Flyash and Effects of Partial Cement Replacement by Fly ash Informational Report", *JTRP Technical Reports*, 82-11.

[3] Soni Neeraj 2011, "Influence of Fly-ash on The Strength and Swelling Characteristics of Bentonite", *Department of Civil Engineering, National Institute of Technology*, Rourkela 769008.

[4] Jonasson Jan-Erik 2005, "Shrinkage Measurements of Mortars with Energetically Modified Fly Ash", *Division of Structural Engineering, Technical Report* 2005:11 ISSN: 1402-1536, ISRN: LTU-TR—05/11—S.

[5] Naik Tarun R, Kraus Rudolph N, and Chun Yoon-moon 2006, "Effect of Different Types of Aggregates on Autogenous and Drying Shrinkage of Concrete", Monterrey Mexico, 2006

[6] ACI Committee Report, "Guide for Modeling and Calculating Shrinkage and Creep in Hardened Concrete", ACI 209.2R-08.

[7] Karim M. R. et.al. 2011, "Strength development of mortar and concrete containing fly ash: A review", *International Journal of the Physical Sciences,* Vol. 6(17), ISSN 1992-1950

[8] Wang Xuhao et.al. "Drying shrinkage behavior of mortars made with ternary blends", *Iowa State University, Civil, Construction and Environmental Engineering* 136 Town Engineering, Ames, IA 50011

[9] South Warren 2009, "A study of the compressive strength and drying shrinkage of cementitious binders prepared using natural pozzolans", University of Wollongong Research Online

[10] Emsiyazici S¸ and Hasan S¸Ahan Arel 2012, "Effects of fly ash fineness on the mechanical properties of concrete", *Indian Academy of Sciences*, Sadhana 37/3,389–403

[11] Kumar Sarath Chandra, Bendapudi 2011, "Contribution of Fly ash to the properties of Mortar and Concrete", *International Journal of Earth Sciences and Engineering* 04(06) SPL, 1017-1023, ISSN 0974-5904

[12] Islam Md. Moinul and Islam Md. Saiful 2010, "Strength Behaviour of Mortar Using Fly Ash as Partial Replacement of Cement", *Concrete Research Letters* Vol. 1 (3)

[13] Rabih S. Khairallah, B. Eng. 2009, "Analysis of Autogenous and Drying Shrinkage of Concrete", Mcmaster University digitalcommons@mcmaster, 9-1-2009

[14] Kiattikomol K et.al. 2001, "A study of ground coarse fly ashes with different fineness from various sources as puzzolanic materials", *Cement and Concrete Composites* 23-2001: 335–343

[15] Gu Erhan , Gesog Mehmet, Mermerdas, "Improving strength, drying shrinkage, and pore structure of concrete using metakaolin", *Materials and Structures*, DOI 10.1617/s11527-007-9296-z.

[16] Ishida Tetsuya et.al. 1997, "Mycro-physical approach to coupled autogenous and drying shrinkage of concrete", *Translation from Proceedings of JSCE* No.578/V-37

[17] IS 4031 (part 10) – 1988 Methods of physical tests for Hydraulic cement (Determination of Drying Shrinkage)

[18] IS 4031 (part 7) – 1988 Methods of physical tests for Hydraulic cement (Determination of Compressive Strength of Masonry Cement)

IOP Conf. Series: Earth and Environmental Science 1084 (2022) 012007 doi:10.1088/1755-1315/1084/1/012007

Eco-Anxiety and Environmental Concern as Predictors of Eco-Activism

Nishtha Jain[1] and Prashasti Jain[2]

[1]Student, Manipal University Jaipur, Jaipur, Rajasthan, India
[2]Assistant Professor, Manipal University Jaipur, Jaipur, Rajasthan, India
Email: nishtha.201016005@muj.manipal.edu; prashasti.jain@jaipur.manipal.edu

Abstract. The earth is under a great threat of destruction due to careless and exploitative anthropological actions. The current research explored the relationship between eco- anxiety (EAn), eco-activism (EAc) and environmental concern (EC) while determining the extent to which eco-activism gets determined by eco-anxiety and environmental concern. Three scales, namely, Hoggs Anxiety Scale, Pro Environmental Behaviour Scale and Taj Environmental Attitude scale were used. Pearsons correlation and regression were used for statistical analysis. A significant (0.01 level) mild positive correlation of 0.22 between eco-activism and eco-anxiety and of 0.33 between eco- activism and environmental concern was found. Regression analysis showed that eco- anxiety and environmental concern significantly predicted eco-activism. The relationship between the variables was further explained in the discussion section.

Keywords: Eco-activism, eco-anxiety, environmental concern

1. Introduction

The unprecedented rate of environmental degradation has left the earth under a great threat of destruction. With a 1.1 degree rise in global temperature compared to pre-industrial levels, catastrophic events are being experienced all over the world. Climate change is not just impacting the survival of the species, but also has dire consequences on the physical and mental health of individuals. Eco-psychology focuses on this relationship between individual's psychological state and nature. Worry, anxiety and ecological grief are commonly experienced by people in response to environmental degradation. APA defines EAn as "the chronic fear of environmental cataclysm that comes from observing the seemingly irrevocable impact of climate change and the associated concern for one's future and that of next generations". EAn is not a disorder in itself in the DSM yet but involves heightened concern about the emergency associated with the climate, which in turn can have psychological consequences on individuals in varying degrees. Symptoms of eco-anxiety involve sleep disturbances, anxiety, stress and nervousness. In severe cases, it is associated with feelings of guilt, PTSD, fatalism and depression. Experience of extreme or chronic levels of anxiety is reported by individuals due to feelings of helplessness about climate change. Experts predict a linear relationship between the rise of climate change and the increased prevalence of eco-anxiety. As per a 2018 study [1], in the USA, 70% of the individuals are worried and 51% of the population feels helpless about climate change. A 2020 study in England reported more than 57% of children and young adults feel distressed about climate change.

Eco activism involves taking direct actions to stop or minimise environmental damage. It can be of various types and includes activism at a local level, individual and political activism, conservational activism and environmental justice activism. The current research focused on environmental activism that is specifically concerned with the protection of the environment and sustainable practices. Eco-activism helps raise levels of environmental awareness and education among individuals about environmental degradation and its conservation. It helps in the formulation of laws and regulations in favour of environmental protection along with promoting pro-environment actions. Environmental concern refers to worry and a sense of anxiousness about the environment. It involves concern regarding the impact of human activity on the natural environment and the planet. Citizens' beliefs around the work done by their governments for environmental preservation were studied for 20 countries [2]. It was found that 26% public median felt that environmental change was affecting where they live, 58% public median felt that the government was doing too little to reduce the effect of climate change and 57% indicated climate change to be a very serious problem. In India, 37% of the population believes that the government is not taking sufficient measures to tackle climate change. Societies gain indirect utility from pollution, without which there would have been no incentive to pollute. Goods and services that cause pollution have some utility for society. Hence making it important to balance the cost of pollution with the utility of the product to achieve efficient outcomes. Innovation and stimulation of low-carbon technologies won't occur unless and until the price of activities associated with the emission of greenhouse gases is increased, i.e. carbon taxes are imposed.

Eco-activism demands a sense of cohesiveness to protect the environment. For instance, the Narmada Bachao Aandolan and Swaach Bharat Abhiyan have been a success given the community involvement and strong enforcement of environmental policies. Environmental activism has now become especially strong on a global level and is fueling pro-environmental behaviour, from large scale platforms like the UN to the very local level in the smallest of communities. The growing concern about environmental preservation has made behavioural and social research on environmental aspects essential. While today APA recognises climate change as a prevalent threat to mental health, there are very few psychologists and mental health professionals trained to deal with eco-anxiety and other adverse effects associated with climate change. Climate change will be the biggest challenge that humanity will have to address over the next decade. Developed nations continue to be the major contributors towards global pollution levels while developing nations are the ones that pay the price. It is crucial that with the COP26 and UN goals in mind, countries act to achieve zero carbon footprint and work towards the preservation of the planet at the earliest. Environmental awareness and eco-anxiety can act as a catalyst to promote eco-friendly behaviours. A threat to the global world is a threat to humans too, making feelings of anxiety and worry about environmental change justifiable.

2. Review of Literature

Climate change has been declared a health emergency by WHO. It threatens the mental well-being of individuals by increasing exposure to social and economic disturbances and gives rise to anxiety associated with existential threats. Eco-anxiety does not affect all individuals equally. It has a greater impact on individuals who are more aware of environmental change and concerned about its impact. It also has a greater impact on the younger generation and on vulnerable populations such as indigenous communities, tribal communities, farmers and fishermen whose livelihoods depend on the environment [3]. The older population and individuals with an existing diagnosis of depression or anxiety are also more affected by it. Disorders like PTSD, depression, anxiety, substance abuse etc. are frequently reported owing to the stronger response to extreme weather events [4]. Knowledge about climate change and the associated mental distress about climatic conditions getting worse manifests as eco-anxiety or solastalgia [3]. It makes individuals feel anxious about their future. Eco-anxiety is different from climate change anxiety and general anxiety. Often, terms like 'ecological stress' and 'climate change anxiety' are used in place of eco-anxiety given the overlap between terms like fear, anxiety, despair and worry. Uncertainty, unpredictability and uncontrollability are the key factors of eco-anxiety. Research also found eco-anxiety to be stemming from rumination, i.e., pondering over climate change and environmental degradation repeatedly [5]. Hoggs defines EAn as anxiety related to a multitude of environmental crisis. The Hoggs scale on EAn found that people are anxious about the negative impact their behaviour has on the planet. Individual's anxiety about various conditions of the environment like a change in climate, environmental degradation and pollution are interrelated.

There is a dearth of research about the variations of eco-anxiety. These variations need to be understood fully in order to address them and help individuals develop resilience for the same. Moreover, feeling of eco-anxiety needs to be explored with communities in diverse lands and especially with members of the indigenous communities. Individuals across continents have been found to experience these EAn, making it a universal phenomenon. In New Zealand, one in three people were found to be worried about climate change and 50% reported being very or extremely concerned about the effect of waste on the planet. 79% of individuals in Australia were found to be concerned about the destruction of native wildlife and 59% saw climate change as a serious threat to safety [6]. Climate change also has adverse impacts on the psychological, spiritual and physical health of children and young adults. It is seen to have similar negative impacts on indigenous people too. In terms of gender, women are more vulnerable to eco-anxiety and have greater experiences of worry and anxiety than men, majorly due to hormonal differences, evolutionary factors and gender-specific trauma. Awareness about the climate crisis often triggers anxiety which in turn gives rise to negative emotions like fear, shame, helplessness, guilt and grief that are commonly associated with EAn. These negative emotions have been found to affect physical health as well and are known to cause insomnia and appetite changes. The desire to avoid these negative emotions gives rise to climate change skepticism and behavioural change resistance, hindering sustainability efforts. While most forms of EAn are non-clinical, few cases of 'pathological' EAn are found as well. In some cases, eco-paralysis happens wherein people become so distressed by the issue that they are unable to act. However, eco-anxiety can also function as 'practical anxiety' and make people re-evaluate their actions and gather information. Various other emotions like fear, guilt, shame and grief that are associated with eco-anxiety can prove to be productive if used constructively. When individuals come together, a sense of empowerment ensues and the belief that individual actions can lead to bigger changes is supported. Group support and sharing of emotions help people come to terms with the problem, take on responsibility towards environmental conservation and deal effectively with difficult emotions. By validating eco-anxiety and creating acceptance for it through the provision of a safe space to discuss these emotions, activities and interventions can help carefully and consciously address the issue. Thus, eco-anxiety can act both as a resource and as a problem.

PBE i.e., 'psychological barriers explanation' is also used to explain human's inactions towards the climate change crisis. It sees human behaviour as being limited and creating psychological barriers in terms of climate actions. The research argued that attributing inaction to psychological reasons leads to human behaviour being misinterpreted as static and disconnected from the environment. While these barriers do exist, not all are linked to psychological factors [7]. A complex interaction between psychological tendencies, social structure and social relations explains the barrier more holistically. Transformative social change needs to be the focus of climate action with psychologists expanding their scope beyond individual consumer behaviour to explain climate (in)action. While there is a strong consensus around humanities responsibility towards conservation, individuals continue to fail to act in line with their environmental attitudes. The gap between attitudes and behaviours can be explained by personal costs, environmental benefits and self- control. When personal costs are low and environmental benefits are high, individual's environmental attitudes more highly predict PEB [8]. Self-control helps individuals act in accordance with their attitudes, thus helping people stick to their long-term environmental goals [8]. Personal experiences and subjective attributions made by individuals regarding climate change impacts their intentions to adopt pro-environmental behaviours.

Pro-environmental behaviour (PEB) typically refers to behaviours that focus on conserving the environment and mitigating acts that damage it. Direct experience with consequences of climate change like wildfires along with personal experiences, environmental concerns and subjective knowledge about climate change impacts an individual's subjective attribution about climate change extremes. These factors cause individuals to believe that climate change is the reason behind these events. These subjective attributions in turn give rise to pro-environmental attitudes and behavioural intentions. Similarly, another research [9] found that individuals believe that human intrusion into the natural environment caused the pandemic, which in turn fostered environmental awareness. It is important to determine what predicts pro-environmental behaviour in order to design interventions that can promote it. Besides attributions and personal experiences, moral emotions also affect pro-environmental behaviour. Guilty conscious i.e., guilt and shame, along with emotions like anger, pride and sadness motivate eco-friendly behaviours and behavioural intentions as amending actions. PEBs are partly determined by economic preferences, explaining the individual differences shown in PEBs. Among seven preference measures (patience, present bias, positive reciprocity, altruism, negative reciprocity, risk-taking and trust) that were studied by researchers, only altruism was found to be significantly associated with PEBs [10]. Other factors that determine PEB include external and internal or personal variables. External variables include societal norms, accessibility of products and costs. Psychological and demographic characters make up the personal domain. A positive and significant relationship exists between 'subjective norms', 'knowledge about the environmental crisis' and 'perceived behavioural control' with individuals' intent to perform pro-environmental behaviour. A sense of personal obligation towards the environment acts as a predictor of individuals intention to adopt PEBs. Individuals also show an increased tendency to engage in environmental behaviour when there is increased harm to the environment.

Environmental crisis impacts the adoption of PEBs by threatening both the well-being and existence of individuals and communities. The covid 19 crisis increased individuals' knowledge about the environment and encouraged PEBs by enhancing their sense of personal responsibility towards the environment and fostering positive attitudes towards environmental conservation. The theory of planned behaviour helped establish links between beliefs about the pandemic, environmental awareness and negative emotions associated with the adoption of pro-environmental behaviours. The Biophilia hypothesis talks about the innate connection that humans have with nature and that they derive psychological wellbeing from this connect. Exposure to nature leads to a sense of positive connect with nature and increases individuals' sense of empathy, belongingness and connectedness with the natural environment (eco-psychology), which in turn enhances the willingness to engage in pro-environmental behaviours (social identification theory). Exposure to nature leads to significant improvement in an individual's mood while helping reduce stress and depression, causing individuals to treat nature with greater kindness. Adverse environmental impacts due to climate change are disturbing this connection and resulting in feelings of loss, i.e., 'ecological grief'. Psychologists play a crucial role in nullifying the mental health impact of climate change by providing support, helping develop active coping skills and by promoting hope and social connectedness as strategies to adapt to climate change. Eco-therapists are now coming up to treat eco-anxiety. Psychotherapeutic practice needs to include ecological concerns and eco-anxiety, along with depth psychology to address the neglected aspects of nature and humanity associated with climate change. No single solution exists for the climate crisis. However, collective action at a community level along with individual efforts can help address the issue at a national and global level.

3. Research Study

3.1 Research Objective:

To study the relationship between eco-anxiety, eco-activism and environmental concern.

3.2 Hypothesis:

The first hypothesis, H1 stated that there is a significant relationship between eco-anxiety and eco-activism. The second hypothesis, H2 stated that there is a significant relationship between eco-anxiety and environmental concern. The third hypothesis, H3 stated that there is a significant relationship between environmental concern and eco activism. The last hypothesis, H4 stated that eco-activism is predicted by eco-anxiety and environmental concern.

4. Methodology

The sample comprised of 157 participants who were chosen randomly ranging from 18-60 years of age for the purpose of data collection. Convenient sampling technique was adopted for the purpose of this study. The study followed an empirical, correlational design. Informed consent was obtained from all participants before administering the three questionnaires. There was no time limit for filling the questionnaire. As per ethical norms, participants were debriefed about the study post completion of the questionnaire.

4.1 Tools.

1. Hogg's eco-anxiety scale (2021): HEAS-13 is a 13-item scale with high reliability and validity along with a unique four-dimensional structure. The items are responded to on a Likert scale ranging from 0 to 3, with 0 corresponding to not at all and 3 corresponding to nearly every day.

   The scale has the highest possible score of 39. It has an excellent internal reliability ($> .82$ for all subscales) and a reliability of Cronbach's alpha: 0.92, $p < 0.001$.
2. Hasten Taj Environmental Attitude scale (section VI on environmental concern) (2001): The TEAS was developed by Hasten Taj in Dr. Haseen Taj in 2001. It comprises of 61 items, divided into 6 dimensions. For the purpose of this study the sixth dimension on environmental concern was used which comprised of 14 items. Highest possible score on the scale was 70. Items on the scale are scored using a Likert scale ranging from 4(strongly agree) to 1(strongly disagree). In case of reverse items, the scoring is reversed as 1(strongly agree) to 4(strongly disagree). Total score is obtained by summating the scores of all items. The scale has a split half reliability of 0.82.
3. Pro-Environmental Behaviour Scale (2013): The PEBS is a 4-dimensional scale comprising of 19 items. The coefficient alpha for the full scale is 0.76 and the coefficient alpha for the sub scales ranges from 0.62 to 0.74. Construct validity for the scale was established using bivariate Pearson correlation. A strong reliability for the test was established through test-retest correlations, with $r = 0.85$, $p<0.01$ for the full scale.

   Factors:

   Conservation: Comprises of 7 items that addresses how often individuals reduce their consumption of heating, hot water, air conditioning and lighting.

   Environmental citizenship: Comprises of 6 items on membership in environmental, conservation and wildlife organisations, along with the frequency of talking to others about their environmental behaviours.

   Food: Comprises of 3 items which address reduction in consumption of beef, pork and poultry in the last one year.

   Transportation: Comprises of 3 items which involves carpooling, use of public transport and walking or cycling in place of driving in the last one year.

5. Results and Discussion

5.1 Results

A total of 165 responses were collected out of which 157 were considered for the final data analysis comprising of 105 females and 52 males. Descriptive statistics was used for data analysis. For the eco-activism scale participants were found to obtain a minimum score of 43 and a maximum score of 87, with a mean score of 64.17 and standard deviation of 9.06. The highest possible score on the scale was 95. For the eco-anxiety scale participants were found to obtain a minimum score of 0 and a maximum score of 39, with a mean score of 15.43 and standard deviation of 8.67. The highest possible score on the scale was 39. For the environmental concern scale participants were found to obtain a minimum score of 38 and a maximum score of 55, with a mean score of 46.89 and standard deviation of 4.00. The highest possible score on the scale was 70.

Table 1. Descriptive statistics

	N	Minimum	Maximum	Mean	Std. Deviation
Eco_Activism	157	43.00	87.00	64.1720	9.06188
Eco_Anxiety	157	0.00	39.00	15.4331	8.67525
Env_Concern	157	38.00	55.00	46.8917	4.00253
Valid N (listwise)	157				

Table 2. Pearson correlation

		Eco_Activism	Eco_Anxiety	Env_Concern
Eco_Activism	Pearson Correlation	1	.222**	.333**
	Sig. (2-tailed)		0.005	0.000
	N	157	157	157
Eco_Anxiety	Pearson Correlation	.222**	1	0.015
	Sig. (2-tailed)	0.005		0.848
	N	157	157	157
Env_Concern	Pearson Correlation	.333**	0.015	1
	Sig. (2-tailed)	0.000	0.848	
	N	157	157	157

**. Correlation is significant at the 0.01 level (2-tailed).

Pearsons correlation was used as the data was found to be normally distributed. Eco-activism was found to be significantly related to eco-anxiety at a significance level of 0.01 with a mild positive correlation of 0.22. Therefore, H1 which states that there is a significant relationship between eco-activism and eco-anxiety was accepted. No significant correlation was found between eco-anxiety and environmental concern, therefore H2 which states that there is a significant relationship between eco-anxiety and environmental concern was rejected. Eco-activism was found to be significantly related to environmental concern at a significance level of 0.01 with a mild positive correlation of 0.33. Therefore, H3 which states that there is a significant relationship between eco-activism and environmental concern was accepted.

Table 3. Regression analysis

Model	R	R Square	Adjusted R Square	Std. Error of the Estimate	F	Sig
1	.398[b]	0.158	0.147	8.36886	14.453	.000[b]

a. Dependent Variable: Eco_Activism
b. Predictors: (Constant), Eco_Anxiety, Env_Concern

Regression analysis and ANOVA show that eco-activism is significantly predicted by eco-anxiety and environmental concern. Therefore, H4 which states that eco-activism is predicted by eco-anxiety and environmental concern was accepted.

5.2 Discussion

Human beings have the power to deplete the environment; however, they also hold the power to save it. Their everyday choices determine the rate of environmental degradation. With rapid urbanisation and globalisation, there has been a massive shift towards a fast-paced life causing negligence towards the environment. This has been witnessed through the immense loss of biodiversity marked by a 70% decline in the population of flora and fauna since the 1970s, recurrent forest fires and other natural disasters along with massive deforestation. 4.4 giga-tons of greenhouse gases are emitted from food wastage alone and air pollution continues to kill 4-7 million people each year worldwide with 9 out of 10 people breathing air filled with high levels of pollutants. These occurrences have led to a rise in the levels of environmental concern and eco-anxiety among individuals.

Given the paucity of research in this domain, the current research helped explore the links between eco-anxiety, eco-activism and environmental concern. It helped determine the extent to which eco-activism is determined by eco-anxiety and environmental concern (14%). Statistical analysis indicated a significant relationship between eco-activism and eco-anxiety and between eco- activism and environmental concern at a 0.01 level. Thus, H1 and H3 were accepted, while H2 was rejected. A mild positive correlation of 0.22 was found between eco-activism and eco-anxiety. The low level of correlation can be explained by the negative emotions that arise due to anxiety and cause behavioural resistance. Low levels of eco-anxiety were found among the population as a significant number of people deny the existence of climate change to avoid the associated negative feelings and worry. Some individuals adopt an egoistic stance and believe that environmental degradation will not affect them. Moreover, the type of coping style adopted by people determine the extent to which they engage in PEBs. Individuals who adopt emotion focused coping style accept the existence of the problem but distance themselves from it. Individuals who adopt meaning focused coping style take concrete steps to deal with the issue.

A mild positive correlation of 0.33 was found between eco-activism and environmental concern. The low level of correlation can be attributed to the lack of awareness about the impact of individual efforts on conservation and to the lack of community support. Individuals tend to feel that the society doesn't care about environmental issues and thus they do not voice their concerns or take measures. Financial restraints, lack of availability of eco-friendly products, ease of access and lack of proper implementation of policies around environmental conservation also explain the gap between environmental concern and action.

Making environmental actions more enjoyable, satisfying and interesting along with a positive and holistic approach towards environmentalism can help promote the switch towards eco-activism. Emotional coaching techniques can be used to promote resilience, create nurturing relationships and promote the development of emotional and behavioural self-regulation. Eco-therapy with its focus on the possible health benefits of interacting with the environment is an effective therapeutic approach to achieve individual health (at the micro-level) and public and environmental health outcomes (at the macro-level). Greater acceptance for environmental psychology and rigorous implementation of laws and regulations around conservation will help achieve the goals set at COP 26 in Glasgow. As the society aims to move towards net zero carbon emission in 2050, this research will aim to pre-empt environmental degradation and its adverse impact on the mental health of mankind.

6. Conclusion

The present study contributes to an understanding of three concepts, namely, eco-anxiety, eco-activism and environmental concern. The study explained the correlation between the three concepts and explored how eco-anxiety and environmental concern together help predict levels of eco-activism shown by individuals. Various studies on eco-anxiety and its impacts on physical and emotional wellbeing were discussed along with ways of addressing it. The study also explained the various factors that cause and hinder eco-activism in addition with ways to promote it. Finally, the study found support for three out the four hypothesis that were stated.

References

[1] Leiserowitz, A. (2006). Climate change risk perception and policy preferences: The role of affect, imagery, and values. *Climatic Change*, *77*(1-2), 45–72. https:// doi.org/10.1007/s10584-006-9059-9

[2] Johnson, C., & Tyson, A. (2020). People globally offer mixed views of the impact of artificial intelligence, job automation on society (Tech. Rep.). Pew Research Center. Retrieved from https://perma.cc/6NYT-DH4M

[3] Hickman, C. (2020). We need to (find a way to) talk about… Eco-anxiety. *Journal of Social Work Practice*, *34*(4), 411-424.

[4] Zhong, S., Yang, L., Toloo, S., Wang, Z., Tong, S., Sun, X., ... & Huang, C. (2018). The long-term physical and psychological health impacts of flooding: a systematic mapping. *Science of the total environment*, *626*, 165-194.

[5] Rusting, C.L., Nolen-Hoeksema, S., (1998). Regulating responses to anger: effects of rumination and distraction on angry mood. J. Personal. Soc. Psychol. 74, 790, https:// doi.org/ 10.1037/0022- 3514.74.3.790

[6] Hogg, T. L., Stanley, S. K., O'Brien, L. V., Wilson, M. S., & Watsford, C. R. (2021). The Hogg eco-anxiety scale: development and validation of a multidimensional scale. *Global Environmental Change*, *71*, 102391.

[7] Schmitt, M. T., Neufeld, S. D., Mackay, C. M., & Dys-Steenbergen, O. (2020). The perils of explaining climate inaction in terms of psychological barriers. *Journal of Social Issues*, *76*(1), 123–135. https://doi.org/10.1111/josi.12360

[8] Wyss, A. M., Knoch, D., & Berger, S. (2022). When and how pro-environmental attitudes turn into behavior: The role of costs, benefits, and self-control. *Journal of environmental psychology*, *79*, 101748.

[9] Daryanto, A., Song, Z., & Soopramanien, D. (2022). The COVID-19 pandemic as an impetus for pro-environmental behaviours: The role of causal attribution. *Personality and Individual Differences*, *187*, 111415.

[10] Lades, L. K., Laffan, K., & Weber, T. O. (2021). Do economic preferences predict pro-environmental behaviour? *Ecological Economics*, *183*, 106977. https://doi.org/ 10.1016/j.ecolecon.2021.106977

Clean drinking water solution for rural India: Portable sip-up

Meraj Ahmad[1], Arti Sharma[2], Quadri S. Mujtaba[1], Saurabh Yadav[1], Ritwik Kulkarni[1], Raj Kumar Satankar[4], Sandeep Gupta[1], Amrita K Nighojkar[3], Meenu Chhabra[2], Ambesh Dixit[5] and Anand Plappally[1]

[1]*Department of Mechanical Engineering, Indian Institute of Technology, Jodhpur, India*
[2]*Department of Bioscience and Bioengineering, Indian Institute of Technology, Jodhpur, India*
[3]*Department of Metallurgical and Materials Engineering, Defence Institute of Advanced Technology, Pune, India*
[4]*Department of Mechanical Engineering, Poornima College of Engineering, Jaipur, India*
[5]*Department of Physics, Indian Institute of Technology, Jodhpur, India*
Corresponding Author Email: meraj.1@iitj.ac.in

Abstract
Ceramic water filters (CWFs) have been used in India due to their low expense, ease of manufacture, and livelihood sustenance features. However, the use and throw character of such systems has not been achieved. The purpose of this study is to develop and investigate the portable CWF variant named Sip-up, which can be attached to any commercially available bottle-based containers. These are porous functional structures made of a distinct volume fraction combination of salty clay, sawdust, and marble powder wet mix, formed into cylindrical shapes and baked at 650°C. The influent and filtrate water samples were analyzed. After filtration, the pH, on average, significantly improves from 8.2 to 7.4. The TDS level, conductivity, and salt content have been reduced by 8.43%, 6.89%, and 5.83%, respectively, which conforms to the WHO limits. A 99% reduction of E. Coli bacteria through CWF was also observed. This portable water filter will be accessible to the people at the lowest economic pyramid.

1. Introduction

Access to clean potable water is a human right [1]. The rising water demand places tremendous stress on the available water resources, especially in rural areas of developing nations. Further, the quality of household water is also questionable in economically backward locations of these nations. Currently, there are scores of water filters available in India's market for household use [2]. However, most of these filtering media are prohibitively costly, energy-intensive, and inaccessible to the economically backward [3]. Sobsey et al. 2008 had identified ceramic water filters (CWF) as one of the sustainable solutions in household water treatment [4]. In the sustainability of point-of-use solutions, operational efficiency, as well as energy consumption, plays a significant role. In this regard, Jamil et al. 2019 comprehensively discussed various portable devices that yield drinkable water using solar energy [5]. Further, the CWF introduced in several countries by NGOs such as Potters for Peace does not use any energy during their operations [6]. Water filtration occurs under the effect of gravity.

The CWF is one of the few kinds of point-of-use (POU) water purification systems being employed to address the issue of water contamination [7]. CWFs are made by combining clay with a combustible or porous substance (sawdust, wheat, or rice husk, and many others) and water [3]. The resultant homogeneous mixture is subsequently formed into the required shape using a mechanical pressing device or manually. The combustible substance burns away during sintering, leaving behind pores [8]. Micro- and nanoscale pores combine to produce a complex porous network [9].

Additionally, the geometry and surface features of the pores play a critical role in the CWFs ability to successfully remove microbiological pollutants to levels below the WHO limit [7]. In a similar way, CWF with similar features was introduced in India in the name of G-Filters as a rural humanitarian initiative [10]. The household manufacturing of G-Filters for providing clean, affordable drinking water is practiced in the states of Rajasthan, Bihar, Kerala, and Tamil Nadu in India. G-Filters are presently available with a volume of 9or 10 liters receptacles and hence, not easy to carry around and are primarily restricted to domestic uses. During COVID 19 times, migration of people became a serious event, and people were required to drink water on the go [11]. During these times, an economically viable low-cost water filter of use and throw thought is required. Although markets contain portable water filters such as Lifestraw, Tata Swach Insta-sip, they are expensive and cannot be afforded by rural Indian communities of the lower economic pyramid [2].
In this paper, a local prototype of a portable G Filter variant named 'Sip-Up' has been developed in Jodhpur, India [12]. Experimental tests were performed to evaluate the performance of Sip Up in improving water quality by measuring pH, Total dissolved salts (TDS), conductivity, salt content, turbidity, and microbial removal. The implications of the statistically significant results are then discussed for a more robust design.

2. Materials and methodology

2.1 Raw materials

The raw materials used to manufacture the prototype of Sip-Up were salty clay, sawdust, and marble processing waste slurry powder, as shown in Figure1(a). Salty clay from Raital, Jalore district, Rajasthan, India, was used during the preparation of the Sip-Up. Sawdust was procured from a timber mart named Rajasthan Timbers in Jodhpur, Rajasthan, India. Using a unit squared millimeter sieve (ASTM 18 mesh size), the salty clay and sawdust were sieved [13]. The marble slurry was provided by a local vendor from Jodhpur. It was air-dried, grounded, and passed through sieves of ASTM 18 size.

2.2 Manufacturing processes

A homogeneous mix of raw materials in equal volumes of 45ml each was used to make the prototype. Additionally, the marble slurry of 10 ml is added along with water (70% of the total volume of clay and sawdust mix). The ingredients were kneaded together using hands and formed into a uniform wet paste Figure 1 (b). The mould was created manually using a cylindrical shaft of diameter 1cm, as shown in Figure 1 (c). This diameter is good enough for a 250 ml water bottle geometry. The diameter of the cylindrical shaft can be varied according to the type of required variant depending upon commercially available water bottles.

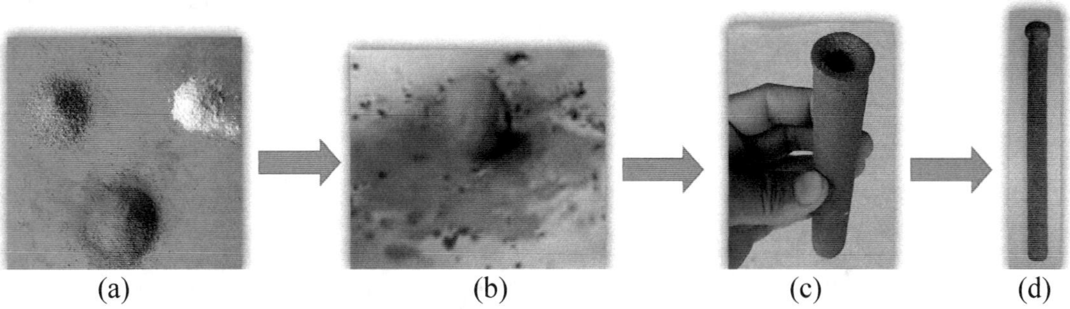

Figure 1. Steps involved during the manufacturing of Sip-Up device: (a) Sample of salty clay, sawdust, and marble powder (b) Uniform kneaded mixture (c) Green mould (d) Final Sip-Up prototype

The green composite was kept for drying at ambient temperature conditions for two days, and the dried sample was then sintered at 650°C in a vertical open-hearth furnace located at Banad village, Rajasthan

[8]. At this temperature, the sintering culminates for the Sip-Up. The weight of the final portable Sip-Up was found to be 68 gm with a radial thickness of 5 mm, as shown in Figure 1(d). The cost incurred for producing such a single prototype is around Rs.3 to the potter. Porosity is one of the significant properties of such CWFs, which plays a crucial role in the rate of water discharge, which can be easily adjusted with the burnout material fraction used in making the filter [9]. The porosity of CWF came out to be 43% which was calculated using a simple method of measuring dry and wet weights [14].

2.3 Working Principle

The Sip-Up device can be used by replacing the cap on commercial water bottles and can be used whenever needed, as shown in Figure 2(c). Filtered water is obtained when a person sucks through the device. The influent water is forced to flow through the porous Sip-up media into an empty space where the filtrate is collected due to the pressure difference created by sucking, as illustrated in Figure 2(b). As the water percolates through the CWF material, it navigates through the pores within the core, which are tortuous in nature [15]. This tortuosity helps in the separation of microscopic non-soluble pollutants [16]. Filtration can occur through physiochemical processes. The used marble slurry is primarily composed of $CaCO_3$, which acts as an activation zone [17]. These zones adsorb the contaminants present in the influent [13]. Also, larger impurities that do not get adsorbed get blocked in the pores due to their size.

Studies have shown that the diaphragm of human beings can create pressures in the range of 11-120 mm of mercury [18]. Thus, a trickle of water equivalent to the pressure applied could be obtained from the device. It is advised to use the device for a single time and throw it away after use.

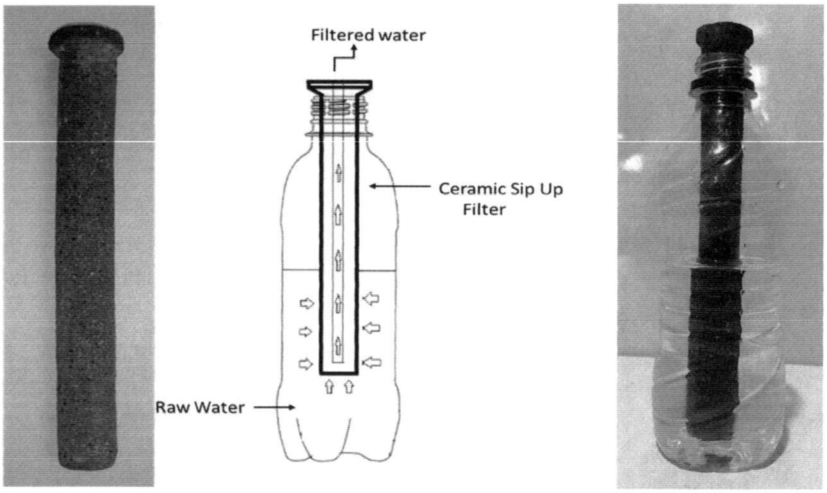

(a) (b) (c)

Figure 2. Depicting working principle of the prototype (a) Sip Up (b) Schematic diagram (c) Actual working model

2.4 Water testing

The raw water used for testing was taken from a household in Banad village, Rajasthan, India, as shown in Figure 3 (a). The quality of influent water can be represented using different electro-kinetic parameters. A Portable water testing device, "Oakton PCS Testr 35" (Eutech PCSTEST35-01X441506/Oakton 35425-10), was used for measuring pH, Electrical conductivity (EC), Total dissolved solids (TDS), and salt content, as shown in Figure 3(b). The turbidity of the samples was measured using a water and soil analysis kit (MT-131™, Mantilab solutions, India). Both the testing equipment was properly calibrated before use as per the instructions provided in the manual. The influent water has the pH, EC, TDS, turbidity as 8.2, 450 µs/cm, 340 ppm, and 50 NTU, respectively. The quality of effluent water was also investigated and compared with WHO guidelines[19]. The difference in the quality of water after filtration was primarily observed through visual inspection, as shown in Figure 3(c).

2.5 Microbial removal study
The bacterial removal capacity of the prepared filters was examined using non-pathogenic *E. coli* Top 10 cells. The *E. coli* cells were grown in Luria Bertani (LB) broth at 37°C for 24 h by stirring at 150 rpm in an incubator shaker (Rivotek, India). The bacterial samples were passed through the filters and filtrates were collected to examine the bacterial growth. Five replicates of the sample along with the control, were investigated for microbial removal study. The *E. coli* in the samples were counted by plate count method as described in the standard methods for the Examination of Water and Wastewater [20].

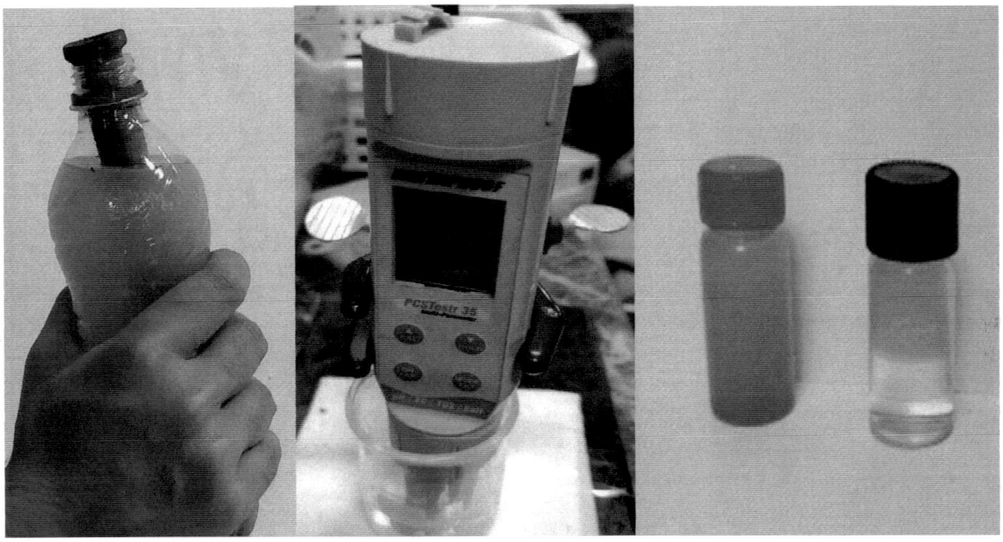

(a) (b) (c)

Figure 3. Water testing (a) Raw water used for testing (b) Oakton PCS Testr35 being calibrated (c) Left and right samples depict influent and effluent water, respectively.

3. Results and discussion
The filtered samples were investigated for the change in quality parameters. A net reduction of 14 % was observed in pH values. The conductivity of filtered water was found to be 419.8µs/cm on average. A drastic decrease of 87.9% was observed in turbidity value. The TDS value for effluent was reduced to an average of 311.3 ppm after passing through the filter. A net change of 8.4% was observed, allowing the filtered water to be very close to WHO limits of 300 ppm, as shown in Table 1. An index of each parameter was also calculated using equation1. The variation of differences between different parameters is shown in Figure 4-9.

$$(X)_{index} = \frac{Influent - Effluent}{Influent} \quad (1)$$

Where X is any electro- kinetic parameter

Table 1. Comparison of electro-kinetic parameters with WHO limits.

Parameter	pH	TDS (ppm)	Turbidity (NTU)	Conductivity (µs/cm)
Influent	8.2	340	50	450
Average Effluent	7.05	311.3	6.02	419.8
WHO limit	6.5-8.5	<300	<5	<400

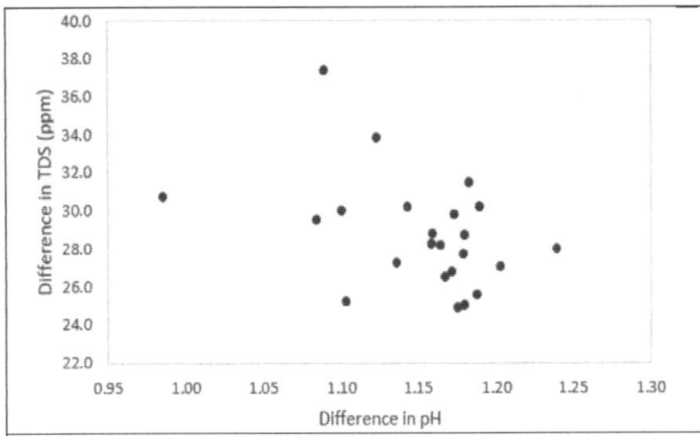

Figure 4. Variation of differences in pH vs TDS

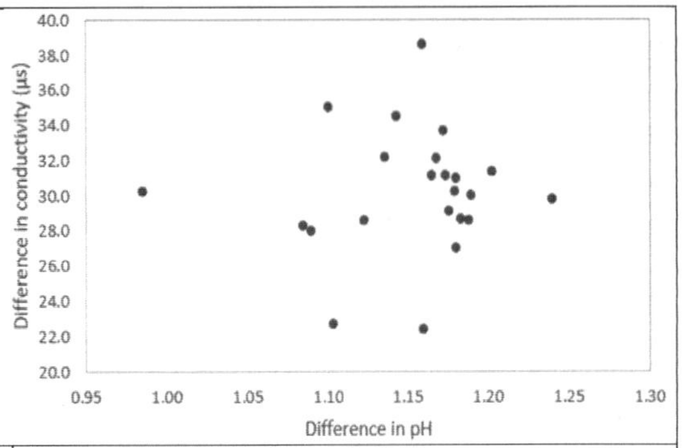

Figure 5. Variation of differences in pH vs Conductivity

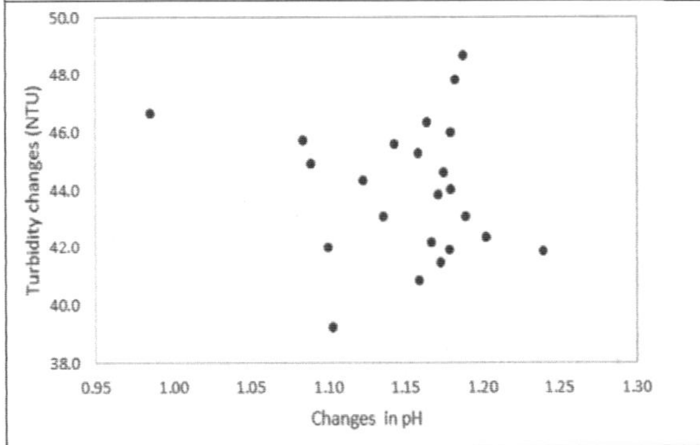

Figure 6. Variation of differences in pH vs Turbidity

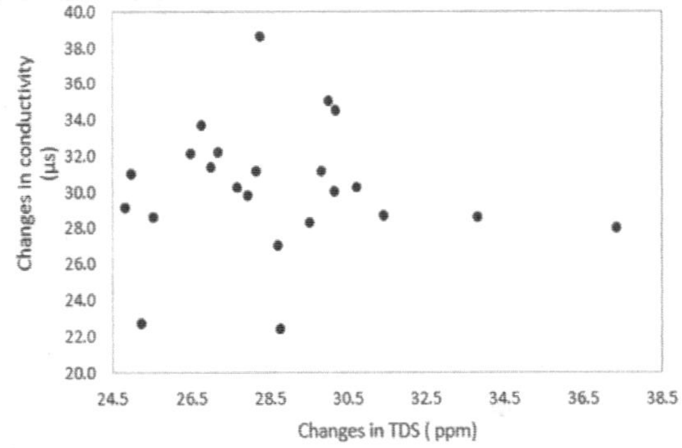

Figure 7. Variation of differences in TDS vs Conductivity

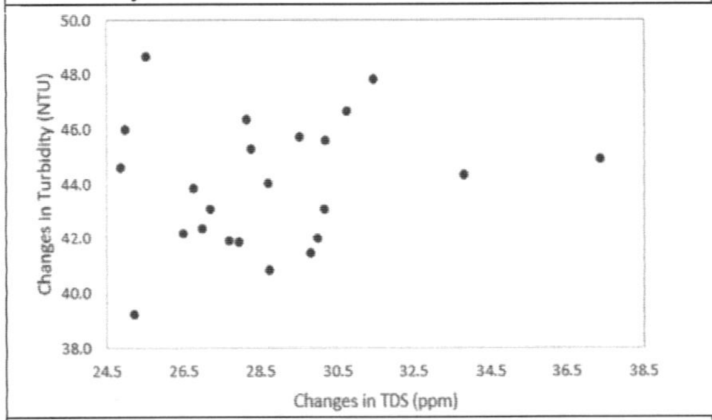

Figure 8. Variation of differences in TDS vs. Turbidity

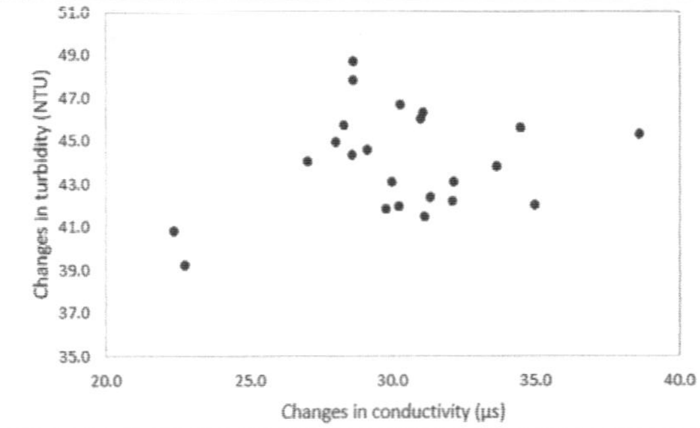

Figure 9. Variation of differences in conductivity vs Turbidity

The influent and filtrate are both dilute solutions in this experiment [21]. The porous media flow treatment can be effectively predicted on the basis of the gradient in electro-kinetic parameters of these dilute solutions as they traverse across the porous clay ceramic cylinders [22]. From Plappally et al. 2010, the characteristic gradient of an electro-kinetic parameter will help predict the treatment provided by the clay ceramic to the water [23][24]. Therefore the gradient in turbidity between the influent and filtrate water from the sip-up can be predicted with an expression in Eq.2 [21,22].

$$\frac{\nabla X_1}{\nabla Y_1} = Z_i = A + B\nabla X_1 + C\nabla X_2 + D\nabla X_3 \tag{2}$$

Where ∇Y_1 is the gradient of turbidity, ∇X_1 is the conductivity gradient, ∇X_2 is the gradient in pH and ∇X_3 is the gradient of total dissolved solids between the influent water to the sip-up and its filtrate. The formulation of the strategy to arrive at Eq. 2 is elaborated by Gupta et al. 2015.

Table 2. The coefficients A, B, C and D for the model and distinct gradient parameters, R^2 and error of the model in Eq.2.

	A	B	C	D	Coefficient of Determination, R^2	Error in the Model, S
Z_i	0.0011	1.032	0.074	-0.052	83.64	0.0039

3.1 Microbial Removal

The *E. coli* removal percentage is shown in the Table 3 and Figure 10 for samples consisting of clay and sawdust and mean value is reported here (n=5). The addition of sawdust in the clay has shown to improve bacterial removal efficiencies. Therefore, it is predicted from the results that the composite material has great surface adsorption activity and biocidal activity which resulted in 99% reduction of bacteria from water [25].

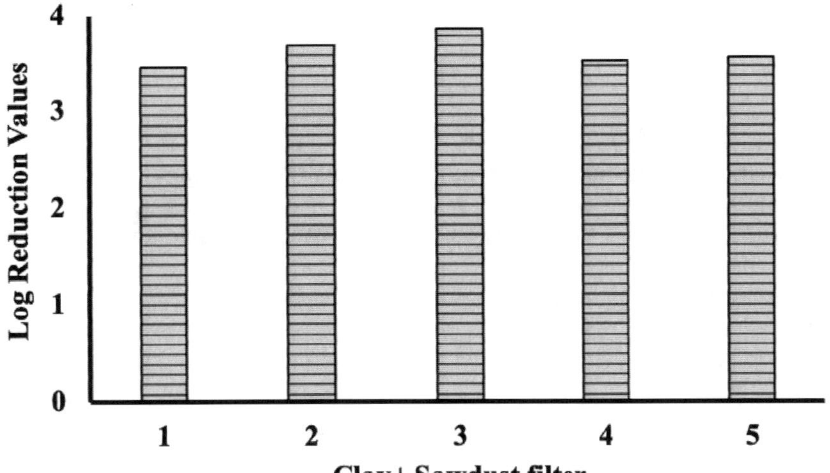

Figure 10. Microbial filtration results for Clay-Sawdust based water filters (n=5).

Table 3. *E. coli* removal data for CWF samples.

Clay-Sawdust composition	Influent Concentration (I)	Effluent Concentration (F)	I/F	Log reduction value (LRV) $Log_{10}(I/F)$	F/I	Filtration Efficiency (1-F/I)
S1 (1)	4.4×10^{10}	1.5×10^{8}	2.933	3.46	0.0034	0.996
S1 (2)	4.4×10^{10}	0.9×10^{8}	4.888	3.68	0.0020	0.997
S1 (3)	4.4×10^{10}	0.6×10^{8}	7.333	3.86	0.0013	0.998
S1 (4)	4.4×10^{10}	1.3×10^{8}	3.384	3.52	0.0029	0.997
S1 (5)	4.4×10^{10}	1.2×10^{8}	3.666	3.56	0.0027	0.997

4. Conclusion

Sip-Up was conceptualized with the goal of creating a low-cost, easy-to-carry ceramic filter to offer a solution towards potable water filtration at point-of-use. However, this is possible only when enough suction pressure is applied by the user within the limit of the human diaphragm pressure or nasal sniff. This device solves the issue of portability by lowering down size and weight with a simple design. The water quality obtained from the filter conformed to the WHO standards. In this work, a new multi-parametric model of electro-kinetic parameters on the basis of Onsager's reciprocal relations is promulgated to predict variation in turbidity through the Sip-up. The model showed a decent coefficient of determination and a very negligible error.

Moreover, the current model demands improvement in its design. One of them would be to design the Sip-Up to be able to fit the standard dimensions of bottles commonly available in the market. In addition, there is a need to make a significant improvement in ergonomics so that the users can satisfactorily meet their requirements. A comprehensive testing on the suitability of the device, its field reliability, and environmental impact assessment of its manufacturing is required. The clay ceramics are fragile and, therefore, more likely to fail when impacted. A strength mechanics investigation is also sought to be done for fail-proof sip ups design. The implication of such a study paves the way for low cost, home-made water filtration devices, functional property introduction and usage in the field of pottery towards water treatment applications.

References

[1] Roaf V, Albuquerque C de and Heller L 2019 The Human Rights to Water and Sanitation *Equal. Water Sanit. Serv.* **1249** 26–43

[2] Plappally A K, Hasija A, Kusins J, Jhaver M, Chee A, Panditrao A, Singh N K, Kumar N, Singh K and Kumar N 2013 Water Use and Related Costs at Households in Western and Northern Parts of India *Hydrol Curr. Res* **4** 2

[3] Plappally A K and Lienhard V J H 2013 Costs for water supply, treatment, end-use and reclamation *Desalin. Water Treat.* **51** 200–32

[4] Sobsey M D, Stauber C E, Casanova L M, Brown J M and Elliott M A 2008 Point of use household drinking water filtration: a practical, effective solution for providing sustained access to safe drinking water in the developing world *Environ. Sci. Technol.* **42** 4261–7

[5] Jamil F and Ali H M 2019 Sustainable desalination using portable devices: A concise review *Sol. Energy* **194** 815–39

[6] Leftwich M C, Yakub I, Plapally A, Soboyejo A B O and Soboyejo W O 2009 Understanding the Filtr´n Ceramic Water Filter *International Ceramic Pot Filter Workshop* vol 28 pp 121–8

[7] Yakub I, Du J and Soboyejo W O 2012 Mechanical properties, modeling and design of porous clay ceramics *Mater. Sci. Eng. A* **558** 21–9

[8] Gupta S, Satankar R K, Kaurwar A, Aravind U, Sharif M and Plappally A 2018 Household Production of Ceramic Water Filters in Western Rajasthan, India *Int. J. Serv. Learn. Eng. Humanit. Eng. Soc. Entrep.* **13** 53–66

[9] Plappally A K, Yakub I, Brown L C, Soboyejo W O and Soboyejo A B O 2009 Theoretical and experimental investigation of water flow through porous ceramic clay compositewater filter *Fluid Dyn. Mater. Process.* **5** 373–98

[10] Venugopal V 2019 The G-filter: A simple high-tech solution to India's water pollution *MRS Bull.* **44** 914–5

[11] Bhagat R B, Harihar R R S, Archana S, Roy K and Govil D 2020 The COVID-19, Migration and Livelihood in India: A Background Paper for Policy Makers 1–40

[12] Mujtaba Q S, Yadav S and Kulkarni R 2019 *Long Term Study of Flow and Electrokinetic Parameter of Thin Ceramic Water filter* (Jodhpur)

[13] Nighojkar A K, Vijay A, Kumavat A, Gupta S, Satankar R K and Plappally A 2019 Use of marble and iron waste additives for enhancing arsenic and E. Coli contaminant removal capacity and strength of porous clay ceramic materials for point of use drinking water treatment *Desalin. Water Treat.* **157** 290–302

[14] Dotson B J, Slobod R L, McCreery P N and Spurlock J W 1951 Porosity-Measurement Comparisons by Five Laboratories *J. Pet. Technol.* **3** 341–6

[15] Simonis J J and Basson A K 2011 Evaluation of a low-cost ceramic micro-porous filter for elimination of common disease microorganisms *Phys. Chem. Earth* **36** 1129–34

[16] Gupta S, Kaurwar A, Satankar R K, Usha K, Sharif M A . and Plappally A K 2016 Flow, Microbial Filtration And Petrophysical Properties Of Ceramic Plate Ware Gravity Water Filter During Cyclic Water Loading Events *From Pollution to Purification* (Kottayam, Kerala.: IUIC, ASCEED & School of Environmental Sciences, Mahatma Gandhi University)

[17] Mehta D, Mondal P and George S 2016 Utilization of marble waste powder as a novel adsorbent for removal of fluoride ions from aqueous solution *J. Environ. Chem. Eng.* **4** 932–42

[18] Gibson G J, Whitelaw W, Siafakas N, Supinski G S, Fitting J W, Bellemare F, Loring S H, Troyer A De and Grassino A E 2002 ATS/ERS Statement on respiratory muscle testing. *Am. J. Respir. Crit. Care Med.* **166** 518–624
[19] Edition F 2011 Guidelines for drinking-water quality *WHO Chron.* **38** 104–8
[20] Nineteenth T and Editions E 2000 Standard methods for the examination of water and wastewater: 20th ed *Choice Rev. Online* **37** 37-2792-37–2792
[21] Plappally A, Soboyejo A, Fausey N, Soboyejo W and Brown L 2010 Stochastic modeling of filtrate alkalinity in water filtration devices: Transport through micro/nano porous clay based ceramic materials *J. Nat. Environ. Sci.* **1** 96–105
[22] Gupta S, Singh R, Chakraborty P R, Sharma R K, Soboyejo A B O, Wei X and Plappally A 2015 Multi-variable approach to determine treatment efficiency of wetland: size effect and electro-kinetic effects *Desalin. Water Treat.* **55** 3576–86
[23] Churaev N ., Lyklema J and Galwey A 2000 *Liquid and Vapour Flows in Porous Bodies : Surface Phenomena*
[24] Onsager L 1931 Reciprocal Relations in Irreversible processes *Phys. Rev.* **37**
[25] Jakubczak M and Jastrzębska A M 2021 A Review on Development of Ceramic-Graphene Based Nanohybrid Composite Systems in Biological Applications *Front. Chem.* **9** 1–21

Experimental investigation of direct injection diesel engine using waste plastic oil as fuel

G S Jatadhara[1], T K Chandrashekhar[2], N R Banapurmath[3], S B Nagesh[4] and N Keerthi kumar[5]
[1] Department of Mechanical Engineering, Siddaganga Institute of Technology, Tumakuru, Karnataka, India and Research Scholar, Department of Mechanical Engineering, Mangalore Institute of Technology and Engineering, Moodabidre.
[2] Department of Mechanical Engineering, RNSIT, Bangalore, Karnataka, India.
[3] KLE Technological University, BVB Campus, Hubballi, Karnataka, India.
[4] Department of Mechanical Engineering Channabasaveshwara institute of Technology, Gubbi, Karnataka, India.
[5] Department of Mechanical Engineering, BMS Institute of Technology and Management, Bangalore, Karnataka, India.
Email: Jatadhara@sit.ac.in

Abstract. Recycling of waste plastics for power generation, automotive traction can provide perfect source of energy due to their energy content. Plastics utilization is growing daily although it is not environment friendly material but can be utilized for different applications. The disposal of after use plastic is a major task and quality of oil obtained from the same has properties similar to fuel. The main advantage of conversion of used plastic to gasoline provides additional fuel generation to be derived besides addressing the disposal issues. The plastic oil obtained from pyrolysis can be added with conventional fuel efficiently and acceptable engine overall performance. Therefore, blends of plastic oil with diesel are considered for the diesel engine applications. In the present study pyrolysis of plastic is done in presence of catalyst at temperature of 380^{0}C to obtain oil with properties similar to petroleum fuels. The plastic oil is mixed with diesel in different volume proportions ranging from 10 to 40% and used as fuel in diesel engine to study its characteristics. Properties of plastic oil blends with diesel obtained were analysed. The lower plastic oil blends P10increased the brake thermal efficiency by 5%and lowered the rate of fuel consumption by 10%. As the blending ratio increased in-cylinder pressures and heat release rates were higher and the longer ignition delay were obtained. The emissions of HC, NOx decreased with lower plastic oil blending as compared to higher blending ratio. The result suggests that the plastic oil P10 can be used effectively as alternative fuel to diesel for engine with optimized operating situations. Utilization of PO with diesel blends of 10 to 30% in conventional engines provide acceptable engine performance and reduction in emission and hence can effectively reduce the dependence on fossil diesel fuel and reduce the foreign exchange.

Keywords: Plastic oil Blends, Diesel Engine, emissions, Pyrolysis, combustion.

1. Introduction

Plastic consumptions is alarmingly increasing to a greater extent for day-to-day usage due to rapid industrialization and urbanization. Search for suitable alternative energy source for diesel engine applications is necessary to not only reduce the alarming pollutions but also to reduce dependence on conventional fuel. In this context the oil extracted from the plastics can be blended with diesel in varied proportions to present increasing demand and as substitute fuel in automotive traction and power generation applications. Further it also reduces the problem of land filling which is a major environmental issue. Uncertainty in the price of depleting fossil fuels has pressurized the point on exploration of alternative energy. Since plastics are not biodegradable, extraction of oil from them by pyrolysis can address both environmental issues as well as high cost of the fossil fuels. The plastic

required for the pyrolysis can be obtained from house old waste. The oil obtained from this can be blended in with diesel in different proportions because the plastic oil has lower viscosity and low self-ignition value [1]. Thermal and catalytic cracking of plastics enables them to transform into liquid and gaseous with 95% efficiency. The fuel cracking and use of catalyst increases the fuel yield. The composition of fuel relies upon working conditions and feedstock available [2]. The properties of oil obtained from plastic reveals their properties being similar to fossil fuels. Quick evaporation of the plastic oil blends provides higher in cylinder pressures, and combustion of fuel blends improves heat transfer characteristics in the combustion chambers. The emissions of engine powered with plastic oil blends are comparable to conventional diesel engine performance [3].
It was observed that engine showed improved performance fuelled with plastic oil blends with increased load. The oil obtained from waste plastic at different pyrolysis temperatures can impact the performance of engine. The oil produced at low pyrolysis temperature has better thermal efficiency and short ignition delay irrespective of the engine load variation. The engine out emissions lowered significantly when the plastic oil is obtained at higher temperature compared to that at lower temperatures. Diesel shows lower emissions comparatively with plastic oil blended operation [4]. The engine performance powered with plastic oil blends has been evaluated and the results are compared with conventional fuels. It is reported that engine can be operated at higher loads using plastic oil blends and the performance was comparable to that with the neat fuel. The thermal efficiency for plastic oil at higher loads decreased marginally but NOx has increased to larger extent. Experiment results indicated that plastic oil is mixed with diesel for optimized operating and working conditions [5].
Properties of fuels obtained from waste polyethylene blends are comparable with diesel gas except with lowered Cetane number and pour point values. The plastic oil blends can be considered as alternative fuel for diesel engines with acceptable performance addressing the environmental concerns as well [6]. The experimental outcomes indicated that the performance with plastic oil blends were comparable to diesel operation while the fuel consumption decreased marginally with WPO and diesel fuel blends [7]. The process of obtaining fuel from the plastics is an important source of energy. In addition, it also provides the solution for land filling and environmental problem. Thermal pyrolysis of plastic leads to oil production in presence of catalyst. The liquid obtained is combination of oil, gasoline and char. The production of char can be reducing by increasing temperature and reaction time. The liquid yield also increases by increasing reaction time but beyond certain time the liquid yield decreases. The liquid fraction obtained in this manner has low fire point [8].
The plastic oil obtained from catalytic pyrolysis process and its subsequent blends are evaluated for different properties and were comparable with direct fuel. The plastic fuel is tested in engine to determine its characteristics. The experiment results of blended fuel with diesel are in close matching at lower blend proportions. As proportions of the plastic oil in diesel increases, it was found that it resulted into lower thermal efficiency while emissions increased. 30% proportion of plastic oil in diesel showed higher NOx emissions. Stability of the engine with increase in the blend proportions was found to be satisfactory. The 10% blend with diesel showed close agreement with diesel alone engine operation and can be easily suggested as a substitute fuel [9]. Engine tests confirmed that 20% blend with diesel showed similar cylinder pressures while increased BTE were obtained as compared to PPO operation. Decrease in fuel consumption compared to diesel and plastic oil blends occurred. NOx emissions reduced approximately by 19% and 26% with the use of diesel compared to PPO blends respectively [10]. The performance of engine powered with biodiesel and plastic oil blends showed improved engine performance with lower emissions [11]. The NOx and smoke can be reduced with slightly reduction in performance by using biodiesel in modified engine [12].
The objective of the present work is to evaluate the suitability of plastic oil blended with diesel as substitute fuel in conventional engine applications and the work basically focuses on the partial substitution of fossil diesel with waste plastic oil for agricultural, power generation and automotive sectors. This thereby addresses both the emission issues as well as the high rising cost of depleting fossil fuels.

2. Preparation of plastic oil and its characterization

2.1Experimental set up of IC engine

The detail of engine used for experiment is shown in table 1. The loading used to do electrically. It has provided with sensors and encoder for the measurements of rise in pressure. The signals from sensors are fed to computer to generate pressure crank angle diagram. Recording of air flow measurement, fuel flow measurements and temperature measurements are done with adequate calibration aspects. Rotameters are used to provide cooling rate of engine components. The setup has software to indicate the performance. The Engine soft software is used for acquiring the performance analysis of the engine. The process involves process of extraction of plastic oil are, trough the pyrolysis unit. The input for the process used for the production of the oil is high density polyethylene which has been crushed in to pieces in crusher machine. Later it is mixed with the suitable catalyst in appropriate proportion. The mixture is fed in to reactor for the process of heating. The temperature of the reactor is monitored. The setup is provided with the condensing unit in order the condense the vapors coming out of the reactor. The oil is collected in a measuring flask from the condenser. The catalyst flash is used in the proportion of 5%. The temperature of the reactor is monitored for the distillation process to takes place. The distillation of oil takes place when the reactor reaches the temperature of 270^{0}C and continues up to 350^{0}C.

The plastic oil obtained from the pyrolysis unit is subject for its physical and chemical characterization. The results suggested that the oil obtained in the process used in the engines as a substitute fuel. The experimental results using plastic oil blends with diesel are carried out by varying the blending ratio from 10% to 40% by volume. The experimental results using plastic oil blends with diesel are carried out. The experiments are carried out for optimizing the blending proportions and study the performance of the engine. Table 2 Properties of plastic oil and Diesel. Table 3 shows the analytical Instruments used in the study and their accuracies. Figure 1 shows the experimental test rig used.

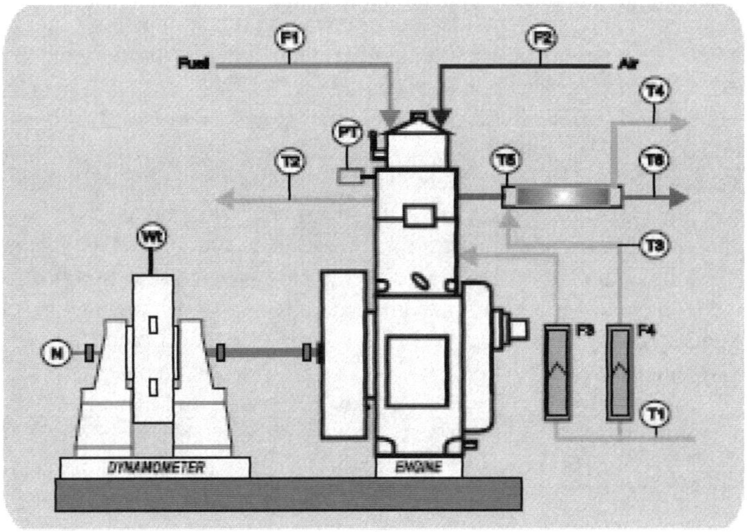

Figure 1. Experimental set-up.

Table 1: Specification of IC engine.

Engine	Make Kirloskar, Model TV1, Type 1
Cylinder power	5.2 kW
Number of cylinders	4 stroke Diesel, water cooled,
Speed	1500 rpm
Stroke	110 mm
Bore	87.5 mm
Compression ratio	17.5

Table 2. Physical and chemical characteristics of plastic oil and diesel.

Property	**Method**	**PPO**	**Diesel**
Density, kg/m3	IS1448	870	830
Kinematic viscosity at 40^0C,cst	IS1448	2.77	3.78
Flash point, ^{0}C	IS:1448	>25	65
Heating Value, MJ/kg	IS1448	35.4	42.46

Table 3. Analytical instruments used.

Gas Analyzer	Range	Accuracy
CO	0-15%	5%
CO_2	0-20%	3%
HC	0-30000 ppm	3%
NOx	0-5000 ppm	1%
Crank angle encoder	720 x 1^0	1^0

3. Results and discussion

3.1 Combustion Characteristics

Figure 2 shows the cylinder pressure against crank angle at full load engine operation. It is noticed that from pressure crank angle diagram for all the blend and diesel similar profile has been obtained. The combustion parameters of blends have been compared with baseline operation. With increasing plastic oil in the blending ratio, it is observed that there is increase in ignition delay. Further increasing the blending ratio beyond certain value results in unstable combustion. The reason for unstable combustion is due to lower cetane number ant the plastic oil contains higher aromatics. The lower value of viscosity also is responsible for the increased ignition delay with increasing the blending ratio. The low value of blending on peak pressure is not that much significant irrespective of engine operating conditions. Higher blending ratio increases peak pressure generally at lower engine load. The reason for this could be due to lower viscosity, cetane value and low flash and fire point. The lower viscosity of oil enhances spray characteristics to increase premixed combustion [1, 2]. The cylinder pressure recorded for diesel is 49.9 bar and with increasing the plastic oil to diesel cylinder pressure decreases and for P20 blend pressure recorded was 48 bar and it was noticed that with increasing plastic oil ignition delay increases.

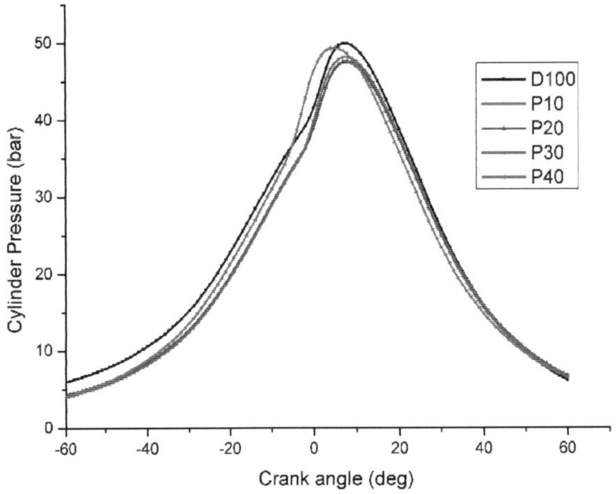

Figure 2. Pressure-crank angle diagram.

As shown in Figure 3 heat release rate variation for different blends, it is observed that in the beginning there is decrease in heat release due to higher ignition delay. The reason for increasing ignition delay is because of increase in fuel burning rates and lower cetane value which results in rapid combustion with increase in heat release value [3,5]. It was also observed that the combustion duration is longer for the diesel fuel compared to the blends from the heat release curve. The increase in heat release is also due to higher air fuel ratio.

By decreasing the loads from 100 to 75%, the effect of ignition delay for increased blending ratio plays very important role. The reason behind increasing heat releases for plastic oil blends is longer ignition delay. This has been confirmed from both pressure crank angle and heat release diagram. The lower viscosity of plastic coil also contributes for better atomization and heat release rate. The heat release rate recorded for diesel is 20.54 J/^{0}C and with increasing the plastic oil heat released decreases but it was less than 1% and for P20 heat release rate recorded was 20.7 J/^{0}C.

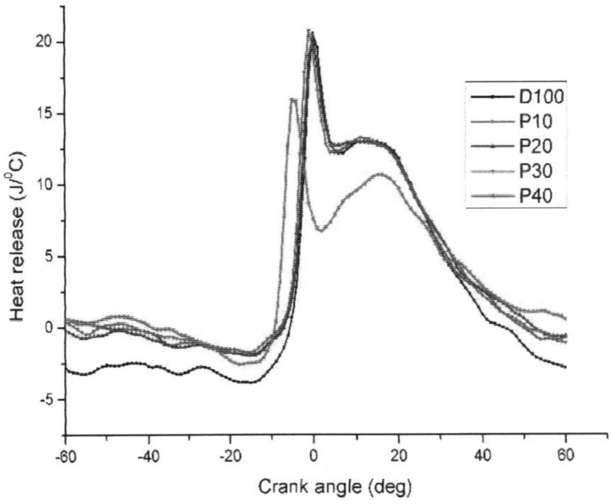

Figure 3. Heat release rate diagram.

3.2 Performance Characteristics

Figure 4 shows variation of brake thermal efficiency against loads. The blending ratio of 10% has close matching with the diesel fuel. As the blending ratio decreases, there is decrease in thermal efficiency. The decrease in thermal efficiency is due to the presence of the aromatic compounds present in the plastic oil. The thermal efficiency does not detoriate; it is marginally lower compared to diesel fuel at average load [3,4]. The brake thermal efficiency recorded for diesel is about 33.6% at full load and at higher load brake thermal efficiency of P20 blend and diesel found to be same about 30.3%.

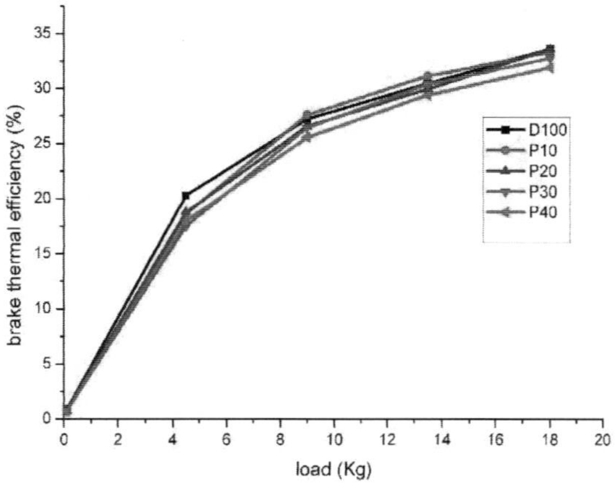

Figure 4. Brake thermal efficiency Vs load.

The specific fuel consumption for all the blending ratio shows reducing trends with engine loading as shown in figure 5. The decrease in fuel consumption for P10 followed by diesel is due to its higher BTE as compared to other blends. The factor contributing lower fuel consumption is lower calorific value of the plastic fuel and their blend as well [5]. For the same load, more quantity of fuel is required compared to diesel. The Specific fuel consumption for diesel decreases with increasing load and found to be 0.25 kg/why. For all the blends of plastic oil specific fuel consumption increases and for P20 blend it was found to be 0.27 kg/KHz.

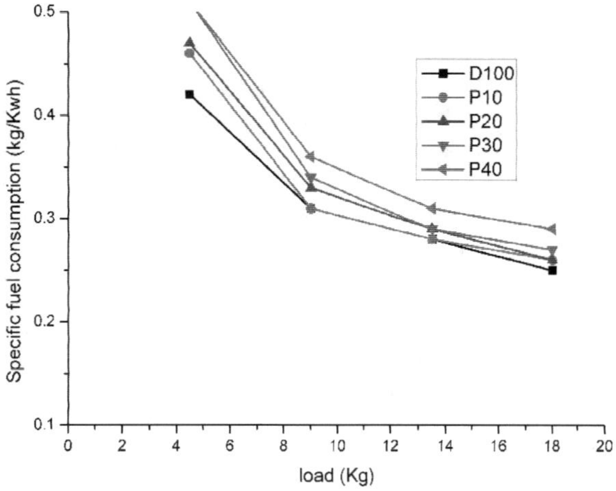

Figure 5. Specific fuel consumption Vs load.

3.3 Emission Characteristics:

CO Emission: It is observed that for all the blends of plastic oil results in incomplete combustion, which results in increase in CO in emission as shown in figure 6. The factors attributing for the increase of CO are lower value of cetane number of oil resulting in lower cylinder pressure. The rise of CO emission does not have impact on the engine performance. CO emissions are released due to incomplete combustion, which may be due to unavailability of oxygen. It is also evident that the duration of combustion for plastic oil blends is more, which also results in incomplete combustion, finally resulting in CO emissions [6]. It has also been observed that higher amount plastic oil injection results in decrease availability of air for combustion [9,10]. The CO emissions can be reduced by improving properties of blended fuel with certain additives.

The CO emissions is low for diesel at full load and with increase in plastic oil CO emissions increases for all the blends and at 75% load P20 blend has CO emissions of about 0.019% and for diesel it is about 0.002%.

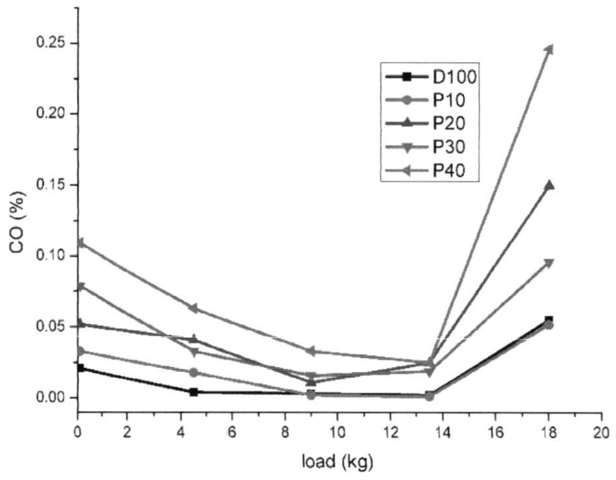

Figure 6. CO emission Vs load.

HC Emission: The Hydrocarbon emissions are because of variation of air fuel ratio. As shown in figure 7, with increasing blending ratio, the hydrocarbon emissions are also increased. The reason for increasing the hydrocarbon emissions is due to higher wall wetting in case of higher blending ratio. It is can be concluded that the possible reason for increasing the hydrocarbons could also be due to aromatic compounds. The other possibility which increases the hydrocarbon emissions are with increasing blending ratio, density of blend increases which stick to the combustion wall without taking place in combustion phenomenon. The factors continuing to this are lower cetane value and longer ignition delay. It is also evident from the properties of plastic fuel, which are responsible for poor combustion inside the cylinder. The hydrocarbon emission can be reduced by improving the properties of plastic oil with higher oxygen content fuel, so that the hydrocarbon emission coming in the exhaust can undergo combustion inside the cylinder [6, 7].

The HC emissions are minimum for diesel and for blend HC emission increases by 17% for P20. At 75%, load HC emissions are low for P20 blend compared to P30.

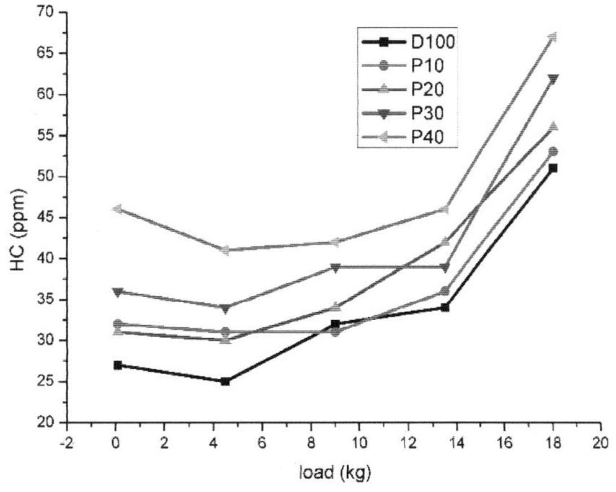

Figure 7. HC emission Vs load.

CO_2 Emission: The product of complete combustion of fuel is carbon dioxide that means fuel is completely oxidizes releasing its energy content. It is required to reduce the emission of carbon dioxide from environment point of view. Carbon dioxide emission in the exhaust revels that the extent of which the fuel has taken place in the process of combustion. The higher value of carbon dioxide for diesel at higher loads is because of higher energy content and better burring properties. It has been seen that higher blending with diesel the carbon dioxide emissions increased since the heating value of blends are comparable to heating value of diesel which increase the combustion performance as shown in figure 8. It is also evident that the ratio carbon to hydrogen is higher for diesel which results in better combustion and thereby releasing carbon dioxide as part of emission [6,7].

The diesel fuel has low carbon dioxide emission at full load and for all the blends of PO carbon dioxide emission increases and for P20 blend the increase in carbon dioxide emission if found to be increased by 5%.

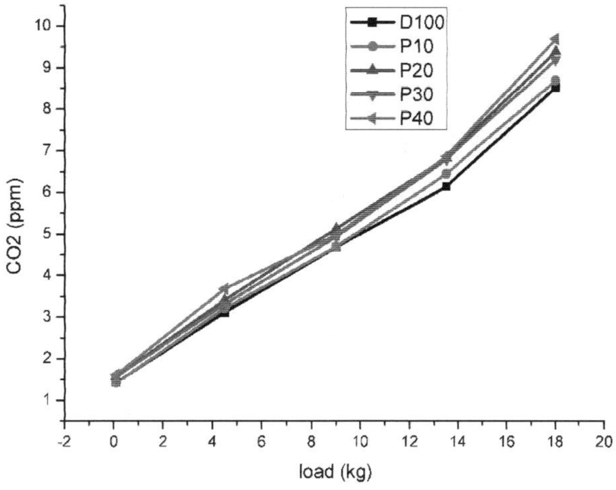

Figure 8. CO_2 emission Vs load.

NOx Emission: Figure 9 shows variation of Nox against load for different blends. In the diesel engines, the NOx formation is due to combustion of fuel with increased temperature in combustion chamber, presence of oxygen and residual time. It has been seen that the NOx emissions for all the blends of plastic oil is higher compared to conventional fuel due to increase in cylinder temperature. The NOx formation can be reduced by using after combustion treatment, which results in reduced combustion temperature with by changing the properties of blended fuel with additives.

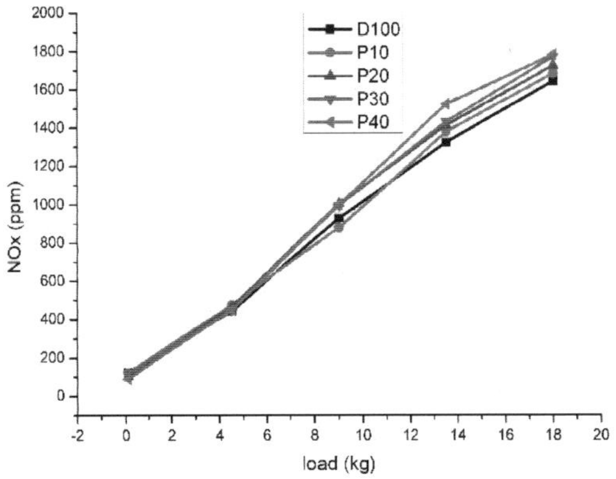

Figure 9. No_x emission Vs load.

4. Conclusions

The following conclusions are made from experiment results.

1. The plastic oil obtained from pyrolysis process showed lower calorific value, lower flash and fire point, viscosity compared to diesel. Hence plastic oil blended with diesel in different proportions.
2. With increase in the blending ratio the cylinder pressure, heat release rate increases up to 20% blend and beyond which lower cylinder pressure and heat release rate were obtained.
3. The peak pressure recorded for the P20 blend were comparable with those of diesel operation.
4. At 75% load the brake thermal efficiency of diesel and P20 blend are comparable.
5. The CO, CO_2, HC and Nox emissions increased for all the blends of plastic oil.

The testing results suggested that P20 can be considered as optimum blend at higher engine load operation with reference to the performance, combustion and emission characteristics obtained. Future investigations on the operation of diesel engine with plastic oil blends is required with improved performance and acceptable emissions with the use of newer additives.

List of abbreviations

1. PO – Plastic oil
2. WPO – Waste plastic oil
3. P10 – 10%Plastic oil+90% Diesel
4. P20 – 20%Plastic oil+80% Diesel
5. P30 – 30%Plastic oil+70% Diesel
6. P40 – 40%Plastic oil+60% Diesel
7. CO – Carbon monoxide
8. CO_2 – Carbon dioxide
9. HC – Hydrocarbon
10. NO_x – Nitrogen oxide

5.Reference

[1] Lee, S., Yoshida, K., & Yoshikawa, K. (2015). Application of waste plastic pyrolysis oil in a direct injection diesel engine: for a small scale non-grid electrification. *Energy and Environment Research, **5(1)**, 18.*

[2] Walendziewski, J. (2002). Engine fuel derived from waste plastics by thermal treatment. *Fuel, **81(4)**,*473-481.

[3] Mani, M., Nagarajan, G., &Sampath, S. (2011). Characterisation and effect of using waste plastic oil and diesel fuel blends in compression ignition engine. *Energy, **36(1)**,*212-219.

[4] Kalargaris, I., Tian, G., &Gu, S. (2017). The utilisation of oils produced from plastic waste at different pyrolysis temperatures in a DI diesel engine. *Energy*, **131,** 179-185.

[5] Kalargaris, I., Tian, G., &Gu, S. (2017). Combustion, performance and emission analysis of a DI diesel engine using plastic pyrolysis oil. *Fuel Processing Technology*, **157,** 108-115.

[6] Güngör, C., Serin, H., Özcanlı, M., Serin, S., & Aydın, K. (2015). Engine performance and emission characteristics of plastic oil produced from waste polyethylene and its blends with diesel fuel. *International journal of green energy*, **12(1),** 98-105.

[7] Syamsiro, M., Saptoadi, H., Kismurtono, M., Mufrodi, Z., & Yoshikawa, K. (2019). Utilization of waste polyethylene pyrolysis oil as partial substitute for diesel fuel in a DI diesel engine. *Int. J. Smart Grid Clean Energy,* **8(1),** 38-47.

[8] Khan, M. Z. H., Sultana, M., Al-Mamun, M. R., & Hasan, M. R. (2016). Pyrolytic waste plastic oil and its diesel blend: fuel characterization. *Journal of environmental and public health*, 2016.

[9] Kumar, R., Mishra, M. K., Singh, S. K., & Kumar, A. (2016). Experimental evaluation of waste plastic oil and its blends on a single cylinder diesel engine. *Journal of mechanical science and technology*, ***30*(10),** 4781-4789.

[10] Singh, T. S., Verma, T. N., & Singh, H. N. (2020). A lab scale waste to energy conversion study for pyrolysis of plastic with and without catalyst: Engine emissions testing study. *Fuel,* **277,** 118176.

[11] Nagesh, S. B., Somesha, G. H., Tilak, S. R., Sachin, C., &Naresh, H. (2021). Performance test on single cylinder diesel engine by lichens biofuel. *Materials Today: Proceedings,* **45,** 161-164.

[12] SB, N., Banapurmath, N. R., TK, C., &Khandal, S. V. (2021). Biodiesels powered in-cylinder common rail direct injection (CRDi) Assisted Homogeneous Charge Compression Ignition (HCCI) engine. *International Journal of Sustainable Engineering*, **14(2),** 147-161.

IOP Conf. Series: Earth and Environmental Science 1084 (2022) 012010 doi:10.1088/1755-1315/1084/1/012010

Integrating Digital Mission and Sustainable Development Goal-9 for Building Technologies: A study of Navratna Companies of India

N Mahajan [1,] ***V Singh*** [2]****N Kaur*** [3]***and A Gupta*** [4]

[1]Associate Professor, AIBS, Amity University, Noida, India ,201303
[2] Professor, AIBS, Amity University, Noida, India ,201303
[3] Assistant Professor, AIBS, Amity University, Noida, India ,201303
[4] Assistant Professor, AIBS, Amity University, Noida, India ,201303

E-mail: navitamahajan07@gmail.com

Abstract: The study focuses on 10 Navratna companies based in India that come under the central public sector enterprises. The study focuses on the Digital Mission by the Government of India ,its impact on the industries operating in India and how technologies have affected the overall productivity of Indian enterprise with their approach to achieve SDG-9. The study finds that the Navratna companies have been successfully working towards achieving the goals of SDG-9 for building resilient infrastructure, sustainable industrialisation and fostering innovation but still they need to use the technologies to their maximum potential in terms of improving their overall productivity and delivery. The data for the study has been collected through secondary resources like websites, annual reports, research materials and company websites. The implication of the study is that other companies operating in India and abroad can learn about implementing the sustainable development goal in their day-to-day practices to increase productivity and improve infrastructure, thereby fostering innovation and growth. This will help India reach its goal of becoming a major manufacturing hub and a digitally advanced nation.

1. Introduction

This research paper deals with the Ninth United Nations (UN) Sustainable Development Goal (SDG 9) which focuses on building and improving the infrastructure of a nation through industrial development measures that are sustainable in nature and that also contribute to the overall technological advancement of a nation. This research paper revolves around SDG from the perspective of Navratna Companies of India, i.e., the major public sector enterprises that are duly recognised by the central Government of India. The paper aims at understanding the contribution of these Navratna companies towards the overall infrastructure and technological growth of India. The paper also emphasises on the due effect of the Digital India Mission on the functioning of the companies. This study also aims to analyse how India is working towards achieving these goals by 2030, as specified by the United Nations. The paper gives special attention to SDG 9 revolving around technological advancement, giving a special emphasis to India's Digital Mission. While the United Nations have several SDGs laid out by it to improve the overall living and earning conditions of nations across the globe, this paper explores the SDG 9 of the United Nations which focuses on the industry, innovation, and infrastructural development of the world. India's overall development as a nation is highly affected by the digital literacy measures taken by the government. The Digital India mission launched in 2015 significantly improved online infrastructure and even increased the accessibility of Internet amongst citizens. Since the Navratna companies are giant industries, which hold a competitive advantage in terms of both resources and manpower in India, this paper will highlight how significant technologies have been introduced and are being implemented by the Navratna companies that have a grip over the manufacturing sector of the Indian economy.

The targets for SD- 9 has 8 dimensions, which include enhancing research and upgrading industrial technologies, sustainable upgradation of industries and infrastructure, easier and increased access of financial services and markets, promoting sustainable and inclusive industrialisation, developing sustainable, resilient and inclusive infrastructure, enabling infrastructure development that is sustainable in nature for countries that are working towards becoming independent and developed, helping in technology development on the domestic front and diversification of industries, along with universal access to information and communications technology. The Digital India mission of India has a significant role to play since it is a core component of the mission of SDG9.Despite increased use of new-age technologies and the IT industry, India had a big divide in terms of people who have access to this digital knowledge and those who don't, and it created economic inequities between those who can and cannot afford technology, harming the country's overall digital growth. To address this gap, the government launched the 'Digital India' effort, which includes a variety of programmes such as e-governance, mobile e-health services, and digital banking for digital inclusion. The 'Digital India' effort is helping all parts of the country; especially small towns and rural areas to improve in terms of digital infrastructure through initiatives.

2. Research Methodology

The study is based on secondary data. For the fulfilment of stated topic of the study , the various official sites of mentioned Navratnas were visited and relevant data was compiled for the purpose of highlighting their various digital emphasis over the time. The detailed study was done on the various mission objectives of SDG-9 and Digital Mission of Government of India to find the common alignment between the two. While studying the official sites of Navratnas companies of India , mention about the technological innovations was specifically taken into consideration for each of them. The study has also mentioned the brief background of each Navratnas, that itself boasts the well-established structures and financial strength of these organizations, hence role in Indian Economy, Digital innovation, and Sustainable Development Goals.

3. India's roadmap for SDG 9 in collaboration with Digital India mission

According to a survey by [1], 60 percent of India's topmost hundred companies comprise Sustainable Development Goals in their long-term strategic plans and primarily the top ten organizations align their business goals with International Goals. Digital India mission by the Government of India is a major component of India's overall mission to be better in terms of Infrastructure and Technology. The Government of India launched various schemes under the Digital India mission, which made lives of the citizens a lot easier when it came to keeping a track of the things, they need to be aware of. Initiatives like Aadhar Card, Digi Locker, My Government application,

The Indian Government has worked significantly over the past few years to improve the condition of the infrastructure and has even worked on the proper allocation of resources. Installed power generation capacity has been steadily increasing during the last five years. Installed capacity in non-fossil-fuel industries has increased by 51.3 percent, while renewable energy capacity has more than doubled (solar, wind, bio- and small hydro power). Furthermore, under its 'Make in India' initiative, India is attempting to become an information technology and manufacturing centre. These efforts have significantly increased FDI inflows and helped the economy uphold annual development proportion of 7.5 percent over the fiscal years 2014-15 and 2016-17. Small-scale business owners were given cheap credit, which helped to develop employment-intensive industrial areas. Furthermore, the 'Start-up India' initiative encourages entrepreneurship and labour-intensive economic growth(A report by NITI Ayog) along with several outcome Indicators, have been set out by for Ministries, at central Level, various schemes at central level along with other government initiatives. Several states have structured their departments and strategies in a similar way. There are nodal ministries at the federal level and nodal departments in several states. Special cells for SDG's

SDG or Excellence Centres have been created in several states to manage SDG implementation. For a periodical evaluation of SDG implementation throughout the country, NITI Aayog also established a special force with contribution from Central and State Ministries.

4. Research Gap

Past research on SDGs has largely focused on their long-term implications for business stakeholders and their significance for future generations [2], however, there has been little research on the role of Public Enterprises ,from developed countries in emerging economies such as India in achieving Global Goals. Furthermore, previous study investigates training and development as an important instrument for achieving strategic success for enterprises [3], however the concept of using this tool to achieve SDGs in an organisation has not been extensively studied. The current study focuses on Navratnas, and how they focus on technology and infrastructural development for the betterment and general progress of the communities they serve.

The research done so far on SDG 9 and its implementation in India has talked about the lack of infrastructural development being the major cause of concern. As per an article released by "*The Borgen Project in October 2020*" the trade industry is critical to a healthy economy that includes job development, strong partnerships, and a larger range of product availability. The standard of commerce and transportation infrastructure has not improved. It has maintained a consistent ranking of 2.91 out of five. Manufacturing has remained stagnant and has shown no signs of expansion. This industry has the potential to contribute to economic success as well. These determinants may be seen in India's industrial growth rate, which has declined by 0.8 percent between 2016 and 2019. India's industries generate a lot of hazardous waste and water waste, which contradicts the concept of sustainability.
As per sources, lack of appropriate funds for research and development contribute to the slow pace of the industrial development and innovative inventions in India. Furthermore, SDG 9 research done so far has not been very inclusive of the Digital India mission launched by the Government of India. Probable reason for this could be the lack of awareness about technologies and the inability of Navratna companies to implement them.

5. Navratnas on the way of SDG-9

5.1 Bharat Electronics Limited (BEL)

Founded in Bengaluru, Karnataka in the year 1954, BEL started with manufacturing equipment for communication in the year 1956, followed by germanium semiconductors, valves, and radio transmitters[4] This government owned entity currently produces advanced electronic devices for aerospace and on-ground applications.
Technological Innovations and Building: BEL has been a strong advocate for the Indian Government's Make in India programme. BEL is concentrating on the fundamental industries in research and development while all the other secondary areas, including MSMEs are subcontracted to Indian companies. The organization has released a long-term outsourcing policy with appointment of Nodal officers for outsourcing and vendor development. Every year, about 800 new indigenous merchants are added. In line with the 'Make in India' concept, BEL continues to carry its indigenization initiatives[5] Missile systems Radars, communication and network-centric systems, tank electronics, gun improvements, electro-optic systems, electronic warfare, and avionics systems all are poised to contribute to BEL's progress. Solar energy, homeland security, smart cards, and telecommunications are all areas where the company is looking for new business.

IOP Conf. Series: Earth and Environmental Science 1084 (2022) 012010 doi:10.1088/1755-1315/1084/1/012010

5.2 Container Corporation of India Ltd

Known as CONCOR, it was established in March in the year 1988 in preview of the Companies Act and began operations in the month of November 1989, taking over the Indian Railways' already established network of 7 Indian container Depots .It has grown from a small venture to become India's indisputable marketplace front-runner, with the largest network of 61 Container Depots (59 terminals and 2 strategic tie-ups).

Technological Innovations and Building: CONCOR has actively participated in the Digital India mission launched by the Government of India. The company launched the "Know your location system" that gives users the location of their containers. There is now a system which enables automation of contractor billing and enables automatic system of invoice savings in Document Management System. CONCOR has taken steps towards paperless offices by replacing physical files with digital ones. CONCOR has recently introduced a Chat-based Application System for Employee Grievance Management System (EGMS) and Customer Grievance Management System (CGMS) to improve employee performance and to help valued EXIM/domestic customers settle their complaints[6] Through its association with the Government of India, CONCOR successfully took measures towards sustainable infrastructural development by taking several green initiatives and constructing classrooms and toilets for children.

5.3 Engineers India Limited (EIL)

Engineers India Ltd, Known as EIL is a renowned organization based on consultancy in engineering and firm based in Engineering, procurement, and construction. The company was founded in 1965 and specialises in technical consulting and EPC services for in the sector of oil, gas, and petrochemical businesses. As of March 31, 2021, EIL's industrial and technical dominance is fuelled by over twenty-four hundred engineers and professionals working as a skilled workforce of over twenty-eight hundred people[7]EIL's design workplaces have around forty-five thousand manpower of technical resources available per year, plus 15 Lakh manpower in construction management services.

Technological Innovations and Building: EIL, which, among other things, builds petroleum refineries and pipelines, had successfully ventured into urban and transportation infrastructure, as well as activities relating to special economic zones. EIL has created over 35 process technologies, with a portfolio that encompasses refining of petroleum products, processing in oil and gas along with aromatics. For the first time in India, EIL's developed DHDT technology and Research and Development by IOCL- has been applied at IOCL's Refinery in Bongaigaon Assam. EIL has 36 active patents along with 31 patent applications which are pending covering a variety of process innovations. The company's most recent revolutionary product was a start-up effort called EngSUI, which has already passed a major milestone with the installation of a 2000 LPD prototype to capture water from the atmosphere at the EIL Gurugram Office Complex.

5.4 Hindustan Aeronautics Limited (HAL)

Hindustan Aeronautics Limited known as HAL has aerospace fully owned by Government and defence enterprise based in Bengaluru, India. HAL is one of the world's oldest and largest aerospace and defence companies, having been founded on December 23, 1940. The President of India, through the Ministry of Defence, appoints a Board of Directors to run HAL. HAL is now working on fighter planes, helicopters, jet engines, and marine gas turbine engines, as well as avionics, software development, spare parts, and refurbishing and upgrading Indian military aircraft. HAL is one of Asia's major aerospace businesses, with an annual revenue of more than US$2 billion. International contracts to manufacture engines for aircrafts, spare parts and other materials for aircraft account for more than 40% of HAL's revenue. HAL has over 20,000 employees with a healthy revenue and consistent performance[8]

Technological Innovations and Building: HAL specialises in the production of aircrafts, helicopters, power plants and aerospace machines among other things. The International Aerospace Community has benefitted from HAL's cost-effective and extensive skills. HAL has demonstrated skills along with offering concerted supplies in the arenas of machine constituent production, metal sheet fabrication, assembly, sub-assembly fabrication, strategy, and expansion (Structural analysis, 3D exhibiting and testing). HAL's efficacy is boosted by refined and updated Information Technology infrastructure consisting of high-end Workplaces, high performance, high dependability scalable Servers, high speed gigabit LANs and high speediness of Internet, ISDN facilities. On November 9th, 2021, HAL and Zero Avia have committed to work on the development of a hydrogen-electric powertrain capable of flying a Dornier 228 aircraft's 19 passengers up to 500 nautical miles. The Aeronautical Development Establishment, DRDO, has placed an order with HAL was given the contract to manufacture, assemble, integrate, test, and provide the ABHYAS High-Speed Expendable Aerial Target (HEAT) System[9] HAL and Elbit Systems have agreed to work together to improve their technological basis and obtain high-end technology for the Digital Overhead HUD System, which is largely utilised in transport aircraft around the world[8]With contemporary optics, the Digital Overhead HUD provides sharp brightness, a broader field of view, and a larger head motion box[9]

5.5 Mahanagar Telephone Nigam Limited (MTNL)

Mahanagar Telephone Nigam Limited, a government-owned telecommunications company that operates in India's metro cities prominently in Mumbai and New Delhi, as well as the African island nation of Mauritius. Till the year 1992, during the time when telecommunication market was allowed to other players in the sector, the business enjoyed a monopoly in Mumbai and New Delhi. The corporation is currently owned entirely by the Indian government. Though it only has presence in two circles, MTNL is the principal internet facility provider in Mumbai and Delhi (in terms of market share) and the third largest internet service provider in the nation. Dialup and DSL are used by MTNL to provide TriBand Internet services. Through TriBand, it offers gaming on demand, demand-based video, and Internet Protocol television(IPTV) services in India. TriBand is aimed towards private residences and small companies.

Technological Innovations and Building: MTNL has been at the forefront of technological advancements, having converted 100 percent of its telephone exchange network to a cutting-edge digital mode. Over the last nineteen years, MTNL has grown fast by modernising the network and adding cutting-edge technologies and a customer-centric approach. Telex, wireless data communication, Telephone, , telematic, and some other similar modes of communication are among the services provided by the company (Internet). MTNL launched its FTTH know as fibre-to-the-home, triple play (video voice, and data along with broadband services of high speed in Delhi. 1 Gbit/s core network speed FTTH was launched in Mumbai apart from IPTV, HDTV, 3DTV, demand-based video, bandwidth on demand, rapid video conferencing, communicating gaming, and a variety of additional value-added services are all available through MTNL FTTH Broadband. Due to massive corruption in the Indian telecom business, MTNL has been trailing in its revenue share and market during the recent years[11]

5.6 National Aluminium Company (NALCO)

National Aluminium Company Limited (NALCO) is a Schedule 'A' Navratna Company. Founded on January 7, 1981, and is headquartered in Bhubaneswar. One of the country's largest integrated Bauxite.Alumina.Aluminium-Power Complex, At the moment, 51% of the equity in company is born by Indian Government. The company owns and operates Asia's largest aluminium complex, in integrated manner which includes bauxite mining, alumina refining, aluminium smelting and casting, power generation, rail and port operations. Nalco, which was established in 1985–87, has emerged as a standout performer in alumina and aluminium production, export, and, more importantly, self–sustained growth[12]

Technological Innovations and Building: NALCO has established a world-class research centre in its pursuit of organisational growth through sustained development in process, product, and technology through research, development, and innovation activity. QEMSCAN (SEM-EDX), ICP-MS, ICP-OES, BET Sedigraph, , UTM, Laser Particle Size Analyzer Analyzer, Potentiometer, OES, CHNS Analyzer, Metallurgical Microscope, Whiteness Index metre, and other advanced equipment are available at the NALCO Research Centre. The testing activities at NRTC began on a regular basis in 2019. On May 29, 2020, the R&D wing of NALCO Research and Technology Centre was recognised by the DSIR, Government of India. As of December 2018, the company has submitted 36 patents, 17 of which had been granted and 5 of which had been commercialised, demonstrating its commitment to R&D[13]. As part of its endeavour to convert waste to wealth, the Company is aiming to rescue iron concentrate from red mud and gallium from discarded liquor. The company has also successfully installed a first-of-its-kind nano-technology-based de-fluoridation process to de-contaminate the Smelter's effluent water, addressing the area's long-standing fluoride pollution problem.

5.7 National Buildings Construction Corporation (NBCC)

National Buildings Construction Corporation is a publicly traded top-class Government of India Navratna enterprise. The current areas of operation of the Company are divided into three major segments, namely Project Management Consultancy, EPC Contracting for Real Estate Development. NBCC (India) Ltd has been designated as a Navratna firm by the Ministry of Housing and Urban Affairs. The company's present fields of business are divided into three primary segments: Project Management Consultancy (PMC), Real Estate Development, and EPC Contracting, and it is ISO 9001:2015 accredited in Project Management and Consultancy by the Bureau of Indian Standards. Since 2012, the company has been listed on both stock exchanges., NBCC has grown at a consistent 20 percent CAGR over the last five years, with over Rs 80,000 crore in order book, the company's unique business model distinguishes it as a leader in the construction sector.

Technological Innovations and Building: The areas of operation of NBCC include three primary segments, Project Management Consultancy (PMC), which includes the execution of landmark projects in a variety of sectors, as well as Government Redevelopment Projects, Engineering Procurement and Construction (EPC), which entails working in a niche market and constructing Chimneys, Cooling Towers for the Power Sector, and Real Estate Development, which entails developing residential and commercial inventories for governments and corporations[14] Some of the notable constructions by the NBCC include construction of 200 Bedded Indira Gandhi Memorial Hospital, Male, Republic of Maldives, New Supreme court building at Maldives and Meer Housing project in Turkey. Recently , NBCC has worked significantly to improve the condition of India, particularly Delhi and NCR by redeveloping several colonies of Central Delhi. NBCC has implemented a policy to incorporate environmentally friendly features into its projects. Zero waste, dual piping, rainwater harvesting, solar energy, smart electricity metering, and LED/energy efficient fixtures are among the features. Green Building standards are followed throughout the company's projects. The company uses ecologically friendly processes and materials such as steel structures, modular construction, pre-cast, prefab components, and light weight concrete slabs. The organisation also maintains rigorous adherence to requirements to reduce air and water pollution at its project sites to contribute to a cleaner environment[15]

5.8 National Mineral Development Corporation (NMDC)

NMDC, National Mineral Development Corporation, formerly known as National Mineral Development Corporation, is a state-owned producer in minerals in India. The Government of India owns 72.43 percent of the company, which is administered by the Ministry of Steel. Iron ore, copper, rock phosphate, limestone, dolomite, gypsum, bentonite, magnesite, diamond, tin, tungsten, graphite, and other minerals are among the minerals explored by the company. With three completely mechanised mines in Chhattisgarh and Karnataka generating more than 35 million tonnes of iron ore, it is India's largest iron ore producer and exporter. In Panna, Madhya Pradesh, it also operates the country's only mechanised diamond mine[16]

Technological Innovations and Building: NMDC's main aim is to develop as a worldwide eco-friendly mining organisation that produces quality steel products. They aim to reach their goals by growing the existing number of mines and by operating mines that are in joint ventures with other companies. Recently, NMDC announced that it has reached a record iron ore production of 40 million tonnes (MT). With production of 40 million MT record iron ore in January 2022[16]Indraprastha Gas Limited Signed Memorandum of Understanding with NDMC, an Indian natural gas distribution company, to establish a waste-to-energy plant and an integrated fuelling station in North Delhi to convert Solid Waste into CBG.

5.9 NLC India Limited (NLCIL)

NLC, formerly known as Neyveli Lignite Corporation Limited is a Central Public Sector Enterprise with Navratna status that reports to the Ministry of Coal. NLC was formed in 1956 under the Companies Act of 1956 with the goal of meeting the electricity demand of India's southern states by excavating lignite for power generation. This is a major company in India's fossil fuel mining sector, focusing on thermal power generation. It produces approximately 30 million tonnes of lignite per year from opencast mines in the southern Indian state of Tamil Nadu and Barsingsar in the Rajasthan district of Bikaner. Its joint venture owns a 1000 MW coal-fired thermal power station. It has recently diversified into renewable energy production, installing a 141 MW solar power plant to generate electricity from photovoltaic (PV) cells and a 51 MW windmill power plant[17]

Technological Innovations and Building: As per a report released in April 2021, NLCIL increased the mining capacity by 66 percent and its generation of green power capacity by 39 percent in the 2020-21 fiscal year when compared to the financial year 2019-20. In fiscal year 2019-20, the company added 515.5 MW of capacity, including a 500 MW lignite-based thermal power plant and a 17.5 MW solar power plant with a battery energy storage system. With this, the total installed capacity of the company increased to 6061.06 MW. The company ventured into the coal mining business launching the activities at its Talabira Mines in Odisha. NLC India Limited has continuously integrated its corporate aims with social goals and made CSR an intrinsic component of its operations at the highest levels. NLC India Limited's various units/departments are working together to attain the best Triple Bottom Line performance. The Production, Maintenance, and Service Units oversee both economic performance and environmental safety, as well as pollution/groundwater control and energy conservation through direct action. Overall Econometric performance is the responsibility of the Projects and Planning Units, Specialized Engineering Units, and Technical Branches/Sections.

5.10 Oil India Limited (OIL)

Oil India Limited, headquartered in Duliajan, Assam, is India's second largest hydrocarbon exploration and production firm. With its operating headquarters in Duliajan, Assam, Oil India Limited (OIL) is India's second largest hydrocarbon exploration and production company. With corporate headquarters in Noida, New Delhi-NCR, the firm is a state-owned Navratna under the administrative jurisdiction of India's Ministry of Petroleum and Natural Gas. OIL has greatly increased its exploration and production efforts in North-East India in recent years. The Northeast Frontier (NEF) project was formed by OIL to expand its exploration activities in the North East's frontier areas, which are logistically challenging and geologically complicated.

Technological Innovations and Building: In the areas of Upper Assam and Arunachal Pradesh, OIL has a long history of hydrocarbon exploration and production. In this regard, the company has continued to adopt new methods and technologies to advance its goal of accurate mapping and imaging of the fold-complicated belt's geology. In this sense, a high density-long offset seismic survey for enhanced illumination of complex geology is a recent innovation. To meet the upcoming challenges of inaccessibility and extreme logistics for the exploration of these frontier areas, OIL will be the first E&P company in India to deploy cable-less seismic data-collecting technology. Oil's in-house field workers are involved in the most efficient acquisition of 2D, 3Dand 4D seismic data using current technology and cutting-edge equipment to satisfy a wide range of exploration and development difficulties. OIL has broadened its exploration efforts to include frontier locations with challenging logistics and geological conditions.

6. Discussion and Conclusion

All forms of transportation have risen greatly since the SDGs were implemented, according to the Voluntary National Review Report on the implementation of the Sustainable Development Goals produced by the Government of India in 2017[18] All villages will have access to roads and power. India intends to connect all village councils in the country to high-speed broadband. The installed capacity of power generation has been steadily increasing. Installed capacity in non-fossil-fuel industries has increased by 51.3 percent, while renewable energy capacity has more than doubled (solar, wind, bio- and small hydro power). Furthermore, with its 'Make in India' initiative, India is attempting to establish itself as an information technology and manufacturing powerhouse. Foreign Direct Investment (FDI) inflows have been considerably expedited because of these efforts. Employment-intensive industrial segments are being strengthened by making small-scale business entrepreneurs eligible for easy borrowing. Furthermore, the "Start-up India" programme encourages entrepreneurship and labour-intensive economic growth[19]

The infrastructure, industry, and innovation pillars of Sustainable Development Goal 9 (SDG 9) are interrelated. The goal of all these pillars is to achieve socially inclusive and environmentally sustainable economic development. SDG 9 will require overcoming resource restrictions, establishing and enhancing developing nations' capacities, and finding innovative methods to handle development difficulties to build resilient infrastructure, promote sustainable industrialization, and boost innovation by 2030. SDG 9 has about 20 targets and indicators connected to its three pillars, and it is closely linked to other SDGs such as job creation, sustainable livelihoods, improved health, technology and skill development, gender equality, food security, green technologies, and climate change.

Increasing investment in digital infrastructure appears to be critical. It is anticipated that India will require US$35 billion in upfront investment every year to be among the top five global digital economies, with such massive investments in digital infrastructure yielding faster growth and long-term economic advantages. To make this happen, we will need an active finance model as well as a robust execution model. Investing in digital infrastructure might follow the model of India's road and bridge building projects, which have considerably contributed to the country's GDP[20]

SDG 9 is linked to many other targets, including industry-specific targets like job development, long-term livelihoods, and food security, to name a few. In addition, innovation is required for the delivery, distribution, and consumption of energy, food, water, and housing, and access to ICTs is critical for achieving SDGs 4 (quality education), 8 (reduced disparities), and 10 (reduced inequalities), among others.

The Navratna companies, through their activities and initiatives, have created an impact on the economy of India. The Public Sector undertakings, be it in the field of mining, construction, or telecommunication service, have made significant progress in terms of the growth and reach of their businesses. The overall impact generated can be seen by their global presence and reach, the number of people they employ, and most importantly how sustainable their initiatives are. Digitization of

Government and its undertakings is also an indicator of the overall progression of the companies. While the Navratnas have adopted certain digital ventures which have increased their overall reach, they still need to use the technologies to their maximum potential in terms of improving their overall productivity and delivery.

7 . Implications of the study

The study will help growing businesses understand the importance of building and innovation themselves in ways that are sustainable in nature, grow infrastructure, develop their companies and the communities they serve and help Indian companies reach their best potential. The study also gives insights on how it is important for Navratna companies to produce in ways that not only benefits them, but also positively impacts the Indian citizens and the economy. SDG 9 implementation is progressing in many areas; nonetheless, developing countries like India will need additional resources and capacity to reach the UN's goals. Furthermore, government funding and support are required so that significant corporations such as Navratnas can take the necessary steps to enhance the country's overall situation. In addition to this, traditional PSUs need to change with the times to adopt the new digital technologies to their benefit to progress with the growing times. Through this study, it is found that the Digital India mission is a significant part of the SDG9, for incorporation of digital enhancement goals is necessary for making a strong foundation for a developed India.

8. References

[1] Futurescape. (2020). India's Top Companies and the SDGs.

[2] Mio, C, 2020 Sustainable development goals and the strategic role of business: A systematic literature review. *Business Strategy and the Environment*, ***29***(8), 3220-3245.

[3] Gusdorf, M. L.,2009, Training design, development and implementation. *Society for Human Resource Management*, 1-38.

[4] Tyagi, A, 2012, A study of the quality circles concept in indian industry *International Journal of Management Research and Reviews*, *2*(9), 1445.

[5] Forbes, N. 2002, Doing business in India: What has liberalization changed. *Economic policy reforms and the Indian economy*, 129-67.

[6] CONCOR / THE COMPANY." *Container Corporation of India Ltd*, "Technologies | Engineers India Ltd." *Engineers India Limited*, https://engineersindia.com/service/technologies.

[7] Kumari, G.et al,2012, Studies on Some Aspects of Job Satisfaction in Engineers India Limited. *Global Journal of Management and Business Research*, ***12*(7)**, 66-74.

[8] Kumar N, 2001, Indian software industry development: international and national perspective. *Economic and Political Weekly*, 4278-4290.

[9] Feb 8,2021 *www.economictimes.indiatimes.com/news/defence/hal-signs-agreement-with-elbit-systems-for-supply-of-digital-overhead-head-up-display-systems/articleshow/80750106.cms.*

[10] Achievement report of HAL up to Q3 Financial Year 2021-22." *PSU Connect*, 9 March 2022, https://www.psuconnect.in/news/hal-achievements-upto-q3-financial-year-2021-22/31641.

[11] Annual Report ,Published on April 8,2022, Department of Telecommunication, India

[12] Chew, I. K, 2004, Human resource management strategies in practice: Case-study findings in multinational firms. *Asia Pacific Journal of Human Resources*, ***42*(1),** 32-56.

[13] Annual Report NALCO INDIA, Published on September ,2021.

[14] https://www.nbccindia.com/webEnglish/overview

[15] Mahesh, D. R., & Ranjithkumar, D. S.2019, Evaluating Financial Strengths of Selected PSU Steel Companies. *International Journal of Management and Business Research*, ***9*(3)**, 16-21.

[16] March 19,2022, www.business-standard.com/article/companies/nmdc-produces-record-40-mt-iron-ore-aims-42-mt-output-in-fy22-122031900770_1.html

[17] www.nlcindia.in

[18] Infrastructure and Industrialization - United Nations Sustainable Development." *the United Nations*, https://www.un.org/sustainabledevelopment/infrastructure-industrialization

[19] Niti Ayog Presentation." | *NITI Aayog*, https://www.niti.gov.in/sites/default/files/2019-01/NITI-Aayog-SDG-Presentation-to-States

[20] Updates on SDG Goal 9 in India." *The Borgen Project*, 23 October 2020, https://borgenproject.org/sdg-goal-9-in-india/

Performance evaluation of the Absorbing Plate of a Hybrid Solar Dryer for Potato Drying

Debashree Debadatta Behera[1*], R.C. Mohanty[1], A.M. Mohanty[1], Shiv Sankar Das[2] and Bijaya Bijeta Nayak[3]

[1]Department of Mechanical Engineering, Centurion University of Technology and Management, Odisha, India
[2]Zenith School of Management, Odisha, India
[3]School of Mechanical Engineering, KIIT University, Odisha, India
*Corresponding author Email: debashree.behera@cutm.ac.in

Abstract. The present study is to develop a long-term livelihood program for drying food products using solar technology and a hybrid type solar dryer. The present study is to analyze of absorbing plate of the dryer for getting maximum temperature by both experimental and simulation procedures. The dryer is made up of aluminium sheet and SS304 grade stainless steel perforated trays and has a capacity of 20 kg/day for drying potatoes. The main objective of this study is to simulate the steady-state thermal analysis using ANSYS software of the absorbing collector plate fabricated with corrugated Aluminium sheet. The maximum temperature, total heat flux, and directional heat flux are found as 88.311°C, 0.131333 W/mm2, and 0.0765 W/mm2 respectively. Experimentally highest temperature of the absorbing plate is reached as 89°C. Performance indicators such as drying efficiency, drying rate, coefficient of performance, and heat utilization factor are evaluated as 43%, 0.607 kg/hr, 0.22, and 0.7948.
Keywords: Absorbing plate, potato drying, total heat flux, directional heat flux, coefficient of performance

1. Introduction

Technology has an impact on us in many ways and many facets of our life. One of these options is to use a solar dryer that runs on renewable energy. Clean energy is defined as energy with no negative externalities. People often find it challenging to make hygienic dry food products. Existing drying techniques are costly (electric drying), unreliable (power is not always available), and unhygienic (open sun drying). Solar drying is a fast and alternative method of drying food in a sanitary way and it is a clean energy product.

Chaudhri et al. [1] have done an experimental investigation on a transverse wedge-shaped roughened absorber plate for delivery of heat air-jet and to increase the heat transfer rate. Maximum exergy efficiency is obtained as 4.95% at a mass flow rate of 0.03 kg/s. Kumar [2] has done an experimental investigation on 3 sided concave dimple roughened duct to derive Nusselt number, friction factor with Reynolds number, and for enhancement of heat transfer rate. Kumar et al. [3] have done an experimental study on spherical ball roughened solar air duct for evaluation of thermal efficiency. Maximum thermal efficiency was found as 81% as compared to 38% for smooth ducts. Kaewkiew et al. [4] performed a comparison study of drying chili in a greenhouse sun drier vs. open drying and concluded that solar drying was more successful. Lingayat et al. [5] conducted an experiment on drying bananas for agricultural purposes using an indirect solar dryer. Chouicha et al. [6] used three different solar drying techniques to examine the hydration of deglet nour dates. Tarigan [7] ran a CFD simulation to look at the average drying temperature and observed that the bottom three treys had a far lower temperature change than the top trey. Rabha et al. [8] tested the drying of ghost chili and sliced ginger at different temperatures. The solar dryer's performance was improved by the addition of insulation and reflectors. El-Sebaii et al. [9] used a blower to circulate hot air to assess the drying effect of indirect sunshine on mint and thymus. Erdem et al. [10] investigated hybrid PVT solar dryers with and without fins in both experimental and modeling experiments. Fins on the absorbent plate and PV panel are used to increase thermal efficiency. Koua et al. [11] designed an indirect type sun dryer with an appropriate air collector, drying chamber, and three treys (wire mesh type created with wooden frames) for drying cocoa beans. Moisture content, desorption isotherms, and thermo physical characteristics including thermal conductivity, thermal diffusivity, and shrinkage were determined with the collector tilted at 17°. For drying mushrooms, Asnaz et al. [12] et al. conducted comparative research on solar dryers independently combined with natural convection, forced convection, and heat pump. The natural convection type consists of a glazed glass collector, a drying chamber, a chimney, and insulating glass wool. Two blowers controlled by a solar panel manage airflow between the collection and drying chamber in a forced convection system. The expansion valve, condenser, evaporator, and compressor, as well as the refrigerant, make up a heat pump sun dryer. Kuan et al. [13] conducted comparison research for drying bananas in four types of continental climatic settings using heat pump conventional, solar dryers, and heat pump assisted dryers. For the creation of energy balanced equations, the heat pump integrated type has been designed for various climate circumstances. The present study is to do both experimental and simulation studies of the Absorbing plate of a solar dryer for getting the highest temperature and also to increase the heat transfer rate of a hybrid type Solar Dryer. Performance indicators such as drying efficiency, drying rate, coefficient of performance, and heat utilization factors are also evaluated.

2. Material and Methods

In this study, several system characteristics such as air intake temperature, humidity, moisture carried by air per minute, and air output temperature were used to build the solar dryer. For design calculations, the features are based on a short examination of the atmosphere in the coastal areas of Bhubaneswar. To offer effective moisture removal, the sun drying device includes an enclosed means, a drying chamber, and a plurality of perforated holders for holding drying material, as well as a collector and a plurality of air blowing, means driven by solar energy.

The solar collector is the dryer's most important component; it's made of glass as a transparent cover plate (3 mm), absorbing plate, and a glass wool (75 mm) acting as an insulator. The food on the trays is heated by the hot air from the solar collector. The drying chamber's saturated air is drawn out by the exhaust fans. By producing convective air currents, air-blowing devices circulate warm air over the drying area. This warm air circulates through the perforations in the trays, drawing moisture from the food placed on top. The trays are aluminium wire mesh. To increase the transitivity, two transparent cover plates made of tempered glass with a 2-inch gap between them were employed. The PV panel, battery, and charge controller were used to power four numbers of DC blowers (6 V, 0.6 A): one at the collector's intake, two at the collector's outlet, and one at the drying chamber's outlet (exhaust fan). The specification of the solar panel is maximum power, voltage and current as 20 W, 17 V and 1.18 A respectively. Battery specification is 12 V, 20 A hour, and charge controller as 12 V, 10 A were connected electrically to run the DC blower. The absorber plate is comprised of a flat 5 mm corrugated aluminium sheet to prevent corrosion and leaking and is black painted. The collector is tilted to match the latitude of the site. The position of the collector is inclined at an angle of 20° and faces towards the south due to the latitude of Bhubaneswar being 20.2961°N and longitude of 85.8245°E.

2.1. Fabrication Process

The dryer was made out of aluminium sheets and angle bars. The dryer's numerous parts were built using a variety of technologies. Welding is used to connecting the frames (electrical arc welding). To hold the trays in place, angle bars are affixed to the dryer's side frame. According to the latitude of the site, the collector is directed directly south and inclined at a 20° inclination. To guarantee proper air movement, a blower is provided at the collector's input. Figure 1 depicts the experimental setup of the hybrid solar dryer with the collector.

Figure 1. Experimental set up of solar dryer with collector

3. Results and Discussion

3.1. Steady-State Thermal Analysis of Absorbing Plate

The steady-state thermal analysis is performed at constant thermal loads (not varying with time).The assumption is taken for making the simulation results easy and accurate. Steady-state thermal study of Absorbing plate (length x width = 1 m×1 m) was simulated using the ANSYS 2016 program. In the meshing section, the entire domain was partitioned into several sub domains, with 86742 and 475179 elements and nodes created, respectively. By applying boundary conditions such as the quantity of heat flow conducted (10^3 W) on the top face and the convective heat transfer coefficient (100 W/mm^2/°C) on the bottom face of the absorbing plate were used to calculate temperature fluctuation, total heat flux variation, and directional heat flux variation. The final result was obtained after applying the aforesaid boundary conditions, namely the temperature variation, heat flow conducted, and the convective heat transfer coefficient. Temperature fluctuation of the absorbent plate, with a maximum temperature of 88.311°C is obtained. The absorbing plate's total heat flux variation was 0.131333 W/mm^2, and the greatest heat flux generated was 0.0765 W/mm^2. Figures 2 and 3 represent the results of the simulation.

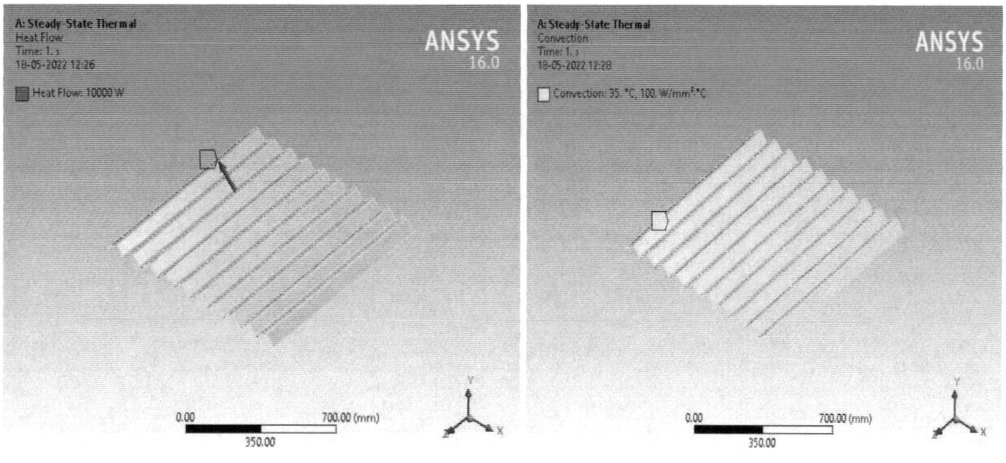

Figure 2. Application of boundary conditions: Heat flow conducted and convective heat transfer coefficient

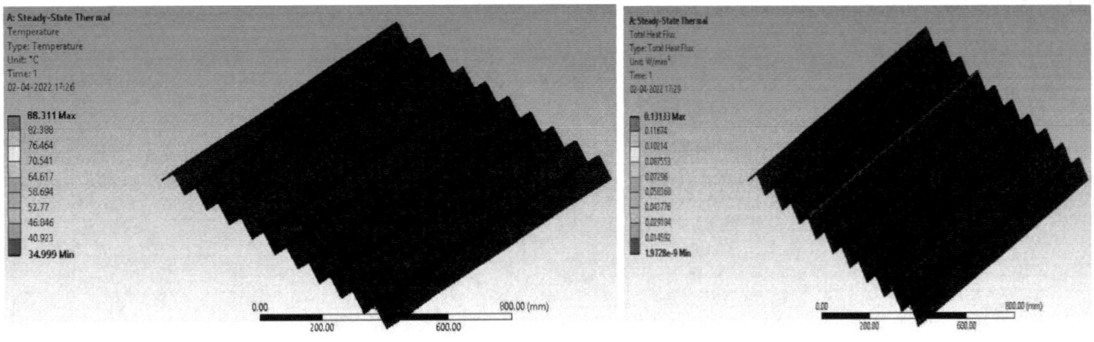

Figure 3. Temperature variation and total heat flux variation of absorbing plate

3.2 Experimental Evaluation

Figure 4 depicts several devices used to measure air velocity, humidity, temperature, and weight of food products during the experiment, including an anemometer, a hygrometer, an infrared thermometer, and a weight machine. Figure 5 depicts the drying of potatoes in the experiment.

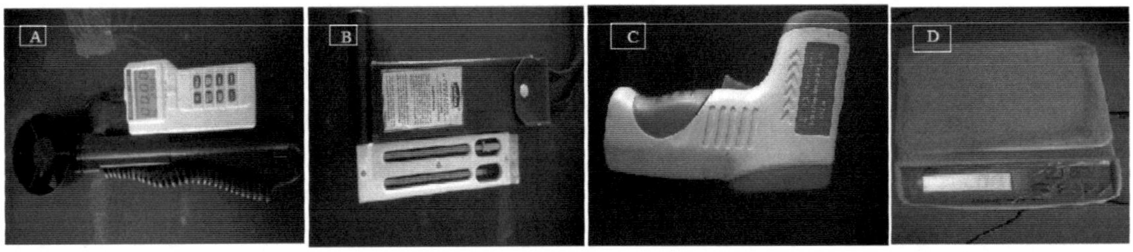

A. Anemometer B. Hygrometer C. Infrared thermometer D. Weight machine

Figure 4. Measuring instruments

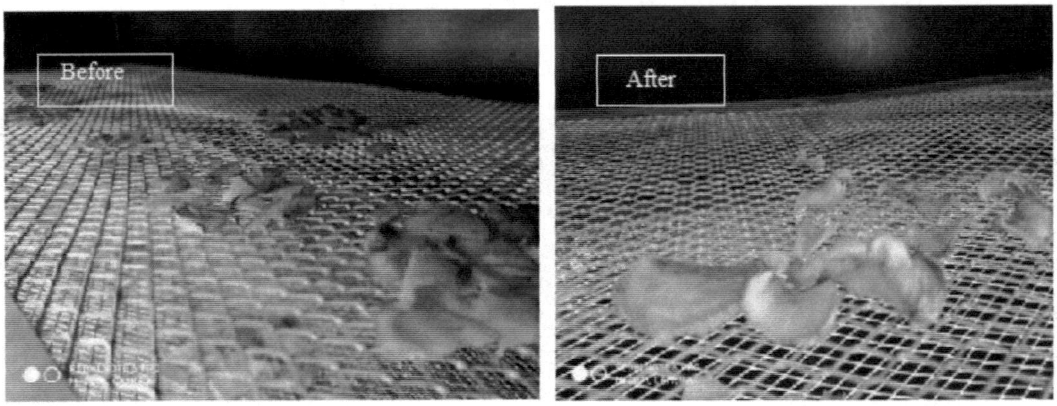

Figure 5. Drying of potatoes inside the drying chamber before and after drying

3.3. Temperature Variation of Solar Collector

The temperatures at different points of the collector such as inlet, glass plate, absorbing plate, and outlet of the solar collector have been measured every one hour time interval from 10.00 A.M. to 4.00 P.M. The temperature variation of the solar collector has been shown in Figures 6, 7 and 8. Also, the instantaneous ambient temperature is recorded. These test readings have been taken for three days considering 6 hours per day. It is observed that as the temperature of the air inside the collector increases, the humidity of air reduces, and the moisture carrying capacity of air increases. The hot air is forced to move from the collector to the drying chamber due to the presence of two blowers at the outlet of the collector. This hot air absorbs water contents from the food product to be dried. The moist air is exhausted into the atmosphere from the top of the dryer by a DC blower. On the first day of the test, it is observed that maximum temperatures of the ambient, absorbing plate and outlet of the collector are recorded as 35° C, 89° C, and 72° C respectively as shown in Figure 6. On the second and third days, a similar test has been conducted and the corresponding highest temperatures are measured as 32.9°C, 87°C, 55.7°C, and 31.3°C, 87°C, and 63.3 °C respectively as shown in the Figures 7 and 8. It is found that temperature gradually increases from 9 A.M. to 1 P.M., after which temperature decreases.

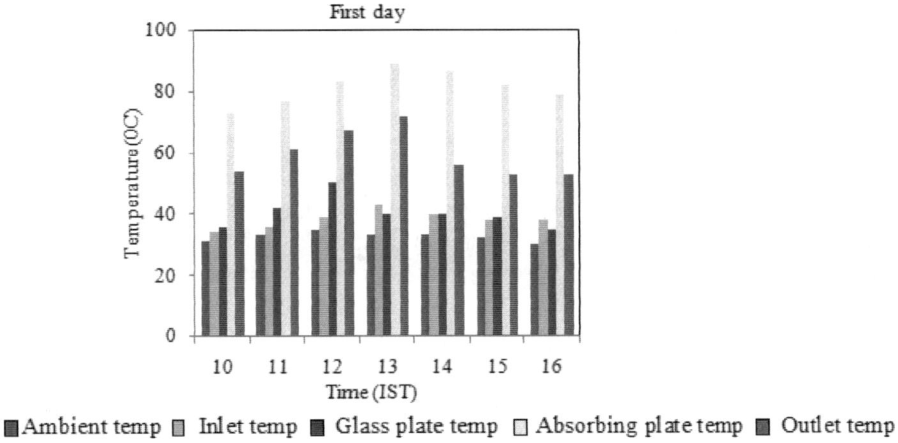

Figure 6. Temperature variation of the solar collector on 1st day vs. Time

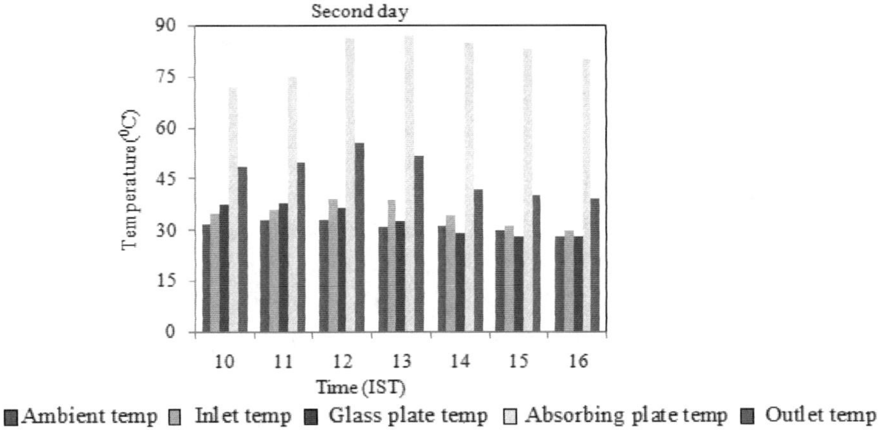

Figure 7. Temperature variation of the solar collector on 2nd vs. Time

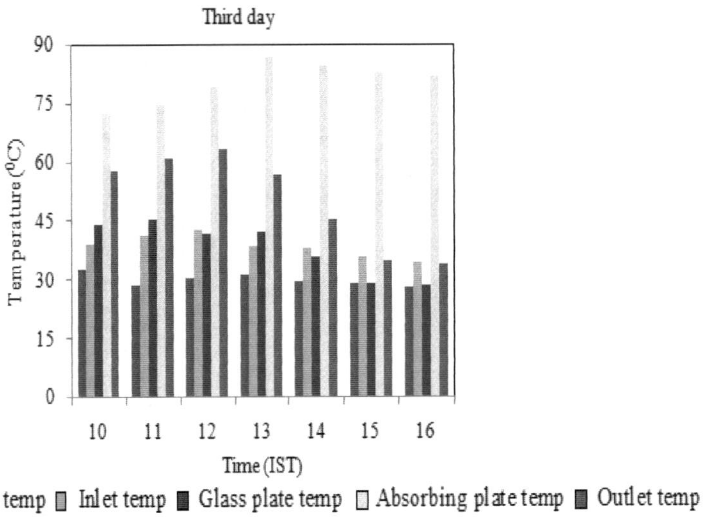

Figure 8. Temperature variation of the solar collector on 3rd day vs. Time

3.4. Validation and Comparison of Experimental and Simulation Results for Absorbing Plate

The maximum temperatures are obtained for the absorbing plate as 89°C, 87°C, and 87°C for the 1st, 2nd, and 3rd day during the experiment. Similarly, by applying the different boundary conditions for 1st, 2nd, and 3rd day, the maximum temperature is obtained as 88.311°C, 86.311°C, and 86.811°C respectively using ANSYS simulation. Therefore, the present study gives a satisfactory result and is valid for getting the optimum temperature of the absorbing plate. Table 1 represents the temperature variation for both experimental and simulation results of the absorbing plate.

Table 1. Temperature variation of absorbing plate from experimental and simulation results

SI No	1st day	2nd day	3rd day
Experimental temperature (°C)	89°C	87°C	87°C
Simulation temperature (°C)	88.311°C	86.311°C	86.811°C

3.5. Evaluation of Performance Indicators of Solar Dryer

Performance indicators such as drying efficiency, drying rate, heat utilization factor, coefficient of performance, moisture content on a dry basis, and moisture ratio have been presented below using Equations 1-4 [14].

[i]. Drying efficiency ($\eta_{thermal}$) is a measure of the overall effectiveness of the drying system. It generally depends on the mass of water evaporated, latent heat of evaporation, the surface area of the collector, and Instantaneous flux incident on a tilted surface. It is defined as the ratio of energy required to evaporate moisture from the food products by supplementing heated air.

$$\eta_{thermal} = \frac{m_{WT} \times L}{A_C \times I} \tag{1}$$

Drying efficiency ($\eta_{thermal}$) has been obtained as 43% by taking the mass of water evaporated m_{WT} as 12.75 kg from 20 kg of fresh food products, Latent heat of evaporation, L as 2260 kJ/kg, collector area Ac as 1 m^2, Instantaneous flux on the tilted surface I as 670 W/m^2 by using equation (1).

[ii]. Drying rate (DR) is defined as the time required for the mass of water to be removed from the food products.

$$DR = \frac{m_{WT}}{t} \tag{2}$$

Drying rate (DR) is evaluated as 0.607 kg/hr by taking the mass of water evaporated as 12.75 kg from 20 kg of fresh food products and the time taken, t as 21 hours by using equation (2).

[iii]. Heat utilization factor (HUF) is calculated from the collector temperature (T_C), drying chamber temperature (T_d), and ambient temperature (T_a) using Equation 3 as presented below.

$$HUF = \frac{T_c - T_d}{T_c - T_a} \tag{3}$$

Heat utilization factor (HUF) is calculated at 0.22 from the collector temperature (T_C) as 72°C, drying chamber temperature (T_d) as 64°C, and ambient temperature (T_a) as 33°C using Equation (3).

[iv]. The coefficient of performance (COP) is calculated as 0.7948 using Equation 4 as presented below.

$$COP = \frac{T_d - T_a}{T_c - T_a} \tag{4}$$

4. Conclusion

According to the design, a solar dryer can replace the traditional method of drying food. The dried food ingredients acquired through this procedure are sanitary and free of contamination. The drying rate increase in case of hybrid solar dryer as compared to open sun drying process. The drying chamber temperature is almost 55.7°C, whereas the ambient temperature is 33°C. The present study gives satisfactory results for getting the optimum temperature of the absorbing plate and are valid for both experimental as well as simulation studies. More research into the diverse structures of absorbent plates and other turbulence methods may be done to minimize heat loss and boost drying efficiency. A CFD simulation of temperature fluctuation inside the drying chamber and the thermal collector may be used to determine the overall loss coefficient.

References

[1] Chaudhri, K., Bhagoria, J.L., Kumar, V. 2022 "Transverse wedge-shaped rib roughened solar air heater (SAH) - Exergy based experimental investigation." Renewable Energy, vol. 184, pp. 1150-1164.
[2] Kumar, V. 2019 "Nusselt number and friction factor correlations of three sides concave dimple roughened solar air heater." Renewable Energy, vol.135, pp. 355-377.
[3] Kumar, V., Murmu, R. 2021 "Experimental Investigation for Thermal Performance of Inclined Spherical Ball Roughened Solar Air Duct." Renewable Energy, vol. 172, pp.1365-1392.
[4] Kaewkiew, J., Nabnean, S., Janjai, S. 2012 "Experimental investigation of the performance of a large-scale greenhouse type solar dryer for drying chilli in Thailand." Procedia Engineering, vol.32, pp. 433-439.
[5] Lingayat, A., Chandramohan, V. P., Raju, V. R. K. 2017 "Design, development and performance of indirect type solar dryer for banana drying." Energy Procedia, vol.109, pp.409-416.

[6] Chouicha, S., Boubekri, A., Mennouche, D., Bouguetaia, H., Berrbeuh, M. H., Bouhafs, S., Rezzoug, W. 2014 "Valorization study of treated deglet-nour dates by solar drying using three different solar driers." Energy Procedia, vol.50, pp.907-916.

[7] Tarigan, E. 2018 "Mathematical modelling and simulation of a solar agricultural dryer with back-up biomass burner and thermal storage." Case studies in thermal engineering, vol.12, pp. 149-165.

[8] Rabha, D. K., Muthukumar, P., Somayaji, C. 2017 "Energy and exergy analyses of the solar drying processes of ghost chilli pepper and ginger." Renewable Energy, vol.105, pp.764-773.

[9] El-Sebaii, A. A., and S. M. Shalaby. 2013 "Experimental investigation of an indirect-mode forced convection solar dryer for drying thymus and mint." Energy Conversion and Management, vol. 74, pp.109-116.

[10] Çiftçi, E., Khanlari, A., Sözen, A., Aytaç, İ., Tuncer, A. D. 2021 "Energy and exergy analysis of a photovoltaic thermal (PVT) system used in solar dryer: A numerical and experimental investigation" Renewable Energy, vol. 180, pp.410-423.

[11] Koua, B. K., Koffi, P. M. E., Gbaha, P. 2019 "Evolution of shrinkage, real density, porosity, heat and mass transfer coefficients during indirect solar drying of cocoa beans", Journal of the Saudi Society of Agricultural Science, vol. 18, pp. 72-82.

[12] Asnaz, M. S. K., Dolcek, A. O. 2021 "Comparative performance study of different types of solar dryers towards sustainable agriculture", Energy. Rep. vol. 7, pp. 6107-6118.

[13] Kuan, M., Shakir, Y., Mohanraj, M., Belyayev, Y., Jayaraj, S., Kaltayev, A. 2019 "Numerical simulation of a heat pump assisted solar dryer for continental climates", Renewable Energy.vol. 143, pp. 214-225.

[14] Chauhan, P. S., Kumar, A., Nuntadusit, C. 2017 "Heat transfer analysis of PV integrated modified greenhouse dryer", Renewable Energy. vol. 121, pp. 53-65.

IOP Conf. Series: Earth and Environmental Science 1084 (2022) 012012 doi:10.1088/1755-1315/1084/1/012012

Approximate climate analysis for assessing the suitability of Indian climate zones for hydroponic fodder applications

Souryadeep Basak[1] and Aviruch Bhatia[1]

[1] Department of Energy and Environment, TERI School of Advanced Studies, Plot No. 10, Vasant Kunj Institutional Area, New Delhi – 110 070

Email: souryadeep.basak@terisas.ac.in

Abstract. Hydroponic fodder has been proposed as a novel low-tech solution to India's fodder scarcity. Hydroponic fodder can reduce water requirements by up to 95% and use up to 70% less land due to their vertically scalable nature. The short crop cycles of approximately 7 days yields a seed to feed ratio of 1:8. This study analyses the external environmental weather conditions in Indian climate zones, assessing their suitability for the application of hydroponic fodder solutions. The weather of each climate zone was approximated by a representative city through parameter selection and optimisation by three-dimensional Euclidean distance minimisation. Simulation and analysis were carried out on the Rhinoceros software platform with the Grasshopper and LadyBug plug-ins. Hourly temperature plots, sun-path diagrams and psychrometric charts with modified predicted mean vote polygons superimposed were generated for each climate zone. The ambient weather suitability of the following representative cities was simulated to be as follows: Ahmedabad (86.70%), Kolkata (81.86%), Bengaluru (94.07%), Lucknow (73.23%) and Srinagar (30.85%). A cost benefit analysis was buttressed by a sensitivity analysis with respect to the selling price of green fodder. A preliminary cash flow analysis yielded a discounted payback period of 4 months for direct fodder sales and 20 months for sale of livestock products as a co-benefit of green fodder production.

1. Introduction

As per the 20th livestock census, India is home to the largest bovine population in the world, with approximately 192.49 million cattle, 109.85 million buffaloes, and 0.45 million mithun and yak. Despite the surplus of cattle in the country owing to, inter alia, the rising demand for milk production and draft power, the Indian cattle is generally considered to be of poorer quality vis-à-vis its global counterparts. Although India has emerged as the largest milk producer in the world, the livestock productivity of India (1538 kg/year) is well below the global average (2238 kg/year) and the principal attributable cause is the deficit in fodder supply [1]. Indian farmers, especially those without access to and the purchasing power to afford green fodder and concentrates, have traditionally relied on raw crop residues such as cotton cake or husks to feed their cattle. Fibrous crop residues have a low nutritive value as feed for ruminants due to their high fiber content, low crude protein and low level of readily available energy and are known to cause lower yields of milk [2].

This problem is exacerbated by the nationwide deficit of green fodder. According to the Indian Grassland and Fodder Research Institute, India faced a 30% shortage of green fodder in 2020. There is currently a net deficiency of 35.6% in green fodder production in India [3]. There is an immediate need to increase the productivity of fodder crops while using marginal acreage to fulfill the fodder requirements of the rising livestock population [4]. To meet the demands engendered by the burgeoning cattle population, the production of green fodder must be enhanced manifold. However, due to the rising competition between other uses for cultivable land, increasing the acreage of fodder crops is not feasible [5]. In this situation, hydroponically grown green fodder has emerged as a viable option to balance the growing demand for green fodder in the country. Not only has this provided succor to debt-ridden, impoverished farmers in the form of an alternative source of income, but it has also increased their resilience against extreme climatic events. Radhakrishnan and Gupta [6] found that despite disruptive weather events that had affected the milk-market chain and other physical infrastructure facilities in Wayanad, innovative practices like hydroponic fodder had increased the adaptive capacity of the region. This study also used livelihood diversification, i.e., the process of building a diverse portfolio of economic activities to improve the standard of living, as one of the components of the Livelihood Vulnerability Index (LVI) constructed on the definition of 'vulnerability' furnished by the Intergovernmental Panel on Climate Change (IPCC).

1.1. Hydroponic fodder
Hydroponic fodder has emerged as a feasible alternative to green fodder for livestock. The distinguishing characteristic of hydroponic fodder is that it requires no soil to grow and due to this unique property, such a system can be vertically scaled [7]. The lack of soil is compensated by the rapid growth of root mass, which provides structural stability to the fodder as well as a buffer for water [8]. The sole requirements for this process are seeds and water. No pesticides or fertilizers are required since the seed contains enough nutrition to sustain the crop for a minimum of 10 days [9]. Several seeds and food-grains may be utilized for growing hydroponic fodder, viz. maize, wheat, barley, oats, millets, and other cereals [10].

The process of production starts with soaking the seeds to soften the seed coat. Germination takes place in a warm and humid environment, commonly gunny bags. Once the sprouts appear, the seeds are placed in a gently sloped tray perforated with holes at the bottom. Both these features prevent standing water, which might lead to mold. The crop cycle of hydroponic fodder is typically around 7 days, with a seed to feed ratio ranging up to 1:8. Mohapatra et al [11] reported highest seed to feed ratio (8.4 kg per kg of seed) and minimum seasonal variability for maize, which was outperformed by wheat during the winters. A low-cost solar prototype has been developed at the TERI School of Advanced Studies campus in New Delhi, as has been shown in figure 1. The cost of the 50 kg fodder unit was Rs. 7640 with the following cost distribution: PVC structure (Rs. 4448), control system (Rs. 1892) and watering system (Rs. 1300). The average seed to feed ratio obtained in the prototype was 1:6.3 for maize fodder during the months of April and May.

1.2. Sustainability of livestock feed
From an environmental perspective, hydroponic fodder adopts a land sparing approach, in contrast to the conventional land sharing one. The land use factor and water use factor for hydroponic fodder are significantly more competitive in comparison to field grown forage, thereby meeting greater demand with lesser land [12]. This is a crucial detail in a country with a rising middle class, and consequently a rising meat and dairy consumption [1]. Due to the lack of pesticides, herbicides or any chemical products, there is no chemical runoff or seepage [3]. Phand and Pankaj [13] have reported a decrease in enteric rumination in cattle fed on a hydroponic fodder diet, which might potentially result in lesser methane emissions from the dairy industry.

2. Literature review

Naik et al [14] reviewed various models of hydroponic fodder production and reported that 1 kg of hydroponic fodder requires about 1.5-3 liters of water across 7 days. This water use efficiency was reported to be significantly high in comparison to an equivalent amount of forage grown under field conditions, in as much as 3-5% of water is utilized in producing a comparable amount of hydroponic fodder [10]. The digestibility of hydroponic fodder was also reported to be higher in terms of total digestible nutrients (TDN) and dry crude protein (DCP). In a typical low-cost hydroponic system, the seed cost accounts for 90% of operational expenses. However, there is an associated increase in livestock productivity of about 13.73%, as reported by Naik et al [15].

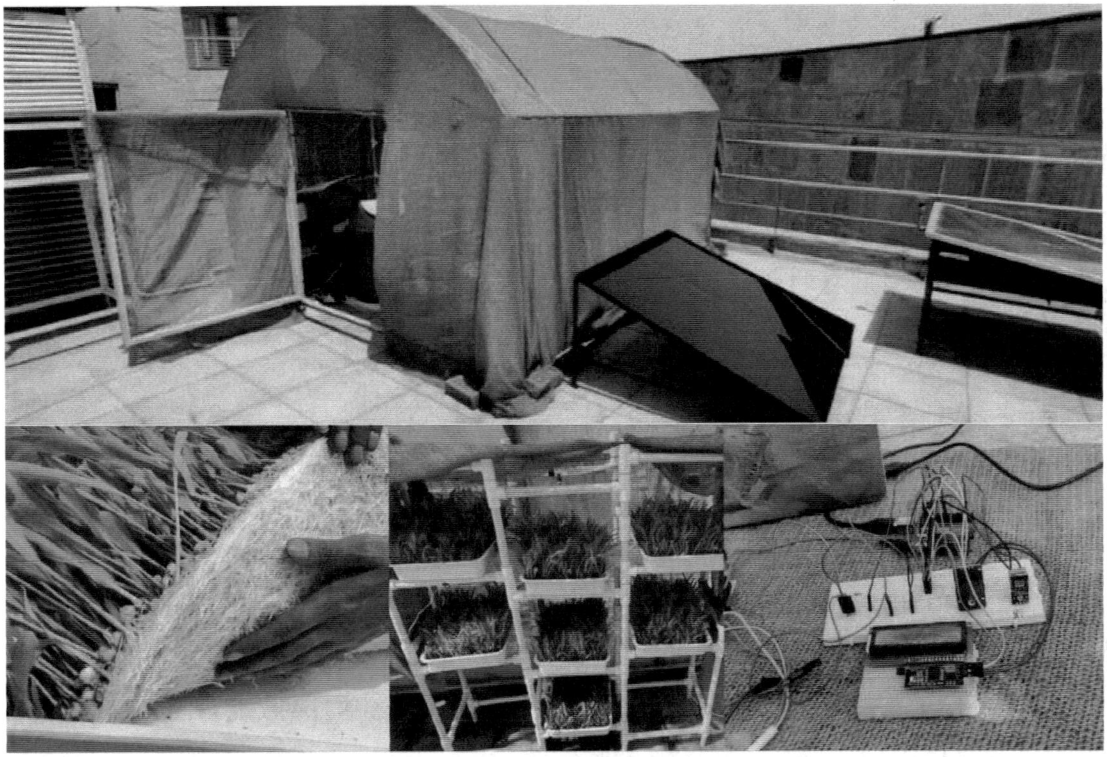

Figure 1. (Clockwise from top) Greenhouse, control system, hydroponic fodder unit, green fodder (Source: Author, TERI School of Advanced Studies, New Delhi).

Roy and Singh [16] have evaluated the Indian fodder availability scenario with respect to the increasing livestock population. Owing to desertification, loss of arable land and the predominantly small size of landholdings in rural India, fodder availability might evolve into a subject of national scarcity. Kumar et al [17] estimated that a 100 kg daily hydroponic fodder unit requires about 6.5 tons of maize seeds annually, to produce which 0.7-1.0 ha of land is required. This is a significantly high land use factor, which is made possible by the vertically scalable nature of the unit. Uddin and Dhar [18] evaluated the economic feasibility of hydroponic fodder in Kishoregonj and Jashore districts of Bangladesh, reporting a benefit to cost ratio of 1.82.

Singh et al [19] developed a temperature and humidity control system to produce hydroponic maize fodder through evaporative cooling and misting, maintaining the daily average temperature and relative humidity in the range of 28.6-30.2°C and 65.2-75.1% respectively against an ambient range of 28.5-41.4°C and 37-77% respectively.

3. Methodology

According to the National Building Code 2016, India is categorized into 5 climate zones, namely hot and dry, warm and humid, temperate, cold and composite. Bhatnagar et al [20] devised a method for approximating climate zones by representative cities. Three meteorological parameters were shortlisted for assessing representativeness of weather, *viz.*, global horizontal irradiance, dry bulb temperature and relative humidity. These three properties of the weather data were plotted on a three-dimensional space and the Euclidean distance was minimized to arrive at the optimally representative city. According to this study, the 5 climate zones of India can be most closely represented by Ahmedabad (hot and dry), Kolkata (warm and humid), Bengaluru (temperate), Srinagar (cold) and Lucknow (composite).

To approximately assess the ambient weather conditions across India, this method has been adopted for reducing the simulation iterations without compromising reliability. The EnergyPlus weather files (.epw) for the 5 cities were imported into the Rhinoceros software platform. The Grasshopper plugin together with the LadyBug toolkit were utilized for the preliminary climate analysis.

An hourly dry bulb temperature plot was generated for all 8760 hours of the year, which allows us to assess the temporal variation of temperature across the year, where the x-axis and y-axis represent respectively the month of the year and time of the day. The sun-path diagram was plotted next, which apart from generating the locus of the sun throughout the year, also codes the hourly sun-points according to the prevalent dry bulb temperature. A joint reading and correlation of the hourly temperature plot with the temperature coded sun-path diagram can enable the design of passive design strategies for hydroponic fodder greenhouses. The simplest application of such a strategy would be the design of overhangs and shades to reduce solar gain during hot spells and permit solar ingress during colder months, when the altitude of the sun is lower in the sky.

Following the hourly temperature plot and the sun-path diagram, the dry bulb temperature and the absolute humidity of all 5 representative cities was plotted on a psychrometric chart. The predicted mean vote (PMV), which is conventionally used as a parameter of human adaptive comfort, has been utilized to identify the percentage of annual hours the ambient conditions permit the cultivation of hydroponic fodder. The percentage of persons dissatisfied (PPD) threshold was increased to 50% from the ASHRAE 55 standard of 20%. The purpose of this exercise is to broaden the range of human tolerance to accommodate growing conditions for hydroponic fodder. Some other modifications to the standard PMV approach include a clothing level of 0.7 clo, low air speed of 0.1 m/s, a metabolic rate of 1.1 met and no absolute humidity constraints.

The next step was to create a comfort polygon to categorize the points on the psychrometric chart into two classes, namely comfortable and not comfortable. Two passive strategies were considered while superimposing the PMV comfort polygon on the psychrometric chart, viz. evaporative cooling and use of fans. The resulting polygon yields the final percentage of annual hours that fall within the desired range of 50% PPD as a PMV parameter.

It is important to acknowledge that the analysis of weather data is carried out in the absence of any glazing material i.e., the above method analyses the outdoor environmental suitability of different climate zones with respect to hydroponic fodder. Furthermore, the weather data of cities has been used as reference for a predominantly rural application due to the urban heat island effect (UHIE), that is associated with impermeable urban environments. Thus, the absence of any glazing material and the added effect of UHIE increase the robustness of the analysis by assuming limiting case conditions. In a practical application of the hydroponic fodder greenhouse, the façade material, and its associated properties such as U-value, solar heat gain coefficient (SHGC) and visible light transmittance (VLT) will facilitate a micro-climate inside and enable desirable temperature and humidity conditions for fodder growth, which have not been considered in this study. Therefore, this study may be considered as a baseline resource assessment before the early design stage.

4. Results

The percentage of annual hours that fall within the desirable range of temperature and humidity and shown in table 1. Four of the five climate zones appear to be suitable for growth of hydroponic fodder throughout the year, except for the 'cold' climate zone, where the ambient conditions meet desired ranges for a meager 30.85% of annual hours. This observation can be first explained and then addressed. Primarily, the 'cold' climate zone is an exceptional climate zone in a tropical country such as India. While other climate zones have a common problem of temperatures soaring above the desired threshold, the 'cold' zone predominantly requires heating strategies, in stark contrast to the other zones. The two low-cost passive strategies that were used for constructing the PMV polygon, namely evaporative cooling and the use of fans are both passive cooling techniques. To increase the percentage of annual hours that are conducive to hydroponic fodder growth, heating strategies such as passive solar gain, north wall insulation and thermal energy storage must be utilized while formulating the PMV polygon.

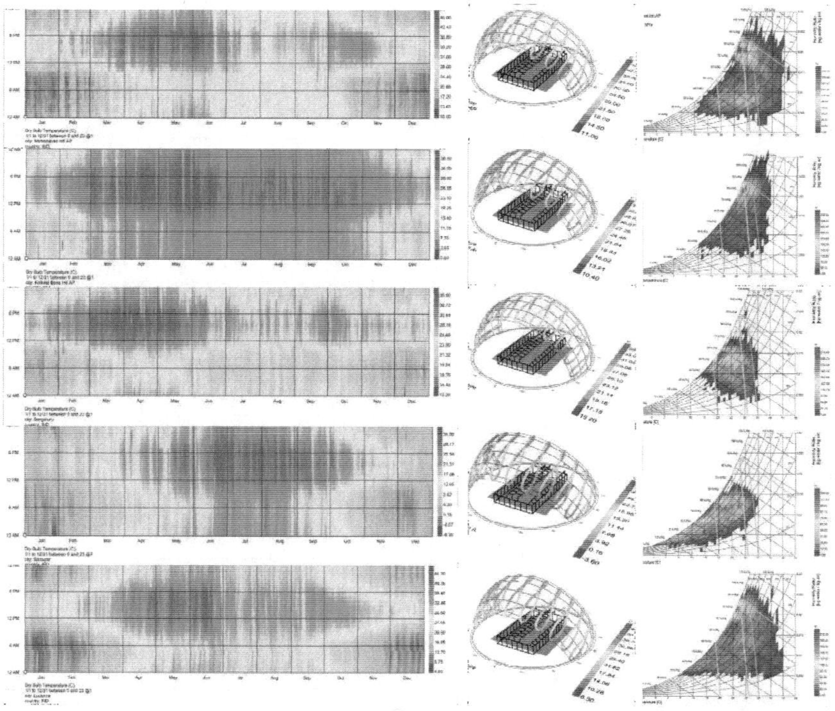

Figure 2. Hourly temperature plot, sun path diagram, and psychrometric chart for Ahmedabad, Kolkata, Bengaluru, Srinagar, and Lucknow

For the rest of the four climate zones, hydroponic fodder emerges as a suitable alternative. The ambient environmental analysis of Ahmedabad (86.70%), Kolkata (81.86%), Bengaluru (94.07%) and Lucknow (73.23%) reveal that the prevalent weather in these climate zones permits the cultivation of hydroponic fodder for a significant portion of the year. The prospect of year-round cultivation is completely achievable in these regions by carrying out fodder operations inside a low-cost shaded greenhouse.

From a joint reading of the hourly dry bulb temperature plots and the temperature coded sun- path diagram in figure 2, it can be intuitively observed that the maximum temperature sun- points are high altitude points in the sky. Therefore, a general design principle for the four climate zones apart from the 'cold' zone would be to integrate shading or radiation attenuating mechanisms to mitigate high-altitude solar radiation as a low-cost passive cooling solar strategy.

Table 1. Indian climate zones - conditions and suitability for hydroponic fodder application

Climate Zone	Mean Monthly Maximum Temperature (°C)	Mean Monthly Relative Humidity (%)	Representative City	Total Comfort Hours (% of 8760)
Hot and dry	>30	<55	Ahmedabad	86.70
Warm and humid	>30 >25	>55 >75	Kolkata	81.86
Temperate	25-30	<75	Bengaluru	94.07
Cold	<25	All values	Srinagar	30.85
Composite	Six months or more does not fall within any of the 4 categories		Lucknow	73.23

4.1. Economics of hydroponic fodder

The cost benefit analysis approach was adopted to evaluate the economic sustainability of hydroponic fodder. The analysis was performed on secondary data of 12 fodder grains provided by the Selco Foundation. The selling rate of green fodder was assumed to be Rs. 5/kg for the cost benefit analysis. The crop cycle duration, feed to seed ratio and seed price data were provided by the SELCO Foundation. The selling rate of fodder was varied from Rs. 3 per kg to Rs. 9 per kg to evaluate the sensitivity of the economic feasibility of different crops.

As can be seen from table 3, maize, sun hemp and barley appear to be the preferable choices for hydroponic fodder. However, the choice of the fodder palette depends on local factors of availability, cost, and logistics. Two economic models have been created for the hydroponic fodder unit. The primary model assumes the direct sale of hydroponic fodder at Rs. 5 per kg. The secondary model assumes the sale of livestock products as a co-benefit of hydroponic fodder. The seed to feed ratio was assumed to be 1:6. A fodder unit with 1000 kg daily green fodder output was compared between the primary and secondary models. The capital cost of such a system would be approximately Rs. 3,00,000. Operating costs were modelled as seed costs and energy costs.

Table 2. Preliminary economic feasibility of 12 fodder crops

Seed	Cycle Duration	Feed:Seed ratio	Seed Price	Revenue	Profit	Profit/month
Maize	8	6.6	15	33	18	68
Cowpea	4	6.6	150	33	-117	-878
Horse gram	4	6	30	30	0	0
Wheat	8	5	26	25	-1	-4
Sun hemp	4	10	30	50	20	150
Pearl Millet	4	3	150	15	-135	-1013
Sorghum	8	3.7	100	18.5	-82	-306
Foxtail Millet	4	4.5	60	22.5	-38	-281
Moth Bean	8	13.5	57	67.5	11	39
Barley	8	5	16	25	9	34
Kodo Millet	8	5.9	55	29.5	-26	-96
Little Millet	8	5.9	35	29.5	-6	-21

Table 3. Sensitivity analysis with respect to selling price of green fodder

Selling Price	3	4	5	6	7	8	9
Maize	43	43	68	92	117	142	167
Cowpea	-927	-927	-878	-828	-779	-729	-680
Horse gram	-45	-45	0	45	90	135	180
Wheat	-23	-23	-4	15	34	53	71
Sun hemp	75	75	150	225	300	375	450
Pearl Millet	-1035	-1035	-1013	-990	-968	-945	-923
Sorghum	-320	-320	-306	-292	-278	-264	-250
Foxtail Millet	-315	-315	-281	-248	-214	-180	-146
Moth Bean	-11	-11	39	90	141	191	242
Barley	15	15	34	53	71	90	109
Kodo Millet	-118	-118	-96	-74	-51	-29	-7
Little Millet	-43	-43	-21	2	24	46	68

Naik et al [21] reported 13.73% increase in quantitative milk yield due to integration of hydroponic fodder in milch cattle diets and Reddy et al [22] reported an income enhancement of Rs. 25-50 per head of cattle per day in the district of Satara, Maharashtra. Based on these findings, the daily average fodder requirement per cattle was assumed to be 12 kg and the income enhancement was assumed to be Rs. 36 per head of cattle per day. The monthly profit from the primary model was Rs. 75,000 and Rs. 14,640 for the secondary model, resulting in a discounted payback period of 4 months and 20 months respectively. The discount rate was assumed to be 4.65%.

5. Further Research

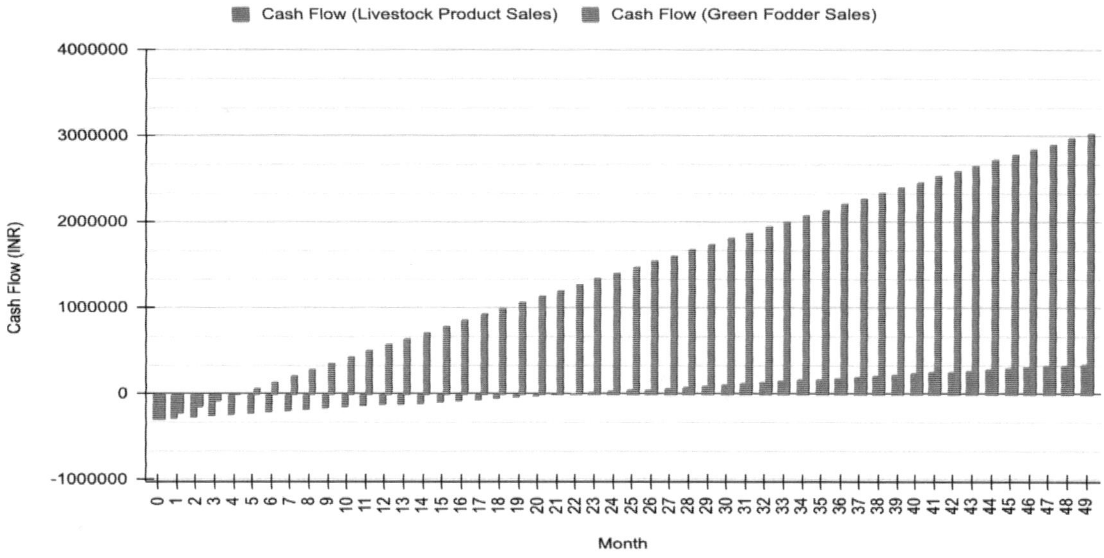

Figure 3. Cash flow analysis for primary and secondary fodder sales

After an approximate analysis of the external environmental suitability of Indian climate zones, further research can be directed at the standardization of the hydroponic fodder cultivation system. While the present study explores the spatio-temporal variation of temperature and humidity, future research may be directed towards greenhouse design, hydroponic fodder structure and affordable micro-climate control systems.

In terms of greenhouse design, façade material selection, greenhouse geometry, shading structures and orientation are a few areas where design optimization can be carried out. Local availability and affordability must be balanced with reliability to achieve a robust standardized design. Furthermore, to facilitate hydroponic fodder cultivation on agricultural land, such greenhouses have to be designed in such a fashion so as to not damage fertile land. Such minimally invasive environmentally benign design can be achieved by using poly-vinyl chloride (PVC) or polypropylene random copolymer type 3 (PPR) pipes as the dominant construction material, which obviates the requirement of sand, cement, steel, and bricks in the greenhouse installation process, thereby preserving the fertility of land and bypassing the need for land use change from a legal perspective.

The hydroponic fodder unit can be standardized with much greater ease in comparison to the greenhouse. Visual cues and diagrammatic manuals can help to overcome the problems of education access and diverse languages during diffusion in rural regions of India. While the fodder unit can be fabricated with locally available materials, such standardization will enable rapid diffusion and easy adoption by agricultural communities due to ease of use. Furthermore, standardization can pave the way for commercial production of such fodder unit structures, which might bring down the price per unit, in accordance with the economies of large scale.

An inexpensive micro-climate control system may be developed using the Arduino microcontroller board and a network of sensors and actuators. Digital sensors such as the DHT- 11 or DHT-22 may be used for monitoring the temperature and humidity, which may be regulated in turn by a system of actuators such as sprinklers and fans connected to the controller via relays. Alternatively, the use of microcontroller boards such as the ESP32 allows for enhanced functionalities such as wireless fidelity (Wi-Fi), Bluetooth and internet of things (IoT) applications. The use of direct current (DC) appliances will also allow for the integration of solar panels without the requirement of an inverter. This is a crucial intervention because minimizing the capital cost of such a standardized system is an objective function when designing for areas with limited paying capacity and without energy access.

6. Conclusion

Hydroponic fodder is a low-tech scalable solution that can boost livestock productivity while providing a secondary income source for agrarian communities. The ambient climate in four out of the five climate zones amply supports the operation of hydroponic fodder units. With the design of standardized greenhouses and low-cost control systems, such solutions can be scaled up to assure import independence in the foreseeable future, which might be economically crippling for the nation with the largest livestock population in the world.

References

[1] Vijay D, Gupta CK and Malviya DR 2018 Innovative technologies for quality seed production and vegetative multiplication in forage grasses *Curr. Sci.* **114** 1 148-154

[2] Harb M and Al-Awawdeh F 2008 Forage: situation, challenges and solutions *Jordanian Agri. Engg. Mag.* **85** 18-23

[3] Singh DN, Bohra JS, Tyagi V, Singh T, Banjara TR and Gupta G 2022 A review of India's fodder production status and opportunities *Grass Forage Sci.* **77** 1 1-10

[4] Dahiya BS and Kharab RPS 2003 Fodder seed production constraints and strategies *Forage Res.* **29** 10-17

[5] Kumar S, Agrawal RK, Dixit AK, Rai AK, Singh JB and Rai SK 2012 *Forage Production Technology for Arable Lands* Indian Grassland and Fodder Research Institute, Jhansi 12

[6] Radhakrishnan A and Gupta J 2017 Vulnerability of Dairy-Based Livelihoods to Climate Variability and Change: A Study of Western Ghats Region in Wayanad, Kerala *Curr. Sci.* **113** 1 123

[7] Emam MSA 2016 The Sprout Production and Water Use Efficiency of some Barley Cultivars under Intensive Hydroponic System *Middle East J. Agric.* **5** 2 161-170

[8] Jan S, Rashid Z, Ahngar TA, Iqbal S, Naikoo MA, Majeed S, Ahmad Bhat T, Gull R and Nazir I 2020 Hydroponics – A Review *Int. J. Curr. Microbiol. App. Sci.* **9** 8 1779-1787

[9] Girma F and Gebremariam B 2018 Review on Hydroponic Feed Value to Livestock Production *J. Sci. Inn. Res.* **7** 4 106-109

[10] Al-Karaki GN and Al-Hashimi M 2012 Green Fodder Production and Water Use Efficiency of Some Forage Crops under Hydroponic Conditions *ISRN Agron.* **2012** 1-5

[11] Mohapatra KK, Mohapatra S, Ekka R, Behera RC and Mohanta RK 2019 Variations in Round-the-Year Fodder Production in a Low-Cost Hydroponic Shed *Natl. Acad. Sci. Lett.* **42** 5 383-385

[12] Kide W, Desai B and Kumar S 2015 Nutritional improvement and economic value of hydroponically Sprouted maize fodder *Life Sci. Int. Res. J.* **2** 2 76-79

[13] Phand S and Pankaj PK 2021 Climate-Resilient Livestock Farming to Ensure Food and Nutritional Security *Clim. Change Resilient Food Syst.* 381-398

[14] Naik PK, Swain BK and Singh NP 2015 Production and Utilisation of Hydroponics Fodder *Indian J. Anim. Nutr.* **32** 1 1-9

[15] Naik PK, Dhuri RB, Karunakaran M, Swain BK and Singh NP 2014 Effect of feeding hydroponics maize fodder on digestibility of nutrients and milk production in lactating cows *Indian J. Anim. Sci.* **84** 8 880-883

[16] Roy MM and Singh KA 2008 The fodder situation in rural India: future outlook *Int. For. Rev.* **10** 2 217-234

[17] Kumar R, Mathur M, Karnani M, Choudhary SD and Jain D 2018 Hydroponics: An alternative to cultivated green fodder: A review *J. Entomol. Zool. Stud.* **6** 6 791-795

[18] Uddin M and Dhar A 2018 Socioeconomic analysis of hydroponic fodder production in selected areas of Bangladesh: prospects and challenges *SAARC J. Agric.* **16** 1 233-247

[19] Singh SK, Patil AK, Kautkar S and Dwivedi PN 2021 Thermal performance for hydroponic maize fodder production *The Bioscan* **16** 1 199-202

[20] Bhatnagar M, Mathur J, Garg V and Iqbal J 2018 Development of a method for selection of representative city in a climate zone *Proc. Building Performance Analysis Conf.* 510-517

[21] Naik PK, Gaikwad SP, Gupta MJ, Dhuri RB, Dhumal GM and Singh NP 2013 Low-cost devices for hydroponics fodder production *Ind. Dairy* **65** 10 68-72

[22] Reddy GVN, Reddy MR and Reddy KK 1988 Nutrient Utilisation by Milch Cattle Fed on Rations Containing Artificially Grown Fodder *Ind. J. Anim. Nutr.* **5** 1 19-22

Acknowledgement

The author would like to extend their gratitude to the IKEA Foundation for providing the funds for prototyping and Engineers Without Borders UK for facilitating the design process during the Efficiency for Access Design Challenge 2020-21. The author would also like to acknowledge Mr. Shingle Sebastian from the SELCO Foundation, who provided operational data on the economics of various fodder crops being prototyped in southern India.

Soil dissolved organic carbon leaching characteristics of the coastal saline seepweed wetlands in the Wuleidaowan national wetland

Yijie Zhang[1], Haotian Zhang[2], Yihang Zhang[2], Tian Xiujun*

Department of Environmental Engineering, College of Civil Engineering, Beijing Jiaotong University, Beijing, China, 100044

*Corresponding author: 1227032587@qq.com

Abstract. As the most active component of soil TOC (Total Organic Carbon), DOC (Dissolved Organic Carbon) is either adsorbed by the soil and used by plants or transported by soil water into the water column, where it participates in the broader carbon cycle. As a substantial "mobile carbon pool" in soils, its leaching process is an essential pathway for the loss of soil TOC. This paper analysed the investigated the leaching characteristics of soil DOC by simulating leaching in an indoor soil column based on the spatial distribution of soil TOC and DOC contents in order to reveal the leaching and migration characteristics of soil DOC at different depths and its main influencing factors. The DOC content of soil leachate varies somewhat under conditions such as PH and soil phosphorous content. However, the overall trend is that the DOC content of soil leachate generally increases with the depth of the soil layer. The DOC content in the soil leachate gradually decreased with increasing leaching time. In the absence of exogenous carbon recharge, the loss of TOC from the upper layers of the soil was significantly more remarkable due to the prolonged leaching. In addition, soil TOC loss declined after leaching and decreased significantly with increasing soil depth. Because the soil DOC is adsorbed and trapped in other layers during leaching, the DOC in the upper layer was always lost during the leaching test. The experiments used highly concentrated DOC solutions as drench solutions. According to the analysis of the intensity of the fluorescence centres of each soil layer in the three-dimensional fluorescence pattern of the drench solution of each soil layer, the ability of different soil layers to adsorb and retain organic matter of different fractions differs. In addition, the selected soil layers planted with Saline seepweed show a general decreasing trend in the ability to adsorb and retain organic matter as the depth increases.

1. Introduction

Terrestrial ecosystems are essential to the global carbon cycle and dominate the global carbon balance. Estimated terrestrial ecosystems store 2,100 Gt of carbon[1], two-thirds of which is stored in soils[2]. Soil carbon stocks in wetland ecosystems are about 450 Pg (Pg = 10 g), accounting for one-third of global soil stocks[3]. Coastal wetland ecosystems, which are included in wetland ecosystems, are in the most active transition zone between land, sea and atmosphere. Their carbon cycle is more complex than other inland wetland types[4]. They are also subject to frequent hydrological fluctuations and frequent leaching processes due to the periodic tidal inundation of the sea[5]. These characteristics impact water infiltration and colloidal material formation, which inevitably makes the leaching process different from other ecosystems[6]. Therefore, studying the mechanisms of carbon sequestration and leaching in coastal wetlands is essential. Dissolved organic carbon (DOC) is the most active component of total organic carbon (TOC) in soils and is characterised by its mobility, ease of decomposition and high biological activity[7]. In addition, DOC sorption on soil can affect not only the carbon cycle but also the soil's nutrient cycle. At the same time, DOC can influence the behaviour of pollutants in soil and groundwater as a carrier of pesticides and heavy metals[8]. Soil DOC leaching has been found to be an essential pathway for soil TOC loss. As an environmental indicator, it is of great significance for studying the carbon cycle and the environment[9]. Especially in coastal wetland ecosystems, which play a significant role in regulating regional climate, the ecological significance of studying the soil DOC leaching characteristics of wetlands is even more critical.

The quantity and quality of DOC in soils vary during soil leaching due to sorption, and the sorption behaviour of DOC varies from soil layer to soil layer depending on the chemical nature of the solution and the soil mineral fraction. Therefore, based on the current focus on content, the study of DOC structure during leaching is a deeper and more complete study of the soil DOC leaching process. The research on soil leaching in coastal wetlands is not yet systematic. This study was carried out to investigate the leaching and transport characteristics of DOC in coastal wetland soils using indoor soil column simulations, with a view to revealing the dynamic changes of DOC content and structural characteristics in different soils, which can help to comprehensively understand the environmental behaviour of DOC content distribution and transport transformation processes in soil profiles, and provide more comprehensive basic data for the study of soil carbon in coastal wetlands.

2. Materials and methods

2.1. Test site overview

The test site is in Weihai Wuweidaowan National Wetland, in Nanhai New District, Weihai City, Shandong Province. The wetland is at the eastern end of the Shandong Peninsula and is bordered by the Yellow Sea to the north, east and south. The area has a temperate monsoon climate, with significant changes in the seasons and monsoon advance and retreat. Compared with inland areas of the same latitude, it is characterised by abundant rainfall, moderate annual temperatures and a mild climate. Due to its proximity to the Yellow Sea, the climate is regulated by the sea, which means that it is cold in spring, cool in summer, warm in autumn and winter, with a small temperature difference between day and night, a long frost-free period, high winds and high humidity. The average temperature in the sampling area is 12.6°C, with extreme highs of 35°C and lows of -13.8°C. The average annual rainfall is 716.1 mm and is unevenly distributed, with 14% in spring, 59% in summer, 23% in autumn and 5% in winter. The sampling area is located in a coastal wetland, i.e. a transitional zone where terrestrial and marine ecosystems intersect. The soil type is saline soil, and the soil vegetation is saline seepweed. This annual herb grows mainly in saline wetlands on seashores, lakesides and deserts and is a typical saline indicator plant.

2.2. Soil Sampling

The experimental soil sampling was carried out on 20 May 2021. Ten soil profiles were selected in the study area using the traditional stratified random sampling method with dense populations of alkali pine vegetation. The ten soil profiles were divided from top to bottom into layer I (0-10 cm), layer II (10-20 cm), layer III (20-30 cm), layer IV (30-40 cm), layer V (40-50 cm), layer VI (50-60 cm), layer VII (60-70 cm), layer VIII (70-80 cm), layer IX (80-90 cm) and layer X (90-100 cm), with three replicates for each soil profile. Soil samples from the same profile were mixed well, and each layer of soil samples collected was picked out of debris, roots and other debris and left to dry naturally in the laboratory. After drying, the samples are sieved and ground. One part is used to determine basic physical and chemical properties; the other is used to fill the soil column.

2.3. Soil column leaching experiments

2.3.1. Preparation of soil columns and high concentration DOC solutions. The soil columns were made of Plexiglas and were 10 cm high with an outer diameter of 5 cm and an inner diameter of 4 cm. A removable circular cap was placed on the top of the column with a 2 cm hole to add soil samples and lysate. In addition, the bottom of the column is sealed with a 2 cm aperture for the leachate collection.

Ten groups of soil samples are placed in ten different columns with 2-3 mm skimmed cotton (after compaction), 2-3 cm quartz sand (70 mesh) and 100 g of soil sample (quartz sand corrected soil, the height of soil sample approx. 6 cm, the density of soil approx. 1.33 g-cm-1), followed by a layer of skimmed cotton and quartz sand (same as at the bottom of the column). To ensure a consistent firmness of the soil inside the column, the soil is filled in 3 stages and compacted appropriately (concerning the original state of the soil).

A high concentration DOC solution was used as a drenching solution for the experiment using soil samples covered with vegetative saline seepweed. The dried saline seepweed was harvested on 31 October 2021 and returned to the laboratory until it was completely dried out. The dried saline seepweed was ground into a powder for use. The powder was added to deionised water and soaked for 24 hours in a ratio of 1:10 (weight: volume). The mixture was correctly shaken to distribute the powder evenly in the solution, added in batches to a centrifuge tube, and centrifuged at 3000 rpm for 15 minutes. The supernatant was passed through a 0.45 μm pore size glass fibre membrane using a vacuum pump to obtain a highly concentrated DOC solution. The high concentration DOC solution was stored in a refrigerator at 4°C.

2.3.2. Simulation of layer-by-layer leaching. Ten soil columns were subjected to leaching experiments. A peristaltic hose is inserted into the top hole of the column and connected to a peristaltic pump, which allows the peristaltic pump to be controlled to add the drench solution at a constant rate while ensuring that each batch of columns is added at the same rate, reducing human error in the experiment. The drench solution is collected through a small hole at the bottom of the column.

Using deionised water as the drench solution, the drench solution is pumped into the column at a constant rate, and collection begins when the column is entirely wet. The drench solution begins to flow, giving 10 ml of drench solution for four hours of operation per day. This is done five times daily, giving 50 ml of leachate. At the end of each column, the soil is removed from the column, air-dried, ground, sieved and stored.

The new soil samples were re-dispensed into the columns, and the drenching experiments were continued using a high concentration DOC solution as the drenching solution. The leachate was pumped into the column using the same pump speed, and the previous operation was repeated to collect the leachate.

2.4. Sample analysis
Twenty sets of collected drench solutions were filtered using a 0.45 um glass fibre membrane, after which the TC content and DOC content of the drench solutions were measured using a total organic carbon analyser. The structural characteristics of DOC in the drench solutions were analysed using a fluorescence spectrophotometer.

3. Result and Discussion

3.1. DOC content variation in leaching solution from various soil depth during deionised water leaching process

3.1.1. Characteristics of DOC content in leaching solution from various soil depth. As can be seen from Figure 1, the DOC content of the leachate obtained with deionised water at different depths varied considerably during the one-week leaching cycle, indicating that the soil layers differed in their sorption of the main constituents of DOC and that DOC was more likely to be lost and migrated. This result is consistent with the study by Xiong Li et al.[10] oon the DOC leaching characteristics of natural forest soil of Quercus Chinensis, i.e. the DOC concentration of leachate from different soil layers increased with the depth of the soil layer. In the present leaching experiment, when comparing the differences in DOC content in the leachate of each soil layer by leaching deionised water, it was found that the increase of DOC content in the leachate of wetland soils with increasing soil depth was not evident after leaching the upper soil layer, i.e. the upper soil layer leached higher DOC content to the lower soil layer. In comparison, the lower soil layer leached less DOC content, which may be related to the upper soil layer's higher TOC and DOC content. This is also consistent with the experimental results of Jiang Hechao[11] using different fertiliser treatments for DOC leaching solutions.

The DOC content in the first 50 cm of the drench solution was high in the seepweed wetland and then suddenly decreased at 50-60 cm depth, mainly due to the effect of DOC sorption by the soil. Due to the relatively short leaching cycle and the air-dried and ground soil, microbial respiration of organic matter mineralisation in the soil had little effect on soil DOC during leaching. It has also been shown that soils preferentially adsorb aromatic and humic substances from DOC[12]. The influence of organic matter on soil DOC sorption is currently controversial. However, there is a consensus that the higher the clay content, the greater the DOC adsorption capacity of the soil. Therefore, the sudden decrease in DOC content of soil leachate at 50cm-60cm depth may be due to the influence of soil physicochemical properties on the sorption-desorption mechanism of DOC in wetland soils. However, the influence mechanism needs to be further analysed and verified.

In a study by Yu Yuanchun *et al.*[13] on the DOC sorption characteristics of fir forest soils, it was noted that phosphorus inhibited DOC sorption by the soil, while Fe and Al in the soil promoted the amount of DOC sorbed by the soil. It can be inferred that soils' phosphorus and soil Fe and Al content differ at different depths. It is also evident from the data that the DOC content of the leachate gradually increases in the depth range of 0cm-50cm and 60cm-100cm, indicating that the phosphorus content of the soil increases again, and the Fe and Al content decreases again with increasing depth. Because when the phosphorus content in the soil is high, and the iron and aluminium content is low, the soil's ability to fix and adsorb doc will become weak. Hence, DOC is easier to flow, resulting in a high DOC content in the leachate. The 50cm DOC content of the leachate shows a sharp decrease in the fault, indicating a sharp decrease in phosphorus and a sharp increase in iron and aluminium in the soil at 50cm.

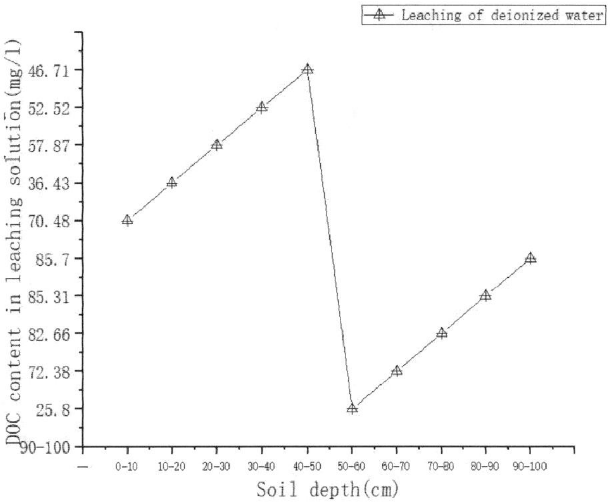

Figure 1. Change of DOC content in leachate of deionized water in different soil layers

3.1.2. Structural and component characteristics of DOC leaching solution

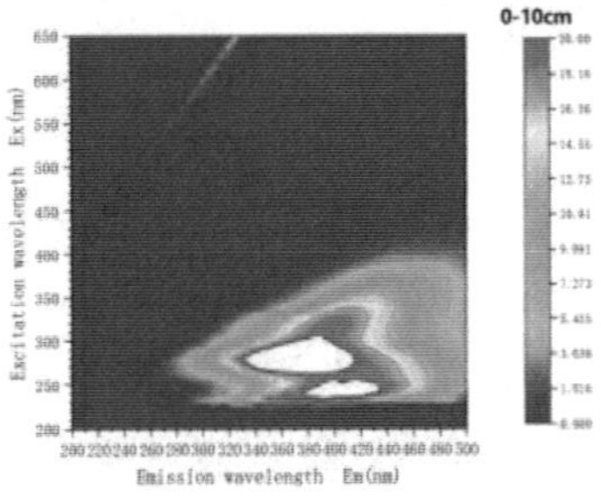

Figure 2.

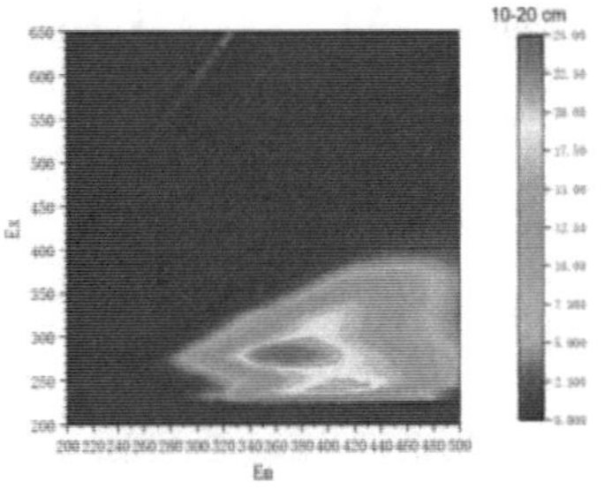

Figure 3.

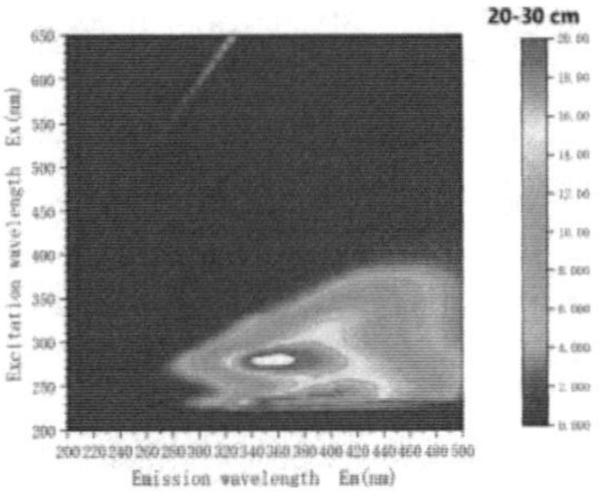

Figure 4.

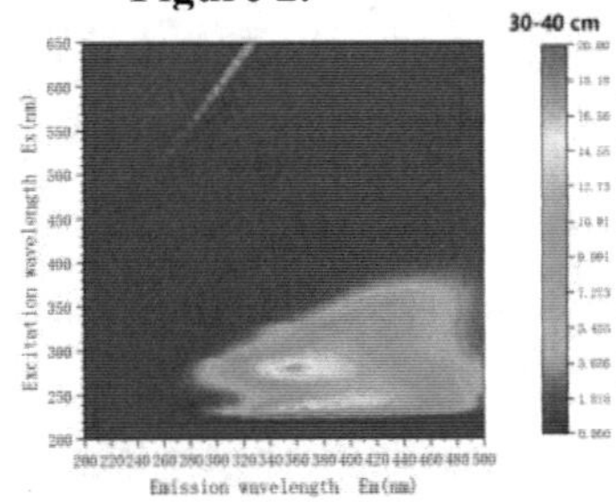

Figure 5.

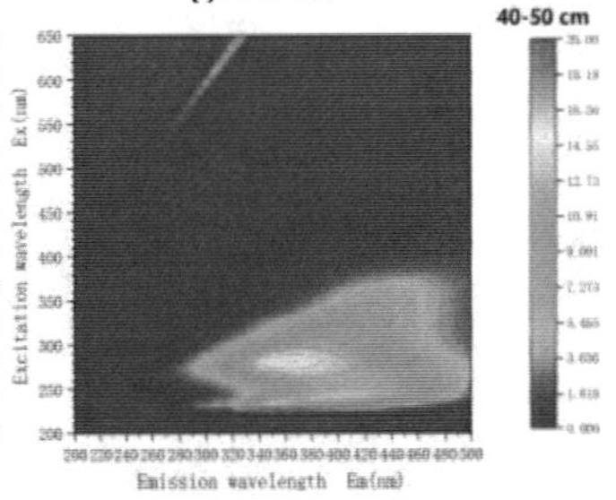

Figure 6.

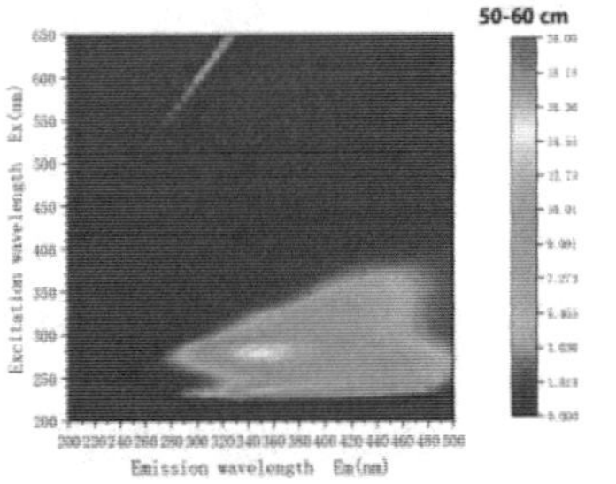

Figure 7.

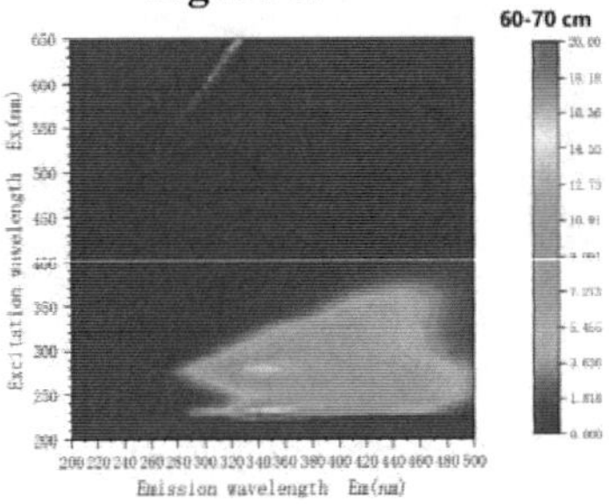

Figure 8.

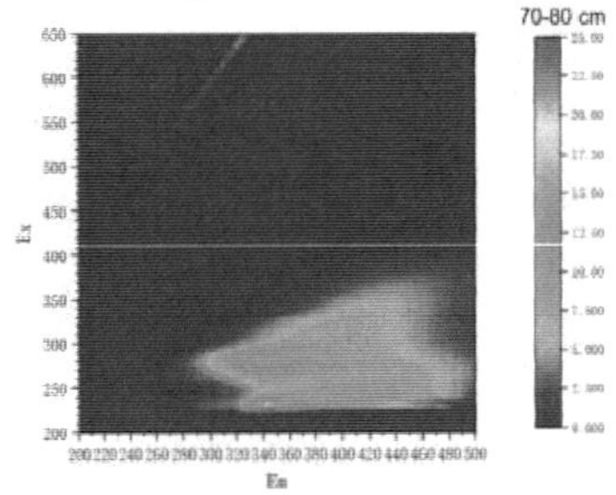

Figure 9.

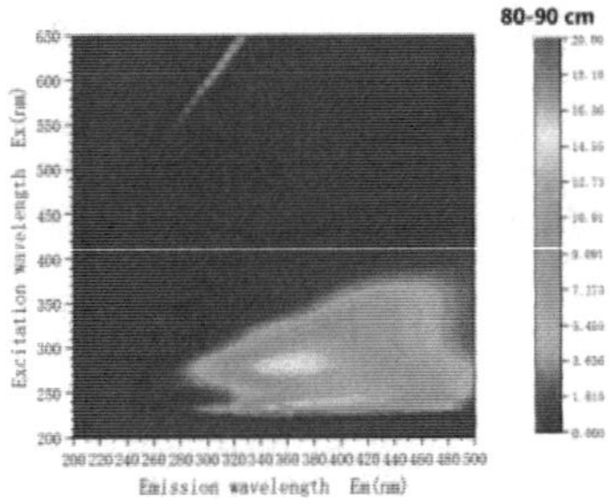

Figure 10.

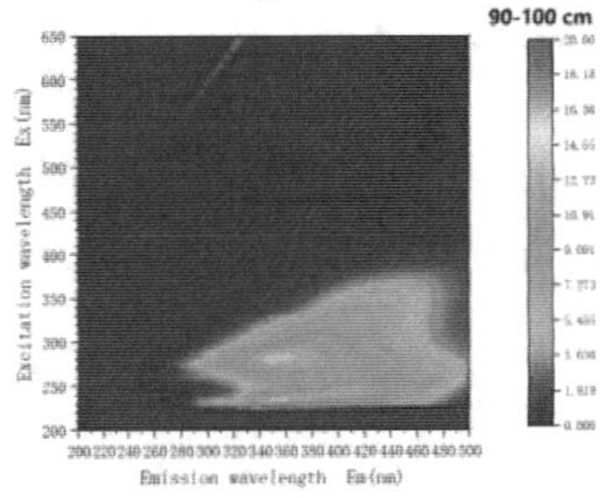

Figure 11.

According to the analysis of Fig 2-11, which is the three-dimensional fluorescence patterns of the drench solutions of each soil layer, the fluorescence intensity of the fluorescence centres in zones 1, 2 and 4 tended to decrease gradually with increasing depth of the soil layer, indicating that the concentration of aromatic proteins, hydrophobic acids, fulvic acids and tryptophan-like substances in the deionised water solution of organic matter gradually decreased. The concentration of organic matter in the deionised water solution gradually decreased. When looking at the fluorescence intensity in zone 4 of the fluorescence spectrum, it can be seen that the intensity in zone 4 is fragile in the 50-100 cm layer, indicating that the hydrophobic and fulvic acid content is low in the 50-100 cm layer.

The fluorescence intensity of the fluorescence centre of the deionised water drench solution is stronger and significantly higher in the 0-30 cm layer than in the other layers, indicating that this layer has a higher content of aromatic proteins, hydrophobic acids, fulvic acids and tryptophan-like substances. These substances have a higher degree of aromaticity and humification and are more easily leached out by the drenching action, which is explained by the organic carbon content and degree of humification in the surface layer of the soil and by the higher content of residues from the decomposition of organic matter. However, as the depth of the soil increases, the amount of organic matter containing higher aromatic compounds decreases, resulting in less organic matter being leached out during leaching. The inference of the organic matter content of the soil layers at each depth is also consistent with the findings of previous experiments on the variation of TOC and DOC with soil depth.

3.2. DOC content variation in leaching solution from various soil depth during leaching process of high-concentration DOC solution

3.2.1. Characteristics of DOC content variation in leaching solution and DOC retention amount from various soil depth

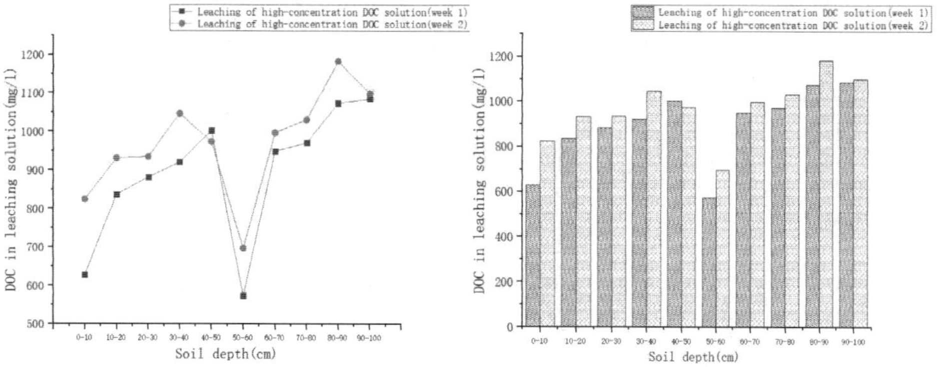

Figure 12. Change of DOC content in leachate of high concentration DOC solution

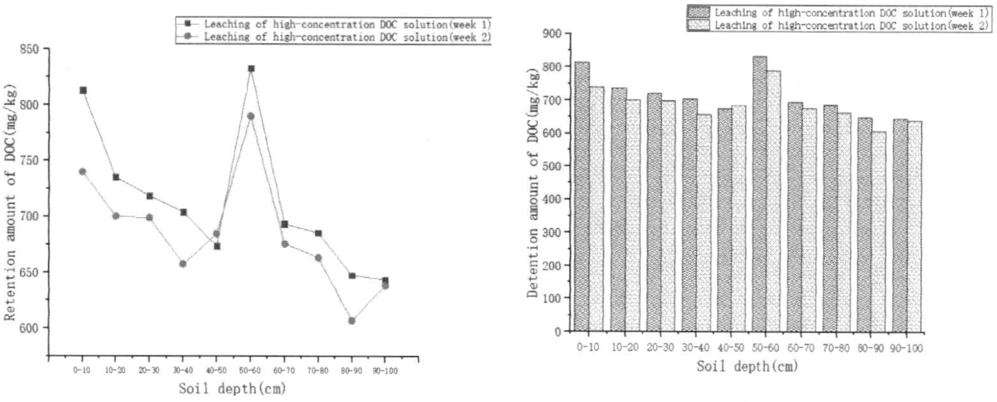

Figure 13. Changes in DOC retention in different soil layers during leaching

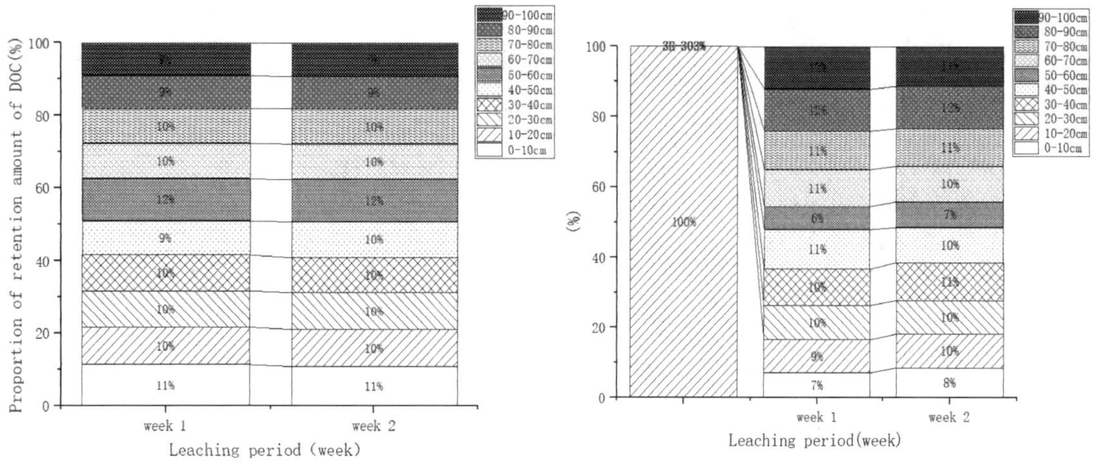

Figure 14. Proportion of retention amount of DOC and DOC content in leaching solution in two weeks

As shown in Fig. 12 and Fig. 13, the DOC content of the leachate decreased during soil leaching when using a high DOC concentration solution as a leachate solution to simulate layer-by-layer leaching experiments, indicating that DOC was trapped in the soil by sorption. In the first week of the leaching experiment, the DOC content of the leachate showed an overall increasing trend with increasing soil depth, i.e. the DOC content trapped in the soil showed an overall decreasing trend with increasing soil depth.

In addition, the DOC content retained in the 0-60 cm layer was significantly higher than that retained in the 60-100 cm layer. This is because DOC is leached from the surface soil downwards, and after being adsorbed layer by layer between the 0 and 60 cm soil layers, it reaches the soil layer below 60 cm with low concentration and less aromatic and humic substances, so it has a weaker affinity for the soil and cannot be effectively adsorbed by the soil, thus limiting the full capacity of the soil below 60 cm.

This indicates that the chemical properties of the DOC itself have a more significant influence on the sorption behaviour of the subsoil than the physicochemical properties of the soil. In addition, it can be seen from the graph that the maximum retention is reached in the 50 to 60 cm soil layer. In Kaiser K's study[14] it was indicated that the lower the organic carbon content, the more favourable the sorption of DOC. As shown in Fig. 13, the maximum retention of DOC during soil leaching in this experiment occurred in the 50-60 cm layer, where the organic carbon content was lowest, which is the same conclusion as in Kaiser K's study.

In the second week of the leaching experiment, the trend in DOC content and soil retention of DOC remained the same as in the first week. However, it is clear that the DOC content of the leachate in the second week was significantly higher than in the first week, i.e. the soil retention in the second week was lower than in the first week. This result indicates that after one cycle of high DOC leaching, the ability of the soil to retain DOC decreases, but it still has a high retention capacity. As shown in Fig. 14, the proportion of DOC retained by different soil layers was not significantly different between the two weeks.

3.2.2. Structural and component characteristics of DOC leaching solution from high-concentration DOC solution. For highly concentrated DOC solutions(Fig.15), there are strong fluorescent centres in zones 1 and 2, with a clear fluorescent centre in zone I at λ ex/ λ em = 275/360, where the fluorescent material is mainly tryptophan-like. In zone II, there is a clear fluorescence centre at λ ex/ λ em = 235/360 and the fluorescent material is mainly aromatic proteins such as tyrosine. There is also some fluorescence intensity in zone 4, with the fluorescence centre at roughly λ ex/ λ em=235/400, demonstrating the presence of some amount of hydrophobic and fulvic acids in the high concentration DOC solution.

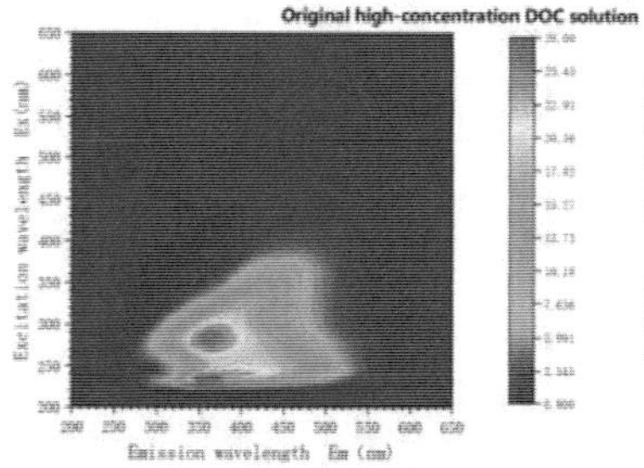

Figure 15

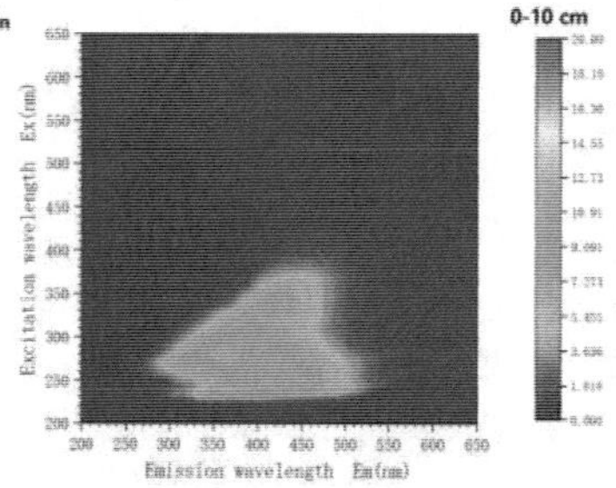

Figure 16

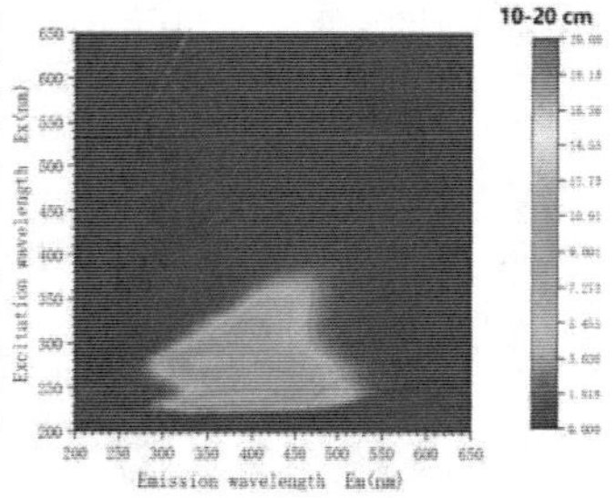

Figure 17

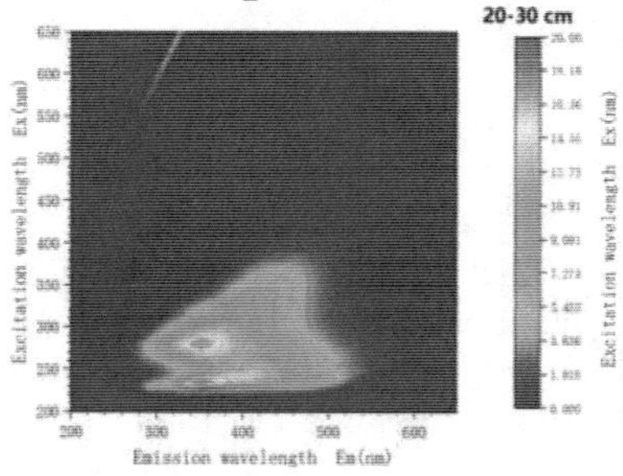

Figure 18.

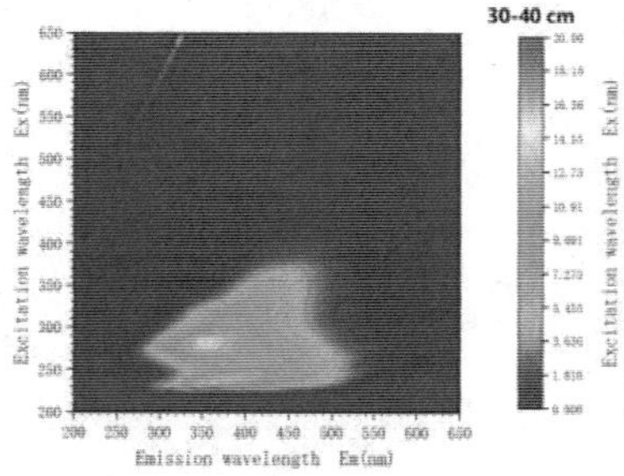

Figure 19.

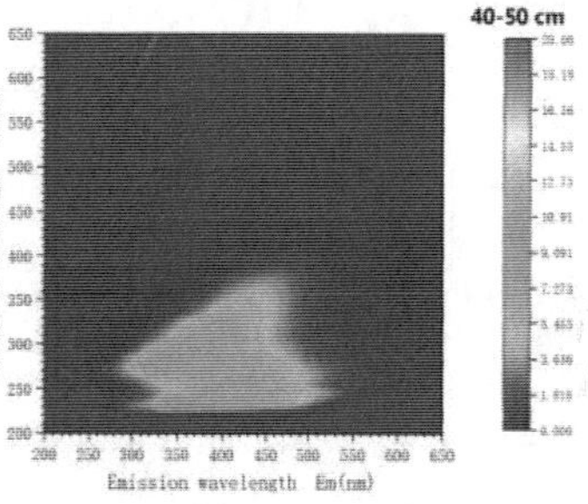

Figure 20.

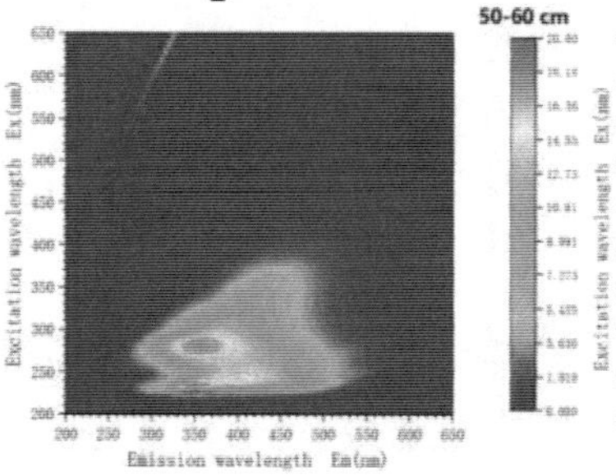

Figure 21.

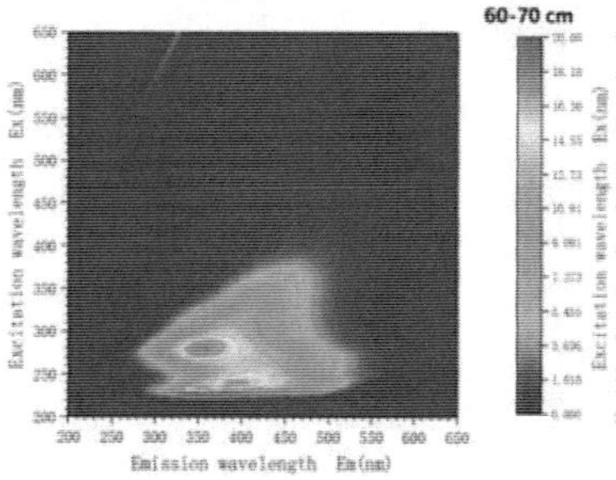

Figure 22.

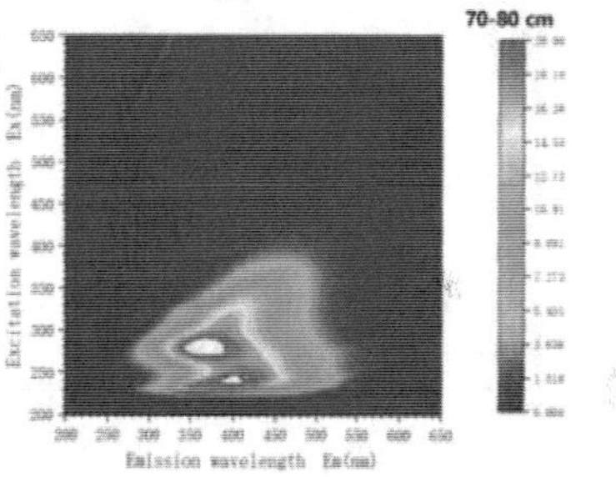

Figure 23.

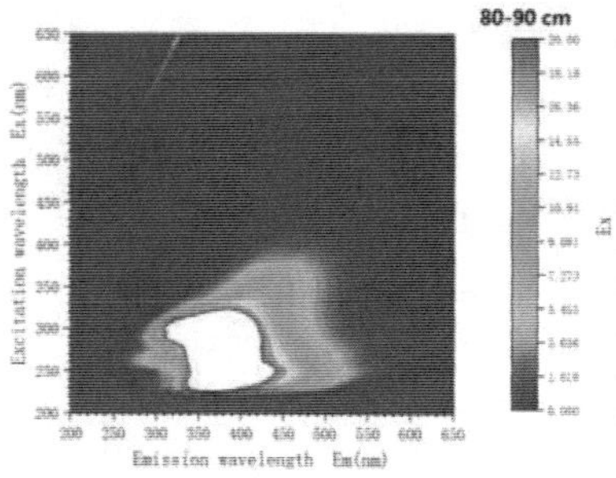

Figure 24.

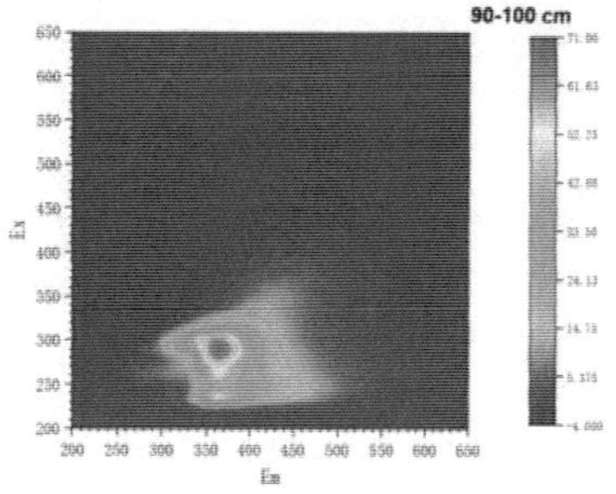

Figure 25.

For the soil leaching solution from 0-10cm (Fig.16), 10-20cm (Fig.17), 30-40cm (Fig.19), 40-50 cm (Fig.20) soil layer. The fluorescence intensity around the four zones was almost the same, but the fluorescence centres of the four zones did not shift significantly compared to the high DOC concentration. Only the fluorescence intensity of each zone decreased a lot. This shows that the concentration of organic matter in the drench solution decreased significantly after passing through the

0-10 cm soil layer, indicating that the 0-10 cm soil layer showed a strong ability in adsorption and retention of aromatic proteins, hydrophobic acids, fulvic acids, tryptophan-like and various humic acids in the organic matter components of the high DOC solution.

For soil leaching solution from 20-30cm (Fig.18),50-60cm(Fig.21), and 60-70cm(Fig.22), the fluorescence intensity of the fluorescence centres in zones 1, 2 and 4 is stronger than in the 0-10, 10-20, 30-40 and 40-50 cm drench solutions. This phenomenon indicates that the 20-30, 50-60 and 60-70 cm layers are weaker than the 0-10, 10-20, 30-40 and 40-50 cm layers in terms of adsorption and retention of aromatic proteins, hydrophobic acids, fulvic acids, tryptophan-like acids and various humic acids from the organic matter fraction.

For soil leaching solution from 70-80cm (Fig.23),80-90cm(Fig.24), and 90-100cm(Fig.25), the fluorescence intensity of the fluorescence centres in zones 1,2,4 was stronger in the 70-80, 80-90 cm drench solution, even compared to the initial high concentration DOC solution. This indicates that the organic matter components (aromatic proteins, hydrophobic acids, fulvic acids, tryptophan-like) in the 70-80, 80-90 cm drench solution showed an increase in concentration. This suggests that the desorption of organic matter components occurred in the 70-80 and 80-90 cm layers, potentially due to the high organic matter content of the 70-80 and 80-90 cm layers or the low content of effective phosphorus and iron and aluminium oxides, resulting in low retention of the organic matter components by adsorption in these layers or even in the sorption phenomenon is not understood.

Therefore, based on the analysis of the 3D fluorescence profiles of the drench solutions of each soil layer, it can be seen that the 70-80, and 80-90 cm drench solutions are consistent with the initial high concentration DOC solution, with both types I and II fluorescence characteristics and three main fluorescence centres. In contrast, the 0-10, 10-20, 30-40, 40-50, 20-30, 50-60 and 60-70 cm drench solutions were of type I fluorescence with two fluorescence peaks, indicating that the hydrophobic acid and fulvic acid represented by the third fluorescence centre decreased significantly after the drenching of these soil layers. And then, based on the intensity of the fluorescence centres in each soil layer, it is possible to indirectly infer the ability of the different soil layers to adsorb and retain organic matter of different composition structures. More specifically, in the 0-50 cm layer (except for 20-30 cm), the soil had approximately the same capacity to sorb and retain aromatic proteins, hydrophobic acids, fulvic acids and tryptophan-like fractions of organic matter with a high adsorption and retention capacity, all resulting in a significant decrease in fluorescence of the drench solution.

However, in the 50-70 cm layer, only zones 1 and 2 showed an increase in fluorescence intensity compared to the 0-50 cm layer, indicating that the ability to adsorb and retain aromatic proteins and tryptophan-like fractions in organic matter decreased in the 50-70 cm layer, but zone 4 did not show a significant increase in fluorescence intensity, indicating that the 50-70 cm layer still has the ability to adsorb and retain aromatic proteins and tryptophan-like fractions in organic matter. However, the fluorescence intensity in zone 4 did not increase significantly, indicating that the soil layer at 50-70 cm still has a strong capacity for retaining hydrophobic and fulvic acids. However, as the depth of the soil layer increases, desorption of organic matter begins to occur.

4. Conclusion

The leaching characteristics of soil DOC vary between different ecological environments. However, the overall trend is that the amount of soil DOC leaching increases with the depth of the soil layer. During the leaching process, the DOC content in the test soil leachate gradually decreased with increasing leaching time because the amount of soil DOC produced in a short period was significantly smaller than the amount of soil DOC lost. In the absence of exogenous carbon recharge, the loss of TOC from the upper layers of the soil was significantly higher due to the prolonged leaching process. In addition, soil TOC loss decreases after leaching, with the depth of the soil layer increasing. The loss of TOC from the upper layer of soil was significantly reduced with increasing depth. The soil DOC was trapped by sorption in all layers during leaching, while the upper layer was lost during the leaching experiment. This also demonstrates that soil TOC is lost in the absence of exogenous carbon input over a short period, which is particularly important for the study of TOC storage and leaching characteristics of soils in the region.

In experiments using highly concentrated DOC solutions as drench solutions, according to the analysis of the intensity of the fluorescence centres of each soil layer in the three-dimensional fluorescence pattern of the drench solution of each soil layer, the ability of different soil layers to adsorb and retain organic matter of different fractions differs, and the selected soil layers planted with saline Seepweed show a general decreasing trend in the ability to adsorb and retain organic matter as the depth increases.

References

[1] Schulze E D 2006 Biological control of the terrestrial carbon sink Biogeosciences **3** pp 147-166
[2] Jobbagy E G, Jackson R B 2000 The vertical distribution of soil organic carbon and its relation to climate and vegetation Ecological Applications **10** pp 423-436
[3] Lal R 2007 Carbon sequestration Philosophical Transactions of the Royal Society B **363** pp 815-830
[4] Xi M, Liu H Y and Lv X G 2006 Research progress on water purification function of watershed wetlands [J] Advances in Water Science **17(4)** pp 566-573
[5] Zi Y Y, Xi M, Kong F L et al 2016 Spatial and temporal distribution and storage of soil organic carbon in coastal wetlands of Jiaozhou Bay [J] Journal of Applied Ecology **27(7)** pp 2075-2083
[6] Xi M, Liu S H, Zhang Y Y et al 2019 Characteristics of soil dissolved organic carbon leaching in coastal wetlands of Jiaozhou Bay [J] Soil and Water Conservation Bulletin **39(1)** pp 16-22
[7] Li L, Qiu S J, Liu J T et al 2012 The role of soil dissolved organic carbon in the carbon cycle of terrestrial ecosystems Journal of Applied Ecology **23(5)** pp 1407-1414
[8] Wang Q K, Wang S L, Feng Z G et al 2005 Soil active organic matter and its relationship with soil quality[J] Ecology Journal of Ecology **25(3)** pp 513-519
[9] Beaumont N J, Jones L, Garbut J D et al 2014 The value of carbon sequestration and storage in coastal habitats[J] Estuarine, Coastal and Shelf Science **137** pp 32-40
[10] Xiong L, Yang Y S, Wang Q Z et al 2014 Leaching characteristics of soluble organic carbon in castanopsis carlesii natural forest soil [J] Journal of Subtropical Resources and Environment **(1)** pp 46-52
[11] Jiang H C 2015 Soil Organic carbon and its leaching characteristics in Jilin Province [D] Jilin Agricultural University(in Chinese with English abstract)
[12] Xiong L, Yang Y S, Wang Q Z et al 2014 Leaching characteristics of soluble organic carbon in castanopsis carlesii natural forest soil [J] Journal of Subtropical Resources and Environment **(1)** pp 46-52
[13] Yu Y C, He S, Li B K et al 2005 Analysis of soil dissolved organic carbon and its influencing factors in Chinese fir forest [J] Journal of nanjing forestry university (natural science edition), **29(2)** pp 14-18
[14] Kaiser K, Kalbitz K 2012 Cycling downwards dissolved organic matter in soils Soil Biology and Biochemistry **52** pp 29-32

Techno-Economic Assessment of LPG operated Porous Radiant Burner based Cook-stoves for Commercial Cooking Applications

Sofia Rani Shaik[1,2], P. Muthukumar[2]* and Pratul Chandra Kalita[1]
[1]Department of Design, Indian Institute of Technology Guwahati, India.

[2]Department of Mechanical Engineering, Indian Institute of Technology Guwahati, India.

email: pmkumar@iitg.ac.in; pmkumariitg@gmail.com

Abstract. The present study is to conduct a Techno-economic Assessment of a 5-7 kW Porous Radiant Burner (PRB) – based LPG cook-stove operating in eateries and restaurants. Process flow study has been performed primarily to understand the technological process involved in the PRB and also to estimate the cost of manufacturing. The cost estimation has been done considering direct costs such as material costs and labor costs, and indirect costs such as factory costs, works costs and administration costs. Sensitivity analysis has been performed considering the production capacity of 50, 250 and 500 cook-stoves each day. An economic feasibility study of PRB cook-stoves has been done by estimating the Pay Back Period, Net Present Value and Internal Rate of Return. For a manufacturing capacity of 250 cook-stoves per day, PRB yields a 51.4% return on the investment with a payback period of 0.68 years. Adopting PRB-based cook-stoves in small, medium and large-scale eateries results in an annual saving of Rs. 77500, Rs. 235000 and Rs. 1170000, indicating that PRB-based cook-stoves are undoubtedly economically viable.

Keywords: Techno-economic Assessment (TEA); Economic sustainability; Porous Radiant Burner; Economic Analysis; Cook-stoves; Process Flow Analysis

1 Introduction

Energy-saving is a global cause and hence efforts are being put worldwide to develop energy-efficient products and technologies that reduce the consumption of non-renewable sources of energy and minimize environmental pollution. Conservation and judicious use of non-renewable energy contribute to sustainable development. The United Nations proposed 17 Sustainable Development Goals (SDGs) in 2015 to provide sustainable and better living conditions to all nations. The 12th SDG is to ensure sustainable production and consumption patterns. The fossil fuel subsidies provided all over the world, threaten the achievement of the 12th SDG by 2030. Energy security concerns can be answered and energy imports can be minimized by adopting energy-efficient technologies. Most of the developing nations import the fuel to meet their energy demands. India is no exception to that and is the second-largest importer of fuel in the world. Of the total fuel requirement, the imported fuel itself accounts for half the total fuel, and cooking consumes a significant amount [1]. Under these circumstances, if energy-efficient cooking technologies are not developed, fossil fuel depletion will be severe. Also, most of the households, especially women and children become victims of harmful emissions. There will also be a concern about global warming, and greenhouse gas emissions. Thus, there is a need for the development of energy-efficient cooking applications. Porous Radiant Burner (PRB) is one such energy-efficient technology.

Studies have been conducted on PRB used for cooking and commercial applications for more than a decade. Research on PRB has been taking place on various aspects of its functioning, fuel consumption, emissions, application, and so on. A PRB is an excess enthalpy burner that predominantly works on the principle of radiation [2]. A PRB is made up of two layers of which one acts as a preheating layer and the other acts as a combustion layer [3]. It has a mixing zone and a mixing tube through which the air-fuel mixture is given as input [4]. The study showed that due to the lean combustion in PRB, there was a reduction of 95% in CO emissions and an 85% reduction in NOx emissions for PRB than its counterparts. In another such study to test the lean combustion of PRB, Li et al. [5] conducted a study on premixed lean-burn on the alumina particles of PRB of cylindrical and cone shapes. The results indicated that for cone-shaped PRB, at the upstream 3 mm diameter alumina particles with equal segment length and at the downstream, 6 mm diameter alumina particles are the best suited for lean-burn with low methane concentration and high flow rate. The same dimensions are best suited for lean-burn for cylindrical PRB too. The research was also done on cluster PRB used in commercial cooking applications to test its stability in a submerged state [6]. The experiment showed that the cluster PRB could combust in a stable state for an 8-20 kW input range. It also yielded 26% higher thermal efficiency and saved around 3.9 kg of 19 kg LPG cylinder, which is a significant saving on fuel.

Focusing on the environmental sustainability of PRB, works on Life Cycle Assessment (LCA) of PRB's impact on the environment have been performed. LCA is performed to analyze the environmental sustainability of a product or a process or a service [7]. It is also performed to compare the environmental impacts of two or more products or processes or services and will act as a decision-making tool to make a choice on the optimal one. Kaushik and Muthukumar [8] performed an LCA and Life Cycle Energy Efficiency (LCEE) to compare PRB to CB functioning at a 5-10 kW input range. Their assessment proved that as PRB has higher thermal efficiency, it led to higher LCEE. In a similar study performed by the authors' research team, LCA was performed to compare the environmental impacts of PRB to the environmental impacts of CB; both operating at 1-3 kW power output and also for their fabrication [7]. The study exhibited that PRB had a 15% lesser negative impact on the environment compared to CB on all the endpoint damage assessment categories. However, testing the environmental sustainability of a product alone is not sufficient for it to be considered a sustainable product.
A product has to be tested for its economic and social sustainability too, besides being an energy-efficient and environment-friendly product. Techno-economic assessment (TEA) tests the economic feasibility of a technology or a product. TEA is significant especially when the viability of a new product or a technology has to be estimated and when there is no ready information available on the industries of the similar technology or product. Wu et al., performed a techno-economic study on the hydrogen storage grid, to analyze the operational flexibility and capability of hydrogen production, storage, compression, and utilization [9]. The authors suggested that as the value streams from bundling grid services sum up to 76% of the total profit, they play a major role in making the hydrogen energy storage project economically feasible. A different assessment was conducted from a land management perspective in Brazil, to assess the financial viability of Bioenergy Livestock Integrated (BLI) systems [10]. It was found that BLI reduces the payback period (PBP) to half the time besides increasing the Net Present Value (NPV) five times the initial investment. In a different study, TEA of building energy-efficient systems was conducted by [11]. The economic assessment was performed for the data collection, analysis and detection of an anomaly, and data visualization of energy-efficient systems. The system proved to be economically viable as the TEA showed up to 28%-68% savings in the consumption of energy. TEA study coupled with LCA was another study performed by Nugroho et al., [12] to assess the environmental and economic viability of

methane supply chains. After the analysis, the team recommended to use an agent-based simulation optimization as this pathway of the technology leads to a higher Internal Rate of Return (IRR), lower PBP, and also lower carbon emissions. However, there is no study performed on the TEA of PRB except for the studies conducted on the PRB-based stove operated with waste cooking oil and LCA and TEA of LPG operated PRB cook-stove for 5-10 kW [2], [13]. The results indicated that PRB is economically feasible when operated with waste cooking oil; and also environmentally and economically feasible even with 5-10 kW LPG. These economic assessments performed for the technology of PRB were focused on the performance of the PRB when used with different fuels. But, no TEA of PRB was performed to assess the economic feasibility of PRB-based cook-stoves working with 5-7 kW; when the industry has to be set up for the mass production of PRB-based cook-stoves. Studies are to be conducted to verify PRB's financial viability, and also the returns the consumers are going to get upon their investment in PRB-based 5-7 kW cook-stove. And so, there is a need to conduct an economic assessment before commercializing the present technology to minimize the cost wherever possible and reduce the financial risks [14]. In the present study, TEA of mass production PRB-based cook-stoves of 5-7 kW is performed. The present TEA not only estimates the returns and cost of production of PRB-based cook-stoves, but it also takes into consideration the socio-TEA of consumers focusing on their returns, savings, and PBP for the investment made by the consumer on the PRB-based cook-stove.

2 Methodology

The methodology of the present study is broadly categorized into three steps viz., (i) design of the concept and procedural flow, (ii) cost estimation and (iii) economic analysis.

2.1 Concept design and procedural flow

Designing the concept and understanding the flow of the process have to be done before performing the economic assessment of any project. This step gives a clear understanding of the material used, equipment set up and the flow of the operation or the procedural flow while the equipment is being operated. Designing the concept and procedural flow are developed into the Process Flow Diagram (PFD). A PFD is a diagrammatic representation of the process, which is normally drawn in a stylized pictorial form using international standards symbols [15]. Fig 1 shows the PFD of the PRB-based cook-stove.

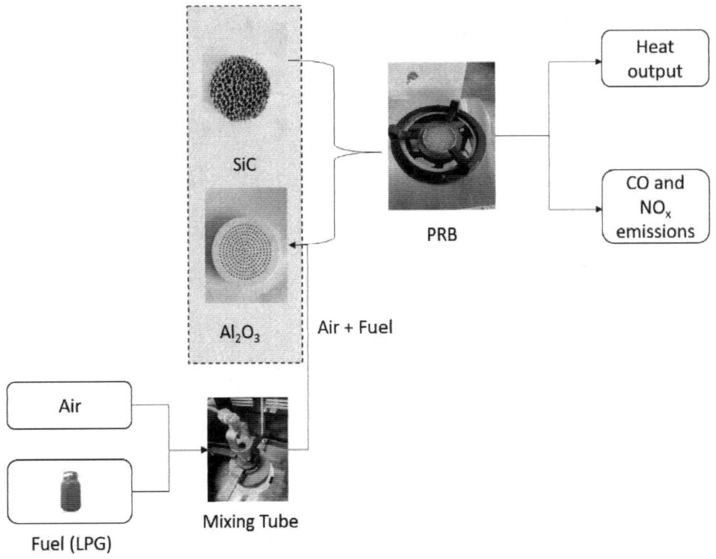

Fig 1 Process Flow Diagram of PRB-based cook-stove

The operational procedure of the PRB-based cook-stove as evident from Fig 1 is that the air and fuel mixture is given as input into the mixing tube, which passes to the preheating zone of the PRB. The preheating zone of PRB is made up of Alumina (Al_2O_3). The combustion takes place in the combustion zone of PRB which is made up of Silicon Carbide (SiC). When the combustion is in progress, heat is given as output along with CO_2, CO and NO_x emissions.

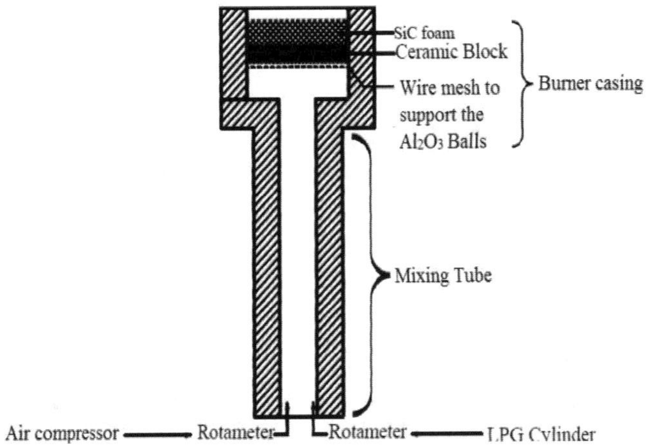

Fig 2 Schematic diagram of PRB

Fig 2 shows the schematic diagram of PRB. The diameter of the burner is 120 mm. The porosity of the preheating zone is 7-8% and the thickness of the combustion zone is 25 mm. Both the preheating zone and combustion zone are held together with the help of casting cement. The casting cement also helps to withstand high temperatures.

2.2 Cost estimation and calculation

In the cost estimation, the total costs required to set up the operational plant such as direct costs, indirect costs, fixed costs, variable costs and works costs are estimated. Cost of investment plays a major role in TEA of any plant involved in the manufacturing of a particular product or providing a dedicated service. A proper cost estimation reduces the risk of financial losses in an organization [16]. In the present study, TEA is conducted to estimate the prime cost, works cost, administration cost, production cost and the cost of sale incurred for a plant that manufactures 250 PRBs of 5-7 kW on daily basis.

Direct costs: Direct costs, also known as prime costs are those expenses that are directly related to producing the goods or services. Whereas indirect costs are those that cannot be directly allocated to the product or service [17]. The more efficient a product, the lesser are the costs incurred [18]. In the present study, the direct costs are calculated by considering raw materials required to manufacture 250 units of PRBs of 5-7 kW capacity for each day. The materials required for the manufacturing of the PRB are SiC, Al2O3, Refractory Alumina casing, Steel frame, etc. Also, the direct costs incurred for PRB-based cook-stove manufacturing includes fitting accessories and the transport to carry the raw material to the manufacturing unit. The manufacturing costs of PRB are represented by MCPRB.

Man-power engaged for the making of PRB is also considered as the direct cost incurred for the making of PRB-based cook-stove. The man-power cost of PRB is indicated as MP_{PRB}. As the cook-stoves manufacturing is not a complicated process and does not require high-end machinery, the man-power requirement is only limited to five staff levels and the direct costs incurred for the man-power are their wages. The prime costs are represented as PrC_{PRB}
Where

$$PrC_{PRB} = MC_{PRB} + MP_{PRB} \quad (1)$$

Factory costs: Factory costs of the PRB-based cook-stove (FCPRB) are indirect costs that include rent of the unit, insurance charges, factory vehicle, expenses incurred for the cleaning and maintenance of the unit.

Administrative costs: The administrative costs for the manufacturing unit of the PRB-based cook-stove (ACPRB) are majorly for advertising.

Production cost: The production cost is the total cost of all the direct and indirect costs. Production costs of the PRB-based cook-stove (PCPRB) is given by

$$PC_{PRB} = PrC_{PRB} + FC_{PRB} + AC_{PRB} \quad (2)$$

Sales cost: The sales cost of PRB-based cook-stove (SCPRB) will analyze the ability of the manufacturing unit to design, procure, source and manufacture the cook-stoves at an affordable and reasonable price for the end-user. The sales cost is given by

$$SC_{PRB} = PC_{PRB} + SP_{PRB} \quad (3)$$

where SP_{PRB} is the selling price of PRB

The sales cost obtained for each PRB-based cook-stove with 50% profit is approximately Rs. 5000.

2.3 Economic Analysis

Economic Analysis is conducted to assess the economic viability of a project to confirm the profitability of the project. In the present study, NPV, IRR and PBP analysis methods have been considered to test the economic viability of manufacturing PRB-based cook-stoves of 5-7 kW operated with LPG fuel input. These analyses are used to make a decision on investing in a process or product or service if viable or rejecting the investment in it [17].

2.3.1 Net Present Value (NPV)

Net Present Value is the present value of all the cash flows calculated at the discounted value of the marginal tax after considering the initial investment. In the present study, net cashflows of PRB (NCF_{PRB}) are estimated for six years and NPV of PRB (NPV_{PRB}) is calculated using the trial and error method of Present Values taken for different costs of capital after discounting the cash flows after-tax ($DCFAT_{PRB}$). Then the initial investment made for the manufacturing of 250 PRB-based cook-stoves is deducted from $DCFAT_{PRB}$ to obtain NPV_{PRB} value. NPV_{PRB} is given by

$$NPV_{PRB} = \sum_{i=1}^{6}(NCF_{PRB} * DCFAT_{PRB}) - Initial\ ivestment \quad (4)$$

The maximum positive NPV_{PRB} is obtained at $DCFAT_{PRB}$ of 50% with the cash flows of Rs. 10171168.

2.3.2 Internal Rate of Return (IRR)

Internal Rate of Return is a financial analysis metric that is used to estimate the profitability of the investments made in a project. IRR of the PRB-based cook-stoves (IRR_{PRB}) present is calculated using the forward method after calculating NPV_{PRB}. The value of IRR_{PRB} is obtained by

IRR_{PRB} = (NPV_{PRB} at the positive percentage value) + ((Difference between NPV_{PRB} at the positive percentage value and negative percentage value)/ Difference in the rate)

The value of IRR_{PRB} obtained is 51.4 %.

2.4 Benefit to the consumer

In the present study, the benefit to the consumer of adapting to PRB-based cook-stoves is obtained by estimating the returns a consumer will get after investing in the PRB-based cook-stove. The returns are estimated by calculating the payback period of the investment on the PRB-based cook-stove (PBP_{LPG}). Also, the cumulative returns are estimated at the end of the life of the stove. In the present study, 10 years of life period is considered.

3 Results and Discussion

3.1 Consumption by PRB-based cook-stove

The consumption of LPG of a PRB-based cook-stove is compared with that of a CB-based cook-stove through the field study in small-scale eateries, medium-scale eateries and large-scale eateries. The capacity of the LPG cylinder is 19 kg, with the operation of the cook-stoves for 6 days a week. Table 1 shows the details of the consumption of LPG cylinders for the cook-stoves with CB and PRB.

Table 1 LPG consumption for PRB-based and CB-based cook-stoves

Factors	Small-scale eateries	Medium-scale eateries	Large-scale eateries
Consumption of cylinders with CB-based cook-stove	2 for a week	1 each day	5 each day
Total cylinders in a year with CB-based cook-stove (312 days)	104	312	1560
Consumption of LPG in kg with CB-based cook-stove (312 days)	1976	5928	29640
Total cylinders in a year with PRB-based cook-stove (312 days)	73	218	1092
Consumption of LPG in kg with PRB-based cook-stove (312 days)	1387	4142	20748
Fuel saved in kg in a year with PRB-based cook-stove (312 days)	589	1786	8892
Total cylinders saved in a year with PRB-based cook-stove (312 days)	31	94	468

The values in Table 1 show that 5-7 kW cook-stoves with PRB save up to 30% on LPG when compared to those cook-stoves with CB of the same capacity.

3.2 Cost savings on LPG by PRB-based cook-stove

Cost savings play a major role in deciding the economic feasibility of a product or a service. The cost savings on the LPG fuel by the small-scale, medium-scale and large-scale eateries when adapted to PRB-based cook-stoves of 5-7 kW is shown in Fig 3.

It is evident from Fig 3 that the total fuel cost spent by the small-scale eateries on 19 kg LPG cylinders annually in India is approximately Rs. 260000 while using CB-based cook-stoves. The fuel cost for medium-scale and large-scale eateries on LPG cylinders is approximately Rs. 780000 and Rs. 3900000, respectively for CB-based cook-stoves of 5-7 kW.

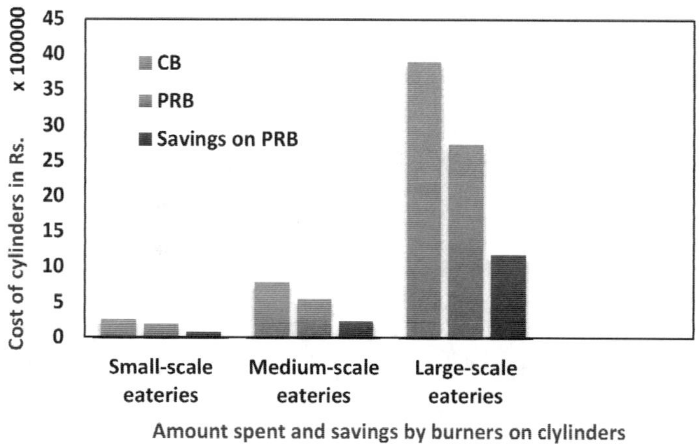

Fig 3 Annual cost spent on LPG cylinders and savings of PRB

Whereas for PRB-based cook-stoves, the annual cost spent on LPG cylinders by the small-scale, medium-scale and large-scale eateries is approximately Rs. 182500, Rs. 545000 and Rs. 2730000, respectively. This indicates that there is a 30% saving on the fuel cost spent on 19 kg LPG cylinders when PRB-based cook-stoves are used [18].

3.3 Cumulative returns at the end of life of PRB-based cook-stove

The returns obtained on the investment of any product or service give assurance to the end-user or the consumer that the investment in the product or process is optimal or not. The cost of the PRB-based cook-stove of 5-7 kW operating is estimated to be Rs. 5000, after performing the assessment. Taking an average investment on 10 stoves for the small-scale, medium-scale and large-scale eateries, the investment is Rs. 50000. This cost is SC_{PRB}, obtained considering a 50% profit margin on PC_{PRB}.

With this cost incurred on PRB-based cook-stove, the end-users PBP of the PRB-based cook-stove (PBP_{PRB}) is estimated as 0.68 years which is a very short PBP_{PRB} assuring PRB-based cook-stove of 5-7 kW operating on LPG fuel to be a commercially feasible and beneficial product. Also, the cumulative return for the consumer at the end of the life of a PRB-based cook-stove is approximately Rs. 72150 on each stove for 10 years in total. Fig 4 shows the cumulative returns of the consumer when invested in a 5-7 kW PRB-based cook-stove calculated based on the fuel-saving of each stove.

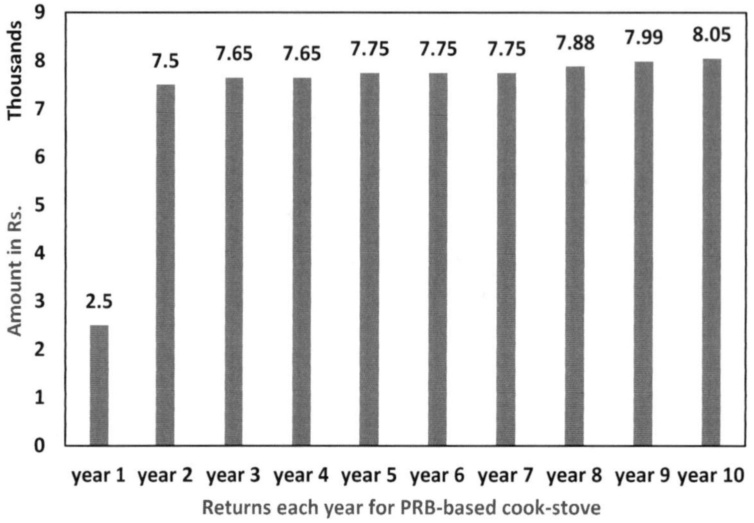

Fig 4 Returns on the investment of 5-7 kW PRB-based cook-stove for its life-period

From Fig 4, it is evident that the investment in a 5-7 kW PRB-based cook-stove made by the consumer will be gained back in less than 1 year approximately, and the for the rest of the life of the cook-stove, it will generate above Rs. 7500 each year.

3.4 Sensitivity Analyses

A sensitivity analysis is performed to analyze the economic feasibility of PRB taking into consideration the possibilities of profit margins of the manufacturing units, PBP_{PRB} of the end-user at different costs that are incurred for the purchase of the 5 -7 kW PRB-based cook-stove.

3.4.1 Profit margin of the manufacturing unit

The sensitivity analysis is done for the estimation of different PBP_{PRB} of the consumers at different profit margins of the manufacturing unit. The different profit margins considered for the sensitivity analysis are 10%, 20%, 30%, 40% and 50%. The respective PBP_{PRB} obtained for the profit margins are shown in Fig 5.

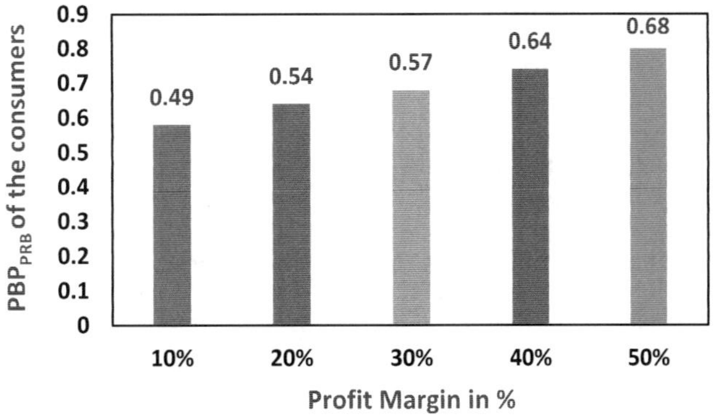

Fig 5 PBP_{PRB} of 5-7 kW PRB-based cook-stoves at different profit margins of manufacturing unit

From Fig 5, it can be understood that the PBP_{PRB} is in co-relation with the profit percentage of the manufacturing unit. However, even for up to a profit margin of 50%, the PBP_{PRB} of 5-7 kW PRB-based cook-stove is less than a year, indicating it to be a highly efficient and economically feasible product.

3.4.2 Cost of the cook-stove

Sensitivity Analysis is done to identify different PBP_{PRB} for the end-user at different costs of the cook-stove. As the PBP_{PRB} plays a key role in determining the economic viability of a product, this data can help in fixing the SC_{PRB}, so that it will be beneficial for both the consumer and the manufacturing unit. Fig 6 shows the details of different PBP_{PRB} at different SC_{PRB}.

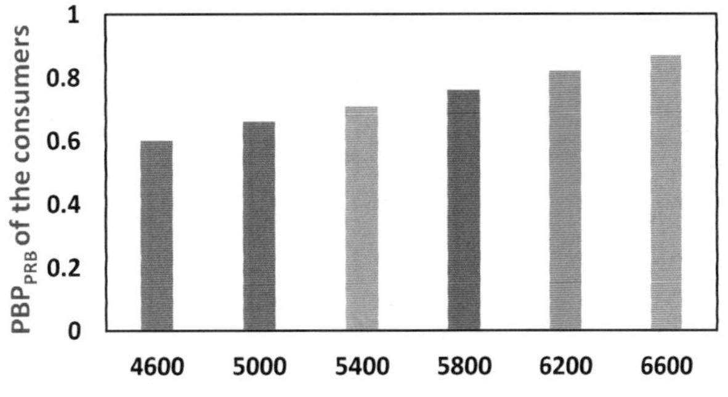

Fig 6 PBP_{PRB} of 5-7 kW PRB-based cook-stoves at different SC_{PRB}

A product is considered to be economically feasible for the consumer or the end-user if the PBP of that product is less than two years. Fig 6 shows that even when the SC_{PRB} of the stove is raised by half of its original SC_{PRB}, the PBP_{PRB} is less than a year which makes the PRB-based cook-stove of 5-7 kW economically feasible for the end-user.

3.4.3 Effect of manufacturing capacity on the PBP_{PRB} of the consumers

Sensitivity analysis is conducted to calculate the PBP_{PRB} for the end-user when the manufacturing unit's capacity is varied. For a manufacturing unit that has the capacity of manufacturing 50 units of PRB-based cook-stove, the SP_{PRB} of each cook-stove even with a 50% margin is estimated to be Rs. 6000. Whereas for the manufacturing units that have the capacities of manufacturing 250 and 500 units of PRB-based cook-stoves, the SP_{PRB} of each cook-stove is estimated to be Rs. 5000 and Rs. 4100, respectively. At these SP_{PRB}, the PBP_{PRB} at different plant unit capacities is calculated. Fig 7 shows the PBP_{PRB} of the consumers at different manufacturing capacities of the plant and the respective SP_{PRB} of the plants with different manufacturing capacities.

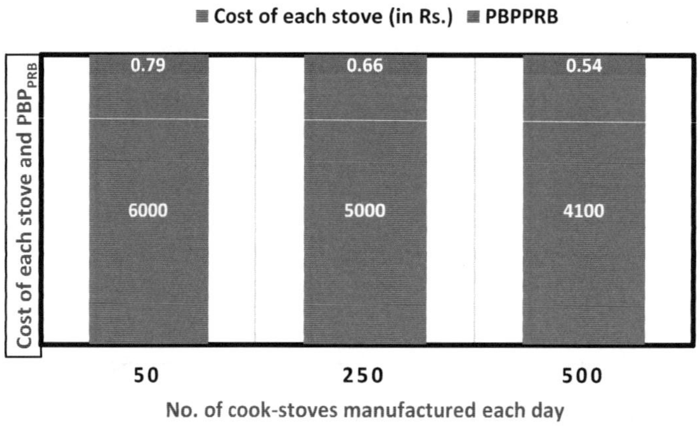

Fig 7 PBP_{PRB} and SP_{PRB} of 5-7 kW PRB-based cook-stoves for different plant capacities

From Fig 7, it can be understood that both the SP_{PRB} and PBP_{PRB} for the end-user will be reduced if the manufacturing plant increases the capacity of its production. The PBP_{PRB} for the end-user if the plant manufactures 500 units per day is estimated to be 0.54 year. Whereas the PBP_{PRB}, when the plant manufactures 50 units and 250 units are estimated to be 0.79 and 0.66 years, respectively. Thus, it can be concluded that a plant with a higher manufacturing capacity proves beneficial for the end-user as it results in lower SP_{PRB} and PBP_{PRB}.

4 Conclusion and future scope

After conducting the TEA of PRB-based cook-stove of 5-7 kW operating with LPG fuel, the results indicate that it is economically a feasible product. The PFD of a PRB-based cook-stove is also not a complicated process and PRB is assembled with easily available materials. Also, as the assembly does not require any complicated procedures and heavy machinery, the direct costs incurred for the material procurement and the manpower is also minimal. Since the TEA is done for a manufacturing unit that manufactures 250 PRB-based cook-stoves per day, other costs incurred like the administration costs and factory costs are also very less. The sales cost of the PRB-based cook-stove accounted for Rs. 5000 for 5-7 kW including 50% profit to the manufacturing unit, which indicates very good returns. The IRR for the manufacturing unit is 51.4% with 50% profit, making a 5-7 kW PRB-based cook-stove a revolutionary and economically feasible product. PRB-based cook-stoves of 5-7 kW when adapted by small-scale, medium-scale and large-scale eateries, will save approximately Rs. 77500, Rs. 235000 and Rs. 1170000, respectively annually on 19 kg LPG cylinders. The consumer when invested in 10 PRB-based cook-stoves, will get the investment back in less than a year, which is a beneficial PBP_{PRB} of 0.68 year. Along with such a short PBP_{PRB}, 5-7 kW PRB-based cook-stoves also provide returns on their investment which account for Rs. 72150 on each stove. With the present TEA, PRB-based cook-stoves are considered highly efficient and economically feasible products as they provide high IRR along with huge savings on fuel, with short PBP_{PRB} and high returns.

- The analysis can be further extended to other aspects like marketing, advertisement and organizational structure for such sustainable products, apart from the economic feasibility of the product to make it easier for the manufacturing units to adapt themselves for the wide manufacturing of PRB-based cook-stoves.
- Training programmes can be provided and implemented to create job opportunities on a wider scale all over the country to promote small-scale industries as the manufacturing of PRB-based cook-stoves is a simple process.
- Policies can be framed to shift from conventional cooking technologies to energy-efficient PRB technologies.

References

[1] P. Oskarsson, K. B. Nielsen, K. Lahiri-Dutt, and B. Roy, "India's new coal geography: Coastal transformations, imported fuel and state-business collaboration in the transition to more fossil fuel energy," *Energy Research and Social Science*, vol. 73, Mar. 2021, doi: 10.1016/j.erss.2020.101903.

[2] L. K. Kaushik and P. Muthukumar, "Thermal and economic performance assessments of waste cooking oil /kerosene blend operated pressure cook-stove with porous radiant burner," *Energy*, vol. 206, Sep. 2020, doi: 10.1016/j.energy.2020.118102.

[3] S. Deb, L. K. Kaushik, M. A. Kumar, S. H. V. Satish, and P. Muthukumar, "Clustered Porous Radiant Burner: A cleaner alternative for cooking systems in small and medium scale applications," *Journal of Cleaner Production*, vol. 308, no. April, p. 127276, 2021, doi: 10.1016/j.jclepro.2021.127276.

[4] S. Devi, N. Sahoo, and P. Muthukumar, "Combustion of biogas in Porous Radiant Burner: Low emission combustion," *Energy Procedia*, vol. 158, pp. 1116–1121, 2019, doi: 10.1016/j.egypro.2019.01.276.

[5] L. Li, Z. Sun, Y. Luo, Y. Liang, X. Wang, and K. Wang, "Experimental study on the premixed lean-burn in alumina particles based porous burner," *Energy Reports*, vol. 7, pp. 7434–7444, Nov. 2021, doi: 10.1016/j.egyr.2021.10.101.

[6] S. Deb, L. K. Kaushik, M. A. Kumar, S. H. V. Satish, and P. Muthukumar, "Clustered Porous Radiant Burner: A cleaner alternative for cooking systems in small and medium scale applications," *Journal of Cleaner Production*, vol. 308, no. November 2020, p. 127276, 2021, doi: 10.1016/j.jclepro.2021.127276.

[7] S. R. Shaik, P. Muthukumar, and P. C. Kalita, "Life Cycle assessment of LPG Cook-stove with Porous Radiant Burner and Conventional Burner – A comparative study," *Sustainable Energy Technologies and Assessments*, vol. 52, p. 102255, Aug. 2022, doi: 10.1016/j.seta.2022.102255.

[8] L. K. Kaushik and P. Muthukumar, "Life cycle Assessment (LCA) and Techno-economic Assessment (TEA) of medium scale (5–10 kW) LPG cooking stove with two-layer porous radiant burner," *Applied Thermal Engineering*, vol. 133, pp. 316–326, Mar. 2018, doi: 10.1016/j.applthermaleng.2018.01.050.

[9] D. Wu, D. Wang, T. Ramachandran, and J. Holladay, "A techno-economic assessment framework for hydrogen energy storage toward multiple energy delivery pathways and grid services," *Energy*, vol. 249, Jun. 2022, doi: 10.1016/j.energy.2022.123638.

[10] N. R. D. de Souza, O. Cavalett, and T. L. Junqueira, "Techno-economic and environmental assessment of bioenergy and livestock integrated systems in Brazil," *Sustainable Production and Consumption*, May 2022, doi: 10.1016/j.spc.2022.05.013.

[11] Y. Himeur *et al.*, "Techno-economic assessment of building energy efficiency systems using behavioral change: A case study of an edge-based micro-moments solution," *Journal of Cleaner Production*, vol. 331, Jan. 2022, doi: 10.1016/j.jclepro.2021.129786.

[12] Y. K. Nugroho, L. Zhu, and C. Heavey, "Building an agent-based techno-economic assessment coupled with life cycle assessment of biomass to methanol supply chains," *Applied Energy*, vol. 309, Mar. 2022, doi: 10.1016/j.apenergy.2021.118449.

[13] L. K. Kaushik and P. Muthukumar, "Life cycle Assessment (LCA) and Techno-economic Assessment (TEA) of medium scale (5–10 kW) LPG cooking stove with two-layer porous radiant burner," *Applied Thermal Engineering*, vol. 133, no. January, pp. 316–326, 2018, doi: 10.1016/j.applthermaleng.2018.01.050.

[14] S. Abdelhady, "Performance and cost evaluation of solar dish power plant: sensitivity analysis of levelized cost of electricity (LCOE) and net present value (NPV)," *Renewable Energy*, vol. 168, pp. 332–342, May 2021, doi: 10.1016/j.renene.2020.12.074.
[15] A. Shah, N. R. Baral, and A. Manandhar, "Technoeconomic Analysis and Life Cycle Assessment of Bioenergy Systems," in *Advances in Bioenergy*, vol. 1, Elsevier Inc., 2016, pp. 189–247. doi: 10.1016/bs.aibe.2016.09.004.
[16] R. Wang, V. Asghari, C. M. Cheung, S. C. Hsu, and C. J. Lee, "Assessing effects of economic factors on construction cost estimation using deep neural networks," *Automation in Construction*, vol. 134, Feb. 2022, doi: 10.1016/j.autcon.2021.104080.
[17] M. Ben-Horin and Y. Kroll, "A simple intuitive NPV-IRR consistent ranking," *Quarterly Review of Economics and Finance*, vol. 66, pp. 108–114, Nov. 2017, doi: 10.1016/j.qref.2017.01.004.
[18] L. K. Kaushik and P. Muthukumar, "TITLE OF THE INVENTION : ECO-FRIENDLY AND ENERGY-EFFICIENT LPG COOKING STOVE WITH NATURALLY-ASPIRATED POROUS RADIANT BURNER FOR COMMERCIAL KITCHENS."

The Uptake of Environmental Management System by Small and Medium Enterprises (SMEs) in India

[1]H P Sharma and [2]K Kumar
[1]Assistant Professor
Institute of Business Management, GLA University, Mathura India
[2]Associate Professor, University School of Business, Chandigarh University, Punjab, India
Corresponding author: hariprapan@gmail.com

Abstract. The purpose of this study to scrutinize the drivers that impact the acceptance and execution of EMP in SMEs companies in India. To identify the various enablers exploratory research method was used. A total of 356 senior executives were surveyed from 89 manufacturing organizations located in NCR, India. Based on factor analysis, four factors were identified responsible for the adoption of environmental management actions in Indian SMEs. These factors were the most important enablers for application of management systems in the organization. These finding will be useful for the practitioners, policymakers and academicians to work upon these enablers to promote application of management system for environment in the organization.

1. Introduction

EMS practices has become one of the most important aspects of overall business operations that aims to mitigate the negative impact on the environment. Currently, the use of environment protection standards and systems is widespread among policy makers, business leaders, practitioners, and regulators [1,2]. EMSs manage the impacts of business operations on the environment through a variety of methods and measures [3]. Implementing an EMS does not only meet the concerns of regulatory agencies, but it mitigates costs in the long run and builds a positive image of company in the market. The implementation of EMS is more prevalent among large companies; however, there is a rising trend of EMS development and execution among small and medium organizations. Acceptance and execution of EMS are no longer restricted to large-cap firms with high margins or high profit margins because of its many benefits [4,5,6]. Due to environmental awareness, a number of small and medium organization are looking EMS for recognizing and dealing bad effects of their economic actions on the atmosphere. EMS provides number of benefits to the SMEs such as: resource wastage, lowers costs, improves efficiency and encourages employees to improve their performance related to environment [7]. Awareness about the environment and reducing cost in long term are the major influencing factors to the SMEs. Due to financial constraints and little of knowledge about the field, developing and implementing an EMS remains a substantial challenge for SMEs in India [8].

Small and medium enterprises sustain and provide livelihood to about 496 million workforces in India and add up to 36% of India's total GDP. A Few research have been undertaken to determine and analyze drivers that contribute to the application of EMS in SMEs in emerging economies [9]. Therefore, it is appropriate to study the factors that lead to the development and application of EMS in SMEs. This study seeks to answer what impact does the application of EMS have on SMEs in India? This research work aims to investigate different factors that enables SMEs to develop and apply EMS [10]. The results of this research have significant implications for policy makers, regulators, industry people and SMEs.

The findings enable to develop an knowledge of the underlying enablers that affect EMS application in organization [11]. The findings signify those concerns about environmental protection are the most important factors influencing firms to identify and control adverse environmental impacts. This study presents an interesting understanding of the four key factors that play a key role in the development and implementation of SME companies in India [12]. These are the features; environment protection, top management commitment, operational support and long-term investment for SD. The whole paper is arranged as follows. The second section discusses important research in the field of environmental protection actions in organizations. The third phase presents the methodology employed in the research work and the fourth section presents the results and findings and the last phase concludes the research.

2. Background

Over the last few decades, companies have had to address environmental considerations into their operations. ISO 14001 is largely implemented by organizations to apply EMS around the world. EMS can signify an integral part of a general management system incorporating planning, process, processes to improve the implementation and maintenance of environmental related policy in an organization. EMS aims to address environmental issues into management planning and provides guiding code for adhering to environmental regulation and contribute in environment protection. The adoption and application of EMS also reflects the management's support to environmental protection [13]. EMS is most effective when combined as part of routine operations as if it were treated as a separate business entity. It assists in promulgating environment related policy and practice in order to achieve the objectives of the firm in the field of risk minimization for the environment and occupational health, compliance with environmental rules, hazard minimization and pollution prevention [14]. Successful implementation of EMS should include the following; a) environmental policy development b) environmental impact assessment due to business performance c) employee roles and responsibilities d) pollution control policies and procedures e) environmental program for awareness f) environmental performance record g) EMS review and research in the business organization [15].

India being the fastest major rising economies of the world with small and medium enterprises (SMEs) often referred to as the integral pillar of the economy. SMEs provide 45 per cent of total economic production and more than 40 per cent of India's total exports. This also provides the 110 million jobs followed by agriculture sector [16]. As SMEs represent the largest share of economic activity as a whole in the economy, there is also a significant negative impact on the environment. SMEs are found mainly in industries that are in dire need of resources and environmentally sensitive as the Indian manufacturing sector. Although the environmental issue at the level of each company may not be obvious, but these companies have a negative impact on the environment. Similarly, SMEs have scarce resources and knowledge to deal with environmental protection issues and challenges. Not only are financial resources scarce and lack of information about their impact on the environment, but SMEs often do not know environmental law in order to comply. More recently, many SMEs have adopted and implemented EMS in their business activities [17]. Previous research suggests that a huge number of companies receiving EMS primarily save costs, regulation, a better public opinion, reduce environmental hazards, comply to the environmental regulations and enhance returns in longer run.

3. Methodology

Both the experimental research design was used in this research to gain insight into the factors influencing the application of EMS in SMEs in India. Data from this study is collected from 356 experienced executives / supervisors at 89 small and medium production companies operating in the, NCR, India. Managers, executives and senior managers working in SMEs were targeted as sample of the study as they are involved in planning and execution of environmental management actions. An objective sample method was used to collect research data. The data was collected in sequential following manners. The questionnaire was handled in person at the company's production facilities and the self-regulatory questionnaire was also posted online via email or linkedln. Pilot study was undertaken initially to test the initial results and identify the content validity of the research instrument.

Further, research data was collected between September 2021 to March 2022. Items were weighed on a 5-point scale and 5 widely agreed upon and 1 strongly disagreed.

A total of 73 questions were handled in person and 146 questions were received through the online forum. PCA employing Varimax rotation has been used to identify the underlying drives sets in a relatively large number of variables. Version 20 of the SPSS was used for analysis the data set. After this, CFA was employed to minimize the formation of variables. During the analysis only 0.7 or more uploaded items have been saved. To test the reliability of the exceptions found in the analysis, the alpha coefficient of Cronbach's coefficient reported reliable from 0.798 to 0.889, and is more than 0.60.

4. Results

The results of the feature analysis applying PCA extraction have extracted four key enablers impacting the application of EMS in SMEs in India. These four factors have been identified as concerns for environmental protection, Top management assurance, operational infrastructure and long-term financial commitment in sustainability. In addition, these features were subject to verification EFA and items with loads of less than 0.7 were removed.

The results are depicted in Table 1.- Factors influencing development and implementation of EMS in small and medium manufacturing companies in India.

	1	2	3	4	Mean	SD
Factor 1	.798				3.3	.45
Environmental	.7339				3.5	.51
Protection Reasons	.897				3.9	.56
Factor 2		.974			3.5	.53
Top Management		.864			3.6	.51
Support		.798			3.2	.48
		.785			3.1	.54
		.767			3.9	.65
Factor 3			.792		3.3	.55
Infrastructural and			.879		3.4	.67
Establishment Support			.922		3.8	.43
Factor 4				.813	3.1	.51
Long-Term Investment				.877	3.2	.59
for Sustainable				.898	3.4	.63
Development				.903	3.1	.48
				.919	3.8	.69

The first element of 'environmental protection reasons' shot three items with loads varying from 0.897 to 0.739 and accounted for 18.23% of total variability. Cronbach's alpha value was 0.883 which suggests considerable reliability of the items. The results support a better correlation between factors such as the negative impact on the environment, environmental protection is crucial and the warranted practices must be adopted to address the environmental considerations. This means that concern for atmosphere and the necessity for environmental protection contributes to the development and application of EMS. The next element of 'top management support' includes five items with feature uploads varying from 0.767 to 0.974. This feature accounts for 17.13 percent of total variance. It shows the availability of qualified employees, exactness of roles and responsibilities among employees, a training activity related to environmental management and appropriate monitoring by senior management. This means that the support management people are essential to the development and application of EMS. Cronbach's alpha value is 0.912 and is considered reliable. The third element of 'infrastructural and establishment support' includes three elements with the feature download from 0.922 to 0.792. It accounts for 14.13 percent of the total amount. This feature represents management help, availability of qualified staff and good organization among all departments for the implementation of EMS. This feature suggests that the support system is one of the most important drivers of application of environmental management system in an organization. Cronbach's alpha value was 0.833, indicated good enough reliability. The final aspect of 'long-term investment for sustainable development' included 5 items between from 0.919 to 0.813. Calculated more than 9.24% of total variance. This feature represents that the application of EMS is taken as a relatively long investment on sustainable development issues of the organization. It involves such things as cost reduction, environmental adherence, improved profit and business development over time. Therefore, companies apply EMS from an investment purpose to get return in long run. The value of Cronbach's alpha is 0.836 and is considered reliable.

5. Conclusion

The originality of this research signifies the determination of enablers that contribute to the application of EMS in Indian SMEs. Following are the key findings:

- The findings of the study raised four key issues namely environmental protection reasons, top management support, infrastructural and establishment sustenance and considerable time-lapsed investment for sustainable development.
- The findings of the study reveal significant repercussions for decision makers, employees, administrators and SMEs.
- The application of EMS to some extent depends on the commitment of senior management, environmental considerations and organizational application support.
- Increased acceptance of EMS in SMEs may be further enhanced by raising knowledge of adverse impact to achieve benefits in longer term.

However, this research work also has some shortcomings. It is worth noting that the present study findings are limited to Indian SMEs. The size of a large multi-sectoral sample can be considered in future studies. This research only determined the views of senior executives about application EMS in the organizations. Opinions may change over time. So, this could be another limit. Future studies with regard to EMS application by SMEs may consider these issues.

References

[1] Sharma, H. P., & Kumar, K. (2021, June). Developing and Implementing Environment Management Practices in Small and Medium Size Manufacturing Companies in India. In IOP Conference Series: Earth and Environmental Science (Vol. 795, No. 1, p. 012022). IOP Publishing.

[2] Hui, I. K., Chan, A. H., & Pun, K. F. (2001). A study of the environmental management system implementation practices. Journal of cleaner production, 9(3), 269-276.

[3] Kumar, K., Kumari, R., Poonia, A., & Kumar, R. (2021). Factors influencing corporate sustainability disclosure practices: empirical evidence from Indian National Stock Exchange. Journal of Financial Reporting and Accounting.

[4] Kumar, K., Sharma, H., Khan, W., & Kumar, R. (2022). Factors influencing adoption of green banking practices: Evidence from commercial banks in India. Journal of Asia Entrepreneurship and Sustainability, 18(1), 41-57.

[5] Lozano, M., & Vallés, J. (2007). An analysis of the implementation of an environmental management system in a local public administration. Journal of environmental management, 82(4), 495-511.

[6] Kumar, K., Kumari, R., & Kumar, R. (2021). The state of corporate sustainability reporting in India: Evidence from environmentally sensitive industries. Business and Society Review.

[7] Ann, G. E., Zailani, S., & Abd Wahid, N. (2006). A study on the impact of environmental management system (EMS) certification towards firms' performance in Malaysia. Management of Environmental Quality: An International Journal.

[8] Hillary, R. (2004). Environmental management systems and the smaller enterprise. Journal of cleaner production, 12(6), 561-569.

[9] Perez-Sanchez, D., Barton, J. R., & Bower, D. (2003). Implementing environmental management in SMEs. Corporate social responsibility and environmental management, 10(2), 67-77.
[10] Zorpas, A. (2010). Environmental management systems as sustainable tools in the way of life for the SMEs and VSMEs. Bioresource technology, 101(6), 1544-1557.
[11] Kumar, K., & Prakash, A. (2018). Developing a framework for assessing sustainable banking performance of the Indian banking sector. Social Responsibility Journal, 15(5), 689-709.
[12] Donnelly, K., Beckett-Furnell, Z., Traeger, S., Okrasinski, T., & Holman, S. (2006). Eco-design implemented through a product-based environmental management system. Journal of Cleaner Production, 14(15-16), 1357-1367.
[13] Khalili, N. R., & Duecker, S. (2013). Application of multi-criteria decision analysis in design of sustainable environmental management system framework. Journal of Cleaner Production, 47, 188-198.
[14] Nee, G. Y., & Wahid, N. A. (2010). The effect of ISO 14001 environmental management system implementation on SMEs performance: An empirical study in Malaysia. Journal of Sustainable Development, 3(2), 215.
[15] Nandy, M., Kumar, K., Kumari, R., & Kumar, R. (2022). Do ownership structures and governance attributes matter for corporate sustainability reporting? An examination in the Indian context.
[16] Hillary, R. (2017). Environmental management standards: what do SMEs think?. In ISO 14001 and Beyond (pp. 333-358). Routledge.
[17] Sharma, H. P., & Chaturvedi, A. (2020). The performance of India in the achievement of sustainable development Goals: A way forward. International Journal of Modern Agriculture, 9(4), 1496-1505.

Sustainable Energy Efficient Green Campuses: A Systematic Literature Review and Bibliometric Analysis

Dr. Chintan Pandya[1], Dr. Saurabh Prajapati[2], Rupam Gupta[3]

[1]Librarian, Anant National University, Ahmedabad. India. Email: pandyachintan@gmail.com | ORCID: 0000-0002-4930-9709

[2]Dy. Librarian, Institute of Rural Management Anand (IRMA), India. Email: saurabh@irma.ac.in | ORCID: 0000-0002-4805-0344

[3]Librarian, Gandhinagar Institute of Technology, Gandhinagar. India. Email: rupam.sikligar@git.org.in | ORCID: 0000-0003-4743-559X

Abstract: A sustainable university campus uses energy that does not produce carbon dioxide, protects natural ecosystems, cuts down on the use of natural resources, and recycles or reuses waste in a useful way. The first steps toward a sustainable campus initiative are to use less energy from fossil fuels, use more energy sources that do not rely on carbon, use more energy-efficient technology, recycle waste, and add more greenery to the campus. The Higher Learning Institutes also have a social duty to help people understand the idea of sustainability better. This paper's primary objective is to study the research and development trends in the realm of sustainable development objectives as they pertain to university campuses. The researchers have undertaken a comprehensive literature study and bibliometric analysis of the pertinent articles. Comprehensive use of the VOSviewer software to map current studies on sustainable energy-efficient green campuses. 707 scientific papers (published between 2012 and 2021) were obtained from SCOPUS for the bibliometric analysis for the influential authors, top-cited articles, and co-citation analysis, while 31 articles published in the United States and India were selected for the systematic literature review to assess the trends and future research directions on this subject. According to the findings, the most researched topics are carbon-neutral campuses, smart technology, food and waste management, transportation, and sustainable campuses, with future research focusing on less explored topics such as energy-efficient campus planning, green infrastructure, and landscape design, and photovoltaics. Since it identifies current research hotspots and new study routes, this paper will be a great resource for academics and practitioners interested in sustainable and energy-efficient green campuses.

Keywords: Sustainability Research, Systematic Literature Analysis, Bibliometric Study, Energy Efficient Campus

1.0 Introduction

Sustainability can be described as a dynamic balance among three mutually interdependent elements: (a) protection and enhancement of natural ecosystems and resources; (b) economic productivity; and (c) provision of social infrastructures such as jobs, housing, education, medical care, and cultural opportunities," according to the Sustainable Development Goals. The notion of a green campus was first proposed during the Stockholm Conference on the Human Environment in 1972. The Brundtland Commission of the United Nations defined sustainability in 1987 as "filling current demands without jeopardising future generations' ability to satisfy their own needs.

"To integrate sustainable development into the teaching and learning process, the United Nations designated 2005–2014 as the "Decade of Education for Sustainable Development." India joined the United Nations' 2030 Agenda for Sustainable Development in 2015. Member nations have begun taking steps to accomplish the United Nations' sustainable development goals. They recognise that ending poverty and other forms of deprivation must be combined with strategies that improve fitness and education, reduce inequality, and stimulate economic growth – all while working to mitigate climate change and safeguard forests and seas.

Higher education institutes play a critical role in research to protect and preserve ecosystems in all environmental challenges. Through educational institutes, to teach future generations about the protection of natural resources such as water, biodiversity, energy efficiency, waste management, and global climate change and sustainability. HEIs all over the world are attempting to instil in students, staff, and faculty a knowledge of the concept of sustainability, its importance, and the necessity for people to progress in an ecological style of living. Universities have geared up with programmes to increase energy conservation and cut carbon emissions, minimise material and food wastes, and train students and staff, all of which are directly or indirectly developing the environment.

A Green Campus is a location where sustainable and eco-friendly activities are combined with education to promote sustainable and eco-friendly practices on campus. The green campus idea allows a university to take the lead in reinventing its environmental culture and building new paradigms by generating long-term solutions to humanity's environmental, social, and economic concerns. As a result of the current global warming situation, there is a greater global commitment to ecological conservation. Global warming and climate change are real environmental challenges that must be addressed as soon as possible.

The purpose of this essay is to examine the various techniques used to implement the Sustainable Campus idea. This study focuses on the techniques used by HEIs to achieve a sustainable campus in areas such as behaviour, learning, and physical facilities. It is expected that this takes into account outlines and educates thoroughly on existing practices in order to create sustainable Campuses.

Therefore, this study aimed to examine the effect of domains of sustainability like education, Carbon Neutral Campuses, Smart Technology, Food and Waste Management, Rainwater Harvesting & Horticulture & Transportation for a sustainable campus. This investigation intended to identify the growth of published documents, keywords used by writers, documents by countries and citations, and the most prolific writers in the field of research related to sustainable energy-efficient green campus viz bibliometric analysis.

2.0 Materials & Methods

To assure the study's quality, the systematic literature review (SLR) approach was used. It is considered to minimise bias by adopting a transparent review approach that fosters repeatability, and it has been recognised as an effective and powerful tool for reviewing and sharing information for this sort of study. Tranfield et al. proposed a three-stage approach to SLRs since it was consistent with other recent comprehensive literature reviews in the field of sustainability. The SLR was conducted from 2012 to 2021.

For a thorough literature assessment and bibliographic research analysis on sustainable and green campus features, we utilized the SCOPUS database. On April 20, 2022, data was obtained from SCOPUS to determine the total research output on the topic of Green Campus and Sustainability. The following is the exact search method we used:

TITLE-ABS-KEY (sustainab*) AND (TITLE (campus* OR institut*OR "universit*"))
TITLE-ABS-KEY (green*) AND (TITLE (campus* OR institut*OR "universit*"))

In the first search strategy, we had retrieved 526 documents and subsequently, for the second search strategy we had retrieved 350 documents. Both the files were extracted in .CSV format and further fine-tune was done by removing duplicate entries from the file. Finally, for an overall understanding of the area, a total of 707 documents have been taken into consideration only for the purpose of bibliometric analysis. For a detailed systematic literature review, we further refined the articles by geographical location i.e. India and the United States of America. Therefore, a total of 31 articles have been taken into consideration for the systematic literature review.

We used VOSviewer bibliometric analysis software to generate figures from the data for the analysis. To create the clusters based on 707 bibliographical data, we used co-authorship analysis of prolific authors, co-authorship analysis of significant countries, co-occurrence of all keywords, country-by-

3.0 Descriptive Literature Review:
The evocative systematic literature review carried out for this research encompasses of total 31 articles retrieved from the SCOPUS database to analyse the research outputs between the two major democratic countries i.e. USA and India. A total of 69 articles were published between 2012-2021 however, 38 articles deliberate on diverse areas related to the universities and campuses in general, and the researchers found them irrelevant for the present study. The authors present a proposition and division of the literature analysis on the main domains of sustainability under the sub-domains of energy-efficient buildings, waste management, effective transportation facilities, horticulture initiatives, training programme, and green campuses.

3.1 Education:
Since India has 1027 universities and the world's largest young population, education is a primary source of addressing Sustainable Development Goals (SDG) for students. In India, private universities have made a number of steps by offering SDG training programmes. Prakasha et al. conducted an interview with students studying at private institutions in India as part of their research, and the results show that important issues, such as ownership, real implementations, leadership, and micro-level efforts, need to be addressed [1]. In their case study on the University of Michigan, Shriberg et al. place a heavy emphasis on the impact of course content and educational programmes on "Sustainability and the Campus." Around 75 teachers and students were polled to determine their attitudes and willingness to support on-campus sustainability. The majority of respondents are eager to embark on problem-based initiatives and participate in the sustainability movement, according to the findings [2].

According to Ryan, students are often flustered through sustainable related training programmes and they slowly adopt practices in the real world. In order to address such demotivating factors, faculty have begun to engage students in the sustainability challenges facing their own campuses. The researcher has conducted a case study on the UG course, Sustainable Green Infrastructure Planning and Design and its curriculum. In the innovative practice apart from classroom teaching, students are engaged in landscape architecture, and green building technology to help develop conceptual designs for the outdoor classroom building [3].

3.2 Carbon-neutral Campuses
The control of climate change and the reduction of greenhouse gas emissions is one of the most crucial problems in enhancing global sustainability. The European Union and the Energy Performance of Building Directive (EPBD) have collaborated on a new building standard with the goal of zero-carbon and net-zero-energy construction. Borghi et al. gathered data on four years of carbon emissions at two

major universities in Europe and the United States, and the results, presented in tonnes of carbon dioxide, show that electricity consumption, particularly in fuel-burning for heating equipment, and commuting are two major sources of carbon emissions at these institutions. The significant initiative "Energia 2020" had a critical role in assisting them in lowering CO2 emissions and operational costs [4]. Heating, ventilation, and air conditioning (HVAC) account for around 44% of total energy use on campus. Lack of maintenance, as well as leakage of pipelines, sensors, and controllers, can cause such equipment to fail. Mchugh et al. set out to figure out what the best degree of detection for such leakage is, as well as how to remedy it [5].

Tonietto et al. highlight the University of Michigan's scalable strategy to carbon-neutrality. They compared multiple remote sensing methods on campus natural lands and urban campuses to find the most accurate and efficient protocol for land assessment and ecosystem service valuation that other institutions could scale as needed. Natural areas and supervised categorization involving designated wetlands were found to be the most accurate and efficient methods [6].

3.3 Smart Technology
Healthcare, public services, transportation, energy consumption, and education are all primarily linked with higher educational institutions in smart city concepts. To make a significant shift in energy use, universities must incorporate greater sustainability. Isabel Mendez et al. conducted a case study of the Mexico campus, examining six different scenarios for energy consumption and carbon footprint, including classroom location, orientation, construction material, schedules, loads, heating setback, and cooling setback, as well as the colling system using thermostats. To reduce energy waste, an artificial neural network (ANN) that operates on smart electricity distribution and temperature management systems has been developed. Adjusting the thermostat setpoint from 22 °C to 24.5 °C saved 50% according to the recommended model [7].

The smart grid is a step forward in grid infrastructure. In Distributed Generation, particularly in the generation of renewable energy, the smart energy system is critical. According to Sharma et al., building a smart microgrid at I.K. Gujral Punjab Technical University will be practical and cost viable because the COE will be greatly decreased. Homer runs a series of simulations and uses sensitivity variables to find the best answer. Only the best choice is recommended when compared to various setup Grids. By using homer optimization, the cost of energy is lowered from INR 7.22/kWh to INR 5.47/kWh [8].

The techniques of sustainable development on major university campuses are quite challenging. An Environmental Management Software (EMS) that allows data to be collected from various parts of the university's infrastructure, such as buildings, labs, and other university infrastructure, and processed for environmental decision-making. Bero et al. analysed data on inventory, transportation, horticulture, greenhouse, electricity/water/fuel usage, carbon emissions, air quality, and solid waste at Northern Arizona University to determine the effectiveness of EMS. The conclusions include goals and processes for resource reduction, as well as an environmental monitoring system that builds an information architecture for acquiring environmental performance data [9].

3.4 Food and Waste Management
Labels on food goods, such as pricing, packing materials, certificate, origin, and so on, have an essential impact on customer purchasing behaviour. Silva et al. performed a consumer survey of 338 students who ate at university dining halls to find out how they felt about buying sustainable food. Students are more interested in purchasing products with culinary, health, and cost features than those with sustainable attributes, according to the survey [10]. On university campuses, municipal solid waste (MSW) mostly consists of papers, plastic, files, cardboards (dry trash), and food waste (wet waste). Because it comprises both biomass and non-biomass combustible elements, MSW is also considered an energy source. Universities may become self-sustaining by recycling their MSW for energy generation [11].

At the University of Cincinnati, Tu et al. analyse three important aspects of waste-to-energy projects, namely waste cooking oil-to-biodiesel, paper-to-fuel, pellets, and food waste-to-biogas. Implementing such a project will increase campus sustainability while lowering carbon emissions from fossil fuels. The data study revealed that by recycling energy created from garbage, the institution can replace 834 gallons of patrolium, 121 tonnes of coal, and 12800 m3 of natural gas per year [12].

Ebrahimi et al. look into purchasing and recycling policies in the implementation of zero-waste management solutions that are sustainable. Many colleges in the United States offer rubbish collection services for those living inside their campus limits. Based on the information gathered, a university should develop an Environmentally Preferable Purchasing Program (EPP) to commit to purchasing items that have a lower effect on human health and the environment than competing products [13].

3.5 Rainwater Harvesting and Horticulture
Water shortage is a major impediment to the development of urban landscapes in cities throughout Iran, including Malayer. In this study, Saeedi et al. address the topic of rainwater collection for horticulture and landscape development on the university campus. The researchers gathered information from campus maps, as well as land, soil, and surface regions where rainfall might be collected. The findings point to the use of cisterns to collect rainwater from rooftops, roadways, and sidewalks [14]. In a case study, Fox et al. explore the relationship between university carbon emissions and about 7000 living trees scattered over 200 species on the University of Georgia main campus. The study focused on measuring diameter, height, and tree species, as well as comparing east-west and north-south blocks and calculating carbon stock with i-Tree Eco V6. The carbon stored in the trees was roughly 3500 tonnes, accounting for only 0.11 percent of the University's overall emissions. The study also discovered that native trees with higher overstory heights store more carbon than non-native trees in general [15].

Birds are essential to Mutualism's beneficial effect on trees and capita reproduction. Thousands of birds travel by the Temple University campus, which is located on a significant path of the Atlantic bird migration flight route. Because birds cannot see the building's glass windows and doors, as well as the reflection of habitat, birds collide with the glass windows, becoming hurt or dying as a result of the impact. In a publication, Elmhurst et al. use the PPBL tool to present a case study of one urban university's attempts to reduce bird-window collisions through independent research, curricular integration projects, and campus awareness campaigns [16].

Chambless et al. created a 4000-square-foot living laboratory in the University of Utah's campus garden using compost derived from the university's food waste and transformed it into a magnificent horticultural laboratory. This was a clear act of collaboration between the biology department and the students, who assisted them in developing gardens and making the campus more sustainable [17].

3.6 Transportation
In a research paper, Verma et al. discuss the importance of vehicle pooling and bike-sharing among university students and employees, and conduct a case study of two major bike-sharing systems, PEDL (smart bicycle sharing system) and Zoom car, on the campus of the Indian Institute of Science in Bangalore. The researchers created a model campus layout with the help of GIS technology that includes parking spaces and access locations in order to make the school more sustainable and minimise carbon emissions caused by vehicle mobility on campus [18].

Universities all around the globe are encouraging their constituents to switch from driving alone to more environmentally friendly modes of transportation such as bicycles, buses, and shared taxis. This has the potential to enhance the environment by lowering pollution, energy consumption, and parking congestion. Schneider et al. performed a travel behaviour study of 3600 University of Wisconsin Milwaukee (UWM) faculty and students to assess significant obstacles to sustainable transportation modes commuting. They discovered that journey time and pick-up/drop-off places are important hurdles to bus commuting, whereas road safety, weather, and distance are major barriers to bicycle commuting [19].

3.7 Sustainable Campus

LEED® certification is an independent verification of a building's highest degree of sustainability during construction and operation. Motwani et al. describe the entire process of renovating an ageing building on a university campus and attaining LEED certification from the ground up. They describe the full process, which includes building redesign, procedure, green score measurement, materials, and sustainability [20]. Tilton et al. conducted a comparison study of LEED-certified vs. non-LEED-certified campuses to assess energy consumption units and building efficiency. The study discovered significant energy performance disparities across residences, office buildings, and classrooms [21].

The notion of "campus as a living laboratory" is demonstrated at California State University Los Angeles, which incorporates its sustainable infrastructure into teaching and research programmes. In their research, Blekhman et al. show how university engineering and technology students contribute through sustainable campus teaching on fuel cells, hydrogen infrastructure, and clean mobility. Solar photovoltaics, hydrogen fuel, smart electric distribution, and car charging systems are among the laboratories established by the institution [22]. The impact of fast population increase and urbanisation on green landscapes and habitats is significant. Rabelo at al. focuses on a qualitative study of the Federal University of Santa Catarina's Water-sensitive urban design (USUD) approaches to promote sustainability and climate change as a model role model (UFSC) [23].

Buildings, roofs, structures, and cooling equipment degrade over time as a result of ageing conditions. A regular examination of a campus building is an important aspect of the infrastructure's upkeep since it detects any problems and requires maintenance. The procedure takes a long time and is somewhat costly. Moore et al. recommended using an unmanned aerial vehicle (UAV) to monitor campus facilities and discover faults as well as possible damages related to water, heat, and the heating, ventilation, and air conditioning (HVAC) system. The study conducted on 30 buildings of the university and the UAV model evaluated different defects using image processing technologies. [24].

The Arabian countries' university campuses are always striving towards sustainability. Koch used textual analysis of multidisciplinary research to perform a case study on three NYUAD, Qatar Foundation, and KAUST to better understand sustainability and renewable energy. KAUST was built by HOK, an American architectural firm, and the campus includes LEED platinum-certified green laboratories. According to the report, the country's authorities are more concerned with the iconic growth of green campuses [25].

Hajrasouliha created and developed the "Campus Score" rating methodology, which uses ten distinct factors to reflect three primary variables: urbanisation, greenness, and campus housing. On 103 research outputs from US institutions, a descriptive literature analysis was conducted. Several elements, including land organisation, connection, campus life, green efforts, and compactness, were primarily debated and assessed [26]. Similarly, Marans et al. discussed the Sustainability Cultural Indicators Program (SCIP), which is used to assess campus sustainability research activities. SCIP data and scores assist university officials in making informed decisions [27].

Libraries have intrinsic sustainability as institutions that lessen the environmental effect of providing informational resources by sharing and reusing materials. Tans discusses the Library Environmental Committee, organising education programmes, collaborating on sustainable projects, and recycling procedures at Michigan State University Libraries. Staff bicycles, a compost programme, recycling, and waste management are all part of the collaborative initiative, which will help the institution attain sustainability [28].

Individuals' pro-environmental activities are influenced by five factors: understanding of concerns, knowledge of processes, social incentives, material incentives, and prompts/reminders. Levy et al. summarise research on environmental behaviour, examine how present campus sustainability initiatives correspond with that research, and explain how they generated evidence-based suggestions to improve environmental sustainability. Researchers have proposed a variety of training programmes to educate university workers on environmental issues [29].

3.8 Other issues

Colleges in the United States have made substantial progress toward energy-efficient, sustainable campuses, but Hudler et al. place a greater focus on social sustainability and social injustice concerns, urging universities to empower students and tackle three topics. i.e. acknowledge campus-related matters, utilisation of common language that encourage university sustainability, and create a venue for a diverse group of students [30]. Sustainable Development (SD) in Higher Education cannot be implemented successfully if knowledge and operational silos exist. The role of social ontology in the implementation framework of Sustainable Development at the University Campus was outlined by Bankole and colleagues [31].

4.0 Bibliometric Analysis

Using the search technique outlined in the methodology, we collected entire bibliographical information and abstracts from the SCOPUS database in order to offer a thorough overview of the specified study field. In addition, in order to comprehend the general trends of the study, we utilised VOSviewer to visualise bibliometric networks on many dimensions, including growth, keyword occurrence, citations, authorship patterns, etc. This section's objective is to offer comprehensive results of the quantitative research output of the chosen study topic.

The findings of the various analyses are shown below:

4.1 Year-wise Growth of Publications

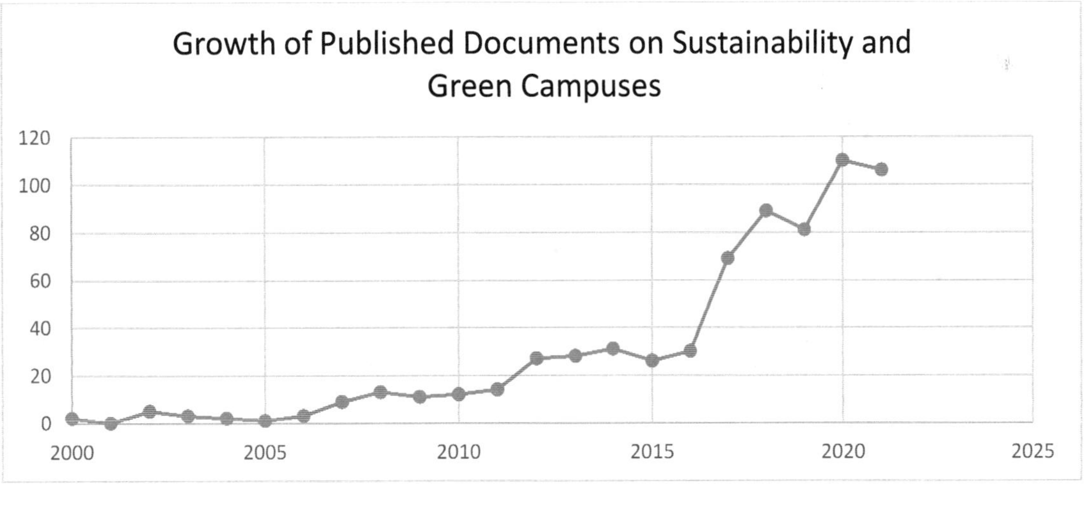

Figure 1. Growth of Documents Published on Sustainability and Green Campuses

Figure 1 shows the significant growth of the publication in the field of sustainability starting from the year 2000 to 2021.

4.2 Keyword Analysis

The network analysis of the co-occurrence of all keywords reported by the Authors is shown in Figure 2. Five was chosen as the threshold value for the minimum number of repetitions of a term. 141 keywords out of 1960 reached the minimum cutoff value of 3. It has been found that researchers' keyword use has evolved through time. 'Environmental sustainability' and 'sustainable development' were the two most often used terms by scholars in this field in the early years. In recent years, study topics such as 'energy efficiency,' 'electric vehicles', sustainable university campuses,' 'landscape design,' 'carbon footprint,' and 'ecosystem services' have been very popular.

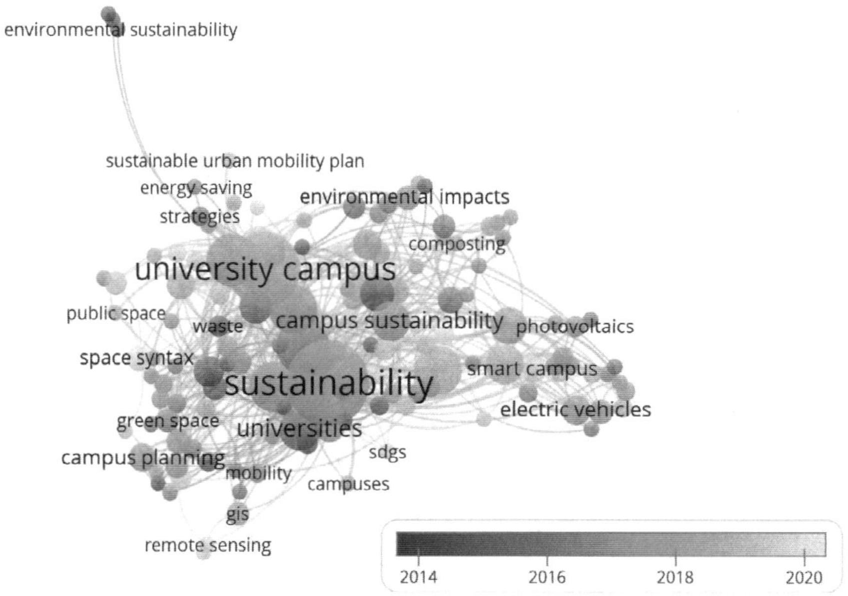

Figure 2. Network Analysis of Author-Keywords

This network analysis is highly essential since it assisted the researcher in understanding the vocabulary used by writers in this specific field.

4.3 Citation Analysis

This section addresses documents by country, citation analysis and reference co-citation analysis based on the 876 articles and their references.

4.3.1 Documents by Country and Citations

The network analysis of authorship and top leading countries was identified and presented in figure 3. These countries were identified based on the rank by a maximum number of documents by particular countries with a maximum number of citations of a country. Out of 85 countries, 57 met the threshold values of 2 countries with 2 citations.

Table 1. Prominent country by documents and citations ranked

Rank	Document by Country	Documents	Rank	Citation by Country	Citations
1	United States	110	1	United States	1070
2	China	68	2	Canada	472
3	Brazil	59	3	United Kingdom	356
4	Malaysia	52	4	Italy	324
5	Italy	43	5	China	307
6	Indonesia	41	6	Malaysia	290
7	Turkey	32	7	Turkey	257
8	Canada	31	8	Brazil	244

9	Spain	31	9	Spain	239
10	United Kingdom	31	10	Mexico	162

Table 1 shows the top 10 countries with the highest graded documents as well as highly cited documents. The top 10 ranked list revealed that, in terms of documents as well as citations, the United States was the first country to research more on Sustainable and Green Campuses, as compared to other countries. This is one of the reasons, we have selected United States documents for a detailed systematic literature review. China was the second-ranked country by documents but when it comes to citations, Canada was the second-ranked country. Similarly, the rest of the counties ranked differently for the number of documents and citations.

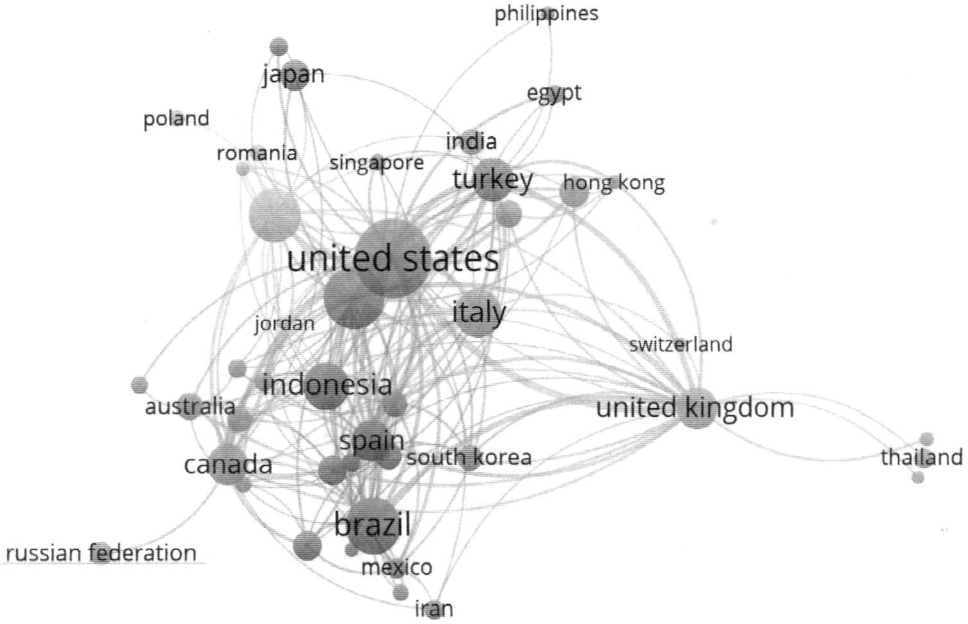

Figure 3. Network Analysis of Citation by Country

Table 2. Top 5 publications based on Citations

Sr.	Author (Year)	Citations
1	Evans J. (2015)	*135*
2	Smyth D.P. (2010)	*109*
3	Armijo De Vega C. (2008)	*106*
4	Anwar N. (2020)	*86*
5	Richardson G.R.A. (2007)	*86*

Table 2 shows the list of the top 5 most cited publications. The most cited study was by Evans J., which focused on how university campuses are being used as a platform for sustainability science. The second is by Smyth D. P., whose main contribution is to reduce solid waste in higher education institutes. These were followed by papers by Armijo De Vega C., Anwar N. and Richardson G.R.A. which focused on solid waste recycling, environment and environmental performance and green buildings on university campuses respectively.

4.4 Most Influential Authors

The most influential authors and co-authorship analysis networks were prepared using the VOSviewer software. Figure 4 and Table 3 depict the list of top 15 most influential authors by documents, as well as another set of top 15 authors whose citations are noteworthy.

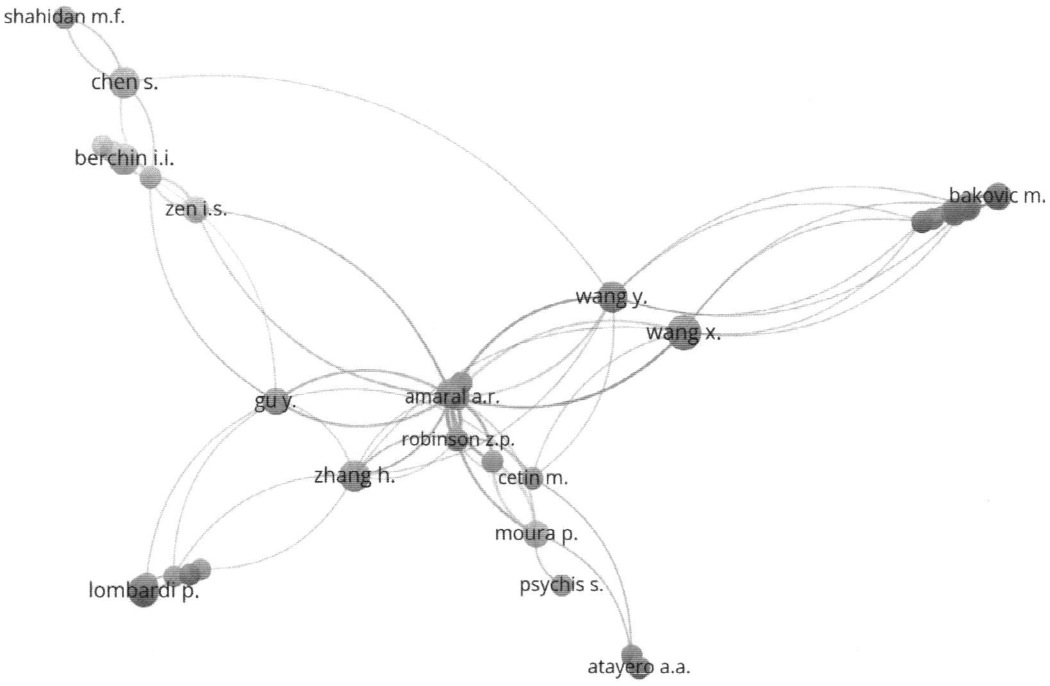

Figure 4. Most Prolific Authors

Table 3. Most Influential Authors

Sr.	Author	Documents	Sr.	Author	Citations
1	Liu J.	6	1	Alves S.	106
2	Longo M.	6	2	Gulwadi G.B.	106
3	Wang X.	5	3	Brooking A.K.	103
4	Attard M.	4	4	Mason I.G.	103
5	Brechin I. I.	4	5	Oberender A.	103
6	Brenna M.	4	6	Wang Y.	84
7	Chen S.	4	7	Lombardi P.	83
8	De Andrade Guerra J.B.S.O.	4	8	Sonetti G.	83
9	Khaustov A.	4	9	Li x.	79
10	Lombardi P.	4	10	Palacios-Pacheco X.	73
11	Papantoniou P.	4	11	Villegas-Ch W.	73
12	Redina M.	4	12	Marans R.W.	72
13	Sonetti g.	4	13	Tan H.	67
14	Wang Y.	4	14	Brandli L.L.	65
15	Yannis G.	4	15	Longo M.	60

Table 3 shows the data about the list of most influencer authors along with the number of documents published and the number of citations received. The first two authors Liu and Longo have (six documents) followed by Wang (five documents). The remaining authors have four documents. Likewise, if talk about the most prolific authors by citations, we saw 103 citations by Brookings, Mason and Oberender followed by Alves and Gulwadi (102 sources) and Wang (81 citations).

5.0 Result and Discussion

Research in the field of sustainable and green campuses continues to expand year after year. Following the announcement of SDG 2030 in 2017, there has been a remarkable growth in the published articles in the field of a sustainable campus. The quantity of literature published in 2021 is 50 times more than that of 10 years ago. Higher Education Institutes have implemented various innovative practices to attain a sustainable campus, but sustainable tools and techniques and their impacts is still a subject under development. The United States is a world leader in university campus research that is both sustainable and energy-efficient. Countries such as China, Canada, Brazil, Malaysia, the United Kingdom, and Italy, on the other hand, are swiftly catching up. "Sustainable development" and "university campus" are the most frequently co-occurring terms in the scientific literature on the sustainability sectors. In this research, we systematically reviewed the most important research articles on diversified areas related to sustainable energy efficient university campuses.

This research contributes value to the current research area in a number of ways, the most important of which is that it gives a systematic summary of the research that has previously been done by scholars all around the world. Furthermore, this paper has attempted to identify the most prolific writers in the chosen field, so that scholars may quickly determine which country is contributing more to this domain. Finally, the evaluation finds that the most prevalent subjects in this research are Carbon-Neutral Campuses, Smart Technology, Food and Waste Management, Transportation, and Sustainable campuses. The findings of this study may be useful for HEIs on their way to redirecting their actions and tackling the sustainability concerns from a practitioner's perspective.

References

[1] G. S. Prakasha, R. Chakraborty, and A. Kenneth, "Energy-saving, waste management, and pollution free steps for university campuses," in AIP Conference Proceedings, Sep. 2021, vol. 2396. doi: 10.1063/5.0066325.

[2] M. Shriberg and K. Harris, "Building sustainability change management and leadership skills in students: Lessons learned from 'Sustainability and the Campus' at the University of Michigan," Journal of Environmental Studies and Sciences, vol. 2, no. 2, pp. 154–164, 2012, doi: 10.1007/s13412-012-0073-0.

[3] R. L. Ryan, "Green infrastructure planning on campus: Case studies from the University of Massachusetts–Amherst," Journal of Green Building, vol. 13, no. 3, 2018, doi: 10.3992/1943-4618.13.3.145.

[4] A. del Borghi, T. Spiegelhalter, L. Moreschi, and M. Gallo, "Carbon-neutral-campus building: Design versus retrofitting of two university zero energy buildings in Europe and in the united states," Sustainability (Switzerland), vol. 13, no. 16, Aug. 2021, doi: 10.3390/su13169023.

[5] M. K. Mchugh, T. Isakson, and Z. Nagy, "Data-Driven Leakage Detection in Air-Handling Units on a University Campus," 2019. [Online]. Available: https://go.gale.com/ps/i.do?p=AONE&u=googlescholar&id=GALE|A616448787&v=2.1&it=r&sid=AONE&asid=b2cc8969

[6] R. Tonietto et al., "Toward a carbon neutral campus: a scalable approach to estimate carbon storage and biosequestration, an example from the University of Michigan," International Journal of Sustainability in Higher Education, vol. 22, no. 5, 2021, doi: 10.1108/IJSHE-05-2020-0188.

[7] J. Isabel Méndez, P. Ponce, T. Peffer, A. Meier, A. Molina, and A. A. Molina Gamified HMI as, "A Gamified HMI as a Response for Implementing a Smart-Sustainable University Campus," p. 10, 2021, doi: 10.1007/978-3-030-85969-5_64ï.

[8] H. Sharma and G. Kaur, "Optimization and simulation of smart grid distributed generation: A case study of university campus," in 2016 4th IEEE International Conference on Smart Energy Grid Engineering, SEGE 2016, Oct. 2016, pp. 153–157. doi: 10.1109/SEGE.2016.7589517.

[9] B. N. Bero, E. Doerry, R. Middleton, and C. Meinhardt, "Challenges in the development of environmental management systems on the modern university campus," International Journal of Sustainability in Higher Education, vol. 13, no. 2, pp. 133–149, Apr. 2012, doi: 10.1108/14676371211211827.

[10] E. Silva et al., "Attitudes of dining customers towards sustainability-related food values at a public University campus," Renewable Agriculture and Food Systems, vol. 35, no. 3, pp. 221–226, Jun. 2020, doi: 10.1017/S1742170519000036.

[11] S. K. Ghosh, Solid waste policies and strategies: Issues, challenges and case studies. Springer Singapore, 2020. doi: 10.1007/978-981-15-1543-9.

[12] Q. Tu, C. Zhu, and D. C. McAvoy, "Converting campus waste into renewable energy - A case study for the University of Cincinnati," Waste Management, vol. 39, pp. 258–265, May 2015, doi: 10.1016/j.wasman.2015.01.016.

[13] K. Ebrahimi and L. A. North, "Effective strategies for enhancing waste management at university campuses," International Journal of Sustainability in Higher Education, vol. 18, no. 7, pp. 1123–1141, 2017, doi: 10.1108/IJSHE-01-2016-0017.

[14] I. Saeedi and M. Goodarzi, "Rainwater harvesting system: a sustainable method for landscape development in semiarid regions, the case of Malayer University campus in Iran," Environment, Development and Sustainability, vol. 22, no. 2, pp. 1579–1598, Feb. 2020, doi: 10.1007/s10668-018-0218-8.

[15] W. Fox, P. Dwivedi, R. C. Lowe, S. Welch, and M. Fuller, "Estimating carbon stock of live trees located on the main campus of the university of georgia," Journal of Forestry, vol. 118, no. 5, pp. 457–465, Sep. 2020, doi: 10.1093/jofore/fvaa025.

[16] K. S. Elmhurst and K. Grady, "Fauna Protection in a Sustainable University Campus: Bird-Window Collision Mitigation Strategies at Temple University," in World Sustainability Series, Springer, 2017, pp. 69–82. doi: 10.1007/978-3-319-47868-5_5.

[17] R. Chambless, A. Parvaz, L. A. Chesson, and J. S. Ruff, "University of utah campus gardens: A living laboratory for sustainability," Sustainability, vol. 5, no. 3, pp. 160–164, Jun. 2012, doi: 10.1089/SUS.2012.9963.

[18] A. Verma, H. Vajjarapu, and M. Thuluthiyil, "Planning and Usage Analysis of Bike Sharing System in a University Campus," in Recent Advances in Traffic Engineering Select Proceedings of RATE 2018, 2018, pp. 339–349. [Online]. Available: http://www.springer.com/series/15087

[19] R. J. Schneider and L. Hu, "Improving University Transportation Sustainability: Reducing Barriers to Campus Bus and Bicycle Commuting." [Online]. Available: www.onsustainability.com,

[20] J. Motwani, N. Afshan, and P. Charath, "Building green infrastructure across university campuses: A research," International Journal of Innovative Technology and Exploring Engineering, vol. 8, no. 11 Special Issue, pp. 498–502, Sep. 2019, doi: 10.35940/ijitee.K1084.09811S19.

[21] C. Tilton and M. el Asmar, "Assessing LEED versus non-LEED Energy Consumption for a University Campus in North America: A Preliminary Study," 2014.

[22] D. Blekhman et al., "Campus Sustainable Infrastructure as Living Lab at California State University Los Angeles," 2019.

[23] V. G. Rabêlo, I. I. Berchin, M. de León, J. H. D. de Toledo, L. R. da Silva, and J. B. S. O. de Andrade Guerra, "University campuses as town-like institutions: promoting sustainable development in cities using the water-sensitive urban design approach," in World Sustainability Series, Springer, 2019, pp. 497–511. doi: 10.1007/978-3-030-15864-4_29.

[24] J. Moore, H. Tadinada, K. Kirsche, J. Perry, F. Remen, and Z. T. H. Tse, "Facility inspection using UAVs: a case study in the University of Georgia campus," International Journal of Remote Sensing, vol. 39, no. 21, pp. 7189–7200, Nov. 2018, doi: 10.1080/01431161.2018.1515510.

[25] N. Koch, "Green Laboratories: University Campuses as Sustainability 'Exemplars' in the Arabian Peninsula," Society and Natural Resources, vol. 31, no. 5, pp. 525–540, May 2018, doi: 10.1080/08941920.2017.1383546.

[26] A. Hajrasouliha, "Campus score: Measuring university campus qualities," Landscape and Urban Planning, vol. 158, pp. 166–176, Feb. 2017, doi: 10.1016/j.landurbplan.2016.10.007.

[27] R. W. Marans and J. Callewaert, "Evaluating Sustainability Initiatives on University Campuses: A Case Study from the University of Michiganâ€™s Sustainability Cultural Indicators Program," in World Sustainability Series, Springer, 2017, pp. 189–199. doi: 10.1007/978-3-319-47889-0_14.

[28] E. D. Tans, "Sustainable Academic Libraries: A Campus Partnership at Michigan State University," in World Sustainability Series, Springer, 2017, pp. 89–100. doi: 10.1007/978-3-319-47895-1_6.

[29] B. L. M. Levy and R. W. Marans, "Towards a campus culture of environmental sustainability: Recommendations for a large university," International Journal of Sustainability in Higher Education, vol. 13, no. 4, pp. 365–377, Sep. 2012, doi: 10.1108/14676371211262317.

[30] K. Hudler, L. Dennis, M. DiNella, N. Ford, J. Mendez, and J. Long, "Intersectional sustainability and student activism: A framework for achieving social sustainability on university campuses," Education, Citizenship and Social Justice, vol. 16, no. 1, pp. 78–96, Mar. 2021, doi: 10.1177/1746197919886860.

[31] A. Bankole, E. Fidelis, and N. Alfred, "Towards a Social Ontology on Sustainable Development in CUT: Understanding Stakeholder Perceptions," in World Sustainability Series, Springer, 2017, pp. 425–439. doi: 10.1007/978-3-319-47889-0_30.

Low frequency Vibration Energy Harvester using Compliant Mechanism driven Rosen-type Piezoelectric Transformer

Sanniva Bhattacharjee[1], Aryan Kanu[2], Arnab Roy Chowdhury[3], Soumyadeep Mondal[4], Nilanjan Chattaraj[5], Nirmal Baran Hui[6]

[1,2,3,4,6] Department of Mechanical Engineering,

[5]Department of Electronics and Communication Engineering

National Institute of Technology Durgapur, India

[5]Email: nilanjan.chattaraj@ece.nitdgp.ac.in

Abstract. This paper represents an energy harvesting technique using a compliant mechanism driven Rosen type piezoelectric transformer from low frequency (3-10 Hz) reciprocal gripping motion. The finite element model of the compliant structure is synthesized using topology optimization implementing the density model and is coupled with a piezoceramic transformer to operate it by human hands. The mechanism can also be operated with the help of external mechanisms. As a case study, the proposed compliant mechanism is modelled considering acrylic plastic material. The mechanism can produce a maximum power of 62.7 μW for a 10 Hz gripping frequency while operating at its optimal conditions. The paper also discusses the resonance condition of the structure to check its marginal state. The harvested energy can be extracted across a load resistances attached at both primary and secondary side of the piezoelectric transformer.

1. Introduction

Energy Harvesting is an emerging research field. The demand for power supply to wearable electronic components has boosted research into the domain of energy harvesting. There are several physics-based approaches associated with energy harvesting, such as (a) Photovoltaic (b) Thermoelectric (c) Vibration (d) Radio-frequency (RF). Batteries have well-known drawbacks in low-power devices, particularly those used in the human body, such as limited lifetime, pollution, and inconvenient maintenance. Piezoelectric energy harvesters provide several advantages while scavenging energy from the surrounding environment [1, 2]. Piezoelectric energy harvesters help to build a smarter world by powering electronic devices with energy harvested from environmental vibrations via the piezoelectric effect. In recent years, numerous researchers have investigated Rosen type piezoelectric transformers [3]. Yang *et al.* [3] reviewed the structural modelling of piezoelectric transformers and stated that transformers made up of piezoelectric materials can be used to increase or decrease a voltage using suitable geometrical parameters. It also stated that in certain low-power applications, piezoelectric transformers outperform electromagnetic transformers. Fukunaga *et al.* [4] analysed that the voltage step-up ratio, efficiency, and power transfer are the functions of the ratio of length to width and thickness of the piezoelectric transformer.

Discovering and examining the promising utility of various mechanisms and structures in energy harvesting is relatively new along the research line [5, 6]. Generally, there are two types of mechanical systems, which are used for energy harvesting applications on the basis of joints: Rigid body mechanism and compliant mechanism. The latter has a number of benefits, including a single structure, less assembly parts, lower manufacturing costs, and reduced friction between the contacts [7]. Compliant mechanisms are used in many applications including MEMS, adaptive structures, surgical tools and the micro-stage [8]. Topology optimization allows designers to determine the best structural architecture for a given performance requirement. Zhu *et al.* [9] presented a technique for optimal topology of compliant mechanism on the basis of continuum type topology optimization approaches. They have demonstrated that the maximum stress level in the mechanism may be governed by confining the permissible displacement at the input side. Maddisety *et al.* [10] proposed a topology optimization technique for designing a piezoelectric actuator for dynamic applications. In recent years, researchers have developed new methods to integrate compliant mechanisms with piezoceramic stack in energy harvesting. There are several articles reported on piezoelectric energy harvesting. However, none of those mentions piezoelectric energy harvesting from bio-mechanical hand-gripping motion.

In this article, we have investigated the scope of energy harvesting from bio-mechanical reciprocal hand-gripping motion. We have presented a novel energy harvester-mechanism comprising of a compliant mechanism and a Rosen type piezoelectric transformer functional at low frequency around 3-10 Hz. We have used topology optimization technique to find the design of the compliant mechanism and have analyzed the performance of the mechanism in integration with a Rosen type piezoelectric transformer to investigate the desired objective. To perform a case study, we have modeled the proposed compliant mechanism assuming acrylic plastic, structural steel and aluminium alloy material. The assembly can produce a maximum power of 62.7 μW for a 10 Hz reciprocal motion frequency at its optimal conditions. We have also discussed the resonance condition of the structure. We have shown that the harvested energy can be pull out across a load resistances, which can be connected at both primary and secondary side of the Rosen type piezoelectric transformer.

2. Design Methodology

The conceptual schematic diagram of the proposed energy harvester comprising of a compliant mechanism and Rosen type piezoelectric transformer is shown in Figure 1.

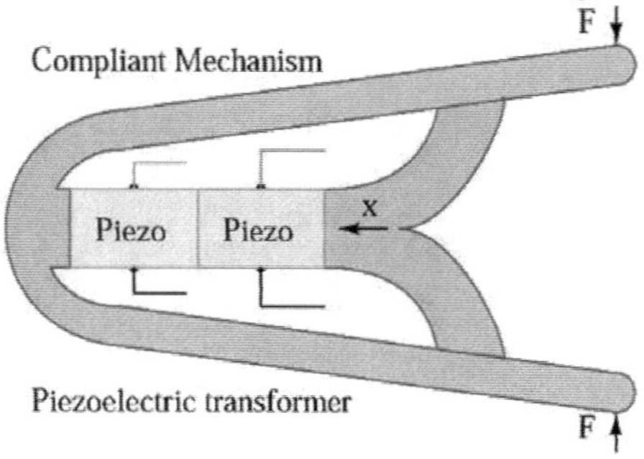

Figure 1. The proposed energy harvester comprising of a compliant mechanism and Rosen type piezoelectric transformer.

2.1. Rosen type Piezoelectric Transformer

A piezoelectric transformer is an electromechanical device that operates at a certain resonant frequency of a vibrating piezoelectric body. Figure 1 illustrates a Rosen piezoelectric transformer with a primary side of length l_1, a secondary side of length l_2, width w, and thickness t. $-l_1 < x_1 < 0$ is considered as the driving portion and is poled along x_3 direction having the electrodes at the top and bottom surfaces of the driving portion. $0 < x_1 < l_2$ is considered as receiving portion and is poled along x_1 direction. The output electrode is at $x_1 = l_2$. The transformer is subjected to a longitudinal vibration and generates a time-harmonic output voltage $V_2(t)$ under a time-harmonic driving voltage $V_1(t)$ regulated with an appropriate frequency. Typically, a load circuit of impedance Z is connected to the output electrodes. The above-mentioned transformer depicted in Figure 2 is the Rosen-type transformer, made of PZT-5H material. The characteristics of the transformer is influenced by linear theory of piezoelectricity [11].

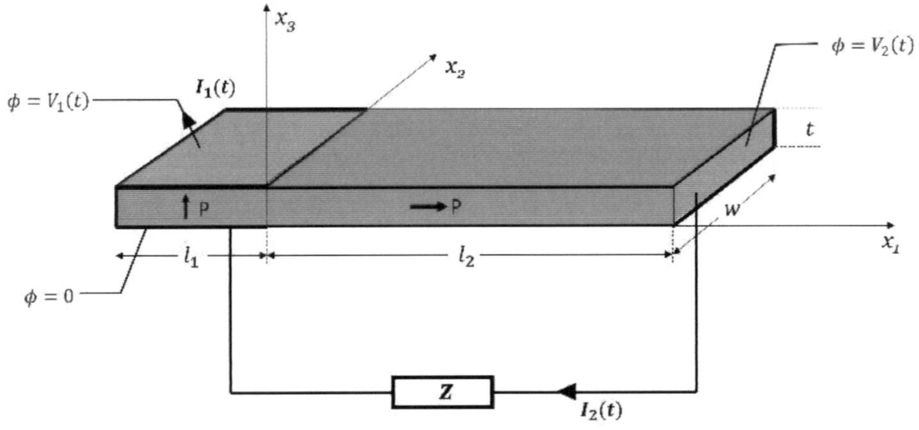

Figure 2. Rosen type piezoelectric transformer.

This paper primarily focuses on symmetric Rosen type transformer where the length of each section is equal i.e., $l_1 = l_2 = l$. A numerical validation study is carried out for the transformer and the results obtained are compared with the reference literature of Jiashi Yang *et al.* [3]. The analysis, which is carried out for validation, includes a) Transforming ratio versus load resistance and b) Transforming ratio versus driving frequency.

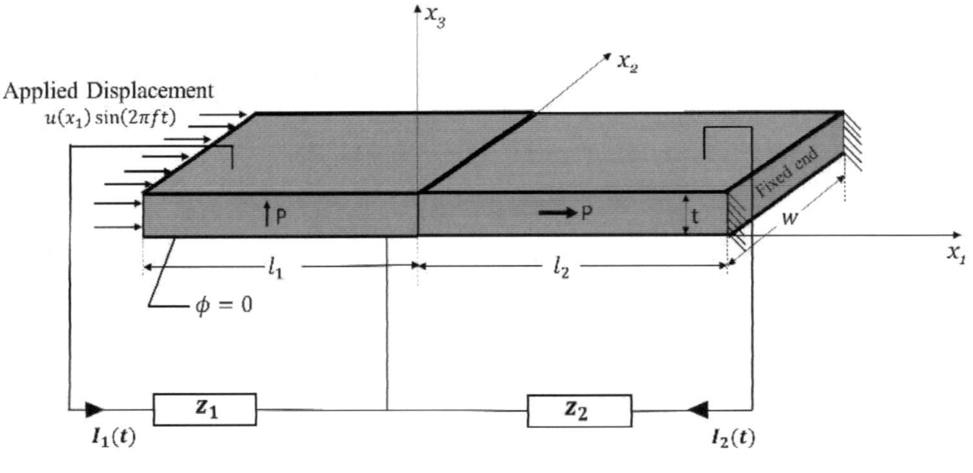

Figure 3. Rosen transformer acting as an energy harvester.

The primary objective of the present paper is to use the piezoelectric transformer integrated with a compliant mechanism for energy harvesting. For this purpose, initially, a lumped model has been structured in such a way that the transformer is acting like a bar and the power is extracted from both primary and secondary side. The bottom surface of both the sections acts as a reference terminal where $\phi = 0$ and the top surfaces of the driving side and the receiving side are considered as terminal 1 and terminal 2, respectively. A schematic of the symmetric transformer is depicted in Figure 3. A harmonic displacement $u(x_1)\sin(2\pi ft)$; (where f is driving frequency) is applied on the driving side considering that the other end is fixed.

2.2. *Selection of the mechanism*

The classification of mechanisms is shown in Figure 4. This paper focuses on compliant mechanisms [12–14] over rigid link mechanisms, since the later have certain limitations that make them less suitable for the targeted application. Some of the limitations are i) Restricted motion of the links, ii) Difficulty in assembling due to a greater number of parts, iii) Friction losses at the joints, iv) Failure of parts. On the contrary, compliant mechanisms can store strain energy during their operation through structural deformation and can release energy while coming back to its original state. This gives a major advantage of less wear, friction and backlash. A compliant mechanism that features distributed elastic deformation can be more reliable and durable. Thus, this paper targets a distributed compliant mechanism for energy harvesting from the targeted application.

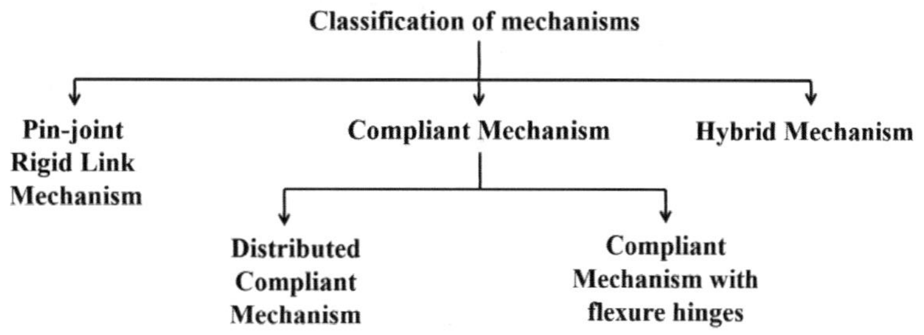

Figure 4 Classification of mechanisms.

2.3. *Synthesis of Distributed Compliant Structure using Topology Optimization*

Figure 5 shows a 2D rectangular geometry of dimension $x_1 \times y_1$ whose area is filled with rectangular mesh elements. The total number of mesh elements along the boundary length and width of the rectangle is considered as N and M, respectively. As a result, the dimension of each rectangular mesh element is $x_1/N \times y_1/M$. Boundary conditions are applied to synthesize the compliant mechanism. A void space is allocated inside the rectangular area in order to accommodate the Rosen type piezoelectric transformer. Fixed boundary conditions are applied at the extreme left boundary of rectangular geometry and the rectangular void space. An input vertical force is applied at the end of the 2D geometry as shown in Figure 5. The prescribed displacement boundary condition is applied at the extreme right end of the void space, such that we get an axial displacement $u \neq 0$ and vertical displacement $v = 0$. This will ensure that the input vertical force will produce a longitudinal axial strain in the Rosen type piezoelectric transformer. The above procedure is depicted in Figure 6 using a flow chart. The SIMP model is used in order to implement the topology optimization. The model uses penalized material volume factor, θ_p which is used for interpolating material parameters. The θ_p for SIMP interpolation is expressed as:

$$\theta_p = \theta_{min} + (1 - \theta_{min})\theta^{p_{\mathrm{SIMP}}} \tag{1}$$

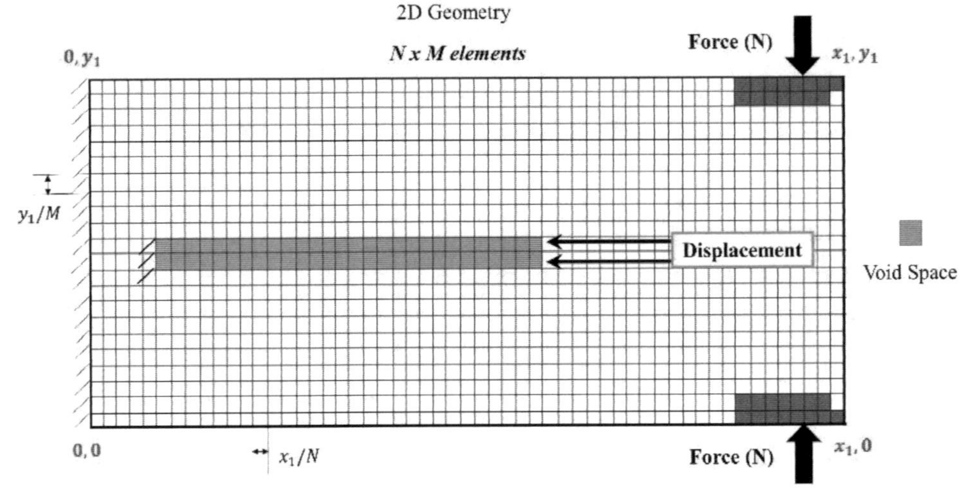

Figure 5. Initial 2D rectangular geometry space with boundary conditions and physical constraints to get the initial design of the compliant mechanism.

Here θ_{min} represents minimum penalized material volume fraction. The p_{SIMP} exponent can be increased to decrease the intermediate stiffness values, so that the purple scale as shown in Figure 6 becomes less favourable. The range of θ_p is [0, 1]. This factor gets multiplied by the density p of solid material to give a new term called penalized density p_p that can be expressed as:

$$p_p = p\theta_p \tag{2}$$

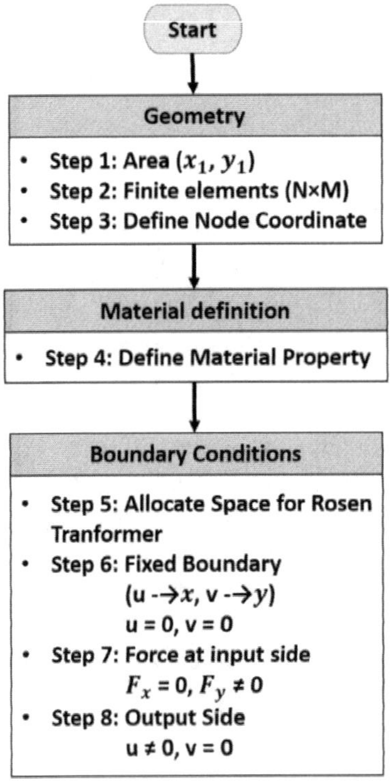

Figure 6. Steps of topology optimization to produce the compliant structure.

The p_p is applied throughout the optimized domains and results in the desired compliant structure as shown in Figure 7. The dark purple elements indicate that θ_p are near 100% and extreme light purple elements correspond to almost 0%. The Rosen type piezoelectric transformer is modelled considering PZT-5H material and the performance of the energy harvester is analysed assuming three different elastic materials for compliant structure i.e., acrylic plastic, structural steel, and aluminium alloy (AL 6061 T-6). The force F as shown in Figure 7 can be applied either by using hands or by using an external mechanism. The bottom electrodes of the Rosen type transformers are acting as reference nodes and the power can be collected across each load resistance as shown in Figure 3. The corresponding results are discussed in the results sections.

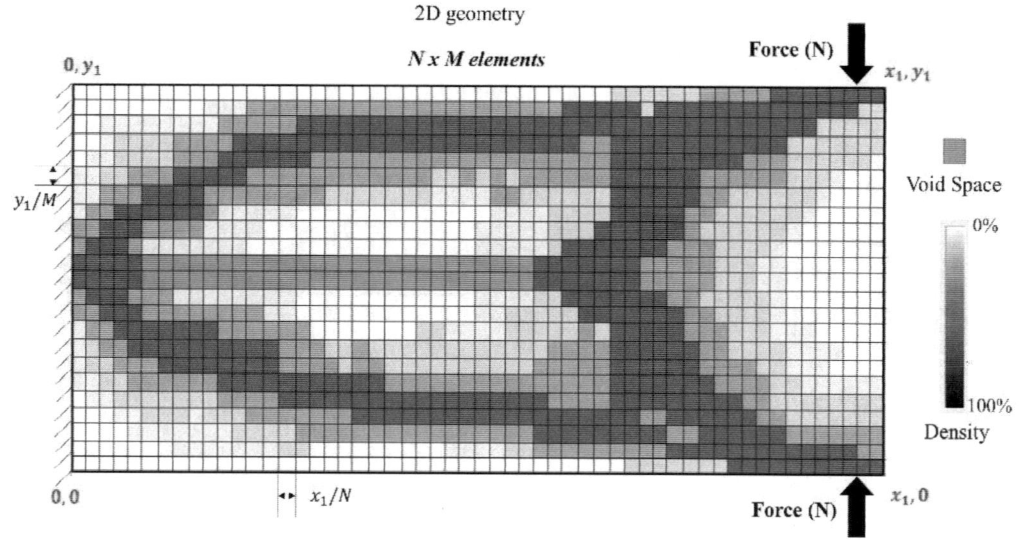

Figure 7. Purple Scale optimized distributed compliant structure.

3. Simulations Results and Discussion

3.1. Numerical Validation of Rosen type Transformer

Initially, the present FE model has been validated by incorporating the design, which is mentioned in [3] and [6, 15]. In FE analysis, the sparse direct solver has been used for structural analysis. Figure 8 depicts the transformation ratio (V_2/V_1) of the PZT-5H transformer at an excitation frequency of 35 kHz and load resistance of 2000 kΩ. The corresponding step-up voltage ratio obtained is around 22, which deviates from the analytical result 20.2 by an error of around 8%. Figure 9(a) depicts a plot of transformation ratio of the transformer against a certain range of load resistance at two different driving frequencies. Here, the primary section receives a 1 V input voltage (V_1), and the output voltage (V_2) is extracted via a load resistance linked to the secondary section. For very large load the transformation ratio is almost constant and exhibits a saturation type of behaviour that is akin to an open circuit ($Z_L = \infty$). From Figure 9(b), it can be inferred that the plot is in good agreement with the reference plot of Jiashi Yang *et.al.* [3]. Here the discrepancies, observed in Figure 9(a) are negligible when small load resistances are considered. However, the difference in values between analytical and FEA results is (~ 8%-10%) when high load resistances are considered.

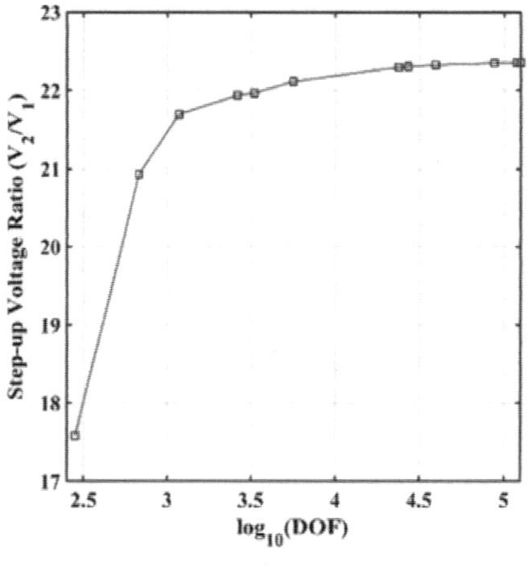

Figure 8. Mesh convergence curve for step-up voltage ratio

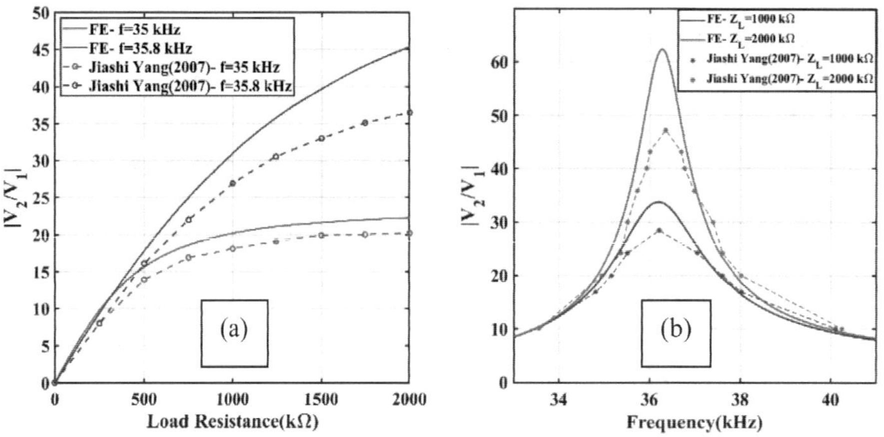

Figure 9. (a) Variation of transformation ratios with load resistance. (b) Variation of transforming ratios with driving frequency.

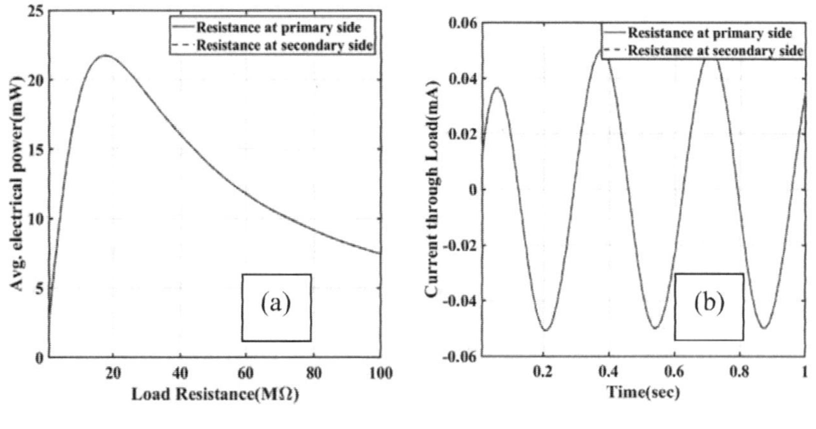

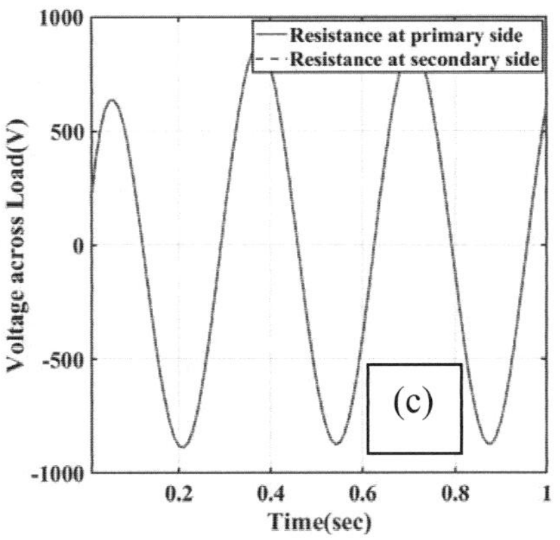

Figure 10. (a) Optimal Load Resistance. (b) Time Variation of current at optimal condition. (c) Time Variation of voltage at optimal condition.

The variation in transformation ratio (V_2/V_1) with the driving frequency at two different load resistances is plotted in Figure 9(b). The resonance behaviour has been checked at two separate load resistances (1000 kΩ and 2000 kΩ), that is later compared with the analytical solution of the Jiashi Yang [3]. The maximum % errors in transformation ratios at resonance frequencies for 1000 kΩ and 2000 kΩ are 18.5 % and 31.8 % respectively. In the further extension, Rosen ceramic transformer is now acting as a bar as discussed in section 2.1 and axial harmonic displacement of amplitude 100 μm is applied at a frequency of 3 Hz. The power is extracted across load resistances connected at both primary and secondary sections. To obtain the maximized power, it is necessary to operate the transformer at its optimal condition. In Figure 10(a) average electrical power has been varied over a range of load resistances and it is observed that at 17.5 MΩ, the maximum power of 21.74 mW is obtained, which is the optimal condition for the present 3 Hz driving frequency.

3.2. Topology Optimization of Compliant Structure

In the present section, the optimized compliant structure is synthesized through the topology optimization technique. As the present system will be operated by hand power only and to accommodate it within this, here we have considered 40 mm × 80 mm rectangular domain to synthesize the topology under a static hand force of 107 N and following boundary conditions as illustrated in Figure 11 and it has been discretized by finite elements having regular triangular areas. All topological operations have been done under the 2D plane strain assumption. Topology optimization with the density method has been implemented during finite element analysis where the control variable field, Helmholtz filter and SIMP penalization are defined according to the present requirements. A normalized objective function to maximize the longitudinal displacement is imposed on the right boundary of the region of the Rosen transformer. Figure 11 shows overall iterative steps, which are performed during topology optimization such as discretization and iteration history, initial and final optimal solution. Finally, a tuned 3D CAD model is prepared with an out-of-plane thickness of 10 mm from the previously obtained 2D topology.

3.3. Dynamic Study of Tuned Topology Optimized Structure

The tuned 3D model is now being used in Comsol Multiphysics to examine the structure's frequency driven dynamic behaviour. The exterior body of the present structure has been tested with three different materials (acrylic plastic [16], structural steel [17], and Aluminium alloy AL 6061 T6 [7]. The structure was subjected to a harmonic hand-driven load of 107 N [18], and the resonance condition (maximum longitudinal deformation of the transformer) was checked and compared with the scenario where there was no piezo transformer in the system. Table 1 shows the resonant displacement amplitudes for all three possible cases. In the presence of a piezo transformer, the maximum resonance displacement for acrylic plastic is 4.7568 mm, as shown in Figure 12(a), and the quality factor is 1000. In the absence of piezo transformers, the resonance frequency has increased significantly (Figure 12(b)). Similar to Figure 10(a), the maximum electrical powers across the output optimal load resistances are now compared for all three aforementioned materials at three different frequencies (5 Hz, 10 Hz, and 100 Hz), which are summarized in Table 2. At the previously obtained resonant frequencies (given in Table 1), the optimal resistance and power have also been shown here in Table 2. For instance, in acrylic plastic, the present compliant structure is capable of extracting a maximum power of 62.7 μW for 10 Hz at frequency from each load resistance affixed in each section of the transformer.

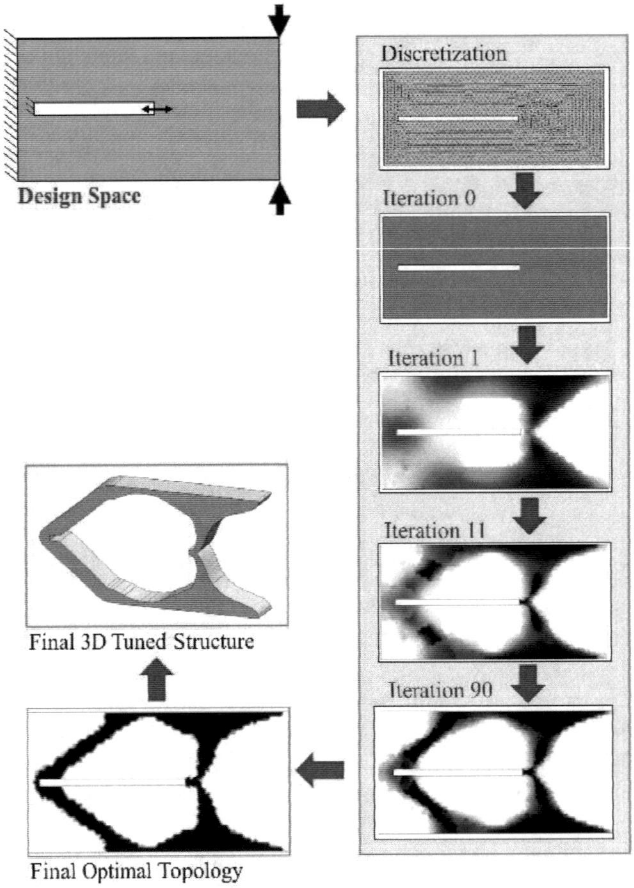

Figure 11. Discretization and iteration history.

Table 1. Displacements amplitudes at resonance.

Material	With Piezo transformer		Without Piezo transformer	
	Resonant Frequency (Hz)	**Displacement (mm)**	**Resonant Frequency (Hz)**	**Displacement (mm)**
Acrylic plastic	4459	4.7568	586	4194.6
AL 6061 T6	5800	0.27753	1851	525.82
Structura l steel	4970	0.02527	1787	170.67

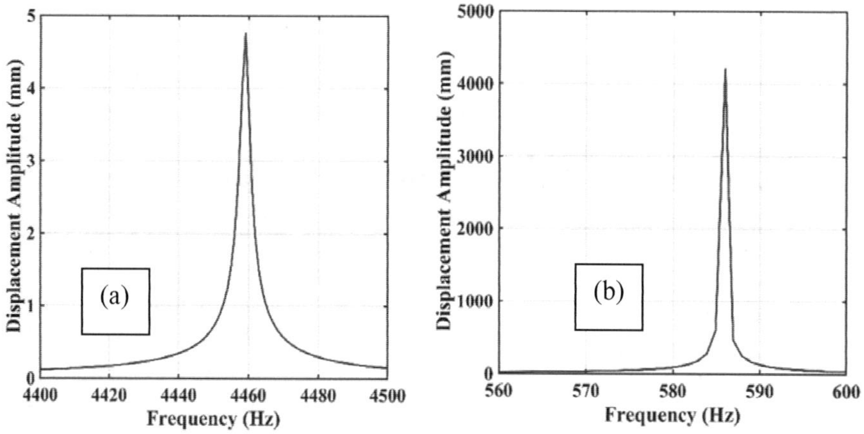

Figure 12. (a) Resonant displacement amplitude (frequency domain study with piezo transformer). (b) Resonant displacement amplitude (frequency domain study without piezo transformer).

Table 2. Power extraction under different input frequency.

Material	Frequency (Hz)	Optimal Resistance (MΩ)	Max. Power developed (μW)
Acrylic plastic	5	10	31.34
	10	5	62.70
	100	0.5	626
	4459 (Resonance)	0.021	198770×10^3
AL 6061 T6	5	10	28.74
	10	5	57.49
	100	0.5	574.90

Material	Frequency (Hz)	Optimal Resistance (MΩ)	Max. Power developed (μW)
	5800 (Resonance)	0.020	67180×10^3
Structural steel	5	10.5	25.28
	10	5	50.53
	100	0.52	505.75
	4970 (Resonance)	0.017	353.16×10^3

4. Conclusion

In this paper, we have investigated the scope of bio-mechanical energy harvesting using compliant mechanism driven Rosen type piezoelectric transformer. The compliant mechanism is synthesized using the density model of topology optimization and is examined at a low frequency around 5 Hz. The harvested energy has been examined considering three distinct elastic materials. For a typical specification, the proposed smart compliant structure is capable of extracting a maximum power of 62.7 μW for 10 Hz reciprocal motion under optimal conditions. The harvested power can be useful to ultra-low-power battery-less IoT devices. Depending on the requirements, the harvested power can be processed through power management circuits in order to ensure the adequacy of power supply in such IoT systems.

5. References

[1] Kim, H. S., Kim, J. H., & Kim, J. (2011). A review of piezoelectric energy harvesting based on vibration. *International journal of precision engineering and manufacturing*, *12*(6), 1129-1141.

[2] Chowdhury, A. R., Mondal, S., Kanu, A., Bhattacharjee, S., Chattaraj, N., & Hui, N. B. (2022). Piezoelectric Energy Harvesting by Utilizing Angular Speed Variation in High-Speed Rotating Beams. *Journal of Vibration Engineering & Technologies*, 1-16.

[3] Yang, J. (2007). Piezoelectric transformer structural modeling-A review. *IEEE transactions on ultrasonics, ferroelectrics, and frequency control*, *54*(6), 1154-1170.

[4] Fukunaga, H., Kakehashi, H., Ogasawara, H., & Ohta, Y. (1998, May). Effect of dimension on characteristics of Rosen-type piezoelectric transformer. In *PESC 98 Record. 29th Annual IEEE Power Electronics Specialists Conference (Cat. No. 98CH36196)* (Vol. 2, pp. 1504-1510). IEEE.

[5] Lee, D. G., Carman, G. P., Murphy, D., & Schulenburg, C. (2007, June). Novel micro vibration energy harvesting device using frequency up conversion. In *TRANSDUCERS 2007-2007 International Solid-State Sensors, Actuators and Microsystems Conference* (pp. 871-874). IEEE.

[6] Roy Chowdhury, A., Saurabh, N., Kiran, R., & Patel, S. (2022). Effect of porous auxetic structures on low-frequency piezoelectric energy harvesting systems: a finite element study. *Applied Physics A*, *128*(1), 1-19.

[7] Patil, V. S., Anerao, P. R., & Chinchanikar, S. S. (2018). Design and analysis of compliant mechanical amplifier. *Materials Today: Proceedings*, *5*(5), 12409-12418.

[8] Khare, P., Madhab, G. B., Kumar, C. S., & Mishra, P. K. (2007, December). Optimizing design of piezoelectric actuated compliant microgripper mechanism. In *The Proceedings of the 13th National*

Conference on Mechanisms and Machines, IISc, Bangalore, India (pp. 12-13).

[9] Zhu, B., Zhang, X., Zhang, H., Liang, J., Zang, H., Li, H., & Wang, R. (2020). Design of compliant mechanisms using continuum topology optimization: A review. *Mechanism and Machine Theory*, *143*, 103622.

[10] Maddisetty, H., & Frecker, M. (2004). Dynamic topology optimization of compliant mechanisms and piezoceramic actuators. *J. Mech. Des.*, *126*(6), 975-983.

[11] Preumont, A. (2006). *Dynamics of electromechanical and piezoelectric systems* (p. 224). Dordrecht, The Netherlands:: Springer.

[12] Linß, S., Henning, S., & Zentner, L. (2019). Modeling and design of flexure hinge-based compliant mechanisms. In *Kinematics-Analysis and Applications*. IntechOpen.

[13] Swan, C. C., & Rahmatalla, S. F. (2004, January). Design and control of path-following compliant mechanisms. In *International Design Engineering Technical Conferences and Computers and Information in Engineering Conference* (Vol. 46954, pp. 1173-1181).

[14] Conlan-Smith, C., Bhattacharyya, A., & James, K. A. (2018). Optimal design of compliant mechanisms using functionally graded materials. *Structural and Multidisciplinary Optimization*, *57*(1), 197-212.

[15] Wang, L., Zhao, L., Jiang, Z., Luo, G., Yang, P., Han, X., ... & Maeda, R. (2019). High accuracy comsol simulation method of bimorph cantilever for piezoelectric vibration energy harvesting. *AIP Advances*, *9*(9), 095067.

[16] Abdelnaby, Mohammed Ali, and Mustafa Arafa. "Energy harvesting using a flextensional compliant mechanism." *Journal of Intelligent Material Systems and Structures* 27, no. 19 (2016): 2707-2718.

[17] Vilorio, C., Stark, B., Hawkins, A. R., Frogget, K., & Jensen, B. (2015). Stress relaxation insensitive designs for metal compliant mechanism threshold accelerometers. *Sensing and bio-sensing research*, *6*, 33-38.

[18] Károly János, B., Ákos, J., & Károly, B. (2010). Force measurement of hand and fingers. *Biomechanica Hungarica*, *3*(1).

Acknowledgments
On behalf of all authors, the corresponding author states that there is no conflict of interest.

Driving through sustainable systems: a Study of Air Quality Index of Delhi during COVID-19 Pandemic

S.Garg[1], S.L. Sahdev[2], N.Mahajan[3] , and N.Goel[4]
[1]Associate Professor, AIBS, Amity University, Noida, Uttar Pradesh, India
[2]Assistant Professor, ISBR, Bangalore, Karnataka, India
[3]Associate Professor, AIBS, Amity University, Noida, Uttar Pradesh, India
[4]Student AIBS, Amity University, Noida, Uttar Pradesh, India

E mail: seemagarg1@ gmail.com

Abstract. The impact of lockdown during covid-19 pandemic on air quality in terms of pollution was investigated in the Delhi/NCR region in this study. The National Air Quality Index (NAQI) uses air quality data for six pollutant measures (PM10, PM2.5, SO2, CO, NO2 and O3) from monitoring stations across the megacity to demonstrate the spatial pattern of air quality before and during the lockdown period due to pandemic. The objective of this study is to investigate the interrelationship among the various parameters the Pearson's correlation analysis was conducted. The results found were able to demonstrate that during lockdown air quality was significantly improved. It was also found that among other pollutants, level of NO2 and CO have also declined during- lockdown phase. It was observed that during this period of lockdown, the air quality index on maximum number of days was found to be good and satisfactory in Delhi that was the clear indication of an improvement in air quality due to a reduction in pollution and emissions of vehicles and industries. It can be implemented to further work on parameters to improve the air quality in future.
Keywords: Air Quality, AQI, Sustainable

1. Introduction

As evidenced by data analysis and correlation for the period prior to the implementation of the lockdown, as well as comparisons to previous years, the cross-country lockdown, which began in March 2020, in response to the COVID-19 pandemic, has resulted in a significant improvement in air quality in the country. The main sources of air pollution are transportation, businesses, power plants, development exercises, and biomass to avoid eating, street dust, and private workouts. Similarly, certain activities add to air pollution, such as DG set activity, cafes, and landfill fires etc. Except for necessary administrations, all vehicle administrations – street, air, and rail – were suspended throughout the cross-country closure. Among other things, modern foundations, development exercises, business, and friendliness administrations were placed on hold. As a result of the cessation of discharges from various sources, air quality improved in a number of Indian towns and urban areas. The current study looks at pre-lockdown and lockdown data of quality of air acquired from surrounding of air quality stations. For key towns and cities in Delhi NCR, the report provides the translation of trends discovered for PM10, PM2.5, NO2, SO2, Benzene and AQI.

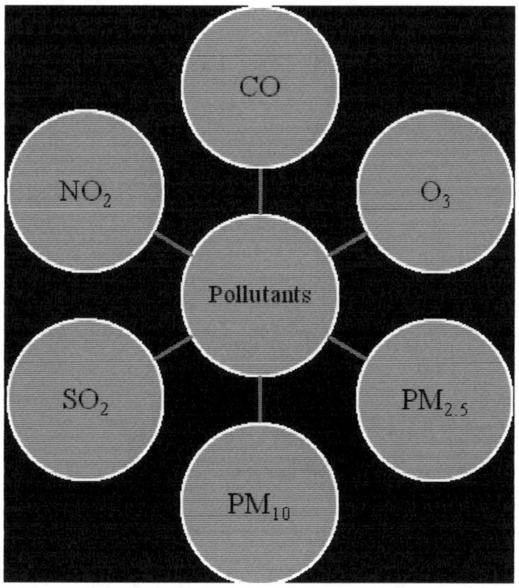

Figure 1. Depicts the major pollutants criteria.

During both phases of the lockdown, critical downfall in PM10, PM2.5, and NO2 levels were seen due to a combination of fewer vehicles on the roads. Significant improvements in air quality were seen as compared to a similar timeframe a year earlier (as shown in Figure2). Air quality improved during the pre-lockdown period in 2020 as well. This could be due to weather conditions, similar to how a few limits on movie theatres, schools, and universities were implemented in the first half of March 2020. In terms of PM2.5 levels, while they reduced by 24% during the pre-lockdown stage, they plummeted by nearly half during both lockdown stages, relative to 2019 levels. During the second period of lockdown, PM10 levels fell by 60% as compared to data taken during a similar timeframe in 2019, with NO2 levels declining by 64%, Benzene 62% and SO2 35%.

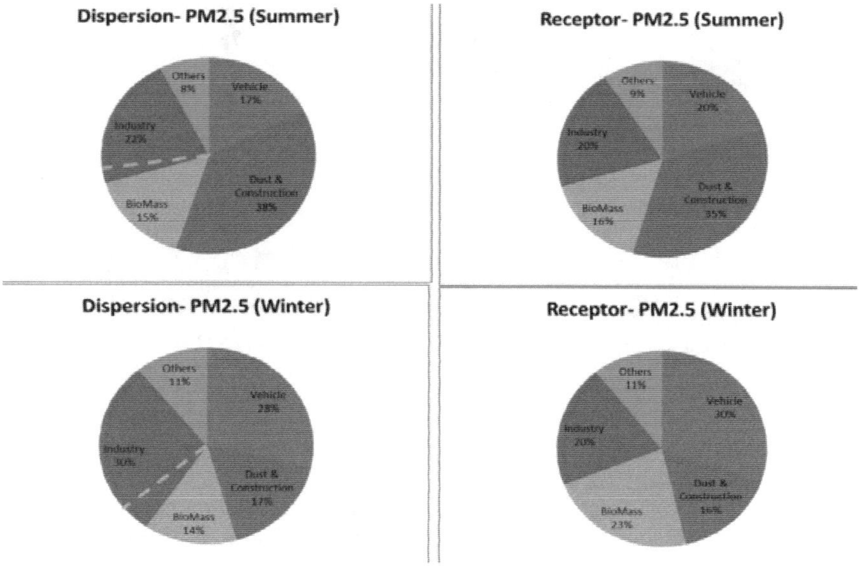

Figure 2. Levels of PM 2.5 and PM 10
Source: Apportionment of PM2.5 & PM10 of Delhi /NCR, in August 2018

PM2 levels had decreased by 38% and 33%, respectively. PM10 levels decreased by 5%, 37%, and 30% , during lockdown stage I and lockdown stage II, compared to pre-lockdown levels. Figure 3 depicts the affect of lockout on ambient quality of air.

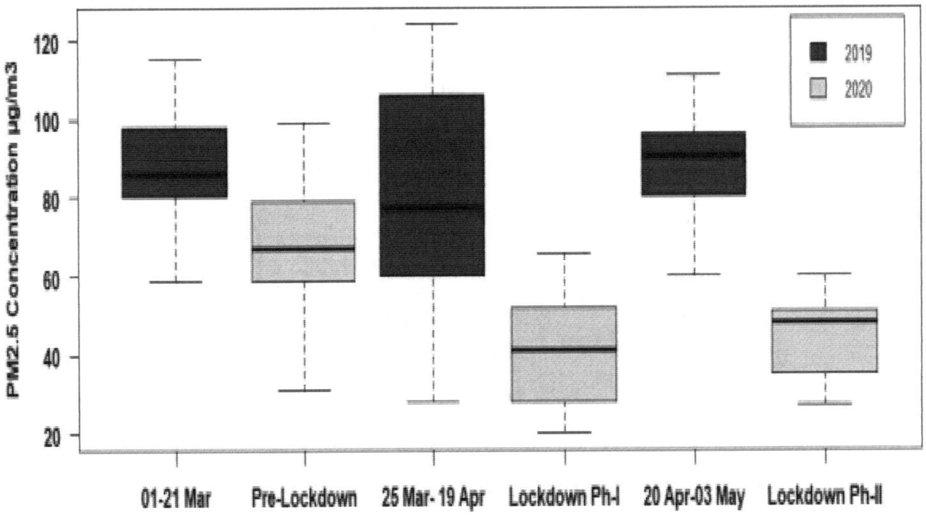

Figure 3 . Depicts the affect of lockout on ambient quality of air.
Source: The Central Pollution Control Board (CPCB) report September 2020.

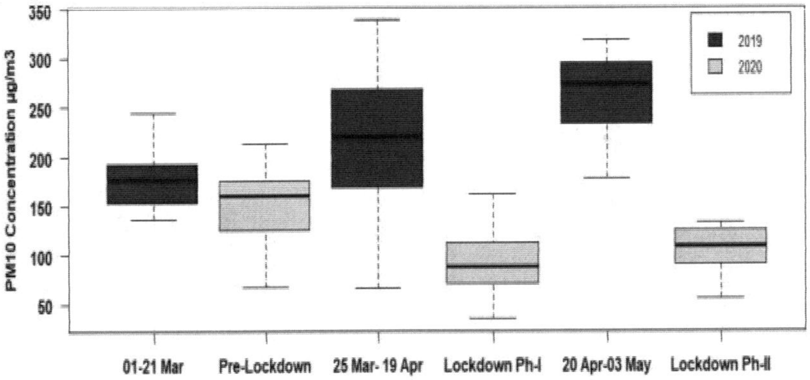

Figure 4. The effect of the Delhi lockdown on ambient air quality, as depicted ina Box Plot of PM10 Concentrations during the dates of 25 March to 3 May 2019 and 2020.(CPCB (Central Pollution Control Board) report, September 2020).

CPCB (Central Pollution Control Board) report, September 2020.
As demonstrated in the above two figures 3 and 4, PM2.5 and PM10 levels appeared to rise in the second seven-day stretch of the stage I lockdown, which may primarily be attributable to changes in climatic circumstances. Temperatures have risen since the start of summer, with an usual temperature of 20.9 degrees Celsius on March 16, 2020 rising to 30.4 degrees Celsius on May 1, 2020, resulting in dry and dusty conditions. On the 14th and 15th of April 2020, a minor residual storm arising from the western part of country and the bay districts also affects Delhi, rapidly raising PM10 levels in Delhi and the NCR. Also, as shown in Figure 5, the range of NO2 esteems during the lockout time frames

has greatly decreased. Because automobile pollution accounts for 81 percent of NO2 in Delhi and the NCR, it is recommended that no visible sources be present. During the lockdown periods, the most extreme 24-hourly value of NO2 was not quite a significant proportion of the peak esteem saw at a similar time in 2019.

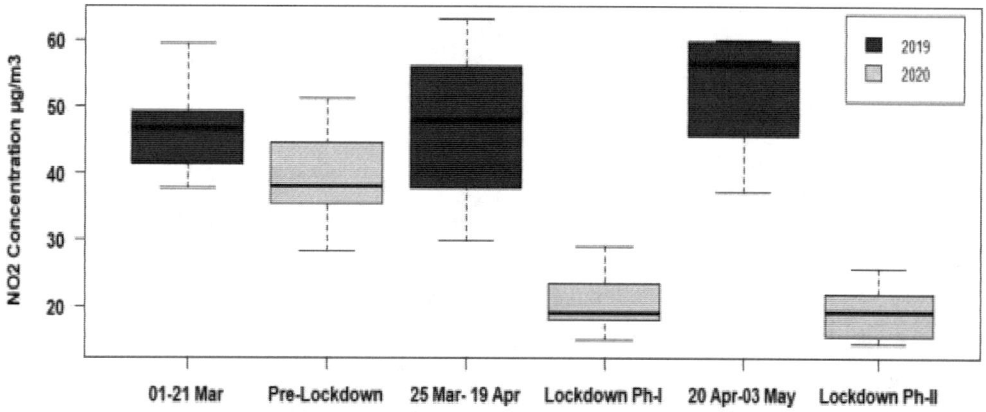

Figure 5. The range of NO2 esteems during the lockout
Source: Report by CPCB, September 2020

Figure 6 shows that SO2 levels diminished by 6% in the stage I of lockdown however stayed like pre-lockdown levels during lockdown stage II, apparently because of activity of enterprises in the encompassing regions. More than 70% of Delhi's SO2 comes from power plants in the surrounding area that were operational amidst lockdown.

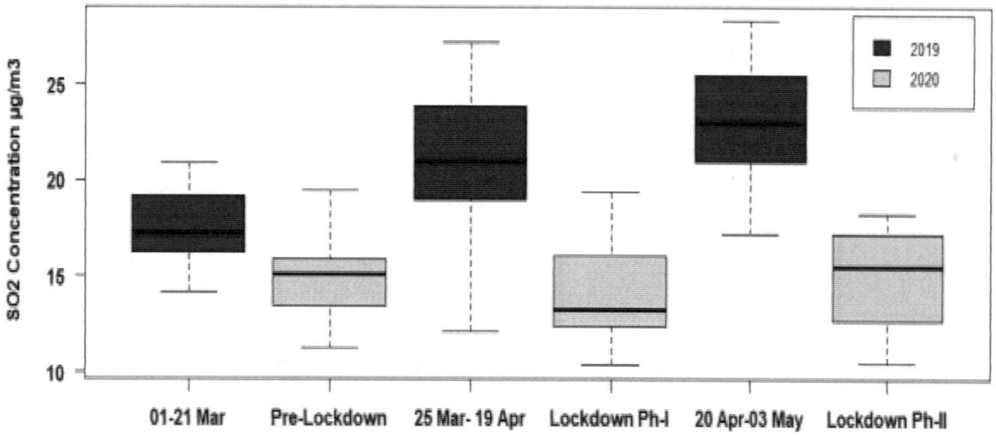

Figure 6. The effect of lockdown in Delhi on ambient air quality represented by Box Plot of SO2 Concentrations for the time duration between 25 March – 3 May 2019 & 2020
Source: Report by CPCB (Central Pollution Control Board), September 2020

1.1 Air Quality Indicators (AQI)

The AQI data as per CPCB for Delhi clearly reveals that air quality had improved, with a higher number of days falling into the Good and Acceptable AQI categories. Because the Public Ambient Air Quality Standards have breakpoints within the Good and Satisfactory classes, it's possible to conclude that Delhi spent more days under public control during the closure than it did during similar periods in 2019.Table 1 shows the AQI data for category wise AQ sub records with concentration of the pollutants

Table1. The AQI data for category wise AQ sub records

AQI Category (Range)	PM_{10} 24 –hr	$PM_{2.5}$ 24-hr	NO_2 24-hr	O_3 8-hr	CO 8-hr	SO_2 24-hr	NH_3 24-hr	Pb 24-hr
Good(0-50)	0-50	0-30	0-40	0-50	0-1.0	0-40	0-200	0-.5
Satisfactory(51-100)	51-100	31-60	41-80	51-100	1.1-2.0	41-80	201-400	0.6-1.0
Moderate(101-200)	101-250	61-90	81-180	101-168	2.1-10	81-380	401-800	1.1-2.0
Poor(201-300)	251-350	91-120	181-280	169-208	10.1-17	381-800	801-1200	2.1-3.0
Very Poor(301-400)	351-430	121-250	281-400	209-748	17.1-34	801-1600	1201-1800	3.1-3.5
Severe(401-500)	430+	250+	400+	748+	34+	1600+	1800	3.5+

For eight toxins (PM10, PM2.5, NO2, SO2, CO, O3, NH3, and Pb) for which current (up to 24-hours) Public Ambient Air Quality Standards are indicated, AQ sub-records and wellness breakpoints have been constructed. It's worth noting that the number of 'Good,' 'Satisfactory,' and 'Moderate' AQI days grew to 40 from March 25 to May 3, 2020, compared to 17 in 2019, while the number of 'Poor,' 'Very Poor,' and 'Severe' AQI days reduced to zero, compared to 23 in 2019.

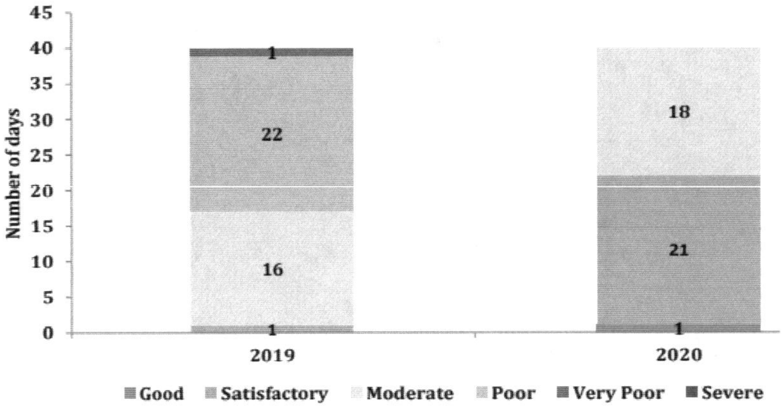

Figure 7. The AQI data for category wise AQ sub records

During the lockout periods, the percentage of Black Carbon in overall PM2.5 composition, which is predominantly provided by biomass burning and vehicle emissions, decreased from 12 to 6-8 percent. The Biomass Burning Factor was shown to be continuously growing after Lockdown. A peak in phase 1 can be connected to an increase in agricultural residue burning around the beginning of May. Despite the fact that concentrations were equal due to a reduction in input from other anthropogenic sources, power plant contribution increased during lockdown phases 1 (25%) and 3 (20%) compared to pre-lockdown phase (10 percent).

2. Review of the Literature

In Delhi, Hyderabad, Kolkata, Mumbai, and Chennai, [1] noted a drop in PM2.5 and AQI (Air Quality Index), as well as a downward trend in troposphere NO2 concentration.

[2], found a significant decrease in ambient AQI (estimated mean reduction of 17.75 percent and 20.70 percent, respectively) and troposphere AQI (estimated mean reduction of 17.75 percent and 20.70 percent, respectively After the lockout, researchers discovered a little increase in O3 density, indicating improved tropospheric air quality. The lockout of COVID-19 resulted in a 72-fold reduction in fatal traffic accidents. COVID-19 cases and fatalities are higher in cities with poor air quality (r =

0.504 and 0.590 for NO2; r = 0.744 and 0.435 for AQI). On the other hand, cities with better air quality had lower mortality rates. These data point to a link between COVID-19-affected areas and the virus's propagation. [3] Discovered that PM10 and PM2.5 concentrations were approximately half of what they were before the lockdown. NO2 and CO levels have also dropped considerably during the lockout. The air quality in transportation and manufacturing areas has improved by more than 60%. The air quality in central and eastern Delhi has significantly improved. On the alternate even days of the lockdown, air quality improved by 40% - 50%.Even if current policies are imposed, PM2.5 concentrations in Delhi will not reach the planned national ambient air quality standards (NAAQS) by 2030, according to Dholakia et al. (2013). Using sophisticated management technologies, PM2.5 levels will be reduced by roughly 60% by 2030, and all-cause mortality will be cut in half.
According to [4] enhancing air quality of Delhi involves coordination with neighbouring areas as well as the inclusion of sources which are less significant in the countries which are industrialized. It was also suggested that several policy measures have numerous other benefits on developed goals in Delhi and its neighbours [5] discovered that air quality improves significantly in lockdown period. When compared to pre-lockdown values, PM2.5 and PM10 concentrations have fallen by more than fifty percent (>50%). It was found that PM10 and PM2.5 levels have decreased by nearly 60% and 39% in comparison to the previous year, respectively, over the same time period [6] looked studied how important air pollutants like PM10, PM2.5, NO2, SO2, and O3 changed throughout two phases: pre-lockdown and post-lockdown. PM10, PM2.5, NO2, and SO2 concentrations reduced by 55 percent, 49 percent, 60 percent, and 19 percent in Delhi and 44 percent, 37 percent, 78 percent, and 39 percent in Mumbai, respectively, during the post-lockdown phase. He also urged stakeholders involved in the establishment of air quality legislation that if enhanced pollution control methods are rigorously applied, considerable improvements in air quality might be realised in the future. On the second and fourth days of the lockdown, the air quality improved by around 40% to 50% Air pollution has become a major cause of concern, posing a threat to inhabitants' health and well-being in cities across India, particularly in Delhi.

3. Research Field

The National Capital Territory (NCT) of Delhi which is the subject area is the world's second-largest megacity and urban agglomeration (United Nations, The World's Cities in 2018, Data Booklet, 2018). India has a population of 1.68 crore people and is growing at a rate of 21% per year [8]. Summer, Monsoon, short Post-monsoon, winter, and Pre-monsoon are the five main seasons in Delhi. Temperatures in the winter range from 4 to 10 degrees Celsius. Temperatures in the summer range from 42 to 48 degrees Celsius [7]. Also more than 80% out of 100% annual precipitation falls during the monsoon season [8]. Nearly 90 % of population who are working in NCT Delhi mostly lives in urban areas, which is much higher than the national average of 31.16 percent. In 2011, there were almost 6.93 million registered automobiles on Delhi's roadways, and by 2020, there were nearly 12 million registered cars, with the number expected to climb to 25.6 million by 2030[9]

4. Research Methodology

The information gathered is secondary information gathered from a variety of sources. To investigate the interrelationship between the six criteria of air pollutants (PM2.5, PM10, NO2, CO, O3, and SO2), and the meteorological parameter which includes the air quality index, relative humidity, temperature, and population density. To find the correlation between various criterions a Pearson correlation analysis was performed on the collected data. The correlation coefficient has a range of -1 to +1. Similarly, the correlation coefficient (r) is a number that ranges from - 1 to 1, with r less than 0.35 indicating a weak connection (less than 0.5), correlation coefficient ranges from 0.36–0.67 indicating moderate correlation, and r equal to greater than 0.68 indicating strong correlation which shows the direction of correlation.

5. Analysis

Table2 shows the descriptive statistics of all the criteria of air pollutants taken into study observed in Delhi, a metropolitan city. PM2.5 concentrations ranged from 98.51 to 34.03 lg/m3 on average (Delhi) In Delhi, there was a decrease in the average concentration of PM10, which was previously high (- 57 percent)

Table 2. The mean concentrations, standard deviations, maximum and lowestvalues for six different air contaminants, as well as the air quality index.

Criterion Pollutants	Mean	S.D	Max	Min
PM10	98.51	34.03	188.43	39.28
CO	0.76	0.15	1.04	0.52
NO2	19.85	3.73	27.38	14.24
03	14.38	2.55	19.29	10.33
SO2	44.19	8.87	60.98	24.06
Air quality index	103.43	28.74	180	45

During the lockdown period in Delhi, there was a decrease in PM10 concentration along with a fall in PM2.5 concentration, indicating that PM2.5 concentration determines a significant portion of PM10 in various aforementioned cities ([60 percent) [10,11]. The main sources of PM10 and PM2.5 pollution are automotive emissions, industrial emissions, dust fallout, and construction-demolition activities. The reduction in PM2.5 and PM10 concentrations was attributable to an 80% decrease in vehicle flux [12] and the suspension of industrial and construction activity during the lockdown period.

Both of these studies support the findings of this study, which reveal that transportation activities contribute significantly to global warming. Particulate matter (PM) refers to a type of particle found in the environment. All activity were ceased during the shutdown. As a result, the amount of PM2.5 in the air has dropped. PM10 is a type of pollutant that can be present in the environment. In this study, the decrease in PM2.5 and PM10 concentrations was observed [13] conducted a study that supports this. During the lockdown phase in 2020, The Pearson correlation coefficient (R) was calculated for the six critical air pollutants (PM2.5, PM10, CO, NO2, SO2, and O3) as well as the air quality index for Delhi during the shutdown and the same period last year. The results of the data analysis for the computation of the correlation coefficient are provided in tables 3 and 4 shown below with pollutant parameters and Air quality index

Table 3. Correlation analysis of six key air contaminants and the Delhi air quality index during the prior year of lockdown period 2019

Delhi Pollutant Parameters	PM2.5	PM10	CO	NO2	SO2	O3	Air quality index
PM10		1	0.54	0.51	0.53	0.29	0.97
CO			1	0.89	0.72	0.18	0.55
NO2				1	0.78	0.28	0.52
SO2					1	0.12	0.47
o3						1	0.38
Air quality index							1

Also table 4 is showing the correlation analysis during lockdown period.

Table 4. Correlation analysis of six criteria air pollutants and air quality index of Delhi during lockdown period (March 25 to May 3, 2020)

Delhi Pollutant	PM2.5	PM10	CO	NO2	O3	SO2	Air

Parameters							quality index
PM2.5	1	1 .86	0.81	0.79	0.1	0.85	0.86
PM10		1	0.81	0.64	0.25	0.76	0.94
CO			1	0.63	0.33	0.77	0.82
NO2				1	-0.17	0.63	0.55
O3					1	0.37	0.37
SO2						1	0.78
Air quality index							1

Both PM2.5 and PM10 were shown to be substantially associated as (R.68) during the pandemic lockdown period in Delhi, and they were similarly strongly correlated over the same period in 2019. During this lockdown period, the significant association between PM2.5 and PM10 promotes to the production of PM10 [11, 14] 15. PM2.5 and NO2 had a strong association (R less than.68) in Delhi, while CO and NO2 and CO and SO2 had a substantial correlation (R.68) over the same period the previous year. During the lockdown in Delhi, the air quality index and PM2.5 were substantially associated (R.68). During the same time period in 2019, it was similarly significantly connected (R.68) in Delhi. During the lockdown, the association between air quality index and PM10 was found to be strong (R.68), and it was similarly strong (R.68) in Delhi during the same period the previous year. During the lockdown and non-lockdown periods of 2019, the substantial positive association between air quality index and PM 2.5, as well as AQI and PM10, revealed that PM2.5, followed by PM10, had the greatest influence on air quality [14]. Also found that PM2.5 was mostly responsible for deteriorating air quality in China and Beijing. In this time, the AQI was highly connected with CO (R =.68). The AQI was found to be substantially linked (R.68) with CO during the lockdown period.

6. Conclusion

The lockdown during COVID-19 has resulted in a significant decline in pollution levels in the Delhi NCR. This was highly required in Delhi, thanks to a significant reduction in local pollution caused by vehicle emissions and industrial operations. This clearly demonstrates that the ambient air quality can be maintained by effectively enforcing regulations, norms, and standards. Criterion pollutants include the concentrations of five pollutants: PM2.5, PM10, CO, SO2, and NO2. To achieve long-term global development, considerable action must be taken to limit pollution exposure.

Before the period of lockdown, the levels of air pollution in Delhi were found to be frighteningly high, well above the World Health Organization's guideline limits. Despite the lockout, air pollution readings were at an all-time low. The improvement observed in the air quality due to lockdown could be a boon to the city's population's health, given Delhi's consistently low air quality for decades and increased sickness and mortalities as a result of poisonous air. The findings show that the composition of gases in the environment, emission sources, and meteorological conditions all influence the amount of contaminants in ambient air. Because the data show that vehicles, industry, and construction activities are the key drivers of air pollution in Delhi and the NCR, it's vital to take aggressive measures to minimise pollution and implement regulations and guidelines that are appropriate for the current circumstances.

References

[1] Singh, R.P. and Chauhan, 2020A. Air Qual Atmos Health **13**, 921–928
[2] Naqvi H R et al,: 2021Society and Environment, Volume **22**, 100473,ISSN2352-9385
[3] Abdollahzadeh S and Sepehr A et al, 2021Society and Environment, Volume **22**, 100503,ISSN 2352-9385
[4] Majumdar D. 2017 Atmos. Environ**. 61**,pg.99–111
[5] Rao S. Et al. Environmental modeling and assessment ; **6830**, 1-10

[6] Mahato S et al. 2020 The Science of the total environment, **730**, 139086.
[7] Kumari P and Toshniwal D,2020. Int J Environ Health Res.**16**,1-8
[8] Kumar P. et al,. 2017Environ. Pollut**. 225**, 20–30
[9] Perrino C et al,2011 Atmos. Pollut. Res**. 2** (4), 418–4279.
[10] Dholakia H.H et al,2013 Atmos. Environ.**75,**241–248
[11] Zhang Y.L and F.Cao, 2015, Scientific Reports, **5**, 14884.
[12] Wang Y et al, Environment International, 2014, **73**, 413–422
[13] Weber J.C and D.R.Lamb,(1970). Statistics and research in physical education. St. Louis: CV Mosby Co, **222**, 59–64.
[14] Sicard P et al,2020, Science of the Total Environment, **735**, 139542.
[15] S Yan et al,2016 Aerobiologia Pollution Research, **23**, 139 996–14007

Strength and durability of concrete by partial replacement of sand by Copper Slag and cement by Egg Shell Powder: A Review

DhirajNaxine[1],Dr.AnitaNag[2]
[1]ResearchScholar,GHRaisoniUniversity,Amravati, India
[2]AssistantProfessor,GHRaisoniUniversity,Amravati, India

Abstract. From some decades the production and use of concrete has been increased. The manufacturing process of concrete as well as the three main ingredients emits lots of harmful gas like CO2 which directly harm the environment. In this study, relevant study done on the fine aggregate and cement replacement with egg shell powder & copper slag were discussed. Also effect of cement and fine aggregate replacement on compressive strength and durability of concrete had been discussed. From study it has been observed that the egg shell powder and the copper slag can't affect much the durability and strength parameters up to 6% and 40% replacement of cement and fine aggregate. Also observed that copper slag and egg shell powder reduced impact of concrete on environment but still need to improve the manufacturing process of copper slag.

Keyword-Concrete , Egg Shell Powder, Copper Slag, Durability, Compressive Strength

1. Introduction

In India most commonly use construction materials are Steel and Concrete (PCC or RCC). The concrete is basically made from three materials such as Cement, Coarse Aggregate and Fine Aggregate. In the concrete, cement is a binding material and fine aggregate assist to produce workability as well as uniformly in mixture. As we know that the pollution rate has been increased in last some decades. Manufacturing of Cement is include different processes begin with the mining then grinding (limestone and clay) raw material to fine powder and then heated to form cement kiln. The aggregates are manufactures by following different processes like blasting and then crushing, screening, washing, shaping and last sorting. But manufacturing of both cement and fine aggregate adversely affect the environment as it includes the processes like combustion and blasting. Combustion and blasting should be avoided or decreased at unnecessary places. To avoid this, there is need to find out alternative material to make eco-friendly concrete. There are many materials are available in the market and the research on different alternative material for fine aggregate and cement is going on. After the literature survey and study, copper slag is using for fine aggregate replacement and egg shell as cement in this particular study. Copper slag is byproduct and egg shell ash is made from the waste material eggshell. The Egg shell will be thrown away from the houses, hotels in big amount as the egg shell is a covering of egg and it is waste. As these both materials are waste, so it might help to reduce the effect on environment.

Fig-1:Egg Shell Powder

Fig-2:Copper Slag

2. LiteratureReview

NahlaHilal*et.al*[1] used self-compacting high strength concrete for experimental work by replacement of cement with egg shell ash along with waste plastic fiber. The cement replaced by egg shell ash by 10%, 20% and 30%. In the first stage, only egg shell ash was added in the concrete. In second stage both egg shell ash and three portion of plastic waste were added. The non destructive test (Ultrasonic pulse velocity test) was performed after 28 days. Also compressive strength as well as flexural test were performed on concrete after 7, 28, 90 days. Result showed that compressive as well as flexural strength were increased when the egg shell ash used with waste plastic fiber.

J. UshaKranti*et.al*[2] has worked on the M40 grade of concrete by fine aggregate replacement by copper slag upto100%. Durability test was performed on concrete with curing period of 90 days. A weight of concrete was decreased during durability test as compared to cube weight placement in acid. The fine aggregate were replaced by 10%, 20% upto 100%. Compressive strength increased up to 40% replacement but after 60%, it was decreased in both acid and sulphate attack cases.

Gopu Ganesh Naidu and KannellaEdukondalu[3] specimen prepared and tested by replacement of cement by eggshell powder along with the fly ash. The different test were performed on concrete such as compressive strength ,slump cone test, splitting tensile strength, modulus of elasticity and flexural strength after 7, 14, 28, 56 days. Results concluded that compressive strength was reduced after replacement of cement by40% of egg shell powder, 50% fly ash by weight. The replacement of cement with egg shell powder and fly ash after 40% and 50%, the compressive strength were reduced.

Khalifa S. Al-Jabri[4] prepared concrete and mortal to find out copper slag effect when partially replaced with fine aggregate. The different test like density ,workability,compressive strength, tensile strength, durability,flexural strength was performed on concrete as well as on mortar. The fine aggregate partially replaced from 0% to 100% by copper slag and it has been found that strength was comparable when copper slag was used in both mortar as well as in concrete. In mortar, strength was enhanced for every mix when fine aggregate replaced by copper slag. A fine aggregate replacement shows 70% increment in strength of mortar while copper slag used as fine aggregate. In concrete there were increment in strength till 40% replacement by copper slag.

Swetapadma Panda ***et.al***[5] investigated concrete with partial replacement of fine aggregate by copper slag.Model was prepared to check copper slag effect on durability. The abrasion resistance. Strength tests, abrasion and durability tests were performed on concrete. The fine aggregate replaced partially and fully with copper slag. The slake durability and abrasion resistance was increased when fine aggregate replaced upto 80%.
BinayaPatnaik***et.al***[6] prepared M20 grade concrete for investigation by partial replacement of fine aggregate by copper slag with 0,10,20,30,40,50 percent .The concrete tested by 7,28,
90 days. Compressive strength test were performed on the cubes and it has been found that compressive strength decreased with increment in copper slag after 40%.It also been observed that cost of concrete by using copper slag was less compared to conventional concrete.
Ali Rezaei LORI ***et.al***[7] investigated pervious concrete with partial replacement of fine aggregate with copper slag by 0, 20, 40, 60, 80, and 100 percent by volume of fine aggregate. The different tests like hydraulic conductivity, split tensile, flexural, compressive strength test, abrasion test were performed after 28 days. The higher strength obtained at optimal replacement of copper slag. The porosity and permeability got affected the concrete with copper slag excluding fine copper slag.
AnkitNainwal***et.al***[8] worked on the concrete by replacing fine aggregate partially. Fine aggregate replaced with the Copper Slag at 0%, 20%, 40%, 60%. The work was investigated for split tensile test, compressive strength as well as durability of concrete. The test result shown that compressive strength and durability were nearer to the conventional concrete at 40% replacement with Copper Slag. The Water absorption decreased with increment of copper slag up to 60%. It was observed that mechanical properties of concrete vary for different percentage of copper slag at different places.
S MohdArif***et.al***[9] performed experiment on the high strength concrete by replacing cement with egg shell ash at 5%, 10%, 15%. The compressive strength and ultrasonic pulse velocity test were performed after 28 days. During experiment it concluded that concrete with partial replacement with egg shell ash at 10% showed more strength as compared to control mix. It has been also found that egg shell ash also affects the workability of concrete as it absorbs more water. The non destructive test such as Ultrasonic pulse velocity was increased after the partial replacement with egg shell ash.
Rokiah Othman ***et.al***[10] investigated the concrete by replacing cement and sand by egg shell powder and waste wire rubber. Both the materials replaced by 5% interval upto15%.Slump cone test were performed to measure rheological properties and compressive strength as well as flexural strength test performed for mechanical properties. From results found that the mechanical properties were optimum and showed minor effect on the slump during the replacement by egg shell powder. Compressive Strength increased till the replacement made with 5% and 10% by egg shell powder [10].
Abhishek Maharishi ***et.al***[11] prepared concrete (copper slag cement) by replacing fine aggregate fully as well as partially with copper slag. The different test was performed on the concrete like split tensile strength, compressive strength, void, acid attack, durability test. GGBS also used by taking 30% fix replacement in every mix. The 100 mm cubes was prepared to perform the compressive strength & durability test.Concrete was tested after 3, 7 & 28 days [11]. Results showed that durability and Compressive Strength were increased when the fine aggregate replaced with copper slag upto 40% but strength fall down after 40% replacement.

ChadalawadaManohar*et.al*[12]prepare and worked on M30 grade concrete with fine aggregate replacement by copper slag. Different tests such as Split tensile,compressive Strength , workability tests performed on concrete. The test on prepared concrete was performed after 3, 7 & 28 days. The different test results showed that the workability, compressive strength, split tensile strength, flexural strength was increased while copper slag used as fine aggregate. Strength was increased by 31% when fine aggregate replaced with copper slag by 21%. It has been seen that the cost of concrete was decreased when copper slag used in concrete [12].

3. Conclusion

The many work has been carried out by using the different material along with the copper slag and egg shell powder to replace fine aggregate and cement. But from the above study it has been observed that many concrete properties were improved when one or two ingredients of concrete replaced with. This particular study focused on the replacement of fine aggregate with copper slag and the cement with the egg shell powder. The above concluded that concrete strength and durability increase when cement replaced by egg shell ash up to 6% and in case of fine aggregate replacement with copper slag up till 40%. Also it has been observed that the workability were increased upto the replacement egg shell by 6% and copper slag by 40%.But in some studies it has been observed that the strength were increased when only one ingredient replaced at a time concrete.So need to do more research on theconcrete with replacement of two or all the ingredients of concrete. The cost of concrete was reduced when the one or two ingredients replaced.

References

[1] N.Hilal,D.M.Saffar,T.Khaleefa,andM.Ali,'Effect of egg shell ash and strap plastic waste on properties of high strength sustainable self-compacting concrete Effect of egg shell ash and strapplastic waste on properties of high strength sustainable self-compacting concrete', no. February,2021,doi:10.1007/s12517-021-06654-x.

[2] J. U. Kranti, A. N. Sai, A. R. Krishna, and K. Srinivasu, 'Materials Today : Proceedings An experimental investigation on effect of durability on strength properties of M40 grade concrete with partial replacement of sand with copper slag', *Mater. Today Proc.*, no. xxxx, 2020, doi:10.1016/j.matpr.2020.09.767.

[3] G. G. Naidu and K. Edukondalu, 'Mechanical Properties of Concrete by Replacing Cement with Egg shell powder Mechanical Properties of Concrete by Replacing Cement with Egg shell powder and Fly ash',no.March,2020,doi:10.35940/ijitee.D1715.029420.

[4] K. S. Al-jabri, A. H. Al-saidy, and R. Taha, 'Effect of copper slag as a fine aggregate on the properties of cement mortars and concrete', *Constr. Build. Mater.*, vol. 25, no. 2, pp. 933–938,2011,doi:10.1016/j.conbuildmat.2010.06.090.

[5] S.Panda, P.Sarkar,and R.Davis, 'Abrasion resistance and slake durability of copper slag aggregate concrete', *J. Build. Eng.*, no. November, p.101987, 2020, doi:10.1016/j.jobe.2020.101987.

[6] B. Patnaik,'An Experimental Investigation on Optimum Usage of Copper Slag as Fine Aggregate in Copper Slag Admixed Concrete An Experimental Investigation on Optimum Usage of Copper Slag as Fine Aggregate in Copper Slag Admixed Concrete', no. January 2011, pp. 5–8, 2016, doi:10.14741/Ijcet/4.5.2014.102.

[7] A. R. Lori, A. Bayat, and A. Azimi, 'Influence of the replacement of fine copper slag aggregate on physical properties and abrasion resistance of pervious concrete',*Road*

Mater.Pavement Des., vol.0,no.0,pp.1–17,2019,doi:10.1080/14680629.2019.1648311.
[8] A. Nainwal, P. Negi, P. Kumar, M. Chandra, and A. Negi, 'Materials Today : Proceedings An experimental investigation to substitute copper slag in concrete with Beas river fine aggregate',*Mater.Today Proc.*,no.xxxx,2021,doi:10.1016/j.matpr.2020.12.481.
[9] E. Science, 'Compressive Strength of Concrete containing Egg shell Powder as Partial Cement Replacement Compressive Strength of Concrete containing Egg shell Powder as Partial Cement Replacement',2021,doi:10.1088/1755-1315/682/1/012031.
[10] R. Othman *et al.*, 'Evaluation on the rheological and mechanical properties of concrete in corporating egg shell with tire powder', *J. Mater. Res. Technol.*, vol. 14, pp. 439–451, 2021, doi:10.1016/j.jmrt.2021.06.078.
[11] A. Maharishi, S. P. Singh, and L. Kumar, 'Materials Today : Proceedings Strength and durability studies on slag cement concrete made with copper slag as fine aggregates', *Mater. Today Proc.*, no.xxxx,2020,doi:10.1016/j.matpr.2020.08.232.
[12] C.Manohar,'Concrete Strength Evaluation by using Copper Slag instead of Aggregates for M30 Grade of Concrete', vol.5, no. September, 2019.
[13] A. Parkash and P. K. Yadav, 'A Review Study of Egg Shell Powder As a Cement Replacing Material in Concrete', *Int. J. Latest Res. Eng. Comput.*, vol. 5, no. 4, pp. 6–9, 2017, [Online]. Available: www.ijlrec.com.
[14] O. Olumide Olu, 'Effect of Saw Dust Ash and Eggshell Powder on the Properties of Cement Blends', *Am. J. Constr. Build. Mater.*, vol. 4, no. 2, p. 88, 2020, doi: 10.11648/j.ajcbm.20200402.16.
[15] S. M.S. and S. K, 'A Review on Impact of Chicken Egg Shell Powder on Strength Characteristics of Cement Concrete', *Int. Res. J. Multidiscip. Technovation*, no. September, pp. 405–410, 2019, doi: 10.34256/irjmtcon55.
[16] Lane AM and Hirst SP., 'Placement of', *J. Nurs. Educ.*, vol. 51, no. 3, pp. 145–149, 2012.
[17] R. Singh and R. Ahmad, 'Mechanical and Durability Characteristics of High Performance Concrete Using Copper Slag as Fine Aggregate', *J. Mater. Eng. Struct.*, vol. 7, no. 2, pp. 277–291, 2020.
[18] T. Mahitha and J. Aswini, 'Effect of copper slag and granite powder on the mechanical properties of reclaimed asphalt pavement aggregate concrete', *Int. J. Sustain. Constr. Eng. Technol.*, vol. 9, no. 2, pp. 1–15, 2018, doi: 10.30880/ijscet.2018.09.02.001.
[19] A. F. Rahman, W. I. Goh, and A. A. Jhatial, 'Flexural study of reinforced foamed concrete beam containing palm oil fuel ash (POFA) and eggshell powder (ESP) as partial cement replacement', *Int. J. Sustain. Constr. Eng. Technol.*, vol. 10, no. 1, pp. 93–100, 2019, doi: 10.30880/ijscet.2019.10.01.009.
[20] K. Murari and R. Siddique, 'Use of waste copper slag, a sustainable material', 2014, doi: 10.1007/s10163-014-0254-x.

Sustainable Rural Development through Renewable Energy Technologies: Identification of Potential Methods to Increase Rural Incomes

Amitabh Patnaik[1], Pallavi Kudal[2], Sunny Dawar[3] and Prince Dawar[4]

[1,2]Dr. D.Y.Patil Institute of Management Studies, Pune, Maharashtra, India
[3*]Faculty of Management & Commerce, Manipal University Jaipur, Rajasthan, India
[4]Poornima Group of Colleges, Jaipur, Rajasthan, India

Email:sunny.dawar86@gmail.com

Abstract. Renewable energy is being promoted in the Organization for Economic Co-operation and Development (OECD)[1] as a potential new source of jobs and rural development. Governments in most nations have poured vast sums of public funds into renewable energy development. But how do these policies and investments affect the economy? Is it true that renewable energy may aid in the development of rural economies? This study examines the answers to these questions. We aim at determining the impact that economically cleaner energy options can play in reducing post-harvest losses in India's rural communities. The food supply chain, energy supplies, and the environment are all being strained by the ever-expanding population. Post-harvest losses, on the other hand, pose a threat to the crop and its products. We have identified how post harvesting losses can be reduced by clean energy options and hence can lead to sustainable and self-sufficient village economies in India.

Keywords. Sustainable energy, clean energy, renewable energy, Organization for Economic Co-operation and Development (OECD), post-harvest loss, cold storage

1. Introduction

India is developing at a rapid pace in the current period. As a result of this growth, both the energy and the electricity demand will increase. Power generation and infrastructure are essential for developing economies like India. With a population of 1.38 billion and an area of 3.29 million km^2, India is the 7th biggest country in the world [1]. In order to serve this large population, India needs to improve energy generation and distribution so that in the future they are able to have sustainable development. India has a GDP of 3000 billion US dollar and the typical GDP growth of seven percent annually [2]. Energy is an indispensable input to the socioeconomic development of our society. Almost every country's energy plan attempts to increase efficiency and security while also offering environmentally friendly access and achieving the best possible mix of primary energy sources. Countries all over the world are shifting their focus on sustainable energy sources and India also need to think on same ground. However, in the next few decades conventional fuels will continue playing a predominant role in the power scenario of India. India has the 5th largest coal reserves in the world, it is also confirmed to be the least expensive source of energy and electricity in India. 75.38 % (1042.5 Terawatt-hour (TWh)) of the total of 1383 TWh generated in 2020 comes from non - renewable fuels (Central Electricity Authority, 2020). In India, as the data shows, the majority of the energy demand is based on coal. These coal reserves may sustain for another 40-50 years, assuming an increased coal production of 5 % each year. The world Institute of Sustainable Energy has prepared a report on coal power in India and predicted that the total capacity of coal power plants would be 400 TWh by the end of 2032. This clearly shows how huge is coal demand in India and this surely is not sustainable by any means. There is dire need for renewable energy sources which can not only help in sustainable development but also helps in rural development. We aim at identifying the need of renewable energy sources in India which can help in rural development and boost rural incomes. The paper further discusses the challenges of clean energy adoption specially in developing economy like India.[1]

In recent years, the international market has been volatile, and the price of crude oil has risen dramatically, reaching above USD 140 per barrel [3]. The sudden surge in crude oil prices is putting a huge strain on the economy. Economic pressures are mounting, particularly in developing countries. India's energy security will be jeopardized only when alternative fuels are created to replace petro-based. The country holds out hope for providing security in biofuels. Biofuels are considered friendly to the environment and their use would deal with global concerns about carbon emissions.

With a population of approximately 1.38 billion people, India is one of the most densely inhabited countries on the globe [4]. India will experience significant energy shortages in the future decades because of rising energy prices and global energy insecurity. India's GDP is expected to increase at a rate of 7-7.5 percent in fiscal year 2022-2023, making it the world's second fastest growing economy after China [5].India's increasing demand for energy is presently satisfied primarily by coal, foreign oil, and petroleum. Non-renewable sources of energy don't provide a lasting solution to the energy crisis and also have negative effects on the environment. Therefore, the time is now right for India to move from fossil fuels to renewable energy sources and achieve greater energy security. The Indian economy is already booming. Renewable sources of energy and technologies are crucial for India's sustainable development. Figure 1 shows that how India's total primary energy supply decreasing over the last two decades and further figure 2 shows that we are still far behind in renewable energy supply production in comparison to other countries.

This sustainable energy revolution can be achieved along with rural development. Rural economy is at the core of India's development. In general, India is an Agro-based country. The agriculture sector comprises nearly one-fifth of India's Gross Domestic Product (GDP). To be able to expand farming, the government has planned a number of programs in rural development in India. The agricultural business and economy are mainly supported by agriculture, handicrafts, fishery, poultry, and dairy products. Rural development seeks to improve rural people's means of livelihood in a sustainable approach, mutually economically and environmentally. This can be accomplished through providing people with better access to natural, physical, human, social, and technological capital, as well as control over productive capital (in economic or financial forms), which allows them to secure their livelihoods in a sustainable and equitable manner. Rural development programmes have mostly aimed to reduce unemployment and poverty.

The government of India has recognized the requirements of rural populations and has implemented development planning methodologies through the five-year plans. One such development plan is introduction of renewable technologies in rural areas. Renewable energy technologies (RTs) are energy-generating technologies that use natural resources as much as possible without depleting the planet's intrinsic energy. These energy sources are renewable in a sense that they may be utilized indefinitely without harming the environment [6]. RTs have a lot of potential for meeting rural communities' energy demands in a sustainable way. Albeit most likely in conjunction with traditional methods. Many RTs are decentralized, allowing them to be customized to meet the needs of different rural communities. In rural areas, RT choices for delivering electricity include wind, solar, small-scale hydropower, biomass, and bio gas. Decentralized RTs are particularly well suited to supplying electricity to remote places. They typically have minimal upfront costs (albeit typical higher prices per KW set up than centralized solutions), and they can also aid

[1] The Organisation for Economic Co-operation and Development (**OECD**) is an international organisation that works to build better policies for better lives.

in the avoidance of excessive transmission and distribution grid costs. Furthermore, given the distinctive success of regular national grid-based rural electrification programs in reaching small, dispersed rural populations in developing nations, the potential for adopting RTs is very important.

Renewable technologies and rural development can go hand in hand and it will be eventually a win-win situation to a developing economy like India [7]. It is important to move up the power ladder. Movement up the power ladder can occur in a variety of ways in rural life [8]. It appears reasonable to believe that improved access to modern energy services, particularly in home cooking, household lighting, and heating, can promote rural development especially in terms of increase in per capita income [9]. The present study is done with an aim to understand how renewable energy technologies can lead to make villages of India self-sufficient in terms of energy requirement and even increase the employment opportunities. However researchers do not deny multitude of factors which pose challenges in rural adoption of renewable technologies. The main objective of this study is to understand role of non-conventional energy sources in rural development to improve rural life and reduce rural people's migration to urban and metropolitan areas in search of employment. A nation like India rich in natural resources, might use these resources to market Agri -based industries, development of green cold chains and creating more mega food parks, thus creating more jobs for rural youth. The novelty of this study is to understand the impact of renewable technologies in rural development. It aims at understanding how sustainable technologies can create employment opportunities for rural youth which would further help in reducing rural migration.

2. Need of renewable energy for rural development

Rural India has dispersed population. Due to the difficulty of installing power grid infrastructure in some distant areas, the cost of grid-connected approaches is higher than the cost of off-grid tactics throughout the rural electrification process [10]. The transmission losses are also high [11]. Hence an effective method for rural electrification in remote locations is to employ dispersed energy obtained from renewable sources rather than low-quality energy to promote off-grid electrification strategies. Rural areas are ideal locations for renewable energy development if the right policies and technologies are in place. Rural areas contain huge open spaces that are appropriate for the building of renewable energy facilities including wind, biomass, and photovoltaic power generating [12]. Thus, the renewable energy development strategy can be rationally planned by fully comprehending the endowment of Renewable Energy resources and the natural environment in rural areas, evaluating the renewable energy development potential, and rationally planning the renewable energy development strategy in accordance with the local economic development level and existing technical level.

Renewable Energy and small-scale non-renewable energy commonly known as hybrid power generation systems have been shown to address the issue of renewable energy power generation's insufficient stability while keeping a clean level in previous research. Renewable energy technology advancements have a huge impact on poverty reduction. On the one hand, renewable energy has aided regional economic development and expanded farmer employment as a self-sustaining business [13]. Rural electrification and the use of renewable energy, on the other hand, have aided regional economic development by increasing non-agricultural employment, ensuring social welfare, improving public service quality, increasing electricity reliability and stability, and, ultimately, reducing multidimensional poverty in rural areas. Advances in renewable energy technology promote the use of energy-efficient stoves, reduce the use of low-quality energy in cooking and heating, and significantly reduce post-harvest losses [14]. Adoption of renewable energy sources can reduce communities' reliance on the forest for firewood, hence improving environmental performance.

2.1. Advantage of Renewable Energy Technologies

Increased home power availability, as well as energy from RTs, can have a substantial impact on the livelihoods of those living in distant places. Cleaner combustion of conventional fuels may improve general health by lowering acute respiratory disease and conjunctivitis, which are typical indoor smog-related ailments [15]. Cooking with more advanced technologies can have broader health benefits, such as making food choices and boiling water more economical or even conceivable .Children and women, in particular, can have a lot more time for education, work, and pleasure [16]. Access to electrical energy can drastically reduce the amount of time spent on household chores. Electric water pumps, for example, can deliver clean water while reducing the amount of labour necessary for collection. In non-urban medical centers, electrical energy can be used to provide refrigeration for operations and vaccines for medical equipment. Access to television and radio can enhance educational opportunities while also providing amusement. Electric lighting provides superior illumination than kerosene lamps, allowing for more study and work time as well as improved security, safety, and comfort.

Improved education and health, as well as much more time to engage in non-energy activities, are critical objectives. On the other hand, modern power provides the added benefit of supporting local residents in launching and running enterprises. The demand for RT solutions can assist in the development of regional economic activity based on these technologies, as well as the management of local industry. RTs are used for a variety of productive tasks, from mechanical wind-powered water pumps to motorized grain milling machines. The availability of power, as well as highway access, can entice Agri-processing companies to set up factories in rural areas, resulting in increased employment opportunities for rural youth [17]. The generation of energy from non-conventional sources has the potential to attract food preservation and processing businesses to rural areas, which might have a substantial impact on the villagers' socioeconomic well-being. Farmers can preserve perishable commodities in their village by establishing a cooperative cold storage facility, which will earn them a higher price during the off-season. Telecommunication services can provide weather forecasts to fishermen and farmers, as well as crop price information to farmers. All of the government's improvement ideas and initiatives can be easily disseminated among farmers through various programs. Better infrastructural development in rural India can help to raise awareness among rural young. These kinds of developments can also attract a variety of private investment in rural India, which is currently concentrating near urban areas due to improved infrastructure. These initiatives may result in the creation of jobs and substantially better living conditions, as well as significant increases in efficiency in remote areas.

2.2. Barriers in Renewable Energy Technologies

At this stage, rural renewable energy production has also experienced various challenges and roadblocks. Renewable energy is highly reliant on the natural climate, has an insecure supply, and is heavily reliant on infrastructure development. The development of renewable energy technologies has been delayed by economic poverty in rural areas, low levels of public services, fragile natural habitats, and poor technological levels. In addition, obstacles such as infrastructure development, energy costs, geographic issues, and household size all impede the promotion of renewable energy in rural India [18]. The imbalanced development of different types of Renewable Energy, which results in distributed Renewable Energy, does not considerably benefit the public amount and monetary growth of rural areas in the long term. As a result, biomass energy sources like as coal, straw, and firewood continue to be heavily used in modern agricultural production and rural living. The failure of the Renewable Energy push can be attributed to immature technology. Residents' discontent with the existing renewable energy promotion process stems mostly from a lack of technical expertise, which leads to equipment breakdowns and excessive servicing expenses [19]. Simultaneously, hydropower facility building will consume farmland and irrigation water, and hydropower operation may have a negative impact on agricultural production due to excessive dam construction regulation and inadequate hydropower station management. Dams built for hydropower could, however, also be utilized to provide irrigation water.

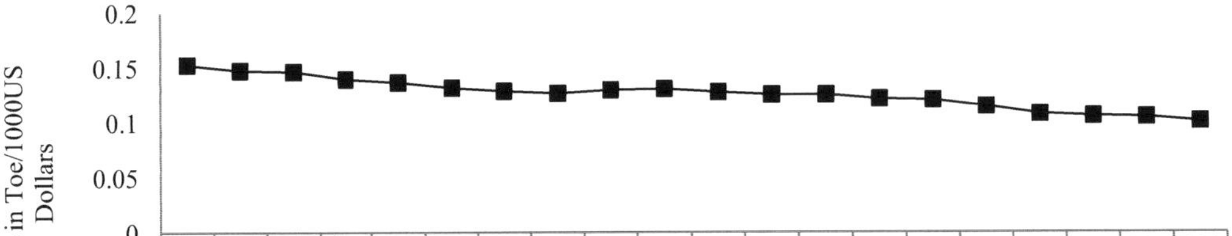

	2000	2001	2002	2003	2004	2005	2006	2007	2008	2009	2010	2011	2012	2013	2014	2015	2016	2017	2018	2019
Series1	0.15	0.15	0.15	0.14	0.14	0.13	0.13	0.13	0.13	0.13	0.13	0.13	0.13	0.12	0.12	0.12	0.11	0.11	0.11	0.1

Figure 1. Primary energy supply Total, Toe/1000US Dollars, 2000-2020
Data source: https://data.oecd.org/energy/primary-energy-supply.htm

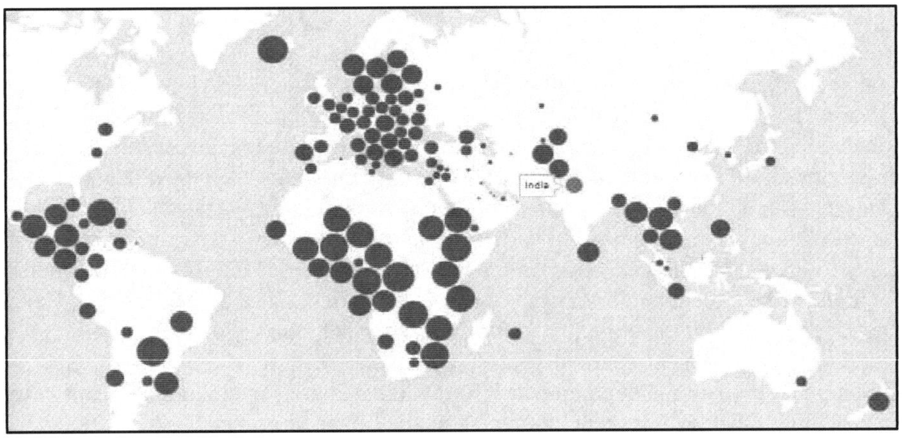

Figure 2. Renewable energy as percentage of Primary Energy supply (2017-2020)
Source: https://data.oecd.org/energy/renewable-energy.htm#indicator-chart

3. Clean Energy Innovations to Boost Rural Incomes

The rural energy gap is a major impediment to increasing smallholder agricultural productivity and adding value after harvest. In rural communities and agricultural value chains, a substantial deployment of renewable clean energy sources is required. There are several ways of achieving this:

3.1 Communal Solar Cold Storage

Farmers may find solar cold storage to be a viable option for storing and preserving perishable foods. This can operate in hybrid mode, meaning, each time there is an uneven distribution of renewable energy, the non-renewable source can compensate. This can help to avoid post-harvest losses. This type of cold storage should be encouraged by the government in collaboration with commercial partners, and a monetary incentive should be offered to encourage more private companies to enter the market.

Another possibility is communal solar cold storage, where towns can step forward to set up this form of cold storage. If the farmer does this, his or her income will go up, and food processing businesses will want to set up shop in areas where raw materials are cheaper.

India's landscape is quite diverse. Because of the lack of grid connectivity, electrification in rural India's distant areas and rugged terrains is not possible. Renewable energy technologies are the only way we can connect to these places. Renewable energy sources are plentiful and unlimited, as well as import-free.

Renewable energy has a low social acceptance in India. Despite the government's generous subsidies for installing solar water heaters and lighting systems, their adoption remains low. As a result, the government must raise public knowledge so that rural populations can accept these technologies. Climate change is the most important issue now being debated around the world. One of the greatest operational strategies to prevent climate shift is to develop renewable energy supplies. Renewable development aids in the overall growth of the country.

3.2 Minimizing Post Harvest loss
India has a variety of climate and geographical zones that produce a variety of fruits and vegetables. According to National Horticulture Board [20], the country is the world's second-largest producer of fruits and vegetables (99.07 million tons) (191.77 million tons). This accounts for 15 percent and 12% of global fruit and vegetable production, respectively [21] Seventy-six percent of the crops are consumed fresh, and 20-22 percent is wasted. India produces around 400 million metric tons of perishables each year, including dairy, meat, poultry, fish, and horticulture products such as fruits and vegetables. However, due to a lack of cold storage facilities and a food processing industry, the waste levels in these perishable commodities are quite high. According to the Ministry of Food Processing Industries, waste levels in horticultural products are nearly 15.9%, 10.5 percent in marine fish, 6.7 percent in poultry, 5.2 percent in inland fish, and 2.7 percent in meat. The yearly value of Agri-product losses is currently estimated to be 92,651 crores. Losses in vegetables and fruits, milk, fish, and meat are projected to be worth 50,473 crores per year [22]. To prevent excessive losses in the perishables supply chain, sufficient and effective cold chain structure is required from the plant to customers.
Cold chain logistics is still in its infancy in emerging countries like India and China [23] . The lack of critical infrastructural facilities such as cold chain, packing, and quality evaluation facilities are barriers in India's food processing industry [24]. There are various issues associated with cold storage, and researchers recently discovered the weak links in the cold chain of fruits and vegetables [25]. The lack of cold storage and agriculture sectors in remote locations is caused by a lack of government support, as well as fluctuations in power availability and poor connectivity. According to data from the Ministry of Food Processing Industries [26].
India currently has 8186 cold storage units with a capacity of 374.25 lakh MT available for storing perishable horticultural produce such as vegetables and fruits, with 75 percent capacity utilization because nearly 70 percent of the available storage is for single commodities, particularly potatoes. Private players participated in 92 percent of cold storage, and these private players charge farmers hefty rental rates. As a result, policymakers should consider increasing their involvement in the development of cold storage facilities, as well as encouraging the start-up of food processing industries by offering a variety of incentives, so that waste food can be preserved or processed into other forms with a longer shelf life. Both of the aforementioned activities necessitate the use of power; hence the government should consider using RTs to generate additional electricity locally.
Power for cold chain infrastructure can be provided by renewable energy sources. [27] Conducted a remarkable study in the field of incorporating renewable energy into the cold chain. They emphasized the importance of incorporating renewable energy into the sector. In the post-harvest period, 25-30% of the crop is lost due to a lack of sufficient storage and transportation facilities. This is due to the inconsistent grid power supply, which necessitated the requirement for self-sustaining, environmentally acceptable renewable energy sources to avoid losses. According to [28] untapped waste energy is having the capacity to generate 439 MW of power from 32,890 TPD of explosive waste. Technologists and policymakers may propose constructing waste-to-energy plants in strategic locations and providing the energy to cold-chain storage facilities via state grids. Storage photovoltaic energy pack, solar photovoltaic system, diesel gen-set hybrid, solar photovoltaic system, phase change material, solar thermal energy vapor absorption machines,

biomass gasifier, and solar or biomass are all renewable technology options for cold storage facilities. These projects can provide both satisfaction and significant financial savings. The hefty initial capital expenditures of creating the plant are a major deterrent. The cost of importing equipment and bringing in installation engineers from overseas is also higher [29].

[30] Discovered that India requires 61 million tons of cold storage capacity, whereas only 26.85 million tons are currently available. Cold storage capacity is required and available in a number of major Indian states, including Uttar Pradesh, West Bengal, Punjab, and Gujarat [31]. Apart from a lack of storage capacity, there are a number of other issues that contribute to post-harvest losses, including inadequate management, excessive costs, unequal distribution, and lack of access to facilities for poor farmers. To alleviate the challenges, the authorities should support the development of collaboration with private partners. Such collaboration can improve public-private partnerships, reducing the government's financial burden. For the growth of the green cold chain, more resources, technology, expertise, and skills should be pooled. Collaborations, on the other hand, are conducted for the private sector's position because they require less risk than independent ventures, the use of clients for secondary business, a competitive advantage, and value chain linkages, and the cost of capital is decreased. These partnerships safeguard the private sector's efficiency, long-term flexibility, and competence, as well as the public sector's social interest.

3.3 Development of Mega Food Park

Mega Food Park was developed by Indian government to provide a framework to connect agricultural output to marketplace by creating collectively producers, processors, and retailers to maximize value supplement, reduce consumption, enhance income of farmer, and create jobs, mostly in rural areas. The Mega Food Park Scheme is based on a cluster concept and comprises the building of cutting-edge backing infrastructure in a well-demarcated Agri-horticultural zone enabling the establishment of recent food processing units in business plots supplied by the park's well-established supply chain. The supply chain infrastructure of a Mega Food Park typically includes collecting centers, a central processing centre, and a primary processing centre. According to ministry of food processing and industries report 2021, there are 24 mega food parks, which are in operation under PPP schemes

The food processing industry employs the most people in compared to other industries. The food industry does not rank in the top five in terms of fixed capital, indicating that it is labour intensive per unit of capital deployed. Food product output remains unaffected despite its low capital intensity, as it is the greatest of all businesses. As a result, every unit of capital financed in the food products industry employs the greatest number of people while also delivering the highest level of output when compared to other businesses. The Food Processing Industry was one of the most labor-intensive industries in the Registered Factory sector in 2017-18, accounting for 12.38 percent of total employment. According to the most recent Annual Survey of Industries (ASI) for 2017-18, the total number of people engaged in the recognized Food Processing Sector was 19.33 lakhs. The unregistered food processing sector employs 51.11 lakh persons, according to the NSSO 73 Round, 2015-16.

The food processing industry is confronted with a set of specific issues that have an immediate impact on the sector's strategy and requires solution. Processors in this industry work with seasonal, perishable resources that must be processed quickly. Because the sector is primarily made up of micro, small, and micro-enterprises, they are unable to create sufficient excess for development or invest in supporting structure. As a result, the industry requires infrastructure development assistance as well as targeted incentives to encourage investment in processing capacity increase. To address these issues, grading and packing centres, testing laboratories, controlled atmosphere facilities, reefer vans, and other subsidiary infrastructure and services such as setting up testing laboratories, research and development, reporting skill training, and marketing provision are all required.

4. Limitations

Understanding the differences between renewables utilized in India and renewables used in the United States, Europe, and other nations is crucial to formulating sustainable policies. The "traditional" renewables challenges in India are still present: (1) variability; (2) location-specific potential (concentrated in regions that are sometimes remote from clients); and (3) higher costs [32]. However, there are significant variations and requirements that warrant a more detailed investigation of renewable energy's long-term viability. It's simple to make renewables profitable for producers by paying them enough but can the rest of the system cope? Can consumer in a developing country like India pay for high-cost renewable resources? All these factors are not considered and are out of the scope of this study, however these are important factors to be considered and surely need in depth research in the same. The study is completely secondary research and on field understanding of the factors affecting the renewable energy resources and its usage in India should surely be considered.

5. Conclusion

The primary goal of this research is to understand how post-harvest losses of perishable agricultural products in rural communities can be reduced with clean energy consumption. This research provides some insight into the energy use in agriculture, rural India's energy consumption pattern, and the growth of rural renewable energy sources. Without a detailed examination, it is impossible to estimate the influence of changes in agricultural production techniques on carbon emissions and the use of agricultural biomass energy. In a developed country, all agricultural activities are carried out on a vast scale, and farming is totally mechanized. As a result, agricultural energy use is comparable to that of the industrial sector. It is not viable to mechanize farming in developing countries like India, where most farmers have small farms. As a result, it will be easier to put renewable energy technologies into practice.

Efforts to reduce the post-harvest losses of Indian-grown fresh fruits and vegetables could assist Indian farmers. In sum, around a third of the world's food production is lost or thrown away. Countless countries in the developing world are suffering from malnutrition, and if we can preserve cultivated produce then we can save many lives by giving them nourishment. Reduced post-harvest losses will improve food availability, safety, and quality. Post-harvest losses and greenhouse gas emissions can be reduced by utilizing renewable energy-enabled devices. A novel method has been created that harnesses renewable energy to reduce post-harvest losses of food crops through the use of efficient drying and cooling. In rural areas, farmers can generate money by storing and processing their agrarian outputs using renewable energy technologies if policymakers establish a favorable and viable strategy for encouraging renewable energy technology. It is possible that both farmers and the government may benefit from this arrangement. With the help of food parks, it is possible to sell these commodities at a higher price, and farmers' incomes can also rise if they are able to profit from selling their products. If farmers' incomes rose, they would be able to spend more on upscale goods, which would help India's GDP grow.

However harnessing renewable energy has many challenges. As this energy is produced in DC and has to be converted to AC. All our relay systems are on AC as result of which we have to create off- grid system for using these current. But there is a great deal of potential for adjusting the quantity of electricity generated to local needs through distribution generation systems. Small amounts of power surplus and demand must be monitored and flexibly re-routed by smart-grid technologies in order to achieve this goal. New grid technology is needed to make building integrated power production a reality, although many agricultural farms may become self-sufficient in their own energy supply from sustainable sources. At the same time, utilities need new financial mechanisms to entice them to fund distribution generation. Renewable energy research and development support is a fundamental flaw in our present energy system since it is inconsistent and sporadic. When it comes to R&D funding, there has been a long history of roller-coaster funding cycles that have harmed more than helped specific technologies [33]. When it comes to solar and fuel cell systems

R&D activities, we've had to justify our expenditures on unreasonably short timeframes rather than on long-term goals that we were dedicated to achieving. The long-term trajectory of our R & D portfolio should be guided by energy production and efficiency targets, rather than specific programmatic or technology subsidies. R & D is required for each of the technologies mentioned above. Thin-film solar technologies, as well as ultra-low-cost organic cells, have a lot of room for improvement in terms of efficiency and cost reductions. Power storage would significantly improve the appeal of both wind and photovoltaic (PV) technology. The costs of fuel cell membranes, and of producing hydrogen, both have an impact on hydrogen's long-term viability. "

References

[1] International Monetary Fund (2021), World Economic Outlook: Adjusting to Lower Commodity Prices; IMF:Washington,DC, USA.

[2] The World Bank Data 2020 GDP growth (annual %) - India. The World Bank.

[3] Moneycontrol.com 2022. Crude oil hits 14-year high: Impact on economy and businesses. Retrieved April 24, 2022, from moneycontrol.com: https://www.moneycontrol.com/news/business/commodities/crude-oil-hits-14-year-high-impact-on-economy-and-businesses-8201721.html

[4] data.worldbank.org. 2022 Retrieved April 24, 2022, from The World Bank Data: https://data.worldbank.org/indicator/SP.POP.TOTL?locations=IN

[5] PTI. 2022, March 12 The Economic Times. Retrieved March 2022, 14, from economictimes.indiatimes.com:https://economictimes.indiatimes.com/news/economy/indicators/india-to-grow-at-7-5-in-fy23-on-strong-investment-to-rise-to-8-next-fiscal-adb/articleshow/90687120.cms?from=mdr

[6] IEA. (2020, November). *IEA*. Retrieved March 05, 2022, from https://www.iea.org/reports/renewables-2020: https://www.iea.org/data-and-statistics

[7] Cherni J, Hill, Y 2009 Energy and policy providing for sustainable rural livelihoods in remote locations – The case of Cuba. *Geoforum,* **40,** 142-158.

[8] Modi V, McDade S, Lallement D, Saghir J. 2005. Energy Services for the Millennium Development Goals. Wshington, D.C. and New York: the World Bank and United Nations Development Programme.

[9] Barnes D, Floor W 1996 Rural Energy in Developing Countries: A Challenge for Economic Development, *Annual Review of Energy and Environment,* **21**, 21-35.

[10] Deichmann U, Meisner C, Murray S, Wheeler, D 2011 The economics of renewable energy expansion in rural Sub-Saharan Africa, *Energy policy*, **39(1)**, 215-227.

[11] Kolhe M L, Ranaweera K I U, Gunawardana A S 2015 Techno-economic sizing of off-grid hybrid renewable energy system for rural electrification in Sri Lanka, *Sustainable Energy Technologies and Assessments*, **11,** 53-64.

[12] Yadav P, Davies P J, Sarkodie S A 2019 The prospects of decentralised solar energy home systems in rural communities: User experience, determinants, and impact of free solar power on the energy poverty cycle, *Energy Strategy Reviews*, ***26***, 100424.

[13] Rahman S M, Ahmad M M 2013 Solar Home System (SHS) in rural Bangladesh: Ornamentation or fact of development?, *Energy Policy*, **63,** 348-354.

[14] Mboumboue E, Njomo D 2016Potential Contribution of Renewables to the Improvement of Living Conditions of Poor Rural Households in Developing Countries: Cameroon's Case Study, *Renewable and Sustainable Energy Reviews*, **61,** 266-279.

[15] Zhang l X, Wang C B, Song B 2013 Carbon emission reduction potential of a typical household biogas system in rural China, *Journal of Cleaner Production*, **47**, 415–421.

[16] Zhen W ,Qin Q, Qian X, Wei Y M 2018, Inequality across China's staple crops in energy consumption and related GHG emissions, *Ecological Economics,* **153**, 17–30.
[17] Murugaperumal K, Raj P A D V 201 Feasibility design and techno-economic analysis of hybrid renewable energy system for rural electrification, *Solar Energy*, **188**, 1068-1083.

[18] Sindhu S, Nehra V, Luthra S 2016 Identification and analysis of barriers in implementation of solar energy in Indian rural sector using integrated ISM and fuzzy MICMAC approach, *Renewable and Sustainable Energy Reviews*, **62**, 70-88.
[19] Leibensperger C, Yang P, Zhao Q, Wei S, Cai X 2021 The synergy between stakeholders for cellulosic biofuel development: Perspectives, opportunities, and barriers. *Renewable and Sustainable Energy Reviews*, **137**, 110613.
[20] NHB N H 2020 *Indian Horticulture Database 2020-21*. Retrieved January 21, 2022, from Ministry of Agriculture, Government of India: http://nhb.gov.in/Statistics.aspx?enc=WkegdyuHokljEtehnJoq0KWLU79sOQCy+W4MfOk01GFOWQSEvtp9tNHHoiv3p49g
[21] APEDA 2020 Retrieved February 2022, from apeda.gov.in: https://apeda.gov.in/apedawebsite/six_head_product/FFV.htm

[22] MoFPI 2021, November 26, *Ministry of Food Processing Industries*. Retrieved March 14, 2022, from www.mofpi.gov.in/: https://www.mofpi.gov.in/Schemes/mega-food-parks.
[23] Xin S, Qiao L, Dawei Z 2013 Evaluation of regional agri-food cold chain logistics competiveness, *Journal of Applied Sciences,* **13(4),** 2670-2675.
[24] Rathore J 2013 Cold chain management for perishable food: Issues and scope for India, *IUP Journal of Supply Chain Management,* **10(1),** 7-14.
[25] Negi S, Anand N 2015 Cold Chain: A Weak Link in the Fruits and Vegetables Supply Chain in India , *IUP Journal of Supply Chain Management,* **12(1),** 48-62.
[26] MoFPI 2020, September 12. *www.mofpi.gov.in*. Retrieved March 2022, 2020, from mofpi.gov.in:https://www.mofpi.gov.in/sites/default/files/OpportunitiesinColdChainSectorinIndia.pdf
[27] Sinha V, Tripathi A 2014 Integrating renewable energy to cold chain: Prospering rural India, *International Proceedings of Chemical, Biological & Environmental Engineering,* **76,** 99-104.
[28] Bag S, Dubey R, Mondal N 2015 Solid waste to energy status in India: a short review, *Discovery*, **39(177),** 75-81.
[29] Joshi R, Banwet D K, Shankar R 2009 Indian cold chain: Modeling the inhibitors, *British Food Journal,* **11(11),** 1260-1283.
[30] Paul V, Ezekiel R, Pandey R 2016 Cold storage in India: Present scenario and future directions, *Process Food Industries,* **19,** 25-28.
[31] Singhal R, Saksena S 2017 Performance assessment of the storage and warehousing industry in India, *Journal of Industrial Statistics,* **6(1),** 15-40.
[32] Tongia R 2014 Growing Complexity and Transformations of the Power Sector: India as an Example of Developing Regions using Enterprise Architecture and Smart Grids. In *A Systemic Perspective to Managing Complexity with Enterprise Architecture* (454-501). IGI Global.
[33] Margolis R M, Kammen D M 1999. Underinvestment: the energy technology and R&D policy challenge, *Science*, **285(5428)**, 690-692.

Numerical Investigation of Concentrated Solar Central Billboard with Hexagonal Tubes.

Sanjay N Havaldar[1], Harsh V Malapur[2], Kaustubh G Kulkarni[3] and Gary A. Anderson[3]

[1,3] School of Mech. Engg, Dr. Vishwanath Karad MIT World Peace University, Pune, India

E-mail: sanjay.havaldar@mitwpu.edu.in (Sanjay N Havaldar)

[2]Student, Department of Elec. and Electronics Eng., The University of Manchester, Manchester, UK.

[2]Professor, Ag and Bio Systems Eng., South Dakota State University, Brookings, SD 57007, USA.

Abstract. The solar receiver is the most important component of any central solar tower power plant (CSP) system. A numerical analysis of four billboard geometry designs of the central tower receiver was undertaken in this study. On the receiver, the surface area, heat transfer fluid (HTF), and intensity of beam radiation were all constant. The mass flow rate of HTF was modified, and the temperature gain was used as a metric to determine the most efficient design based on the analytical results. Four different models of vertical tube receiverswere designed and analyzed. In the analysis, the surface area of the receivers was the same. When compared to the vertical variable circular tube receiver design, the vertical variable hexagonal tube receiver design achieved a marginally higher temperature for the HTF fluid at lower mass flow rates.

Keywords: hexagonal tubes billboard solar receiver, billboard solar receiver, and solarconcentrator

1. Introduction

A special arrangement of mirrors (heliostats) reflects and concentrates solar energy on a receiver on top of a tower in a CSP system. The heat transfer fluid (HTF) that travels through the pipes of the central solar receiver is then heated by the focused solar beam. The usual design of a central receiver for a CST system is shown in Fig. 1. The sunbeam radiation from the field heliostats is collected by multiple panels of tubes along the receiver's periphery. The geometric considerations of this panel of tubes have a considerable impact on the system's central solar receiver's total heat transfer conversion efficiency.

In concentrated solar power systems, the design of CSP receivers is crucial. CSP optically redirects the beam radiation (solar flux) those are received on its surface, the intensity of which is related to the heliostat characteristics in the CSP field. Another crucial element is that as the radiations of the sun are pointed toward the receiver, the beam radiation intensity changes from the center of the receiver to the perimeter. As indicated in Fig. 1, tubes are connected to headers in each receiver panel. To boost heat absorptivity, receiver tubes are coated with Pyromark on the outside [3]. The tubes are arranged in the panel such that they are separated by around 2.5 mm to avoid contact after the expansion due to overheating. The frame prevents the tubes from bowing outwardly and laterally, as well as preventing gaps from emerging between the tubes on the receiver's back.

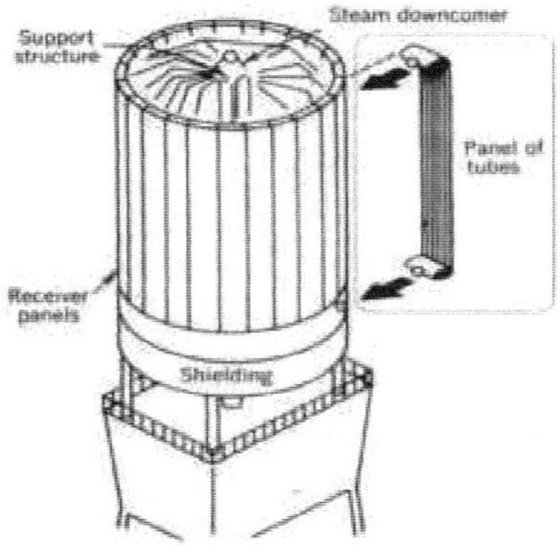

Figure 1: Layout of an external receiver of the Solar One facility in Barstow, CA[1]. The tubes of the receiver are shown in the rectangular box.

2. Literature Review

In concentrated solar power generation systems, CSP receiver design is critical. On its surface, CSP receives optically redirected beam radiation (solar flux), the strength of which is determined by the number of heliostats in the CSP field. The concentrated beam radiation profile of the reflected solar beam on the receiver on top of the tower is visible. The flux valueof the irradiance is larger in the middle and lower towards the outskirts. The temperature distribution in the receiver determines the thermal energy loss. Thus the receiver geometry (tubes of receiver panel) design is at the heart of the CSP system. To avoid overheating of the tubes due to contact, the tubes in the panels are separated by roughly 2.5 mm. Several academics have studied the problem of using solar thermal power in high-temperature manufacturing processes, analyzing high-temperature solar devices with computational fluid dynamics (CFD) to improve designs and the performance of the high-temperature process [4]. Three layers of material make up the solar prototype (refractory, frame, and insulation). The goal of this design method is to create the widest temperature range possible while retaining a more consistent thermal profile [5].

An analysis of the temperature of a heat transfer fluid in a spiral receiver for variable heat flux input and HTF flow rates was undertaken using an analytical framework using "n" reflecting mirrors [6]. The amount of solar irradiation reflected by the heliostats on the receiver determines the amount of heat input to the receiver [7]. A circular solar receiver's pressure drop and heat transfer coefficient are also compared to those of a traditional solar receiver [8]. In a concentrated solar thermal power plant, intense sunlight irradiates receiver tubes, converting solar energy into heat [9]. The heat flux density on the receiver surface could be as high as 2.5 MW/m^2. The targeting approach [10] determines the heat flux densitydistribution on the receiver surface. The heat flux in the receiver (heat exchanger) system is transferred from the receiver tube walls to the heat transfer fluid (HTF) [11]. A higher HTF temperature is associated with a higher receiver-focused solar beam temperature [12, 13]. The thermal energy in the fluid is subsequently converted to steam, which is then used to generate electricity via a turbine [14]. In this computational study using ANSYS Fluent 16, hexagonal tubes of uniform cross-section and variable cross-section are compared to standard circular tubes and variable circular tubes of the receiver panel. By altering the HTF flow rate, the temperature gained in the receiver panel was estimated for each of the designated receiver types. A number of various designs of circular solar receivers were numerically analyzed in [14, 15]. They suggested that a variable tube central receiver has better heat absorbing configuration in their leaf type circular solar receiver.

3. Receiver Modeling

3.1 *Receiver Design Modeling*

For numerical simulation and comparison, four types of receivers were modeled, as illustrated in Figures. 2 (a) to 2 (d) (d). Figure 2 depicts the simplest model of a vertical uniformtube diameter billboard solar receiver (VUTDR) (a). The cold HTF enters from the bottom header, while the heated HTF leaves from the top header after passing through the modeled tube receiver. The receiver tube was 20 mm in diameter; with HTF entering at point 1 and exiting at point 2 (see Fig. 2(a)).

2(a) VUTDR 2(b) VVTDR 2(c) VUXDR 2(d) VVXDR

Figure 2: Various design s of receivers analyzed in the simulation. (a) VUTDR–A conventional vertical uniform tube diameter receiver, (b) VVTDR–A vertical tube variable diameter receiver, (c) VU XD R–A vertical uniform hexagonal diameter receiver, (d) VVXDR–Avertical variable hexagonal diameter receiver. In Figs. 2(a) to 2(d), 1 is HTF fluid inlet and 2 is HTF fluid outlet.

Figure 2(a) shows the proposed vertical tube variable diameter billboard solar receiver (VVTDR) of constant diameter. Because the temperature intensity of the reflected solar beam is higher in the middle than its perimeter, a vertical tube variable diameter receiverwas designed, as illustrated in Fig. 2 (b). In receiver model VVTDR, HTF enters centrally from the bottom header and the higher temperature HTF leaves centrally from the top header in this model (VVTDR). The vertical uniform hexagonal tube billboard solar receiver (VUXDR) was proposed for the third type of design, as shown in Fig. 2 (c). The cold HTF enters from the bottom header, while the hot HTF exits from the top header via the billboardreceiver. HTF entered from point 1 and exited from point 2 in the receiver hexagon tube, which had a diameter of 20 mm (diagonal width) (see Fig. 2(c)). Figure 2 (d) depicts the designed model of a vertical variable hexagonal tube billboard solar VVXDR.

The variable tube diameter receiver was designed because the temperature intensity of the sun rays was greater at the center than around the sun ray perimeter. Second, heat absorption may be higher in hexagonal tubes due to their larger surface area. All of the models (Figs. 2(a) to 2(d)) were created with the same surface area so that they would be subjected to the same amount of solar beam radiation in this study. Table 1 lists the parameters utilized in the design of the four circular receiver models.

Table 1: Design parameters for receiver tubes.

Sr. No	Parameters	Value
1	Receiver area (m^2)	0.6
2	Thickness of Tube (mm)	1
3	Flow Rate (LPM)	0.1 to 0.5
4	Heat Flux (w/m^2)	1000 to 3000
5	Tube material	Stainless Steel

3.2 *Numerical Modeling*

All of the modeled solar receivers follow continuity, momentum, and energy laws in the numerical analysis. The continuity, momentum and the energy equations are represented by Eqns. (1) to (3), respectively.

$$\frac{\partial u_i}{\partial x_i} = 0 \quad (1)$$

$$\frac{\partial u_i u_j}{\partial x_i} = -\frac{1}{\rho}\frac{\partial p}{\partial x_i} + \frac{\partial}{\partial x_i}\left((v + v_t)\left(\frac{\partial u_i}{\partial x_j} + \frac{\partial u_j}{\partial x_i}\right)\right) \quad (2)$$

$$\frac{\partial u_i T}{\partial x_i} = \rho \frac{\partial}{\partial x_i}\left(\left(\frac{v}{P_r} + \frac{v_t}{P_r}\right)\frac{\partial T}{\partial x_i}\right) \quad (3)$$

Table 2: Boundary conditions settings

Boundary Conditions	
Inlet Conditions	Mass flow inlet
Outlet Conditions	Pressure Outlet
Initial Pressure (gauge)	Zero Pascal Inlet
Temperature	300 K
Grid Type	Fine, Tetrahedral
Number of Nodes	18000 to 23000 (as per model)
Number of Elements	60000 to 73000 (as per model)

Table 3: Simulation parameters settings in ANSYS Fluent

Parameter	**Setting**
Heat transfer fluid	Water
Modeling	Energy
Receiver Pipe Make	Stainless Steel
Pressure-Velocity Coupling	SIMPLE Spatial
Spatial gradient	Green Gauss, Cell Based
Initialization method	Standard
Computed for zones	all the zones

Table 2 shows the domain and boundary conditions set in simulation software ANSYS Fluent 16 which was used for numerical simulations. To solve for convergence in the simulation, the SIMPLE algorithm was utilized using a finite-volume formulation (R). For the pressure term, a standard scheme was employed. The setup was modeled for steady state simulation. In the software a First order upwind setting was used for solving the Eqns. (1) to (3) iteratively for convergences. All of the models were simulated with the same boundary conditions, mass flow rate, and irradiance flux on the receivers' surfaces. The test was performed, and the analytical result was recorded and examined. The simulation settings are listed in Table 3.

3.3 *Analysis Procedure*

For each of the modeled receivers, initial, boundary, and convergence conditions were defined. To solve for convergence in the simulation, the SIMPLE technique was used with a finite- volume formulation. The pressure term was solved using a standard scheme while the governing equations were solved using a second-order upwind scheme. The output from ANSYSFluent was recorded and used to perform various calculations using the mathematical model.

3.4 *Mathematical Modeling*

The sun's focused energy is absorbed by the central receiver, which then transfers the heat to the working fluid (HTF). This is how a receiver's mathematical modeling is explained. The available receiver tube area that receives flux is expressed as,

$$A = L * \pi * \frac{D_o}{2} \tag{4}$$

Heat flux on the circular receiver can be given as,

$$Q_t / A_r = I_B \gamma_b \rho \tag{5}$$

The receiver energy balance can be written as:

$$Q_t = Q_a + Q_c + Q_\gamma \tag{6}$$

The density used for the SS receiver was 9078 kg/m^3, C_p was 481 J/kg-K, and thermal conductivity was 487 W/m-K. The analysis software employed an absorptivity fraction of 0.8, resulting in a reflectivity of 0.2. Q_a (Absorptivity) was 80% of the total Q_t in Eqn. (6), Q_c (stainless steel) was 20%, and Q_γ (reflectivity) was 20%. Convective losses reduce the heat transfer coefficient. If V is the velocity of the surround air then we get convective heat losses are reduced and given as:

$$h = 5.7 + 3.8V \tag{7}$$

The temperature of HTF was measured at the output by,

$$T_i + \left(\frac{I_{b*}\gamma_{*b}*\rho*\alpha*A_m - h*A_0(T_w - T_a) - \varepsilon*A_0*(T_\omega^4 - T_a^4)}{m*_p}\right) = T_0 \tag{8}$$

The heat absorbed by HTF was calculated using q as the flux input (solar irradiation). The efficiency (η) was calculated by dividing the amount of absorbed heat by HTF by the solar flux input amount.

4. Analysis of results and discussion

Temperature, pressure, and velocity measured output were used to evaluate the performance of all modeled billboard type vertical solar receivers. The CFD output for various characteristics for HTF (temperature, pressure, and velocity) for the modeled VUTDR solar receiver is shown in Figs. 3a to 3d as the mass flow rates was varied. As seen inFig. 3a, the HTF temperature for the VUTDR model peaked at the upper left section of the header. This was due to the fact that there was no or very little HTF flow in the area. The HTF velocity curve is as shown in Fig. 3c. The HTF had an average temperature level in theVUTDR's center zone. The pressure fluctuation of the HTF for the VUTDR (Fig. 3b) indicates that there was lower pressure in right upper part (blue shade) and greater pressure in the bottom right part (red shade). This shows that the flow rate was lower in the sections farthest from the HTF entrance point into the VUTDR. The temperature distribution in the blue area is lowered as a result of this impact, which occurs on a continuous basis. As observed by the yellow color, the flow rate of heat transfer fluid on the borders of the tube has decreased, suggesting that the temperature from radiation wasnot active on the top right side of the model.

Fig. 3d shows the change of HTF temperature gain (*To -Ti*) versus mass flow rate of HTF for three different solar irradiations (1000, 2000, and 3000 W/m^2) in the VUTDR model for three different solar irradiations (1000, 2000, and 3000 W/m^2). Lower mass flow rates are observed to result in higher HTF temperatures.

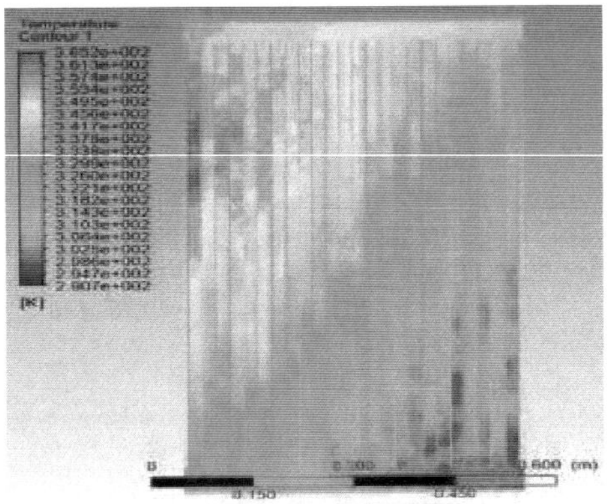

Figure 3a: Typical HTF temperature map in the VUTDR.

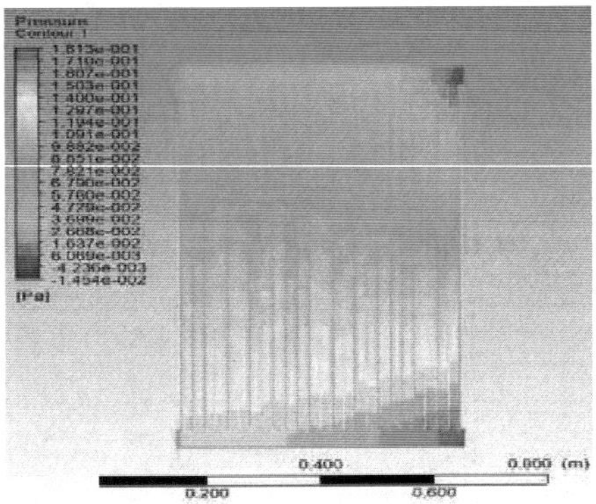

Figure 3b: Typical HTF pressure map in the VUTDR.

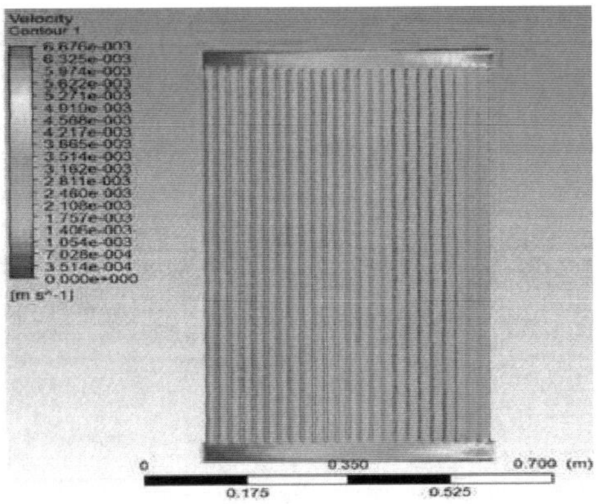

Figure 3c: Typical HTF velocity map in the VUTDR.

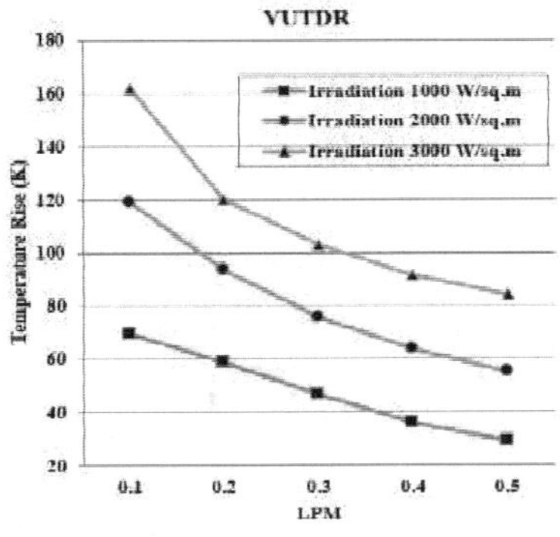

Figure 3d: HTF Temperature versus flow rate in VUTDR.

The CFD output of HTF temperature, pressure, and velocity, as well as the plot of temperature gain (To – Ti) with mass flow rate in LPM for the modeled VVTDR solar receiver,are shown in Figs. 4a to 4d. The HTF temperature zone in the center tubes of the VVTDR variant was lowered. The pressure fluctuation of the HTF for the VVTDR (see Fig. 4b) shows a greater pressure in the upper receiver zone (as represented by the yellow shade). There is less pressure in the lower receiver zone (as represented by the blue shade) than in the upper receiver zone (green shade). The cold HTF entered the intake by the bottom-center header, and the flow took the shortest course through the middle larger diameter tubes to reach the exit header (top header). As shown in the VVTDR center zone, this effect happens on a continuous basis, leading to decreased temperature dispersion in the blue color area. This means that the mass flow rate in the center tubes is greater, but there is little or no HTF fluid in the VVTDR tubes in the periphery tubes. The temperature fluctuation for the peripheral tubes is depicted in red in Fig. 4b. In Fig. 4d, the fluctuation of HTF temperature gain (*To – Ti*) versus mass flow rate of HTF for three different solar irradiations (1000, 2000, and 3000 W/m^2) for the VVTDR model is shown for three different solar irradiations (1000, 2000, and 3000 W/m^2). Higher HTF temperatures are attained at lower mass flow rates in a VVTDR model.

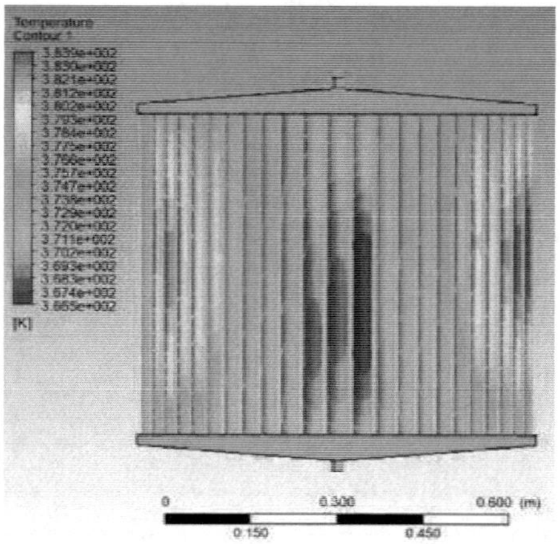

Figure 4a: Typical HTF temperature map in the VVTDR.

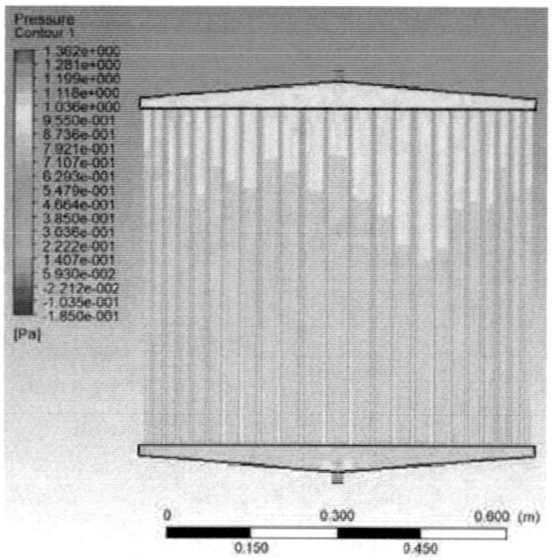

Figure 4b: Typical HTF pressure map in the VVTDR.

Figure 4c: Typical HTF velocity map in the VVTDR.

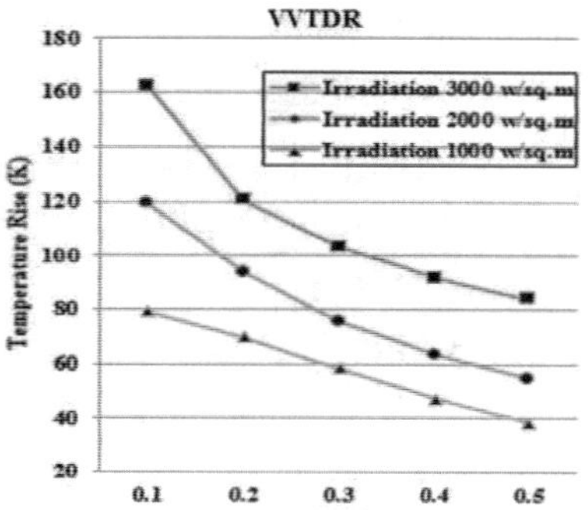

Figure 4d: HTF Temperature versus flow rate in VVTDR.

Because the hexagonal shape offers a higher surface area to transfer heat to the HTF flowing inside it, a hexagonal tube model for the solar receiver was proposed. The temperature distribution, pressure, and velocity of the vertical uniform hexagonal solar receiver (VUXDR) are depicted in Figs. 5a to 5d. The hexagon pipes had a diameter of 20 mm. The hexagonal pipes experience larger temperature rises, as there is little or no mass movement of HTF in the hexagonal tubes, as seen in Fig. 5a.

At the entry of a cold HTF, the pressure is lower, and at the departure zone of the VUXDR model, it is higher. This indicates that there was no mass or only a small flow of HTF into hexagonal tubes from the receiver's initial half (from the input point). Figure 5d shows a visualization of the gainin temperature of the HTF fluid (*To Ti*) for the VUXDR model versus mass flow rate. It can be shown that the gain in HTF temperature is greater for lower mass flow rates than for larger mass flow rates. Because the hexagonal tubes are assaulted from outside with heat flux, at lower mass flow rates, the HTF flows slower. As a result, as compared to HTFtravelling at a quicker speed in hexagonal tubes, the heat absorbed is greater by the heat transfer fluid.

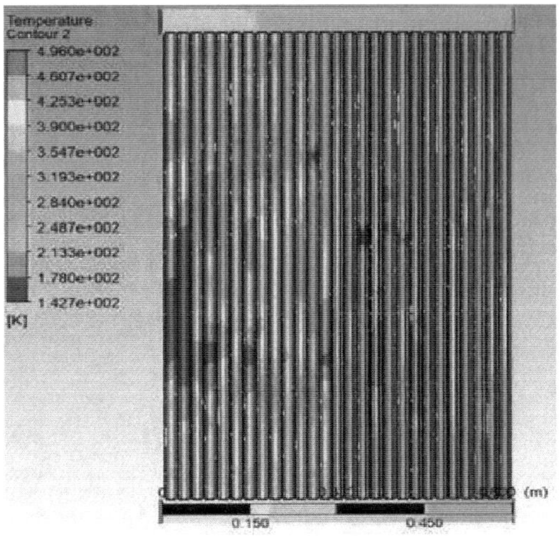

Figure 5a: Typical HTF temperature map in the VUXDR.

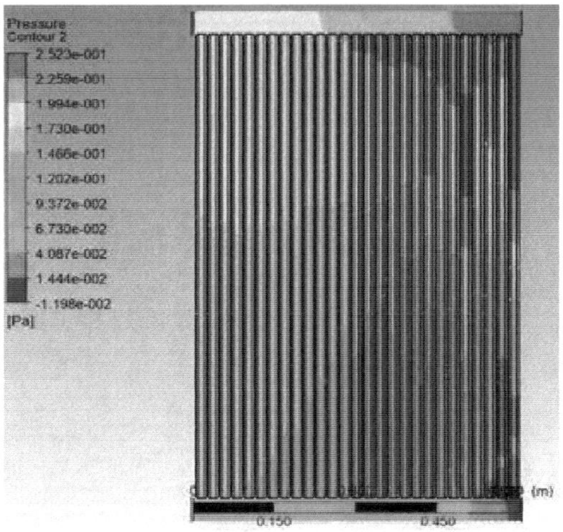

Figure 5b: Typical HTF pressure map in the VUXDR.

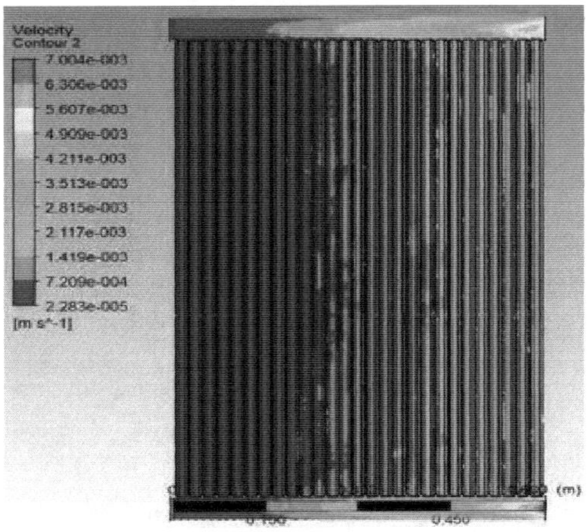

Figure 5c: Typical HTF velocity map in the VUXDR.

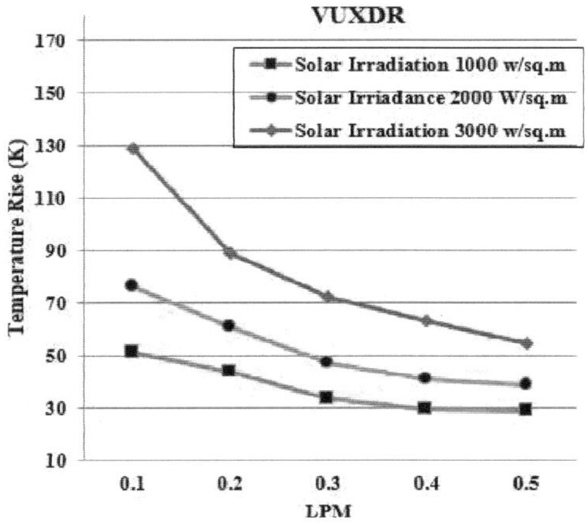

Figure 5d: HTF Temperature versus flow rate in VUXDR.

The VVXDR model output of the analysis for the gain in HTF temperature, pressure, and velocity in variable hexagonal tubes vs mass flow rates is shown in Figures. 6.a to 6.d. Figure 6.a shows that the periphery hexagonal tubes have little or no HTF flow, resulting in greater temperatures, as opposed to the center tubes, which have the most HTF flow and hence the shortest path to the receiver's outlet. The pressure map versus mass flow rate (Fig. 6.b) shows that the outer hexagonal tubes achieve higher pressure. This means that the core hexagonal tube has a lower pressure and a higher velocity than the smaller hexagonal tubes on the periphery, which have a higher pressure but a lower velocity.

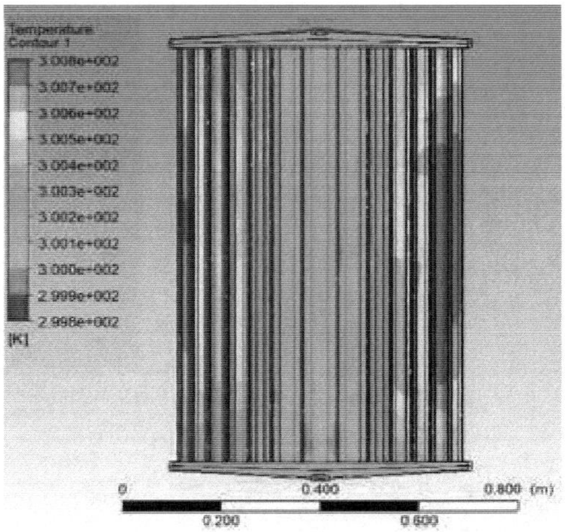

Figure 6a: Typical HTF temperature map in the VVXDR.

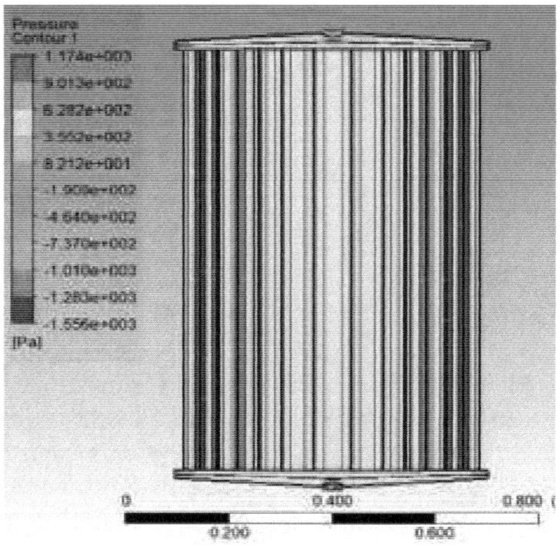

Figure 6b: Typical HTF pressure map in the VVXDR.

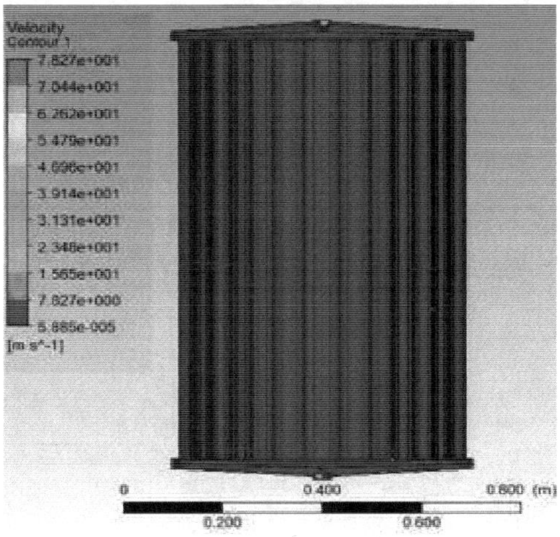

Figure 6c: Typical HTF velocity map in the VVXDR.

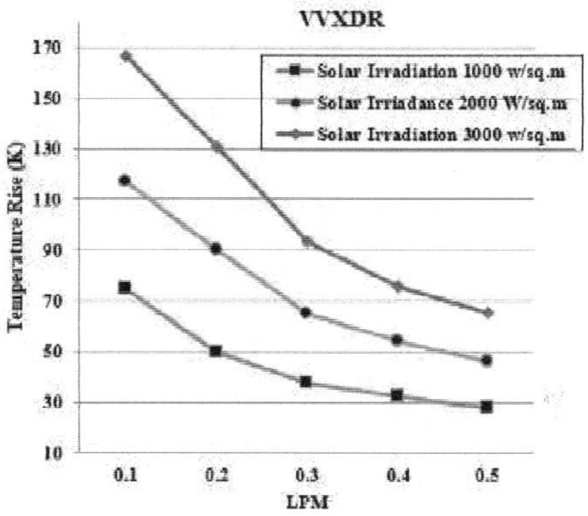

Figure 6d: Heat transfer fluid temperature versus flow rate in VVXDR.

For all of the modeled solar receivers, plots of the gain in temperature of the HTF fluid at varying irradiation intensities vs mass flow rates are shown in Figures 7 to 9. For lower mass flow rates, it was discovered that the Vertical Variable Hexagonal Tube Receiver (VVXDR) achieves a marginally higher temperature than the Vertical Variable Tube Receiver (VVTDR). This is due to the arrangement of cross-sections of circular tubesor cross-sections of hexagonal tubes in the design, which allows the HTF to take longer to migrate from the receiver's entry to exit under solar irradiation. As the steam moves from the receiver's entry to exit, the HTF acquires a greater temperature.

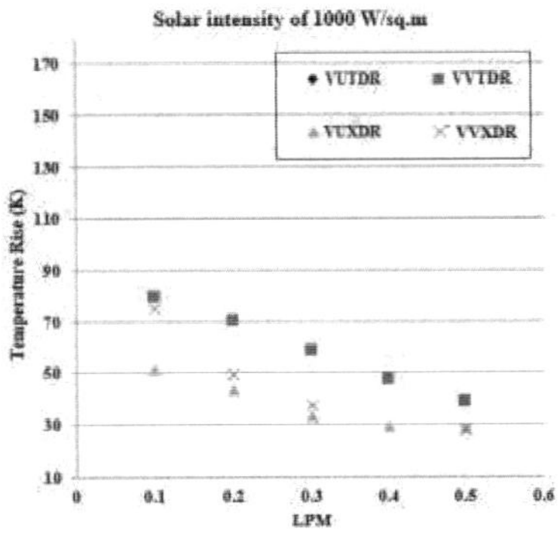

Figure 7: Variation of heat transfer fluid temperature (To – Ti) with flow rate (LPM) 1000 W/m 2 solar intensity in receivers.

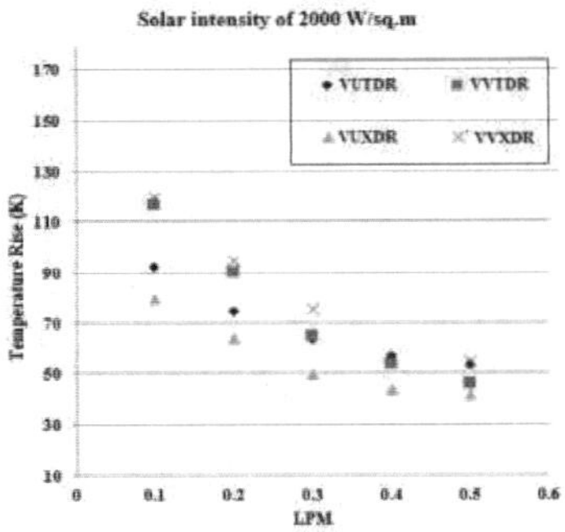

Figure 8: Variation of heat transfer fluid temperature (To – Ti) with flow rate (LPM) 2000 W/m 2 solar intensity in receivers.

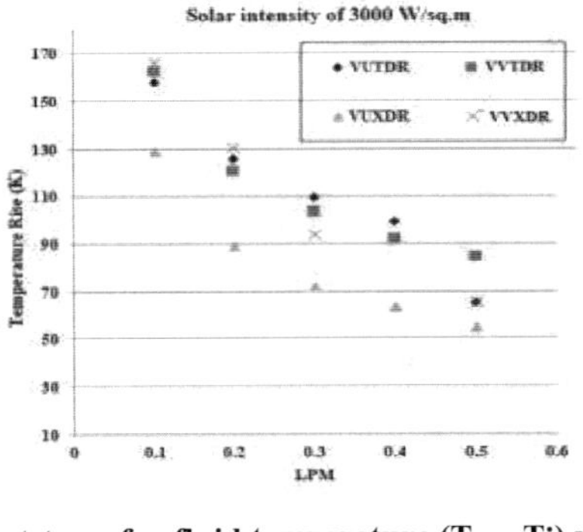

Figure 9: Variation of heat transfer fluid temperature (To – Ti) with flow rate (LPM) at 3000 W/m 2 solar intensity in receivers.

5. Conclusions

ANSYS Fluent 16 was used to analyze the velocity, temperature and pressure distribution of heat transfer fluid (water) for 4 distinct vertical billboard solar receiver modeled geometries. The mass flow rate of water was varied in the receiver, and also the intensity of the sun beam in three steps. The billboard tubes receiver surface area was nearly constant for4receiver modeled geometries. It was discovered that the heat gained orabsorbed by the HTF (*To – Ti*) in the vertical variable hexagonal tube receiver was marginally higher than the heat gained or absorbed by the HTF (*To – Ti*) in the vertical variable tube receiver. The flow of the HTF in the VVXDR was much more uniform than that in the VVTDR. Moreover, the velocity distribution of the HTF in the VVXDR was slightly more linear than the velocity distribution in VVTDR. The pressure of the HTF in VVXDR was also consistent across the tubes, unlike in the VVTDR, where the pressure slightly decreases mid length.

Nomenclature

D_o - Receiver tube outside diameter
D_i - Receiver tube inside diameter
L - Receiver vertical tube length
N - Receiver tubes number of turns
Q - Theoretical rate of heat transfer
I_B - Solar beam radiation (W/m^2)
I_G - global radiation (W/m2)
I_D - diffuse radiation (w/m2)
Q_t - heat absorbed (W)
Q_a - heat absorbed fluid (W)
Q_c - loess due to convective (W)
Q_γ - losses due to radiation (W)
H - Coefficient of heat transfer (W/m^2 K)
T_a - the ambient temperature of air (K)
E - Emissivity of tube material

T_w - mean temperature of tube surface (K)
T_0 - outlet temperature of working fluid (K),
T_i - the Inlet temperature of working fluid (K),
m - mass flow rate (Kg/sec)
C_p - specific heat (kJ/Kg k)
V - velocity (m/s)

References

[1] William B, Stine M and Geyer.
[2] Fritsch A, Uhlig R, Marocco L, Frantz C, Flesch R and Hoffschmidt B 2017 *SolarEnergy,* 155 259–66.
[3] Zavoico A 2001 *Solar power tower design basis document* (Sandia National Labs)
[4] Abuseada M and Ozalp N 2020 " *Solar Energy* **197** 396–410
[5] Lachhab E, Saad A and Bliya E 2019 AIMS Energy **7** 743–59 [6] Yu Q and Fu P 2020 Energy **200** 117505–117505.
[6] Kaustubh Kulkarni, Snjay Havaldar and Nabharun Bhattacharya, IoP Conf. Series: Materials Science and Engineering, 1136 (2021) 012018, doi: 10.1088/1757-899X/1136/1/012018
[7] Zhu G and Libby C 2017 *Review and Future Perspective of Central Receiver Design and Performance* (Abu Dhabi, United Arab Emirates) URL https://doi.org/10.1063/1.4984395
[8] Tawfik M, Tonnellier X and Sansom C 2018 *Renewable and Sustainable Energy Reviews* 90 802–815.
[9] Schiel W, Geyer M and Carmona R 1987
[10] Vignarooban K, Xu A X, Arvay K, Hsu A M and Kannan 2015 " *Applied Energy* **146** 383–96[11] Ozalp N and Jayakrishna D 2010 *Int J Hydrog Energy* **35** 6248–60
[11] Yu, Q., Fu, P., Yang, Y.H., et al. (2020). Modeling and parametric study of molten salt receiver of concentrating solar power tower plant. Energy, https://doi.org/10.1016/j.energy.2020.117505
[12] Rodat S, Abanades S, Sans J L and Flamant G 2009 *Sol Energy* **83** 1599–610
[13] Jamel M S, Rahman A H A A and Shamsuddin 2013 *Renewable and Sustainable EnergyReviews* 20 71–81.
[14] Kulkarni, K.G., Havaldar, S.N., Malapur, H.V. 2022, Numerical analysis of central solar receivers with various geometries. International Journal of Heat and Technology, Vol. 40, No. 1, pp. 339 - 346. https://doi.org/10.18280/ijht.400141.
[15] Kulkarni, K.G., Havaldar, S.N., Malapur, H.V. 2022, Numerical analysis of concentrated beam solar circular receivers. International Journal of Heat and Technology, Vol. 40, No. 2, pp. 375-382. https://doi.org/10.18280/ijht.400203.

Acknowledgements
The simulation work was carried out at the CFD Lab, School of Mech. Eng., Dr. VishwanathKarad MIT-World Peace University, Pune, INDIA

An Experimental Behavior on Green Concrete utilizing Granite Slurry Waste (GSW) as Partial Substitute of Cement

Sachin Sharma[1*], Priyanka Pandey[2], Pooja Gupta[3], Abhay M. Shende[4] and Rakesh Bhandari[5]

[1]Research Scholar, Department of Civil Engineering, Sangam University, Bhilwara, India

[2]Assistant Professor, Department of Civil Engineering, Sangam University, Bhilwara, India

[3]Professor, Department of Civil Engineering, Poornima College of Engineering, Jaipur, India

[4]Principal, Priyadarshini J. L. College of Engineering, Nagpur, India

[5]Professor, Department of Mechanical Engineering, Sangam University, Bhilwara, India

Email: sachin2410s@gmail.com

Abstract. Granite Granite materials are widely used by the construction industries. Granite slurry waste is a fine powder similar to cement that is causing climate and environmental problems around the world today. Individuals' respiratory and sensitivity issues are exacerbated by the disposal of this granite waste. Concrete seems to be the most versatile building material. The usage of GSW in concrete could help alleviate a variety of issues, including CO_2 emissions in the environment. Various molds were formed in this study to investigate the durability properties of concrete (water permeability, abrasion resistance, UPV test, and modulus of elasticity (static) for green concrete to be w/c 0.4 and 0.5. More workable green concrete is obtained at the varied replacement of granite slurry waste. The result has been shown to increase the depth of water penetration and modulus of elasticity with the addition of GSW for two w/c 0.5 and 0.4. The maximum water penetration depth was reported to be 85 mm at a 30% substitution level of GSW. The abrasion loss was step by step increased with an increasing of different substitution ranges (5% to 30%) of GSW as a partial substitute of cement (OPC and PPC). The highest abrasion loss was likewise observed to be less than 1 mm. The modulus of elasticity (Static) of concrete was increase at the substitute range of 5%. The UPV test results showed that the quality of GSW concrete is good. These studies show that the inclusion of granite slurry waste in green concrete will show that it will be cheap, protect the resources, and be more sustainable.

Keywords. Aggregates, Concrete, Elasticity, Abrasion, Workability, UPV

1. Introduction

Granite stones are the most well-known plutonic rocks found in mountain ranges and mainland safeguards. India is one of the Earth's main mining of granite and exporting countries, with abundant granite reserves. Granite deposits abound in the country's southern and eastern areas geologically. In Tamil Nadu, Karnataka, Assam, Maharashtra, Bihar, Meghalaya, Rajasthan, Odisha, and Madhya Pradesh, different hues of granite can be found in abundance. Granite stone is a widely used building material that generates a significant quantity of waste, accounting for over half of finished granites, during extraction and processing [1,2]. The quantity of waste generated in the granite sector in India contributes to around 20 to 30 percent of global production, implying that millions of tonnes of waste are freely available from cutting, processing, polishing, etc. Granite slurry waste (GSW) is made up of very fine, and it is causing environmental difficulties all over the world. Granite garbage disposal puts people's health at risk, causing respiratory and allergy difficulties [3]. Granite slurry waste is suitable for use in the building sector since it is easily available, has the necessary physical and chemical properties, all, is a quality waste material [4, 5].Concrete has the more demand construction material in infrastructure and due to the depletion of natural resources, innovation in the ingredients (cement, fine aggregates, coarse aggregate) is urgently required. Cement manufacture necessitates natural resources, and a lot of energy, and emits roughly 5-7 percent carbon dioxide [6, 7, 8]. Concrete advancements could lower the consumption of other energy sources and natural resources, reducing pollution burdens on the environment even further [10, 11]. They stated that cold, heat, moisture, stains, cracks, and scratches are all better tolerated by granite particles [9]. Previously, researchers [4] employed granite powder waste as a fine aggregates substitute in normal vibrated concrete (NVC) and obtain that adding granite slurry waste (GSW) lowered the slump value (workability) of the concrete mixes. By-products are currently occupying about 35% of the total area of the industry's premises, which are discarded in open yards [12]. Some companies, such as granite stone processing and mining industries, produce enormous volumes of GSW, which is distributed by the surrounding area and has negative environmental consequences [13, 14]. Granite slurry waste (GSW) can be mixed with cement concrete cubes to produce a material with high compressive strength and workability. And if the construction costs are decreased, it can aid in the overall solution to the environmental issue [15]. Abubaker et al. [6] conducted a research experiment on the mechanical strength qualities of design concrete constructed with granite slurry waste instead of cement. Divakar et al. [3] obtain the effects of using granite slurry dust as a partial substitute for fine particles in concrete at various levels of substitution. Rubanincheran and Ganeshan carried out the durability test on fiber concrete for curing times of 7 & 28 days containing partial substitute of OPC by granite slurry waste. In this investigation, M20 grade concrete has been prepared using OPC 43 grade cement. It can be observed from the preceding experiment investigations that using GSW could be used to design green concrete as a cement substitution. The use of GSW in fresh concrete will solve several issues linked to waste formation, natural resource consumption, and CO_2 emissions [16, 17]. This increased demand for cement will result in greater CO_2 emissions, which will harm the environment [18].

2. Experimental Programs

2.1 Ingredients (Materials)

In this research work, fine aggregates, granite slurry waste, coarse aggregates, and Ordinary Portland cement were used to manufacture the green concrete. To find the physical properties of ingredients in the laboratory have been classified in the following paragraph:

2.1.1Cement

Mostly available types of OPC of 53 grade (Shree Ultra) were utilized for the investigation. We used pozzolana Portland cement, and PPC for comparison purposes, and some specimens were also cast by PPC. In the laboratory, the Physical Properties of this OPC have been performed using a testing procedure as per the relevant IS code. The normal consistency and specific gravity of OPC were 33% and 3.11.

Table 1. Physical properties of cement

S.No.	Physical Properties/Test	Result
1.	Normal Consistency	33%
2.	Initial Setting Time	110 minute
3.	Final Setting Time	180 minute
4.	Specific Gravity	3.11

2.1.2 Berach River Sand (Fine Aggregates)

Berach River Sand (Fine Aggregates) collect from nearby Udaipur City, Rajasthan for this exp0eriment. We observed the sieve analysis and specific gravity of fine aggregates as follows as per IS 383-1970, IS 2386 (part III)-1963 in the laboratory. The water absorption and specific gravity of Berach River sand were 0.81 and 2.65 observed and fine aggregates belong to zone-II.

Table 2. Sieve analysis result for Berach river sand (fine aggregate)

Sieve size	Quantity of retained (gm)	Percentage of retained	Cumulative % retained	Cumulative percentage passing
10 mm	2	0.02	0.02	99.98
4.75 mm	26	2.6	2.62	97.38
2.36 mm	19	1.9	4.52	95.48
1.18 mm	207	20.7	25.22	74.78
600 micron	386	38.6	63.82	36.18
300 micron	158	15.8	79.62	20.38
150 micron	193	19.3	98.92	1.18
75 micron	8	0.08	99	1
Pan	3	0.03	99.03	0.97

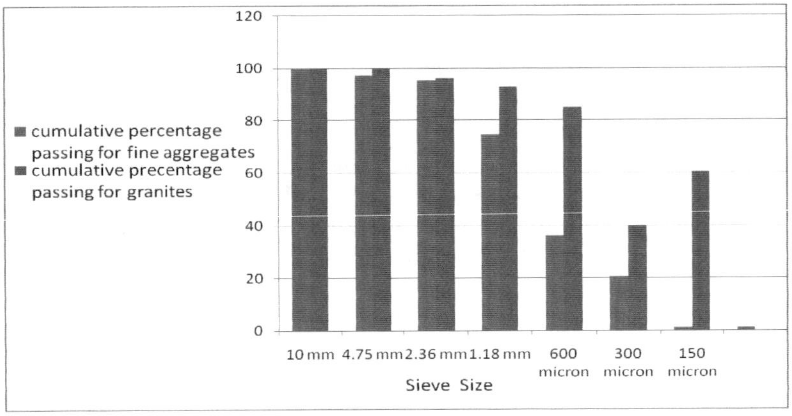

Figure 1. Graphical representation of fine aggregates and granite slurry waste sieve analysis

Table 3. specific gravity of (Berach river sand) fine aggregate

Wt. of Vessel + Sample + W^r (A) (in gm.)	Wt. of Vessel + W^r (B) (in gm.)	Wt. of Saturated & dry surface sample (C) in gm.	Wt. of Oven dry sample (D) in gm.	Sp. Gravity (S)	W^r Absorption
1743	1430	500	496	2.65	0.81

2.1.3 Coarse Aggregates

For this research work, the size of coarse aggregates was used as 20 mm and 10 mm. The sieve analysis test for coarse aggregate was investigated in the laboratory as per IS 383:1970. The water absorption and specific gravity of coarse aggregates were observed at 0.91 and 2.71.

Table 4. Sieve analysis for coarse aggregate

Sieve size	Quantity of retained (gm)	Percentage of retained cumulative % Retained	Cumulative % retained	Cumulative percentage passing
20 mm	17	0.02	0.02	99.98
16 mm	532	2.6	2.62	97.38
12.5 mm	148	1.9	4.52	95.48
10 mm	101	20.7	25.22	74.78
4.75 mm	202	38.6	63.82	36.18

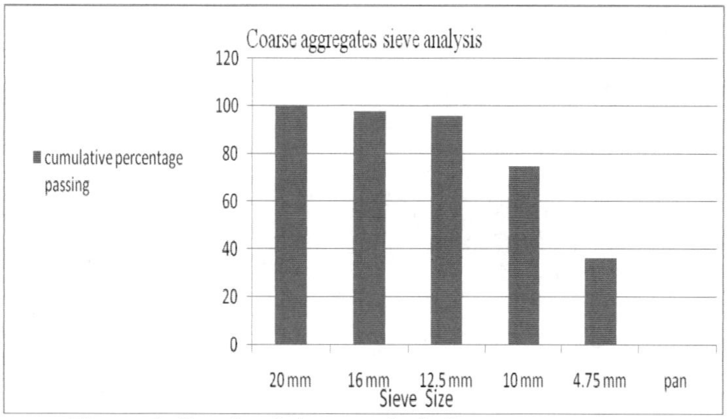

Figure 2. Result of sieve analysis for coarse aggregate

Table 5. Specific gravity of coarse aggregate

Wt. of Vessel + Sample + W^r (A) (in gm.)	Wt. of Vessel + W^r (B) (in gm.)	Wt. of Saturated & dry surface sample (C) in gm.	Wt. of Oven dry sample (D) in gm.	Sp. Gravity (S)	W^r Absorption
3848	3218	994	985	2.71	0.91

2.1.4 Granite Slurry Waste

The powder of granite waste collected from the grinding unit of Ravi Marble & Granites, Udaipur, Rajasthan was used for the research work. The test on chemical composition was identified at Mines & Geology Department, Udaipur, and obtained the identified physical properties of Granite slurry waste from CTAE, Udaipur.

Figure 3. Granite slurry waste

Table 6. Chemical composition of granite slurry waste

Composition	Percentage
SiO_2	71.24%
Fe_2O_3	0.58%
Al_2O_3	12.50%
TiO_2	0.17%
CaO	1.45%
MgO	0.78%
LOI	1.15%
Na_2O	6.19%
K_20	4.58%

Table 7. Physical characteristics of granite slurry waste

S. N.	Characteristics	Results
1.	Porosity%	0.87
2.	Colour	Black
3.	Water Absorption%	0.38
4.	Specific Gravity	2.17
5.	Form	Fine Powder

2.2 Methodology

We have discussed ingredients such as granite slurry waste, fine aggregates, cement, and coarse aggregate in Clause 2.1 and all properties of the ingredient have been tested in the laboratory.

2.2.1 Design of Mix Concrete Preparation

M25 concrete was designed to use the methodology of IS 10262:2009 and IS 456:2000 for this experimental study to determine the proportion of ingredients satisfying the desired condition at different two w/c proportions (0.5 and 0.4). Three series (Trial Mix) were cast in this study. In the first series 0.5 of w/c ratio is used whereas in this second series mixes. The third series mix was cast utilizing PPC instead of OPC with w/c 0.4.

Table 8. Various tests performed

Test	Shape of the Specimen	Dimensions of the Specimens	Curing Period (in days)
Ultra Sonic Pulse Velocity	Cube	(150 mm X 150 mm X 150 mm)	28
Water Permeability	Cube	(150 mm X 150 mm X 150 mm)	28
Abrasion Resistance	Cube	(100 mm X 100 mm X 100 mm)	28
Modulus of Elasticity (Static)	Cylinder	(150 mm X 300 mm)	28

Table 9. Mix proportion of concrete (Grade of concrete: M25)

rial Mixes	Specimen No.	w/c ratio	Cement type	Mix proportion by weight				
				Cement Kg/m^3	% of Replacement cement by granite slurry waste	GSW as a partial substitute for cement Kg/m^3	Berach River Sand (Fine Aggregates) Kg/m^3	Coarse Aggregates (CA) Kg/m^3
Series-I	D1	0.5	OPC	370.0	0	0.0	585	1280
	D2	0.5		351.5	5	18.5	585	1280
	D3	0.5		333	10	37.0	585	1280
	D4	0.5		314.5	15	55.5	585	1280
	D5	0.5		296	20	74.0	585	1280
	D6	0.5		277.5	25	92.5	585	1280
	D7	0.5		259	30	111	585	1280
Series-II	E1	0.4	OPC	370.0	0	0.0	585	1280
	E2	0.4		351.5	5	18.5	585	1280
	E3	0.4		333	10	37.0	585	1280
	E4	0.4		314.5	15	55.5	585	1280
	E5	0.4		296	20	74.0	585	1280
	E6	0.4		277.5	25	92.5	585	1280
	E7	0.4		259	30	111	585	1280
Series-III	F1	0.4	PPC	370.0	0	0.0	585	1280
	F2	0.4		351.5	5	18.5	585	1280
	F3	0.4		333	10	37.0	585	1280
	F4	0.4		314.5	15	55.5	585	1280
	F5	0.4		296	20	74.0	585	1280

3. Result and discussion

For fresh concrete, the Workability test of Concrete was used to be obtained by Slump cone Apparatus, and durability was carried out. For the ductile behavior of concrete, the modulus of elasticity (Static) has been carried out and for the quality of concrete, a non-destructive evaluation test (UPV) has been studied. The results for the slump test, water permeability test, abrasion test & modulus of elasticity test (static), and ultrasonic pulse velocity meter test (UPV) are presented. tests were performed on hardened concrete.

3.1 Workability of concrete

In the laboratory, To examine the workability from the Slump cone test for both fresh control and design concrete as per IS: 1199–1959. The slump test value for mixing fresh control concrete or design concrete using granite slurry waste at w/c fractions of 0.5 and 0.4 is shown in Table 10.

Table 10. Workability of concrete (slump result)

S.No.	Specimen Number of Series	W/C Ratio	% Substitute of Granite Slurry Waste (GSW)	Slump Result (mm)
1	D1	0.5	0	40
2	D7	0.5	30	75
3	E1	0.4	0	5
4	E7	0.4	30	20

3.2 Water permeability of concrete

On further increase in the level of granite as a cement substitute, a gradual increase in water penetration depth of concrete was noted. The value of water permeability in series I-III as compared to controlled mix concrete contained GSW as partial substitution of cement in w/c proportion (0.4 and 0.5) increases.

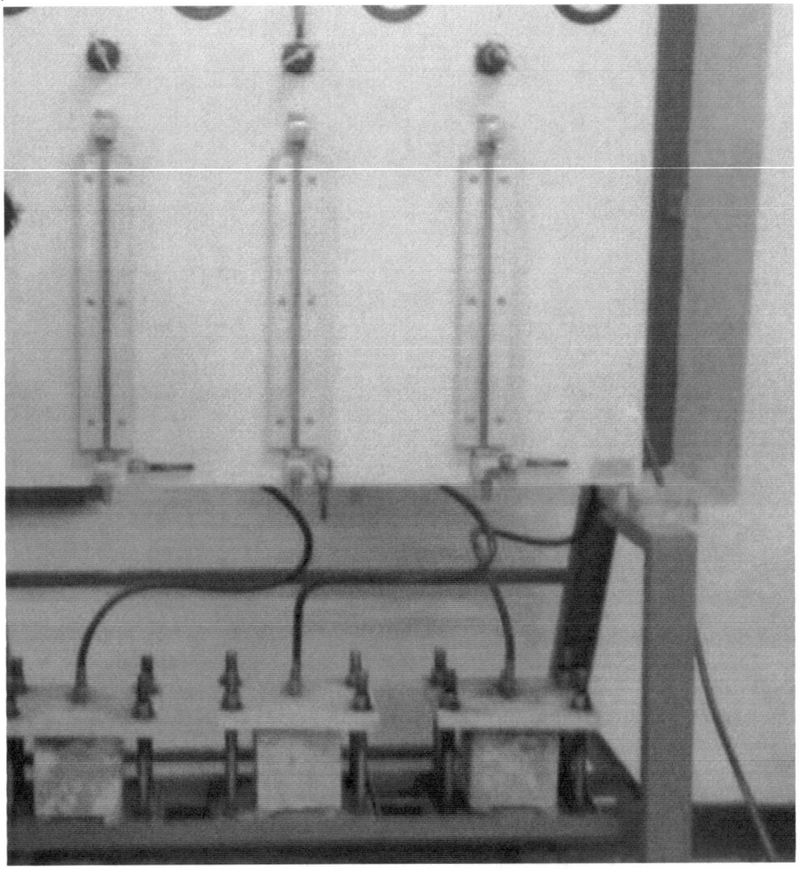

Figure 4. Water permeability test apparatus

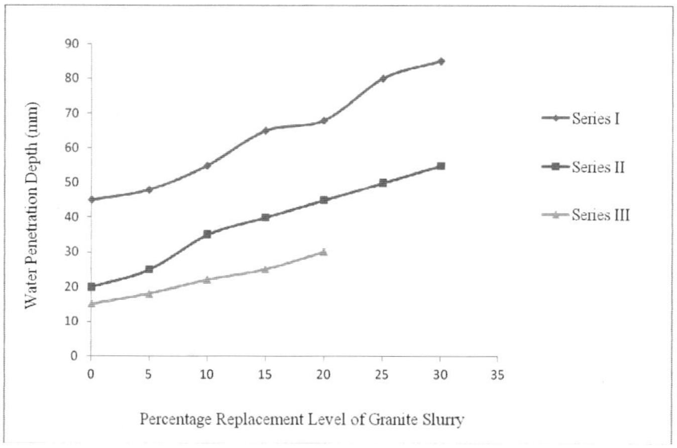

Figure 5. Variation in water penetration depth of concrete using granite slurry waste

3.3 Abrasion Resistance of Concrete

The abrasion resistance test has been determined for the surface wear's depth. The graph represents showing the increased result in the wear's depth concerning the different varied percentages (0% to 30%) with the interval of 5% of granite slurry waste as a substitute of cement at w/c fraction of 0.5 and 0.4 in Figure. 7. On further increase in the level of granite as cement Substitution, a gradual increase in abrasion loss of concrete was noted. As can be observed from the above data, the control sample gets the lowest amount of thickness loss (abrasion), but the largest loss occurs when the granite slurry waste substitute level is set to 30%. It is also observed from the above-cited Figures that the thickness loss for green concrete with substitute level of granite slurry waste from the range 5-30% is less than 1 mm.

Figure 6. Loss angles abrasion test apparatus

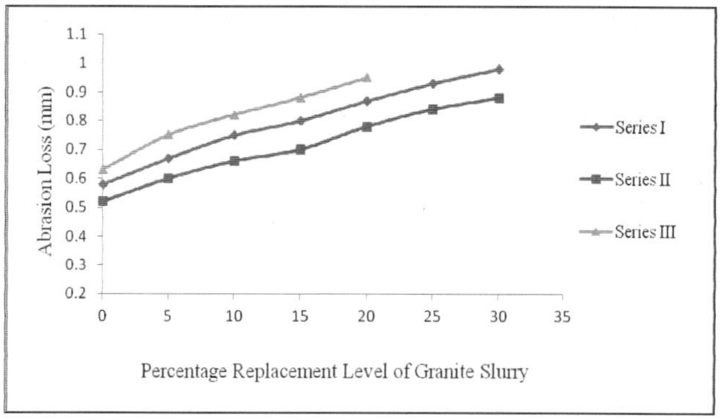

Figure 7. Variation in loss angles abrasion test of concrete using granite slurry waste

Table 11. Results of abrasion resistance test for concrete specimen

Trial Mixes	Specimen No.	W/C	Cement	% Substitute of granite slurry	W_1 (gm) Initial wt	W_2 (gm) Final wt. after abrasion	Wt. Loss (W_1-W_2)	Abrasion Loss (T mm)
Series I	D1	0.5	OPC	0	2667	2652	15	0.58
	D2	0.5		5	2566	2548	18	0.67
	D3	0.5		10	2538	2519	19	0.75
	D4	0.5		15	2604	2582	22	0.80
	D5	0.5		20	2582	2560	22	0.87
	D6	0.5		25	2502	2478	24	0.93
	D7	0.5		30	2578	2553	25	0.98
Series II	E1	0.4		0	2692	2676	16	0.52
	E2	0.4		5	2751	2734	17	0.60
	E3	0.4		10	2579	2562	17	0.66
	E4	0.4		15	2594	2575	19	0.70
	E5	0.4		20	2709	2688	21	0.78
	E6	0.4		25	2657	2634	23	0.84
	E7	0.4		30	2678	2655	23	0.88
Series III	F1	0.4	PPC	0	2693	2675	18	0.63
	F2	0.4		5	2611	2593	18	0.75
	F3	0.4		10	2657	2638	19	0.82
	F4	0.4		15	2661	2640	21	0.88
	F5	0.4		20	2597	2575	22	0.95

3.4 Modulus of Elasticity (Static) of Concrete

The Modulus of Elasticity (Static) of concrete mold specimens was evaluated after 28 days of the curing time. The graphical variations of the modulus of elasticity for two w/c ratios 0.4 and 0.5 have been also shown in Fig. 9. It can be obtained that the test of modulus of elasticity for GSW concrete for series-I more improve up to 5% substitute level whereas increase value of modulus of elasticity for GSW green concrete for series-II is substitute level of 10% or less. When PPC is used to replace OPC for w/c ratio is 0.4, the modulus of elasticity for GSW concrete is gradual increase up to 5% substitute level of cement. The modulus of elasticity of concrete is reduced after a 5% partial cement substitute at a w/c ratio of 0.5, as seen in Fig. 9. It can be also shown from series-II, that the modulus of elasticity for GSW green concrete is reduced after a 10% partial substitute of cement in Fig. 9. It has been shown that the reduction in modulus of elasticity (Static) of modified concrete is increased after the substitute of 5% cement at a w/c proportion of 0.4.

Figure 8. Modulus of elasticity test apparatus

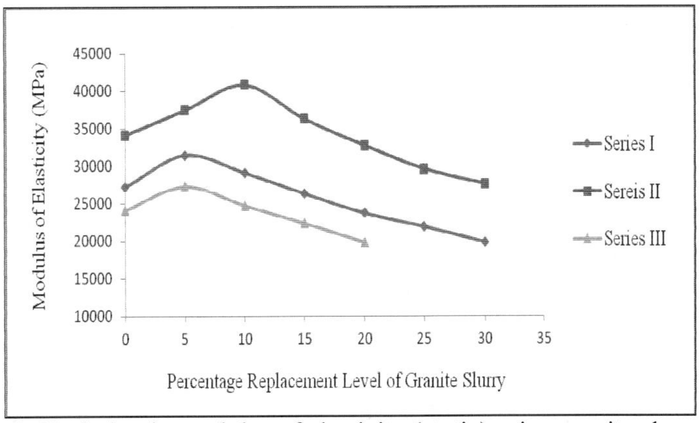

Figure 9. Variation in modulus of elasticity (static) using granite slurry waste

3.5 Ultra-Sonic Pulse Velocity (UPV) of Concrete

This test is done to examine the nature of modified green concrete by UPV process according to IS: 13311 (Part 1) - 1992 [30]. UPV of modified green concrete has been observed at w/c (0.5 and 0.4) for various substitution levels (ranges from 5% to 30%). When compared to the UPV value achieved for the Control mix, the UPV value of concrete from series I through series III, including series II, decreases (level 0 percent). The minimum value of UPV for series I too series III including series II is 3230 m/s, 3523 m/s, and 3950 m/s. As per IS 13311 (Part 1), concrete's quality is good for various results obtained in series-I (5% to 30% substitute level), and the quality of concrete is also good for series II and series- III.

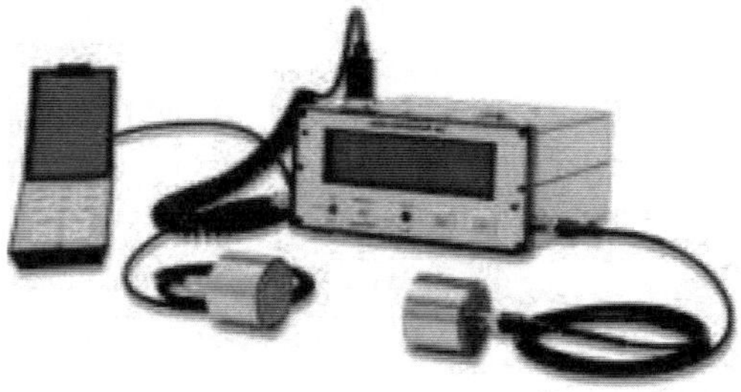

Figure 10. Ultra sonic pulse velocity meter test apparatus

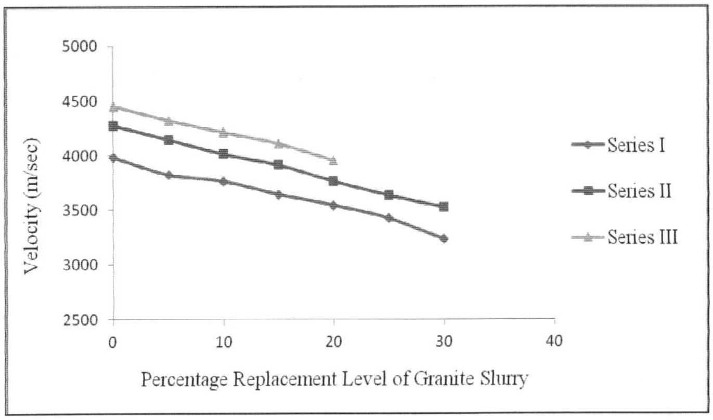

Figure 11. Graphical representation in pulse velocity of concrete using granite slurry waste

4. Conclusion

In this experimental research work, the suitability of modified design green concrete using GSW (Granite Slurry waste) has been evaluated. Workability tests for fresh concrete and durability properties and modulus of elasticity have been performed for hardened concrete.

- The value of the slump test of modified green concrete was raised when OPC and PPC cement was substituted by GSW at a substitution range of 30% level as compared to the fresh concrete mix (substitution range 0%) for w/c 0.4 & 0.5. More workable green concrete is obtained at the varied replacement of granite slurry waste.
- For the w/c ratios 0.5 and 0.4 (Series I/II/III), the obtained amount of water penetration depth expanded slowly when compared to control fresh concrete. The maximum water penetration depth was reported to be 85 mm at a 30% substitution level of granite slurry waste as well as a w/c of 0.5.
- The value of abrasion loss was step by step expanded with an increase of different substitution ranges (5% to 30%) of GSW as a partial substitute of cement (OPC and PPC). It is additionally seen that the most extreme loss of Abrasion is under 1 mm. This sort of concrete can be utilized in numerous advantageous applications (Modified concrete blocks, concrete paver blocks, etc).
- For a w/c proportion of 0.5, the modulus of elasticity of GSW green concrete has expanded at a substitute ranges of 5% while for a w/c proportion of 0.4, the relating increment at the substitute range up to 10%. At the point when PPC was utilized instead of OPC, the expansions in modulus of elasticity (Static) of concrete have seen at the substitute range of 5%.
- Ultra-sonic pulse velocity (UPV) results show that more durable GSW green concrete (IS 13311 (Part 1)) was obtained for w/c ratios of 0.5 and 0.4 (Series I/II/III).
- Subsequently, this green concrete will be monetarily less expensive as GSW material and openly accessible. Furthermore, the cost savings in cement will ensure the usual assets, making this type of green concrete much more realistic and sustainable.
- The use of GSW will take care of the removal issue. Additionally, this alternate green concrete will decrease the CO_2 outflow in light of less utilization of cement which understudy will save the pollution.

References

[1]. Elyas Asadi Shamsabadi, Mansour Ghalehnovi, Jorge de Brito, and Ali Khodabakhshian. "Performance of Concrete with Waste Granite Powder: The Effect of Superplasticizers", *Applied Science (MDPI),* 2018, 8, 1808; doi:10.3390/app8101808

[2]. Mendoza, J.; Feced, M.; Feijoo, G.; Josa, A.; Gabarrell, X.; Rieradevall, J. Life cycle inventory analysis of granite production from cradle to gate. Int. J. Life Cycle Assess. 2014, 19, 153–165.

[3]. Divakar Y., Manjunath.S and Aswath M.U.,"Experimental Investigation on Behavior of Concrete with the Use of Granite Fines", *International Journal of Advanced Engineering Research and Studies*, Vol. I, No. IV, (2012), 84-87.

[4]. S. Singh, S. Khan, R. Khandelwal, A. Chugh, R. Nagar, Performance of sustainable concrete containing granite cutting waste, J. Cleaner Prod. 119 (2016) 86–98.

[5]. Abhishek Jain, Rajesh Gupta, Sandeep Chaudhary. "Sustainable development of self-compacting concrete by using granite waste and fly ash", *Construction and Building Materials* (2020)- https://doi.org/10.1016/j.conbuildmat.2020.120516

[6]. Abubakar K.A. and Sasi D., "Strength Properties of Concrete with Partial Substitute of Cement by Granite Quarry", *International Journal of Engineering Research & Technology*, Vol. 3, No. 9, (2013), 344-348.

[7]. Devangan A. and Pawar M., "The Significance of Partial Substitute of Cement With Waste Marble Powder", *Periodical Research,* Vol. 3, No. 1, (2014),1-6.

[8]. Sharma S., Gupta T., and Sharma R.K., Assessment of Mechanical Properties of Concrete Containing Slurry Waste, *International Journal of Engineering*, 29,2016, 599-605.
[9]. A. Karmegam, A. Kalidass, D. Ulaganathan, Utilization of granite sawing waste in self-compacting concrete, *Gradevinar* 66 (11.) (2014) 997–1006.
[10]. Anwar A. and Ahmad J. "Study of Compressive Strength of Concrete by Partial Substitute of Cement with Marble Dust Powder", *International Journal on Mechanical Engineering and Robotics (IJMER)*, Vol 2, No. 3, (2014),1-4.
[11]. G.Latha, A.Suchith Reddy, K. Mounika, "Experimental Investigation on Strength Characteristics of Concrete Using Waste Marble Powder as Cementitious Material" *International Journal of Innovative Research in Science, Engineering and Technology,* Vol. 4, No. 12, (2015), 12691-12698.
[12]. Kala, T. Effect of granite powder on strength properties of concrete. *International Journal of Engineering and Science*. 2013, 2, 36–50.
[13]. S. Ghorbani, I. Taji, J. de Brito, M. Negahban, S. Ghorbani, M. Tavakkolizadeh, A. Davoodi, Mechanical and durability behaviour of concrete with granite waste dust as partial cement substitute under adverse exposure conditions, *Construction Building Material*, 194 (2019) 143–152.
[14]. S. Ghorbani, I. Taji, J. de Brito, M. Negahban, S. Ghorbani, M. Tavakkolizadeh, A. Davoodi, "Simultaneous effect of granite waste dust as partial substitute of cement and magnetized water on the properties of concrete exposed to NaCl and H2SO4 solutions", *Construction Building Material,* 288 (2021) 1–11. https://doi.org/10.1016/j.conbuildmat.2021.123064
[15]. Koneti Vamsi, Rajarapu Bhushaiah, Gangavarapu Manohar, Kakani Krupasagar, Shaik Mohammad, Rama Harshitha. "Partial Substitute of Cement in Concrete with Granite Powder and Fine Aggregate with Saw Dust", *International Research Journal of Engineering and Technology (IRJET),* (2019) *V*olume: 06 Issue: 03 PP: 3210-3222.
[16]. Kumar Ranjan and Kumar Shyam, "Partial Substitute of Cement with Marble Dust Powder", *International Journal of Engineering Research and Applications*, Vol. 5, No. 8, (2015), 106-114.
[17]. Singh Jashandeep and Bansal R.S., "Partial Substitute of Cement with Waste Marble Powder with M25 Grade", *International Journal of Technical Research and Applications*, Vol. 3, No. 2, (2015), 202-205.
[18]. Digvijay S. Chouhan, Yash Agrawal, Trilok Gupta, and Ravi K. Sharma. "Utilization of Granite Slurry Waste in Concrete: A Review", *Indian Journal of Science and Technology,* Vol 10(6), DOI: 10.17485/ijst/2017/v10i6/88279, 2017

Life Cycle Costing Methodology for Sustainable Construction

Mustafa S Sadliwala[1] and Nivedita G Gogate[2]

[1]M.Tech (CEM) student, MIT World Peace University, Pune, India.
[2]Associate Professor, Civil Engineering, MIT World Peace University, Pune, India.

Email: msadliwala97@gmail.com

Abstract. There has been a pressing need to consider lifelong building costs and spread awareness about the life cycle cost of a building in the construction sector. Before 1970s, most of the clients', contractors' and developers' decisions were solely based on the initial capital cost with no consideration of the entire life cycle cost of the building. If the cost of a building is estimated considering its entire life, approximately 70% portion is required for operational aspects. This approach of focusing only on the initial cost has failed to give the best value for money to the client in the long term. In this research, questionnaire-survey method is utilized to understand the awareness level and importance of different tangible and intangible costs of the construction over its useful life amongst the construction community. List of critical factors affecting life cycle cost of a building was extracted through an extensive literature survey. The questionnaire survey was prepared by incorporating the most significant parameters and circulated amongst the stakeholders consisting of respondents working at various hierarchical positions in the construction industry. The factors were ranked based on weightages obtained through Relative Importance Index (RII) methodology. The data analysis showed that the top 5 costs which should definitely be considered in life cycle costing are construction cost, Land Acquisition and associated cost, Design team fees and associated cost, Operation & Maintenance Cost and Taxes. Respondents also agreed that the life cycle costing methodology would provide a better and comprehensive estimate of the building cost as a whole and would be crucial in decision making. Incorporating life cycle costing in the design and planning phase would significantly help engineers as well as contractors to focus on the most important cost parameters. Furthermore, the analysis indicated that a lower emphasis is still given to cost factors related to environment. It needs to be emphasized that a comprehensive cost estimate for a project can only be obtained by incorporating sustainability aspects. Thus, a LCC framework considering all important cost parameters would further improve the process of selection of the most cost-effective alternative, as well as significantly increase sustainability in construction projects.

Keywords: Life Cycle Costing, Sustainable Construction, Life cycle analysis, Initial cost, Awareness

1. Introduction

Life cycle costing (LCC) aims to evaluate the cost-effectiveness of alternative design strategies by considering the potential initial and operational costs incurred over a specified period of time. Values which can be expressed in monetary terms are considered in LCC calculations; thus, intangible impacts such as comfort and environmental load are neglected. Rising awareness of the importance of buildings' operational and maintenance costs and the growing "value for money" trend has significantly increased the application of life cycle costing methodology in construction projects.

"Life Cycle Costing" is an essential design process for controlling the initial and the future cost of building ownership" (Tim, 1999). The LCC method is used mostly to determine the lowest results or the most cost-effective choice among several alternatives. Although the initial cost increases when the LCC method is applied, due to inclusion of maintenance, operation, and repair or replacement costs in the initial stages, savings could still be realized. In contrast, a lower initial cost will probably increase the running costs and cancel out the initial savings through the project's life span. While, LCC analysis provides the most economically feasible choice in most of the cases, it may not be required in case of a building where both initial as well as running costs are low (ASTM, 2005). Making a decision is a common task in which the most desirable solution from a set of alternatives must be chosen. Many decision-making processes in the real world are based on inaccurate or inadequate data. A significant amount of the total cost of infrastructure assets such as highways is incurred after its completion. As a result, a whole-life cost approach can improve the effectiveness of final judgments (Sinhal et al. 2001; Kirk and Dell 1995). In most cases, life-cycle cost (LCC) analysis is utilized to compare competing alternative investment options. The term "life-cycle cost" refers to an evaluation method that considers all costs that arise during a project's life cycle. The method is generally used to simplify the selection process and provides an easier way to choose a better alternative amongst the available options. It's best done in the early stages of viability, feasibility, and conceptual design, when most, if not all, choices are on the table.

The LCC technique provides a mechanism for assessing the economic viability of various available options, with a measurable scale for ranking the alternatives. LCC is used to select the most appealing option throughout all life cycle stages. The LCC method can benefit from budgeting, cost control, project feasibility studies, preliminary design, and asset or product assessments (Zhang, 1999). Realizing the advantages of LCC methodology, building expenses are increasingly being considered in the cost analysis, and financial methodologies for evaluating their life cycle costs are being developed. Many clients, developers, and consultants made investment decisions purely based on the initial capital expenses before 1970 (A1-Hajj and Homer, 1998). Several investigations, including those by Egan (1998) and Latham (1994), have emphasized the need to consider the long-term cost of a project. The current rule for projects financed under the Private Finance Initiative (PFI) route allows the use of LCC methodologies precisely because they assess project long-term costs (Jones, 2000). Hence, a comprehensive framework for analysing the long-term cost of a construction project and for taking appropriate decisions is the need of the day.

Sustainable development has gained appeal in several areas, including the building industry, over the previous few decades. However, because of the globalization of the business environment, demographic change, environmental sustainability, climate change, new materials and technologies, and governance and regulation, the practice of sustainability in the construction industry continues to face challenges. It is necessary to increase economic value while utilising natural resources in a sustainable manner and contributing to society's social purposes and objectives for the business sector to contribute to sustainable development.

Despite the growing importance of the sustainability concept in infrastructure development, many infrastructure owners are still concerned about financial constraints due to the large capital requirements for project development. Understandably, cost estimation is an important aspect of the development process and a significant component of the decision-making toolset. According to List (2007), life-cycle cost analysis (LCCA) aids in the achievement of these goals. Decision-makers may use LCCA to compare competing proposals and determine the best sustainable growth route for the shared infrastructure.

Life Cycle Costing (LCC) technique which has 50 years of history, is able to provide a very good estimate of the total cost of ownership in construction assets. In spite of its proven utility, its successful application and diffusion within the construction domain is scarcely seen (Manewa et al. 2021). According to Olubodun et al (2010), only 50% projects have implemented the life cycle costing approach in the construction industry in UK. Moreover, the authors observed that this technique is not clearly known to the construction professionals and there is lack of a standard methodology for implementation. A study conducted in Nigeria highlights the fact that the data on factors affecting applicability of LCA approach in construction, particularly in developing countries, is limited

(Opawole, et al. 2020). It is also observed that non-availability of reliable data and standardized approach for LCC implementation in construction industry further inhibit the adoption of this model (Hingham et al. 2015). A study conducted for Indian residential projects reveals critical barriers in implementation and acceptance of LCC approach (Guru and Mohibullah, 2021). This clearly indicates the necessity to understand the current status of LCC implementation for construction and infrastructure projects in India. The literature review points out that there is limited data availability for implementation of LCC methodology in construction. Hence, it is necessary to assess the level of knowledge amongst construction professionals regarding the relative importance of various LCC factors. This analysis would help in prioritizing the LCC factors and further utilize their weightages in formulating a standard LCC framework. The present work attempts to check the level of awareness regarding LCC methodology in construction professionals. It further attempts to identify the critical cost factors being considered currently for implementation of LCC methodology in the building sector. A questionnaire survey was designed and circulated amongst construction industry stakeholders. The response was analysed and weightages were calculated based on RII technique. Moreover, the cost factors were ranked to select the most critical ones from the list.

2. Research Methodology

The following is an overview of the research methodology: The cost factors for life cycle costing were extracted from an extensive literature review and informal discussions with stakeholders, and grouped into seven sub-categories. A questionnaire to capture the perspective of stakeholders regarding LCC methodology awareness and implementation was created. Another questionnaire was developed to select critical LCC parameters from a pool of cost factors. Both the questionnaires were sent to stakeholders working in the construction sector at various hierarchical levels. Table: 1 and Table: 2 present the LCC factor list considered for ranking and RII values obtained regarding the awareness level respectively.

Following the collection of responses, a Relative importance index value was calculated.

The formula used for calculating the **Relative importance index** is given below:

$$RII = \frac{\Sigma W}{(A*N)} \qquad 0 \leq RII \leq 1 \tag{1}$$

W = weighting given to each factor by the respondents.

A = Highest weight

N = Total no. of respondents

Table 1. Life cycle costing parameters.

Criteria	Sub Criteria	Parameters
Life Cycle Costing	Initial cost	Land acquisition & associated cost Design team fees & associated cost Construction cost
	Operation & maintenance, Repair and Replacement cost	Operation & Maintenance Cost Repair cost Replacement cost
	Energy & water cost	Energy Consumption Cost Gas consumption Cost Water usage cost
	Environmental impact cost	Green construction measurement cost Virtual abetment Cost (Air Pollution, Noise pollution etc) Environmental degradation cost
	Waste management cost	waste disposal cost waste segregation cost

Residual/ Salvage value	Salvage value Disposal/Demolition cost
Other cost	Insurance cost Interest rate cost Taxes

3. Results and Discussion

According to the results of the survey, 76.9% of those who replied work as contractors, while 15.4% work as real estate developers.

Table 2. RII of awareness about LCC.

Questions	RII	Rank
Do you take into account building initial Investment cost during planning stage?	0.82857	1
Do you take into account building O&M, Repair and Replacement cost during planning stage?	0.71429	2
Do you take into account building occupancy cost during planning stage?	0.71429	2
Do you take into account building end of the life/disposal cost during planning stage?	0.4	4

Table 3. Parameter Ranking.

Questions	RII	Rank
Land Acquisition and associated cost	0.91429	2
Design team fees and associated cost	0.87143	3
Construction cost	0.95714	1
Operation & Maintenance Cost	0.85714	4
Repair cost (major and minor repair cost)	0.72857	11
Replacement cost (major and minor replacement cost)	0.65714	15
Energy Consumption Cost	0.82857	6
Gas consumption Cost	0.78571	7
Water usage cost	0.78571	7
Green construction measurement cost	0.67143	14
Virtual abetment Cost (Air Pollution, Noise pollution etc.)	0.6	16
Environmental degradation cost	0.58571	17
Waste disposal cost	0.7	12
Waste segregation cost	0.68571	13
Salvage value	0.54286	19
Disposal/Demolition cost	0.58571	17
Insurance cost	0.77143	9
Interest rate cost	0.75714	10
Taxes	0.85714	4

The analysis of survey results regarding LCC awareness levels shows that construction and land acquisition costs are the ones which are majorly being considered currently. Professionals are now including O&M as well as Design team costs in the initial stages. However, there is lack of awareness regarding costs incurred for environmental impact abatement and end of life disposal.
Construction costs, land acquisition and related costs, Design team fees and associated costs, Operation & Maintenance Costs, and Taxes are the top five parameters according to the RII values obtained from data analysis of LCC parameters, as shown in Table 3. It can also be observed that costs regarding environmental impact, degradation as well as waste management occupy lower positions in the ranking.
The responses revealed that when undertaking life cycle costing, stakeholders almost always evaluate initial costs and seldom consider end-of-life costs. This indicates that the initial cost is still given a higher priority as compared to other LCC parameters. Nevertheless, the high RII values of O&M as well as building occupancy cost reveal that the other LCC factors have started gaining importance and indicate a positive trend towards adoption of LCC methodology in near future.

4. Conclusion.

The advantage of LCC is its capacity to produce more comprehensive and accurate cost projections, since the relevance of considering operation and maintenance costs rather than capital expenditures throughout the life of an asset is becoming increasingly apparent. As a result, life cycle costing of a building must be incorporated at the conceptual design evaluation stage. The purpose of this study was to assess the awareness levels of LCC methodology and its implementation among construction industry experts in India. Professional responses were collected using questionnaires, and data was analysed using RII (Relative importance index).
As can be observed from the responses, the cost focus still revolves around initial cost. However, there is increasing trend of considering the operational costs for evaluation of alternatives. Further insights into the LCC approach and implementation were captured through informal communications with some of the construction industry professionals. The respondents strongly agreed that operation and maintenance costs, as well as other life cycle costs should be incorporated into LCC and should be considered at the planning stage itself. Respondents also agreed that LCC serves as a management and decision-making tool and is suitable for calculating the total costs of facilities.
However, the data analysis regarding awareness of LCC factors reveals that the cost factors representing waste, pollution abatement and environmental degradation occupy lower positions in ranking. This clearly depicts that sustainability factors are seldom being considered for life cycle costing. Since this analysis was done by capturing stakeholders' perspectives, currently working in the construction sector, there is still a strong need to raise the awareness level regarding sustainability aspects in construction sector. It needs to be emphasized that a comprehensive cost estimate for a project can only be obtained by incorporating sustainability aspects. While there is a marginal increase in today's expenditure owing to the inclusion of environmental cost factors, it will provide substantial savings in the long run. The analysis of LCC factors has provided the data regarding current status of factor wise awareness levels and their inclusion in costing. It is observed that the estimates are still based on construction, O&M, land acquisition and design team costs. The factors like waste management, environment degradation and abatement costs, end of life cost are neglected. These costs ultimately affect the sustainability of the project throughout its design life. These factors further affect the liveability, building occupancy costs, health & well-being of the residents and the impact of the

structure on the surrounding environment. Moreover, if such sustainable measures are adopted, it would improve the social perception along with the additional tangible benefits in terms of lower O&M as well as water and energy costs and tax concessions and rebates. The LCC factor rankings would also help in developing a framework for effective implementation of LCC in current projects by suitably simplifying the process. The LCC factors can be incorporated in the framework successively based on the weightages. This would result in an integrated framework which can be practically implemented in the planning stage. Thus, adoption of LCC framework considering all important cost parameters at planning stage would substantially raise value for money as well as significantly increase sustainability in construction projects.

References

[1] Aghili, N., Hosseini, S. E., Bin Mohammed, A. H., & Zainul Abidin, N. (2019). Management criteria for green building in Malaysia; relative important index. *Energy Sources, Part A: Recovery, Utilization, and Environmental Effects*, *41*(21), 2601-2615.

[2] Alshamrani, O. (2012). *Evaluation of school buildings using sustainability measures and life-cycle costing technique* (Doctoral dissertation, Concordia University).

[3] Altaf, M., Alalaoul, W. S., Musarat, M. A., Hussain, A., Saad, S., Rabbani, M. B. A., & Ammad, S. (2022). Evaluating the awareness and implementation level of LCCA in the construction industry of Malaysia. *Ain Shams Engineering Journal*, *13*(5), 101686.

[4] Ammar, M., Zayed, T., & Moselhi, O. (2013). Fuzzy-based life-cycle cost model for decision making under subjectivity. *Journal of Construction Engineering and Management*, *139*(5), 556-563.

[5] Atia, N. G., Bassily, M. A., & Elamer, A. A. (2020). Do life-cycle costing and assessment integration support decision-making towards sustainable development?. *Journal of Cleaner Production*, *267*, 122056.

[6] Goh, B. H., & Sun, Y. (2016). The development of life-cycle costing for buildings. *Building Research & Information*, *44*(3), 319-333.

[7] Guru, A.; and Mohibullah (2021). Exploring the Acceptance of Life Cycle Cost for Residential Projects in India. In *Recent Developments in Sustainable Infrastructure* (pp. 855-863). Springer, Singapore.

[8] Heralova, R. S. (2017). Life cycle costing as an important contribution to feasibility study in construction projects. *Procedia engineering*, *196*, 565-570.

[9] Higham, A., Fortune, C., & James, H. (2015). "Life cycle costing: evaluating its use in UK practice", Structural Survey, Vol. 33 (1), pp. 73-87. DOI 10.1108/SS-06-2014-0026

[10] Manewa, A., Siriwardena, M. and Wijekoon, C. 2021. Life cycle costing in construction: Current trends and emerging directions. In: Sandanayake, Y.G., Gunatilake, S. and Waidyasekara, K.G.A.S. (eds). Proceedings of the 9th World Construction Symposium, 9-10 July 2021, Sri Lanka. [Online]. pp. 403-412. DOI: https://doi.org/10.31705/WCS.2021.35.

[11] Oduyemi, O. I. (2021). *Life cycle costing methodology for sustainable commercial office buildings*. University of Derby (United Kingdom).

[12] Oduyemi, O., Okoroh, M., & Dean, A. (2015). Developing an artificial neural network model for life cycle costing in buildings. *Management*, *843*, 852.

[13] Olubodun, F., Kangwa, J., Oladapo, A. and Thompson, J. (2010), "An appraisal of the level of application of life cycle costing within the construction industry in the UK", Structural Survey, Vol. 28 No. 4, pp. 254-265. https://doi.org/10.1108/02630801011070966

[14] Opawole, A., Babatunde, S.O., Kajimo-Shakantu, K. and Ateji, O.A. (2020), "Analysis of barriers to the application of life cycle costing in building projects in developing countries: A case of Nigeria", Smart and Sustainable Built Environment, Vol. 9 No. 4, pp. 503-521. https://doi.org/10.1108/SASBE-11-2018-0057

[15] Rahim, F. A., Muzaffar, S. A., Yusoff, N. M., Zainon, N., & Wang, C. (2014). Sustainable construction through life cycle costing. *Journal of building performance*, *5*(1).
[16] Sayed, S. S. M., & Sawant, P. H. (2015). Life-cycle cost and financial analysis of energy components in mass housing projects–A case project in sub-urban India. *International Journal of Sustainable Built Environment*, *4*(2), 202-221.

Urban Organic Waste as a Source of Bioenergy for Electricity Generation in Bhopal, Madhya Pradesh

Nirjhar Gupta[1], Jitendra Patel[2], Nitesh Gupta[3] and Amit Vishwakarma[4]

[1]Ph.D. Research Scholar, Department of Civil Engineering, UIT RGPV, Bhopal (M.P), 462033 India
[2]Post Graduate Student, Department of Civil Engineering, SATI Vidisha (M.P), 464001 India
[3]Assistant Professor, Department of Civil Engineering, Institute of Technology, Nirma University, Ahmedabad (Gujarat), 382481 India
[4]Professor, Department of Civil Engineering, UIT RGPV, Bhopal (M.P), 462033 India

Email id. nirjhargupta070@gmail.com

Abstract. Municipal Solid Waste Management (MSWM) is essential as it impacts health, the environment, and aesthetic society if it is not managed correctly. Hence improving the quality and standard of living. There is an urgent requirement to neutralize the waste and the best possible solution for organic waste. It can be used to generate biofuels and utilized to generate electricity. The production of energy from waste depends on organic waste, so the energy potential of Bhopal city has been categorized into different sectors such as Urban Organic Solid Waste (UOSW), Pulp & Paper, Cattle, and Poultry, in which UOSW is 42% and, the Urban Organic Liquid Waste (UOLW) 25%, Cattle 23%, Poultry 2% and Pulp & Paper 8%. The UOSW has been analysed, and it has been categorized in Slaug (1%), Organic Solid waste (33%), Poultry (3%), and Cattle (63%), and the UOLW in Slaug (1%), UOLW (98.9%), and paper and pulp (0.10%). Total waste generated in Bhopal per day is 1123 TPD of which, 54.3% is organic waste that we can use to generate biofuel, i.e., 609 TPD, which can generate 165MW energy per day.

1. Introduction

Communities used to live in peace with nature in ancient times, and hence the formation of Solid Waste (SW) was minimal. At the same time, people's lifestyles have evolved dramatically during the past several decades worldwide. For our comfort, we are plundering natural resources. Consequently, the volume and pattern of material and commodities consumption have expanded dramatically, thus affecting and increasing the creation of SW [1]. Furthermore, unbelievable population expansion, and unrestrained, and unmonitored urbanization have boosted waste creation by many orders of magnitude in the past several decades. Various human activities produce SW at the residential, commercial, agricultural, and industrial levels, both organic and inorganic. SW can be divided into several categories, such as hazardous waste, biological waste, municipal solid waste (MSW), and electronic waste, according to the materials that make up the waste and the location where it was generated. If solid waste is not correctly handled or disposed of, it may cause significant pollution. Unlike affluent nations, emerging countries are experiencing faster urbanization and economic expansion, making them more vulnerable to SW concerns. MSW is a substantial component of SW [2–4]. MSW has different definitions in different countries. Hence it differs from nation to country. In India, however, it is often referred to as waste or trash, and it is generated by houses, workplaces, hotels, restaurants, marketplaces, commercial buildings, and institutions, among other places [5]. Packaging the durable and non-durable items, yard garbage, and solid and semi-solid kitchen waste make up most of it, as illustrated in Figure 1.

A rigorous bottom-up analysis of the global availability of sustainability for biogas and biomethane revealed that the technological potential to create these gases is enormous and virtually unexplored. Agriculture, animal, and MSW residues for direct biomethane [3]. Only a small fraction of the whole potential was realised in 2018, when the total output of biogas and biomethane was approximately 35 million tonnes of oil equivalent (mton). If the sustainable potential were fully utilised, it would be able to cover around 20 percent of the world's gas consumption in the present day [6,7]. Recovery of energy from municipal, industrial, and agricultural wastes or residues is the goal of a Waste to Energy (WtE) Program. This energy can be recovered in the form of biogas, bio-compressed natural gas (CNG), or power depending on when it is needed. In India, the WtE program, through a grant of Central Financial Assistance, supports the Sustainable Alternative Towards A- portable Transportation (SATAT) initiative of the Ministry of Petroleum and Natural Gas, wherein it is envisaged to set up 5000 CBG plants by 2023-24 with a production target of 15 MMT of Bio-CNG. It will facilitate new employment opportunities and enhance farmers' income, thereby helping realize the vision of "Annadaata se Urjadaata" and invigorate the rural economy. The WtE program is also aligned with the Swachh Bharat Mission by promoting Municipal Solid Waste (MSW) to power projects. Under the Waste to Energy program, as of 31st Jan 2021, projects with a cumulative capacity of 74.7 MW power from Municipal Solid Waste (MSW) have been given in-principal approval by the Ministry for grant of Central Financial Assistance (CFA). These projects are estimated to utilize around 6000 Tons/day of MSW to generate electricity (CPCB report 2022) [8,9]. To make this program effective and well implemented, we have carried out the research in our local area to analysis the potential of Bhopal city to generate Biofuel for future [10–12]. Madhya Pradesh is a state in India with a population of 73 million people and 9.5 percent of the total land area of the country is contained within this region (308,000 sq. Km.). The state's urban population, spread among 378 ULBs, rose by 2.2 percent from 16 million in 2001 to 21 million in 2011 [13–15]. The increasing pace of urbanization has resulted in a greater emphasis concerning the delivery of municipal services and urban infrastructure, which has raised the need for investment in urban development. However, as a consequence of increasing rate of urbanisation, enormous quantities of municipal solid waste are being produced in both large cities and small villages. As a result, to enhance environmental conditions and make cities livable for all citizens, each city and town in the state must prepare a time-targeted action plan for managing Municipal Solid Waste (MSW) in line with population developmental growth.

As a fast-rising country with a big population, India has numerous environmental concerns, including a pressing need to enhance both living standards and ecosystems. MSW management is one of the most serious issues threatening India's environmental quality and urban development [16–18]. MSW created millions of tonnes of waste in Madhya Pradesh over a 20-year period (2001-2020). The incorrect and wasteful disposal of MSW has caused serious environmental problems. Traditional methods, such as landfill, are confronting a land scarcity dilemma. At the same time, waste to energy (WtE) technologies like incineration and gasification have shown to be successful around the world. As a result, an MSW management system with the added benefit of recovering energy from a waste source is a safe alternative with the potential to solve Madhya Pradesh's MSW disposal problem. WTE is gaining popularity in India, owing to its potential to reduce the amount of MSW that must be disposed of in landfills.

The research work aims to emphasize the municipal solid waste specially generated in the Bhopal, city of Madhya Pradesh to develop a comprehensive, integrated, sustainable, environmentally friendly, and implementable municipal solid waste. The best possible solution to encounter the MSW is to produce a Bioenergy from it. This study examines the amount of municipal solid garbage created per inhabitant per day in ULBs as well as the different forms of waste. for various sources and types of waste developed to encourage the establishment of projects for the recovery of energy from urban, industrial, and agricultural waste in the form of Biogas/ Bio-CNG/ Power, as well as captive power and thermal utilization in industries through gasification. The survey of selected site has been conducted in some prime-area of Bhopal to know the volume and type of waste generated with the help of Central Pollution Control Board (CPCB) data. The established methodology has been analysed, and the potential of electricity generated from MSW is calculated by using gasifier mass balance method.

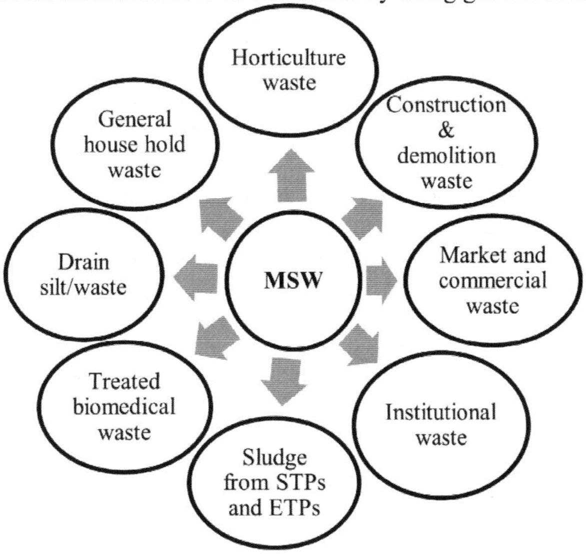

Figure 1. A schematic of MSW generated from different sector.

2. Study area

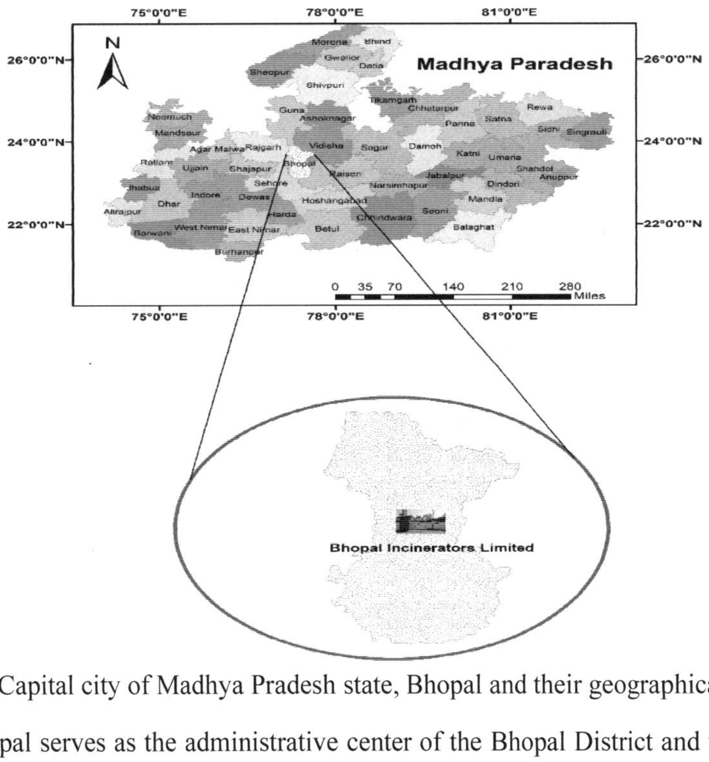

Figure 2. Capital city of Madhya Pradesh state, Bhopal and their geographical locations.

The city of Bhopal serves as the administrative center of the Bhopal District and the Bhopal Division in addition to being a city located in the Madhya Pradesh region of India that serves as the state capital. The name "Bhopal" literally translates to "City of Lakes". It is the 131st largest city in the

globe and ranks 17th among the major cities in the Country. The city is situated in India's central region, with a latitude of 23°16' and an east longitude of 77°36'. Bhopal city's administration is handled under the Bhopal Municipal Council (BMC), and the total land area that is managed by the BMC is 413 square meters. Old Bhopal and New Bhopal are the two distinct sections that make up the city.
On average, there are around 6956 people crammed onto each square kilometer of land in the city of Bhopal. However, the town's inhabitants are not dispersed in a uniform manner throughout the town. Figure 2 is the selected study area with which bio-based plants only, i.e., Bhopal Incineration Limited.

3. Data collection and analysis

Essential needs for the development of strategies for the management of solid waste include information on the types of wastes, as well as their composition, physical and chemical features, and the amounts of trash that are produced [19,20]. In order to effectively manage solid waste, it is required to investigate the characteristics of the waste material, in addition to determining where the trash came from. As a result, we can put them into one of the following categories:

- Biological agents like bacteria are able to break down biodegradable organic substances into their component parts. Examples include food waste, waste from fruits and vegetables, waste from gardens (also known as plant waste), and so on.
- Plastic, paper, and metals are all examples of recyclable materials.
- Combustibles are defined as materials that are relatively dry and have a high calorific value. Examples of combustibles include paper, plastic, rags, cardboard, and other similar items.
- The term "hazardous waste" refers to a collection of materials that, due to their chemical or pathogenic composition, pose a threat to human or animal health as well as to the environment. Some examples of hazardous trash include waste from hospitals and particular industrial waste.
- Dust, cinder, grit, and other forms of debris are all examples of inert debris.

During the process of collecting samples of municipal solid trash, primary collection sites are determined, which allows for coverage of a broader population. The study area is divided into different types, such as residential, commercial, and industrial, as well as market and slum areas, and sampling stations are spread out evenly across the entire region. There is a consistent distribution of sampling locations over the whole study area. The following is a list of the seven sampling places that are being considered: AIIMS Zone, Badbada, Beraghat, Bhanpura Kunti, Danapani, DIG Bungalow, Kolar Road and Sehore road.

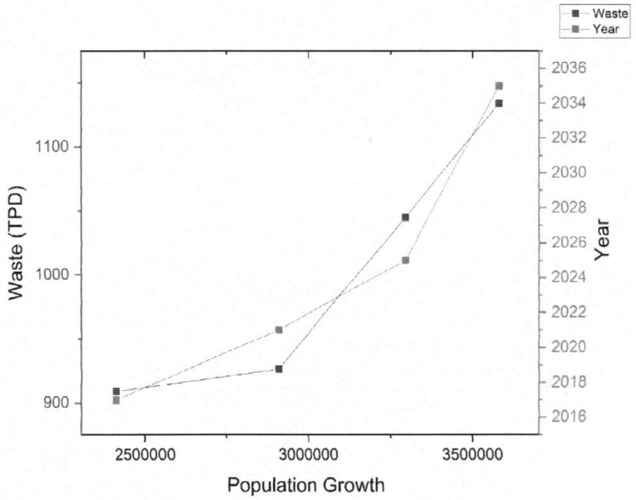

Figure 3. Relationship of generated waste with growth of the population.

Both the residential and the business regions contribute a large amount to the generation of solid trash. Table 1 presents the per capita waste generation rates specified in the CPHEEO manual. This table can be found following the previous one. The population growth in the future and the waste degeneration have been projected to 1100 (TPD) in the year 2035. Figure 3 illustrates the projected numbers i.e., relationship of waste generated with respect to the growth of the population.

Table 1. Waste generation rates per capita.

Population range (in million)	Average per capita (kg/capita/day)
01 to 0.5	0.23
0.5 to 1.0	0.27
1.0 to 2.0	0.28
2.0 to 5.0	0.37
>5.5	0.56

3.1 Sectoral share analysis of energy

The production of energy from waste depends on organic waste, so the energy potential of Bhopal city has been categorized into different sectors such as Urban Organic Solid Waste, Urban Organic Solid Waste, Pulp & Paper, Cattle, and Poultry, as demonstrated in Figure 4. in which Urban Organic Solid Waste is 42%, Urban Organic Liquid Waste 25%, Cattle 23%, Poultry 2% and Pulp & Paper 8%.

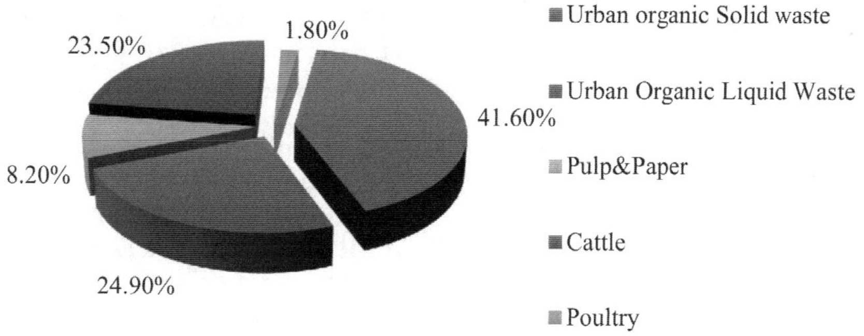

Figure 4. Energy potential share from various sectors in Bhopal.

3.2 Sectoral share in solid waste generation and liquid waste generation

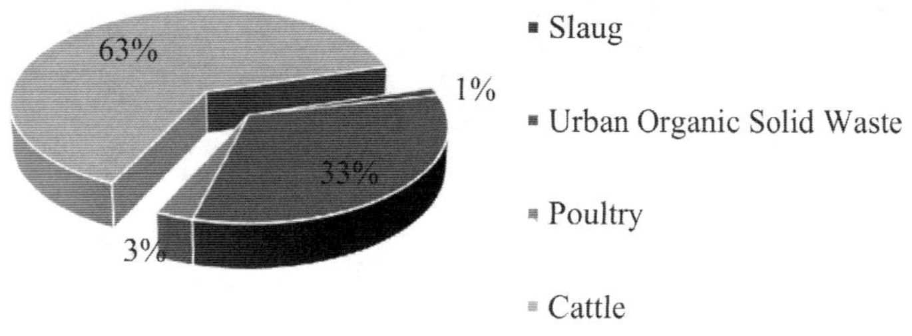

Figure 5. Solid Waste Generation share from various sectors.

Sectoral share of solid waste generation has been analysed. It has been categorized in Slaug (1%), Organic Solid waste (33%), Poultry 3%, and Cattle (63%), as illustrated in Figure 5, and the Liquid

Waste Generation has been categorized as Slaug (1%), Urban organic Liquid waste (99%), and Pulp and Paper (0.10% approx. zero) as illustrated in Figure 6.

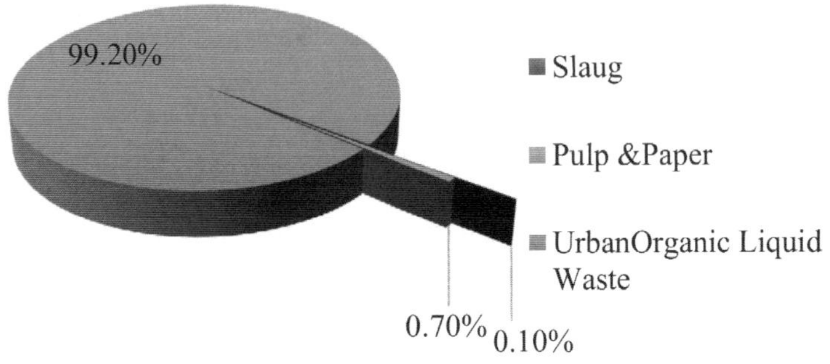

Figure 6. Liquid Waste Generation from different sectors.

3.3 Sector-wise utilizations of energy

Biogas was utilized for electricity and heat around two-thirds of the time in 2018 (with about equal percentages of electric-only and co-generation plants). The remaining 30% was converted to biomethane and mixed into gas networks or utilized as a transportation fuel. With the balance being restored to biomethane and incorporated into gas networks. Transitioning from technology-neutral renewable energy auctions (such as power purchase agreements) to technology-neutral feed-in tariffs might restrict future chances for electricity-only biogas projects in several nations. Unlike wind and solar P.V., biogas plants may operate flexibly and provide balancing and other grid services. Recognizing these benefits might help biogas plants gain appeal in the future. Co-generation with biogas is more cost-effective than an electricity-only generation if local heat off-take is available. This is because co-generation may use up to 35% biogas energy to create electricity and up to 40% waste heat to generate electricity. Wet waste with a high organic content is produced by several industrial subsectors, such as food and beverage and chemicals, and is ideal for anaerobic digestion. Biogas may be utilized as a waste treatment method and a heat and energy source on-site. Only a small fraction of worldwide biogas is converted to biomethane. This business has a lot of room for expansion, but it depends on the strength and design of decarbonization programs in different world regions.

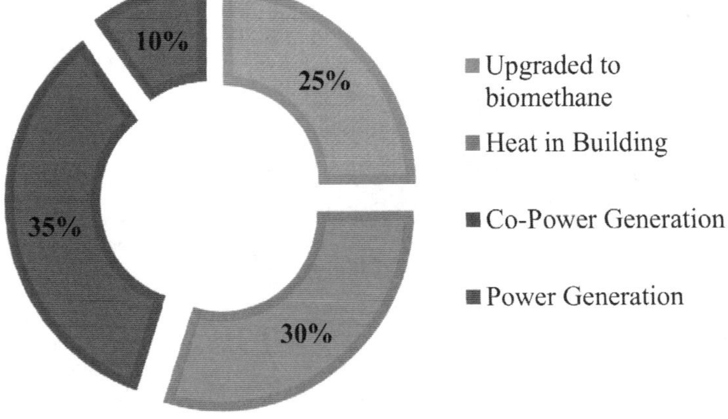

Figure 7. Utilization of energy from different sector.

4. Results and discussion

Approximately, 18 percent of the raw gas produced by the gasifier is composed of carbon monoxide, whereas only 6 percent is composed of carbon dioxide (30 percent moisture in the MSW after drying). The feedstock, once dried, has a moisture content that is equal to 30 percent by weight, and the quantity of oxygen that is provided is equal to 0.3 mol/mol of dry waste. The final feedstock analysis is (C: 54.8, H: 8, O: 33.4, N: 3.8) [10].

The final waste analysis in Bhopal is (C: 55.75, H: 7.54, O: 34.57, N: 1.87, S: 0.27), which is quite comparable. The mass balancing is carried out under the same conditions (after drying, the moisture content is 30% w/w). Given the feedstock and moisture content homogeneity, CO and CO_2 percentages in the raw gas are estimated to be around 18% and 6%, respectively. Carbon balance is employed before and after the gasifier to compute mass balance.

Using 1 kg of waste, the reaction before and after the gasifier may be represented in the equation (1) below, with all units being mole:

$$CXHYOZ + aH_2O + bO_2 + 3.71bN_2 = n1H_2O + n2H_2 + n3N_2 + n4CO + n5CO \quad (1)$$

According to the ultimate analysis and the values of the elements (C = 46.46, H = 75.4, and O = 21.6), the moles X, Y, and Z in the equation above represent the elements that may be found in MSW.

- a, equal to the number of moles of water in 1 kg of municipal solid waste. Since the moisture content is 30 percent, a = 1000 g * 30 percent / 18 g/mol = 16.67 moles.
- b, is the number of moles of O_2 from air that were added during the process. You can think of air as being made up of 78 percent nitrogen and 21 percent oxygen. So, 3.71 billion moles of N_2 from the air are also added to the process.

Following the reaction, the moles of different gas compositions are denoted by the numbers n1, n2, n3, n4, and n5, respectively.

Before reaction: X = 46.46

n4 + n5 = 46.46 after reaction because the CO and CO_2 percentages in the raw gas are 18 and 6%, respectively, n4: n5 = 3: 1. The numbers n4 = 34.845 and n5 = 11.615, respectively, can be calculated. After the reaction, the amount of oxygen produced by CO and CO_2 is 46.46 * 3 / 4 + 46.46 * 1 / 4 *2 = 58.075 moles.

4.1 Nitrogen balance

Before reaction: 3.71b * 2 = 7.42b

After the reaction, as CO and CO_2 account for 24% of the raw gas, the remaining gases account for 76% of the raw gas, implying that moles from the remaining gases equal (moles of CO + moles of CO_2) / 24% * 76% i.e., 46.46 / 24 % *76 % = 147.12 moles.

As a result, the nitrogen amount after reaction is 2 * (147.12 – n1 –n2). We derive b = (147.12 - n1 - n2) / 3.71 using nitrogen balance.

4.2 Oxygen balance:

Before the reaction: 21.6 + a + 2 * (147.12 - n1 - n2) / 3.71

After the reaction: n1 + 58.075, based on oxygen balance, we can say: 220.77 = 5.71n1 + 2n2

4.3 Hydrogen balance:

75.4 + 2a = 108.74 before reaction 2n1 + 2n2 after response

We get the following results based on hydrogen balance: 108.74 = 2n1 + 2n2

The gas composition of the raw gas is presented in Table 2 by solving two linear equations in two unknowns from oxygen and hydrogen balance:

Table 2. Gas compositions and volumes after gasifier.

Gas	Vol %
CO_2	6
N_2	47.91
CO	18
H_2	12.49
H_2O	15.90

Total waste generated in Bhopal per day is 1123 TPD, out of which 54.3% is organic waste that can be used to generate Biomethane Energy, i.e., 609 TPD

The plant's capacity is 609 TPD, 25.37 tons per hour.

Dulong Formula can be used to compute the MSW's high heating value (HHV).

HHV in MJ/kg = $33.86 \times C + 144.4 \times (H - O / 8) + 9.428 \times S$

The mass fractions determined through ultimate Analysis are C, H, O, and S.

The HHV of the MSW in Bhopal, according to the preceding Eq. and the final analysis, is 23.55 MJ/kg.

As a result, the energy input from MSW can be computed as follows:

Energy input = 23550 * 103 J/kg * 25.37* 103 kg / 3600 S = **165 MW**

Bhopal generates 1123 TPD of total trash per day, of which 54.3 percent is organic waste that may be used to generate biomethane energy, i.e., 609 TPD energy input from MSW can be generated at 165MW per day.

5. Conclusion

The following conclusion has been drawn for this study.

- The production of energy from waste depends on organic waste, so the energy potential of Bhopal city has been categorized into different sectors such as Urban Organic Solid Waste, Urban Organic Liquid Waste, Pulp & Paper, Cattle, and Poultry in which Urban Organic Solid Waste is 42%, Urban Organic Liquid Waste 25%, Cattle 23%, Poultry 2% and Pulp& Paper 8%.
- Sectoral share of solid waste generation has been analyzed. It has been categorized in Slaug (1%), Organic Solid Waste (33%), Poultry (3%), and Cattle (63%), and the Liquid Waste Generation has been categorized in Slaug (1%), Urban Organic Liquid Waste (98.9%) and Pulp and Paper (0.10% approx. zero).
- Total waste generated in Bhopal per day is 1123 TPD, out of which 54.3% is organic waste that can be used to generate biomethane energy, i.e., 609 TPD energy input from the MSW can be generated at 165MW per day.

References

[1] Anon 2019 Solid Waste Management in Urban India: Imperatives for Improvement *J. Contemp. Issues Bus. Gov.* **25**

[2] Parihar R S, Ahmed S, Baredar P and Sharma A 2017 Characterisation and management of municipal solid waste in Bhopal, Madhya Pradesh, India *Proc. Inst. Civ. Eng. Waste Resour. Manag.* **170** 95–106

[3] Kumar A and Agrawal A 2020 Recent trends in solid waste management status, challenges, and potential for the future Indian cities – A review *Curr. Res. Environ. Sustain.* **2** 100011

[4] Patel M L and Jain R 2010 A Glance at the World *Waste Manag.* **30** 2396–400

[5] Schneider S H (Centra. P H and E E O (CPHEEO)) 2017 Municipal solid waste management manual *Energy Conversion, Second Ed.* 73–9

[6] United Nations, Change C, Report A and 2020 2021 United Nations Change, Climate Report, Annual 2020

[7] Kumar M, Kumar S and Singh S . 2021 Waste Management By Waste to Energy Initiatives in India *Int. J. Sustain. Energy Environ. Res.* **10** 58–68

[8] Rong L, Zhang C, Jin D and Dai Z 2017 Assessment of the potential utilization of municipal solid waste from a closed irregular landfill *J. Clean. Prod.* **142** 413–9

[9] Cole-Hunter T, Johnston F H, Marks G B, Morawska L, Morgan G G, Overs M, Porta-Cubas A and Cowie C T 2020 The health impacts of waste-to-energy emissions: A systematic review of the literature *Environ. Res. Lett.* **15**

[10] CPHEEO-Part I 2016 CPHEEO (Central Public Health and Environmental Engineering Organization), Swachh Bharat Mission- Municipal Solid Waste Management Manual Part I: An Overview *Minist. Urban Dev.*

[11] Planning Commission 2014 Report of the Task Force on Waste to Energy: Volume I prepared by Planning Commission (PC) *Task Force Waste to Energy* **I** 1–178
[12] Thakur A, Kumari S, Sinai Borker S, Prashant S P, Kumar A and Kumar R 2021 Solid Waste Management in Indian Himalayan Region: Current Scenario, Resource Recovery, and Way Forward for Sustainable Development *Front. Energy Res.* **9** 1–18
[13] Katiyar R B, Suresh S and Sharma A K 2013 Characterisation of municipal solid waste generated by the city of Bhopal, India *Int. J. ChemTech Res.* **5** 623–8
[14] Scribe C D and We C D 2011 Lecture 4 Lecture 4 **1** 1–13
[15] Pfadt-Trilling A R, Volk T A and Fortier M O P 2021 Climate Change Impacts of Electricity Generated at a Waste-to-Energy Facility *Environ. Sci. Technol.* **55** 1436–45
[16] Millati R, Cahyono R B, Ariyanto T, Azzahrani I N, Putri R U and Taherzadeh M J 2019 *Agricultural, industrial, municipal, and forest wastes: An overview* (Elsevier B.V.)
[17] Muthudineshkumar R and Anand R 2019 *Anaerobic digestion of various feedstocks for second-generation biofuel production* (Elsevier Ltd.)
[18] Sangeetha T, Rajneesh C P and Yan W M 2020 *Integration of microbial electrolysis cells with anaerobic digestion to treat beer industry wastewater* (INC).
[19] Awasthi M K, Zhao J, Soundari P G, Kumar S, Chen H, Awasthi S K, Duan Y, Liu T, Pandey A and Zhang Z 2019 *Sustainable management of solid waste* (Elsevier B.V.).
[20] Szilágyi Á, Bodor A, Tolvai N, Kovács K L, Bodai L, Wirth R, Bagi Z, Szepesi Á, Markó V, Kakuk B, Bounedjoum N and Rákhely G 2021 A comparative analysis of biogas production from tomato bio-waste in mesophilic batch and continuous anaerobic digestion systems *PLoS One* **16** 1–17.

Experimental Study on Strength Behaviour of Geogrid Reinforced Subgrade Soil

Sateesh Pisini, Swetha Thammadi, Monish Nadan, Naadirah Edward, Shaniza Nisha

Department of Civil Engineering, Fiji National University, Fiji

Email: sateesh.pisini@fnu.ac.fj

Abstract. This paper presents the summary of experimental program to evaluate the potential of KOBM (Klockner Oxygen Blown Maxhutte) slag to stabilize the geogrid-reinforced subgrade soil in Suva, Fiji. This KOBM slag is a by-product of steel manufacturing process in New Zealand. The standard California bearing ratio (CBR) laboratory tests were performed on the soil samples with various combinations of geogrid and KOBM slag, along with other geotechnical tests to determine the strength behaviour of subgrade soil. The laboratory experiments were undertaken to investigate the impact of using KOBM slag (0%, 5%, 10%, 15%, 20%) with single and double layers of geogrid reinforcement placed at H/2 and H/4 (where H is height of CBR mould) in the soil sample. The effect of geogrid reinforcement with respect to the position and number of layers with varying contents of KOBM on the strength behaviour of subgrade soil has been examined. Results of the current study indicate that there is a significant increase in CBR values for subgrade soil stabilized with KOBM and reinforced with geogrid. Maximum CBR value was observed in the soil sample reinforced with single geogrid layer placed at H/2 with 20% KOBM slag. This is the best outcome amongst all the combinations of KOBM, position/no. of layers of geogrid reinforcement in the current experimental study consisting of 50 experiments.

1. Introduction

The strength of weak subgrade soil for paved and unpaved roads can be improved using geosynthetics. Although various mechanical and chemical methods for soil stabilization are available, geosynthetics have demonstrated to be an effective means of soil improvement. Giroud and Noiray [1], established an analytical treatment for geotextiles that can be used as reinforcement. Geosynthetics can be incorporated in pavement design to enhance tensile strength and stabilize weak soil. For separation and stabilization of weak soil, geogrids and geotextiles reduce the aggregate thickness utilized in pavement design while increasing durability. Numerous studies demonstrated the benefits of using geosynthetic reinforcement for subgrades in unpaved roads to enhance durability and increase the load bearing capacity of weak soil in a cost effective manner [2]. On granular soil, CBR tests were carried out using geotextiles and geogrid. The impact of geogrid and geotextile on unpaved roads was investigated using soft subgrade clayey soil with low to medium compressibility. Numerous field and laboratory tests illustrated that the geosynthetics used to reinforce unpaved roads on soft soil reduces rut depth, increases the load bearing capacity, service life increases as well as reduces the amount of fill material needed [1]. The thickness of the pavement is proportional to the amount of aggerated used. Palmeira and Antunes [2] demonstrated that geosynthetic reinforcement reduces the cost of maintenance of unpaved road thus lowering the overall cost of road construction. Since the exploitation of natural/traditional construction materials is restrained by environmental regulations, alternatively using geogrid as reinforcement is a sustainable approach in road construction and other civil engineering projects [2, 3].

The strength of the subgrade is determined using the California Bearing Ratio (CBR) test, which at specific penetration, measures the ratio of test load to standard load by a standard plunger. The test is frequently used to determine the mechanical strength of materials used in roadway construction. The CBR test is the ultimate parameter used in design of unpaved highways due to its simplicity [4]. Steel slag (KOBM Binder™) is an industrial waste that is made from the steel refining procedure where one ton of steel creates about 100kg of slag (KOBM) therefore roughly 130 million tons of steel slag (KOBM) is generated internationally per year and 43% are used on road project [5]. KOBM binder is a sustainable material that has been tested for stabilizing clay soil for use in pavement constructions which also decreases the need for natural aggregates [6]. Adding steel slag and fly ash improves the strength characteristics and stabilization of soft soil [5]. A study used soil-cushion system with cushion positioned on an expansive soil bed was used to conduct the CBR test. There was substantial rise in the soaked CBR values for both the cases where the cushioning materials were utilized with the by-product of thermal power and steel plants i.e., granulated blast furnace slag and fly ash [7]. Hussien and Omar [8], concluded that the addition of fine steel slag has a straight effect on the Free Swell and Plasticity Index of the soil as observed through an experimental investigation carried out on the stabilization of highly plastic subgrade soil using fine steel slag. Conclusively, 20% addition of fine steel slag by weight resulted in the increase of CBR value, maximum dry density, and unconfined compressive strength of the soil. Also, fly-ash was utilized to thicken the fine-grained subgrade and expanding soft soil in South Wisconsin [9].

There is no published research on the use of KOBM Binder™ steel slag as a stabilizing material in geosynthetic reinforced clay soil. Therefore, the aim of current research is to evaluate the suitability of ACEgrid geogrid GG100 as a reinforcement material for soil-KOBM mix samples by using laboratory compaction tests to improve the behaviour of the soil-KOBM mix samples and their compaction characteristics. The core objective of this study is to examine the strength performance of unreinforced and geogrid reinforced subgrade soil using KOBM Binder™ steel slag. Also, determining the optimal location for positioning geosynthetic either at H/2 or H/4 (from top and bottom) in single layer and double layers (where H is the height of CBR mould). Therefore, this study displays the outcomes of an experimental program to analyse the effect of KOBM Binder™ steel slag as a stabilizing material on geotechnical characteristics of geogrid reinforced subgrade soil.

2. Materials and methods

For the present research, soil for laboratory testing was collected from Ravalevu, Nausori, Suva in the Central division, Fiji. Soil sample for this study can be categorized as a representative soil sample since it is suitable for determining the engineering properties like Atterberg limits and specific gravity of the subgrade soil [10]. The detailed engineering properties of soil sample are listed in Table 1. The soil samples were prepared for drying using two approaches which include oven drying and air drying in accordance with AS1141.1-2015 and AS 1289.0- 2014 standards. The particle size distribution of subgrade soil after dry sieve analysis is shown in Figure 1.

2.1 Geogrid

Geogrids are generally used for reinforcement and separation, particularly in soils with very large particle sizes. In the current study, ACEgrids were used to understand the performance of geogrid reinforced subgrade soil. Use of Geogrid as reinforcement is dependent on its stiffness or high tensile modulus, as well as its open geometry, which accounts for its great capacity for interlocking with soil particles. Engineering properties of geogrid material used in this study are mentioned in Table 2. It has been noted that for geogrids, the interlocking of soil particles through the geogrid apertures creates an efficient interlocking effect [11] [12].

Table 1. Engineering characteristics of subgrade soil

Properties	**Subgrade soil**
Gravel %	4
Sand %	56
Silt and clay %	40
D_{10} (mm)	0.0206
D_{30} (mm)	0.1839
D_{60} (mm)	0.9118
C_c	1.80
C_u	44
Specific Gravity	2.68
(%) Liquid limit	56
(%) Plastic limit	36
(%) Plastic index	20
Cohesion, c (kPa)	10
Friction angle (°)	20
Maximum Dry Density, MDD (g/cm^3)	0.5
Optimum Moisture Content, OMC (%)	35
USCS classification	SC

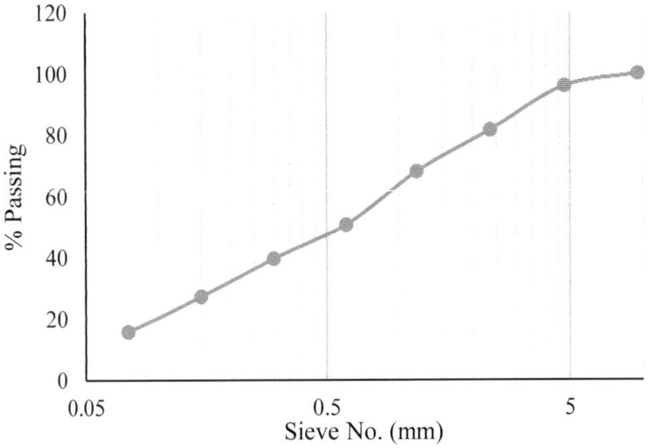

Figure 1. Particle size distribution of subgrade soil

2.2 KOBM steel slag

KOBM Binder™ steel slag is an extremely alkaline material; therefore it is recommended to treat KOBM with extra caution and store in a covered area absent from moisture and water. Likewise, KOBM is corrosive and hydroscopic to few substances. Furthermore, it is important to prevent direct contact with the material and proper ventilation should be available for working areas which generate high amount of dust. Light metal with alloy or galvanized materials should be avoided [13]. Moreover, Carbonation can be used to activate ground KOBM steel slag to increase the strength up to 80 MPa in 2hrs and 109 MPa after subsequent hydration of 28 days [6]. The characteristics and particle size distribution of KOBM used in the current study are detailed in Table 3 and Figure 2. Typical Oxide composition of KOBM slag is shown in Figure 3. As KOBM slag consists of free Calcium and Magnesium oxides, it results in volumetric instability thus limiting the use in geotechnical engineering applications beyond a certain limit.

Table 2. Engineering Features of Geogrid Material [13] [14][15].

Mechanical Characteristics	MD Values	XMD Values
Rib Pitch, mm (in)	32 (1.26)	32 (1.26)
Rib Breadth, mm (in)	3.3 (0.13)	1.2 (0.05)
Rib Width, mm (in)	3.5 (0.14)	2.0 (0.08)
Shape of the Aperture	Rectangular	
True Initial Modulus, kN/m (I/ft)	550 (37,700)	350 (24,000)
Stability of the Aperture, N-m/deg	0.45	
Overall Flexural Rigidity, mg-cm	2,100,000	
Radial Stiffness at low strain @ 0.5% Strain, kN/m (I/ft)	321 (22,000)	
Apparent Coefficient of Friction Soil	1.25	
Resistance to Installation Damage, %SC/%SW/%GP	100/100/100	
Resistance to Long Term Degradation, %	100	
Resistance to UV Degradation, %	100	
Declared Service Life (years)	100	

Table 3. Characteristics of KOBM [13]

Common name	KOBM Binder™
Appearance	Granular
Bulk Density	1500kg/m^3
Particle Size	9.5mm minus
Colour	Dark Grey
Specific Gravity	2.88
Maximum pH value @ 25°C	12

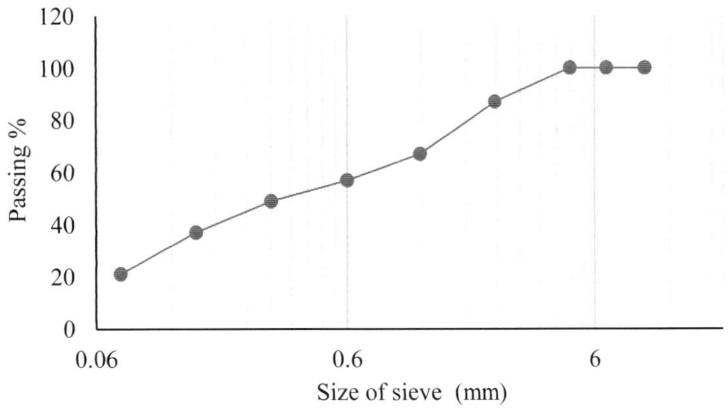

Figure 2. Particle size distribution of KOBM slag

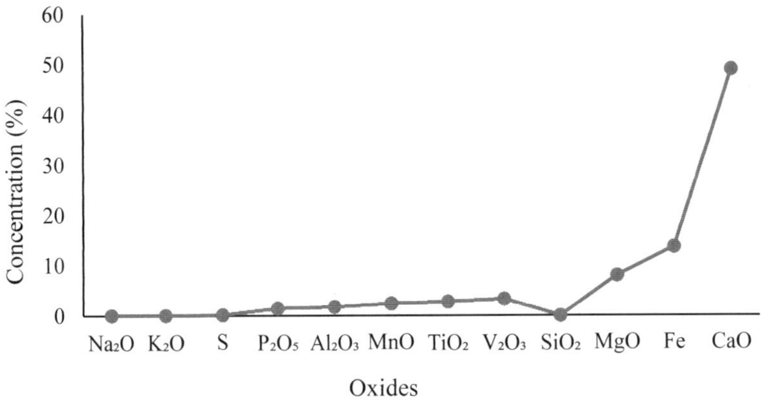

Figure 3. Typical Oxide composition of KOBM slag

3. Experimental Investigation

All the experiments conducted in this study were strategized in advance to make the investigational procedure and the interpretation of results effective. Dry sieve analysis of soil was performed initially to determine the classification of soil (Table 1). Subsequently, Atterberg limits, specific gravity, optimum moisture content and maximum dry density though standard proctor test were determined for soil sample. Direct shear test was conducted to determine the value of cohesion and internal angle of friction. Further, CBR tests were carried out for the soil samples blended with KOBM – with/without geogrid reinforcement positioned at different heights in five scenarios: Scenario 1 - CBR mould with unreinforced soil. Scenario 2 - Single layer of geogrid reinforced soil i.e, SGRS at H/2 in CBR mould with 5%, 10%, 15% and 20% KOBM slag. Scenario 3 - SGRS at H/4 of CBR mould with 5%, 10%, 15% and 20% KOBM slag. Scenario 4 - Double layers of geogrid reinforced soil i.e, DGRS at H/2 of CBR mould with 5%, 10%, 15% and 20% KOBM slag and Scenario 5 - DGRS at H/4 of CBR mould with 5%, 10%, 15% and 20% KOBM slag.

4. Results

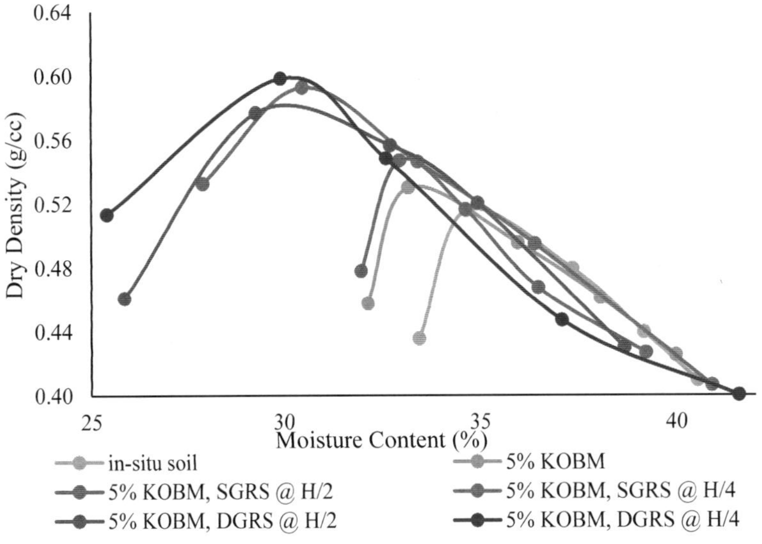

Figure 4. Compaction curves at 5% KOBM with/ without geogrid reinforcement.

The Compaction curves with/without geogrid reinforced layer at 5%, 10%, 15% and 20% KOBM are represented in Figures 4 – 7. The reinforcement efficiency (η) is used to measure the change in CBR value of subgrade soil using reinforcement. It is defined as a ratio of the CBR value of soil using reinforcement to that of the original soil.

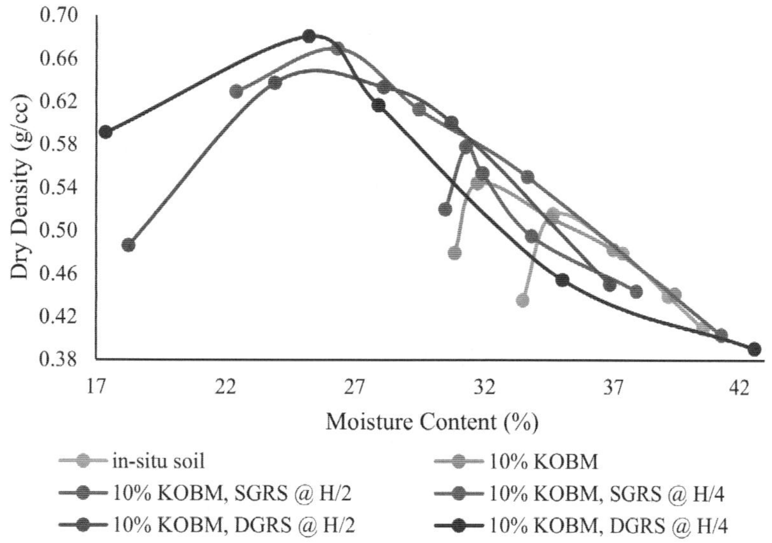

Figure 5. Compaction curves at 10% KOBM with/ without geogrid reinforcement.

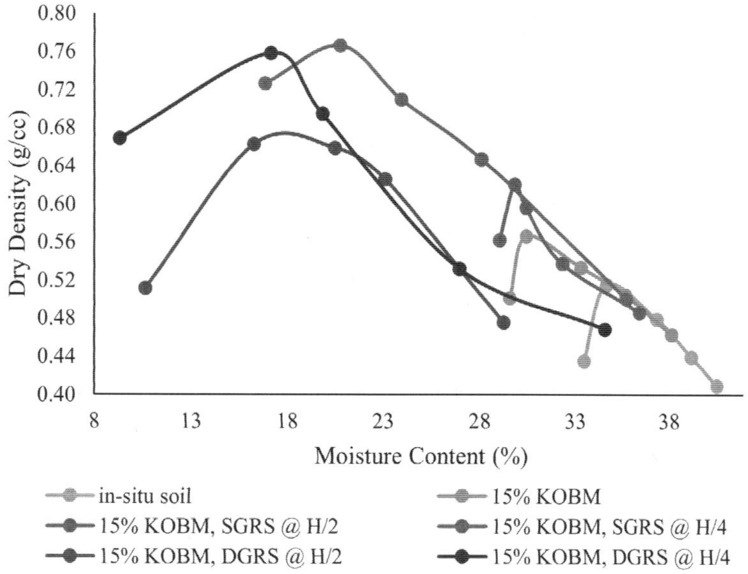

Figure 6. Compaction curves at 15% KOBM with/ without geogrid reinforcement

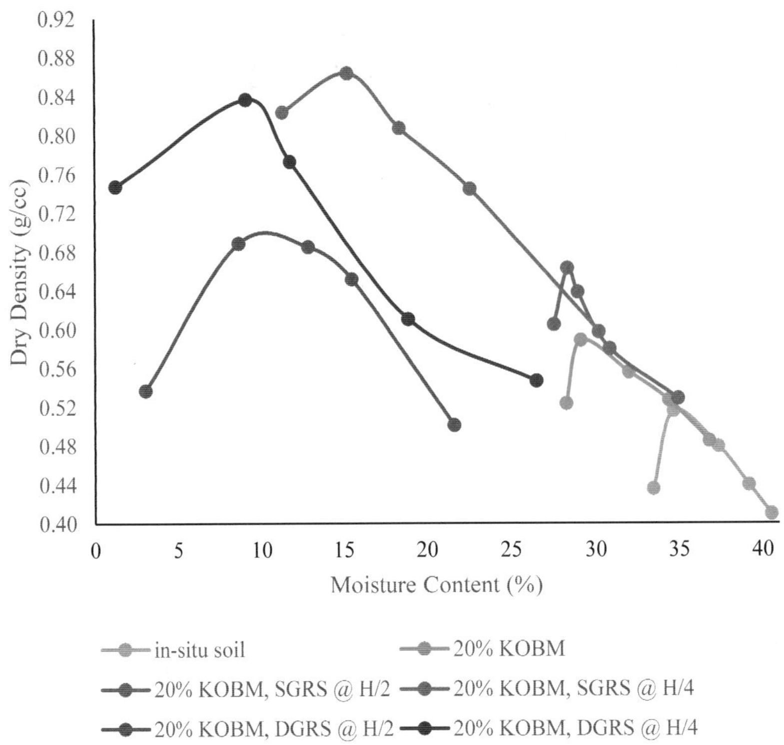

Figure 7. Compaction curves at 20% KOBM with/ without geogrid reinforcement.

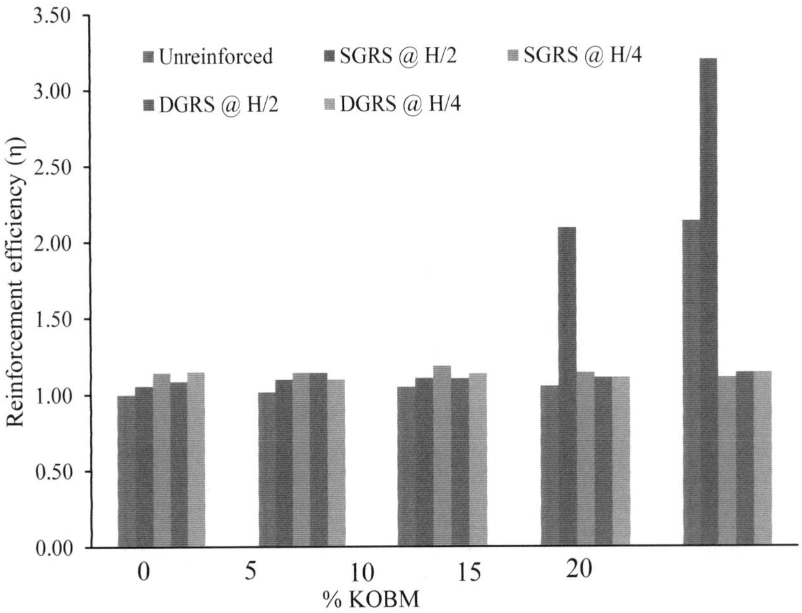

Figure 8. Variation in reinforcement efficiency with position of geogrid at varying KOBM content

Table 4. Variation in Reinforcement efficiency (η) with varying KOBM content

Reinforcement	% KOBM	CBR (%)	η
	0	1.92	1.00
	5	1.95	1.01
Unreinforced	10	2.02	1.05
	15	2.03	1.05
	20	4.11	2.14
SGRS @ H/2	0	2.03	1.06
	5	2.11	1.10
	10	2.13	1.11
	15	4.02	2.09
	20	6.15	3.20
SGRS @ H/4	0	2.19	1.14
	5	2.20	1.14
	10	2.28	1.19
	15	2.19	1.14
	20	2.13	1.11
DGRS @ H/2	0	2.08	1.08
	5	2.19	1.14
	10	2.12	1.10
	15	2.13	1.11
	20	2.20	1.14
DGRS @ H/4	0	2.21	1.15
	5	2.11	1.10
	10	2.18	1.13
	15	2.13	1.11
	20	2.20	1.14

5. Discussion

Table 4 demonstrates the variation in reinforcement efficiency for varying contents of KOBM. Based on CBR values, it can be concluded that 20% KOBM yields best results in unreinforced case, SGRS @ H/2 (CBR value at 6.15% which is the best combination amongst all the trials), and DGRS @ H/4 from top and bottom. However, 5% KOBM yields commendable results for SGRS @ H/4 from top and bottom, as well as DGRS @ H/2.

Figure 8 deduced from Table 4 illustrates the variation in reinforcement efficiency with position of geogrid for varying contents of KOBM from 0 to 20% for each position. It is evident that DGRS @ H/2 displays maximum strength at 5% KOBM while DGRS @ H/4 from top and bottom exhibits maximum strength at 20% KOBM. On the other hand, SGRS @ H/4 from top and bottom displays highest CBR value at 5% KOBM; while SGRS @ H/2 yields the best performance at 20% KOBM with highest CBR value of 6.15% which is the best combination amongst all the trials.

Maximum CBR values for single- and double-layers geogrid reinforcement for each composition of KOBM content (i.e., 0%, 5%, 10%, 15% and 20%) have been compared. In the case where no KOBM is added, DGRS @ H/4 from top and bottom demonstrates slightly greater strength when compared to SGRS @ H/4 from top and bottom. At 5% KOBM, SGRS @ H/4 from top and performs like DGRS @ H/2. At 10% KOBM content, DGRS @ H/4 from top and bottom performs slightly better than SGRS @ H/4 from top and bottom. At 15% KOBM content, SGRS @ H/2 performs significantly better than DGRS @ H/4 from top and bottom. Finally at 20% KOBM content similar to 15% KOBM, SGRS @ H/2 performs notably better than DGRS @ H/4 from top and bottom.

Compaction curves determined using Standard Proctor Tests for unreinforced and reinforced soil samples are presented in Figures 4 – 7. In unreinforced case, MDD values increase with increase in percentage of KOBM. In case of SGRS and DGRS placed at H/2 and H/4, there is a steady increase in MDD values with increase in KOBM percentage. Also, there is a decrease in the OMC values with increase in percentage of KOBM in unreinforced case. Moreover, in case of SGRS and DGRS placed at H/2 and H/4, there is a steady decline in OMC values with rise in KOBM percentage. The reduction in OMC values with rise in KOBM percentage can be attributed to hydrophobic nature of KOBM and water absorption capacity while reduction in MDD values with rise in KOBM percentage can be attributed to higher specific gravity of KOBM.

6. Conclusion

In the current study, CBR values have been used to evaluate the strength behaviour of different combinations of KOBM slag and effort has been made to enhance the strength of subgrade soil using geogrid reinforcement by varying the position and number of layers. In addition, reinforcement efficiency has been used to evaluate the performance of subgrade soil with the inclusion of geogrid reinforcement and KOBM slag. It is to be noted that the CBR value increases with geogrid reinforcement at varying depths with few exceptions at 15% and 20% KOBM for SGRS @ H/4, DGRS @ H/2 and DGRS @ H/4. On the other hand, in the unreinforced case and SGRS @ H/2; CBR values increase with increase in KOBM content. For SGRS @ H/2, maximum CBR was observed at 20% KOBM which is observed to be the highest CBR value amongst all the combinations of KOBM and position of geogrid in the current experimental investigation (two trials for each test) of 50 experiments.

Overall, it can be summarized that there is a significant amount of improvement in the strength of subgrade soil stabilized with KOBM and reinforced with geogrid material. The level of improvement is subjected to the characteristics, type, depth, and number of layers of geogrid reinforcement. Nonetheless, imperatively it is to be noted that all the findings from this study are derived from the small-scale experimentation in the laboratory. Therefore, possibilities of different findings can be expected if carried out in the field. Therefore, for better reliability in results, definite field tests must be conducted.

References

[1] Madhu S N Sanjay S 2020 *Indian Highways* pp. 15–21
[2] E M Palmeira 2016 *Soils and Rocks* **39** 2 pp. 113–135
[3] M Singh, A Trivedi and S K Shukla 2019 *Transp Geotech* **19** pp. 54–60
[4] Bernardo C and Cristhian M 2015 *Transportation Geotechnics* in *Fundamentals to Applications in Geotechnics* D. Manzanal and A.O. Sfriso (Eds.) pp 95 - 100.
[5] D R K T Laxmikant Yadu 2013 *J Adv Sch Res Allied Educ* **15** 2 pp. 414–418
[6] Ghouleh Z Guthrie R I L Shao Y 2015 *Constr Build Mater* **99** pp. 175–183
[7] Dordi C M Vyasa A N Santhanam M 2013 *3rd Int conf on Sustain Constr Mater Technol.*
[8] Aldeeky H Hattamleh O A 2017 *Adv Civ Eng* https://doi.org/10.1155/2017/9230279
[9] Senol A Edil T B Bin-Shafique M S Acosta H A Benson C H 2006 *Resour. Conserv. Recycl* **46** 4 pp. 365–376
[10] Venkatramaiah C 2006 *Geotechnical Engineering 3rd ed* Tirupati India: New Age International(P) Limited Publishers.
[11] David A C L 2018 Guide to Pavement Technology Part 4G: *Geotextiles and Geogrids* Sydney: Austroads Ltd
[12] Shukla S K 2016 *An Introduction to Geosynthetic Engineering*. Leiden: CRC Press/Balkema.
[13] Julie R 2018 *KOBM Analytical Laboratory Report* Glenbrook.
[14] Tenax 3D *Triplanar Geogrids* Baltimore.
[15] Material Safety Data Sheet- KOBM BINDER *Hiway Stabilizers* NZ.
[16] Shi J L Subhashini A Yi S T Mavinakare E R 2020 *Minerals* **10** 863 pp. 1 - 13.

Acknowledgement

The authors thank Mr. Peter Watts, General Manager and Allen Browne, Executive manager, Technical and Innovation at Hiway Stabilizers Fiji for their technical assistance and support in providing KOBM material.

Durability Characteristics of Recycled Aggregates for Physical and Chemical Properties of Concrete

Rohit Choudhary[1], Satish Jangid[2], Ramdayal Meena[3] and Shailesh choudhary[4]

[1]Assistant professor, MBM Jodhpur, India
[2]Research Scholar, MBM Jodhpur, India
[3]Research Scholar, MBM Jodhpur, India
[4]Associate Professor, MBM Jodhpur, India

Email.rohit10choudhary@gmail.com

Abstract. In structural concrete, it becomes essential to perform the comprehensive research work for recycled aggregates (RA) in order to check their material properties as well as assess the influence the second-generation concrete quality. This work has the objective to assess and suggest the technical standards to check the performance and behaviour of concrete as compressive strength developed from recycled aggregates. The demolished crushed waste concrete can provide the RAs. Such aggregates are heterogeneous in composition from a quality point of view as being resultant from diverse adhered mortar and materials. Such properties of RAs must be assessed, as these are utilized in concrete; consequently make an effort to further investigate the RAs characteristics that are needed to be implemented in concrete mixes. This paper presents the characteristics of RAs as physical and chemical properties with further examination on the concrete properties prepared with RAs. It also presents the comprehensive strength to show the durability properties in order to set the standard guidelines. The developed nations such as Japan have even laid down standards for RA usage in concrete. Thus, it becomes essential to make guidelines for the material utilization in construction, especially to match with the local Indian conditions.

1. Introduction

The sustainable development in the construction waste management and concrete industry, the recycled aggregate utilization certainly becomes a significant phase. Recycled aggregate (RA) becomes an alternative solution of natural aggregate helping in the ecological conservation. The variations in aggregate properties is one of the crucial aspect that influence the RA usage. The collected and delivered material can also affect the recycled aggregate quality in the recycling plants. Thus, it is difficult to obtain the recycled aggregate production at a satisfactory quality and cost because of existing recycling plants limitations. Such challenges are creating issues with customers about the inconsistent aggregate properties and production stability. The main goal of this research is to examine the inconsistency of aggregate properties with their influence on the production of concrete. Some of the mechanical and physical characteristics influencing the concrete's durability and strength are texture, shape, specific gravity, moisture content, absorption, gradation and aggregate strength. Hence, it becomes essential to assess such characteristics before using the aggregate. This research paper offers a procedure to evaluate the recycled aggregate properties from an unidentified source composed during the six months from a recycling plant. Furthermore, it shows the investigation on concrete properties i.e. made with 100 % recycled aggregates.

The ecological issues developing from indiscriminate as well as unempirical municipal solid waste (MSW) disposal i.e. the actual threat for the entire society. Such MSW are growing gradually because of rise in industrialization, urbanization and population. The MSW properties illustrates that it includes nearly15–20% of inert material, 25–35% of recyclable/dry materials, and 55–65% of compostable material [1]. The annual rate of growing the MSW amount produced per capita is evaluated as 1–1.33% in India [2, 3]. The science and technology play a significant role in the successful employment of waste management as the land filling provision for MSW management is not a key solution. It requires a pattern alteration in managerial concept vision to concentrate more on MSW minimization or comprehensive waste recycling instead of getting rid of land filling. The step of landfill has to be eliminated from the procedure by delivering additional prominence on 100% waste recycling. Processing and treatment as the recovery/recycling can be a significant part of solid waste management. The demolition activities and extreme construction can produce the inert waste majorly involves the stone pebbles, bricks, concrete, and so forth. According to central pollution Control Board (CPCB) of India predicted the production of MSW as nearly 48 million tonnes per annum i.e. the one-fourth of the construction industry [1].

The total waste production is projected to be approximately 12–14.7 million tonnes per annum. The growing quantity of demolition waste has made a substantial influence on society as well as ecology and its management becomes a great concern for worldwide. Aspects like good quality of raw material, lack of land for disposal, economy and sustainability for construction make it authoritative that demolition/construction waste appropriately recycled. According to the previous studies, the demolition waste earlier thought as non-usable, now has been validated that it could be used entirely as a recycled aggregate for ecological friendly concrete production with its several applications in construction [4–8]. Recycled aggregate usage can sort out the disposal issue as well as also lessen the natural resources utilization in construction industry as well as therefore it contributes to the ecology protection by plummeting.

2. Background

The key aspects towards the stable utilization as well as introduction of RA concrete are availability and quality to the construction industry. In this research work, the crushed stone aggregate was acquired from a recycling plant that was build and made efforts towards the reduction of waste production from the construction industry. It offers an effective solution to recycle the aggregate by reusing it [4-6]. The procedure includes aggregate classification into different grades according to particle size, manual removal of impurities such as wood, plastic, etc., metal separation by a magnet and crushing. The portion of RA in concrete mixes can influence the concrete properties such as durability and stability. The construction industry development presents numerous challenges with respect to the natural aggregate resources availability as these are being promptly exhausted [7]. Current figures depicted the enhancing demand of construction aggregate to attain the mark of 58.3 billion metric tons up to 2025 with the maximum consumption in Asia (The Freedonia Group 2012). Such growing demand is followed by the growth of construction waste. For instance, European Union nations produce the construction wastes refers approximately 31 % of the total waste production per annum. Likewise, the waste generation in Hong Kong was approximately 20 million tons in 2011 establishing nearly 50 % of the worldwide waste production [8]. The landfills disposal is the usual approach to handle the construction waste that develops a large deposits of demolition waste and construction sites. Making the attempts to restrict such practice and to support the demolition waste/construction recycling in several construction applications resulting as the usage up to 10 % of RA in various construction applications. Thus, recycling has the capability to lessen that waste materials amount that preserve the natural resources as well as landfills deposition [9-10].

3. Method and material

The development of a concrete mix design is a technique of choosing appropriate constituents for concrete with estimating their ratio that would generate a concrete that fulfils the demand of acceptable durability/workability and least compressive strength. The entire aspects that impact on the compressive strength is called normal aggregate concrete, although it would be necessary to know the substantial parameters affecting the mix ratios with RAs. According to the previous work, the RAs become the attribution of lower specific gravity and greater water absorption with respect to normal aggregates. Thus, the mix design approach could be implemented in developing concrete mixes with RA requires to be detected since the material characteristics varies with respect to the normal aggregates.

3.1 Features Affecting the Mix Proportions Selection

i. Grade Designation of Concrete

It provides the required characteristic of concrete as compressive strength parameter. According to IS: 456-2000, the characteristic compressive strength refers the value lower that is not greater than 5% of the test outcomes are projected to drop. It becomes the key aspect affecting the design of mix. The concrete mix has to be made for a target mean compressive strength according to the degree of control at the site. The value is nearly greater than the characteristic strength. Nevertheless, IS: 456- 2000 has no provision for the grade designation of RA concrete. Therefore, investigators made effort to attain the M20 grade concrete by utilizing RAs.

ii. Grade/Type of Cement

A cement shows minimum standard deviation or least variations as it has consistent quality. It can be stated as its compressive strength that can estimate the most cost-effective ratio of cement needed to achieve a certain concrete mix grade. It can be done only by varying the proportion of fine to coarse aggregates. With the high water absorption by RAs, 53 grade cement utilization may result the cracking in concrete because of high hydration heat, therefore, this work utilizes the 43 grade of cement in proportioning the RA concrete.

iii. Coarse Aggregate with Optimum Nominal Size

The coarse aggregate with optimum nominal size is estimated by conducting sieve assessment as well as is selected by the higher sieve size against the largest size, where ≤15% of the RA is taken. The RAs with optimum nominal size to be utilized in concrete is regulated by the reinforcement spacing as well as section size. Based on IS: 1343-2012 and IS: 456 -2000 the optimum nominal size of aggregate should not be greater than 25% of the least member thickness. Nominal optimum aggregate size within the limits may be as greater as possible, as the greater maximum aggregate size can be resulted as lower requirement of cement for a certain proportion of water and cement. With an upsurge in the optimum normal aggregate size, the workability also increases. The smaller size aggregates in case of RAs on account of greater water absorption negatively impact on compressive strength and workability of concrete. Furthermore, the extra connected mortar on the RA also influences the mortar matrix along with bonding characteristics.

iv. Combined Aggregates Grading

The coarse and fine aggregates with their relative proportion in a concrete mix become the significant aspects influencing the concrete strength and workability. It becomes also necessary that the fine and coarse aggregates are well graded for dense concrete. Well graded aggregates produces concrete that has smaller size particles such as sand fill the voids among greater particles, lessening the space. Water cement paste can fill this space and therefore enhance the density. This outcomes show lower cost, minimum shrinkage and enhanced strength of the concrete.

Generally, locally present aggregates has the incapability to provide the confirmation on standard grading. Such aggregates require to be integrated in an appropriate ratio that offers the desired grading with the integrated or resultant grading approximates. The integrated aggregate procedure is focused at achieving a grading nearby to the coarsest grading of standard grading curves. It provides the most efficient mixture with greatest permissible aggregate cement proportion.

v. Water Cement Proportion

The concrete with compressive strength at a definite age under standard temperature relies on the water cement proportion. The comprehensive strength is greater with the lesser water cement ratio and vice versa. However, it is also valid for RA concrete, as the water cement proportion selection

IOP Conf. Series: Earth and Environmental Science 1084 (2022) 012026 doi:10.1088/1755-1315/1084/1/012026

for a 28 days aim to compressive strength if RA may not follow the similar empirical correlations according to guidelines.

vi. Workability

The concrete workability for suitable compaction as well as location is regulated by the shape/size of the concreted container. It involves the reinforcement spacing/quantity, implemented approach for transportation as well as compaction/placing of concrete. The condition requires to be analysed to reach at the anticipated workability. The objective to involve the least conceivable workability consistent with suitable location and concrete compaction. Inadequate workability leads to the inadequate compaction that may relentlessly impact on the surface finish, durability and strength of concrete. Therefore, it becomes wasteful in the long run. Though, RAs have the greater water content adjustment that is to be developed to attain the essential workability. By considering such aspects that would impact on the mix design of RA concrete as an appropriate mix design approach that has been suggested in this work to reach at M20 grade concrete.

3.2 *Compressive Strength Test*

The concrete cubes is going through the compressive strength test as it depicts about the entire concrete characteristics. This single test can provide the appropriateness of the concreting. Compressive strength relies on several aspects like quality control during concrete production, quality of concrete material, cement strength and water-cement proportion etc. Compressive strength test is performed with the help of cube and cube moulds having the dimensions of 15cm x 15cm x 15cm. Several guidelines have been followed with many standard codes for the testing of concrete cube.

Figure 1. Water Absorption and Specific Gravity of Fine and Coarse Aggregates

This concrete is poured in the mould and tempered as well as vibrated appropriately in order to remove any voids. Such moulds are eliminated after 24 hrs. As well as test specimens have been cured by placing in water. These specimens should have smooth and even top surface. This can be performed by placing and spreading the cement paste smoothly on the entire specimen area. After 7 days and 28 days, such specimens are examined by compression testing machine. Till the specimens fails, load should be applied progressively at the rate of 140 kg/cm^2 per minute.

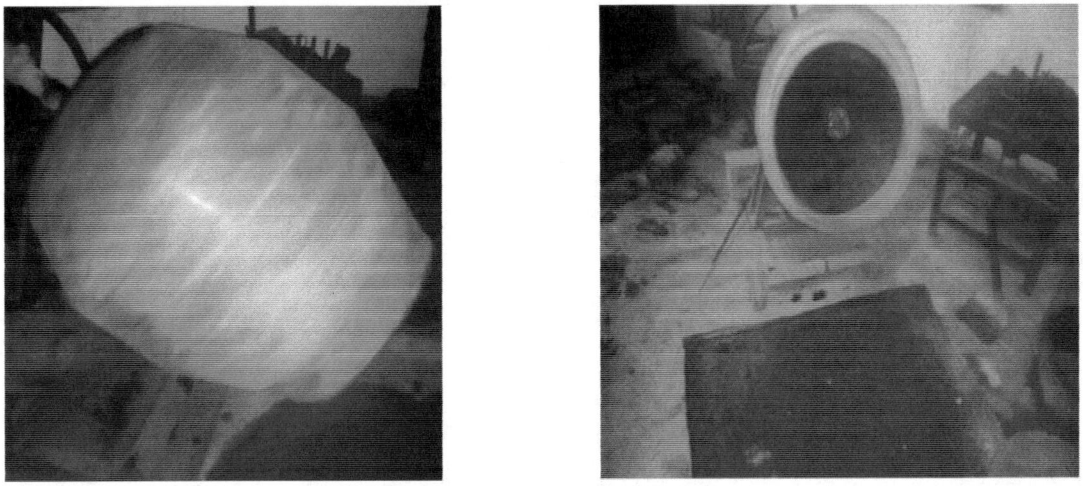

Figure 2 Mixing of concrete by tilting mixer

The compressive strength of concrete can be achieved by measuring the load at the failure divided by area of specimen. The comprehensive strength is reduced by 5-25% by replacing 50-100 % virgin aggregates with RA. Although, RA concrete can replace the virgin aggregate up to 30% without any impacts on concrete strength. For the first 7 days, the strength gain for RCA concrete is lesser with respect to normal aggregate concrete (NAC). While, fine RA has a more harmful impact on compressive strength than coarse RA

4. Results & Discussion

For this experiment the design mix was done with the mix M20. Various properties were determined of list material already mentioned to do the mix design. The list of IS codes used for doing this mix design are;

- IS:10262 : 2009 - Concrete Mix Design Ratio
- IS:456 : 2000 – Reinforced and Plain Concrete
- IS:383 : 1970 – Fine and Coarse Aggregates Specifications from Natural Sources for Concrete
- IS:8112 : 2013 – Ordinary Portland Cement 43 Grade
- IS: 2386 (Part 3): 1963 – Testing approaches for Aggregates of Concrete.

Using these various codes of Indian Standards tests were done to find out the various properties required of the components and then using these properties the mix design was done accordingly.
There are some parameters which are set before the calculations of the mix design being done as per the requirement and conditions in which the mix is used. The parameters used for the following mix design are illustrated in Table 1.

Table 1. Technical Specifications of concrete

Variables	Values
Compressive Strength Characteristics after 28 days	20 N/mm
Type of Cement Used	43 grade OPC
Minimum cement Content	300 kg/m^3
Degree of Workability	50mm Slump
Type of Exposure	Moderate
Maximum Size of Aggregate Used	20 mm
Maximum Water Cement Ratio	0.45
Quality Control	Good

Several other properties of material used in Cement are also important such that type of cement is 43 grade OPC cement, Cement brand is J K Lakshmi Cement with the specific gravity of 3.05. While, the considered fine Aggregate has the water absorption of 1.16%, specific gravity is 2.560 and Silt Content is 0.15%. The coarse Aggregate has the features of specific gravity is 2.67 and water absorption is 0.21

The test was conducted for water-cement ratio of .45 and .43 and results were noted for compressive strength test at 7 days, 28 days and slump test for its workability.

Table 2. Compressive strength at 7 days for w/c ratio 0.45

% Replacement of RCA	Compressive Strength at 7 days	(%) Variation	W/C Ratio
0%	17.4	0.00	
10%	15.71	9.71	0.45
20%	14.14	18.74	
30%	13.05	25.00	

For the compressive strength test once the cubes were left over for cuing wait was done for the cubes to attain the age of 7 days. After completing 7 days the cubes were taken out from curing tank and they were wiped up with a cloth so as to make sure that there is no outside material left onto the surface of the cube and then they were allowed to get dry until we can't see any moisture onto the surface of the cubes.

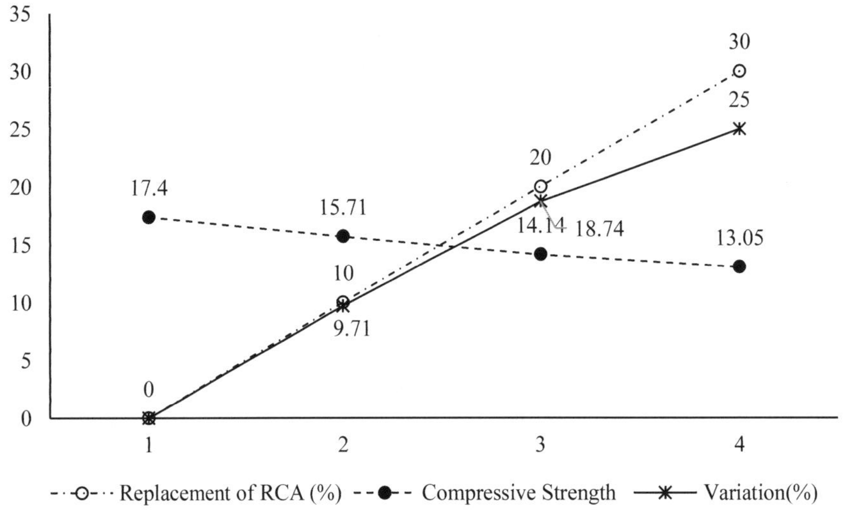

Figure 3. Graphical representation of compressive strength obtain after 7 days with the w/c ratio of 0.45

Now bearing surface of the cube testing machine was properly cleaned and the cubes were located into the machine in such a way that the loads get applied to the opposite sides of the casting of the cubes. It was made sure that the cube was aligned into the center of the base plate of the cube testing machine. The movable portion was rotated slowly to get the cube to fix inside the machine.

Table 3. Compressive strength at 28 days for w/c ratio 0.45

% Replacement of RCA	Compressive Strength at 28 days	(%) Variation	W/C Ratio
0%	24.96	0.00	
10%	22.08	11.54	0.45
20%	21.92	12.18	
30%	20.31	18.63	

The machine was then started and the load was applied progressively without any shock at the rate of 140 kg/cm /minute until the cube failed and stops taking any kind of load. The reading was recorded for and noted down. It was kept in mind that if the strength for any cube is less than 15% of the average strength then that cube should be rejected. So in this way the cube test results for the 7 days and 28 days strength was done.

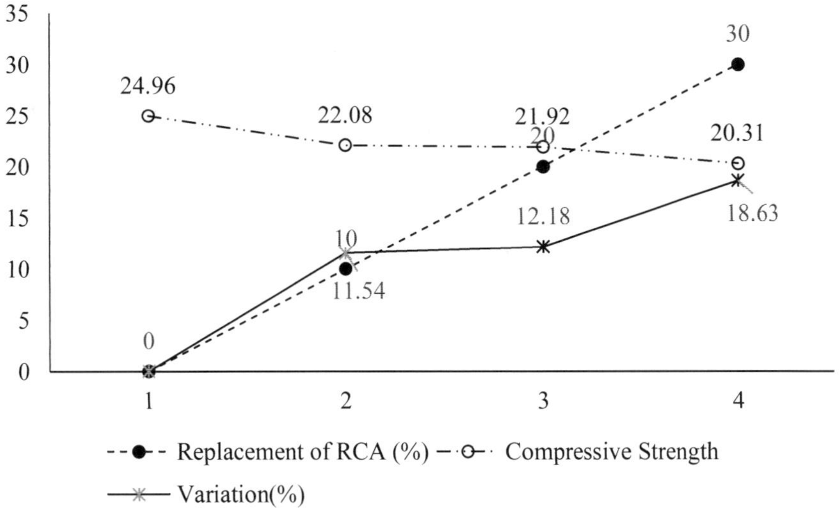

Figure. 4 Graphical representation of compressive strength obtain after 28 days with the w/c ratio of 0.45

5. Conclusion

Worldwide, the properties of RA and their applications have been studied from the mid-seventies onwards. In this study we are using coarse aggregate from Demolition and Construction waste as fractional replacement of coarse aggregate in concrete by the observation of workability we found the concrete is workable with replacement of coarse aggregate in concrete. By the observation of compression test we found that the structural properties in concrete with partial replacement of coarse aggregate in concrete, By the observation of compression test we found the maximum compressive strength at 10% replacement of coarse aggregate in concrete of 22.08 MPa and minimum at 30% replacement of coarse aggregate 20.31MPa with at 0.45 W/C ratios. It shows that the strength reduces with the upsurge in the replacement of coarse aggregates of concrete. By the observation of compression test we found the maximum compressive strength at 10% replacement of coarse aggregate in concrete of 23.14 MPa and minimum at 30% replacement of coarse aggregate 21.58 MPa with at 0.43 W/C ratios. Also, it depicts that the strength reduces with the enhancement the replacement of coarse aggregates in concrete. We observed by results of compressive test with the compression of compressive strength is decreasing with increasing the replacement of recycled coarse aggregates in percentage variation at 10%, 20% and 30% are 11.54, 12.18 and18.63 respectively at 0.45 W/C ratio.

References

[1] Abbas, A., Fathifazl, G., Isgor, O. B., Razaqpur, A. G., Fournier, B., & Foo, S. (2009). Durability of recycled aggregate concrete designed with equivalent mortar volume method. *Cement & Concrete Composites*, 31(8), 555–563.

[2] Abdelfatah, A., Tabsh, S., & Yehia, S. (2011). Utlilization of recycled coarse aggregate in concrete mixes. *Journal of Civil Engineering and Architecture*, 5(6), 562–566.

[3] Akbarnezhad, A., Ong, K. C. G., Tam, C. T., & Zhang, M. H. (2013). Effects of the parent concrete properties and crushing procedure on the properties of coarse recycled concrete aggregates. *Journal of Materials in Civil Engineering*, 25(12), 1795–1802.

[4] ASTM C29/29 M. (2009). Standard test method for bulk density (unit weight) and voids in aggregate. *West Conshohocken, PA: ASTM Internationa*l.

[5] ASTM C78/C78 M-10e1. (2010). Standard test method for flexural strength of concrete (using simple beam with thirdpoint loading). *West Conshohocken, PA: ASTM International.*

[6] De Brito, J., & Saikia, N. (2013). Recycled aggregate in concrete: Use of industrial, construction and demolition waste. 445 p., London, UK: Springer.

[7] Eisa, A. (2014). Properties of concrete incorporating recycled post-consumer environmental wastes. *International Journal of Concrete Structures and Materials*, 8(3), 251–258.

[8] Fathifazl, G., & Razaqpur, A. G. (2013). Creep rheological models for recycled aggregate concrete. *ACI Materials Journal*, 110(2), 115–126.

[9] Garg, P., Singh, H., & Walia, B. S. (2013). Optimum Size of Recycled Aggregate. GE-*International Journal of Engineering Research*. pp. 35–41, ISSN:2321-1717.

[10] Knaack, A. M., & Kurama, Y. C. (2013). Design of concrete mixtures with recycled concrete aggregates. *ACI Materials Journal*, 110(5), 483–493.

Review on Recycling of Microplastics in Cigarette Butts

Mary George[1] and Rupali Khadtar[2]

Anjuman-I-Islam's Abdul Razzak Kalsekar Polytechnic
Email-mary.george@aiarkp.ac.in[1],rupali.khadtar@aiarkp.ac.in

Abstract. It's a lesser-known fact that the cigarette filter is made of cellulose acetate, a kind of non-biodegradable plastic and it takes almost 10 years for this ingredient to completely decompose. Cigarette smoking is on the rise throughout the world. Ever since tobacco smoking began there are both opponents and proponents for the same. Instead of getting into a discussion as to whether one should smoke or not; would like to focus on cigarette butts which a micro plastic that is harmful to the environmentand needs to be disposed of in an appropriate manner to conserve the environment and human health. As the trend is today, cigarette smoking is on the increase. And in light of this event recycling cigarette butts give rise to a sustainable material.
This paper will cover how cigarette butts can be collected and recycled to tackle the unfriendly and unsightly remains of cigarettes after smoking.
Keywords: Cigarette butts, Solutions to handle cigarette butt waste, Awareness regarding probable health and environmental hazards

1. Introduction
Filters were initially used from 1860 to 1920 to prevent particles of tobacco from entering the mouth. Inthe 1950s after published studies began to reveal the health threats of tobacco, the cigarette industry propagated the use of cigarettes with filters. The use of filters gave cigarette smokers an illusion that they are safe from the ill effects of cigarette smoking and started inhaling as muchor more tar, nicotine, and noxious gases as smokers of unfiltered cigarettes. The fast life of cities and the increased pressure in day-to-day life due to the industrial revolution causing men to competewith machines has caused many men to smoke to relieve stress and tension. Consequently, there is an increased litter of cigarette butts.

2. In what way are cigarette butts harmful?
2.1. What are micro plastics?
Micro plastics comprise tiny fibers from nylon clothes and other synthetic textiles. Plastic particles smaller than 5 millimeters are considered to be micro plastics. Well, it is likely that microplastics continue breaking down even past 10nm. Plastics do not blend with the soil easily but remain in the soil in a particulate manner, probably just becoming into even smaller pieces. The tinier the particle, the easier it is for them to pass cell membranes and cause tissue damage. For example, it is observed that micro plastics can absorb toxic chemicals and then release them into an animal's digestive system. This would definitely be harmful to our health. [1]
Research has shown these nano plastics can enter the intestines during digestion. Whether they can intermingle with the blood is not clear, however, another rather unnerving study demonstrated thatnano plastic particles lodged in the brains of fish affected their behavior. The study found that

plastic particles made fish eat slowly and more lethargic. Research has not yet established Nano plastics penetrate brain tissue in humans, let alone affect behavior. The filter of cigarette is made of micro plastic. Plastics cannot be digested or decomposed and once produced, one cannotget rid of them. They are mistaken for food by animals and ingested blocking the digestive systemof animals due to which there are low levels of oxygen and consequently reduced energy levels inanimals. Very tiny micro plastics get embedded in the animal tissues. They enter human beings through the food chain. Micro plastics find their way to humans through ingestion or respiration.

2.2 What are micro plastics in cigarettes made up of?

The majority of cigarette filters are made of cellulose acetate, which is a type of plastic. The cigarette filter is not cotton, but a plastic that does not disintegrate similar to other types of plasticin the environment. Cigarette butts (CB) are one of the most common types of litter in cities, withan estimated 4.5 trillion thrown each year, accounting for22-46 percent of visible litter. 76 percentof cigarettes smoked in public in metropolitan areas are littered rather than disposed of properly. Butts are then carried to coastal locations by rain or river water, where they are typically the mostabundant part of litter collected. The chemical makeup of cigarette butts waste has an impact on the environment. Because cigarette butts are made up of a highly twisted network of fibers combined with plasticizers, the breakdown is sluggish and difficult. Furthermore, toxic metals and nicotine have been found to seep out of cigarette butts. The Lethal Concentration 50 for test fish exposed to cigarette butts' leachates is one cigarette butt immersed in a liter of water for 96 hours.When this trash is released into the environment, it poses a significant risk to living species and ecosystem health. As a result, cigarette butts pose a threat to the environment and public health. [2]

3. Where are cigarette butts found?

Cigarette butts are tossed on the street or littered in parks, and beaches, they may be thrown in street gutters too. Since indoor smoking is banned by many organizations smokers have been pushed outdoors. Cigarette butts are mostly found near corporate offices, hospitals, colleges, and tourist destinations. Also, outside bars cigarette butts are found in large quantities.

4. How can these cigarette butts be effectively collected?

Countries are beginning to put the burden of cleaning up these butts onto cigarette companies themselves. People who collect cigarette butts for the purpose of recycling are termed as heroes because common men due to lack of education and ignorance do not like to endeavor in this field for prestige sake. One of the ways of collecting cigarette butts is to provide special fire-safe receptacles. It would be wise to charge a percentage of the cost of the cigarettepack as a clean-up cost. Attractive crucibles could be placed on the street to promote the appropriate disposal of cigarette butts, as shown in Figures 1a and 1b.

Figure 1. a) California b) Seattle

At present companies who recycle cigarette butts in India pay a collection fee of Rs. 700 for everykilogram and Rs. 80 for every 100 grams of cigarette waste.

5. What is harmful about cigarette butts?

Cigarette butts take anywhere from 18 months to ten years to decompose.

Chemicals and heavy metals, the same things that kill smokers, leak into the environment whenbutts are discarded. They frequently wash into the gutters, polluting our rivers and endangering fish and other creatures.

The most serious issue is the filters, which are mostly comprised of cellulose acetate, a type ofplastic. That means the filters take a long time to biodegrade, even when they do, thousands of tiny plastic fibers wind up in the environment, streams, and wildlife. Arsenic, benzene, and lead are among the substances found in the fibers.

Cigarette butts are hazardous plastic fragments that are frequently mishandled and contain over15,000 separate strands of plastic fiber. Cigarette butts can be carried into rivers and lakes, eventually ending up in the ocean. This source might potentially release 300,000 tons of micro-plastic fibers into the aquatic ecosystem each year.

Furthermore, harmful compounds such as nicotine, carcinogenic tar, and polycyclic aromatichydrocarbons have a highly toxic effect on aquatic creatures, causing serious damage.

According to the WHO, tobacco product waste contains over 7,000 toxic chemicals. Cellulose acetate, the major component of cigarette butts, along with paper and rayon, pollutes land and water. A report by Truth Initiative, a U.S.-based NGO says that the chemicals that leach from a single cigarettebutt can release enough toxins to kill 50% of saltwater and freshwater fish exposed to it for 96 hours. [3,4]

There is also the fear of birds, fish, and animals eating cigarette butts that can be disastrous," viii. As a result, it is vital to comprehend the mechanism by which cigarette butts release micro Plastic fibers, the potential environmental impact, and viable mitigation techniques for cigarette butt litter. [5]

6. How is the cigarette butt processed for recycling?

The cigarettes get shredded and separated into 2 key parts, the organic matter which is the ash, tobacco, and paper composted and the inorganic component which is the filter is melted.

The paper is ground in an industrial grinder and treated with an organic binder. The ground material is spread into sheets and dried. They are later cut and sold in local shops as mosquito repellant. The paper with remains of nicotine act as a mosquito repellant. To use the mosquito repellant, it just has to be burnt.

The plastic in the cigarette butt is also ground in the industrial grinder, they are then soakedin sterilizing chemicals for 24 hours which leaves them looking like cotton. Later the plastic fibersare fluffed and made to the required specifications.

It is not necessary that the plastic from cigarette butts be recycled separately. It can be combined with other plastics such as polyethylene and polypropylene. Cellulose acetate is found in objects like combs, toothbrushes, playing cards, tool handles, pen barrels, etc. made from plastic. Hence all these plastics can be recycled together. [6]

7. What can be made from cigarette butts?

- Composted organic waste from cigarette butts is used as fertilizers.
- At present in India cigarette butts are recycled and used to make mosquito repellants and soft toys.
- Butts is also used to make packing and building material as well as plastics for key chainsand toys and cushion. [7]
- In France the fibers from cigarette butts are used to make jackets, here the plastics obtained from cigarette butts can be used as an insulating material, as shown in figure 2 a & 2 b. [8]

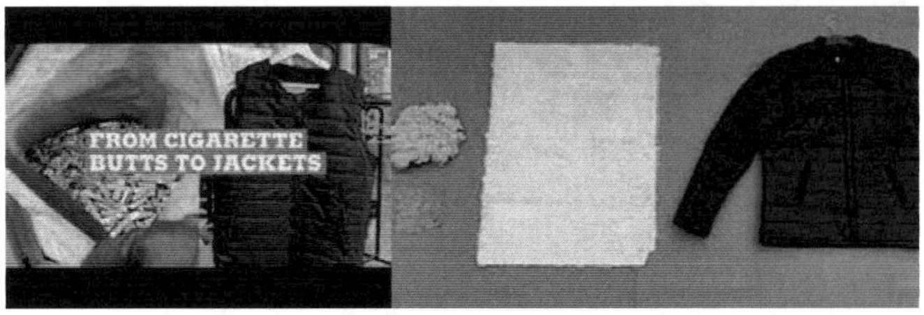

Figure 2. Insulating material from CB used for jackets in France

If the cigarette butt is not recycled into an industrial or commercial product it is converted into pellets which can be used in any plastic industry Figure 4.

Figure 3. Pellets of plastic from cigarette butts

Also, France manufactures benches and furniture from plastic obtained from cigarette butts figure 5.

Figure 4. Benches made from CB in France

8. How cigarette butts can be used in the construction industry.

Cigarette butts with leftover tobacco had an energy value of 16.5 MJ/kg. If just 2.5% of all bricks produced annually worldwide included 1% CB content, all of the CBs currently produced could be recycled in bricks, and it is estimated that global firing energy consumptioncould be reduced by approximately 20 billion MJ (mega joules). This is roughly equal to the annual energy consumption of one million houses in Victoria, Australia. "Firing butts into bricks is a reliable and practical way to deal with this terrible environmental problem, whileat the same time cutting brickmaking production costs," Mohajerani said.
Cigarette butt bricks are also lighter and provide superior insulation, resulting in lower heating and cooling expenditures in the home. Mohajerani's research team discovered that once cigarette butts were employed on the pavement, they could not only withstand high traffic, butit could also minimize heat created by the sun hammering down on black asphalt, aphenomenon known as the urban heat-island effect. It was discovered that butts encapsulatedin bitumen and combined with class 170 bitumen performed better in heavy traffic than paraffin-encapsulated butts mixed with c170 and c320 bitumen after completing various testson the samples. The butts encased in paraffin were barely able to withstand light traffic. [9]
Insulating material that can be used in the construction industry can be made from the plastic in cigarette butts.

9. Is recycling cigarette butts a small business?

About 6 trillion cigarettes are produced every year, leading to more than 1.2 million tons of cigarette butt waste. These figures are expected to increase by more than 50 percent by 2025,mainly due to an increase in the world population.

It is estimated that 4.5 trillion cigarette butts are disposed of annually. An estimated 766,571metric tons of cigarette butts make their way into the environment every year.

10. Has the litter of cigarette butts reduced wherever an attempt has been made?

By the financial year 2019-2020 a whopping 10,000 kg of cigarette waste, of which 40 percent has been recycled and utilized. Each kg of cigarette waste comprises 4,000 cigarette butts. A company by the name CODE has pioneered in India recycling cigarette butts. Having started in 2016 in Noida so far the duo, Vishal Kanet and Noman Gupta have successfully collected 500-600 kilograms of cigarette waste and recycled 40-50 kgs. CODE stands for Conservation of the depleting environment.

11. Who should give a go-ahead to the project?

Non-smokers should give a go-ahead for this method of recycling micro plastics. When smokers are careless about the harm they do to their own bodies by smoking, non-smokers who strongly oppose this habit should propagate the ill effects of the micro plastic environment and come up with innovative methods of freeing the environment of this nuisance.

12. Scope of this innovative method

Before anything can be done technically to recycle cigarette butts, education has to play a major rolein this field. Non-smokers who understand the harm of cigarettes should take the lead in recycling cigarette butts. Social stigma should not be feared and a bold and confident approach should be taken to recycling this notorious element. Smokers should be educated about the harm of cigarette butts and should be encouraged to dispose of cigarette butts in the proper manner. The misconception that cigarettebutts are made of cotton which is biodegradable should be changed and people need to know that cigarette butts are ma of non-biodegradable micro plastic. Fines may be imposed on those who arecareless in this matter. Government should finance start-ups in this area and motivate them towards a cleanIndia and world.

13. Conclusion

Like every drop makes an ocean, every cigarette butt adds up to the litter which amounts to 4.5 trillion individual butts polluting our global environment. Due to a long history of the tobacco industry targeting, low-income communities of color, they have greater smoking rates and confront systemic disparities that make it difficult for them to access health information and resources. Furthermore, they are disproportionately served by inequitable waste management, which can exacerbate the consequencesof cigarette butt pollution. Hence recycling cigarette butts makes a sustainable innovative solutionto the problem.

References

[1] Puls, J.; Wilson, S.A.; Hölter, D. Degradation of Cellulose Acetate-Based Materials: A Review. J. Polym. Environ. 2010, 19, 152–165. 8. Tchounwou, P.B. Retraction: Herndon J.M. Evidence of Coal-Fly-Ash Toxic Chemical Geoengineering in the Troposphere: Consequences for Public Health. Int. J.Environ. Res. Public Health 2015, 12, 9375–9390.
[2] Novotny, T.E.; Slaughter, E. Tobacco Product Waste: An Environmental Approach to Reduce Tobacco Consumption. Curr. Environ. Health Rpt. 2014, 1, 208–216.
[3] Green, A.L.R.; Putschew, A.; Nehls, T.; et al. Littered cigarette butts as a source of nicotine in urban waters. J. Hydrol. 2014, 519, 3466–3474
[4] Potentially toxic elements leachate from cigarette butts into different types of water: A threat for aquatic environments and ecosystems? Akhbarizadeh, R., Dobaradaran, S., Parhizgar, G., Schmidt, T.C., Mallaki, R. 2021, Environmental Research
[5] Recycling Waste Cigarette Butts in Dense Graded Asphalt Tareq Rahman, M., Mohajerani, A.2021 Journal of Materials in Civil Engineering Smoked cigarette butts: Unignorably source for environmental micro plastic fibers, Shen, M., Li,Y., Song, B., (...), Gong, J., Zeng, G.2021, Science of the Total
[6] Environment Converting the cigarette butts into valuable products using the pyrolysis process, Hazbehiean, M., Mokhtarian, N., Hallajisani, A.,2022, Global Journal of Environmental Science and Management
[7] Murugan, K.; Suresh, U.; Panneerselvam, C.; Rajaganesh, R.; Roni, M.; Aziz, A.T.; Hwang, J.-S.; Sathish Kumar, K.; Rajasekar, A.; Higuchi, A.; et al. Managing wastes as green resources: Cigarette butts synthesized pesticides are highly toxic to malaria vectors with little impact on predatory copepods.Environ. Sci. Pollut. Res. 2017, 25, 10456–10470.
[8] A second life for cigarette butts? A review of recycling solutions, Marinello S., Lolli F., GamberiniR., Rimini B., (2020) *Journal of Hazardous Materials*, 384, art. no. 121245
[9] Selmar, D.; Radwan, A.; Abdalla, N.; Taha, H.; Wittke, C.; El-Henawy, A.; Alshaal, T.; Amer, M.; Kleinwächter, M.; Nowak, M.; et al. Uptake of nicotine from discarded cigarette butts—A so far unconsidered path of contamination of plant-derived commodities. Environ. Pollut. 2018, 238, 972– 976.

THE NEED FOR LEGISLATION ON RENEWABLE ENERGY FOR SUSTAINABLE DEVELOPMENT IN INDIA

A Bhargava[1]and K Kochhar[2]
[1]Assistant Professor, Faculty of Law, Banasthali Vidyapith, Rajasthan, India
[2]Assistant Professor, Faculty of Management Studies, Banasthali Vidyapith, Rajasthan, India

Abstract: Energy is required for each and every activity since the beginning of this universe. With the advancement of human beings their activities have developed and have become more and more complex with time. As the world developed at a very high pace energy requirements also increased and this resulted in more utilization of conventional energy sources. Use of conventional energy sources (fossil fuels) has resulted in environmental degradation which has become a major problem worldwide.Development is indispensible but at the same time it should not be at the cost of environment so with this came up the concept of sustainable development. To achieve this we have to switch over to non conventional sources of energy. Considering the geographical conditions, India has huge potential regarding renewable energy. For systematic development of renewable energy sector we need a comprehensive legislation. If we look at international level there are many countries which have made separate and ancillary legislations regarding renewable energy and their growth is remarkable. Therefore, this study has focused to examine the need of separate renewable energy law for sustainable development.
Keywords: Renewable Energy, Environment degradation, sustainable development, legislation.

1. Introduction

"The four elements of life - earth, water, air, energy - are common property of the human race. The management and use of such portions thereof as are vested in or assigned to particular ownership, private or corporate or national or regional, of definite or indefinite tenure, of individualist or collectivist economy, shall be subordinated in each and all cases to the inherent interest of the common good."[1]

Energy development is considered to be barometer of economic growth.[2] As the world developed at a very high pace energy requirements also increased and this resulted in more utilization of conventional energy sources. As per a 2007 report of the Government of India, 'Energy is the sine qua non of development'.[3] Till now, human beings have used conventional sources of energy for economic growth and development, which has resulted in environmental degradation. So, what we require is not just development but sustainable development.

[1]*See,* Preliminary Draft of World Constitution, 1948.
[2]*See,*B. Sudhakara Reddy, Gaudenz B. Assenza, Dora Assenza, and Franziska Hasselmann, *Energy Efficiency and Climate Change: Conserving Power for a Sustainable Future* (New Delhi: SAGE Publications, 2009), p. 1.
[3]*See*, Ministry of Environment and Forest and Bureau of Energy Efficiency under Ministry of Power, Government of India, 'India: Addressing Energy Security And Climate Change" (2007), p.1, available http://www.ceeindia.org/cee/pr/publication.pdf,accessed on 20th February 2019.

Therefore, the problem before the world is switching to other energy options without hampering economic growth which cannot be compromised. The experts say, switching to renewable energy and other alternative forms of clean energy is not just the best choice under present circumstances, but it is our 'only option'.[4] Moreover it should be affordable as high cost will again hamper economic growth.
India is a party to Paris Agreement where as a good global citizen it has set targets regarding production of energy from renewable energy sources. India has raised its solar target from 20 GW (pre-Paris) to 100 GW by 2022. Given that India's output levels are between 3-4 GW, meeting, meeting the 100 GW will require more than a 50 percent compound annual growth rate for the next few years.[5]In addition to that India's target for overall renewable of 175 GW is ambitious. For achieving these targets and to meet its ever increasing energy demand for keeping growth rate intact India must have proper legal and administrative mechanism in place.

2. Literature review

M. Afzal Wani (2018) explains the interplay between Humanity, Energy & Law give suggestions as to how a policy has to be framed by keeping balance between these three and importance of alternative energy sources in this regard.[6]
Klaus Bosselmann and Pushpa Kumar Lakshmanan (2018) explains the legal concept and doctrines related to energy, why there is a need for the world to switch to renewable energy, how renewable energy is perceived at international level and what is the approach of different countries in this regard.[7]
Megha Kaladharan (2015) does a critical analysis of renewable energy sector in India. The lacunas that exist in legal and administrative set up in India. The approach of executive and judiciary regarding renewable energy in India. [8]
Armin Rosencranz, Rajnish Wadhera, Neelakshi Bhadauria& Pranay Chitale (2018) explain the potential India has in renewable energy sector. What legislative and policy initiatives India has undertaken. How has been growth of this sector and the financial and legal measures required to promote investments in this sector.[9]
Sanjay Upadhyay (2018) explains the current legal position of renewable energy in India, the lacunas which exist in legal and administrative mechanism and as to why a separate legislation is required for renewable energy[10]

---

[4]*See,* 'The Energy Report: 100% Renewable Energy by 2050(WWF, Ecofys, OMA, 2011), p.13.
[5]*See,*V.Sivaram, 'Do India's Renewable Energy Targets Make Sense?', Council on Foreign Relations (11 March2015), available http://blogs.cfr.org/levi/2015/03/11/do-indias-renewable-energy-targets-make-sense/,accessed on 3rd April 2019.
[6]*See,*Wani, M. Afzal. 2018. Humanity, Energy and Law Urgencies and Challenges In Usha Tandon (Ed.), Energy Law and Policy (pp.348-378). New Delhi, Delhi: Oxford University Press.
[7]*See,*Bosselmann, K. and Lakshmanan, P., 2018. The Atmosphere as a Global Commons and Cleansing It with New Energy Options. In: U. Tandon, ed., *Energy Law And Policy*, 1st ed. New Delhi: Oxford University Press, pp.3-28.
[8]Kaladharan, Megha (2015) *India Renewable Energy in India: An Analysis of the Regulatory Environment and Evolving Policy Trends.* Retrieved fromhttps://www.cprindia.org:

[9]Rosencranz, A., Wadhera, R., Bhadauria, N. and Chitale, P., 2018. Clean Energy in India: Supply and Prospects. In: U. Tandon, ed., *Energy Law And Policy*, 1st ed. New Delhi: Oxford University Press, pp.121-154.
[10]*See,*Upadhyay, S., 2018. Renewable Energy Development in India: The Need for a Robust Legal Framework. In: U. Tandon, ed., *Energy Law And Policy*, 1st ed. New Delhi: Oxford University Press, pp.155-180.

Sushanta K. Chatterjee (2017) explains what should be the approach regarding renewable purchase obligation (RPO), fixed contract Feed-in-Tariff (FIT) and the market-based Renewable Energy Certificate (REC) mechanism in India.[11]
Erimma Gloria Orie (2018) explains the remarkable growth of China in renewable energy sector. The legal framework, policy framework and administrative framework that exists in China and what we can learn from China.[12]

3. Research Objectives

a) To understand the role of renewable energy in sustainable development.
b) To analyze the current position of renewable energy laws in India and in other countries.
c) To understand the importance and significance of separate and comprehensive legislation for systematic development of renewable energy sector.

4. Methodology

The present study is descriptive and analytical in nature. The sources of data collection are purely secondary based. Data has been extracted from books, journals, articles and different websites.

5. Analysis

a) Role of renewable energy in sustainable development
b) Current position of renewable energy laws in India and in other countries
c) Importance and significance of separate and comprehensive legislation for systematic development of renewable energy sector

Role of Renewable Energy in Sustainable Development

Energy is required for each and every activity since the beginning of this universe. With the advancement of human beings their activities have developed and have become more and more complex with time. Industrial revolution is considered as a major milestone in human civilization. With industrial revolution came huge mechanization and to drive machines energy was required which was generated through fossil fuels particularly coal. The use of these conventional sources of energy has resulted in pollution. This pollution had reached to that extent that it became a matter of global concern which was addressed at various international forums.
It was realized that development is important but not at the cost of environment. Such development is known as sustainable development. Thus, the problem before the world is switching to other energy options without hampering economic growth which cannot be compromised. "The experts say, switching to renewable energy and other alternative forms of

[11]Chatterjee,Sushanta.K. (2017) *The Renewable Energy Policy Dilemma in India: Should Renewable Energy Certificate mechanism compete or merge with the Feed-in-Tariff Scheme*? www.hks.harvard.edu/mrcbg:
[12]*See*, Orie,Erimma. Gloria. 2018. Examining the Legal Impediments in the Development of Renewable Energy in
Nigeria. In: U.Tandon ed., *Energy Law and Policy*, 1st ed. New Delhi: Oxford University Press, pp.181-214.

IOP Conf. Series: Earth and Environmental Science 1084 (2022) 012028 doi:10.1088/1755-1315/1084/1/012028

clean energy is not just the best choice under present circumstances, but it is our ‘only option’.”[13]

Law Relating to Renewable Energy in India

The constitution of a country is the basic and supreme law of the land. It is the basic source from which other laws emanate. When we go through the seventh schedule of the Indian Constitution which comprises of three list that is the Union, State and Concurrent list, we find that renewable energy has not been specifically mentioned in any of these lists. But we see that electricity is mentioned in the concurrent list at entry 38[14]which means that centre and state governments are empowered to legislate on this matter. The Constitution of India lists the subject matter of laws which can be made by Parliament and by the legislatures under Article 246.[15]Under this Article the Parliament has plenary powers to make any law including the law to amend the Constitution subject to the limitations laid down therein.[16]Further, the power to legislate on a topic of legislation carries with it the power to legislate on an ancillary matter which can be said to be reasonably included in the power given.[17] The State under Article 48-A read with Article 51-A (g), Article 14 and Article 21 of the Constitution, must endeavour to protect and improve the environment in addition to Article 39(b) where the State has a duty to direct its policy towards ensuring that the ownership and control of the material resources of the community are so distributed as best to sub-serve the common good.[18]Entry 33 of List III relates to trade and commerce in and the production, supply, and distribution of the products of any industry where the control of such industry by the Union is declared by Parliament by law to be expedient in public interest and supply and distribution includes the power to control the price of the commodities coming under this entry.[19]As we see that renewable energy has not been specifically mentioned in the seventh schedule so as per the scheme provided under Article 248 read with Entry 97, List I of Seventh Schedule of the Indian Constitution the residuary power of legislation has been given to the Parliament where it can legislate on those matters which are not mentioned in the Concurrent List or State List and can impose tax on such matters which are not mentioned in any of these lists. Through the Constitution (Seventy Third Amendment) Act, 1992 Part IX which deals with Panchayats was added to the Indian Constitution and through this amendment Eleventh Schedule was also added in which non-conventional energy resources have been included as 15th item. So,we see that renewable energy finds place in the Eleventh Schedule of the Indian Constitution.

Besides the scheme provided in the Indian Constitution we have legislations like The Electricity Act, 2003 and The Energy Conservation Act, 2001 which have provisions for renewable energy. As per Section 3 of the Electricity Act, 2003 “the Central Government shall, from time to time, prepare the National Electricity Policy and tariff policy, in consultation with the State Governments and the Authority for development of the power system based on optimal

[13] ‘The Energy Report: 100% Renewable Energy by 2050(WWF, Ecofys, OMA, 2011), p.13.

[14]Schedule VII, Concurrent List, Entry 38, The Constitution of India

[15]Upadhyay, S., 2018. Renewable Energy Development in India: The Need for a Robust Legal Framework. In: U. Tandon, ed., *Energy Law And Policy*, 1st ed. New Delhi: Oxford University Press, pp.155-180.

[16]Golak Nath v. State of Punjab, (1967) SC 1643

[17]State of Rajasthan v. G. Chawla, *(1959) AIR SC 544*

[18]Upadhyay, S., 2018. Renewable Energy Development in India: The Need for a Robust Legal Framework. In: U. Tandon, ed., *Energy Law And Policy*, 1st ed. New Delhi: Oxford University Press, pp.155-180.

[19]Jain v Union of India, (1970) 2 S.C. W.R 59 (India)

utilization of resources such as coal, natural gas, nuclear substances or materials, hydro and renewable sources of energy."[20]

As per Section 4 of the Electricity Act, 2003 "the Central Government shall, after consultation with the State Governments, prepare and notify a national policy, permitting stand-alone systems (including those based on renewable sources of energy and other non-conventional sources of energy) for rural areas."[21] Analyzing this we find that these sections as such do not create any substantive legal mandate apart from bringing renewable energy also as a player into the centre-stage of policy formulation. To give real boost to this sector it is important to have a clear legal mandate to attract investors.[22]

The Appropriate Commission i.e. Central Regulatory Commission or State Regulatory Commission shall be guided by the principle for promotion of co-generation and generation of electricity from renewable sources of energy and the National Electricity Policy and tariff policy while specifying the terms and conditions for the determination of tariff.[23] Analyzing this we find that tariff fixation is also the responsibility of the State Electricity Regulatory Commissions (SERCs) and tariffs have to be fixed separately in different category of renewable energy, that is solar, biomass, wind, hydro, and other sources where within these categories there are a lot of differentiation based on technology used and the size of the project.[24]

The other issues that emerge include the time line within which the tariff has to be fixed and there is no guarantee of payment of fixed tariff for fixed number of years which is decided primarily keeping in mind the life of a project and profitability of the project among other factors.[25] Take for example in Germany where it extends up to 20 years for solar and winds based projects.[26] The tariff fixation does not take into account the actual gradual technology advancement and market development in consequence of technology becoming cheaper in determining the tariff for the later years of project life; there should be a specific provision which takes into account the gradual technological advancements as well as market development; tariff fixation is of critical importance for any renewable energy project and is highly technical and nuanced process which has to take into account a multiplicity of factors- technical, financial, and market related.[27] In addition to this other factors like the location of the project and the quality of the resource play a very important and significant role where one set of tariff for wind energy cannot fit all the projects of wind energy.[28] This section gives all

[20]Section 3 (1), Electricity Act, 2003

[21]Section 4 (1), Electricity Act, 2003

[22]Upadhyay, S., 2018. Renewable Energy Development in India: The Need for a Robust Legal Framework. In: U. Tandon, ed., *Energy Law And Policy*, 1st ed. New Delhi: Oxford University Press, pp.155-180.

[23]See Section 61, Electricity Act, 2003

[24]Upadhyay, S., 2018. Renewable Energy Development in India: The Need for a Robust Legal Framework. In: U. Tandon, ed., *Energy Law And Policy*, 1st ed. New Delhi: Oxford University Press, pp.155-180.

[25]*Supra*

[26]'Feed-In Tariffs in Germany', available https://en.wikipedia.org/wiki/Feed-in_tariffs_in_Germany, accessed on 3 July 2021.

[27]Upadhyay, S., 2018. Renewable Energy Development in India: The Need for a Robust Legal Framework. In: U. Tandon, ed., *Energy Law And Policy*, 1st ed. New Delhi: Oxford University Press, pp.155-180.

[28]*Supra*

IOP Conf. Series: Earth and Environmental Science 1084 (2022) 012028 doi:10.1088/1755-1315/1084/1/012028

the powers regarding this to the State Electricity Regulatory Commission but there should be a provision in the Act itself about the manner in which this process is carried out.[29]

As per Section 86 of the Electricity Act, 2003 the State Commission shall promote co-generation and generation of electricity from renewable sources of energy by providing suitable measures for connectivity with the grid and sale of electricity to any person, and also specify, for purchase of electricity from such sources, a percentage of the total consumption of electricity in the area of a distribution licensee[30] which means that it is mandatory for the State Commissions to provide Renewable Power Purchase Obligations (RPPO). Renewable Power Purchase Obligations creates an environment which is conducive in nature for the development of renewable energy sector in the state by generating confidence among developers as it is mandatory for the distribution licensee to purchase the electricity produced by them from the renewable energy sources.[31]But we see that the arrangement provided by the legislation is still not free from the challenges as the mandate of the Electricity Act, 2003 does not fix any time line within which this renewable power purchase obligations (RPPO) has to be fixed; the Electricity, Act, 2003 does not venture into the minimum percentage of renewable energy that the State Electricity Regulatory Commission have to fix; the Electricity, Act, 2003 is also silent on how a gradual increase in percentage of RPPO can be ensured by SERCs; there is no mandate to provide a National RPPO; the orders/regulations specifying the percentage of renewable power procurement lack the monitoring and enforcement mechanism at present.[32] In addition to this the SERCs have no control over the renewable energy industry development, availability of renewable energy equipment, availability of and access to latest technology, availability of tailor made or suitable finances to ensure that generation of electricity from renewable energy increase over a period of time as these are some of the critical areas without which renewable energy generation in the country cannot take off and these are also the areas which have not been dealt with in the Electricity Act, 2003.[33]

The other legislation that deals with renewable energy is the Energy Conservation, 2001 which has been enacted with the objective to provide for efficient use of energy and its conservation and for matters connected therewith or incidental thereto.[34]According to the definition of 'Energy' given in Section 2(h) of Energy Conservation Act, 2001 'Energy' includes electricity generated from renewable sources of energy or bio-mass connected to the grid but other applications of renewable energy has not been discussed within the ambit of 'energy'.[35] The expansion of the word 'energy' in context of renewable energy by including any form of energy derived from any renewable source will widen the ambit of the scope of functions and powers of the Bureau of Energy Efficiency (BEE) enabling it to take steps to increase the efficiency of energy generated from renewable energy sources.[36]

---

[29]*Supra*

[30]See Section 86(1) (e), Electricity Act, 2003

[31]Upadhyay, S., 2018. Renewable Energy Development in India: The Need for a Robust Legal Framework. In: U. Tandon, ed., *Energy Law And Policy*, 1st ed. New Delhi: Oxford University Press, pp.155-180.

[32]*Supra*

[33]*Supra*

[34]Preamble, Energy Conservation Act, 2001.

[35]Upadhyay, S., 2018. Renewable Energy Development in India: The Need for a Robust Legal Framework. In: U. Tandon, ed., *Energy Law And Policy*, 1st ed. New Delhi: Oxford University Press, pp.155-180.

[36]*Supra*

After examining the pertinent provisions from the Electricity Act, 2003 and the Energy Conservation Act, 2001 regarding renewable energy, we see that there are legal issues which are not addressed[37]:

1. The statement of objects specifies 13 main features of the Act but at the same time it does not focus on development of RE, with an exception where it allows for the setting up of freestanding systems for generation and distribution in rural and remote areas. The Electricity Act does not focus on the promotion and development of renewable energy.
2. Holistic approach is required for the development of renewable energy sector in India covering various sources and their applications ensuring their efficient use for generation of energy not only in the form of electricity but other forms of energy as well. These aspects are missing in the existing legal framework.
3. For the development of renewable energy sector, resource mapping is highly required and the existing legal framework does not address this important and significant aspect.
4. The technology development in the field of renewable sector has so far happened in a fragmented manner thus it is yet to be made mandatory by law so as to avoid being too discretionary.
5. To ensure the development and promotion of renewable energy sector the use of renewable energy resources has to be made legally mandatory, particularly at government and commercial establishments. The Electricity Act has not made any provisions in this regard as it only deals with electricity generation.
6. The provisions so as to provide incentives to the people for generating their own electricity by installing solar water heaters or solar panels are not made in the legislation.
7. The existing legal framework regarding renewable energy does not include resources like Biofuels rather it should also include the full economic chain right from the farm to the market.
8. The existing legal framework does not include fuel cell development, hybrid cars and other technological developments within its ambit.

Here we can see that the provisions in the Electricity Act[38] dealing with renewable energy are enabling provisions thus lacking efficiency and a separate legislation which is comprehensive in nature providing clear mandate is highly required.

The Requirement for Separate Legislation

We need separate legislation regarding renewable energy for the following reasons[39]

1. Legislation provides more certainty than a collection of policies as there is a probability that policies change with the shift in political environment.

[37] *See,* Upadhyay, S., 2018. Renewable Energy Development in India: The Need for a Robust Legal Framework. In: U. Tandon, ed., *Energy Law And Policy*, 1st ed. New Delhi: Oxford University Press, pp.155-180.

[38] *See,* Upadhyay, S., 2018. Renewable Energy Development in India: The Need for a Robust Legal Framework.

In: U. Tandon, ed., *Energy Law And Policy*, 1st ed. New Delhi: Oxford University Press, pp.155-180.

[39] Upadhyay, S., 2018. Renewable Energy Development in India: The Need for a Robust Legal Framework. In: U. Tandon, ed., *Energy Law And Policy*, 1st ed. New Delhi: Oxford University Press, pp.155-180.

2. Legislation removes inconsistencies and provides clarity thus making things easier for the foreign investors which eventually leads to increased foreign investment ensuring financial viability.
3. The provisions regarding renewable energy sector in the Electricity Act, 2003 are enabling provisions which results in their ineffective enforcement while for the development and promotion of renewable energy sector mandatory provisions are required.
4. Currently the renewable energy sector in India is governed through the Electricity Act, 2003, the Energy Conservation Act, 2001 and the various policies dealing with tariff and electricity etc. which do not touch all the issues pertinent to this sector, thus it is insufficient and fragmentary eventually becoming ineffective in reality.
5. Various aspects pertinent to renewable energy sector will be covered in the comprehensive legislative framework thus ensuring a holistic approach which is indispensable for the systematic development of renewable energy sector in India.
6. The countries where separate legislative framework has been provided for the renewable energy have made remarkable and significant development so it is the most effective means to achieve desired results which is evident when one goes through the legal set up created in those countries.

Legal Framework Created through Legislation(s) in Various Countries

China

In China the development of renewable energy sector has been remarkable, perhaps it has set an example for the world. Here also a huge credit can be given to the comprehensive legal framework which has been developed in China regarding renewable energy sector. The first law to regulate renewable energy sector the Renewable Energy Law came into force in the year 2006 (amended in December 2009) along with other legislations which were enforced covering ancillary matters like the Land Administration Law (1999) Cleaner Production Promotion Law (2003), Energy Conservation Law (2007) and Circular Economy Promotion Law (2009). The law has been made so as to promote the development and utilizing the renewable energy by increasing the supply of energy, improving the structure of energy, safeguarding the safety of energy, protecting environment and realizing a sustainable economic and social development.[40]Under this law "it shall be the duty of state to give priority to the development and utilization of renewable energy in energy development and promote the establishment and development of the renewable energy market by setting an overall target for the development and utilization of renewable energy and adopting corresponding measures and it shall be the duty of state to encourage economic subjects of different ownership to participate in the development and utilization of renewable energy and to protect the legitimate rights and interests of those who develop and utilize renewable energy".[41]

[40] Article 1 Renewable Energy Law of People's Republic of China.
[41] Article 4 Renewable Energy Law of People's Republic of China.

Germany
Germany over the years has come up with a series of legislations which are conducive to the changing needs and requirements of renewable energy sector like Electricity Feed-in Act (1991), Renewable Energy Sources Act (2000), PV Interim Act (2003, Renewable Energy Sources Act (2004), PV Act (2010), PV Interim Act (2011), Renewable Energy Sources Act (2012), PV Act (2013), Renewable Energy Sources Act (2014), Renewable Energy Sources Act (2017) and Renewable Energy Act(2021). We can see that Germany has made remarkable growth in the renewable energy sector adopting 'energy U-turn' popularly known as *Energiewende* where it has set a goal to reduce its green house gases to 95% by 2050.

United Kingdom
In United Kingdom the Sustainable Energy Act, 2003 was enacted which deals with the provisions regarding the promotion and development of a sustainable energy policy.

Australia
In Australia through the Renewable Energy (Electricity) Act, 2000 a mandatory national renewable energy target has been set.

6. Conclusion

Energy is required for every activity thus it is indispensable for the further growth and sustainable development. This can only be achieved through energy generated from renewable energy sources. Considering the various geographical and climatic conditions like big geographical area having vast coast line, receiving plenty of sunlight throughout year, India has huge potential to become an international hub in renewable energy sector thus providing employment opportunities to millions. The performance of India in this sector has been good but it is not up to its potential. So far, this sector has been majorly regulated through various policies and the legislative efforts made concerning this sector have proved to be piecemeal measures. While the countries which have made exclusive legislation(s) regarding renewable energy and other ancillary matters have made significant and remarkable development. The energy sector involves various aspects like installation, generation, transmission and distribution. The systematic development requires holistic approach which can only be achieved through various exclusive legislations covering not only the various aspects of renewable energy but other ancillary matters as well. Thus, what we need is a separate legislative framework.

References
[1] B. Sudhakara Reddy, Gaudenz B. Assenza, Dora Assenza, and Franziska Hasselmann, *Energy Efficiency and Climate Change: Conserving Power for a Sustainable Future* (New Delhi: SAGE Publications, 2009), p. 1.
[2] Ministry of Environment and Forest and Bureau of Energy Efficiency under Ministry of Power, Government of India, 'India: Addressing Energy Security And Climate Change" (2007), p.1, available http://www.ceeindia.org/cee/pr/publication.pdf,accessed on 20th February 2019.

IOP Conf. Series: Earth and Environmental Science 1084 (2022) 012028 doi:10.1088/1755-1315/1084/1/012028

[3] 'The Energy Report: 100% Renewable Energy by 2050(WWF, Ecofys, OMA, 2011), p.13.
[4] V.Sivaram, 'Do India's Renewable Energy Targets Make Sense?', Council on Foreign Relations (11 March2015), available http://blogs.cfr.org/levi/2015/03/11/do-indias-renewable-energy-targets-make-sense/,accessed on 3rd April 2019
[5] Wani, M. Afzal. 2018. Humanity, Energy and Law Urgencies and Challenges In Usha Tandon (Ed.), EnergyLaw and Policy (pp.348-378). New Delhi, Delhi: Oxford University Press.
[6] Bosselmann, K. and Lakshmanan, P., 2018. The Atmosphere as a Global Commons and Cleansing It with New Energy Options. In: U. Tandon, ed., *Energy Law And Policy*, 1st ed. New Delhi: Oxford University Press,pp.3-28.
[7] Kaladharan, Megha (2015) *India Renewable Energy in India: An Analysis of the Regulatory Environment and Evolving Policy Trends.* Retrieved fromhttps://www.cprindia.org:
[8] Rosencranz, A., Wadhera, R., Bhadauria, N. and Chitale, P., 2018. Clean Energy in India: Supply and Prospects. In: U. Tandon, ed., *Energy Law And Policy*, 1st ed. New Delhi: Oxford University Press, pp.121-154.
[9] Upadhyay, S., 2018. Renewable Energy Development in India: The Need for a Robust Legal. Framework. In: U. Tandon, ed., *Energy Law And Policy*, 1st ed. New Delhi: Oxford University Press, pp.155-180.
[10] Chatterjee,Sushanta.K. (2017) *The Renewable Energy Policy Dilemma in India: Should Renewable EnergyCertificate mechanism compete or merge with the Feed-in-Tariff Scheme*? www.hks.harvard.edu/mrcbg:
[11] Orie,Erimma. Gloria. 2018. Examining the Legal Impediments in the Development of Renewable Energyin Nigeria. In: U.Tandon ed., *Energy Law and Policy*, 1st ed. New Delhi: Oxford University Press, pp.181-214.
[12] 'The Energy Report: 100% Renewable Energy by 2050(WWF, Ecofys, OMA, 2011), p.13.
[13] Schedule VII, Concurrent List, Entry 38, The Constitution of India
[14] Golak Nath v. State of Punjab, (1967) SC 1643
[15] State of Rajasthan v. G. Chawla, *(1959) AIR SC 544*
[16] Jain v Union of India, (1970) 2 S.C. W.R 59 (India)
[17] Section 3 (1), Electricity Act, 2003
[18] Section 4 (1), Electricity Act, 2003
[19] See Section 61, Electricity Act, 2003
[20] 'Feed-In Tariffs in Germany', available https://en.wikipedia.org/wiki/Feed-in_tariffs_in_Germany, accessed on3 July 2021.
[21] Section 86(1) (e), Electricity Act, 2003.
[22] Preamble, Energy Conservation Act, 2001.
[23] Article 1 Renewable Energy Law of People's Republic of China.
[24] Article 4 Renewable Energy Law of People's Republic of China.

Utilization of Silica Fume and Fly ash in Cement and Partial Substitution of Fine Aggregate Using Quarry Dust in Concrete

SudhirKapgate[1] and Anjul Tomar[2]
[1]Research Scholar, G H Raisoni University, Amravati, India
[2]AssistantProfessor, G H Raisoni University, Amravati, India

Email: kapgate.sudhir68@gmail.com

Abstract-Thecemententerprisebetheessentialhotspotintendedfornaturalcontamination because it admissions a tremendous measure of energy during the creationas well as deliveries a ton of CO_2which adds to an unnatural weather change. Reducethese impacts, concrete is to some degree subbed among the blend of Silica Fume (SF) and Fly Ash (FA) and sand by Quarry Dust (QD) with addition of steel fibre. This paper is centered on theconduct of substantial blend of M30 grade with particular of mechanical properties. Various mixes of rate substitutions of concrete by SF (5%) and FA (5%,10%,15%, 20%, 25%) and quarry dust (5%, 10%, 15%,20% , 25%) with addition of steel fibre (1 %) were embraced for the making of mixed substantial blend. Test results indicated to facilitate the accumulation of silica fume and fly ash in cement and quarry dust in sand improved the compressive strengths of concrete.
Keywords: Silica Fume, Fly Ash, Quarry Dust, Compressive strength.

1. Introduction

A dangerous atmospheric devation and incessant changes in environment are moving issues all over the planet. The expanded interest for the assembling and use of cement in concrete is related with ecological contamination and prompting gigantic discharge of green house gases. Creation of 1 ton of folio produces CO_2 equivalent amounts into the environment, accounting for 5 to 7% of entire CO_2 discharged worldwide. The creation of Portland concrete arrived at 85.4 million tons internationally in 2018, which is expanding by 12% beginning around 2013. This figure is probably going to increment considerably more in the next few decades as numerous nations are focusing on quick up grade towards frame work in fulfilling the requests for the developing population [1]. Regular sand has been the standard fine aggregate inside substantial creation used for a long time.Notwithstanding, there has been broad examination into other materials that could be used toward succeeds and in concrete. The necessity to monitor sand trade originates starting the mode so as to there is developing worry in many areas of the planet about the exhaustion of sand reserves, as well as the environmental and financial risks associated with sand extraction from riverbanks,beachfront areas, and homestead lands. Fly ash, silica stone and recycled fine aggregate are a portion of the elective materials that have been investigated for use as a sand substitute [2-5].

Quarry dust (QD) appears to be the most cost-effective of the several materials investigated, as it is widely accessible in enormous amounts in many regions of the globe. Quarry dust has been used as a structural substance in high-level countries for the past thirty years. The level consumption is the result of funded research into boosting the utilize of quarry fines in total. The mechanical qualities of concrete are enhanced by cementitious materials, for example, silica fume and fly ash. The growth of fly ash in

concrete has numerous advantages [2].
The lack of suitable amounts of Ca(OH)2 to respond among the fly residue particle be the fundamental reason for cement's slow early strength progress . Thus, working on the pozzolanic response of fly ash by adding silica fume is a possibility. Another explanation for fly ash concrete's poor early strength advancement is that the responsive silica in the fly ash collects in the interior piece of its particles, which is gradually degraded in the hydrated concrete paste climate. Silica fume, with a larger surface area and higher SiO2 concentration has been seen as more responsive than fly ash and has thus been mixed with fly ash in low degrees of Portland cement substitution. As a result, an inquiry was conducted on the effects of silica fume and flyash in concrete taking place the early development of its strength and characteristics. [6-8].

1.1 Research significance
The principal focal point of the current exploration is to limit the amount of cement used substance concrete production. This assists with decrease in arrival of ozone harming substances into the environment. The utilization of these materials likewise maintains a strategic distance from the removal furthermore, landfill issues which are going to cause serious natural issues. Use of these materials as a choice to solidify in a productive manner is principal significant in keeping up with maintain ability. Consequently the current exploration is pointed toward creating practical cement with the utilization of modern squanders (SF and FA).

2. Objectives and methodology
The current study's main goal is to accomplish the best ternary significant blend by adding SF, FA, QD and steel fibre. To achieve the purpose for the review, an exploratory program in unmistakable changed into planned. This study is based on to find the most excellent mix of SF and FA which be able to supplant concrete to some degree. Various mixes of substitutions of cement by FA (5%, 10%, 15%,20%, and 25%) and SF (5%) and also fine aggregate by QD (5%, 10%, 15%, 20%, and 25%) to increase the strength steel fibre (1%) be adopted. The mechanical properties of every ternary concrete blended mixes were set up as well as contrasted and the regular blend in with attain the most excellent mix for ternary blended significant mix [2].

3. Materials and mix proportions
The OPC grade utilized in this examination was I.S:12269- 1987, among a particular gravity of 3.15.The early and finishing setting time were 52 and 330 minutes, separately. The fine aggregates utilized in this study has a particular gravity of 2.45 along with a mass density of 1460 kg/m^3, separately, and is brought from the Bhandara River. As coarse aggregates of 20 mm adjusting to IS: 383-1970 was utilized, which were locally accessible. It had a particular gravity of 2.67 and a mass density of 1530 kg/m^3. The FA for this test was acquired from Koradi, Nagpur. It has a 2.4 particular gravity. I.S. 15388-2003 indicates particular gravity of 2.23 for SF, which is received from Pachgaon, Nagpur. It had a particular gravity of 2.5 and a mass density of 1800 kg/m^3. Restoring and blending were finished with consumable water. The synthetic arrangements of concrete, SF, FA plus QD which are utilized inside continuous audit are introduced in Table1.and the mix proportion of concrete are introduced in Table 2.

Table 1:- Oxides compositions of OPC, silica fume, fly ash and Quarry dust

Oxides	Portland cement (%)	Silica fume (%)	Fly ash (%)	Quarry dust (%)
SO_3	3.55	1.37	0.051	0.12
SiO_2	18.76	85.15	62.51	50.12
Al_2O_3	4.25	1.82	14.30	10.54
Fe_2O_3	1.5	3.15	5.32	16.06
CaO	68.64	2.49	2.46	11.17
MgO	1.45	2.39	0.39	4.42
Cl	0.2	0.075	0.048	0.055
Loss of Ignition	1.4	3.07	0.63	2.91

Table 2. Mix proportion of M 30 concrete by Indian Standard as per IS 10262:2019 for $1m^3$concrete

Sr.No.	Items	Volume for $1m^3$	Ratio
I	OPC	0.121	1
II	Water	0.192	0.5
III	Sand	0.274	1.75
IV	Coarse Aggregate	0.412	2.87

3.1. Preparation of specimens

The blend extents of the concrete pastes without and with adding silica fume by (5% by weight of OPC) and fly ash as a cement substitution (5%, 10%, 15%, 20% and 25 % by weight of OPC) and fine aggregates supplant with quarry dust(5%, 10%, 15%, 20%, and 25 % by weight of FA) are given in Table 3. For paste blends, the binders (OPC, SF and FA) were first mixed in the blender for 3 minutes prior to blending in with water, utilizing a water/cement proportion 0.5.From that point onward, the mixes were filled oiled moulds (150×150×150 mm) , compacted and covered for 24 hours. The sample was then taken out from the mould sand restored into water at a steady room temperature. Three specimens of each blend were ready for compressive strength tests at 28 days in the wake of relieving. Comparing tests for slump value, compaction factor and compressive strength were likewise performed. Tests were tried for compressive strength after curing in water at 7, 14, 21 and 28 days.

Table3. Obtain blend quantity for 1meter cube of green concrete

Mixes	Cement	Silica Fume	Fly Ash	Fine Aggregate	Quarry Dust	Coarse Aggregate	Water
M_1	383.2	-	-	673.26	-	1100.57	191.6
M_2	344.88	19.16	19.16	639.6	33.66	1100.57	191.6
M_3	325.72	19.16	38.32	605.94	67.32	1100.57	191.6
M_4	306.56	19.16	57.48	572.28	100.98	1100.57	191.6
M_5	287.4	19.16	76.64	538.61	134.65	1100.57	191.6
M_6	268.24	19.16	95.8	504.95	168.31	1100.57	191.6

M1, M2, M3, M4, M5, M6 :- Concrete with cement substitution by 0% and 0% and 5% by silica fume and 0% , 5% , 10% , 15% , 20% and 25% by fly ash and fine aggregates by 0% ,5% , 10% , 15% , 20% and 25% by quarry dust

4. Results and discussion

In these experimental programs studied on the conventional concrete mix with partial substitution of cement with 5% of silica fume and various percentages of fly ash and fine aggregates with various percentage of quarry dust in concrete are discussed. The parameters such as Slump value, Compaction factor and Compressive strength comparisons between the various mixes are represented.

4.1. Slump value

To investigate the impact on new significant properties when silica fume and fly ash were added to concrete materials as well as fine aggregates were replaced with quarry dust for various mixes (M1, M2, M_3, M_4, M_5, and M_6). The slump test results for several mixtures by a steady water/cement proportion (0.50) were 100,125,160,180,185, and 187 mm, respectively. Figure 1 shows the data expressed as a slump value [9]. It showed that the Portland-fly ash cement' utility was greater than that of the OPC control blend. This is because of the way that the volume of blinder in fly ash increased. Slump values were lower in Portland-fly ash concrete mixes with consistent silica fume content than in blends with similar fly ash percentage but no silica fume [10]. This is because the bigger the surface area of silica fume particles, the more water is required.

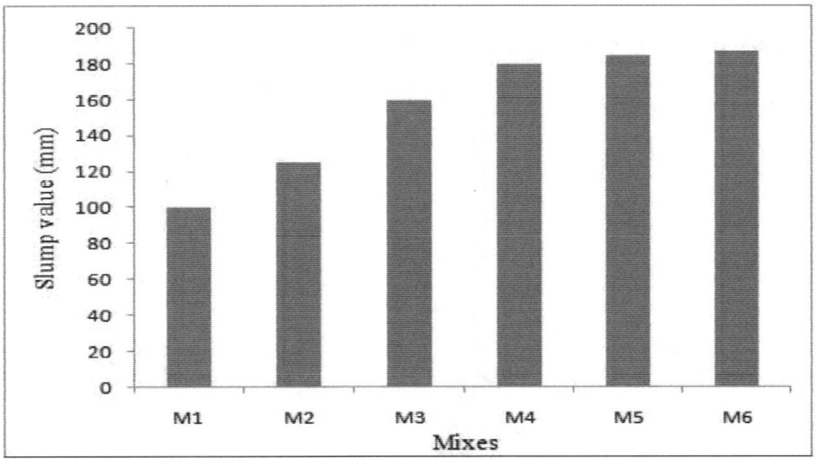

Figure 1. Slump value

4.2. Compaction factor

With a fixed w/c ratio, workability is assessed in terms of compaction factor (0.5). The values obtained for various mixtures (M_1, M_2, M_3, M_4, M_5, and M_6) are 0.75, 0.78, 0.83, 0.85, 0.90, and 0.92, as shown in figure 2.

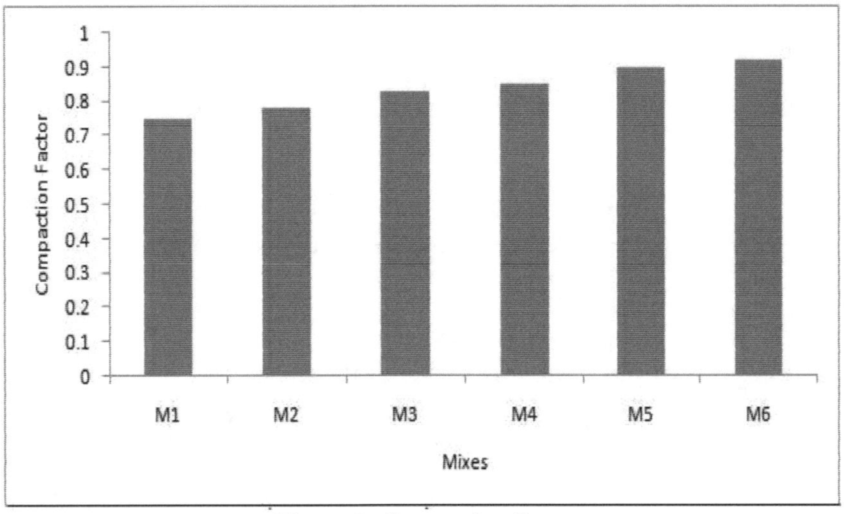

Figure 2. Compaction factor

4.3. Compressive strength

The concrete's compressive strength was evaluated during Portland-fly residue, silica smoke, and quarry dust. Compressive strength as expressed in Figure 3 and relative quality-after immersing the blends in water for 7, 14, 21, and 28 days. The compressive strength of this mixture simplified among the addition of fly ash and quarry dust at all ages studied, although it decreased beyond 20 percent. At 7 days, considerable blends with and without fly residue and silica smoke concentration were found to have a pressure of 16.3 MPa for mix M1 and 28.2 MPa for mix M5. After 14 day so relieving, most extreme and least compressive strength for the blends M5 and M1 was found 32.8 MPa and 22.7 MPa. At 21days, the compressive strength of portland silica fume with different measure of fly residue seen to be for the most part higher than the solidified cement [11-14]. Additionally, the drawn out strength as recognized at 28 days restoring, the compressive strength advancement of fly residue blends was likewise found to have higher strength than the portland concrete mix. In any case, when compared to hardened cement, it was deemed to have stronger strength in all SF blends than fly ash blends. These findings clearly demonstrate that combining silica smolder with fly debris can result in a ternary mix concrete with increased compressive strength.

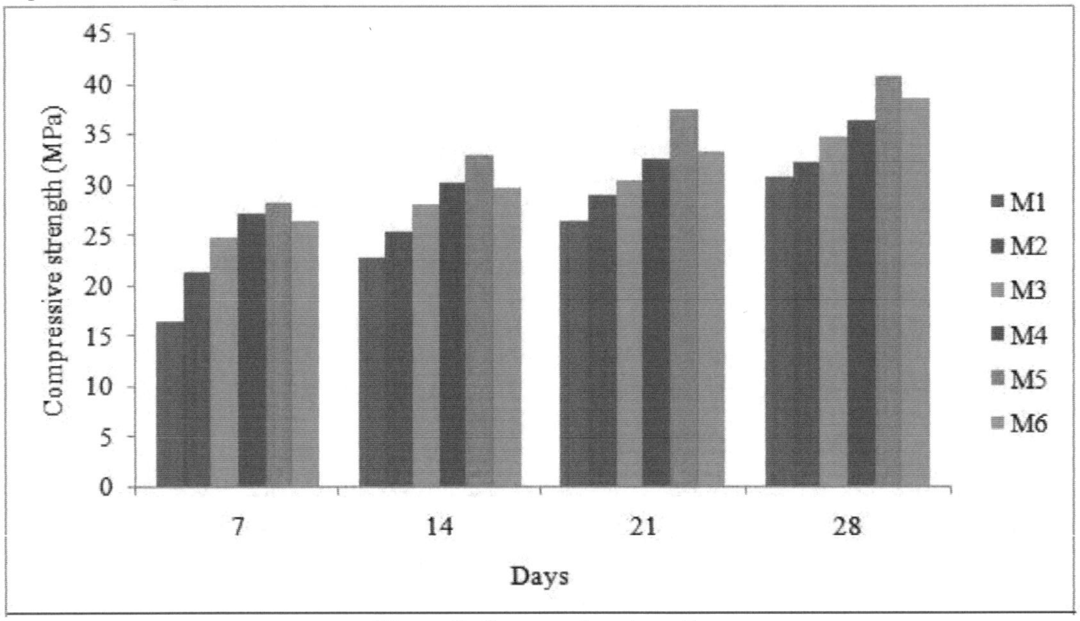

Figure 3. Compressive strength

5. Conclusions

It is reasonable to conclude from this study that the water requirement of fly ash with silica fume has a greater water interest than blends without silica fume and fly ash. The addition of silica smoke to concrete with fly residue increased the compressive strength of considerable mixtures.The ability to increase then formed concrete was discovered. The high compressive strength of combined portland-fly residue and silica smoke cements was due to both the filler effect and silica fume's pozzolanic response. The use of fly ash with silica fume not only improves the strength of the material, but it also allows to used for a better result (silica fume) inside combination with fly ash, allowing for more combinations to be used while maintaining fantastic new substantial qualities. Additionally, combining both side-effects would provide a natural benefit by reducing the amount of portland used in solidification while also improving the characteristics of fly debris concrete.

6. References

[1] Padavala Anjaneya Babu, Potharaju M, KodeVenkata Ramesh,2021, *Mechanical properties of ternary blended mix concrete of fly ash and* Proceedings,**43,**2198–2202.

[2] Kankam Charles K., Meisuh Bismark K., Sossou Gnida ,Buabin Thomas K,2017, *Stress-strain characteristics of concrete containing quarry rock dust as partial replacement of sand,* Case Studies in Construction Materials **7,**66–72.

[3] Aditya G., Lakshmayya M.T.S., 2016,*Effective utilization of various industrial waste in concretefor rigid pavement construction—a literature review*, IJIRST– International Journal for Innovative Research in Science & Technology,**3,**65–73.

[4] Kou S.-C., Poon, C.-S.2009, *Properties of concrete prepared with crushed fine stone, furnace bottom ash and fine recycled aggregate as fine aggregates*, Construction and building materials,**23**,2877 -2886.

[5] Siddique R.,2003, *Effect of fine aggregate replacement with Class F fly ash on the mechanical properties of concrete*, Cement and Concrete Research,**33**,539–547.

[6] Lee Chin-Lai, Huang Ran, Lin Wei-Ting Weng Tsai-Lung,2012, *Establishment of the durable indices for cement-based composite containing supplementary cementitious materials*, Materials and Design,**37**,28-39.

[7] Prasanna P.K., Srinivasu K., Murthy A. Ramachandra,2021, *Strength and durability of fiber reinforced concrete with partial replacement of cement by Ground Granulated Blast Furnace Slag,* Materials Today:Proceedings,**47** (6).

[8] Nisnevich M,Sirotin G,Eshel Y,2003,*Light weight concrete containing thermal power station and stone quarry waste*, Magazine of Concrete Research,**55,**313–320.

[9] Naik T R, Singh S S, Hossain M M,1995, *Properties of high-performance concrete systems Incorporating large amounts of high-lime fly ash*, Construction and BuildingMaterials,.**9**(4), 195–204.

[10] Kumar V.V. Praveen , Prasad D. Ravi ,2019, *Study on strength and durability characteristics of lime sludge based blended cement concrete* ,Advance Concrete Construction **7**,2, 75-85.

[11] Shannag M.J.,2000, *High strength concrete containing natural pozzolan and silica fume*, Cement and Concrete Composites**, 22** (6) 399–406.

[12] Thomas M.D.A., Shehata M.H., Shashi prakash S.G., Hopkins D.S., Cail, K.,2019, *Use of ternary cementitious systems containing silica fume and fly ash in concrete* ,Cement Construction Res.**29**,1207–1214.

[13] Megat Johari M.A., Brooks J.J., Kabir S., Rivard P.,2011,, *Influence of supplementary cementitious materials on engineering properties of high strength concrete*, Construction building materials, **25** 2639–2648.

doi:10.1088/1755-1315/1084/1/012030

Biofuel Production through Algae Route using Solar Assisted Carbon Capture Plant

M Samartha[1], R K S Pippal[2] and V K Sethi[3]

[1] Associate Prof. Bio-Science RKDF University, Bhopal, MP, India
[2] Dean & Prof., Vedica Institute of Technology, RKDF University Bhopal, MP, India
[3] DG (Research), RKDF University, Bhopal, MP, India

Abstract. In present scenario when there is great hike in energy demand which results in increased prices of petroleum products and the big matter of concern is that the resources are diminishing so rapidly, these diminishing fuels should be replaced by biofuels. Other sources of biofuels are limited and difficult to cultivate therefore microalgae has comes out as an important source of renewable biofuel. Microalgae grow in the presence of sunlight and CO_2, and those rich in lipids are excellent source of biofuels (biodiesel) after processing. The present study was aimed at production of biofuel through microalgae route using carbon capture pilot plant and exploring feasibility of scaling-up of bio-fuel production on a mega scale coal fired thermal power plant through analysis and selection of species that can grow in ash-pond area on mass scale.

1. Introduction

Climatic change abatement strategies on the 'Anvil' include accelerated deployment of renewable energy technologies like solar, wind, biomass and biofuels. Depleting trend of fossil fuels and increasing CO_2 emission partially from combustion of fossil fuels has targeted focus on Carbon Capture and Sequestration (CCS) on fossil fuel based power plants. The CO_2 so captured can be put to use for production of multi-purpose fuels such as hydrogen, methane, methanol and biodiesel through 'Algae- route'. Microalgae biodiesel production can be an option of easily accessible and cost-effective [1, 2] by usingmethanol solvent and two-step process, microalgae oil, can be converted efficiently and biodiesel of Fatty Acid Methyl Acid (FAME) yield of 95.1% can be achieved [3]. Fertile land is not required by microalgae to grow and thus has no competition with food production [4, 5].

1.1. Biodiesel from microalgae-Methodology

The solvent extraction method recovers almost all the oils and leaves behind only 0.5% to 0.7% residual oil in the raw material.

The trans-esterification is the process involving removal of waxy portion from algal oil using methanol as process chemical and KOH as a catalyst. At the project site, a bio-diesel production unit of capacity of 50 liters per batch is operational which can use jatropha, karanja and algae as feed stock. The present study was aimed at on production of biofuel through microalgae route using carbon capture pilot plant and exploring feasibility of scaling-up of bio-fuel production on a mega scale coal fired thermal power plant through analysis and selection of species that can grow in ash-pond area on mass scale.

2. Pilot scale production of bio-diesel through algae route

The solar integrated carbon capture and sequestration pilot plant is provided with CO_2 sequestration arrangement from the Reactor/Re-boiler to a specifically designed Algal pond of area 1500 sq. meter. The selected species of Algae were grown through direct solar flux, pH control and CO_2 injection for a period of 4 months. The high lipid algae so produced was dried and subjected to trans-esterification in the bio-diesel production unit using methanol and KOH as input chemicals to set algal oil.

2.1. Cultivation of algae

The production process and application of algal biomass is given below at Figure 1.

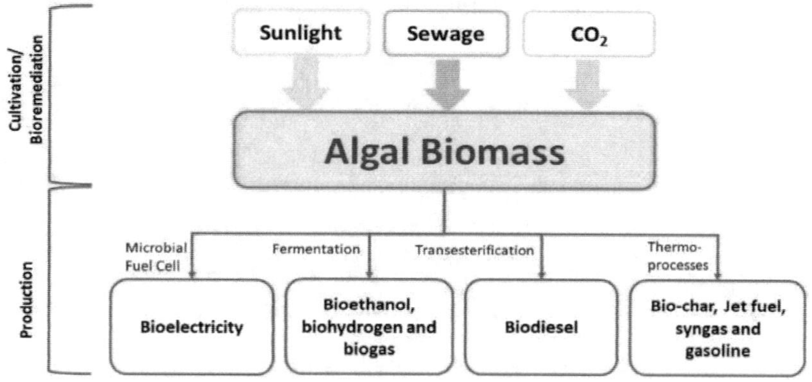

Figure 1. Carbon sequestration through algae route.

There are several factors to determine the growth rate of algae.

- Solar flux for the photosynthesis process.
- Temperature- in the range of 25- 35 degree Celsius
- Medium/Nutrients – water parameters, salinity, and pH (between 7 - 9)
- Algae Type – Selected algal strain high lipid content.
- Aeration – BOD of water in acceptable range
- Mixing – Using paddler device

2.2. Production methodology

The experiments were carried out at biodiesel production unit at RKDF University, Gandhi Nagar Bhopal. The major steps are; CO_2 sequestration, algal biomass production and biodiesel/biofuel production through algae route. CO_2 is selectively captured from the flue gas by MEA solvent and released in to the algal pond, where CO_2 will expedite growth of algae in pond water and in turn sequester in biomass. Algal biomass so produced may be utilized as biofuel. Main production techniques of synthesis from microalgae oil to biodiesel consists of crushing of dried algae followed by trans-esterification process of production of biodiesel using KOH as catalyst and methanol.

A 500 MW pulverized coal fired unit would require a CCS plant of 310 tons/hr. capacity. This interdisciplinary project is expected to resolve certain frontline issues in CO_2 sequestration such as energy intensive process optimization in terms of cost of generation and development of effective catalyst for methane, hydrogen and biodiesel recovery through algae route.

3. Production of multi-purpose fuels from captured CO_2 – Some test result

- Using flue gas containing CO_2 and treated lignite the carbon monoxide is obtained having a Low Heat Value (LHV) of 2417 kCal/kg. This would reduce the fuel required by the boiler and also reduce the net emissions, once recycled back to the boiler using separate short flame burners.
- For hydrogen production, water-gas shift reaction was used and hydrogen so formed was evaluated to the extent of 1.43%-1.79% by weight equivalent to hydrogen by volume 17.44% to 21.83% creating the opportunity to the production of multipurpose fuels from captured CO_2.

3.1 Test results on algae produced from the pilot plant

Figure 2 & 3 provides glimpses of CCS pilot plant and algal pond where the pilot study was carried out as shown below:

Figure 2. Solar integrated carbon capture plant at RKDF University, Bhopal.

Figure 3. (A) Algal pond of carbon capture pilot plant (B) Trans-esterification unit at RKDF University, Bhopal.

The 'Growth' of selected micro algae was found to be increased by 2.4 to 2.7 times by weight with CO_2 injection as shown below:

*[i]*Dry mass of algae *chlorella vulgaris*

Before release of CO_2: 5.24 gm.

After release of the CO_2: 14.22 gm.

[ii] Dry mass of algae *Monoraphidium minutum*

Before release of CO_2: 4.71 gm.

After release of the CO_2: 11.17 gm.

3.2 Analysis of micro-algal biomass in CO_2 sequestration process

3.2.1 Carbon sequestration on CCS pilot plant

The analysis is based on inputs like:

- Average carbon content assumed in coal is 40% (*NCL annual report 2021-22*)
- Carbon to CO_2 as per molar ratio is 3.6 kg CO_2 with per kg Carbon

Using above the sequestration of CO_2 from boiler of coal capacity 40kg/hr. works out to 104 kg/hr. algal production which is of the order of 37-40 tons/month.

3.2.2 Carbon sequestration on scaled up plant.

On a scaled up plant, the proportionate algal production per acre works out to about 1200 tons/month/acre. A 500 MW plant having ash pond decantation area of 15 acres for algal production would give about 50 tons per hour.

In this feasibility study, the scaling-up has been carried out to 30% CO_2 capture & sequestration to produce multi-purpose fuels on a 500 MW unit at selected thermal plant in Singrauli coal fields region of Northern India. Efforts have been made with the Government of UP in collaboration with TOSHIBA corporation; the EPC contractor and supplier of the Anpara B TPS (2x500 MW Plant) for establishment of CCS plant on one of the pit head thermal unit number 4 of 500 MW at ANPARA in Singrauli Region of UP and MP border.

A conceptual view of the scaled-up 100 tons per hour CCS Plant on a 500 MW Unit is shown below (refer Figure 4):

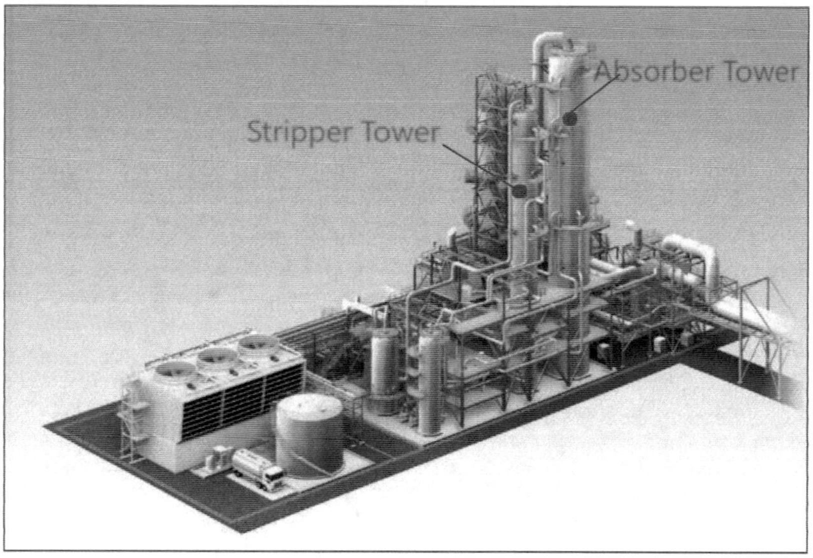

Figure 4 Scaled-up 100 tons per hour CCS Plant on a 500 MW Thermal Power Plant

Since a 500 MW unit emits about 8000 tons of CO_2 per day (about 330 tons per hour), the 30% capacity scaled-up carbon capture and sequestration plant will be of capacity 100 tons per hour. No compression and storage of CO_2 is presently envisaged.

The strategy of scaling-up of the pilot plant of the proposal is briefly given below:

- Optimization of scale of carbon capture (range 30% to 100%)
- Algae plant stage of implementation for species to grow in Ash decantation well.
- Feasibility study of power plant modification required for production of multi- purpose fuels like H2, CH4 etc.
- Analysis to utilization of CCS plant produced H2 in fuel cell for lighting application
- CO recycling option to boiler, using short flame burner design
- Energy penalty reduction through various option of solar thermal- technologies
- Sequestration to coal seams of depleted mines

3.3 Some anticipated results on application of captured CO_2 on the Scaled-up plant

- Production of Algal Bio-mass @55 tons per hour using selected species of algae viz. *Monoraphidium minutum* and *chlorella vulgaris* in the specially designed bio-reactor for culture preparation and algal pond. These varieties have been tested to have growthin the Ash decantation water also. Bio-diesel shall be produced using trans- stratification process.
- Incorporation solar thermal plant of capacity 70 tons per hour for steam production both for MEA solvent regeneration and water gas shift reaction with a make-up water consumption of 4.5%, i.e. 3 tons/hr. This novel concept of using renewable energy will reduce energy penalty from a level of 15-17% to about 4-5%
- Our feasibility study also showed that for a 500 MW Unit presently the best solution would be to produce algae-based bio-diesel and later use the depleted 'KAKRI coal field' for initial sequestration and later use 'Bina coal field' for CO_2 sequestration for long-term utilization of captured CO_2.

4. Concluding remark

The 'Growth' of selected micro algae found to be increased by 2.4 to 2.7 times by weight with CO_2 injection. The study conducted for CO_2 capture from flue gases & release and production of hydrogen using water-gas shift reaction and algal biodiesel has prompted us to design and develop the 'Scaling-up of CCS Plant on a 500 MW Coal-fired Unit of Thermal Power plant' in Singrauli region with facility to produce bio-diesel through algae route in which the selected species of algae was capable of growing in decantation ash pond of the Power Station. The average yield in a 500 MW unit for selected species for algae was calculated at about 50 tons/hr. for the plant running at full capacity.

5. References

[1] Demirbas A 2016 Optimization of Energy Systems. In: Waste Energy for Life Cycle Assessment *Green Energy and Technology, Springer*, Cham.

[2] Sharma Y C and Singh, V 2017 Microalgae biodiesel: A possible solution for India's energy security *Renewable and Sustainable Energy Reviews, Elsevier* 67 p72.

[3] Chang J S, Show P L, Ling T C, Chen C Y, Ho S H, Tan C H, Nagarajan D and Phong W N 2017 Photobioreactors In: Current Developments in Biotechnology and Bioengineering *Bioprocesses, Bioreactors and Controls, Elsevier*, p 313.

[4] Bibi R, Ahmad Z, Imran M, Hussain S, Ditta A, Mahmood S and Khalid A 2017 Algae bioethnol production technology: A trend towards sustainable development *Renewable and Sustainable Energy Reviews, Elsevier* 71 p 976.

[5] Bhagea R, Bhoyroo V and Puchooa D 2019 Microalgae: the Next Best Alternative to Fossil Fuels after Biomass-A Review *Open acess, Page Press. Microbiology Research* 10 p 7936.

Acknowledgments

The Authors gratefully acknowledge the Central Power Research Institute (CPRI), Ministry of Power (Govt. of India) for sponsoring the pilot project on Carbon Capture & Sequestration integrated with Solar Thermal plant for solvent regeneration. The authors are also thankful to Ministry of New & Renewable Energy (MNRE) for sponsoring parallel project to the University on Solar Thermal Energy Storage (TES). Contribution of Dr. Partha S. Dutta Professor of RPI, NY, USA is also acknowledged for providing technology transfer on TES.

Assessment of Fixed Rack and 1–Axis Tracking Crystalline Silicon Photovoltaic Array on Their Ideal Energy Performance

Prakhar Duggal[1], Ravinder Kumar Tomar[1], Vigneshwaran S[2], Vegesna Rohith Varma[3], and Vekhohu Puro[4]

Department of Civil Engineering, Amity School of Engineering and Technology,

Amity University, Noida, Uttar Pradesh

Email: vigneshwaranabc7@gmail.com

Abstract. An energy assessment is examined in ten different cities spread across India on a 50 kWp crystalline silicon photovoltaic system comparing the performance of Fixed Rack and 1–Axis Tracking Array. Simulation is done on the photovoltaic systems. The varying methods of installation used for this study are fixed positioning and sun tracking. The software used for simulating the radiation condition is NREL PV Watts. The data is gathered from available data using appropriate conditions for all the considered cities in this paper. The output results show the study of comparison between Fixed rack and 1–Axis Tracking array based on the monthly data and yearly data. The overall potential ability of simulated PV devices for every place taken in the study with unique configuration strategies is also compared and evaluated.

1. Introduction

Energy is the most essential basic need of everyday life. An uninterrupted energy supply is what makes dailylife run smoothly. The sustenance of the economy all over the world could be in great danger without a steady supply of electricity. With increasing demand, countries are looking for alternative resources of energy to narrow the supply-demand gap. Depletion of conventional energy resources and the increasing rate of environmental exploitation, the need for investment in renewable sources of energy is at a new high. Hydropower, Solar power are some of the renewable energies that the world is growing into. While hydropower cannot be equipped in all places, solar energy is suited to all places. The Sun for billions of years has helped in the making of life, now the same energyis used to generate electricity through the help of Photovoltaic cells. Solar energy has great potential in terms of wide applicability but due to the lack of suitable arrays, the efficiency is limited. The conversion efficiency of thelatest array is not more than 20%, which shows that there is still room for immense improvement. This research paper shows that energy performance and assessment of 50 kWp Crystalline Silicon array taken using two diverse specifications (fixed rack & 1-axis tracking installation types) under comparable field conditions is considered. This study is done with the help of NREL (National Renewable Energy Laboratory) PV Watts software where thesimulation considers array performance and losses in the system.

2. Methodology

Identification of cities that covers the whole spectrum of climatic conditions in India was done. Then, Photovoltaic array type of Fixed Open rack Crystalline Silicon array and 1-Axis tracking Crystalline Silicon array was considered. To analyse the required parameters, a software named NREL PV Watts was used. NREL PV Watts is open-source software that in response to the geographical coordinates provides us with the solar radiance and AC energy for the given module type, array type, DC system size, System loss, Tilt angle, Azimuth angle, Inverter efficiency, ground coverage ratio, DC to AC size ratio. NREL PV Watts calculates the weather data from a background operation. Solar irradiance, monthly average energy, and annual energy for both installations were obtained. The data is exported to an excel file where it is analysed and visualized through a graph for better understanding. Implications for the graphs were provided with a detailed discussion which provide insight into the study.

3. Explanation of fixed open rack array and 1-axis tracking array

Solar energy is one of the easy and simplest energy production methods lacking any complicated steps unlikewhen compared to hydropower, thermal power plant, or geothermal power plant. The process involved in the production of electric power is very simple and remains the same from watts to megawatts. The installation process can be done in any geographical location which has a descent to an adequate amount of solar irradiance and the demand can be provided. In this comparison, a solar fixed open rack array and the 1-axis tracking array are taken as the source for the comparison and the photovoltaic module that is used in the array is a standard crystalline silicon cell with an approximate efficiency of 15% which is covered on glass. An array refers to a stringof solar modules connected on a series and those arrays are connected in parallel with connecting wires and connectors.

3.1. *Fixed Open rack Crystalline Silicon array*

The Fixed open rack array is a highly effective ground-mount system that has good airflow on and under it forthe cooldown of its photovoltaic cells. Its temperature provides the modules to work on reduced cell operating temperatures. (The output in a system increases as the module cells operate at a decreased temperature, providedat a given incident solar irradiance.)

3.2. *1-Axis tracking Crystalline Silicon array*

1-axis tracking array works on the disadvantages of the fixed open array in that the angle of incidence of solarrays is not always 90 degrees to the solar photovoltaic module's aperture, so to improve the collection of sunlightby the array a sun solar tracker is introduced to the sun's radiance on a single axis. solar tracker helps to keep thesolar array's Aperture that is 90 degrees to the array when the solar radiation is maximum to improve the overallperformance of the arrays.

4. Designing of a photovoltaic system for the simulation in NREL PVWatts

Estimating the size of a Photovoltaic power producing system is usually done by a suitable requirement of the site wherever the solar electricity generation is going to be installed. So as mentioned, a photovoltaic estimation in thecase is done by Firstly, selection of 10 cities in India that are dissimilar in the geographic conditions and could experience a variety of climates around the year, these cities were taken under a detailed observation, after the selection all ten cities, specific location coordinates were collected and given to than solar database for collectionweather forecasting in a specialized data measurement source. after the collection of information, the requirementof the photovoltaic system for the desired output becomes easier. This can also be done in a mathematical methodto analyze all of these parameters which are battery capacity, demand load, size of the inverter, the capacity of thecharger controller, and photovoltaic panels. Besides mathematical dimensioning, other ways to model such projects are based on the energy project modeling software.

Table 1. locations in India

Name of the city	Latitude (°)	Longitude (°)	Nearest location of theGrid (Satellite Data) (km)
Amritsar	31.65	74.85	1.8
Chennai	13.09	80.30	1.4
Delhi	28.65	77.25	1.5
Hyderabad	17.37	78.46	1.1
Kochi	9.97	76.26	0.6
Kohima	25.65	94.1	0.8
Kolkata	22.57	88.34	0.4
Mumbai	18.95	72.85	1.1
Nagpur	21.13	79.10	1.9
Srinagar	34.05	74.85	2.4

4.1. Solar radiance data for the selected locations

Data required for the analysis of the performance of the Solar Photovoltaic systems for a location needed an extensive collection of radiation data. In this case, there is a long-run average solar radiation data is been given by the NREL PV Watts software which is a web-based software mainly used for solar radiance data collection, International Weather for Energy Calculations (IWEC)) was used to estimate the feasibility of the investigation of photovoltaic systems in 10 locations, i.e. Delhi, Chennai, Hyderabad, Nagpur, Mumbai, Kolkata, Kochi, Amritsar, Srinagar, and Kohima shown in Table.1, and the monthly average radiation data for each city is provided in the below Table.

4.2. National Renewable Energy Laboratory (NREL) PV Watts.

PVWatts is one easiest and most well-equipped web-based software tools that can be used in the estimation ofthe power output of a PV array at a given location based on the solar radiation database. Based on the sites surveyed in this study, PVWatts will determine the incident solar irradiance to a photovoltaic array and the temperature of the environment, and the temperature of a panel cell for every given hour in the year. The data provided is of much use to the general public to know alternative estimations.

Table 2 Annual and Monthly average Solar radiation data for ten cities

Solar radiation data of ten cities ((kWh/Sq m/Day)													
City	annual solar radiance	Monthly average solar radiance											
		Jan	Feb	Mar	Apr	May	Jun	Jul	Aug	Sep	Oct	Nov	Dec
Amritsar	5.94	3.71	4.80	6.39	7.62	8.12	7.21	6.37	6.54	6.54	5.71	4.39	3.85
Chennai	6.32	7.19	7.59	8.83	8.57	7.14	6.10	5.35	5.20	5.65	5.25	3.84	5.09
Delhi	5.99	4.26	5.55	7.07	7.94	7.57	6.74	5.64	6.13	6.41	5.82	4.68	4.12
Hyderabad	6.39	6.64	7.52	8.10	7.70	7.82	6.03	5.01	4.93	4.54	5.52	6.63	6.20
Kochi	5.17	6.89	7.07	6.81	5.76	5.83	4.67	2.24	2.21	4.56	5.66	5.81	4.49
Kohima	5.33	5.67	6.16	6.71	5.83	5.41	4.42	4.38	4.59	4.51	4.87	5.89	5.51
Kolkata	5.90	5.32	6.48	7.56	8.06	7.48	5.50	4.98	4.92	4.86	5.97	5.05	4.87
Mumbai	6.70	6.64	7.60	8.35	8.53	8.39	6.05	4.85	4.92	6.06	6.84	6.24	5.89
Nagpur	6.29	5.88	6.75	7.51	8.87	8.68	5.27	4.50	4.06	5.32	6.28	6.41	5.91
Srinagar	6.14	3.46	3.49	5.61	6.47	8.11	8.78	8.07	7.50	7.25	6.48	4.92	3.56

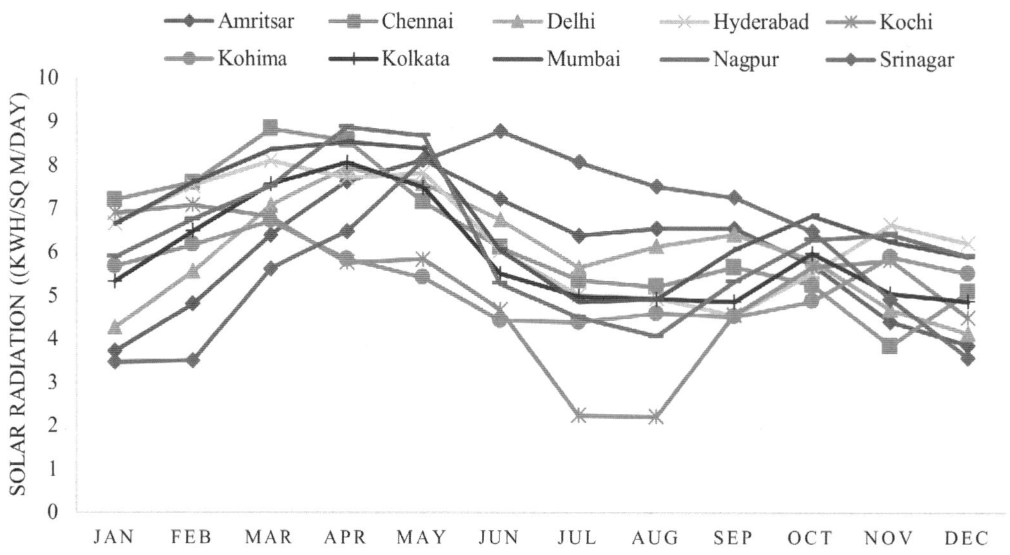

Figure 1. Monthly solar radiation over ten cities in India

The PVWatts system has the flexibility to provide the site's specific requirement of the Photovoltaic system that needs to be installed. So, in that several specifications can be mentioned, firstly Photovoltaic Module type which includes crystalline and thin-film, array type that consists of open fixed rack, 1- axis tracking, 2-axis tracking, and 1-axis backtracking, DC system size, System loss, Tilt angle, Azimuth angle, Inverter efficiency, ground coverage ratio, DC to AC size ratio. After providing the suitable data, Then the simulation process beginsand the output results are visible on the Results column of the software window. The simulation starts in the system as a background simulation of the weather data and solar radiance in the solar database, followed by the calculation the of DC output per hour from the Photovoltaic array DC rating and the incident irradiation, and which can be selected about the Photovoltaic cell temperature. The product of DC power to the DC-AC ratio provides the system AC output and then can be adjusted by inverter efficiency, which is the operating load. The PV system that produces an AC output recording on an hourly basis for over a month and then a year is designed. In this case, a 50 kWp Photovoltaic system is provided in PV watts making use of a standard crystalline silicon cell, Fixed Open rack Crystalline Silicon array, and 1-Axis tracking Crystalline Silicon array, with a size ratio of DC-AC being 1.1, and with an efficiency of inverter value being 96 %.

5. Result and Discussion

After the process of simulation, the output result data that were obtained from the PV watts are exported to theMS Excel Workbook. the data that were obtained from the software are exported and to analyse those data, Tables,and graphs were made it makes more understandable. Based on the analysis and comparison of output energy performance, the Fixed Open rack Crystalline Silicon array and 1-Axis tracking Crystalline Silicon array, the results obtained denote the energy potential of two different array methods annually and monthly for ten cities, shown in Figure.2

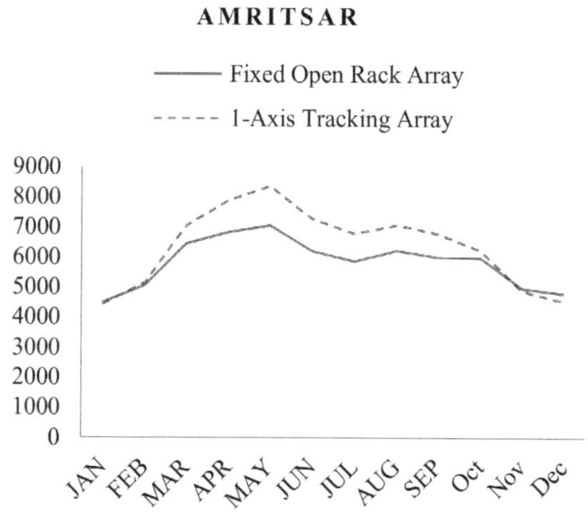
AMRITSAR
Fixed Open Rack Array
1-Axis Tracking Array
9000
8000
7000
6000
5000
4000
3000
2000
1000
0
JAN
FEB
MAR
APR
MAY
JUN
JUL
AUG
SEP
Oct
Nov
Dec

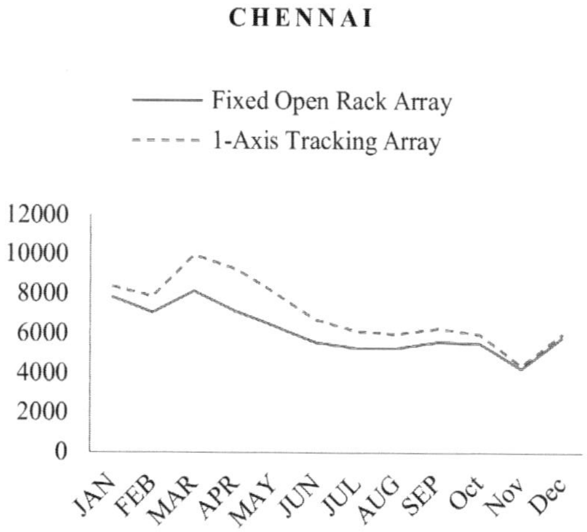
CHENNAI
Fixed Open Rack Array
1-Axis Tracking Array
12000
10000
8000
6000
4000
2000
0
JAN
FEB
MAR
APR
MAY
JUN
JUL
AUG
SEP
Oct
Nov
Dec

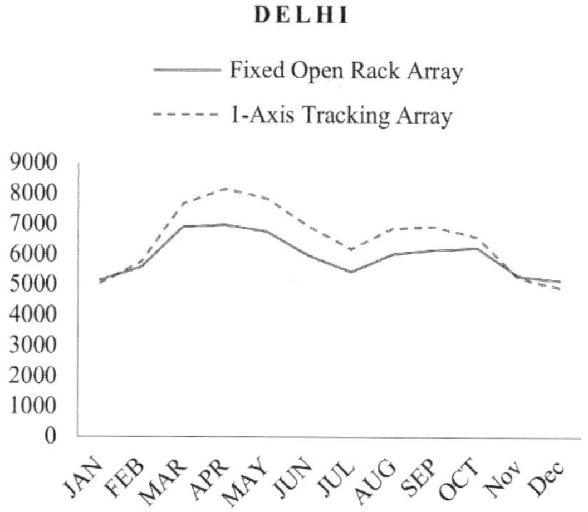
DELHI
Fixed Open Rack Array
1-Axis Tracking Array
9000
8000
7000
6000
5000
4000
3000
2000
1000
0
JAN
FEB
MAR
APR
MAY
JUN
JUL
AUG
SEP
OCT
Nov
Dec

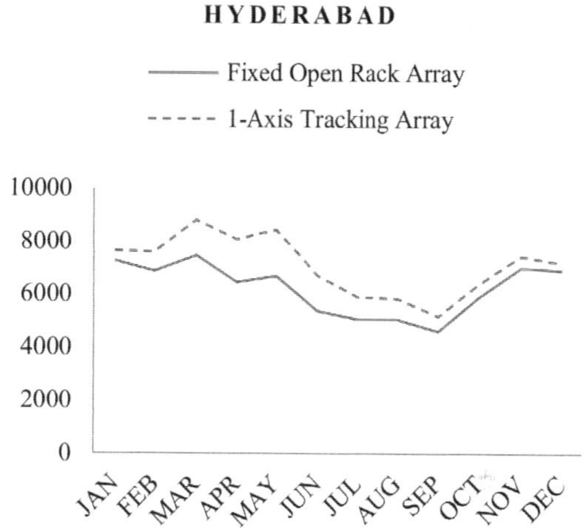
HYDERABAD
Fixed Open Rack Array
1-Axis Tracking Array
10000
8000
6000
4000
2000
0
JAN
FEB
MAR
APR
MAY
JUN
JUL
AUG
SEP
OCT
NOV
DEC

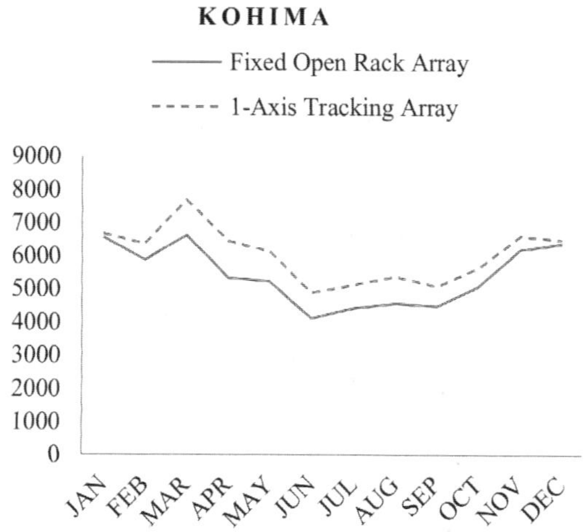
KOHIMA
Fixed Open Rack Array
1-Axis Tracking Array
9000
8000
7000
6000
5000
4000
3000
2000
1000
0
JAN
FEB
MAR
APR
MAY
JUN
JUL
AUG
SEP
OCT
NOV
DEC

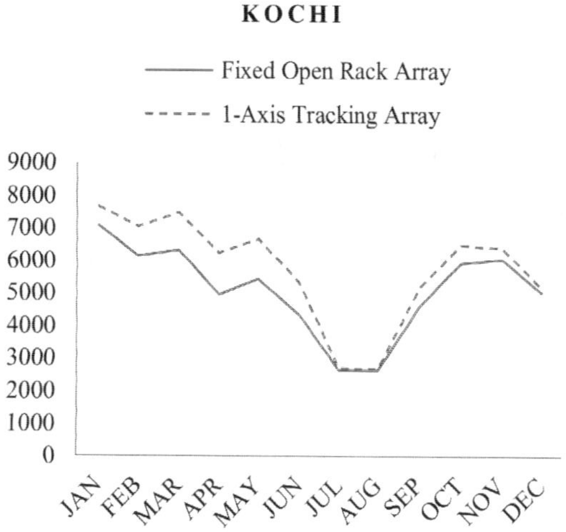
KOCHI
Fixed Open Rack Array
1-Axis Tracking Array
9000
8000
7000
6000
5000
4000
3000
2000
1000
0
JAN
FEB
MAR
APR
MAY
JUN
JUL
AUG
SEP
OCT
NOV
DEC

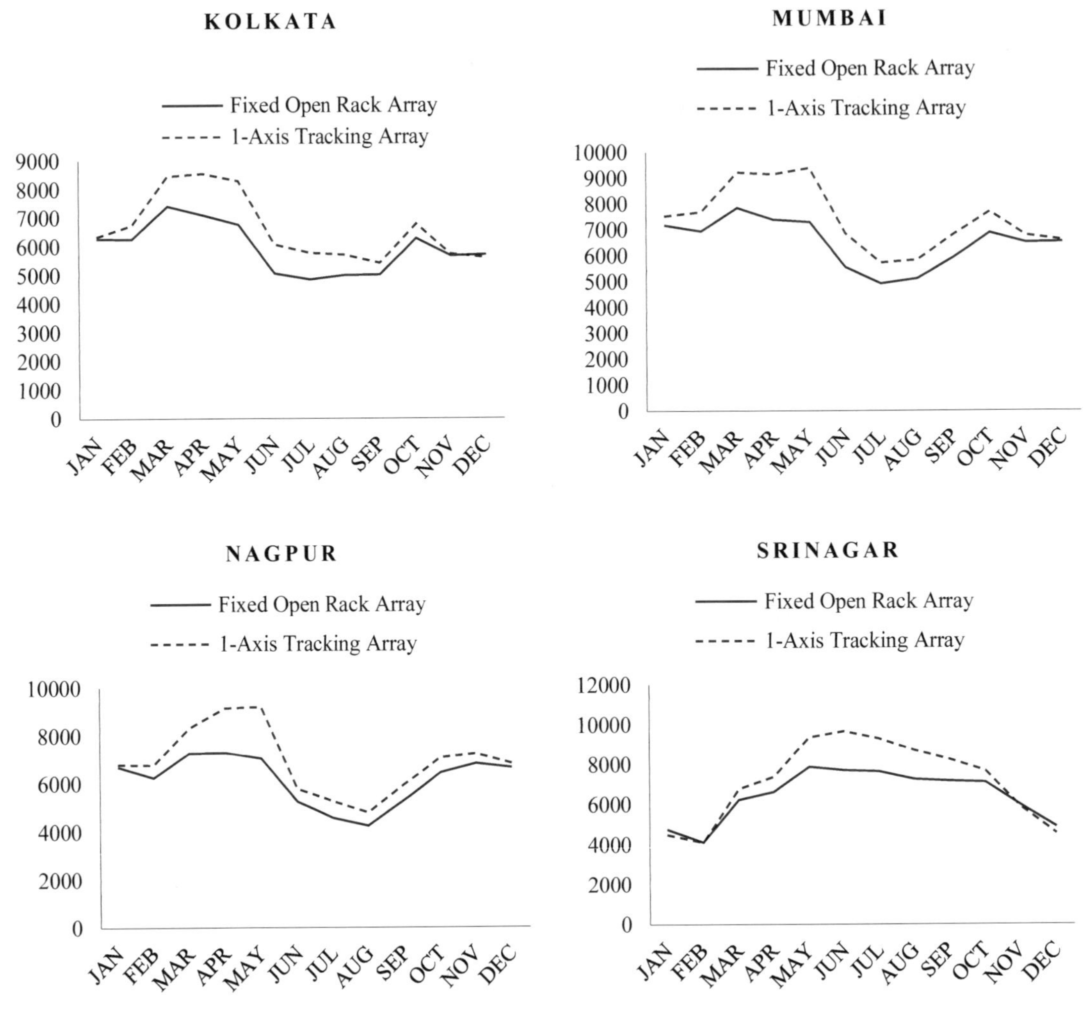

Figure 2. Monthly energy potential over ten cities in India: Open rack vs. 1-axis tracking

5.1. *Energy potential from considered locations*

Amritsar is located in the North-west of India, which is warm and temperate with a latitude of 31.65°N and a longitude of 74.85 °E. The temperature average annually is 23.4 °C in Amritsar. Precipitation here is about 786mmThe study which is carried out on 50 kWp on the array system in the area shows that the highest energy records in May with the Fixed Open rack array producing 7058 kWh and 1-Axis tracking array producing 8374 kWh and the lowest in January with the fixed open rack array producing 4496 kWh and 1-axis tracking array producing 4436 kWh.

Chennai is in the extreme south part of India, with a latitude of 13.09 °N and a longitude of 80.30 °E. This coastal city has a tropically wet and dry climate that lies near the thermal equator. The average annual temperature is 29 °C in Chennai. Precipitation here is about 1382 mm. The study which is carried out on 50 kWp on the array system in the area shows that the highest energy records in March with the fixed open rack array producing 8152 kWh and 1-axis tracking array producing 9970 kWh and the lowest in November with the fixed open rack array producing 4266 kWh and 1-axis tracking array producing 4415 kWh.

Delhi the capital of India is at the north-centre at a latitude and longitude of 28.65 °N and 77.25 °E respectively. It has dry winters and hot semi-arid summers. The annual temperature average is 25.2 °C in Delhi. Precipitationhere is about 762 mm. The study which is carried out on 50 kWp on the array system in the area shows that the highest energy records in April with the fixed open rack array producing 6974 kWh and 1-axis tracking array producing 8155 kWh and the lowest in December with the fixed open rack array producing 5141 kWh and 1- axis tracking array producing 4914 kWh.
Hyderabad is situated in the southeast of India at a latitude of 17.37 °N,and a longitude of 78.46 °E. It has a tropical wet and dry climate on a hot semi-arid climate. The average annual temperature is 26.8 °C in Hyderabad. Precipitation here is about 854 mm. The study which is carried out on 50 kWp on the array system in the area shows that the highest energy records in March with the fixed open rack array producing 7467 kWh and 1-Axis tracking array producing 8810 kWh and the lowest in September with thefixed open rack array producing 4618 kWh and 1-axis tracking array producing 5180 kWh.
On the southwest coast of India in Kochi, at a latitude of 9.97 °N, and a longitude of 76.26 °E. It has a tropical monsoon climate. The average annual temperature is 26.5 °C in Kochi. Precipitation here is about 3014 mm. The study which is carried out on 50 kWp on the array system in the area shows that the highest energy records in January with the fixed open rack array producing 7270 kWh and 1-axis tracking array producing 7631 kWh and the lowest in August with the fixed open rack array producing 2647 kWh and 1-axis tracking array producing 2717 kWh.
Kolkata has a humid subtropical climate and is located in the east part of India at a latitude of 22.57 °N, and a longitude of 88.34 °E. The average annual temperature is 26.8 °C in Kolkata. Precipitation here is about 1836 mm. The study which is carried out on 50 kWp on the array system in the area shows that the highest energy records in April with the fixed open rack array producing 7401 kWh and 1-Axis tracking array producing 8444kWh and the lowest in July and September with the fixed open rack array producing 4845 kWh (July) and 1- axis tracking array producing 5416 kWh (September).
Kohima is located in the southern part of Nagaland at a latitude of 25.57 °N, and a longitude of 88.34 °E. It has a subtropical highland climate. The average annual temperature is 18.2 °C in Kohima. Precipitation here is about 1756.9 mm. The study which is carried out on 50 kWp on the array system in the area shows that the highest energy records in January with the fixed open rack array producing 6580 kWh and 1-axis tracking array producing 6688 kWh and the lowest in June with the fixed openrack array producing 4122 kWh and 1-axis tracking array producing 4892 kWh.
Mumbai is located in thewest-central part of India at a latitude of 18.95 °N, and longitude of 72.85°E. It has a humid tropical climate that records seven months of dryness in a year. The average annual temperature is 27.7 °C in Mumbai. Precipitation here is about 2184 mm. The study which is carried out on 50 kWp on the array system in the area shows that the highest energy records in May with the fixed open rack array producing 7282 kWh and 1-axis tracking array producing 9202 kWh and the lowest in August with the fixed open rack array producing 4239 kWh and 1-axis tracking array producing 4788 kWh.
Nagpur is located in the central part of India at a latitude of 21.13 °N, and a longitude of 79.10 °E. It has a tropical savannah climate. The average annual temperature is 23.4 °C in Nagpur. Precipitation here is about 1100 mm. The study which is carried out on 50 kWp on the arraysystem in the area shows that the highest energy records in April and May with the fixed open rack array producing 7293 kWh (April) and 1-axis tracking array producing 9202 kWh (may) and the lowest in the monthof august with the fixed open rack array producing 4239 kWh and 1-axis tracking array producing 4788 kWh.
Srinagar is at a latitude of 34.05 °N, and a longitude of 74.85 °E. It has a humid subtropical climate. classificationwith seven months of dryness in a year. The average annual temperature is 13.7 °C in Srinagar. Precipitation here is about 721.8 mm. The study which is carried out on 50 kWp on the array system in the area shows that thehighest energy records in May and June with the fixed open rack array producing 7876 kWh (May) and 1-axis tracking array producing 9662 kWh (June) and the lowest in

February with the fixed open rack array producing4112 kWh and 1-axis tracking array producing 4101 kWh.

5.2. Comparative analysis of annual energy potential for the fixed rack and 1-axis tracking Array

Table 3. Energy potential over a year for the fixed rack and 1- axis tracking over ten Indian cities

City	System size (kW)	Photovoltaic Array Specification			
		Fixed Rack		1-Axis Tracking	
		Annual energy potential (kWh)	Capacity factors (%)	Annual energy potential (kWh)	Capacity factors (%)
Amritsar	50	69976	16.0	76509	17.5
Chennai		74066	16.9	85193	19.5
Delhi		71537	16.3	78040	17.8
Hyderabad		74760	17.1	85323	19.5
Kochi		61215	14.0	69155	15.8
Kohima		64983	14.8	72618	16.6
Kolkata		71311	16.3	79412	18.1
Mumbai		78011	17.8	89235	20.4
Nagpur		73860	16.9	83198	19.0
Srinagar		77334	17.70	86118	19.7

As shown above, the data of energy output taken annually using a 50 kWp array for the ten cities taken acrossIndia using the installation of a fixed rack Crystalline Silicon array and 1-axis tracking Crystalline Silicon array.Mumbai annually records around 78011 kWh for a fixed rack and 89235 kWh for 1-axis tracking which is the highest among the cities taken with a capacity factor of 17.8 % and 20.4 % correspondingly. Hyderabad records 74760 kWh for the case of fixed rack array and 85323 kWh of energy for 1-axis tracking with 17.1 % and 19.5 % As plant capacity factors respectively. The simulation of energy potential for Chennai also records around 74066 kWh for 1-axis tracking and 85193 kWh for fixed rack PV systems. The same Photovoltaic array taken at appropriate conditions for the region of Delhi shows a very lower energy output than Srinagar, Hyderabad, Chennai,and Mumbai. Delhi showed an energy potential of 71537 kWh for the fixed rack PV array, whereas the case of 1-axis tracking shows 78040 kWh of energy with a capacity factor of 16.3 % and 17.8 %. Placed under the similarInstallation of PV array, Kochi records the least annual output over the comparison with all the ten cities taken in this study with an annual energy potential of 61215 kWh for a fixed rack and 69155 kWh for 1-axis tracking with capacity factors of 14.0 % and 15.8 %. Srinagar shows better performance values similar to that of Mumbai,being the highest ineffectiveness. The recorded energy potential value annually for the fixed rack array is 77334kWh and for the 1-Axis Tracking model is given to be 86118 kWh. The factor of capacity taken for a 50 kWp model in the Srinagar region is 17.70 % and 19.7 % for a fixed rack and 1-Axis Tracking configuration respectively. The annual energy potential for Nagpur has a better average of 73860 kWh for a fixed rack array and 83198 kWh for 1-axis tracking, whereas the capacity factors are 16.9 % and 19.0 % for a Fixed rack, and 1-axis tracking respectively. The city of Amritsar and Kohima has an energy potential of 69976 kWh and 64983 kWh with a capacity factor of 16.0% and 14.8% for Fixed rack arrays respectively whereas it is 76509 kWh and 72618 kWh for 1-axis tracking with a capacity factor of 17.5 % and 16.6% respectively. Kolkata records an annual energy potential of 71311 kWh with a capacity factor of 16.3% for the Fixed rack array and 79412 kWh for 1-Axis Tracking with a capacity factor of 18.1%. From the analysis, it is concluded that the 1-axis tracking array produces better performance for all ten cities on comparing to the Fixed rack array. The reason behind this is an installation of a solar tracker which tracks the sun's radiance on a single axis. Solar tracker helps to keep thesolar array's Aperture at a 90-degree angle to the incident light for maximizing the solar radiation, so its miles are a key aspect to enhancing the overall performance of solar panels. In some cases, fixed open arrays perform betterby a small value but these occur due to the shading of panels and solar tracker's misjudged tracking or unpredictable solar radiance position. Overall, on average 1-axis tracking array displays highly effective performance for all the ten cities taken under this study.

6. Conclusion

In this paper, the study disbursed is, that leads to finding the photovoltaic systems that produce the best energyoutput potential. Energy performance is another factor that is studied on Crystalline Silicon arrays which shows varying radiation data collected on-site. It has also been found that the installation method of the Crystalline silicon array also plays an important role in energy performance. Ten cities in India are the study locations takenfor this research. On simulating the energy results it is concluded that the 1-Axis tracking Crystalline silicon arraydisplays high performance in comparison to the Fixed rack Crystalline element array for all the study locations taken. This paper gives a better understanding of the pre-feasibility and practicability of PV systems to scholars, researchers, analysts, etc., who are interested in improvement in this sector.

References

[1] Aghaei, Eskandari, Vaezi, S. Chopra, 2020, '*Solar PV power plants*, Photovoltaic Solar Energy Conversion, https://doi.org/10.1016/B978-0-12-819610-6.00010-7.

[2] Spanos, Simons, Holmes,2010, '*Cost savings by application of passive solar heating*, Structural Survey, DOI 10.1108/02630800510593684.

[3] Brunoro, 2012, '*An assessment of energetic efficiency improvement of existing building envelopes in Italy'*, Management of Environmental Quality: An International Journal, DO 0.1108/14777830810904939.

[4] Agarwal1, Gautam, Tarique, 2021, '*Assessment of appropriate areas for the establishment of Solar Thermal Power Plant in Rajasthan'*, IOP Conf. Series: Materials Science and Engineering, doi:10.1088/1757-899X/1149/1/012030.

[5] Liu, Bimenyimana, Asemota,2019,' *Assessment of power generation performance characteristics using different solar photovoltaic technologies'*, IOP Conf. Series: Journal of Physics: Conf. Series, doi:10.1088/1742- 6596/1311/1/012004.

[6] Pedrero J, Hermoso N, Hernández P, Muñoz I, 2019, '*Assessment of urban-scale potential for solar PV generationand consumption*', IOP Conf. Series: Earth and Environmental Science, doi:10.1088/1755-1315/323/1/012066.

[7] Park. Kim,2016, '*Building information modeling-based energy performance assessment system*', Construction Innovation

[8] Praveen, Nallapaneni, Kumar, 2017,' Optimal energy performance and comparison of open rack and roof mountmono c-Si photovoltaic Systems', 1st International Conference on Power Engineering, Computing, and control Duffie, Beckman,2012,' *measure and estimation of the solar resource*', Solar Engineering of Thermal Processes

[9] Vares, Savolainen, Häkkinen,2020,' *Concepts and type building for carbon neutral construction in arctic Finlandbased on tradition*', World Sustainable Built Environment conference, doi:10.1088/1755-1315/588/2/022032.

[10] Crosbie, Dawood, Dean,2016,' Energy profiling in the life-cycle assessment of buildings, Management ofEnvironmental Quality: An International Journal, DOI 10.1108/14777831011010838.

[11] Yadav, Hema, Shivakumara, 2020,' *Environmental impact assessment on renewable energy: A review*, IOPConference Series: Earth and Environmental Science, doi:10.1088/1755-1315/573/1/012048.

[12] Martinez-Gracia,2019,' *Solar energy availability'*, Solar Hydrogen Production. https://doi.org/10.1016/B978-0-12-814853-2.00005-9.

[13] Kumar, Sharma,2014, *'Review and analysis of solar photovoltaic software'*, International Journal of Engineeringand Technology.

[14] Sreenath, Sudhakar, Yusop,2021, *'Performance assessment of conceptual bifacial solar PV system in varyingalbedo conditions'*, IOP Conf. Series: Materials Science and Engineering, doi:10.1088/1757-899X/1078/1/012033.

[15] meur, Arechkik & Berrada, Asmae & Loudiyi, Khalid & Aggour, Mohammed. (2020). Forecast modeling and performance assessment of solar PV systems. Journalof Cleaner Production. 267. 122167. 10.1016/j.jclepro.2020.122167.

[16] Bravo, Ruben & Ortiz, Carlos & Chacartegui, Ricardo & Friedrich, Daniel. (2021). Multi-objective optimization and guidelines for the design of dispatchable hybrid solar power plants with thermochemical energy storage. Applied Energy. 282. 10.1016/j.apenergy.2020.116257.

[17] Li, T. & Roskilly, Tony & Wang, Yaodong. (2017). Life cycle sustainability assessment of grid-connected photovoltaic power generation: A case study of Northeast England. Applied Energy. 227. 10.1016/j.apenergy.2017.07.021.

[18] Yadav, Somil & Panda, S.K. & Hachem, Caroline. (2020). Method to improve performance of building integrated photovoltaic thermal system having optimum tiltand facing directions. Applied Energy. 266. 114881. 10.1016/j.apenergy.2020.114881.

[19] Zarzo, Manuel & Marti, Pau. (2011). Modeling the variability of solar radiation data among weather stations using principal components analysis. Applied Energy.88. 2775-2784. 10.1016/j.apenergy.2011.01.070.

[20] Gao, Yuan & Dong, Jianfei & Isabella, Olindo & Santbergen, Rudi & Tan, Hairen &Zeman, Miro & Zhang, Guoqi, 2019. "Modeling and analyses of energy performances of photovoltaic greenhouses with sun-tracking functionality," Applied Energy, Elsevier, vol. 233, pages 424-442.

Performance Evaluation of Water Treatment Plant of Kekri, Rajasthan, India

R S Sinsinwar[1] and M Verma[2]

[1]Department of Civil Engineering, Bhagwant University, Ajmer-305023, India
[2] Department of Mechanical Engineering, Bhagwant University, Ajmer-305023, India
E-mail id: rsroopsing81@gmail.com

Abstract. Drinking water is a sole element of human life. The rapidly growing population is adding to the excessive demand for water. Increasing demand for drinking water is rising the over-exploitation of water. Raw water sources require water treatment with a high production capacity of treatment. Present research provides a case study of the water treatment plant of Kekri, Rajasthan, India. In these two water treatment plants water purification efficiency is good as compared to conventional water treatment plants. Adding poly aluminum Chloride as a coagulant increases water treatment efficiency. Many parts of the water treatment plant add coagulation-flocculation, filtration, and disinfection by using poly aluminum chloride. Two water treatment plants with the capacity of 142 MLD and 132 MLD are having the net purifying capacity to remove the turbidity and Color. In this research, it has been found that turbidity removal of 142 MLD and 132 MLD WTP was found about 83.4 and 74.1% respectively. Total dissolved solid of surface reservoir Bisalpur dam found within acceptable limits. Total dissolved solids in Bisalpur dam surface water such as Chloride, Fluoride, and Nitrates were found of 60 mg/l, 0.3 mg/l, and 2 mg/l respectively. The pH value of surface water was found of 7.90 and remains almost constant with the seasonal variation impact. Turbidity and color reduction were found within the acceptable limit according to Indian standard 10500-2012.

Keywords: Water treatment plant; Coagulant; Poly-aluminum Chloride; Turbidity; pH Value; Coagulation; Flocculation

1. Introduction

Natural water is available in various sources like dams, rivers, lakes, springs, and streams. Nowadays water consumption capacity is increasing with the development of civilization. This increasing demand for water has also increased the use of filter plants. The utility of the filter is to purify the contaminated water and deliver it to the consumer. Chlorine is used to kill bacteria and viruses and the water treatment plant removes the taste odor, color, Turbidity, and metals [1]. The water born disease like cholera, typhoid, and dysentery spreading concept was adopted by the year 1990 [2]. Particulate suspended matters such as sand, clay, loam, chemicals, and minerals are dissolved in water. Moreover, surface water contains the water bacterial load, E Coli [3]. The surface water quality available in the Bisalpur dam requires some conventional water treatment. This conventional water treatment plant contains four stages, coagulation-flocculation, sedimentation, filtration, and disinfection. Sludge generated from water treatment plants is of environmental impact adversely. Sludge discharge in water bodies pollutes the groundwater and water bodies due to high concentrations of heavy metals likely Al_2O_3; FeO_3, K_2O etc. [4].

In Kekri water treatment plants 142 MLD WTP have recirculation concrete lined diggi open to the sky. Two reservoirs contain for collection and recirculation of backwash water. Settling of the suspended sand particles from backwash water is reused for water treatment with again discharge it to receiving chamber along with surface water at 142 MLD. Another 132 MLD WTP has an unlined open pond for backwash water recirculation. Turbidity should be less than acceptable 1.0 NTU and permissible limit up to 5.0 NTU [5]. The world health organization also recommended this range 5-7 NTU so corrective measures should therefore be pursued [6,7]. Poly-Aluminum Chloride is one of the most common coagulants for wastewater treatment [8]. The Poly-Aluminum Chloride (PAC) contains about 70% Al base and the aluminum concentration was 0.35 mol/l [9,10].

This research paper describes the comprehensive performance evolution (CPE) phase conducted in all two water treatment plants. The turbidity, color, and pH were selected as parameters that are essential tools in the evaluation of performance. Other parameters of surface water such as total dissolved solids, chloride, fluoride, and Nitrates were also analyzed. Performance evaluation was conducted in these two water treatment plants under the guidance and use of a manual for the optimization of water treatment.

2. Material and Methods

2.1 Description of the Project

This research work is based on Bisalpur Ajmer water supply project of Rajasthan (India). This project consists of two water treatment plants. The old project was commissioned in 1994 to produce 132 million Liter/day and the new water treatment was commissioned in 2011 to produce 142 million/liter/day. Due to the increasing water demand in all towns and Ajmer city province, six numbers vertical turbine pumps were installed at the intake well (Figure-1 shows the site of water treatment plant Kekri (Raj.) India. Details of water treatment plants dimensions are shown in Table 1.

Figure1. Water treatment plant Kekri (Raj.) India

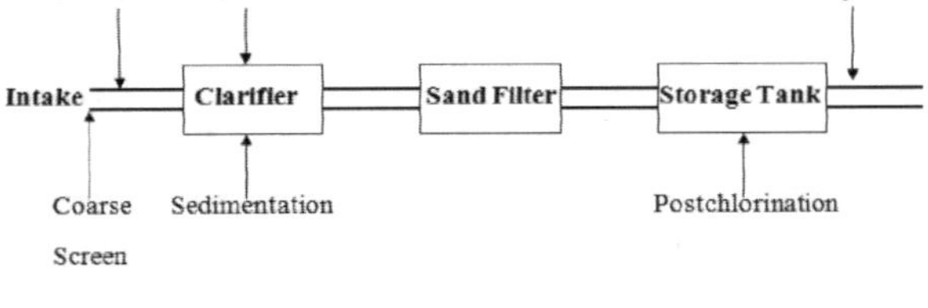

Figure2. Flow diagram of conventional water

2.2 Purification Process

Figure 2 shows the general layout of a conventional water treatment plant which consists of raw water receiving chamber treatment and storage facilities. This figure gives the description of the water treatment unit's components facilities, including clarifier, flocculator, filter beds, and clear water reservoir for storage of treated water. Figure 3 shows the intake which is located on the Banas River on the downstream side of the Bisalpur dam. This dam is the source of drinking water and irrigation purpose. There are six vertical turbine pumps. Although four vertical turbine pumps set are in running position and two vertical turbine pumps are on standby. The total discharge of each pump is 4236 m3/hour and the head of the pump is 45 m.

Figure3. Image of intake at BanasRiver on the downstream side of Bisalpur dam

2.3 Coagulant Dose

A Jar test was conducted to find out an optimum dose of poly-aluminum chloride. Both water treatment plants have an equal dose of coagulant 30 mg/l of PAC and both WTP treat the same surface water of the Bisalpur dam. The optimal dose of PAC coagulant was obtained by conducting the Jar test and its value was found of 30 mg/l. Poly aluminum Chloride (PAC) was added to the raw water and then distributed the raw water into the four clarifiers. Poly-aluminum Chloride and pre-chlorination were applied across a 500 mm diameter UPVC pipe inside the coagulation tank. Flex mixture was used for flocculation inside the coagulation and it was the second step of water treatment. Which is discharged from the concentrator and central chamber of each clarifier, and flows under gravity to the sludge tank. The dose of PAC changes in weather conditions and in cases according to turbidity and pH test. According to Indian standards of drinking water quality, after coagulation-flocculation, sedimentation process completion the water goes to the opening as (V-Notch) shape to the filtration during the time period (detention period) which is shown in Figure 4. New WTP contains 12 numbers of filter beds and in the old WTP 10 numbers of beds.

Figure 4. Image of filtration process

Table 1. Details of water treatment plants at Kekri (Raj.)

Particulars	132 MLD WTP	142 MLD WTP
Output Capacity	132 MLD	142 MLD
Working hours	24	24
Inlet & Clarification works	4 & 4	4 & 4
Inlet chamber size detention period	3.00x4.80x5.50 m. 45 Second's	6.60x3.30x3.75 m. 30 Second's
Size of Raw water Channel	3.00 m. wide x 25 m. long	3.00 m. wide x 25 m. long
Distribution chamber size detention period	3.00x4.80x5.50 m. 45 Second	7.2x3.50x3.00 m. 30 Second
Distribution chamber size detention period	3.00x4.80x5.50 m. 45 Second	7.20x3.50x3.00 m. 30 Second
Flash mixture numbers size detention period	4 No. 2.30x2.3x2.5 m. 30 Second	4 No. 3.05x3.05x3.00 m. 60 Second
Clariflocculator Numbers Type	4 No. Trapped velocity gradient	4 No. Trapped velocity gradient
Flocculators zone Size	39.5 m. dia x3.15 side LD 2hours	43.0 m. dia x v3.5 m side LD 2.5hours
Detention Period Surface loading Weir loading Launder size Sludge outlet pipe	40 cum/sqm/day - 650x750 mm. 250 mm C.I.	30.8 cum/sqm/day 12.32 cum/m/hr. 1000x650 mm. 250 mm C.I.
Clarifier sludge with drawl system	Auto/Manual	Auto/Manual
Size of each bed	twin and each section 4.1m.x4.40m.	twin and each section 5.35 m.x12.40 m.
Rate of Filtration Content head over sand	6 cum/sqm/hr.2 m.	4.76 cum/m^2/hr. 2 m.
Filter media thickness:		
• Sand	600mm.	700 mm.
• Grit (2.5 – 6mm)	150 mm.	200mm.
• Grit (6.0-12 mm)	100 mm.	-
• Gravel (12-37 mm)	100 mm.	-
• Gravel (17-50 mm)	100 mm.	250 mm(50mm. size)
Chemical House: Chemical for clarification	Poly aluminum chloride chlorination (pre+post)	Poly aluminum chloride chlorination (pre+post)
Disinfection method	NA	1 No.
Wastewater Recirculation: • Sludge-well • DIGGI • No. • Size	2 2.50m. x1.5 m.x2.0m.	1 68.63x56.27x1.90m.

Filter media consist of graded sand and gravel. The rate of filtration of old WTP 132 MLD and new WTP 142 MLD are 6 cum/sqm/hr and 4.76 cum/sqm/hr respectively. The water after filtration through filter beds is pure but not sterilized. Washing of filters is accordingly due to removing mud balls on the filter sand beds. Backwash water of filter plant collected in the backwash water tank. The backwash tank is shown in Figure 5 and the jar test apparatus is shown in Figure 6.

Figure5. Image back wash water collection tank

Figure6. Image of Jar Test Apparatus

3. Disinfection of Filtered Water

Chloride is mechanically added as a disinfection matter. Chlorine was put at the rate of 2 ppm dose after post chlorination in the accepted rainy season. In the rainy season, chlorine dose is applied according to turbidity and microorganism so that 0.5 ppm of chlorine is reached the consumer after the disinfection is completed and the water has become drinkable. The mechanism of pre and post-chlorination is shown in Figure 7.

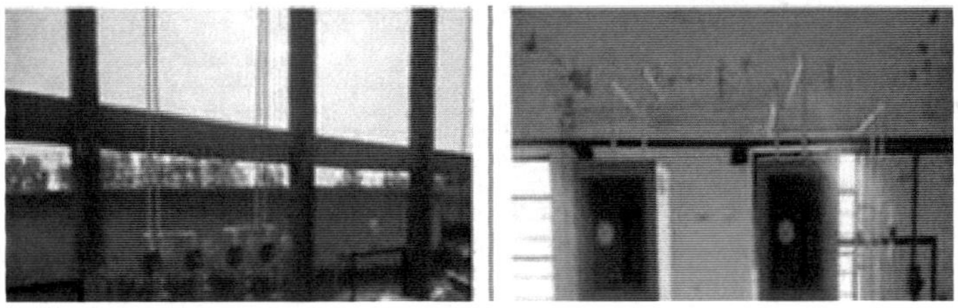

Figure7. Chlorination System

Two water samples of raw and treated water have been taken for analysis at different stages of treatment. All the water samples were taken in the month of November 2021. The aim of the daily laboratory test is to ensure that potable water confirming according to Indian standards 10500-2012. A physical test was concluded to determine the quality of water. Water treatment is properly done during each stage to examine whether the treated water confirms Indian standards. Turbidity, pH, and color testing data were collected from 1-11-2021 to 20/11/2020. The adopted dose of chlorine used on the testing date is shown in Table 3.

Table 2. Indian specification (IS) of drinking water IS10500-2012

Characteristic	Requirement (Acceptable limit)	Permissible limit in the Absence of Alternative source
Color Hazen unit Max	5	15
Odour	Agreeable	Agreeable
pH value	6.5-8.5	No Relaxation
Taste	Agreeable	Agreeable
Turbidity, NTU Max	1	5
Total dissolved solid mg/l max	500	2000
Calcium (as Ca) mg/l, max	75	200
Chloride (ascl), mg/l, max	250	1000
Fluoride (asF) mg/l, max	1.0	1.5
Free residual Chlorine mg/l, min	0.2	1
Iron (as Fe), mg/l, max	0.3	No relaxation
Sulphate (as SO_4) mg/l, max	200	400
Nitrate (NO_3), mg/l, max	45	No relation
All water intended for drinking		
E-coli or thermo tolerant Coli form bacteria	Shall not be detectable in any 100M sample	

4. Result and Discussion

4.1 Turbidity

It is indicated in Tables 2 and 3 that it is obvious not that the turbidity of the water in the plant had peaked on 20/11/21 of old WTP 132 MLD and New WTP 142 MLD during the study period with a value of 6.69 NTU and 6.21 NTU respectively. The lowered turbidity is found on 11/11/2021 for the old WTP 132 MLD and 10/11/2021 for the new NTU 142 MLD with a total value of 5.95 NTU and 5.83 NTU respectively. The turbidity of treated water peaked on 3/11/2021 of old WTP 132 MLD and in new WTP 142 MLD on 08/11/21 with a value of 2.0 and 1.27 NTU respectively. Its value was lowered on 11/11/2021 with a value of 1.27 NTU for old WTP 132 MLD and on 01/11/21 for New WTP 142 MLD value was 0.98 NTU. Whereas the overall rate of turbidity of treated water was found of 1.74 NTU for old WTP and 1.22 NTU for new WTP. The removal efficiency of turbidity reached its optimum magnitude of old WTP on 11/11/2021 with about 79.00% and new WTP was 83.90% on 01/11/2021 while it is lowered on 3/11/2021 with about 67.21% for old WTP and on 8/11/2021 with about 78.69% for new WTP. The turbidity increased in the rainy season in July & August month. The relationship between turbidity and the date of the test is reported in Table 4.

4.2 pH Value

pH values found ranged between 7.50-8.5. Which are among the determinants of global and do not have a significant impact on other water features. The data clearly indicates that all pH values of treated water were found appreciable within Indian standard limits of 6.5-8.5. Table 3 indicates the data on pH value.

Table 3. pH value of raw water, treated water of old and new water treatment plant

Date	142 MLD WTP (New Plant)				132 MLD WTP (New Plant)			
	Raw Water pH value	Treated Water pH value	Pre Chlorine dose mg/l	Post Chlorine dose mg/l	Raw Water pH value	Treated Water pH value	Pre Chlorine dose mg/l	Post Chlorine dose mg/l
1/11/2021	7.76	7.59	2.00	2.00	7.87	7.55	2.00	2.00
2/11/2021	7.87	7.56	2.00	2.00	7.85	7.56	2.00	2.00
3/11/2021	7.77	7.52	2.00	2.00	7.95	7.56	2.00	2.00
4/11/2021	7.89	7.60	2.00	2.00	7.95	7.65	2.00	2.00
5/11/2021	7.86	7.57	2.00	2.00	7.94	7.54	2.00	2.00
6/11/2021	7.89	7.54	2.00	2.00	7.76	7.69	2.00	2.00
7/11/2021	7.78	7.51	2.00	2.00	7.94	7.72	2.00	2.00
8/11/2021	7.75	7.54	2.00	2.00	7.96	7.66	2.00	2.00
9/11/2021	7.92	7.55	2.00	2.00	7.84	7.66	2.00	2.00
10/11/2021	7.87	7.55	2.00	2.00	7.88	7.64	2.00	2.00
11/11/2021	7.73	7.53	2.00	2.00	7.92	7.57	2.00	2.00
12/11/2021	7.84	7.54	2.00	2.00	7.87	7.62	2.00	2.00
13/11/2021	7.86	7.58	2.00	2.00	7.87	7.63	2.00	2.00
14/11/2021	7.89	7.56	2.00	2.00	7.88	7.54	2.00	2.00
20/11/2021	7.91	7.54	2.00	2.00	7.94	7.64	2.00	2.00

4.3 Add Chloride

The dose of Chlorine was within the specifications of Indian standards ranging between 1-16 mg/l. It is evidenced that the Chlorine dose increased in the rainy season. This is because of the high value of turbidity which leads to the rising number of contaminants. The data of added Chloride of old and new WTPs are shown in Table 3.

4.4 Color

It is obviously not that the color of the water of the plant peaked on 20/11/2021 for old WTP 132 MLD and new WTP during the study period with a value of 40.39 Hazen and 40.14 Hazen respectively. The lowered color was found for the old WTP 132 MLD on 8/11/2021 and for the new WTP 142 MLD on 1/11/2021 with a total value of 25.40 and 24.10 Hazen respectively. The color of treated water peaked on 20/11/2021 for old WTP 132 MLD and in five dates 4/11/2021, 10/11/2021, 12/11/21,13/11/2021 to 20/11/2021 and on 20/11/2021 for old WTP and its value was found of 5 Hazen. lowered on three dates on 7/11/2021, 8/11/2021, and 11/11/2021 of old WTP 132 MLD and new WTP 142 MLD found 3Hazen and 3 Hazen respectively. The minimum color reduction value was found at 3 Hazen in both WTPs. The removal efficiency of color reached its optimum magnitude of old WTP on 11/11/2021 with about 90.19% and new WTP was 90.89% on 10/11/2021. The relationship between color between removal efficiency of old and new WTPs shown in Table 5.

Table 4. Turbidity and removal efficiency of 142 MLD & 132 MLD water treatment plant

	142 MLD WTP (New Plant)			132 MLD WTP (New Plant)		
Date	Raw Water NTU	Treated Water NTU	Removal Efficiency (%)	Raw Water NTU	Treated Water NTU	Removal Efficiency (%)
01/11/21	6.09	0.98	83.90%	6.10	1.65	72.95%
02/11/21	6.21	1.02	83.57%	6.15	1.66	73.00%
03/11/21	5.92	1.14	80.74%	6.10	2.00	67.21%
04/11/21	6.18	1.17	81.06%	6.15	1.93	68.61%
05/11/21	6.07	1.00	83.52%	6.27	1.91	69.29%
06/11/21	5.93	1.02	82.80%	6.17	1.85	70.07%
07/11/21	5.89	1.10	80.06%	5.96	1.54	74.16%
08/11/21	5.96	1.27	78.69%	6.12	1.62	73.53%
09/11/21	5.90	1.22	79.32%	5.96	1.82	69.46%
10/11/21	5.83	1.10	81.13%	5.95	1.54	74.11%
11/11/21	6.09	1.21	80.13%	6.05	1.27	79.00%
12/11/21	6.15	1.25	79.67%	6.14	1.87	69.54%
13/11/21	6.17	1.23	80.00%	6.23	1.91	69.34%
14/11/21	6.01	1.22	79.70%	6.10	1.83	70.00%
20/11/21	6.21	1.26	79.71%	6.69	1.70	73.99%

4.5 Filter Media

New water treatment plant 142 MLD performance is found good due to using 50 mm size aggregated with 5-70 mm well graded round and smooth surface pebbles water worn in 250 mm thick in the bottom of the filter beds. This is an advanced technology that improves the removal of color, turbidity, and pH reduction. The Ideal sand media for intermitted sand filters is coarse sand with an effective size between 0.3–0.5 mm. The media sand grains were found relatively uniform in size having a low uniformity coefficient value (less than 4.0) to promote the movement of water and prevent clogging. Both the filter plants are running at the same Poly aluminum Chloride dose of 30mg/l (part per million). There is a difference in the quality of the purification water capacity of both water treatment plants 132MLD and 142MLD. The sand media of the new WTP 142 MLD was found to be good according to Indian standards. In the old WTP 132 MLD, the sand media of the water treatment plant filter beds were sand sampled at a depth of 150 mm below.The filter sand media of the water treatment plant was tested. The mud balls were found inside the filter media. Sand media is working under old WTP 132 MLD for about 15 years. The quality and flow rate of water purification was found to be low due to the presence of clay particles and loss in the ignition was found 1.0% more than the 0.7% permissible limits. Defects in filter sand media may result in reduced water purification. Head loss of old filter plant 132MLD WTP has been found more than new WTP 142 WTP MLD. The water

IOP Conf. Series: Earth and Environmental Science 1084 (2022) 012032 doi:10.1088/1755-1315/1084/1/012032

purification capability of filter beds of the old water treatment plant 132 MLD is lower than the new WTP 142 MLD. Head loss of the water treatment plant is shown in Figure 9.

Table 5. Color and removal efficiencies of 142 MLD and 132 MLD water treatment plant

Date	142 MLD WTP (New Plant)			132 MLD WTP (New Plant)		
	Raw Water Color	Treated Water Color	Removal Efficiency (%)	Raw Water Color	Treated Water Color	Removal Efficiency (%)
01/11/21	24.19	3.00	87.60%	29.60	4.0	86.48%
02/11/21	25.10	3.00	87.65%	25.40	4.0	84.25%
03/11/21	29.85	3.00	89.83%	25.50	4.0	84.31%
04/11/21	31.30	3.00	90.41%	27.60	5.0	81.88%
05/11/21	29.70	4.00	86.53%	26.60	4.0	84.07%
06/11/21	29.40	3.50	88.09%	29.50	4.0	86.44%
07/11/21	26.34	3.00	88.61%	24.12	3.0	87.50%
08/11/21	26.59	3.00	88.71%	24.10	3.0	87.50%
09/11/21	30.50	3.00	90.16%	27.10	4.0	85.24%
10/11/21	32.95	3.00	90.89%	31.45	5.0	84.10%
11/11/21	30.98	3.00	90.32%	30.60	3.0	90.19%
12/11/21	32.90	4.00	87.84%	29.20	5.0	82.87%
13/11/21	28.95	3.00	89.64%	30.51	5.0	83.61%
14/11/21	29.54	3.00	89.84%	29.95	4.0	86.64%
20/11/21	40.14	4.20	89.53%	40.39	5.0	87.62%

Figure8. Image of 142 MLD and132 MLD WTP filter bed

4.6 Testing Effectiveness of Filter Media Sand

The sand media of both the filter plants132MLD and 142MLD was tested to increase the water purification efficiency of the plants. Impurities, such as clay, loam, silts, etc. were found at 13% in old WTP 132 MLD and 6.5% in new WTP 142 MLD. Acid soluble matters were found of 10% of old WTP 132 MLD and 5% of new WTP 142 MLD. Results are shown in Table 6.

Table 6. Filter media testing results of water treatment plants

Sand media testing Parameters	As per IS code 8419 (Part-I) 1977	Old WTP 132 MLD	New WTP 142 MLD
Specific gravity	not less than 2.50	2.90	3.05
Impurities such as clay loam silt	not more than 5%	13%	6.50%
Acid solubility	not more than 5%	10%	5%
Loss on Ignition	not more than 0.7%	1.0%	0.61%

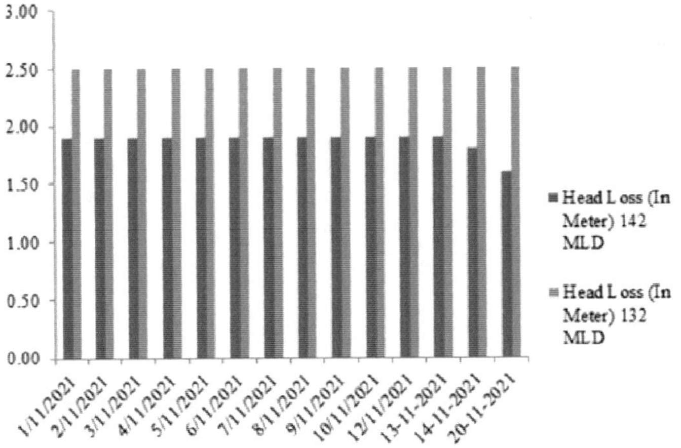

Figure9. Head losses in filter beds of 132 MLD & 142 MLD

5. Conclusions

The conclusions that can be drawn from the laboratory investigation are summarized as follows:

- Turbidity, color removal, and pH reduction of new WTP 142 MLD were found too high. The filter media sand has been used the effective size is suitable between 0.3 mm and 0.5 mm, uniformity coefficient was found more than 4.0.
- Excessive amount of sticky clay, loam, and organic matter was found under the sand media of the 132 MLD filter plant. The purification capacity old water treatment plant, 132 MLD is comparatively less than that of the new water treatment plant 142 MLD.
- The purification capacity of the new water treatment plant 142 MLD has increased by adding 5-7 mm round shaped filter media with 50 mm size aggregates in the bottom of filter beds.
- Almost pH value of the dam is constant with a simple difference that results from the suspended particles in it.
- The water temperature varies with the air temperature without affecting the other water properties.
- The average turbidity removal was relatively acceptable. Its efficiency was raised by reducing flow velocity and increasing the amount of PAC to promote flocculation.
- The Amount of added chlorine is relative preciously within the limits.

6. References

[1]. Al-Samawi AA, Al-Bayati SA and Al-Obaidi AH 2008 Aluminum Concentrations in Baghdad Water Supplies During 2004,*Engineering and Technology Journal*, **26(11)**, 1301-1315.

[2]. World Health Organization report, 1963.*Website:http://whqlibdoc.who.int*

[3]. Pandey MK, Jenssen PD2015 Reed Beds for Sludge Dewatering and Stabilization,*Journal of Environmental Protection*, **6(4)**, 341-530.

[4]. Black AP and Willems DG 1961 Electroscopic studies of Coagulation for removal of organic Color, *Journal of American Water Works Association*; **53(5)**, 589-604.

[5]. Amirtharajah A, Wetstein and Daniel P 1980Initialdegradation of effluent quality during filtration,*Journal of American Water Works Association*, **79(9)**, 518-524.

[6]. A Report of Indian standard of drinking water quality IS 10500-2012. Website:https://www.indiawaterportal.org

[7]. Selman HM, Wahid AAA and Selman GM 2015 Evaluating the Performance of Water Treatment Plant (Case Study: Al-Rumaitha Treatment Plant, Al-Muthanna, Iraq), *Civil Engineering,* 82, 32086-32093.

[8]. Albaggar AKA 2021 Investigation of some physical, chemical, and bacteriological parameters of water quality in some dams in Albaha region, Saudi Arabia, *Saudi Journal of Biological Sciences,***28(8)**, 4605-4612.

[9]. Varma VG, Jha S, Raju LHK, Kishore RL, Ranjith V 2022 A review on decentralized wastewater treatment systems in India, *Chemosphere*, 300, 134462.

[10]. Repot of Indian standards 8419 (Part I-1977) requirements for filter sand and for gravel used in filtration water.*Website: https://www.cracindia.in*

Framework for freight movement in Bangalore city, India

Seelam Srikanth[1*], Syed Omar Ballari[2] and Sala Eswar[3]

[1]Associate Professor, School of Civil Engineering, REVA University, Yelhanka-560064, Bangalore, Karnataka, India.
[2]Professor, Dept. of Civil Engineering, Guru Nanak Institutions Technical Campus, Ibrahimpatnam, Telangana, India.
[3]Assistant Professor, Department of Civil Engineering, Gudlavalleru Engineering College, Andhra Pradesh, India.

*corresponding author's e-mail: srikanths.reddy@reva.edu.in

Abstract. The movement of goods throughout the country ensures need and availability across the countries. The process and methodology of movement of goods are termed Freight. From the process of identification of the source, production, storage, and delivery of goods to respective customers there are logistics involved in the process, known as freight logistics. The market size of the logistics sector in India is estimated to be between USD 90-125 billion. Freight is also the major user of roads, rails, and other modes of transport. Despite holding promise the logistics sector in India remains mired in several complexities which have the potential of holding it back. As a solution to this Freight Logistics needs a framework and Directive policies, in this regard this documentation was carried out with the help of the Directorate of Urban Land Transport to review various Directive policies of freight around the world. The present study will map the types of goods, different stakeholders participating in freight, Infrastructures involved in freight, Logistics or Supply Chain processes, different types of Logistic Service Providers, etc. Finally in this paper, the draft framework for Freight Logistics Policies is worked out under mainly three sub-headings Institutional Landscape, Infrastructure, and Logistics or Supply Chain Process.

1. Introduction

A simple definition of urban transportation is the movement of goods into, out of, in, or out of the city by small or heavy vehicles, which includes service delivery and vehicle demolition as well as waste and procurement. Home shopping trips are not considered part of urban transportation as these are considered to be passenger transports [1]. Logistics by definition is a detailed process for handling large complex tasks, varying from business to business. Freight Logistics manages the movement of goods and other resources between the place of origin and the place of use to meet the needs of consumers. Logistics includes a variety of end-to-end supply chain management features, be it information integration, transportation, inventory, warehouse, inventory management, and packaging. Supply chain (Logistics) (SCM) management represents a set of management processes, emphasizing the integration of supply chain development as it involves a variety of processes and interdependent stakeholders [2].

The chain of Indian shops is suffering from the presence of many links. There are up to five intermediaries in a typical supply chain of perishable goods such as fruits and vegetables. GPS technology and tracking have already been developed for advanced IT tools that are being processed and used by senders but there are approximately 2% of LSPs follow these processes. Transportation practices often lead to unsafe practices such as overcrowding, which disrupts road safety for both truck drivers and their passengers — more than 20% of the 1.4 lakh deaths in 2014 were truck drivers (According to Road accidents in India, 2018). It is estimated that 40% of the time lost on the road is due to stopping at checkpoints at government borders only. The industry has led to problems related to good organization, lack of leadership, incompetent skills and positions, and a lack of process-driven IT systems.

To encounter these challenges through Policy interventions the study was designed, which can be listed as follows: Review of Directive policies of International organizations, review of Directive policies of National organizations, review of City logistics frameworks of various cities across the world, and draft Framework for the Freight Management Policy for Indian cities-Challenges, Structures, and Reforms.

2. Review of Directive Policies

Various policies framed by International organizations were reviewed and the contents of those policies that were relevant to the Indian context were briefly listed below: Fleet renewal schemes are to be adopted which will improve road safety and also protect the environment. It also suggests the adoption of technology in the fleet to improve their efficiency. A Professional Certificate of Performance should be issued to road transport operators and drivers to comply with complex laws and regulations and to be acquainted with the latest technologies and standards. Recognizing the Informal Sector with incentives to encourage the carriers to operate as legal entities by providing fiscal advantages, facilitation in accessing credit, etc [3]. GIZ document suggests the adoption of Route Planning, scheduling applications, and GIS tracking all help improve logistic efficiency by 20%. It recommends keeping the distance between consolidation centers and the inner city delivery area to be as low as possible, to avoid traffic congestion inside the city [4]. OECD promotes the adoption of Public-Private Partnerships to achieve common objectives of freight management [5].

Amongst many nations, few undertaken reviews were found to have some of the best freight management practices in the world. German policies suggest operators step up their investment in innovative technologies for optimized control of logistics processes and the removal of bottlenecks in the transport system is essential to ensure that avoidable journeys do not take place [6]. UAE policies talk about setting up an Integrated Logistics Platform where all transport modes, logistics, and value-added services, including light manufacturing and assembly, shall be made available in a single custom bonded and Free Zone environment area. It promotes the setting up of Institutions for collaboration (IFC) in an attempt to facilitate interactions between the government and the business sectors [7]. The environmental zone should protect the interior of the central road network around the city center from traffic pollution. BAG's IT department has a central computer center with a storage capacity of 100 terabytes and houses about 140 IT applications [8]. Seattle advisory Committee: This committee is responsible for Doing Neighbor freight studies. Off-Hours Delivery (OHD) program +Trusted Vendor Program (TVP) is done by coordinating with the major players of the city, to avoid traffic congestion during peak hours [9].

3. Freight Management Policy for Indian Cities

The draft Framework for Freight Management Policy can be broadly classified into an institutional setting, infrastructure, and logistics efficiency. First, the existing trends/challenges in cities are analyzed concerning the above categories and then, the most fitting solutions from best practices around the world can be adopted to bring about reforms in India's Freight Management. After studying various cities Like Bangalore, Delhi, Mumbai, and Chennai, general challenges faced in freight management by these cities are Poor Infrastructure, Improper Balance, fierce Competition, Unsafe Practices, and Time Loss. Non-optimal intermodal distribution of freight traffic is estimated to have cost the Indian economy 16% of the total transport cost in 2017. The top 20 logistics companies in India together constitute a mere 2% of the industry. India where fleets of 5 trucks or less own 75% of the market share. About 60-70% of freight flow is through the road. Around 45 billion US dollars are lost every year due to inefficiencies in the network. 40% of the fruits and vegetables grown in India get wasted. Only 2% of Logistic Service Providers adopt the technology. Inefficient organization structures and lack of leadership As many as five intermediaries in a typical supply chain for perishable commodities. After analyzing the challenges, the Structure of Freight and reforms to be brought about in these categories are suggested below.

The local authorities will list the various stakeholders operating in the city and aim at facilitating better freight management in the city. It will also frame policies that lead to a reduction in environmental nuisances as well as in increasing the safety of road traffic. It will adopt best practices in cities from around the world and tailor them to meet the needs of their city. The resource supply stakeholders are responsible for the infrastructure set available for freight transport as shown in figure 1. These stakeholders and their investments determine the possibilities for urban freight transport. These stakeholders consist of Infrastructure Providers, Infrastructure Operators, and Land Owners [10]. The supply chain actors who are responsible for sending, carrying, and receiving goods are shown in figure 2. Freight resource from end to end is transported and stored using these physical infrastructure facilities. These are the few infrastructure facilities available in Indian cities as shown in figure 3.

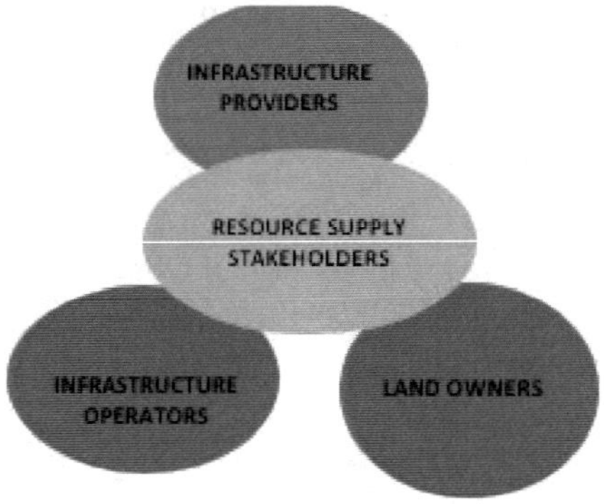

Figure 1. Resource Supply Stakeholders [11]

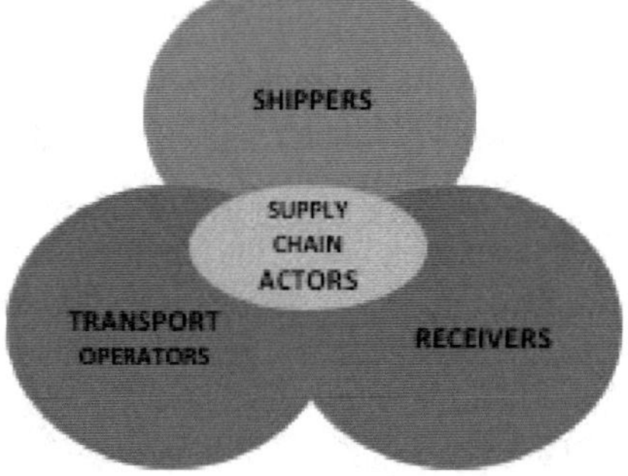

Figure 2. Supply Chain Actors [10]

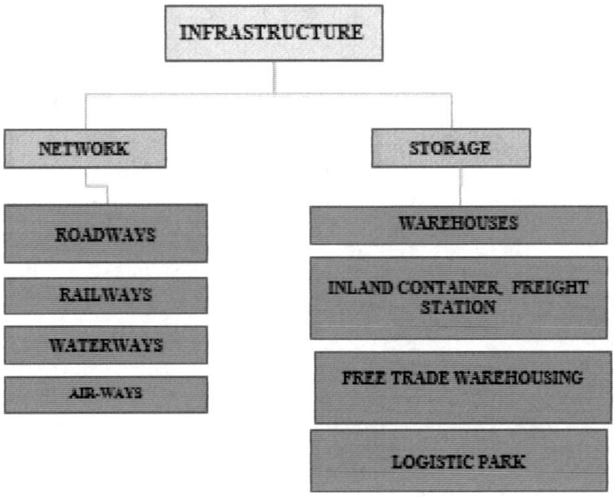

Figure 3. Structure of Infrastructure in Indian Cities [11]

From the procurement of materials to the delivery of final products to the consumer's infrastructure plays the main role in the freight supply chain [11]. Networks - for an end-to-end flow of goods. Infrastructure for production and manufacturing of goods like farms, factories, etc. Storage facilities- different types of goods demand different types of storage facilities, thus preservation of goods. Terminals facilities to receive and export the goods. Ex- Bus terminal, rail terminal, intermodal terminal, etc. Distribution facilities like distribution centers for cross-docking, packaging, and as a common platform for import and export. Though there are many reforms to be brought and researches are in progress, a few can be summarized as follow: zone-wise mapping to enable sustainable freight movement, Integration of transport modes, neighbor-hood friendly policies and facilities, planning, design, creation, and strengthening of freight network, creating awareness about existing and upcoming infrastructures and public-private partnership programs and policies to work freight network and freight movement. Enhancing Skills for transport operators to encourage safe and efficient sustainable practices within the freight transport sector. To adopt IT into freight management to increase logistics efficiency. Study various markets in urban freight and the challenges faced by them and propose measures to tackle the same. Conferences with Businessmen and academicians to review the existing policies and facilities and discuss the future strengthening of the freight network [12]. The management of the Freight network and movement needs more attention, though the government and private institutions are striving for it. The integration and efficiency of the network might make Freight more seamless than the existing conditions. The topic demands more discussion and research, yet the scope of the paper is framed under a few topics.

4. Freight Management Policy for Bengaluru

Following along the lines of a policy framework for Indian Cities, the Policy framework for Bengaluru is prepared. Perishable goods along with a controlled environment for storage also need the same for transport. As the name suggests they perish very vast thus need to reach the customer on time. Existing facilities such as 19 Mega Food Parks are operational, Plans for the agro-processing cluster, operational greens-MFP, Food Processing Training Centres (FPTC), Investments from, Global e-commerce, and training at recognized institutions at State/National level are available. The Main challenges are Wastage of goods- though India ranks among one of the major producers of goods, the lack of logistics leads to wastage of resources. Storage facilities- The lack of knowledge about storage methods and construction techniques leads to the wastage of produced goods. The Proposed solution is the collection of real-time data, it's processing, and making it available to logistics players and stakeholders so that the fresh and edible goods move around. Maintaining the right temperature-cold chain logistics- to extend the life of perishables.

Construction materials are the major industry involved in building an urban environment, the controlled and timely transport of these materials increases the rate of development of the country. Cement constitutes 90% of the total goods handled at the terminal 70% of orders are vacated immediately to construction sites or other warehouses, while 30% are stored at the CRWC warehouses at the terminal. Railways handle major construction materials such as cement, proper warehouse demands must be assessed and provided. Challenges Faced include Lack of data about the number of vehicles, sources of construction materials, and insufficient warehouse storage. Increase the Construction Consolidation Centers to supply the construction sites with the construction materials and equipment on a just-in-time basis which can bring down the pollution levels in the city. Application of reverse logistics, to avoid the empty trucks/vehicles moving in and around the cities.

Courier and cargo, one of the oldest freight which fetches people their day-to-day needs through cargo and connects people socially from various parts of land through posts. Due to a lack of logistics, improper importing and exporting techniques may damage goods. The lack of knowledge about storage methods and construction techniques may lead to casualties. Loss in freight structure due to lack of economic support to minor LSPs. The collection of real-time data, it's processing, and making it available to logistics players and stakeholders so that the goods reach the consumer without delay. Technology and logistics avoid the freight traffic interfering with present traffic. Good Storage Facilities help in delivering quality goods.

5. Conclusions

After the review of various policies, case studies, and literature, some of the methods suitable for improving the logistics supply chain can be listed as follows. A single interface for information on freight between public authorities and freight operators to ease supply based on estimated demand, for parking facilities, licensing, etc. This portal should aim to simplify a complex list of information and requirements from various operator-facing sources. Users of the portal include freight operators and drivers, businesses, freight associations, and local government. This will help the freight community to efficiently and effectively plan journeys, increase journey time reliability and reduce emissions. Encouraging cargo operators to handle raw vessels and use best practices and increase the sustainability of the supply chain must be adopted. Drivers will join the scheme as members, with membership categories indicating the success of the property operators. This will promote Best practices in improved safety, emissions reduction, fuel efficiency, and adoption of electric vehicles. Following sustainability and efficiency levels rewards like bronze level, silver level, gold level, etc. will be awarded to freight operators. A trusted vendor program will enable connecting highly rated shippers with potential clients to expand the adoption of Off-Hours Delivery and promote more efficient usage of the freight network. Institutions for funding freight has to be set up to encourage start-ups in freight, for providing loans/recovery fund for the causalities that occurred during freight. Employment opportunities and the scope of the structure should be established and educated among people. Employment benefits as in other secured jobs can be provided, to remove the fear of freight employment. And make it approachable and accessible to all. To protect the cities from pollution, the areas can be zoned as green areas and restrict the movement of freight in chosen areas.

References

[1] Cossu P et al 2014, *Clean Last mile transport and logistics management for smart and Efficient local Governments in Europe*, C-LIEGE Project.

[2] Urzawska A 2020 Towards Responsible and Sustainable Supply Chains – Innovation, Multi-stakeholder Approach and Governance *Philosophy of Management* **19**, 267–295.

[3] Ahmed SS, Hulme KF, Fountas G, Eker U, Benedyk IV, Still SE and Anastasopoulos PC 2020 The Flying Car—Challenges and Strategies Toward Future Adoption *Front. Built Environ* **6**, 106.

[4] Pandey A, Venkataraman C 2014 Estimating emissions from the Indian transport sector with on road fleet composition and traffic volume *Atmospheric Environment* **98**, 123-133.
[5] Jens K R, Michael A L, Gerard G 2014 Are public–private partnerships a healthy option? A systematic literature review *Social Science & Medicine* **113**, 110-119.
[6] Kersten, W *et al* 2017 Digitalization in Supply Chain Management and Logistics: Smart and Digital Solutions for an Industry 4.0 Environment *Proceedings of the Hamburg International Conference of Logistics (HICL)*, No. 23, ISBN 978-3-7450-4328-0, Berlin.
[7] Szymonik A 2012 *Logistics and Supply Chain Management*, text book.
[8] Zhang K, Batterman S 2013 Air pollution and health risks due to vehicle traffic *Sci Total Environ* 450-451, 307-316.
[9] Hau L L, Zuo J S 2020 Supply chain and logistics innovations with the Belt and Road Initiative *Journal of Management Science and Engineering* **5**, 77-86.
[10] Astrid B, Hanne S, Terje K, Kelly P 2019 The potential for coordinated logistics planning at the local level: A Norwegian in-depth study of public and private stakeholders, *Journal of Transport Geography* **76**, 34-41.
[11] Xiang L 2014 Operations Management of Logistics and Supply Chain: Issues and Directions *Discrete Dynamics in Nature and Society* **2014**, Article ID 701938, 7 pages.
[12] Mitra S *et al* 2020 Developing multimodal logistics parks in India, *ADB briefs* , ISBN: 978-92-9262-263-3, 10 pages

Energy, geopolitics and the dying arctic ice fields: an enviro-political perspective

A Thangaraj[1] and A Chowdhury[2]

[1] Jindal School of International Affairs, O.P. Jindal Global University, Sonipat-131001, Haryana, India.

Corresponding author: [2] Jindal School of Environment and Sustainability, O.P. Jindal Global University, Sonipat-131001, Haryana, India. abhiroop.chowdhury@gmail.com;

Abstract
In view of climate change's effects across the world and the present global escalation of conflicts it is important to assess potential future conflict zones in order to pre-emptively place measures to avoid damage and loss of life. It is identified that Arctic can be a potential future conflict zone, given its geopolitical complexities and abundance of energy resources. One of the several precursors of conflict at present is the rapid militarization of the region. Recent geopolitical instability due to Russian invasion on Ukraine and resulting sanction on Russia can accelerate the energy extraction process at Arctic's, if Russia wants to compensate its EU energy markets by connecting with hydrocarbon starved economies of East Asia. Arctic's biodiversity and abiotic environment is deteriorating rapidly. The recommendations thus put forth advocate for stronger cooperation between Arctic states as well as pushing for the ratification of international law specific to the Arctic region.

Keywords: Arctic; Geopolitics, Environment, UNCLOS; Energy, Russia-Ukraine war

1. Introduction

The Arctic has become a hot topic recently, with climate change transforming it from an icy no man's land to a piece of fresh real estate with plentiful resources beneath the ice. This paper will address three key factors in Arctic expansion:

- Resources and their viability
- The current laws and regulations concerning the Arctic
- The current evolving geopolitical state of the Arctic.

The Arctic has plentiful resources beneath the ice but their extraction, until recently, was economically unfeasible due to the difficult and challenging weather and landscape of the region. The Arctic's resources can be broadly classified into four groups:

- Living resources
- Oil and gas reserves
- Mineral resources

In terms of living resources fish are plentiful in the area and the "Arctic Five" (Russia, US, Denmark, Canada and Norway) have large communities dedicated to the exploitation of this resource. This has already resulted in environmental problems in the area and over fishing of several species has led to a depreciation of the catch in recent years. The Arctic fisheries working group (AFWG) has published a report in 2020 [1], has shown that the catchment of several varieties of fish has been on a relatively steady decrease.

The problem in the Arctic arises from several countries' claims as to their 'continental shelf'. According to United Nations Convention on the Law of the Sea (UNCLOS), a country's Exclusive Economic Zone extends to 200 nautical miles off their baseline. The EEZ of a country allows them sovereign right over any resources in the water column and sea bed. There exists a provision in UNCLOS that a country's rights need not end at their EEZ by virtue of a geographic feature known as a continental shelf. A country has rights over exploitation over any and all resources on the *sea bed* till the extent of their continental shelf. This presents a problem in the Arctic in particular as demarcating the continental shelf in these icy areas is a problem of capability.

This becomes important when it comes to Natural gas resource exploitation due to their proximity to the sea bed. This is a strategic resource and conflicting claims are a driver for future conflict in the area. Overlapping territorial disputes in the Arctic region will be a driver for conflict in the near future. Oil and natural gas resources in the Arctic region are well documented and are the primary driver of tension in the area. According to present estimations, there are about 90 billion barrels of oil in the Arctic circle, equal to 110% of Russia's reserves, and 339% of US's reserves. In terms of natural gas, there are approximately 1669 trillion cubic feet of natural gas reserves, equal to 24.3% of the world's total natural gas reserves. The objective of this paper is to assess the policies governing the Arctic at present and thereby understand future energy, environment and security scenarios in the Arctic.

Table 1. Petroleum reserves by region [2]

Petroleum Province	Crude Oil (Billion Barrels)	Natural Gas (Trillion Cubic Feet)	Natural Gas Liquids (Billion Barrels)	Total (Oil Equivalent in Billions of Barrels)
West Siberian Basin	3.66	651.50	20.33	132.57
Arctic Alaska	29.96	221.40	5.90	72.77
East Barents Basin	7.41	317.56	1.42	61.76
East Greenland Rift Basin	8.90	86.18	8.12	31.39
Yenisey-Katanga Basin	5.58	99.96	2.68	24.92
Amerasian Basin	9.72	56.89	0.54	19.75
West Greenland-East Canada	7.27	51.82	1.15	17.06

2. Methodology

The author will be conducting an intensive literature review, primarily focusing on official policy papers published by government departments as well as resource reviews. The primary literatures considered for this study were policy papers published by governmental agencies. The author has refrained from considering opinion pieces or research papers instead focusing on official policy stances by governments in the Arctic region. Policy papers published by the US Army, *Regaining Arctic Dominance,* policy stances by Canada, security analysis by Denmark's Security and Intelligence Service, policy papers published by Norway's government and policy stances published by Russia are the main sources of information in the paper.

3. Why is the Arctic a Future Conflict?

The Arctic has been barreling towards conflict for a long time. As one of the only physical meeting points between three superpowers (USA, Russia and China) is could be argued that it was inevitable that it would be a geopolitical conflict zone, especially considering the long standing conflicts between these three countries. There have been conflicting claims on the Arctic for a long time but they have been more or less resolved bilaterally (E.g. Russia-Norway dispute in the Barents Sea). It should be noted that the example of the Barents Sea took 40 years to reach a resolution, only settling in 2010. Even during the dispute collaboration in the Arctic, particularly scientific collaboration did not decrease but was strengthened. So the question arises as to why experts believe the Arctic is going to become a zone of conflict in the future. The answer is climate change.

According to the latest estimates [3] the Arctic will have an ice-free summer by 2050. This would remove a significant barrier to both oil exploration, drilling as well as shipping. This would prove to be a boon for any country that can exploit the resources of the Arctic. Here arises the problem, as we have seen in countless conflicts before, as superpowers prepare to exploit the ever more attainable Arctic there will be conflict as to who has the rights over such resources.

3.1 Who are the main agents in the Arctic?

To understand if there will be a conflict on Arctic we must understand the present policies and players in the Arctic. First, in a geophysical way there are the Arctic Five: United States of America, Canada, Norway, Denmark and Russia. This present situation is further complicated by China's unilateral declaration of being a 'Near Arctic State.' Then there is the Arctic Council, a high level intergovernmental multilateral body presiding over policies in the Arctic. The member states of the Arctic Council are as follows: US, Canada, Denmark (representing Greenland), Iceland, Finland, Russia, Sweden and Norway. Although full membership is limited to those countries geologically close to the Arctic observer status is open to countries that are vetted by the Arctic Council members. Present observers of the Arctic Council are: Germany, Netherlands, Poland, UK, France, Spain, China, India, Italy, Japan, South Korea, Singapore and Switzerland.

3.2 Current Laws in the Arctic

It must be understood that the Arctic is not as 'unclaimed' as most think it to be. In fact only a small portion of the Arctic is as yet unclaimed by one of the Arctic Five right at the pole. The problem arises due to conflicting claims and border disputes among the members, particularly Russia. Russia currently has the longest Arctic coastline and is the most dependent on Arctic resources for its economy. It must be understood that, at present, the most significant statute of 'law' in the Arctic is the United Nations Convention on the Law of the Sea (UNCLOS). But UNCLOS, unfortunately, does not have anything significant to say about the Arctic specifically (the word 'Arctic' is not mentioned anywhere in UNCLOS rather Arctic 234 alludes to 'Ice Covered Areas'). This has left the nature of Arctic policy up to the states that are stakeholders in the region.

First, we will analyze the present policies for the Arctic Five, both for brevity as well as that they are currently the most significant players in the Arctic.

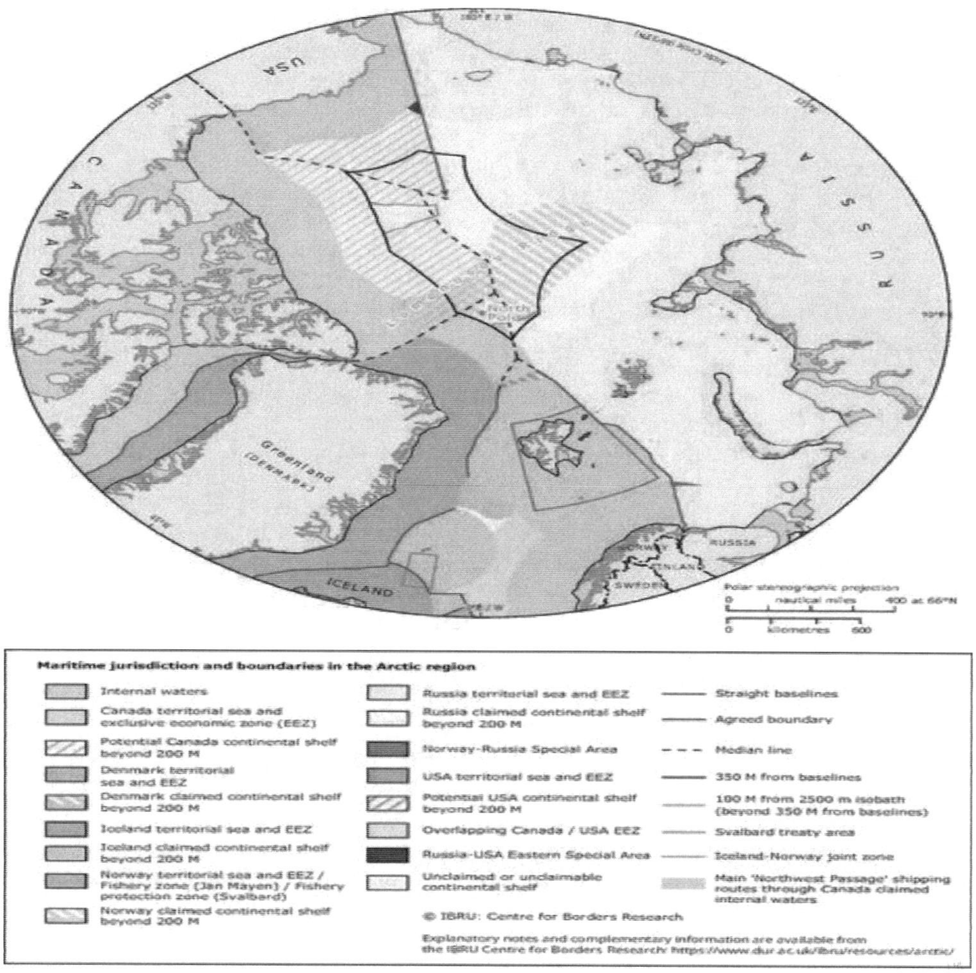

Figure 1. Conflicting Claims on Arctic [4]

3.2.1 United States of America

Since 2021 US has viewed the Arctic as a realm of increasing geostrategic importance, with a particular emphasis on the militarization of the Arctic. In the policy paper published by the US army Regaining Arctic Dominance [5] the US has laid out in an official capacity that the Arctic is going to be a focal point for conflict as well as the fact, through the implication of the title itself it shows that the US perceives that it has 'lost' dominance in the Arctic as well as the fact that they intend to regain it. The report details that Washington is pushing for a larger military presence in the Arctic. One of the main reasons for this push is the present thaw of the permafrost in the Arctic making the race for control of the Northern Sea Route, amongst other routes that are becoming more viable due to climate change more and more important. In this way US is pushing for more and more infrastructure building in the Arctic shipping thermally durable materials, long term durability and resistance to freeze-thaw cycles.

The Arctic provides the US (and others) an opportunity for power projection and the aforementioned report detailed the placement of bases and weaponry. Fort Wainwright, Joint Base Elmendorf-Richardson and Fort Greenly are the US's main encampments in the Arctic. These forces participate in several missions most importantly the Army Space and Missile Defence which provide space and ballistic missile capabilities. The US, under the Obama administration has since placed anti-ballistic missile systems in the Arctic supposedly to counter 'rouge' states like North Korea and Iran but Russia has been wary of this military expansion, enhancing its own capabilities in the region. This is a particular concern in the Arctic given its fragile ecosystem.

The report clearly details the requirement for Arctic-capable soldiers, trained for the extreme cold of the region, with appropriate equipment and capacity behind them. This is very important not just to US but to India as well as these extreme-cold conditions would allow the US to assess the requirements of Indian troops fighting in the Himalayan regions. The report details the need for military presence to secure the resources in the Arctic region, as it is a treasure-trove of natural resources, particularly Liquid Natural Gas which is a fossil fuel with a much smaller carbon footprint than petroleum. The US is also committed to 'cooperation' in the Arctic in terms of preservation of the ecosystem, scientific research and climate action.

3.2.2 Canada

Canada's policy on the Arctic has been historically aligned with the US, pushing for the self reliance of the native peoples of the Arctic (the Inuit People), respect for the land and environment of the Arctic regions, economic and social development and mainly, the "protection of Canada's Sovereignty over its Northern Regions was its first and 'Non-negotiable priority' in Arctic policy [6]." Canada also has extended continental shelf claims over the Arctic which may incur future conflict with other countries.

The main crux of Canada's claim on the Arctic at the international level is their claims over sovereignty over the Northwest Passage, the second most important trade route after the Northern Sea Route. Canada does not seek to interfere with international trade in the region but only states that all vessels travelling the Northwest Passage to comply with Canadian laws. US has disputed this claim, saying that the Northwest Passage is part of international waters and thus Canada has no right to claim sovereignty or jurisdiction over these waters.

3.2.3 Denmark

Denmark has historically stressed 'low tension in the Arctic' as a goal and a philosophy. Historically wary of NATO presence in the region, after the Russia-Ukraine war, Denmark is now welcoming NATO presence especially given a report on alleged Russian cyberattacks on Danish intelligence services [7]. Denmark has recently allocated 235 Million USD for drone and satellite surveillance in the Arctic region.

Denmark has been wary of Chinese interests in the area, strategically declining an offer from China for the establishment of a Port and mine in the Arctic region as early as 2018. Denmark has not increased its 'boots on the ground' in the Arctic instead trying to deescalate tensions by placing drones and satellites over the region.

3.2.4 Norway

Norway, like Denmark, is wary of increased Russian involvement in the Arctic. A recent report published by the Norwegian Government cites Russia's increased military capabilities in the Arctic region [8]. Since the days of the Soviet Union, Norway's security policy has been based on the maintenance of sovereignty through their domestic military as well as assurances of military reinforcements from Allies in the case of a conflict. After the violation of international law that was the Ukraine war, Norway has suspended all bilateral military ties with Russia.

To add to this they have participated in Operation Cold Response, a multilateral military exercise by NATO and allies testing their military capacity in extreme cold weather on land, sea and air [9]. Norway's military has also become more closely allied with US as their keep Russia on an even ground diplomatically. Recently Norway acquired five ice capable frigates outfitted with the US Aegis Combat System. Norway has also tried and failed to convince NATO to make the Arctic a priority but this would again prove to be a flashpoint with Russia and hence NATO has refused to do this [10].

3.2.5 Russia

Russia's claims over the Arctic are both historic and ambitious. Especially considering the Ukraine war and its devastating effects on the Russian natural resource-dependent economy, Russia has strengthened its claims over the Arctic regions, boosting exploration as well as mining in the region. Losmonov ridge in particular is of high geostrategic and economic importance to Russia as petroleum prices increase day by day. Russia's wariness of NATO will also lead to a security dilemma in the region, especially given Washington's ambitions of Arctic dominance. Russia has a particular advantage over the rest of the countries as they have a larger coastline, a more trained military force adapted to the region as well as a deeper understanding of the region. Russia also has a great technological advantage over the region, as one of the largest producers of ice-breaker hulled ships capable of traversing the Arctic. Russia's main source of insecurity in the Arctic is the presence of US military capabilities in the region. US's strategic placement of anti-ballistic missile systems with the capacity to intercept Russian ICBMs as well as US's own ICBM systems in the Arctic are leading to a security dilemma and arms race in the region.

4. Environmental impact

To understand the relevance of such a study there must be an understanding of the ecological impact of any anthropogenic activity particularly as it comes to resource extraction, shipping and military activity. Already there is documentation of the impact of shipping in the Canadian Arctic as it increases the concentration of pollutants like $PM_{2.5}$ and increases the mixing ratio of O_3 [11]. This increase in O_3 increases the effects of climate change and therefore pushes the already fragile ecosystem to the brink. Natural gas extraction as well is known to pollute water resources and resuspend sediment through the laying of pipes [12]. This would devastate the unique marine ecosystem of the Arctic. Cod (*Gordus morhua*) catch as measures in tonnes peaked in 2008 with 9889 tonnes and reached its lowest point in 30 years by 2019 with a meager 2965 tonnes caught [13]. There have been studies indicating that these fish are moving away from the Arctic due to climate change and increased anthropogenic activity and this will only get worse with warmer temperatures making it easier for humans to conduct various activities in the area.

The melting of the Arctic ice is not an easily reversible process and most of the plant and animal life there is dependent on the temperatures. Recent studies have highlighted that local species extinctions have direct relation with the rising temperature in the Arctic's [14]. The freshwater quality and chemistry is also changing rapidly across the regions that can destabilize the functionality of the ecosystem [14]. There is a projection that infers that the temperate boreal ecosystems will advance poleward if the current trend of climate change continues, reducing the tundra biodiversity [15]. Arctic plants are going to suffer local extinction with loss of snow cover in the region [16]. With rapid changes in abiotic environment, invasive species are more likely to show up in the melting Arctic region causing immense pressure to the endemic species [17]. There is a need for cooperation between the countries to conduct regular monitoring or abiotic as well as biotic components of Arctic ecosystem to formulate a plan to save the endemic species from locale extinction [18].

5. Recent international developments and the Arctic's future

The international arena and global geopolitics have shifted significantly in the recent months. At 24th February 2022, Russia, the great energy giant, with 167 billion U.S. dollars' worth of energy exports in 2020, invaded Ukraine [19,20]. This destabilized the entire geopolitics in the region with sanctions hitting the country's exports. Russia have the highest share in the Arctic which is also known for its hidden hydrocarbon reserves. EU economic giants such as Germany is heavily dependent on Russian oil and gas supplies accounting for 40% of Berlin's total gas and 25% of its total oil imports. Russia have immense reserves of hydrocarbons in their Caspian- Volga, Yenisey-Khatanga, Western Siberia, Pechora, Vilyuy- Angara- Lena (Eastern siberia), Anadyr – Shakhalin (Far eastern gas fields) basins. They export these reserve through soviet era pipeline networks to EU- Ukraine route, Yamal-Europe pipelines crossing Belarus and Poland to Germany, and the Nord Stream 1 pipeline passing below the Baltics. With sanction hitting hard and EU nations looking for alternative hydrocarbon supplies from Azerbaizan via Trans-Anatolian Natural Gas Pipeline (TANAP) through Turkey, African nations via Trans Adriatic Pipeline, supplies from the vast Dutch Groningen gas reserves; Russia's business with EU can vaporize within 2030. So the next market alternative will be in the eastern Asiatic economic hydrocarbon starved giants, such as China and India. For which the Arctic reserves will be nearer and easy to transport. The Russian energy companies- Gazprom and Novatek is already exploring the gas reserves at Arctic [21]. The Russian gas transport networks at Arctic are depicted in Figure-2.

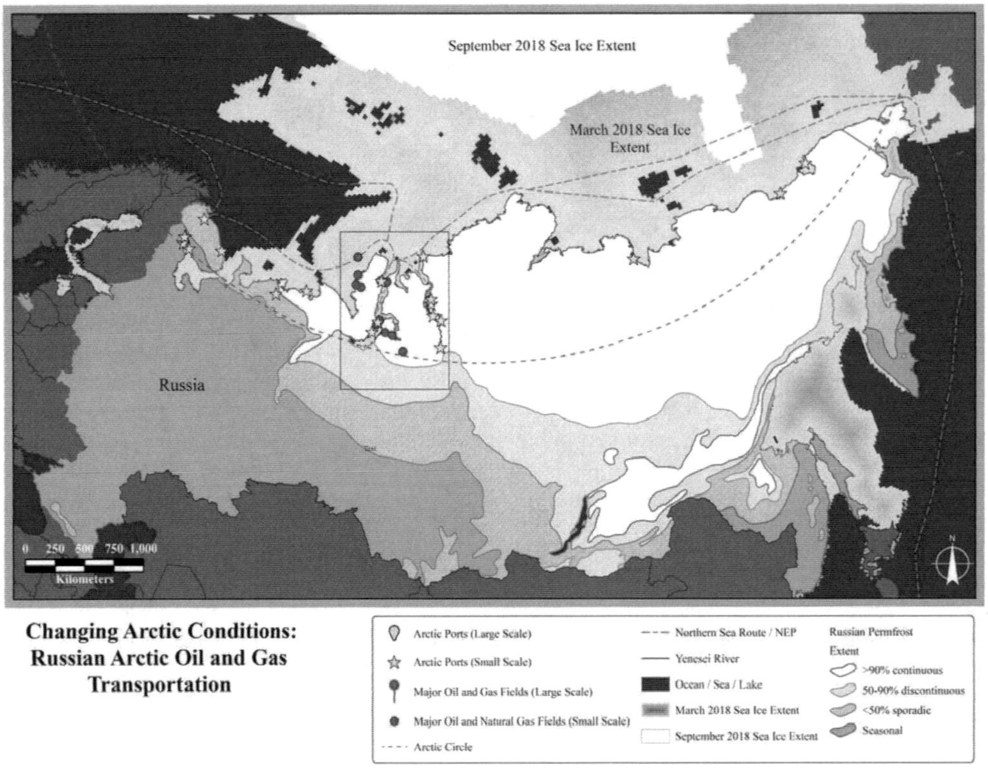

Figure 2. Russian hydrocarbon field and transport routes at the Arctic. (Adapted from McGee, 2020 [21])

If sanction made EU look elsewhere for their energy needs, Russia may lose market worth of 108 billion dollars as per estimates of 2021, and it will make the situation of Arctic even bleak. With tensions running all time high between Russia and the west, since the invasion of Ukraine, soft power may not be work to dissuade Russia from exploring the dying Arctic ice fields, which can signal the doom's day for the dying Arctic.

6. Conclusion and Recommendations

Ratification of UNCLOS is needed in the Arctic's to better govern the region with special provisions for exploration and exploitation in the region. These ratifications should include, but not be limited to, strict time frames for continental shelf claims, delimitation of conflicting claims with a legally binding resolution between the countries. The overlapping claims must not be zones of conflict but of joint cooperation between the countries [22] (Kadir, 2014). Enforcement of Canadian sovereignty over the Northwest Passage and its waters need to be discussed with all parties. A treaty similar to the Milan Protocol need to be implemented to avoid an ecological disaster in the Arctic with increased resource mining. Any future protocol over the Arctic must mandate strictly limiting security forces and capabilities, determined by an external geopolitically neutral agency. Increased Climate Change mitigation action is required to achieve the UN Sustainable Developmental Goal-13 (Climate action), to slow the thawing of Arctic and deescalate growing security tensions in the region. In view of recent geopolitical instability in the backyard of the Arctic's, a consolidated global efforts and soft power diplomacy are required to manage the looming threat over the region.

References

[1] Johannesen, J. A. (2020). *Arctic Fisheries Working Group.* Copenhagen: International Council for Exploration of the Sea.
[2] Circum-Arctic Resource Appraisal: Estimates of Undiscovered Oil and Gas North of the Arctic Circle: Kenneth J. Bird and others, United States Geological Survey, Fact Sheet 2008-3049. July 2008.
[3] Arctic Sea Ice in CMIP6. Dirk Nortz, SIMIP Community, Geophysical Research Letters Volume 47, Issue 10, 28 May 2020, e2019GL086749
[4] Artic Continental Shelf Claims (2017) European Environment Agency. https://www.eea.europa.eu/data-and-maps/figures/arctic-continental-shelf-claims
[5] Department of the Army. (2021). *REGAINING ARCTIC DOMINANCE.* United States Army. Harper, S. (2010). Prime Minister of Canada.
[6] Danish Secuirty and Intelligence Service. (2021). *ASSESSMENT OF THE ESPIONAGE THREAT TO DENMARK: The threat from foreign state intelligence activities targetting Denmark.* Danish Secuirty and Intelligence Service.
[7] Norwegian Government. (2021). *The Norwegian Government's Arctic Policy.* Norwegian Ministry.
[8] Kenneth J. Bird, R. R. (2008). *Circum-Arctic Resource Appraisal: Estimates of Undiscovered Oil and Gas North of the Arctic Circle.* US Geological Survey.
[9] Norwegian Armed Forces. (2022). Operation Cold Response 2022. https://www.forsvaret.no/en/exercises-and-operations/exercises/cr22
[10] Brook A. Smith Windsor. Putting the "N" back in NATO: A High North Policy Framework for Arctic Alliance? (July 2013) 94 *NATO Defence College Research Paper*
[11] A. A. Aliabadi, R. M. Staebler, and S. Sharma. Air quality monitoring in communities of the Canadian Arctic during the high shipping season with a focus on local and marine pollution, Atmospheric Chemistry and Physic, March 2015
[12] Erik E. Cordes, Daniel OB Jones, et al. Environmental Impacts of the Deep-Water Oil and Gas Industry: A Review to Guide Management Strategies, Frontiers in Environmental Science September 2016
[13] Niittynen, P., Heikkinen, R. K., & Luoto, M. (2018). Snow cover is a neglected driver of Arctic biodiversity loss. Nature Climate Change, 8(11), 997-1001.
[14] Heino, J., Culp, J. M., Erkinaro, J., Goedkoop, W., Lento, J., Rühland, K. M., & Smol, J. P. (2020). Abruptly and irreversibly changing Arctic freshwaters urgently require standardized monitoring. Journal of Applied Ecology, 57(7), 1192-1198.
[15] Alabia, I. D., Molinos, J. G., Saitoh, S. I., Hirata, T., Hirawake, T., & Mueter, F. J. (2020). Multiple facets of marine biodiversity in the Pacific Arctic under future climate. Science of the Total Environment, 744, 140913.
[16] Phoenix, G. (2018). Arctic plants threatened by winter snow loss. Nature Climate Change, 8(11), 942-943.
[17] Goldsmit, J., McKindsey, C. W., Schlegel, R. W., Stewart, D. B., Archambault, P., & Howland, K. L. (2020). What and where? Predicting invasion hotspots in the Arctic marine realm. Global change biology, 26(9), 4752-4771.
[18] Burgass, M. J., Milner-Gulland, E. J., Stewart Lowndes, J. S., O'Hara, C., Afflerbach, J. C., & Halpern, B. S. (2019). A pan-Arctic assessment of the status of marine social-ecological systems. Regional Environmental Change, 19(1), 293-308.
[19] Chowdhury, A., Rosencranz, A. (2022). Who Is Winning the Pyrrhic War? Economic & Political Weekly, 57(20),4-5. https://www.epw.in/journal/2022/20/letters/who-winning-pyrrhic-war.html
[20] BBC report (2022). Kharkiv fightback and jets roar in rehearsal - round-up. https://www.bbc.com/news/world-europe-61362553 (accessed on 5th June 2022)
[21] McGee, R. (2020) Mapping Russia's Arctic Hydrocarbon Development Scheme. https://www.thearcticinstitute.org/mapping-russia-arctic-hydrocarbon-development-scheme/ (accessed on 5th June 2022)

[22] Rizal A. Kadir. (2014) Is joint development possible in the Arctic? Polar Oceans Governance in an era of Environmental Change. New Horizons in Environmental and Energy Law. Published by Edward Elgar Publishing Limited

Application of eco-friendly and smart materials in geotechnical engineering for subgrade stabilization: a review

A I Dhatrak[1] and P V Kolhe[2*]

[1]Associate Professor, Department of Civil Engineering, Government College of Engineering Amravati, Maharashtra, India
[2]Research Scholar, Department of Civil Engineering, Government College of Engineering Amravati, Maharashtra, India

*Corresponding author email: piyushkolhe007@gmail.com

Abstract: The performance of road pavement mainly depends upon the properties of subgrade soil. The failures occur in case of road pavement is due to poor strength characteristics and high volume change behaviour of subgrade soil. Thus, considering the performance criteria, subgrade soil needs to be treated so as to make it suitable for the construction of road pavement. It was observed that, soil stabilization using various admixtures is more effective and commonly used practice for enhancing the properties of subgrade soil. The previous research shows that subgrade soil stabilization can be achieved with the application of various environmental friendly (Bio – Enzymes) and smart materials (Agricultural and Industrial waste) in required quantity to enhance the physical, strength and volume change characteristics of natural soil. It was also observed that, the combined use of such materials makes the subgrade soil more suitable for the construction of road pavement. From the experimental investigations of various available research works on subgrade stabilization, it was observed that strength and volume change properties such as California Bearing Ratio (CBR), Unconfined Compressive Strength (UCS) and Swell Pressure of subgrade soil can be effectively changed with the help of use of such materials. It was also observed that use of conventional methods has negative impact on environment. Thus, subgrade can be stabilized with eco – friendly and cost effective materials to achieve both economic and serviceability criteria.

Keywords: bio – enzymes, agricultural waste, industrial waste, road pavement, stabilization

1. Introduction

The expansive soil mostly encountered with problem of low bearing capacity and high swell potential which itself make it unsuitable for the construction of road pavement [1, 2]. The high swelling and shrinkage behaviour of clayey soil leads to formation of cracks on the surface of road pavement. The soil stabilization was proven to be most effective and commonly used method among all ground improvement techniques to enhance the subgrade soil properties. The soil stabilization mainly focuses on the formation of cementatious material when treated with different proportion of stabilizing agent which will further become the main reason of soil improvement [5, 6]. In general, the stabilized soil is the composite material results from the chemical reaction takes place between the natural soil and stabilizing agent [4]. Generally, subgrade stabilization can be carried out with the help of conventional methods such as use of lime, cement and flyash [3]. In recent years, use of chemical additives such nano chemicals and nano materials was found to be more effective in case of subgrade stabilization.

The various agricultural waste such as rice husk ash, bio polymers and industrial waste can also be used for the soil stabilization. The eco - friendly materials (bio – enzymes) prepared from vegetable and plants extract are widely used to stabilize the subgrade soil [7, 8]. The main advantage of use of bio – enzymes over conventional methods is the environmental friendliness and cost effective. Now a day the combined use of such eco-friendly and smart materials in the area of geotechnical engineering will help to enhance the soil properties to make it useful as subgrade material for the construction of road pavement. [9]

1.1 Eco-friendly material (bio-enzymes)

The eco-friendly bio-enzymes are the substances (organic catalyst) which accelerate the rate of a chemical reaction when added into the soil without becoming a part of the end product [5]. The quantity of eco-friendly material required for the entire improvement process will be very less which makes it more cost effective. Based upon the available literature, the bio-enzymes are found to be more effective in case of soil having more clay content. The addition of bioenzyme improve the soil particle chemical bonding resulting in a more permanent structure which is more resistant to weathering, water penetration, wear and tear [13]. In the enzymatic treated soil, enzyme catalyses the interaction between the clay and the organic cations that speed up cationic exchange without becoming part of the final product. Figure 1 shows the mechanism of soil stabilization using bio-enzymes where the reduction in thickness of double layer can be achieved with the help of cat-ionic exchange [14].

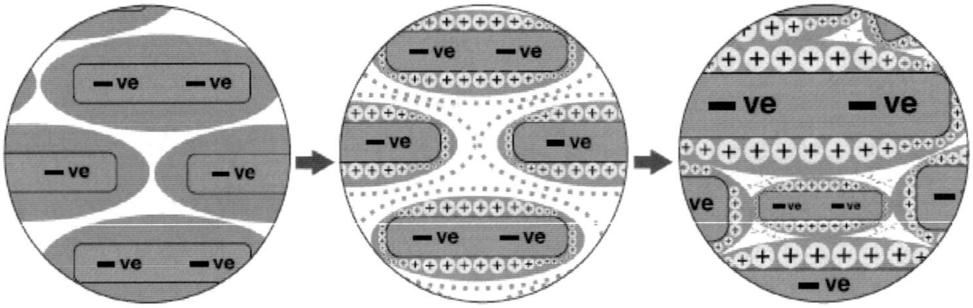

Figure 1. Cat-ionic exchange mechanism using bio-enzyme [Renjith R, Robert D 18]

1.2 Smart/cost effective material (agricultural/industrial Waste)

The cost effective materials such as bio polymers, cement, lime, flyash, micro silica, chemical additives, rice husk ash and rubber tyre shread are found effective in the improvement of subgrade soil properties [9]. The available literature shows that a lot of work has been performed on treating soil with various agricultural and industrial wastes. The treatment includes the chemical reaction which occurs when these materials are added and compacted at optimum water content to achieve the desired values. The optimum dosages of all these stabilizing materials can be obtained by reviewing the available research work. Therefore, a combined use of eco-friendly (Bio-enzymes) and cost effective materials (Agricultural and Industrial waste) will be an innovative approach for subgrade stabilization of road pavement [1, 10].

2. Review of Application of Eco-Friendly and Smart Materials for Subgrade Stabilization

The review of available literature based on experimental studies carried out with the use of eco-friendly (Bio-enzymes) and smart material (Industrial and Agricultural waste) in geotechnical engineering to treat the subgrade soils are classified as follows:

2.1 Experimental investigation of lime, rice husk ash, rubber tyre shread, bio - polymers and micro – silica treated soil

The experimental studies carried out by various authors to investigate the effects on properties of soil treated with lime, rice husk ash, rubber tyre shread, bio - polymers and micro Silica with their concluding remarks are as shown in Table 1.

Table 1. Study based on lime, rice husk ash, rubber tyre shread, bio - polymers and micro – silica treated soil

Sr. No.	Author	Type of soil treatment	% of stabilizer	Concluding remarks
1	Thakare and Chauhan [1] (2016)	Clay (CH) + Lime + Micro Silica	Micro silica - 0, 2.5, 5, 7.5, 10, 15, and 20 % Lime - 0, 3, 6, and 9 %	Optimum percentages of stabilizers were found as Micro Silica – 5% and lime - 3%. CBR (soaked and unsoaked) was found to be 6.5 times and 1.6 times respectively in comparison to natural soil.
2	Tiwari and Satyam [2] (2019)	Black Cotton (CH) + Lime + Micro Silica + Coir Geotextile	The coir geotextile fibres were 80 mm length, 20 µm diameter and L/B as 4 which were treated first with lime and micro silica suspension.	Upward swelling pressure due to lime treated coir geotextile was found to be approximately 52.19 %, 81.89 % in single and double layer respectively. The improvement in CBR value for the same case (lime treated) was found to be 399 % and 435 % in single and double layer respectively.
3	Ghavamshirazi and Bilsel [3] (2020)	Soil (CH) + Lime + Micro - Silica	CH + Micro – Silica (10 %and 20 %) CH + Lime (3 - 5 %) CH + Lime + Micro - Silica	The values of 10 % for micro - silica and 3 % for lime as optimum values in reduction of swell potential by 92 % however 10 % micro- silica and 5 % lime increase compressive strength by 81.5 %. The best mixture used for stabilization not only reduces the possibility for swelling but also reduces the amount of lime used.
4	Thomas Karatai *et al.* [4] (2016)	Clay + Rice Husk Ash (RHA) + Lime	Rice Husk Ash – 2.5, 5, 10, 20 % and Lime - 2, 4 and 6 % by weight of dry soil	The values of 20 % of RHA and 2 % of lime were found to be optimum for the treatment. The improved CBR of expansive clay was 800 % with decrease in free swell by approximately 70 % for optimum values of stabilizer.
5	Kolhe and Langote [5] (2018)	Clay + Rubber Tyre Shread	Rubber Tyre Shread – 2, 4, 6, 8 and 10 % by weight of soil	Maximum CBR was achieved for 8% of rubber tyre shread. The shear strength parameters, cohesion decreases and angle of friction increases with percentage of tyre shreads.

From above mentioned study based on use of smart materials such as lime and micro for stabilization of subgrade soil (clayey soil) shows that in case of clay of high compressibility clay (CH), optimum percentages of lime and micro silica can be considered as 3 % and 5 % respectively. Both strength values i.e. California bearing ratio (CBR) and unconfined compressive strength (UCS) can be improved with the application of smart materials. In few cases, it was found that combined use of lime and micro silica gives better results compared with individual one. The use of smart material in subgrade stabilization significantly reduced the swelling potential of highly expansive soil. The main reason of strength gain was the chemical reaction which occurs between clay particles, lime and micro silica with curing period.

2.2 Experimental investigation of soil treated with bio – enzyme

The experimental studies carried out by various authors to investigate the effects on properties of bio-enzyme treated soils with their concluding remarks are as shown in Table 2.

Table 2. Study based on bio - enzyme treated soil

Sr. No.	Author	Type of soil treatment	Dosages of stabilizer	Concluding remarks
1	Safana and Athira [6] (2017)	Soil (Clay content 21% + TerraZyme)	The samples were tested for CBR and UCS values with different dosages of Terrazyme with 0, 7, 14, 21 and 28 days of curing period	Both the strength parameters (CBR and UCS) were improved for treated soil. The UCS and CBR values of soil was increased by 281.5 % and 139.32 % respectively when compared to natural soil for 28 days of curing period.
2	Das and Maharana [7] (2017)	Soil (Clay content 42 %) + TerraZyme	Terrazyme as 200 ml per (3.5 m^3, 3 m^3, 0.5 m^3, 0.25 m^3, 0.15 m^3 and 0.075 m^3) of soil	UCS value increases more than 138 % when compared to natural soil. The cohesion value was increased by 100% and friction angle was normally about 1^0- 4^0 higher that of untreated soil
3	Aboukhadra *et al.* [8] (2018)	Soil (Fine and coarse grained) + PermaZyme + TerraZyme	Concentration ratio 2, 4, 6, 8 and 10 g/l for both stabilizers	Permazyme was more efficient. The both types of enzymes showed a small decrease in dry density in sandy soil. The CBR values of fine grained soils improved 5 times the untreated soil. It was found that 18% of clay content and 0.5 % concentration of Permazyme was the optimum combination to achieve maximum CBR Value.
4	Sodhi and Kumar [9] (2018)	Soil (Silt and clay 59%) + Alkazyme	200 ml/1.5 m^3, 200 ml/1 m^3 and 200 ml/0.5 m^3	UCS value by increased by 358.33% and CBR value increased by 180.73% when compared to untreated soil for the dosage of 0.06 ml/kg. However, the strength characteristic decreases after the increase in dosage of Alkazyme beyond optimum.
5	Sinha *et al.* [10] (2018)	Soil (Silt and clay 68.70 %)+ DZ-1X	Treated soil samples were tested for consistency limits, Free swell and CBR values with different dosages of DZ 1X	Consistency limits (Liquid and Plastic limit) due to addition of DZ – 1X product in the soil. The free swell index reduced from 72.80 % to 58.81 % and unsoaked CBR increased from 3 .93 to 8.03.
6	Muguda and Nagraj [11] (2019)	Soil (SC) (clay 21 % and sand 60 %) + TerraZyme	200 ml/3.5 m^3, 200 ml/2.5 m^3 and 200 ml/2.0 m^3 and 200 ml/1.5 m^3	Plasticity properties (Liquid, Plastic and Shrinkage limits) of treated soil were changes from 42.2 % to 38.2 %, 22.8 % to 26.0 % and 21.3 % to 24.3 respectively for the optimum dosages of

				133.33 ml/ m^3 and curing period of 60 days. The structure may lead to more flocculated when compared to natural.
7	Khan *et al.* [12] (2020)	Soil (Clay content 36 %) + 10 % Bentonite + TerraZyme + EarthZyme and DZ – 1X	Three dilution ratios (DR) of 5/1000, 1/1000 and 1/1000 for DZ-1X, EarthZyme and TerraZyme respectively with application ratios (AR) of 1/7035, 1/55275 and 1/41875. The dosages were selected as 1, 2, 5 and 10 times of AR.	Samples were tested at water content of OMC – AR/DR + 1 % and cured for four months. The swell potential was maximum for (EAR – D5) samples where it reduced from 6.63 % to 6.48 %. The maximum increase in UCS of soil was observed 5 % for EAR – D5 and TER – D5. The overall improvement was found with ageing period.
8	Reddy *et al.* [13] (2021)	BC, Lithomargic and locally available soil + TerraZyme	200 ml for 2.0, 2.5 and 3.0 m^3 of soil	The highest UCS value of BC soil was attained for 15 days of curing for all dosage levels, and it is observed to be double that of untreated soil, according to the experimental findings.

From the above mentioned studies, it was found that, bio - enzymes are eco-friendly materials when compared with lime and micro silica. Thus, use of such bio – enzymes in different proportion with varying curing period for stabilization of expansive soil provides innovative and environment friendly approach. The bio - enzymes when introduce into soil mass can act as a catalyst which accelerate the rate of cationic exchange and further reduces the adsorbed layer. The optimum for various bio enzymes was found different as clay content play an important role in bio enzymatic treatment. The swelling potential can be effectively reduced even for small dosages of bio – enzymes. The improvement in CBR and UCS values with varying curing period is due to precipitation occurs on the surface of sealed soil samples. The use of such eco - friendly materials is really a good alternative for admixtures which not environmental friendly and are energy consuming too.

2.3 *Experimental investigation of soil treated with combination of bio – enzyme, cement, lime, biopolymer and micro – silica*

The experimental studies performed by various researchers to investigate impact on characteristics of soil treated with addition of Bioenzyme, Lime and Micro - Silica with their concluding remarks are as shown in Table 3.

Table 3. Study based on combination of bio – enzyme, cement, lime, biopolymer and micro silica treated soil

Sr. No.	Author	Type of soil treatment	% of stabilizer	Concluding remarks
1	Singh and Garg [14] (2015)	Silty Sand (SM) + Renolith + Cement	The samples were prepared with (2%, 4%, 6%, 8% and 10%) cement and (1%, 2%, 3%, 4% and 5%) Renolith	The improvement in CBR and MDD values of silty sand (SM) has been observed with addition of Renolith also the permeability values decreases with use of Renolith. The usage of Renolith was shown to reduce the cost of pavement construction by roughly 15 to 28 percent.

2	Chandrakaran *et al.* [15] (2017)	Kaolinite, Clay fines (40 % and 60%) + Lime + TerraZyme	The samples were prepared with varying dosages of hydraulic lime and enzyme	Optimum enzyme lime mixture was determined to be 70 mL/m^3 of enzyme and 1.75 percent of lime. CBR values increased by upto 5 times when treated with lime alone and by upto 3 times when treated with enzyme alone whereas increase of more than 6 times was observed when treated with enzymatic lime. The result of 2 weeks curing demonstrates that with lime and enzyme stabilization, the improvement was less than 2 times, whereas it is approximately 4 times in case of combination of both.
3	Kumar and Amarnath [16] (2019)	Red, Black Soil + polymer + Renolith	(2.5 %, 5 %, 7.5 % and 10 %) of Renolith by the weight of cement	Stabilized UCS and unsoaked CBR values shows maximum improvement in strength when cement was 4.0 % and Renolith was 7.5 % for black cotton soil whereas for red soil it was 2.0 % and 5.0 %.
4	Thomas and Rangaswamy [17] (2021)	Soft Clay + Nano Silica + TerraZyme + Cement	The soil samples were treated with optimum of cement and Terrazyme as 1 percent and 220 ml/2 m3 of dry soil respectively and results were compared with natural soil	Cyclic triaxial tests were performed at optimum dosage of nano silica in enzymatic cement-treated clay. In the compressive strength of soft clay, the combination of cement and enzyme with nanoparticle produces a realistic performance.

A very few studies are available on combined use of smart material (lime and micro silica) and eco-friendly (bio – enzymes) for subgrade stabilization of clayey soil. It was found that, the combined use of all these three materials in right proportion can lead to achieve overall economy when compared with individual use. The CBR and UCS values for combined use increases upto 6 times when compared with untreated soil. The use of bio enzyme is soil specific therefore a good quality of research is needed before its use in the field. The chemical reactions occur in case combined use can shows high strength enhancement with less curing period.

3. Conclusions

From the above mentioned published research work based on use of eco-friendly (Bio-enzyme) and smart materials (Agricultural and Industrial Waste) in geotechnical engineering to modify the characteristics of soil subgrade, following concluding remarks can be done:

- The individual and combined use of bio-enzymes, various industrial and agricultural waste materials can improve the characteristics of subgrade soil to make it more suitable for road pavement.
- The bio-enzymes are the reformulations of other products so it is important to check its effect on the properties of soil before its actual use in the field.

- The different types and various dosages of enzymes will enhance the characteristics such as CBR, UCS along with compaction characteristics (OMC & MDD) of subgrade soil.
- The curing period and clay content will play significant part in improving the strength characteristics (CBR and UCS) of subgrade soil.
- The optimum dosages for micro silica and lime were found to be 5 % and 3% respectively for clayey subgrade. The optimum dosages of bio - enzyme will depend upon the clay content of treated soil.
- The swell potential of high compressible clay can be reduced substantially with the application of small dosages of bio-enzymes.
- The use of eco-friendly and smart materials in stabilization of subgrade soil will achieve the overall economy and becomes eco-friendly approach for soil stabilization compared to conventional methods.
- The use of conventional stabilizing agent such as cement and flyash are energy consuming and hence its use in the filed can be replaced by smart and innovative materials.
- The agricultural waste (Rice husk ash) can be effectively used to enhance the properties of subgrade soil.
- Lime and micro silica when introduce to treat expansive soil, chemical reactions will occur which results into formation of calcium silicate hydroxyl gel.

References

[1] Thakare S, Chauhan P 2016 *Journal of Engineering Research* **5** 9-13
[2] Tiwari N, Satyam A, 2019 *Journal of Engineering Science and Technology Elsevier* **8** 1-9
[3] Ghavamshirazi S, Bilsel H 2020 *Journal of Acta Geotechnica Elsevier* **5** 1-14
[4] Thomas K, James K, Charles K 2016 *Journal of Construction Engineering Management* **5** 1-5
[5] Kolhe P, Langote R *Journal of geotechnical Studies, MAT Journals* **2** 1 - 8
[6] Safana B, Athira S, Sabu K 2017 *Journal of Engg. Research & Tech.* **6** 547 – 549
[7] Das S, Maharana M 2017 Journal of Recent Scientific Research 8 19029-19034
[8] Zidan A, Gaber Y, Aboukhadra A 2018 *Journal of Cogent Engineering* **5** 1 – 11
[9] Sodhi P, Kumar Y 2018 *Journal of Engineering and Technology* **5** 1681-1684
[10] Sinha K, Waghmare S 2018 *Journal of Scientific Research & Development* **6** 90 -92
[11] Muguda S, Nagaraj H 2019 *Journal of Geo-engineering Springer* **10** 1-14
[12] Khan T, Taha R, Khan M 2020 *Journal of Minerals, MDPI* **10** 1 – 16
[13] Reddy R, Nandini N 2021 *Journal of Advanced Science and Technology* **29** 5200 – 5209
[14] Singh A, Garg P 2015 50th Indian Geotechnical Conference, IGC 1 - 8
[15] Eujine G, Chandrakaran S, Sankar N 2017 *Journal of Materials in Civil Engineering* **29** 1-6
[16] Kumar C, Amarnath M 2019 *Journal of Engineering Research & Technology* **6** 1097-1100
[17] Thomas G, Rangaswamy K 2021 *Journal of Soil Dynamics and Earthquake Engg.* **137** 1-10
[18] Renjith R, Robert D 2021 *Journal of Materials in Civil Engineering ASCE* **32** 1-12

Four-year Spatiotemporal Distribution & Analysis of $PM_{2.5}$ and its Precursor Air Pollutant SO_2, NO_2 & NH_3 and their Impact on LST in Bengaluru City, India

G Suthar, R P Singhal, S Khandelwal, N Kaul, V Parmar and A P Singh
Department of Civil Engineering, Malaviya National Institute of Technology, Jaipur, Rajasthan, India
Abstract. Bengaluru is one of the most populated cities in India. The rise in urbanization has caused an increase in air pollutant concentrations and land surface temperature. The present study explored the four-year (2019-2022) seasonal relationship between $PM_{2.5}$ and its precursor's air pollutants SO_2, NO_2, NH_3, and meteorological parameters (WS, WD, RH, SR). Another relationship was explored to find the combined impact of air pollutants ($PM_{2.5}$ SO_2, NO_2, NH_3) and meteorological parameters on LST. Results from correlation analysis show that $PM_{2.5}$ is strongly and very strongly correlated with NH_3 during 2022 and moderately correlated with SO_2 in 2021 and NO_2 in 2022, respectively, during both seasons. LST is positively associated with air pollutants except for $PM_{2.5}$ in 2020 and NO_2 in 2022. The multiple linear regression analysis shows that $PM_{2.5}$ strongly correlates with its precursors and meteorological parameter in both summers (0.66 to 0.74) and winters (0.59 to 0.93). Similarly, LST is strongly associated with all four air pollutants and meteorological parameters in summer (0.66 to 0.81) and winter from 0.60 to 0.79 from 2019 to 2021. The result shows that multiple parameters are responsible for the formation of secondary $PM_{2.5}$ and change in LST.

1. Introduction

Urbanization and industrialization are the most critical issues in cities. Due to urbanization, the green and barren lands are converted into impervious surfaces. It also caused a rise in air pollution in urban areas. Impervious surfaces absorb solar radiation, making their temperature higher than the surroundings. This phenomenon is known as the Urban Heat Island (UHI) effect. Urbanization is responsible for increased anthropogenic activities such as fossil fuel combustion, air pollutant emission from traffic, open waste burning, etc. These activities contribute to primary air pollutants, i.e., particulate matter ($PM_{2.5}$ or PM_{10}), Sulfur dioxide (SO_2), Nitrogen dioxide (NO_2), and Carbon monoxide (CO). The higher concentration of these air pollutants is majorly responsible for bad air quality in a city. The presence of air pollutants also affects the UHI. It is derived from the difference between urban and rural land surface temperature (LST).

Air pollutants absorb and scatter solar radiation from the earth's surface and affect the amount of solar radiation absorbed by impervious surfaces. LST was mainly related to land use and cover, waterbody, vegetation, etc. [1-4]. However, some researchers have studied the relationship between air pollutants and LST. Their results show that a significant relationship exists between air pollutants and LST. Wen and Yang (2006) showed that dust particles, NO_2, and SO_2 positively correlate with LST [5]. Guo et al. (2018) found that NO_2 is negatively associated with LST [6]. Another study by Ngarambe et al. (2021) shows that PM, SO_2, and NO_2 positively correlate with LST [7].

Due to anthropogenic activities, both primary and secondary air pollutants are formed. The secondary $PM_{2.5}$ is formed due to its precursors of air pollutants SO_2, NO_2, and NH_3. It is formed from the photochemical reaction of changing SO_X and NO_X into aerosol sulfate and nitrate. Their contribution depends upon the locality and activities that generate $PM_{2.5}$ and solar radiation. Due to the process of gas to particle conversion, many inorganic ammonium salts of ammonium sulfate ($(NH_4)_2SO_4$), ammonium chloride (NH_4Cl), and ammonium bisulfate (NH_4HSO_4) are formed [8]. Secondary $PM_{2.5}$ formed from the various gas and aqueous phase oxidation reactions of SO_2 and NO_2 and NH_3 hydrolysis [9, 10]. A study conducted by Wu et al. (2016) found that NH_3 is more strongly associated with $PM_{2.5}$ than SO_2 and NO_X [11]. Another study by Jiang and Xia (2017) shows that NH_3 can indirectly enhance new particle formation ($PM_{2.5}$) by converting SO_2 and NO_2 [12]. Lee et al. (2018) conducted a study in six cities in Taiwan to find the long-term (2005-2015) trend of precursor air pollutants SO_2 and NO_2. It has been found that NO_2 and SO_2 are correlated with the secondary $PM_{2.5}$ in the range of 0.58–0.93 and 0.32–0.66 [13].
Indian cities are the most populated cities globally, with a higher increase in population. The population in urban areas in 2011 was 377 million, which is expected to increase to 469 and 595 million in 2026 and 2036, respectively [14]. The increase in population creates two significant issues in the urban area: air pollution and LST. Globally, many studies show the relationship of $PM_{2.5}$ with its precursors, but their studies are minimal for developing nations like India. Also, no research shows the relationship between air pollutants and LST. Thus, this study has been carried out by selecting Bengaluru as one of the study areas. The study's primary aim is to find the four-year (2019 to 2022) spatiotemporal relationship of $PM_{2.5}$ with its precursors SO_2, NO_2, and NH_3 and their impact on LST during the summer and winter seasons of Bengaluru.

2. Methodology

2.1. Study area
Bengaluru, located at 12.97°N latitude and 77.56°E longitude, is among India's top five fastest-growing cities. This city has been chosen based on the population density, urbanization/urban sprawl, and presence of CAAQMS. Bengaluru is the third most populous city in India. 8.4 million people live in the city [15], expected to be 10.9 and 11.8 million in 2026 and 2031, respectively. The city's climate is tropical savanna with wet and dry seasons throughout the year. Figure 1 shows the location of the study area. Bengaluru is located on the Deccan Plateau of southern India, at over 3000 ft above sea level (the highest elevation of a city in India). Bengaluru is globally known as the Silicon Valley of India because of the leading information technology sector. The maximum and minimum temperature in the city is noted to be 38.9°C and 7.8°C, respectively, and humidity ranges from 35% to 80% [16]. The city has a summer season from March to May and a winter season from December to January

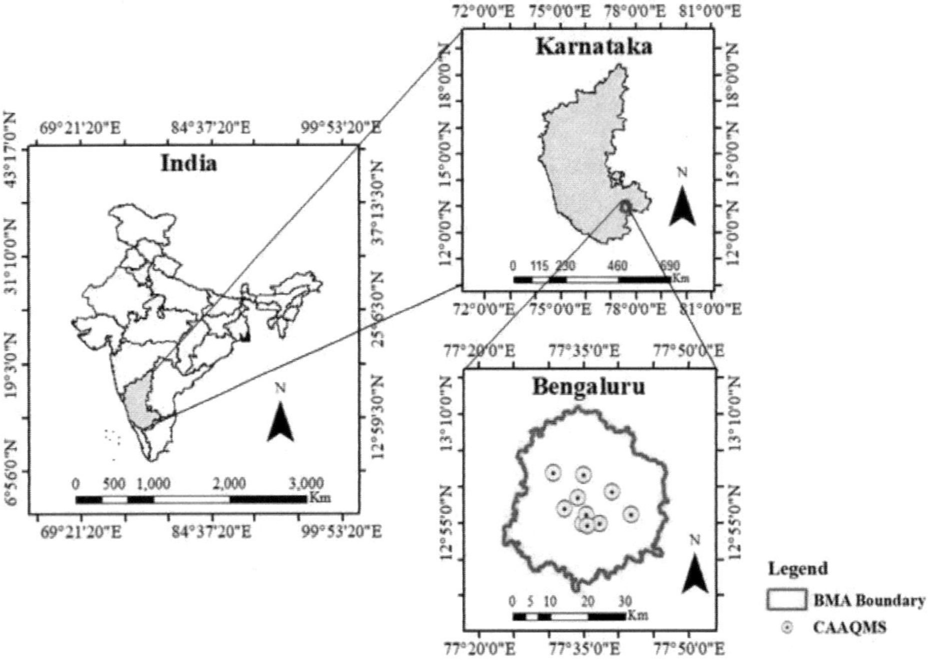

Figure 1. Location of the study area

2.2. Data collection

2.2.1. Air pollution and meteorological data
The city has obtained the 24-hour daily ambient air quality and meteorological data from ten CAAQMS. For the present study, the air pollutant (i.e., PM, SO_2, NO_2, & NH_3) and meteorological parameters (WS, WD, RH, SR) have been collected to analyze and explore the association of $PM_{2.5}$ with its precursor air pollutants (SO_2, NO_2, & NH_3) and to see their impact on LST in summer and winter seasons during 2019-2022.

2.2.2. LST data
LST data has been obtained from MODIS product MOD11A1 v.6. Moderate Resolution Imaging Spectroradiometer (MODIS) is a sensor onboard the Terra and Aqua platforms, managed by National Aeronautical Space Agency (NASA), USA. Terra and Aqua sensors collected data in 36 spectral bands, including thermal, visible, and shortwave (Wan, 2019). MOD11A1 v.6 gives daily daytime and nighttime LST data with 1 kilometer by 1-kilometer spatial resolution. Further, a quality flag has been applied to obtain high-quality LST images, including a control check of 1-bit unsigned integers. Nighttime daily LST images have been collected from MODIS because of its capacity to emit longwave radiation compared to daytime.

2.2.3. Data processing
The daily nighttime LST images are obtained in HDF-EOS format using the Sinusoidal Projection System, which changes to UTM Projection System with WGS 1984 as a datum. The HDF-EOS format changes to Geotiff format. The higher temperature pixel images have been selected using Wan's quality check details [17]. The LST images with more than 92% higher temperature pixels have been selected for the study having errors less than 1K. It has been observed that January and March have shown higher temperature pixels in LST images for four years. Thus, January and March represent the winter and summer seasons.

The air pollutants and meteorological data from CAAQMS were collected for the days when higher temperature pixel LST images were obtained. Ten CAAQMS are located in different parts of the city. The data monitored at each station is its surroundings' representative air quality and meteorological parameters. CAAQMS are installed at some height at the location surrounded by various air pollution sources such as construction, industries, transportation, etc. The spot measurement of air pollutants and meteorological data from CAAQMS have been converted into raster images. Raster images show the whole city's air pollutants and meteorological data using the inverse distance weighted (IDW) technique. Each station is joined by its neighbouring station by a line to a form polygon represented as the study area (SA-I), as shown in figure 2. A buffer of 2km (SA-II) around SA-I was also created to check the suitability of the interpolation. The spot measurement of air pollutants and meteorological parameters at each station have been interpolated using the IDW technique to obtain the data for the whole city with a spatial resolution of 1kilometer by 1kilometer. The IDW interpolation has been carried out for SA-I and SA-II having 157km^2 and 274km^2, respectively. Both SA-I and SA-II are separately analyzed. The air pollutant, meteorological, and LST data have been extracted from raster images using the raster to ASCII feature in ArcMap software. The relationship of $PM_{2.5}$ with its precursor's air pollutants SO_2, NO_2, and NH_3, along with meteorological parameters, have been investigated. The impact of air pollutants and meteorological parameters on LST is also explored. The statistical analysis has been performed using Pearson correlation and regression analysis using Minitab Software (v.19).

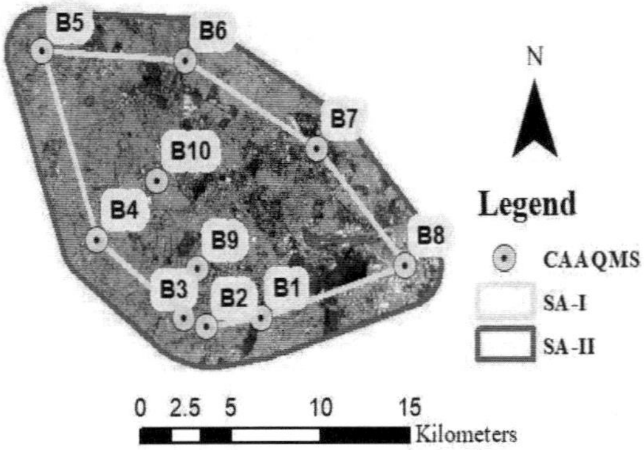

Figure 2. Study areas and Urban Sprawl

3. Results and discussion

3.1. Bengaluru air quality

The daily 24-hour air pollutants and meteorological data have been collected from the interpolated images for the summer and winter seasons from 2019 to 2022. Figure 3 And table 1 show the trend of air pollutants concentrations and meteorological parameters, respectively. Figure 3 shows that NO_2 exceeds the 24-hours standard limit prescribed by WHO, 25 μg/m^3 [18]. $PM_{2.5}$, SO_2, and NH_3 show an increasing trend, whereas NO_2 shows a decreasing trend in 2022. The reduction in NO_2 may be due to the implication of government policies on transportation by introducing good quality fuel in 2021, reducing the emission of NO_2 from vehicles. At the same time, the increasing trend in other air pollutants may occur due to the increase in population and industrial activities in the city.

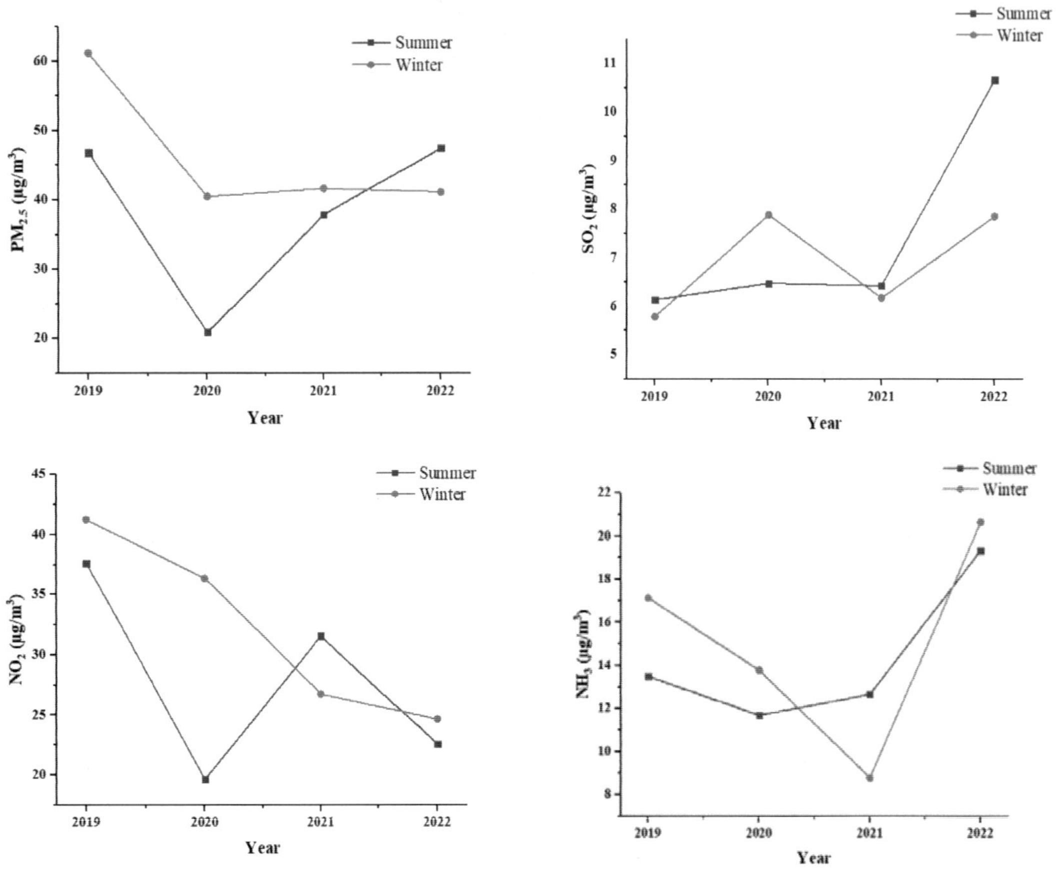

Figure 3. Yearly and seasonal trend of air pollution trend during 2019-2022

The interpolated data of meteorological parameters show that the four-year averaged wind speed is mainly in the south and east. RH and SR range from 39 to 41% in summer and 190 to 262 W/m2. In winter, RH and SR range from 55 to 64% and 155 to 214 $W/m^{2,}$ respectively.

Table 1. Daily 24-hour averaged (mean±sd) of meteorological parameters during summer and winter seasons from 2019 to 2022

Parameters	Summer				Winter			
	2019	**2020**	**2021**	**2020**	**2019**	**2020**	**2021**	**2020**
WS (m/s)	1.1±0.1	0.8±0.7	1.0±0.2	1.1±0.2	1.3±0.2	1.2±0.2	0.9±0.2	1.1±0.1
WD (deg)	131±11	102±32	143±26	121±10	108±12	110±28	88±15	151±16
RH (%)	39±2	40±7	43±3	41±3	55±3	55±2	60±7	64±6
SR (W/m^2)	206±25	190±10	231±54	262±85	155±12	162±2	189±31	214±40

The air pollutants' pollution distribution (PD) maps have been prepared from the IDW technique. The air pollutant concentration of the study area has been designed by interpolating spot data at each CAAQMS. The PD maps have been compared to see the change in the air pollutants distribution change over the study area in 2019 and 2022. PD maps show the estimated value of each air pollutant for both study areas, i.e., SA-I & SA-II. As shown in figure 4, the maximum concentration of $PM_{2.5}$ and NO2 levels was recorded to be higher in 2019 than in 2022. The maximum concentration of $PM_{2.5}$, SO_2, NO_2, and NH_3 are represented by dark green, blue, red, and Gray colors, respectively. The maximum concentration of $PM_{2.5}$, SO_2, NO_2, and NH_3 is characterized by light green, blue, red, and Gray colors, respectively. In 2019, the maximum concentration of $PM_{2.5}$ ($88\mu g/m^3$) in the winter and NO_2 ($67\mu g/m^3$) in the summer season was recorded at B2 and B4 stations, respectively higher than their concentration in 2022. In 2022, the maximum concentration of SO_2 (22µg/m^3) and NH_3 (68µg/m^3) were recorded at B7 and B4 stations. The maximum $PM_{2.5}$ and NO_2 recorded in the summer and winter seasons of 2019 and 2020 exceed the 24-hour WHO and national ambient air quality standards (NAAQS) [18, 19].

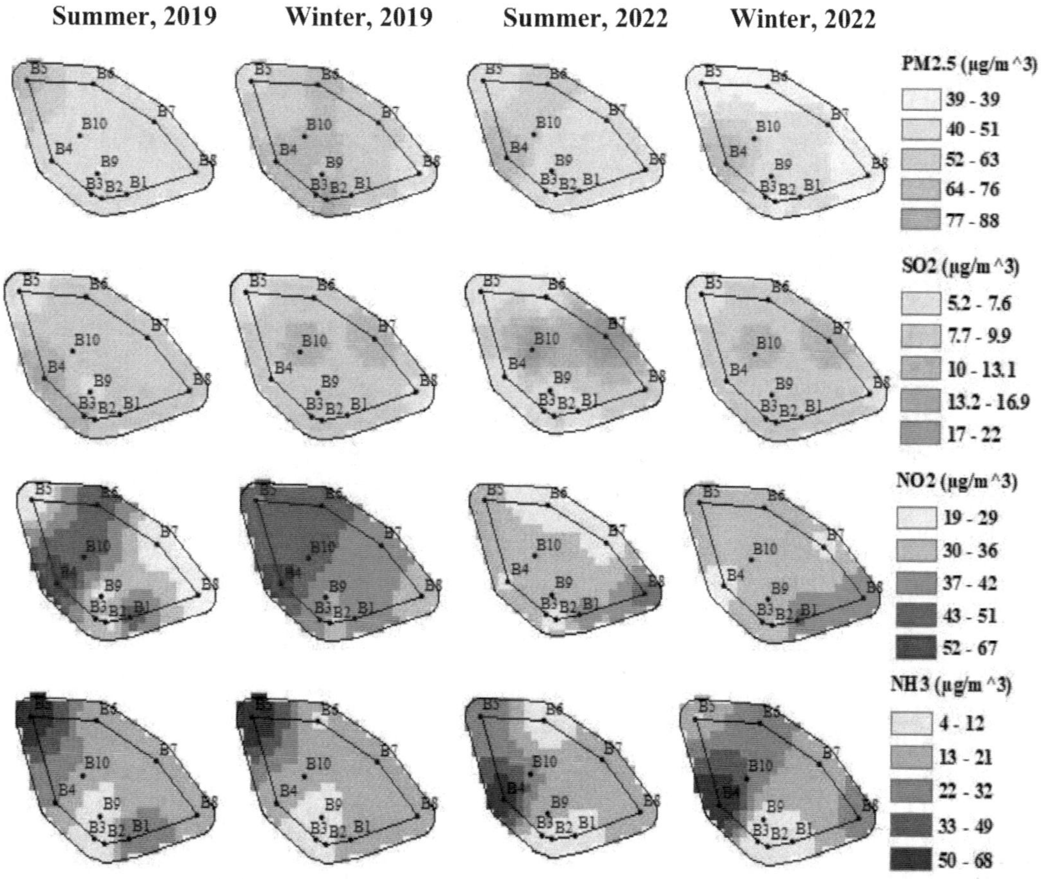

Figure 4. Pollution Distribution maps of 2019 and 2022 during summer and winter

3.2. LST in Bengaluru

The best quality nighttime LST images with higher temperature pixels are collected for the summer and winter seasons from 2019 to 2022. LST images summer and winter seasons of 2019 and 2022 are shown in figure 5. The figure shows higher temperature pixels in red color, moderate LST pixels indicated in yellow color, and lower temperature pixels represented in blue color. LST values are noted higher for the complete study area in each season except for the Southern-Eastern part, where lower temperature pixels are observed. It can be seen from the figure that red pixels are found mainly in the southwest and northwest part of the study areas, and blue pixels are primarily observed in the southeast part during both the summer and winter seasons, respectively. However, the variation of LST values can be observed. In summer, the maximum and minimum LST values in 2019 are 27°C and 21°C, higher than in 2022.

Similarly, in 2022 the LST values for the winter season are noted as lower than in 2019. All stations are located within the built-up area of the study area except for B8, which is located near the boundary in the southeast part, where blue color pixels are present. The blue pixels in any urban area represent the barren land, water body, and vegetation. Here at B8, the lower temperature pixels are observed due to the presence of Bellandur Lake (area of 3.61 km^2)

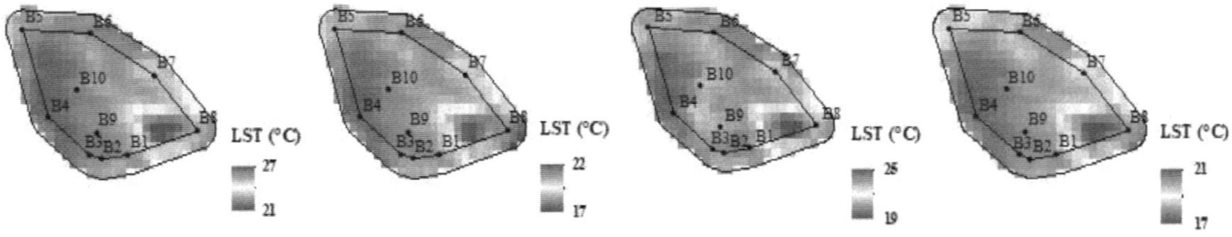

(a) Summary, 2019 **(b) Winter, 2019** **(c) Summer, 2022** **(d) Winter, 2022**

Figure 5. LST during the summer and winter seasons of 2019 and 2022

3.3. Correlation between air pollutants, meteorological parameters, and LST

Pearson correlation analysis has been carried out for interpolated data obtained from LST images and PD maps to find the relationship between air pollutants ($PM_{2.5}$, SO_2, NO_2, NH_3), meteorological parameters (WS, WD, RH, SR), and LST in summer and winter season during 2019-2022. The correlation plots are shown in figure 6. $PM_{2.5}$ has a very weak correlation with SO_2 and NO_2 in the summer and winter of 2019. But it has a very strong correlation with NH_3 in the summer season. LST is positively correlated to all air pollutants in both summer and winter. Though, variation in the relationships has been observed. Except for NO_2, all other air pollutants are very weakly correlated with LST in summer. In the summer season, LST strongly correlates with $PM_{2.5}$ and NO_2. The result obtained for 2020 for the relationship between $PM_{2.5}$ and its precursor are quite different: all three air pollutants are weakly correlated with $PM_{2.5}$ in both seasons. LST is weakly correlated with $PM_{2.5}$, NO_2, and NH_3 in summer and moderately correlated with $PM_{2.5}$ and SO_2 in the winter. In 2021, $PM_{2.5}$ had a positive correlation with all three air pollutants. Moderate relationships of $PM_{2.5}$ with SO_2 in summer, NO_2, and NH_3 in the winter season, respectively.

Similarly, LST is also positively correlated to all air pollutants. A moderate correlation has been obtained for NO_2 and NH_3 in summer and NO_2 in winter. In 2022, a strong correlation was obtained for NH_3 in the summer and winter, respectively. $PM_{2.5}$ negatively correlates with NO_2 in both seasons, and their relationship is moderate. For the same year, LST strongly correlates with NO_2 despite its lower concentration in the atmosphere. The relationship between $PM_{2.5}$ and LST is moderate for both seasons. It can be observed from the correlation results that LST is positively correlated with $PM_{2.5}$ in the summer and winter of all years except 2020.

The air pollutants in the atmosphere can scatter or absorb the upcoming solar radiation. $PM_{2.5}$ can scatter and absorb solar radiation; it is also deposited on the earth's surface. $PM_{2.5}$ originates from different sources such as industrial, residential, traffic, burning of fossil fuel, etc., getting deposited under particular meteorological conditions [20]. The surface temperature increase due to the deposition of $PM_{2.5}$ on urban surfaces. The deposition of air pollutants depends on physical properties. $PM_{2.5}$ also scatters solar radiation, which reduces the upcoming solar radiation and thus, plays a significant role in reducing LST. NO_2 positively affects LST in both seasons in all years except for 2022. NO_2 in the atmosphere can reduce the SR reaching the earth's surface due to scattering.

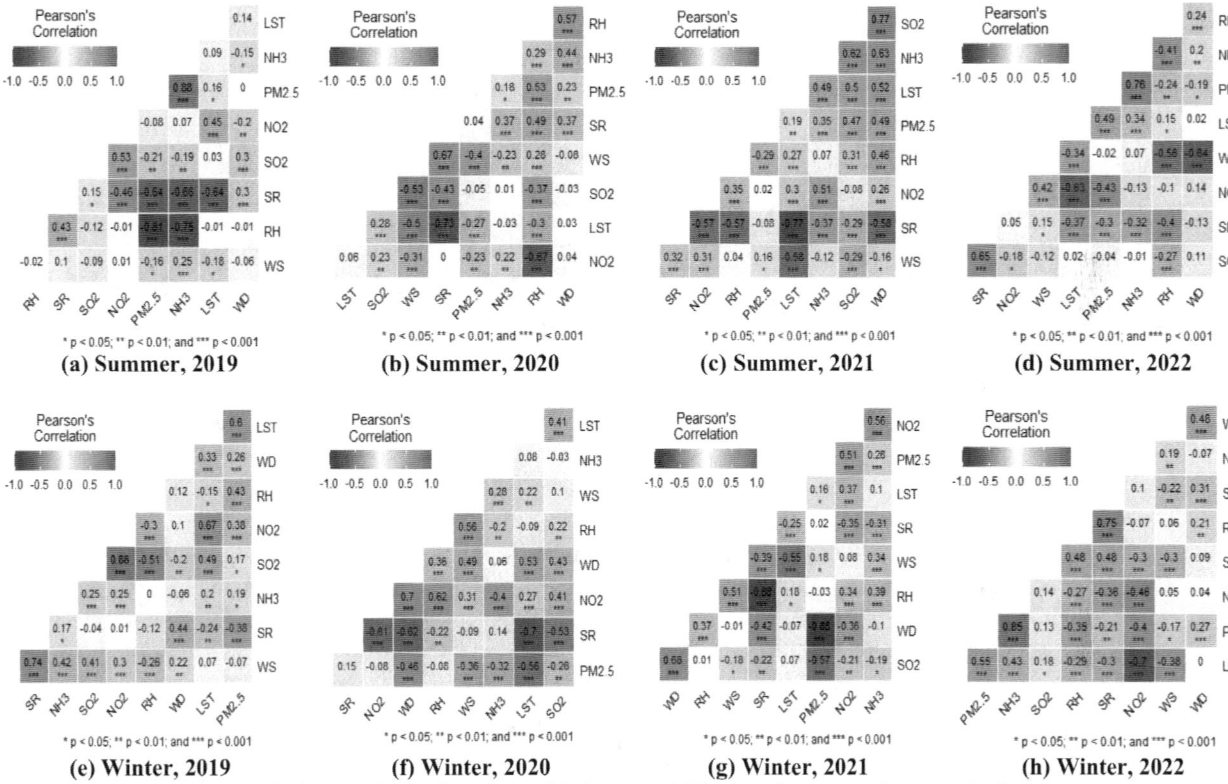

Figure 6.Correlation matrix for the relationship between LST, air pollutants, and meteorological parameter

3.4. Impact of air pollutants and meteorological parameters on LST

The results shown in the above section indicate that the change in a single parameter cannot be attributed to the change in LST. The explanation of the air pollution trend cannot be complete with meteorological parameters. Air pollutants' transfer, dispersion, and deposition depend on meteorological parameters. Thus, the associations of air pollutants, meteorological parameters, and LST have been carried out using multiple linear regression analyses. Table 2 shows the strength of the relationship between LST and air pollutants & meteorological parameters in the summer and winter seasons during 2019-2022. The table shows that the relationship is almost strong in summer and winter except for 2022. In 2022, a lower association was observed in summer for all air pollutants than in the winter. Lower R^2 indicates that the change in LST may not be attributed to air pollutants; other factors, such as vegetation, also play an essential role. At the same time, the strong relationships indicate the impact of air pollutants on LST.

Table 2. Strength of relationship of LST with individual air pollutants and meteorological parameters in SA-I

Strength of relationship	`Summer				Winter			
	2019	2020	2021	2022	2019	2020	2021	2022
LST with $PM_{2.5}$, WS, WD, RH, SR	0.73	0.77	0.75	0.41	0.60	0.79	0.72	0.59
LST with SO_2, WS, WD, RH, SR	0.76	0.79	0.77	0.43	0.70	0.75	0.71	0.69
LST with NO_2, WS, WD, RH, SR	0.66	0.67	0.74	0.5	0.65	0.76	0.74	0.59
LST with NH_3, WS, WD, RH, SR	0.66	0.67	0.81	0.42	0.54	0.76	0.72	0.61

3.5. Relationship of $PM_{2.5}$ with its precursor air pollutants

The relationship between $PM_{2.5}$ and its precursor air pollutant (SO_2, NO_2, and NH_3) has shown in table 3. As shown in figure 6, the relationship between $PM_{2.5}$ and individual air pollutants is relatively low, and many variations have been observed over the four years. Thus, to better understand the relationship, meteorological parameters are included. The strength of the relationships has been enhanced under favourable meteorological conditions. The strength of the relationship is strong and very strong for 2019 to 2021, but the relationship will be moderate in 2022. During 2019-2021, R^2 values for the relationship between $PM_{2.5}$ and precursor air pollutants in summer range from 0.66 to 0.74 and in winter from 0.59 to 0.93, respectively. In 2022, a strong relationship exists between $PM_{2.5}$ and NH_3 during the winter. The result obtained in 2022 shows that the $PM_{2.5}$ present in the atmosphere is not entirely generated from the chemical reaction of its precursor. The result also indicates that the precursor air pollutants have a moderate relationship with $PM_{2.5}$, partially involving secondary $PM_{2.5}$ formation. The substantial R^2 value obtained in the summer and winter seasons shows that the $PM_{2.5}$ not only from primary emission sources but also from various heterogenous and homogenous chemical reactions of its precursor air pollutants in Aqueous and gas-phase oxidation reactions [9, 21]. The air pollutants SO_2 and NO_2 form $PM_{2.5}$ due to the photochemical reaction of SOx and NOx, which are converted into sulfate (SO_4^{-2}) and nitrate (NO_3^{-}) aerosol [22]. The contribution of sulfate and nitrate in secondary $PM_{2.5}$ formation depends on various factors such as particulate episode events, SR, seasonal changes, and geographical changes [23, 24].

Table 3. Strength of relationship (R^2) of air pollutants with their precursor and meteorological parameters in SA-I

Strength of relationship	Summer				Winter			
	2019	2020	2021	2022	2019	2020	2021	2022
$PM_{2.5}$ with SO_2, WS, WD, RH, SR	0.66	0.67	0.81	0.42	0.59	0.82	0.94	0.46
$PM_{2.5}$ with NO_2, WS, WD, RH, SR	0.66	0.67	0.74	0.5	0.59	0.76	0.93	0.48
$PM_{2.5}$ with NH_3, WS, WD, RH, SR	0.76	0.79	0.77	0.43	0.68	0.78	0.93	0.94

The relationship of $PM_{2.5}$ with SO_2, NO_2, and NH_3 has been further explored by taking the ratios $PM_{2.5}/SO_2$, $PM_{2.5}/NO_2$, and $PM_{2.5}/NH_3$, which are shown in table 4. The highest ratio value has been obtained for $PM_{2.5}/SO_2$, ranging from 4.9 - 7.8 during summer and 5.4 - 10.8 during winter, respectively, over four years. These results indicate that high SO_2 concentration in certain meteorological conditions can increase the secondary $PM_{2.5}$ concentration. It can also be observed that the ratios are higher in winter compared to summer, which may be due to the variation in meteorological conditions and the emission of SO_2 from vehicles. The $PM_{2.5}/NO_2$ and $PM_{2.5}/NH_3$ flow the same trend as $PM_{2.5}/SO_2$, i.e., the ratio is higher in summer and lower in winter. However, the lowest ratio is noted for $PM_{2.5}/NO_2$, ranging from 1.1 - 2.2 during summer and 1.2 - 1.7 during winter. This lower ratio in both seasons may suggest that the lower concentration of NO_2 in the atmosphere leads to lower $PM_{2.5}/NO_2$.

Table 4. Seasonal average data (mean±sd) of air pollutant rations $PM_{2.5}/SO_2$, $PM_{2.5}/NO_2$, and $PM_{2.5}/NH_3$ in MSA

Pollutants	Season	2019	2020	2021	2022
$PM_{2.5}/SO_2$	**Summer**	7.8±1.7	3.5±1.7	5.9±0.6	4.9±1.6
	Winter	10.8±0.2	5.4±0.1	7.3±3.1	5.5±1.5
$PM_{2.5}/NO_2$	**Summer**	1.3±0.3	1.1±0.5	1.2±0.2	2.2±0.5
	Winter	1.5±0.3	1.2±0.3	1.6±0.3	1.7±0.4
$PM_{2.5}/NH_3$	**Summer**	3.6±0.7	2.0±1.0	3.4±1.7	3.0±1.4
	Winter	3.7±0.7	3.0±0.6	4.9±1.4	2.2±0.8

Conclusions

This study analyzed the four-year (2019 to 2022) relationship between $PM_{2.5}$ and its precursor air pollutants SO_2, NO_2, and NH_3. Also, the impact of these four pollutants on LST is explored. From the literature, it was found that only a few studies are available that show this kind of relationship. Due to the rise in population and the availability of ten CAAQMS, Bengaluru city has been selected for the study. The city has a greater number of CAAQMS as compared to other cities. In this study, the spot data from each location has been interpolated to obtain the data for the complete study area. The air pollutant ($PM_{2.5}$, SO_2, and NO_2) and meteorological parameters (WS, WD, RH, & SR) from each station have analyzed the associations. The correlation between $PM_{2.5}$ and precursor air pollutants show that $PM_{2.5}$ is almost weakly correlated with individual precursor air pollutants in both seasons. However, a strong and very strong positive correlation has been obtained for NH_3 in the summer and winter seasons of 2022. LST positively correlates to all air pollutants in both seasons from 2019 to 2022 except $PM_{2.5}$ in 2020 and NO_2 in 2022, respectively. The correlation analysis of LST and air pollutants shows that the impact of air pollutants is also low. It is a well-known fact that air pollution dispersion depends upon meteorological parameters. Thus, a multiple linear regression shows a strong association between $PM_{2.5}$ and its precursor in summer ranging from 0.66 to 0.74 and in winter from 0.59 to 0.93 from 2019-to 2022. The result obtained for 2022 is moderate except for NH_3 in the summer (0.68) and winter (0.94) seasons. LST strongly correlates with air pollutants and meteorological parameters in summer ranging from (0.66 to 0.81) and in winter from 0.60 to 0.79 from 2019 to 2021. Similar to the relationship of $PM_{2.5}$ with its precursor in 2022, LST has a lower relationship with all four air pollutants during the summer and winter seasons. This result shows that the impact of individual parameters could not explain the formation of secondary $PM_{2.5}$ and change in LST. Thus, multiple parameters have been considered in this study.

References

[1] Pal, S., & Ziaul, S. K. (2017). Detection of land use and land cover change and land surface temperature in English Bazar urban center. *The Egyptian Journal of Remote Sensing and Space Science*, **20(1)**, 125-145.

[2] Khandelwal, S., Goyal, R., Kaul, N., & Mathew, A. (2018). Assessment of land surface temperature variation due to change in elevation of area surrounding Jaipur, India. *The Egyptian Journal of Remote Sensing and Space Science*, **21(1)**, 87-94.

[3] Mukherjee, F., & Singh, D. (2020). Assessing land use–land cover change and its impact on land surface temperature using LANDSAT data: A comparison of two urban areas in India. *Earth Systems and Environment*, **4(2)**, 385-407.

[4] Mathew, A., Sarwesh, P., & Khandelwal, S. (2022). Investigating the contrast diurnal relationship of land surface temperatures with various surface parameters represent vegetation, soil, water, and urbanization over Ahmedabad city in India. *Energy Nexus*, **5**, 100044.

[5] Weng, Q., & Yang, S. (2006). Urban air pollution patterns, land use, and thermal landscape: an examination of the linkage using GIS. *Environmental monitoring and assessment*, **117(1)**, 463-489.

[6] Guo, Z. Wang, Y., Du, H., Xu, Y., Lu, D., & Wang, X. (2018). Temporal and spatial variation relationship and influence factors on surface urban heat island and ozone pollution in the Yangtze River Delta, China. *Science of the Total Environment*, **631**, 921-933.

[7] Ngarambe, J., Joen, S. J., Han, C. H., & Yun, G. Y. (2021). Exploring the relationship between particulate matter, CO, SO_2, NO_2, O_3 and urban heat island in Seoul, Korea. *Journal of Hazardous Materials*, **403**, 123615.

[8] Tsai, J. H., Chang, L. P., & Chiang, H. L. (2014). Airborne pollutant characteristics in an urban, industrial and agricultural complex metroplex with high emission loading and ammonia concentration. *Science of the total environment*, **494**, 74-83.

[9] Sarwar, G., Fahey, K., Kwok, R., Gilliam, R. C., Roselle, S. J., Mathur, R., ... & Carter, W. P. (2013). Potential impacts of two SO2 oxidation pathways on regional sulfate concentrations: Aqueous-phase oxidation by NO2 and gas-phase oxidation by Stabilized Criegee Intermediates. *Atmospheric Environment*, **68**, 186-197.

[10] Zhang, R., Wang, G., Guo, S., Zamora, M. L., Ying, Q., Lin, Y., ... & Wang, Y. (2015). Formation of urban fine particulate matter. *Chemical reviews*, **115(10)**, 3803-3855.

[11] Wu, Y., Gu, B., Erisman, J. W., Reis, S., Fang, Y., Lu, X., & Zhang, X. (2016). $PM_{2.5}$ pollution is substantially affected by ammonia emissions in China. *Environmental pollution*, **218**, 86-94.

[12] Jiang, B., & Xia, D. (2017). Role identification of NH_3 in atmospheric secondary new particle formation in haze occurrence of China. *Atmospheric Environment*, **163**, 107-117.

[13] Lee, C. S., Chang, K. H., & Kim, H. (2018). Long-term (2005–2015) trend analysis of $PM_{2.5}$ precursor gas NO_2 and SO_2 concentrations in Taiwan. *Environmental Science and Pollution Research*, **25(22)**, 22136-22152.

[14] MoHFW (Ministry of Health & Family Welfare). (2020). Population projection for India and state (2011-2036). Report on technical group of population projections. National commission on population, Delhi. Retrieved from https://main.mohfw.gov.in/sites/default/files/Population%20Projection%20Report%202011-2036%20-%20upload_compressed_0.pdf. Accessed on April, 2022.

[15] Census of India. Bengaluru city population 2011-2012. Retrieved from https://www.census2011.co.in/census/city/448-bangalore.html. (Accessed on March, 2022).
[16] Rajashekara, S. (2020). Monthly and Annual Variation of Temperature in Urban Habitats of the Bengaluru Region, India. *Transactions on Science and Technology*, **7(2)**, 29-34.
[17] Wan, Z. (2019). Collection6.1 MODIS land surface temperature products users' guide. *ICESS, University of California, Santa Barbara*.
[18] WHO. (2021). The new 2021 WHO air quality guideline limits. Breeze Technologies. Retrieved from https://www.breeze-technologies.de/blog/new-2021-who-air-quality-guideline-limits/. Accessed on April 2022.
[19] CPCB. (2009). National Ambient Air Quality Standards, Central Pollution Control Board. Ministry of Environment, Forests and Climate Change (MoEF&CC), Government of India, New Delhi, India. Retrieved from http://www.indiaenvironmentportal.org.in/files/file/Permissible%20Level%20for%20Pollutants.pdf. Accessed on March 2022.
[20] Zhu, L., Liu, J., Cong, L., Ma, W., Ma, W., & Zhang, Z. (2016). Spatiotemporal characteristics of particulate matter and dry deposition flux in the Cuihu wetland of Beijing. *PloS one*, **11(7)**, e0158616.
[21] Zhang, R., Wang, G., Guo, S., Zamora, M. L., Ying, Q., Lin, Y., ... & Wang, Y. (2015). Formation of urban fine particulate matter. *Chemical reviews*, **115(10)**, 3803-3855.
[22] Shandilya, K., Gupta, A. B., & Khare, M. (2011). Formation of atmospheric nitrate under high Particulate Matter concentration. *World Review of Science, Technology and Sustainable Development*, *8*(2-4), **148-165**.
[23] Cheng, Z., Wang, S., Fu, X., Watson, J. G., Jiang, J., Fu, Q., ... & Hao, J. (2014). Impact of biomass burning on haze pollution in the Yangtze River delta, China: a case study in summer 2011. *Atmospheric Chemistry and Physics*, **14(9)**, 4573-4585.
[24] Zhao, M., Wang, S., Tan, J., Hua, Y., Wu, D., & Hao, J. (2016). Variation of urban atmospheric ammonia pollution and its relation with PM2.5 chemical property in winter of Beijing, China. *Aerosol and Air Quality Research*, **16(6)**, 1378-1389.

Acknowledgments
The authors thank Central Pollution Control Board (CPCB) for providing air pollutants data and meteorological parameters. The authors are also grateful to Land Processes Distributed Active Archive Center (LP DAAC), United State Geological Survey (USGS), and Earth Resources Observation and Science Center (EROS) for providing LST data.

Urban environmental challenges associated with dynamics of slum settlement

Rashmi Kumari

Assistant Professor, Architecture Department, National Institute of Technology Hamirpur, India

rashmi@nith.ac.in

Abstract. The urban environmental quality is one of the major issues which need immediate attention as it has direct impact on the human life around the globe. The growing population in the urban areas has put a pressure on the housing sector, which leads to slum formation, and the decreased environmental quality due to it. The urban system cannot be imagined without the slums as these are inexorable part of urban areas, especially in the developing nations. There is a direct association between the complex spatial arrangement and the socio-economic fabric of the slum dwellers. In the present study, the statistical analysis has been performed to establish the interrelation between the determinants of social and economic status of the slum dwellers, and the air quality. Further, the space syntax analysis on the selected case has been performed to understand the spatial dynamics and its interrelation with the socio-behavioural pattern of the slum in the study area of Patna city. The present investigation advocated to improving the quality of slum settlements for improving the quality of the environment.

Keywords: Slum; Environment; Air Pollution; Regression; Space Syntax

1. Introduction

The frequency and severity of slum formation in the urban areas of developing nations are increasing more rapidly as compared to the developed nations [1]. In general, the slum and squatter settlements formed gradually in the bleak urban areas, which are severely lacking in the provision of basic infrastructure facilities. These areas can be characterized by substandard living conditions in terms of amenities, health, education, hygiene, safety, overcrowding, poverty, etc. [1,2]. The economic disparity; dense population; migration; housing shortage; poverty; unemployment; rental inconsistency; and inefficient planning policies; are few major factors contributing towards slum formation [3,4]. The planning and management deficiencies magnify the severity of degradation in city's livability and environmental quality due to slum and squatter settlements [4]. The slums exacerbate environmental problems with inadequate water and sanitation facilities [5]. The increased level of air pollutants in urban slums are mainly due to dust and burning of trash. Most of the slums are situated near industries or on the road side, which causes air pollution due to vehicular and industrial emissions [6]. The poor ventilation results in infiltration of pollutants, raising the levels of indoor air pollution too. Thus, the air pollution in deprived urban areas is one of the foremost research motivations [6,7,8,9] .

In past, slums were considered as jeopardy and malignance for an urban society. However, based upon contemporary research findings, it is evident that besides being an environmental threat, slum is a necessary fuel to accelerate an urban growth engine [6,8]. Slums are provider of goods and services demanded by the urban population. Hence, the studies are being conducted to find ways of improving the slum conditions. There is great need of study the slum dynamics in order to provide optimum spatial layout in regard of achieving overall improvement of the slum condition as well as improvement of the whole urban environment [9].

The studies suggest that statistical analysis provides systematic comparative knowledge about a particular settlement case. Based upon which, the detailed spatial analysis can be performed on each settlement to benchmark the present scenario and prioritize areas for intervention, and also identify the methods and level of intervention required. The space syntax method has been adopted by many researches to get more relevant results in these cases [10,11].

2. Research Method

The 'Case Study Research Method' among the varied range of methods, has been selected to be implemented in the present research which involves elaborate explanatory research study of a definite spatial extent and specific social group. The present investigation relies on both subjective and objective data. The problems associated with the slums have been identified in the delineated study area of city Patna, India. The relevant data regarding identified variables have been collected from primary as well as secondary sources. The survey at grassroots level has been conducted to collect the aggregate information from seven different identified slums in the study area. The responses to different questions were summarized and analyzed with detailed approach after data processing, which include data feeding and data vetting. The statistical analysis tools like regression and correlation have been utilized to analyze the data. The extensive exploration of one of the selected slums has been performed using space syntax method, to analyze the scenario in depth. The results of the present investigation provide evidences about the association of slum conditions and pollution level in their surrounding environment. Hence, the research provides the conclusion regarding provision of planning policies for slum upgradation which in turn will heal the environment, definitely.

3. About the study area

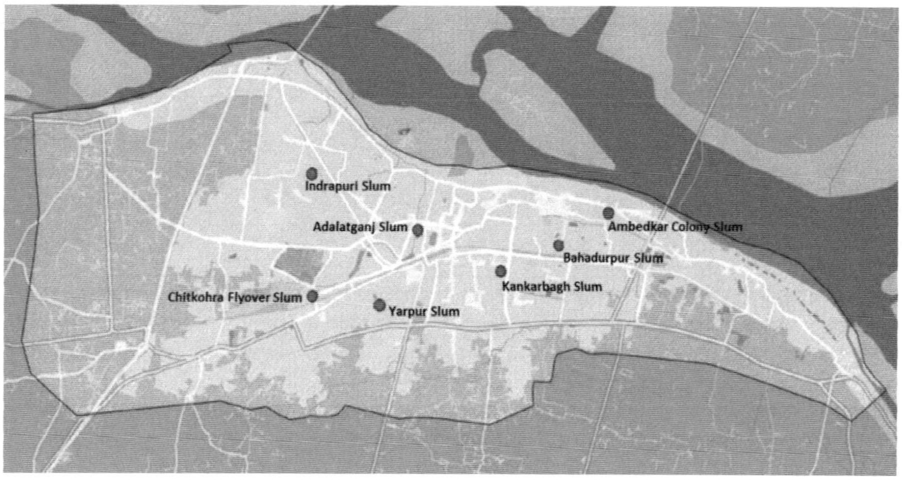

Figure 1. Slum locations shown on the map of Patna city, Bihar, India

The study area of city Patna, India, historically known as Pataliputra, is one of the oldest continually inhabited places in the world. As per the data released by United Nations, the population of Patna is 2.35 million (year 2018), which makes it the 19th largest city in India [12]. It is the capital and largest city of the state of Bihar in India with an area of 250 sq km. The GDP (2011-12) per capita of Patna is ₹1,08,657 and its GDP growth rate is 7.29 per cent. The seven slums identified for the present investigation are Yarpur slum, Chitkohra Flyover Slum, Adalatganj Slum, Ambedkar Colony Slum, Bahadurpur Slum, Indrapuri Slum, and Kankarbagh Slum [13]. The locations of these slums are shown in the map (figure 1 & 2). These slums came into existence about 20 to 100 years ago. The number of households in these slums varies from only 122 in Kankarbagh area to 1800 in Adalatganj slum. Similarly the population varies in each slum from 610 to 9000 (table 1) [14].

Table 1. Slums identified for present investigation

S. No.	Name of Slums	Ward no	years of existence	Number of Households	Population
1	Yarpur Slum	19	100	1560	7800
2	Chitkohra Flyover Slum	9	20	760	3800
3	Adalatganj Slum	21	60	1800	9000
4	Ambedkar Colony Slum	51	50	850	4250
5	Bahadurpur Slum	36	70	600	3000
6	Indrapuri Slum	7	70	250	1250
7	Kankarbagh Slum	45	100	122	610

Source: City Development Plan (2010-30)

Table 2. Slum condition score with respect to identified indicators

Sl No.	Name of Slums	Average Area of dwelling (in m^2)	Average house height (m)	Roof material	Cleanliness of surrounding	Average monthly income per household (in INR)	Sanitation (toilet facility)	Sanitation (drainage)
1	Yarpur Slum	7	2	0.6	1.5	9000	1.8	1.2
2	Chitkohra Flyover Slum	6.2	2.1	0.5	1.4	8000	1.2	1.1
3	Adalatganj Slum	6	2	1.1	1.3	7700	1.1	1
4	Ambedkar Colony Slum	6	2.2	1.2	1	6800	1	0.9
5	Bahadurpur Slum	5.9	2	1	0.7	7000	0.8	0.8
6	Indrapuri Slum	5.1	2.2	1.2	0.8	6000	0.9	0.5
7	Kankarbagh Slum	4.6	2.2	1.5	0.5	5000	0.5	0.3

Roof material Score: tarp=0, thatch/tin=1, cement sheet=2, brick=3, concrete=4
Cleanliness of surrounding (waste quantity, odour, view) Score: intolerable=0, very poor=1, poor=2, clean=3, very clean=4
Sanitation (toilet) Score: none=0, shared public=1, shared private=2, private=3
Sanitation (drainage) Score: none=0, open rough=1, open sturdy=2, closed=3
Source: Survey by the researcher

Figure 2. Images showing seven slums under investigation

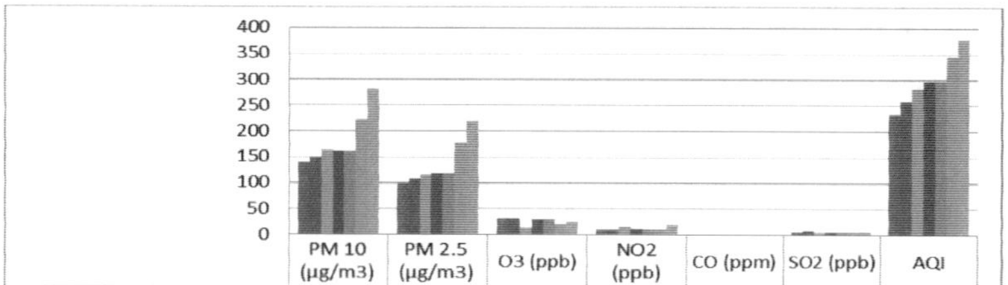

	PM 10 (µg/m3)	PM 2.5 (µg/m3)	O3 (ppb)	NO2 (ppb)	CO (ppm)	SO2 (ppb)	AQI
Yarpur Slum	140.07	99.95	31	10.46	0.88	5.65	233
Chitkohra Flyover Slum	149.21	107.62	32	9.67	1.22	7.66	259
Adalatganj Slum	163.11	115.38	13	15.91	1.08	5.56	283
Ambedkar Colony Slum	161.61	119.03	30	11.38	1.38	5.81	297
Bahadurpur Slum	161.61	119.03	30	11.38	1.38	5.81	297
Indrapuri Slum	222.38	179.03	21	11.93	1.6	6.09	345
Kankarbagh Slum	281.24	219.9	25	19.22	1.83	7.01	377

Figure 3. Chart showing pollution level in the slums under investigation

The parameters selected for depicting the socio-economic condition of slum are dwelling area; height of houses; roof material; cleanliness of the surrounding area; average monthly income; and sanitation facility [15,16]. The scores have been assigned for assessment of parameters. The lowered the score, poorer the condition of each indicator. The higher score denotes better condition of the particular indicator. The overall values and scores calculated from the result of surveyed household in each slum area have been tabulated (table 2). As per Bihar State Disaster management authority, the pollution level in terms of quantity of PM 10 (μg/m3), PM 2.5 (μg/m3), O3 (ppb), NO2 (ppb), CO (ppm), SO2 (ppb), and AQI, in these areas have been analyzed (figure 3).

4. Result and Discussion

The regression and correlation analysis have been performed between the AQI (Air Quality Index) and score values of each indicator of slum condition. The results have been analyzed rigorously and presented in the form of graphs (figure 4).

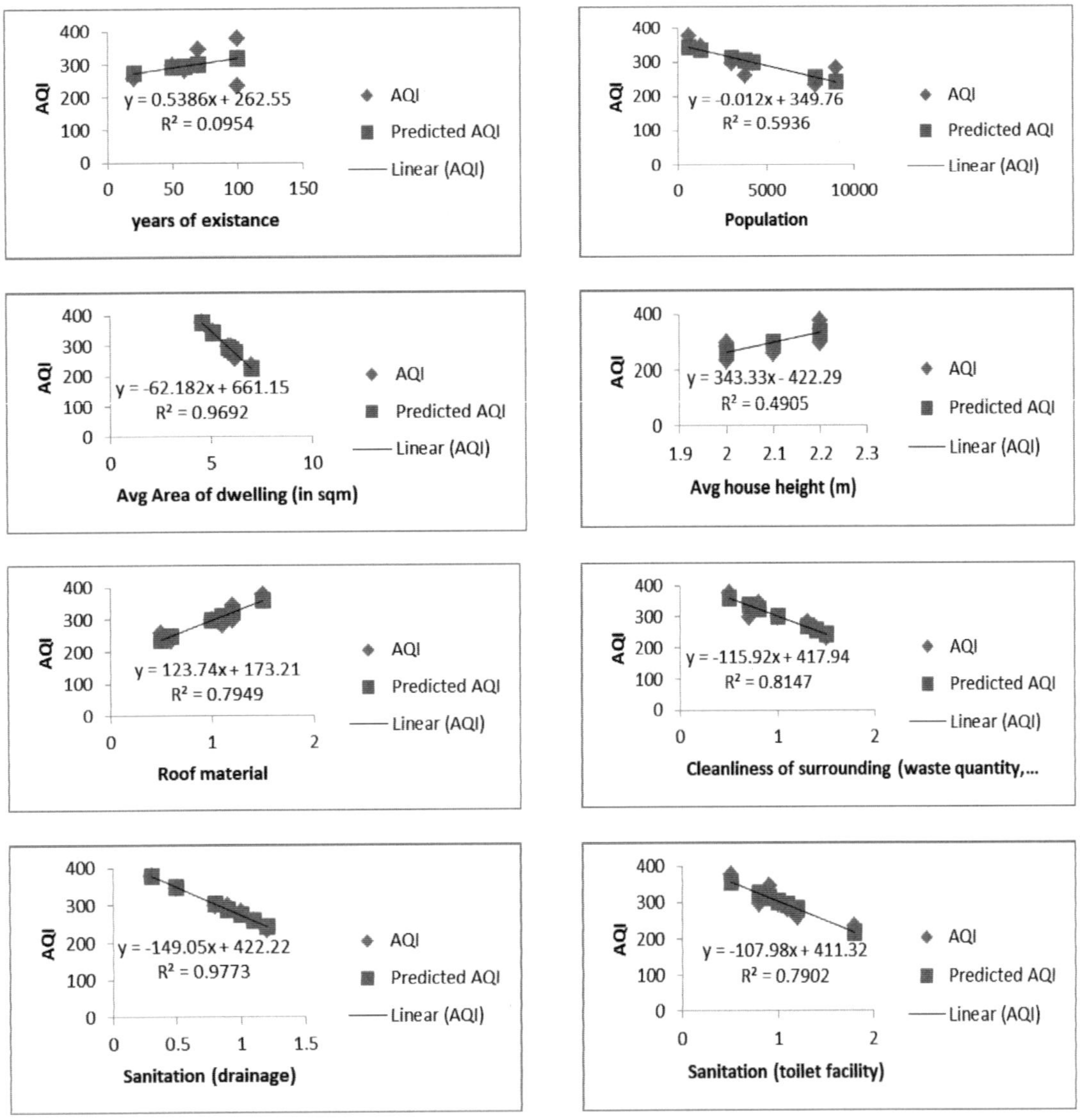

Figure 4. Charts showing results of regression and correlation analysis

The results suggest there is strong correlation (R^2 value 0.9778) between AQI and Average monthly income of the slum. The year of existence shows very less or no correlation (R^2 value 0.0954) with the AQI. The population of the slum has less correlation (R^2 value 0.5936) with the AQI. The average area of dwelling, an indicator of living condition, has strong correlation (R^2 value 0.9692) with the pollution level in terms of AQI. The height of house has not much variation and is almost same among the surveyed slums and hence does not show correlation with the AQI. The material used for roof construction has been found as tarp, thatch and bamboo, tin, cement corrugated sheet, brick and concrete. The material of roof construction is a strong indicator of standard of living. It displays correlation (R^2 value 0.7949) with the pollution level in the surrounding environment. Another important parameter describing the socio-economic condition is the cleanliness of surrounding environment in terms of waste disposal, insects, flies, odor, muddy patch, etc. The cleanliness has strong correlation (R^2 value 0.8147) with the AQI pollution level. The drainage facility type is very much associated with the pollution level, and shows strong correlation (R^2 value 0.9773) with the AQI. Further, the toilet facility, an essential requirement, shows correlation (R^2 value 0.7902) with the AQI of the surrounding environment. Hence, it can be established that the quality and standard of living of the slums impacts the surrounding air quality too. This finding can be precursor towards further research on this subject. The planning policy guidelines may be formulated keeping in view of the slum upgradation in order to achieve environmental improvement.

4.1. Case Study

The Bahadurpur slum area, one among the seven selected slums in the present investigation, is prototype substandard bleak area in the study area of Patna city. Hence, the detailed study of the Bahadurpur slum area has been conducted to analyze the problems and their possible planning solutions in the study area. The Bahadurpur slum area with a population of 3000 people, is situated in the ward no 36 of the study area of Patna city. It exists since past 70 years near the Moinul Haq Stadium (figure 5).

Figure 5. Map showing Bahadurpur Slum Area

The spatial layout of the selected case has been analysed using space syntax method to understand the interrelation between the spatial organisation and the society along with their development pattern. The models have been generated to objectively analyse the ease of access and passing flow within the slum area. The result of complex spatial relations of the slum is represented as graphs (figure 6). The axial line graph shows the maximum extension of a point of space is less in the left half of the slum area, which is more dense as well as poorer in condition, and higher in the right half of the slum. The connectivity graph shows that the number of spaces immediately connecting a space of origin is almost high at all spaces within the slum area, making the area lacking in private spaces. The segment length graph shows that the metric lengths of the segments are less in the slum area, making it more haphazard and vulnerable. The choice graph shows that it is less likely a street segment it is to be used on all shortest routes choices between the spaces in the entire slum. It makes the segments usable by less population making it more unsafe. The integration graph suggests that distance from any space of origin to all others in the slum is neither very less nor very high. The sum of the topological depth from any node to all the others is observes as less in the total depth graph. The correlation between angular connectivity and global integration is less, hence the understanding the whole slum area from looking at its part is very difficult, making the intelligibility frailer (figure 7). The synergy graph shows that the internal structure of alum relates at higher degree to the larger-scale slum in which it is embedded, making the whole slum area monotonous (figure 8). Hence, the space syntax analysis gives an idea about the spatial characteristics in relation to social and behavioral pattern of the slum in the study area.

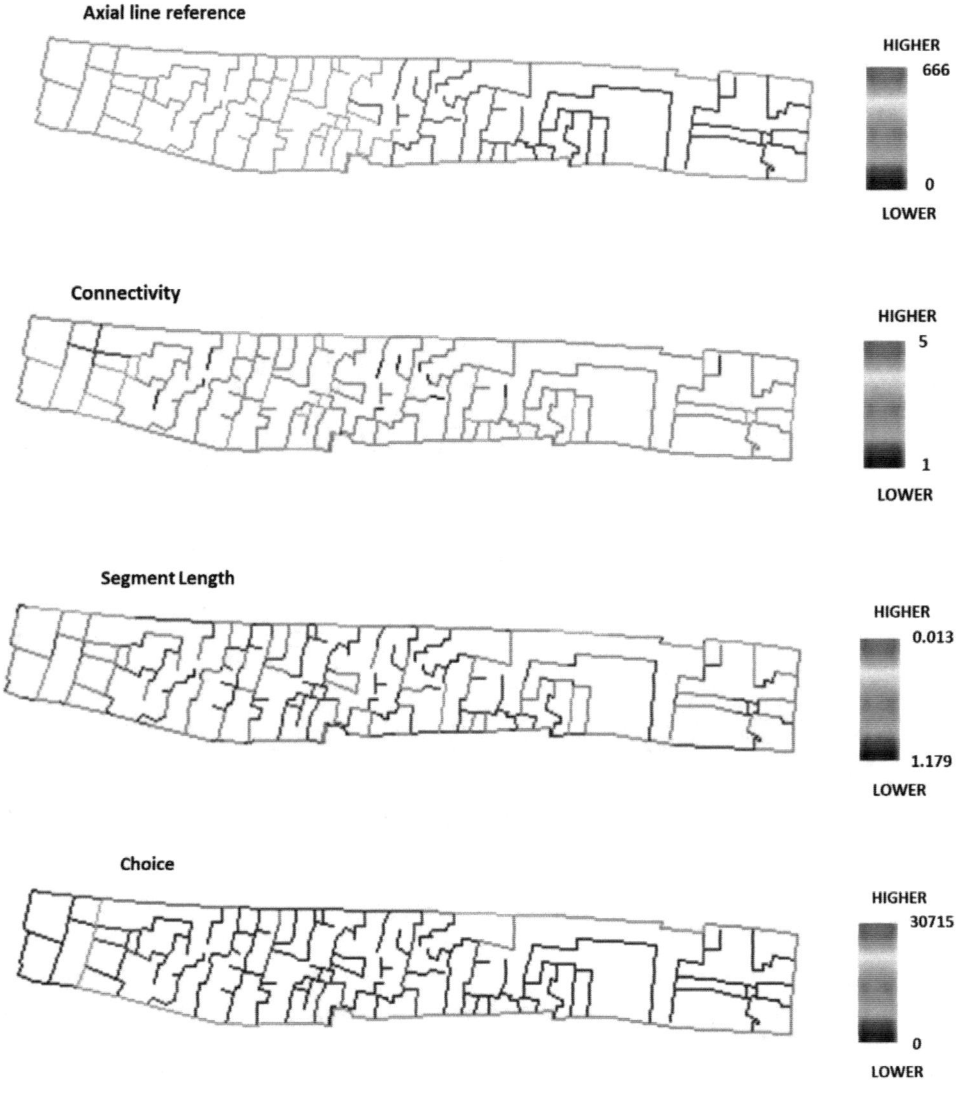

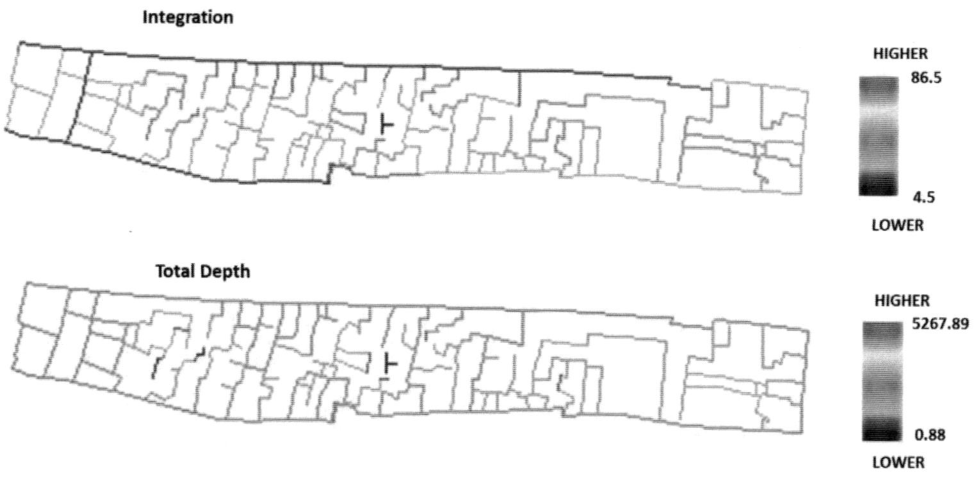

Figure 6. Maps showing results of space syntax analysis of Bahadurpur Slum Area

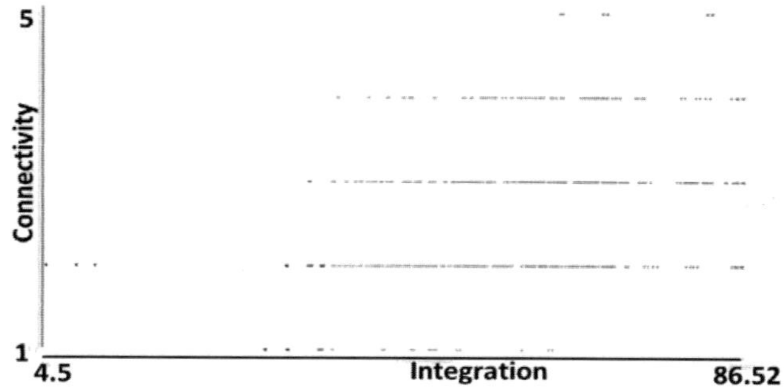

Figure 7. Intelligibility Graph

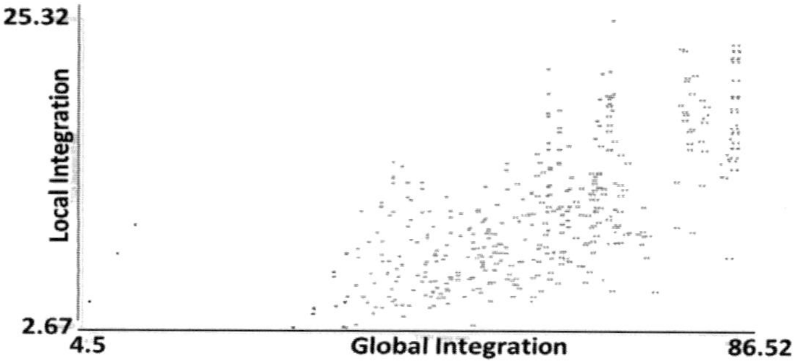

Figure 8. Synergy Graph

5. Conclusion

The result of the analysis explains that the formation of slum in an urban area have a distinct pattern and features guided by deep routed socio-economic characteristics of the slum dwellers, which in turn impacts the social and behavioral phenomenon as well as the environmental quality. It is imperative to find the planning strategies to interpose the dynamics of slum development and improve the quality of life of the slum dwellers for overall improvement of the urban environment. There can further, more in-depth, research be conducted to study the environmental impacts of slums at large scale.

References

[1] Anderson N 1959 The Urban Community: A World Perspective (Henry Hall and Company, New York) pp 190-93
[2] National Sample Survey Organisation 2003 *Conditions of Urban Slums-2002 Report no 486* Ministry of Statistics and Programme Implementation. Government of India
[3] Kumari R and Devadas V 2017 Modelling the dynamics of economic development driven by agricultural growth in Patna Region, India *Economic Structures* **6** 15 https://doi.org/10.1186/s40008-017-0075-x
[4] Clinard M 1966 Slums and Community Development *Experiments in Self Help* (The Free Press, Glencoe, New York) pp 24-42
[5] Singh O et.al.1988 *India's Urban Environment: Pollution, Perception and Management* (Tara Book Agency, Varanasi)
[6] Clinard M 1981 Slums and Community Development *Journal of the Institute of Town Planners* Special Issue **108** pp 15-24
[7] Egondi T, Kyobutungi C, Ng N, Muindi K, Oti S, van de Vijver S, Ettarh R, and Rocklöv J 2013 Community perceptions of air pollution and related health risks in Nairobi slums *International journal of environmental research and public health* **10** 10 pp 4851–4868 https://doi.org/10.3390/ijerph10104851
[8] Stores C J 1962 A Theory of Slums *Land Economics* **38** 188
[9] Fermo P, Comite V., and La Russa M F 2022 The interaction between environmental pollution and cultural heritage: from outdoor to indoor "MetroArcheo2020" *Environ Sci Pollut Res* **29** pp 29382–29384 https://doi.org/10.1007/s11356-022-19233-8
[10] Kumari R 2021 ADAPTNESS ASSESSMENT OF PEDESTRIAN STREET DURING CRISES LIKE COVID-19 *Journal of Urban and Environmental Engineering* **15** 1 pp 50-57
[11] D K Putri et al 2020 The use of space syntax method in mapping passive contacts in slum area of Semanggi, Surakarta *IOP Conf. Ser.: Earth Environ. Sci.* **447** 012003
[12] The World's cities in 2018 Data Booklet *United Nations*
[13] Census of India 2001 *Metadata and Brief Highlights on Slum Population* available at http://www.censusindia.net/results/slums/metadata_highlights.pdf
[14] City Development Plan 2010-30 (Patna) *Urban Development and Housing Department, Government of Bihar*
[15] Kumari R 2021 Impact of Aspect Ratio of Floor Plan on the Energy Performance of Office Rooms in New Delhi, India *IOP Conf. Ser.: Earth Environ. Sci.* **795** 012035 p 7
[16] Kumari R 2022 Contemplating the traditional rural architecture of Nalanda: A case study *IOP Conf. Ser.: Mater. Sci. Eng.* **1212** 012007 p 8

Acknowledgements

The support from National Institute of Technology Hamirpur and Ministry of Education, Government of India, is acknowledged. Further, I would like to thank the anonymous reviewers for their valuable comments.

Delay models for Signalised Intersections with Vehicle Actuated Controlled system in Heterogeneous Traffic Conditions

Borigarla Barhmaiah[1], AChandrasekar[2], TanalaRamya[3] andS Moses Santhakumar[4]

[1]Research Scholar, Transportation Engineering and Management, Department of Civil Engineering, National Institute of Technology Tiruchirappalli, Tiruchirappalli, Tamil Nadu, India.
[2,3,] Student, Department of Civil Engineering, Vignan's Institute of Information Technology Visakhapatnam, Andhra Pradesh, India.
[4]Professor Transportation Engineering and Management, Department of Civil Engineering, National Institute of Technology Tiruchirappalli, Tiruchirappalli, Tamil Nadu, India.
E-mail: brahmamg44@gmail.com

Abstract. Delay is a critical parameter used to assess the performance of signalised intersections. Non-deterministic arrival and departure rates of vehicles at intersectionseffect the determination of delay. Furthermore, driver behaviour, vehicle characteristics, and type of signal control system at intersectionwere also influencing parameters. Previous literature has offered analytical and empirical models to estimate the average delay using hourly traffic flow data and a fixed cycle time. In India, major signalised intersections on urban arterial roads are controlled with vehicle actuated controlled system. In the present study, Simpson'sone third rule was applied to estimate the overall delay by using arrival rate and departure rate of vehicles in everycycle. In addition, the current studydevelopedregression-based delay model and modified the Indo-HCM model by changing incremental delay term in themodels. Thisensuredaccurateestimation of the delays at signalised intersections in both the VAC and fixed signal control systems. The proposed modelswere statistically significant, with R^2values of 0.89 and 0.51, respectively.

Keywords: *Signalized intersection, Simpson's one third rule, Average delay, Vehicle actuated controlled system.*

1. Introduction

Approach delay is a critical parameter used to assess the performance of signalised intersections. It is the average delay a vehicle experiences owing to the presence of signal control; it includes stopped delay, queued delay and acceleration–deceleration delay. It is difficult to calculatebecause of the non-deterministic arrival and departure processes of vehicles at intersections.Generally, in India, through and turning vehicles are allowed in the same green phase, and the light vehicles try to surge to the front of queues by using possible lateral and longitudinal gaps without following lane discipline. As a result, there are numerous conflicts,reducing the capacity of intersections and increasing the delays.It is due to the driver behaviour, vehicle characteristics, and type of signal control system at intersection. Signal control systems can be classified as either fixed-time or vehicle-activated control (VAC) systems. In a fixed-time signal control system, a predetermined signal cycle time repeatedfor all phases. The VAC programming ensures that the CCTV cameras are installed in the system and measures the length of traffic at all approachesas well as estimates the length of green and red time for each approach.The traffic demand in each approach varies from time to time. Hence, the VAC system signal timings are always being calculated and continuously adjusted for every cycle. Therefore, proper constructed traffic signals can minimise the delays by periodically offering the same space to vehicle traffic from various approaches.

Theprevious researchers have offered analytical and empirical models to estimate the average delay of vehicles at signalised intersections under homogeneous traffic conditions. Out of these models Webster's classical delay formula is most frequentlyadopted for unsaturated conditions (Webster, 1958) [1]. Numerous researchers have revised Webster's delay model to integrate saturated flow conditions, such as the Robertson model (Hurdle 1984) [2], the Akcelik model (Roess et al., 2006) [3], and the control model (HCM 2010) [4]. Although these models produced good results for homogeneous traffic situations, the same could not be replicated in heterogeneous traffic conditions. At a signalised intersection Mousa [5] estimated the average deceleration and acceleration delay to be 11.7 sec/veh at a signalized intersection, The study ignored the delay caused by non-stopped vehicles (7% of the total delay) compared to that of stopped vehicles (93% of the total delay). Dion et al. [6] also evaluated the delays at signal-controlled intersections at a fixed signal time and traffic flow ranged from unsaturated to highly saturated conditions. The study reported similar results for all delay models for signalised intersections with low traffic demand, but increasing differences were observed at high saturation flow. Darma et al. [7] identified the factors that contribute to control delays at signalised intersections. SIDRA and TRANSYT-7Fsoftwareswere used to calculate the delays which were found to be strongly associated with the cycle time,amber time, number of phasings, and number of lanes. Hoque and Imran [8] added an adjustment term to Webster's delay model to fit Bangladeshi traffic conditions.Raval and Gundaliya (2012) have adjusted Webster's method to account for traffic in Ahmadabad, Gujarat. Su et al. [9] estimated delays for heterogeneous traffic at signalised intersections. The findings revealed that delays are mainly affected by the proportion and position of heavy vehicles in the queue. Kumar and Dhinakaran [10] also estimated delays for mixed traffic conditions and found there was no good association between observed and projected delays. Jian Huang et al. [11] developed an analytical delay model, based on observations of queue formation and dispersal. Murat et al. [12] proposed Multiple Regression delay model to establish the relationship between cyclic vehicle queues and vehicular delays. Wang Z et al. [13] compared various delay models such as HCM 2000, Shanghai adjusted, Webster model, and deterministic queuing modes. HCM 2000 model and the Shanghai adjusted model perform satisfactorily and show similar outcomes at different volume-capacity (v/c) ratios. Whereas the Webster model performs poorly when the v/c ratio is close to one, and the deterministic queuing paradigm is ideal for high volume to capacity (v/c) ratios. Fawaz and Khoury [14] proposed a model to estimate uniform delay accurately at under-saturated signalised intersections. Lee Vien Leong and Jong Hui Lee [17] investigated the queuing behaviour of motorbikes and found that if motorbikes stopped at the stop line there is no considerable impact on delays. Ashish Verma et al. [18] modified Webster's delay model for Indian conditions. They observed an error between 4–14% in the modified Webster's model and 14–35% in the basic Webster's delay model.

Although few researchers have modified Webster's and HCM delay models for estimating delay in heterogeneous traffic conditions with poor lane discipline. Preethi P et al. [15] modified Webster's delay model using an Artificial Neural Network (ANN) approach and suggested a correction factor for signalised intersections with varying control conditions. Arpit Saha et al. [16] modified random delay (existing HCM model) using Simpson's rule, with an estimating error of 3.9 percent, the modified delay model produces the best results.The models proposed in the previous studies are for a fixed-time signal control system and unsaturated traffic flow conditions. In a vehicle-activated signal control system, the delay of vehicles is dependent on queue accumulation and departures. The queue accumulation depends on the arrival rate and time of arrival. The queue departure is influenced by (v/c) ratio, vehicle composition, and lane discipline. Hence, the direct applications of proposed models may be doubtful to find the delay accurately in a VAC system. Therefore, the current study aims to develop delay models for signalised intersections with the VAC system.

2. Methodology

As part of the Smart City Project in India, major signalised intersections in Visakhapatnam were equipped with VAC systems. According to the literature, previous studies estimate the average delay using hourly traffic flow data and a fixed cycle time. Therefore, the current study developed mathematical models to estimate the average delay at signalised intersections with the VAC system. The overall delay of vehicles in each cycle was estimated by using Simpson's one-third rule. Furthermore, the variables that influence the delay were identified and a regression-based delay model was developed and modified the Indo-HCM model by changing the incremental delay term in the model.

2.1 Data Collection and Analysis

This study examined four signalised intersections. The data from Satyam, Gurudwara, and Fourth Town intersections were used to construct the models, and the data from the Akkayyapalem intersection was used to validate the developed models. At all intersection approaches, straight, and right turn movements are allowed in the same phase, and the left turn is treated as a free left. The study section comprises three types of data: geometric, traffic, and traffic control data. The geometric data was collected manually. The traffic and traffic control data were collected using the videography techniqueduring morning peak hours (8:00 am to 10:30 am) at North Bound (NB) approach and evening peak hours (4:00 pm to 06.30 pm) at South Bound (SB) approach. The methodology described by Arpit Saha et al. [16] was used in this study to collect the arrival and departure of the vehicles. To capture the arrival and departure of the vehicles at an intersection, two cameras were placed on the same approach, of which one was placed on the upstream side at 80 m from the stop line to record vehicle arrival rates and signal timings, while the other was fixed on the downstream side of the intersection to capture vehicle departure rates. Vehicles entering and exiting at an intersection were detected by making longitudinal traps on selected approaches as shown in Figure 1. Table 1 shows geometric data, traffic flow, and traffic control characteristics of signalised intersections.

(a) (b)

Figure1.Trap at Satyam junction from the stop line to queue end (a) start of the queue; (b) end of the queue

Table1. Geometric, traffic, and traffic controlling characteristics signalized intersections

Intersection Name	Approach	Approach Width (m)	Range of Green Time (Sec)	Range of cycle time (Sec)	Number of observed cycles	Traffic Volume (PCU/hr)
Satyam Junction	NB	12	30-50	130-195	44	1722

	SB	12	30-50	140-195	40	1977
Gurudwara	NB	12	20-50	85-165	51	1234
Junction	SB	12	20-50	100-165	49	1685
4th Town	NB	12	20-50	60-125	49	1516
Junction	SB	12	15-50	60-135	47	1545
Akkayyapalem Junction	SB	10.2	30-40	95-125	10	1483

From the recorded video, both the arrival and departure rates of vehicles were extracted every five seconds for all the signal cycles, with further saturation flow estimated for every cycle by using equation 1.The number of vehicles in the queue was calculated by adding the number of vehicles currently in the queue and the vehicles arriving during every five-second interval, excluding those that crossed the stop line during that time interval (TRB 2010). All vehicles in this study are divided into six categories: two-wheelers (2W), three-wheelers (3W), standard cars (SC), buses (BS), light commercial vehicles (LCV), and heavy commercial vehicles (HCV). Figure 2 depicts the composition of vehicles for the considered approaches during peak hours. During peak hours, the 2W had the highest traffic composition, ranging from 51 to 69 percent, among the other classes of vehicles.

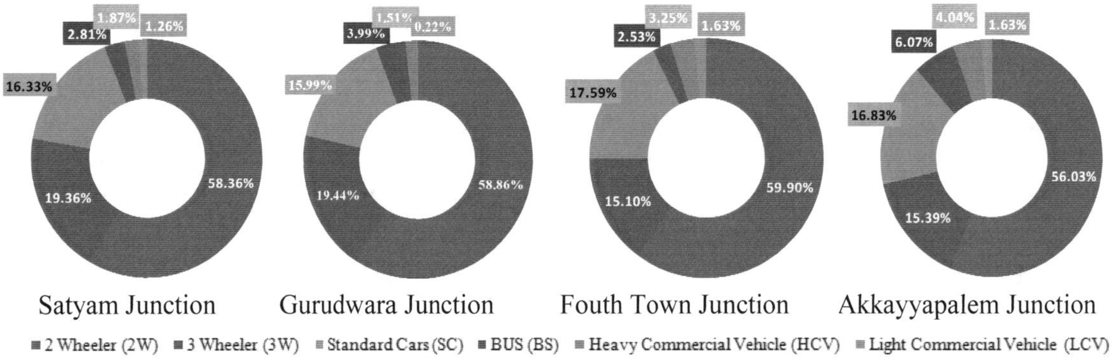

Figure2. Composition of traffic volume at South Bound (SB) approaches of selected Intersections to develop simulation models.

2.2 Estimation of Saturation flow (SF) and Capacity

In this study, traffic volume, capacity, and SF were analysed in terms of the number of vehicles (N). The SF is calculated by plotting the discharge at the stop line every five seconds. At the Satyam intersection, Figure 3 shows the saturation flows for three sample signal cycles. The departure rate changes every 5 seconds, and the variations during the beginning of green time are ascertained by traffic composition. The departure rates decreased with an increase in the heavy vehicle composition. The effective green time was calculated as the length of the green period during which at least three vehicles could discharge. When there were a lot of vehicles moving quickly together, a minimum of three vehicles in the queue were considered. SF is defined as the total number of vehicles that pass through the intersection during the saturated green time, which is calculated using Equation 1.

$$S = [\frac{N}{g_e} * CT] \quad (1)$$

Where; S is the saturation flow, N is the number of vehicles and g_e is the effective green time (s) and CT is the cycle time.

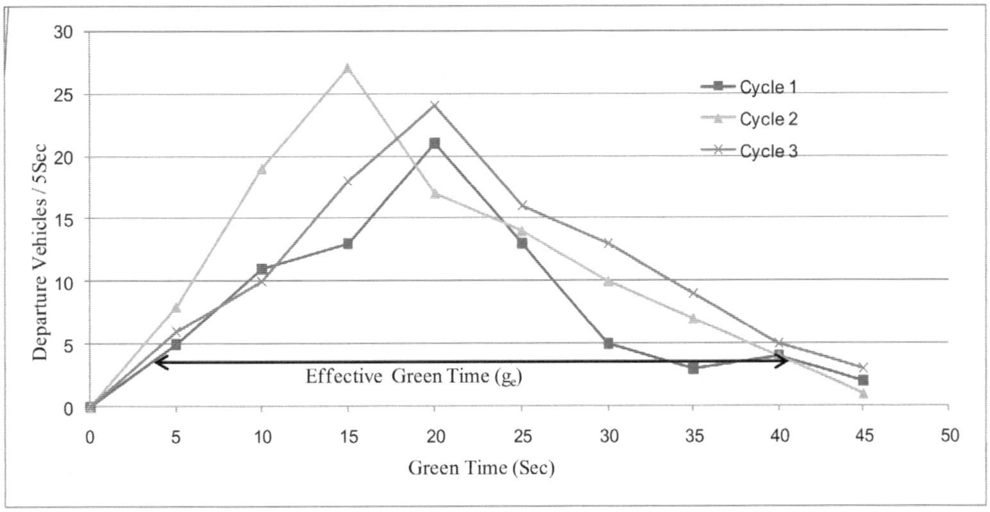

Figure 3. Measurement of saturation flow

The Indo-HCM proposed an equation to estimate the capacity of the approach for each cycle, and the same equation was used to estimate the capacity as shown in equation 2.

$$C = S*[\frac{g_e}{CT}] \quad (2)$$

2.3 Estimation of Queue length andOverall delay

Vehicle arrivals and departures data were monitored at five-second intervals and estimated the queue length (QL) of each cycle. TRB's (2010) simple approach wasused to calculate the queue length for a five-second interval. Arpit Saha et al. 2017 [16] also used the same approach to determine the queue length and total delay per cycle. The QL is estimated by subtracting the number of vehicles that passed the stop line during the five-second interval from the total vehicles in the approach. The area between the length of the queue and the corresponding time was then estimated by using Simpson's one-third rule, which is shown in Equation (3). The area was taken into account while calculating the overall delay for that cycle.

$$\int_o^c f(Q_L)d_q = \frac{h}{3}[(q_0 + q_n) + 4(q_1 + q_3 + ... + q_{(n-1)}) + 2(q_2 + q_4 + ... + q_{(n-2)})] \quad (3)$$

$$h = \frac{(CT-0)}{n} \quad (4)$$

Where C-0 is the time interval between the beginning and end of the queue; Q_L is the queue length; CT is the cycle time; n is the number of five-second intervals; and $q_0, q_1, q_2 \dots q_n$ is the queue length at the 0th, 1st, 2nd... n^{th} intervals.

Equation 3 calculated the total delay of the approach vehicles in each cycle. The average vehicle delay was calculated by dividing the total delay by the number of vehicles queued. HCM (2010) recommends visualising a trapezoid or triangle approach between queue length and cycle time to compute the overall delay of an unsaturated cycle. On the diagram, each triangle or trapezoid shows the overall delay of a single cycle. The main assumption for building a triangle or trapezoid is that the arrival and departure rates were constant over time. At the Satyam intersection, Figure 4 shows how the queue length changes every five seconds. The surcharge and excess zones on both sides of AB and BC are not the same, as shown Figure 4.Moreover, inaccurate results are possibly predictedwith the change in the slopes of triangle sides. Although rectangular area elements are considered under queue and cycle length in the trapezoidal rule, their horizontal line tops are replaced with sloped line. This is done by transforming each element's area into a trapezoid, then which is used to determine overall delay. In this case, the error is similar to that of the triangular rule. So, to minimize this error, Simpson's one-third rule was used in this study.

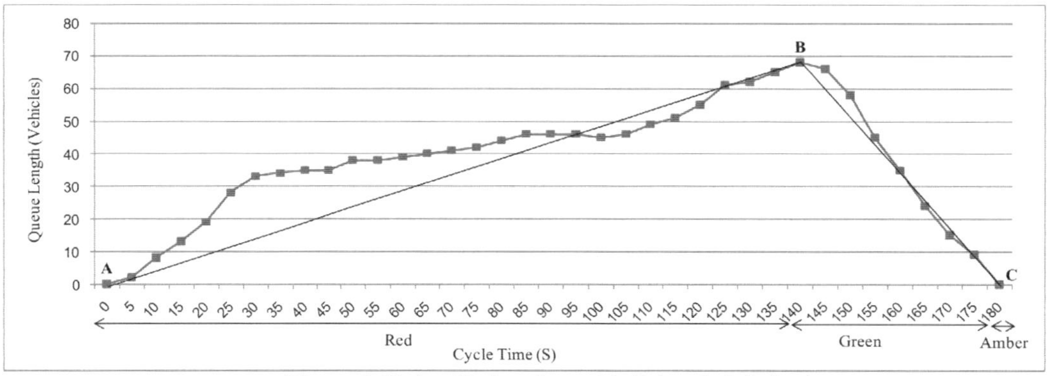

Figure4. Delay Estimation with Triangle Method

3. Formulation of the Delay Model

According to experts, the length of the cycle time, the green time to cycle time (g/C) ratio, and the v/c ratio are the main factors that influence delays at signalised intersections (Varia et al. 2001; Dion et al. 2004). All of these investigations used HCM or Webster's technique todetermine the effect of these variables on control delay. However, these methods produced accurate results at unsaturated and with fixed cycle timeconditions. But in aVAC system, cycle time varies from cycle to cycle based on approach volume, resulting in these methods not producing accurate results. As a result, this study presents a more accurate mathematical model for delay at saturated flow and activated signal control systems. The data for variables (v/c ratio, g/C ratio, traffic volume, capacity, SF, vehicle compositions, and platoon ratio(R_P) were extracted and found platoon ratio by using equation 5.

$$R_p = \frac{PVG}{PTG} \tag{5}$$

Where R_P isthe Platoon ratio; PVG is theproportion of vehicles arriving during the green time; PTG is the proportion of time green.

The platoon ratio is a metric for assessing the effectiveness of an approach's progress. It is the ratio of vehicles that arrive during the green time to the percentage of green time in that overall cycle.HCM (2010) stated that avalue between 0.5 and 2.0 is employed in the estimation of approach delays and capacity. The estimated $R_p$values are ranging from 0.6 to 1.6

3.1 Regression model.

A multiple linear regression (MLR) delay model was constructed by using the variables $R_{p,}$ v/c ratio, g/C ratio, and CT. The regression coefficient indicates how much the dependent variable is predicted when one of the independent variables is changed while the other independent variables remain constant. The estimated coefficients of the proposed model wereshown in Table 2. With an R squared value of 0.89 and a standard error of 5.3, the proposed model is statistically significant. The MLR delay model is shown in equation 6.

$$Average\ Dealy = 26.9—9.35R_p + 1.72 * \frac{v}{c} - 47.5 * \frac{g}{C} - 0.32 * CT \tag{6}$$

Table2. Coefficients and Statisticalvalues of delay Models

Variable	Coefficients	t Stat	p-Value
Intercept	26.9	9.52	<0.005
R_p	-9.35	-9.32	<0.005
v/c ratio	1.72	1.82	<0.005
g/C ratio	- 47.5	-10.6	<0.005
Cycle Time	0.32	33.3	<0.005

3.2.Modification ofIndo-HCM model

Central Road Research Institute (CRRI) conducted extensive research with the help of leading academic institutions in the country and found that the control delay model of HCM (2010) from the United States is very close to the observed control delay at selected intersections in India. The same theoretical form has been used in Indo-HCM [30] as shown in equation 7.

$$d_c = 0.9 * d_1 + d_2 + d_3 \tag{7}$$

Where,

d_c is the control delay (s/veh); d_1 is the uniform delay assuming uniform arrivals (s/veh) as shown in equation 8; d_2 is the incremental delay to account for the effect of random arrivals and oversaturation (s/veh), as shown in equation 9; d_3 is theinitial queue delay (s/veh), in this study assumed as no initial queue at the start of analysis period;

$$d_1 = 0.5 * C \frac{[(1-\frac{g}{c})^2}{\{1-\left[\min(X, 1.0) * \left(\frac{g}{c}\right)\right]\}} \tag{8}$$

$$d_2 = 900\mathrm{T}(\mathrm{X} - 1 + \sqrt{[(\mathrm{X}-1)^2 + (\frac{4\mathrm{X}}{\mathrm{CT}})]} \tag{9}$$

Where;

g is the effective green time (s); c is the capacity of lane group (veh/h); C is the cycle length; T is the evaluation time.

Field measurements of delay automatically take into account all of the above-listed parameters. Uniform delay (d_1) has not changed because of these three considerations. d_2 is the additional delay term considered to account for random arrivals, and it is varying with platoon ratio.d_1 was determined for each cycle to estimate the second component of the delay equation (d_2). Since the overall delay is equal to the sum of d_1, d_2, and d_3, the expression (d–d_1) shows that delay d_2 is viewed as negative. Even though every possible influencing factor, such as the percentage of 2W, g/c ratio, and the v/c, was investigated to correlate the modulus of incremental delay, only the platoon ratio was found to have a good relationship with incremental delay. Video footage was used to estimate the platoon ratios for each cycle.As illustrated in Figure 5, the polynomial relationship was the best observed, with an R squared value of 0.512. Equation 10 shows the modified delay model.

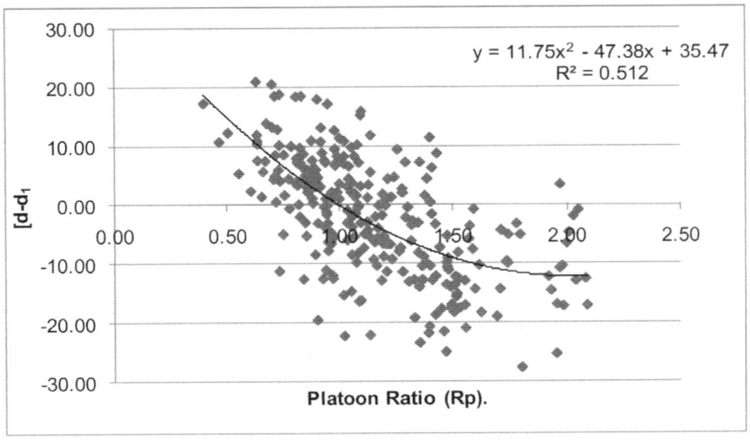

Figure 5. Relation of platoon ratio with Incremental Delay [d − d_1]

$$d = 35.47 + 0.5 * C \frac{[(1-\frac{g}{c})^2}{\{1-[\min(X,1.0)*(\frac{g}{c})]\}} + 11.5R_p^2 - 47.38R_p (R^2 = 0.512) \quad (10)$$

3.3. Validation of Delay Models

The proposed delay models were validated using data collected at the Akkayyapalem intersection. As shown in Figure 6, delays were measured using field data and compared to the delays estimated using equations (6) and (10). When compared to the Indo HCM modified delay model, the MLR delay model produced closed results with field delay. The maximum difference estimated by MLR and Indo HCM modified models was 8.2 sec and 14 sec, respectively. This shows that the MLR delay model accurately predicts the average delay.

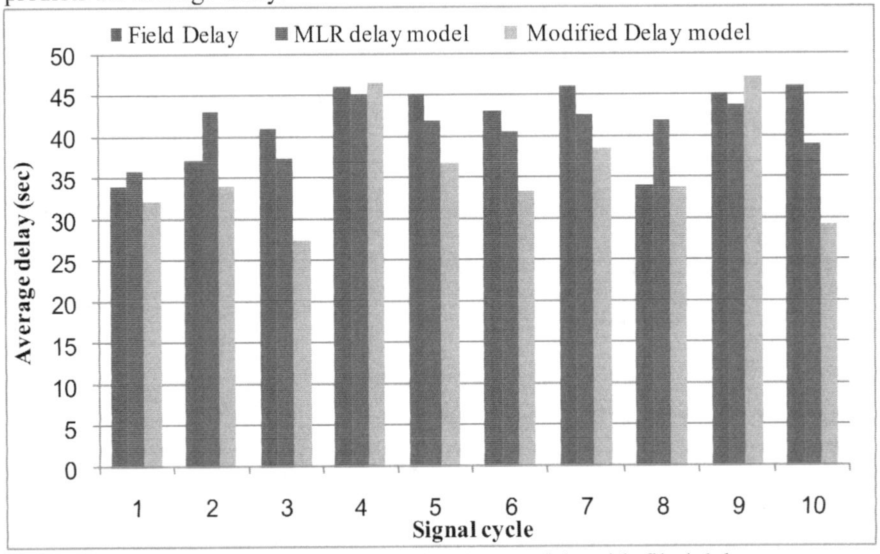

Figure 6. Comparison of Delay models with filed delay.

4. Conclusions

The proposed delay models accurately estimate average delay at signalised intersections in both the VAC and fixed signal control systems.

1. When compared to the Indo HCM modified delay model, the MLR delay model predicted the closed results with field delay.

2. The MLR and Indo HCM-modified models estimated a maximum difference in delay of 8.2 and 14 seconds, respectively, when compared to field delay.
3. The proposed MLA and modified delay models were statistically significant, with R^2values of 0.89 and 0.51, respectively.

References

[1] Webster, F.V. (1958). "Traffic Signal Settings. Department of Scientific and Industrial Research", Road Research Technical Paper No. 39, Her Majesty's Stationary Office, London, England.
[2] Hurdle, V.F.(1984). "Signalized Intersection Delay Models- a primer for the Uniformed", Transportation Research Record, 971, 96-105.
[3] Roess R P., McShane, W.R., Prassas, E.S. (2006). Traffic Engineering 2nd Edition, Prentice-Hall, New Jersey.
[4] HCM *(2010)* "HCM 2010: Highway Capacity Manual", Special Report No. 209, 5th Edition,
[5] Mousa, R. M. (2002). "Analysis and modelling of measured delays at isolated signalized intersections." J. Transp. Eng., 10.1061/(ASCE)0733-947X(2002)128:4(347), 347–354.
[6] Dion, F., Rakha, H., Kang, Y (2004). "Comparison of delay estimates at under-saturated and over-saturated pre-timed signalized intersections', Transp. Res. B, Method. 2004, 38, (2), pp. 99–122.
[7] Darma, Y., Karim, M., Mohamad, J., and Abdullah,S. (2005). "Control Delay Variability at Signalized Intersection Based on HCM Method". Proceedings of the Eastern Asia Society for Transportation studies. Vol 5, 945-958.
[8] Hoque, S., Imran, A. (2007). "Modification of the Webster's Delay Formula under Non–lane Based Heterogeneous Traffic Condition. Journal of Civil Engineering". 81-92.
[9] Y Su, Z Wei, S Cheng, D Yao, Y Zhang and L li (2009) "Delay Estimates of Mixed Traffic Flow at Signalized Intersections in China", Tsinghua Science and Technology ISSN 1007-0214, 02/18llpp157-160.
[10] Prasanna Kumar, R., and Dhinakaran, G. (2013). "Estimation of delay at signalized intersections for mixed traffic conditions of a developing country." International Journal of Civil Engineering, 11(1), 53–59.
[11] Jian Huang, Ge Li, Qi Wang, Haitao Yu (2013) "Real Time Delay Estimation for Signalized Intersection Using Transit Vehicle Positioning Data" 2013 13th international conference on ITS Telecommunications.
[12] Murat, S.Y., Kuthuhalan, S., Cakici, Z. (2014). "Investigation of cyclic Vehicle Queue Delay and delay Relationship for Isolated Signalized Intersections". Procedia- Social and Behavioural Sciences, 111,252-261.
[13] Wang, Z., Chen, S., Yang, H., Wu, B. & Wang Y. 2016. "Comparison of delay estimation models for signalized intersections using field observations in Shanghai", IET intelligent transport systems, 10(3): 165- 174.
[14] Fawaz, W., and El Khoury, J. (2016). "An exact modelling of the uniform control traffic delay in under saturated signalized intersections." *Journal of Advanced Transportation*, 50(5), 918–932.
[15] Preethi.P, Aby Varghese, Ashalatha.R (2016) "Modelling Delay at Signalized Intersection under Heterogeneous Traffic Condition''.
[16] ArpitaSaha, Satish Chandra, and Indrajit Ghosh. (2017), "Delay at Signalized Intersections under Mixed Traffic Conditions" Journal of Transportation Engineering, Part A: Systems, 2017, 143(8): 04017041.
[17] Lee Vien Leong and Jong Hui Lee(2017) "Impact of Motorcyclists' Travel Behaviour on Delay and Level-of-Service at Signalized Intersections in Malaysia" © Springer International Publishing Switzerland 2017.
[18] Ashish Verma, G. Nagaraja C. S. Anusha Sai Kiran Mayakuntla (2018). "Traffic Signal Timing Optimization for Heterogeneous Traffic Conditions Using Modified Webster's Delay Model." Transportation in Developing Economies (2018) 4:13.

Geo-statistical analysis of ground water fluctuations for sustainable management of groundwater resources within the chaliyar basin, southwest India

M I Ansari[1], L N Thakural[2] and Q Hassan[3]

[1,3]Department of Civil Engineering, Jamia Millia Islamia, New Delhi-110025, INDIA
[2]Surface Water Division, National Institute of Hydrology, Roorkee-247667, INDIA

[1]izharansari17@gmail.com

Abstract. Ground water is the primary source for potable water and it forms the fundamental connecting link amongst the Climates, Society & Natural Ecosystem. Phenomenon like climate change has a marked influence on ground water recharge giving rise to spatial variation in ground water levels. Today, fluctuations in ground water level have become a major problem and their study is of prime importance from water security & sustainable development point of view. To meet such challenges, geospatial technologies have proven to be an effective enabler i.e. spatial interpolation mechanism using geo-statistical analysis has become very popular. The Kriging method is an example of such techniques. Hence, this paper attempts to study & analyze the spatio-temporal variations of ground water levels pertaining to 29 observation wells located in Chaliyar basin, Southwest India for which Geo-statistical analysis was performed using ordinary Kriging method in Arc GIS environment. Finally, the spatial maps showing the ground water level fluctuations were generated for the Chaliyar basin which could be used for sustainable management & planning of Ground water resources within the basin.

1. Introduction

During the previous century, global water demand has increased exponentially. In addition to hydrological stress, territories around the globe are facing increasing challenges associated with water scarcity caused by climate change. A study conducted by the council on Energy, Environment & water in 2021, more than 80% Indian are living in districts vulnerable to climate risks. Just a single degree rise in temperature, could end up costing the country 3% of GDP. As per the data available on United Nations website, population living in water-stressed countries is around 2.3 billion people (as of 2018). Regions such as northern Africa, western & southern Asia and are worst affected, presenting alarming water stress levels higher than 70%.

Today, the world is lagging to achieve the Sustainable development Goal No. 6 of United Nations which talks about ensuring sustainable availability & management of water resources. In order to attain this target by 2030 (deadline set by United Nations for attaining SDGs), we need to focus on the Key Natural Resource i.e. Ground water.

Ground water is the primary source of potable water & it forms connecting link amongst climates, Society & Natural ecosystem. Today, majority of the world's population depends primarily at groundwater for drinking, industrial, domestic & agricultural purposes. However, the rate of groundwater extraction is continuously increasing with the escalation of the population, agriculture, urbanization & industrial development. Excessive extraction of groundwater, especially in metropolitan cities & industrial towns, is responsible for severe groundwater fluctuations & corresponding groundwater level declination in many areas [1]. This depletion in groundwater level has negative effects on water pollution, salinity intrusion, groundwater contaminant & ecosystem imbalance, etc.

Climate change, with respect to temperature variations & more importantly in intensity & quantity of precipitation [2] has substantial effects over water volume that consists of recharge to aquifer [3], i.e. phenomenon like climate change has a marked influence on ground water recharge giving rise to spatial variation & fluctuations in ground water levels [4].

Determining how natural water recharge gets affected by these variations or fluctuations is a question of utmost importance, as this constitutes an important part in hydrological cycle. Assessment & knowledge of this parameter is mandatory for effective management of ground water resources [5]. Today, Ground water level fluctuations space to space (Spatial) and time to time (Temporal) has emerged as severe issue, hence their study is of prime importance from water security & sustainable development point of view.

2. Study Area & Data Collection

Area of study is Chaliyar River Basin, Southwest India. Majority of the basin area lies in state of Kerala (minor portion falls in Tamil Nadu also). Chaliyar is the third largest river of Kerala state [6]. Kerala, also Known as Gateway to Monsoon in India, has found to be extremely sensitive from Climate change point of view, hence, evaluations of Ground water fluctuations in these circumstances is of greater significance for this basin.

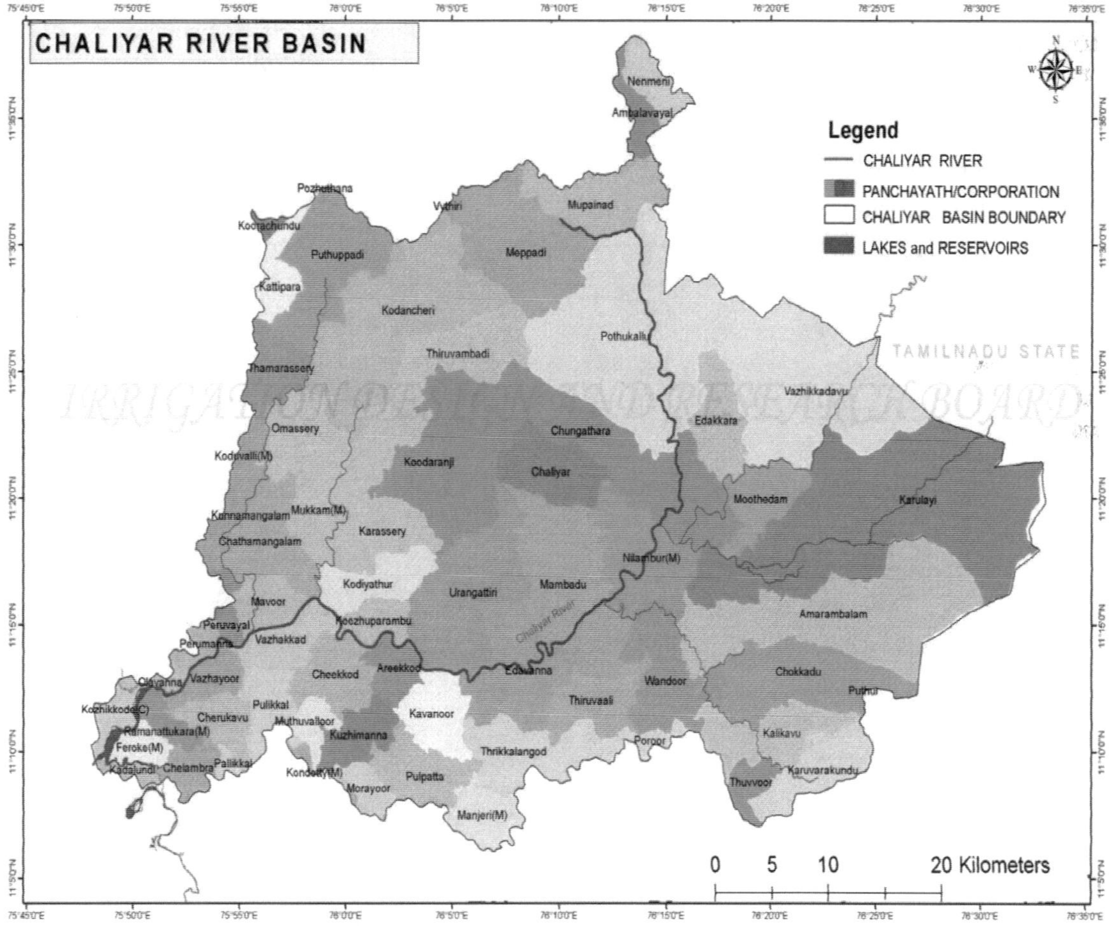

Figure 1. Chaliyar River Basin, Kerala, India.

In the study area shown in Figure 1, 35 observation wells (fairly evenly distributed) were located to record variation of ground water levels of Chaliyar basin. GPS was used in locating sample collection locations coordinates, so as to carry out the purpose of continuous monitoring. Although, samples of groundwater collected only by 29 observation wells were used in our study, as remaining observation wells were showing discontinuity in their data.

Table 1. Well stations with their Geographical coordinates & Districts.

S. No.	Station	Longitude	Latitude	District
1	Mavoor-1	75.955	11.257	Kozhikode
2	Beypore	75.806	11.171	Kozhikode
3	Ramanattukara1	75.858	11.175	Kozhikode
4	Koduvally	75.909	11.358	Kozhikode
5	Tamarasseri	75.938	11.421	Kozhikode
6	Puduppadi	75.984	11.49	Kozhikode
7	Chelavur	75.882	11.299	Kozhikode
8	Mavoor-2	75.942	11.25	Kozhikode
9	Malayamma	75.933	11.35	Kozhikode
10	Vaduvanchal	76.225	11.555	Wayanad
11	Ambalavayal	76.229	11.606	Wayanad
12	Trippanachi	76.044	11.158	Mallapuram
13	Kalikavu	76.32	11.169	Mallapuram
14	Buliyampadam	76.25	11.41	Mallapuram
15	Karulai	76.327	11.283	Mallapuram
16	Kadalundi	75.827	11.136	Mallapuram
17	Areakode	76.041	11.233	Mallapuram
18	Kavanur	76.055	11.196	Mallapuram
19	Manjeri	76.12	11.118	Mallapuram
20	Karulayi	76.297	11.281	Mallapuram
21	Chokkad	76.33	11.233	Mallapuram
22	Olavattur	75.95	11.21	Mallapuram
23	Edavanna1	76.15	11.208	Mallapuram
24	Thiruvali	76.178	11.19	Mallapuram
25	Wandoor	76.233	11.2	Mallapuram
26	Edakkara	76.293	11.373	Mallapuram
27	Kondotty	75.964	11.146	Mallapuram
28	Nilambur	76.241	11.283	Mallapuram
29	Maruda	76.325	11.416	Mallapuram

The data was collected from CWRDM, Kerala [7] for these 35 observation wells in the format of pre-monsoon ground water level & post-monsoon ground water level for the duration of 1996 to 2014. Groundwater levels measured on seasonal basis (pre- monsoon & post-monsoon ground water levels) were investigated for groundwater fluctuations of the year. The well stations with their geographical coordinates & districts are given above in table 1.

Details such as well location, depth to static water level, etc. were collected & tabulated for each well. Thereafter, using the GIS environment, locations of observation wells were transferred to base map. Representation of wells, according to their locations, is shown below as Figure 2 on Chaliyar basin Map with the name of their respective districts.

Then the ground water level data (pre monsoon, post monsoon & fluctuations) & the spatial data (coordinates of observatory wells) were joined with the help of Arc GIS. After linking the spatial & non-spatial data, groundwater quality point layer was generated for further analysis. Afterwards, remaining analysis was performed using Kriging Technique in the Arc GIS 9.3 software.

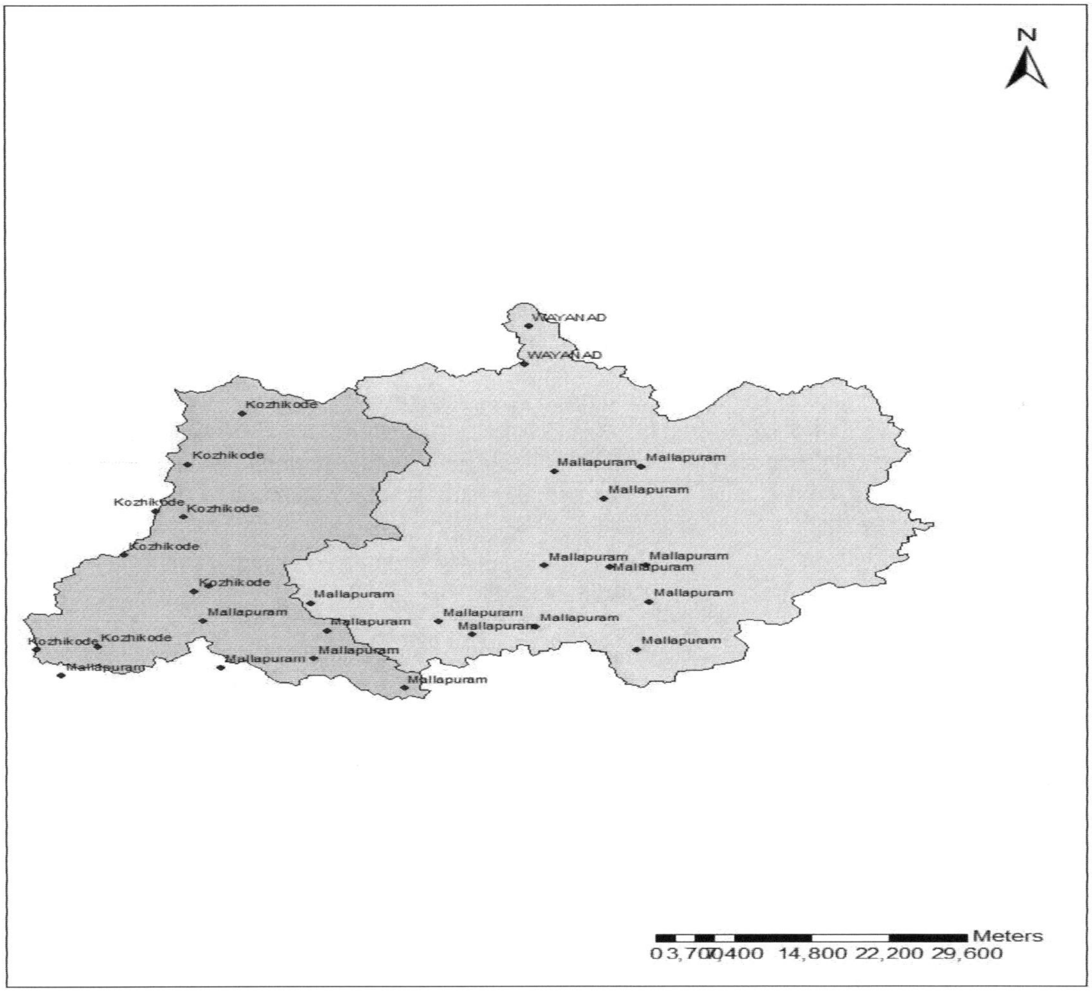

Figure 2. Geographical representations of different wells on study area map districtwise.

3. Methodology

Regular monitoring of groundwater levels is quite important to cope with the current and future water demand. However, remotely placed observation wells do not provide detailed information about spatial coverage of water level. To meet such challenges, geospatial technologies have proven to be an effective enabler i.e. application of Geospatial Technology can be used for management of such Ground water Hydrology problems. Recently, spatial interpolation mechanism through geo statistical analysis has emerged as a popular technique to assess & analyze such spatiotemporal characteristics of water level fluctuations.

Geostatistical analysis

It has been widely used in hydrology to determine various important properties of the aquifer. The geo-statistical methods are capable of creating a prediction surface & in fact accuracy predictions & certainty measures can also be obtained from it [8]. For a particular point, value of the variable can also be calculated from this method, using weighted sum of available point observations. Selection of the weights for the data should be done in a way such that the interpolation is free from biasing and the variance attains a minimum value [9]. The appropriate example of such techniques is Kriging method.

Kriging Method

The Kriging method is perfect reflection for cluster of geo-statistical techniques used for interpolating random field value. A continuous (prediction) surface can be produced using sampled point values from this Interpolation technique. Sample measurements at all locations (from raster dataset) are used for making predictions, irrespective of the fact; measurements have been taken at the location or not [10].

Matheron (1971) [11] formalized & named this method in honor of veteran South African mining engineer, Daniel G. Krige. He is the pioneering father of geo-statistics. Statistical models incorporating autocorrelation is the basis for Kriging method. Statistical relationships between measured points referred as Autocorrelation.

In this method, weighted sum of available point observations is used for calculating value of the variable at a particular point. Data weights are selected in a manner such that the interpolation is free from biasing & the variance attains a minimum value [12].Generally, Kriging should be Authorized, Linear, Unbiased & Optimal. The first method of interpolation that takes into account spatial dependence structure of data is Kriging technique.

Kriging is of multiple types, (a) simple Kriging (SK): assumption is that mean is constant & known (b) ordinary Kriging (ORK): where the mean is constant but unknown (c) universal Kriging (UNK): assumption is that mean will exhibit a polynomial function of spatial coordinates.

In our present analysis, we use ordinary Kriging Interpolation technique to prepare ground water fluctuation map for our study area at different time intervals & further to interpret the spatial variation of ground water levels in GIS environment.

In brief methodology adopted in our study is as follows:

1. Arrange the data in excel sheets.
2. Review the GWL table in Excel.
3. Add table to Arc Map.
4. Interpolate and prepare the ground water level fluctuation maps using ordinary Kriging interpolation technique in Arc GIS.
5. Study & interpret the spatial Ground water fluctuations for Chaliyar river basin.

4. Results & Discussion

Ground water level fluctuations primarily depend on the difference between water level of pre-monsoon & post-monsoon periods, which is correlated with recharge & discharge of groundwater.

Computations regarding this problem are performed for 35 wells located within the basin. To compare spatial & temporal variations of groundwater level in the study area, five Ground water depth maps were exported for the year 1996, 2000, 2004, 2011 & 2014. Furthermore, water table fluctuation maps were generated incorporating difference between pre & post monsoon water level data. Interpolation (Kriging) technique of Arc GIS software is applied for preparation of fluctuation maps. The resultant spatial variability maps showing different groundwater levels for the year 1996, 2000, 2005, 2011 & 2014 are shown below in Figure 3, Figure 4, Figure 5, Figure 6 & Figure 7. Substantial spatial variations in ground water levels for different locations can be noticed from these maps drawn for different years over a uniform time interval. Distribution of Groundwater level in different classesis evident from each map which is a clear indication of variation in ground water level over different locations within the Chaliyar basin. These maps are not only providing us spatial variation in Ground water level within the basin but interpretation can be made regarding temporal variation also, using these maps, as these are drawn for different years maintaining uniform time interval, hence spatial variations may also be compared year wise with the help of these maps. We can observe substantial fluctuations & corresponding variations in ground water level with the help of these figures. Considerable variations could be noticed for few years, which are of grave concern.

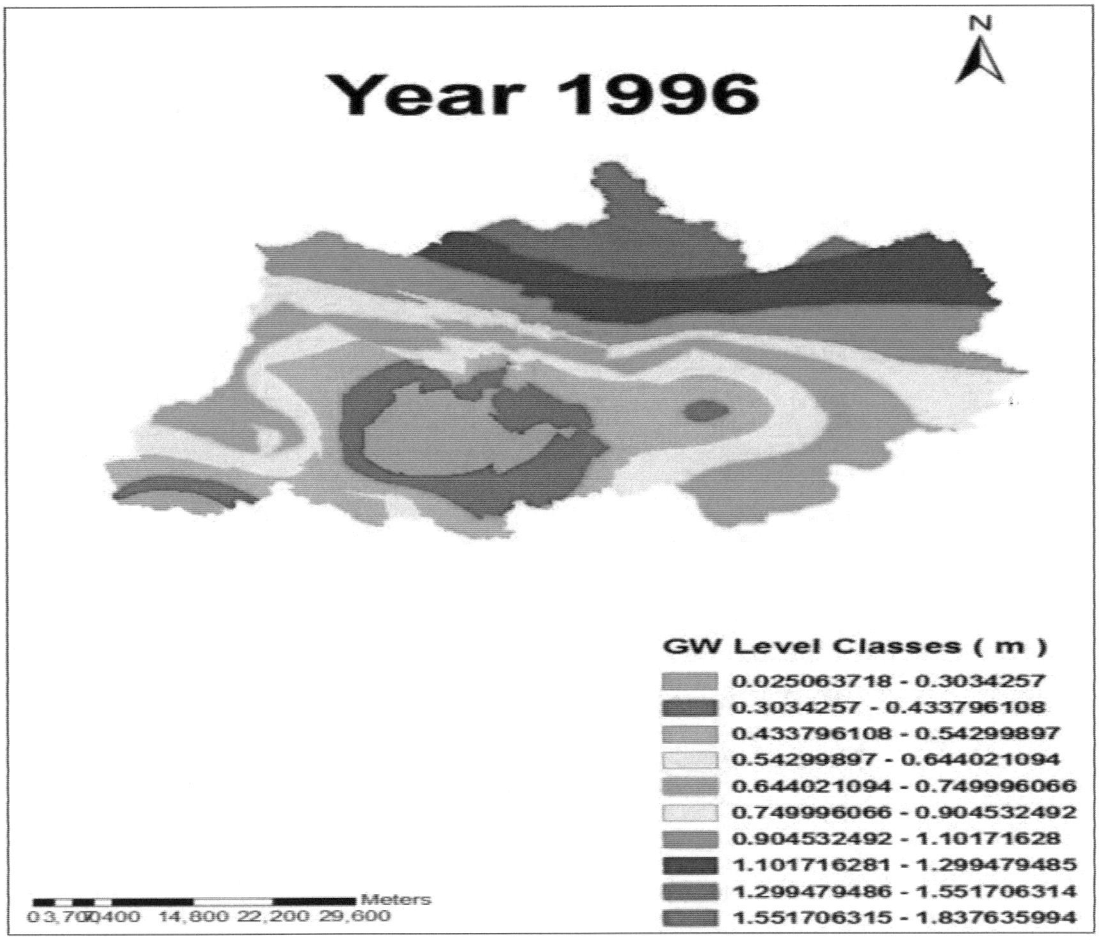

Figure 3. Spatial variability map of groundwater levels for the year 1996.

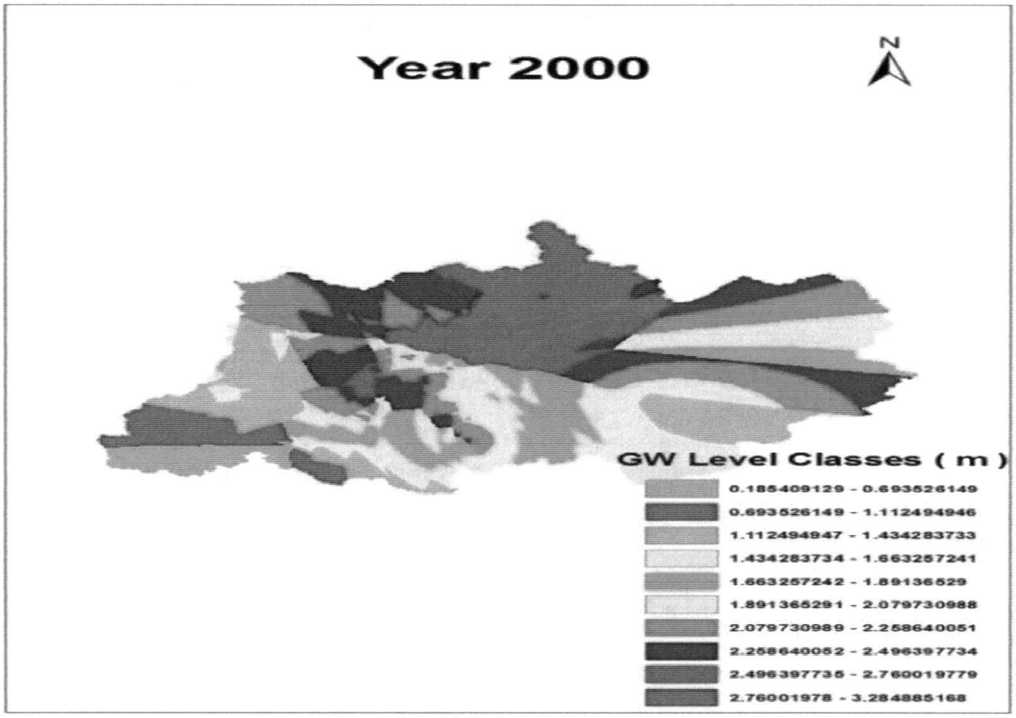

Figure 4. Spatial variability map of groundwater levels for the year 2000.

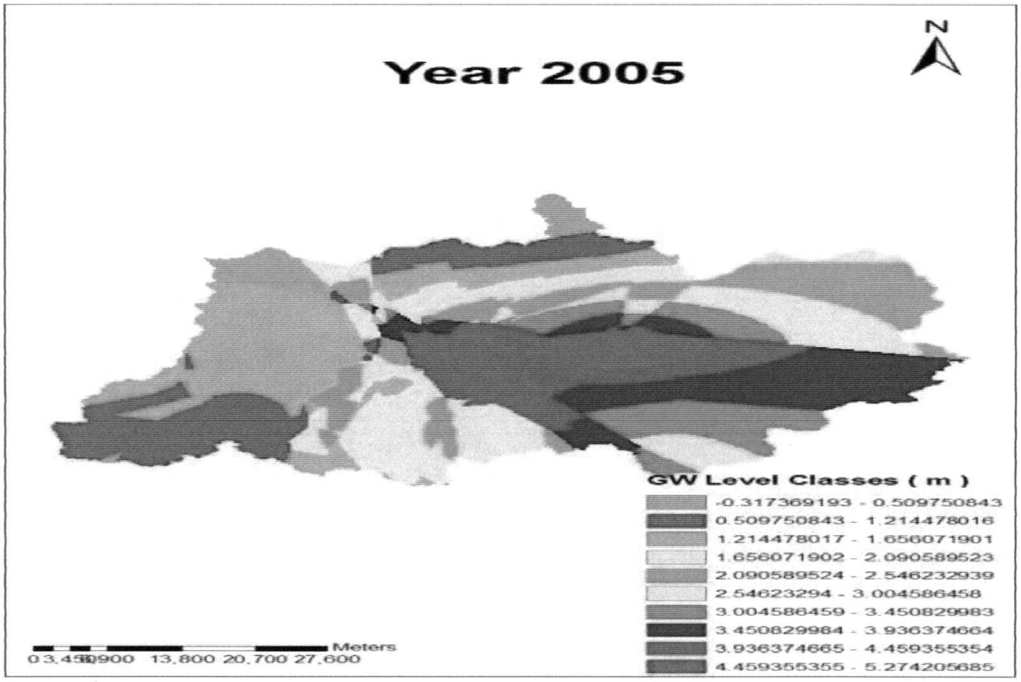

Figure 5. Spatial variability map of groundwater levels for the year 2005.

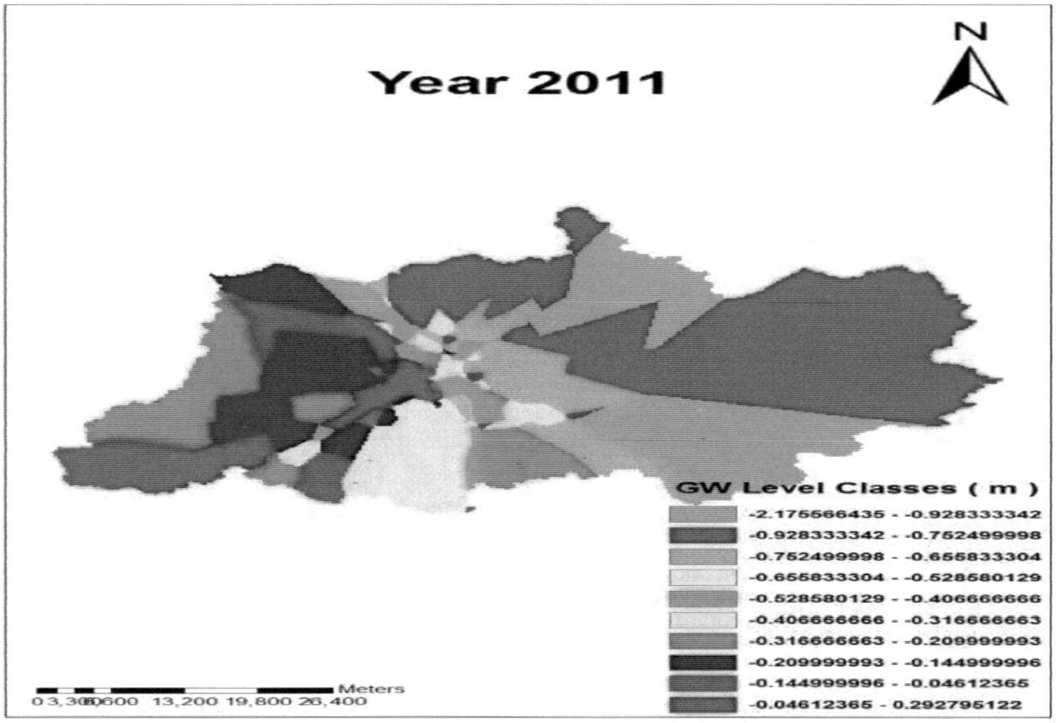

Figure 6. Spatial variability map of groundwater levels for the year 2011.

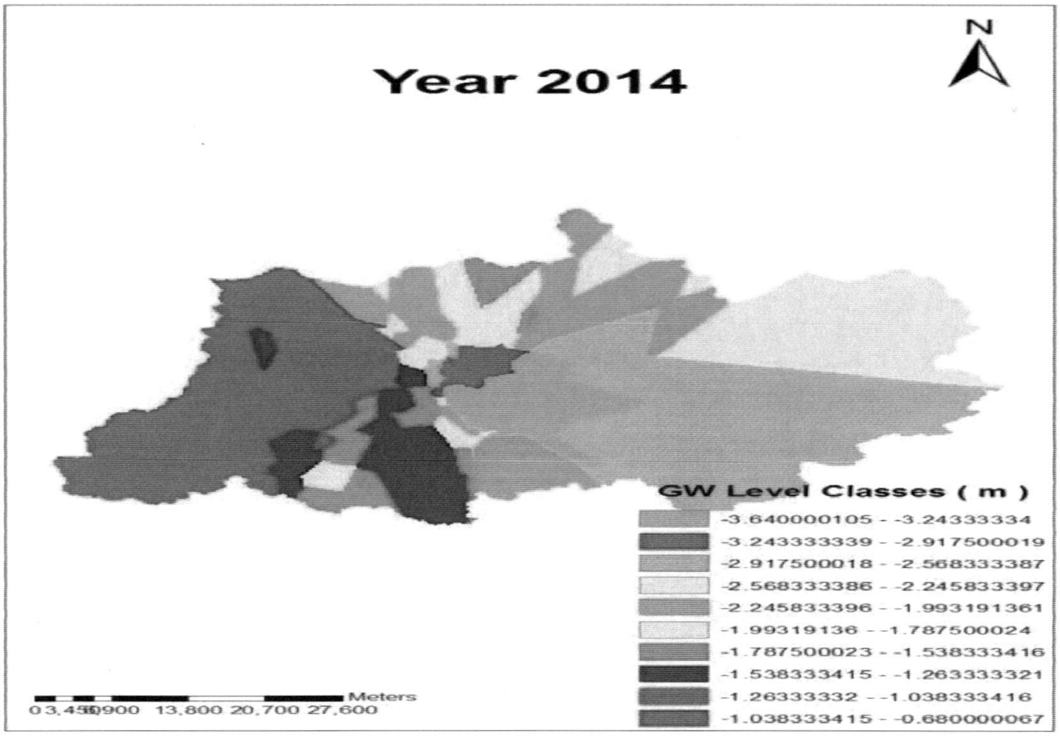

Figure 7. Spatial variability map of groundwater levels for the year 2014.

5. Conclusion

The aim of United Nations Sustainable Development Goals is to execute integrated water resources management at all levels, ensuring sustainable supply & withdrawals of freshwater to address water stress issue & decreasing the percentage of population fighting with water scarcity by the year 2030. A dramatic acceleration or doubling the rate of implementation is required globally as per the current progress rate and in integrating the holistic approaches to water management.

Moving in this direction, spatial variation of ground water levels were investigated in our study for Chaliyar basin by analyzing the fluctuation data obtained for the wells which were fairly evenly distributed across the basin. The spatial variability maps of groundwater level for years 1996, 2000, 2005, 2011 & 2014 were prepared for the basin using spatial interpolation technique (ordinary Kriging) in Arc GIS. Substantial spatial variations in ground water levels can be noticed from these maps, which are of great concern.

Although, Climate change could be one probable reason behind these fluctuations but other factors such as Climatic extremes (Droughts & Floods), Socio-Economic factors, geological & geographical features etc; need to be explored to arrive at final conclusion. More Participation of stakeholders is required to address the above issue; especially participation by local communities should be promoted. As per the precedent set by recent COVID 19 Pandemic, online meetings & online consultations may also be adopted sometimes, for wider & instant stakeholder engagement in water resources management.

In the end, it is suggested to develop & disseminate technologies, monitor & implement national policies effectively regarding sustainable management of India's Ground Water Resources, including their assessment, conservation, exploration & augmentation. The application of the principles of ecological & economic efficiency with equitable distribution is the need of the hour.

Finally it can be concluded that more work needs to be done in the direction of this ground water fluctuations & corresponding ground water depletion problem, but this analysis may be used by water resources development planners, scientists & engineers as a preliminary study, as so as to designwater resources schemes and plans for the region.

References

[1] Shahana, Thakural L N, Prasad J, Prasad H J S and Ansari M I 2020 Spatio-temporal variability and trends of rainfall for a Peninsular Basin. *ESSENCE Int. J. Env. Rehab. Conserv.*, **XI** (SP2), 268—274.

[2] Eckhardt K and Ulbrich U 2003 Potential Impacts of Climate Change on Groundwater Recharge and Streamflow in a Central European Low Mountain Range. *Journal of Hydrology*, **284**(1), 244-252.

[3] Scibek J and Allen D M 2006 Modeled Impacts of Predicted Climate Change on Recharge and Groundwater Levels. *Water Resources Research*, **42**(11), 1-18.

[4] Thakural L N, Anbu Kumar S and Ansari M I 2018 Trend Analysis of Rainfall for the Chaliyar Basin, South India. *International Journal for Research in Applied Science and Engineering Technology*, **6**(7), 91-100.

[5] Marechal J C, Dewandel B, Ahmed S, Galeazzi L, Zaidi F K 2006 Combined estimation of specific yield and natural recharge in a semi-arid groundwater basin with irrigated agriculture. *Journal of Hydrology*, **329**(1-2), 281-293.

[6] Ansari M I, Thakural L N, and AnbuKumar S 2020 StreamflowModelling for a Peninsular Basin in India In S. Ahmed et al. (eds.), Smart Cities-Opportunities and Challenges. *Lecture Notes in Civil Engineering*, **58**, 645-659.

[7] CWRDM: Centre for Water Resources Development and Management, Kozhikode, Kerala, India.

[8] Kumar P, Herath S, Avtar R and Takeuchi K 2016 Mapping of groundwater potential zones in Killinochi area, Sri Lanka, using GIS and remote sensing techniques. *Sustain. Water Resour. Manag.*, **2**, 419–430.

[9] M.M. Hussain, Bari S H, Tarif M E, Rahman M T and Hoque M A 2016 Temporal and spatial variation of groundwater level in Mymensingh district, Bangladesh. *Int. J. Hydrology Science and Technology*, **6**(2), 188-196.

[10] Tiwari A K, Singh P K, Chandra S & Ghosh A 2015 Assessment of groundwater level fluctuation by using remote sensing and GIS in West Bokaro coalfield, Jharkhand, India. *ISH Journal of Hydraulic Engineering*. DOI: 10.1080/09715010.2015.1067575

[11] Matheron, G 1971 *The Theory of Regionalized Variables and Its Applications. Les Cahiers du Centre de Morphologie Mathematique in Fontainebleu* vol 5 (Paris: École Nationale Supérieure des Mines).

[12] Ahmadi S H & Sedghamiz A 2007 Geostatistical Analysis of Spatial and Temporal Variations of Groundwater Level. *Environ Monit Assess*, **129**, 277–294.

Trend Analysis and Forecasting of Long-Term Temperature Data to Anticipate Local Warming: A Case Study of Jawa Timur, Indonesia

Abdur Rohman[1], Retno Utami Agung Wiyono[2], and Fitriani[1]

[1] Department of Environmental Engineering, Faculty of Engineering, Universitas Jember, Jl. Kalimantan 37, Jember, Jawa Timur, Indonesia 68121
[2] Civil Engineering Master Program, Faculty of Engineering, Universitas Jember, Jl. Kalimantan 37, Jember, Jawa Timur, Indonesia 68121

Email/ORCID: abdur@unej.ac.id[1], retnoutami@unej.ac.id[2], fitrianii191201@gmail.com[3]

Abstract. Numerous studies have confirmed the significant rise in the mean land surface air temperature in recent decades at the global scale. However, studies on the land surface air temperature at local scales in Indonesia in recent decades are rare. Such studies are necessary to comprehend the imminence of the threat of climate change in regions with intensive agricultural industry, such as Jawa Timur, Indonesia. The first objective of this study is to analyse the trend of air temperature in six locations in Jawa Timur based on at least 30-year air temperature data retrieved from NOAA Global Summary of the Day data. After the significance or insignificance of the temperature trend is verified, the second objective is to compare the accuracies of two forecasting methods, namely Exponential Smoothing and ARIMA, on the long-term air temperature data in each location. The Mann-Kendall test along with sieve-bootstrapping is used in the air temperature trend analysis. Mean Absolute Error (MAE) metric is used to measure the accuracy of the methods. The trend analysis shows that statistically significant monotonic rises in air temperature are found in four locations, namely Bawean, Kalianget Madura, Surabaya Perak, and Banyuwangi. It is also found that ARIMA (3,1,3) is able to forecast the temperature data for 2021 with MAE < 1.30 degrees Celsius and it outperforms the Exponential Smoothing method.

1. Introduction

A number of studies have confirmed the significant increase in the mean land surface air temperature in recent decades at the global scale, as reported by IPCC [1]. Some studies also confirmed the temperature rise at local scales. An increase in temperature was found at two contrasting sites for the period of 1931 to 2014 in Azul and 1952 to 2014 in Marcos Juarez of the Argentinean wheat region [2]. The authors of this study analysed temperature and precipitation factors by using regression and Mann-Kendall test of the trend in the data and made future forecasting. They concluded that at both locations temperature significantly increased in mean and extreme values. Another study was carried out on the data of monthly, seasonal, annual temperature during 1901 to 2017 over thirty subdivisions and seven regions of India [3]. For the trend analysis they also made the use of Sen's slope and Mann-Kendall test. The net changes in mean temperature indicates temperature rise in majority of the days in the year. Another study was also been performed on annual temperature trends at nine states of the Northeastern United States [4]. They used a modified Mann-Kendall test to analyse the data of the years 1900 to 2011.

Statistically significant positive trends were found in all the states, with the linear trend of temperature ranging from 0.00006 to 0.02 degrees Fahrenheit per year.

Few studies on temperature trends in Indonesia have been conducted. A study analysed surface air temperature variations and trends in Jawa Island for 30 years (1990 to 2019) [5]. In this study, seven stations of BMKG (Indonesian Agency for Meteorological, Climatological and Geophysics) under analysis are located in Jawa Barat (West Jawa) province, while four stations are located in Jawa Tengah (Central Jawa) province, and five stations are located in Jawa Timur (East Jawa) province. The statistical test results show that except for Banyuwangi station, observed temperature data in all the BMKG stations showed increasing trends, ranging from 0.017 to 0.048 degree Celsius per year. Another study analysed and compared homogeneity test results on average temperature and relative humidity in Medan City in Sumatera Island for 30 years, namely 1991 to 2020 [6]. The results showed positive trends existed on average temperature and air humidity Medan City for the last 30 years. The sea surface temperature has also been actively studied. A study evaluated 28 years (1982-2009) sea surface temperature data using Mann-Kendall test over Indonesia, which ranges from 15° N-15° S and 90° E-145° E [7]. The results showed that the sea surface temperature in Indonesian water was correlated with the monsoon season. The sea surface temperature in the northern region was found warmer during the Asian summer monsoon, whereas in the southern region the sea surface temperature was found warmer during the Australian summer monsoon. In general, the sea surface temperature is increasing in Indonesian waters, with a likelihood of over 95%. Despite the previously mentioned studies, considering the vast area of the lands of Indonesia, studies on the land surface air temperature at local scales in Indonesia in recent decades are rare. Such studies are necessary to comprehend the imminence of the threat of climate change in regions with intensive agricultural industry.

Jawa Timur is a province in Indonesia that has essential roles in the agricultural industry, indicated by the existing operation of 34 plantation fields with total area of 81, 278.4 hectares owned by PTPN XII, a state-owned plantation company [8]. In addition, according to the 2020 National Census of Indonesia, the population of Jawa Timur province was 40.67 million, which accounted for 15% of the total population of Indonesia [9]. Compared to the population in the 2010 National Census, which was 37.48 million [10], on average the population of this province has been increasing with the rate of 319 thousand per year. The rapid growth of the population directly relates to the increasing need for food and fresh water, of which the quality and quantity are highly dependent on the climate condition. The objective of this study is to analyse the past long-term temperature data recorded in climate stations in Jawa Timur. The result of this study is expected to benefit the agriculture industry in developing strategies to deal with the climate change that has potentially adverse effects in this region.

2. The Temperature Datasets

The datasets used in this research are subsets of the *Global Summary of the Day* data provided by NOAA (the United States National Centers for Environmental Information, National Oceanic and Atmospheric Administration) [11]. The Global Summary Data is collected every day from over 9000 stations located in many countries. Each dataset consists of 12 daily elements of climate entities including temperature, sea level pressure, wind speed, and precipitation amount. Because this study focuses on analysis of temperature, only daily temperature data are analysed in this study. All steps of the data collection, data cleaning, and analysis and visualisation are conducted in the R [12] and RStudio [13].

2.1 Data selection and retrieval

The datasets are retrieved from the NOAA webpage [11] by using the *rnoaa* package [14]. To get the data of stations located in Indonesia, the country code of "ID" is used, which resulted in 164 stations. Subsequently, to extract the data from stations in Jawa Timur province, the data are filtered by setting the latitudes to less than -5.43° and more than -8.80°. The longitudes are set to less than 114.07° and more than 111.00°. The coordinate points are derived from the geographic coordinates of Jawa Timur borders published by the Statistic Bureau of Jawa Timur [15]. However, the northernmost latitude of

IOP Conf. Series: Earth and Environmental Science 1084 (2022) 012040 doi:10.1088/1755-1315/1084/1/012040

Jawa Timur Province, which is Bawean island, is set to -5.43° based on the study of [16]. Ten stations are selected based on the coordinates. The final selection is the availability of long-term data. We choose 30 years as the minimum period of data coverage to be eligible for time-series analysis. Only six of ten stations meet the criteria. Table 1 shows meta-data of the six stations selected for further analysis.

2.2 Data availability and missingness

Figure 1 shows the line graph of the published daily data from January 1, 1982 to December 31, 2021 for the six selected stations. Figure 2 shows the distribution of missing data over the years and stations. The number of complete data points varies, namely 11915, 11267, 13089, 10121, 12256, and 12941 data points for Bawean, Surabaya/Perak, Surabaya/Juanda, Surabaya/Gedangan, Kalianget, and Banyuwangi stations, respectively. The daily climate data that cover 40 years are available for download for five stations, namely Bawean, Surabaya/Perak, Surabaya/Juanda, Kalianget, and Banyuwangi. As for Surabaya/Gedangan, the data of the year 1991 are completely missing, hence the available data is 1982 to 1990 and 1992 to 2021. As seen in Figure 2, missing data are found in all stations in all years except in the year 2012.

Table 1. Six selected stations in Jawa Timur Province

USAF Code	Station Name	Latitude	Longitude	Elevation (m)	Begin Date	End Date
969250	Sangkapura (Bawean Is.)	-5.85	112.63	3	19530101	20220206
969330	Surabaya/Perak	-7.22	112.72	3	19490101	20220206
969350	Surabaya Juanda	-7.38	112.79	3	19640909	20220206
969370	Surabaya/Gedangan	-7.23	113.72	3	19801001	20220206
969730	Kalianget (Madura Is)	-7.05	113.97	3	19750918	20220206
969870	Banyuwangi	-8.22	114.38	5	19560101	20220206

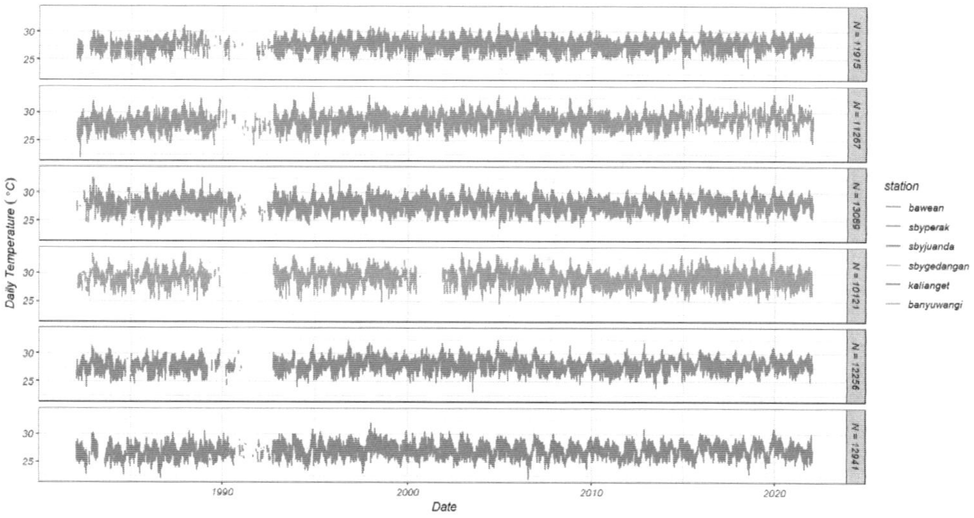

Figure 1. Line plots of daily temperature data on the six stations during 1982 to 2021. N indicates the number of available data points.

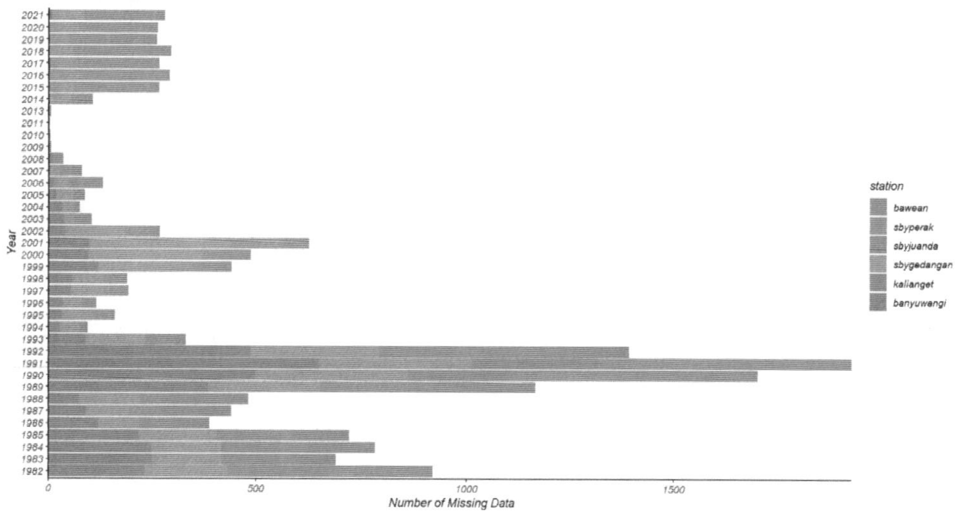

Figure 2. Distribution of missing data

2.3. Autocorrelation detection

Detection of autocorrelation is important for data modelling. The presence of significant autocorrelation implies that the observations are not independent, which violates the assumption in many statistical tests. It also shows the necessity of considering autocorrelation as a part of any model used to make predictions. Hence, we suggest that the use of ARIMA (Autoregressive Integrated Moving Average) models is appropriate for the data in which significant autocorrelation is present.

Strong, significant autocorrelations are found in the data of all stations under analysis, indicated by large correlation coefficients in numerous first lags values. To save space, Figure 3 shows the autocorrelation function plots only for the temperature data of three stations, namely Bawean, Surabaya/Perak, and Banyuwangi.

Table 2. Summary of available temperature data in 1982 – 2021

Station	Minimum (°C)	Mean (°C)	Median (°C)	Std.Dev (°C)	Maximum (°C)
Bawean	23.6	28.0	28.0	1.1	31.7
Surabaya/Perak	21.9	28.5	28.4	1.3	34.0
Surabaya/Juanda	23.4	28.0	28.0	1.2	32.6
Surabaya/Gedangan	24.6	29.2	29.2	1.3	34.0
Kalianget	23.2	28.1	28.1	1.1	32.3
Banyuwangi	22.1	27.1	27.1	1.2	31.9

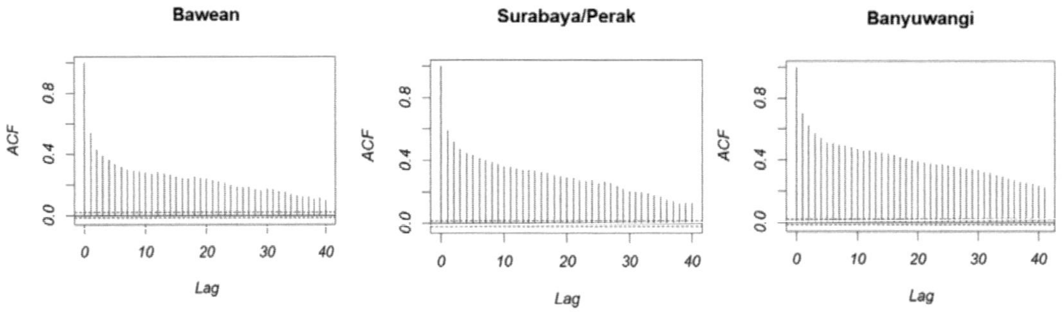

Figure 3. Autocorrelation function plots of temperature data of Bawean, Surabaya/Perak, and Banyuwangi stations.

3. Trend Analysis of Temperature Data

3.1. Mann-Kendall Test with sieve-bootstrapping

The Mann-Kendall test is applied to determine the existence of trends in temperature data in all stations. The statistic of this test is presented in equations (1), (2), and (3)

$$T_{MK} = \frac{S - sign(S)}{V(S)^{1/2}} \quad (1)$$

$$S = \sum_{i=1}^{n-1} \sum_{j=i+1}^{n} sign(y_j - y_i) \quad (2)$$

$$V(S) = \frac{1}{18}[n(n-1)(2n+5) - \sum_{m=1}^{n} t_m m(m-1)(2m+5)] \quad (3)$$

where

- y_i is the time-series datum at at timestamp i,
- y_j is the datum at at timestamp j,
- *sign* is the (positive or negative) sign of the subtraction $y_j - y_i$,
- n is the length of observation data being analysed,
- t_m is the length of ties in the observations data of rank m.

The statistic of this test, T_{MK}, follows the standard Normal distribution asymptotically [17].

The assumption used for the validity of Mann-Kendall test is that the observation data are independent. Because strong autocorrelation is present in the temperature data of all stations, the Mann-Kendall test combined with sieve bootstrapping as proposed in [17] is preferred. The sieve bootstrapping concept consists of pre-whitening the observed data that are correlated temporally. The pre-whitening is conducted by using an Auto Regressive filter, AR(p(n)), of which the order, p(n), has possibility to increase when the sample size n is increasing. After the pre-whitening, resampling is conducted on the resulted residuals so that newly constructed time series data are obtained. Then, simulation is conducted on the resampled data with a noise generated from the resampled residuals [17]. In this study, the Mann-Kendall test combined with sieve bootstrapping that is implemented in **funtimes** package in R [18] is used.

Table 3. Statistics and p-values of Sieve-bootstrap Mann-Kendall's test applied to temperature data of the stations

Station	Number of Data (n)	Years Covered	Mann-Kendall Test		T-test	
			Statistic (T_{MK})	p-value	T-value	p-value
Bawean	11915	1982 to 2021	**0.107**	< 0.001	14.201	0.009
Surabaya/ Perak	11267	1982 to 2021	**0.158**	< 0.001	23.780	< 0.001
Surabaya/ Juanda	13089	1982 to 2021	0.056	0.090	9.416	0.304
Surabaya/ Gedangan	10121	1992 to 2021	**– 0.108**	< 0.001	–13.766	0.014
Kalianget	12256	1982–2021	**0.072**	0.016	10.395	0.122
Banyuwangi	12941	1982–2021	**0.066**	0.042	15.389	0.060

The results of the Sieve-bootstrap Mann-Kendall test show that statistically significant positive trends are found in the temperature data of four stations, namely Bawean, Surabaya/Perak, Kalianget, and Banyuwangi. On the other hand, a statistically significant negative trend is found in the temperature data of Surabaya/Gedangan station. A positive trend is also found in the temperature data of Surabaya/Juanda station, but the trend is not statistically significant at 0.05 significance level. The results of t-tests show that statistically significant positive trends are found in the temperature data of two stations, namely Bawean and Surabaya/Perak. On the other hand, a statistically significant negative trend is found in the temperature data of Surabaya/Gedangan station, similar to the result of Mann-Kendall test for this station.

These results imply that in the last 40 years, the temperature has been increasing in Bawean and Surabaya/Perak areas. The significance of t-test statistics for these areas show that the trends are linear, implying relatively constant rate of temperature increase.

3.2. Sen's slope analysis

To estimate the rate of temperature increase, Sen's slope analysis is applied to the data in these two stations by means of **trend** package in R [19]. The slope is calculated as in equation (4):

$$d_m = \frac{y_k - y_j}{k - j} \quad (4)$$

for ($1 <= j < k <= n$).

in which d denotes the slope, y represents the data point, n represents the length of the observation data, and each of m, k, j is an index. The Sen's slope is computed as the median of all d_m [17,19].

The results, as presented in Table 4, show that the rate of temperature increase is 4.01×10^{-5} degree Celsius per day in Bawean area and 8.62×10^{-5} degree Celsius per day in Surabaya/Perak area. Both slopes are statistically significant on the 0.05 significance level.

Table 4. Sen's slope and confidence interval of the temperature data in three stations

Station	Sen's slope (° C /day)	z-score	95% CI (° C /day)
Bawean	4.01×10^{-5}	14.467	$[3.46 \times 10^{-5}, 4.56 \times 10^{-5}]$
Surabaya/ Perak	8.62×10^{-5}	22.651	$[7.88 \times 10^{-5}, 9.35 \times 10^{-5}]$
Surabaya/ Gedangan	-5.96×10^{-5}	-13.590	$[-6.82 \times 10^{-5}, -5.10 \times 10^{-5}]$

On the other hand, the significant negative linear trend in Surabaya/Gedangan implies that in the last 30 years, the temperature in this area has been decreasing. Based on the Sen's slope analysis, the rate of decrease is -5.96×10^{-5} degree Celsius per day. This is unexpected because other stations in Surabaya show positive trends. A possible explanation is that this station is located nearer to sea than other stations in Surabaya.

4. Analysis of Forecasting Models

The exponential smoothing and ARIMA are two models that are highly popular in time series forecasting and analysis. The exponential smoothing models are generated on the basis of the trend and seasonality found in the time-series data. ARIMA models, on the other hand, are mainly generated based on the autocorrelations in the data [20, 21]. In this study, an exponential smoothing model and an ARIMA model are built for each station. The data are split into two parts, namely training data and testing data. The training data are the temperature data before January 1, 2021. The test data are the temperature data beginning from January 1, 2021 up to December 31, 2021.

4.1. Holt's Exponential Smoothing

This method is considered because it is suitable for time-series data that have a clear trend [21]. This method involves three equations, namely the forecast equation, the level equation, and the trend equation, represented in equations (5), (6), and (7), respectively:

$$\hat{y}_{t+h|t} = \lambda_t + hb_t \tag{5}$$

$$\lambda_t = ay_t + (1-a)(\lambda_t - 1 + b_t - 1) \tag{6}$$

$$b_t = c(\lambda_t - \lambda_t - 1) + (1-c)b_{t-1} \tag{7}$$

where λ_t denotes an estimate of the series level at time *t*, b_t denotes an estimate of the time-series slope at time *t*, *a* denotes the smoothing parameter for the level, and *c* denotes the smoothing parameter for the trend. Each of *a* and *c* should have the value not less than 0 and not more than 1. In this study, the exponential smoothing method implemented in the **smooth** package in R [22] is used. The best fit parameters for each station are presented in Table 5.

4.2 Autoregressive Integrated Moving Average (ARIMA) models

There are three components of ARIMA, namely: the autoregressive (AR) components, the moving average models (MA) ones, and the integrated (I) ones. A general expression of an ARIMA model is ARIMA (p, d, q), in which p expresses the *autoregressive* order, d expresses the *integrated* order, and q expresses the *moving-average* order [20, 21]. An autoregressive (AR) model of the order p is the one in which the current observation y_t is regressed on previous observations, represented in equation(8)

$$y_t = c_1 + b_1 y_{t-1} + b_2 y_{t-2} + \ldots + b_p y_{t-p} + \varepsilon_t \tag{8}$$

where ε_t is white noise. Eq. (9) is known as an AR(p) model, which is an autoregressive model of order p.

A moving-average (MA) model incorporates past forecast errors as presented in equation (9)

$$y_t = c_2 + \varepsilon_t + d_1\varepsilon_{t-1} + d_2\,\varepsilon_{t-2} + \ldots + d_q\varepsilon_{t-q} \tag{9}$$

where ε_t is white noise. This is called an MA(q) model, a moving average model of order q. An ARIMA model is constructed from the combination of differencing, autoregression, and a moving average model. The integration is the inverse of differencing. The entire model can be written as in equation (10) [20]:

$$y_t{}' = c_3 + b_1 y_{t-1} + \ldots + b_p y_{t-p} + \varepsilon_t + d_1\varepsilon_{t-1} + d_2\,\varepsilon_{t-2} + \ldots + d_q\varepsilon_{t-q} \tag{10}$$

where y_t' denotes the truncated series. Both lagged y_t values and lagged errors are included in the predictors on the right-hand side. This is known as an ARIMA(p,d,q) model as previously explained.

4.3 Forecasting Results

Table 5 summarizes the best fit parameters for exponential smoothing models applied to temperature data up to year 2020. Table 7 summarizes ARIMA models applied to temperature data of the stations up to year 2020. Table 6 summarizes the accuracies, represented by Mean Absolute Error values, for the forecasting of both models to 2021 daily temperature data.

Table 5. Best fit exponential smoothing models applied to temperature data up to year 2020.

Station	A	b	AIC
Bawean	0.2627	0.000	29869.93
Surabaya/ Perak	0.2702	0.0000	31919.52
Surabaya/ Juanda	0.3656	0.0000	33208.12
Surabaya/ Gedangan	0.2891	0.0000	27967.93
Kalianget	0.3571	0.0000	29092.69
Banyuwangi	0.357	0.0000	30719.74

Table 6. Mean Absolute Errors in forecasting 2021 data with Exponential Smoothing and ARIMA models

Stations	Exponential Smoothing (°C)	ARIMA (°C)
Bawean	1.23	1.02
Surabaya/Perak	1.14	1.11
Surabaya/Juanda	1.64	1.26
Surabaya/Gedangan	1.59	1.28
Kalianget	1.18	1.03
Banyuwangi	0.75	0.75

The results show that ARIMA (3,1,3) is able to forecast the temperature data for 2021 with MAE < 1.30 degree Celsius in all stations and it outperforms the Exponential Smoothing method. However, it is worthy to note that the relatively small errors in the annual mean values do not guarantee good agreement in daily mean values.

Table 7. Best fit ARIMA models applied to temperature data of the stations up to year 2020.

Station	ARIMA(p,d,q)	lags	AIC
Bawean	(3,1,3)	AR(1) 0.3619, AR(2) 0.1719, AR(3) -0.0050, MA(1) -0.9907, MA(2) -0.0761, MA(3) 0.1283	29360.43
Surabaya/ Perak	(3,1,3)	AR(1) -0.2371, AR(2) 0.5752, AR(3) -0.0220, MA(1) -0.4073, MA(2) -0.8179, MA(3) 0.3344	31040.99
Surabaya/ Juanda	(3,1,3)	AR(1) 0.6085, AR(2) 0.1170, AR(3) -0.0643, MA(1) -1.1684, MA(2) 0.1217, MA(3) 0.1070	32596.12
Surabaya/ Gedangan	(3,1,3)	AR(1) -0.1139, AR(2) 0.6664 AR(3) -0.0874, MA(1) -0.4988, MA(2) -0.8513, MA(3) 0.4385	27624.25
Kalianget	(3,1,3)	AR(1) 0.3897, AR(2) 0.2394 AR(3) -0.0669, MA(1) -0.9440 MA(2) -0.1476, MA(3) 0.1672	28460.97
Banyuwangi	(3,1,3)	AR(1) 0.7900, AR(2) -0.2251 AR(3) 0.0211, MA(1) -1.3611 MA(2) 0.5816, MA(3) -0.1435	30151.40

The zero values of b in Table 5 shows that the exponential smoothing employed in this study failed to find the positive slope of the time-series temperature data of all stations, which has been found statistically significant in three stations as demonstrated by the results of Mann-Kendall test with the sieve bootstrapping in Table 3. A possible cause of this failure is the presence of missing values in almost all year, as shown in Figure 3. In addition, despite statistically significant, the trends are very small so the exponential smoothing cannot capture them well. On the contrary, the ARIMA models captures the serial correlation well so that their predictions are more accurate than the predictions of the exponential smoothing models.

5. Discussion

The significant positive trends in temperature found in Bawean, Surabaya/Perak, Kalianget, and Banyuwangi based on Mann-Kendall test show that the temperature rise has been occurring in different parts of this province. One of the expected impacts of this rise is the acceleration of water cycle in this region, which possibly results in lesser rain intensity and lower river discharge in the future, similar to the predicted condition for a neighbouring province discussed in [23, 24]. Coupled with the rise of population in this region, this condition poses an imminent threat to the agricultural sustainability in local scale. The local government and the local community in this region should consider applying relevant practical mitigation measures must be considered, such as rainwater harvesting, water conservation led by the community, and improving water regulation facilities, as suggested in [23, 24].

6. Conclusions and Recommendations

In this study, the past long-term temperature data of six stations in Jawa Timur province, namely Bawean, Surabaya/Perak, Surabaya/Juanda, Surabaya/Gedangan, Kalianget, and Banyuwangi stations are analysed. It is found that there are statistically significant positive trends in temperature in Bawean, Kalianget Madura, Surabaya Perak, and Banyuwangi, and statistically significant decreases in temperature in Surabaya/Juanda. Therefore, mitigation measures should be immediately implemented in order to anticipate the climate change in these areas. It is also found that ARIMA models outperformed the exponential smoothing models in all stations. By using the models built on pre-2021 temperature data to forecast the 2021 temperature data, it is found that the Mean Absolute Errors of ARIMA (3,1,3) forecast data are smaller than 1.30 degrees Celsius in all stations. It shows that incorporating serial errors in the data alone with no external variable can result in relatively good prediction for annual mean temperature data. We recommend future studies to aim at finding the relationship between temperature data and other climate variables, such as the rainfalls, to help comprehend the dynamics of climate change in this region. We also recommend to incorporate various practical considerations (such as seasons and holidays) in the forecasting models.

Acknowledgement

The authors are grateful to the reviewers of the ICSEEGT 2022 for their valuable comments, and to Aditya Wiratama for his correction to the draft of this article.

References

[1] IPCC, 2022: *Climate Change 2022: Impacts, Adaptation, and Vulnerability.* Contribution of Working Group II to the Sixth Assessment Report of the Intergovernmental Panel on Climate Change [H.-O. Pörtner, D.C. Roberts, M. Tignor, E.S. Poloczanska, K. Mintenbeck, A. Alegría, M. Craig, S. Langsdorf, S. Löschke, V. Möller, A. Okem, B. Rama (eds.)]. Cambridge University Press. In Press.

[2] S. M. L. Basile, J. A. Tognetti, M. L. Gandini, and W. J. Rogers, "Climate change in the temperature and precipitation at two contrasting sites of the Argentinean wheat region," *Theor. Appl. Climatol.*, 2022, doi: 10.1007/s00704-022-03936-6.

[3] A. Mondal, D. Khare, and S. Kundu, "Spatial and temporal analysis of rainfall and temperature trend of India," *Theor. Appl. Climatol.*, vol. 122, no. 1–2, pp. 143–158, 2015, doi: 10.1007/s00704-014-1283-z.

[4] K. Neha, "Trend Detection in Annual Temperature & Precipitation using the Mann Kendall Test – A Case Study to Assess Climate Change on Select States in the Northeastern United States," *Mausam*, vol. 66, no. 1, pp. 1–6, 2015, [Online]. Available: http://repository.upenn.edu/mes_capstones/47.

[5] S. Prasetyo, U. Hidayat, Y. D. Haryanto, and N. F. Riama, "Variasi dan Trend Suhu Udara Permukaan di Pulau Jawa Tahun 1990-2019," *J. Geogr. Media Media Inf. Pengemb. dan Profesi Kegeografian*, vol. 18, no. 1, pp. 60–68, 2021, doi: 10.15294/jg.v18i1.27622.

[6] B. P. Syahputra, D. F. Sidik, H. Akbar, M. Rivaldo, and N. Sagita, "Perbandingan Uji Homogenitas Periode 30 Tahun Data Suhu Rata-Rata dan Kelembaban Udara sebagai

Representasi Analisis Trend Iklim," National Conference of Mathematics and Science Universitas PGRI Banyuwangi, 2022.

[7] M. N. Habibie and T. A. Nuraini, "Karakteristik Dan Tren Perubahan Suhu Permukaan Laut Di Indonesia Periode 1982-2009," *J. Meteorol. dan Geofis.*, vol. 15, no. 1, pp. 37–49, 2014, doi: 10.31172/jmg.v15i1.171.

[8] Company Profile of PTPN XII, https://ptpn12.com/2019/07/09/profil/

[9] BPS-Statistics of Indonesia. 2020. 2020 National Census: 2020 Indonesia Population based on Region, Age Group, and Sex (*in Bahasa Indonesia*) https://sensus.bps.go.id/topik/tabular/sp2020/86/175748/0 (accessed May 10, 2022).

[10] BPS-Statistics of Indonesia. 2010. 2010 National Census: 2010 Indonesia Population based on Region, Counting Documents, and Sex (*in Bahasa Indonesia*) https://sp2010.bps.go.id/index.php/site/tabel?tid=337&wid=0 (accessed May 10, 2022).

[11] National Climatic Data Center NOAA US Department of Commerce, "Gobal Summary of the Day NOAA," 2021. https://www.ncei.noaa.gov/data/global-summary-of-the-day/access/ (accessed Apr. 20, 2022).

[12] R Core Team, "R: A language and environment for statistical computing. R Foundation for Statistical Computing," *Vienna, Austria*, 2022. https://www.r-project.org/. (accessed Apr. 20, 2022

[13] RStudio Team, "RStudio: Integrated Development for R," *PBC, Boston, MA*, 2020. http://www.rstudio.com/ (accessed Apr. 20, 2022).

[14] S. Chamberlain, "rnoaa: 'NOAA' Weather Data from R," 2021. https://cran.r-project.org/package=rnoaa (accessed Apr. 20, 2022).

[15] BPS-Statistics of Jawa Timur Province, *Jawa Timur Province in Figures 2022*. Surabaya: BPS-Statistics of Jawa Timur Province, 2022.

[16] M. F. Aziri and A. Wahyudi, "Become Boyan: The Arrival and Development of Baweanese in Singapore in the 19-20th Century," *J. Marit. Stud. Natl. Integr.*, vol. 3, no. 2, p. 54, 2019, doi: 10.14710/jmsni.v3i2.6082.

[17] Noguchi, K., Gel, Y. R., & Duguay, C. R. (2011). Bootstrap-based tests for trends in hydrological time series, with application to ice phenology data. *Journal of Hydrology*, *410*(3–4), 150–161. https://doi.org/10.1016/j.jhydrol.2011.09.008

[18] Lyubchich,V. and Gel, Y.R. (2022). funtimes: Functions for Time Series Analysis. R package version 8.2. https://CRAN.R-project.org/package=funtimes

[19] Thorsten Pohlert (2020). trend: Non-Parametric Trend Tests and Change-Point Detection. R package version 1.1.4. https://CRAN.R-project.org/package=trend

[20] Rahayu, W. S., Juwono, P. T., & Soetopo, W. (2019). Analisis Prediksi Debit Sungai Amprong Dengan Model Arima (Autoregressive Integrated Moving Average) Sebagai Dasar Penyusunan Pola Tata Tanam. *Jurnal Teknik Pengairan*, *10*(2), 110–119. https://doi.org/10.21776/ub.pengairan.2019.010.02.04

[21] Hyndman, R.J., & Athanasopoulos, G. (2021) *Forecasting: principles and practice*, 3rd edition, OTexts: Melbourne, Australia. https://otexts.com/fpp3 . Accessed on 20 April 2022.

[22] Ivan Svetunkov (2022). smooth: Forecasting Using State Space Models. R package version 3.1.6. https://CRAN.R-project.org/package=smooth

[23] Kuntoro, A. A., Cahyono, M., & Soentoro, E. A. (2018). Land cover and climate change impact on river discharge: Case study of upper citarum river basin. *Journal of Engineering and Technological Sciences*, *50*(3), 364–381. https://doi.org/10.5614/j.eng.technol.sci.2018.50.3.4

[24] Yoshida, K., Azechi, I., Hariya, R., Tanaka, K., Noda, K., Oki, K., Hongo, C., Honma, K., Maki, M., & Shirakawa, H. (2013). Future Water Availability in the Asian Monsoon Region : A Case Study in Indonesia. *Journal of Developments in Sustainable Agriculture*, *8*(1), 25–31. https://doi.org/10.11178/jdsa.8.25

Emission performance assessment of in-use diesel-driven cars

Abhinav Pandey[1], Rajeev Kumar Mishra[2], Govind Pandey[3]

[1]Department of Environmental Engineering, Delhi Technological University, India
[2]Department of Environmental Engineering, Delhi Technological University, India
[3]Civil Engineering Department, Madan Mohan Malaviya University of Technology, India)

Email: navabhigkp@gmail.com

Abstract. The overall fleet of light-duty vehicles in most of the developing countries is characterized by various manufacturers of diesel-driven passenger cars. There is a need of a thorough assessment of the actual conformance levels of such cars towards in-use emission norms, particularly the diesel-driven ones, in consideration of their high proportion in the overall privately-owned passenger cars fleet. In this context, the study conducted on 400 number of light-duty diesel cars reported that the compliance levels for BS (Bharat Stage) III diesel cars are better than that for BS IV, although not varying significantly (90 % in case of former vs. 80 %). Based on the findings, there is a need to include vehicle-related parameters during emission certification, such as, age, mileage and maintenance records along with emission measurements while deciding upon phasing-out of vehicles rather only based upon age (10 years of allowable registration life for diesel cars in the state of Delhi). Based on the results, the inclusion of mass emission-based testing in the overall transport emission assessment and reduction policy of the developing countries is highly recommended.

Keywords: Emission performance, In-use emission norms, Inspection/Maintenance program, Maintenance category, Tailpipe emission, Environmental policy.

1. Introduction

The exponential and advertent surge in the motor vehicle density (number of vehicles per thousand of human population) in most of the urban settlements worldwide looks to have surpassed the expansion rate of the cities itself. Such expansion is more manifest in the urban areas of the developing countries which are often marked by the relatively faster growth in few cities compared to the remaining parts. Although, there has been an apparent decline in the sales of diesel cars considering the preceding decade, a substantial chunk of the 'in-use vehicles' is still characterized by such cars (Pandey et al. 2022a). Most of the times, this high proportion of diesel cars is aggregated in the megacities (including suburbs) in contrast to several other urban areas letting such megacities emit a larger fraction of air pollutants at global scale (Lawrence et al. 2007; Butler 2013; Pandey et al. 2022a). The worldwide passenger car (PC) sales have witnessed a moderate-to-very high growth of 21.16 % (in 2019 compared to 2010) and 17.97 % (in 2019 against that in 2005) while African nations recorded the lowest sales volume. Compared to a total of 45.25 million passenger cars sold globally in 2005, 2019 witnessed the numbers as 90.42 showing almost doubling of numbers in only 14 years' span (OICA 2019; Pandey et al. 2022b). The infrastructure development, unfortunately, has not been able to equalize with this exponentially rising degree of motorization, aggravating the problem of poor urban air quality (Wang et al. 2019; Jaikumar et al. 2017), unplanned growth and changing meteorology (Bai et al. 2017).

Until recently, driven by their advantages like better thermal efficacy, higher fuel mileage, and relatively higher notch of consistency, higher magnitude of torque compared to other engines, the diesel engine remained a prime choice. On the other hand, as they also, tend to emit harmful exhaust emissions, such as CO, unburned hydrocarbons HC, NO_X, SO_2 and smoke or PM, more or less similar to the other extant engine type. Further, as the NO_X and SE (smoke) are relatively higher in tailpipe exhaust from the diesel engines compared to petrol engines, their environmental performance has been subject to strict policy actions as well (Karthikeyan and Prathima 2016; Sharma and Marechal, 2020; Pandey et al. 2022a).

According to the global health report, over 4.2 million deaths annually in developing countries alone, *inter-alia* several cases of pulmonary ailments are being attributable to the deteriorating urban ambient air quality (WHO 2019). Moreover, both long-term and short-term (chronic and acute in other terms) exposures to vehicular exhaust emission (VEE) has been triggering various epidemiological consequences, ranging from cardiovascular disease, waning pulmonary function, oncological cases and even mortality in extreme instances (Ogunseye et al. 2018; Rice et al. 2015; Khreis et al. 2016). Various researches in the past have evidenced vehicular emission (VE) to be contributing to approximately 70–80% of air pollution in mega cities of developing nations. Such high apportionment is being credited not only to the increasing sales of passenger cars, but also to the presence of significantly large number of older cars coupled with poor maintenance history, cars having clocked higher mileage, inadequate access and circulation infrastructure and inferior fuel quality (Singh et al. 2007; Pandey et al. 2016; Pandey el al. 2022a, b). Among the criteria (air) pollutants, CO is the major pollutant emanating from the transport sector and contributing to approximately 90% of overall emissions with Hydrocarbons following closely. Further, the vehicular emissions are known to contribute to the small air particles such as black carbon (BC), nitrogen oxides (NOx), and particulate matter (PM_{10}, $PM_{2.5}$ and PM_1) as prominent air pollutants (Rao et al. 2018).

As part of overall environmental policy, the I/M (Inspection and Maintenance) program in India mandates all the in-use diesel-driven passenger cars to acquire and possess the Pollution Under Control (PUC) certificate at a fixed periodicity. Such certificate is issued in the name of the owner/registered car if it is found to comply to the relevant smoke (exhaust) emission norm conducted using the smoke testing under free acceleration mode (FAST). These emission checks must be carried out only at the government-authorized centres (PUC centres) which use approved smoke meters (CPCB 2010). Previous relevant studies have pointed towards the need to explore and characterize the emission quality of vehicles with a fairly larger and heterogeneous dataset with more model years, and in the context of more parameters concerning vehicles, such as, age, mileage clocked, maintenance history and records, in-use emission norms and various engine-related traits as the key ones (Pandey et al. 2016).

Governments, policymakers, and managers have been affecting and trying to strengthen various policy measures to improve urban air quality by maintaining ambient air pollution levels below the specified national and international standards. At-the-source control of vehicular pollution is among the most promising practices for pollution abatement, but this approach requires a very stringent administration system, policy revisits and re-strengthening, public awareness and support, and monitoring for implementation (Gulia et al. 2020).

Except US, EU and other developed countries plus China and Brazil, the emission certification policies for in-use passenger cars in most developing countries are struggling to meet the overall vehicular emission reduction goals. There is lack of a comprehensive exploratory analysis of exhaust emission compliance of the diesel-driven light-duty passenger cars to the in-use emission norms required covering a large and heterogeneous make-wise dataset. Therefore, in the present study, 400 numbers of private registration type diesel cars appearing at PUC (Pollution Under Control) centres for emission recertification, were assessed for their tailpipe smoke emission (SE) in HSU (Hartridge Scale Unit) terms in a collaborated testing program. The testing criteria was in line with CVMR (Central Motor Vehicle Rules) prescribed by the Ministry of Road Transport and Highways (MoRTH), Government of

India. The protocol, at present, requires the diesel vehicles to be tested under stationary and no-load conditions and the PUC certificate is issued in case of conformance to the in-use emission limits.

2. Materials and methods

2.1. Vehicular emission testing

The vehicle testing schedule covered 400 numbers of diesel-driven light-duty passenger cars. The cars appearing for emission compliance certification at respective PUC (Pollution-Under-Control) centres in the NCT (National Capital Territory) of Delhi, India were checked for smoke emission from the tailpipe in FAST mode. The compliance level of the diesel cars was checked against respective in-use norms applicable in India **(Table 1)**. The analysis was carried out for three dataset scenarios – whole, top 5 makes and top 13 models.

Table 1. Sample characteristics at a glance

Emission norm	**HSU @ FAST**
BS (Bharat Stage) IV	50
BS (Bharat Stage) III	65

2.2. Instrumental set-up and data analysis tool

A smoke meter of the Ozone make (model no. Oz-Gas-02) was deployed to measure smoke emitted by the diesel vehicles (in HSU terms) under the FAST mode from the tailpipe of the cars. Further, the SPSS package (version 23) was used in the present study for the emission data compilation, analysis and processing along with the MS-Excel program as a support (input) tool.

3. Results and discussion

The status of compliance of diesel cars for different dataset scenarios is presented in Figure **1**, followed by **Fig. 2** and **Fig. 3** depicting the same for make and model-wise status, respectively.

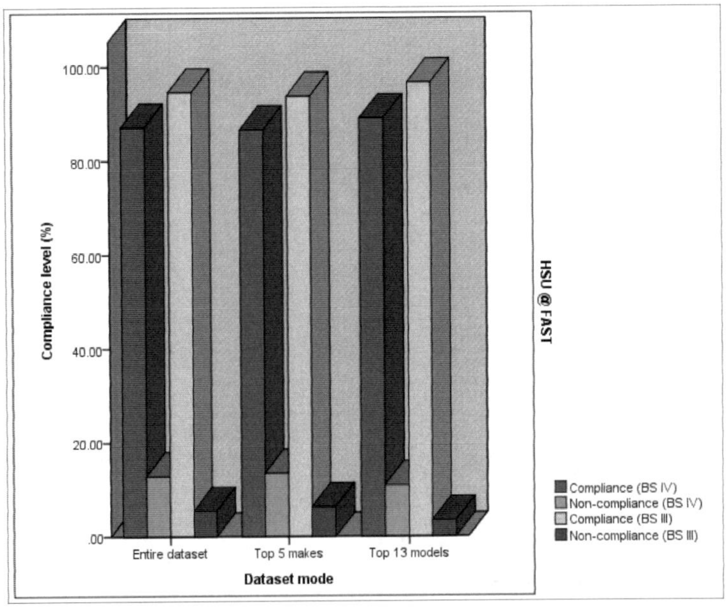

Figure 1. Emission compliance levels for diesel-driven cars (all datasets)

As evidenced, all the datasets reveal a very high level of compliance to both BS IV and BS III norms for HSU in the FAST mode of testing. As the BS III benchmark is higher than that of BS IV (65 HSU against 50 HSU), compliance of diesel cars towards BS III is significantly and logically so, higher in all dataset conditions.

In terms of makes, the MML has lowest degree of compliance amongst the 5 makes closely followed by TKMPL whereas HMIL depicts the highest degree of compliance (over 90%), Figure 2.

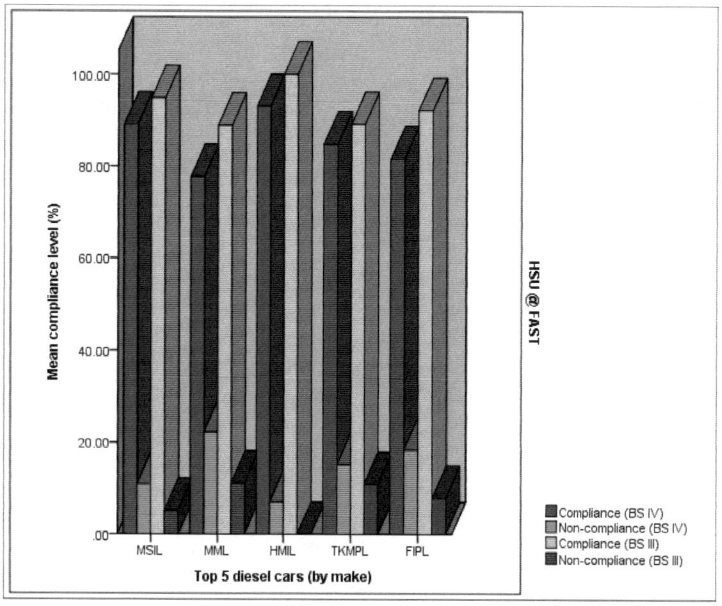

Figure 2. Emission compliance levels for diesel-driven cars (top 5 makes)

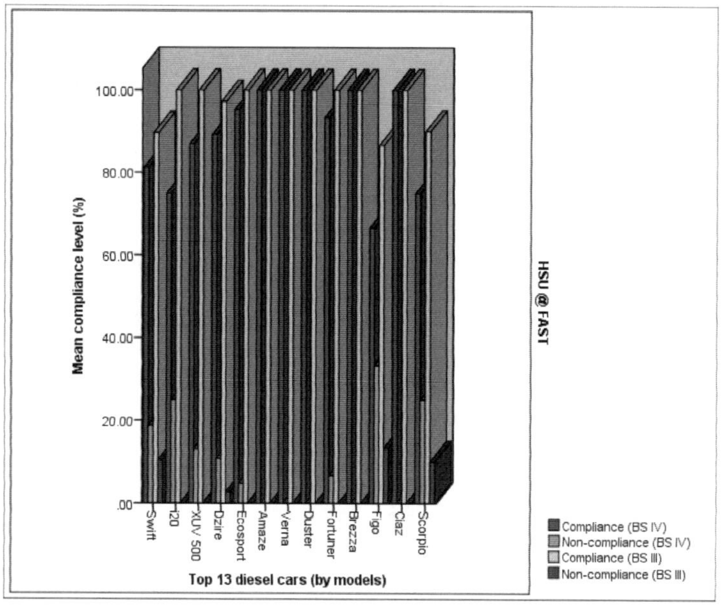

Figure 3. Emission compliance levels for diesel-driven cars (top 13 models)

The top 13 models' dataset scenario fares better than other datasets in terms of non-compliance to both BS III and BS IV norms returning the lowest values. Amaze, Verna, Duster, Brezza and Ciaz depict the highest degree of compliances for both BS IV and BS II emission norms of HSU under FAST showing almost 100% compliance. Figo leads the list of lowest levels of compliance followed by Scorpio, i20 and Swift models (Figure 3).

4. Conclusion

The present study is an attempt to investigate the compliance levels of smoke emission from 400 diesel-driven light-duty passenger cars towards in in-use emission norm. The cars tested were registered in the different regional transport offices (RTOs) across the state of Delhi, India. The study reported that the diesel cars reveal a very high level of compliance to both BS IV and BS III norms for HSU in FAST mode of testing. While make-wise, HMIL performs the best towards both the emission standards, followed by MSIL and TKMPL, MML and FIPL have higher percentage of non-compliance, especially for BS IV. Amaze, Verna, Duster, Brezza and Ciaz depict the highest degree of compliances for both BS IV and BS II emission norms of HSU under FAST showing almost 100% compliance. In model terms, Figo leads the list of lowest levels of compliance followed by Scorpio, i20 and Swift models. Although such compliance levels seem to be encouraging for the car manufacturers in justification of the fact that the improved vehicle and engine technologies have contributed to the less-emitting engines, the increasing ambient levels of both gaseous and particulate pollutants in Delhi and in other major economies worldwide present a contrasting picture. This issue or the gap needs to be dealt by the future studies in the concurrent area. Nevertheless, the policymakers should take into account the more realistic real-time driving tests to correlate the results reported using the in-situ laboratory-based emission tests. It is emphasized that in view of the introduction of and insistence over the stricter emission standards worldwide (mostly for the new vehicles), the future research works need to be carried out with vehicle model or fuel-based emission datasets. Such research works also need to keep in mind the real-world driving conditions and emission generation thereof.

MSIL – Maruti Suzuki India Limited; MML – Mahindra Motors Limited; HMIL – Hyundai Motors India Limited; TKMPL – Toyota Kirloskar Motors Private Limited; FIPL – Ford India Private Limited; NOx – Oxides of Nitrogen; SO_2 – Sulphur Dioxide; SE – Smoke Emission; PM – Particulate Matter; CPCB – Central Pollution Control Board; EU – European Union; US – United States (of America); HSU – Hartridge Smoke Unit; I/M – Inspection & Maintenance; IBM – International Business Machines; SPSS – Statistical Package for Social Sciences; VEE – Vehicular Exhaust Emission; VE – Vehicular Emission; FAST – Free Smoke Acceleration Testing

References

[1] A. Pandey, G. Pandey, and R.K. Mishra, An in situ exploratory analysis of diesel cars' emission: way forward on policy evaluation, Environmental Science and Pollution Research, 2022.
[2] M.G. Lawrence, T.M. Butler, J. Steinkamp, B.R. Gurjar, and J. Lelieveld, Regional pollution potentials of megacities and other major population centres, Atmospheric Chemistry and Physics, 7, 2007, 3969–3987.
[3] T. Butler, Urban and global effects of megacity pollution. In: Barnes I., Rudziński K. (eds) Disposal of dangerous chemicals in urban areas and mega cities, 2013 NATO Science for Peace and Security Series C: Environmental Security. Springer, Dordrecht.
[4] OICA (International Organization of Motor Vehicle Manufacturers) 2019, *Global sales statistics of passenger cars,* https://www.oica.net/category/sales-statistics/ Accessed 10 January 2020.
[5] A. Pandey, R.K. Mishra; and G. Pandey, Investigating exhaust emissions from in-use passenger cars: exploratory analysis and policy outlook, Journal of Environmental Engineering (American Society of Civil Engineering), 2022, 148(7), 04022035.

[6] J. Wang, Q. Wu, J. Liu, H. Yang, M. Yin, S. Chen, P. Guo, J. Ren, X. Luo, W. Linghu, and Q. Huang, Vehicle emission and atmospheric pollution in China: problems, progress, and prospects. PeerJ, 16,7, 2019, e6932 PMID: 31143547; PMCID: PMC6526014
[7] R. Jaikumar, S.M. Shiva Nagendra, and R. Sivanandan, Modal analysis of real-time, real world vehicular exhaust emissions under heterogeneous traffic conditions. Transportaiton Research Part D Transport and Environment, 54, 2017, 397–409
[8] X. Bai, T. McPhearson, H. Cleugh, H. Nagendra, X. Tong, T. Zhu, and Y.G. Zhu, Linking urbanization and the environment: conceptual and empirical advances. Annual Review of Environment and Resources, 42, 2017, 215–240.
[9] S. Sharma, and F. Marechal F, Carbon dioxide capture from internal combustion engine exhaust using temperature swing adsorption. Frontier in Energy Research 7, 2019, 143.
[10] S. Karthikeyan, and A. Prathima, Environmental effect of CeO_2 nano-additive on biodiesel, Energy Sources A: Recovery, Utilization and Environmental Effects, 38(24), 2016, 3673–3679.
[11] WHO (World Health Organization) 2019, *The global health observatory*, https://www.who.int/data/gho/data/themes/theme-details/GHO/air-pollution Accessed 7 January 2020.
[12] O.O. Ogunseye, G.R.E.E. Ana, D.C. Uhiara, and D.G. Shendell, Carboxyhaemoglobin levels among traders exposed to vehicular emissions in three motor parks in Ibadan, Nigeria, Journal of Environmental and Public Health, 2018, 1–8.
[13] M.B. Rice, P.L. Ljungman, E.H. Wilker, K.S. Dorans, D.R. Gold, J. Schwartz, P. Koutrakis, G.R. Washko, G.T. O'Connor, and M.A. Mittleman, Long-term exposure to traffic emissions and fine particulate matter and lung function decline in the Framingham Heart Study, American Journal of Respiratory and Critical Care Medicine, 191, 2015, 656–664.
[14] H. Khreis, K.M. Warsow, E. Verlinghieri, A. Guzman, L. Pellecuer, A. Ferreira, I. Jones, E. Heinen, D. Rojas-Rueda, N. Mueller, P. Schepers, K. Lucas, and M. Nieuwenhuijsen, The health impacts of traffic-related exposures in urban areas: understanding real effects, underlying driving forces and co-producing future directions, Journal of Transport and Health, 3(3), 2016, 249–267
[15] A.K. Singh, H.K. Gupta, K. Gupta, P. Singh, V.B. Gupta, and R.C. Sharma, A comparative study of air pollution in Indian cities, Bulletin of Environmental Contamination and Toxicology, 78, 2007, 411–416.
[16] A. Pandey, G. Pandey, and R.K. Mishra, Tailpipe emission from petrol driven passenger cars, Transportation Research Part D Transport and Environment, 44, 2016, 14–29.
[17] X. Rao, J. Zhong, R.D. Brook, and S. Rajagopalan, Effect of particulate matter air pollution on cardiovascular oxidative stress pathways, Antioxidants & Redox Signaling, 28(9), 2018, 797–818.
[18] CPCB (Central Pollution Control Board), 2010, Status of the vehicular pollution control programme in India. Central Pollution Control Board, Ministry of Environment, Forests and Climate Change, Delhi.
[19] S. Gulia, S.S. Nagendra, J. Barnes, and M. Khare, Urban local air quality management framework for non-attainment areas in Indian cities, Science of the Total Environment, 619, 2018, 1308–1318.
[20] S. Caserini, C. Pastorello, P. Gaifami, and L. Ntziachristos, Impact of the dropping activity with vehicle age on air pollutant emissions, Atmospheric Pollution Research, 4, 2013, 282–289.

Acknowledgements
The authors would like to express their thanks to the PUC centres in Delhi and the Advance Air and Acoustics Research Lab, Delhi Technological University (DTU) for their support in the emission measurement program, data collection and processing thereafter.

Environmental Impact Assessment of Maize Production in Northern India

Rohit Kumar[1], Arvind Bhardwaj[1], Lakhwinder Pal Singh[1], and Gurraj Singh[1]

[1]Department of Industrial and Production Engineering, Dr BR Ambedkar National Institute of Technology Jalandhar, Punjab, India

Email: rohitk.ip.19@nitj.ac.in

Abstract. Agricultural production has been recognized as a one of the major contributors to environmental degradation and a source of several forms of pollution that have far-reaching consequences for ecosystems, water supplies, soil quality, air quality, and human health. Maize cultivation has also placed a major environmental cost, particularly when considering resource utilization and the environmental impacts of fertilizer, insecticides, and electricity production. With the goal of measuring the impacts and identifying the major environment hotspot, a life cycle assessment was performed using the cradle-to-farm-gate approach to evaluate the environmental effects of maize cultivation. The environmental impact of the inputs and outputs studied by life cycle assessment was estimated using the Sima Pro version 9.0 and the CML 2 baseline 2000 technique. According to the findings, the fertilization phase had the most detrimental influence on the ecosystem, followed by the harvesting period. The current research makes a number of recommendations for agricultural practices to be used in conjunction with maize production, based on agri-environmental indicator scores and the correlations between them for eco-friendly maize production.

Keywords: Impact Assessment; Life Cycle Assessment; Sustainable Agriculture Production; Maize production; CML 2 baseline 2000

1. Introduction

Agriculture is a thriving area that offers a vast variety of products that are beneficial to society. India produces 7.39% of the world's total agricultural output, making it the second largest agricultural economy after China's. About half of India's population relies on farming as their main means of living, making it the country's most important industry. In addition, the agricultural, forest, and fisheries sector's estimated FY20 Gross Value Added was Rs. 19.48 lakh crore, or $276.37 billion. [1]. However, as a result of recent technological advancements and population expansion, agriculture is increasingly under pressure to fulfill society's rising need for feed, food, fiber, and bioenergy [2], [3]. In light of increasing land scarcity, the FAO (food and agriculture organization) of the United Nations states that agricultural production intensification is the most viable option [4]. Furthermore, agriculture yields have also grown as a consequence of greater usage of synthetic agrochemicals (for example, fertilizers and pesticides), additional field-work methods (for example, irrigation), and mechanization in recent decades [2], [5], [6].

However, the modern practice of growing, plant genetics, inorganic chemicals like pesticides and fertilizers, and advancements in technology have many unfavorable repercussions on the natural environment [7]. Moreover, Agriculture production needs massive quantities of capital such as water, fossil fuels and agriculture chemicals, whose application destroys the ecosystem in various ways [8]. In addition, the overuse of agricultural chemicals contributes to the release of greenhouse gases and the contamination of water sources [8]. In India, freshwater water bodies are being polluted through drainage and leaching of nitrates from soil and mismanagement of nutrient application [9]. As per most recent evaluation by the National Institution for Transforming India (NITI), nearly 70% of the nation's fresh water supply, whether it is in the ground or on the surface, is polluted. The extent of these environmental effects and the utilization of energy in various ways largely depend on agricultural management activities, soil properties and agricultural ecosystem conditions [10]. In addition, natural resources are also heavily reduced by agricultural production, which imposes a strong environmental burden on the planet. In reaction to the growing environmental awareness, decision makers are likely to steer future advancements toward more environment friendly agricultural practices. Therefore, it is crucial to conduct an in-depth analysis that provides the environmental profile of the agricultural production system [11].

1.1 Life Cycle Assessment

Life-cycle assessment (LCA) across the agriculture production system is one of the most useful methodologies for quantitatively comparing the environmental performance of various agriculture strategies among the plethora of measuring approaches. Among the plethora of measurement approaches, Among the many methods available for assessing the environmental performance of different agricultural strategies quantitatively, LCA across agriculture production systems is widely regarded as one of the most illuminating [12]. LCA has typically been used for the study of industrial production processes [13], but has been updated to determine the environmental effects of agriculture over the past 15 years. To account the environmental effect involved with agriculture production has a number of different challenges, that included a diverse range of farming categories as well as challenges related to field (e.g., different geographical conditions, crop rotation, and seasonal fluctuations). LCA is a comprehensive technique for assessing the environmental implications of a product across its full life cycle: from raw material extraction through processing, manufacturing, distribution, use, and disposal or recycling (Figure 1). It is an effective tool for quantifying and analyzing the potential environmental impact of various agricultural production processes [14]. LCA of agriculture products allows for better understanding of the impact of food on the environment or ecosystem [15], [16]. Life cycle evaluation also allows farmers to maximize productivity by applying alternatives proposed by the LCA model by product comparison.

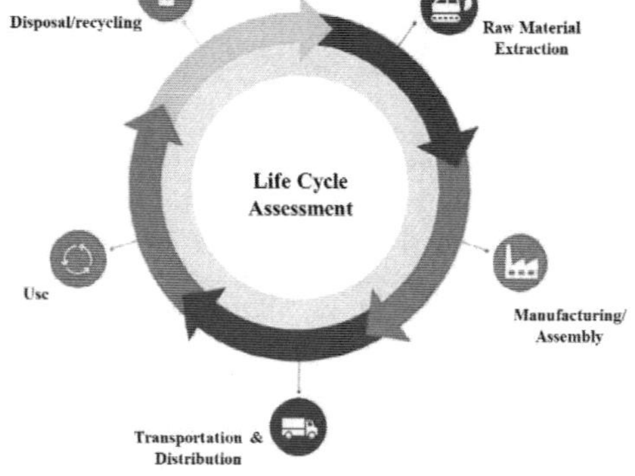

Figure 1. Schematic of product life cycle by LCA

1.2 Maize as a crop

Maize is one of the most versatile growing crops, with a wide range of adaptability under a variety of agro-climatic conditions. Because it has the highest genetic yield potential of any grain, maize is referred to as the "Queen of Cereals". It is cultivated on approximately 150 million hectares (ha) in over 160 nations, with a broad range of soil, environment, ecology, and management activities, and accounts for 36% (782 million tonnes) of global grain production. Maize contributes to roughly 9% of India's national food basket [17] . Maize is a primary raw material utilized in industrial commodities at a larger scale, including starch, oil, protein, alcoholic beverages, sweeteners, medicinal, cosmetic, film, clothing, gum, box, and paper industries, among others [18]. However, the crop has its own set of issues for society and the environment as a result of the recent increase in demand [19]. Figure 2 shows the product life cycle of maize production including various agriculture practices. This study aims to analyze the possible environmental impact indicators w.r.t to maize cultivation. In order to mitigate or make actionable statements to reduce environmental burden, it is necessary to analyze those negative environmental impacts that are damaging to society. There are a number of LCA studies available which shows the environmental burden of agriculture burden but there are very few studies has been carried on, especially maize production (cradle to farm gate) [20]–[23]. However, there is also scarcity of studies on the environmental performance of maize production, especially in the context to Indian climate conditions. Therefore, a comprehensive study that includes more information on the environmental impact of various agricultural practices is needed to establish more sustainable maize production systems. In order to fill in the gaps and give a more comprehensive evaluation of the environmental impact of maize production practices, the current research used the life cycle methodology to identify the environmental consequences of maize production.

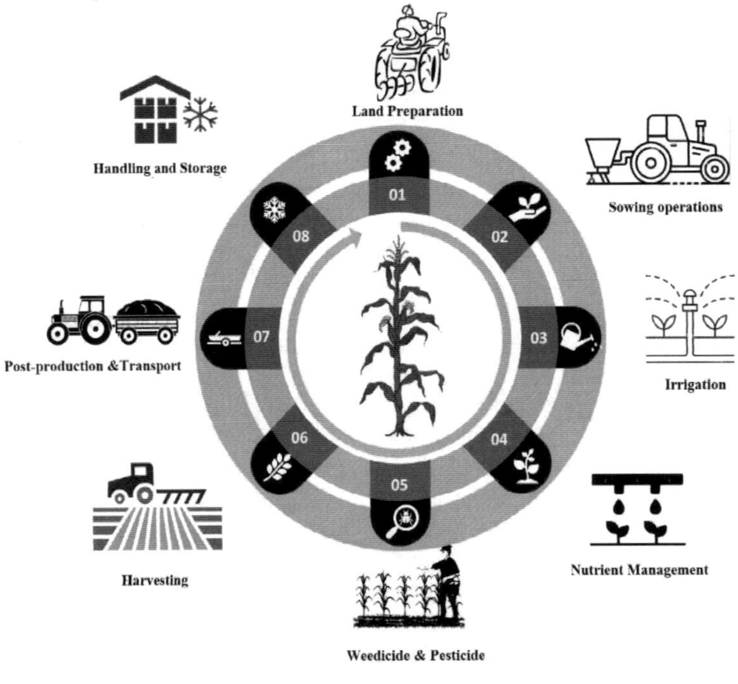

Figure 2. Product life cycle of maize production from cradle to farm gate

2. Materials & Methods

The analysis was carried out on primary data gathered from maize farming systems in the Punjab province of India, around the Jalandhar district region (as this region is ideal for maize production). A total of twelve medium-sized maize farms (less than five hectares) were chosen and surveyed for the purpose of data collection for 2021-22 crop session. Additionally, the yield of maize production per hectare was also recorded at the end of the crop cycle (Table 2). To assess the environmental impact of per hectare maize production LCA methodology was used; which included four phases: Goal and scope of study; Life cycle inventory; Impact assessment and interpretation of results (ISO 2006) [24].

2.1. Goal & Scope of the Study

The goal of this research is to evaluate and identify the major hotspot in the production chain of maize for farmland over its life cycle period. The findings of the study may be used for the development of environmentally friendly maize production practices, as well as economic and environmental analyses. Such findings can be highly assistive in policy framing by providing useful data to governments & other similar agencies. The functional unit of the current study has been taken as per hectare of maize production. Additionally, cradle to farm gate system boundary has been taken in the study which includes various agricultural practices for instance land preparation, sowing, nutrient management, weeding & pesticide, irrigation, and harvesting. The whole system boundary of the maize production system is shown in Figure 3. The findings of the study may be used to develop eco-friendly maize production practices, as well as for agriculture government policies.

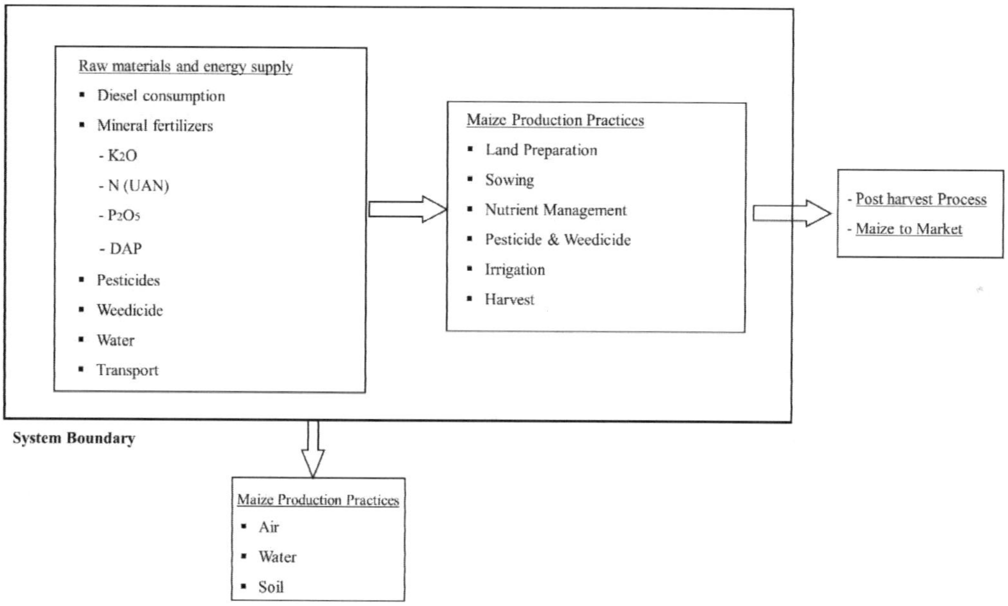

Figure 3. System boundary of one hectare maize production

2.2. Life Cycle Inventory

The environmental impacts due to production of maize not only made through from agricultural activities but also from the direct and indirect inputs that is manufacturing and transportation (fuel, electricity, fertilizers and pesticides). As a consequence, to conduct a comprehensive review, all the factors must be taken into consideration. All aspects of maize production were extensively examined in order to acquire the required data. The life cycle inventory was done using primary data gathered via a field survey and expert judgment. As indicated in Table 1, each field activity and usage of inputs was recorded by means of a resource category, weight, equipment usage time, and fuel consumption. To data consistency during the analysis, it was made to ensure that all the units were identical to acquire all forms of data.

The inventory data was gathered from twelve farms (primarily in Punjab, India), and the average of all input data was computed. For seed emergence and root growth, maize requires a well-pulverized, fine, and smooth field. For good field preparation and early season weed control, a deep tillage operation with a mould board plough, two harrowing's, and one planking is necessary. The diesel consumption during land preparation activities is taken in this study. A total of 20 kg per hectare of seed has been noted during the inventory collection of maize production. Furthermore, in the fertilization phase, a nutrient requirement of the crop is quite high and the application of fertilizers is considered essential to obtain economic and high yields. In the Nutrient management phase inventory of fertilizer and pesticide as well as crop management activities (weeding management) have been taken in this study. As per the field survey maize production needs 400–700 mm of water per growing season. Rainfall meets only a portion of the water demand during the kharif season. As a result, 1-4 irrigations are expected, which can vary based on rainfall level. All the selected farmers used electricity as a resource to run the water pumps. So, an average electricity consumption has taken in the study from all the farms. Despite of these practices, When the husk has dried and turned brown, maize can be harvested. Aside from hand-picking, a combined harvester may be used to reap quickly. Diesel consumption has been taken during the harvesting of maize. The background data was taken from the world-leading database Ecoinvent v3.7 which is accessible in SimaPro software.

Table 1. Life cycle inventory taken for one hectare maize production

Input	Unit	Total	Land Preparation	Sowing	Nutrient Management	Weedicide & Pesticide	Irrigation	Harvesting
Diesel Consumption	Litre	26	16					10
Water	m^3	1070			30	20	1020	
Urea as N	Kg	300.00			300.00			
Potassium fertilizer, as K2O	Kg	39.00			39.00			
Phosphate fertilizer, as P2O5	Kg	39.00			39.00			
Di-ammonium Phosphate	kg	350.00			350.00			
Nitrogen, as N	Kg	15.00			15.00			
Atrazine	Kg	1.25				1.25		
Pendimethalin	Kg	2.22				2.22		
Seed	Kg	20.00		20.00				
Transport	km	50.00		10	40.00			
Electricity	KWh	140.00					140.00	
OUTPUT								
Yield	Kg	9000.00						9000.00

2.4. Impact Assessment & Interpretation

For the Impact assessment methodology, SimaPro version 9.1.1 application coupled with ecoinvent database version 3.7 was used to conduct a life cycle impact assessment, and CML 2000 [24] was used to analyze the environmental impact. A CML baseline 2000 methodology has been adopted as this has enough range of impact indicators that can show a clear environmental profile with possible environmental indicators. The technique is separated into baseline and non-baseline impact categories, with the baseline being the most often utilised effect category in LCA.

4. Results & Discussion

The goals of this study were to identify the factors that might have a major influence on the LCA outcome of maize production in order to build a link between different agricultural practices that have a lower negative impact. The result shows the ten environmental impact indicators including Abiotic depletion, Acidification potential, Freshwater and marine aquatic ecotoxicity potential Eutrophication potential, Global warming potential, , Human toxicity potential, Terrestrial ecotoxicity potential, Ozone layer depletion and Photochemical oxidation potential.

Table 2. Characterization results of one hectare maize production

Impact category	Unit	Total	Land Preparation	Sowing	Nutrient Management	Weedicide & Pesticide	Irrigation	Harvesting
Abiotic depletion	kg Sb eq	15.903	1.375	0.247	10.067	0.176	1.705	2.333
Acidification	kg SO2 eq	11.050	1.052	0.256	6.769	0.089	0.834	2.048
Eutrophication	kg PO4--- eq	3.416	0.290	0.249	1.731	0.039	0.399	0.706
Global warming (GWP100)	kg CO2 eq	1974.480	158.836	44.376	1235.996	22.268	199.703	313.302
Ozone layer depletion (ODP)	kg CFC-11 eq	0.000	0.000	0.000	0.000	0.000	0.000	0.000
Human toxicity	kg 1,4-DB eq	2049.837	233.081	35.031	1078.079	42.524	91.317	569.804
Fresh water aquatic ecotoxicity	kg 1,4-DB eq	599.091	43.956	24.332	291.897	5.515	101.319	132.073
Marine aquatic ecotoxicity	kg 1,4-DB eq	1285961.672	74810.698	27915.839	763750.949	12222.392	174259.072	233002.722
Terrestrial ecotoxicity	kg 1,4-DB eq	13.887	0.425	0.219	11.094	0.063	0.712	1.375
Photochemical oxidation	kg C2H4 eq	0.559	0.066	0.010	0.270	0.006	0.031	0.175

A closer look at the data in Table 2 shows that the Marine aquatic ecotoxicity followed by Global warming potential and Human toxicity indicators stands out amongst the other 7 impact indicators. Out of these indicators, nutrient management appears to be the dominant position with more than fifty percent shares in each category. The major contribution was coming due to the production of phosphorus and potassium, two finite mineral resources in fertilizer productions that, unlike fossil fuels, have no substitutes. The high amount of synthetic fertilizer application also led to an increase in ecotoxicity (both marine and human) in the environment [25]. Despite of these a significant contribution also came from global warming which was mainly attributed to Nitrous oxide releases from synthetic fertilizer applications [26]. The influence of various agricultural techniques on the ten impact categories is widely diverse (Figure 4); for example, fertilization is responsible for about 80% of terrestrial ecotoxicity, nearly 60% of acidification and global warming potential but less than 50% for the others impact categories. In one-hectare maize production, agriculture practices that have a substantial impact on the environment are nutrient management, harvesting, irrigation, land preparation, sowing, and crop protection (weedicide & pesticide). It is also confirmed with the historical data available, that highlights the higher contribution of the nutrient management phase for other similar agricultural processes [23], [27], [28]. Aside from irrigation and nutrient management, other impact categories are directly and indirectly characterized by the usage of diesel which led to environmental load; however, it is virtually unachievable to do high-yield maize production without these agriculture practices. Nutrient management phase in maize production, owing to the use of synthetic fertilizer in production. Longo et al. 2016 also stated that the use of agricultural chemicals, as well as the use of diesel by agricultural equipment, account for a considerable portion of the impact on the environment [27].

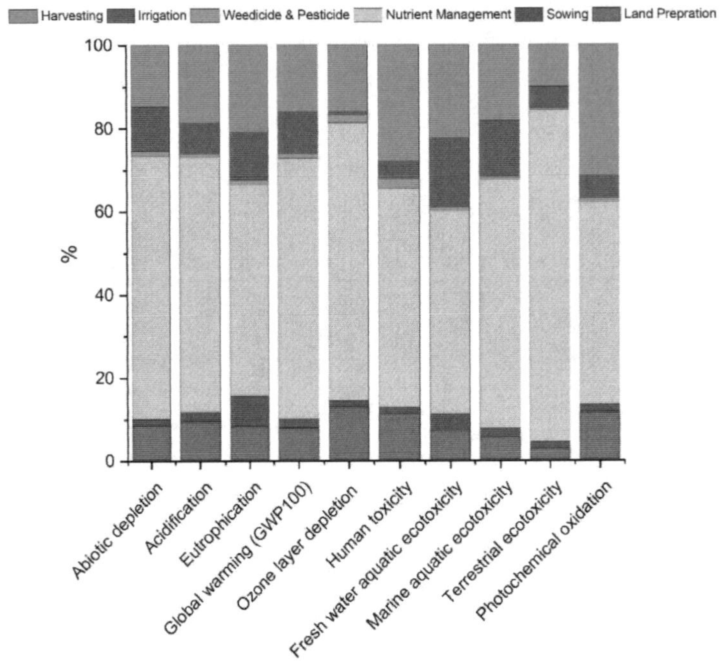

Figure 4. The distribution of the environmental impact of the maize production using CML baseline 2000 methodology

A negative environmental contribution made by irrigation practice due to high amount of freshwater consumption in order to hydrate the crop as well coal-based electricity usage which increases in resources usage. The rest of the practice's contribution is apparently squeezed in comparison to other practices due to fewer inputs from nature and techno sphere. Furthermore, figure 5 depict the cumulative point bar graphs for maize production. The normalization of the results is given in order to bring them on a common relative scale. It can be seen that Marine and freshwater ecotoxicity are a category that has higher scores among others. In these indicators, again nutrient management takes precedence position over others. Increased pesticide use resulted in prevalent nutrient pollution, which contributed to acidification, eutrophication, toxicity-related effects, and even climate change [2], [29]. Moreover, the detrimental impact of nutrient management is not only coming from the application of synthetic fertilizers but also from the production of these fertilizers [28] as well as transportation of these fertilizers from point of sale to field [28].

5. Conclusion and Recommendations

Overall, our LCA revealed obvious environmental advantages associated with switching from conventional to more sustainable agricultural practices related to maize production. In light of the fact that nutrient management is the most significant contributor to all environmental impact categories, it may be interesting to explore optimizing fertilizer use in order to minimize consumption and, therefore, reduce environmental impacts. It is necessary to set limits on the use of fertilizers and pesticides up to a particular point in order to protect the environment. It was found that nutrient management contributes more than 50 percent in almost in each impact category. So, if we limit the excessive use of fertilizer while maintaining crop yield, we may lessen the environmental impact and then focus on other procedures to correct the excessive use of natural resources and safeguard the environment for future generations. Moreover, by switching flood irrigation with modern technologies like pressured irrigation may reduce water consumption, which in turn reduces power consumption and burden on the environment.

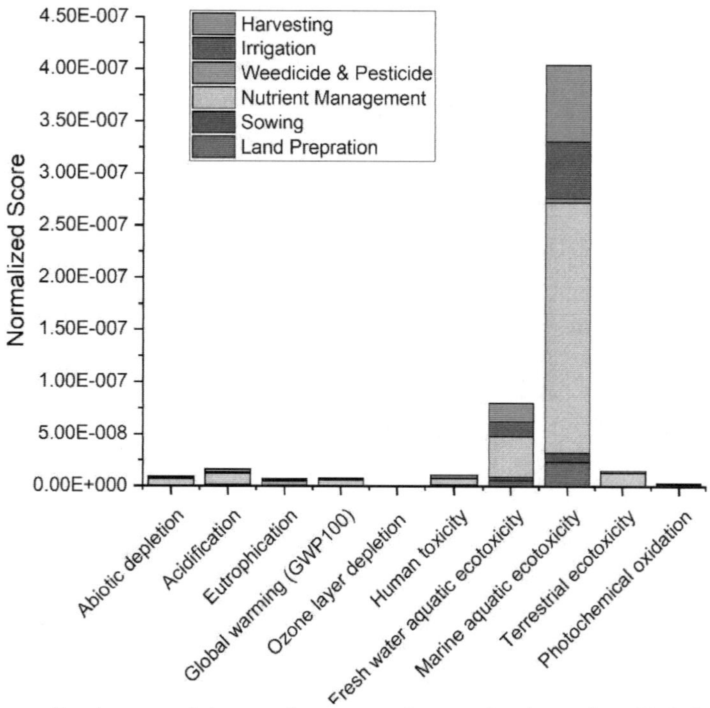

Figure 5. Normalized score of the one-hectare maize production using CML baseline 2000 methodology

References

[1] "IBEF," *Indian Brand Equity Found.*, vol. 8, no. 3, pp. 729–736, 2020, [Online]. Available: https://www.ibef.org/industry/agriculture-india.aspx

[2] J. A. Foley *et al.*, "Solutions for a cultivated planet," *Nature*, vol. 478, no. 7369, pp. 337–342, 2011, doi: 10.1038/nature10452.

[3] Kslat, "The global need for food, fibre and fuel," vol. 151 (4), p. 96, 2012.

[4] R. Pérez-Escamilla, "Food security and the 2015-2030 sustainable development goals: From human to planetary health," *Curr. Dev. Nutr.*, vol. 1, no. 7, pp. 1–8, 2017, doi: 10.3945/cdn.117.000513.

[5] J. Bürger, A. Günther, F. de Mol, and B. Gerowitt, "Analysing the influence of crop management on pesticide use intensity while controlling for external sources of variability with Linear Mixed Effects Models," *Agric. Syst.*, vol. 111, pp. 13–22, 2012, doi: https://doi.org/10.1016/j.agsy.2012.05.002.

[6] S. Huysveld *et al.*, "Resource use assessment of an agricultural system from a life cycle perspective – a dairy farm as case study," *Agric. Syst.*, vol. 135, pp. 77–89, 2015, doi: https://doi.org/10.1016/j.agsy.2014.12.008.

[7] F. Brentrup, J. Küsters, J. Lammel, and H. Kuhlmann, "Methods to estimate on-field nitrogen emissions from crop production as an input to LCA studies in the agricultural sector," *Int. J. Life Cycle Assess.*, vol. 5, no. 6, p. 349, 2000, doi: 10.1007/BF02978670.

[8] M. Shafiq and T. Rehman, "The extent of resource use inefficiencies in cotton production in Pakistan's Punjab: An application of Data Envelopment Analysis," *Agric. Econ.*, vol. 22, no. 3, pp. 321–330, 2000, doi: 10.1111/j.1574-0862.2000.tb00078.x.

[9] A. Agrawal, R. S. Pandey, and B. Sharma, "Water Pollution with Special Reference to Pesticide Contamination in India," *J. Water Resour. Prot.*, vol. 02, no. 05, pp. 432–448, 2010, doi: 10.4236/jwarp.2010.25050.

[10] A. T. M. A. Choudhury and I. R. Kennedy, "Nitrogen Fertilizer Losses from Rice Soils and Control of Environmental Pollution Problems," *Commun. Soil Sci. Plant Anal.*, vol. 36, no. 11–12, pp. 1625–1639, Jun. 2005, doi: 10.1081/CSS-200059104.

[11] M. S. Meier, F. Stoessel, N. Jungbluth, R. Juraske, C. Schader, and M. Stolze, "Environmental impacts of organic and conventional agricultural products – Are the differences captured by life cycle assessment?," *J. Environ. Manage.*, vol. 149, pp. 193–208, 2015, doi: https://doi.org/10.1016/j.jenvman.2014.10.006.

[12] M. de Vries and I. J. M. de Boer, "Comparing environmental impacts for livestock products: A review of life cycle assessments," *Livest. Sci.*, vol. 128, no. 1, pp. 1–11, 2010, doi: https://doi.org/10.1016/j.livsci.2009.11.007.

[13] E. Audsley *et al.*, "Harmonistion of environmental life cycle assessment for agriculture," *Final report. Concert. action AIR3-CT94-2028*, p. 107, 2003, [Online]. Available: http://randd.defra.gov.uk/Default.aspx?Menu=Menu&Module=More&Location=None&Completed=0&ProjectID=11442

[14] J. Martínez-Blanco *et al.*, "Compost benefits for agriculture evaluated by life cycle assessment. A review," *Agron. Sustain. Dev.*, vol. 33, no. 4, pp. 721–732, 2013, doi: 10.1007/s13593-013-0148-7.

[15] S. Clune, E. Crossin, and K. Verghese, "Systematic review of greenhouse gas emissions for different fresh food categories," *J. Clean. Prod.*, vol. 140, pp. 766–783, 2017, doi: 10.1016/j.jclepro.2016.04.082.

[16] S. Foteinis and E. Chatzisymeon, "Life cycle assessment of organic versus conventional agriculture. A case study of lettuce cultivation in Greece," *J. Clean. Prod.*, vol. 112, pp. 2462–2471, 2016, doi: https://doi.org/10.1016/j.jclepro.2015.09.075.

[17] APEDA, "Apeda products," p. 1, 2021, [Online]. Available: http://apeda.gov.in/apedawebsite/SubHead_Products/Dairy_Products.htm

[18] T. Rouf Shah, K. Prasad, and P. Kumar, "Maize—A potential source of human nutrition and health: A review," *Cogent Food Agric.*, vol. 2, no. 1, p. 1166995, Dec. 2016, doi: 10.1080/23311932.2016.1166995.

[19] J. E. Cairns *et al.*, "Chapter one - Maize Production in a Changing Climate: Impacts, Adaptation, and Mitigation Strategies," in *Advances in Agronomy*, vol. 114, D. L. B. T.-A. in A. Sparks, Ed. Academic Press, 2012, pp. 1–58. doi: https://doi.org/10.1016/B978-0-12-394275-3.00006-7.

[20] D. Dressler, A. Loewen, and M. Nelles, "Life cycle assessment of the supply and use of bioenergy: Impact of regional factors on biogas production," *Int. J. Life Cycle Assess.*, vol. 17, no. 9, pp. 1104–1115, 2012, doi: 10.1007/s11367-012-0424-9.

[21] J. Bacenetti, A. Fusi, R. Guidetti, and M. Fiala, "Life Cycle Assessment of maize cultivation for biogas production," *J. Agric. Eng.*, vol. 44, no. 2s, pp. 579–582, 2013, doi: 10.4081/jae.2013.s2.e114.

[22] L. Boone *et al.*, "Environmental life cycle assessment of grain maize production: An analysis of factors causing variability," *Sci. Total Environ.*, vol. 553, pp. 551–564, 2016, doi: https://doi.org/10.1016/j.scitotenv.2016.02.089.

[23] V. Fantin, S. Righi, I. Rondini, and P. Masoni, "Environmental assessment of wheat and maize production in an Italian farmers' cooperative," *J. Clean. Prod.*, vol. 140, pp. 631–643, 2017, doi: https://doi.org/10.1016/j.jclepro.2016.06.136.

[24] J. B. Guinée *et al.*, "Life cycle assessment: An operational guide to the ISO standards," *III Sci. Backgr.*, no. May, p. 692, 2001, [Online]. Available: http://www.gbv.de/dms/hbz/toc/ht013470560.pdf

[25] A. Agrawal, "Water Pollution with Special Reference to Pesticide Contamination in India," *J. Water Resour. Prot.*, vol. 02, pp. 432–448, Jan. 2010, doi: 10.4236/jwarp.2010.25050.

[26] I. Mukumbuta, M. Shimizu, and R. Hatano, "Mitigating global warming potential and greenhouse gas intensities by applying composted manure in cornfield: A 3-year field study in an Andosol soil," *Agric.*, vol. 7, no. 2, pp. 1–20, 2017, doi: 10.3390/agriculture7020013.

[27] S. Longo, M. Mistretta, F. Guarino, and M. Cellura, "Life Cycle Assessment of organic and conventional apple supply chains in the North of Italy," *J. Clean. Prod.*, vol. 140, pp. 654–663, 2017, doi: https://doi.org/10.1016/j.jclepro.2016.02.049.

[28] K. Hasler, S. Bröring, S. W. F. Omta, and H.-W. Olfs, "Life cycle assessment (LCA) of different fertilizer product types," *Eur. J. Agron.*, vol. 69, pp. 41–51, 2015, doi: https://doi.org/10.1016/j.eja.2015.06.001.

[29] F. Brentrup, J. Küsters, J. Lammel, P. Barraclough, and H. Kuhlmann, "Environmental impact assessment of agricultural production systems using the life cycle assessment (LCA) methodology II. The application to N fertilizer use in winter wheat production systems," *Eur. J. Agron.*, vol. 20, no. 3, pp. 265–279, 2004, doi: https://doi.org/10.1016/S1161-0301(03)00039-X.

GIS Mapping for Distribution of Ground Water Quality in Udaipur

Sangeeta Choudhary[1*], Denis Jangeed[2], Prithvi Raj Rankawat[3], Sweety Tank[4] and Nikhil Sen[5]

[1*] Associate Professor, Department of Civil Engineering, Geetanjali Institute of Technical Studies, Udaipur (Rajasthan), E-mail: sangeeta27apr@gmail.com

[2] Assistant Professor, Department of Civil Engineering, Geetanjali Institute of Technical Studies, Udaipur (Rajasthan)

[3,4,5] UG Students of B.Tech., Department of Civil Engineering, Geetanjali Institute of Technical Studies, Udaipur (Rajasthan)

Abstract. Ground water supplies drinking water and important component in many industrial processes and irrigation use. Groundwater resources have been exposed to excessive risk due to overexploitation and inadequate management, making groundwater quality assessment and suitability determination vitally important. The current study has been undertaken with an aim to analyse the ground water quality for drinking by utilizing GIS mapping and graphical presentation of various chemical constituents. Water sample values were taken from 'Ground Water Year Book 2020-21' of Central Ground Water Board, Ministry of Jal Shakti. This study has been included 22 locations in Udaipur district with 8 water quality parameters which are Cl, SO4, NO3, Ca, Mg, F, TDS and Alkalinity. The GIS environment was used to create the Water Quality Indices maps. The findings suggest that the calcium, sulphate, and chloride hazards are not a problem. About 72.7% samples of calcium, 95.50% samples of sulphate and 86.40% samples of chlorine are within acceptable limit of BIS. In case of fluoride only 18.20% samples are within the permissible limit 1.0 – 1.5 mg/l, 68.20% are less than 1 mg/l and 13.6% are greater than 1.5 mg/l. In case of nitrate only 31.8% samples are within the permissible limit. The findings show that the ground water in the studied area requires special treatment before it can be used.

1. Introduction

The most difficult difficulty is delivering safe and clean water for residential, agriculture, and industrial uses due to declining ground water resources. Monitoring of ground water quality plays important role for safe and clean water supply [1, 2, 3]. It is required to determine the water quality parameters for identifications of danger zone. Ground water is an important resource in Rajasthan. So this study represents the different zones with ground water quality distribution through GIS mapping. There are 8 water quality parameters of 22 stations in present study. Cl, SO_4, NO_3, Ca, Mg, F, TDS, and Alkalinity are the water quality parameters which includes in present study. Udaipur, the city of lakes, lies in the 24.5854^0 N latitude and 73.7125 E^0 longitudes and this study includes the nearby villages in Udaipur district. The annual rainfall of the Udaipur is 817 mm. Tables 1 show the all sample locations with their latitudes and longitudes. Geographic Information System (GIS) is a dynamic and growing field with possibilities that increase as the world becomes more connected. Advancements like crowd-sourced location data and drone technology have paved the way for incredible touch and specificity in spatial analyses, providing easier and more comprehensive solutions in areas like city planning, conservation, transportation and much more. GIS can also help address some of the most urgent challenges like flood, earthquake, covid19 etc.

2. Methodology

This study included to analyse the status of ground water quality through GIS mapping and graphical presentation [4, 5, 6]. Total 22 water samples of Udaipur district were collected from 'Ground Water Year Book 2020-21' of Central Ground Water Board, Ministry of Jal Shakti. The Central Ground

Water Board has a system in place of 1266 stations in Rajasthan. Table 1 shows all the locations and parameters for this study work. Four times a year, the water levels in these NHS are checked. Water samples for ground water quality are taken once a year in the month of May, when chemical constituent concentrations are predicted to be at their highest [7, 8, 9]. There are different colour codes in GIS mapping for showing the distribution of various chemical constituents according to the permissible range as per IS Drinking Water Specifications (IS 10500:1991). Table 2 shows the different colour codes used in GIS mapping for the current study [10].

Table 1. Values of Chemical Constituents with Locations of the Study area

S. No.	Village	Long	Lat	Cl	SO_4	NO_3	Ca	Mg	F	TDS	Alkalinity
1	Bhatewar	24.617	74.013	92	5	4	28	31.616	0.97	501	346.67
2	Bhinder	24.508	74.167	206	26	68	116	21.888	0.5	1027	699.84
3	Gadoli	24.75	73.978	752	435	1050	156	308.864	0.3	3068	230
4	Ghori Mari	24.306	73.388	149	32	60	48	36.48	0.5	533	340.16
5	Intalikhara	24.006	73.959	191	66	60	64	39	0.5	793	169.67
6	Jaswantgarh	24.8	73.468	142	24	55	60	46	1.17	670	640.16
7	Kalyanpura	24.001	73.758	177	40	125	48	51	0.75	982	230.16
8	Kanod	24.433	74.267	177	98	200	52	75	0.97	1040	670.16
9	Kathari	24.89	73.57	149	9	240	76	34	0.45	1034	629.84
10	Kheroda	24.58	74.059	149	58	9.8	12	39	3.2	923	140.16
11	Kholri	24.072	74.063	35	22	80	68	24	1.5	462	959.84
12	Luniyara	24.356	73.386	142	27	55	60	24	1	533	134.43
13	Mavli	24.783	73.983	64	12	15	48	27	0.25	332	250
14	NeechaTalab	24.075	73.433	184	10	1.8	40	27	1.5	670	340.16
15	Padawali	24.613	73.393	213	96	150	80	90	0.92	975	204.92
16	Punawali	24.878	73.429	142	34	160	120	73	0.95	930	629.51
17	Salumber	24.128	74.047	411	17.5	500	84	68	1.25	1859	259.84
18	Semari	24.065	73.85	355	78	13	24	61	2.5	1216	163.93
19	Sisarma	24.591	73.659	163	33.12	65	40	71	0.3	787	329.51
20	Som	24.171	73.343	78	36	60	64	34	0.7	514	109.67
21	Srimai kikaria	24.776	73.633	71	24	0.2	60	15	0.5	384	336.07
22	Undri	24.496	73.613	135	33.6	8.8	32	32	0.97	585	214.75

Table 2. Permissible Ranes og the chemical constituents as per BIS

S. No.	Color Codes for GIS Mapping	Green Color Code	Blue Color Code	Red Color Code
	Chemical Constituents	(Desirable limit as per BIS	In the absence of an alternative source, the permissible limit	Beyond the limits
1	Calcium	≤ 75	75 -200	> 200
2	Alkalinity	≤ 200	200-600	> 600

3	Chloride	≤ 250	-	> 250
4	Fluoride	< 1	1 -1.5	> 1.5
5	Magnesium	≤ 30	30-100	> 100
6	Nitrate	≤ 45	45-100	> 100
7	Sulphate	< 200	200-400	> 400
8	TDS	≤ 500	500-2000	> 2000

3. Result and discussions:

GIS Mapping of ground water samples are main tool for analysing the distribution of chemical constituents in the study area. Figure 2 to 5 show the distribution of various chemical constituents in GIS mapping with their permissible ranges and also graphical presentation has been included in these figures on the basis of Table 2. There are following three ranges for the distribution pattern:

i. Desirable limit as per BIS
ii. In the absence of an alternative source, the permissible limit
iii. Beyond the limits of extreme permissible

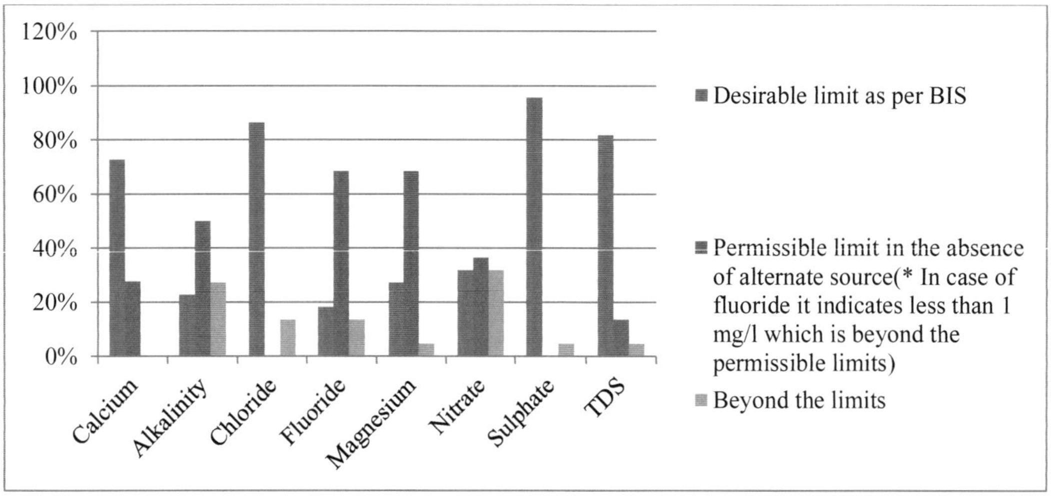

Figure 1. Distribution of chemical constituents in the study area

Alkalinity, Fluoride, Magnesium and Nitrate are exceed the permissible limits of drinking water in this study area. Ion exchange, reverse osmosis and distillation methods can be used to removed the nitrate and fluoride and alkalinity from water [11].

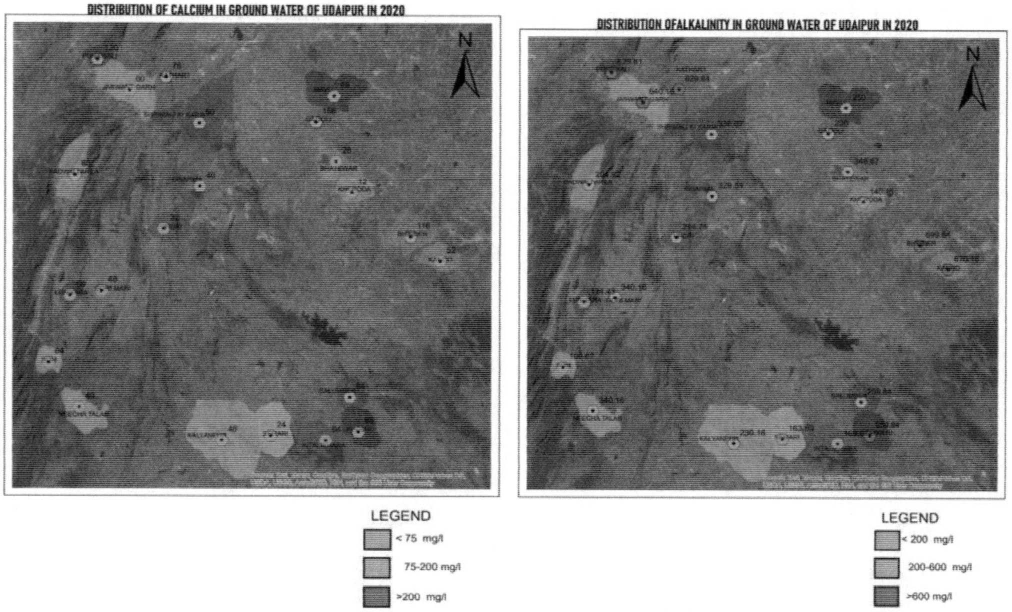

Figure 2.GIS mapping of calcium and alkalinity for study area

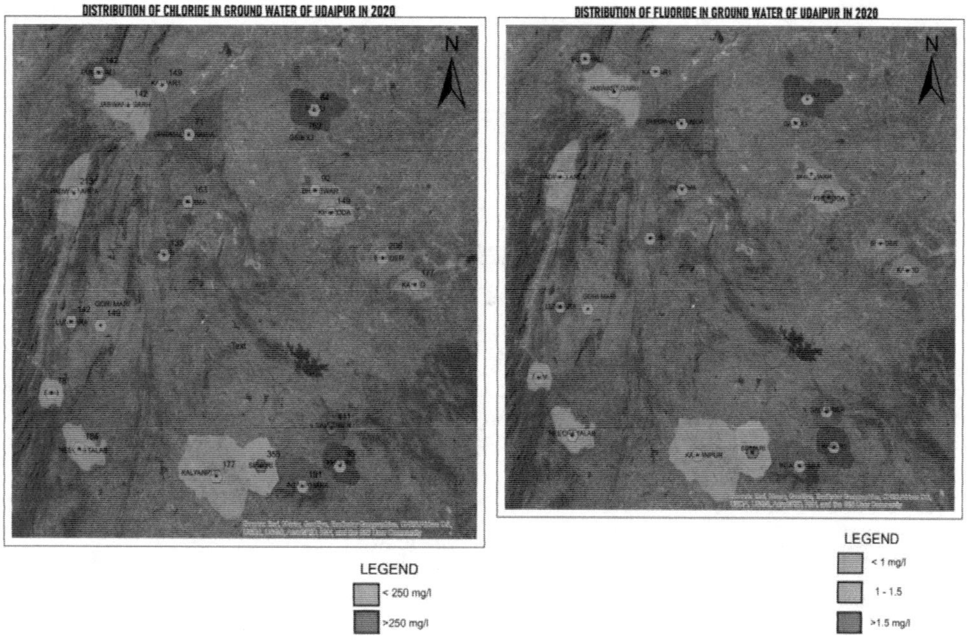

Figure 3. GIS mapping of chloride and fluoride for study area

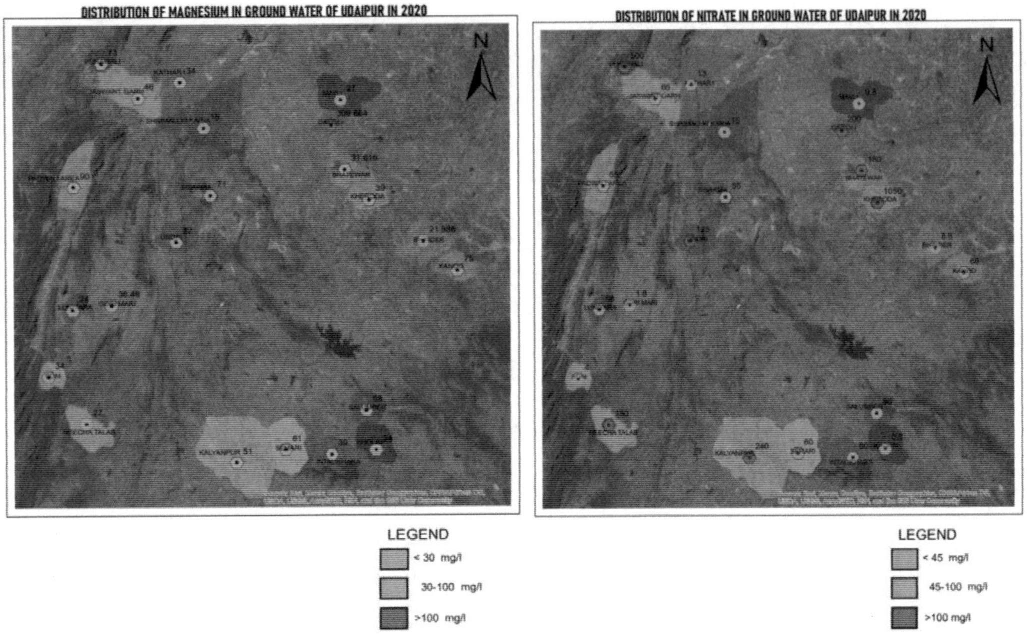

Figure 4. GIS mapping of magnesium and nitrate for study area

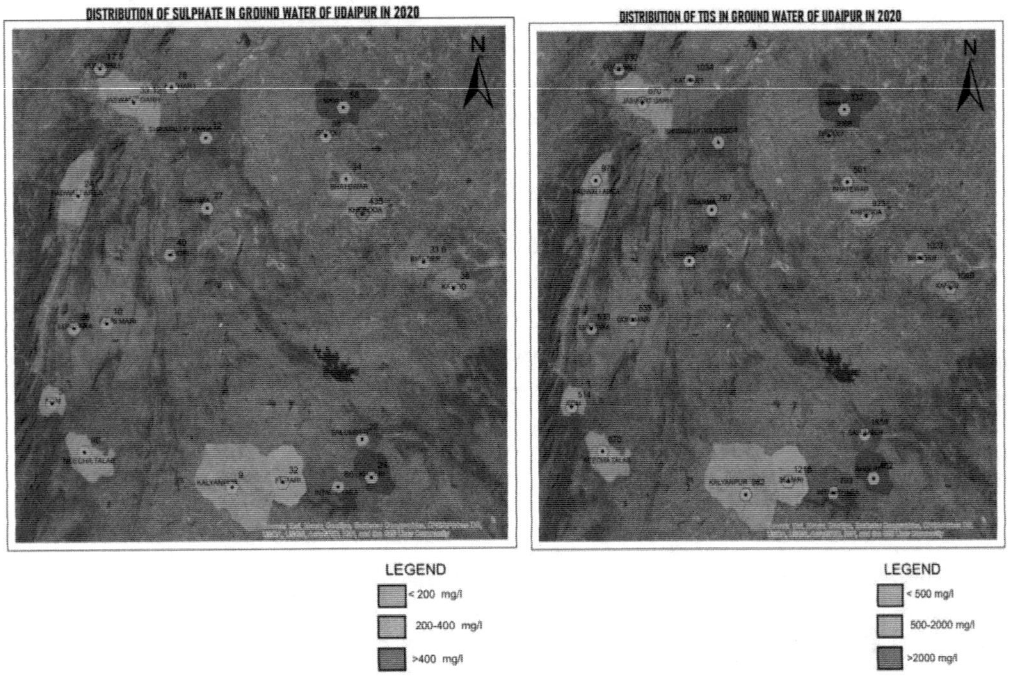

Figure 5. GIS mapping of magnesium and nitrate for study area

After analysing the GIS maps and graphs it was found that there is no bad cause of concern about the Calcium, Sulphate and Chloride hazard. About 72.7% samples of calcium, 95.50% samples of sulphate and 86.40% samples of chlorine are within acceptable limit of BIS. There is very less percentage of beyond the permissible limits.

A fluoride concentration of less than 0.8-1.0 mg/l may be harmful and may cause dental carries due to the formation of excessive cavities in the teeth of young children during calcination of their permanent teeth. High fluoride concentrations, greater than 1.5 mg/l may again be harmful, causing spotting and discolouration of our teeth and excessive fluoride consumption may even cause deformation of bones. Only 18.20% samples are within the permissible limit 1.0 – 1.5 mg/l wherage 68.20% are less than 1 mg/l and 13.6% are greater than 1.5 mg/l. Figure 1 shows the comparitive distribution in percentage of chemical constituents in the study area as mentioned in Table 2. Distribution of alkalinity in permissible range is very low.

4. Conclusion

Groundwater is one of the most valuable natural resources, and its use in various industries has made it an indispensable component of millions of people's lives. The continual and rapid increase in groundwater demand has put enormous strain on groundwater supply and appropriateness. N.S. Magesh has done work on 'Ground water quality assessment using WQI and GIS techniques, Dindigul district, Tamil Nadu, India'. The ground water samples were collected from 59 bore wells covering the entire study area and analysed using standard methods. Groundwater contamination is a global trouble and it is essential to identify the danger zones of high intensity contamination. This study represents the range of groundwater quality of Udaipur district in Rajasthan. There is no bad cause of concern about the Calcium, Sulphate and Chloride hazard as there is very less percentage of beyond the permissible limits. Alkalinity, Fluoride and Nitrate concentration in ground water exceeds the limits in high percetage. Mainly the distribution of fluoride less than 1 mg/l is around 68%. A fluoride concentration of less than 0.8-1.0 mg/l may be harmful and may cause dental carries due to the formation of excessive cavities in the teeth of young children during calcination of their permanent teeth. GIS mapping plays important role for decision making plans for treatment of ground water.

To minimise overexploitation and pollution of this vital resource, the most critical aspect of using groundwater is to strike the correct balance between removing and allowing the aquifer's level to recover. The durability and chemical properties of ground water can be studied in depth. Satellite remote sensing allows for a more thorough examination and systematic analysis of various geomorphic units and lineament features, which can then be combined with the use of a Geographical Information System to demarcate groundwater potential zones. Pollution hazard zones in water bodies and agricultural lands next to water bodies can be mapped using GIS. The research can be expanded to find acceptable artificial recharge sites and technologies for augmenting groundwater resources in moderate and poor prospective zones. Provides information to local governments and planners regarding places appropriate for groundwater exploration in order to assure well yield sustainability and combat water table depletion.

References

[1] Bamiker I., Mohamed A.&Hiyama T. 2007 assessing the groundwater quality by using GIS. Water Resour.Manage.21,699-715
[2] Choudhary, S., & Sharma, J. (2021). Surface Water Quality Trends and Regression Model through SPSS in Udaipur, Rajasthan. Iarjset, 8(10), 1–8. https://doi.org/10.17148/iarjset.2021.81030
[3] Data on assessment of groundwater quality for drinking purpose and irrigation in rural area Sarpol-e Zahab city, Kermanshah province , Iran
[4] Data Brief , 17(2018),pp. 148-156, 10.1016/j.dib.2017.12.061
[5] Kumar S.,Kumar P., Babu R &RaoC. 2010 an Assessment and mapping of groundwater quality by using GIS Int.J.Eng.Sci.Technol.2(11),6035-6046
[6] Nayak, T.R.,Bhar ,A.K. 2003. Monitoring Groundwater Fluctuation in a Part of Bundelkhand Hydrology Journal, 26(1-2),45-53
[7] Groundwater quality assessmwnt using water quality index and GIS technique in Modjo River Basin, centalethiopia

[8] Assessment of Groundwater quality for irrigation and drinking purposes and identification of hydrogeochemical mechanisms evolutioninnortheastern, Tunisia.
[9] A.K Batabyal , S. Chakraborty, Hydrogeochemistry and water quality index in the Assessment of groundwater quality for drinking uses Water Enviorn.Res.,87(2015),pp, 607-617
[10] Alam.W.Singh, K.S.Gyanendra, Y.,Laishram, R.J. and Nesa. N.(2020). Hdrogeochemical assessment of groundwater qualityfor few habitationsofChandel district, Manipur (India). Applied Water Science ,10(5) 10-13
[11] Ojha, Shrikant and Choudhary, Sangeeta (2017). Environmentally sustainable sand mining it's based on GIS based sediment yield estimation. Engineering & Technology in India, 8 (1&2): 49-57; DOI: 10.15740/HAS/ETI/8.1&2/ 49-57.

Measuring the Level of Road Safety through Road Safety Audit and Star Rating Method – A Case Study from Himalayan Expressway

Saurabh Kumar S[1] and Sunil Sharma[2]

[1]MTech Scholar, Department of Civil Engineering, National Institute of Technology Hamirpur, India
[2]Assistant Professor, Department of Civil Engineering, National Institute of Technology Hamirpur, India

E-mail: 20mce216@nith.ac.in

Abstract. National Highways (NH) carry the major portion of the total traffic in India. Road safety on NH is therefore of paramount importance. In the present study, a road safety audit has been performed for a 4-lane NH using Star Rating method. ViDA software developed by International Road Assessment Programme (iRAP) is used for finding the Star Rating of the road sections. This safety rating has been determined for four road user types: Vehicle occupant, Motorcyclist, Pedestrian and Bicyclist. The purpose of the study is to measure the level of road safety and lower the severity level by proposing countermeasures. The case study is a five Kilometre (km) section of NH-5 from Sanwara toll plaza to Jabali in the state of Himachal Pradesh in India. Encroachment on the land, roadside activities, insufficient infrastructure, imperfections in the geometric design, and non-motorized access were some of the severe safety flaws identified through the study process. The Star Rating results indicated that among all the road user types, the vehicle occupant is much safer than the others.

1. Introduction

Road safety not only saves lives but also boosts the country's economy. As India is a developing country, many roads have been built or in the process of being built in recent years. Road safety records of past years also indicates that there is a need for more safety improvement programs, which is necessary from time to time [1]. According to the World Road Statistics, 2018, India tops the 199 countries in terms of road accident mortality. According to the World Health Organization's 2018 Global study on road safety, around 11% of all deaths caused by accidents occur in India. On the National Highways (NHs) of India which account for 2.03% of the entire road network, a disproportionate share of 35.7 percent of road accident deaths occurred in 2019. Among various types of vehicles, two-wheelers account for 35% of all road accident deaths on NHs, followed by Light Motor Vehicles (18.6%), pedestrians (14%), trucks (10.7%), buses (4.9%), and bicycles (2.9%). In 2018, India's total estimated socio-economic cost of road accidents was Rs. 1,47,114 crores, equal to about 0.77 percent of the country's GDP [2]. Typically, four variables cause road accidents: driver/rider behaviour, vehicle issues, road factors, and environmental factors. One of these, the road factors may be significantly improved by applying engineering principles [3,4]. The present study demonstrates how road factors can influence the road safety and why their accurate measurement can significantly improve the overall safety on NHs. The RSA is first carried out to identify the road features that can cause road accidents. This is followed by rating the highway sections based on their safety features for occupants of automobiles, motorbikes, pedestrians and bicyclists. Based on the findings of RSA and the safety rating, suggestions are proposed to fill the gaps and improve the overall safety for all types of road user.

The first manual on RSA was published in India by Indian Roads Congress (IRC) in 2010 which was later revised in 2019 to include all types of urban and rural roads [5]. Several RSA-based studies have been carried out in various parts of the country [6,7]. Roy and Chakraborty [8] predicted the number of future accidents in the city of Kolkata in India. The developed model predicted the number of total accidents, fatality, and injury by applying the regression analysis technique using the available 15-years data. Another study by Budania et al. [9] established that value of maintenance requirements has a negative and significant connection with road accidents, which means if the cost of repair work is kept low, the number of accidents will be high. It was also observed that rainfall had positive relationship with accidents, which means that if there is more rain, the number of accidents will also be more. Singh [10] analysed the traffic fatalities in India at national, state, and urban city level. The study presented the proportion of road accidental deaths and injuries with respect to age, gender, month and time. The analysis showed that there were more accidents in the months of May & June and December & January than in the rest of the months. This was attributed to the bad weather conditions during these months of rainy and foggy season respectively. Kurumlarin [11] studied the severity of accidents with respect to the geometric design standards. Kumar and Chowdary [12] conducted RSA at a section of NH-65 connecting the cities of Vijayawada and Hyderabad in the state of Andhra Pradesh in India. They established that absence of crash barriers, poor road markings, bad condition of carriageway pavement and shoulders were the major reasons of accidents. The study suggested to provide service lane for segregating the slow-moving traffic from the fast-moving traffic and closing the unauthorized gaps in the carriageway median. The study also suggested to provide foot-over bridges for the pedestrians to cross the highway safely along with the sidewalk for pedestrians near the industries and educational institutions. Chatterjee et al. [13] carried out RSA on the full stretch of two 2-lane NHs number 117 and 60. The audit was performed with respect to three elements: geometry of the road, road infrastructure and road users. Critical safety deficiencies were identified, and appropriate countermeasures were suggested.

iRAP (International Road Assessment Programme) was formed in 2006 and registered as a charity in England in 2011 to help tackle the devastating social and economic cost of road crashes worldwide. ViDA is iRAP's free online data processing engine for star ratings [14]. By 2030, the yearly figure of road deaths worldwide is projected to increase to 2.4 million. Most of this will occur in low-income and developing countries, which already suffer nine out of ten of the world's road fatalities. Nearly half of those killed will be vulnerable road users: motorbikes, bicyclists and pedestrians [15]. A standardised evaluation of major roads in India was performed by iRAP through the locally developed programme, IndiaRAP, in cooperation with the Ministry of Road Transport and Highways, National Highways Authority of India, the World Bank, Bloomberg Philanthropies, Private Concessionaires and State Governments [16]. The study demonstrated a significant shortage of adequate facilities for road users, especially non-motorized vehicles and pedestrians. The star rating of the roads contributed the authorities know the safety lapses and suggest precise solutions. A study on two roads situation was conducted, one in the state of Gujarat and the other in Karnataka in India. The results of the study showed that after implementing the targeted road safety treatment plan, the rating of the roads improved substantially. As per the star rating method, 5 stars indicate the safest while 1star indicates a highly unsafe road. The methodology in the present study is discussed in the subsequent section.

2. Study method

The task of identifying the section's deficiencies by performing a road safety audit and implementing a star rating system to assess the present level of safety provided to road users. Finally, measures for improving the rating and reducing the deficiency have been discussed.

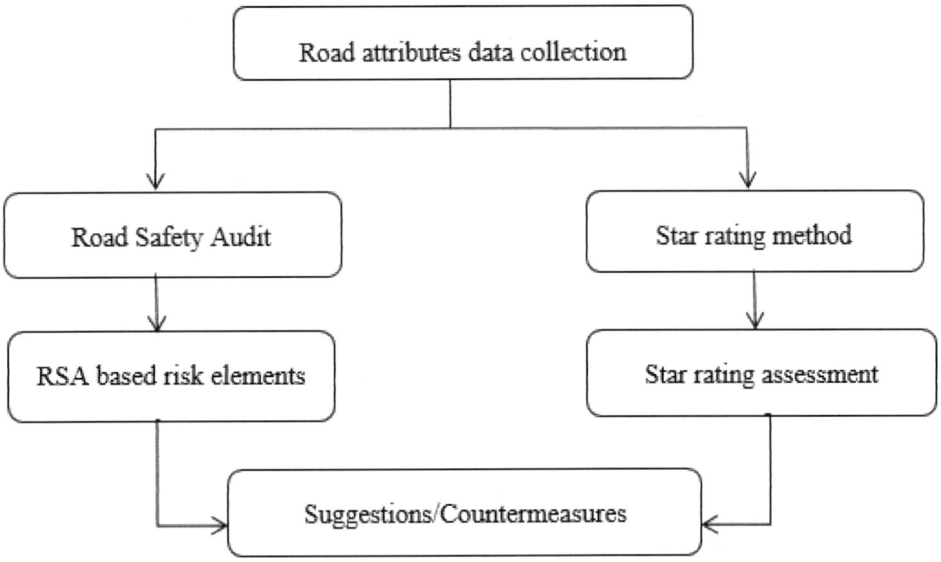

Figure 1. Road safety audit and star rating plan process

2.1. *Road attributes and data collection*

The existing 5-km segment accommodate significant volume of traffic and the deficiency on it was almost similar across the National Highway 05. This segment was chosen in view of the fact that major accidents occurred over the section as compared to the overall length of NH 05. A survey of the road section was conducted through visual inspection as well as videography. Various road attributes and traffic characteristics were collected in the process.

2.2. *Road safety audit*

The locations of accident sites were also obtained from the National Highway Authority of India (NHAI) office. The process of RSA as outlined in the manual on road safety audit called IRC SP 88 – 2019 is an exercise to improve road safety and reduce accidents by involving checks on various stages of highway development such as planning, designing, construction, at handover, and during operation [5,17]. The present study has performed RSA on an operational 4-lane highway. The survey consisted of obtaining information about the following: Sight Distances, Intersections, Interchanges, Crossing Segments, Roadside Obstacles, Drainage, Signposts, Pavement Markings, and Delineation, Vulnerable Road Users, Access to Property and Advancements, Issues with Lighting and Night-time troubles, General Road Safety Considerations.

The evaluation and subsequent rating of all NHs is to be carried out twice a year in the months of October and April, before and after the rainy season. It has been observed that the road conditions in India vary to some extent after the monsoon season. The highway rating is to be done based on three broad parameters: efficiency, safety, and road users' perspective. The present study is focused on safety.

2.3. *Star rating method*

Star Ratings involve an inspection of road infrastructure attributes that are known to have an impact on the likelihood of a crash and its severity. Based on some input information, the software assigns a star rating to the road. The most secured roads with a star rating of 5 have road safety features that are sufficient for the prevailing traffic speeds. Some of the road infrastructure elements on a safe road are division of opposite traffic by a broad median or barrier, good lane-marking and design of junctions, providing extra lane width, sealed (paved) shoulders, clear of vulnerable risks on the roadside such as poles, and adequate provision for bicyclists and pedestrians such as footpaths, bicycle lanes and pedestrian crossings [15,18]. The software computes and displays the star rating for a road user type based on the input road attributes.

Star Rating is obtained through ViDA for each of the four road user categories that account for the majority of road users on Indian roads [19]: vehicle occupants, motorcyclists, bicyclists, and pedestrians. The types of crashes that can occur for each type of road user is provided in Table 1.

Table 1. Crash types for each type of road user [20]

Vehicle occupants	**Motorcyclists**	**Bicyclists**	**Pedestrians**
• Run-off road • Head-on • Collision on intersections and access points	• Run-off road • Head-on • Collision on intersections and access points • Travelling along the road	• Riding along the road • Collision on intersections • Run-off road	• Walking along the road • During crossing the road

2.4. Case study

A 4-lane dual carriageway NH built recently has been selected as a case study. This 5-km section from Sanwara toll plaza to Jabli market is a part of 120-km Parwanoo-Solan NH which is managed by NHAI. It is currently a partial access-controlled expressway but is expected to be fully access-controlled in near future. The first phase of the project, about 28km in length from Zirakpur in the state of Punjab to Parwanoo in the state of Himachal Pradesh became functional in April 2012. The other phase of the project, about 39km in length from Parwanoo to Solan in Himachal Pradesh began operation in April 2021.

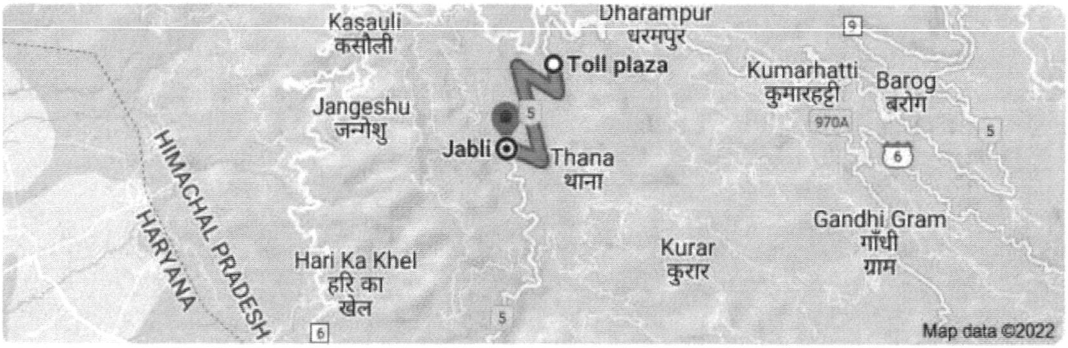

Figure 2. Google image of 5km road length of selected Highway

3. Results and discussion

3.1. RSA findings

The major audit findings are discussed along with site photographs in Table 2 and Table 3 contain a detailed list of risk elements and potential solutions or countermeasures to enhance safety.

3.1.1. Inadequate sight distance. A large number of curves did not have enough sight distance due to insufficient set back distance.

Figure 3. Inadequate sight distance on horizontal curves

3.1.2. Median opening. Delineators are not installed on some median openings due to which the drivers may have trouble in visibility during the night and the availability of sight distance was also less in such locations.

Figure 4. Median opening without delineators and broken median

3.1.3. Shoulder condition. The shoulder condition was poor across the length, which may lead to accidents involving automobiles and the pedestrians.

Figure 5. Shoulder condition at curves and shoulder at straight section

3.1.4. Drainage. Although side drains were available, they were obstructed at many places.

Figure 6. Inadequate surface drainage along the side of highway

3.1.5. Illegal vehicle parking. Illegal parking in front of the hotels and restaurants was frequent resulting in reduced road width and hindered visibility.

Figure 7. Illegal parking

3.1.6. Dumped materials. Construction material has been dumped at many places along the highway.

Figure 8. Dumped construction material near crash barrier and on service lanes

3.1.7. Broken barriers. Crash barriers were damaged in several locations and installation finishes were generally poor.

Figure 9. Damaged crash barrier

3.1.8. Lighting issues. Populated areas lacked sufficient lighting which can put pedestrians at risk during night time.

Figure 10. Poor visibility

3.1.9. Pedestrian sidewalk and crossing. Populated areas did not have sidewalks or railings or safe places for crossing.

Figure 11. Pedestrian facilities

3.1.10. Road marking. Road markings had faded at many places. Due to overgrown vegetation, certain road signs were also not visible.

Figure 12. Invisible road sign and faded road marking along the stretch

3.1.11. Roadside hazards. Large sized stones were noticed at many places, especially along the curves due to landslides.

Figure 13. Falling boulders at hill sides

3.1.12. Service lanes. In certain locations, the service lane was not maintained and had damaged at some places.

Figure 14. Damaged service lanes

3.2. Star rating

Star rating was performed at three locations which were selected based on accident data and RSA findings. The road attributes such as roadside details, mid-block details, intersections details, flow details, vulnerable road users (VRU) facilities and land use, speed Details that were entered into the ViDA to generate Star Rating.

All ViDA users have access to the Demonstrator, which may be used to create Star Ratings for single locations or small portions of road. The Demonstrator is especially useful for figuring out how certain road elements affect Star Ratings. To access the Star Rating Demonstrator, first step is to select the 'Demonstrator' button on the ViDA Dashboard. After that, all of the road attributes were entered into the Demonstrator, and the rating is presented by clicking on the star rating tab. The Star Ratings for the specified road attributes are displayed in the Star Ratings display. the star rating result has been obtained as shown in Figure 15. This particular location has a 1-star rating for bicyclists and pedestrians, 3-star rating for motorcyclists, and 4-star rating for vehicle occupants. Based on the results, this particular highway is quite safe for vehicle occupants, but quite unsafe for pedestrians and bicyclists. The star rating results obtained for all the three locations is shown in Table 2. Suggestions to mitigate risks identified through RSA are given in Table 3.

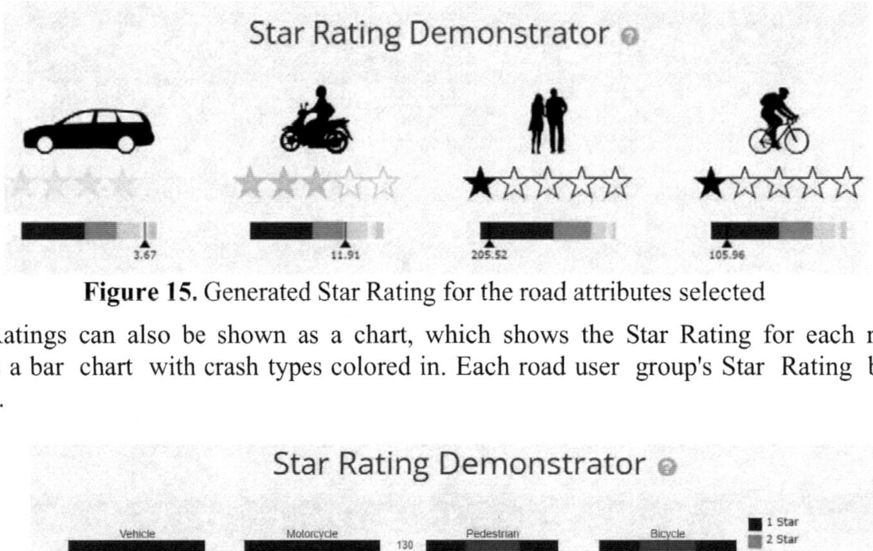

Figure 15. Generated Star Rating for the road attributes selected

The Star Ratings can also be shown as a chart, which shows the Star Rating for each road user category as a bar chart with crash types colored in. Each road user group's Star Rating bands are also shown.

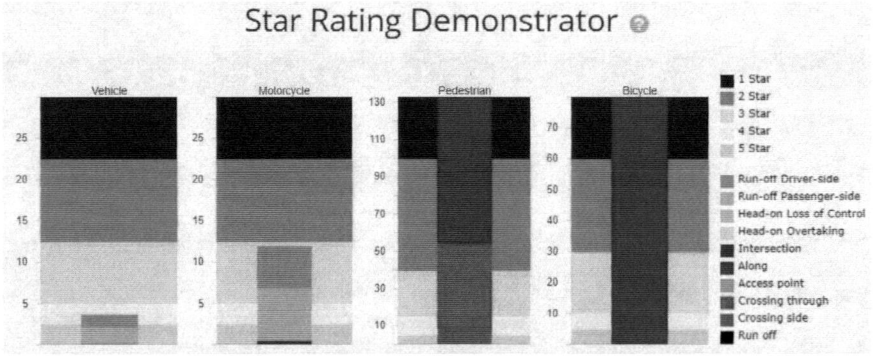

Figure 16. The Star Rating for each road user group as a bar chart

Table 2. Star Rating result for the selected three critical locations

Sr. No.	Selected locations	Generated Star Rating	
1.	Location 1 -Chainage (80+040) km to (79+940) km of 100 metres length	Vehicle occupants Motorcyclists Pedestrians	★ ★ ★ ★ ★ ★ ★ ★

		Bicyclists	★
2.	Location 2 - Chainage (76+450) km to (76+350) km of 100 metres length	Vehicleoccupant	★ ★ ★ ★
		Motorcyclists	★ ★
		Pedestrians	★
		Bicyclists	★
			★ ★ ★ ★
3.	Location 3 - Chainage (74+200) km to (74+100) km of 100 metres length	Vehicleoccupant	★ ★ ★
		Motorcyclists	★ ★
		Pedestrians	★
		Bicyclists	

Table 3. Risk elements identified through RSA and proposed solution/suggestions

S No.	Risk Elements	Suggestions/Countermeasures
1.	Inadequate Sight distance on curves	• Installation of Rumble strips along the centre and edge lines ofhorizontal curves to warn the traffic • There should be clear road markings and warning signals
2.	Median Opening with right turning	• Prior to merging, usage of speed humps to slow down the speed onthe minor road • At large and small road intersections, flexible pole barriers shouldbe utilized to segregate traffic
3.	Shoulder condition	• Paved shoulder should be reconstructed, throughout the length ofthe highway
4.	Side trains	• Over the shoulders, temporary sloppy side drain structures shouldbe removed • In waterlogged areas, drains should be adequately maintained
5.	Illegal Vehicle Parking	• For truck and bus parking, provide wide paved shoulders • Edge rumble strips should be installed to warn drivers if they aremoving away from the road
6.	Dumped construction materials	• Building materials dumped on the side of the road should be forbidden, and penalties should be imposed by authorities and police
7.	Barrier condition and its requirements	• Broken metal barriers should be replaced immediately • Metal barrier end should be driven into the concrete
8.	Lighting and Night-time Issues	• At uncontrolled median opening and built-up regions, high mastlighting should be provided
9.	Pedestrian facility	• Pedestrian guard rails or sidewalk should be provided in high-riskregions to separate people from high-speed vehicles • Close to markets and schools, wide paved shoulders should beprovided • At regular intervals, adequate cross walks should be given • Vehicle speeds should be limited to a maximum of 20 km/h aroundmarkets and densely populated region
10.	Road marking and signs	• Remove all long grass, shrubs, or tree branches that obscure thesign's visibility • Put a guard rail around the sign to protect it from accidents, orplace it far away from the hazard zone

IOP Conf. Series: Earth and Environmental Science 1084 (2022) 012044 doi:10.1088/1755-1315/1084/1/012044

11.	Roadside hazards	• The boulders that are lying over the road should be cleared • Checking should be done on a regular basis
12.	Service Lane	• Damaged service lanes must be repaired • It is important to clear the vegetation that has grown along the sideof the service lane

The following suggestions are proposed to improve star rating scores:

- Segregate the Bicyclists and Motorcyclists from interactions with larger, heavier vehicles and the risk for conflicts by providing separate Bicycle and Motorcycle Lane
- Shoulder should be maintained, along the entire length of the road
- Streetlights must be installed in habitat areas, median openings, sharp curves, and other spots where accidents are likely to occur
- Development of a walkway in the habitat area, as well as crossing facility at junctions
- Increasing the visibility of road markings and traffic signs

4. Conclusion

A large number of NHs have been recently constructed or upgraded in India. The owner, NHAI desires to carry out a periodic assessment of these roads for safety. Based on the parameters identified by NHAI and safety standards defined by NHAI, the present study was undertaken to identify the safety deficiencies on a four-lane NH. Along with RSA, ViDA software was used to assign safety-based star rating to the selected highway sections. The RSA findings show deficiencies such as lack of appropriate sight distance at horizontal curves, poorly illuminated median openings and built-up areas, missing pedestrian facilities, illegal parking, dumped construction material along the roadside, and damaged shoulders. The outcome of the star rating through ViDA software indicates that although the highway is safe for vehicle occupants, it is highly unsafe for bicyclists and pedestrians who are at significant risk. For motorcyclists, the highway scores average rating for safety. The following are the key outcomes from the RSA-based inspection and Star rating.

- Due to landslides debris on the road, encroachment, and overgrowth of vegetation in the median, the deficiency of appropriate sight distance was constantly noted at the curves and at the junction entrances across the study road, despite the fact that sight distance is one of the most important needs on high-speed routes like NHs.
- The lack of adequate infrastructure such as a walkway, crossover bridges, and bicycle lane had exposed bicyclist and pedestrians to high-speed vehicles.
- Road users were found to lack the fundamental understanding of road safety. This is because drivers aren't used to high-speed facilities in densely populated places. The majority of drivers were found to be driving faster than the posted speed limit.
- Road users can be influenced, and their current attitudes and habits can be changed by awareness and education. Implementing course-based road safety instruction can help young people to develop safe mindsets early on, reducing the likelihood of future traffic crashes. Better drivers can be produced by increasing public awareness of traffic regulations and proper road driving.

References

[1] McLeod, S., & Curtis, C. (2022). Integrating urban road safety and sustainable transportation policy through the hierarchy of hazard controls. *International Journal of Sustainable Transportation*, *16*(2), 166-180.

[2] Ministry of Road Transport and Highways, Road Accidents in India 2019, 2020, [Online]. Available: https://morth.nic.in/sites/default/files/RA_Uploading.pdf.

[3] Zheng, L., Sayed, T., & Mannering, F. (2021). Modeling traffic conflicts for use in road safety analysis: A review of analytic methods and future directions. *Analytic methods in accident research*, *29*, 100142.

[4] Job, R. S., & Brodie, C. (2022). Road safety evidence review: Understanding the role of speeding and speed in serious crash trauma: A case study of New Zealand. *Journal of road safety*, *33*(1), 5-25.

[5] IRC:SP:88-2019, Manual on Road Safety Audit, 2019, [Online]. Available: https://law.resource.org/pub/in/bis/irc/irc.gov.in.sp.088.2019.pdf.

[6] Alonso, F., Useche, S. A., Valle, E., Esteban, C., & Gene-Morales, J. (2021). Could Road Safety Education (RSE) help parents protect children? Examining their driving crashes with children on board. *International Journal of Environmental Research and Public Health*, *18*(7), 3611.

[7] Lawrence, B., Fildes, B., Cairney, P., Davy, S., & Sobhani, A. (2021). Evaluation of raised safety platforms (RSP) on-road safety performance. *Sustainability*, *14*(1), 138.

[8] Chakraborty, S., & Roy, S. K. (2005). Traffic accident characteristics of Kolkata. *Transport and Communications Bulletin for Asia and the Pacific*, *74*, 75-86.

[9] Budania, P., Chaudhary, M., Kumar, S., & Anand, V. (2014). Highway Safety in India (A Case Study).

[10] Singh, S. K. (2017). Road traffic accidents in India: issues and challenges. *Transportation research procedia*, *25*, 4708-4719.

[11] Dadashova, B., Ramírez, B. A., McWilliams, J. M., & Izquierdo, F. A. (2016). The identification of patterns of interurban road accident frequency and severity using road geometry and traffic indicators. *Transportation research procedia*, *14*, 4122-4129.

[12] Kumar, B. T., & Chowdary, J. C. (2018). Road safety audit: a case study on NH-65. *International Journal of Engineering & Technology*, *7*(2.1), 69-74.

[13] Chatterjee, S., Bandyopadhyaya, P. S., & Mitra, S. (2020). Identifying Critical Safety Issues on Two-Lane National Highways in India–A Case Study from NH 117 and NH 60. *Transportation research procedia*, *48*, 3908-3923.

[14] ViDA online software - iRAP. [Online]. Available: phttps://vida.irap.org/en-gb/home.

[15] International Road Assessment Programme (iRAP), iRAP Methodology Fact Sheet #1, pp. 1–3, 2014.

[16] Bhavsar, J., Tharakan, A., Rogers, L., Smith, G., & McInerney, R. (2019). Star Ratings for life-saving road improvements in India. *Transport Commun. Bull. Asia Pac*, *2030*, 1-24.

[17] Nautiyal, A., & Sharma, S. (2022). Methods and factors of prioritizing roads for maintenance: a review for sustainable flexible pavement maintenance program. *Innovative Infrastructure Solutions 2022 7:3*, *7*(3), 1–15. https://doi.org/10.1007/S41062-022-00771-6.

[18] Zhang, Y., Li, H., Sze, N. N., & Ren, G. (2021). Propensity score methods for road safety evaluation: Practical suggestions from a simulation study. *Accident Analysis & Prevention*, *158*, 106200.

[19] ViDA User Guide, p. 29, 2020, [Online]. Available: https://resources.irap.org/Specifications/ViDA_User_Guide.pdf.
[20] R. A. P. Methodology and F. Sheet, Crash types, 2013.

Acknowledgments
The authors are grateful to National Highways Authority of India and Himachal Pradesh Public Works Department for providing accidents data and facility to survey the selected stretch.

Psychological predictors of Sustainable Transportation Behaviour: Implications for Low Carbon Economy and Society

B Arya[1] and S Chaturvedi[2*]

1 Manipal University Jaipur, Jaipur
2 St. Xavier's College, Jaipur

*shiromichaturvedi@gmail.com

Abstract. There's been a consistent increase in carbon emissions globally in the recent decades, primarily because of a consistent increase in personal transportation needs. Thus sustainable transportation behaviour has become a pertinent variable for investigation. The current study thus aimed to study the impact of environmental consciousness, green identity, values (biospheric, altruistic, egoistic), and consideration of future consequences on sustainable transportation behaviour such as usage of public transport and carpooling. The study employed a quantitative survey design with 351 participants in total and (M=67% F=33%) who were daily commuters to offices or universities. Results indicated that environmental consciousness, green identity and consideration of future consequences to be positively correlated with sustainable transportation behaviour. Biospheric and altruistic values were also found to have a positive correlation with sustainable transportation behaviour while egoistic values had a negative correlation with sustainable transportation behaviour. These findings have implications for a low carbon economy and society since they highlight important psychological variables that can promote sustainable transportation behaviour at an individual level and reduce overall carbon emissions.

1. Introduction

One of the greatest concerns of the current times is climate change. Transportation is a major source of pollution, and it is the greatest contributor to air pollution after energy generation and industrial processes. Daily commuting behaviour accounts for around 8% of total carbon emissions, and this ratio is steadily increasing. Many countries' recent economic progress has enabled an increasing number of individuals to afford personal vehicles increasing the threat of carbon emissions of each household. Automobile travel is still on the rise within developing nations as well, and it is quickly becoming the most significant producer of carbon dioxide from passenger transportation [1]. Furthermore, a large proportion of the world's transportation system is still reliant on the use of automobiles such as cars, buses, motorbikes, which is responsible for three-fourth of transport related greenhouse gas emissions [2]. Consequently, how we commute every day may have a noteworthy impact on each individual's carbon footprint. Thus, sustainable transport for our daily transportation has become a pertinent topic of study and research.

Many people believe that taking public transport is the most apparent and obvious way to reduce their transportation emissions, and that it is especially important in the world's most congested cities.

If public transportation is unavailable, carpooling remains one of the most efficient ways to decrease one's individual emissions from driving. Experts believe that one way that would be incredibly helpful is if individuals could just adopt a simple motto of never driving alone; if a car has to be used it is imperative to ensure that at least one other person accompanies them in the car [2]. Internet based mobile applications such as Uber, Ola, BlaBlaCar, and Lyft also encourage carpooling.

The decision to carpool is influenced by an individual's attitudes, their demographic characteristics, and certain socioeconomic considerations among others. Neoh et al. [3] did a meta-analysis study to determine the elements that might impact a person's decision to use carpooling as their method of transportation. Findings indicated that there were two important elements impacting carpooling namely "internal" psychological factors and "external" factors including interventions and policies. According to several experts, saving money is the primary motivation for carpooling, making it a low-cost means of transportation. Thus, it is pertinent to understand the psychological variables that impact the usage of sustainable transport for daily commuting like car-pooling or civil transportation in order to promote their usage and reduce overall carbon emissions. Understanding people's attitudes toward carpooling situations will further aid policy makers in devising techniques for successfully executing this eco-friendly system of daily transport.

2. Review of Literature

According to the International Energy Agency, the transportation industry plays a significant part in people's everyday lives all over the world, accounting for more than twenty two percent of carbon dioxide emissions. Research [4] has recently focused on the need to work towards the reduction of the transportation sector's current emissions. Moreover, the Department of Economics and Social Affairs [5] states that sustainable transportation is critical to achieving the broader sustainability goals of universal access and decreased environmental and climate impact, alongside ensuring enhanced efficiency. As a result, it is critical element for the Agenda for Sustainable Development and the Paris Climate Change Agreement to be realised by 2030. Furthermore, transformative action is urgently needed to hasten the transition to sustainable transportation at a global and personal scale to reduce carbon dioxide emissions.

According to the Federal Transit Association (USA), public transportation has a significant impact on reducing overall carbon emissions, by assisting cities in meeting national ambient air quality standards (NAAQS) by lowering total vehicle emissions and smog-causing pollutants and reducing the amount of fuel used and subsequently the amount of greenhouse gas emissions. Thus, sustainable transportation behavior such as using public transport and carpooling is a key component of realizing climate change goals and must be worked towards by individuals, communities, and governments.

A study by Semenescu and Alin Gavreliuc [6] adds to a more thorough understanding of automobile usage behavior by including structural and psychological variables in addition to environmental variables. Such findings encourage the inclusion of psychological aspects in promoting sustainable behaviour as well as intervention design and policy making.

A primary psychological variable that impacts pro-environmental behavior including but not limited to sustainable transportation behaviour, is environmental consciousness also known as environmental concern or environmental awareness. Environmental consciousness is the individual's knowledge regarding their impact on the environment. Environmental awareness is a strong predictor of environmentally friendly behavior [7]. Knowing that there is a problem with the environment and being aware about what should be done about it are crucial prerequisites for pro-environmental behavior such as pro-environmental consumption tendencies, buying eco-friendly products and engaging in sustainable fashion.

Another factor which plays a significant role in people engaging in environment-friendly behavior, especially from a psychological perspective, is that of green identity. Self-identity has been defined as "the label used to characterise oneself in relation to a certain behavior" [8]. The green identity approach explains how group membership to 'green groups' as well as identifying oneself as 'environmentally conscious' influences a person's attitudes, emotions, and pro-environmental behaviours. Green identity has been associated with conservation behaviour [9], as well as other environment-friendly behaviours like recycling, purchasing fair-trade goods, and not flying on vacation [9]. Thus the construct of 'green identity' has a lot of potential to explain sustainable behaviour at large and sustainable transportation behaviour as well.

Values provide an important framework to understand sustainable or pro-environmental behavior of people since a range of pro-environmental activities have been demonstrated to be strongly influenced by values [10]. According to Steg et al., [10], biospheric values (referring to a persons' s concern for nature and environmental conservation), altruistic values (referring to an individual's concern for other people's well-being), and egoistic values (referring to a person's primary concern for personal resources and protection) have all been closely associated with pro-environmental behaviour. An individual's biospheric values have been favourably and significantly connected with environment friendly behaviour when compared to other values. Studies have also found that a strong association exists between biospheric values and green identity and both are also associated with various kinds of environment-friendly behaviors like using energy efficient equipment and sustainable transport [11].

Individuals also differ in their expectations of the consequences of their acts, with some focusing on long-term consequences and others failing to see beyond the immediate ones. On this foundation, the notion of future consequences consideration (CFC) was born. Individuals with high CFC scores were shown to be more concerned about environmental concerns, to exhibit pro-environmental attitudes, and to follow or wish to follow "eco-friendly" patterns of behaviour to a greater extent. In existing literature, the importance of CFC in association with environment-friendly behavior has been researched but needs exploration in greater detail with regards to sustainable transportation for daily use.

3. Current Research

Past studies have looked at the effects of a variety of factors on the rise in carbon emissions in the transportation industry at large. However, few studies have examined the psychological aspects that impact carbon emissions due to household transportation or daily transportation behaviour for individuals. The current study aims to understand how various psychological variables such as green identity, consideration of future consequences, and values impact the use of public transport and carpooling as part of sustainable transportation behavior.

Based on the existing literature and conceptualization of variables the following hypotheses were framed:

Figure 1. Conceptual Model

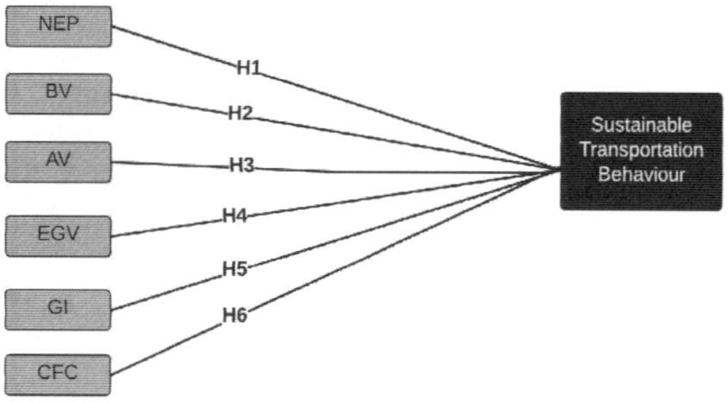

4. Research Methodology

4.1. Sample

The target population of the study was daily commuters to office or university. To be eligible to participate, all participants were required to own a car, be able to drive a car, have access to a car and have a driving license. Prior to participation, the aim of the study was shared with the subjects and they were required to read and sign a consent form. Once they agreed, they were directed to an online form. Cross-sectional survey design was used to collect data through anonymous Google Forms using purposive sampling. Purposive sampling method helps to select respondents who fulfill the inclusion criteria. The online survey was publicised on social networking platforms (Facebook, LinkedIn, WhatsApp, Instagram, twitter etc.) and shared through emails. Participation was voluntary and anonymous. The questionnaire was filled in by 438 participants. Researchers were left with 351 responses for final analysis after data cleaning. The final sample comprised 67% men and 33% women in the age range of 21 to 68 years (M= 38.63 years, SD=8.42 years). 59% commuted for work and 41% were students who commuted to attend various types of classes. 71% of the respondents had a six-day week and the remaining 29% worked for five days in a week.

4.2. Tools

Questionnaire used for the study had two sections. Section one consisted of filter questions to ensure the appropriateness of the sample. Section two had questions pertaining to the variables of the study. The following scales were administered:

Consideration of Future Consequences (CFC) Scale [12] The CFC has 12 items, with a five-point Likert scale;

Values: Values were measured using the adapted version of Schwartz's Value Scale [13] having a 7-point Likert scale in order to assess their orientation for egoistic values (EGV), altruistic values (AV) and biospheric values (BV);

New Ecological Paradigm Scale (NEP) [14]: This 15-iten questionnaire is designed to test environmental awareness through a seven-point Likert scale regarding raising awareness of people's effect on the ecosystem and their capacity to disrupt the equilibrium within the ecosystem;

A green-environmental self-identity scale (GI) was developed which consisted of four items, measured on a five-point Likert type rating scale, with the aid of measures from previous studied [8];

Sustainable Transportation Behaviour. Measured by responses given to four items pertaining to transportation behaviour on a daily basis on a five-point self-developed Likert scale.

4.3. Analysis

Data analysis of the study followed two steps and was done using SPSS 21. Firstly, factor loadings were identified by exploratory factor analysis (EFA) followed by the Kaiser-Meyer-Olkin (KMO) test. Secondly, hierarchical multiple regression was performed. The dependent variable was sustainable transport behaviour. At the first step, environmental consciousness was entered, values as well as green identity were added in the second step and consequences of future consequences was entered at the final step. Results have been presented in detail.

5. Results

Descriptive statistics pertaining to the primary variables under study are listed in table 1. Preliminary analysis that sustainable transportation behaviour is positively associated with environmental consciousness ($r= .411, p<.05$), altruistic ($r= .517, p<.01$) and biospheric ($r= .588, p<.01$) values, consideration of future consequences ($r= .562, p<.01$) and green self-identity ($r= .633, p<.01$). People with egoistic value orientations are less likely to opt for sustainable transportation choices ($r=-.462, p<.05$)

Table 1. Descriptive Statistics

S.N.	Variable	Mean	SD	1	2	3	4	5	6	7
1.	Environmental Consciousness	59.07	9.28	1	.634**	.496**	-.636**	.502**	.540**	.411*
2.	Altruistic Values	11.42	3.12	-	1	.554**	-.464**	.515**	.532**	.517**
3.	Biospheric Values	10.79	3.39	-	-	1	-.521**	.590**	.655**	.588**

4.	Egoistic Values	11.31	3.57	-	-	-	1	-.385*	-.579**	-.462**
5.	Consideration of Future Consequences	34.07	8.28	-	-	-	-	1	.604**	.562**
6	Green Self-Identity	9.73	3.61	-	-	-	-	-	1	.633**
7	Sustainable Transportation Behaviour	8.86	2.95	-	-	-	-	-	-	1

Hierarchical regression analysis was done in order to examine all the hypotheses. The dependent variable sustainable transportation behaviour was regressed onto the predictor variables. Results are presented in Table 2. Environmental consciousness was entered at first step and was a significant predictor of sustainable transportation behaviour (R2=.421, p<.000). Addition of values and green identity the second step significantly increased the proportion of variance accounted for (R2=.613). Total variance accounted for at step 3 was 65.4% (R2=.654).

Environmental consciousness, biospheric values, consideration of future consequences and green self-identity emerged as significant positive predictors, whereas egoistic value orientation emerged as significant negative predictor of sustainable transportation behaviour in the final step.

Table 2. Regression Analysis

Step	Predictor	R Square	Adjusted R Square	R Square Change	F Change	Step 1β	Step 1β	Step 1β
1	Environmental Consciousness	.498	.421	.498	134.031***	.581**	.462*	.321**
2	Biospheric Values	.648	.613	.150	17.053**		.333**	.213**
	Altruistic Values						.051	.065
	Egoistic Values						-.301**	-.247*
	Green Self-Identity							.301**
3	Consideration of Future Consequences	.687	.654	.039	6.84*			.241**

6. Discussion

The findings of the current research indicated that environmental consciousness has a significant, positive correlation with sustainable transportation behavior. This means that as an individual's environmental consciousness or awareness increases, they engage in more sustainable transportation behavior such as using public transport or carpooling. These findings support past studies that have looked into the link between environmental concern and various antecedents and outcomes and have found that persons who are concerned about the environment are more likely to engage in pro-environmental action [15]. These studies also showed that government policies on environmental protection, can elicit environmental concern, which can increase usage of sustainable transportation modes.

Results also indicated that biospheric values and altruistic values, both have a significant, positive correlation with sustainable transportation behavior i.e. if people are high on biospheric and altruistic values then they are more likely to demonstrate sustainable transportation behavior. This finding is also supported by past studies that showed that people are motivated to act pro-environmentally by biospheric values [16] and altruistic values [17] even if such behavior is relatively expensive.

On the other hand, results indicated that egoistic values such as protecting and promoting personal assets have a significant, negative correlation with sustainable transportation behavior i.e. if persons place importance on their personal needs and resources due to high levels of egoistic values then their likelihood of demonstrating sustainable transportation behavior is reduced. Egoistic values, on the other hand, have long been a source of concern in the literature on pro-environmental behaviour because they are aimed at maximising individual benefits while minimising individual costs [10], which is likely to reduce sustainable transportation behaviour because many of these behaviours entail some personal costs.

The study also found that green identity and consideration of future consequences (CFC) have a significant, positive correlation with sustainable transportation behavior which implies that if an individual has a "green identity" and is high on their CFC score then they will demonstrate sustainable transportation behavior such as using public transport and carpooling. The findings of past research also suggest that green-identity is an important behavioural driver for carbon offsetting behavior [9].

References

[1] Schafer A, Victor DG 1999 Global passenger travel: implications for carbon dioxide emissions *Energy* **24**(8):657-79

[2] Timperley J 2020 How our daily travel harms the planet [Internet] *BBC* Available from: https://www.bbc.com/future/article/20200317-climate-change-cut-carbon-emissions-from-your-commute

[3] Neoh JG, Chipulu M, Marshall A 2017 What encourages people to carpool? An evaluation of factors with meta-analysis *Transportation* **44**(2):423-47

[4] Brand C, Goodman A, Rutter H, Song Y, Ogilvie D 2013 Associations of individual, household and environmental characteristics with carbon dioxide emissions from motorised passenger travel *Appl. Energy* **104**:158-69

[5] UNDESA 2021 World Social Report I DISD [Internet] *www.un.org* Available from: https://www.un.org/development/desa/dspd/world-social-report/2021-2.html

[6] Semenescu A, Gavreliuc A 2021 Structural, psychological and contextual predictors of car use *Frontiers in Psychology* **12**

[7] Arya B, Chaturvedi S 2020 Extending the Theory of Planned Behaviour to Explain Energy Saving Behaviour *Environmental And Climate Technologies* **24**(1):516–28

[8] Cook AJ, Kerr GN, Moore K 2002 Attitudes and intentions towards purchasing GM food *J. Of Economic Psychology* **23**(5):557-72

[9] Whitmarsh L, O'Neill S 2010 Green identity, green living? The role of pro-environmental self-identity in determining consistency across diverse pro-environmental behaviours *J. Of Environmental Psychology*. **30**(3):305-14

[10] Steg L, De Groot JI, Dreijerink L, Abrahamse W, Siero F 2011 General antecedents of personal norms, policy acceptability, and intentions: The role of values, worldviews, and environmental concern *Society And Natural Resources* **24**(4):349-67

[11] Balundė A, Perlaviciute G, Steg L 2019 The relationship between people's environmental considerations and pro-environmental behavior in Lithuania *Frontiers In Psychology* 2319

[12] Strathman A, Gleicher F, Boninger DS, Edwards CS 1994 The consideration of future consequences: weighing immediate and distant outcomes of behavior *J. Of Personality And Social Psychology* **66**(4):742

[13] De Groot JI, Steg L 2008 Value orientations to explain beliefs related to environmental significant behavior: How to measure egoistic, altruistic, and biospheric value orientations *Environment And Behavior* **40**(3):330-54

[14] Dunlap R, Liere KV, Mertig A, Jones RE 2000 Measuring endorsement of the new ecological paradigm: A revised NEP scale J. Of Social Issues **56**(3):425-42.

[15] Minton AP, Rose RL 1997 The effects of environmental concern on environmentally friendly consumer behavior: An exploratory study *J. Of Business Research* **40**(1):37-48

[16] Loebnitz N, Aschemann-Witzel J 2016 Communicating organic food quality in China: Consumer perceptions of organic products and the effect of environmental value priming *Food Quality and Preference* **50**:102-8

[17] Thøgersen J, Ölander F 2002 Human values and the emergence of a sustainable consumption pattern: A panel study *J. Of Economic Psychology* **23**(5):605-30

Modeling and optimization of the anaerobic digestion system for domestic waste water treatment with the development of monitoring techniques

Rajshree Mathur[1], Meena Kumari Sharma[2]
[1,2]Department of Civil Engineering, Manipal University Jaipur, India

Abstract. Water scarcity and wastewater management are two most serious environmental problems faced by developing countries like India. Recycling and reuse of the treated wastewater for non-potable purpose may significantly reduce the stress on fresh water requirement. Anaerobic digestion methods are used to treat wastewater discharge from various fields. Software tools provide an opportunity for engineers to create models which represent activities, resources, and the environment in powerful, affordable and user-friendly way. Modeling and simulation results serve as an important guideline in the optimization and process control for plant operators and other specialists in the field of engineering. Model-based development of software using tools such as Math Works Simulink has become common in the engineering. The Anaerobic Digestion Model (ADM1) from the International Water Association might be used for modeling and optimization of digestion operations. This research reveals the study of domestic wastewater treatment using ADM1. The process equations are solved using the input data, and output is generated. The results show that the proposed method is capable of predicting process parameters of excellent stream with high degree of accuracy (R^2=0.945).
Keywords:
Wastewater, anaerobic digestion, software tools, modeling, optimization, simulation.

1. Introduction

Hazardous chemicals are often released in the municipal wastewater resulting in large scale of pollution. Wastewater treatment plants are there for mandatory. Reliable model is necessary not only to estimate the excellent performance but also for controlling the complex biological processes involved in treatment of wastewater [1]. The most commonly used parameter defines organic pollution in both wastewater and surface water is chemical oxygen demand (COD). COD is used in decisive actions such as determining size and efficiency of the treatment plant [2]. The effect of the effluent discharge on the water body can also be determined with COD. Higher levels of COD indicate list of oxidations in the organic compound (effluent). This decreases the level of dissolved oxygen and it affects the entire ecosystem [3]. It can be stated that maintaining effluent quality in wastewater treatment plants is extremely important and should not be overlooked or compromised [4].

Despite the technological advancement in sensing and modeling technologies, reliable measurements of effluent COD are still a strenuous task. This is because the characteristics of wastewater treatment plant are quite dynamic and several nonlinear and complex processes are involved. These processes are inherently complex to predict or explain using a mathematical or statistical model. The processes represent the general interaction taking place in the system [5]. In the present time water is polluted at an elevated scale this makes provision of water quality management very hard and resource consuming [6]. The advanced technologies must be used not only for improving the excellent quality but also for reducing the operation and maintenance cost of the plant. Most of the deterministic models fail to assess the excellent quality [7]. Soft computing techniques have been applied in this domain to assess the uncertainty, but they also fail to capture dynamic nature of the process [8]. The reason for the above mentioned that most of the machine learning techniques for forecasting methods are based on static data and does not include information regarding process parameters and variables and their dependencies on both external and internal factors. In this research work a mathematical model for ADM representation using differential equations is developed. The biological equations are modeled and solved. This model is easy to calibrate (can be used to represent processes at different locations) and is not computationally demanding. The significance of this model is that it can be used in conjunction with data-intensive forecasting on machine learning techniques to reduce the uncertainty and predict the effluent parameters with maximum efficiency.

The last decade has witnessed enormous development not only in the generation and storage of data but also in the field of data handling and analysis [8]. An immediate effect is that the complex industrial processes can be easily modeled, analyzed and controlled. Moreover, the cost and computational resources required are very economical [9]. Anaerobic digestion of sludge, which is a well-known method and conventional process for managing municipal solid sludge, has historically been one of the principal gas emission sources in a typical wastewater treatment plant. Hydrolysis, acidogenesis, acetogenesis, and methanogenesis are the four stages of digestion [10]. Complex compounds including carbohydrates, proteins, and lipids are broken down into fatty acids with long hydrocarbon chains, amino acids and sugars during the hydrolysis stage. Hydrogen, volatile fatty acids and CO_2 are produced during the acidogenesis step (VFAs). Modeling and analysis of wastewater treatment plants (WTP) have benefited from Programmable Logic controllers, data acquisition system and supervisory controlled techniques [11]. The research community has extensively studied the industrial excellent source and parameters. Previous studies have used biochemical, biological, and physicochemical processes to build models that simulate various anaerobic digestion processes, including municipal wastewater treatment plants (MWTP)[12]. Artificial neural networks (ANNs) and fuzzy models are widely used to explain relationships inside complex structures and processes, making them suitable for WTP modelling. An adaptive network-based fuzzy inference system (ANFIS) is a hybrid of an ANN and a fuzzy inference system (FIS) that has gained popularity as a method for describing complex processes [13]. Other MWTP facilities have employed both the ANN and ANFIS techniques in the past to anticipate biogas production rates.

Given the wide variation in operation and geographic location among WTPs, facility-specific models must be addressed for reliable forecast capabilities. Because of its location in Indian Institute of Technology Kharagpur, India, the currently researched facility, for example, is subjected to a wide range of temperatures. Each WTP has its own set of process parameters, necessitating the development of distinct models based on these parameters. Total solids (TS), total volatile solids (TVS), Sludge loading rate (SLR), sludge retention time (SRT), organic loading rate (OLR), volatile solids (VS), digestion time (DT), alkalinity (ALK), temperature (T), pH, and carbon to nitrogen (C/N) ratio were previously common selected model input variables for anaerobic and biological digesters [12; 14;15]. It's vital to remember that models are frequently constrained by the availability of monitored parameters that can be employed in simulations. Furthermore, input data pre-processing to reduce input parameters (e.g., principal component analysis (PCA)) has not been adequately established. This pre-processing can be effective in restricting model input data that is strongly correlated, resulting in a greatly reduced model processing time and, as a result, a more efficient model.

Wastewater treatment plant analysis can be viewed as a sophisticated control problem with nonlinear dynamic behaviour [16]. Analyzing the wastewater treatment plant demands knowledge from different domains and incorporation of microbial chemical and physical features [17]. A large number of machines learning and forecasting techniques can be found in the literature. The focal point of these techniques is using that are acquired in the past for predicting future outcomes. Plethora of research works using artificial neural networks, regression analysis, and artificial intelligence on wastewater treatment plants can be found in the literature. Machine learning techniques have redefined the analytical techniques to solve elaborate problems. However, it can be observed that it cannot handle the uncertainty involved in treatment of wastewater. Hybrid model which is capable of learning from the previous data as well as drawing conclusions from the present environmental conditions can be a milestone in predicting effluent parameters. The significance of this research work is development of a mathematical model can represent the process parameters of an anaerobic digestion model as closely as possible.

2. Materials and Methods

2.1 Model Description

ADM1 was released by the International Water Association (IWA) in 2002 [18]. ADM1 examines the anaerobic system's biochemical and materialization processes in depth. Table 1 lists the biological reactions used in this work, including acidogenesis, acetogenesis, and methanogenesis.

Table 1. Biochemical Reactions in the ADM1

Reaction Process	Reaction Equation
Acidogenesis from glocose	$C_6H_{12}O_6 + 2H_2O = 2CH_3COOH+2C+O_2+4H_2$ $3C_6H_{12}O_6 = 4CH_3\ CH_2COOH + 2CH_3COOH + 2CO_2+ 2H_2O$ $C_6H_{12}O_6 = CH_3\ CH_2\ CH_2COOH + 2CO_2+ 2H_2$
Acetogenesis from butyric acid	$C_4H_8\ O_2+2H_2O = 2CH_3\ COOH+ 2H_2$
Acetogenesis from propionic acid	$CH_3\ CH_2COOH+2H_2O = CH_3\ COOH+ 3H_2+CO_2$
Acetoclastic metanogenesis	$CH_3COOH = CO_2+CH_4$
Hydrogenotrophic methanogenesis	$4H_2+CO_2= CH_4+2H_2O$

The ADM1 is a structured mathematical model involving 19 metabolic rate processes and 32 dynamic state concentration variables. The ADM1 model has the following set of differential equations (DE): 10 (DE) to describe the evolution of soluble matter concentrations in liquid phase, and two (DE) to model the levels of inorganic carbon (IC) and inorganic nitrogen (IN) in liquid phase. Twelve (DE) to represents the dynamic behaviour of particulate matter and biomass concentrations in liquid phase; two (DE) to model cations and anions levels in liquid phase; and six (DE) for acid–base reactions to determine the pH of effluent and calculate ionised forms of VFA, free ammonia nitrogen, and carbon dioxide concentration [18].

2.2 Model Implementation

The general layout of an ASMADM interface-based wastewater treatment plant is shown in Figure 1, with the help of a variety of microbes, organic stuff in the water is broken down into simpler chemicals. The microorganisms change with the changes in operating conditions such as temperature and pH. For this reason, predictive analysis is used for COD and consistent decision making. The datasets are received from a wastewater treatment plant and 5 variables are collected from 13 different locations. The following were the process of primary variables: Flow, COD of influent water, and suspended particles in influent water (SS), Nitrogen (N), pH. Programming platform of MATLAB is used to form the differential equations and solve them using the input data.

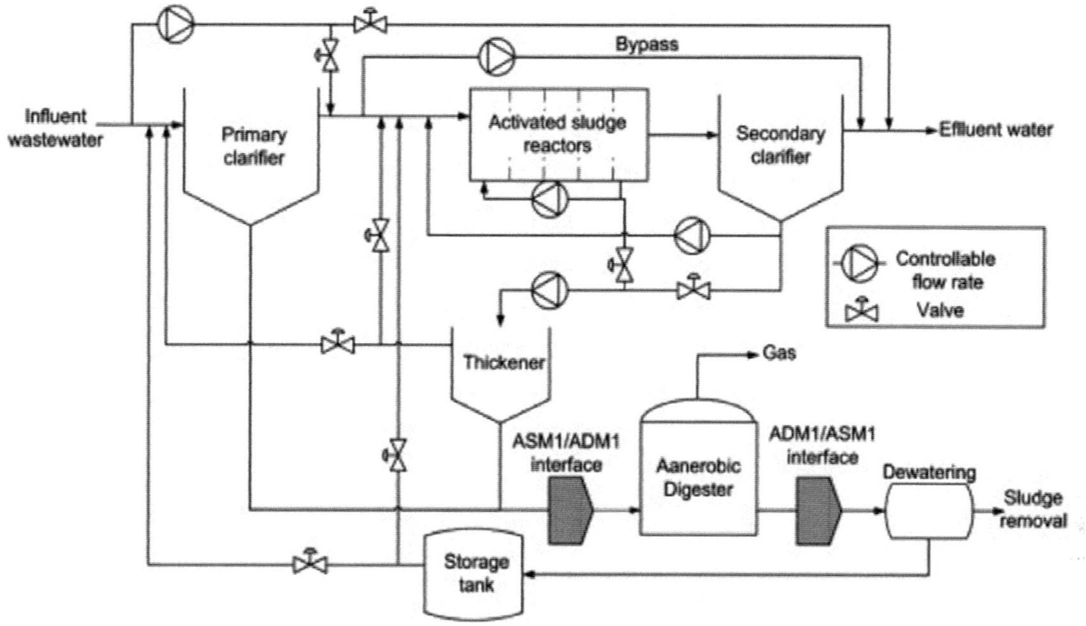

Figure 1. Schematic representation of ADM1 model.

The different stages involved in anaerobic digestion model are shown in Figure 2.

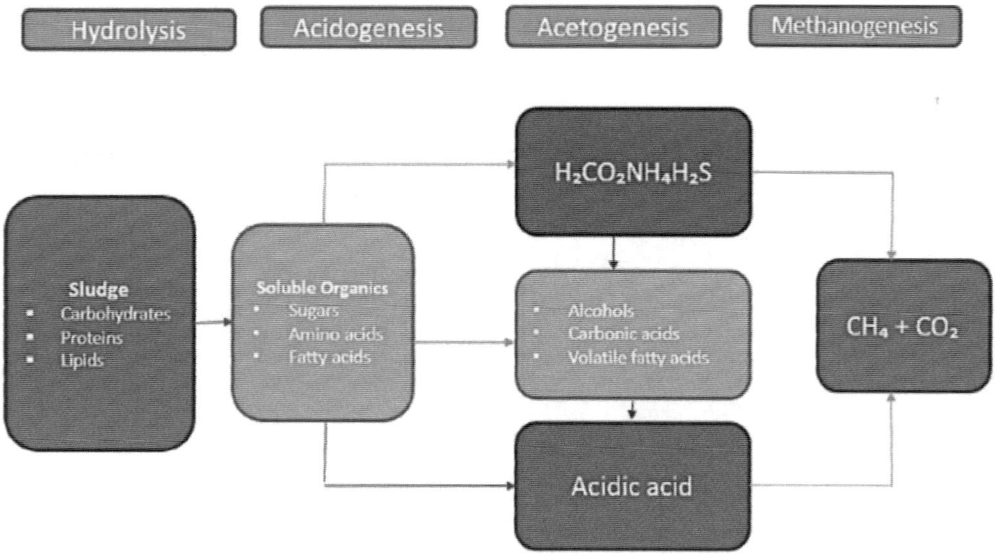

Figure 2. Stages involved in ADM.

A detailed description about these stages and the processes involved can be found in [19] Figure 3, represents mass balance for a simple and anaerobic digester. The mass balances cannot be obtained by direct measurement. The conversion from one state to another can be studied. The processes are dependent on each other, error in the earlier stages manifests into later stages as well [20]. The final calculations of COD are represented in the equations (1-4).

$$COD_s = S_{aa} + S_v + S_{sua} + S_{bu} + S_{pro} + S_{fa} + S_{ac} + S_{h2} + S_i\ S_{ch4} \quad (1)$$

$$COD_p = X_{su} + X_{fa} + X_i + X_{aa} + X_{pro} + X_{c4} + X_{ac} + X_{h2} + X_{ch} + X_{li} + X_c + X_{pr} \quad (2)$$

$$COD_t = COD_p + COD_s \quad (3)$$

$$TKN_t = N_{xc}X_c + N_{bac}\,(X_{su} + X_{aa} + X_{fa} + X_{c4} + X_{pro} + X_{ac} + X_{h2}) + N_i\,(S_i + X_i) + S_{in} + N_{aa}\,(X_{pr} + S_{aa}) \quad (4)$$

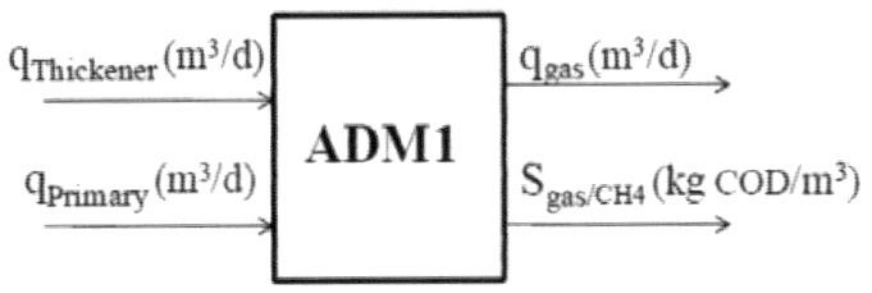

Figure 3. Mass balance of ADM bioreactor.

The list of ADM state variables is given in Table 2. These equations are implemented in the programming platform and solved using the concept of differential equations. The resultant output in terms of effluent COD is assessed and compared with the measured values.

Table 2. ADM1 State variables.

State variable Description	Symbol	Units
Long chain fatty acids (LCFA)	Sfa	
Amino acids	Saa	
Monosaccharides	Ssu	
Soluble Inerts	Si	
Total valerate	Sva	
Hydrogen gas	Sh2	
Methane gas	Sch4	
Total butyrate	Sbu	
Total acetate	Sac	
Total propionate	Spro	KgCOD m^{-3}
Proteins	Xpr	
Composites	Xc	
Particulate inerts	Xi	
Lipids	Xli	
LCFA degraders	Xfa	
LCFA degraders	Xfa	
Carbohydrates	Xch	
Sudar degraders	Xsu	
Amino acid degraders	Xaa	
Propionate degraders	Xpro	
Acetate degraders	Xac	
Hydrogen degraders	Xh2	
Inorganic carbon	Sic	Kmole Cm^{-3}
Inorganic nitrogen	Sin	Kmole Nm^{-3}
Cations	Scat	Kmole m^{-3}
Anions	San	

3. Result and Discussion

The Chemical Oxygen Demand (COD) of a two-stage anaerobic digestion system of domestic waste water from different locations was simulated using Anaerobic Digestion Model No.1(ADM1). As input variables, onsite measurements of home wastewater influent and effluent parameters were used. and benchmark respectively. After running the ASM interface we got the ASM and ADM input variables that are: SI g COD/m^3, SS g COD/m^3, XI g COD/m^3, XS g COD/m^3, XBH g COD/m^3, XBA g COD/m^3, XP g COD/m^3, SO g COD/m^3, SNO g N/m^3, SNH g N/m^3, SND g N/m^3, XND g N/m^3, SALK mol HCO3/m^3, TSS mg SS/l.

ADM output variables using ODE Solver have been achieved after initializing ASMADM interface. These output variables are: Spro ,Saa, Sva , Sbu , Sfa , Ssu, Sh2 , SI, SIC, Sch4, SIN , Xpr , Xc , Xch , Xli , Xpro, Xsu , Xaa ,Xfa , Xc4, Xh2 , Xac COD (Kg/m^3), XI , Scat+, San- Flow rate (m3/d), Temperature (degC). After doing the sum of aforementioned 27 variables the COD got out then validate the data using the curve fitting method. The final COD was calculated by summing all the output variables of ADM1 model. The comparison between the experimental and simulated results can be seen in figure 4. After regression analysis it can be observed that the simulated values are in close proximity to the experimental values (R^2 = 0.945).

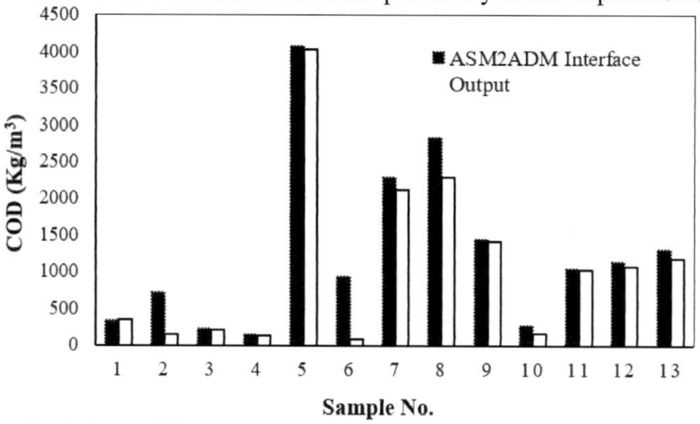

Figure 4. Variation of COD values from experiments and ASM2ADM interface

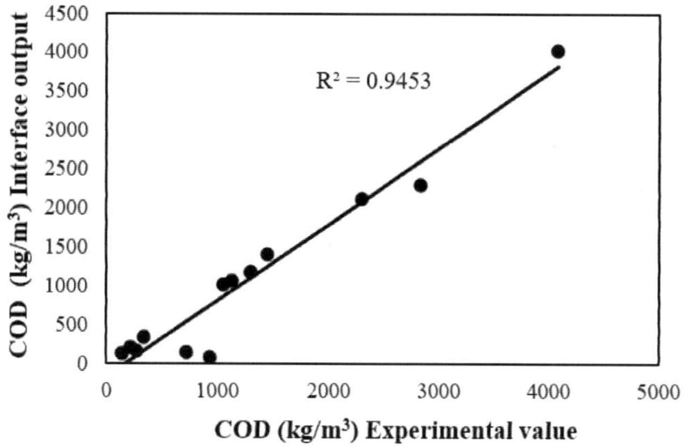

Figure 5. Validation of COD values from Experiments and ASM2ADM Interface

4. Conclusion

In this paper ADM1 model are used for predicting effluent characteristics in wastewater treatment plant. The measured data can contain error and result in ambiguous outputs. The development of process design, evaluation parameters and benchmarking on modeling approaches is hindered by this problem. Approach of numerical analysis of ADM1 process is presented. Complex equations involved in all four stages of anaerobic digestion are designed and solved, to acquire the effluent parameters. Effluent characteristics can be predicted for whole plant by this method. Results obtained from ASM2ADM interface is nearby experimental value (R^2=0.945). The proposed approach is validated using input obtained from in situ two stage digestor. These technologies demonstrate how mathematical modelling can aid in the improvement of biotechnological processes in general.

5. References

[1] Copp J B, Jeppsson U and Rosen C 2003 Towards an ASM1-ADM1 state variable interface for plant-wide wastewater treatment modeling *Proc. 76th Annual WEF Conf* p517
[2] Alisawi H A O 2020 Performance of wastewater treatment during variable temperature *J. Appl Water Sci* 10, 89
[3] Shrestha A, Chan T K, Aibinu A A and Chen C 2017 Efficient risk transfer in PPP wastewater treatment projects *Elsevier,* Volume 48, p 132-14
[4] Silva M I and Bortoli D 2021 Development of a Model for the Process of Anaerobic Digestion and Its Solution by the Modified Adomian Decomposition Method. *Int. J. Appl. Comput. Math*
[5] Madhura L, Singh S., Kanchi S. et al. Nanotechnology-based water quality management for wastewater treatment. *Environ Chem* Lett 17, 65–121
[6] Regmi P, Stewart H, and Amerlinck A 2018 The future of WRRF modelling - Outlook and challenges *Water Science and Technology*. 79. 10.2166/wst.2018.498
[7] Dürrenmatt, D J and Gujer, W 2012 Data-driven modeling approaches to support wastewater treatment plant operation. *Environ. Model. Softw*., 30, 47-56.
[8] Huang Y, Lan Y, Thomson S, Fang A, Hoffmann, C and Lacey R 2010. Development of soft computing and applications in agricultural and biological engineering. *Computers and Electronics in Agriculture* 71. 107-127. 10.1016
[9] Dieu B 2001 Application of the SCADA system in wastewater treatment plants. *ISA transactions* 40. 267-81. 10.1016/S0019-0578(00)00053-7.
[10] Tchobanoglous G, Stensel H D, Tsuchihashi R. and Burton F 2014 *Wastewater Engineering: Treatment and Resource Recovery* 4th edition, Mc.Graw Hill
[11] Pang J, Yang S, He L, Chen Y and Ren, N 2019 Intelligent control/operational strategies in WWTPs through an integrated Q-learning algorithm with ASM2d-guided reward. *Water*, 11, 927.
[12] Najafi B and Ardabili S 2018 Application of ANFIS, ANN, and logistic methods in estimating biogas production from spent mushroom compost (SMC) *Resources Conservation and Recycling* 133 DOI:10.1016/j.resconrec.2018.02.025.
[13] Ghiasi A M, Lotfan S, Fallah M and Sadeghi H 2016 ANN-based modeling and reducing dual-fuel engine's challenging emissions by multi-objective evolutionary algorithm NSGA-II. *Applied Energy* Volume 175, 1 p 91-99.
[14] Qdaisa A H, Hani K B and Shatnawia N 2010 Modeling and optimization of biogas production from a waste digester using artificial neural network and genetic algorithm *Resources, Conservation and Recycling* Volume 54, Issue 6, p359-363.
[15] Murugan S and Debabrata B 2015 Assessment of sustainable biogas production from de-oiled seed cake of karanja-an organic industrial waste from biodiesel industries. *Fuel* 148:25-31DOI:10.1016/j.fuel.2015.01.072.

[16] Li D, Yang H Z and Liang X F 2013 Prediction analysis of a wastewater treatment system using a Bayesian network. *Environ. Model.Softw.* 40, 140–150
[17] Haggege J, Benrejeb M and Borne P 2005 On the design of a neuro-fuzzy controller—Application to the control of a bioreactor. *J. Syst. Sci. Syst. Eng.* 14, 417–435.
[18] Batstone D, Keller J, Angelidaki I, Kalyuzhnyi S, Pavlostathis S, Rozzi A, Sanders W, Siegrist H and Vavilin V 2002 Anaerobic digestion model No 1 (ADM1) *Water science and technology* a journal of the International Association on Water Pollution Research 45. 65-73.
[19] Nadiri A A, Shokri S, Tsai F T and Asghari M 2018 A. Prediction of effluent quality parameters of a wastewater treatment plant using a supervised committee fuzzy logic model. *J. Clean. Prod.* 180, 539–549.
[20] Han H, Zhu S, Qiao J and Guo M 2018 Data-driven intelligent monitoring system for key variables in wastewater treatment process. *Chin. J. Chem. Eng.* 26, 2093–2101.

IOP Conf. Series: Earth and Environmental Science 1084 (2022) 012047 doi:10.1088/1755-1315/1084/1/012047

Assessment and Identification of drought prone zone in a Low Laying Area by AHP and MIF method: A GIS based study

Nameirakpam Momo Singh and Thiyam Tamphasana Devi*

Department of Civil Engineering, National Institute of Technology, Manipur – 795004, India

*thiyam85@gmail.com

Abstract. The study location, the Imphal West district of Manipur state, is known for its flood-prone character due to its low-lying topography in India's north-eastern region. Unexpectedly, in the year 2019, a drought-like scenario with extremely limited available surface water arose, severely affecting agricultural activity in the region, which is primarily covered by agricultural land. Drought-prone zones for 2019 were discovered using GIS (Geographical Information System) tools and satellite data, employing both the AHP (Analytic Hierarchy Process) and MIF (Multi Influencing Factor) methods. Seven parameters (rainfall, temperature, slope, drainage density, soil, landuse/landcover, and groundwater yield) were taken as input parameters. Drought zones were simulated and classified as mild (22.82% according to AHP and 39.42% according to MIF), moderate (60.10% and 54.71%), severe (16.16% and 5.55%), and extreme (0.92% and 0.32%). Drought data (ground data) from the State Government Department (2019) shows that 43.71% of drought zones are mild, while 51.32% are moderate, indicating that the MIF approach can more accurately simulate the drought region than the AHP method.

1. Introduction

Drought is the lack or deficiency of precipitation below a normal range over a period of time in a region. It is one of the natural disasters that occur all over the world and can be caused by a variety of factors such as decreased rainfall, less surface runoff, and groundwater table depletion [1-2]. Their effects can have a short- or long-term impact on humans as well as the environment, including decreased food production, decreased forest productivity, decreased water level, and harm to wildlife habitat and migration activities, to name a few [3-4]. It is a complex natural phenomenon occurring at different spatial and temporal scales as meteorological drought (less rainfall, less intensity of rainfall), hydrological drought (less surface runoff in rivers and streams, drying up of ponds, lakes, rivers and reservoirs), and agricultural drought (reduced soil moisture necessary for sustaining plant and crop growth). Therefore, it is important for the identification of drought prone zones to take the necessary proper actions and mitigation measures.

Satellite remote sensing (RS) products are useful for observing natural occurrences and disaster-related events in any part of the world, both before and after they occur. With the combinations of geographic information systems (GIS) and RS products, different studies related to drought monitoring and assessment of drought conditions are conducted [5-9]. The GIS technique provides a simple and easy method for the assessment of drought zones compared to conventional methods with its capability to integrate and analyse different data sources, including RS datasets [10]. Integration of GIS with different multi-criteria decision-making methods such as Analytic Hierarchy Process (AHP) and Multi-influencing Factor (MIF) methods was used for the identification or mapping of drought zones and groundwater potential studies were conducted [11-20]. Studies utilising both the AHP & MIF methods in drought assessment are very rare. However, in recent years, studies on drought assessment utilising GIS and AHP methods have been observed and the method has proved to be useful and reliable as well [21-23]. The AHP method developed by Saaty [24–25] helps in deciding or prioritising decision criteria in the pairwise comparison matrix when multiple alternatives are available for obtaining or deciding the best objective or goal. Using several criteria in a pairwise comparison matrix, the identification of drought zone areas was analyzed. Using the weights from the AHP & MIF method, GIS operations were conducted using the weighted overlay analysis tool for the identification of drought prone zones in the study area, and its validation of the result was conducted using field data (report on drought affected portion) from the agriculture department, Government of Manipur, India.

The objective of this study is to identify or map the drought prone zones through GIS and AHP & MIF methods in the Imphal West district of Manipur (North-Eastern region of India). The study's findings are expected to assist and raise awareness of the drought-prone region among decision-makers, people, and communities involved in agricultural activities, water resource management, and water conservation measures.

2. Study Area

Imphal-West is located in the valley district of Manipur (India's North East region), surrounded by sixteen other districts of plain and hilly terrain. The Imphal-West district lies between the coordinates of $24^{0}30'00"$ N to $25^{0}0'00"$ N latitude and $93^{0}40'00"$ E to $94^{0}10'00"$ E longitude (figure 1), surrounded by four districts from all corners, i.e., the north (Senapati), the east (Imphal-East & Thoubal), the south (Thoubal & Bishnupur), and the west (Senapati & Bishnupur) respectively.

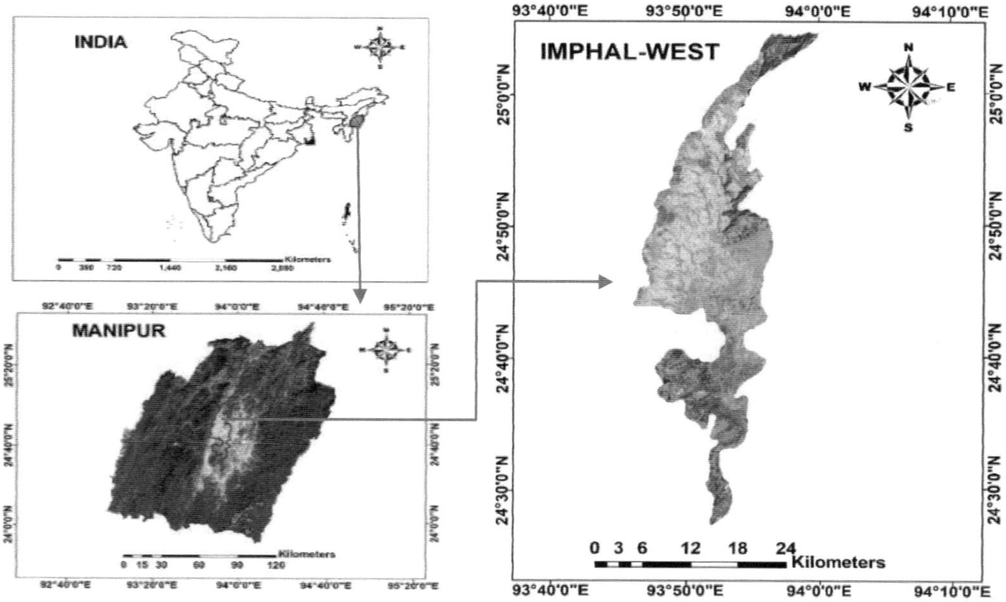

Figure 1. Location of study area (Imphal-West District)

It has an area of 528 km2 with a dense population of 5,17,992 [26] in the state. The district lies at an average elevation of 790m above mean sea level, experiencing a sub-tropical climate in the region and enjoys comfortable weather throughout the year. The district is under the influence of hot and humid rainy seasons during the summer and cool and dry seasons during the winter, with an annual rainfall of around 1581 mm and a temperature range from 10°C to 34°C.

3. Materials and Method

In this study, spatial and non-spatial data have been used. Spatial data includes Shuttle Radar Topography Mission (SRTM) digital elevation model (DEM) and Landsat 8 images, whereas non-spatial data includes rainfall, temperature, soil and groundwater yield data. These data are downloaded and collected from a variety of sources, including SRTM DEM & Landsat 8 (2019) image (30m resolution) OLI/TIRS C2 Level-1 (United States Geological Survey, USGS), rainfall & temperature data (2019) (Directorate of Environment and Climate Change, Manipur), soil (National Bureau of Soil Survey and Land Use Planning, NBSS & LUP, Nagpur), and groundwater yield data (1986-2014) (Public Health of the geographic information systems (GIS) was used for the integration and analysis of spatial and non-spatial datasets in the GIS environment. In the analysis of the drought prone zone region, the weighted overlay analysis tool in ArcGIS was used by integrating all the raster layers of drought influencing factors in the GIS platform. The roles of RS and GIS were involved in the preparation of these raster layers such as slope, drainage density, soil type, landuse/landcover (LULC) and auxiliary data such as rainfall, temperature, and groundwater yield data. Lastly, for validation purposes, drought affected paddy regions (sub-division & taluks) of the study area were collected from the Imphal-West Agriculture Department, Government of Manipur. The coordinates of these sub-divisions and taluks are plotted over the raster layer of the resultant drought-prone zone regions.

Table 1. Data Source used in the study

Sl.No.	Data	Data source	Generated/Extracted data type
1	SRTM DEM (30m resolution)	USGS	Slope & drainage density map
2	Landsat 8 (2019) Collection 2 Level 1	USGS	Landuse/landcover map
3	Meteorology	DoECC, Manipur	Rainfall & Temperature map
4	Soil type	NBSS & LUP, Nagpur	Soil type map
5	Groundwater yield	PHED, Manipur	Groundwater yield map

The AHP method developed by Saaty [24] was used as a function of multi-criteria decision making for the identification of drought prone zones in the Imphal-West district. Using the methods, the weights of the different criteria used for the identification of drought prone zones were calculated through a pairwise comparison matrix and normalized weight. These weights are based on expert knowledge and literature reviews, including suggestions from government officers, farmers, and surrounding communities. Further, these weights are used in the weighted overlay analysis, including their degree of influence on the scale (1–5), to give the final result of the drought zone in ArcGIS®. The weights are given on the scale of importance using a pairwise comparison matrix in the range of 1 to 9. The AHP matrix used in the study is given in table 1. The lambda maximums, consistency index (CI) and consistency ratio (CR) of the AHP matrix obtained are 7.635, 0.106, and 0.08 respectively. Since *CR* (0.08) 0.1, it indicates the matrix assigned is consistent and can be used in the analysis. The equations used in finding the consistency index (*CI)* and consistency ratio (*CR*) are given as:

$$CI = \frac{\lambda_{\max} - n}{n - 1} \tag{1}$$

$$CR = \frac{CI}{RI} \tag{2}$$

Where, λ_{max} = maximum eigen value; n = number of parameters in matrix; RI = random index value (for $n = 7$; $RI = 1.32$)

3.1. *AHP Method*

An AHP method is a multi-criteria decision-making method based on the analysis of a pairwise comparison matrix of parameters used for defining an objective. The parameters used in the method are ordered in a hierarchical manner of influence on the objective. Based on the objective of the study problem or analysis, the parameters are placed descending on top of the hierarchy order. A comparative analysis between the parameters (from most important to least important) was performed for a comparative judgement. The comparative judgement is based on the scale factor from 1 (equal importance or least important) to 9 (extreme importance) as given in table 2 [24].

Table 2. AHP scale of importance & its description (Saaty, 1977)

Scale of importance	Definition	Justification
1	Equal importance	Two activities contribute equally to the objective
3	Moderate importance of one over another	Experience & judgment slightly favour one activity over another
5	Essential or strong importance	Experience & judgment strongly favour one activity over another
7	Very strong importance	An activity is strongly favoured & its dominance is demonstrated in practice
9	Absolute importance	The evidence favouring one activity over another is of the highest possible order of affirmation
2,4,6,8	Intermediate values between two adjacent judgements	When compromise is needed

The pairwise comparison matrix and corresponding weights assigned for each parameter are by the AHP method tabulated in table 3.

Table 3. Pairwise comparison matrix using AHP method

Parameters	a	b	c	d	e	f	g	Normalized AHP weights
Rainfall (**a**)	1	2	3	5	5	3	5	0.332
Temperature (**b**)	1/2	1	3	5	5	2	5	0.267
Slope (**c**)	1/3	1/3	1	2	3	2	1/2	0.104
Drainage density (**d**)	1/5	1/5	1/2	1	3	2	2	0.094
LULC (**e**)	1/5	1/5	1/3	1/3	1	1	1/3	0.043
Soil type (**f**)	1/3	1/2	1/2	1/2	1	1	1/2	0.065
GWY (**g**)	1/5	1/5	2	1/2	3	2	1	0.096

3.2. *MIF Method*

In the multi-influencing factor method, seven parameters influencing the effect of drought events are considered for drought zone mapping of the study area. The seven parameters include rainfall, temperature, slope, drainage density, landuse/landcover, soil type, and groundwater yield potential. The interrelationships and effects of one parameter among each parameter are studied and a chart is developed (figure 2). Each relationship has been given a weightage in terms of "rank" according to its hydrologic characteristics (table 4). The representative rank of a factor in the potential zone is the sum of all weights from each factor. A factor with a higher weight value shows a larger impact, and a factor with a lower weight value shows a smaller impact on groundwater recharge potential. A major effect represents the direct influence of one factor over another, and a minor effect represents indirect influence. The major and minor effects are classified based on their ability to carry or reach water to a larger portion of the study area and their roles in causing drought events.

Table 4. MIF weight for mapping drought prone zone region

Parameter	Major effect (A)	Minor effect (B)	Relative weights (A + B)
Rainfall	1 + 1	0.5 + 0.5	3
Temperature	1 + 1	0.5	2.5
Slope	1 + 1 + 1	0.5	3.5
Drainage density	1	0.5	1.5
LULC	1	0.5	1.5
Soil type	1	0	1
GWY	0	0.5	0.5

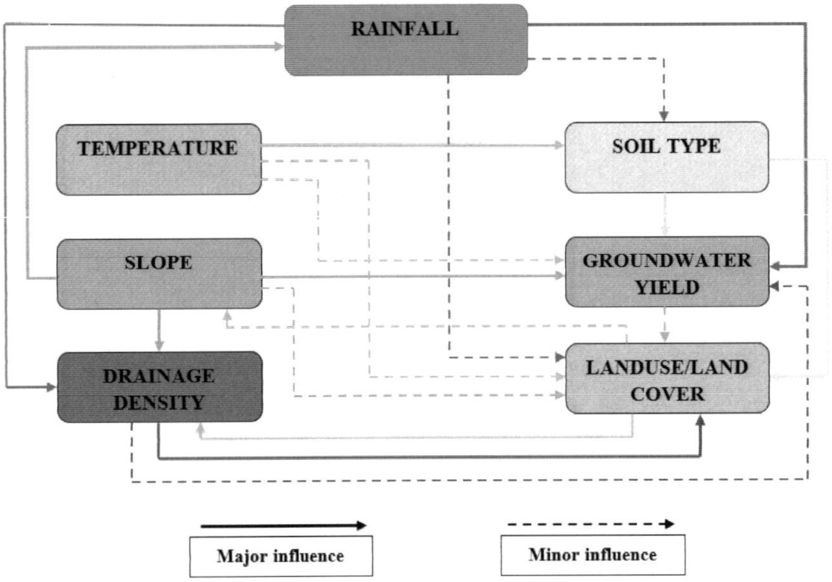

Figure 2. Multi influencing factors and their interrelationship used for mapping drought prone zone

The effect of each major and minor factor is assigned a value of 1.0 and 0.5 respectively. The proposed score of each influencing factor has been calculated as:

$$\text{MIF Score } (S) = \left[\frac{(A+B)}{(\Sigma\, A+B)} \right] * 100 \tag{3}$$

Where, A is major influencing factor and B is minor influencing factor.

4. Results and Discussion

In the weighted overlay analysis in GIS, the weights and ranks of each parameter used in AHP and MIF are shown in table 5. The weights of each parameter are multiplied by each sub-class of the parameters, which is again ranked according to the nature of the influence of each sub-class on the parameter. The rank of each sub-class influence is ranked from 1 to 5.

Table 5. AHP and MIF rank and weight for mapping Drought prone zone

Sl. No.	Parameter	Rank	Vulnerability	AHP weight	MIF weight
1	Rainfall			0.332	22
	61-69	5	very high		
	70-75	4	high		
	76-80	3	moderate		
	81-85	2	low		
	86-93	1	very low		
2	Temperature			0.267	19
	30.7-31.4	1	very low		
	31.5-31.9	2	low		
	32-32.3	3	moderate		
	32.4-32.8	4	high		
	32.9-33.4	5	very high		
3	Slope			0.104	26
	0-3	1	very low		
	3.1-8.5	2	low		
	8.6-17	3	moderate		
	18-27	4	high		
	28-58	5	very high		
4	Drainage density			0.094	26
	0-1.2	5	very high		
	1.3-2.6	4	high		
	2.7-4.1	3	moderate		
	4.2-6.2	2	low		
	6.3-11	1	very low		
5	Landuse/landcover			0.043	11
	Vegetation	4	high		
	Built-up	3	moderate		
	Agriculture	5	very high		
	Barren	2	low		
	Waterbody	1	very low		
6	Soil type			0.065	7
	Clay	1	very low		
	Clayey loam	1	very low		
	Marshy land	1	very low		
7	Groundwater yield			0.096	4
	8-100	5	very high		
	110-230	4	high		
	240-400	3	moderate		
	410-600	2	low		
	610-1100	1	very low		

4.1. Climatic Parameters (Rainfall & Temperature)

Among the parameters compared, rainfall is the main parameter directly affecting the drought condition of a region. The drought condition is also controlled by the amount of rainfall, intensity of rainfall, and rainfall seasons. Generally, rainfall may be normal or above average in a year if it feeds the reservoirs, groundwater aquifers, and allows for efficient production of crops. With 15 weather station location data points, the rainfall map of the study area has been interpolated using the Inverse Distance Weightage (IDW) tool in ArcGIS®. The annual monthly average rainfall distribution map is classified into five classes with a maximum average rainfall of 61 mm to 93 mm (figure 3a). On the scale of the impact of drought, the region with low rainfall is assigned a higher rank and a lower rank to the areas receiving maximum rainfall. The AHP weight of rainfall given is 0.332 (approximately 34%) and the MIF weight of 22%.

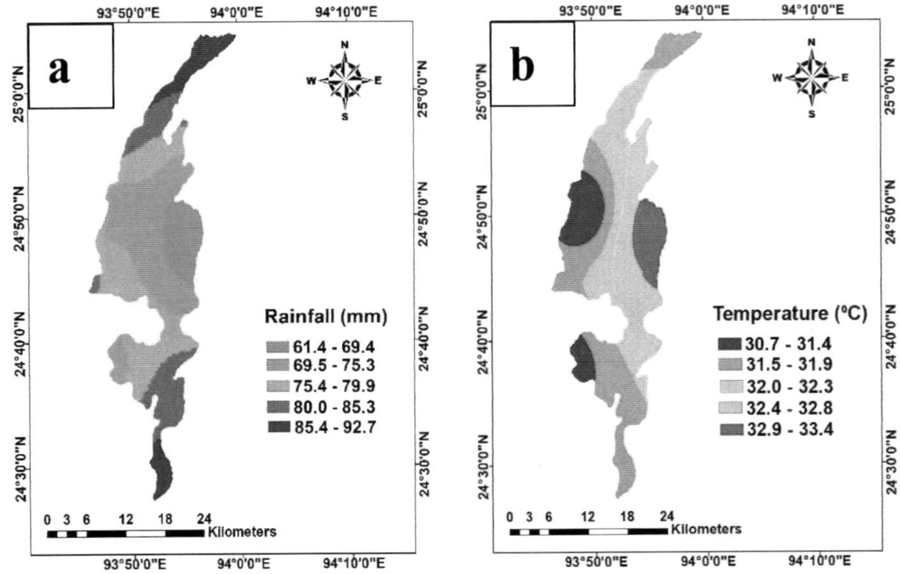

Figure 3. Climatic parameters (annual monthly average) for the year 2019 (a) rainfall and (b) temperature

The annual monthly average temperature of each rainfall station was used for the preparation of the temperature raster map. Temperature in a region also plays a significant role in the assessment of drought zones. It is more likely to observe severe drought zones in areas with higher temperatures and high evapotranspiration. With 15 weather station locations, the temperature map of the study area has been interpolated using the IDW technique. The annual average monthly maximum temperature of the study area ranges from 31°C to 33°C. The temperature ranges are classified into five classes (figure 3b). On the scale of the impact of drought, the region with high temperatures is assigned a higher rank, and the areas experiencing low temperatures are given a lower rank. The AHP weight of the temperature given is 0.267 (approx. 27%) and the MIF weight of 19%.

4.2. Hydrological parameters (slope, drainage density & groundwater yield)

The slope map (figure 4a) generated using SRTM DEM (30 meter) resolution data acts as a decisive factor for the rainfall-runoff and rainfall-recharge process. A high-slope region experiences high runoff and therefore less infiltration capacity. The study area is surrounded by a few steep slopes and more flat and gentle slopes, experiencing good runoff quantity and a high infiltration rate in the less sloping area. Thus, the region with a high slope is more vulnerable to drought and less vulnerable to low-slope regions. The slope class of the region is divided into five classes, ranging from 00 to 580. The AHP weight of the slope given is 0.104 (approx. 10%) and the MIF weight of 26%.

The drainage density is defined as the ratio of the total length of the stream channel per unit area of the region. A region with highly permeable terrain has low drainage densities and hence less runoff potential and vice-versa. Also, a region with high drainage densities has the capacity to carry large volumes of surface runoff to different stream channels. The drainage density map (figure 4b) of the region is classified into five classes. In the weighted overlay tool analysis, the highest scale of drought effect is given to the low drainage density region, where the chances of drought prone zone are higher, whereas a value of low scale is given to the high drainage density region. The AHP weight of drainage density given is 0.094 (approx. 9%) and the MIF weight of 11%.

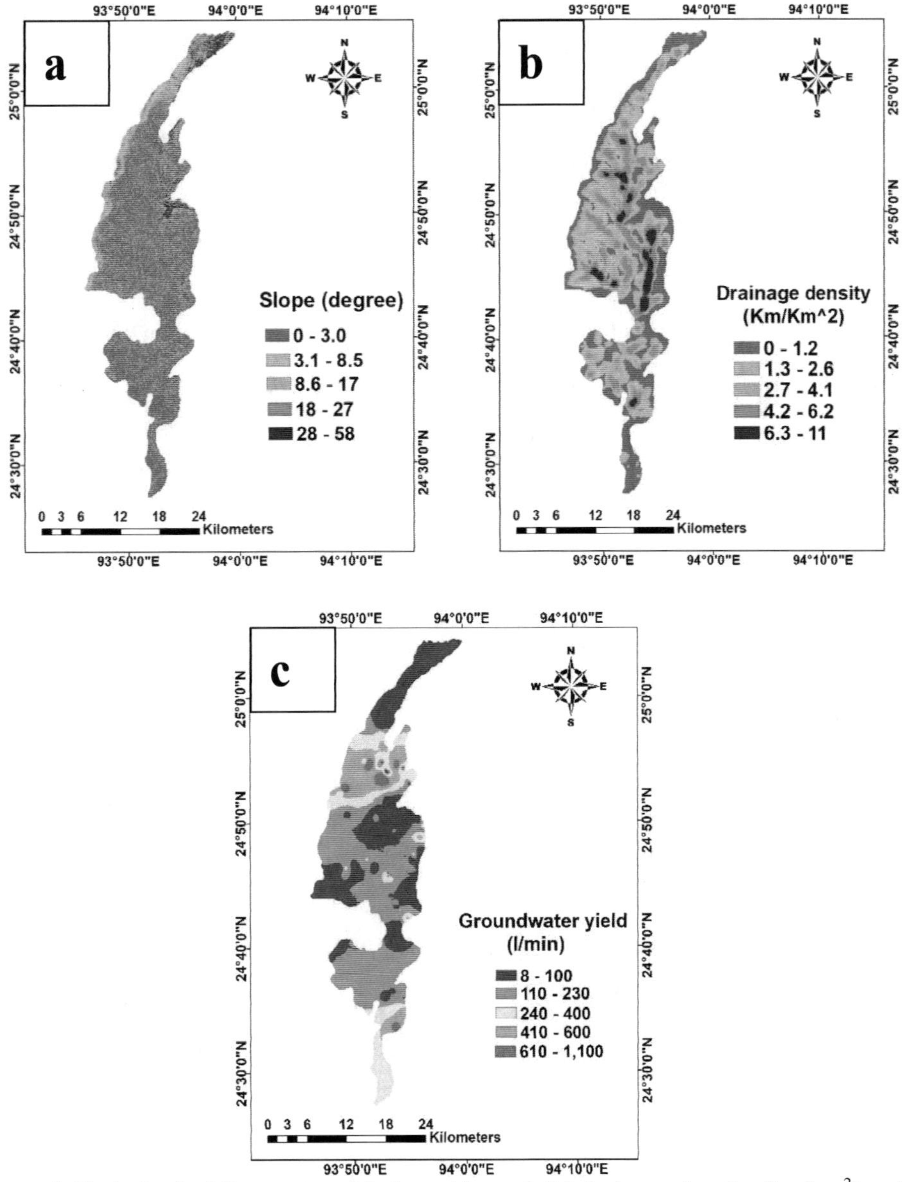

Figure 4. Hydrological Parameters (a) slope (degree) (b) drainage density (km/km^2) and (c) groundwater yield (l/min)

The groundwater yield capacity can act as an indicator of the severity of drought in a region. The groundwater in a region is affected by various reasons such as rainfall variation, soil type and texture, infiltration capacity of the soil, soil moisture, temperature and climate conditions of the region. With 117 tube well location data, the groundwater yield map of the study area has been interpolated using the IDW technique. Regions with high groundwater yield are a good indication of well-drained surface water and base flow to the region and hence less likely to be affected by drought. Thus, a lower rank is assigned to a region with a high yield capacity on the vulnerability scale. The groundwater yield capacity (litres per minute) of the study area is classified into five classes (figure 4c). The AHP weight of the groundwater yield given is 0.096 (approx. 10%) and the MIF weight of 4%.

4.3. Soil type and Landuse/landcover (LULC)
The type of soil available in a region greatly influences the capacity of water recharge, type of plant/crops, and growth of the plant. Soil-water interaction plays a huge role in the growth of plants and crops. The texture of soil also greatly influences the permeability and water-holding capacity in a region. The soil map (figure 5a) of the region is classified into three types: clay, clay loam, and marshy land. All three soil types have a good capacity to hold or store water; they are less likely to pose a threat to drought conditions. However, considering other factors discussed in the study may affect the quality and quantity of the soil condition in the study area, The AHP weight given to the soil type is 0.065 (approx. 6%) and the MIF weight of 7%.

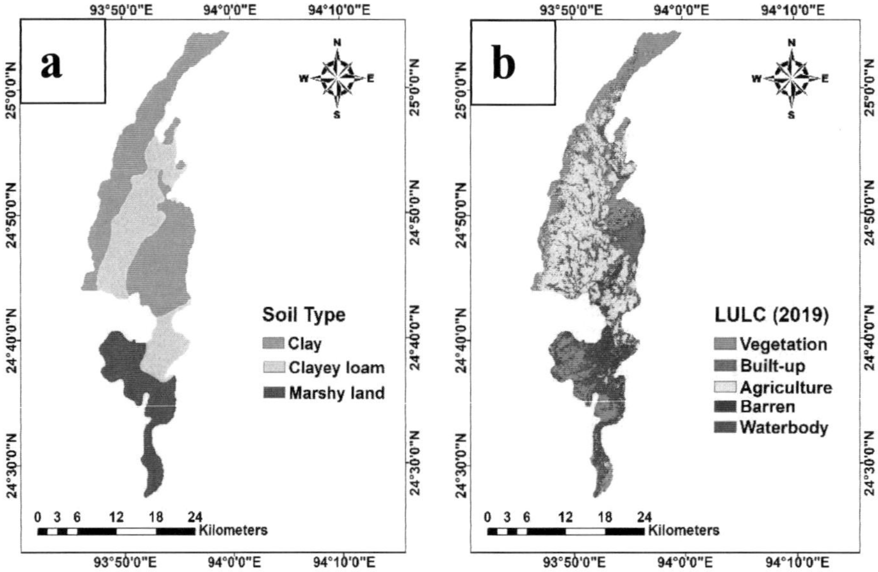

Figure 5. (a) Soil type and (b) LULC (2019) maps

The landuse/landcover pattern is concerned with the type of land cover such as vegetation, agriculture, waterbodies, built-up, etc. on the surface of the earth and how this landcover is utilised and managed. It has a significant effect on the hydrological cycle of that region. Its role includes runoff, evapotranspiration, and soil erosion in a region. In the study, the LULC maps (figure 5b) are obtained using supervised classification, and five classes of landcover are classified as vegetation, agriculture, built-up, barren, and waterbodies. The impact of drought is more likely to be in agriculture and vegetation (forest, marshy), than in built-up, barren land and waterbodies (ordered in high to low vulnerability). Accordingly, the scale of impact was given on the weighted overlay analysis. The AHP weight of the LULC given is 0.043 (approx. 4%) and the MIF weight of 11%.

5. Identification of drought prone zone and validation
The drought zone map of the study area is obtained using two methods, namely the AHP and the MIF method, as shown in figure 6. The drought regions (local names) are sekmai, lamsang, salam, konthoujam, patsoi, hiyangthang, wangoi, lilong chajing, and mayang Imphal, which are indicated in figure 5. The drought zone is classified into four classes: mild, moderate, severe, and extreme drought. In addition, the simulated drought prone zones have been verified using the data from the Department of Agriculture, Government of Manipur [27] for the drought affected locations (figure 7).

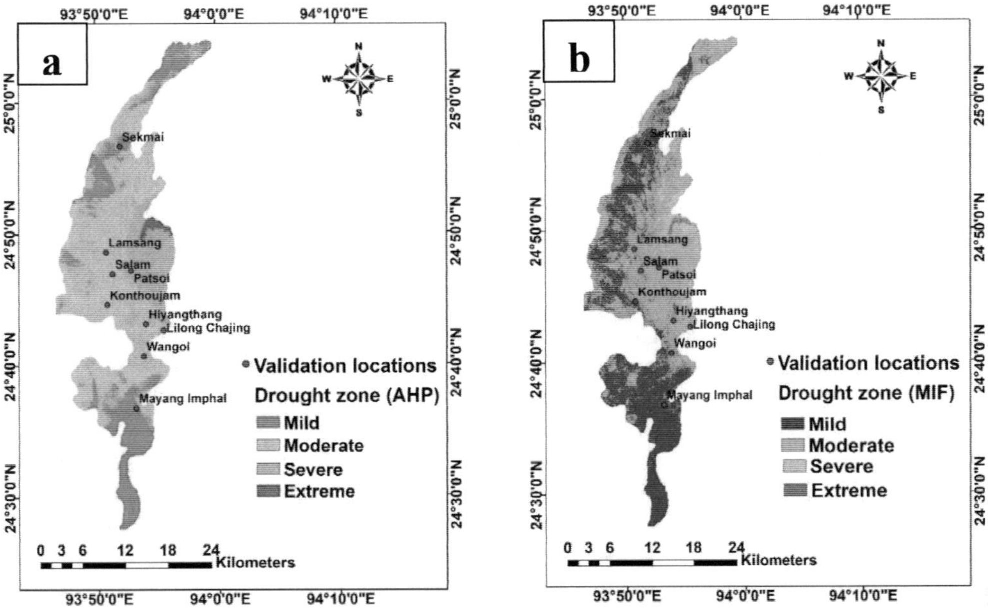

Figure 6. Drought prone zone map of study area (Imphal-West District) obtained using (a) AHP method and (b) MIF method

As depicted from figure 7, the percentage (%) of drought affected area simulated by AHP as mild is 22.82%, moderate is 60.10%, severe is 16.16%, and extreme is 0.92%, whereas by MIF it is 39.42% mild, 54.71% moderate, 5.55% severe, and 0.32% extreme. Ground data from the State Government indicates that 43.71% mild and 51.32% moderate drought in the study area, which shows that MIF is more effective in accurately simulating the drought affected area than the AHP method.

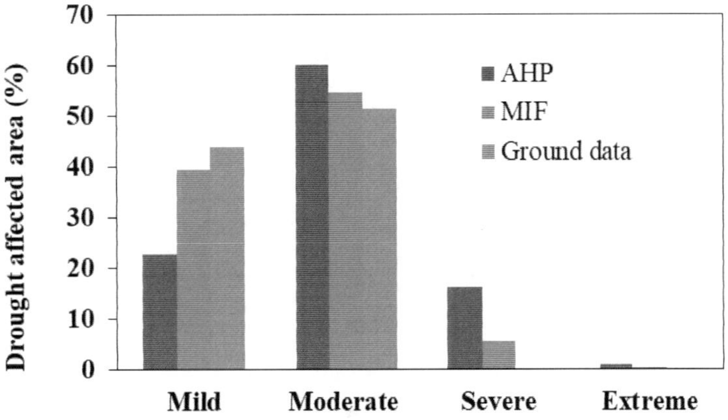

Fig.7: Area (%) of drought pone zone in the study area

It is important to note that the data was assessed for paddy crops affected by the drought that occurred in the Imphal-West district in the year 2019. Since rice is the main crop and mostly cultivated crop in Manipur, assessments of drought conditions are the most essential factor for food security and socio-economic concerns.

6. Conclusion

A study on the identification of drought prone zones in the Imphal-West district, Manipur was conducted. Through this study, the efficacy of RS & GIS integrated with the AHP method and MIF method has shown the huge potential of defining drought zones in the region. Using GIS and remote sensing products has considerably reduced the necessity of large field inspections and operations in reachable as well as inaccessible regions. The method, in addition to the field data, aims to improve or provide other alternatives to study or identify drought zones when limited resource and time constraints are pushed. Thus, policy and decision-makers can make a swift response to any drought condition in a short period of time. Also, the AHP method has eased up in assessing the drought prone with its pairwise comparison matrix among the parameters, which helps in ranking the influencing factors for the identification of drought zones. The normalised weights obtained from the method have efficiently solved the hierarchical relationships among the parameters such as rainfall, temperature, slope, drainage density, soil type, landuse/landcover, and groundwater yield where these weights are used in the weighted overlay analysis. Similarly, the MIF method has helped in identifying the major and minor influencing factors for mapping the drought prone zone areas. The output obtained from the weighted overlay analysis shows zones with mild, moderate, severe, and extreme drought cover over the study area. Further, these drought zones are validated or supported by the government documents collected from the Imphal-West Agriculture Department, Government of Manipur. The data thus collected validates the region affected during the year 2019 in which the government officially announced/acknowledged and also reported in the news articles the need to take necessary precautionary measures and actions to relieve the farmers as well as the communities affected by the drought conditions. The present study only highlights the alternative and simple methods for the identification of drought zones in the region. Later in future studies, studies conducted on drought indicators using indices such as normalised difference vegetation index (NDVI), standard precipitation index (SPI), vegetation condition index (VCI), etc. are suggested. Researchers incorporating future climate models and predictions on drought severity conditions can improve and help in alerting the government, policy and decision-makers, farmers, and communities at large to necessary future measures.

References

[1] Thomas T, Jaiswal RK, Galkate R, Nayak PC and Ghosh NC 2016 *Nat. Hazards* **81**(3) 1627-52
[2] Pei W, Fu Q, Liu D, Li T, Cheng K and Cui, S 2019 *Water Resour. Manag* .**33**(6) 2033-47
[3] Zhang Q, Sun P, Li J, Xiao M and Singh V P *Theor. Appl. Climatol.* **121**(1) 337-47
[4] Pei Z, Fang S, Wang L and Yang W 2020 *Water* **12**(7) 1925
[5] Shahid S and Behrawan H 2008 *Nat. Hazards* **46**(3) 391-413
[6] Shamsipour AA, Zawar-Reza P, Alavi Panah SK and Azizi G 2011 *Int. J. Remote Sens.* **32**(24) 9559-69
[7] Dalezios NR, Blanta A, Spyropoulos NV and Tarquis AM 2014 *Nat. Hazards Earth Sys. Sci.* **14**(9) 2435-48
[8] Paparrizos S, Maris F, Weiler M and Matzarakis A 2018 *Theor. Appl. Climatol.* **131**(1) 259-70
[9] Bhardwaj K and Mishra V 2021 *Water Secur.* 14100104
[10] Pandey RP, Pandey A, Galkate RV, Byun HR and Mal BC 2010 *Manag.* **24**(15) 4199-217
[11] Palchaudhuri M and Biswas S 2016 *Nat. Hazards* **84**(3) 1905-20
[12] Das S, Gupta A and Ghosh S 2017 *Spat. Inf. Res.* **25**(6) 749-56
[13] Thomas R and Duraisamy V 2018 *Egypt. J. Remote Sens. Space Sci.* **21**(2) 121-37
[14] Abijith D, Saravanan S, Singh L, Jennifer JJ, Saranya T and Parthasarathy KSS *Hydro Res.* **3** 1-14
[15] Taheri K, Missimer TM, Taheri M, Moayedi H and Mohseni PF 2020 *Nat. Resour. Res.* **29**(2) 1163-91
[16] Serele C, Pérez-Hoyos A and Kayitakire F 2020 *Geosci. Frontiers* **11**(4) 1403-13
[17] Pandey P, Tiwari SK, Pandey HK, Chaurasia AK and Singh S 2021 *Water Conserv. Sci.*

Eng. **6**(3) 105-25
[18] Kumari S, Poddar A, Kuma, N and Shankar V 2021 *Model. Earth Syst. Environ.* 1-12
[19] Pande CB, Moharir KN, Panneerselvam B, Singh SK, Elbeltagi A, Pham QB and Rajesh J *Appl. Water Sci.* **11**(12) 1-20
[20] Zagade ND and Umrikar BN 2021 *Nat. Hazards* **105**(2) 1165-88
[21] Pandey S, Pandey AC, Nathawat MS, Kumar M and Mahanti NC 2012 *Nat. Hazards* **63**(2) 279-303
[22] HoqueM A A, Pradhan B and Ahmed N 2019 *Sci. Total Environ.* **705** 135957
[23] Hoque M, Pradhan B, Ahmed N and Alamri A 2021 *Sensors* **21**(20) 6896
[24] Saaty TL 1977 *J. Math. Psychol.* **15**(3) 234-81
[25] Saaty TL 1990 *Eur. J. Oper. Res.* **48**(1) 9-26
[26] Census report 2011 *District Census Handbook for Imphal West* (Government of Manipur, Manipur)
[27] Drought report 2019 *Damage to agriculture crops due to drought in Imphal West* (Agriculture Department, Government of Manipur, Manipur)

Standard Operating Procedure for Assessment and Rating of National Highways based on Highway Efficiency

Agrita Singh[1] and Sunil Sharma[2]

[1]MTech Scholar, Department of Civil engineering, National Institute of Technology Hamirpur, India
[2]Assisstant Professor, Department of Civil engineering, National Institute of Technology Hamirpur, India

Email: 20mce203@nith.ac.in

Abstract. The transportation network and ways in which it is maintained are critical to the development of a nation. National Highways (NHs) constitute only 2.3 per cent of total road network but carry 40 per cent of road traffic in India. National Highway authority of India (NHAI) and Ministry of Road Transport and Highways (MoRTH) of India have framed Standard Operating Procedure (SOP) for assessment and rating of divided 4 to 6-lane NHs based on highway efficiency, highway safety, and user services for a total of 100 marks. The present study has assessed and rated a 4-lane NH based on highway efficiency carrying 45 marks. A manual survey along with videography was conducted on the Kalka-Shimla NH No. 5 to collect road features and traffic data. There are 15 parameters which will give the overall efficiency of the highway enabling one to assess where and how much each parameter is inefficient, dangerous, and deficient in certain functions. The highway got 35.92 marks out of 45 marks which is approximately 80% score indicating excellent performance of various parameters on the selected highway. The efficiency assessment of highways will not only help in prioritizing highways for maintenance and upgradation but also identify accurately the missing points.

Keywords: National Highway, Standard Operating Procedures, Highway Efficiency

1. Introduction

Although the construction and development of new roads and transportation infrastructure has been emphasized in the past, the maintenance and rehabilitation of existing facilities has become increasingly prominent in recent years [1]. As of December 2019 the length of NHs in India is 1, 32,500km [2]. NHs through improved road quality provides comfort, speed, safety, access and lower vehicle operating costs. A highway efficiency assessment could be defined as the comparison between current status of roads in terms of road capacity, traffic volume, geometric design, etc. and standard of roads prescribed for well-functioning of roads. It is first of its kind initiative in the world. This 'Benchmarking and Rating of Corridors' program will aid us in developing standards for outstanding highway services, and improved highway quality will provide road users with better value for their money. Furthermore, increased highway quality in India will result in a safer driving experience. For this experience several road network evaluation systems have gained popularity in recent years, including: CEEQUAL, Greenroads, Envision, STARS, Green Guide for road, INVEST, GreenLITES, BE2ST-In-Highways etc. In India, initiatives to develop/implement a rating system have been made. Both the construction of new road sections and the upkeep of existing ones require an effective road rating system. Road rating systems are designed to establish standards that may be used to identify areas of success and areas for improvement. For highways, a rating framework with 56 indicators and 170 credits has been proposed, divided into six categories: site selection and planning, energy conservation, water conservation, material and resource conservation, environmental quality, and design innovation [3]. Another rating system has been presented that may assess the level of safety of any road network. For vehicle occupants, motorcyclists, pedestrians, and bicyclists, the iRAP model was utilized to create a baseline star rating for the existing route network [4]. A technique for establishing a green highway rating system in India has been presented in order to determine if a road can be classified as greener in terms of its use of green technologies [5]. Similarly, several efforts were made to improve the efficiency of the road network. A data envelopment study and the Malmquist productivity and efficiency index, which combine a single financial framework and institutional data for each Portuguese concession, were used to examine the efficiency of seven highway projects in Portugal [6]. The elements impacting prioritization and the relative relevance of each were discovered through an expert survey in order to propose a scientific technique for prioritizing pavement maintenance. The Analytic Hierarchy Process (AHP) was used to determine the weight (importance) of each component based on its relative importance to other factors in a scientific manner [7]. Using AHP, a similar study was conducted to produce a prioritization model or priority index (PI) for urban pavement section maintenance (analytical hierarchy process) [8]. It may be observed that the various road rating systems in use around the world differ in nature due to their pre-defined set of criteria. The efficiency and safety of roadway networks are both influenced by highway geometry. Highway authorities should focus strongly on improving highway geometry since it is an important component of the road network [9]. Having this in mind NHAI and MoRTH of India have framed Standard Operating Procedure (SOP) for assessment and rating of divided 4 to 6- lanes NHs based on highway efficiency, highway safety, and user services for a total of 100 marks. The present study has assessed and rated a 4-lane NH based on highway efficiency carrying 45 marks. A manual survey along with videography was conducted on the Kalka-Shimla NH No. 5 to collect road features and traffic data. There are 15 parameters which will give the overall efficiency of the highway enabling one to assess where and how much each parameter is inefficient, dangerous, and deficient in certain functions as well as accordingly categorize the highway section as excellent, good or poor in efficiency parameters.

2. Methodology

The methodology adopted to carry out the study is shown with help of a flowchart which explains in detail what work was carried out during the site visit. Figure. 1 shows the flowchart of methodology adopted.

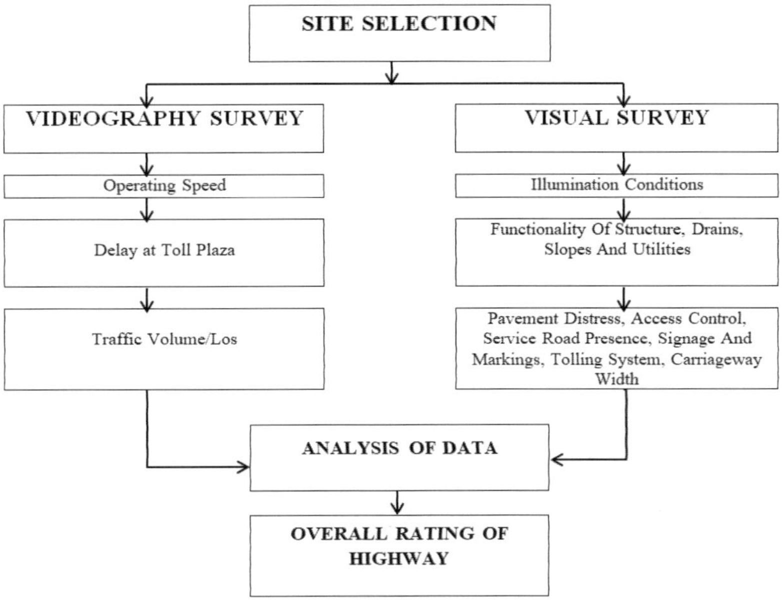

Figure 1. Methodology flow chart

2.1 Study area

NHs is maintained by NHAI. NH-5 is a 660.2 km long highway connecting the towns of Shimla and Solan in the state of Himachal Pradesh in India. A 5 km stretch on this highway starting from Sanwara toll plaza to Jabli market (80+540 to 76+430 chainage) has been selected as a case study. The toll plaza is selected in order to extract data like traffic volume, toll collection etc. Figure. 2 shows the site location from Sanwara toll plaza to Jabli market a 5 km stretch taken from Google Earth.

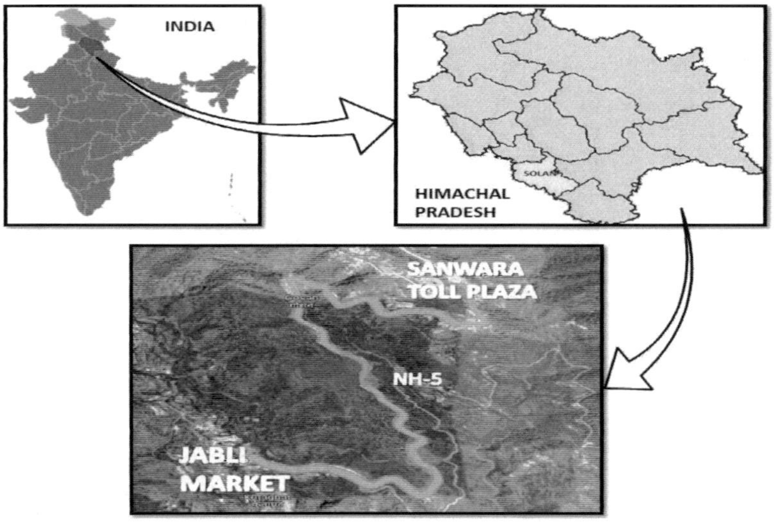

Figure 2. Selected site location (**Source**: Google Earth)

3. Data analysis and procedure

3.1 Sub parameters of highway efficiency

3.1.1. Operating speed

The speed survey was conducted with video camera installed at both ends of survey location and data was extracted from it. The difference in the times was noted and for every vehicle, taken as running time. So every vehicle's average speed has been estimated. by considering distance of stretch taken in consideration divided by the running time. Figure. 3 shows camera installation point showing the movement of vehicles before and after toll plaza crossing at the survey location Table 1 shows the weightage to be given for operating speed. Weightage chart is given in procedure of NHAI guidelines [10].

Table 1. Weightage chart given for evaluation of marks for operating speed

Weightage - 17 Marks		
Full marks	Intermediate/half marks	Zero marks
>50kmph	0.85 marks deducted for every 1 kmph reduction in speed	<40 kmph

Figure 3. IN and OUT movement of vehicle recorded during videography for calculation

3.1.2. Delay at toll plaza
Delay was calculated using videography on the toll booth and data was extracted for each vehicle. Data was collected only through the FASTag lanes. At the toll plazas, an In and Out survey was conducted, and the average delay was computed. Table 2 shows the weightage to be given for delay studies.

Table 2. Weightage chart given for evaluation of marks for delay at plaza

Weightage - 4 Marks		
Full marks	Intermediate/half marks	Zero marks
<1 minute	0.4 marks deducted for every 12 second delay	>3 minutes

3.1.3. Traffic volume/Capacity Ratio/Level of Service (LOS)
The volume of the highway was determined using the Annual Average Daily Traffic (AADT) for the current fiscal year that was gathered from ATCC/IHMCL at the toll plaza. The designed capacity of the 4 lane to be taken as 20,000 for mountainous and steep terrain. The Actual Traffic Volume/Design traffic for Level 'B' Service ratio of less than or equal to 1.0 shall be taken as LOS "B" and 1.0-1.5 should be taken as LOS "C" and higher than 1.5 shall be taken as LOS "D". LOS was calculated using V/C ratio. Table 3 shows the weightage to be given for Traffic volume/capacity ratio/level of service.

Table 3. Weightage chart given for evaluation of marks for LOS

Weightage – 2.5 Marks		
Full marks	Intermediate/half marks	Zero marks
B	C (1.25 marks)	D

3.1.4. Percentage illumination in habitation/built-up area
Every single location, including built-up areas, toll plaza areas, truck lay-bys, rest areas, bus bay and bus shelter locations, interchanges, flyovers, underpasses, overpasses, and any other place specifiedas per clause 1.19.2 IRC SP 84 was identified as locations ideally to be illuminated as per desired lux level and taken as Ideal Value [11]. The number of locations that actually have the desired lighting as measured in reality. Figure. 4 shows the illumination conditions at the site location present solely due to road side amenities. Table 4 shows the weightage to be given for illumination condition of highway.

Table 4. Weightage chart given for evaluation of marks for illumination on highway

Weightage – 2 Marks		
Full marks	Intermediate/half marks	Zero marks
100%	0.02marks deducted for every 1% decrease in illumination availability	0%

Figure 4. Highway lighting during night

3.1.5. Functionality and adequacy of structures: major/minor bridges, guide bunds, river training structures etc.

The length of all structures (bridges, culverts, river trains, and works) taken as a whole and added to determine their ideal functional length. The total of all the good lengths to be added and taken as actual functional length. The actual functional length expressed as percentage of ideal functional length. Figure. 5 shows the damaged pitching works and safety barriers which are in very bad conditions. Table 5 shows the weightage to be given for adequacy of structures.

Table 5. Weightage chart given for evaluation of marks for adequacy of structures

Weightage – 1 Marks		
Full marks	Intermediate/half marks	Zero marks
100%	0.01 marks deducted for every 1% decrease in functionality	0%

Figure 5. Damaged pitching works and safety barriers at site

3.1.6. Percentage uniformity of carriageway width (road with structure)

All of the locations (Bridges, Grade Separators, and Culverts) where the main roadway changes into a structure were examined properly. Section where the width of the approach road without paved shoulders was less than the width of the structure, then zero marks was given otherwise full marks. Table 6 shows the weightage to be given for uniformity of carriageway.

Table 6. Weightage chart given for evaluation of marks for uniformity of carriageway

Weightage – 1 Marks	
Full marks	Zero marks
Width of all structures > width of corresponding approach roads	Width of even a single structure < width of approach road

3.1.7. Functionality of drains on main carriageway

Both physical condition and sanitation of the drains with in built-up areas and some other significant site-specific places were assessed visually. For each stretch of drain with each built-up area/important location, observations were made at a minimum of 5 locations spaced no more than 200m apart to check for blockages and lining breaks. Figure. 6 shows the drainage system present at the site location at regular interval. Table 7 shows the weightage to be given for functionality of drains.

Table 7. Weightage chart given for evaluation of marks for functionality of drains

Weightage – 1 Marks		
Full marks	Intermediate/half marks	Zero marks
100%	0.01 marks deducted for every 1% decrease in functionality	0%

Figure 6. Drainage system is at regular intervals

3.1.8. Functionality and appearance of slope projection

The embankments were checked for a randomly selected 1m in every 100m of the embankment's length. Video filming was done in addition to the panel's ocular inspection. Table 8 shows the weightage to be given for Functionality and appearance of Slope Projection.

Table 8. Weightage chart given for evaluation of marks for functionality and appearance of slope projection

	Weightage – 0.5 Marks	
Full marks	Intermediate/half marks	Zero marks
100%	0.005 marks deducted for every 1% decrease in normal slope	0%

3.1.9. Existing utilities laid as per MoRTH norms

The length of the utilities installed along or across the highway was determined and regarded as an ideal value. The facilities along the highway must be installed at least 15 metres beyond the carriageway's centerline and 0.6 metres below the surface of the earth. Actual value with standard deviation represented as a percentage of the ideal value. Table 9 shows the weightage to be given for Existing utilities.

Table 9. Weightage chart given for evaluation of marks for existing utilities

	Weightage – 0.5 Marks	
Full marks	Intermediate/half marks	Zero marks
100%	0.005marks deducted for every 1% decrease in utilities laid as per MoRTH guidelines	0%

3.1.10. Pavement quality rating value

Various defects like potholes, cracks, shovel etc. were detected using visual survey. The measurements were taken with help of measuring tape; wheel meter and metre scale. Table 10(a) shows the weightage to be given for quality of pavement. Table 10(b) shows the nature of defects along with their weightage. Figure. 7 shows the pavement distress showing patching and cracking present at the location.

Table 10 (a). Weightage chart given for evaluation of marks for quality of pavement

	Weightage – 6 Marks	
Full marks	Intermediate/half marks	Zero marks
2.1-3.0	1.1-2.0 (3 marks)	1.0

Table 10 (b). Nature of defects and their weightage

Defects	Range of Distress			Weight over a scale of 5.75
	I	II	III	
Cracking (in %)	>10	5 to 10	<5	1
Raveling (in %)	>10	1 to 10	<1	0.75
Potholes (in %)	>1	0.1 to 1	<0.1	0.5
Shoving (in %)	>1	0.1 to 1	<0.1	1
Patching (in %)	>10	1 to 10	<1	0.75
Depression (in %)	>5	1 to 5	<1	0.75
Marks	1	1.1-2.0	2.1-3.0	

Figure 7. Pavement distress (Patching, Cracking)

3.1.11. Access control

Expressways are to be considered completely access controlled. Partial Access controlled highways shall be defined as the highways with no median opening and no right turning movement at any of the at-grade intersections. Table 11 shows the weightage to be given for access control.

Table 11. Weightage chart given for evaluation of marks for access control

Weightage - 2 Marks		
Full marks	Intermediate/half marks	Zero marks
Complete	Partial(1 marks)	None

3.1.12. Percentage availability of service road in built-up area

The ideal value was noted throughout the corridor as the length of the actual part of the habitation area as of the date. It was measured and the actual length for service road that was offered in the habitation zone part was used. the percentage of the actual value over the ideal value. The term "built up area" has been defined as the area used for habitation as per clause 1.19.2 IRC SP 84 [11]. Table 12 shows the weightage to be given for Availability of service road in habitation area.

Table 12. Weightage chart given for evaluation of marks for availability of service road in habitation area

Weightage - 3 Marks		
Full marks	Intermediate/half marks	Zero marks
100%	0.03marks deducted for every 1% decrease in service road availability	0%

3.1.13. Availability of closed tolling

Closed tolling (toll collection based on actual distance travelled) to be checked in the influence length under consideration. Table 13 shows the weightage to be given for Availability of closed tolling.

Table 13. Weightage chart given for evaluation of marks for Availability of closed tolling

Weightage – 0.5 Marks	
Full marks	Zero marks
Yes	No

3.1.14. Percentage of substandard signage
According to IRC 67, every area where road signs were to be installed was identified, and the total number of these locations was regarded as the ideal value. [12]. The number of appropriate traffic signs was counted and assigned actual value. The actual value to be expressed as a percentage of the ideal value. The percentage of incorrect road signs is equal to 100% less the percentage of true value. Figure. 8 shows the improper road signs installed. Table 14 shows the weightage to be given for Percentage of substandard signage.

Table 14. Weightage chart given for evaluation of marks for Percentage of substandard signage

	Weightage – 2 Marks	
Full marks	Intermediate/half marks	Zero marks
0%	0.01 marks deducted for every 1% increase in improper signage	100%

Figure 8. Sign board too close to carriageway and broken

3.1.15. Percentage of substandard markings
According to IRC 35, the lengths wherein road markings should be erected were identified, and the total length of these sites was considered as the optimal value [13]. The length of proper road markings identified, and taken as actual value. The actual value expressed as a percentage of the ideal value. The percentage of incorrect road markings is equal to 100% less the percentage of true value. Figure. 9 shows the road markings not being visible and the retro-reflector is installed away from the carriageway. Table 15 shows the weightage to be given for Percentage of substandard markings.

Table 15. Weightage chart given for evaluation of marks for percentage of substandard marking

	Weightage – 2 Marks	
Full marks	Intermediate/half marks	Zero marks
0%	0.02marks deducted for every 1% increase in improper marking	100%

Figure 9. Improper road markings and retro-reflector

4. Result

A score sheet is prepared to represent the data collected and marks are given to each sub category in tabular form. From the table it can be concluded which parameter requires special attention. Table 16 shows marks allotted to each parameter and Sub-total efficiency and percentage efficiency calculation. Table 17 shows the classification of highway section as per efficiency percentage.

Table 16. Sub-total efficiency and percentage efficiency calculation

Sub parameters	**Marks obtained**
Operating Speed (50%Cars+50% Trucks) on main carriageway(17)	15.8
Delay at toll plaza (4)	3.91
Traffic Volume/Capacity Ratio/Level of Service (2.5)	2.5
Percentage illumination in habitation/built-up area (2)	0.0
Functionality and adequacy of structures: major/minor bridges, guide bunds, river training structures etc. (1)	0.98
Percentage uniformity of carriageway width (road with structure)(1)	1.0
Functionality of drains on main carriageway (1)	1.0
Functionality and appearance of slope projection (0.5)	0.5
Existing utilities laid as per MORTH norms (0.5)	0.49
Pavement quality rating value (6)	3.0
Access control (2)	2.0
Percentage availability of service road in built-up area (3)	1.14
Availability of closed tolling (0.5)	0.5
Percentage of substandard signage (2)	1.84
Percentage of substandard markings (2)	1.95
Sub-Total Efficiency	**35.92/45**
Efficiency Percentage (%)	**79.82**
Classification*	**Excellent**

Table 17. Classification of highway section as per efficiency percentage

Marks	35-45	25-35	15-25	< 15
*Classification	Excellent	Very Good	Good	Poor

5. Conclusion

The vast network of NHs in India carries majority of the traffic day and night. A high efficiency of this network is critical to India's economy. Periodic assessment and rating of these highways will help in monitoring their performance and suggesting improvements. The efficiency of a highway is dependent on parameters like pavement condition, shoulder width, traffic signs, traffic flow, adequacy of structure etc. A detailed study was carried out to assess these parameters. Following conclusions were drawn from manual and video graphic survey such as the overall highway efficiency rating of 5 km stretch came out to be 79.82 per cent falling under the category of 'excellent'. Even though classified as 'excellent', the highway lacked certain parameters like, there were illegal parking and boulder falling making it difficult for free and safe movement of vehicles, and by visual survey it was seen that delay was mostly due to machinery defects rather than FASTag recharges. Illumination was insufficient affecting the desired lux and was not in working condition at many places due to which it scored a zero. The pavement was constructed recently, so the quality of pavement was good with very few distresses present at long gaps. Traffic flow scored full marks indicating excellent traffic conditions on the highway. The primary reasons can then be intervened, recommendations, and corrective actions can be proposed to minimize them. This effort will aid in the development of benchmarking the assessment standards for Indian highway corridors.

References

[1] W. Version, "Vol. 11-No.1, Jan-Jun, Web Version," vol. 11, no. 1, 2021, doi: 10.13140/RG.2.2.15461.12006.

[2] Ministry of Power, "Year End Review 2019 – Ministry of Power," pp. 1–11, 2020, [Online]. Available: https://pib.gov.in/newsite/PrintRelease.aspx?relid=197411.

[3] Singh, J., & Jain, A. K. (2019). Development of Framework for Rating System for Indian Green Highways. *American Journal of Engineering Research: Stamford, CT, USA*, 250-259.

[4] Rogers, L., Smith, G., Devika, K. R., Kumar, H. S., & Prakash, G. S. (2012, September). Star rating road designs performance indicators for roads in India. In *ARRB Conference, 25th, 2012, Perth, Western Australia, Australia.*

[5] T. Bhattacharya and A. Shukla, " Procedure for Green Highways Rating System in India (GHRSI)," Vol. 78, no. 1, pp. 45-56, 2017

[6] Sarmento, J., Renneboog, L., & Matos, P. V. (2017). Measuring highway efficiency by a DEA approach and the Malmquist index. *European Journal of Transport and Infrastructure Research, 17.*

[7] Nautiyal, A., & Sharma, S. (2021). Scientific approach using AHP to prioritize low volume rural roads for pavement maintenance. *Journal of Quality in Maintenance Engineering.* https://doi.org/10.1108/JQME-12-2019-0111/FULL/PDF

[8] Prakasan, A. C., Tiwari, D., Shah, Y. U., & Parida, M. (2015). Pavement Maintenance Prioritization of Urban Roads Using Analytical Hierarchy Process. *International Journal of Pavement Research & Technology*, *8*(2).

[9] Shawky, M., & Hashim, I. (2010, June). Impact of horizontal alignment on traffic performance at rural two-lane highways. In *4th International Symposium on Highway Geometric Design* (Vol. 20, pp. 1-19)

[10] H. R. Report, "Standard Operating Procedure (SOP) for Assessing and Rating for divided carriageway (4 / 6 lane National Highways) based on : Highway Efficiency (45 Marks) Highway Safety (35 Marks) User Services (20 Marks)," no. i, 2021.

[11] IRC, "IRC SP: 84-2019 Manual of Specifications and Standards for Four Laning of Highways." p. 219, 2019.

[12] IRC, "IRC: 67-2012 Code of Practice for Road Signs".2012.

[13] Indian Road Congress, "IRC 35.pdf." 2015.

Strength Behaviour of Expansive Subgrade Soil Stabilized with Coronus Material

Sateesh Kumar Pisini and Swetha Priya Darshini Thammadi

Department of Civil Engineering, Fiji National University

Email: sateesh.pisini@fnu.ac.fj

Abstract. Expansive subgrade soil needs to be re-engineered to enhance the load-bearing capacity as it lacks the capacity to sustain traffic and pavement load. Although various chemical modification techniques have proven to be effective for stabilization, they are expensive and unsustainable. Coronus aggregate material is an uplifted coralline deposit abundantly available in the Pacific. The current study focuses on the strength behavior of fine-grained soil notably clayey soil when blended with coronus material as a potential alternative to the usual soil stabilization methods employed in the construction industry. To successfully stabilize the clayey soil sample with the coronus material, the particle size distribution, coefficient of uniformity, and curvature of the clayey soil sample and coronus material were determined. The results indicate that the coronus material is well-graded. The consistency of the soil sample was determined using the four-point Casagrande method for Atterberg limits i.e, the liquid limit, plastic limit, and shrinkage limit of the clayey soil. The load-bearing capacity of the clayey soil stabilized with coronus material was determined using standard compaction tests to calculate the maximum dry density and optimum moisture content of clay before and after blending with 10% increments of coronus material. The California Bearing Ratio (CBR) test results of all the samples were determined using standard CBR laboratory tests. It has been observed that the addition of coronus material has improved the load-bearing capacity of clayey soil. Hence, it is recommended that, wherever possible, coronus material be utilized as a stabilizing agent to improve the geotechnical properties of expansive subgrade soil.

1. Introduction

Soil Stabilization is a fundamental aspect of all Civil engineering applications as it is the foundation on which all construction projects are built. It is therefore critical that foundation soil has the strength and ability to support external loads acting on the soil. Traditionally, soil stabilization techniques have been effective in stabilizing the clayey soil thus improving the characteristics of soil and its behavior [1]. Clayey soils have often been the bane of civil engineers in general due to it's difficult characteristics. Low strength, high compressibility, and volumetric changes in the presence of water has given rise to the failure of structures due to settlement when the clay is consolidated. The necessity of altering clay soil with additives by geotechnical engineers has been adapted fora variety of structures such as embankments, roads, and structures that require a large load-bearing capacity from the soil. Clayey soils which are fine-grained soils, change their volume depending on the moisture content, that is, they swell when they get wet and shrink as they dry up. Fine-grained soils are more susceptible to moisture compared to coarse-grained soils since they have smaller air voids between the soil particles which reduces the ability of water to dissipate or pass through them easily thus causing them to be easily prone to swelling and shrinkage. Therefore, engineers need to take these challenges into account when constructing on clayey soils.

The Pacific basin is geologically characterized by deep ocean trenches associated with their tectonic activity. Both dead and living coral formations characterize atolls, which are weathered residues of past volcanic structures. Geology and Geography affect the availability of locally available pavement material [2]. The material obtained from the live or dead offshore reefs is known as coralline material. Coralline material is subdivided into coral, cascajo and coronus material [3]. The uplifted coralline deposit, inland limestone, is called coronus material [4]. It is abundantly available in the Pacific and is locally available in most Pacific Island Countries (PICs). Coral-based materials (natural limestone) have been utilized largely for road pavement structures since World War II in the Pacific [2]. It has been identified that including non-expansive clays in many coral-derived aggregates does allow for the material to perform satisfactorily as a surface aggregate [3, 5]. So far, lime (hydrated or quick) and cement are the most frequently used stabilizing agents in the Pacific while other options include foamed bitumen/cement, fly-ash, granular replacement additives, polymer additives, recycled glass, plastic and tires, and geosynthetic reinforcement [2].

Investigation of airport runways in Vanuatu and Tonga revealed that the overlying bitumen-bound surface was distressed, however, the coronus base still provided good support. It shows that with proper water drainage from the coronus base, it can last for a long time before it requires maintenance. In Tonga, for example, the base coralline limestone at the international airport displayed good support to the surface layer [2]. Public Works Department (PWD) Vanuatu states that coronus is one of the most widely spread materials in the country. It is used as a wearing course and as a concrete aggregate in Vanuatu's road works. It has varied properties in its natural state (grading, hardness, plasticity etc.), thus requires careful selection, handling, construction procedures, and equipment in order to produce the best outcome for a particular work [6].

Kingdom of Tonga which is a small island nation in the South Pacific looked at the successful application of otta seals using locally available aggregates. It has provided sustainable road maintenance for the country and provides promising solutions for other PICs faced with similar issues [7]. Another study focused on Solomon Islands and Vanuatu investigated the upgrade of concrete pavement for unpaved low-volume roads to improve connectivity and the overall quality of life associated with these roads [8]. However, no research was conducted to investigate the use of coralline limestone as a stabilizer for clay soil. In the PICs, larger islands like PNG, Fiji etc. coral and volcanic materials are abundantly available. Coral rock is the prevalent local aggregate source in Tuvalu, Kirbati, and small island atolls across the Pacific. However, pavement design and construction specifications are largely based on empirical methods [2]. The successful use of empirical design methods has proven to provide sustainable means of conducting pavement investigation, design construction and maintenance works in PICs provided that the design engineer understands the assumptions behind the use of these design methods. There is a need to find cost-effective ways to use locally available coronus material to stabilize fine-grained soil while considering factors such as the limited availability of funds, the local technical capacity, environmental sustainability, etc in the Pacific. The purpose of the current research is to determine the Geotechnicalproperties of clay-coronus soil mix and report the optimum combination of coronus material for pavement aggregate applications thus ascertain the suitability of coronus material for stabilization of clayey soils.

2. Methodology

Materials used in this study consist of Soil and Coronus material. The coronus material was extracted as huge chunks of rocks and manually crushed. The Coronus material was sieved through the 19mm sieve and the passing materials were used throughout all the tests. Experimental investigations were conducted to determine the geotechnical properties of soil-coronus mix samples by adding coronus material to the soil at incremental rates. The experiments were conducted by increasing the percentage of coronus material in soil-coronus mix to prepare the 100% sample (e.g. 10% coronus: 90% clay soil) and determine the optimum soil-coronus mix with maximum strength. Geotechnical laboratory tests were performed in accordance with Australian Standard AS 1289 series [9].

The Atterberg limits determine the critical points of moisture content of fine soils such as clay and silts. The procedures outlined in the current study complies with the AS 1289.3.3.1:2009, AS 1289.3.4.1:2008 and AS 1289.3.1.1:2009 for determination of Liquid limit, Plastic limit and Shrinkage limit. Particle size distribution method was used to determine the size and the range of the particles in both the soil (clay) samples and the Coronus material. The procedure was in accordance with AS1289.3.6.1:2009 [9]. Standard Proctor Compaction Test established the maximum unit weight using a controlled compact force at an optimum water content. California Bearing Ratio (CBR) test is a penetration test used to evaluate the strength of subgrade soil which eventually influences the thickness of road pavement and thus impacts the construction cost [10].

3. Results and Discussion

3.1 Atterberg Limits

Results of Atterberg limit test show the liquid limit (LL) recorded at 81.38% for subgrade soil. Since LL $\geq$ 50%. The plastic limit (PL) and shrinkage limit (SL) calculated were 10.61% and 17.41% respectively, thus the plasticity index (PI) determined from the LL and SL was 70.77%. Thus, soil can be classified as clay with high plasticity according to USCS classification. Since PI depicts the amount of clay in soil unlike the LL and PL which depend on the amount of clay contained in the soil, high PI implies that the clay content is very high in the soil. Also, it was observed that the experimental specific gravity formulated using the pycnometer test method for clay (2.36) is less than the specific gravity of coronus material (2.44). However, specific gravity for soils usually ranges between 2.60 to 2.80 [11]. The specific gravity of soil below this range is probably due to the inorganic material contained in the soil.

3.2 Particle size distribution

The results for the particle size distribution of clay soil as shown in Figure 1 indicate that 83% of the sample passes through the 75 microns (0.075mm) sieve which indicates that the amount of fine particles contained in the soil is very high. However, according to the soil classification table of the AS 1289.3.6.1:2009, the amount of soil passing includes both silt and clay since silt particles are 0.002mm to 0.06mm in size and clay is less than 0.002mm.

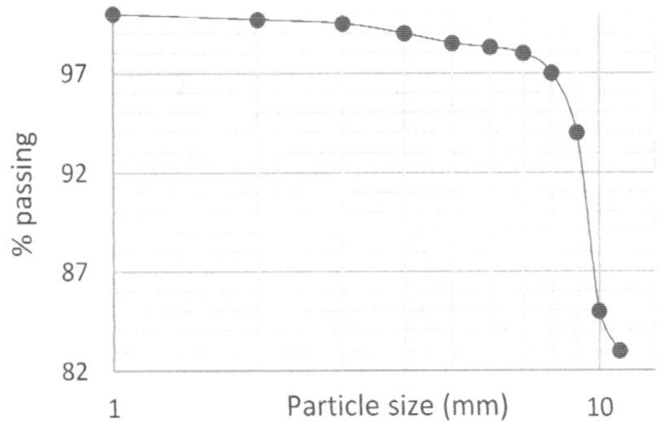

Figure 1. Particle size distribution of clay soil

Particle size distribution of coronus material illustrated in Figure 2 indicates the percentage sample passing through the 75 microns (0.075mm) sieve is 0%. It implies that the sample may be classified as coarse-grained soil according to the soil classification table of the AS 1289.3.6.1:2009. This is evident from the specific gravity values as well as physical examination. The coefficient of curvature (C_c) and the coefficient of uniformity (C_u) determined for Coronus material were 1 and 5.6 respectively. USCS soil classification is well-graded gravel.

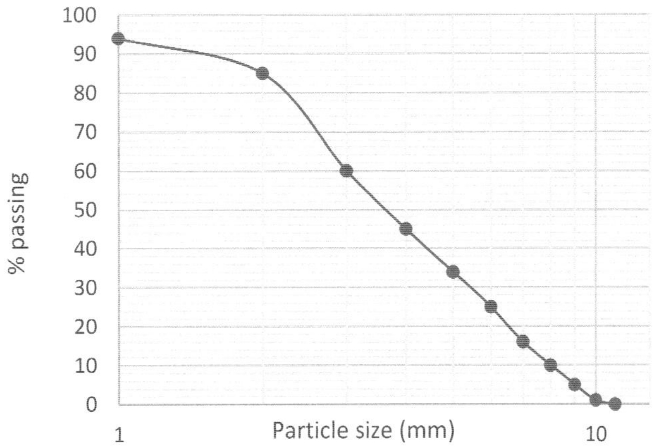

Figure 2. Particle size distribution of coronus material

3.3 Compaction test results

As per the Standard Proctor Test (SPT) results in Figure 3, the clayey soil reaches its maximum dry density (MDD) of 1.06 g/cm^3 at 52.96% optimum moisture content (OMC) while the coronus material reaches its MDD at 1.81 g/cm^3 at 16.03% OMC. Therefore, it is evident that the coronus material has commendable mechanical properties than clay.

Therefore, when clayey soil was stabilized with coronus material, the MDD increases as the amount of coronus material increases while OMC values decrease except for 10% coronus blended with 90% clay, which may be due to highest affinity towards water upon initial addition of water to the coronus material. Moreover, SPT results illustrate that the addition of coronus material improves the properties of clay soil by increasing its MDD.

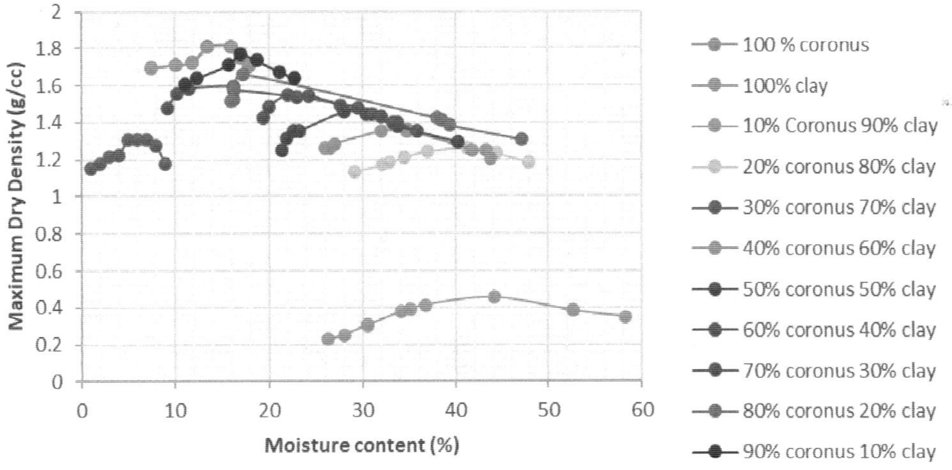

Figure 3. SPT results for the soil-coronus samples

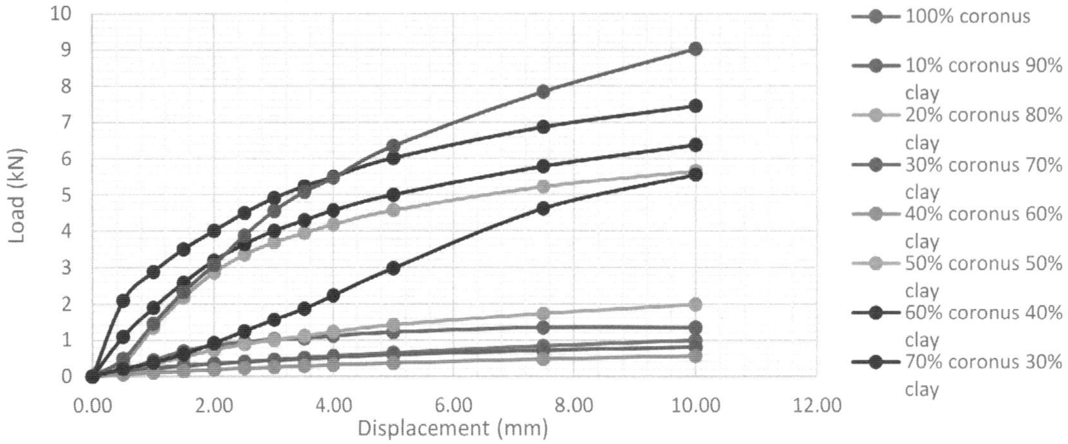

Figure 4. CBR test results of load-penetration curves for the clay-coronus samples

Soil samples stabilized with coronus material may have lesser voids and require less moisture to achieve MDD when compared to 100% clay. Experimental CBR test results shown in Figure 4 illustrate that a combination of 40% clay and 60% coronus material has the highest CBR value. Also, it is evident that the CBR value of coronus material is greater than the CBR value of clay. The CBR for 100% clay and 100% coronus material were determined to be 3.23 and 40.05 respectively. Coarse grained soils have higher CBR values than fine grained soils, thus CBR values for coarse grained soils range from 20-80, whereas CBR values for fine grained soils are less than 5 [11]. For coarse grained soils, well graded soils have the CBR value ranging from 40-80, therefore, the coronus material which is similar to coarse grained soil has a CBR value of 40.05. It can be deduced that coronus material is well graded and is a good material for pavement applications. When the coronus material is stabilized with clay, the results show that the maximum CBR value (of 34.09) was achieved when the soil sample comprised of 60% coronus material and 40% clay. However, the other blending ratios show slight changes when the addition of coronus material is less than 50% but from 50% to 90% the CBR values increase significantly. The fluctuations in the CBR values may be due to the variations in the OMC and MDD achieved during the standard proctor test.

4. Conclusions

Sustainable Development Goals (SDGs) are designed to achieve a clearer and more sustainable future for all by 2030. In line with the sustainable agenda, there is a need to tackle the challenges of climate change due to human-induced activities through reducing greenhouse gas (GHG) emissions contributed by construction sector. Although the use of conventional additives for subgrade stabilization such as lime (hydrated or quick) and cement, yield technically sound response, they have adversely impacted the environment in terms of water bodies pollution, GHG emissions, ecosystems damage and natural resources obliteration during the production of cement. Using locally available materials for subgrade stabilization, as demonstrated in this study is a greener way of resource conservation, economical, sustainable and effective alternative in enhancing the engineering properties of expansive subgrade soil. The study reveals the potential of using coronus material in road pavement construction for subgrade stabilization and would contribute towards achievement of SDG 9 i.e, Industry, Innovation and Infrastructure and SDG 13 i.e, Climate Action.

Following recommendations may be proposed based on the findings of the experimental investigations performed in the current research.

- In terms of soil stabilization, it can be concluded that coronus material increases the strength of clay soil. Due to the heterogenous texture of coronus material, coronus composites are relatively soft and make it easy to breakdown.

- Coronus material can blend evenly with expansive subgrade soil reducing air voids in clayey soil and minimize its susceptibility to shrinking and swelling or a sudden change in volume when exposed to moisture thus making it more stable. It is to be ensured that the locally available material is sourced sustainably.
- As demonstrated in this study, locally available coronus material can be used in the Pacific for pavement subgrade stabilization ensuring acceptable local pavement design and construction practices comprehending that the variation in material properties depends on the source, method of stabilization and robust sealing.
- Subgrade stabilization using locally available coronus material is advisable if the cost of importing and employing the stabilizing agent is less than the alternative already being used locally.
- Further research can progress with the current understanding of the potential of coronus material in the subgrade stabilization applications of road pavement structures around the Pacific.

References

[1] Amakye S Y Abbey S J Booth C A Mahamadu A M 2021 Enhancing the Engineering Properties of Subgrade Materials Using Processed Waste: A Review *Geotechnics* **1** 2 307-329 https://doi.org/10.3390/geotechnics1020015

[2] PRIF Pacific Research Infrastructure Facility 2016 Road Pavement Design for the Pacific Region

[3] Bullen F and Heaton B 1986 Evaluation of coronus pavement materials *Conference proceedings of the 13th Australia Road Research Board* **13** 3 pp. 63 - 75

[4] Belkum K V and Djou S K 1982 Effects of Tropical Soils on Palau Airport Design *ASCE Geotechnical Proceedings*

[5] Bullen F 2003 Use of Coral-Derived Aggregates for Construction of Low-Volume Roads. *Transportation Research Record: Journal of the Transportation Research Board* **1819** 1 134-142

[6] Anon 2001 Using lime for soil stabilization and modification: A proven solution https://www.lime.org/documents/publications/free_downloads/fact-soilstabilization.pdf

[7] Sharif Y G Yang Z Ali N Halil C Sunghwan K Ali A Charles T J Charles Ø 2019 Design, Construction, and Preliminary Investigations of Otta Seal in Iowa *Transportation Research Record: Journal of the Transportation Research Board* **2673** 12 pp. 821-833

[8] Johnson S Faiz A and Visser A 2019 Concrete Pavements for Climate Resilient Low-Volume Roads in the Pacific Island Countries. World Bank

[9] AS 1289 Australian standard 1289: Methods of Testing Soils for Engineering Purposes:

[10] Amakye S Y Abbey S J Booth C A Oti J 2022 Road Pavement Thickness and Construction Depth Optimization Using Treated and Untreated Artificially-Synthesized Expansive Road Subgrade Materials with Varying Plasticity Index *Materials* **15** 8 2773 https://doi.org/10.3390/ma15082773

[11] CBR Testing UK 2020 https://www.cbrtesting.com/california-bearing-ratio-typicalvalues

IOP Conf. Series: Earth and Environmental Science 1084 (2022) 012050 doi:10.1088/1755-1315/1084/1/012050

Soil Erosion mapping using Remote Sensing and GIS - A case study of Rewa River Watershed, Naitasiri province, Fiji

Swetha Priya Darshini Thammadi and Sateesh Kumar Pisini

Department of Civil Engineering, Fiji National University

Email: swetha.thammadi@fnu.ac.fj

Abstract. Rewa river watershed covers approximately up around 33% of the Viti Levu's total Land area in Fiji's central division. Due to anthropogenic development activities, this region has undergone soil erosion and flooding. Soil erosion leads to a variety of environmental issues, with a direct impact on agriculture and infrastructure development. An evaluation of soil erosion in the Rewa river watershed can be used to plan conservation activities. Modeling provides a quantitative and consistent technique to estimating the quantity of soil loss under various situations. The goal of this research is to estimate average annual soil erosion in the Rewa river watershed in Naitasiri province, Fiji, using a GIS (Geographic Information System) combined with the RUSLE (Revised Universal Soil Loss Equation) model. According to the findings, a maximum annual soil loss occurs near to developing regions and river tributaries in areas with a high risk of erosion. To avoid potential concerns in places prone to moderate to high erosion risk, erosion control and watershed protection measures must be implemented immediately. In this context, the RUSLE model can be used to analyse soil erosion in small watersheds in a quick, cost-effective, and less labor-intensive manner, hence assisting in sustainable development and supporting healthy land use practices.

1. Introduction

Soil erosion occurs naturally due to factors such as steep slopes, frequent high-intensity rainfall, and tropical cyclones. These factors have a significant impact on land profile and use [1]. Increased soil erosion leads to reduced soil fertility, land loss, increased runoff due to rainfall, heavy sediment flow, poor drinking water quality, reduced groundwater cultivation, increased flooding, economic loss and repair along with increased cost of rehabilitation. Human induced activities have also contributed to soil erosion over the years by degrading valuable farming lands through changes to the topsoil [2]. As a result of these land use changes, ecosystems such as rainforests, agrarian and infrastructural development activities tend to experience a decline in their quality and efficiency. In watershed areas, runoffs from precipitous slopes greatly harm surface areas of arable lands to the extent that sustainable land management and conservation becomes difficult. This also leads to pollution due to heavy sediments flowing into rivers or reservoirs [3]. Satellite remote sensing and Geographic Information Systems (GIS) applications' ability to extract, delineate, and manipulate land characteristics, as well as their integration with the Revised Universal Soil Loss Equation (RUSLE), make them important tools for estimating spatial soil erosion [4 - 7]. The RUSLE model is widely used and has been validated over the years, in addition to well-documented shortcomings [8,9].

Due to the spatio-temporal variability of soil erosion due to water, factors such as rainfall, topography, land use land cover and soil characteristics play a prominent role [10]. Numerous studies on estimating and assessing the rate of soil erosion using RUSLE model have been reported for Africa focusing on river watersheds [10], People's Republic of China including the effect of urbanization in inner Mongolia [11] and India with attention to watersheds [12], hilltop mining [13] etc amongst others. From the late 1980s to the early 1990s in Fiji, research and academic work was conducted on soil erosion where the investigators used USLE (Universal Soil Loss Equation) technique to determine soil losses [14]. However, there is a lack of research on mapping potential erosion using GIS and modeling soil erosion risk in Fiji. Between 1981 till date, there are no recorded studies of soil erosion assessment using RUSLE (Revised Universal Soil Loss Equation) model being carried out in Fiji.

Over the years, there have been discussions on situation of Viti Levu, Central division of Fiji in terms of soil losses and flooding [15] with subsequent sediment deposition from surface runoffs and erosion reducing the efficiency of agrarian activities through poor soil fertility. Additionally, the alarming rate of soil erosion in Fiji has led to the need for a viable soil erosion model that can consider the climate and landscape factors that trigger soil erosion. The Rewa River basin, which occupies about 33% of the total area of Viti Levu is prone to soil erosion and flooding. It is a catchment area with heavy rainfall and the highest runoff coefficient of major rivers in Fiji but has not been investigated so far [15]. Therefore, the Rewa River basin in Fiji's Viti Levu, Naitasiri province is the area of interest for this study. A working methodology into the Rewa waterway watershed flood issues is significant in understanding soil erosion or soil loss issues and land degradation. Thus, the theory of developing a Soil Erosion map using RUSLE and GIS was applied to identify the areas prone to erosion risks in the Naitasiri province. An erosion model using RUSLE is used considering Advanced Spaceborne Thermal Emission and Reflection Global Digital Elevation Model (ASTER G-DEM) and LANDSAT 8 satellite image. A map of susceptibility to soil erosion is classified into five categories from high to low soil erosion risks.

2. Materials and Methods

2.1. Study area

Naitasiri Province (Figure 1) is located north of the capital city of Suva and is home to the Medrausucu mountain ranges and the five river systems that feed into the Rewa River. The province arguably holds all of the Rewa River's watershed and is a very important contributor to Fiji's economy with high yield of agricultural products such as Dalo and Cassava. Data relevant to the modelling of soil erosion along with description and source of data are detailed in Table 1.

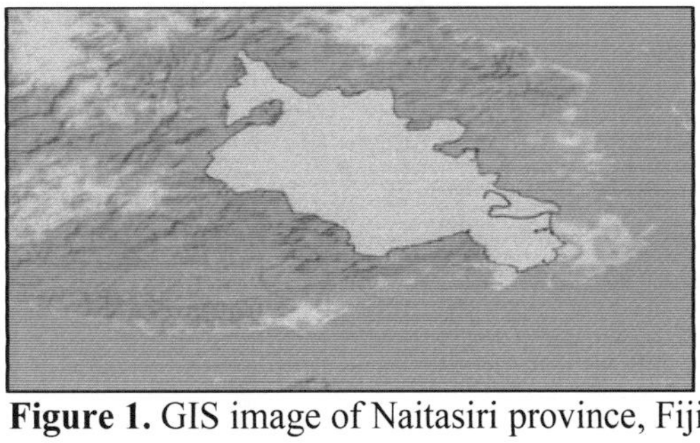

Figure 1. GIS image of Naitasiri province, Fiji

2.2. Data processing

RUSLE model is mathematically categorical as shown in Equation 1:

$$A = R \times K \times LS \times C \times P \tag{1}$$

where A is the mean year soil loss (t/ha/yr), R is the runoff erosivity factor, K is the soil erosivity factor, LS is gradient length or slope steepness (topography factor), C is the cover management factor & P is conservation practice factor. The methodology adopted in the current study is outlined in Figure 2.

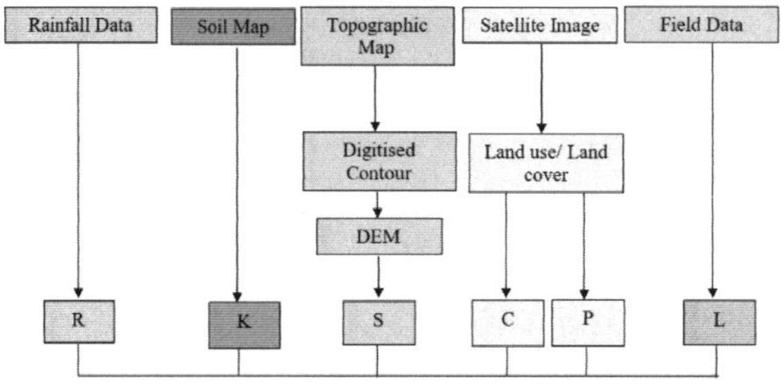

Figure 2. Flow chart for mapping soil erosion risk using RUSLE, Remote sensing and GIS

Table 1. Data relevant to the study and respective sources.

Data	Description	Source
Soil factors & Soil Attributes	Derived from Fiji LUP Guidelines. Data source year: 2015	Fiji Ministry of Agriculture
Rainfall factors	Derived from Fiji Meteorology Department Data source year: 2019	Fiji Meteorological Department
Land Use/Zoning/Built infrastructures	Derived from Fiji Department of Town & Country Planning. Data source year: 2018	Fiji Department of Town and Country Planning
Slope	Derived from SPC. Data source year: 2018	Fiji Mineral Department
Landsat 8 ETM & satellite image (30m spatial resolution – 2017)	Downloaded from https://libra.developmentseed.org/ for verification purposes. Data source year 2019.	PNG University of Technology
Geology (rock type classification)	Derived from Fiji Geology map. Data source year: 1936, 1959, 1990, 1994 & 2015.	Fiji Mineral Resources Department

2.2.1 Rainfall erosivity factor

Rainfall is one of the primary factors that causes soil erosion. The rainfall erosivity factor (R) simply displays the relationship of the effects of rainfall to the study area. Alternatively, it is also a reflection of the effect's runoff posed to the study area due to the intensity of rainfall over a given period. Rainfall erosivity is estimated as the measure of kinetic energy occurring because of rainfall intensity [16]. Rainfall data extracted from Fiji Meteorological department is shown in Figure 3.

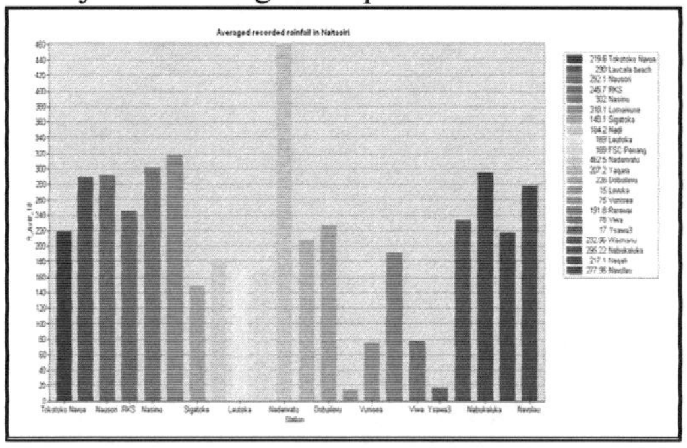

Figure 3. Graph representing rainfall data of Viti Levu

The R value changes based on the orographic and climatic highlights of each location. The Rainfall erosivity factor map for the study area of Naitasiri province primarily involved the incorporation of 22 rainfall stations of the whole Viti Levu area having key continual precipitation data. Kriging interpolation method was adopted to compile the data obtained from the rainfall stations. This method of interpolation arguably is the best tool for spatial distribution precipitation values for the Naitasiri province since the stations expanded across the whole Viti Levu region and not only of the province itself. The R values corresponding to various locations in the study area were then obtained after application of kriging method.

2.2.2 Soil erodibility factor

Nature of soil is a prominent parameter in erosion since weathering down of particles relies upon the properties of soil and earth. The soil erodibility factor (K) represents the susceptibility of a soil type to disintegration.

Table 2. K factor for various soil textures

Soil Texture	**K factor (ton-acre-hr/ hundreds of acre.ft.tons.in)**	**K factor (t. ha.h/ha.MJ.mm)**
Sand	0.02	0.0026
Loamy sand	0.04	0.0053
Coarse sandy loam	0.07	0.0092
Fine sand	0.08	0.0105
Loamy fine sand	0.11	0.0145
Sandy clay	0.12	0.0158
Sandy loam	0.13	0.0171
Fine sandy loam	0.18	0.0237
Sandy Clay loam	0.2	0.0263
Clay	0.22	0.0289
Silty clay	0.26	0.0342
Clay loam	0.3	0.0395
Loam	0.3	0.0395
Organic	0.3	0.0395
Silty clay loam	0.32	0.0421
Silt loam	0.38	0.0500

Soil data set collected for Naitasiri study area were grouped in hydrological soil group as class A, B, C and D. K values proposed by Stone and Hilton (2000) as shown in Table 2 have been assigned to soil texture under each hydrological soil class and the average K value was obtained to represent each hydrological soil group (class A, B, C and D).

2.2.3 Topographic factor

Topographical data is the richest source of data in the soil erosion process. The length slope factor is the total of the slope length (L) and the slope gradient (S) for a given geographical profile. This factor determines the effect of the topographical layout in soil erosion. After the flow accumulation and slope raster were created from the DEM, following empirical formula was used to calculate the LS factor using the raster calculator function in the Equation 2.

$$LS = \text{Power}\left(\frac{\text{Flow Accumulation} * \text{cell size}}{22.13}, 0.4\right) \text{ x Power}\left(\frac{\sin\text{(slope)} * 0.01745}{0.09}, 1.4\right) \text{x } 1.4 \qquad (2)$$

2.2.4 Land cover management factor

The cover management factor (C) considers the land cover attributes such as the type of vegetation, the facilities available on the grounds and the agricultural practices done in the study area. The agriculture and fallow land experience high rate of erosion, whereas intensive vegetation displays less erosion [17]. The NDVI data was obtained from the PNG University of Technology and exported into the ArcGIS

IOP Conf. Series: Earth and Environmental Science 1084 (2022) 012050 doi:10.1088/1755-1315/1084/1/012050

environment as shown in Figure 4. Once the NDVI data is exported into ArcMap, the raster calculator is used to derive the C factor using the Equation 3.

$$C = \exp\left[-a\frac{NDVI}{b-NDVI}\right] \tag{3}$$

where a & b are dimensionless variables used to define the shape of the curve in relation to NDVI and cover management.

2.2.5 Soil conservation practice factor

The Soil Conservation Practice factor, P is a dimensionless ratio of soil loss under a given support practice upon the soil loss if crops were planted up and down the slope [18]. In the current study, we used the slope gradient generated from the DEMs in ArcGIS to obtain the P factor. The Conservation practice factor map depicts the type of conservation practices of land being used within the study area and typically ranges from 0.0 to 1.0. P factor values adopted from Chadli, 2016 [19] are shown in Table 3.

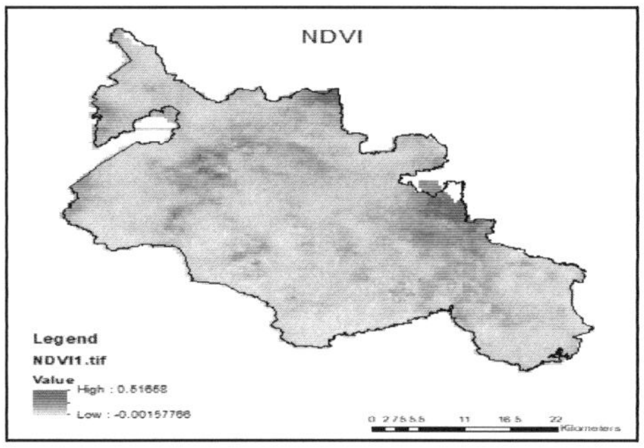

Figure 4. NDVI data for the Naitasiri Province

Table 3. P factor values

Land Use	Slope (°)	P factor
Agriculture	1 - 4	0.11
	5 -10	0.09
	11 - 20	0.18
	21 - 40	0.00
	>40	0.2
Plantation	all	0.13
Other	all	1.00

3. Results and Discussion

3.1 Factor maps

The rainfall erosivity factor map was generated using data collected from 22 rainfall stations on Viti Levu Island. As evidently displayed on the map in Figure 5, the western part of the region has a higher rainfall while the rainfall values' decreases towards the eastern part. The R factor is in the range of 220 to 424 MJ mm ha^{-1} $year^{-1}$.

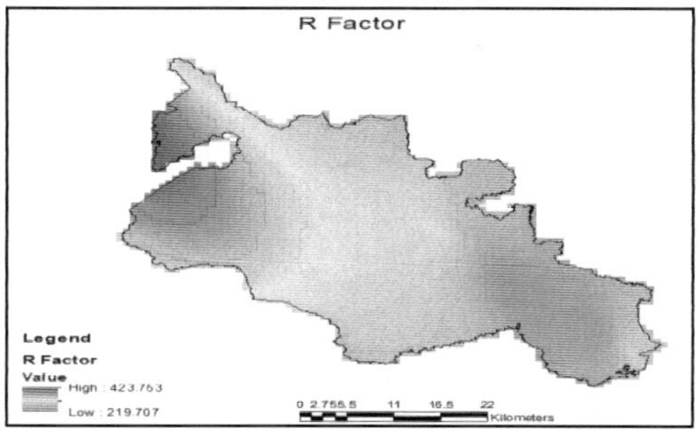

Figure 5. Rainfall Erosivity factor map

The soil erosivity map (Figure 6) shows rate of erosions based on soil types in the study area, Naitasiri Province. It represents the organic matter and textural properties of soil [20]. From the map, it can be observed that the soil erosivity factors of Naitasiri Province varies from 0.008 to 0.029 t ha h ha^{-1} MJ^{-1} mm^{-1}.

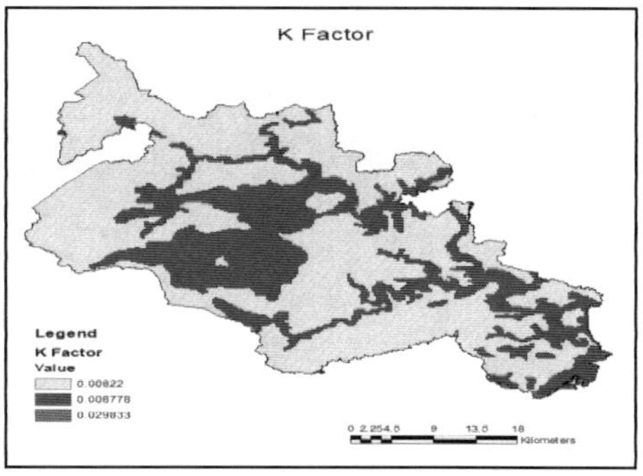

Figure 6. Soil erosivity factor map

The lowest K value (0.008 t ha h ha^{-1} MJ^{-1} mm^{-1}) represents the dominant soil type; class A which consists mostly of sand and gravel which characterizes perfectly drained condition i.e., allows water to penetrate the soil resulting in less runoff and hence less erosion. The moderate K value (0.008 t ha h ha^{-1} MJ^{-1} mm^{-1}) was found mostly in the central region of the study area. This value represents the class C soil which comprises of mostly clay, clay loam and silty clay loam texture soil. This type of soil is poorly drained soil which allows more runoff thus, accounting for slightly higher K values when compared to class A soil. The highest K value (0.029 t ha h ha^{-1} MJ^{-1} mm^{-1}) as per the soil erosivity factor map was found in a small portion of the study area. Soil found in this area were mostly clayey texture soil and silt. Since clayey texture soil are very fine, they possess a waterlogged nature and has low infiltration rate, making it susceptible to running water, thus accounting for high K factor value of (0.029 t ha h ha^{-1} MJ^{-1} mm^{-1}).

The slope length and steepness factor map (Figure 7) indicates LS factor values vary from 0 to 16,716. The lowest LS values indicate plain and flat areas while the highest LS factor highlights the peak or summit of hills in the study area. Moderate LS factor represents areas that have the topographical characteristics of plain and hill.

The cover management factor was generated from the NDVI (Normalized Difference Vegetation Index) raster file of the study area which is an indicator of soil erosion [21]. It can be noted that lower value of C factor, indicates greater vegetation cover, thus leading to low soil erosion occurrences and higher value of C indicates higher soil erosion occurrences.

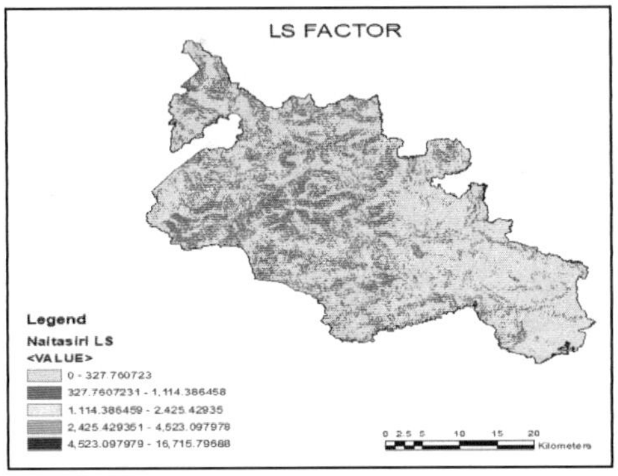

Figure 7. Slope Length and Steepness factor map

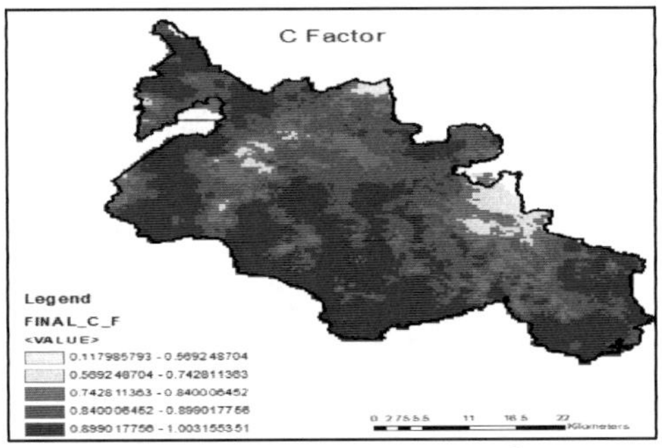

Figure 8. Crop Management factor

As can be seen in Figure 8, scattered layers of yellow flecks can be seen in the centre and north-western part of the province with C-factor values ranging from approximately 0.1 –0.6, which indicates that there is high vegetation cover in these areas. However, majority of the purple and blue layers can be seen covering most of the province indicating a low vegetation cover with C-factor values roughly ranging from 0.6-1.0. This is supported by the presence of river watersheds, farmlands, rocky mountainous regions, and urban centres with developed sites and no vegetation covers.

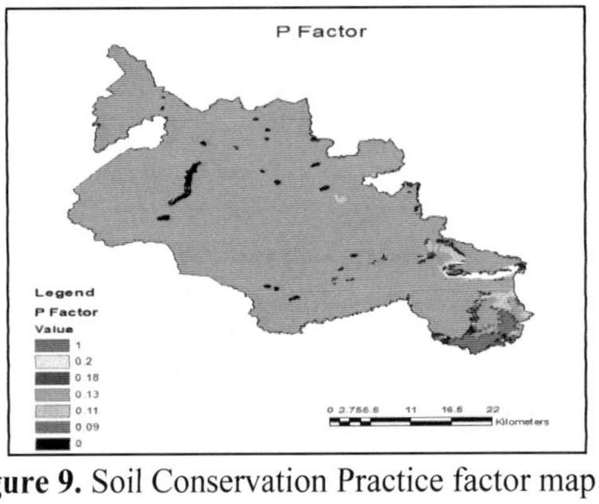

Figure 9. Soil Conservation Practice factor map

The P-factor map represents poor conservation practices where the lowest value represents a good level of conservation practice being done in the study area [19]. As evident from the map in Figure 9, forest land covered majority of the map with a P factor of 0.13 which slightly resembles a good conservation practice. However, the forest land shows lower P-factor value evident due to undistributed land. Agricultural land has P factor values ranging from 0.0 to 0.20 hence indicating a corrupt conversation practice with its peak value at 0.20. For the others category, a high P factor value of 1.0 can be seen on the map indicating a very poor level of conservation practice. The colored layers representing layers of the others category can be seen as situated right and surrounding areas of the town centre. It illustrates that much of the development activities occurring in towns proved poor land conservation practices hence demonstrating substantial evidence as to why the P factor value is higher within the vicinity of town when compared to those areas further away from it.

3.2 Potential soil loss map

The potential soil erosion map for the province of Naitasiri can be seen in the Figure 10. As portrayed in the figure, the colored spots represent erosion rates in ascending order from light red to dark red with the latter being the highest erosion prone area and the former areas having the lowest rate of erosion. The erosion rate of the study area range between 0-4400 tonnes/hectare/year (t/ha/yr) and helps to identify the areas of high and low erosion.

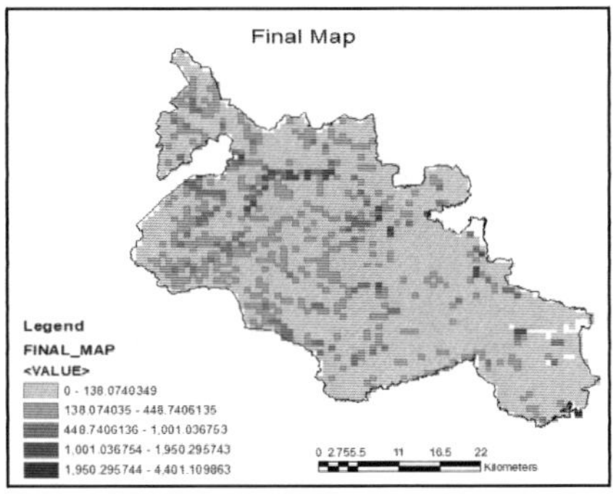

Figure 10. Potential Soil Loss map of the study area

Upon observation of the map (Figure 10), most of the province's land mass are covered with lighter shades of red which is an indication of little to no soil erosion occurrences, i.e., between 0 – 138 t/ha/yr as compared to inland areas of the province where the erosion rate can be found up to 4400 t/ha/yr. A simplification of the varying rates of soil erosion on the map can be aided by the preceding discussion on five factors of the slope length and steepness, rainfall erosivity, soil erosivity, cover management and conservation practice factors. With reference to the five factor maps (Figure 5 – Figure 9), nearly 80% of the land mass in the study area displays a light red on the potential soil loss map and as being lowly rated with regards to erosion, have high vegetation cover as observed from both the cover management and conservation practice factor maps. Very little cover concentration is evident from areas with high erosion rates as seen in dark red spots of Figure 10.

From the cover management factor map (Figure 8), there is high concentration of vegetative cover overlying majority of the land mass with most of it being forest area as supported from observations of the conservation practice factor map. An increase in canopy cover can have significant contributions to the resistance capabilities erosion prone areas have against the forces of erosion itself and this could be one major contributing factor to the low erosion rates being displayed in the potential soil loss map (Figure 10). Moreover, an in-depth study of the conservation practice factor map showed that the areas being prone to high erosion actions as depicted above in dark red spots are in fact agricultural areas. As it may be, poor agricultural practices bring about the higher rate of erosion if inadequate land management techniques are used. In consideration of the observations from the factor maps, an area of steep slope feature associated with poor management of agricultural activities would consequently exhibit high erosion as seen in the centralised high erosion rate shown with distinct features of red spots (Figure 10).

Also, as observed from the soil erosivity factor map (Figure 6), most of the land mass prone to low erosion rate also has the dominant 'soil class A' comprising primarily of sand and gravel and therefore allowing the penetration of water and exhibiting good drainage. This explains the low erosion rates in much of the areas displayed in light red in the potential soil erosion map (Figure 10) with those of high erosion rates being of clay loam and silty clay loam. Clay loam and silty clay loam texture soil have the tendency of being easily washed away by surface runoffs due to their poor drainage traits. The clay loam and silty clay loam texture soil are mostly found in the central region hence evidently showing the high erosion rates are more focused at the central areas of the study area. The two factor maps being of slope length and steepness (LS) (Figure 8), and of rainfall erosivity (Figure 6) presented the level of steepness within the areas of high erosion rates whilst the latter shows a distributed range of rainfall occurrences ascending from southeast region of the map towards the north-western region. The rate at which soil erodes is a direct proportion of the steep slope feature of a particular area. From the LS factor map (Figure 8), the central region of the map has steeper slopes than areas outlying the map thus further agreeing to the evident mountainous feature of the province itself. However, the rainfall erosivity factor map displayed higher rainfall pattern at the north-eastern region of Naitasiri province whilst descending towards the south-eastern part of the province. This supports the higher erosion rates being centralised on parts with steep slope of the province inland than on the outlying areas of the map.

Additionally, steep slope lands with high occurrence of rainfall are liable to high erosion rates on areas of poor drainage soil such as those of clay loam and silty clay loam found in the same parts of the map corresponding to steep slope lands. Since there are less slope areas moving outwardly towards the boundary of the map with better soil composition, heavy rainfalls do not necessarily erode the soil thus adding to the cause of low soil erosion rates towards the edges of the map than at the centre. The potential soil loss map was used to classify the study area into five categories (Figure 11). They are (i) very low (ii) low (iii) moderate (iv) high and (v) very high susceptibility to erosion. Approximately 80% of the land area in Naitasiri province displayed low to very low rates of erosion risk.

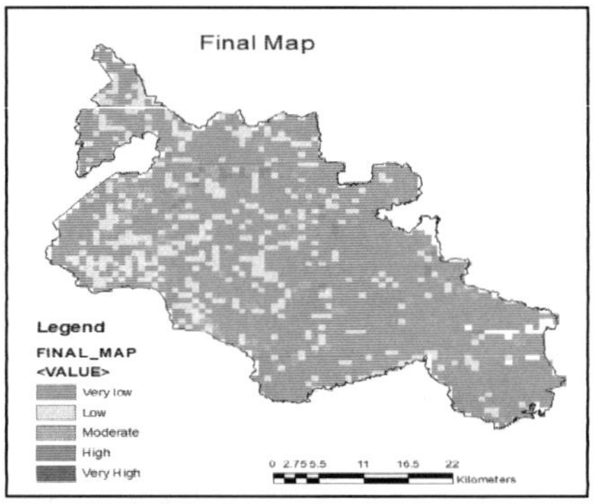

Figure 11. Soil erosion risk map in five categories

3.3 Data validation

To validate the Potential Soil Loss map generated using the RUSLE model integrated with ArcGIS, the final potential soil loss map in Figure 11 was overlayed with all the major rivers in the Naitasiri Province. It is evident that the areas that are prone to higher erosion rate are located along the major rivers or the areas surrounding it. This clearly shows the impact the rivers have on the rate of soil erosion. Although the areas have thick vegetation cover, but due to the frequent rainfall and river flooding, the soil is easily vulnerable to be washed away by the strong flow of water. The other areas where the erosion rates are higher but are not located near the major rivers are steep terrains and slopes in the mountainous regions. The validation of our results was also performed using site observation of geographical features such as terrain, soil and vegetation cover of five chosen sites in the study area – Navatuvula (clay soil – low erosion), Waibau (quarry site – high erosion), Waimanu (riverbank – high erosion), Lomaivuna (light vegetation – moderate erosion) and Vuniniudrovu (dense vegetation – low erosion). GPS wave points were plotted for each site and later exported into the final ArcGIS map to be verified if the field observations correlated with the model generated outputs or results. The field results were found to be analogous with the modelling results indicating that the model works well for the study area.

4. Conclusion

Soil erosion is a global environmental concern that has a detrimental influence on agricultural output and poses a flood risk to infrastructure. This paper demonstrates the usefulness of remote sensing and GIS in deriving inputs to the empirical soil erosion RUSLE model to estimate the potential soil erosion risk map in the Rewa river watershed region, Naitasiri province, Fiji. The potential soil loss map was classified into five categories (very low, low, moderate, high and very high) to identify areas with moderate to very high erosion risk. It has been observed that erosion rates vary with land use/land cover and terrain. Nearly 20% of the land area in Naitasiri province displayed moderate to very high rates of erosion risk. The results of the study indicate that all the factors—namely, R, K, LS, C and P—play a significant role in assessing the erosion potential of the region. The results and analysis conclude that the average annual soil loss is about 4400 t/ha/yr in the Rewa river watershed of the study area. Soil erosion control measures to prevent soil erosion helps protect vulnerable areas at risk of erosion. Overall, the development of soil erosion risk map can be a vital tool for watershed development and assist with local and regional planning. This study can further be extended to the entire Viti Levu for better understanding the interrelation between Land use and soil erosion and to further plan developmental activities.

References

[1] Kadam K A, Umrikar B N and Sankhua R N 2018 *Remote Sensing of Land* **2** 1 pp 65-75
[2] Pimentel D and Burgess M 2013 *Agriculture* **3** 3 pp 443-463
[3] Tadesse L Suryabhagavan V K Sridhar G Legesse G 2017 *International Soil and Water Conservation Research* pp 85-94
[4] Wang G Gertner G Fang S 2020 *Photogramm Eng Remote Sens* ***69*** pp 889–898
[5] Lu D Li G Valladares G S Batistella M 2004 *Land Degrad Dev* ***15*** 499–512
[6] Li X S Wu B F Wang H Zhang J 2011 *J Remote Sens* ***15*** 372–387
[7] Chalise D Kumar L Shriwastav C P Lamichhane S 2018 *Enviro. Earth Sci 77* 685
[8] Renard K G Foster G R Weesies G A McCool D K Yoder D C 1997 Predicting Soil Erosion by Water: A Guide to Conservation Planning with the Revised Universal Soil Loss Equation (RUSLE) United States Government Printing Office: Washington DC, USA
[9] Shamshad A Azhari M N Isa M H Hussin W M A W Parida B P **2008** *Catena 72* 423–432
[10] Andrew K M Brigitta S 2020 *Sustainability* **12** 8531
[11] Li-Yan W Yi X En-Ming R Ling J Yang X Zhi-Yun O 2018 *Int. J. Environ. Res. Public Health* **15** 550
[12] Ganasri B P Ramesh H 2016 *Geoscience Frontiers* **7** 953-961
[13] Mhaske S N Pathak K Dash S S Nayak D B 2021 *J. Env. Mngt.* **294** 112987
[14] Printemps J 2008 *Noumea: CRISP*
[15] Tamata B U Comley J and Tokalauvere L 2012 *The Journal of Pacific Studies* **32** pp 11–24
[16] Sujatha E R and Sridhar V 2018 *Water* **10** 11 pp 1–17
[17] Mercer K 2019 Faculty of Science, Medicine & Health - Honours Theses | University of Wollongong Thesis Collections | University of Wollongong. Available at: Ro.uow.edu.au.
[18] Mondal A Khare D and Kundu S 2016 *Geocarto International* **33** 1 pp 89-103
[19] Chadli K 2016 *Modeling Earth Systems and Environment* **2** 2
[20] Schwab G Frevert R Dminster T Barnes K 1966 *The Fergunson Foundation Agricultural Engineering Series* John Wiley & Sons: New York, NY, USA p 683
[21] Magesh N S and Chandrasekar N 2016 *Environmental Earth Sciences* **75** 16 pp 1–17

IOP Conf. Series: Earth and Environmental Science 1084 (2022) 012051 doi:10.1088/1755-1315/1084/1/012051

Chambal River Water Quality Assessment at Kota Metropolis Through the Irrigation Water Quality Indices from 2017-2021

Aparna Sharma[1], Kuldeep[2] and Anil K Mathur[3]

[1] M. Tech Scholar, Department of Civil Engineering, University Department, Rajasthan Technical University, Kota, Rajasthan, India
[2] PhD Research Scholar, Department of Civil Engineering, University Department, Rajasthan Technical University, Kota – 324010, Rajasthan, India
[3]Professor, Department of Civil Engineering University Department, Rajasthan Technical University, Kota – 324010, Rajasthan, India

Email: kuldeep.kambo44@rtu.ac.in

Abstract. The Chambal river plays an essential role in satisfying all water demands for the two million population of Kota and nearby districts of Rajasthan (India). This research aims to analyse the Chambal river water quality for irrigation in the upstream and downstream of Kota dam, Rajasthan (India), for five years, from 2017 to 2021. The irrigation water quality of the Chambal river at Kota is estimated with the help of five indices, namely, the permeability index (PI). Kelly ration (KR), magnesium hazard (MH), soluble sodium percentage (SSP), and sodium absorption ratio (SAR). Calcium, magnesium, potassium, sodium, and bicarbonate were analysed from 2017 to 2021 to complete irrigation water quality indices (IWQIs) for the Chambal river upstream and downstream of the Kota dam. Akalgarh is situated upstream of the Kota dam, while Rangpur, SRRT, and Keshoraipatan in the downstream. The irrigation water quality indices (IWQIs) are estimated to access river water quality for agricultural use. SAR, SAP, KR, MH, and PI had higher values in downstream sampling points than upstream during the observation period of five years. Hence, upstream river water is more suitable for use in agricultural activities. The evaluated IWQIs are the best rating tool for Chambal river water quality. Different characteristics of Chambal river water may be assessed through these IWQIs. The best management program and pollution control strategies can be determined, which ultimately helps stack holders for decision making.

1. Introduction

Water is an essential component required for the survival of flora, fauna, and human beings on the Earth, and hence, it is listed in sustainable development goals (SDGs) [1], [2]. The population of India is primarily dependent on agriculture for the survival of their lives [3]. Water is the basic need for all agricultural activities, and it is regularly satisfied by surface water bodies around the world [4]. The fitness of available water for specific purposes like irrigation, industrial and domestic uses guarantees sustainable development [5]. However, impaired river water quality is a limiting factor in crop production and yield and can cause salinity and alkalinity. Irrigated agriculture contributes to nearly 38 % of global food production and consumes 60–80 % of the total water usage worldwide [6].
Anthropogenic activities such as agricultural runoff, surface water logging, domestic, and commercial waste, wastewater streams, leachates from solid waste sites, and Surface runoff are the cause of

several environmental menaces [7], [8]. These activities may change the physical, chemical, and bacteriological activity, making them unsuitable for irrigation[9]. Poor-quality irrigation changes soil's physical and chemical properties, reduced soil productivity and yield. Therefore, the river water quality needs to be continuously monitored to safeguard long-term sustainability [10]. Many studies have been conducted on surface water quality for irrigation purposes in different parts of the world and India. Kota, an educational city of Rajasthan (India), is greatly dependent on the Chambal river to fulfill its all-water demand, such as domestic, commercial, industrial, municipal, and fire.
This study aims to analyse the Chambal River Water Quality for irrigation in the upstream and downstream of Kota Dam, Rajasthan, India, for the period of five years from 2017 to 2021. Sodium absorption ratio (SAR), soluble sodium percentage (SSP), kelly ratio (KR), magnesium hazard (MH), and permeability index (PI) were calculated to determine Irrigation water quality with the help of 5 water quality parameters, namely, Calcium, Magnesium, Potassium, Sodium, and Bicarbonate.

2. Study area and observation period

Kota is a fast-growing smart city in Rajasthan, India, with a population of more than 2 million [11]. It is globally recognised as the educational hub of India and situated in the southeast of Rajasthan State, and its distance from the State capital, Jaipur, is 250 km. The city and nearby areas are greatly dependent on Chambal River for its commercial, domestic, industrial, fire, and agricultural demand.

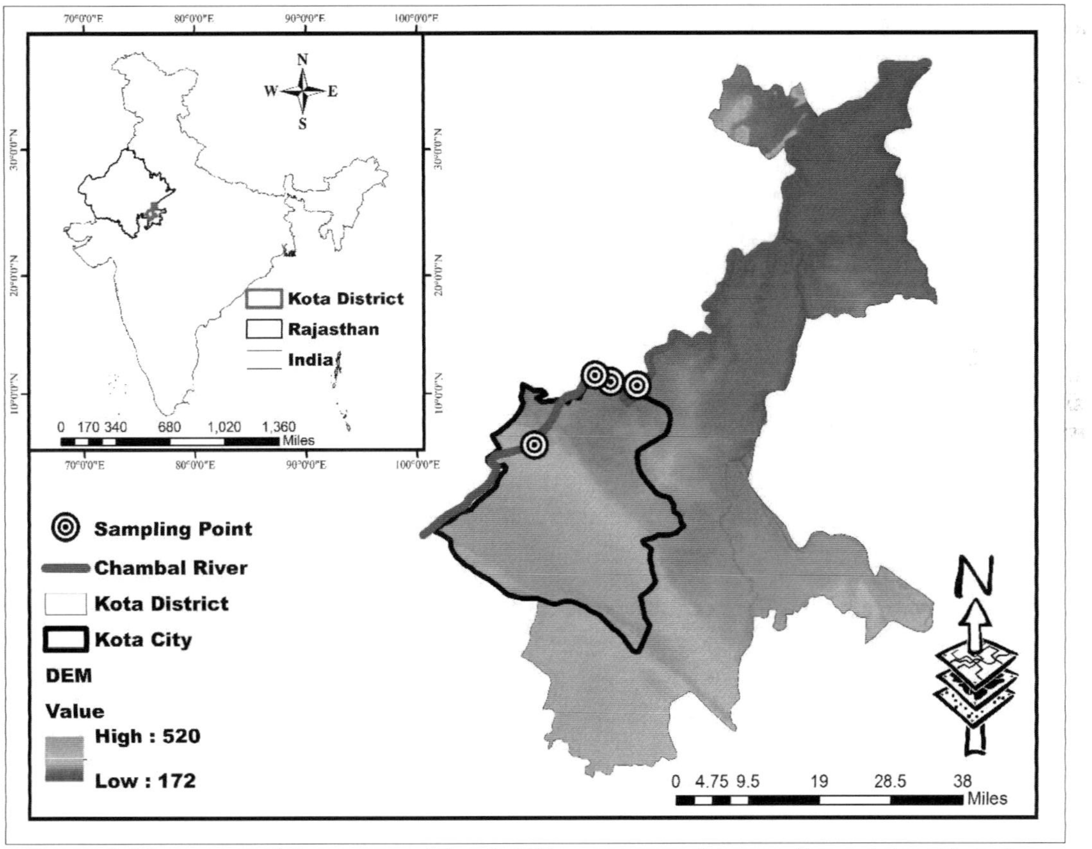

Figure 1. Upstream and downstream Chambal River water quality sampling points at Kota dam, Rajasthan, India.

Rapid urbanisation, population growth, improper town planning, and illegal encroachment on both banks of the Chambal river have an adverse effect on Chambal river water quality in downstream of the Kota dam. The Kota farmers and nearby districts rely on the Chambal river for irrigation.

Hence, Five water quality parameters, namely, Calcium, Magnesium, Potassium, Sodium, and Bicarbonate, were analysed regularly during the study period of five years from 2017 to 2021 at four sampling locations, namely, Akelgarh in Upstream and Rangpur, SRRT, and Keshoraipatan in the Downstream to analysis river water quality to use in agricultural activities. The testing of samples was performed in the laboratory of the Regional Office, RSPCB, Kota. The latitude and longitude of each sampling point are mentioned in Table 1. The study area map with sampling points is illustrated in Figure 1.

Table 1. GPS Coordinates for upstream and downstream of Kota Dam.

Sampling Point	Latitude	Longitude	Name	Location
1.	25.151	75.812	Akelgarh	Upstream
2.	25.267	75.955	Rangpur	Downstream
3.	25.260	76.005	Keshoraipatan	Downstream
4.	25.256	75.990	Shree Rajrajeswar Mahadev Temple	Downstream

3. Irrigation water quality index

IWQIs can be determined using five parameters: SAR, KR, SSP, MH, and PI. All are selected from the literature, and their governing equation and standard limits are mentioned in Tables 2 and 3 [12]–[14]. Seasonal and annual values of SAR, SSP, KR, MH, and PI at each sampling point are shown in Figure 2, Figure 3, Figure 4, Figure 5, and Figure 6, respectively. Irrigation water containing high SAR may adversely affect the soil structure and reduce the hydraulic conductivity of soil in fine-textured soil. The extent of sodium absorbed by soil depends on the amount of sodium to divalent cations of calcium and magnesium. Leaf burns of almond, avocado, and stone fruit are examples of negative impacts of high sodium concentration [15].

Table 2. Various irrigation water quality indices.

Index	Equation
Sodium Absorption Ratio	$SAR = \frac{Na^+}{\sqrt{\frac{Ca^{2+} + Mg^{2+}}{2}}}$
Soluble Sodium Percentage	$Na\% = \frac{Na^+ + K^+}{Ca^{2+} + Mg^{2+} + Na^+ + K^+} \times 100$
Kelly Ratio	$KR = \frac{Na^+}{Ca^{2+} + Mg^{2+}}$
Magnesium Hazard	$MH = \frac{Mg^{2+}}{Ca^{2+} + Mg^{2+}} \times 100$
Permeability Index	$PI = \frac{Na^+ + K^+ + \sqrt{HCO_3^-}}{Ca^{2+} + Mg^{2+} + Na^+ + K^+} \times 100$

The excess amount of sodium in irrigation water is determined through the kelly ratio. The KR values bigger than one (KR > 1) determined the excess sodium, while the KR values lower than one are suitable for irrigation [4], [13]. The value of SSP less than 50 is an indication of good water quality for irrigation use, while an SSP greater than 50 indicates that the water is unsafe for irrigation.

Excess sodium ions in irrigation water make it saline or alkaline depending upon its occurrence in the water body [16]. High sodium content in irrigation water makes soils relativity impermeable to air and water. The soil becomes hard when dry difficult to till as plastic, and sticky when wet. These negative impacts affect plant growth [12]. Long-term exposure to irrigation water with high content of Na^+, Ca^{2+}, Mg^{2+}, and alkalinity influenced the appropriateness of irrigation water and is determined by the permeability index. The physical properties of soils are greatly affected by the high magnesium content in irrigation water [6].

Table 3. Standard limits for irrigation water parameters.

Index	Range	Quality
SAR	0 – 10	Excellent
	10 -18	Good
	18 – 26	Doubtful
	> 26	Unsuitable
SSP	< 50	Good
	> 50	Unsuitable
KR	< 1	Suitable
	>1	Unsuitable
MH	< 50	Suitable
	> 50	Unsuitable
PI	< 80	Good
	80 - 100	Moderate

4. Observation

4.1. Akelgarh

The annual SAR values for Akelgarh were 10.11, 13.05, 9.77, 6.57, and 7.26 in 2017, 2018, 2019, 2020, and 2021, respectively. The annual SSP values for Akelgarh were 56.93, 61.83, 55.86, 45.37, and 44.22 in 2017, 2018, 2019, 2020, and 2021, respectively.

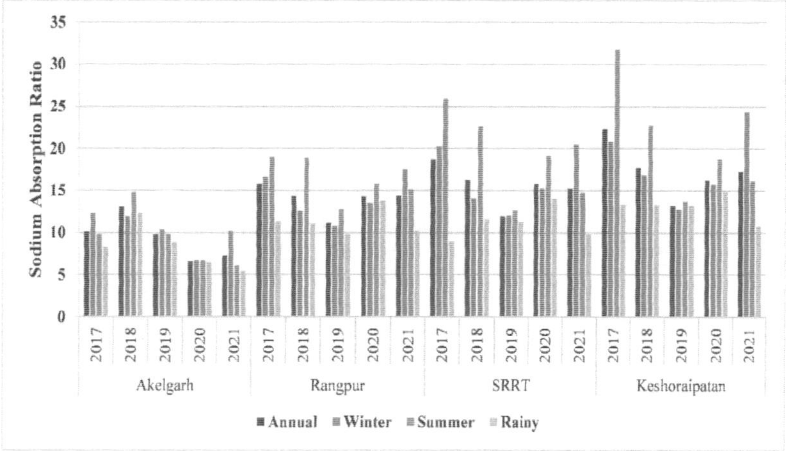

Figure 2. Seasonal and annual values of sodium absorption ratio (SAR) at each sampling point.

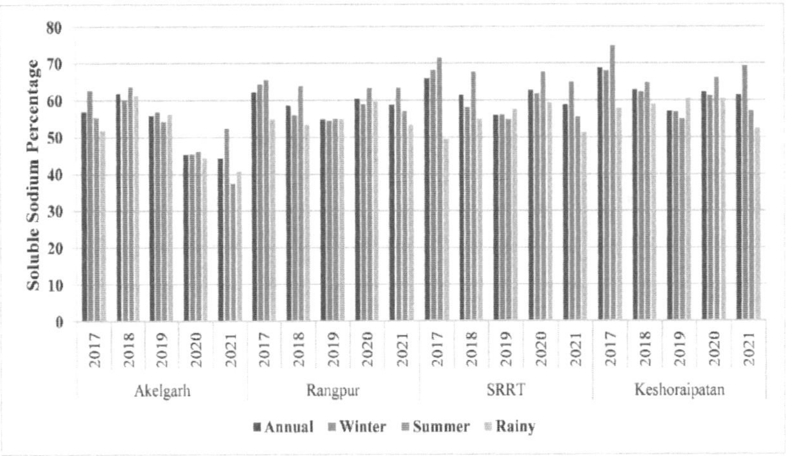

Figure 3. Seasonal and annual values of soluble sodium percentage (SSP) at each sampling point.

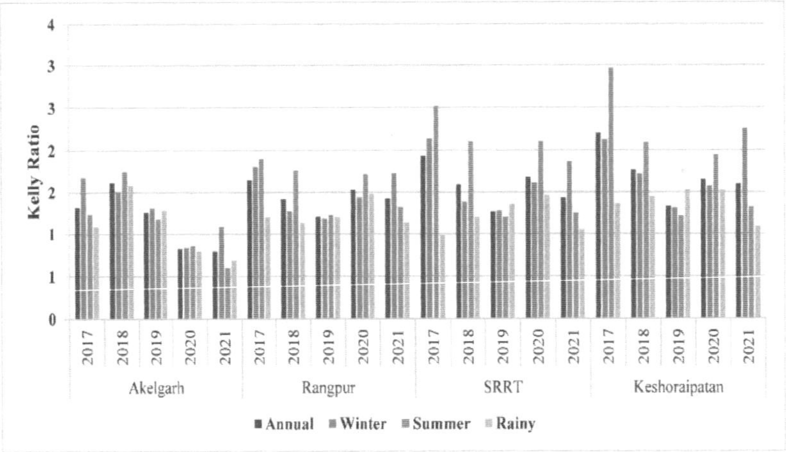

Figure 4. Seasonal and annual values of Kelly ratio (KR) at each sampling point.

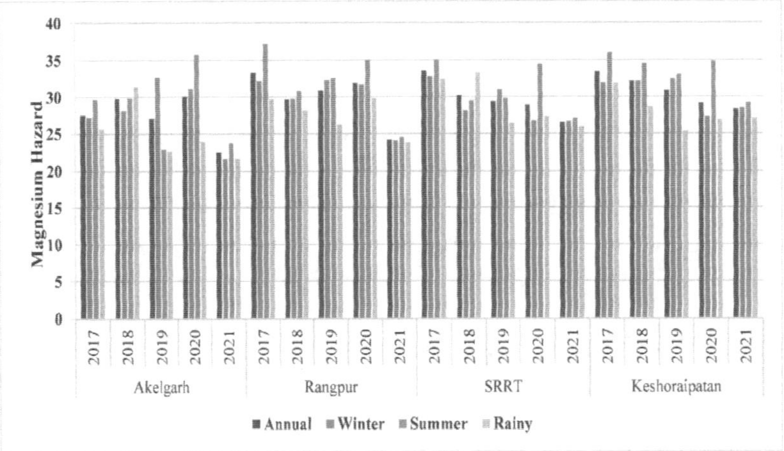

Figure 5. Seasonal and annual values of magnesium hazard (MH) at each sampling point.

The annual KR values for Akelgarh were 1.32, 1.62, 1.27, 0.83, and 0.79 in 2017, 2018, 2019, 2020, and 2021, respectively. The annual MH values for Akelgarh were 27.53, 29.80, 27.14, 30.14, and 22.48 in 2017, 2018, 2019, 2020, and 2021, respectively. The annual PI values for Akelgarh were 71.14, 73.95, 69.31, 59.76, and 54.81 in 2017, 2018, 2019, 2020, and 2021, respectively.

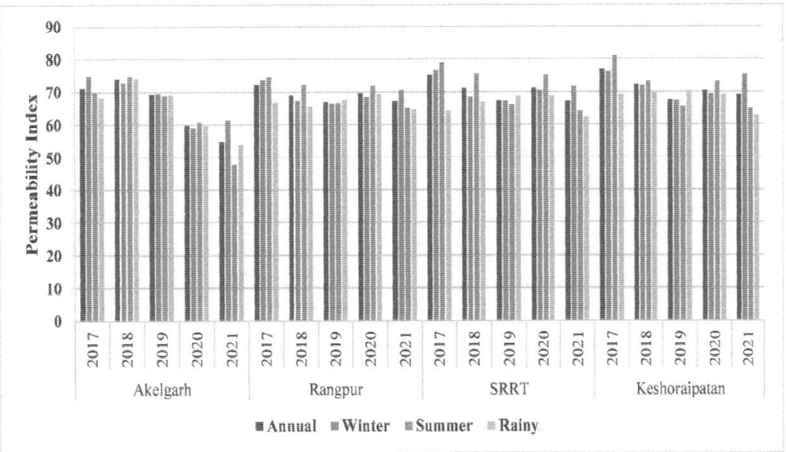

Figure 6. Seasonal and annual values of permeability index (PI) at each sampling point.

4.2. Rangpur

The annual SAR values for Rangpur were 15.75, 14.32, 11.17, 14.23, and 14.37 in 2017, 2018, 2019, 2020, and 2021, respectively. The annual SSP values for Rangpur were 62.22, 58.65, 54.74, 60.44, and 58.72 in 2017, 2018, 2019, 2020, and 2021, respectively. The annual KR values for Rangpur were 1.65, 1.42, 1.21, 1.53, and 1.42 in 2017, 2018, 2019, 2020, and 2021, respectively. The annual MH values for Rangpur were 33.25, 29.67, 30.90, 31.92, and 24.17 in 2017, 2018, 2019, 2020, and 2021, respectively. The annual PI values for Rangpur were 72.23, 68.86, 66.92, 69.66, and 67.23 in 2017, 2018, 2019, 2020, and 2021, respectively.

4.3. Shree Raj Rajeswar Temple (SRRT)

The annual SAR values for SRRT were 18.66, 16.27, 11.96, 15.82, and 15.29 in 2017, 2018, 2019, 2020, and 2021, respectively. The annual SSP values for SRRT were 62.22, 58.65, 54.74, 60.44, and 58.72 in 2017, 2018, 2019, 2020, and 2021, respectively. The annual KR values for SRRT were 1.65, 1.42, 1.21, 1.53, and 1.42 in 2017, 2018, 2019, 2020, and 2021, respectively. The annual MH values for SRRT were 33.25, 29.67, 30.90, 31.92, and 24.17 in 2017, 2018, 2019, 2020, and 2021, respectively. The annual PI values for SRRT were 75.16, 71.18, 67.37, 71.20, and 67.11, in 2017, 2018, 2019, 2020, and 2021, respectively.

4.4. Keshoraipatan

The annual SAR values for Keshoraipatan were 22.28, 17.73, 13.19, 16.26, and 17.29 in 2017, 2018, 2019, 2020, and 2021, respectively. The annual SSP values for Keshoraipatan were 68.79, 62.93, 57.07, 62.29, and 61.44 in 2017, 2018, 2019, 2020, and 2021, respectively. The annual KR values for Keshoraipatan were 2.20, 1.77, 1.33, 1.65, and 1.59 in 2017, 2018, 2019, 2020, and 2021, respectively. The annual MH values for Keshoraipatan were 33.44, 32.17, 30.91, 29.13, and 728.34 in 2017, 2018, 2019, 2020, and 2021, respectively. The annual PI values for Keshoraipatan were 76.84, 72.21, 67.57, 70.37, and 69.00 in 2017, 2018, 2019, 2020, and 2021, respectively.

5. Results and Discussion

The SAR, SSP, KR, MH, and PI values are higher in the downstream river water than upstream of the Kota dam. The maximum values of irrigation water quality indices (SAR, SSP, KR, MH, and PI) are found at Keshoraipatan, followed by SRRT, Rangpur, and Akelagrh during the study period of 5 years. The SAR values improved for Akelgarh in recent years, and maintaining its level below 10 indicates the excellent quality of river water for irrigation. While the values for Rangpur, SRRT, and Keshoraipatan are between 10-32, i.e., river quality varies from good to unsuitable category in different years. The SSP values are less than 50 for Akelgarh, confirming the suitability of river water for irrigation. While the values for Rangpur, SRRT, and Keshoraipatan are more than 50, i.e., river quality is unsuitable in downstream for irrigation. The KR values are greater than 1 for Akelgarh, Rangpur, SRRT, and Keshoraipatan, exhibiting the unsuitability of river water for irrigation. The MH values are less than 50 for Akelgarh, Rangpur, SRRT, and Keshoraipatan, ensuring the suitability of river water for irrigation at all places. The PI values are lower than 8 for Akelgarh, Rangpur, SRRT, and Keshoraipatan, exhibiting the good quality of river water for irrigation.

The Chambal river flows through a large and deep gorge starting from Chaurasigarh and ending at Kota Dam. Both ends of the Chambal river between it, are free from pollution sources, and it is the main reason behind the minute pollution level in the upstream river water of Kota Dam at Akelgarh. Hence, the Chambal river water quality is suitable for irrigation of agricultural lands as SAR, SSP, KR, MH, and PI values indicate that river water is within permissible limits. The scenario is different in the downstream Chambal river water after Kota Dam. Wastewater streams originating from domestic and commercial activities are directly dumped into the river without treatment through unaccountable drainage channels (nallahs).

Wastewater streams contain a wide range of contaminants such as suspended solids, biodegradable organics, pathogens, nutrients, and refractory organics. Agricultural runoff and surface runoff also enhance the nutrients load on river water, leading to the degraded river water quality of drinking and irrigation. Excessive contamination load on the river also reduces the self-purification capacity of the river. Hence, river water quality worsens in the Kota dam downstream than upstream (Akelgarh).

The seasonal variation has also been seen in the values of IWQIs. The highest values of these indices were in Summer, followed by Winter and Rainy seasons, with some exceptions. The high concentration of water quality parameter used for the calculation of IWQIs in the Summer season is due to two leading causes, (i) evaporation of river water to raise the concentration of dissolved salts, (ii) less discharge of river water from the Kota dam due to maintaining the river water level in upstream. While in the Rainy season, the concentration of these contaminants is significantly less due to the frequent discharge of river water from the Kota dam.

This study revealed no significant variation in the values of different IWQIs at each downstream sampling point over the last five years. i.e., Covid-19 restrictions do not significantly affect the contamination load during the lockdown period in 2020 and 2021.

6. Conclusion

This study concludes that the water quality of downstream river water for irrigation purposes is less suitable than upstream as different IWQIs (SAR, SAP, KR, MH, and PI) had higher values in downstream. Wastewater streams, agricultural runoff, surface runoff, commercial and domestic wastes, and the civic sense of inhabitants towards river water pollution are the main reasons behind the higher downstream river water pollution. The water quality parameters of the Chambal river are greatly influenced by temperature, river water discharge from the Kota dam, and the amount of rainfall that occurs in the region. Lockdown restrictions were not significant to controlling or reducing the downstream river water pollution.

Wastewater streams should be collected through different drainage channels at a common effluent treatment plant (CETP), where effluent should be treated before discharging into the river or using it for different purposes. Public awareness should be created because every system may fail one day if people don't change their bad habits to throw waste into the river water. The provision of heavy lavy should be strictly followed to change people's habits suddenly. This study may be used as a baseline study for future studies on the Chambal river.

References

[1] Kuldeep, S. Sisodiya, and A. K. Mathur, "An assessment of water quality indices of Chambal river in Kota City for drinking and irrigation Purposes," WEENTECH Proceedings in Energy vol. 7, no. February, pp. 118–133, 2021. https://doi.org/10.32438/WPE.122021

[2] G. Asadollahfardi, A. Hemati, S. Moradinejad, and R. Asadollahfardi, "Sodium Adsorption Ratio (SAR) Prediction of the halghazi River Using Artificial Neural Network (ANN) Iran," *Curr. World Environ. J.*, vol. 8, no. 2, pp. 169–178, 2013, DOI: 10.12944/cwe.8.2.02.

[3] B. Singh, C. K. Arya, J. Singh, and K. K. Mourya, "Analysis of rainfall data for storage and irrigation planning in humid south-eastern plain of Rajasthan in India," *J. Appl. Nat. Sci.*, vol. 6, no. 1, pp. 214–219, 2014, DOI: 10.31018/jans.v6i1.403.

[4] V. Satish Kumar, B. Amarender, R. Dhakate, S. Sankaran, and K. Raj Kumar, "Assessment of groundwater quality for drinking and irrigation use in shallow hard rock aquifer of Pudunagaram, Palakkad District Kerala," *Appl. Water Sci.*, vol. 6, no. 2, pp. 149–167, 2016, DOI: 10.1007/s13201-014-0214-6.

[5] V. Wagh, S. Mukate, A. Muley, A. Kadam, D. Panaskar, and A. Varade, "Study of groundwater contamination and drinking suitability in basaltic terrain of Maharashtra, India through pig and multivariate statistical techniques," *J. Water Supply Res. Technol. - AQUA*, vol. 69, no. 4, pp. 398–414, 2020, DOI: 10.2166/aqua.2020.108.

[6] A. O. Ogunfowokan, J. F. Obisanya, and O. O. Ogunkoya, "Salinity and sodium hazards of three streams of different agricultural land-use systems in Ile-Ife, Nigeria," *Appl. Water Sci.*, vol. 3, no. 1, pp. 19–28, 2013, DOI: 10.1007/s13201-012-0053-2.

[7] S. Madhav, A. Ahamad, A. Kumar, J. Kushawaha, P. Singh, and P. K. Mishra, "Geochemical assessment of groundwater quality for its suitability for drinking and irrigation purpose in rural areas of Sant Ravidas Nagar (Bhadohi), Uttar Pradesh," *Geol. Ecol. Landscapes*, vol. 2, no. 2, pp. 127–136, 2018, DOI: 10.1080/24749508.2018.1452485.

[8] R. Sharma *et al.*, "Analysis of Water Pollution Using Different Physicochemical Parameters: A Study of Yamuna River," *Front. Environ. Sci.*, vol. 8, no. December, pp. 1–18, 2020, DOI: 10.3389/fenvs.2020.581591.

[9] S. Yıldız and C. B. Karakuş, "Estimation of irrigation water quality index with development of an optimum model: a case study," *Environ. Dev. Sustain.*, vol. 22, no. 5, pp. 4771–4786, 2020, DOI: 10.1007/s10668-019-00405-5.

[10] N. Adimalla, "Groundwater Quality for Drinking and Irrigation Purposes and Potential Health Risks Assessment: A Case Study from Semi-Arid Region of South India," *Expo. Heal.*, vol. 11, no. 2, pp. 109–123, 2019, DOI: 10.1007/s12403-018-0288-8.

[11] Kuldeep, S. Sisodiya, and A. K. Mathur, "Noise Descriptors For Kota Metropolis, Rajasthan (India)," Pollution research, vol. 16, pp. 1–14, 2021.

[12] S. Sisodiya, R. K. Mehar, and A. K. Mathur, "Datasets for Assessment of Water Quality Indices for Irrigation and Drinking for Hadoti Lakes, Rajasthan," vol. 13, no. 2, pp. 121–130, 2020.

[13] Kuldeep, S. Sisodiya, and A. K. Mathur, "Assessment of the Chambal River Quality at Kota Metropolis Through the Drinking Water Quality Index and Irrigation Water Quality Index," *Lect. Notes Mech. Eng.*, vol. I, pp. 559–573, 2022, DOI: 10.1007/978-981-16-5371-1_49.

[14] S. Sisodiya, R. K. Mehar, and A. K. Mathur, "Statistical Analysis for Urban lakes of Hadoti Region, Rajasthan," vol. 13, no. 1, pp. 35–42, 2020.

[15] S. Shil, U. K. Singh, and P. Mehta, "Water quality assessment of a tropical river using water quality index (WQI), multivariate statistical techniques and GIS," *Appl. Water Sci.*, vol. 9, no. 7, pp. 1–21, 2019, DOI: 10.1007/s13201-019-1045-2.

[16] K. K. Singh, G. Tewari, and S. Kumar, "Evaluation of Groundwater Quality for Suitability of Irrigation Purposes: A Case Study in the Udham Singh Nagar, Uttarakhand," *J. Chem.*, vol. 2020, 2020, DOI: 10.1155/2020/6924026.

Acknowledgments
We are grateful to Rajasthan State Pollution Control Board (RSPCB) for providing particulate matter data for the study. The authors are obliged for the financial support to perform this research work from University Departments, Rajasthan Technical University, Kota , India.

Organic Farming in the Tropical, Disaster-Prone Coasts: Interventions, Major Challenges and a Success Story from Sundarban Biosphere Reserve, India

A Chowdhury[1, 2], A S Iyer[1], S Bhattacharyya[2], S U Khan[3] and A Naz[4, 2*]

[1]Jindal School of Environment and Sustainability, O.P. Jindal Global University, Sonipat-131001, Haryana, India.
[2]Tagore Society for Rural Development, Kolkata- 700005, West Bengal, India
[3]Amity School of Communication, Amity University Rajasthan, Jaipur- 303002, Rajasthan, India.
[4]Jindal School of Liberal Arts and Humanities, O.P. Jindal Global University, Sonipat-131001, Haryana, India.

*E-mail: naazaliya6@gmail.com

Abstract. Organic agricultural practices are sustainable alternatives to chemical methods. This article highlights the major challenges faced by environmentalists, social workers in popularizing organic agriculture as well as its positive field application. The primary soil parameters (pH, alkalinity, electrical conductivity, soil, organic carbon, soil texture, and available N, P and K) are compared between global coastal agricultural soils. Organic farming practices adopted in different regions were also focused in this article. Sundarbans is the world's largest contiguous mangrove ecosystem, largest delta, abode of mangrove dwelling tigers and Indian part is the home of 4.6 million natural disaster vulnerable populations. This work showcases success story of organic farming by 84 families of Indian Sundarbans since 2012, with use of vermicompost, biopesticides (neem oil) and natural microbe augmentations (*Trichoderma sp*), as a part of a socio-environmental intervention project. This intervention resulted in increased yield in both kharif season (10-12%) and rabi season (5-10%) by 2017.

Keywords: Organic agriculture; Sundarbans, Delta, Coast; Self Help Groups, Sustainability

1. Introduction

With increasing climate change problems like inconsistent rainfall and worsening soil quality, organic agriculture is being promoted as a sustainable solution to ensure food security and maintain the agricultural economy [1]. In developing countries, the rising population demands more agricultural yield. India is known for its large agricultural output [2]. At the same time, the country is also facing numerous problems from soil degradation. This is caused by poor agricultural practices and natural factors like salinization, waterlogging and droughts. This threatens the future food security of countries like India and also causes huge economic losses.

The repetitive use of chemical fertilisers and pesticides has rendered acres of soil infertile [3]. This also leads to water pollution, loss of plant biodiversity and is unhealthy for consumption due to the bioaccumulation of chemicals in the crops. As the production of food increased since the Green

Revolution, the nutrient deficiency in the soil also increased simultaneously. Cultivating crops beyond the capacity of the land has depleted the groundwater in many fields (water used for irrigation). Nitrates enter the groundwater because of excessive nitrogen fertilisers. Burning residual crop waste returns some of the nutrients to the soil, but it is lesser when compared to the nutrients used up to increase the yield and it cause air pollution. Thus, the nutrient imbalance leads to soil infertility. Added to this, poor agricultural practices like extensive tilling worsen soil texture. It decreases the soil organic matter, which is essential for soil productivity [2]. Overall, continuous application of chemical fertilisers for a long period of time leads to acidification degrades soil structure and makes the land vulnerable to soil erosion [3].

Sundarbans is The United Nations Educational, Scientific and Cultural Organization (UNESCO) world heritage site, largest RAMSAR site across India and is largest mangrove forest of the globe. Indian Sundarbans covers 102 islands, in which 54 are inhabited with human and rest are reserved forest. In these inhabited islands there are about 4.6 million people live. The major source of income is farming, aquaculture, tourism and honey collection. The agricultural land is continuously decreasing due to frequent cyclone surge [4]. In the 17 years from 2010 to 2017 the agricultural land decreased about 3.1% from 52,148 ha to 50,508 ha [5]. Climate change, sea water intrusions, and depleting fresh water resource are other major challenges in these regions. The uses of chemical fertilizers and pesticides are ruining the conditions as well. The increased soil salinity inhibits the ammonia oxidizing bacteria, and thus decreases the ammonia sorption capacity of soil. This phenomenon resulted in the conversion of ammonium to free ammonia in form of N_2O [6]. N_2O is the one of the greenhouse gas, which leads global warming. Thus the organic farming can be effective for the farmers in order to reduce soil salinity and improve soil fertility of these unique coastal lands.

Organic agriculture is being promoted by environmentalists to remediate manage soil degradation and infertility. This method refers to farming without the use of chemical pesticides and fertilisers. There is no excessive irrigation because only crops suited to the region (endemic crops) and soil type are cultivated. It has numerous environmental and social benefits like remediating degraded soil, sustaining agricultural economy, and food security. Farmers become more self-reliant and use their indigenous knowledge [7]. Organic farming still remains unpopular as farmers feel discouraged by the lower yield. They find it complex to maintain the fields without using pesticides or fertilisers. This also involves high labour costs and organic farming techniques may not be widely known. However, the global economic share in organic food production is rising because of the health and environmental awareness of consumers, which has increased the demand for organic food. Organic agriculture has been successful in developing countries like Bangladesh and some parts of India, like Sikkim [7]. Thus the research question arises to what extent is organic farming a sustainable and economically viable solution to recover coastal, saline soils?

The objective of this study is to understand the importance of organic farming in soil recovery and the challenges of organic farming in coastal areas. The latter will be explored through the case of Sunderban delta, where organic farming was promoted, to generate livelihood and food security in the disaster-prone area.

2. Agricultural soil parameters

In order to measure whether the soil fertility is improving after agriculture, soil parameters are the major indicators. These indicators are based on the physical and chemical properties of the soil necessary for plant growth. For coastal soil, the natural conditions will be different from inland soil, so the indicators help us compare coastal soil to fertile soil. The soil health indicators used in this paper are pH, alkalinity, and electrical conductivity (EC), nitrogen, phosphorous and potassium (NPK), soil organic carbon (SOC) and soil texture.

2.1. pH

Soil pH affects microbial activity and the nutrient solubility. The pH of a soil conveys whether there are enough nutrients to support the growth of plants. It depends on the native soil conditions, because some crops in tropical regions prefer slightly acidic soils while others might prefer alkaline soil [8]. We can see that the coastal soil in Bangladesh is slightly acidic, with a pH of 5.39-6.02 (Table 1). The pH values in fertile soils were found range of 6-8 referred in Table 1.

2.2. NPK

Nitrogen (N) is the most important nutrient for plant growth- found in different forms like nitrates. Phosphorus (P) is essential for agricultural yield and determines the soil quality [8]. Phosphorous play a crucial part in the enzymatic pathways of the crops. This is very important for the photosynthesis, biosynthesis process such as ADP, ATP, NADP. Potassium (K) takes part in the enzymatic pathways and very important for the plant growth. These nutrients are measured as Total and Available N, P, K.

2.3. Soil structure

Soil structure influences the soil texture. It was found that upon long-term recovery of coastal soil in China, the soil particle size gradually decreased and the soil had higher percentage of clay particles and lesser sand [9]. Coastal soil retains a lot of moisture, which was reduced upon introducing agriculture in China; because seawater intrusion was prevented using dikes [9]. In saline soils, excess sodium ions cause clay particles to separate and repel. This results in a loose texture that cannot retain much water [10]. Similarly, in Bangladesh, coastal areas had higher percentage of sand and silt, when compared to its inland soil (Table 1). Soil structure is also quantified through bulk density. Lower bulk density indicates better soil texture.

2.4. Alkalinity

Coastal soils are alkaline due to the presence of salts. The pH and salt content in the soil decreases with agriculture [9]. In Egypt, it was seen that alkalinity reduced when the soil is farther from the coast [11].

2.5. Electrical conductivity (EC)

Electrical conductivity is a measure of soil salinity [10]. When agriculture is practised in saline soils, the cations are displaced because other nutrients in the soil increase, and electrical conductivity seems to decrease over time [9]. Adding too many nutrients could lead to acidification of soil, as is the case in many agricultural lands in India. Overuses of chemical fertilisers have caused this acidification [3]. Table 1 compares the soil parameters between coastal soil and fertile soil across various regions. The information is based on experiments conducted in different regions of tropical, developing nations.

Table 1. Comparison of soil parameters of agricultural soils with special focus to coastal zones.

Parameter	Countries	Values		Remarks	Reference
		Coastal soil	Inland soil		
pH	Bangladesh	Topsoil= 5.39-6.02; Deeper soil sections= 8.1	6.5 - 7.5	Coastal topsoil was slightly acidic and became more alkaline in the deeper sections. Higher pH indicates lesser nutrients while acidic pH is caused by soil	[10]

				pollution through chemicals in Bangladesh.	
	India (Bihar)	8.38 (coastal sandy soil)	6.0 - 6.5	Indian coastal soils are slightly alkaline. This could lead towards nutrient deficiency. Alluvial soil in Bihar is very fertile, and is extensively cultivated. However, excessive uses of chemicals in certain parts of Bihar have led to soil acidification with pH 4.4.	[24-25]
	Egypt (45km from sea)	8.14	8.11	Throughout Egypt, there seems to be less variation in pH between inland and coastal soil.	[11]
Alkalinity	Egypt	12.76	8.86	Alkalinity decreases when the land elevation increases- because this raises the level of groundwater and lowers the capillary intrusion of saltwater. So, inland soil is less alkaline.	[11]
EC	Bangladesh	2.1 - 15 dSm^{-1}	2.4 dSm^{-1}	Salinity increases EC. A value that exceeds 4 dSm^{-1} is considered very saline. It increases during the dry season due to tidal intrusion or poor agriculture practices.	[10]
	India	1.12 dSm^{-1}	0 - 1 dSm^{-1}	According to the study, 0 - 1 dSm^{-1} is considered the ideal range for Indian coastal soils. EC that is too low indicates the lack of nutrients in the soil.	[24]
	Egypt	Topsoil: 5.61dSm^{-1} Deeper section: 13.26dSm^{-1}	2.94 dSm^{-1} (45km from sea)	The high salinity is caused by saltwater intrusion due to low land elevation. The water level is low, so EC is higher in deeper sections of the soil in both coastal and inland soil.	[11]
Soil Organic Carbon (SOC)	Bangladesh	0.55% (at Kalapara)	0.85-2.8% (at Assasuni)	SOC indicates the health and productivity of the soil. Inland soil seems to be healthier because of high SOC concentration.	[10]
	China	5g kg^{-1} (North China)	8.58g kg^{-1} (South China)	Wang et al.'s study applied organic amendments to the saline soil to improve its fertility. Compost, sludge peat, etc. were added to the coastal soil in 2010-2012. The SOC increased with the supplements. The SOC of 8.58g kg^{-1} is of an agricultural area in South China.	[26-27]
	Aceh, Indonesia	<12g kg^{-1}	>12g kg^{-1}	SOC level for crop production should ideally be greater than 12g kg^{-1} for the tropical soil in Aceh. However, according to the study, most of the	[28]

				samples showed SOC levels below the critical value. Aceh is seen to have low fertility and overall, an acidic soil.	
Soil Texture	China	Bulk density = 1.55g cm^{-3} Soil porosity = 0.38% (North China)	Bulk density = 1.15g cm^{-3} (East China)	High soil porosity and low bulk density indicate fertile soil. In Wang et al.'s study, the same organic amendments were used to improve coastal soil texture. Soil recovery in the coastal soil in East China reduced the BD to 1.15g cm^{-1}, suited for crops.	[18, 26]
	Bangladesh	Clay- 37% Silt- 51% Sand- 12% (Kalapara)	Clay- 42% Silt- 52% Sand- 6% (Assasuni)	Clay minerals help in holding soil particles together, which increase when salinity reduces. Assasuni has more clay particles than coastal Kalapara and half the concentration of sand particles. This indicates that Assasuni soil is more fertile.	[10, 26]
Available N	China	44.33mg kg^{-1}	79mg kg^{-1}	In Wang et al.'s study, the values for available N, P and K significantly increased in two years (2010-2012) after the organic inputs. These are initial values.	[26-27]
Available P		5.38mg kg^{-1}	10.8 mg kg^{-1}		
Available K		143.33mg kg^{-1} (North China)	122mg kg^{-1} (East China)	The soil in east China had an ideal amount of NPK after recovery from organic farming.	
Total N	Aceh, Indonesia	< 2.0g kg^{-1}	2.0g kg^{-1}	According to McLeod et al, the amounts of nutrients in the soil in Aceh varied widely, but were below the critical amount needed for cultivation.	[28]
Total P		< 8.3mg kg^{-1}	8.3mg kg^{-1}		
Total K		< 166mg kg^{-1}	166mg kg^{-1}		
Total N	India	0.10-0.35g kg^{-1}	141-474kg ha^{-1}	The soil in Punjab, an intensely cultivated region had turned acidic and infertile due to overuse of fertilizers.	[25-29]
Total P		0.03-0.084g kg^{-1}	2.2-68.2kg ha^{-1}	The alluvial soil in Bihar is still suitable for cultivation and is important to India's agricultural economy.	
Total K		1.118-1.436g kg^{-1} (Punjab)	107-903kg ha^{-1} (Bihar)		

Less NPK is available for plants in saline conditions. At Kalapara (coastal Bangladesh) this could be due to salt intrusion, over cultivation, improper soil management. At Assasuni (inland Bangladesh), N was much higher than in coastal soil. The NPK can be maintained if organic additions like manure are used as required.

3. Organic and Conventional Farming

According to Roos et al. [1], increasing crop yield through unsustainable inorganic processes has several negative impacts on ecosystem, and human health [1]. Nowadays, farmers are highly dependent on inorganic fertilizers, for increasing short term yield. However, there are many processes which can increase crop yield through organic agriculture. This work also sheds light on the strategies for sustainable agriculture through organic farming and other risk free opportunities to increase crop yield. The advantages of organic agriculture over conventional methods are debated over environmental, economic and social factors. There is not enough research in this field to conclude that one method is better than the other. However, we can weigh the advantages and disadvantages of the two methods to decide which is best suited for the location and people.

There is more demand for organically grown food given the spread of awareness regarding its expected health benefits. But, there is also a rising demand for food due to the growing world population. Organic agriculture has proved to be more profitable and sustainable as compared to conventional farming methods. Though the yield is lower in this method, the high economic value for organically produced food compensates for the low yield and brings in profit. It also consumes lesser resources because chemical fertilisers/ pesticides are not used.

3.1. Feasibility

However, given that land resources are scarce and demand for food is high, the sustainability per unit product is questioned because more land is needed to maintain or increase the yield [12]. It is estimated that yield is lower by 19-25% [13-14]. This varies according to the crop type and growing conditions. The yield is lesser in legumes but higher in cereals [1]. If organic products did not have such high economic value, farmers would incur losses due to low productivity of organic farms [1].

However, according to a study conducted by Schrama et al., the yield gap between organic and conventional farms reduces over time. In the 13 year study, lower yield was recorded from organic farms at the beginning [15]. Then, yield increased to almost the same level as conventional farms, with an average gap of 13% across the 13 years [15]. In fact, the yield was achieved even with lower N inputs.

Crops in an organic farm are more susceptible to pests and diseases due to the absence of chemical pesticides [1]. This again puts yield at risk. Organic fertilisers are low in nitrogen as plant wastes do not contain much nitrogen in available state. To overcome this, farmers rotate crops- legumes can fix nitrogen in the soil for the crops grown in the next season. However, this might not be feasible if certain soils are not favourable to grow legumes. Organic farms are vulnerable to weeds. Farmers need to invest in weed removal technology or manually remove weeds, which increases the labour costs. Excessive tillage to remove weeds can damage the soil structure and loss of organic matter [13].

In spite of these drawbacks, organic agriculture can have numerous benefits on the soil, if organic fertilisers and farming methods are managed properly.

3.2. Sustainability

Studies have shown that microbial diversity is higher on organic farms. The soil ecosystem is healthier due to the presence of these microbes. Soil Organic Carbon (SOC) is also higher in organic farms; potentially contributing to carbon sequestration [1].The organic amendments used in this type of farming makes the soil fertile in the long-term. The soil texture allows for water retention as well as prevents soil erosion. So, crops have more probability to survive extreme conditions like droughts [13]. Using organic manure improves the SOC from the decomposing matter. There is also a lower risk of nutrient leaching because of lower fertiliser use. This ensures that the soil ecosystem thrives for a long duration [15].

Organic farms have higher yield stability. This means that production can be sustained over a longer time period for the same patch of land. Crop diversity ensures nutrient recycling to maintain a balance of nutrients in the soil [13]. One concern is nutrient losses (N and P, for example) during crop rotation because the rate of uptake varies between plants [1]. Also, organic farming is heavily dependent on external inputs of organic matter. Farms could suffer from a shortage of organic inputs because of the slow decomposition [15]. These two concerns can be overcome by undertaking more research on this topic to implement necessary agricultural techniques.

3.3. Cost-effectiveness and social aspects

Organic agriculture is becoming more commercialised as it is seen as a 'commodity for the elites' rather than a basic food product because of the high purchase cost associated with them. 75% of the organic farms are located in developing countries while most of the produce is exported to developed nations in Europe and North America, causing food security concerns [13]. This process involves large corporations and certification processes which may be tedious and expensive for small scale farmers. However, there are cases in developing countries where farmers and consumers form their own network to buy/sell organic food. This has positively helped local farmers [13]. Organic agriculture is also reducing a farmer's economic dependence on a single crop [13]. This way, the negative environmental impacts of monoculture, like unsustainable use of water, can be mitigated [12].

In Bangladesh, self-help groups were formed between farmers to share information and resources for organic agriculture. This has increased the organic farming percentage in the country and helped farmers find social and financial support when required [16]. Thus, organic farms create better working conditions for farmers where the exposure to chemicals significantly decreases. For customers too, though higher nutrient content is not proven in organic food, there is lesser risk of ingesting chemicals through plants grown using synthetic pesticides and fertilisers [13]. This reduces the negative externalities posed by conventional farming. One drawback is that, to increase yield, sometimes, farmers use high-yielding variety crops. This makes the processes similar to conventional agriculture, so the environmental benefits will not be long-term [12].

Government policies play a huge role in promoting organic agriculture. They set targets to increase the percentage of farms practising organic agriculture. They can incentivise organic agriculture and enter into trade agreements to exchange organically produced food, due to the profits involved. It is also important that they regulate labour laws so that small farmers are not exploited for manual labour in certain organic farms [12].

4. Environmental and Economic impacts

Organic farming has been implemented in many regions across the world. The results have been diverse, depending on various factors like location, crop choices, socio-economic conditions, etc. While many cases have shown positive results (Such as Egypt and Indonesia Table 1, organic farming does not seem viable in certain agricultural practices. It is common for coastal farmlands to include aquaculture beside their farms for additional economic gain [17, 18]. Table 2 provides a review of some organic farming initiatives.

Table 2. Adopted organic farming methods across various regions and their impacts on soil and crop production

Region	Organic farming method	Impact on soil and livelihoods	References
Karnataka, India	The state government decided to introduce organic farming to counter agricultural problems. Manure is obtained from	<u>*Economic Impact:*</u> Dry area: Food crops like millet and rice had higher yield under organic methods, so higher profit. Low input costs helped farmers reduce their debts. During crop failures, net loss was lesser by upto 40%.	[21]

	cows and crop residues. If this is not sufficient, farmers purchase vermicompost from the market. Crop rotation was practiced. The soil type in this region is red, sandy and loamy. The dry areas in the state were impacted positively by organic farming but wetter, more fertile areas had a slightly negative impact.	Wet area: Commercial crops had lower yield. Farmers had to spend on vermicompost and additional labour. Profits were lower for most crops. *Environmental Impact:* Dry area: N leaching was reduced by about 50%. Soil water retention capacity had improved. However, there was nutrient imbalance, which could cause deficiency in future. Wet area: Not much change was observed apart from lower yield. There was no nutrient imbalance.	
Eastern, Coastal China	Reclamation of coastal saline soil was done over 30 years. *The process followed:* - Initially, abandoned coastal soil is put under mariculture. Or, plants that grow in saline conditions are allowed to grow, to reduce the salinity. - After 10 years, salt tolerant crops like cotton, corn and wheat are grown. - Mariculture gradually changes to freshwater culture because salinity decreases. The soil also becomes conducive for crops. - Currently in this soil, rapeseed, wheat, paddy, corn etc. are grown along with practising aquaculture. Changing the land use patterns was an important process for the recovery- whether land was abandoned, used for aquaculture, or under organic agriculture, type of crops like rapeseed or beans, etc.	The climatic conditions in the eastern coast of China are extreme. It is humid, receives a high amount of rainfall and sunshine throughout the year. These conditions make it ideal for agriculture, but its proximity to the sea makes the soil saline with marine deposits. This also makes this soil sensitive to over cultivation beyond its capacity. *Environmental impacts:* Bulk density decreased over the years to reach 1.15g cm^{-3}. The soil water content did not change significantly. The pH decreased from an alkaline state to 7.97. EC ranged from 0-1dSm^{-1} after recovery. The SOC, total N and total P followed a trend of increase and decrease based on the crops grown. Overall, there was an increase. Aquaculture farms showed the opposite trend- higher soil water content, pH and EC, lower in nutrients and SOC. However, it suited the requirement for aquaculture farms.	[18]
Andaman Islands	Raised bed and furrow system, along with organic inputs.	The Andaman Islands are located near the equator. The land has a low elevation, and is vulnerable to climate change. High evaporation during the summer	[17,22]

	Raised beds were created on coastal farmland to prevent tidal intrusion and improve the soil's drainage. The dug-out area surrounding the raised bed is the furrow, used for rainwater harvesting or aquaculture. Organic supplements prepared from previous crop residues and other farm waste were applied to the soil, to improve its fertility. Importing fruits and vegetables makes it expensive for the residents of the island due to the transportation costs involved. Increasing profit from organic farming could help the locals increase their household income.	causes water scarcity and high rainfall during monsoon causes waterlogging. This method helped improve soil fertility and agriculture in this region. Adding organic matter increased the SOC concentration. The soil texture improved as the bulk density decreased. The initial EC of 10.9dSm^{-1}, indicating very saline soil, was brought down by 87%, to 1.40 dSm^{-1}. The nature of the coastal soil and limited land area hinder any major crop cultivation. Even subsistence needs are not met, as grains are imported from mainland India. However, plantation crops like coconut can grow organically and can be exported for economic gain. The residues can be used for vermicomposting.	
Bangladesh	Due to the global market for organic products, and increasing popularity for organic food among Bangladeshi customers, government and NGOs are promoting organic farming in this country. Farmers are given training programs. Women are also trained separately to raise awareness and to convert household waste into compost. Integrated pest management is a part of their training. NGOs plan to set up more support groups among farmers for any technical, social or economic difficulties they face. As per literature organic farming seems to work well	0.1% of cultivable land in Bangladesh is under organic farming, and its global importance in this field is growing. Generally, farmers welcome organic farming as it has fetched them more profits. However, certain small farmers have not benefited much by organic farming. There are lesser support groups in rural areas. They are unable to cope with food insecurities during extreme conditions as organic inputs are not as efficient as chemical fertilizers for their small plots. They can't afford vermicomposting equipment either. Livestock possession seems to improve their economic situation because of the organic waste for farms and dairy products they can sell.	[7, 23]

	in Bangladesh's humid, saline and tropical conditions.	

5. Case study

This is the successful case study on the organic farming practice at Indian Sundarbans (Figure 1) initiated in 2012. Where, 84 families were facilitated to using organic method of agriculture with vermicompost, biopesticides like neem oil and natural microbe augmentations such as Trichoderma, Phosphate Solubilizing Bacteria (PSB). This was an achievement to motivate them to adapt the organic method, leaving their age old process using eco-destructive chemical fertilizers and pesticides. The improved seeds with bio-pesticides have proven benefited for the farmers and yield is significantly increased for the both Kharif season (10-12%) and Rabi season (5-10%) by 2017.

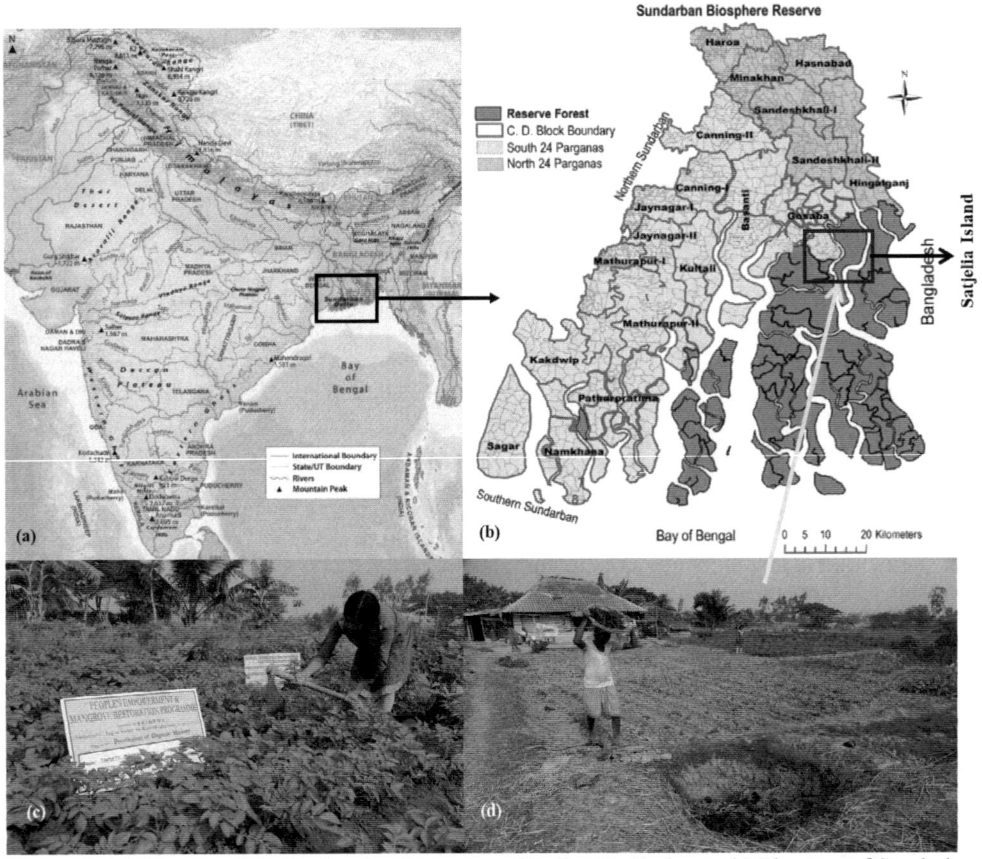

Figure 1. (a) The map of India showing the region of Indian Sundarbans, (b) The area of Sundarbans zoomed out focusing on Satjelia Island where the intervention has taken place and the sustainable agricultural practices are implemented by local community as a part of the project. Two pictorial representations of the agro-practices from the project are represented by the black coloured arrow, (c) Beneficiary has been seen to be working in the kitchen garden cultivated in their back yard through organic sustainable agricultural practices, (d) Beneficiary is seen to be collecting cow dung for composting to be used later at fertilizer in their agricultural fields.

Organic farming and plantation of indigenous variety trees helped to save nature as well as to reduce environmental pollution. Thus, out of 84 families, 10 families have started producing Vermicompost, after receiving adequate training. Two families of Sudhangsupur and Hamilton Abad Hamlet villages have trained to prepare demonstration plot using only organic methods and indigenous seed, and they stand as an example amongst the entire populace of the Satjelia Island, practically highlighting the benefits of Organic Agriculture. The other beneficiary families have trained for household Plantation programme focused on the production of fodder, fuel, small timber, minor forests produce to meet the local need [18]. Dry timber used as fuel, fruit trees produced a fruit which adds up the nutrition value of the household, where mal-nutrition is a reality and this burden is borne by the children and the women.

Awareness, organic farming skill and perspective building among the target people have proven beneficial for the locals and helped them to become resource person, developing healthy attitude & practice. This helped them to take leadership of the society for betterment. Infant mortality rate were also reduced through nutritional fruits and vegetables cultivated through organic methods. This project also helped in the renovation of ponds which further being used by beneficiaries' in plantation works and aid in agricultural activity as intended in the project objective.

This was an example of Sustainability and financial independence of the beneficiaries, which was the prime target of the project. Farmer were also encouraged to create 'Farming companies' where they can directly dictate the sale of the products to the local market without getting dependent on middle man.

5.1. Challenges

One of the major challenges faced by the farmers in the region is their socio-economic vulnerability [20]. Organic farming practices takes times to make the soil fertile whereas chemical based methods give immediate productivity boose [14]. In case of Indian Sundarbans, most of the farmers are marginalized with very limited land holdings. Hence, they are reluctant to dedicate whole of their farming land holdings for organic agriculture. Only through training, awareness and initial monitory support the issue can be addressed.

Organic certification process is a time taking matter with lot of bureaucratic red tape regulations. These make the certification process difficult for the farmers to comply or pursue. Though such certifications can improve the selling price of the product and indirectly motivate farmers to go for organic practices. Organic agriculture in the case study has been supported by the social initiative in the region. Otherwise, both the certification and organic processes are economically taxing for a marginalized farmer which makes it unpopular to marginalize.

6. Conclusion

Organic farming is attracting the attention nowadays. The chemical based farming practices not only destroy the human health but also proven unsustainable for soil fertility. Coastal areas of India are facing huge challenges in the soil fertility as well as the lower yield. To achieving high yield through organic farming process can improve the soil health with providing sufficient nutritional crops. This work shed light on the feasibility of organic farming in the vulnerable, saline and infertile coastal agricultural belt. This work also highlighted the challenges faced by the farmers in the way of organic farming in remote coastal belts of tropics. This work concludes with the way to overcome with the major challenges in organic farming in the coastal belt of India. Case study of the Indian Sundarban villages, where 84 families of different coastal villages were provided financial support through a foreign funding explains the successes story of skill development, and organic farming.

References

[1] Roos E, Mie A, Wivstad M, Salomon E, Johansson B, and Gunnarsson S *et al.* 2018 Risks and opportunities of increasing yields in organic farming: A review *Agron. Sustain. Dev.* 38 1-21 doi: 10.1007/s13593-018-0489-3

[2] Bhattacharyya R, Ghosh B, Mishra P, Mandal B, Rao C, Sarkar D and Franzluebbers A J *et al.* 2015 Soil Degradation in India: Challenges and Potential Solutions *Sustainability* **7**(4) 3528-3570 doi: 10.3390/su7043528

[3] Bhat S A, Dar M U D and Meena R S 2019 Soil erosion and management strategies. In Sustainable Management of Soil and Environment pp. 73-122 Springer, Singapore.

[4] Chowdhury A, Naz A, Iyer A S and Bhattacharyya S 2021 Ecosystem based disaster risk reduction at Indian Sundarbans: A lesson learned from AMPHAN supercyclone In IOP Conference Series: Earth *Environ. Sci.* **796** 012042. IOP Publishing.

[5] Thakur S, Maity D, Mondal I, Basumatary G, Ghosh P B, Das P and De T K 2021 Assessment of changes in land use, land cover, and land surface temperature in the mangrove forest of Sundarbans, northeast coast of India *Environ. Dev. Sustain.* **23** 1917-1943.

[6] Li Y, Xu J, Liu S, Qi Z, Wang H, Wei Q and Hameed F *et al.* 2020 Salinity-induced concomitant increases in soil ammonia volatilization and nitrous oxide emission *Geoderma* **361** 114053.

[7] Ferdous Z, Zulfiqar F, Datta A, Hasan A and Sarker A 2020 Potential and challenges of organic agriculture in Bangladesh: a review *J. Crop Improv* **35** 403-426 doi: 10.1080/15427528.2020.1824951

[8] Cardoso E, Vasconcellos R, Bini D, Miyauchi M, Santos C and Alves P *et al.* 2013 Soil health: looking for suitable indicators. What should be considered to assess the effects of use and management on soil health? *Sci. Agric* **70**(4) 274-289 doi: 10.1590/s0103-90162013000400009

[9] Li J, Pu L, Zhu M, Zhang J, Li P and Dai X *et al.* 2014 Evolution of soil properties following reclamation in coastal areas: A review *Geoderma* **226** 130-139 doi: 10.1016/j.geoderma.2014.02.003

[10] Naher N, Uddin M and Alam A 2011 Impacts Of Salinity On Soil Properties Of Coastal Areas In Bangladesh *Agrivita,* **33** 161-173. Retrieved on 21 July 2021.

[11] Ding Z, Koriem M A, Ibrahim S M, Antar A S, Ewis M A, He Z and Kheir A M S 2020 Seawater intrusion impacts on groundwater and soil quality in the northern part of the Nile Delta, Egypt *Environ. Earth Sci.* **79** 1-11 doi:10.1007/s12665-020-09069-1

[12] Eyhorn F, Muller A, Reganold J, Frison E, Herren H and Luttikholt L *et al.* 2019 Sustainability in global agriculture driven by organic farming *Nat. Sustain.* **2** 253-255. https://doi.org/10.1038/s41893-019-0266-6

[13] Seufert V and Ramankutty N 2017 Many shades of Gray—The context-dependent performance of organic agriculture *Sci. Adv.* **3**(3). https://doi.org/10.1126/sciadv.1602638

[14] Lakner S and Breustedt G 2017 Efficiency analysis of organic farming systems a review of concepts, topics, results and conclusions. German Journal of Agricultural Economics, 66(670-2020-978), 85-108.

[15] Schrama M, De Haan J J, Kroonen M, Verstegen H and Van der Putten W H 2018 Crop yield gap and stability in organic and conventional farming systems. Agriculture, ecosystems & environment, 256, 123-130.

[16] Ferdous J 2019 Rural Development in the Era of Globalization: An Assessment of the Performance of Bangladesh Academy for Rural Development (BARD) (Doctoral dissertation, University of Dhaka) http://localhost:8080/xmlui/handle/123456789/1518

[17] Velmurugan A, Swarnam T P, Ambast S K and Kumar N 2016 Managing waterlogging and soil salinity with a permanent raised bed and furrow system in coastal lowlands of humid tropics Agric. Water Manag. 168 56–67. doi:10.1016/j.agwat.2016.01.020

[18] Xie X, Pu L, Wang Q, Zhu M, Xu Y and Zhang M 2017 Response of soil physicochemical properties and enzyme activities to long-term reclamation of coastal saline soil, Eastern China Sci. Total Environ. 607-608 1419–1427. doi:10.1016/j.scitotenv.2017.05.1

[19] Chowdhury A, Maiti S K and Bhattacharyya S 2016 How to communicate climate change 'impact and solutions' to vulnerable population of Indian Sundarbans? From theory to practice SpringerPlus 5(1), 1-17.

[20] Sahana M, Rehman S, Paul A K and Sajjad, H 2021 Assessing socio-economic vulnerability to climate change-induced disasters: evidence from Sundarban Biosphere Reserve, India Geol. ecol. landsc. 5 40-52.

[21] Patil S, Reidsma P, Shah P, Purushothaman S and Wolf J 2014 Comparing conventional and organic agriculture in Karnataka, India: Where and when can organic farming be sustainable? Land Use Policy 37 40–51. doi:10.1016/j.landusepol.2012.01.

[22] Bandla G, Kundu A, Murugan A, Pandey G, Poddar R, Navin K, Sar U 2017 Basic information of agriculture and organic farming in Andaman & Nicobar

Islands https://www.researchgate.net/publication/352361639_Basic_information_of_agriculture_and_organic_farming_in_Andaman_Nicobar_Islands (Retrieved on 13 August 2021)

[23] Barua P and Rahman S H 2021 Livelihood Security Determinants Of The Organic Farm Household In South-Eastern Coast Of Bangladesh Food Agribusiness Manag. 2 01-05.

[24] Singaravel R, Prasath V and Elayaraja D 2005 Role Of Bio-resources In Improving The Fertility Of Coastal Sandy Soils For Sustainable Groundnut Production Management Of Tropical Sandy Soils For Sustainable Agriculture. ISBN: 978-974-7946-96-3

[25] Reza S K, Nayak D C, Mukhopadhyay S, Chattopadhyay T and Singh S K 2017 Characterizing spatial variability of soil properties in alluvial soils of India using geostatistics and geographical information system *Arch. Agron. Soil Sci.* **63** 1489-1498. DOI: 10.1080/03650340.2017.1296134

[26] Wang L, Sun X, Li S, Zhang T, Zhang W and Zhai P 2014 Application of Organic Amendments to a Coastal Saline Soil in North China: Effects on Soil Physical and Chemical Properties and Tree Growth *Plosone* **9** e89185. https://doi.org/10.1371/journal.pone.0089185

[27] Zhang W, Xu M, Wang B and Wang X 2008 Soil organic carbon, total nitrogen and grain yields under long-term fertilizations in the upland red soil of southern China *Nutr. Cycl. Agroecosystems* **84** 59-69. https://doi.org/10.1007/s10705-008-9226-7

[28] McLeod M, Sufardi S and Harden S 2020 Soil fertility constraints and management to increase crop yields in the dryland farming systems of Aceh, Indonesia *Soil Res* **59** 68. https://doi.org/10.1071/sr19324

[29] Bhatti S, Kumar V, Singh N, Sambyal V, Singh J, Katnoria J and Nagpal A 2016 Physico-chemical Properties and Heavy Metal Contents of Soils and Kharif Crops of Punjab, India *Procedia Environ. Sci.* **35** 801-808. https://doi.org/10.1016/j.proenv.2016.07.096

Acknowledgement

We acknowledge the support and funding of KKS (Karl Kübel Stiftung für Kind und Familie, Germany) and BMZ (Federal Ministry for Economic Cooperation and Development, Germany) through project entitled "Peoples Empowerment towards Restoring Mangrove Vegetation and Resource Conservation" implemented in the island of Satjelia by Tagore Society for Rural Development (TSRD) between 2012 and 2017. First author was the 'Programme coordinator' of this project in its inception years and continued as a consultant for the project till its completion.

Perpetual Flexible Pavement vs. Rigid Pavement: An Economic and Environmental Cost Comparison

Rahul Kumar[1] and Sunil Sharma[2]

[1]MTech Scholar, Department of Civil Engineering, National Institutes of Technology Hamirpur, India
[2]Assistant Professor, Department of Civil Engineering, National Institutes of Technology Hamirpur, India

Email: 20mce212@nith.ac.in , rahulnith96@gmail.com

Abstract. India needs roads that can handle a large volume of traffic and have a long service life due to the scarcity of natural resources and their rising cost. Over the last decade, type. As a result, the notion of a perpetual flexible pavement with a life of more than 50 years and great structural capacity has emerged. Perpetual flexible pavement has a direct competition with cement concrete-based rigid pavement. The comparison of life cycle cost (LCC) and life cycle environmental cost (LCEC) is an analytical method for determining the long-term viability of any pavement. This paper examined the overall sustainability of each pavement by comparing (1) the life cycle cost with the help of net present value (NPV) method and (2) life cycle environmental cost considering greenhouse gas emissions from various activities in construction and maintenance of pavements over a 50-year analysis period. The cross-sections of the Delhi-Mumbai expressway with perpetual flexible pavement and the Mumbai-Pune expressway with conventional rigid pavement, both of which have been designed for a long service life and a good structural capability were chosen for the investigation as a case study. The findings of the present study reveal that the per kilometre life cycle cost of perpetual flexible pavement is 2.12 lakh rupee less than the rigid pavement. Also, perpetual pavement emits 4.5 times less greenhouse gases than rigid pavement over their design life. In comparison to a rigid pavement during a 50-year design life, the study concludes that perpetual flexible pavement is not only economically effective but also environmentally sustainable

1. Introduction

The second-largest road network in the world is reported from India with 54, 82,000 km, but just 2% are national highways, carrying 40% of road traffic. The government wants to expand freight traffic on national roads and expressways from 40% to 80% [1]. The expense of building and operating a highway pavement network is not only enormous, but it also has a considerable social and environmental effect. The transportation industry accounts for around 14 % of worldwide greenhouse gas (GHG) emissions, with road building, rehabilitation, maintenance, service, and consumption accounting for 72 percent [2]. According to the Intergovernmental Panel on Climate Change (IPCC), the global mean surface temperature is expected to climb by 1.5 degrees Celsius or more over the next 20 years [3]. It is anticipated that if global warming induced by greenhouse gases continues, about 28.6 million people will be vulnerable to coastal flooding, and assets subjected to flooding will be valued at around USD 4 trillion [4].Cost and environmental effects become impediments to meeting the goal of building a national highway and expressway. In the highway construction industry, sustainability includes optimizing the efficient use of natural resources and energy, as well as creating long-lasting, cost-effective pavement with minimal carbon emissions. With the advice of industrialists and academics, the government has made several initiatives toward sustainable development. Alternative pavement designs and materials, pavement performance model[5], condition-based planning and scheduling[6]&scientific approach[7], and factor of prioritizing road for new road construction and existing road maintenance[8], recycling in raw material consumption processing in the pavement layer and many more are the factor that contribute to pavement sustainability. Generally rigid pavement is advantageous where high structural capacity and cost-effectiveness are required because its life cycle cost is lower than conventional flexible pavement, but it requires a high initial investment and results in the production of unnecessarily extra cement as a binding material to pave the roads, which has an overall negative impact on the environment. Flexible pavement promotes sustainability in construction by employing bitumen as the binding medium, which is the most recyclable. The hefty bituminous pavement of United Kingdom (UK) roads fared significantly better, Similar works in the United States validated the concept, earning it the moniker "perpetual pavement" [9]. Perpetual pavement can be built to survive for more than 50 years without requirement of significant rehabilitation and rebuilding work, requiring simply periodic surface renewal maintenance in response to the pavement's distress confined top. Milling, recycling, and hot mix plant (HMA) technological advancements over the last few decades have made pavement better, more durable, and cost-efficient.

The pavement kinds used should be competent in terms of performance and cost-effectiveness when selecting pavement types, for that life cycle evaluation, and environmental impact are also considered. Research comparing the LCC (initial construction cost, maintenance & rehabilitation (M&R) cost, and user cost) of conventional flexible pavement versus perpetual flexible pavement under varied traffic and weather situations found that perpetual flexible pavement reduces LCC by 4-20%[10]. According to research that compared the life cycle costs of conventional Portland cement concrete (PCC) and conventional bituminous pavement, the initial cost of conventional bituminous pavement is 30% cheaper than conventional PCC pavement, but the life cycle cost in terms of net present value is 46.05% greater than PCC pavement[11].HMA has a considerably higher ecological cumulative consumption ratio than PCC in terms of computation of energy and commercial energy usage, making it even more sustainable, and PCC generates roughly 2.6 times more pollutants than HMA[12].Flexible pavements, according to one research, have a lower carbon impact than rigid pavements when CO_2 generated in the production of raw materials and mixtures is examined, it is significantly less than the quantity generated in the production of PCC, whereas the cement manufacturing has the biggest carbon generator in it[13].

The rigid pavement had found several disadvantages, but its construction is still widespread for high-traffic volume roads and perpetual flexible pavement will be the best alternative to provide an acceptable solution. In India, the notion of perpetual flexible pavement is still in its infancy, with little concrete data available. In this study, a comparison of pavements for long service periods and significant traffic loads will be examined from an economic and environmental standpoint. The objectives of this study are to examine and compare the life cycle cost (LCC) of perpetual flexible pavement and conventional rigid pavement from an economic standpoint and to conduct an life cycle environmental cost (LCEC) in terms of carbon dioxide equivalent (CO_2e) from various activities that occur during the pavement's life cycle in order to point out the environmental impact due to construction of pavement, so that practitioners can easily evaluate more sustainable pavement with the help of LCC and LCEC.

2. Methodology

The methodology's goal is to investigate the performance of pavement options while considering economic and environmental issues, and to determine the most sustainable pavement option from a decision-making position.

2.1. Computation of life cycle cost

Although bituminous pavement is cheaper than PCC pavement to construct but it involves much more maintenance cost than rigid pavement. The idea of NPV is applied to calculate the "time worth of money" using discount rate. Hence, the framework for life cycle cost analysis using NPV as shown in Figure 1 is a suitable method to compare the overall cost of two pavements. The equation term as (1) used to calculate net present value was derived from the study report [14].

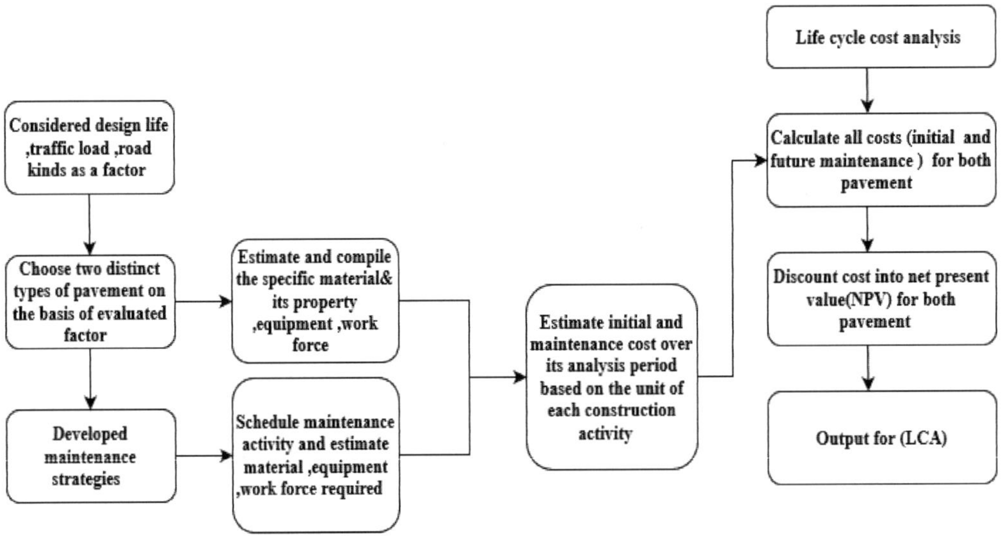

Figure 1. Proposed framework for life cycle cost (LCC) of pavement

NPV= initial cost +Σ up keeping $cost_k$ * $[1/ (1+i)^{n}{}_{k}]$ (1)

Where, i= rate of discount, n =number of the year from initial construction at which expenditure occurs, k= individual M&R activities

2.2. Computation of life cycle environmental cost

It is a method for evaluating environmental implications at various activities of a project's life cycle. The carbon footprint computation throughout its design life is depicted in the Figure 2.The formula term as (2) for calculating greenhouse gas emissions in terms of carbon dioxide equivalent (CO_2e) is adapted from a research paper [15]

$KgCO_2e$= (Amount of gas in kg) * (GWP of the gas) (2)

Where, $kgCO_2e$ = carbon dioxide equivalent in kilogram

GWP= green warming potential is determined over a certain time period measured by the ability of trapped heat in the atmosphere.

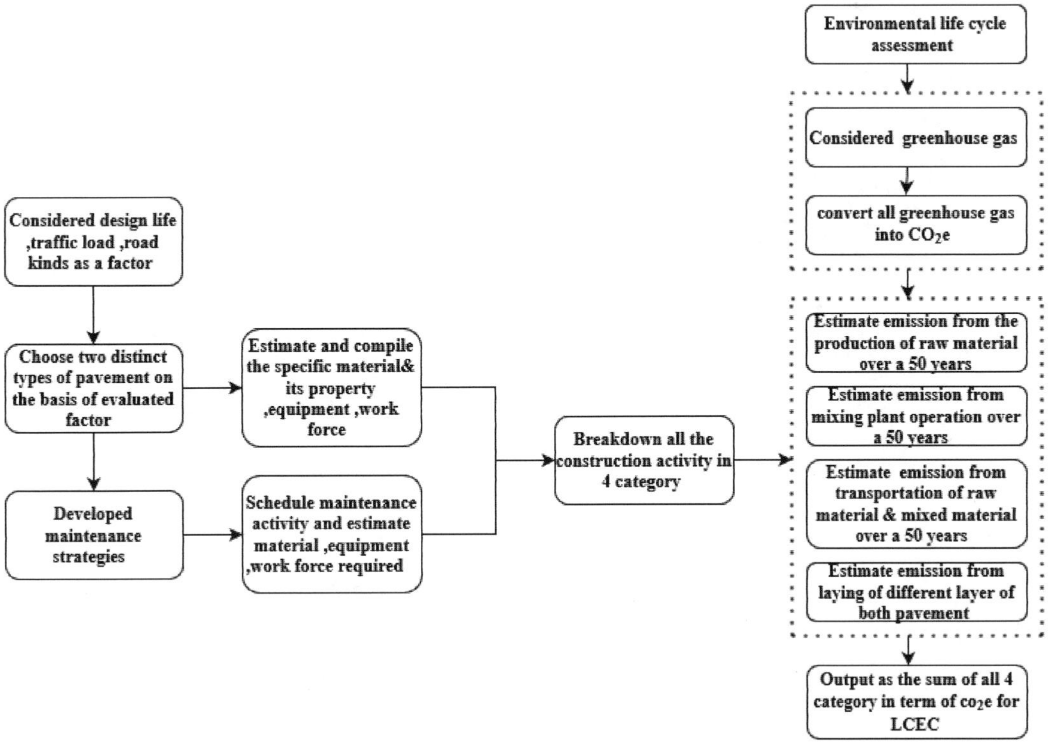

Figure 2. Proposed framework for life cycle environmental cost (LCEC) of pavement

3. Data collection

This case study uses life cycle cost and life cycle assessment of greenhouse gas emissions from raw material extraction to final pavement paving during its analysis period on the Delhi-Mumbai expressway with perpetual flexible pavement and the Mumbai-Pune expressway with the rigid pavement as shown in Figure 3. The life cycle cost study comprises the initial construction costs of the main carriageways of 1000m length and 14m width, as well as its significant preventive maintenance costs, rehabilitation costs, and rebuilding costs throughout a 50-year analysis period. Environmental life cycle evaluation in terms ofCO_2efrom raw material manufacturing, raw material transportation, raw material mixing, mixed material transit, and final paving throughout a 50-year analysis period. The Indiana Department of Transportation's (INDOT) pavement design procedure [16] and a research paper [17] are used to determine rigid pavement maintenance activity and scheduling over a 50-year periods shown in Figure 4 which includes both new and existing pavement maintenance activity and scheduling. Maintenance activity and scheduling of perpetual flexible pavement over a 50-year period as shown in Figure 4 are done with the reference of the study article "Design of Optimal Perpetual flexible pavement Structure". According to a total of 3,213 Mechanistic-Empirical Pavement Design Guide (MEPDG) simulations, the deployment of perpetual pavements should include a resurfacing plan every 10 years to remove rutting in the top surface [18]When comparing the two pavements, just that layer with a different value is chosen for quantifying cost and emission.

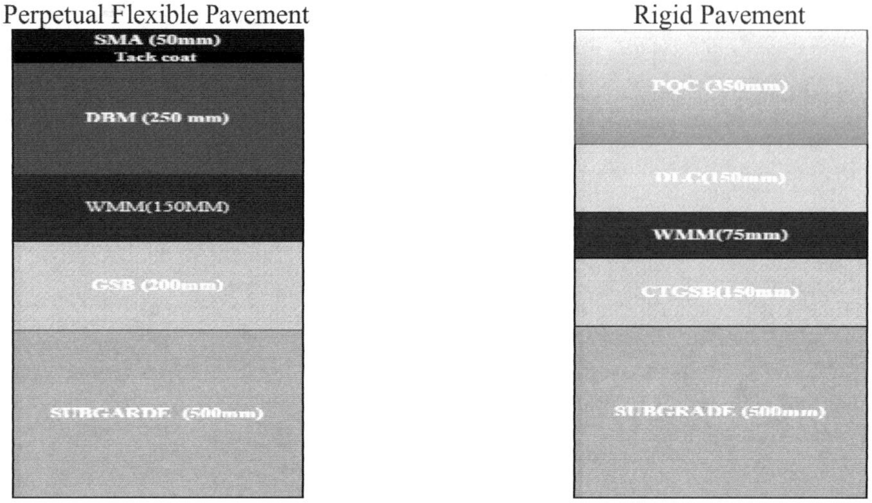

Figure 3. Cross Section of Pavement

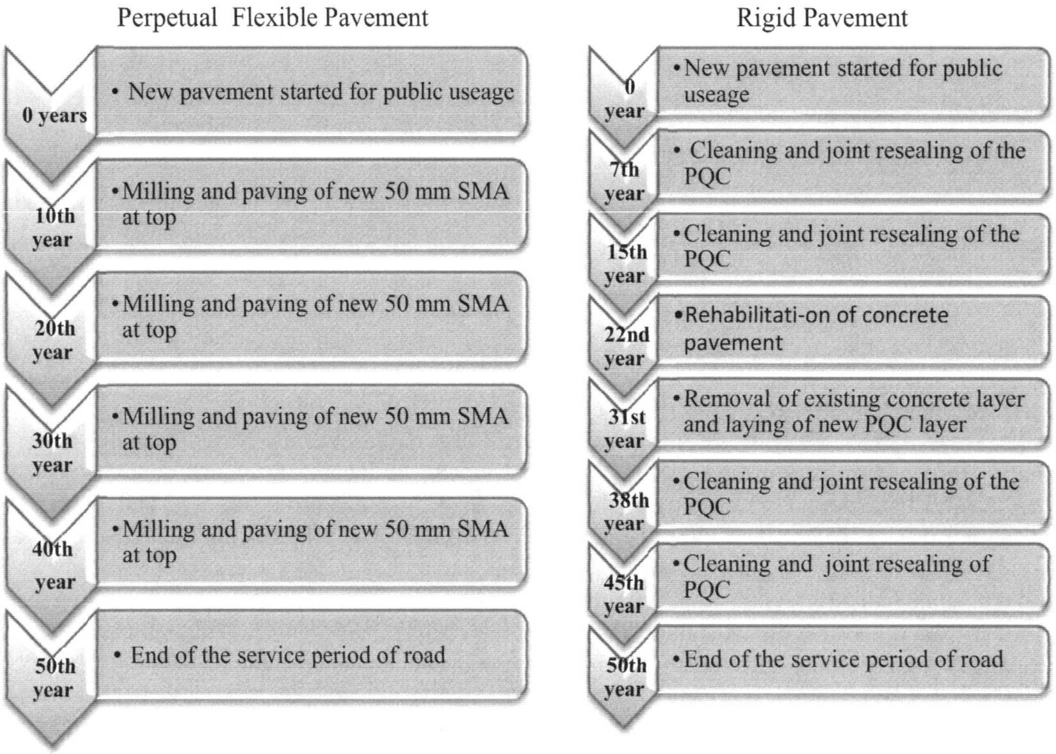

Figure 4. Major Preventive Maintenance Activities & Its Scheduling of Pavement

3.1. Input data for computation of life cycle cost

To compute the LCC for both pavements, reference prices as per ministry of road transport& highways (MORTH) specifications from Haryana state in India as shown in Table 1 is used. The NPV for pavements is calculated in Indian rupees (₹) using a discount rate of 12 % and an inflation rate of 5%, according to a recent study [15].

Table 1. Schedule of rate (2019-2020) as per MORTH specification

S.No.	Item with description	Unit rate
1	Stone matrix asphalt having NMAS 13 mm	9337.5 ₹/cum
2	Dense bituminous macadam	6190 ₹/cum
3	Granular sub-base (grade 1)	1786₹/cum
4	Tack coat	9₹/cum
5	Wet mix macadam	1930₹/sqm
6	Pavement quality concrete of M40	5606 ₹/cum
7	Drainage layer (CTGSB)	2229₹/cum
8	Dry lean concrete of M10	2612₹/cum
9	Milling of top renewable surface	56 ₹ /sqm
10	Joint resealing	45 ₹/m
11	Concrete restoration with epoxy mortar	530 ₹/m
12	Removal of existing PQC layer	1062 ₹/cum

3.2. Input data for computing environmental cost

To calculate total emissions, all activity over a 50-year period is divided into four categories: raw material production, mixing plan operation, material movement, and final layer laying is taken from different paper as shown in Table 4. The NHAI provides a layer with defined properties, and the job mix formula is drawn from the kandhal book [19] as shown in Table 2 to estimate the material necessary for both pavements. For material's movement, distance between sources to destination as shown in Table 3 is taken from real geographical location and recent study [15] .

Table 2. Job mix formula of different layers of pavements

S. No.	Layers name	Density (kg/m^3)	Job mix formula
1	GSB (grading V)	2185	40mm-40%,10mm-20%,6mm-30%, river sand-10%
2	WMM	2260	40mm-30%,10mm-20%,6mm-30%, stone dust -30%
3	DBM (grade I)	2460	O.B.C-4.0%, (40-24mm)-35 %, (24-14mm)-5%, (14-6)-10%, (<6mm)-50
4	DBM (grade II)	2450	O.B.C -4.5%, (40-24mm)-16 %, (24-14mm)-20%, (14-6)-24%, (<6mm)-40%
5	SMA (NMAS 13mm)	2450	O.B.C -6 %, (13.2-9.5mm)-35%, (9.5-4.75 mm)-41%(4.75-1.18mm)- 9 %, (<1.18mm)-5%, Celluloid fiber by wt. of total mix -4%
6	PQC (M40)	2350	W/c ratio -0.38, Cement content (kg/m^3) by wt. of total mix-394, C.A having NMAS 25mm (kg/m^3) by wt. of total mix-1276, F.A (kg/m^3) by wt. of total mix -644
7	DLC (M10)	2550	W/c ratio-0.45, Cement content (kg/m^3) by wt. of total mix -236, C.A having NMAS 20mm (kg/m^3) by wt. of total mix -1386, F.A(kg/m^3) by wt. of total mix -739
8	CTGSB	2316	2316 Cement content by % of total mix-2%, W/C ratio-4.0,40mm-40%,10mm-20%,6mm-25%, river sand -8%

Table 3. Material's movement distance

S.No	Material's source to destination	Transportation distance (km)
1	Coarse aggregate to mix plant	100
2	Fine aggregate to mix plant	25
3	Bitumen to mix plant	150
4	Cement to mix plant	150
5	Steel to mix plant	500
6	Mixing plant to the construction site	5

Notes. Capacity of truck used for transportation material is 14 tones

Table 4. Equivalent carbon dioxides (CO_2e) emission from different material

S.No.	Component	Value	Reference
1	Production of coarse aggregate	0.0028 $kgCO_2e/kg$	[20]
2	Production of bitumen	0.426 $kgCO_2e/kg$	[20]
3	Production of fine aggregate	0.0025 $kgCO_2e/kg$	[20]
4	Production of cement	0.8207 $kgCO_2e/kg$	[20]
5	Production of steel	4.67 $kgCO_2e/kg$	[20]
6	Truck (14 tonnes) for Material movement	1.1$kgCO_2e$/(vehicle*km)	[21]
7	Milling of existing Pavement for rap	0.8 $kgCO_2e$/Ton	[22]
8	Paving of cement concrete pavement	0.2$kgCO_2e$/Ton	[22]
9	Laying of asphalt Layers	0.6 $kgCO_2e$/Ton	[22]
10	Laying of untreated granular material	0.4 $kgCO_2e$/Ton	[22]
11	Laying of cold mix asphalt	0.4 $kgCO_2e$/Ton	[22]
12	Material mixing value of cold mix plant	1.0 $kgCO_2e$/Ton	[22]
13	Material mixing value of PQC	0.000056($kgCO_2$ eq./kg)	[23]
14	Material mixing value of DLC (M10)	0.000083($kgCO_2$ eq./kg)	[23]
15	Material mixing value of hot asphalt	0.0194 (kgCO2e/kg)	[24]
16	Tack Coat	0.0205 ($kgCO_2e/m^2$)	[25]

4. Result & discussion

4.1. Life cycle cost

The schedule rate is used to calculate the starting construction cost of perpetual and rigid pavement layers per layer, as indicated in Table 1. The cost of paving DBM in perpetual flexible pavement and PQC in rigid pavement as shownFigure5 is to be the highest in their category. It is necessary to do research in this layer in order to minimize its cost without compromising its strength.

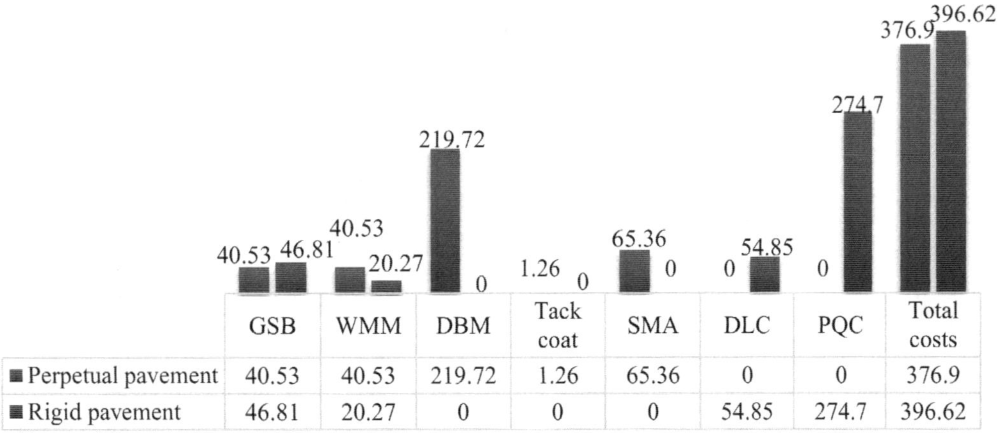

	GSB	WMM	DBM	Tack coat	SMA	DLC	PQC	Total costs
■ Perpetual pavement	40.53	40.53	219.72	1.26	65.36	0	0	376.9
■ Rigid pavement	46.81	20.27	0	0	0	54.85	274.7	396.62

Figure 5. Comparison of the initial construction cost (in lakh ₹) of pavement layers per kilometre

The cost of maintaining a perpetual flexible pavement varies consistently during its service life and only requires resurfacing every 10 years, but the cost of maintaining a rigid pavement varies up and down throughout its analysis period. Perpetual flexible pavement has lower overall maintenance cost as shown in Figure6 but also has a low maintenance frequency, which saves users money.

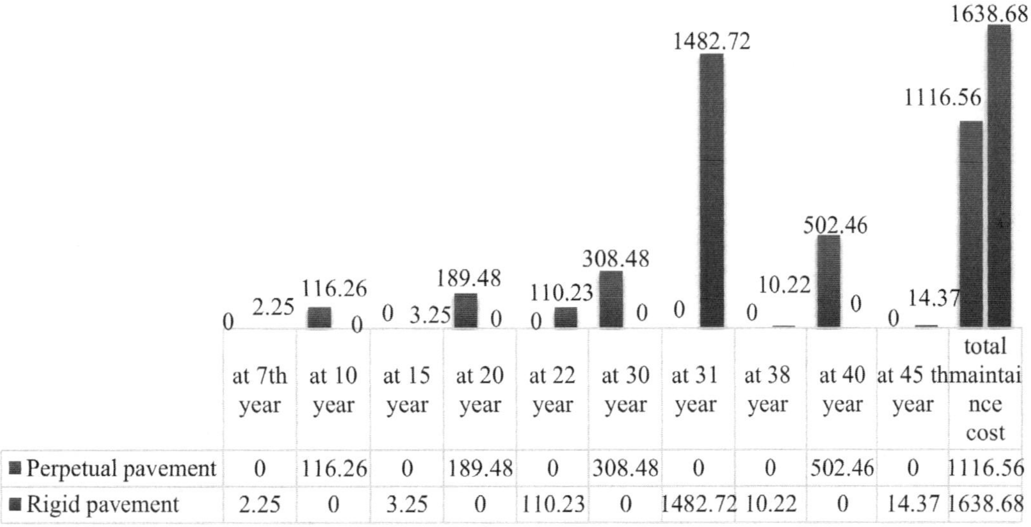

	at 7th year	at 10 year	at 15 year	at 20 year	at 22 year	at 30 year	at 31 year	at 38 year	at 40 year	at 45 th year	total maintaince cost
■ Perpetual pavement	0	116.26	0	189.48	0	308.48	0	0	502.46	0	1116.56
■ Rigid pavement	2.25	0	3.25	0	110.23	0	1482.72	10.22	0	14.37	1638.68

Figure 6. Comparison between per kilometre maintenance cost of pavements over 50 years

NPV is used to compare the money spent on construction/ maintenance /rehabilitation of pavement over time of 50 year into a particular year i.e., at 0th year as presented in Table 5 and Table 6.

Table 5. Life cycle cost of perpetual flexible pavement by NPV method

Years	Inflated cost @5% in₹	Inflated cost @5% in $	Discount cost @ 12% in ₹	Discount cost @ 12% in $
0th year (initial cost)	3,76,88350	4,98391	3,76,88350	4,98391
10th year	1,16,25828	1,53740	37,43205	49501
20th year	1,89,37249	2,50427	19,63163	25961
30th year	3,08,46783	4,07918	10,29602	13616
40th year	5,02,46159	6,64456	5,39985	7141
Total NPV over 50 years			**4,49,64305**	**5,94295**

Table 6. Life cycle cost of conventional rigid pavement by NPV method

Years	Inflated cost @5% in₹	Inflated cost @5% in $	Discount cost @ 12% in₹	Discount cost @ 12% in $
0th year (initial cost)	3,96,62000	524214	3,96,62000	5,24214
7th year	2,25100	2976	1,01823	1346
15th year	33,25576	43955	60760	803
22nd year	1,10,23260	1,45695	9,10990	12041
31st year	14,82,72272	1,959718	44,18769	58402
38th year	10,21517	13502	13771	182
45th year	14,37377	18998	8766	116
Total NPV over 50 years			**4,51,76879**	**597104**

Notes. 1 dollar ($) =75.66 rupees (₹)

From the method of net present value (NPV), it is found that NPV of the perpetual flexible pavement has cost of ₹ 4, 49, 64305 whereas rigid pavement has cost of ₹ 4, 51, 76879 during the analysis period of 50 years. Per kilometer cost of perpetual pavement is 2.12 lakhs ₹ lower than rigid pavement during analyzing period of 50 years.

4.2. Life cycle environmental cost

A life cycle methodology is a strategy for estimating a product's environmental effect from cradle to grave. Greenhouse gas emissions for various activities throughout the processes of raw material extraction, transportation from material's source to plant, mixing at the plant, shipping from mixing plant to site, and paving at the site, including maintenance and rehabilitation as shown in Figure (7.a-7. d). The carbon dioxide emissions from traffic throughout this pavement's design period were not taken into account in this analysis. Emissions of greenhouse gases from various activities occurring over a 50-year period are measured in units of CO2e (in tons).

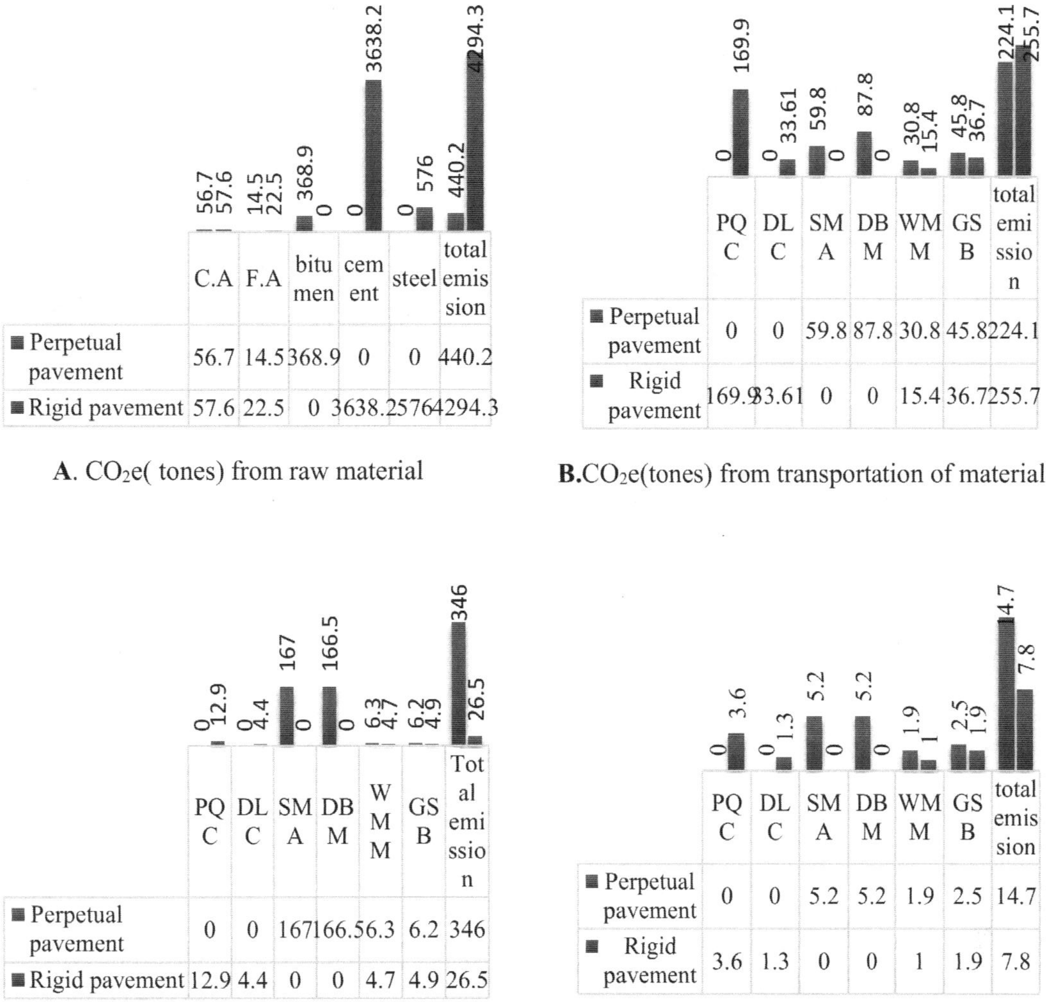

	C.A	F.A	bitumen	cement	steel	total emission
Perpetual pavement	56.7	14.5	368.9	0	0	440.2
Rigid pavement	57.6	22.5	0	3638.2	576	4294.3

A. CO_2e(tones) from raw material

	PQC	DLC	SMA	DBM	WMM	GSB	total emission
Perpetual pavement	0	0	59.8	87.8	30.8	45.8	224.1
Rigid pavement	169.9	33.61	0	0	15.4	36.7	255.7

B.CO_2e(tones) from transportation of material

	PQC	DLC	SMA	DBM	WMM	GSB	Total emission
Perpetual pavement	0	0	167	166.5	6.3	6.2	346
Rigid pavement	12.9	4.4	0	0	4.7	4.9	26.5

C. CO_2e (tones) from mixing plant operation

	PQC	DLC	SMA	DBM	WMM	GSB	total emission
Perpetual pavement	0	0	5.2	5.2	1.9	2.5	14.7
Rigid pavement	3.6	1.3	0	0	1	1.9	7.8

D. CO_2e(tones) from laying of layers

Figures 7: LCEC in terms of CO_2e (in tons) from the different activities of pavement construction over a 50-year period

Comparison and depicture of overall carbon footprints C02e (in tons) from the development of perpetual flexible and conventional rigid pavement layers throughout a 50-year analysis period is shown in Figure8.

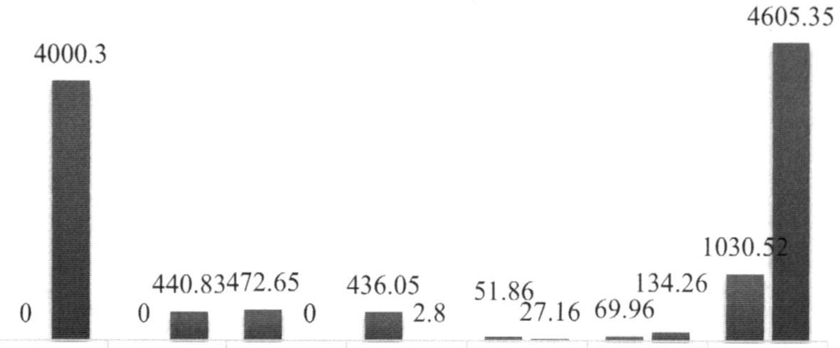

	PQC	DLC	SMA	DBM	WMM	GSB/CTG SB	total emission
■ Perpetual pavement	0	0	472.65	436.05	51.86	69.96	1030.52
■ Rigid pavement	4000.3	440.83	0	2.8	27.16	134.26	4605.35

Figure 8. LCEC in terms of CO_2e (in tones) from the building of pavement layers over a 50-year period.

5. Conclusion

Sustainability is a key factor that is considered across the world when selecting new building projects such as pavements. The goal is for the product to last a long time, be cost-effective, and environmentally beneficial. The present study evaluated the life cycle cost and environmental cost of perpetual flexible pavement and rigid pavement during the initial construction and maintenance phases over a 50-year analysis life considering 1km length and 14 m width. The initial cost of perpetual flexible pavement is 5% less than rigid pavement, and the total maintenance cost over 50 years period is 31.9 % less. This cost benefit allows spending more money on public welfare. The costs of perpetual flexible pavement and rigid pavement over a 50-year analysis period were converted into net present value (NPV) using a 5% inflation rate and a 12% discount rate. The perpetual flexible pavement has a NPV of 449.64 lakh rupees while the rigid pavement has a NPV of 451.76 lakh rupees, a difference of 2.12 lakh ₹. The perpetual flexible pavement emits 1030.52 tons of greenhouse gas during construction and maintenance over a 50-year period, whereas rigid pavement emits 4605.35 tones. It shows that rigid pavement emits 4.5 times more greenhouse gas during its life cycle, with most of its emissions coming from the manufacturing and extraction of cement, the chief raw material in rigid pavement. Hence, for long-life, high traffic-volume roads, perpetual flexible pavement outperforms rigid pavement in terms of life-cycle cost as well as environmental sustainability. Also, perpetual flexible pavement uses bitumen as a binding ingredient which is one of the most recycled substances. This makes it a better choice than cement which is almost non-recyclable

A comprehensive life cycle analysis may be performed between rigid pavement and perpetual flexible pavement that considers all other aspects, like user cost factor & social cost factors. We should also look for alternative pavement that have the same specific properties but are less expensive and emit less greenhouse gases

References

[1] MoRTH, "Bharatmala Pariyojana," 2017. New Delhi

[2] Albuquerque, F. D., Maraqa, M. A., Chowdhury, R., Mauga, T., & Alzard, M. (2020). Greenhouse gas emissions associated with road transport projects: current status, benchmarking, and assessment tools. *Transportation Research Procedia*, ***48***, 2018-2030.

[3] Skea, J., Shukla, P., Al Khourdajie, A., & McCollum, D. (2021). Intergovernmental Panel on Climate Change: Transparency and integrated assessment modeling. *Wiley Interdisciplinary Reviews: Climate Change*, ***12***(5), e727.

[4] Puja' 'awasthi, "28.6 million people in 6 Indian cities to face flooding if sea levels rise: Report,"

The week, Sep. 21, 2021

[5] Nautiyal, A., & Sharma, S. (2019). A model to compute service life of rural roads using present pavement condition and pavement age. *Compusoft*, ***8***(7), 3261-3268

[6] Nautiyal, A., & Sharma, S. (2021). Condition Based Maintenance Planning of low volume rural roads using GIS. *Journal of Cleaner Production*, ***312***, 127649

[7] Nautiyal, A., & Sharma, S. (2021). Scientific approach using AHP to prioritize low volume rural roads for pavement maintenance. *Journal of Quality in Maintenance Engineering*

[8] Nautiyal, A., & Sharma, S. (2022). Methods and factors of prioritizing roads for maintenance: a review for sustainable flexible pavement maintenance program. *Innovative Infrastructure Solutions*, *7*(3), 1-15.

[9] Harvey, J., Monismith, C., Horonjeff, R., Bejarano, M., Tsai, B. W., & Kannekanti, V. (2004). Long-life AC pavements: A discussion of design and construction criteria based on California experience. In *International Symposium on Design and Construction of Long Lasting Asphalt Pavements, 2004, Auburn, Alabama, USA*.

[10] Amini, A. A., Mashayekhi, M., Ziari, H., & Nobakht, S. (2012). Life cycle cost comparison of highways with perpetual and conventional pavements. *International Journal of Pavement Engineering*, ***13***(6), 553-568.

[11] Adanikin, A., Falade, F., Olutaiwo, A., Ajibade, T., & Adeoye, I. (2021). Life-Cycle Cost Analysis (LCCA) comparison of pavements (Flexible, rigid and rigid-admixed with cow bone ASH). *ITEGAM-JETIA*, *7*(30), 75-80.

[12] Kucukvar, M., & Tatari, O. (2012). Ecologically based hybrid life cycle analysis of continuously reinforced concrete and hot-mix asphalt pavements. *Transportation Research Part D: Transport and Environment*, ***17***(1), 86-90.

[13] Kristjánsdóttir, Ó., Muench, S. T., Michael, L., & Burke, G. (2007). Assessing potential for warm-mix asphalt technology adoption. *Transportation Research Record*, ***2040***(1), 91-99.

[14] Papagiannakis, A. T., & Masad, E. A. (2008). *Pavement design and materials*. John Wiley & Sons.

[15] Siva Rama Krishna, U., & Naga Satish Kumar, C. (2020). A case study on maintenance of bituminous concrete pavement considering life cycle cost analysis and carbon footprint estimation. *International Journal of Construction Management*, 1-9.

[16] Lamptey, G., Ahmad, M. Z., Labi, S., & Sinha, K. C. (2005). Life cycle cost analysis for INDOT pavement design procedures.

[17] Basu, C., Das, A., Thirumalasetty, P., & Das, T. (2013). Perpetual pavement–A boon for the Indian roads. *Procedia-Social and Behavioral Sciences*, ***104***, 139-148.

[18] Tarefder, R. A., & Bateman, D. (2012). Design of optimal perpetual pavement structure. *Journal of transportation engineering*, ***138***(2), 157-175.

[19] Prithvi' 'Kandhal, *Bituminous road construction in India* . PHI learning New Delhi, 2016

[20] White, P., Golden, J. S., Biligiri, K. P., & Kaloush, K. (2010). Modeling climate change impacts of pavement production and construction. *Resources, Conservation and Recycling*, ***54***(11), 776-782.

[21] Stripple, H. (2001). Life cycle assessment of road. A pilot study for inventory analysis.

[22] Chappat, M., & Bilal, J. (2003). The environmental road of the future: Life cycle analysis. *Colas SA, Paris*, ***9***(2003), 1-34.

[23] Singh, A., Vaddy, P., & Biligiri, K. P. (2020). Quantification of embodied energy and carbon footprint of pervious concrete pavements through a methodical lifecycle assessment framework. *Resources, Conservation and Recycling*, ***161***, 104953.

[24] Copeland, A., D'Angelo, J., Dongre, R., Belagutti, S., & Sholar, G. (2010). Field evaluation of high reclaimed asphalt pavement–warm-mix asphalt project in Florida: Case study. *Transportation Research Record*, ***2179***(1), 93-101.

[25] Sreedhar, S., Jichkar, P., & Biligiri, K. P. (2016). Investigation of carbon footprints of highway construction materials in India. *Transportation Research Procedia*, ***17***, 291-300.

Prediction of Rainfall Using Hybrid SVM-HHO Model

L Moharana[1], A Sahoo[2*], and D K Ghose[3]

[1] M.Tech, Department of Civil Engineering, National Institute of Technology Silchar, Fakiratilla, Silchar - 788010, Assam, India
[2] Research Scholar, Department of Civil Engineering, National Institute of Technology Silchar, Fakiratilla, Silchar - 788010, Assam, India
[3]Assistant Professor, Department of Civil Engineering, National Institute of Technology Silchar, Fakiratilla, Silchar - 788010, Assam, India

Email: bablusahoo1992@gmail.com

Abstract. Rainfall modeling is considered a need of the hour in order to understand, control, and monitor the quality and quantity of water resources. Modeling rainfall is one of the key components of the meteorological engineering process. Hydrological and climatological studies require accurate rainfall estimation in order to effectively manage water resources. Thus, adopting an advanced, reliable, and intelligent model for forecasting will be of great value in water resources engineering. The present study tried to establish a hybrid model with the combination of Support Vector Machine (SVM) and Harris Hawks Optimization (SVM-HHO) for predicting the rainfall time series of Cachar district located in Assam, India. The predictive performance of models is examined based on statistical analysis. Statistical measures like root mean squared error (RMSE) and coefficient of correlation (CC) is used to evaluate the considered hybrid model. It is observed from obtained results that proposed hybrid model exhibited least RMSE value of 20.29, and highest CC value of 0.9714, during testing period. Findings of this study confirm that proposed hybrid optimisation strategy can be regarded as a powerful forecasting tool for achieving better generalisation capability and higher forecasting accurateness.

Keywords: Cachar, Rainfall, SVM, SVM-HHO

1. Introduction

Rainfall is a major environmental component, which has a significant impact on various sectors, including hydroelectric power, agronomy, and water resource management, and it depends on space and time. Temperature, relative humidity, wind speed, and elevation influence the occurrence of precipitation in different areas. Artificial intelligence (AI) models encompass multilayer perceptron (MLP) [1], artificial neural networks (ANNs), long short term memory (LSTM) [2, 3], adaptive neuro-fuzzy inference system (ANFIS) [4, 5] and radial basis function network (RBFN) [6, 7] have grown increasingly popular in the creation of models.

In last few years, machine learning algorithms for hydrological models have made substantial improvements. Machine learning models make it easier to simulate complex (hydro-meteorological) systems and dynamic connections between multiple sub-processes. ANNs have gotten a lot of attention in rainfall forecasting because of their low variance, plasticity, and computation retraining without any previous knowledge of the underlying dynamic processes. Kim et al. [8] observed monthly rainfall from 10 climatological stations and proposed a new model in the Han River basin, South Korea. Moreover, interesting findings are reported by Samantaray et al. [9] with implementation of Adaptive Network Fuzzy Inference System (ANFIS), SVM, and Recurrent Neural Model (RNN) techniques for rainfall analysis in Bolangir district of state Odisha, India. Likewise, In China, Wang et al. [10] employed an RNN model to execute statistical downscaling and demonstrated its advantages by comparing it to a standard artificial neural network (ANN) model. There have been several applications of different neural network models in rainfall-runoff modelling [11, 12].

Over the last decade, ANN has been frequently used in rainfall and runoff prediction research. The risk of being stuck in local optimal solution, overtraining, uncertainty in establishing model parameters, random initialisation of weights in every simulation, and constituents of its sophisticated structure are all downsides of ANN models. A new kernel-based approach known as a support vector machine has emerged in the last decade (SVM). It has become prominent in modelling studies to overcome the above complications by ANN. Wu et al. [13] used HPSOGA (integration of particle swarm optimisation into genetic algorithm) to determine constraints of RBFN for predicting rainfall. Their findings revealed that hybrid optimisation algorithm could be utilised as a promising alternate prediction tool. Bui et al. [14] used a spatial model to predict precipitation events and shallow landslides in Central Vietnam. They introduced differential evolution optimization and Least-Squares Support Vector Machines, two heterogeneous database techniques.

Yang et al. [15] used a rapid elitist Semi Sorting Metaheuristic Optimization to propose a predictor assortment strategy for building SVM-based typhoon rainfall prediction model (NSGA-II). A two-step hybrid neural network (HNN) prediction model was proposed for predicting rainfall in Southern portion of West Bengal [16]. Proposed HNN model was compared with MLP and found that HNN model outperformed MLP in rainfall prediction. To predict daily precipitation in Hoa Binh province, Vietnam, Pham et al. [17] compared ANFIS optimised with Particle Swarm Optimisation (ANFIS-PSO), ANN, and SVM. To create a rainfall-runoff model, Tikhamarine et al. [18] used advanced machine learning algorithms like HHO and PSO. Malik et al. [19] used hybrid SVM-HHO and PSO algorithms to estimate meteorological drought. In order to calculate the longitudinal dispersion coefficient of natural streams, Azar et al. [20] adopted LS-SVM and ANFIS optimised using HHO algorithm method. Paryani et al. [21] evolved a groundwater spring potential mapping model using SVM and metaheuristic algorithms. This study's main goal is to see how well the hybrid SVM-HHO model can predict monthly rainfall intensity in Cachar, Assam, based on quantitative statistical performance indicators and graphical interpretations.

2. Study area

This research was carried out in Cachar district, situated in central portion of the Barak valley in Assam, India as shown in Figure 1. Geographically, the study area is bounded between 24°22’ to 25°8' N North latitudes, and 92°24' E, to 93°15' East longitudes. It acquires a total area of 3,786 sq. Km. Cachar receives heavy annual rainfall of more than 3,000 mm [7]. The maximum rainfall occurs during monsoon period between May to August. The district is part a of the sub-Himalayan area zone II, which was later sub-regionalized into the Barak Valley zone climatic zone, with hot and wet summers and cool winters.

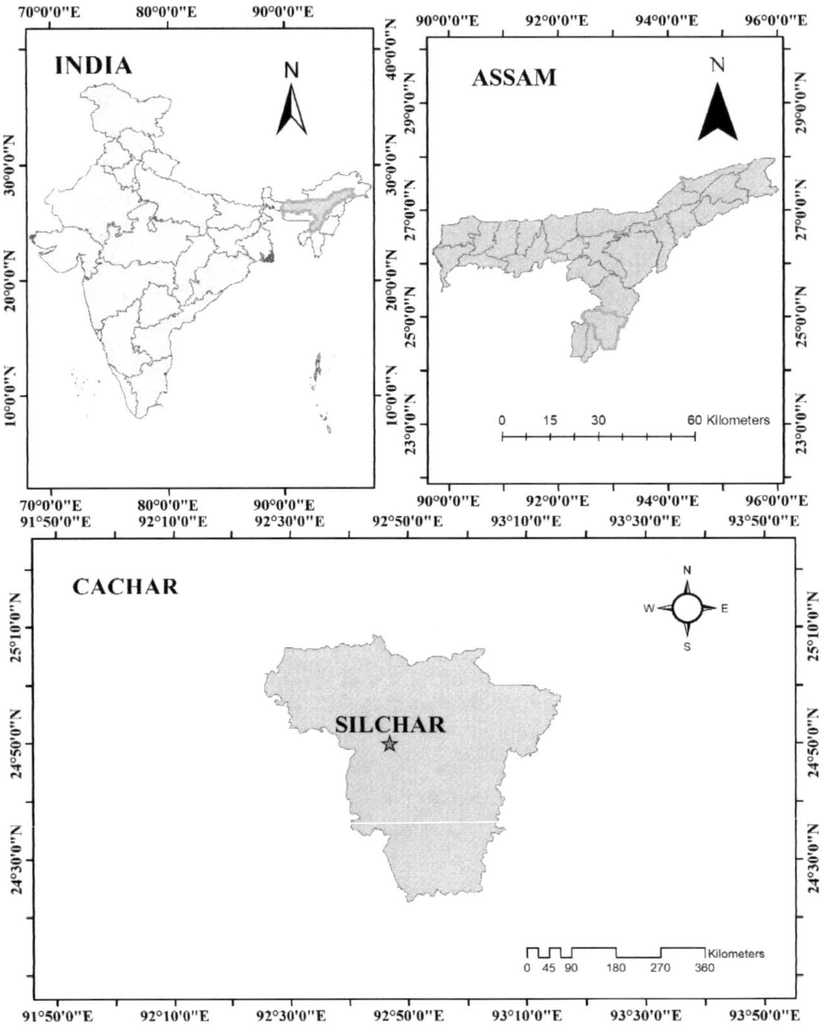

Figure 1. Study area map of Cachar watershed

3. Material and Methodology

3.1. SVM

SVM is a supervised artificial intelligence technique to solve regression and classification issues [22]. It has a broad variety of usefulness and can handle both non-linear and linear problems. It receives data by means of input and prints a line separating classes. SVM method was used for finding points from both classes adjoining to reference line [12]. Names specified to these points are called support vectors. Margin is the distance amid support vectors and a line that have been determined. Ideal hyper plane is the one with biggest margin. Nonlinear regression function is defined using subsequent equation:

$$f(x) = \sum_{i=1}^{N}(\alpha_i^* - \alpha_i)K(x, x_i) + B \quad (1)$$

Where, $\alpha_i^*, \alpha_i \geq 0$- Lagrange multipliers; B- bias; $K(x, x_i)$- Kernel function

3.2. HHO

HHO is a nest optimization method that mimics Harris' hawks' surprise pounce pursuit style [23]. It is a revolutionary metaheuristic algorithm that is used to solve complex issues in a competitive manner. It has a straightforward and time-varying structure that aids in a smooth transition amid primary stages. To design an optimisation method, this approach mathematically establishes dynamic patterns and behaviours [24]. When compared to other eminent published approaches, this model's efficacy yields highly promising and occasionally competitive results. This approach is mathematically described and capable of tackling difficult issues with computational implementation characteristics.
In HHO, Harris' hawks perch at random places and wait for prey using one of two approaches. (i) Exploration phase

$$X(t+1) = \begin{cases} X_{rand}(t) - r_1 |X_{rand}(t) - 2r_2 X(t)| & q \geq 0.5 \\ (X_{rabbit}(t) - X_m(t)) - r_3 \left(LB + r_4(UB - LB)\right) & q < 0.5 \end{cases} \tag{2}$$

$X(t+1)$ = Hawk's location vector in subsequent iteration t, $X_{rabbit}(t)$ = Location of rabbit,

$X(t)$ = Hawk's present location vector, r_1, r_2, r_3, r_4, and q are arbitrary numbers within (0-1), UB and LB display upper and lower boundaries of variables, which are updated in each iteration. $X_{rand}(t)$ = Hawk drawn at arbitrary from current population, and X_m= Average position of current population of hawks.

Using Eq. (2), the average hawk position is:

$$X_m(t) = \frac{1}{N} \sum_{i=1}^{N} X_i(t) \tag{3}$$

$X_i(t)$ = Every hawk's position in t iteration, N = Number of hawks.

(ii) Exploration to exploitation transition

$$E = 2E_0(1 - \frac{t}{T}) \tag{4}$$

E = Escape energy of target, T = Highest number of iteration, and E_0 = Preliminary stage of its energy. The detailed methodology is presented in Figure 2.

3.3. Evaluating Constraints

Quantitative statistical measures considered for finding the best performing model are RMSE and CC. These can be mathematically articulated as:

$$RMSE = \sqrt{\frac{1}{l} \sum_{i=1}^{n} (f_i - a_i)^2} \tag{5}$$

$$CC = \frac{\sum_{i=1}^{n} (a_i - \bar{a})(f_i - \bar{f})}{\sqrt{\sum_{i=1}^{n} (a_i - \bar{a})^2 * \sum_{i=1}^{n} (f_i - \bar{f})^2}} \tag{6}$$

Where, a_i and f_i= Estimated actual and rainfall values; $\bar{a}$ and $\bar{f}$= Average of actual and estimated rainfall values.

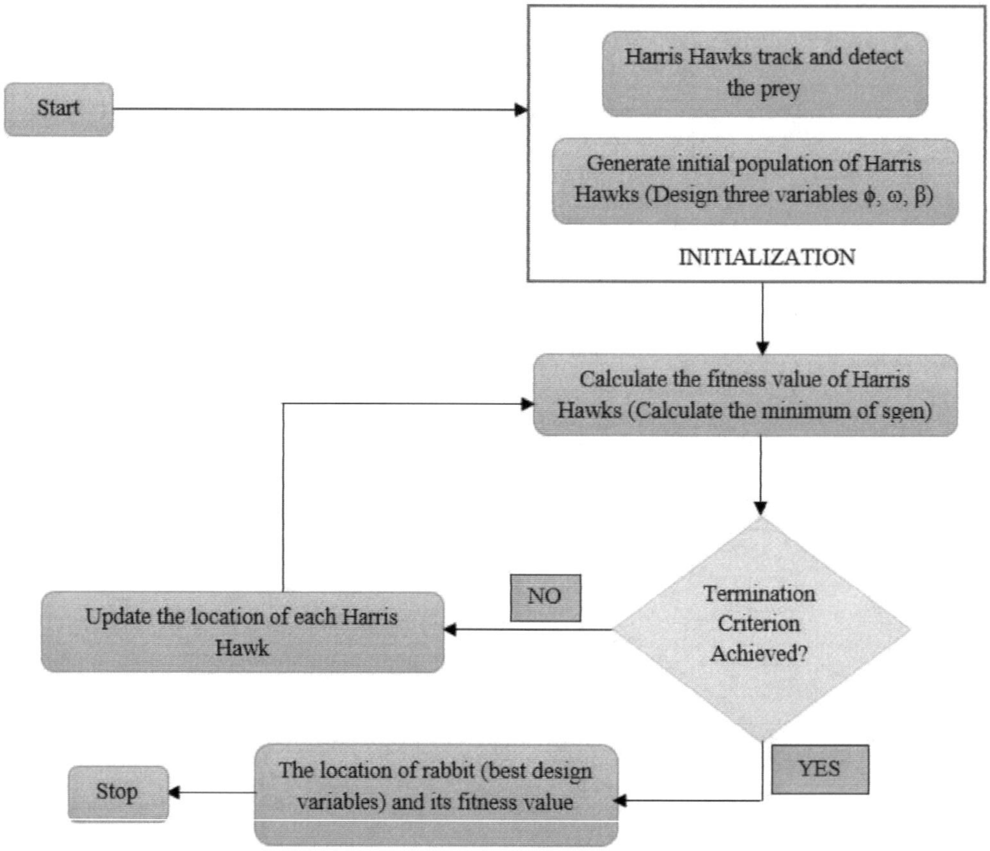

Figure 2. Methodology for rainfall modelling using hybrid algorithm

4. Result and Discussion

For the present study SVM and SVM-HHO models are established in estimation of monthly rainfall. The performance of these models is evaluated with consideration of statistical indices such as RMSE and CC for monthly rainfall during testing phase, which is summarized in Table 1. Moreover, it can be inferred from the figures and table that the presently investigated SVM-HHO model outperformed the conventional model. The scatter plots are established between the observed and estimated rainfall by the proposed SVM and SVM-HHO models for the best scenarios during testing phase. The same has been demonstrated in Figure 3. Figure 3 shows that for SVM model the scatter points are mostly dispersed while, on the other hand, the scatter points are grouped together near the line 'y=x' in SVM-HHO model. An interesting result achieved between observed and estimated monthly rainfall during testing phase is reported here. Additionally, the time-series of observed and predicted monthly rainfall data during the testing phase are investigated using the proposed models depicted in Figure 4.

Table 1. Statistical performance evaluation of investigated models

Station	Model	Input	Training		Testing	
			RMSE	R	RMSE	R
	SVM*I	I(t)	56.57	0.9203	58.48	0.9115
	SVM*II	I(t), I(t-1)	52.24	0.9235	54.92	0.9202

	SVM*III	I(t), I(t-1), I(t-2)	48.82	0.9319	50.41	0.9213
Silchar	SVM*IV	I(t), I(t-1), I(t-2), I(t-12)	42.84	0.9526	44.23	0.9387
	SVM-HHO*I	I(t)	31.06	0.9671	34.93	0.9526
	SVM-HHO*II	I(t), I(t-1)	27.47	0.9747	29.81	0.9625
	SVM-HHO*III	I(t), I(t-1), I(t-2)	22.91	0.9805	24.37	0.9678
	SVM-HHO*IV	I(t), I(t-1), I(t-2), I(t-12)	18.25	0.9926	20.29	0.9714

From Figure 4, it is clearly observed that there is a close agreement between observed and estimated monthly rainfall data in case of proposed hybrid SVM-HHO model compared to those achieved by SVM model. The result shows that the SVM-HHO model exhibited more satisfactorily than that of SVM model in estimating both the time-series and peak values of monthly rainfall data. Considering the statistical analysis performance of RMSE, the SVM-HHO model attained the best result (20.29 mm), whereas SVM model executed with the worst result (44.23 mm) in testing phase. Besides these, SVM-HHO model performed the best in terms of CC (0.9714), while the SVM model couldn't perform well, resulting in a poor value of CC (0.9387).

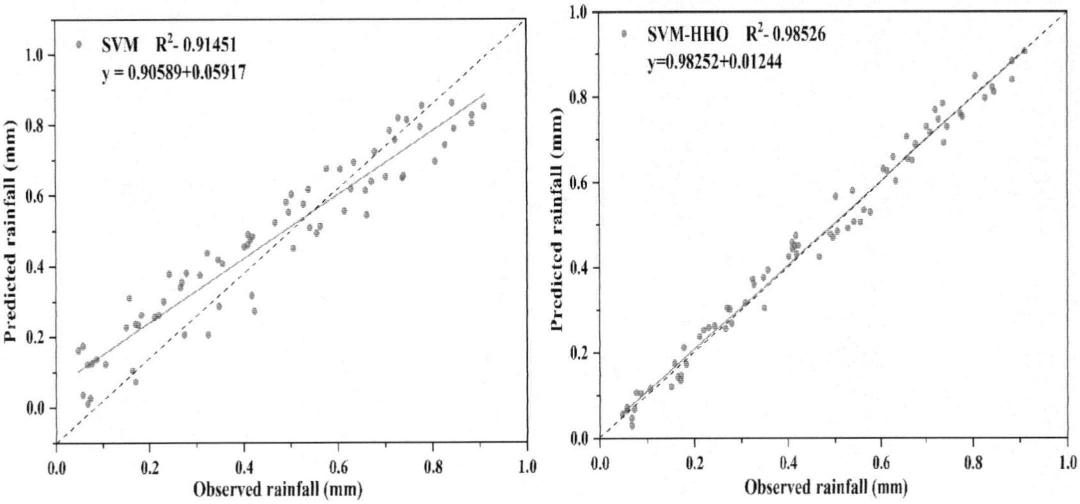

Figure 3. Scatter plot showing prediction of rainfall data using SVM and SVM-HHO models

Furthermore, it is noticeably manifested in R^2 values demonstrated in Figure 3 that the SVM-HHO has the highest predictive performance compared to others discussed in the present study. It is further confirmed by depicting actual and predicted rainfall values in form of a box plot during both the training and testing phase for the studied models. The predicted values from SVM and SVM-HHO in comparison with actual rainfall data during training and testing phase are shown in Figure5. It is noticeably evident that the SVM-HHO model outperformed well with an acceptable range of error during training and testing period compared to SVM model for the present study. The predictive rainfall results of SVM–HHO model have almost equal median, minimum, mean, and interquartile ranges.

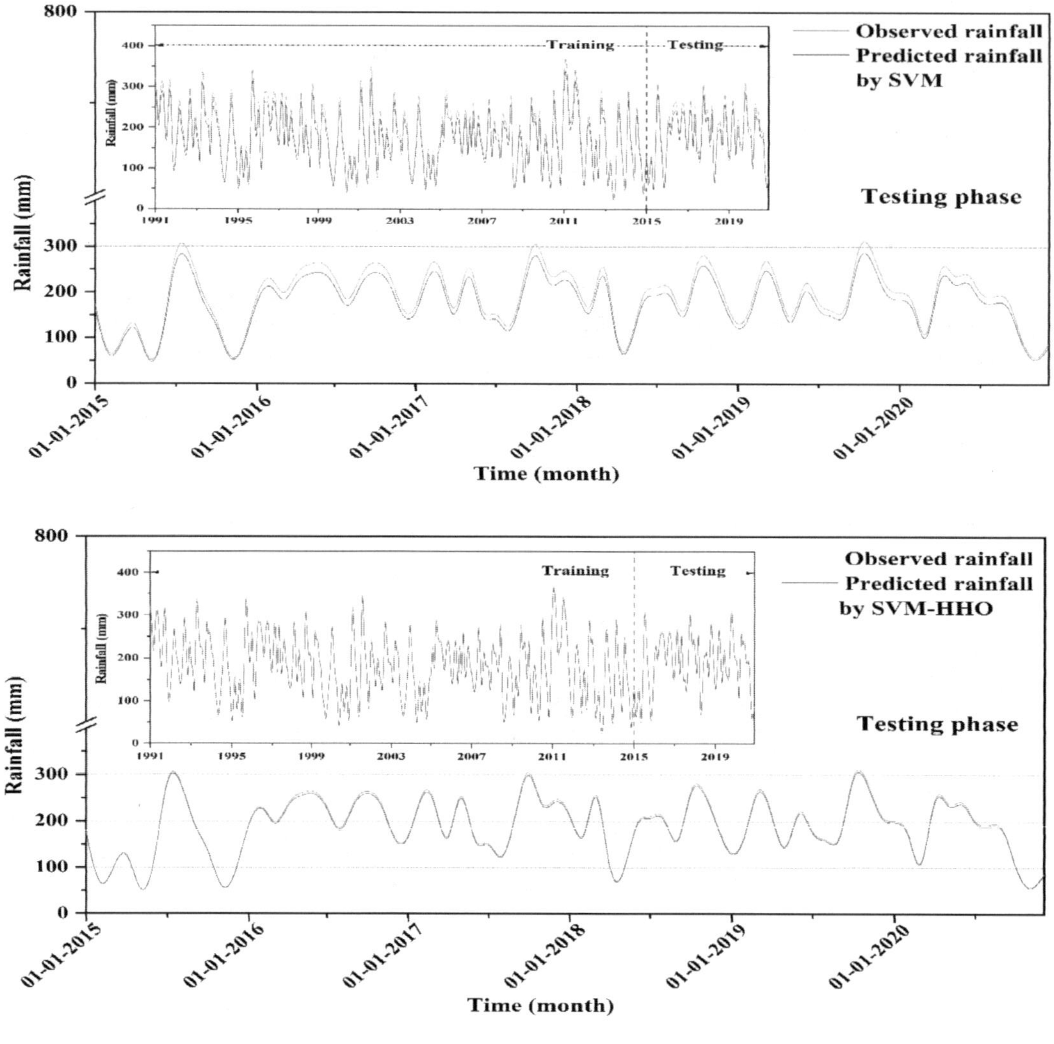

Figure 4. Prediction of rainfall for the best scenarios using SVM and SVM-HHO models

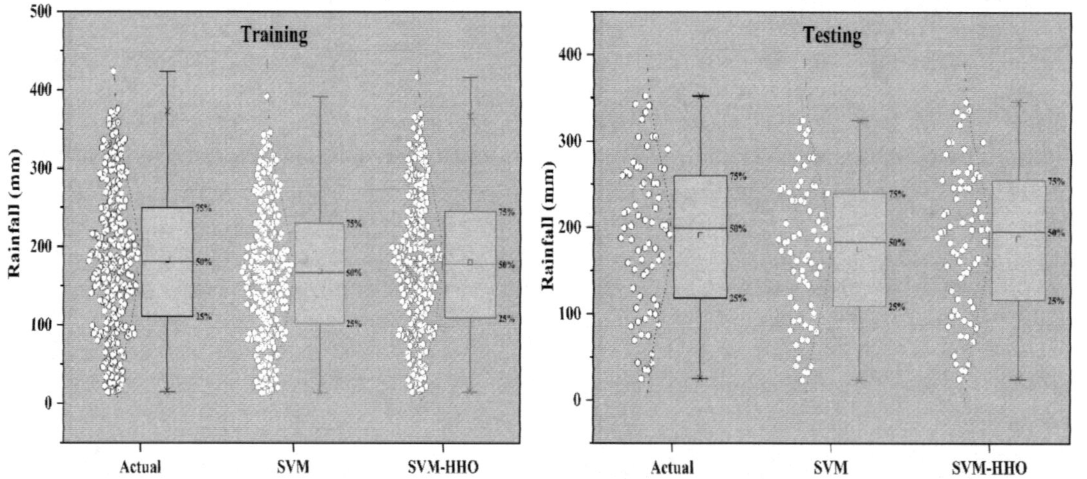

Figure 5. Box plot using investigated models (SVM and SVM-HHO) during training and testing phases

The proposed model delivers an augmented, improved, and advanced gateway in assessing the trends or monthly rainfall data with the implementation of one factor or two and three published approaches. The findings of these novel approaches provide guidance on optimizing the use of natural resources, land, labour and creating awareness for emergencies months ahead by following valuable feedbacks continuously.

Now-a-days climate change is an emerging field of study as it greatly disrupts the established seasonal rainfall trends [25]. Hence advanced models need to be employed quickly within a short time span so that the severe increase in variability can be controlled. This leads to an uncertainty in short- and long-term forecasting events, most probably by incorporating the data from remotely sensed platform into time-history analyses using artificial intelligence as well as genetic algorithms. Therefore, model calibration and validation must be conducted on a priority basis as the scenario becomes more adverse and challenging.

5. Conclusion

Rainfall forecasting is a challenging task in recent times due to its complexities, erratic nature, and persistent applications. The approach that uses artificial intelligence in conjunction with time-series data are being explored as the most promising approach to overcome these difficulties. This paper investigated the performance of a hybrid model which uses the combined approach of SVM and HHO in estimating monthly rainfall time-series data by employing different statistical indicators, namely RMSE and CC. The result findings clearly inferred the superiority of hybrid model (RMSE- 20.29, CC- 0.9714) compared to conventional SVM model (RMSE- 44.23, CC- 0.9387). The graphical representations also proved the high performance of SVM-HHO than the SVM model. However, there exists a model limitation in capturing extreme events of rainfall patterns where the noise analysis needs to be carried out. This suggests that the potential of hybrid model can be reviewed more rigorously to assess the downscaled rainfall in future. Moreover, its performance can be applied as a predictive model for budget-wise rainfall forecast applications. The model can be verified in downscaling rainfall in other climatic conditions, especially in regions where rainfall seems very erratic and scarce.

References

[1] Sahoo A and Ghose DK 2022 Application of Hybrid MLP-GWO for Monthly Rainfall Forecasting in Cachar, Assam: A Case Study. In *Smart Intelligent Computing and Applications, Volume 1* (pp. 307-317). Springer, Singapore.

[2] Afzaal H, Farooque, A.A., Abbas F, Acharya B and Esau, T 2020 Computation of evapotranspiration with artificial intelligence for precision water resource management. *Applied Sciences* **10** 1621.

[3] Ghose DK, Mahakur V and Sahoo A 2022 Monthly Runoff Prediction by Hybrid CNN-LSTM Model: A Case Study. In *International Conference on Advances in Computing and Data Sciences* (pp. 381-392). Springer, Cham.

[4] Ghose DK, Tanaya K, Sahoo A and Kumar U 2022 March. Performance Evaluation of hybrid ANFIS model for Flood Prediction. In *2022 8th International Conference on Advanced Computing and Communication Systems (ICACCS)* (pp. 772-777). IEEE.

[5] Agnihotri A, Sahoo A and Diwakar MK 2022 Flood prediction using hybrid ANFIS-ACO model: a case study. In *Inventive computation and information technologies* (pp. 169-180). Springer, Singapore.

[6] Sahoo A, Samantaray S and Ghose DK 2021 Prediction of flood in Barak River using hybrid machine learning approaches: a case study. *Journal of the Geological Society of India* **97** 186-198.

[7] Poonia V and Tiwari HL 2020 Rainfall-runoff modeling for the Hoshangabad Basin of Narmada River using artificial neural network. *Arabian Journal of Geosciences* **13** 1-10.

[8] Kim T, Shin JY, Kim H, & Heo JH 2020 Ensemble-based neural network modeling for hydrologic forecasts: Addressing uncertainty in the model structure and input variable selection. Water

Resources Research **56** e2019WR026262.
[9] Samantaray S, Tripathy O, Sahoo A and Ghose DK 2020 Rainfall forecasting through ANN and SVM in Bolangir Watershed, India. In *Smart intelligent computing and applications* (pp. 767-774). Springer, Singapore.
[10] Wang Q, Huang J, Liu, R, Men C, Guo L, Miao Y, Jiao L, Wang Y, Shoaib M and Xia X 2020 Sequence-based statistical downscaling and its application to hydrologic simulations based on machine learning and big data. *Journal of Hydrology* **586** 124875.
[11] Feng Q, Wen X, and Li J 2015 Wavelet analysis-support vector machine coupled models for monthly rainfall forecasting in arid regions. Water resources management **29** 1049-1065.
[12] Yu P S, Yang T C, Chen SY, Kuo CM and Tseng HW 2017 Comparison of randomforests and support vector machine for real-time radar-derived rainfall forecasting. Journal of Hydrology **552** 92-104.
[13] Wu J, Long J and Liu M 2015 Evolving RBF neural networks for rainfall prediction using hybrid particle swarm optimization and genetic algorithm. *Neurocomputing* **148** 136-142.
[14] Tien Bui D, Nguyen, QP, Hoang ND and Klempe H 2017 A novel fuzzy K-nearest neighbor inference model with differential evolution for spatial prediction of rainfall-induced shallow landslides in a tropical hilly area using GIS. *Landslides* **14** 1-17.
[15] Yang TC, Yu PS, Lin KH, Kuo CM and Tseng HW 2018 Predictor selection method for the construction of support vector machine (SVM)-based typhoon rainfall forecasting models using a non-dominated sorting genetic algorithm. *Meteorological applications* **25** 510-522.
[16] Chatterjee S, Datta B, Sen S, Dey N and Debnath NC 2018 January. Rainfall prediction using hybrid neural network approach. In *2018 2nd International Conference on Recent Advances in Signal Processing, Telecommunications & Computing (SigTelCom)* (pp. 67-72). IEEE.
[17] Pham BT, Le LM, Le TT, Bui KTT, Le VM, Ly HB, and Prakash I 2020 Development of advanced artificial intelligence models for daily rainfall prediction. Atmospheric Research **237** 104845.
[18] Tikhamarine Y, Souag-Gamane D, Ahmed, AN, Sammen SS, Kisi O, Huang YF and El-Shafie A 2020 Rainfall-runoff modelling using improved machine learning methods: Harris hawks optimizer vs. particle swarm optimization. *Journal of Hydrology* **589** 125133.
[19] Malik A, Tikhamarine Y, Sammen SS, Abba SI and Shahid S 2021 Prediction of meteorological drought by using hybrid support vector regression optimized with HHO versusPSO algorithms. Environmental Science and Pollution Research **28** 39139-39158.
[20] Azar NA, Milan SG and Kayhomayoon Z 2021 The prediction of longitudinal dispersion coefficient in natural streams using LS-SVM and ANFIS optimized by Harris hawk optimization algorithm. Journal of Contaminant Hydrology **240** 103781.
[21] Paryani S, Neshat A, Pourghasemi HR, Ntona MM and Kazakis N 2022 A novel hybridof support vector regression and metaheuristic algorithms for groundwater spring potential mapping. *Science of The Total Environment* **807** 151055.
[22] Vapnik V 1995 The nature of statistical learning theory, Springer, New York.
[23] Heidari AA, Mirjalili S, Faris H, Aljarah I, Mafarja M and Chen H 2019 Harris hawks optimization: Algorithm and applications. *Future generation computer systems* **97** 849-872.
[24] Houssein EH, Hosney ME, Oliva D, Mohamed WM and Hassaballah M 2020 A novel hybrid Harris hawks optimization and support vector machines for drug design and discovery. *Computers & Chemical Engineering* **133** 106656.
[25] Ohba M and Sugimoto S 2019 Differences in climate change impacts between weather patterns: possible effects on spatial heterogeneous changes in future extreme rainfall. *Climate Dynamics* **52** 4177-4191.

Performance Rating of National Highways Based on Road User Perspective – A Case Study of Kalka–Shimla Highway

Pratik Solanki[1] and Sunil Sharma[2]

[1]MTech Scholar, Department of civil engineering, National Institute of Technology Hamirpur, India
[2]Assistant Professor, , Department of civil engineering, National Institute of Technology Hamirpur, India

Email: 20mce211@nith.ac.in

Abstract. The road network plays the most important role in the development of India's infrastructure. Through the initiative of rating national highways, we will be able to set standards for quality highway services, and as a result, road users will receive excellent value for their money. Moreover, better quality Indian Highways will lead to safer travel experiences and consequently result in fewer road accidents and fatalities. The effort for evaluating the roads was done by US Department of Transportation in 1950s as has been highlighted in the Highway Sufficiency Rating Manual of 1952. However, these ratings were based only on efficiency and safety parameters, completely neglecting the user perspective and services. As the road users gain access to information about products and services in real time over the internet, they are more empowered than ever before. Thus, the performance evaluation of a highway stretch must include additional aspects associated with the user experience. This paper examines the real-time problems they face along the journey. National Highway authority of India (NHAI) and Ministry of Road Transport and Highways (MORTH) of India have framed Standard Operating Procedure (SOP) for assessment and rating of 4-6 lanes divided carriageway NHs based on road user perspective. Road user perspective carries a 20 per cent weightage in overall highway efficiency. A questionnaire survey was taken as part of the road user perspective study on the Kalka-Shimla NH No. 5 in the state of Himachal Pradesh in India. The highway scored 17.1 marks out of 20 based on users' response, representing an 85.5% score, and indicating minor deficiencies regarding the selected highway facilities.

1. Introduction

It is important to construct and maintain a highway that can withstand the merits of users as their money is used to do so. An overview of the national highway between Kalka and Shimla (NH-05), also referred to as the Himalayan expressway, is included in this paper. Prior to this, rating was based primarily on technical aspects such as safety and efficiency and completely overlooked the perspective of road users. However, it is road users who are most impacted by road conditions, whether they are concerned about comfort, speed, travel times, or road conditions. The national highway length increases, but India's performance in "the logistic performance index" did not increase significantly. In addition to technical expectations, road user assistance gives an overall picture of the road from the perspective of users. The information does not consider traffic parameters but includes toll plaza conditions, visibility, safety, etc. and many other features that can contribute to a road's overall quality. The road user rating gives a fundamental idea that what are the real time problems faced by user along the journey time like availability if motor repair, parking, phone booth rest area etc. some of the studies based on rating did at Motorway Route 7. This study used at least 800 questionnaire samples to explore and understand road users' satisfaction on Motorway Route 7 to survey road users' satisfaction and to know road users' demands. The major elements that most affected road users' comfort, according to the multiple regression results, were traffic volume, road surface cleanliness, features of the Motorway Route 7, and other drivers' compliance, Suanmali et al. [1].

In another study performance of current road infrastructure was studied and found the result as In Satellite City, the overall user satisfaction index for three road sections is still acceptable and this study is done by Resdiansyah et al.[2]. This study shows that level of service provided to users by the observed road segment is worsened by the grades of road quality, transportation facilities, lateral road signs, and road maintenance this study is done by Subotić et al.[3]. Another study done in Europe signifies that the proportion of Unprotected Road Users death and injury crashes is predicted to rise, owing to the population's ageing, urbanization, and despite ICT/ITS advancements in EU. This study was carried out by Rob Methorst et al.[4]. Alexander Paz examined Nevadans' [5] satisfaction with their highway transportation system and the Nevada Department of Transportation's accompanying expenses (NDOT). The findings show that customer satisfaction with traffic safety is declining on highways in Northwestern and Southern Nevada.

Eliza rosmaya puri [6] study recommended the five physical road conditions as a new set of conditions for roadways in Indonesia's city network. The scorecards can be used to compare results. Customer satisfaction was studied in terms of service quality, restroom facilities, and perceived price at Toll Road of Cipali by Reka Harjuna Pratama and Mudji Sabar [7]. The dimension of road safety should be considered, and indices of road safety have been developed By Abeer Khudhur Jameel and Harry Evdorides[8] .Mohammad Mahmudul Hasan et al. [9] determined the total satisfaction level in Rajshahi City Bypass Road based on five major elements such as convenience, safety, utilities, roadside signpost, and rescue services. NHAI and MORTH of India have made standard operating procedure (SOPs) for the rating of a divided carriageway 4 to 6-lanes NHs based on highway efficiency, highway safety, and user services. Out of a total of 100 marks, 20 marks are assigned to the ROAD USER FACILITIES which is sub divided into two major categories as follows; wayside amenities, satisfaction index, which deals with assessing different sub parameters under them. The area of study is a 5-kilometer stretch of Kalka-Shimla NH-5 from Sanwara toll plaza to village Jabli bus stop. The NH-5 was widened from a 2-lane to 4-lane highway between Sanwara toll plaza to Parwanoo to raise transport capacity and alleviate traffic congestion. The main aim of this study is to carry out study of highway rating on 4-lane highway as per standard operating procedure (SOP) of NHAI as well as categorizing the section as excellent, good, or poor to help in improving highway performance standards.

2. Methodology

2.1. Data collection

In this study, two research methods have been employed. Rating of a road according to guidelines (SOP) provided by NHAI and MORTH - Adopting a procedure for assessing the functionality of the road (wayside amenities) in relation to the facilities that it offers its users. In the second method, questionnaires were distributed to gather data, which asked respondents to rate their satisfaction levels with the road facilities based on their perspective as a road user. The questions are answered based on a 5-point Likert Scale ranging from “most satisfied” to “most dissatisfied “To meet real time problems faced by road users and know the demands for improving the highway facilities.

2.2. Site study

The road stretch is selected by keeping in mind to full fill all the requirements of the study which includes a bus stop, a ROB, two habitual areas of 300 meters, toll plaza a total of 5 different places. The area for research starts from Sanwara Toll Plaza to Jabli market which is a four lane Kalka-Shimla National Highway (NH-5) i.e., also known as Himalayan expressway with a total length of about 5 km.

Considering the, NHAI decided to enlarge NH-5 from Solan to Shimla from a two-lane highway to a four-lane highway. The route chosen is a tourist route connecting important cities. Due to apple dealing and tourists, the road is extremely congested throughout the winter months. As a result, there was a pressing need to widen the existing route to boost the area's economic potential. From the Sanwara toll plaza to the Jabli market, Fig. 1 depicts the site position.

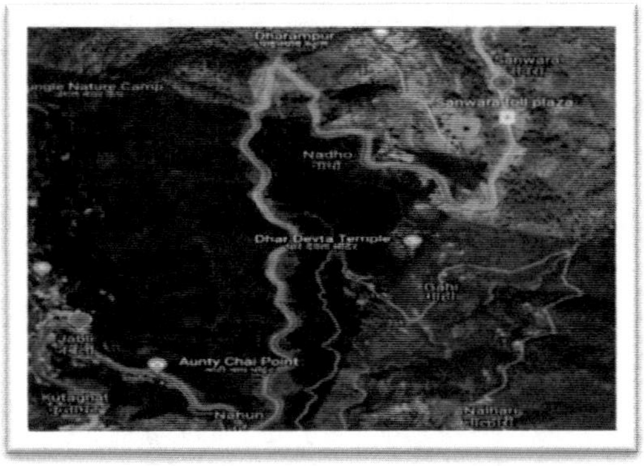

Figure 1. Site location (Source: Google Earth)

Figure 2. Actual site location

3. Data extraction and analysis

3.1. Examples of the questionnaire based on which the study is carried out

Table 1. statements of surveyed people for Satisfaction Index

Please select the box according to people's statements				
Gender	Age	Education Level	Occupation	Mode of travel
Male/Female	18-27	High school or Vocational certificate	Student	car
Male/Female	28-37	associate degree or Vocational Diploma	Teacher/Professor	jeep
Male/Female	38-47	bachelor's degree	Government officer	van
Male/Female	48-57	Master's Degree	State enterprise employee	LCV
Male/Female	58 and above	Above Master's Degree	House husband/ Housewife	bus
			Driver	Truck
			Other	3 axel truck

Table 2. Satisfaction parameters questions for Satisfaction Index

services	Very satisfied (5 marks)	Satisfied (4 marks)	Neutral (3 marks)	Dissatisfied (2 marks)	Very dissatisfied (1 marks)

Quality of roads
Signs and Road markings
Toll Plaza experience
Rest Areas/ WSA
Light and visibility
Congestion
Accidents

Table 3. Process for evaluating and rating for four- or six-lane national highways with divided carriageways [10]

S.No.	Sub parameters	Weightage marks	Procedure for assessing the parameter and marking
1	% Length median plantation and avenue plantation	3 marks	For median a total no. of 666 plants to be planted in 2 rows 333 each. (Only 4.5m wide will be considered) and for avenue total 583 plants to be planted 333 trees in first row, 166 trees in second row, 84 trees in third row. (Only considered where ROW is enough to accommodate the same) and 0.03 marks deducted for every 1% decrease in plantation numbers.
2	Cleanliness along Highways	1 mark	Whole ROW is to be observed for cleanliness regarding dumping of garbage, condition of paint on the structures, open sewage drains. Either you should give full marks or zero marks.
3	Functionality of Wayside amenities	4 marks	At least one operational (WSA) along the corridor either privately owned or Govt. owned should be which include drinking water, paved parking, eating facility, proper illumination, security personnel to guard the premise and if there is 1 WSA is fulfilling all the conditions you provide full marks or either zero.
4	Functioning of Toilet & availability of drinking water at Toll Plaza	2 marks	All toilets and water ATM near the concerned toll plaza should be inspected for cleanliness and water availability. The water ATM should be assessed for availability of potable water and if there is only 1 either toilets or water ATM is available then provide half marks.
5	Hinderance free from speed breaker	1 mark	The total no. of speed breakers be identified along the influence length and provide half marks if there is 1 speed braker is present along the full influence length of toll plaza give zero.
6	Hinderance free from traffic barrier	1 mark	The no. of traffic barriers either police check posts or forest check posts etc, shall be recorded within the influence length of the toll plaza and provide half marks if there is 1 speed braker is present along the full influence length of toll plaza either give zero.

7	Availability of motor repair workshop	1 mark	Availability of a motor repair shop within legal access permission shall be observed within the influence length of the toll plaza. There is no intermediate marking if workshop is fulfilling all the conditions provide full marks or either zero.
8	Encroachment and illegal hoardings:	2 marks	Area under encroachment and area of illegal hoarding be calculated and taken as actual value in SQM (square meter). 0.035% marks will be deducted for all illegal hoardings present on ROW in influence length of toll plaza.
9	Noise Quality Index	1 mark	Monitoring stations near built up areas to be set up for measurement and CPCB guidelines be used for any assistance and Average sound intensity across the corridor be calculated and taken as actual value. If avg. noise intensity is between 65db-75db then deduct half marks if intensity exceeds 75db provide zero marks.
10	Satisfaction Index	4 marks	50 surveys per day on any one weekday and any one weekend to be conducted, and 0.04 marks be deducted for every 1% decrease in average percentage marks from all the surveys

3.2. Analysis and the results of data collected from the face-to-face interviews

3.2.1. Assessment of the satisfaction level of road users/citizens in terms of the road conditions.

The survey found that road users/citizens were satisfied in five distinct locations on weekends and weekdays. Weekly day satisfaction was higher than weekend day satisfaction, indicating that weekly days have better road conditions than weekend days. The road conditions survey was used to confirm this assumption.

Table 4. The satisfaction level of road users based on road conditions

Satisfaction	Weekly days		Weekend days	
	Number of Road Users	Total marks	Number of Road Users	Total marks
Very satisfied (5 marks)	58	290	64	320
Satisfied (4 marks)	38	152	32	96
Neutral (3 marks)	4	12	0	0
Dissatisfied (2 marks)	0	0	4	8
Very dissatisfied (1 marks)	0	0	0	0
Average score	100	4.54	100	4.4

3.2.2. Identification of reasons for the satisfaction/dissatisfaction of road users in terms of the road. Satisfaction scores were linked to physical road condition and drainage characteristics, as well as their functions. Road users were satisfied when the physical conditions of the roads and drainage were good and well-functioning; nevertheless, road users were dissatisfied when the physical conditions of the roads and drainage were unstable. According to the survey results, respondents connected these issues with flaws in maintenance efforts. The outcomes were as follows:

- Road distress and potholes — This poor road quality reduces comfort as well as speed because users must constantly apply the brake, increasing journey time. Road users indicated that they are unable to drive peacefully and smoothly because they are constantly afraid of an accident or collision with another car due to potholes, while anguish causes discomfort in driving since the vehicle pendulous when driven at high speeds.

- Uneven shoulder - many consumers expressed dissatisfaction with their shoulder condition while participating in the study. They stated that the elevation of the shoulder and pavement edge changes abruptly. As this should be present in any condition along the highways, the driver tends to shift away from the edge, which leads to reduced utilisation of the full width of the road, which leads to a drop in speed and, eventually, congestion. Because of the height difference, laden tucks and LCV may become unbalanced and slide in the valley because NH-05 is on hilly ground. Because this driver drove their vehicle in the centre of the carriageway, other vehicles were unable to overtake them easily. This results in the establishment of a moving queue on the road.
- Deferred maintenance/faulty maintenance- because of late and improper road maintenance. The user notices abrupt changes in surfaces as they pass unintentionally through a little hump that runs the entire width of the road. This results in an uncomfortable driving stance, which limits speed.
- Drainage conditions — drains along the highway are either clogged or battered. The effect of improper drainage eventually leads to pavement surface as the edges of the road disappear user drive slowly specially loaded vehicle's and in case of cars user don't find the opportunity to overtake as they can't judge road width due to logging of water on pavement surface and slippery pavement surface.
- Natural disaster/land slide- the NH-05 segment between Sanwara toll plaza to Jabli bus stop is being enlarged by cutting hills. During the rainy season, as water rushes down the hill, there will be a sudden landslide in various spots, coupled with mud and rock, which will occasionally obstruct the road. As a result, traffic jams develop along the route. A summary of the overall reasons given by road users are shown in the following table.

Table 5. Reasons behind satisfaction and dissatisfaction of road users

Reasons for satisfaction: good and well-functioning physical conditions of roads/drainage. Reasons for dissatisfaction: unstable road conditions.		
Caused by	Description	Physical conditions of road/drainage
Distress and pothole	A slow response of the road agency to repair	A small volume of pavement distress in a scattered area and Moving Road deterioration
Uneven shoulder	Improper maintenance of road	There is sudden height difference in road and shoulder
Drainage conditions	Improper construction	Water logging along roadside and missing of road marking
improper maintenance	A slow process of repair and unskilled workers	Rapid decay of a newly paved road and construction of new without levelling
Natural disasters	Land slide with muddy water	Blocked road and missing of road markings

Figure 3. shows the poor repair and maintenance

Figure 4. Shows land slide (natural disaster)

3.2.3. Assessment of the satisfaction level of road users in terms of the Sings and road markings. While conducting the questionary poll, some road users believe that the quality of road markings is adequate, while others believe that there should be more road markings present. Some road users argue that traffic signs situated near habitual areas should be surrounded by unlawful hoardings so that they cannot see the sign while entering the habitual region. That is rather distressing, given that location is both crowded and contains unlawful parking. The results reveal that road users are less satisfied with the road marking on weekdays than on weekends. On weekdays, locals are in a hurry to get to their destination. If they discover incorrect signs and markings, they will lose a lot of time travelling since they will be misled by the road information provided.

Table 6. shows the satisfaction level of road users based on signs and road markings

	Weekly days		Weekend days	
Satisfaction	Number of Road Users	Total marks	Number of Road Users	Total marks
Very satisfied (5 marks)	68	340	86	230
Satisfied (4 marks)	32	128	4	16
Neutral (3 marks)	0	0	2	6
Dissatisfied (2 marks)	0	0	2	4
Very dissatisfied (1 marks)	0	0	0	0
Average score	100	4.68	100	4.86

3.2.4 Assessment of the satisfaction level of road users in terms of the toll plaza experience. The sorts of fundamental service quality that must be attained in implementing toll road operations are as follows: a) Toll Road conditions; b) Average speed; c) Accessibility; d) Mobility; e) Safety; f) Rescue unit and relief service; g) Environment; and h) Rest area. On weekends and weekdays, the road user gets a distinct experience. They are likely to be dissatisfied on weekends since they experienced traffic congestion at the Sanwara toll plaza once the FAST-TAG service was launched later. Road users find it simple because contactless payment is used in this procedure. They are pleased with the services provided at the toll plaza but are disappointed with the accessibility, mobility, and toll road conditions.

Table 7. shows the satisfaction level of road users based on services provided by the toll plaza

	Weekly days		Weekend days	
Satisfaction	Number of Road Users	Total marks	Number of Road Users	Total marks
Very satisfied (5 marks)	76	380	48	240

Satisfied (4 marks)	12	48	16	64
Neutral (3 marks)	4	12	16	48
Dissatisfied (2 marks)	8	16	8	16
Very dissatisfied (1mark)	0	0	12	12
Average score	100	4.36	100	3.8

3.2.5 Assessment of the satisfaction level of road users in terms of the rest area/wayside amenities (WSA). The study is carried out by asking 200 road users questions in real time. According to the findings of the study, there is a need for a rest place for families because the current rest area is utilised by truck drivers or buses. The hygienic condition of such rest place is deplorable. People want to see an increase in the number of rest areas and the availability of drinking water. They come up with ideas such as, 1] An increase in the number of public restrooms. 2] The cleanliness of public restrooms. 3]The ease of access to the service area/fuel stations. 4]The safety while spending time in the service area's vehicle park is preferable to stealing. According to the findings of the study, the service of the rest area is unpleasant on weekends.

Table 8. shows the satisfaction level of road users based on rest area/wayside amenities (WSA)

	Weekly days		Weekend days	
Satisfaction	Number of Road Users	Total marks	Number of Road Users	Total marks
Very satisfied (5 marks)	80	400	52	260
Satisfied (4 marks)	14	56	28	112
Neutral (3 marks)	6	18	8	24
Dissatisfied (2 marks)	0	0	12	24
Very dissatisfied (1 marks)	0	0	0	0
Average score	100	4.74	100	4.20

3.2.6 Assessment of the satisfaction level of road users in terms of the light and visibility. When distances are considered, night-time driving is up to three times riskier than daylight driving. [11] When questioned about traffic lights and the visibility quality of the road length. Drivers are "very displeased" with the traffic signal situation over the whole 5km route. There are no streetlights in functional order, and there are no retroreflectors (road studs) throughout the whole length of the road. Despite all the road maintenance, the worst situation is one of light and visibility since fog conditions grow in steep places throughout the winter season. The road user considers the light conditions to be worse on weekdays rather than weekends.

Table 9. shows the satisfaction level of road users based on light and visibility

	Weekly days		Weekend days	
Satisfaction	Number of Road Users	Total marks	Number of Road Users	Total marks
Very satisfied (5 marks)	32	160	32	160
Satisfied (4 marks)	30	120	36	144
Neutral (3 marks)	8	24	18	54
Dissatisfied (2 marks)	26	52	8	16
Very dissatisfied (1 marks)	4	4	6	6
Average score	100	3.72	100	3.80

3.2.7 Assessment of the satisfaction level of road users in terms of the congestion. The NH-05 is being expanded from a two-lane undivided to a four-lane divided carriageway. Because the concerned road section has a steep inclination, road users are pleased with the road enlargement. The jam state is worse when accidents or landslides, and the risk of congestion is greatly increased owing to widening. However, road users discover that the trip time is not as short as they had hoped. because they anticipate being able to move at a faster speed owing to the widening Route users discover that the road is more clogged on weekends than on weekdays owing to tourist vehicles.

Table 10. shows the satisfaction level of road users based on congestion

	Weekly days		Weekend days	
Satisfaction	Number of Road Users	Total marks	Number of Road Users	Total marks
Very satisfied (5 marks)	72	360	36	180
Satisfied (4 marks)	18	72	42	168
Neutral (3 marks)	6	18	4	12
Dissatisfied (2 marks)	4	8	16	32
Very dissatisfied (1 marks)	0	0	2	2
Average score	100	4.58	100	3.94

3.2.8 *Assessment of the satisfaction level of road users in terms of the accidents.* The most common measure of road unsafety is the number of fatalities. When we talk about stay, we count two types of road users: unprotected road users (URU) who are not protected in their vehicles: pedestrians, cyclists, and powered two-wheelers, and protected road users (PRU). The main issue on NH-05 is PRU because all road users go by bus, truck, or privately owned car. Road users are extremely happy with the safety service offered to them since the authority checks for problematic locations where the chance of an accident is higher across the whole influence length on a regular basis. They have two patrol vehicles, an ambulance service, and a phonebooth located in that location.

Table 11. shows the satisfaction level of road users based on accidents

	Weekly days		Weekend days	
Satisfaction	Number of Road Users	Total marks	Number of Road Users	Total marks
Very satisfied (5 marks)	94	470	90	450
Satisfied (4 marks)	6	24	10	40
Neutral (3 marks)	0	0	0	0
Dissatisfied (2 marks)	0	0	0	0
Very dissatisfied (1 marks)	0	0	0	0
Average score	100	4.94	100	4.90

4. Result

4.1 To grasp the perspective of a road user, a list of sub parameters is provided with the obtained marks.

Table 12. shows the marking based on user perspective of wayside facilities and satisfaction index provided by the road to the user.

S.No.	Sub parameters	Weightage marks	Marks obtained
1	% Length median plantation and avenue plantation	3 marks	2.66 marks
2	Cleanliness along Highways	1 mark	1 mark
3	Functionality of Wayside amenities	4 marks	4 marks

4	Functioning of Toilet & availability of drinking water at Toll Plaza	2 marks	2 marks
5	Hinderance free from speed breaker	1 mark	1 mark
6	Hinderance free from traffic barrier	1 mark	1 mark
7	Availability of motor repair workshop	1 mark	0 marks
8	Encroachment and illegal hoardings:	2 marks	1.93 marks
9	Noise Quality Index	1 mark	0 marks
10	Satisfaction Index	4 marks	3.51 marks
11	Total marks	20 marks	17.10 marks

5. Conclusion

This study investigated road users' satisfaction level and rated NH-05, Kalka-Shimla highway (Himalayan expressway) by using at least 200 samples of questionnaire survey. Road users were asked to indicate what aspects of road facilities they would like to see improved. As shown by the results, the significant factors that negatively affected road users' experience were streetlight and visibility as only 72.4% of road users were satisfied with this service. In addition, congestion was the third most significant factor that contributed to unsatisfied road users, even though the road was recently widened from 2 to 4 lanes. The services such as quality of the pavement, rest area/wayside amenities (WSA), signs and road markings, and accident response system exceeded expectation of road users. Users were pleased with the quality of services provided in these aspects. It was also observed from the study that safety equipment and aesthetics along the highway were of prime concern to the road users. Following the NHAI highway rating system, the Kalka-Shimla National Highway received 17.1 out of a possible 20 marks, or an 82.5 percent score, indicating shortcomings in relation to the specified road user facilities.

References

[1] Suanmali, S., Chankao, K., Korbsanthia, P., & Ammarapala, V. (2015). Factors affecting road users' satisfaction: the case of Motorway Route 7. *Songklanakarin Journal of Science & Technology*, *37*(2).

[2] Adikesuma, T. N., & Nahdiyah, N. (2020, February). Performance of Urban Infrastructure: Road User Satisfaction Index in Satellite City. In *2nd International Symposium on Transportation Studies in Developing Countries (ISTSDC 2019)* (pp. 107-111). Atlantis Press.

[3] Subotić, M., Tešić, M., & Vidović, N. Analysis of the level of satisfaction of road network users-case review of road section Koprivna–Modriča (r-465). *JOURNAL TTTP-TRAFFIC AND TRANSPORT THEORY AND PRACTICE*, *2*(1-2)..

[4] Methorst, R., Eenink, R., Cardoso, J., Machata, K., & Malasek, J. (2016). Single unprotected road user crashes: Europe we have a problem!. *Transportation research procedia*, *14*, 2297-2305..

[5] Paz, A., de la Fuente-Mella, H., Singh, A., Conover, R., & Monteiro, H. (2016). Highway expenditures and associated customer satisfaction: A case study. *Mathematical Problems in Engineering*, *2016*..

[6] Puri, E. R. (2017). Development of Parameters of Road Conditions based on Road User/Citizen Satisfaction. *IPTEK Journal of Proceedings Series*, *3*(5).

[7] Pratama, R. H., & Sabar, M. Analysis of the Effect of Service Quality, Rest Area Facilities, and Perceived Price on Customer Satisfaction at The Cipali Toll Road. *vol*, *4*, 1285-1289.

[8] Jameel, A. K., & Evdorides, H. (2021). Developing a safer road user behaviour index. *IATSS research*, *45*(1), 70-78.

[9] Hasan, M. M., Alam, A., Mim, A. M., & Das, A. (2020). Identifying User Satisfaction Level of Road services: A Focus on Rajshahi City Bypass Road, Bangladesh. *Transportation Research Procedia*, *48*, 3132-3152.

[10] H. R. Report, "Standard Operating Procedure (SOP) for Assessing and Rating for divided carriageway (4 / 6 lane National Highways) based on : Highway Efficiency (45 Marks) Highway Safety (35 Marks) User Services (20 Marks)," no. i, 2021. https://morth.nic.in/sites/default/files/circulars_document/Highway%20Rating%20Circular.pdf

[11] Wood, J. M., Isoardi, G., Black, A., & Cowling, I. (2018). Night-time driving visibility associated with LED streetlight dimming. *Accident Analysis & Prevention*, *121*, 295-300..

Mapping Groundwater Potential Zone and Flood Risk Zone in the Visakhapatnam District, India.

P. Vamsi Krishna Prasad[1], M.Leela Priyanka[2], R. Sarath[3], B.Raghupathi Naidu[4] and T.Ravi[5]

[2]Assistant Professor, Vignan's Institute of Information & Technology, Duvvada, Visakhapatnam, Andhra Pradesh, India

[1,3,4,5]Graduate student, Department of Civil Engineering, Vignan's Institute of Information & Technology, Duvvada, Visakhapatnam, Andhra Pradesh, India

Email: pvkprasad029@gmail.com ; viitpriyanka09@gmail.com

Abstract. Groundwater is a vital resource that significantly contributes to the annual supply. On the other hand, Overexploitation has resulted in a significant reduction in groundwater supplies and, in some cases, soil sinking. Therefore, it's critical to assess the possible zone of groundwater rejuvenation to safeguard water quality and manage subsurface water systems. Potential groundwater zones are detected using RS and GIS technologies. GIS technologies were used to build the composite map. Accurate data is required to determine the criteria used to estimate the potential groundwater zone. DEM, soil, Rainfall, and LULC data are all generated by the satellites, and, at a scale of 1:50000, topo sheetsare collected from the Survey of India (SOI). All the input data is combined with a weighted overlay in ArcGIS. Each of these criteria is given an appropriate ranking. The ability of various geomorphic units to store groundwater is assigned weight values. This procedureis repeated for each of the remaining layers, with the resulting layers being categorized. Extremely poor, poor, moderate, good, and exceptional are the five categories of potential groundwater zones. The method is used in a particular research area in the Visakhapatnam district. This data will help you figure out where you can get water, where water can be recharged, and where groundwater is limited.

1. Introduction

Groundwater is a primary natural supply that promotes human health, economic prosperity, and biological variety. It has become a crucially important object due to its various intrinsic properties. Water supply can be considered a crucial aspect both in rural and urban settlement regions irrespective of their climatic conditions. Groundwater is a sort of water that fills every vacancy in a geological layer. Water-bearing crustal formations serve as both transmission channels and reservoirs for storing water.

The creation of porosity determines the existence of groundwater in a geological formation and the capacity for its extraction. Runoff often increases with increasing terrain slope, and infiltration is typically encouraged by confined contours like as valleys in topography. A region with a high drainage density has more surface runoff than a low drainagedensity area. Recharge zones can be rivers, ponds, and other pools of surface water. Because of the rising significance of groundwater due to consumption, illicit groundwater mining has occurred,resulting in a water stress situation. This risky condition necessitates a low-cost, quick-to-implement approach for analyzing and managing groundwater supplies [1]. A groundwater development program needs to collect massive data from many sources. As a result, identifying and

measuring thesecharacteristics is crucial for constructing a groundwater potential model of a research region. Groundwater has become a critical water resource due to its potential to supply water under drought situations, deliver water to large-scale urban populations, and low infrastructure construction costs. However, groundwater study has remained hazardous despite substantial research and technological improvement owing to the absence of a direct way of viewing water below the surface. As a result, its presence or absence may only be deduced indirectly by examining geologicaland surface factors.
Potential zones of groundwater can be identified using various themes of hydrogeography of any study region. Remote sensing and Geographic information system (GIS) techniques can open up newvenues for water resource study. The study of remote sensing data, coupled with topographical sheets and ancillary information from the Survey of India (SOI) and the requisite ground-truth verifications, contributes to developing baseline information for groundwater targeting. Using remote sensing data, identifying groundwater occurrence locations is based on an indirect examination of immediately observable terrain elements such as geological formations, geomorphology, and associated hydrologic properties. Lineaments are also crucial in groundwater research across all terrain types [2].
GIS and remote sensing may also be utilized for multi-criteria resource evaluation and hydro geomorphological mapping in managing water resources. For example, the Visakhapatnam district's remote sensing and GIS techniques to extract precise drainage, slope, and geomorphic characteristics propose potential groundwater zone research strategies.

2. Study area
Visakhapatnam District is one of Andhra Pradesh's North-Eastern Coastal districts, located between the latitudes of 17° 15' and 18° 32' in the north and 18° 54' and 83° 30' in the east. It has its headquarters in Visakhapatnam and spans an area of approximately 11,342.84 square kilometres. It shares borders with state of Orissa on the north and the west (partially with Vizianagaram on the north), the Bay of Bengal on the east, and the district of East Godavari on the south. The district consists of 3265 villages, 15 cities, which comprise one municipal organization, two municipalities, and 12 census cities.Visakhapatnam town is only Municipal Corporation and two municipalities are Elamanchili, Narsipatnam present in study area.
The district is divided into two major geographical sections. The lowlands division is the strip of land along the coast and interior. At the same time, the Agency Division is the mountainous territory of the Eastern Ghats surrounding it on the north and west. The Agency Division has hilly lands sheltered by the Eastern Ghats with around 900 meters, interspersed with several peaks above 1200 meters. Total Rainfall received was 1098.4 MM during the year 2017-18, and its deficiency was -9 % of average Rainfall

3. Materials & Methodology
Numerous datasets were utilized to identify potential groundwater areas, including earlier maps, satellite data, Landsat 8, Quick Bird, updated in July 2020, and SRTM DEM. ArcGIS 10.8 is used to create thematic maps of lineament density, drainage density, slope, and soil. Additionally, Landsat TM 30 m was utilized to create a land use and land cover thematic layer. The Supervised classification methodology is used to classify land use and cover. The source geological map of Visakhapatnam district at a 1:50000 scale survey of India (SOI) toposheet was used to create the area's geological map.The soil type map is based on the FAO's India soil map. Annual Rainfall for 2019-2020 was retrieved from the CRU website [3].

3.1.Methodology
The suggested technique for the study includes several tasks such as base map preparation, LULCmap preparation, digitization and image processing with software, and output interpretation. The first stage comprises establishing a spatial database utilizing a 1:50000 scale survey of India (SOI) toposheet and Land sat TM 30 m satellite data. Then, GIS and remote sensing technologies create groundwater-related thematic maps such as drainage density, contour, and stream length. Furthermore, the Land Utilization Survey Database, geologic maps, and on-site inquiry are utilizedto statistically and qualitatively define the hydro-geo-logical conditions in the area. In the second stage, a digital elevation model (DEM) was downloaded by interpolating a contour map digitized from an SOI toposheet. Slope, aspect, flow accumulation, and stream order are all calculated usingDEMs. This technology is commonly used to create a runoff potential map for a small to medium- sized involved drainage basin.

The third stage entails georeferencing and geometric rectification using digital image processing of satellite data. The supervised classification approach is then used to generate multiple topic layers.The properties from the acquired data were then combined to construct the buffer map for the agricultural and settlement regions. It is then followed by the generation of additional critical data that will be utilized to evaluate groundwater potentials later on, such as a land use/land cover map,soil map, and rainfall map [4]. Finally, the DEM map generates the slope map and the drainage density. In the last stage, all of the themes mentioned above are further processed and analyzed in the overlay, and a score is given to evaluate appropriate groundwater potential zones. In the currentstudy, all thematic layers will be overlaid using GIS to obtain the final integrated output of potentialgroundwater zones. DEM, slope, drainage density, LULC, soil, and rainfall maps were employed inthe research to determine potential groundwater zones.

3.2. Factors influencing ground water potential zone maps

3.2.1. Drainage density map

A drainage basin is a natural entity that collects the runoff water and directs it to a recurring location. Water bodies, rivers, tributaries, perennial and ephemeral streams, and ponds are all depicted on the drainage map. The drainage is prepared by analyzing the DEM of the research region. Hydrologytools are used to extract hydrologic data from a DEM. The density of drainage is defined as the tight spacing of stream channels.

Table 1. Drainage density map

Class	Km/Km	Drainage density
1	0.12-0.68	Very Low
2	0.68-0.95	Low
3	0.95-1.15	Moderate
4	1.15-1.33	High
5	1.33-1.78	Very High

Table 2. Slope map

Class	Slope (%)	Slope category
1	0-5.14	Very Low
2	5.14-12.35	Low
3	12.35-20.33	Moderate
4	20.33-29.08	High
5	29.08-65.63	Very High

3.2.2 Slope map

It is calculated using the total time taken by the flow section of all orders per unit area. Permeability is proportional to drainage density in the inverse direction. The less porous the rock, the less rainwater penetrates, resulting in runoff concentrating on the floor. The slope affects the drainage density and varies according to location. The drainage map outputs a density map using the spatial analyst's line density tool. The drainage density of the studied region is divided into four categories: very low, low, medium, high, and extremely high.

3.2.3 DEM map

A digital elevation model (DEM) is a digital representation of the topography or terrain of the ground surface. It is usually limited to using a raster grid of elevation values. Remote sensing techniques are commonly used to create DEMs, but land surveying can also be used. DEMs are the most common basis for digitally-produced relief maps and are frequently used in geographic information systems. To create a Dem map, we download the study area part in suitable standard from the world Dem in Aster format from earth data, one of the NASA websites. It will be mosaic to a single image and other mandatory settings are applied. Then clipped using study area shape file.

3.2.4 LULC map

The LULC factor influences groundwater occurrence, reduces runoff, increases infiltration water, and increases recharge rate. In addition, it determines the amount of rain that percolates to the groundwater; for example, in a densely vegetated urban environment, intercepting precipitation over the ground is accessible for groundwater recharge [5]. Radiance calibration and atmospheric correction create the land use/land map from Landsat-8 pictures. Erdas imagine software was used for this research, and some points were inserted as training samples before supervised classification using the maximum likelihood technique. As a result, the research area was divided into seven categories: water, settlements, agriculture, fallow land, natural vegetation and barren land [Figure.4]

Table 3. DEM map

CLASS	Meters	DEM
1	0-222	Very Low
2	222-536	Low
3	536-820	Moderate
4	820-1055	High
5	1055-1681	Very High

Table 4. LULC map

CLASS	Land use /Landcover
1	Water
2	Settlements
3	Agriculture
4	Fallow Land
5	Natural Vegetation
6	Barren Lan

3.2.5 Soil map

Soil maps are primarily used to identify soils and their qualities. Still, they are also required for more particular objectives, such as identifying a soil's suitability for specific crops or the land drainage capacities. In digital mapping, a Geographical Information System (GIS), which employs a computer to store and manage data on soils and other related topics, is becoming more popular. The Food and Agriculture Organization of the United Nations provides a global soil map, then used to exact the research area using a shapefile. By matching soil codes in both files, data relating to each soil was found utilizing a Microsoft access file from the same organization. SNum present in the map is used to find soil properties from the access file [6].

3.2.6 Rainfall map

Rainfall is a significant source of floor water replenishment. It is critical to the hydrologic cycle, which controls groundwater potential. It also controls how much water may percolate into the groundwater system. Even though the region receives much rain each year, the unequal distribution and uncontrolled use of water resources regularly lead to floods. In contrast to steep slope places where the water goes as quickly as runoff, flat areas can absorb Rainfall and allow groundwater recharging. Therefore, rainfall measurement is a one-time occurrence that may not represent the studied area.

Therefore, achieving a sufficient uniform Rainfall intensity in the examined place is essential to get greater dependable and representative results. It is downloaded from The Climatic Research Unit's website, a component of the University of East Anglia. First, 10 years of rainfall data from the study area is downloaded. Then, it is separated using the bands of the required year by converting the raster to grid points. Finally, the points are used in interpolation to create a rainfall map [7].

Table 5. Soil classes and Types

Class	Types of soils present in
1	Clay, Loam
2	Loam
3	Loam
4	Sand, Loam
5	Loam
6	Sand, Clay, Loam
7	Sand, Loam
8	Sand, Loam
9	Clay

Table 6. Rainfall classes and intensity

S.No	Class	Rainfall
1	Very Low	1246.56-1357.08
2	Low	1357.08-1420.47
3	Moderate	1420.47-1467.61
4	High	1467.61-1530.99
5	Very High	1530.99-1661.02

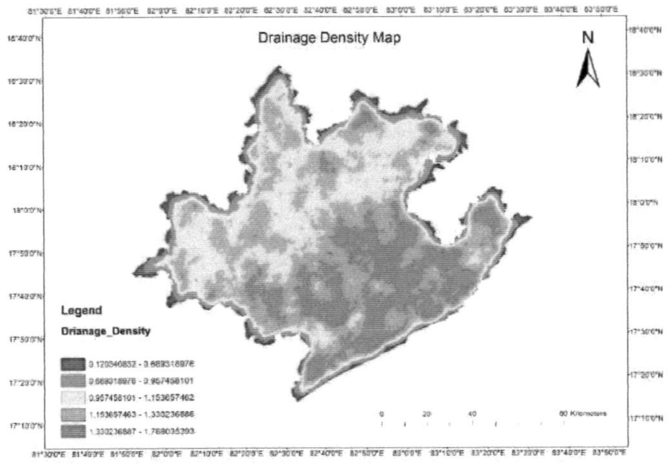

Figure 1. Drainage density map of Visakhapatnam district,India

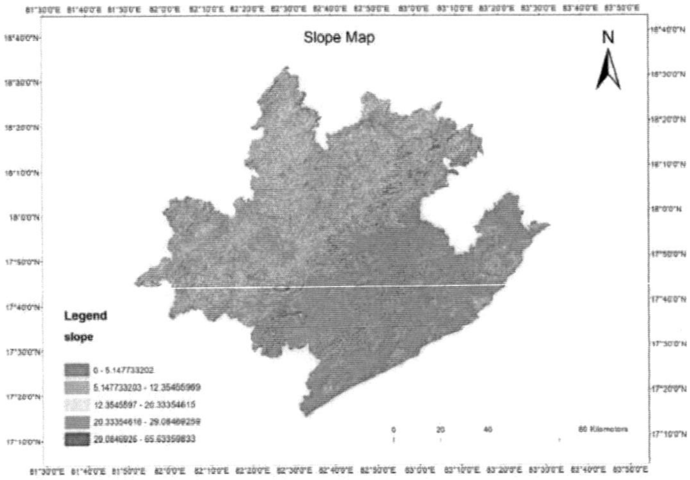

Figure 2. Slope map of Visakhapatnam district,India

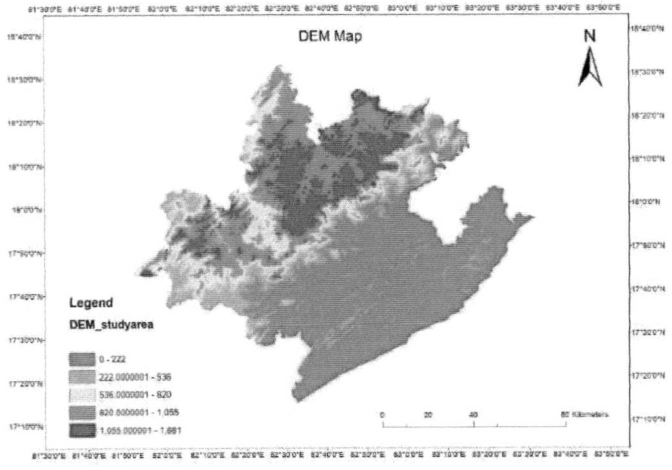

Figure 3. DEM map of Visakhapatnam district,India

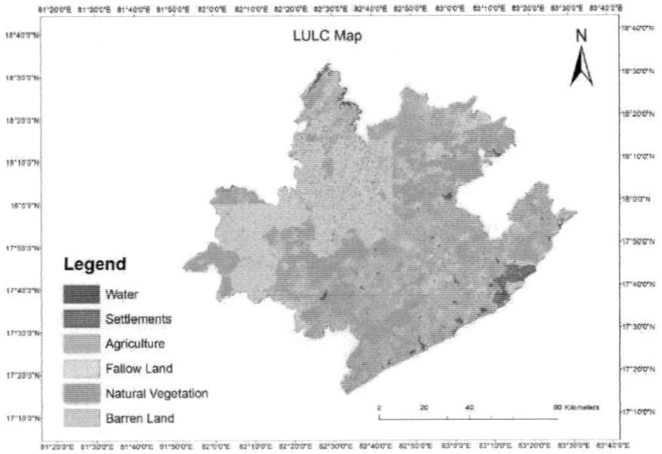

Figure 4. LULC map of Visakhapatnam district,India

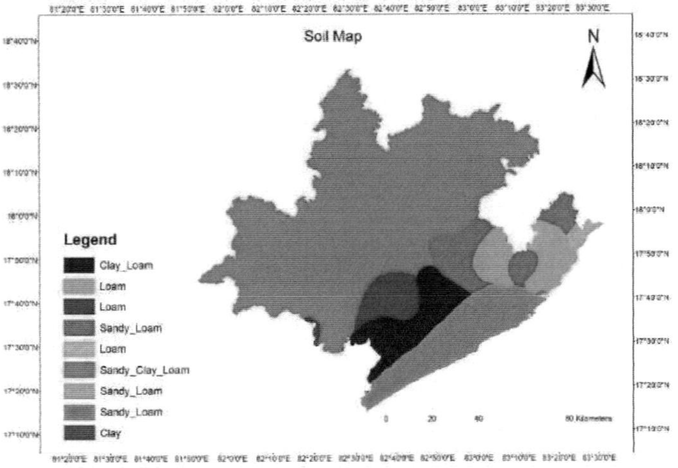

Figure 5. Soil map of Visakhapatnam district,India

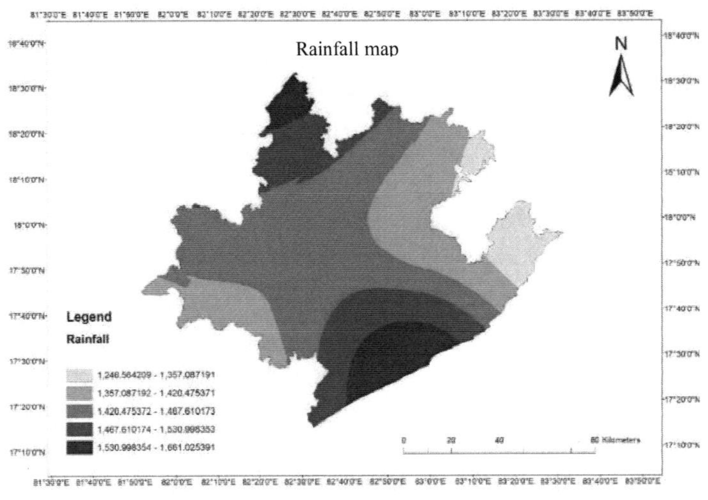

Figure 6. Rainfall map of Visakhapatnam district,India

Table 7. Weightage and ranking of the parameters of groundwater potential zones

Parameter	Classes	Weight(%)	Rank
Drainage Density	0.12-0.68	30	1
	0.68-0.95		2
	0.95-1.15		3
	1.15-1.33		4
	1.33-1.78		5
Slope	0-5.14	5	5
	5.14-12.35		4
	12.35-20.33		3
	20.33-29.08		2
	29.08-65.63		1
DEM	0-222	30	5
	222-536		4
	536-820		3
	820-1055		2
	1055-1681		1
LULU	Water	10	5
	Settlements		4
	Agriculture		3
	Fallow Land		2
	Natural Vegetation		2
	Barren Land		1
Soil (SNum represent Soil Properties)	3714	10	1
	3729		2
	3754		2
	3795		3
	3829		3
	3830		4
	3845		4
	3848		5
	3864		5
Rainfall	0-222	15	1
	222-536		2
	536-820		3
	820-1055		4
	1055-1681		5

Table 8. Groundwater potential zones of the study area

Sr. No	Type of zones	Area in km²	Area in percentage
1	Very Good	771.75	6.67
2	Good	4062.92	35.14
3	Moderate	5419.76	46.88
4	Poor	1306.19	11.29
5	Very Poor	0.0128	0.000111

4. Results & Discussion

As the name suggests, weighted overlay analysis is an analysis process done in the overlay method. Overlay means overlapping one map on another map to present all data in one map. Multiple maps are used in overlay analysis. So the thematic maps used in the analysis are the maps whose properties influence groundwater levels and their recharge [8].The first step is to open the overlay tool in the toolbox, and a popup of the overlay will appear. The next step is to add maps prepared in advance to analyze an overlay window.

Generally, the map's appraisal (property) is based on the domination of one property that influences groundwater recharge over others. Therefore, the sum of the weight should be 100. Finally,

it is assigned based on the above conditions and assumptions over the study area. Now we give ranking to the attribute table of each map. The two ranking options are 1to 5 by 5 and 1 to 9 by 9.

Generally, we use one option as it is suitable for most analyses. As we said, it is ranking. In the attribute table, which influences more, our analysis will be assigned rank 1, and less will be rank 5. Same ranks may be assigned in one attribute table based on assumptions and facts of study of the area that we analyze [5].

It is a supervised analysis where we have seen analyst inputs the weightage and ranks are given based on assumptions and facts of the study area. Now the ranks are allocated. Next, click on the analysis option. The system will take some time to analyze and generates a groundwater potential map. Finally, we will complete it by furnishing with naming the zones from top to bottom in the attribute table of that map as Very poor, Poor, Moderate, Good, and Very Good, as shown in Figure.7.

Overlay analysis can use to create flood maps, drought maps, etc. can be created by assigning weights and ranks based on how these factors influence the subject and purpose. For example, the flood Risk zones map was produced in the same method as shown in Figure.8.

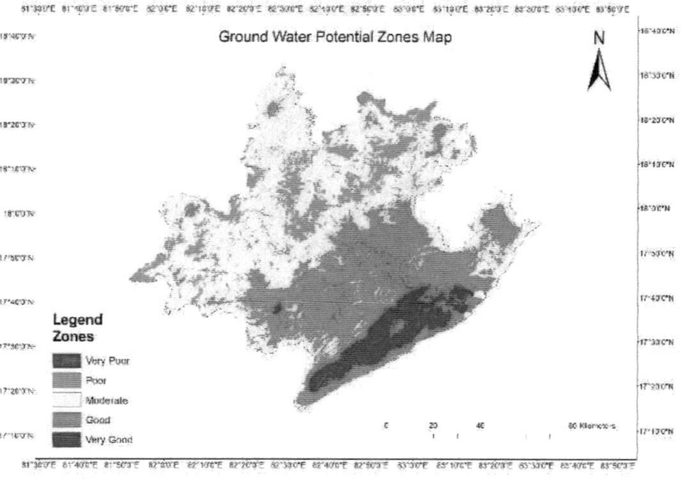

Figure.7 Ground Water Potential Zones map of Visakhapatnam district, India

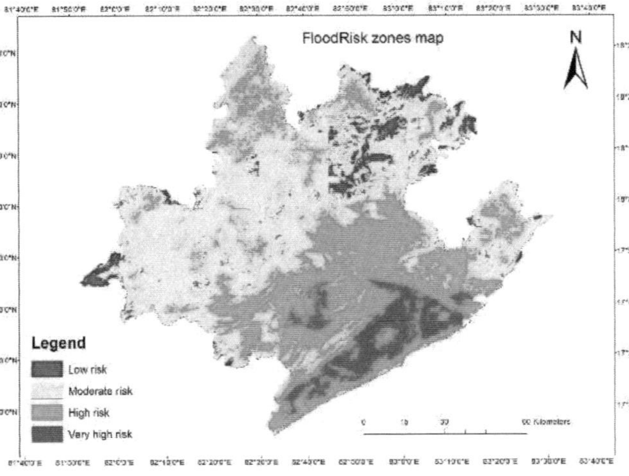

Figure.8 Flood Risk Zone map of Visakhapatnamdistrict, India

5. Conclusion

Remote Sensing and GIS are used for this analysis using satellite data. By unified, six theme maps are Drainage density, slope, Dem, soil, LULC, and Rainfall. Overlay analysis is used where the importance of maps is given based on their influence on groundwater recharge. Using drainage density exhibits the streams and flow in the study area basin. The slope shows the slopes which influence the flow path of water. Dem manifests the elevation of ground without topography [9].soil map depicts the different soil types and their qualities in the research region. The land use and the land cover map show how much land is used or covered in the research region. The rainfall map depicts the Rainfall of the research region and ranges using the kriging interpolation method. The weighted overlay analysis has been completed, and a map of groundwater potential zones has been generated. It is divided into five zones: (1) very poor, (2) poor, (3) moderate, (4) good, and (5) very good.

Groundwater cannot be seen directly so people don't know the condition of groundwater. Groundwater potential zones map will give people and local authorities about the groundwater situation of Visakhapatnam district and use suitable application of recharge. In addition, it will help plan and design water conversation programs and structures to preserve groundwater levels in the future. This map will help people better install new bores, agricultural crop choices based on groundwater zone, groundwater recharge methods, future research, etc.

A Flood Risk map is also created by repeating the process. In which the process of making six theme maps is repeated as above mentioned. After the maps are ready. Overlay analysis will begin by inputting the maps. Now weights and ranks are assigned based on these factors which will influence the flood. After overlay analysis, a map is generated showing the flood risk zone of the study area from low to high-risk zones. Flood Risk Zone Map can be used by authorities in town planning, disaster management, awareness programs, emergence response, and excavation planning. It furnishes necessary information also usefulin future studies and assessments.

6. References

[1] Magesh NS, Chandrasekar N, Soundranayagam JP. Delineation of groundwater potential zones in Theni district, Tamil Nadu, using remote sensing, GIS and MIF techniques. Geosci Front.2012;3(2):189–96.

[2] Shimpi S, Rokade VM, Upasani K. Application of Remote Sensing and GIS for Groundwater Potential Zonation: A Case Study of Bori-Chikli Watershed, Maharashtra, India. Bull Pure Appl Sci Geol. 2019;38f(1):114.

[3] Kattimani J, et al. Identification of Ground Water Potential Zone Using Remote Sensing And Gis Techniques. Int J Adv Res. 2018;6(5):948–53.

[4] Agarwal E, Agarwal R, Garg RD, Garg PK. Delineation of groundwater potential zone: An AHP/ANP approach. J Earth Syst Sci. 2013;122(3):887–98.
[5] Nag SK. Application of lineament density and hydrogeomorphology to delineate Groundwater potential zones of Baghmundi Block in Purulia district, West Bengal. J Indian Soc Remote Sens. 2005;33(4):521–9.
[6] Das B, Pal SC, Malik S, Chakrabortty R. Modeling groundwater potential zones of Puruliya district, West Bengal, India using remote sensing and GIS techniques. Geol Ecol Landscapes [Internet]. 2019;3(3):223–37.
[7] Reddy YVK, Lakshmi DSV. Identification of Groundwater Potential Zones Using Gis and Remote Sensing. Int J Pure Appl Math. 2018;119(17):3195–210.
[8] Allafta H, Opp C, Patra S. Identification of groundwater potential zones using remote sensing and GIS techniques: A case study of the shatt Al-Arab Basin. Remote Sens. 2021;13(1):1–28.
[9] Ganapuram S, Kumar GTV, Krishna IVM, Kahya E, Demirel MC. Mapping of groundwater potential zones in the Musi basin using remote sensing data and GIS. Adv Eng Software. 2009;40(7):506–1.

Role of ADR methods in environmental conflicts in the light of sustainable development

J Alkhayer[1], N Gupta[1] and C M Gupta[1]

[1] School of law, Shoolini University, Solan, Himachal Pradesh, India

E-mail: jaffaralkhayer89@gmail.com

Abstract. Sustainable development could not be envisaged regardless of protection of the environment and natural resources and achieving economic growth detached from resource use, in other words, sustainable development relies on environmental sustainability. On the other hand, Goal 16 of the 2030 Agenda for Sustainable Development, calls for peace, justice, and strong institutions to promote rule of law and access to justice at a global level. Settlement of environmental conflicts is an integral part of environmental sustainability and a part of accessing justice as well. In the light of increasing environmental challenges as a result of the continuous increasing rate of population and lack of global use of the renewed resource of energy has had a substantial influence on the environment, resulting in a corresponding increase in environmental conflicts. With the significantly varied court systems of nations throughout the world and the less efficiency of most of these systems, the question of rethinking and rebuilding effective alternative dispute resolution and further its role in environmental conflicts could be basic and urgent. But on the other hand, alternative dispute resolutions (ADR) have also barriers and negative aspects. The present paper examines and studies the role of ADR methods in environmental disputes and evaluates its efficiency and research on factors that lead to its success or its failure by studying real environmental conflicts from different countries to provide recommendations and suggestions to boost ADR methods in environmental disputes.

Keywords: *ADR methods, Environmental conflicts, Sustainable development.*

1. Introduction

The twentieth and 21 years of the twenty-first centuries saw accelerated growth in the rate of population from 1.65 B to 7.9 B people according to the Last World Population scale(1), as well as a series of actors that ushered in substantial variations in the world, such as the effect of two World Wars, technological developments, and, awareness of danger around the environment. All of these series of changes confirmed the necessity of an international partnership that works to control and make them revolve around serving the world and people by achieving sustainable development and avoiding the dangers that result from it. The Earth Summit in Rio de Janeiro, Brazil, in June 1992 was the beginning when over 178 nations endorsed the Agenda of 2021, a comprehensive strategy that involves global cooperation, aims to boost sustainable development and people's lives and protect the environment(2). the next major step in the stream of Sustainable Development was the World Summit on Sustainable Development in South Africa in 2002, that adopted The Johannesburg Declaration on Sustainable Development and the Plan of Implementation, that reconfirmed the universal society's environmental obligations while emphasising the need of multilateral collaborations(3). In January 2015 the process came to a climax with the General Assembly launching the process of negotiation around the post-2015 development agenda. In September 2015, the 2030 Agenda for Sustainable Development has been accepted at the UN Sustainable Development Summit. The 2030 Agenda for Sustainable Development in 2015 lays out a mutual vision among member states for accomplishing, prosperity and perpetual peace, for human beings and the world now and forthcoming. The agenda emphasises the Objectives of Sustainable Development which were considered as the heart of the agenda, and an emergency invitation to hold the responsibility of all nations, developed and developing, to work together in a universal framework. The Objectives of Sustainable Development are the outcome of years of discussions and negotiations by governments all over the world, as well as the UN. the 2030 Agenda had been embraced by all members of the United Nations(4). Boosting rule of law and making access to justice available to the public at the universal level is one of the objectives of sustainable development which is dedicated to goal 16th of the 2030 Agenda, that call for achieving peace, justice, and strong institutions(5). Goal 16th will be an essential concern of this paper due to it is a significant link between sustainable development and alternative dispute resolution methods (hereinafter ADR methods).

ADR methods as alternative mechanisms for accessing and achieving justice cannot be ignored anymore, especially it is looked at them as cost and time-effective and successful methods(6). Achieving justice which is embodied in Goal 16 of the 2030 Agenda, interacts with some other objectives in the field of environmental sustainability, Which are (6th),(7th),(11th),(12th),(13th),(14th),(15th), and (17th) goals of 2030 Agenda, which concern respectively about: accessing to safe water (Goal 6th), accessing to sustainable energy(Goal 7th), addressing the environmental consequences of urban development (Goal 11th), detaching between the growth of the economy and non-sustainable resource (Goal 12th), climate change(Goal 13th), protection of resources of marine, seas, and oceans and usage of these resources based on sustainable base (Goal 14th), protection and sustainable use of terrestrial ecosystems(Goal 15th), finally, the cooperation among all entities, governmental or non-governmental, for supporting and achieving of the 2030 Agenda objectives(Goal 17th) (7). It is pertinent to note that the environmental sustainability as a part of the sustainable development cannot be met in isolation from other goals of sustainable development, therefore the interaction among the Sustainable Development objectives is essential for accomplishing the environmental sustainability(8). Resolution of environmental conflicts in a rapid and effective method is a part of environmental sustainability and a part of 16th Goal of 2030 Agenda for sustainable development as well, and to achieve peaceful resolution in environmental issues, it is required to reinvent dispute resolution systems(9).

2. Adoption of ADR in environmental conflicts by the international community

The Rio Declaration of 1992, could be considered a start point for promoting the channel of access to environmental justice. Principle 10 of the Rio Declaration formats three basic rights when it describes environmental challenges, its emphasis was on three points that form pillars of environmental governance: information related to dangers to the environment should be available to the public, the public should participate in the decision-making process, methods and channels of accessing to justice should be available to all individuals(10). The Aarhus Convention 1998, is a further important step, article 16 of the Aarhus Convention supported the general standards of the Rio Declaration of 1992 and encouraged solving disputes by amicable methods like conciliation and mediation or any other type of resolving methods, as well as it permits to pursue binding methods like arbitration and adjudication if the amicable methods have failed (11). Resolution 65/283 of the United Nations General Assembly invites to consolidate the use of mediation as a peaceful method for settling the disputes, prevent and resolve the conflicts, The resolution focuses on the importance of contributions of all key parties and entities, including Member States, regional, national and international organizations and institutions and activities of civil society, and for supporting new perspectives and ideas for promoting and boosting adaptation of mediation and amical methods to contemporary conflicts(12). In 2013 the Minamata Convention on Mercury was signed, and in 2017 went into force, which confirmed the importance of ADR in conflict resolution. Article 25 of the Minamata Convention, took the same position as Article 16 of the Aarhus Convention and supported amicable resolution of disputes first through peaceful means chosen by the parties such as negotiation and mediation, before resorting to mandatory methods such as arbitration or adjudication(13). It was observed, that a great theoretical value was given to ADR by the international community as a peaceful method for the resolution of contemporary disputes and conflicts in an extensive frame of cases, whether it was civil disputes or environmental cases among others.

3. The Role of alternative dispute resolution in accessing and achieving justice

As mentioned above that accessing and achieving the justice is one of the most important goals of the sustainable development stream, alternative dispute resolution has proved as reasonable and successful methods may play an alternative role to traditional methods, especially if it is put into the consideration, the enormity of problems and obstacles which from the official judicial authorities and national courts suffer. Therefore, ADR as an alternative method for settlement disputes and an indispensable approach for accessing and achieving justice cannot be ignored.

ADR comprises a group of non-traditional dispute resolution methods, such as negotiation, conciliation, mediation, and arbitration. These mechanisms play a role is similar to the role of the official judiciary system for resolving the disputes between different entities, physical or legal, but the ADR methods may are distinguished with a more cost-effective, time-efficient, flexible, and consensual solving process, and at the same time, some types of ADR methods have enjoined with the same advantages that the judicial process have, in term of binding awards and ability of official authorities to interfere to enforce the parties to dispute to implement the binding content of ADR awards. In other words, the ADR awards produce a legally enforceable result as judicial decisions, with one more feature has come from the nature of ADR as a consensual process that may make the ADR methods more convenient to the parties of disputes.

3.1 ADR types and their characteristics

As a result, going in-depth research and studying the role of ADR in sustainable development as part of access to justice especially its role in environmental challenges and disputes, necessitate first of all understanding each ADR method(14).

3.1.1 Negotiation

Negotiation is the most simple and consensual method of alternative dispute resolution; this type of ADR method does not entail a neutral third party to make a final and binding decision(15), that binding agreement will be a result of the negotiation process between the parties to dispute themselves based on the ground that as the parties make negotiation process, without the intervention of a third party, for signing a contract which from a dispute has emerged, it will be reasonable to make similar negotiation process for reaching to an agreement for solving this dispute without intervention from a third party also. It is pertinent to note that the negotiation process cannot consider a judicial or arbitral process, in other words, the result of negotiation cannot be considered an arbitral award or judicial decision but it can be considered a new agreement or a new contract that will be binding based on the principle of compulsory nature of contracts, that means that the parties of dispute can change the result of negotiation any time if each party has agreed to these changes. It can be said that result of the negotiation is binding but not final, while the judicial decisions or arbitral awards are binding and final and cannot be changeable after its issue.

3.1.2 Conciliation and mediation

Conciliation and mediation are other types of ADR methods. Wherein a collaborative technique in which an impartial and unbiased third party who is called a "conciliator" or "mediator" manages dialogue sessions among the parties of dispute in which they practise a constructive discussion about their contentious issues for reaching a solution. The conciliator or mediator performs a consulting and advisory role, they suggest and offer potential and reasonable resolutions and remedies to the conflicts. The conciliation and mediation process lead to a settlement between the parties which is binding and final upon the parties. The conciliation and Mediation processes are voluntary processes, depending on the will and consent of the parties of the dispute. Confidentiality is a basic component of the process, whereas the conciliator or mediator is not permitted to unfold or disclose anything associated with the proceedings to outside. In the conciliation and mediation process, the parties' interests and relationships are given greater weight than legal standards, resulting in its popularity in delicate, confidential, and/or private conflicts(16).

The distinction between conciliation and mediation is sometimes obscured, and the two concepts are occasionally used interchangeably. However, it's vital to remember that facilitation only improves dialogue between parties, whereas mediation aims to establish an agreement(15).

In the principle, the settlement agreement which resulted from the mediation and conciliation should be binding and final for the parties of disputes, but it cannot be said that this agreement upgrade to the level of the arbitral award or judicial decision, in connection with the power of enforcement, therefore the question will be open, what a party can do when another party refuses to implement the terms of the agreement. In some jurisdiction such as India, they consider the conciliation settlement agreement enjoys the same value and effect as if it is an arbitral award issued by an arbitral tribunal(11). In other jurisdictions the settlement of conciliation and mediation are not enforceable or supported by the intervention of official authorities, therefore the settlement will consider a new agreement or contract which is binding based on the principle of the compulsory nature of contracts. Any dispute that will emerge from an implementation of a conciliation or mediation agreement may need an arbitral or judicial process to legitimate the official authority's intervention. But in the other hand, that does not reduce the importance of conciliation and mediation as ADR methods, especially when it comes to the most sensitive contentious issues whereas the parties will be reluctant to pursue binding and enforceable methods such as arbitration which produce a find and binding award, giving them no option to change or amend its terms(16).

3.1.3 Arbitration

Arbitration is one of the ADR methods which is considered the most similar to the judicial process, due to being adjudicative rather than consensual, or it can be said, Arbitration is a combination of litigation and alternative dispute resolution. It involves an arbitrator/s who should be a neutral and unbiased third party, the arbitrators render a binding, final and enforceable award to the disputants as it is issued by a judicial court(18). But on the flip face, arbitration gives the parties of disputes the ability to appoint their arbitrators and define the frame of their competence as well as the parties can select the applicable and procedure law, the seat of arbitration and the language which should be used during the process(15).

It became necessary, more than any time before, to be aware of the dangers surrounding the environment, the same should create the awareness of environmental justice which must be an integral part of justice, and accessing justice as a goal of sustainable developments must include the accessing to the environmental justice, especially when it comes to the natural resource governance, whereas the environmental justice is a significant component helping to manage successful policies. Sustainable development cannot be envisaged if the saving of the environment will be ignored. The same will shed the light on promoting the channels of access to environmental justice. ADR methods as types of these channels cannot be hidden(19).

4. Practical application of ADR methods

UN member states differ in the implementation of ADR methods, due to various factors such as the nature of the dispute and the most suitable procedure to be followed for reaching its solution. It can be seen that it sometimes takes the form of a local panel for solving complaints and grievances and sometimes takes the form of an institutional panel and is sometimes held secretly or publicity.
Regardless of the nature of procedures, some cases have shown a successful experiment of the ADR method when the parties have the willingness to negotiate and compromise their disputes and provide some concessions for achieving an agreement, in general, the participation of all concerned parties and stakeholders and their diligently and real intention to reach to an agreement and their eagerness to protect the environmental public interests is considered as the most assistant factor to success ADR methods. But in other cases, on the other hand, it has been shown that participation of many parties with the absence of their influence hampers the reaching of a solution, in general, the lack of trust between parties and their reluctance to agree on pursuing a specific type of ADR or their unwillingness to make concessions for reaching a whole party- satisfactory settlement, these factors increase the difficulties of reaching to a final settlement by ADR. The following paragraph shed the light on some cases in different countries to see how the mentioned factors influence ADR mechanisms.

4.1 Cases study

4.1.1 IPPC permit for the kunda pulp plant factory (Estonia)

ICCP Permit for the Kunda Pulp Plant Factory in Estonia, An example of successful mediation between NGO and developers. An enforceable agreement regarding the conditions of an IPPC permit was reached.
In the case at hand, AS Estonian cell intends to build a Greenfield aspen pulp plant in the eastern part of Kunda Laane-virumaa, Estonia. An environmental impact assessment (EIA) of the proposed development was conducted following the Estonian environmental impact assessment and was subjected to the environmental auditing act 2000 as well. The concern EIA was approved by the ministry of environment on July 29, 2002. To establish a pulp factory at Kunda, Estonia's northern shore, AS Estonian Cell sought an integrated environmental permit (IPPC). The pollution prevention measures were prescribed by the IPPC. The Estonian Fund for Nature (ELF) was the first to challenge the IPPC, citing its shortcomings as a severe threat to the Baltic Sea's marine ecology. A Negotiation process had launched among the stakeholders. The parties had reached an agreement to change the conditions, and a new IPPC was issued. Though the technique was useful in showing the likelihood of key actors reaching an agreement, the end outcomes involved some substantial concessions on both sides and no side was entirely satisfied with the final results. In addition, lack of time and experience especially from the side of NGOs contributed to increasing the complicity of the process(20) (14).

4.1.2 Returning the protected status to natural areas in the Lviv region (Ukraine)

In the Lviv Region, Returning Protected Status to Natural Tracts, was a great experience that shows the ADR method may be useful to save time while conserving important forest lands that would otherwise be chopped down. Several environmental NGOs and specialists banded together to challenge the Regional Council's decision to declassify the region as a protected area without a scientific basis. Media campaigns played the role of "facilitator" informally and virtually, supporting the discussions that led to an emerging panel of key parties entrusted with examining and determining the condition of protected areas(20) (14).

4.1.3 Znesinnia regional landscape park versus electric power supplier (Ukraine)

Another case is Znesinnia Regional Landscape Park v. Electric Power Supplier illustrates how public interest might persuade parties to reach an acceptable solution through Negotiation. The public council at the Lviv Oblast State Administration on Environment and Natural Resources used ADR processes to resolve the problem from the start of a request for the destruction of a considerable magnitude of trees to simplify Electricity Supply Networks Maintenance. Instead of focusing on the immediate crisis, the conversations yielded long-term answers to the larger issue(20) (14).

4.1.4 Saaremaa deep harbour (Estonia)

The instance of the Saaremaa Deep Harbour, on the other hand, demonstrates how ADR is not a viable choice when either side is reluctant to reach a solution, leaving all parties dissatisfied with the outcome. In this case, the construction of a port on Saaremaa Island inside a gulf designated as a Special Protected Area for birds was rejected by environmentalists. The environmental impact assessment and water usage permission issued by the ministry were challenged by NGOs. Even though the NGOs attempted to negotiate a possible resolution with the construction company, the Ministry refused to participate and the effort was rejected(20) (14).

4.1.5 Szentgal regional landfill (Hungary)

The Northern Lake Balaton Regional Waste Disposal Facility was a project of a group of projects under a comprehensive system to renew solid waste disposal in the country, the concerned project was given an environmental permit. The neighbouring communities have challenged the permission. attempts to facilitate the conflicts among the parties have been conducted by two NGOs, the NGOs tried to provide information about suitable measures to the parties for resolving the conflicts, but no party have agreed to initiate one of them, in the light of disagreement between the parties on a particular method, the NGOs initiated facilitated negotiation, but due to the hesitation of some of conflicting parties and absence of others, no outcomes were reached, this case sheds the light on the fact lack of the trust between the conflicting parties is the main reason leads to failure of ADR methods in environmental conflicts(20) (14).

4.1.6 Route 10 (Hungary)

In the case of Route 10, it was depicted how the severality of conflicting parties, with contradictory agendas and little actual influence, impede the success of ADR methods in settling the disputes. Several environmental non-governmental organisations (NGOs) have demanded dismissing the Route 10 environmental protection permit, arguing that the project will worsen the region's traffic and environmental problems. In preparation for direct or aided discussions, many parties had a series of meetings and conciliatory conversations. However, none of the sessions succeeded in altering the stakeholders' thoughts, the lack of involvement of a large number of players and absence of cohesive entities and their contradictory aims and dearth of their influence led to failing ADR in settling this case(20) (14).

The mentioned cases show that pursuing ADR processes in the environmental conflicts field serves to boost public and NGOs' access to justice and at the same time have proved time and cost-efficient. Furthermore, the conflicting parties have a greater grasp of the problem and are happier with the solution. In any incidental or consequential concerns, this allows for long-term advantages. However, low levels of commitment have hampered adoption, particularly among public authorities, who are afraid of losing power and hesitant to change owing to an ingrained litigation culture. The inadequacy of ADR measures as a general concept is further hampered by the intricacy of some issues or the rising number of players. Despite these hurdles, the assessment of advantages and disadvantages leads to the extraction that ADR methods are promising mechanisms in the future(15) (22).

Table 1. Brief explanatory of circumstances and outcomes of each case study

S. No.	Name of case study	Country	Type of ADR	Topic area	Outcome of process	Notes
1-	IPPC Permit for	Estonia	Negotiation	Industry, trade,	The parties had reached	The end outcomes

	the Kunda Pulp Plant Factory			enterprises, nature conservation	an agreement to change the conditions, and a new IPPC was issued.	involved some substantial concessions on both sides and no side was entirely satisfied with the final results.
2-	Returning the Protected Status to Natural Areas in the Lviv Region	Ukraine	Mediation-like	Nature conservation	The discussions that led to an emerging panel of key parties entrusted with examining and determining the condition of protected areas	Media campaigns have played a facilitating role that led to ease the success of process.
3-	Znesinnia Regional Landscape Park Versus Electric Power Supplier	Ukraine	Negotiation	Land use planning, nature conservation, power industry	The conversations yielded long-term answers to the larger issue	Impact of public interest might persuade parties to reach an acceptable solution through mediation
4-	Saaremaa Deep Harbour	Estonia	Negotiation	Land use planning, traffic, transportation, nature conservation	The parties had failed to reach to a possible resolution.	ADR is not a viable choice when either side is reluctant to reach a solution
5-	Szentgal Regional Landfill	Hungary	Cooperative Discourse	Waste management	No outcomes were reached	lack of the trust between the conflicting parties is the main reason leads to failure of ADR methods in environmental conflicts
6-	Route 10	Hungary	Mediation-like	Urban and land use planning, traffic, trade and enterprises, neighbourhood conflict	None of the sessions succeeded in altering the stakeholders' thoughts, therefore no agreement was reached.	Involvement of a large number of players and absence of influential entities led to failing ADR in settling this case

4.2 Role of ADR methods in environmental issues in India

India is one of the countries that were liberated from locus standi rule in terms of environmental issues. This rule states that courts are not allowed to allow anyone to participate in a forum unless they can show that some of their rights were violated. This means that only those who have been injured can participate in the court proceedings. The liberalisation of the classical locus standi rule offered environmental NGOs and civil society groups a huge advantage in approaching the Court in public interest cases including environmental cases when the injured parties were marginalized or difficult to prove or they can participate in the case as a third party or jointly with the injured party. As a result, environmental administration bodies by pursuing ADR methods can help to mitigate the harsh critiques directed at the concept of public interest litigation(23).

In fact, administrative ADR, which lets public interest groups bring up environmental issues, is meant to ease critics' worries about the courts' ability to choose whether or not to accept these actions. This takes a lot of pressure off the courts and makes environmental justice easier for the public to access. A precedent for including non-legal specialists from the scientific community in environmental lawsuits was set in 1986 in India. There are 21 to 41 members of the Indian National Green Tribunal, which was founded on October 18, 2010. The chairman (judicial), 10 to 20 full-time judicial members and 10 to 20 experts are all selected by the Central Government. Strictly competent judges and technical specialists are needed to provide a proper balance(23)(24).

It is mandatory in India to file a claim for relief and compensation under the National Green Tribunal Act 2010 no later than five years from the initiation of the original action of supposed damage. Because environmental concerns entail intricate biochemical and ecological processes, the National Green Tribunal in India has a special feature, it is allowed to design its own procedures to circumvent the rigour of the Civil Procedure Code and the Indian Evidence Act while collecting and recording evidence, which may not be appropriate for making a reasonable and fair evaluation of environmental damage in such complicated procedures. There is a Civil Court-like authority over the National Green Tribunal when it comes to summons, attendance, testimony on affidavit, and examination under oath, as well as the ability to give interim and ex parte injunctions. Through stare decisis, the National Green Tribunal Act included strict responsibility, precautionary, and polluter pay as part of Sustainable Development management. As a rule, civil courts respect and follow the National Green Tribunal's rulings(24).

5. The standard of justice, administration and representativeness of environmental ADR

Acceptance of ADR methods as mechanisms for solving environmental challenges contingent on two pillars, efficiency and justice, perhaps the efficiency of ADR methods does not need so much debate, whereas ADR methods have proved their efficiency for addressing issues including environmental issues, but on the other hand, ADR methods were accused of not meeting the standard of justice as the courts apply or at least concern about efficiency more than justice. That may be rational for some authors, whereas the ADR process depends on the will of parties who define how this process will go on. Owen Fiss was one of the first to raise concerns about the extent of justice of ADR methods in his article, “Against Settlement”(25).

In other words, the will of parties on which ADR methods are based may reduce the standards and conditions of justice, while the courts or judicial authorities cannot ignore these standards due to the courts cannot invoke the pretext of the will of the parties. In addition, the absence of accountability of third parties in private conflict settlement may will led to the loss of several legal values and principles which are indispensable in public litigation, such as the neutrality of third parties.

Another issue revolves around the administration of the process of ADR methods, whereas the consensual and private nature of ADR methods will increase the non-ability of inspection by official authorities and the public, in other words, the level of transparency in ADR method is less compared with judicial methods. In environmental conflicts, the issue of representativeness of all parties is considered the most important element, especially if it is put in mind the subsequent influence of these conflicts, it is necessary to secure public and representative stakeholder engagement in environmental

disputes. In the field of sustainable development, this element should expand to include the issue of future generations' representation(26).
To discuss these issues, first of all, it should be mentioned, that ADR methods are not completely based on the will of parties, ADR methods are recognized and established by laws and regulations issued by legislature authorities in most countries, these laws and regulations administrate the ADR methods and put forth some limitations and restrictions on ADR process, in other words, not everything the parties have agreed upon, will be admissible, for instance, ADR method will not be applicable if violate the public order or public interest, as well as not all issues and disputes, can be arbitrated.
Putting into consideration what has been mentioned above, it can be said that mentioned problems are not concentrated in ADR methods themselves, but in the laws and regulations which manage and regulate ADR methods. The extent of freedom of the ADR method can be expanded or shrunken by these laws. These laws can impose more limitations and restrictions on the ADR process to make it more identical to the justice standard for example it can be set forth that third parties should be specialized in the type of disputes which they mediate or arbitrate. Most ADR laws in different countries already state some measures to ensure the impartially and neutrality of thirds parties and regulate some repercussions if has been corroborated that arbitrators or mediator were biased, the most important consequence is annulment or set aside of final agreement or award. That can be seen in the most important international arbitration laws such as Article 34 of UNICITRAL(27) and Article 52 of the ICSID Convention(28).
With its adoption as a part of the judicial culture, the issue of the administrative system of ADR processes is crucial in the present time. The administrative side of the procedures has grown more systematic since most member states have adopted ADR as a legal resort and many organisations and institutions have been established to facilitate the administrative aspects of ADR methods, such as Arbitration Centres and Mediation Councils. In addition, in this contemporary time, there is a stream forward institutionalising ADR methods, in India for example, the 2019 amendment law(29) states that tasks that were entrusted to the supreme or high courts should be entrusted to an arbitral institution, the same will ease the inspection function of judicial authorities and public.
Scholars believe that the solution to the representativeness issue should comprise all its aspects, in the light of environmental issues, the fact is that any change in the environment will hold an effect on the upcoming generation, it could be said the limitation of the effect of environmental change cannot be defined or familiar precisely, this fact lies the responsibility all parties to a dispute, including the neutral, to consider the interests of all those who may be harmed. Especially if it is put in mind, the settling of environmental disputes is not just a resolving dispute, in many cases, settling of environmental disputes leads to new regulations and establishing new approaches to deal with issues of protecting the environment(22). To ensure that the representativeness in settling environmental disputes by the ADR mechanisms will reach the largest limit, that needs to fill the lack of knowledge and acceptance of ADR methods in terms of public, ADR methods need to more spread by education and training and activate the role of media campaigns.

6. Conclusion

Although the fact that the nature of environmental conflicts is complicated and intricate. Arbitration, mediation, conciliation, and negotiation, the four primary types of ADR, have all been used successfully by various countries in the settlement of environmental conflicts. Even when ADR methods have been failed in reaching a final settlement of some disputes, some positive results were achieved, such as convergence of the points of view and reduce the range of conflicts, furthermore, in the most complicated conflicts, adopting ADR methods unfolds the obstacles and barriers of the signing of a settlement, such as the lack of trust between the parties, lack of participation of key stakeholders and concerned parties, or participating of many parties with the absence of active ones. And as it is known, identifying the reasons for failure is a step toward success. As a result, it can be said that the adoption of ADR methods in environmental conflicts will be successful, if not fully, at least partially. Therefore, the main problem is the obstacles that hamper the adoption of the ADR mechanism. The most crucial obstacle to ADR adoption may be brought back to lack of acceptability which results from a lack of awareness and knowledge. In light of the fact, Settling environment conflicts are a part of the environment maintenance approach which in its role is an essential base of sustainable development on one hand and accessing justice and further the rule of law is another essential base on another hand, that makes clear the use of ADR methods will be indispensable, but the issue does not only stand on this limit, it should be extended to open the prospects for boosting the current ADR methods and creating new ADR methods as a result of the principle "constant change entails constant creation and innovation".

References

[1] Awuah KGB, Abdulai RT. Urban Land and Development Management in a Challenged Developing World: An Overview of New Reflections. L 2022, Vol 11, Page 129. 2022 Jan 14;11(1):129.
[2] Dodds F. Maurice Strong – From Poverty to UN Under-Secretary General and Father of Sustainable Development – The Rio Earth Summit. Heroes Environ Dipl. 2022 May 3;111–32.
[3] Hens L, Nath B. The world summit on sustainable development. springer. 2005.
[4] Kostetckaia M, Hametner M. How Sustainable Development Goals interlinkages influence European Union countries' progress towards the 2030 Agenda. Sustain Dev 2022.
[5] Hope KR. Peace, justice and inclusive institutions: overcoming challenges to the implementation of Sustainable Development Goal 16. https://doi.org/101080/1478115820191667320. 2020 Jan 2;32(1):57–77.
[6] Mutuku B, Alaii J, Krumeich A, den BH van, Westerndorp I. FORMAL JUSTICE AND ALTERNATIVE DISPUTE RESOLUTION ON LAND BASED CONFLICTS IN KENYA. Int J Gend Stud. 2022 Apr 1; 7(1):1–19.
[7] Lindsey I. An overview of Sustainable Development Goal 17. Routledge Handb Sport Sustain Dev. 2021 Dec 8;429–41.
[8] Mensah J. Sustainable development: Meaning, history, principles, pillars, and implications for human action: Literature review. Cogent Soc Sci. 2019 Jan 1; 5(1).
[9] Cohen H, Alberstein M. Multilevel Access to Justice in a World of Vanishing Trials: A Conflict Resolution Perspective. Fordham Urban Law J. 2019;47.
[10] UNEP Implementing Principle 10 of the Rio Declaration (2016, August 19). https://www.unenvironment.org/news-and-stories/story/unep-implementing-principle-10-rio-declaration. Available from: https://www.unep.org/news-and-stories/story/unep-implementing-principle-10-rio-declaration
[11] Jerzy Jendrośka, Lorenzo Squintani MR. The Courts as Guardians of the Environment – New Developments in Access to Justice and Environmental Litigation — the University of Groningen research portal. university of groningen. 2020.
[12] Secretary-General UN. "Strengthening the role of mediation in the peaceful settlement of disputes, conflict prevention and resolution" (2012, June 25). https://peacemaker.un.org/sites/peacemaker.un.org/files/SGReport_StrenghteningtheRoleofMediati. Available from: https://peacemaker.un.org/node/81.
[13] Minamata Convention on Mercury (n.d.) http://www.mercuryconvention.org/Portals/11/documents/conventionText/MinamataConvention on Mercury_e.pdf.
[14] Ruis B. Alternative Dispute Resolution in Environmental Cases United Nations Environment Programme. https://www.unece.org/fileadmin/DAM/env/pp/a.to.j/TF6-2013/4_Ruis_ADR.pdf.
[15] M. van der Bank CM van der B. Alternative Dispute Resolution in the Settlement of Environmental Disputes in South Africa. World Academy of Science, Engineering and Technology International Journal of Humanities and Social Sciences Vol:11, No:11, 2017.

2017.
[16] Maser C, de Silva L. Resolving Environmental Conflicts : Principles and Concepts. Resolv Environ Conflicts. 2019 May 6 .
[17] India Code: Arbitration and Conciliation Act, 1996. https://www.indiacode.nic.in/handle/123456789/1978?sam_handle=123456789/1362
[18] Artamevia D, Putri E. Conflict and conflict management. IOSR J Bus Manag. 2013. www.iosrjournals.org
[19] Crawhall N, Silverman A, Crawhall# N. ACCESS TO JUSTICE AND THE RIGHT TO SUSTAIN NATURE. Int UNION Conserv Nat Comm Environ Econ Soc POLICY. 2016.
[20] Handler M, Purker L, Dana Carmen Romanescu, Tingas K. promoting environmental mediation as a tool for public participation and conflict resolution. from: www.oegut.at
[21] Rev. JS-PEL, 2007 U. Alternative dispute resolution in environmental enforcement cases: A call for enhanced assessment and greater use. Pace Environ Law Rev. 2007.
[22] Tharakan RJ, Lahoti K. Conquest of Sustainable Development through Reformative Mechanisms of ADR. Eur J Sustain Dev. 2019 Oct 1 ;8(5):254–254.
[23] Abdullah NC. Going Green in Urbanisation Area: Environmental Alternative Dispute Resolution as an Option. Procedia - Soc Behav Sci. 2015 Jan 27;170:401–8.
[24] Bakshi, P. & Yadav M. New judicial roles and green courts In India. Paper presented at the Ninth International Conference on Environmental Compliance and Enforcement, British Columbia, Canada. 2011.
[25] Brown JG. Ethics in Environmental ADR: An Overview of Issues and Some Overarching Questions. Valparaiso Univ Law Rev. 2000;34.
[26] MENKEL-MEADOW C. Ethics in Alternative Dispute Resolution: New Issues, No Answers from the Adversary Conception of Lawyers. South Tex Law Rev. 1997.
[27] 34 - UNCITRAL Model Law on International Commercial Arbitration (1985). https://www.jus.uio.no/lm/un.arbitration.model.law.1985/34.html.
[28] 52 - Convention on the Settlement of Investment Disputes Between States and Nationals of Other States - International Centre for Settlement Of Investment Disputes, Washington 1965.
[29] The Arbitration and Conciliation (Amendment) Bill, 2019. https://prsindia.org/billtrack/the-arbitration-and-conciliation-amendment-bill-2019.

A Study of Cost Comparison of Flexible Pavement Using Different Soils

Sridhar Bhupathi[1], **Jayasree Kuchu**[2] **and Sastri M V S S**[3]

[1,2,3] VasaviCollege of Engineering (A), Hyderabad-500031, India.

E-mail. mvss.sastri@staff.vce.ac.in

Abstract. India is home to the world's second-largest road network. It has a diverse range of soil types, each with its own set of conditions and characteristics, on which the road network is built. IRC: 37-2018 was used for the Design and Cost Analysis of Flexible Pavement for five distinct types of soils: Black cotton soil, Alluvial soil, Sandy soil, Red Laterite soil, and Murram soil. The Resilient modulus which depends on CBR value of the soil has been found to influence the pavement thickness, strength and performance. The project's cost is determined by the pavement thickness, which again is determined by the soil's CBR value. By employing a soil with a high CBR value, the project's pavement thickness and cost will be decreased. On poor or problematic soils, soil stabilisation techniques should be employed to improve CBR value and save project expenses. According to the findings, employing the CBR value improves the pavement's strength and hence saves time when planning a pavement.

Keywords: Resilient Modulus; CBR value; Different Soils; Flexible Pavement; Cost;

1. Introduction

A pavement structure is composed of various layers superimposed on each other designed to take loads acting on the pavement and able to resist on the existing subgrade. Improper design causes early pavement failure, lowering the quality and performance of roadways. The performance of the pavement has been assessed over a life time and the failure in the design aspects would overall impact the pavement. Flexible pavements are preferred over rigid pavements, because they may be strengthened and upgraded in stages as traffic increases, as well as being able to grind and recycle their surfaces for rehabilitation [1].

The literature provides various approaches to flexible pavement design [2] that are empirical, mechanistic, and mechanistic-empirical approaches. Seven distinct pavement design methods were investigated, with each method being utilized to determine the flexible pavement thickness. The authors employed the Fuzzy PROMETHEE decision approach to find the best way for creating the pavement and discovered that Road Note 29 (RN29) is the best option, whereas HD26/01 will be suggested for new road design. Layered system concept has been adopted by many approaches in which the uppermost layer is of superior quality with high modulus of elasticity to withstand the heavy loads from the surface. The authors [3] discussed various methods for development of design charts based on finite element analysis varying the thickness of pavement. The pressure and elastic modulus of subgrade were also taken as input parameters. The charts are presented in a manner such that, if any two variables are known other can be determined. The aim of the study is to limit developed stresses caused by wheel loads to the underlying layers not to exceed the bearing capacity of the sub-grade. A review of recycled materials that were used in different layers of flexible pavements was presented by [4]. Despite having reduction in cost suggested possible ways to promote recycled materials in flexible pavement. A correlation with soil moisture content and subgrade resilient modulus changes was observed in the pavement [5]. A comparative evaluation of pavement suitability based on several characteristics such as material, loading, longer life, and cost efficiency was presented [6] and [7]. The IRI (International Roughness Index) approach was utilized to study the damage to flexible pavement [8] and it was compared to the Surface Distress Index method adopted in Indonesia. When employing statistical analysis to compare and examine the data, the IRI approach is also recommended. [9] investigated the feasibility of combining fly ash and oil shale ash to modify silty clay in subgrade as a result of the combination was investigated [9]. The tests revealed that fly ash and oil shale ash considerably improve the strength parameters of silty clay.

1.1. Objectives
The main objective is to compare the variation in pavement thickness with alternative materials and Cost of the pavement for five chosen types of soils along with soil stabilization methods of subgrade soil for five different chosen soils based on resilient modulus in turn depends on CBR value.

2. Methodology of study
The data was collected for the traffic parameters for a two-lane road with carriageway and also CBR values of five chosen types of soils. The criteria selected for estimating design traffic as per IRC 37 – 2018 was based on post construction commercial traffic with a consideration of annual growth over the design life.
The following parameters for the design of flexible pavement thickness was considered

- A two-lane road with 7.5m
- Design period of (n= 20 Years)
- Moderate traffic with P=1000 PCUs
- Growth rate, r =6%
- The Lane distribution Factor is taken as 50% in each direction (D=0.5).
- Vehicle Damage Factor (F=5.2%)
- Construction period (x=2 years)
- Projected number of axles for the design life = 42 msa, and
- CBR value was determined for different soils.

Since the determination of CBR is easy as compared to Resilient Modulus as specified in [10], this value was determined from the equations proposed in the code. From the cumulative number of standard axles, the pavement thickness of flexible pavements was found using IIT PAVE software. It is s an improved version of FPAVE which is developed by research scheme R-56 of MORTH. This is a multilayer analysis programme used for design, analysis of the flexible or bitumen pavement using IRC: 37-2018 guidelines. In this software we enter thickness of pavement layers, loads applied over the surface of pavement, tyre pressure, spacing between the wheels and Poisson´s ratio as inputs. After running the software actual horizontal tensile strain and vertical compressive strains at critical points of the pavement are found. Before carrying out IITPAVE analysis, the flexible pavement is designed as per IRC: 37-2018 guidelines and the thickness of each layer in pavement structure are determined. The strains obtained are compared with the allowable strains. If the actual strains obtained from IITPAVE software is less than the allowable strains obtained from IRC: 37-2018, then the given pavement thickness is said to be safe and satisfied.
The relation between CBR and resilient modulus as per the [10] is

$$M_{RS} = 10.0 \times \text{CBR for CBR} \leq 5\% \tag{1}$$

$$M_{RS} = 17.6 \times (\text{CBR})^{0.64} \text{ for CBR} > 5\% \tag{2}$$

Design resilient modulus of granular layer= $M_{RG} = 0.2(h)^{0.45} \times M_{RS}$ (3)

where M_{RS}= Resilient modulus of subgrade soil (in MPa)
CBR= California bearing ratio of subgrade soil (%)
h=thickness of granular layer in mm
The strains, stresses and deflections caused at different locations in a pavement by a uniformly distributed loads over the circular contact area at the surface of the pavement was computed using this software and a sample calculation sheet was shown in Figure.1. The pavement inputs required to analyze were:

- Cross section of pavement layers – number of layers
- All layers' resilient modulus or elastic modulus values are in MPa.
- The value of the Poisson's ratio for all layers.
- Thickness of all layers (in mm) except subgrade and Tyre pressure is in MPa.
- Wheel load in Newton.

2.1. Fatigue failure check
The induced horizontal tensile strain (ε_t), taken should be considered at the bottom of the bituminous layers computed at two radial distances of '0' and '155with the maximum among of the tangential and radial strains.Resilient modulus of 3000MPa with effective binder volume of 11.5% and air void content of 3% is considered in the fatigue failure analysis. Allowable horizontal tensile strain at the bottom of bituminous layer was computed using equation.

$$N_f = 0.5161 \times C \times (10)^{-4} \left(\frac{1}{\epsilon_t}\right)^{3.89} \times \left(\frac{1}{M_{Rm}}\right)^{0.854} \quad (4)$$

2.2. Rutting failure check
The vertical compressive strain (ε_v) was taken as the maximum of the two strain values from the between granular layer and subgrade. Allowable vertical compressive strain on subgrade is computed for the calculated design traffic of 42 msa. 90% reliability has been assumed for the analysis

$$N_R = 1.41 \times 10^{-8} \left(\frac{1}{\epsilon_v}\right)^{4.5337} \quad (5)$$

The induced vertical compressive strain and horizontal tensile strains must be less than the allowable vertical compressive strain and horizontal tensile strains for safe design.

3. Soil Classification
Results of the various index property tests conducted on different soil samples are given in Table 1. The disturbed soil samples were collected from five different locations in Telangana state. Lab tests were conducted on the soil samples to determine its index properties. They are used to classify the soil as per the Indian Standard Soil Classification System (ISSCS). Based on the index properties of soil such as particle size distribution curve, soaked CBR values and plasticity characteristics, soil samples were utilized in the experimental work. The picture of the soil sample collection and its tests are shown in Figure 1.

Figure 1. Soil sample collection and its testing in lab

3.1. Types of tests

The various types of tests conducted, and their lab results are shown in Table 1.

Table 1. The different types of tests conducted on the soil samples

Name of the test	Black cotton soil	Alluvial soil	Sandy soil	Red Laterite soil	Murram soil	IS code
Specific gravity	2.4	2.6	2.67	2.69	2.65	[11]
Liquid limit	65%	26.60%	-	76%	28%	[12]
Plastic limit	30%	24%	-	40%	18%	[12]
Plasticity index	37%	3.6%	-	36%	10%	[12]
OMC	17%	13%	-	26%	11%	
Maximum dry unit weight (kN/m^2)	17.4	21.4	16.99	14.98	19	[13]
Unconfined compressive strength (kPa)	53	46.8	-	70	-	[14]
CBR % (Soaked)	3	4.83	8	10	15	[15]

The Strains computed in output results were checked with allowable strains for fatigue and rutting conditions to know whether the designed pavement thickness was sufficient to withstand the loads passed over it.

Table 2. Comparison of pavement thickness and different soils

S.No.	Parameter	Black Cotton soil	Alluvial soil	Sandy soil	Red Laterite soil	Murram soil
1.	The Effective CBR value %	3	5	8	10	15
2.	Cumulative number of standard axles (msa)	42	42	42	42	42
3.	Resilient modulus of soil (MPa).	30	49.3	66.6	76.8	99.58
4	Resilient modulus of subbase granular layer (MPa)	130	154.10	208.185	240.131	278

Variations in Pavement Thickness with Alternative materials have been used within the layers of pavement. The alternatives considered for analysis are granular subbase, Cement treated sub base and base, Foam bitumen and RAP and Cement treated (CT) sub base and CT sub base with crack relief layer. The addition of alternative materials improves the strength of the subgrade [4,9]

Table 3. Variation in PavementThickness with Alternative materials

S. No	Alternative materials	Black Cotton soil	Alluvial soil	Sandy soil	Red Laterite soil	Murram soil
1	Granular sub-base	810	625	600	590	570
2	Cement treated sub-base and base with crack relief layer	505	500	500	500	500
3	Cement treated sub-base and base with **SAMI** layer	450	450	435	435	425
4	CT subbase treated with RAP and foam bitumen	430	430	410	400	400

The quantity of materials required and the total cost of pavement for flexible pavements designed for five different soils was calculated and a sample calculation for black cotton soil was enclosed in Table 4 and Abstract estimate of flexible pavement for different soils is shown in Table 5. The Schedule of rates of 2020-21 has been considered for the study.

4. Design Cross section of the pavement

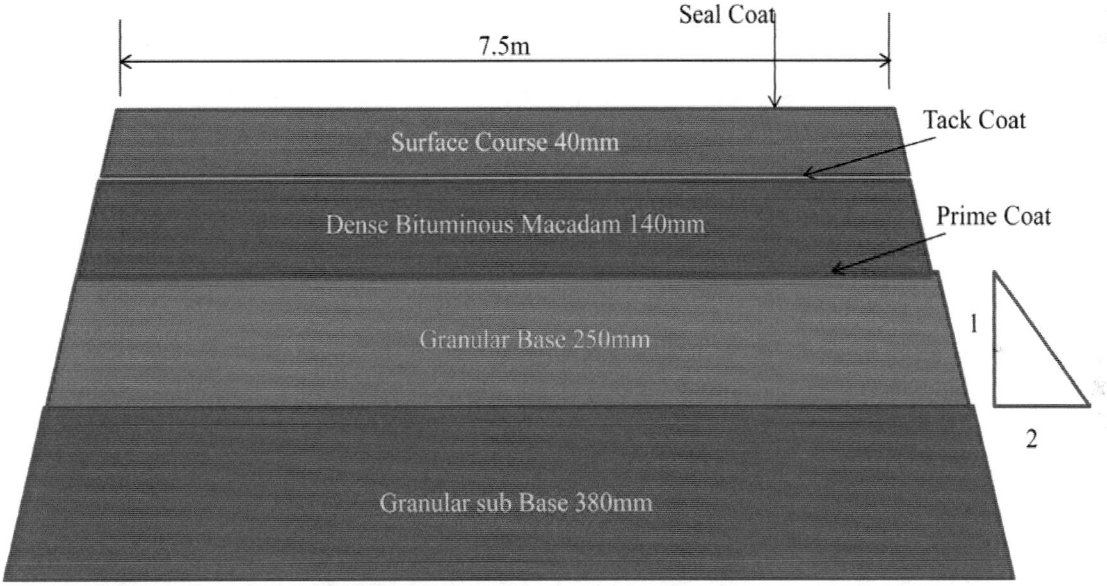

Figure 2. Cross section of the pavement

Table 4. A sample detailed estimation of flexible pavement for black cotton soil

S. No	Description of item	no	Length (m)	Breadth (m)	Depth (m)	Quantity	Rate (Rupees)	Total cost (Rupees)
1.	Excavation	1	1000	11.34	0.95	10773m^3	107	11,52,711
2.	Compaction of sub-grade	1	1000	11.34	0.3	3402m^3	450	15,30,900
3.	Granular Sub-base (laying, spreading and compaction of stone aggregate)	1	1000	9.98	0.38	3792.4m^3	1649	62,53,667

4.	Granular Base (laying, spreading & compaction of stone aggregate)	1	1000	8.72	0.25	2180m^3	1870	40,76,600
5.	Application of Prime coat with bitumen emulsion	1	1000	8.22	0.14	8220m^2	29.40	2,41,668
6.	Dense bituminous macadam (laying, spreading and compacting stone aggregate)	1	1000	7.94	0.14	1111.6m^3	7370	81,92,492
7.	Application of Tack coat with bitumen emulsion	1	1000	7.66	0.04	7660m^2	10.13	77,595.8
8.	Surface course	1	1000	7.58	0.04	303.2m^3	7560	22,92,192
9.	Application of seal coat with bitumen emulsion	1	1000	705	0.04	7500m^2	10.13	75,975
Total cost (Rupees)								2,38,93800.8

Table 5. Abstract estimate of flexible pavement for different soils

S.No.	Abstract of the estimate	Cost (Rupees)
1.	Abstract estimate of flexible pavement for Black cotton soil (in Rupees)	2,38,93800.8
2.	Abstract estimate of flexible pavement for Alluvial soil (in Rupees)	2,00,68,378.8
3.	Abstract estimate of flexible pavement for Sandy soil (in Rupees)	1,84,06,213.8
4.	Abstract estimate of flexible pavement for Red Laterite soil (in Rupees)	1,77,06,781.8
5.	Abstract estimate of flexible pavement for Murram soil (in Rupees)	1,02,04,552.8

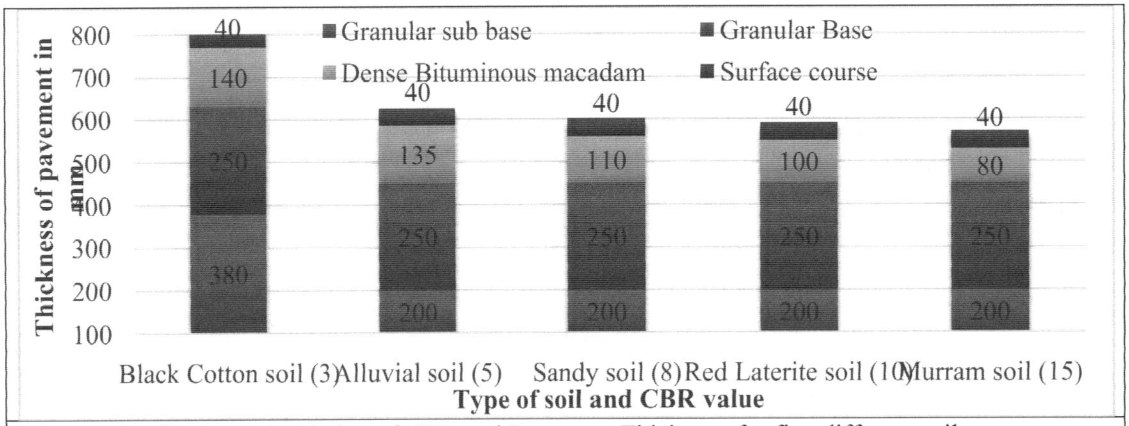

Figure 3. Variation of CBR and Pavement Thickness for five different soils

From Figure 3 it has been discovered that when the CBR value of the soil increases, the thickness of the pavement decreases. A well-known factor is as the value increases, the soil's strength also increases. As a result, the value of pavement thickness decreases.

From Figure 4 it is observed that the pavement thickness has decreased and hence, the cost of the pavement decreases. In Black cotton soil, lime stabilization enhances the CBR value and enhances the strength of the sub-grade, decreases the pavement thickness and also increase the strength of the pavement.

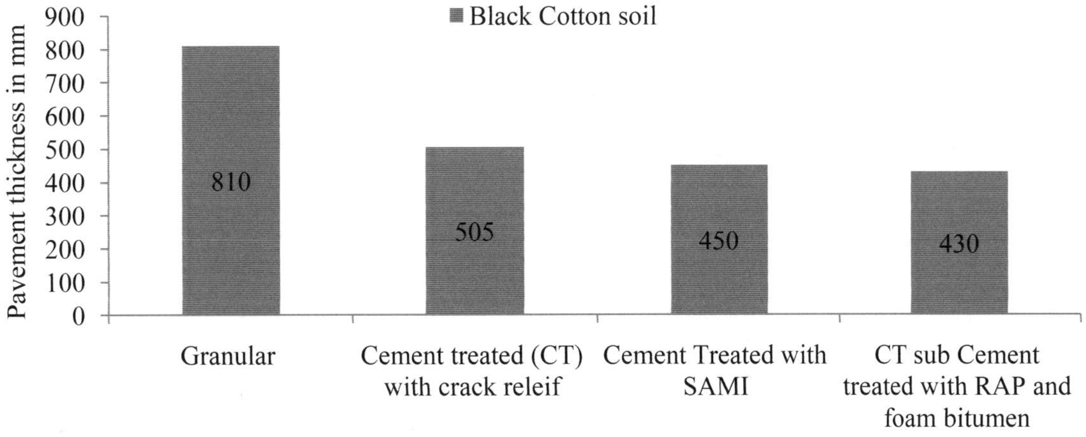

Figure 4. Graph between Pavement thickness and Alternative materials for Black cotton soil

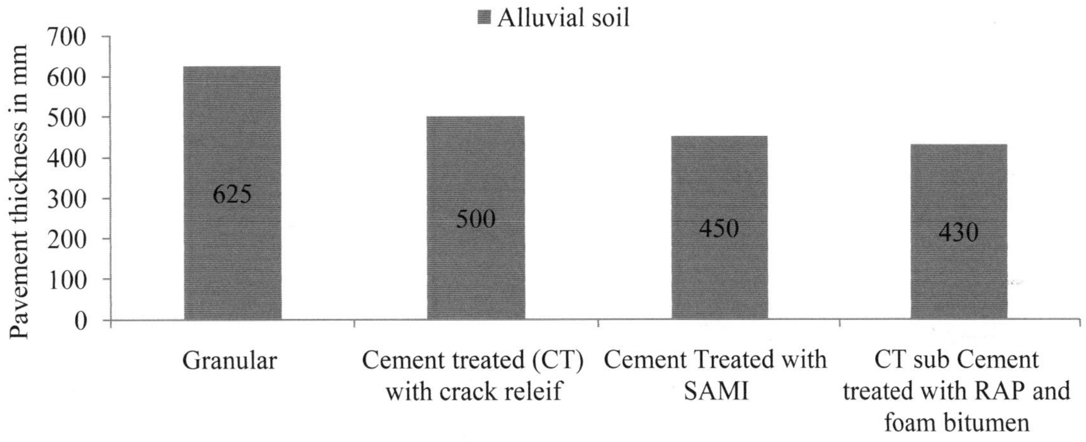

Figure 5. Graph between Pavement thickness and Alternative materials for Alluvial soil

From the above Figure 5 graph, it is observed that the pavement thickness decreased. Hence, the cost of the pavement decreases. Soil stabilization with Oil shale ash (OSA) and Fly ash (FA) in alluvial soil improves the CBR value, increases the strength of sub-grade, decreases the pavement thickness and also increases the strength of the pavement.

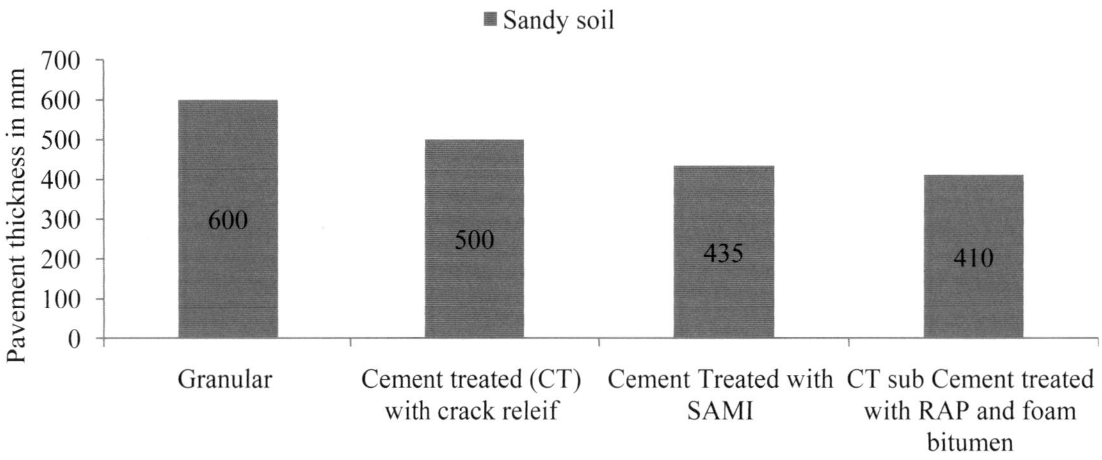

Figure 6. Graph between Pavement thickness and Alternative materials for Sandy soil

From Figure 6 it is observed that the pavement thickness decreases with the Alternative materials usage within the pavement layers and increases the performance of the pavement, the pavement cost decreases. Soil stabilization with Low Density Polythene sheets (LDPE) in Sandy soil improves the CBR value of soil, strength of sub-grade and decreases the pavement thickness and also strength of pavement increases.

Referring to the Figure7 it is observed that there is a reduction of pavement thickness. Hence, the cost of the pavement decreases. Soil stabilization with Cement and Sodium bentonite in Red Laterite soil improves the CBR value of soil, increases the strength of sub-grade, decreases the pavement thickness and also increases the strength of the pavement.

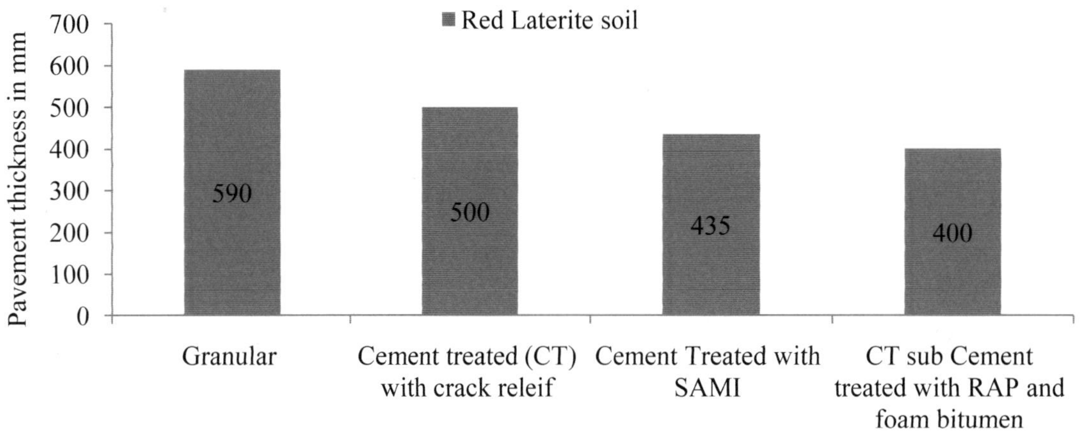

Figure 7. Graph between Pavement thickness and Alternative materials for Red Laterite soil

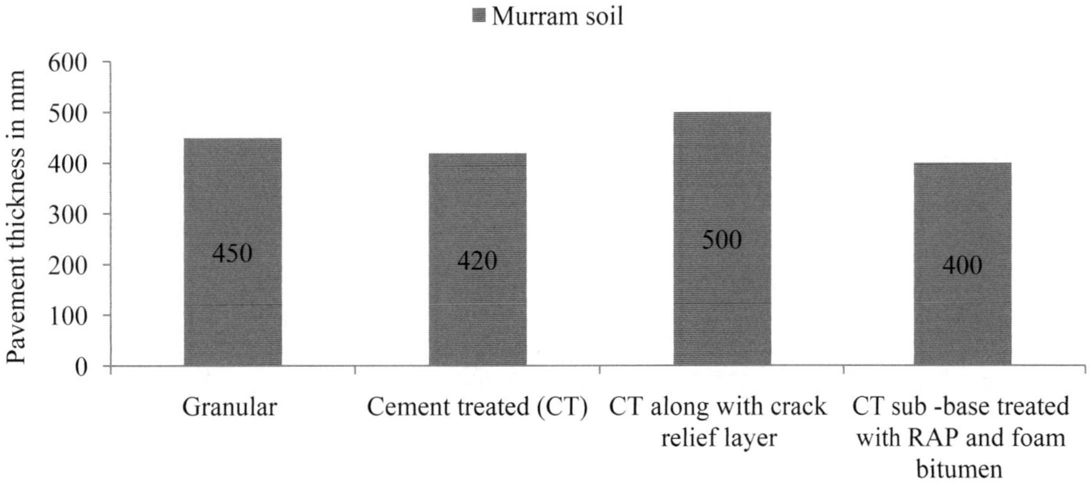

Figure 8. Graph between Pavement thickness and Alternative materials for Murram soil

From the Figure 8, it is observed that the pavement thickness decreases which helps in reducing the cost the pavement. Soil stabilization with Cement in Murram soil improvise the CBR value of soil, thereby increases the strength of sub-grade and decreases the pavement thickness, which increases the pavement strength.

Table 6. Comparison of Pavement thickness, Resilient modulus and Cost of Pavement

S.No.	Type of soil	Resilient modulus (MPa)	% Improvement in Resilient modulus	Cost reduction (%)
1.	Black cotton soil	30	0%	0%
2.	Alluvial soil	49.3	64%	-16%
3.	Sandy soil	66.6	122%	-23%
4.	Red Laterite soil	76.8	156%	-26%
5.	Murram soil	99.58	232%	-57%

Table 6 shows that as the resilient modulus are increased the cost required to lay the flexible pavement also decreases.

Table 7. Variation in Pavement Thickness and cost by Soil stabilization techniques

S. No	Type of soil	Stabilization technique	Improved resilient modulus (%)	Reduced pavement thickness (mm)	Cost reduction (%)
1.	Black cotton soil	Lime stabilization	15	450	44.31
2.	Alluvial soil	Oil shale ash & Fly ash	20	380	45.6
3.	Sandy soil	Low density polythene sheets	18	420	27.27
4.	Red Laterite soil	Cement& Sodium bentonite	22	350	26.91
5.	Murram soil	Cement stabilization	25	300	40.5

From the Table 7, if the soil was stabilized using the techniques discussed, the thickness of the pavement will decrease further it reduces the cost of construction thereby saving material as well as economy.

5. Conclusion

In the present investigation, design of flexible pavements and their cost estimation was carried out for five different soils i.e., Black cotton soil, Alluvial soil, Sandy soil, Red laterite soil and Murram soil. The following conclusions were drawn:

- Comparing the Resilient modulus value of soil and pavement thickness, it can be observed that Flexible pavement thickness is higher for poor or weak soils with lower resilient modulus and less for soils with good or higher resilient modulus values for same traffic conditions.
- The Pavement thickness can be reduced by introducing alternative materials within the layers of flexible pavement and strength of the pavement can be improved.
- The Cost of the pavement depends on the Pavement thickness which intern depends on type of soil and its resilient modulus value.
- Soil stabilization improves the strength of soil and resilient modulus value which reduces the pavement thickness and cost of pavement.
- Black cotton soil when stabilized with lime has an improved resilient modulus value of 15%, pavement thickness has been reduced to 450mm and total cost has reduced to 44.31%.
- Alluvial soil when stabilized with oil shale ash (OSA) and fly ash (FA) has an improved CBR value of 20%, pavement thickness has been reduced to 420 mm and total cost has reduced to 34.9%.
- Sandy soil when stabilized with Low density polythene (LDPE) has an improved CBR value of 18%, pavement thickness has been reduced to 430mm and total cost has reduced to 27.27%.
- Red Laterite soil when stabilized with cement and Sodium Bentonite has an improved resilient modulus value of 22%, pavement thickness reduced to 400mm and cost has reduced to 26.91%.
- Murram soil when stabilized with cement has an improved CBR value of 0% pavement thickness reduced to 350mm and cost has reduced to 40.5%.

References

[1] Y. H. *Huang* 1993*Pavement analysis and design*. Prentice-Hall, Incorporated.
[2] I. K. Umar, H. Gökçekuş, and D. UzunOzsahinş, "Comparative Analysis of Flexible Pavement Design Methods Using Fuzzy PROMETHEE," in *Application of Multi-Criteria Decision Analysis in Environmental and Civil Engineering*, D. UzunOzsahin, H. Gökçekuş, B. Uzun, and J. LaMoreaux, Eds. Cham: Springer International Publishing, 2021, pp. 173-182.
[3] S. Gill and D. K. Maharaj 2015. "Comparative study of design charts for flexible pavement," *Int Res Engineering Technology,* Vol. **2**, pp. 339-348.
[4] P. K. Gautam, P. Kalla, A. S. Jethoo, R. Agrawal, and H. Singh 2018. "Sustainable use of waste in flexible pavement: A review," *Construction and Building Materials,* Vol. **180**, pp. 239-253.
[5] F. U. Rehman and R. Balyan 2018. "Pavement Material Characteristics and its Influence on Stress and Strain Pavement Layer," *International Journal for Research in Applied Science & Engineering Technology* Vol. **6**, no. 7, pp. 346-355.
[6] S. Gill and D.K.Maharaj 2018. "Comparative study of flexible and rigid pavements by finite element method," *International Journal of Latest Research in Science and Technology,* Vol. **4**, no. 6, pp. 44-50.
[7] S. Jain, Y. P. Joshi, and S. S. S. Goliya 2013. "Design of Rigid and Flexible Pavements by Various Methods & Their Cost Analysis of Each Method," *International Journal of Engineering Research and Applications,* Vol. **3**, no. 5, pp. 119-123.

[8] H. A. Rahardjo and Suparman 2017. "The Effectiveness of IRI Compared To SDI System for Assessing the Quality And Performance Of Materials Used In Flexible Pavement In Java, Indonesia," *IOP Conference Series: Materials Science and Engineering,* Vol. **241**, p. 012007.
[9] H. Wei, Y. Zhang, J. Cui, L. Han, and Z. Li 2019. "Engineering and environmental evaluation of silty clay modified by waste fly ash and oil shale ash as a road subgrade material," *Construction and Building Materials,* Vol. **196**, pp. 204-213.
[10] IRC:37 2018. "Code of guideline for the design of flexible pavement," New Delhi, India.
[11] IS:2720-3 1987. ":(reaffirmed 2002):Methods of test for Soil: Determination of specific gravity: Section 1 Fine grained soils," Bureau of Indian Standards, New Delhi, India.
[12] IS:2720-5 1985. ":(reaffirmed 2006):Methods of test for Soil: Determination of Liquid and Plastic limit," Bureau of Indian Standards, New Delhi, India.
[13] IS:2720-7 1985. ":(reaffirmed 2006):Methods of test for Soil: Determination of water content-dry density relation using light compaction ", Bureau of Indian Standards, New Delhi, India.
[14] IS:2720-10 1991. "Methods of test for Soil: Determination of Unconfined Compressive strength," Bureau of Indian Standards, New Delhi, India.
[15] IS:2720-16 1987. "(reaffirmed 1997): Methods of test for Soil : Laboratory determination of CBR," Bureau of Indian Standards, New Delhi, India.

Acknowledgements
The authors express their sincere thanks to Principal and Management, Vasavi College of Engineering (A), Hyderabad for the help provided in giving the laboratory support and financial help.

Traffic Count Analysis during COVID-19: Case Study of Toll Plazas under Project Highway along Nagpur- Jabalpur National Highway (NH44) near Nagpur, India

Kalpana N Kawathekar (Thakare)[1] and V K Bakde[2]

[1]Research Scholar at Department of Architecture and Planning, Visvesvaraya National Institute of Technology, Nagpur andFaculty at Kavikulguru Institute of Technology and Science, Ramtek, Nagpur, Maharashtra, India.
[2]Associate Professor, Department of Architecture and Planning, Visvesvaraya National Institute of Technology, Nagpur, Maharashtra, India.

Email. [1]kalpana@students.vnit.ac.in; [2]vkbakde@arc.vnit.ac.in

Abstract. The Corona virus is originated in Wuhan, China, and has spread throughout the world since December 2019. Despite being close to China, India took two months to detect its first case. In months' time it reached major countries of the world. Since then, the social life of the people has been disrupted because of the lockdown situations. Senior members were shocked to see such a situation for the first time in their lives, as all activities have been affected by the restrictions on both domestic and international travel and transport.
This case study is an attempt to analyze the vehicle traffic counts at four toll plazas namely Khumari, KKBP, NBP and Borkhedi Toll Plaza located along Nagpur-Jabalpur National Highway (NH44) near Nagpur city from Maharashtra state of India during Session 2020 (during January to December) as an effect of COVID -19 under lockdown and unlocking situations. Tolls at Nagpur have been selected for being the national geographic centre with a hub of all activities. The study critically analyses the vehicle traffic count along with the comparative study of four toll plazas thereby making the analytical remarks with conclusions.

Keywords. Corona Virus, COVID- 19 effects, Socio-Economy and Health, Toll Traffic.

1. Introduction

The Corona viruses (CoV) are a large family of viruses that cause illness ranging from the common cold to more severe diseases. It was recognized as a new virus family in 1968 and a novel corona virus, a severe acute respiratory syndrome coronavirus-2 *'SARC-CoV-2'*, is also known as Corona Virus Disease 2019, 'COVID-19' for being detected in 2019. SARC CoV infection was first time identified in 2003 and has known to be originated in bats and transmitted to humans in China[1],[2],[3],[4]. The Wuhan seafood market in China was implicated as the epicentre of the COVID-19 pandemic identifying seven cases of pneumonia in Wuhan from 8th to 18th December 2019. With increasing number of cases through international transport communication in neighbouring countries and worldwide, the disease spread to several countries and it was declared a Public Health

Emergency of International Concern by the World Health Organization (WHO) on 31[st]January 2020[5]. The epidemic intensified and spread without boundaries and affected more than 215 countries and a few international conveyances, which led the WHO to declare COVID-19 as a pandemic on 11[th]March 2020[2]. Since then the human life has seen the restrictions as a participatory measure to COVID-19. Under COVID-19, the minimal preventative measures were social separation and mask wearing.

1.1COVID- 19 Effects

The overall number of corona positive cases in the world as of November 14th, 2021 was 252, 826, 597, with a total of 5,092,761 deaths due to COVID. With first and second places, the United States and India were at the top with corona deaths climb by 4,136 in the first week of November 2021, according to the World Health Organization [6].COVID 19 had resulted in a dramatic drop in social health over the world. The frequency of corona positive cases and deaths has increased dramatically after December 2019. For the interest of social safety and to prevent the spread of the corona virus, all socio-physical and economic activities have been halted locally and worldwide, with the slogans "Stay home, stay safe" and "Work from home" [7]. Economic activity has come to a halt, which has had a tremendous impact on state economies. Many people have lost their jobs as a result of the economy, and others' salaries have been cut as a result of the unlocking of employment activities[8]. Based on the amount of fatalities and health complications caused by its after effects, each new wave of Corona has been observed to be more deadly and effective. Thousands of children lost their parents and homes as a result of fatalities during all of these periods, and millions of families were destroyed [9]. Those that have succeeded are still struggling with the economical and health effects of the COVID-19 situation.

Transportation has also been affected by COVID-19. Transportation infrastructure is a source of worry for all citizens in COVID-19. During COVID-19, all public transit (airway, waterway, railway, subway, and bus transportation) in the world's countries has been completely shut down, and everyone be obliged to stay at home for their own safety resulting in a drastic decline in ridership volumes of up to 90% due to the lack of public transportation, dubbed a "lockdown," resulting in a 60% reduction in air pollution. In the event of an emergency health scenario, private autos with a small number of passengers were allowed to travel with social distance[10]. In the absence of public transportation, India and the world have experienced the greatest mobility, thanks to Indian workers/ labourers who used ancient transportation techniques such as walking and cycling to go up to 1700 kilometres to their homes [11]. India's state governments also offered public transit to assist them while returning home. Thus all traffic movements have been significantly reduced as a result of the COVID 19 lockdown.

1.2Need of the Study

The need of this study was identified to understand traffic flow in 2020, from January to December at tolls where international, interstate, business and local traffic is usually seen. For this, the four toll plazas on NH44 (also known as NH7) on the Nagpur-Jabalpur Highway in India near Nagpur were chosen for the case study which satisfied the requirement of study for its analysis for wise reasons as Nagpur is the natural logistics hub for the Indian Subcontinent to allow movement of international and domestic cargo. It hosts a “Zero Mile” and is an India’s geographic centre. It is the converging point of the nations’ railway and road network linking a vast hinterland. It has Asia’s largest industrial area, Butibori and the Hingna Industrial Estate located on the western fringes of the city. Nagpur hosts the Themal Power Stations at Koradi, Khaparkheda along with the National Thermal Power Plant at Mouda and operated by MSPGCL. The major commercial market areas in Nagpur are Sitaburdi, Itwari, Gandhibag and Mahal. Kalamna is one of the largest wholesale markets for oranges and grains in Asia. Nagpur has also established main centres for thousands of small and medium size businesses and industries (Corporation) and Multi-International Hub Airport at Nagpur, MIHAN, a Socio-

Economic Zone, SEZ project for overall development of Nagpur. In addition to these all, Nagpur also witnesses urbanization due to rurban migration (Census data , 2011)(GoI, 2021).

With the above establishments it is obvious that approaching highway roads have significance for Nagpur and its economic development. Also Nagpur is well connected by its Ring Road to the National Highways and Asian Highways. The highways have tolls at the intervals. The data of tolls on these highways give the direct information on the traffic and transport throughout the year. During COVID- 19 the traffic transport activities were stopped and released after instructions in stages. Thus study need was found to understand the sudden stop and later on gradual growth the traffic at the nearest four tolls on NH-44 which analyses the traffic at tolls with different situations viz. international traffic, traffic near industries zone, interstate boundaries and urban and rural residential localities.

2. Aim of the study

Due to lock down Traffic counts on all roads and at tolls have fallen as a result of the COVID- 19 traffic legislation. The research' purpose was to analyse traffic flow when activities were shut down and reopened in stages in 2020. Therefore, a case study on highway traffic and transport is done at four toll plazas on NH-44 Nagpur Jabalpur road in India during the January – December 2020 peack session of COVID-19.

3. Methodology

During the lockdown and unlock periods of COVID-19, analysis of traffic at four toll plazas on NH-44 (formerly NH-7) Nagpur-Jabalpur road is performed to understand the traffic count characteristics at tolls namely Khumari, KKBP, NBP and Borkhedi Toll Plaza. Comparative case analysis method is used for understanding the traffic count differences at selected four tolls on NH44 near Nagpur city during session 2020.

4. Significance and limitation

Socio economic growth of Nagpur city has a significant relation with roads and highways development. The highway development under the leadership of honourable Minister of Transport Shri. Nitinji Gadkari has made the distances shorter and accessible for the Nagpur city and its public. Therefore the selection of Nagpur city and the four tolls is significant for the study analysis. The study of four selected tolls give the overview of all situations in the region and can help understanding the effect of COVID-19 situation on traffic and hereby economy and the societal health. But the study is limited to traffic count at these four tolls and not intervened into economic and social factors further.

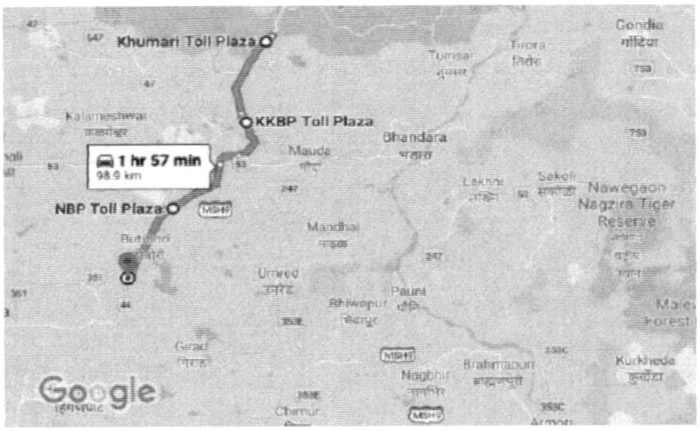

Figure1. Google Map of Nagpur showing NH 44 as outer ring road passing around the city along with the other Highways viz. Asian Highways, National Highways and Maharashtra State Highways [14].

5. Traffic count Study at four tolls under Project Highway at Nagpur on NH 44 (also known earlier as NH7)

The traffic count study was conducted on the Project Highway with divided carriageway on NH-44 (also known as NH-7) with part of Jabalpur - Nagpur Road, i.e. Nagpur - Kamptee - Kanhan bypass and part of Nagpur - Hyderabad Road (Total Length 117.078 Km)[13]. The project highway contains the four toll plazas positioned on NH44 in the part near Nagpur city as described above considering the vehicular movement from different directions namely Khumari Toll Plaza, KKBP Toll Plaza, NBP Toll Plaza and Borkhedi Toll Plaza from North to South (Refer figure 1). The data is based on five types viz. Car / Jeep / Van or Light Motor Vehicle are classified as Category 1; Light Commercial Vehicles, Light Goods Vehicles, or Mini Buses are classified as Category 2; Buses and Trucks (2 Axles) are classified as Category 3; Heavy Construction Machinery (HCM), Earth Moving Equipment (EME), or Multi Axle Vehicles (3 to 6 Axles) are classified as Category 4; and Oversized Vehicles are classified as Category 5. (7 or more axles).

5.1. Khumari Toll Plaza Traffic Count Study

The northernmost toll plaza, Khumari Toll Plaza, is located near Mansar hamlet on the Maharashtra–Madhya Pradesh state line. The Khumari toll Plaza monitors traffic on the Nagpur-Jabalpur and Nagpur-Mansar-Tumsar routes between Maharashtra and Madhya Pradesh and Maharashtra and Chhattisgarh.

Table 1. Vehicles category wise Monthly Traffic Count at Khumari Toll Plaza during January to December 2020[15].

Vehicle Cate gory	Jan-20	Feb-20	Mar-20	Apr-20	May-20	Jun-20	Jul-20	Aug-20	Sep-20	Oct-20	Nov-20	Dec-20
1	39,147	42,100	33,927	1,163	16,666	16,420	20,877	24,322	30,160	42,126	63,784	75,663
2.	6,142	6,656	5,038	1,065	9,286	2,380	2,451	2,382	2,962	3,535	3,658	4,154
3.	17,455	20,014	14,793	1,561	11,818	1,049	1,702	2,712	4,537	5,493	5,360	6,400
4.	43,515	43,936	32,830	9,773	34,870	2,486	3,373	5,501	11,058	17,109	17,259	17,451
5.	47	42	18	10	18	1	0	2	0	1	7	4
Total	1,06,306	1,12,748	86,606	13,572	72,658	22,336	28,403	34,919	48,717	68,264	90,068	1,03,672

5.2. Counting Traffic at the KKBP Toll Plaza

On NH 44, near Kanhan village, there is a toll called KKBP Toll Plaza. The KKBP toll Plaza monitors traffic flow from Khumari Toll to Nagpur and vice versa, as well as local traffic flow due to the presence of numerous villages nearby.

Table 2. Vehicles category wise Monthly Traffic Count at Khumari Toll Plaza during January to December 2020[16].

Veh. Cate Gory	Jan-20	Feb-20	Mar-20	Apr-20	May-20	Jun-20	Jul-20	Aug-20	Sep-20	Oct-20	Nov-20	Dec-20
1	14,046	14,163	10,258	786	8,921	9,681	10,371	13,096	13,548	15,672	22,358	26,692
2.	4,274	4,481	3,219	788	5,835	2,271	2,039	2,454	2,687	2,619	2,427	2,897
3.	9,390	10,140	8,055	1,462	10,029	2,924	3,334	3,986	4,079	4,952	4,895	5,060
4.	41,498	43,159	34,719	10,808	37,332	12,035	14,650	17,858	17,108	24,201	25,229	26,105
5.	85	35	34	8	28	22	12	12	48	14	19	35
Total	69,293	71,978	56,285	13,852	62,145	26,933	30,406	37,406	37,470	47,458	54,928	60,789

5.3. NBP Toll Plaza Traffic Count Study

NBP Toll Plaza is a toll plaza on NH 44 near Wardha Road crossings. Because of the large number of towns nearby and the National and Asian Highways crossing near the NBP Toll, the Plaza monitors traffic travelling locally, nationally, and internationally from and through Nagpur.

Table 3. Vehicles category wise Monthly Traffic Count at Khumari Toll Plaza during January to December 2020[17].

Veh. Cate Gory	Jan-20	Feb-20	Mar-20	Apr-20	May-20	Jun-20	Jul-20	Aug-20	Sep-20	Oct-20	Nov-20	Dec-20
1	24,399	25,885	18,637	1,703	19,071	20,640	22,745	27,828	33,905	34,514	43,263	48,117
2.	15,396	15,188	12,217	3,540	16,196	9,181	10,598	10,511	12,502	12,507	10,896	13,695
3.	29,314	31,553	25,708	4,604	29,785	20,487	23,296	24,356	27,645	28,607	25,970	28,255
4.	1,24,636	1,27,074	95,944	22,414	87,098	86,812	95,513	91,366	98,575	1,02,742	96,537	1,13,951
5.	151	138	94	20	50	109	97	53	87	81	73	120
Total	1,93,896	1,99,838	1,52,600	32,281	1,52,200	1,37,229	1,52,249	1,54,114	1,72,714	1,78,451	1,76,739	2,04,138

5.4. Borkhedi Toll Plaza Traffic Count Study

Borkhedi Toll Plaza is a toll station on NH 44 near the Butibori Industrial Area in Borkhedi village. It monitors traffic flow between the Asian and National Highways via Nagpur Road, as well as local traffic flow due to the large number of urban and rural populations connected to MIHAN and SEZ.

Table 4. Vehicles' type wise Monthly Traffic Count at Borkhedi Toll Plaza during January to December 2020[18].

Veh. Cate Gory	Jan-20	Feb-20	Mar-20	Apr-20	May-20	Jun-20	Jul-20	Aug-20	Sep-20	Oct-20	Nov-20	Dec-20
1	95,254	1,07,933	69,694	2,811	31,879	62,859	64,962	78,074	96,500	1,13,109	1,35,313	1,40,797
2.	18085	19,427	13,436	2,712	17,215	15,642	14,593	14,112	15,338	16,678	15,955	17,416
3.	29099	30,725	22,815	2,813	18,359	21,752	23,115	24,357	30,146	36,093	37,265	40,410
4.	70409	71,657	54,345	14,365	54,784	65,903	65,710	67,463	72,068	76,899	73,524	78,272
5.	130	162	88	1	31	73	42	58	70	51	43	45
Total	2,12,977	2,29,904	1,60,378	22,702	1,22,268	1,66,229	1,68,422	1,84,064	2,14,122	2,42,830	2,62,100	2,76,940

6. Analysis of Traffic Count at four toll Plazas

Based on the traffic count it is noticed that the total number of tolls collected in Session 2020 is 55,26,647 at the studied four toll plazas on NH44 highway near Nagpur city.

Table 5. Total number of vehicles through four toll plazas on NH44 at Nagpur during 12 months in session 2020.

Month - Session 2020	Total Vehicles at Khumari Toll Plaza	Total Vehicles at KKBP Check Toll Plaza	Total vehicles at NBP Check Toll Plaza	Total Vehicles at Borkhedi Toll Plaza (New Location)	Monthwise Total Count at all 4 plazas
Jan-20	1,06,306	69,293	1,93,896	2,12,977	5,82,399
Feb-20	1,12,748	71,978	1,99,838	2,29,904	6,14,468
Mar-20	86,606	56,285	1,52,600	1,60,378	4,55,869
Apr-20	13,572	13,852	32,281	22,702	82,407
May-20	72,658	62,145	1,52,200	1,22,268	4,09,271
Jun-20	22,336	26,933	1,37,229	1,66,229	3,52,727
Jul-20	28,403	30,406	1,52,249	1,68,422	3,79,480
Aug-20	34,919	37,406	1,54,114	1,84,064	4,10,503
Sep-20	48,717	37,470	1,72,714	2,14,122	4,73,023
Oct-20	68,264	47,458	1,78,451	2,42,830	5,37,003
Nov-20	90,068	54,928	1,76,739	2,62,100	5,83,835
Dec-20	1,03,672	60,789	2,04,138	2,76,940	6,45,539
Total	7,88,269	5,68,943	19,06,499	22,62,936	55,26,647

IOP Conf. Series: Earth and Environmental Science 1084 (2022) 012059 doi:10.1088/1755-1315/1084/1/012059

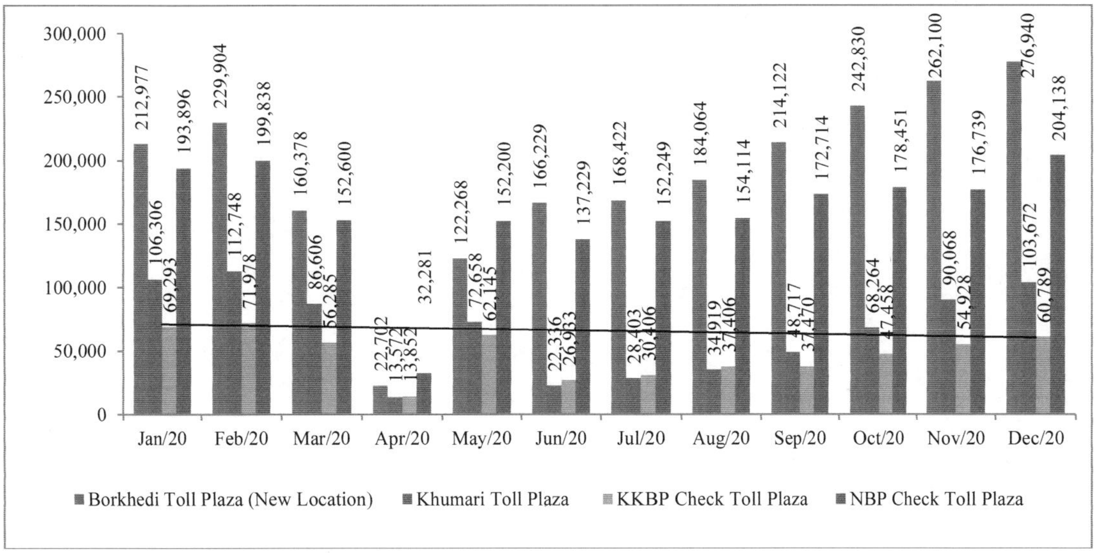

Figure 2. Monthwise number of vehicles crossing four Toll Plazas on NH44 at Nagpur during 2020.

The figure 2 shows that April 2020 month has seen the lowest count of tolls in all four tolls with the minimum count as 13,572 vehicles at Khumari toll plaza. Whereas, December 2020 month has seen the highest count of tolls for light motor vehicles and Heavy construction machinery vehicles but for light commercial and goods vehicles it is still to resume traffic count of pre-COVID months period.

Table 6. Toll Plaza ranking based on Maximum Total Vehicle Count.

Sr. No.	Toll Wise Total Vehicles count Rank in Session 2020	Toll	Vehicles Count	Remarks
1	Rank 1	Borkhedi Toll Plaza	22,62,936	Industrial Zone
2	Rank 2	NBP Toll Plaza	19,06,499	City and other Highways crossroad
3	Rank 3	Khumari Toll Plaza	7,88,269	Inter State Boundary
4	Rank 4	KKBP Toll Plaza	5,68,943	Local Villages nearby

Table 7. Overall Total Month count (Minimmum and Maximum) of Vehicle Traffic out of four Toll Plazas.

S. No.	Overall Month' Total count of Vehicles Traffic(Minimum and Maximum)	Month	Vehicles Count	Remarks
1	Overall Total Minimmum Month count of Vehicles Traffic	April 2020	82,409	Complete Lockdown nationally and internationally
2	Overall Total Maximum Month count of Vehicles Traffic	December 2020	6,45,539	Activities started slowly

Overall Total Month count of Vehicles Traffic Minimum is observed in the month of April 2020 with total vehicle count of 82,409. Overall Total Maximum Month count of Vehicles Traffic is seen in the month of December 2020 with the total count of vehicles as 6,45,539 numbers.

Table 8: Toll Plazawise Maximum and Minimum Vehicles traffic count month of 2020.

Toll Plaza	Maximum Vehicles traffic count	Maximum Vehicles traffic Count Month	Minimum Vehicles traffic Count	Minimums Vehicles traffic Count Month	Remarks
Khumari Toll Plaza	1,12,748	February 2020	13,572	April 2020	Inter State Boundary
KKBP Toll Plaza	71,978	February 2020	13,852	April 2020	Local Villages nearby
NBP Toll Plaza	2,04,138	December 2020	32,281	April 2020	City and other Highways crossroad
Borkhedi Toll Plaza	2,76,940	December 2020	22,702	April 2020	Industrial Zone

It shows that, Khumari toll plaza counted minimum tolls in April 2020 with 13,572 numbers and maximum in February 2020 with 1,12,748 numbers. KKBP toll plaza give the count of minimum tolls in April 2020 with 13,852 numbers and maximum in February 2020 with 71,978 numbers. NBP toll plaza counted minimum tolls in April 2020 with 32,281 numbers and maximum in December 2020 with a count of 2,04,138 vehicles. Borkhedi toll plaza gives minimum tolls count in April 2020 with 22,702 numbers and maximum count in December 2020 with 2,76,940 numbers.

Table 9. Comparative Analysis of Vehicles' Category wise minimum and maximum count month for individual toll plaza.

Vehicle Category	Min/ Max Vehicle Count (Month)	Khumari Toll Plaza	KKBP Toll Plaza	NBP Toll Plaza	Borkhedi Toll Plaza	Remarks
1	Minimum Vehicle Count (Month)	1,163 (April 2020)	786 (April 2020)	1,703 (April 2020)	2,811 (April 2020)	KKBP (April 2020)
1	Max Vehicle Count (Month)	75,663 (December 2020)	26,692 (December 2020)	48,117 (December 2020)	1,40,797 (December 2020)	Borkhedi (December 2020)
2	Minimum Vehicle Count (Month)	1,065 (April 2020)	788 (April 2020)	3,540 (April 2020)	2,712 (April 2020)	KKBP (April 2020)
2	Max Vehicle Count (Month)	9,286 (May 2020)	5,835 (May 2020)	15,396 (Jan 2020)	19,427 (February 2020)	Borkhedi (February 2020)
3	Minimum Vehicle Count (Month)	1,049 (June 2020)	1,462 (April 2020)	4,604 (April 2020)	2,813 (April 2020)	Khumari (June 2020)
3	Max Vehicle Count (Month)	20,014 (February 2020)	10,140 (February 2020)	43,936 (February 2020)	40,410 (December 2020)	NBP (February 2020)
4	Minimum Vehicle Count (Month)	2,486 (June 2020)	10,808 (April 2020)	22,414 (April 2020)	14,365 (April 2020)	Khumari (June 2020)
4	Max Vehicle Count (Month)	43,936 (February 2020)	43,159 (February 2020)	1,27,074 (February 2020)	78,272 (December 2020)	NBP (February 2020)
5	Minimum Vehicle Count (Month)	0 (July & September 2020)	8 (April 2020)	20 (April 2020)	1 (April 2020)	Khumari (July & September 2020)
5	Max Vehicle Count (Month)	47 (January 2020)	85 (January 2020)	151 (January 2020)	162 (February 2020)	Borkhedi (February 2020)

Note: Vehicles Category 1.: Car / Jeep / Van or Light Motor Vehicle
Vehicles Category 2.: Light Commercial Vehicles, Light Goods Vehicle or Mini Bus
Vehicles Category 3. : Bus / Truck (2 Axles)
Vehicles Category 4. : Heavy Construction Machinery (HCM), Earth Moving Equipment (EME) or Multi Axle Vehicle (MAV) (3 to 6 Axles)
Vehicles Category 5. : Oversized Vehicles (7 or more axles)

6.1. Analytical Remarks

The analytical remarks on the four toll plazas after studying the part of Project Highway on NH 44 (NH7) are as follows:

1. The total number of vehicles passing through all toll plazas during the 2020 session is 55,26,647.
2. During the main lockdown month of April 2020, in all 82,409 cars are crossed at four tolls, marking it as an overall Total Month Count (minimum) of vehicle traffic in all Toll Plazas. Due to the unlocking of activities, the Total Month Count (Maximum) of Vehicle Traffic in all Toll Plazas is noted in December 2020 with 6,45,539 vehicles counted.
3. Vehicle Count Toll Ranking in Session 2020: Borkhedi Toll Plaza recorded the largest vehicle traffic during the 2020 session, with a total of 22,62,936 cars, placing it top. It could be owing to the presence of an industrial zone near the toll plaza and the proximity of other highway highways to the Borkhedi toll plaza. In the 2020 session, NBP Toll Plaza was ranked second, with 19,06,499 vehicles. Its location at the intersection of city and other Asian and National Highways links could be the explanation. Due to its location near the state line, the Khumari Toll Plaza records a vehicle count of 7,88,269 in 2020. The KKBP toll plaza is located near Nagpur city and has both urban and rural populations. It ranks fourth in terms of vehicle count with 5,68,943.
4. Maximum and Minimum Vehicle Traffic Counts by Toll Plaza' in a Month of 2020 show that: Khumari and KKBP toll plazas saw similar vehicle counts in April and February 2020, with minimum counts of 13,572 and 13,852 and maximum counts of 1,12,748 and 71,978 respectively. However, NBP and Borkhedi Toll Plaza traffic have comparable characteristics, with minimum counts of 32,281 and 22,702 in April 2020 and maximum counts of 2,04,138 and 2,76,940 in December 2020, respectively. It shows that traffic at the Khumari and KKBP Toll Plazas has not returned to normal after COVID-19 until December 2020, although traffic at the NBP and Borkhedi Toll Plazas has grown by December 2020.

6.2. Vehicles' Category wise minimum and maximum count month for individual toll plaza gives the Analysis as follows

a. In the month of April 2020, the KKBP toll plaza saw the least number of vehicles 786 in category 1, which includes cars, jeeps, vans, and light motor vehicles; and
b. In the month of April 2020, the KKBP toll plaza saw the least number of vehicles 788 in category 2, which includes light commercial vehicles, light goods vehicles, and mini buses.
c. In the month of June 2020, the Khumari toll plaza had the lowerest number of vehicles (1,049) in category 3, which includes buses and trucks (2 Axles).
d. In the month of June 2020, the Khumari toll plaza had the lowest number of vehicles (2,486) in category 4, which comprises Heavy Construction Machinery (HCM), Earth Moving Equipment (EME), or Multi Axle Vehicle (MAV) (3 to 6).
e. In the months of July and September 2020, Khumari toll plaza saw no cars, however Borkhedi toll plaza saw 1 vehicle crossing in April 2020 under category 5, which includes Oversized Vehicles (7 or more axles).
f. In the month of December 2020, the Borkhedi toll plaza saw a total of 1, 40,797 vehicles in category 1, which comprises cars, jeeps, vans and light motor vehicles.
g. In the month of February 2020, the Borkhedi toll plaza saw a total of 19,427 vehicles in category 2, which includes Light Commercial Vehicles, Light Goods Vehicles, and Mini Buses.
h. In the month of February 2020, the NBP toll plaza saw a total of 43,936 vehicles in category 3, which includes buses and trucks (2 Axles).
i. In the month of February 2020, the NBP toll plaza saw the highest number of vehicles (127,074) in category 4, which comprises Heavy Construction Machinery (HCM), Earth Moving Equipment (EME), and Multi Axle Vehicle (MAV) (3 to 6 Axles).
j. In February 2020, the Borkhedi toll plaza had a maximum of 162 cars crossing under category 5, which includes Oversized Vehicles (7 or more axles).

7. Conclusions

The COVID 19 lockout has certainly had an impact on traffic movement along NH 44 in Nagpur city of the Maharashtra state, India in 2020. April and May 2020 were the months of total lockdown owing to the COVID 19 pandemic, with significantly limited traffic throughout these two months. The traffic data study provides that count cannot be regarded typical in both social and economic contexts. From June to December 2020, the number of vehicles on road has gradually increased. With the exception of the few cases where the usual traffic of February 2020 is still to be realized the traffic count, has seen the gradual increase to the regular minimum by December 2020. Reduced vehicular travel has had a significant impact on the states and nations economic operations.

During the second wave after January 2021, the COVID-19 scenario in India has got disastrous. COVID positive cases and deaths have continued to rise in the months of March, April, and May 2021, with a reduction at the end of May 2021. From June 2021 onwards, all operations in major districts, including Nagpur, have been unlocked, subject to certain limits imposed by the local administrative authority from time to time. The COVID- 19 waves are continued further. Traffic movements were not restricted further as in 2020 session. Though Third and now fourth wave was expected to have an impact overall but COVID cases were within control. Further, it will also be a time to wait and watch while executing our tasks maintaining social distancing and sanitization during these situations.

References

[1] R. Rana, A. Tripathi, N. Kumar, and N. K. Ganguly, "A Comprehensive Overview on COVID-19: Future Perspectives," *Front. Cell. Infect. Microbiol.*, vol. 11, no. September 2021, pp. 1–16, 2021, doi: 10.3389/fcimb.2021.744903.
[2] K. J. Goh *et al.*, "Rapid Progression to Acute Respiratory Distress Syndrome: Review of Current Understanding of Critical Illness from Coronavirus Disease 2019 (COVID-19) Infection," 2020.
[3] U. WHO Director, "WHO Director-General's opening remarks at the media briefing on COVID-19 - 11 March 2020," 2020.
[4] WHO update 31 october 2021, "Global overview."
[5] Y. S. Malik *et al.*, "Coronavirus disease pandemic (Covid-19): Challenges and a global perspective," *Pathogens*, vol. 9, no. 7. MDPI AG, pp. 1–31, Jul. 01, 2020, doi: 10.3390/pathogens9070519.
[6] WHO update 14th Nov 2021, "Global overview."
[7] M. Chaudhary, P. R. Sodani, and S. Das, "Effect of COVID-19 on Economy in India: Some Reflections for Policy and Programme," 2020, doi: 10.1177/0972063420935541.
[8] H. Chung, H. Seo, S. Forbes, and H. Birkett, "WFH DURING THE LOCKDOWN: Changing preferences and the future of work UK," *J. Univ. Kent*, 2020.
[9] The Hindu, "Krishnadas Rajagopal NATIONAL," Jun. 2021.
[10] M. Vannoni, M. McKee, J. C. Semenza, C. Bonell, and D. Stuckler, "Using volunteered geographic information to assess mobility in the early phases of the COVID-19 pandemic: A cross-city time series analysis of 41 cities in 22 countries from March 2nd to 26th 2020," *Globalization and Health*, vol. 16, no. 1. BioMed Central Ltd, Sep. 23, 2020, doi: 10.1186/s12992-020-00598-9.
[11] Hindustan Times, "India under lockdown_ Migrant labourer pedals 1,700 km in 7 days to reach home _ Latest News India - Hindustan Times," Apr. 19, 2020.
[12] T. Bhawan, P. Street, N. Delhi, and S. Sl, "Government of India Ministry of Road Transport & Highways The List of National Highways in the Country is as under : Government of India Ministry of Road Transport & Highways," pp. 1–42, 2019.
[13] E. Plaza and B. K. Complex, "Oriental infratrust," no. 8, 2020.
[14] Google maps, "Map data ©2021 2 km," 2021.
[15] O. I. for P. Highway, *KHUMARI*. 2021.

[16] O. I. for P. Highway, *KKBP*. 2021.
[17] Oriental Infrastructure for Project Highway, *NBP*. 2021.
[18] Oriental Infrastructures for Project Highway, *BORKHEDI*. 2021.

[19] Census data . (2011). Retrieved 2020, from Census of India : www.censusofindia.gov.in/
[20] Corporation, N. M. (n.d.). Retrieved 6 2022, from NMC local Governemnt : https://www.nmcnagpur.gov.in/city-with-latent-economic-potential
[21] GoI. (2021). India Statistics . Retrieved May 2022, from https://www.indiastatdistricts.com/maharashtra/nagpur/demographics/growth/data-year/all-years

Formulation of Parking Demand Estimation Model in Transit Oriented CBDs: A Review

Debasish Das[1], Debargha Banerjee[2], Dibyendu Das[3], Sharmistha Bhowmik[3], Ishita Roy[3], Tanmoy Sarkhel[3], Suhel Raj[3] and Pritam Das[3]

[1] Assistant Professor and Head, Civil Engineering Department, JIS College of Engineering, Kalyani, WB
[2] Assistant Professor, Civil Engineering Department, JIS College of Engineering, Kalyani, WB
[3] Graduate Students, Civil Engineering Department, JIS College of Engineering, Kalyani, WB

Email. debasish.das@jiscollege.ac.in

Abstract. Parking shortages are a regular occurrence in the urban Central Business District (CBD). In most Indian cities, the demand for parking outnumbers the supply. Managing parking demand is a critical aspect of transportation planning, particularly in metropolitan areas. Due to a shortage of off-street parking options in the CBD, tourists are compelled to park on the street, causing unnecessary congestion. The goals of this paper are to review existing research, identify key parameters, and provide models for forecasting on-street parking demand. To predict on-street parking demand in metropolitan CBDs with public transit (PT) service, a cost parameter (CF) model is developed. Some surveys, such as in-out surveys, license plate surveys, questionnaire surveys, video graphic survey methods, and so on, are likely to be utilized to collect data for the suggested model.

1. Introduction

The rise in population in metropolitan areas has led in an increase in travel demand. To accommodate the increased demand, vehicle ownership is expanding. The desire to own and operate personal vehicles has resulted in a significant increase in parking demand in major cities [1]. On-street parking is a difficulty in almost all of the metropolitan areas. The high concentration of urban activity in the CBDs has resulted in a significant increase in parking demand. To reduce the walking distance between the parking spot and the destination, every car owner wishes to park their vehicle as near to the destination as possible, resulting in parking cruising [2]. High concentrations of urban activity in India's CBDs have resulted in more job opportunities. The parking demand can be addressed by improving PT service customer satisfaction [3, 4]. CBDs that are transit-oriented are likely to have less traffic congestion. Parking and transit service have a significant impact on the transportation system, both individually and in combination. The majority of emerging countries' historic cities are poorly planned. They built up in an unplanned manner, with insufficient road space and no adequate planning. Finding the necessary characteristics to estimate the parking demand is critical. In order to lower on-street parking demand, some parking management measures must be enacted.

Land usage, land value, parking price, parking supply, search time, easy time, transit accessibility, total trip time, manoeuvrability space, mode choice, age, family income, current climatic condition, in-situ appeal, and other factors all influence parking demand. For CBDs with transit, Equation 6 is proposed. A literature survey was undertaken and reported in the next part to better understand the various elements affecting on-street parking demand and estimation.

2. Literature review

A review of the literature on parking demand estimation was conducted. On this topic, a broad literature review has been attempted. However, developing countries, particularly India, have received special attention. The goal of the study is to identify key factors that influence on-street parking demand and to choose a model for forecasting demand.

In real life, users' expectations and expectations that are not met might influence any constructive planning job. Though this notion is commonly utilised in marketing research, it is also used to implement/improve PT services in order to reduce on-street parking demand. For demand estimation, Tiexin et al. [5] used criteria such as average turnover rate, parking occupancy, service level, parking fees, and automotive growth rate. The basic parking generation rate model was modified using these parameters. Because most of the key parameters were taken into account, the model was reliable. The basic parking generation rate model as given (Equation 1) below:

$$Y = \left(\sum_{i=1}^{n} a_i \times r_i\right) \quad i=1,2,3.....,n \tag{1}$$

Where,

Y refers to parking demand in certain CBD area during rush hour (in lot)

The improved rate model is given in Equation 2 of the same paper. It includes some more parameters. The model is as follows:

$$Y = \left(\sum_{i=1}^{n} \frac{a_i}{\mu_i} \times \frac{r_i}{\gamma_i}\right) \times \delta \times L \times \beta \tag{2}$$

Where,

Y is as mentioned above

L = the parking price impact coefficient

δ = parking level of service

The improved parking generation rate model (Equation 2) will be trying to improve further in this study.

The pleasure of users is the most important factor in any effective planning. If a PT system can be implemented in accordance with the wishes of the users, the profit of the company will be maximised, and the demand for on-street parking will be decreased. When implementing Mode Shift (MS), users' happiness or dissatisfaction has to be clearly known [6, 7, 8, 9, 10, 11]. As previously said, failed re-affirmation is not always achievable, and Steven E. Polzin [12] provided the best example of this. The relationship between transportation and land use was detailed in detail. This link has been influenced by factors like as development momentum, agglomeration economics, and so on. The choice of land for transportation purposes is extremely important. This study also revealed the detrimental consequences of taking the wrong path. Stradling et al. [13] looked into the pull and push policy and discovered that a motorist cannot be forced to switch from a private car to public transportation. It was discovered that pulling rather than pushing the motorist was more efficient. The elderly and disadvantaged urban people were also discovered to be more vulnerable to push measures. They identified critical parameters in the pull and push approach, such as comfort in PT service, travel cost of PT service, frequency of PT service, and so on, based on these pull and push methods. Friman et al. [14] shown how to execute the PT service by taking into account the frequency of negative critical occurrences and customer satisfaction. The frequency of significant occurrences cannot be manipulated for a good transit system. The only method to provide a quality transit system was to eliminate negative critical incidents or change important criteria intended to improve transit service, such as new vehicles, new schedules, and new routes. Friman et al. [15] used a stated preference survey to extend his earlier study in order to determine the most important factors among service reliability, personnel behaviour, information ease, and vehicle design. They discovered that an employee's behaviour is the most sensitive factor in determining whether they accept or reject PT. The MS effect for work travels was explored by Kingham et al. [16]. They discovered that commuters were willing to switch from private transportation to bicycle, public transportation, or car sharing. The biggest constraint, however, was the travel distance. Although customers were concerned about pollution and traffic congestion caused by the increasing use of private automobiles, they were obliged to use PV since PT was too time consuming. The PT in Indian cities has been fully swamped, according to Pucher et al. [3]. Most public transportation modes, such as bus and train, are overcrowded, slow, inconvenient, and unsafe, forcing people to rely on their personal vehicles for everyday use, resulting in a high demand for parking. Immediate activity is required to obtain a high-quality PT service that will naturally attract users. In Kolkata, Phanikumar et al. [4] assessed the characteristics of urban buses. They calculated how much users were willing to pay for various quantitative and qualitative aspects of the bus system. They also discovered that, in addition to quantitative factors, qualitative factors influence the decision to use an urban bus service. Travel time, waiting time, comfortable seating, comfortable standing, nice look, noise level, and other factors were deemed to be important. Hollander et al. [17] used game theory to find the optimal number of parking in 2006. They assumed the government wanted to encourage people to use public transportation. To improve the transit service in the city core, a quantitative analysis was conducted. The travellers' choice distribution was also calculated using the Logit model. However, game theory was a theoretical concept that needed to be tested in a real-world setting to be accepted. The integration of the Logit model and game theory required further investigation. A. Nelessen Associates [18] investigated on-street parking and concluded that transit-oriented construction in the CBD is preferable to reduce on-street parking congestion. Previous studies have also shown that adding high-quality PT can help manage on-street parking. Each parking lot should be positioned between significant offices or commercial centres, according to the research. The parking lot should also be within walking distance, i.e. it should be located in the heart of any CBD area. Cantwell et al. [19] discovered that 'commuting stress levels' have an impact on PT commuting enjoyment. Every person has a limit to how much stress they can take. In their research, they identified the stress limit. They discovered a significant effect of this parameter, which can be employed in other comparable investigations. Qin et al. investigated the impact of PT accessibility on parking demand in Beijing business buildings [20]. To determine the link under various PT accessibility, they simply looked at one parameter. Because PT accessibility is a qualitative phrase, they introduced the transit accessibility index to turn it into a quantifiable measure. Eriksson [21] looked on the important factors in MS. He found that people were forced to utilise the PV for long distance travels because PT was time consuming. However, Eriksson noted that increasing the PV trip cost could make the PT more effective. For MS of PV to PT, the service dependability and frequency of the PT were critical elements [22]. Reduced rates and other habit-breaking measures are also attracting PV customers to PT. The study's major goal was to determine the factors that attract the most PV users to the PT. However, for each sort of sample population, a distinct set of parameters was utilised. Combining cycling and walking with PT may reduce automobile commuting [23]. The "perceptions of captive (i.e., those who do not have a car) and choice riders (i.e., those who do have a vehicle) about bus service features in Kolkata, India" [24] was also investigated. A willingness to pay (WTP) survey was done to gather information on a variety of factors. The obtained data from multiple trip makers was examined using the Logit model in a stated choice experiment. According to the WTP poll, captive and choice users have distinct preferences for bus service qualities. The WTPs of choice riders were much higher than captive riders for most of the criteria. The detailed outcomes of the above studies and the novelty of this study have been listed in Table 1.

Table 1. Outcomes from the studies.

Author (s) of the articles / report	Parameter (s) considered for estimating mode shift/Major finding (s)
Oliver (1980), Churchill et al. (1982), Cadotte et al. (1987), Oliver et al. (1988), Tse et al. (1988), Fornell (1992)	The concept of users' satisfaction or dissatisfaction.
Steven E. Polzin (1999)	The relation between transportation and land use
Stradling et al. (2000)	Pull and push policy
Friman et al. (2001)	Effect of negative critical incident on PT service

Friman et al. (2001)	Employee's behavior found to be most vulnerable variable which controls the preference of the users' for accepting or rejecting public transit.
Kingham et al. (2001)	Distance between origin and destination
Pucher et al. (2004)	Existing situation of PT in India
Phanikumar et al. (2006)	i. Travel time, ii. Waiting time, iii. Comfortable seating, iv. Comfortable standing, v. Good appearance, Noise level
Cantwell et al. (2009)	Level of stress causing by commuting PT
Qin et al. (2010)	Public transit accessibility index
L. Eriksson, (2011)	Travel cost
Lauren et al. (2013)	i. The service reliability and Frequency of the public transport
Hamre et al. (2014)	Effect of cycling and walking
Overall selected parameters to estimate the mode shift in previous studies	
Distance	Travel Time
Waiting Time	Comfortable Standing
Comfortable Seating	Good appearance
Noise level	Street causing due to commuting public transit
Frequency of public transit	Travel cost
Novelty of this study	
The above-mentioned parameters [Phanikumar et al. (2006)]. This study aims to propose a new model by incorporating maximum number of parameters. The new model is proposed using 13 number of parameters in a later section (Equation 3) to estimate the probable mode shift and the effect on on-street parking demand due to MS	

As per the previous study from various researchers the following methodology is proposed for the current study.

3. Proposed Model

It is found from the previous studies that many variables, like land use, floor area, parking fees, age, annual income, ease time, search time, transit facility, LOS etc. plays a vital role in estimating parking demand. All parameters are initially considered in developing the model. CF model are proposed to manage the on-street parking demand. Cost factor is the primary parameter to control the parking demand in CF model. CF model estimate the probable MS. Comfort Index (CI) plays a vital role in the MS. A regression analysis may be carried out with SPSS to estimate the parking model. $a_{i,} r_{i,} \mu_i, FI$, AI_p, AI_t, β, α, CF, accessibility index of parking at present LOS (AIP_L) and γ are considered to formulate CF model (Equation 3) along with $a_{i,} r_{i,} \mu_i$, AI_t and β.

CFmodel

It is observed from previous study [38] that CI is also an important parameter for MS. The CI is used to estimate the U_f (Equation 5) along with other variables like A, D, AI. A regression analysis needs to be carried out to obtain U_f (Equation 5). γ is calculated using Equation 4. A new parameter CF is also used for MS (Equation 3). 'Travel cost by PT' needs to be maximized to control the parking demand. This may be increased up to the travel cost of private vehicles and in this condition the value of CF will be 1 which will provide the optimal parking demand. With the increased travel cost of PT, the CI will increase, which in turn will reduce the parking demand. The CF model is shown below:

$$Y = \left[\sum_{i=1}^{n} \frac{a_i}{\mu_i} \times r_i\right] \times CF \times \frac{AIP_L}{AI_t} \times \beta \times \gamma \quad i=1,2,3,\ldots,n \tag{3}$$

Where,

$Y, a_i, r_i, \mu_i, AI_t, \beta$ = same as Equation 2

$$CF = \frac{Travel\ cost\ by\ private\ car}{Travel\ cost\ by\ PT}$$

$$AIP_L = \frac{\text{Std.Travel time by vehicle + std.Search time + std.Ease time}}{\text{Actual travel time by vehicle} + \left(\frac{\text{Std. Search time}}{\text{Level of serive(LOS)of parking lot}}\right) + \left(\frac{\text{Std.Ease time}}{\text{Level of serive(LOS)of parking lot}}\right)}$$

$$\gamma = 1 - \left[\frac{e^{U_f}}{1 + e^{U_f}}\right] \tag{4}$$

Where,

$$U_f = P_1 \times A + P_2 \times AI + P_3 \times D + P_4 \times CI + Q \tag{5}$$

Where,

P_1, P_2, P_3, P_4 = Coefficient of A, AI, D and CI respectively
Q = Constant

4. Proposed survey(s) for data collection

The propose surveys required for collecting the data for the parameters used in CF Model are indicated in Table 2.

Table 2. Type of surveys required for the study.

S. No.	Parameter(s)	Required survey	Type of survey
1	Empty parking space available for a particular time duration, Occupancy	In-out survey	
2	Parking duration, Parking turnover	License plate survey	
3	Parking price, Drivers' choice, Trip generation, Walking time, Age, Family annual income, Travel distance, Service level, Travel time, Search time, Ease time, Travel cost, Accessibility index of parking at present LOS, Transit accessibility, Search time, Ease time, Travel distance, Comfort index	Questionnaire survey	Primary
4	Road capacity	Traffic survey	
5	Land use, Parking supply, Population growth, Vehicular growth		Secondary

It is observed from Table 3 that, the parking accumulation, parking volume, parking lot, average parking duration, parking turnover, etc. can obtain from various types of parking survey. In-out survey, license plate method of survey and questionnaire surveys are the primary surveys for conducting the study. Detailed of these surveys are discussed below.

4.1. In-out survey

This survey is used to figure out how full the parking lot is. The parking lot gets taken for the first time. For a certain time, interval, the number of cars entering and exiting is counted. Finally, that lot is occupied. This survey necessitates relatively little labour. This survey can be completed by only one person.

A parking lot must first decide. The lot will then be divided into pieces to make the survey easier to complete. For each sector, parking spaces will be labelled 1, 2, 3,...., n. The survey will be completed both manually and graphically after it has been set up. The data will be sent to an excel document, from which the parking accumulation and occupancy will be calculated.

4.2. License plate survey
The data from this survey is the most accurate and realistic. This survey estimates parking length, parking turnover, and parking fee. Each parking bay will be watched for 15 minutes at a time, and the licence plate number will be recorded. The odds of missing short-term parkers are reduced when the time interval is reduced. However, this procedure is time consuming. It is necessary to estimate the length of time a vehicle will be parked in a bay. The parking price will be calculated based on the length of time you will be parked.

4.3. Questionnaire survey
All of the survey locations must conduct the questionnaire survey. To comprehend the current parking and mode choice scenario, a qualitative and quantitative investigation was required. A willingness to pay survey should also be done to determine the consumers' desire to switch from private vehicles to public transportation. Personal Details, Trip Characteristics, Parking Characteristics, and Parking Location Choice are the four key groups in the questionnaire. The data can be organised in Excel, and the analysis can be done with statistical software such as SPSS.

5. Conclusion
13 numbers of parameters are found out and presented in Table 1. CF model (Equation 3) is proposed to evaluate the on-street parking demand in the CBDs with transit service. The estimated demand will be compared with the present supply and necessary actions are to be taken accordingly. The proposed surveys for data collection are briefly described in Section 5. The parameters affecting on-street parking demand have been identified through literature survey. Basic model (Equation 1) for estimating on-street parking demand has been identified through literature survey. The model has been modified by incorporating various parameters obtained and shown in Equation 3.

Reference
[1] Yao, G., Guo, H., Li, C. and Sun, H. 2016 Analyzing the elements related to parking demand: An empirical study in Beijing. Advances in Mechanical Engineering, Vol. 8 (4) 1–9, DOI: 10.1177/1687814016645064.
[2] Shoup, D.C. 2003 Truth in transportation planning. University of California Transportation Center.
[3] Pucher, J., Korattyswaroopam, N. and Ittyerah, N. 2004 The crisis of public transport in India: overwhelming needs but limited resources. Journal of Public Transportation, 7(3), p.5. DOI: http://dx.doi.org/10.5038/2375-0901.7.3.5.
[4] Phanikumar, C.V. and Maitra, B. 2006 Valuing urban bus attributes: an experience in Kolkata. Journal of Public Transportation, 9(2), p.4., DOI: http://dx.doi.org/10.5038/2375-0901.9.2.4.
[5] Tiexin, C., Miaomiao, T. and Ze, M. 2012 The Model of Parking Demand Forecast for the Urban CCD. Energy Procedia, 16, pp.1393-1400. DOI:10.1016/j.egypro.2012.01.221.
[6] Churchill Jr, G.A. and Surprenant, C. 1982 An investigation into the determinants of customer satisfaction. Journal of marketing research, pp.491-504.
[7] Cadotte, E. R., Robert B. W., and Roger L. J. 1987 Expectations and mode shift in models of consumer satisfaction, Journal of Marketing Research, 24: 305–314.
[8] Fornell, C. 1992 A national customer satisfaction barometer: The Swedish experience. the Journal of Marketing, pp.6-21.
[9] Oliver, R.L. 1980 A cognitive model of the antecedents and consequences of satisfaction decisions. Journal of marketing research, pp.460-469.
[10] Oliver, R.L. and DeSarbo, W.S. 1988 Response determinants in satisfaction judgments. Journal of consumer research, pp.495-507.

[11] Tse, D.K. and Wilton, P.C. 1988 Models of consumer satisfaction formation: An extension. Journal of marketing research, pp.204-212.
[12] Steven E. P. 1999 Transportation/land-use relationship: Public transit's impact on land use. Journal of urban planning and development, 125(4), pp.135-151.
[13] Stradling, S.G., Meadows, M.L. and Beatty, S. 2000 Helping drivers out of their vehicles integrating transport policy and social psychology for sustainable change. Transport policy, 7(3), pp.207-215, DOI: 10.1016/S0967-070X(00)00026-3.
[14] Friman, M., Edvardsson, B. and Gärling, T. 2001 Frequency of negative critical incidents and satisfaction with public transport services. I. Journal of Retailing and Consumer Services, 8(2), pp.95-104, DOI: 10.1016/S0969-6989(00)00003-5.
[15] Friman, M. and Gärling, T. 2001 Frequency of negative critical incidents and satisfaction with public transport services. II. Journal of Retailing and Consumer Services, 8(2), pp.105-114., DOI: 10.1016/S0969-6989(00)00004-7.
[16] Kingham, S., Dickinson, J. and Copsey, S. 2001 Travelling to work: will people move out of their vehicles. Transport policy, 8(2), pp.151-160. DOI: 10.1016/S0967-070X(01)00005-1.
[17] Hollander, Y., Prashker, J.N. and Mahalel, D, 2006 Determining the desired amount of parking using game theory. Journal of urban planning and development, 132(1), pp.53-61.
[18] Nelessen A. Associates 2008 Parking study report Journal Square, Available:http://www.thejcra.org/jcra_files/File/development_projects/journal_square/Parking_Study.pdf.
[19] Cantwell, M., Caulfield, B. and O'Mahony, M. 2009 Examining the parameters that impact public transport commuting satisfaction. Journal of Public Transportation, 12(2), p.1. DOI: http://dx.doi.org/10.5038/2375-0901.12.2.1.
[20] Qin, H., Xiao, Q., Guan, H. and Pan, X. 2010 Analysis on the Parking demand of the Commercial Buildings Considering the Public Transport Accessibility- Commercial Buildings in Beijing as an Example, Nature and Science. 2010; 8(3):63-68.
[21] Eriksson, L. 2011 Vehicle Users' Switching to Public Transport for the Work Commute.
[22] Lauren, R., Friman, M., Gärling, T. and Hartig, T. 2013 Quality attributes of public transport that attract vehicle users: A research review. Transport Policy,25, pp.119-127.
[23] Hamre, A. and Buehler, R. 2014 Commuter mode choice and free vehicle parking, public transportation benefits, showers/lockers, and bike parking at work: evidence from the Washington, DC Region. Journal of Public Transportation, 17(2), p.4.
[24] Maitra, B., Dandapat, S. and Chintakayala, P. 2014 Differences between the Perceptions of Captive and Choice Riders toward Bus Service Attributes and the Need for Segmentation of Bus Services in Urban India. Journal of Urban Planning and Development, 141(2), p.04014018, DOI: 10.1061/(ASCE)UP.1943-5444.0000205.

Acknowledgments
Authors wishing to acknowledge assistance or encouragement from colleagues, special work by technical staff, parent institute and nearby Medical College & Hospital for their support and cooperation for the completion of the current study.

Correlation Study of Physico-Chemical Parameters and Water Quality Index Around Kodungaiyur Dumping Ground

Aishwarya R[1], Faizuneesa A[1], Hemamathi A[1], Ramshankar P[1], Yuvaraj M[2]

[1]Assistant Professor, Department of Civil Engineering, R.M.K Engineering College, Tamil Nadu, India
[2]Propreitor, Skyrider Aerosolutions, Chennai, India

Email.raa.civil@rmkec.ac.in

Abstract. The amount of waste that enters the Kodungaiyur dumping ground, in Tamil Nadu, India increases day by day due to the urban sprawl and growth of the metropolitan limits. The leachate infiltration in the groundwater accounts for the contamination and it is a critical issue to be addressed. The survey conducted among the residents of Kodungaiyur represents a disagreement on the water quality among the residents of their area. The go-to rack and ruin health condition of the people in the locality makes them prone to frequent hospitalisations. Physico-chemical parameters from 12 bore wells around the dumpsite are tested. On average, the conductivity, Total hardness, Chlorides, COD, BOD, Total Dissolved Solids, Calcium, Magnesium, Sodium, Potassium and Total alkalinity are more than the BIS prescribed limits of drinking water standards. On analyzing the water quality index (WQI) using the weighted arithmetic method, it is clear that all the bore wells have a WQI of more than 100, which is the deprecatory indicator for the category of "Totally unsuitable for drinking". The worrisome situation indicates that the groundwater is polluted and unsuitable for drinking and domestic purposes. The WQI has negatively correlated to pH as the pH of all the samples has values ranging between 6.5-8.5. Also, the WQI is high-positively correlated to Chlorides, Sulphates, COD, BOD, Total hardness, Calcium, Magnesium, Alkalinity, Nitrate and Sodium with a correlation factor of more than 0.900. This indicates the excessive presence of physical, chemical and biological parameters contributing to high values of WQI in the water samples. After evaluating the data, it can be concluded that the drinking water of Kodungaiyur is not potable and a direct demand to revamp the groundwater status in this region is the need of the hour to prevent the population from facing adverse health effects.

Keywords: Water quality index, BIS, weighted arithmetic method, BOD, COD, Groundwater, dump yard.

1. Introduction

Kodungaiyur dumping ground has experienced several issues of groundwater pollution, air pollution, soil pollution and socio-economic issues by handling large quantities of the legacy waste for almost 40 years. According to Greater Chennai Corporation, the dumping ground vast 258 acres and receives 3000-

3600 tonnes of legacy waste that only increases in quantity exponentially day by day. The gauged magnitude of debris uncharacterized for disposal is around 64 lakh cubic metres spread across four dumps. [1] Laasya Shekhar (2019) has registered in her blog that tests conducted for heavy metals for many sites of leachate, water, soil and air confirmed the presence of eruptive organic compounds accounting for frequent fires in the trash yard. It was also mentioned that the quantities of eco-toxic metals are more than the values prescribed by the US Environmental Protection Agency (USEPA) and the Bureau of Indian Standards (BIS) for certain metals. [2] Saravanan et al. stated that it is an open trash dump, which does not have a sound lining to control leachate percolation into the surrounding groundwater. There are no appropriate actions to oversee the leachate generation and percolation from the dumpsite. [3] Aishwarya R (2021) mentioned bout the ill effects of leachate migration on groundwater quality to comprehend the effects of the Kodungaiyur trash yard. The values of the Physico-chemical parameters and heavy metals were illustrated in interpolated maps. [4] Navneet Kumar and Sinha D.K (2010) performed a statistical analysis of the Physico-chemical parameters of water and documented them from different parts of the World and India. The study reveals that the statistical parameters are one of the important tools for exploring the variation and interrelation of the parameters of the water. In another study for the analysis of groundwater quality for Al-Mukalla city, Hadhramout, Yemen, Sami G. Daraigan et al. mentioned that all the Physico-chemical parameters are more or less correlated with each other in linear correlation analysis of drinking water quality [5]. Water-borne diseases are transmitted through the water of substandard quality owing to public health concerns. Hence, many countries have exacted water quality safeguarding assays and monitoring by-laws [6,7].
The suitability of water can be evaluated through Water Quality Index (WQI). It is also a means for disseminating details on the overall quality of water to the citizens and policymakers [8, 9]. WQI is a dimensionless numeral which unifies numerous water-quality parameters into a single number by standardizing weights [10]. The factors enclosed in the WQI model could differ counting on the specific uses of the said water and local preferences. Water quality indices (WQIs) are devised to integrate water quality variables [11,12 and 13] and qualify water quality data into simple terms (e.g., excellent, good, bad, etc.) for documenting which are used by the administrators and the people [14]. [15] Vinod Kothari (2021) performed statistical analysis to find the correlation coefficient of WQI and physico-chemical parameters. The study showed a significant linear interrelation and a high correlation coefficient between different water quality parameters. Therefore, the present work focuses on leachate percolation and its impact on groundwater by analysing the Water Quality indices and performing correlation analysis.

2. Study area

The study's main aim is to quantify and qualify groundwater pollution because of the presence of unacceptable limits of basic minerals due to leachate penetration from the dumping ground through Water Quality Indexing (WQI).
Quoting to the newsletter by [17]Afroz Khan (2020) Kodungaiyur dumping ground in North Chennai was once 400 acres of pleasing alluvial land with huge and rampant grassland. But our mismanagement and inadequate processing of littering started generating a huge volume of waste and pose serious dangers to the environment. It was an intermittent flood-prone alluvial meadowland of the Kosasthalaiyar river in Kodungaiyur. Multiple species of flora and fauna peacefully co-existed alongside a small residential conglomeration of Kaviarasu Kannadasan Nagar, Krishnamurthy Nagar and MKB Nagar. Until the 1980s, Kodungaiyur and its surroundings were a retirement retreat, a rightful recline, and a sanctuary for nature enthusiasts. There was no water problem, air smog or any other environmental issues. But the imprudent invasion of post-consumer waste and sewage treatment plants into the alluvial grassland has made its testing times for the people of Kodungaiyur. It comes under Perambur taluk of Chennai city surrounded by educational institutions and industrial sectors such as Madras fertilizer limited, Indian oil corporation and Madras refinery limited etc. Many studies are done to assess the concentration of Physico-chemical parameters in the groundwaters around the dumpsite. But there has been very little research done on Water Quality pollution indexing around the Kodungaiyur dump yard. Hence, this investigation determines the groundwater quality in the Kodungaiyur residence around the dumping ground aimed at determination of WQI and analysing the correlation of Physico-chemical parameters and WQI.

Figure 1. Study area

3. Physico-chemical parameters

The groundwater contamination study is initiated by conducting a survey of the residents around the dump yard and water samples from 12 were collected, numbered and sealed along with the latitude and longitude noted using GPS (Figure 1). The parameters such as pH, COD, BOD, Total hardness, Conductivity, TDS, Chlorides, Sulphates, Calcium, Magnesium, Alkalinity, Nitrate and Sodium were determined for all 12 samples and represented in Table 1. L1,L2, … L12 represent the locations of the samples. Table 2 shows the mean, median, Standard Deviation (SD) and the prescribed limits of the parameters. It is seen from table 2 that many parameters exceed the limits prescribed by the Tamil Nādu Pollution Control Board (TNPCB) and Bureau of India Standards (BIS) of drinking water.

It is evident from table 2 that the mean values of all the parameters exceed the BIS which is also represented in the graph (figure 2). The red bars represent the mean values of water quality parameters in Kodungaiyur while the yellow bars represent the BIS prescribed limits for the same parameters. Although there are many authorities to fix the thresholds for the water quality parameters, the mean value exceeds in comparison with all CPCB, BIS, WHO, ICMR prescribed limits. The mean values are compared to the permissible limits stated by various regulatory bodies and it is declared that the groundwater is safe or unsafe for human consumption, polluted or contaminated.

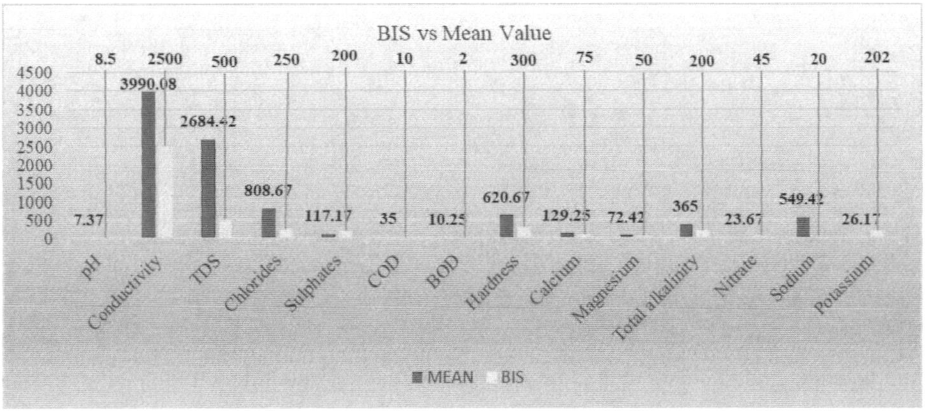

Figure 2. Comparison of BIS and mean values

Table 1. The concentration of Physico-chemical in water samples

Location	pH	COD (mg/L)	BOD (mg/L)	Conductivity (µs/cm)	Total hardness (mg/L)	Chlorides (mg/L)	Sulphates (mg/L)	TDS (mg/L)	Calcium (mg/L)	Magnesium (mg/L)	Sodium (mg/L)	Potassium (mg/L)	Alkalinity(mg/L)	Nitrate (mg/L)
1	7.37	498.304	527.793	888	557.282	260	468.815	425	586.771	616.26	704.727	4	645.749	675.238
2	6.99	109 11.908	116 56.811	16820	12 401.714	3949	10 167.005	12443	131 46.617	13891.52	161 26.229	38	14636.423	15381.326
3	7.3	4984.26	5176.67	9120	5369.08	1475	4791.85	6641	5561.49	5753.9	6331.13	51	5946.31	6138.72
4	7.25	273.5	273.125	647	272.75	110	273.875	335	272.375	272	270.875	4	271.625	271.25
5	7.46	24	5	1350	388	180	19	811	133	14	95	22	477	5
6	7.31	28	6	1697	344	310	9	962	97	25	174	28	456	6
7	7.34	20	4	609	142	120	40	350	34	14	50	2	138	2
8	7.11	12	4	3130	374	550	216	1861	101	29	598	38	541	8
9	7.53	8	5	2540	718	625	112	1556	162	76	260	19	254	12
10	7.25	24	5	5410	880	1225	254	3253	186	101	898	44	721	13
11	7.48	12	6	3460	324	575	372	2109	65	39	772	45	392	11
12	8.04	8	4	2210	596	325	307	1467	170	42	298	19	329	4

Table 2. Statistical analysis of Physico-chemical in water samples

Parameters	Mini Mum	Maxi Mum	Mean	Median	Sd	TNPCB Tolerance Limits		BIS Standards	CPCB	WHO	ICMR	Effects Of Undesirable Limits
						Min	Max					
pH	6.99	8.04	7.37	7.33	0.26	6.5	8.5	6.5-8.5	8.5	8.5	9.2	Taste
Conductivity	609	16820	3990.08	2375.00	4706.77	-	-	2500	2000	-	-	-
TDS	335	12443	2684.42	1511.50	3535.83	-	500	500	-	-	-	Gastrointestinal effect
Chlorides	110	3949	808.67	437.50	1077.69	-	250	250	1000	200	1000	Taste affected
Sulphates	9	372	117.17	39.50	133.38	-	400	200	400	-	400	Laxative effect
COD	4	236	35.00	20.00	63.78	-	-	10	10	-	-	-
BOD	4	71	10.25	5.00	19.15	-	-	2	10	-	-	-
Hardness	69	2113	620.67	381.00	597.88	-	300	300	600	500	600	Scale formation
Calcium	16	324	129.25	117.00	91.24	-	200	75	200	75	200	Scale formation
Magnesium	7	317	72.42	34.00	93.66	-	100	50	100	50	-	Scale formation
Total alkalinity	64	721	365.00	387.00	194.83	-	-	200	600	-	-	Boiled rice turns yellow

Nitrate	2	140	23.67	7.00	41.84	-	20	45	100	-	100	Blue baby syndrome
Sodium	50	1649	549.42	279.00	582.36	-	-	20	-	-	-	-
Potassium	2	51	26.17	25.00	17.21	-	-	-	-	-	-	-

Table 3. Water Quality Index calculation

Parameters	pH	Conductivity	TDS	Chlorides	Sulphates	COD	BOD	Total hardness	Calcium	Magnesium	Alkalinity	Nitrate	Sodium	Potassium	WQI
BIS(Sn)	8.5	2500	500	250	200	10	2	300	75	30	200	45	200	10	
1/Sn	0.117647	0.0004	0.002	0.004	0.005	0.1	0.5	0.003333	0.013333	0.033333	0.005	0.022222	0.005	0.1	
Sum1/Sn								**0.911269**							
k= 1/(sum(1/Sn))								**1.09737**							
w=k/Sn	0.129102	0.000439	0.002194	0.004389	0.005487	0.109737	0.548685	0.003658	0.014632	0.036579	0.005487	0.024386	0.005487	0.109737	
ideal value	7	0	0	0	0	0	0	0	0	0	0	0	0	0	
concentration value of site 1	7.37	888	425	260	13	4	4	69	16	7	64	3	109	4	
Q= (V-Vi)/ (VSn-Vi)	0.246667	0.3552	0.85	1.04	0.065	0.4	2	0.23	0.213333	0.233333	0.32	0.066667	0.545	0.4	
Q*100	24.66667	35.52	85	104	6.5	40	200	23	21.33333	23.33333	32	6.666667	54.5	40	
W*Q	3.184526	0.015591	0.18655298	0.456506	0.035665	4.389482	109.737	0.084132	0.312141	0.85351	0.175579	0.162573	0.299033	4.389482	**124.2818**

Table 4 Groundwater categorization based on WQI [17-19]

WQI	Category	Possibility of usage
<25	Excellent	Drinking, Irrigation, Industrial
26-50	Good	Drinking, Irrigation, Industrial
51-75	Poor	Irrigation, Industrial
76-100	Very Poor	Irrigation
>100	Totally unsuitable	Required proper treatment before usage

4. Water Quality Index

The weighted arithmetic water quality index (WQI) takes the water quality variables and gives a rating into a single number. WQI has been developed and formulated [17] Brown et al. 1972 as given in equation 1.

$$\mathrm{WQI} = \frac{\sum_{i=1}^{n} Qiw_i}{\sum_{i=1}^{n} w_i} \quad (1)$$

Qi is called as the sub-index and Wi is called as unit weightage of the ith parameter while n is the total number of parameters. Fourteen parameters are considered in this study. The sub-index (Qi) is calculated by equation 2.

$$\mathrm{Qi} = 100 \times \frac{\sum_{i=1}^{n}[V-Vi]}{\sum_{i=1}^{n}[Vsn-Vi]]} \quad (2)$$

The standard value is given as Sn for the ith parameter as dictated by BIS. V is the observed value and Vi denote the ideal value of the ith parameter. Bis has prescribed an ideal value for pH alone which is 7 denoting neutral water. The other 13 parameters do not have any ideal value. Then overall pollution level in the sample locations can be evaluated be determining the weighted arithmetic WQI model. Table 3 shows the WQI calculation for one site and similar calculations were carried out for the other 11 sites, the values of which are given in table 4. It is seen that the WQI exceeds 100 for all the samples and is regarded as unsafe for drinking. The WQI for site 2 is 2322 which is very near to the dumping ground and the lowest value 124.28 is for site 1 which is far above the standard value.

Table 5. Water Quality Index

Site	WQI	Status
1	124.2818	
2	2321.991	
3	235.7984	
4	173.3255	
5	199.6335	
6	240.9971	Totally Unsuitable for drinking
7	143.7254	
8	185.5453	
9	187.8318	
10	245.2338	
11	244.5822	
12	155.2258	

Based on the critical index presented by [17-19] given in Table 4, the WQI as calculated in table 5, all the samples have a WQI of more than 100 and are classified as Totally unsafe for drinking purposes [20]. Therefore, a correlation matrix is drawn to analyze the relationship between each variable and the WQI as given in table 6. It is seen that the WQI is high-positively correlated to Chlorides, Sulphates, COD, BOD, Total hardness, Calcium, Magnesium, Alkalinity, and Sodium and negatively correlated with the pH of the samples as it falls within the BIS prescribed limits for all the twelve sites. The WQI is low and positively correlates with potassium and it describes the WQI and Potassium concentration are less interrelated. Therefore, it can be concluded that the groundwater is not polluted by pH. The highest positive correlation exists between COD and BOD exemplifying that BOD and COD of a water sample vary in the same manner and they are highly interrelated. The lowest positive correlation lies between potassium and nitrate indicating that these two salts do not have much impact on each other.

IOP Conf. Series: Earth and Environmental Science 1084 (2022) 012061 doi:10.1088/1755-1315/1084/1/012061

Table 6. Correlation analysis of Physico-chemical parameters and WQI

	pH	Conductivity (µs/cm)	TDS (mg/L)	Chlorides(mg/L)	Sulphates(mg/L)	COD (mg/L)	BOD (mg/L)	Total hardness (mg/L)	Calcium (mg/L)	Magne Sium (mg/L)	Alkali Nity (mg/L)	Nitrate (mg/L)	Sodi Um (mg/L)	Pota ssium (mg/L)	WQI
pH	1.000														
Conductivity (µs/cm)	-0.465	1.000													
TDS (mg/L)	-0.456	0.998	1.000												
Chlorides (mg/L)	-0.494	0.985	0.983	1.000											
Sulphates (mg/L)	-0.462	0.958	0.972	0.950	1.000										
COD (mg/L)	-0.476	0.950	0.965	0.944	0.999	1.000									
BOD (mg/L)	-0.476	0.950	0.964	0.946	0.999	1.000	1.000								
Total hardness (mg/L)	-0.459	0.965	0.977	0.964	0.997	0.997	0.997	1.000							
Calcium (mg/L)	-0.472	0.952	0.966	0.952	0.999	0.999	1.000	0.998	1.000						
Magnesium (mg/L)	-0.477	0.951	0.965	0.951	0.998	0.999	1.000	0.998	1.000	1.000					
Alkalinity (mg/L)	-0.483	0.959	0.972	0.961	0.997	0.997	0.998	0.999	0.999	0.999	1.000				
Nitrate (mg/L)	-0.478	0.949	0.963	0.951	0.998	0.999	0.999	0.998	1.000	1.000	0.999	1.000			
Sodium (mg/L)	-0.484	0.964	0.975	0.967	0.996	0.996	0.996	0.998	0.998	0.998	0.999	0.998	1.000		
Potassium (mg/L)	-0.252	0.605	0.573	0.512	0.402	0.376	0.372	0.401	0.373	0.367	0.394	0.360	0.402	1.000	
WQI	-0.473	0.878	0.886	0.932	0.908	0.910	0.914	0.926	0.923	0.925	0.934	0.929	0.937	0.271	1.000

5. Conclusion
The present investigation appraises the pollution of the groundwaters of the Kodungaiyur Dumping yard by evaluating WQI. The average value of WQI is 371.51, a figure above 100 and classified as totally unsafe for drinking. The results point out the impact of leachate percolation and anthropogenic activities in the study area reveals tremendous groundwater contamination, thus making it of water unsuitable for drinking and domestic purposes. The denizens of the Kodungaiyur region have been making use of the water constantly and are prone to several health hazards. It is thus vital that groundwater pollution needs to be dealt one step at a time by reinforcing forthright control measures such as biomining and obeying with legitimate standards towards littering. Thorough pollution inspection efforts are to be instituted via strict enforcement of laws to assure good functioning and maintenance of the biomining and sewage treatment plants in the dumping ground. The health issues are scaled up by consuming the said water and hence it is crucial to minister the groundwater to ensure its potable nature.

References
[1] Laasya Shekhar, 2019, "Nearly a third of Kodungaiyur residents have difficulty breathing, but no action on waste dump", Citizen consumer and Civil Action group, Blogs.
[2] Saravanan. R, Lenin. V, Navaneetha. A "Modeling of Groundwater Pollution due to Solid WasteDumping – A Case Study," Int. J. Earth Sci. Eng., vol.03, no. 02, pp. 176–187, 2010.
[3] R Aishwarya, 2021, "Assessment of Spatial Distribution of Physico-Chemical Parameters of Groundwater around Kodungaiyur Dump yard", J. Phys.: Conf. Ser. 2070 012215
[4] Navneet Kumar and Sinha D.K (2010), "Drinking water quality management through correlation studies among various physicochemical parameters: A case study" Int J. Environ Sci,1(2), 253-259
[5] Sami G. Darigan, Ahmed.S Wahdain, Ahmed S. Ba. Mosa and Manal H Obid (2011), "linear correlation analysis study of drinking water quality data for Al Mukalla City, Hadhramout, Yemen", Int. J Environ. Sci, 1(7), 1692-1701.
[6] Bermel, S., Damour, M., Ludwig, R. & Rodriguez, M. J. "Water quality monitoring strategies - A review and future perspectives". Sci. Total Environ. 571, 1312–1329 (2016).
[7] Romero, E. et al. Long-term water quality in the lower Seine: "Lessons learned over 4 decades of monitoring". Environ. Sci. Policy 58, 141–154 (2016).
[8] Asadi, S.S., Vuppala, P., Anji, R.M, "Remote sensing and GIS techniques for evaluation of groundwater quality in the municipal corporation of Hyderabad (Zone-V)", Indi. Int. J. Environ. Res. Public Health, Vol (4), No (1), (2007),45-52.
[9] Buchanan, S. Triantafilis, J. "Mapping water table depth using geophysical and environmental variables", Groundwater, Vol(47), No(1),(2009),80-96.
[10] Astel, A., Biziuk, M., Przyjazny, A. & Namiesnik, J. "Chemometrics in monitoring spatial and temporal variations in drinking water quality". Water Res. 40, 1706–1716 (2006).
[11] Cude C. "Oregon water quality index: A tool for evaluating water quality management effectiveness". Journal of the American Water Resources Association, Vol (37), No (1), (2001), 125–137.15.
[12] Liou, S, Lo, S, and Wang, S. "A generalized water quality index for Taiwan." Environmental Monitoring and Assessment, Vol(96), No (1-3), (2004) 35–52.16.
[13] Said, A, Stevens, D, and Selke, G. "An innovative index for evaluating water quality in streams" Environmental Management, Vol (34), No (3), (2004), 406–414
[14] Saksena D.N, Garg R.K, and Rao R.J. "Water quality and pollution status of Chambal River in National Chambal Sanctuary", Madhya Pradesh. J. Environ. Biol., Vol (29), No (5), (2008), 701-710.
[15] Vinod Kothari, Suman Vij, SuneshKumar Sharma, Neha Gupta, "Correlation of various water quality parameters and water quality index of districts of Uttarakhand, Environmental and Sustainability Indicators", Volume 9, 2021, 100093, ISSN 2665-9727, https://doi.org/10.1016/j.indic.2020.100093.
[16] Afroz Khan, 2020, "Destruction of marshlands due to garbage dumping", Citizen consumer and Civil Action group, Newsletter

[17] Brown R.M, Mcclelland N.J, Deininger R.A, Oconnor M.F.A, "Water quality index- crossing the physical barrier" Proc. Int. Conf. on water pollution research, Jerusalem, Vol(6), (1972), 787-797.
[18] C. Chatterji and M. Raziuddin, "Determination of water quality index of a degraded river in Asansol industrial area, Raniganj, Burdwan, West Bengal," Nature Environment and Pollution Technology, vol. 1, no. 2, pp. 181–189, 2002.
[19] G. Asadollahfardi, "Water Quality Management Assessment and Interpretation", Springer-Verlag, Berlin, Germany, 1st edition, 2015.
[20] A.Faizuneesa, S.P Kanniyappan "Analysis of Groundwater Quality in Thiruvallur" Turkish Online Journal of Qualitative Inquiry (TOJQI)Volume 12, Issue 8, July, 2021:7387 – 7397

Acknowledgements
The residents around Kodungaiyur dump yard are thanked for their kind cooperation in the collection of water samples and for filling out the questionnaire.

Spring Water Management to Ensure Long Term Sustainability in North-Eastern Regions of India

Prem Ranjan[1], Pankaj Kumar Pandey[2], Vanita Pandey[3] and Pema Tshering Lepcha[4]

[1,4]Research Scholar, Department of Agricultural Engineering, North Eastern Regional Institute of Science and Technology, Nirjuli, Arunachal Pradesh, 791109, India.
[2]Associate Professor, Department of Agricultural Engineering, North Eastern Regional Institute of Science and Technology, Nirjuli, Arunachal Pradesh, 791109, India.
[3]Assistant Professor, Department of Agricultural Engineering, North Eastern Regional Institute of Science and Technology, Nirjuli, Arunachal Pradesh, 791109, India.

Corresponding Author E-mail: prem009ranjan@gmail.com

Abstract. Spring water is the lifeline of the Himalayan communities. They depend on it for drinking and household use and farming. In recent decades, the region has been facing acute water shortage in the lean period due to climatic variabilities, especially rainfall. The rainfall pattern has changed from low-intensity, longer duration to high-density, shorter duration, which causes the “too little and too much water” syndrome. Rainwater does not have sufficient time to infiltrate due to high reliefs and undulating terrains. As a result, quick runoff dominates the region, reducing groundwater recharge and health. Springs also contribute to the base flow of many rivers in the region. The changing land use and land cover is another issue in water management, and the different studies indicate that deforestation, forest fire, mounting tourist spots, rapidly growing towns, and development activities reduce the infiltration capacity. In this region, most rural populations are completely dependent on perennial springs and streams for water needs such as drinking, agriculture, livestock feeding, and other household consumption. It is noticed that sudden water scarcity and demand increased during the non-monsoon period, which is uncertain for the long-term sustainability of the human population in the region. Hence, different water-saving techniques must be adopted to sustain this water scarcity condition at the household level and in irrigation practices. This can be achieved by bringing awareness of different water minimizing and saving techniques at the community level. Also, implementing a highly efficient irrigation system during the lean period instead of a flood irrigation system such as gravity-fed drip irrigation, bamboo drip irrigation, and pitcher irrigation is recommended.

Keywords: Water Demand, Gravity-fed, Scarcity, Structure, Tapping.

1. Introduction

Humans naturally tend to camp near potable and perennial sources [1]. In hill region communities living in rural and urban areas, they depend on springs to meet their water needs. Thus, springs are the lifeline of the Himalayan regions [2]. It provides drinking water to local habitants, helps maintain the river's base flow during the dry season, and supports the ecology [3]. Now it has been observed that these perennial water sources are drying due to climate change. Human activity such as deforestation and the cutting of hills for new construction in spring catchments is also responsible for declining spring discharge [4]. The majority of spring discharge is concentrated in monsoon months only, which shows that spring occurs only in a particular season [5]. Spring discharge depends on rainfall pattern, and in recent times, the rainfall has changed from lower intensity-long duration to high intensity-short duration fashion [6,7]. The change in precipitation pattern and undulating topography, steep slope, and faulty land management coupled with high-intensity storms result in high runoff and soil loss [8,9]. The knowledge of springs science is still developing, especially the concrete relation between precipitation and spring discharges has not been well-developed yet [10]. The springs come into existence when the combined geological, hydrologic, and human interventions disrupt the underlying soil and rock layers where subsurface water and groundwater are in movement (shown in Figure 1).

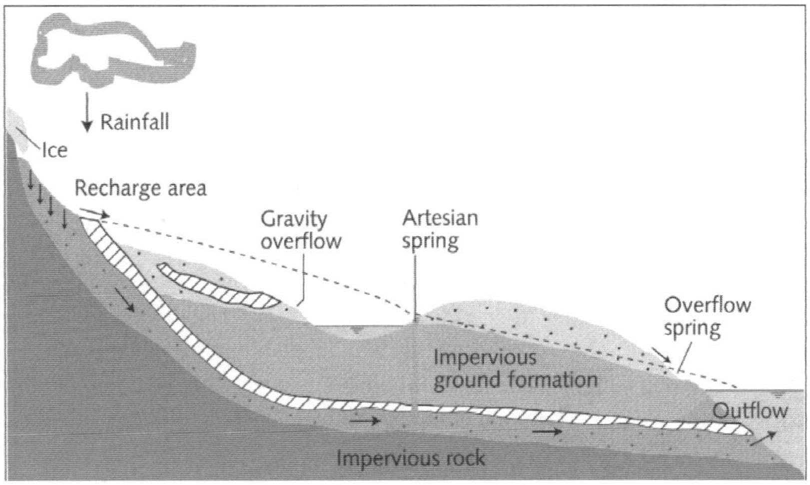

Figure 1. Occurrence of spring [26]

Most mountainous parts of Sikkim, Arunachal Pradesh, Meghalaya, and Nagaland, depend upon springs for drinking water and irrigation practice. These springs are also lifelines for wildlife and other ecosystems, not just humans. The decline in spring discharge and its increasing demand depict the future water crisis. Different water management solutions must be developed to tackle these challenges. It is well understood that different water management tactics like reviving, augmenting recharge area of traditional water sources by constructing trenches, CCTs, ponds, pits, check dams, gabion structures, bunds, etc. across the slope, development of seepage springs, water harvesting structures [11] and techniques, and well-managed distribution of water could help in meeting the water demand for inhabitants especially residing in remote and inaccessible terrain of Himalayas. When the concern is to meet the drinking water requirements of people living in hilly terrain in a cost-effective and eco-friendly manner, the strategy should be made to restore and conserve existing springs [12]. When effectively executed, such an approach would solve the water-related issues permanently.

1.1 Major challenges of water management practices in the Indian Himalayan region

a) Steep slopes and rugged topography are two major challenges that affect the aquifer restoration and water harvesting,
b) Poor terrain's water retention capability
c) Water crisis increases from mid-November to mid-March or the onset of the next monsoon,
d) Carrying water from distant sources for farm irrigation is hectic and laborious
e) Women and children walk long distances to fetch water, affecting women's empowerment and children's education.
f) Due to the rough landscape, ground channelling is also impractical
g) Lack of historical data for analysis

2. Spring inventory, water harvesting and distribution

2.1 Spring inventory
Primarily, locate a spring already in use and collect information from the users about its seasonality and other related information. Also, collect GPS locations and measure discharge rate and quality. Secondarily, identify the new locations with the help of farmers, hunters, and grazers, as they have good knowledge of the location of springs [13]. Also, following the animal's footprints, small streams in valleys and dense green vegetation in the dry area indicate spring source.

2.2 Distributing water at the community level
In a hilly region, spring water is conveyed using bamboo pipes, plastic pipes, or iron pipes from the spring eye to storage tanks and then to the community level (shown in Figure 2). The major reason for pipe supply is to avoid any contamination during transportation. Steps for tapping spring water:

- Survey the site: First, survey the spring eye location and make it tappable. Check the elevation and distance from the spring to the community point or storage structure.
- Source reliability: Before tapping, it is necessary to ensure the spring's reliability. It must provide water throughout the year to the community. Hence, investigate the discharge rate of spring or estimate the yield of spring during the lean period. It helps to design the supply system and storage tank.
- Water quality: Measure the water quality before distributing it to the community as per standard provision. If water quality is under the permissible limit by WHO and BIS, then distribute it to the community. Also, measure water quality periodically to ensure it is potable water.
- Install storage tank and supply system: If communities manage the water supply system, install a storage or settling tank at the spring site and then distribute it to different houses or community points.
- Maintenance of supply system: Sometimes pipes get blocked due to entrapped air, plant and animal debris, suspended particles, pipes breakage, and leakage problems [6]. Communities must appoint a person regularly to maintain the system.
- Protect catchment: When spring water is used for household purposes, the springshed must be protected. Its catchment areas must be a no-go zone for humans and animals, and no cultivation should be permitted. To avoid entry, we fenced the catchment area and made some laws for communities.

2.3 Spring water harvesting
To sustain the water scarcity, conserve it when it is excess, especially during the monsoon or rainy season. Promote water harvesting structure among communities to check the water demand during dry periods. Adopting conventional water harvesting techniques may have limited options due to the soil type and depth, steep, slopes, and rainfall characteristics. So, it is essential to develop a proper strategy for water resource management [24]. Generally, overhead storage tank size is decided based on one- or two-days family demand. It would become a very difficult task if we decided to store water for a completely dry period, extending up to 150 to 200 days. The size of the water harvesting structure depends upon the yield of springs and the number of users, and their demands. Also, the structure size directly increases as per the reduction of spring water during the dry period [6, 25].

2.3.1 *Estimate yield of spring*
Spring yield is defined as the volume of water collected per unit, expressed in various ways. Spring yield may vary from a few litres per minute to several thousand per minute. Spring discharge is measured using the time and volume method by a container and stopwatch in liters per second (l/s). Collect reading three or five times in the day and repeat it every week. Once the data is collected, estimate the total discharge of the spring. Spring yield can be evaluated by summing up daily discharge values for the considered period [6]. A general mathematical equation for estimating spring yield (V*)* shown for the recession period with one exponential component can be expressed as Equation 1:

$$V = \sum_{t_i=0}^{T_i} Q_{t_i} = \sum_{t_i}^{Ti} Q_{0_1} . e^{\alpha i.(\tau - 0)} \qquad (1)$$

Where Q_{0i} and Q_{ti} represent the spring discharge rates for the ith exponential component of the recession curve at time t = 0 and at time t = t_i, respectively, in contrast, T_i and α_i denote the period and recession coefficient of the ith exponential component, respectively [6].

2.3.2 Estimate the theoretical and actual water demand

A water supply plan for household purposes can be only based on perennial sources such as spring. Thus, the availability of spring determines the adequacy of water availability to the area's people, particularly for domestic needs.

- Theoretical domestic water requirements: It can be estimated through the number of persons and animals multiplied by the standard norms of water use. It can be calculated as If the total number of populations is assumed as P_N and daily domestic water demand is D litres per capita per day [6]. The total daily demand can be calculated in Equation 2:

$$T_D = P_N \times D \qquad (2)$$

Also, T_D is known as a daily basic water requirement of P_N Persons.

- Actual domestic water use: It can be estimated based on information from each family by conducting a questionnaire survey such as drinking, cleaning, bathing, cooking, animal use, etc. These individual survey data can be used to estimate the water demand.

3. Spring water use

3.1 Drinking and household use

IHR and other hilly regions use springs and seep springs as the main water source for drinking and household use [14]. So, spring has different religious and cultural values. The reason is that peoples are well aware of the importance of water; they treat spring as sacred and perform many rituals around it. Communities worshipped spring sites as devithan or secret groves. People in the region have developed their hydraulic technologies, which are unique in their uses.

3.1.1 Benefits of using spring water

It has high mineral content such as potassium, magnesium, calcium, iron, zinc, and sodium, which is beneficial for human health and enhances its taste. These minerals come in ionic form, so the human body and rich easily absorb them in oxygen. Also, hot springs have been beneficial for skin and joint disorders.

3.1.2 Public awareness of water usage

In the hilly region, the acute water shortage has been faced during the non-monsoon season due to a decline in winter rains. In this scenario, water resources should be properly managed to meet the demand needs. So, estimate water demand and create awareness among communities for various water-saving techniques such as changing day-to-day behaviour (keeping tap close when not in use) or equipment modifications as discussed below.

- In the Kitchen: Clean vegetables in containers or bowls instead of running water.
- In Bathroom: Take short showers, install minimum uses shower head, close taps when not in use,
- In Laundry: Run full loads of laundry.

Figure 2. Tapping of spring water (a) water directly collected at spring site, (b) water is stored and conveyed with the help of PVC pipe to end-user

3.2 Water-saving irrigation practices

Shortage of potable water has a serious impact on hilly communities. People will collect water from distant sources or pay high prices for water to meet their household needs. But gratifying irrigation demand is a very difficult task, and it seriously affects the farming system and the economic condition of communities. Although, the allocation or accessibility of water for agricultural use cannot be ignored. So, also use water precisely in farming using efficient irrigation systems such as drip irrigation instead of traditional flood irrigation.

3.2.1 Gravity-fed drip irrigation

Modern irrigation methods, particularly drip irrigation, have high efficiency (90%). This irrigation system is suitable and cost-efficient for hilly farmers [15,16]. Based on operating pressure, it is classified into two types: (a) gravity-fed low head drip irrigation system (GLHDIS) and (b) high-head drip irrigation system (HHDIS). The implementation of HHDIS requires a pumping unit to build up a high pressure of 15 m or more. It requires electricity and fuels to operate the pumping unit, which is costly for small farmers; also, it may be impractical to have electric power available or carry engines to remote locations. But GLHDIS, a gravity-fed drip irrigation system (GDIS), requires neither any pumping equipment nor electric power unit to drive the pump (refer Figure 3). The operating pressure for GDIS is 0.5 - 0.3 m, which is naturally available in hilly areas due to the elevation difference between traced field and the overhead tank. The major advantage of GDIS is that it can be adopted in various types of soil under undulating topography. It minimizes soil erosion by maintaining the soil moisture, and conserving water, is cost-effective, saves energy, increasing crop productivity and farmer's income. GDIS consists of a storage tank, stand, mainline, sub-mainline laterals, valves, tubing adapters, fittings, drip tubing, emitters, and end caps.

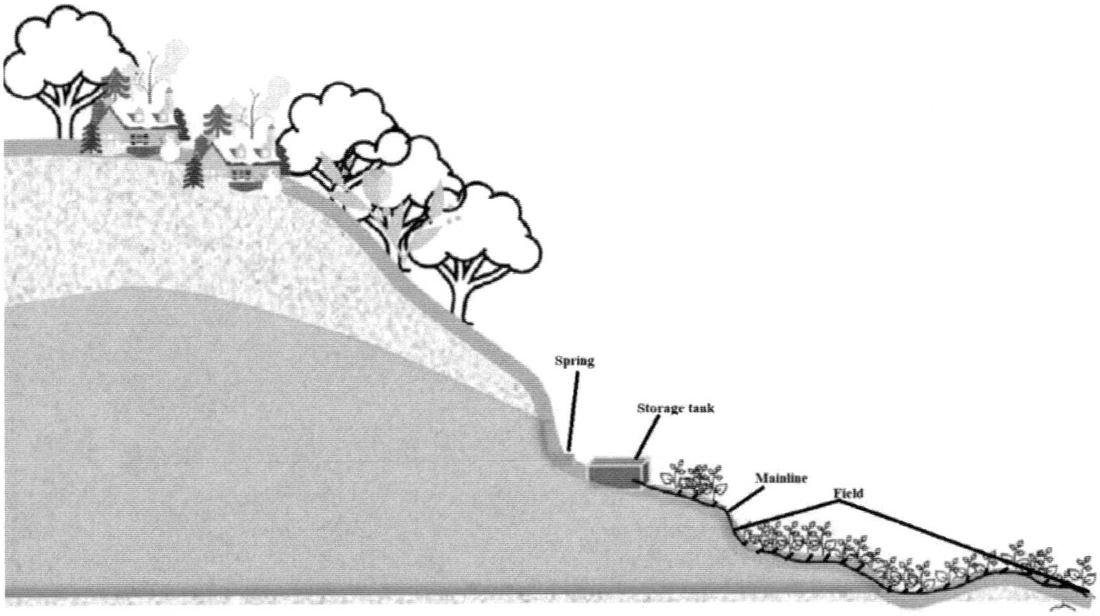

Figure 3. Gravity-fed drip irrigation system

3.2.2 Bamboo drip irrigation

Bamboo drip irrigation (shown in Figure 4) is the traditional irrigation system of Meghalaya, Mizoram, and the other North East States. In this indigenous system, spring and stream water are tapped for irrigation using bamboo channels. In this region, soil's water retention capacity is low due to rocky terrain and steep slopes where ground channelling is impractical. So, the farmers of Khasi and Jaintia hills have been using bamboo drip irrigation systems for more than 200 years [17]. They employ a bamboo drip irrigation system to irrigate betel leaf and black peppers in mixed orchards during the lean period. The different diameters of holed bamboo shoots divert spring water in terraced cropland. This complex design of the irrigation system has been perfected over 200 years of practice, which deliver water at a rate of 18 to 20 lpm [18] and can gradually deliver water over more than 100 meters [19, 20]. Finally, it is decreased to 20 - 80 drops per minute at the end of its length.

The Channel section of bamboo drip irrigation diverts and conveys water to the cropland and distributes it into branches without leakage. The flow of water in lateral pipes can be monitored by manipulating the positions intake pipe. About four to five distribution phases are involved from the water diversion point to the implementation point. In the last phase of water implementation, reduced channel parts and diversion units are used. The last segment of the channel allows the water to drop near the plant's roots. Channels are placed with bamboos of variable diameters. Approximately 1/3rd of the outer shell in length and bamboo piece internodes need to be separated while the system is being manufactured. Other parts are small pipes and channels of variable dimensions used for the main channel water diversion and distribution [17].

Materials required: Dao for cutting and smoothing bamboo shoots, various size bamboo strands, forked branches, smaller bamboo shoots for channel diversions, and two trained workers. Approximately in 15 days, two workers can assemble an irrigation system covering 1 hectare of cropland [17,19]. From the point of the water source to the endpoint of water application, about four to five-stage zig-zag channels are constructed (shown in Figure 3). Ultimately, 18 to 20 lpm water spread at 20-80 drops per minute [21].

Steps to installing bamboo drip irrigation: The first step is to locate nearer water resources such as springs or streams. Then choose an area of at least 30 m slope variation. Then cut the bamboo shoots and their forked branches and fix larger shoots in the first channel and smaller ones in the last stage (installation plan for 4 or 5 stages). Puncture and space a sequence of holes in the shoots. Water can be dropped near the plant roots due to a gradually dropped ground clearance in the last section. To strengthen the structure, use fibre-rich twine to bind pipes and forked branches. Use smaller bamboo shoots at diversion points to divert water [22].

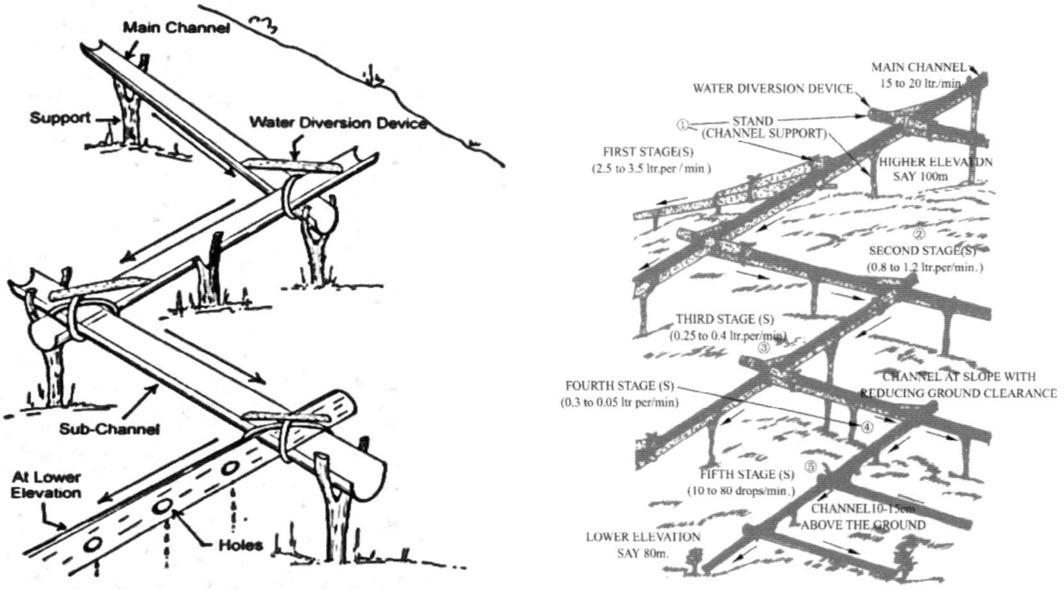

Figure 4. Design of bamboo drip irrigation [21,22]

Maintenance of irrigation system: Farmers maintain the irrigation system by themselves or in communities. Repair work shall be performed as and when necessary. Generally, materials used in bamboo drip irrigation last for about three years after installation. So, maintenance is limited to cleaning and strengthening. The major reason behind using of bamboo drip irrigation system in Meghalaya is:

- It is a faster-growing renewable resource, making it a reliable supply for farmers in the next decades.
- There is no need for harmful fertilizers and pesticides to grow, and it has antibacterial features. It also generates approximately 35% of the oxygen that keeps the air clean.
- It's easy to repair and bio-degradable. It can naturally turn into humus that can maintain soil fertility.
- It increases crop yield with less water, prevents leakage, and is naturally available at a low cost.
- When the worldwide temperature alteration, exhaustion of assets, and deforestation compromise the equalization of the fragile regular world and its different biological systems, bamboo is a reasonable arrangement and asset that is surprisingly valuable and eco-accommodating. It is a supportable, sustainable asset.

3.2.3 Bottle irrigation/pitcher irrigation

Both bottle and pitcher irrigation (shown in 5) irrigation techniques are similar to drip irrigation but less expensive to install. Also, no need for high-tech gadgets. It is ideal for small landholdings and gardens. In pitcher irrigation, it requires the round clay containers used for water storage ranging from 10-20 litres and buried in the soil. The natural pores of the pot allow water to seep into the soil, providing moisture for crop growth. It is good for deep-rooted plants [23]. Bottle irrigation requires one or two or more bottles, punch holes per plant requirement and hang withstand. It releases water slowly and keeps plants hydrated. Slow watering helps plant roots develop deeply. The major advantage of pitcher and bottle irrigation are (a) the area near the pot and bottle gets irrigated, (b) evaporation of water is minimum, (c) reduce expenses, and (d) saves water.

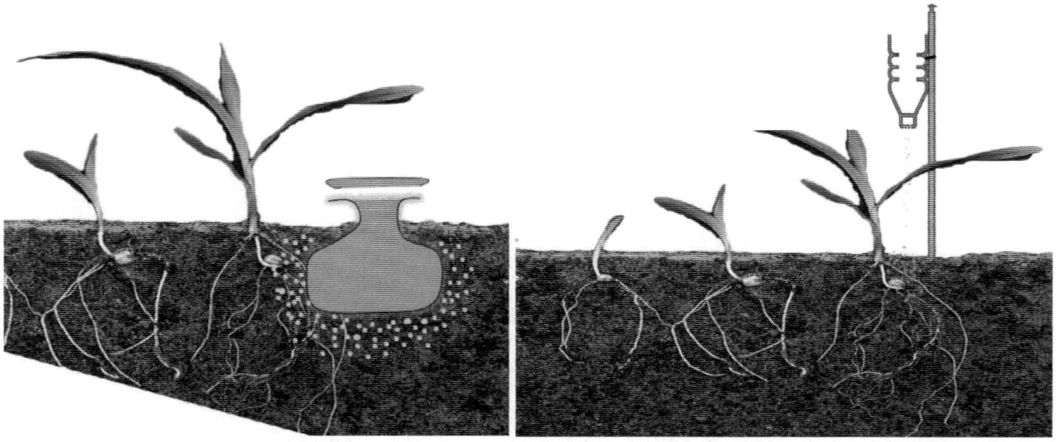

Figure 5. (a) Pitcher irrigation and (b) Bottle Irrigation

4. Conclusion

A well water management strategy is necessary to sustain water scarcity conditions. Springs are the primary water sources of the hilly region, so in this climate change scenario, spring water management practices are necessary. Planning proper storage, distribution, and utilization at the community level is required. Also, water transfer from excess to shortage areas through gravity flow is required. Springwater harvesting and tapping using the gravity-based technique to convey water to the community can easily end the drudgery of the tribal women and children. Before harvesting spring water, estimate the total demand in the dry period and then construct an efficient water harvesting structure. This is based on gravity-fed and no need for energy to convey water, so this system is a financially and technically sustainable solution for communities. An efficient irrigation system is highly recommended for the region to minimize spring water use. Such traditional bamboo drip irrigation systems widely used in the north-eastern state, especially in Meghalaya, for a long back are eco-friendly practices that help in conservation, protection, and better use of natural resources. Also, adopt modern irrigation methods such as drip irrigation, sprinklers, and locally suitable or modified methods with higher efficiency than traditional irrigation practices.

References

[1] Vashisht AK, 2015 Adaption of seepage spring development technique to manage the water scarcity in Himalayan region–a practical approach. *Water Utility Journal* **11** 93-98.

[2] NITI Aayog, 2017 Inventory, and Revival of Springs in the Himalayas for Water Security. Dept. of Science and Technology, Government of India, New Delhi.

[3] Tambe S, Kharel G, Arrawatia ML, Kulkarni H, Mahamuni K, and Ganeriwala AK, 2012 Reviving dying springs: climate change adaptation experiments from the Sikkim Himalaya. *Mountain Research and Development*, **32** 1 62-72.

[4] Ranjan P, Pandey PK, Pandey V, and Lepcha PT, 2021 Development of seepage spring for rural water security. In *IOP Conference Series: Materials Science and Engineering* **1070** 1 012047. IOP Publishing.

[5] Mall RK, Gupta A, Singh R, Singh RS, and Rathore LS, 2006 Water resources and climate change: An Indian perspective. *Current Science* 1610-1626.

[6] Vashisht AK, 2017 Use of mathematical techniques for determining the reliability of springs. In *Development of Water Resources in India* 113-124. Springer, Cham.

[7] Tambe S, Arrawatia ML, Bhutia NT, and Swaroop B, 2011 Rapid, cost-effective and high resolution assessment of climate-related vulnerability of rural communities of Sikkim Himalaya, India. *Current Science* 165-173.

[8] Lepcha PT, Pandey PK, and Ranjan P, 2021 Hydrological significance of Himalayan surface water and its management considering anthropogenic and climate change aspects. In *IOP conference series: materials science and engineering* **1020** 1 012013. IOP Publishing.

[9] Tambe S, and Arrawatia ML, 2012 Climate Change initiatives in Sikkim. *Climate Change in Sikkim Patterns, Impacts and Initiatives. Information and Public Relations Department, Government of Sikkim, Gangtok.*
[10] Sharma B, Nepal S, Gyawali D, Pokharel GS, Wahid S, Mukherji A, Acharya S, and Shrestha AB, 2016 Springs, storage towers, and water conservation in the midhills of Nepal. *International Centre for Integrated Mountain Development.*
[11] Ranjan P, Pandey PK, Pandey V, and Lepcha PT, 2021 Water Resources Assessment Issues and Application of Isotope Hydrology in North East India. In *Water Resources Management and Reservoir Operation* 103-116. Springer, Cham.
[12] ICIMOD 2015 Reviving the Drying Springs Reinforcing Social Development and Economic Growth in the Midhills of Nepal.
[13] Ranjan P, and Pandey PK, 2020 Reviving, development and protection of springs to increase water security in the Himalayan Region. Computer research network (SSRN) 1-7. *ICCM 2019.*
[14] Vashisht AK and Sharma HC, 2007. Study on hydrological behaviour of a natural spring. *Current Science* 837-840.
[15] Bhatnagar P, and Srivastava R, 2003 Gravity-fed drip irrigation system for hilly terraces of the northwest Himalayas. *Irrigation science*, **21** 4 151-157.
[16] Rigve R, Bhardwaj A, Satpute S, Singh J, and Garg S, 2018 Standardization of Gravity-Fed Low Head Drip Irrigation System under Variable Operational Head for Small Farms.
[17] Ghosh SN, 2016 Eco-technological Practices for Sustainable Development. In *Environmental Hydrology and Hydraulics* 255-392.
[18] Jain SK, Agarwal PK and Singh VP, 2007 *Hydrology and water resources of India* **57**. Springer Science & Business Media.
[19] Kumar A, and Madhukar AK, 2019 Management of traditional water system and their conservation in North Eastern Region through local traditional wisdom. In *AIP Conference Proceedings* **2142** 1 210001). AIP Publishing LLC.
[20] Jeeva SRDN, Laloo RC and Mishra BP, 2006 Traditional agricultural practices in Meghalaya, North East India.
[21] Singh RA, and Gupta RC, 2002 Traditional land and water management systems of North-East hill region.
[22] Dabral PP, 2002 Indigenous techniques of soil and water conservation in North Eastern Region of India. In *12th ISCO Conference, Beijing, China* 26-31.
[23] Pal A, Adhikary R, Karbera M, and Nayak R, 2020 Pitcher irrigation: a water-saving technique in arid region. *International Journal of Engineering, Science and Mathematics* **9** 2 50-57.
[24] Kusre BC, Bora PK, Rai D, Adhikari K, Khuman NC, and Tabing J, 2018 Developing water resources management strategies for South Sikkim District, India. *Water Science and Technology: Water Supply* **18** 2 648-659.
[25] Vashisht AK, 2016 Development of an analytical function for optimizing the capacity of spring water storage structure. *Sustainable Water Resources Management*, **2** 4 337-352.
[26] Tayong A, 2003 8 Spring water tapping. *Small Community Water Supplies: Technology, people and partnerships.*

Application of circular economy in wastewater treatment using biochar based adsorbent derived from sewage sludge

Pooja Kenchannavar, Anupama Surenjan*

Department of Civil Engineering, National Institute of Technology- Karnataka, Surathkal, Mangaluru-575025

*anupama@nitk.edu.in

Abstract. The Circular Economy intends to reuse wastewater and recover resources and energy from the sludge in the wastewater sector. The conversion of sewage sludge into biochar by pyrolysis technique is a sustainable, environmentally friendly option for sewage sludge management. In the current work, biochar was produced by co-pyrolyzing sewage sludge and coconut shell, and it was then utilized as an adsorbent to eliminate methylene blue (the model pollutant) from wastewater. The sewage sludge is abundant in functional groups and metal ions, and the addition of coconut shells increased the porosity of the adsorbent. The biochar produced by mixing coconut shell and sewage sludge in the ratio1:1 showed higher adsorption capacity compared to biochar obtained from pure sewage sludge; hence the co-pyrolyzed biochar was used for further studies. The morphology and functional groups present on biochar were examined using SEM and FTIR analysis. The kinetics and mechanism of methylene blue adsorption was adequately explained by the pseudo-second-order kinetic model. Using the Langmuir isotherm model, the maximal adsorption capacity of the methylene blue dye on biochar was observed to be 31.64 mg/g.

1. Introduction

During the last couple of decades, a substantial quantity of sewage sludge has been generated in sewage treatment plants worldwide. The management and disposal of sewage sludge are crucial due to toxic compounds detrimental to the environment. Developing countries like India lacks sufficient wastewater treatment plants and sludge treatment. According to the CPCB 2021 report, India's sewage generation is estimated to be 72,368 MLD. But there are only 1631 STPs with a total capacity of 36,668 MLD [1]. The current waste management situation requires sustainable ways to switch toward the recovery of resources from waste. Hence, the shift towards sustainable sewage sludge management is a foremost factor for zero waste generation. In this regard, the shift toward a Circular Economy (CE) is essential since it is an economic model that strives to preserve the utility of products as much as possible, reduces waste generation, and promotes material reuse. The CE integrates economic prosperity and environmental quality and is developed as another possibility for the linear economic model. The linear economy model has usually followed the 'take-make-use-dispose' approach, in which the collected raw material is transformed into useful products, used for particular purposes, and disposed of as waste finally. The CE follows the 4R's framework: Reduce, Reuse, Recycle and Recover [2].

Sludge is a solid or semi-solid residue produced during wastewater treatment as a by-product. An enormous volume of sewage sludge is generated at WWTPs globally. The proper disposal and management of sludge are difficult due to its harmful effects on the environment. Circular Economy finds application in recovering nutrients, proteins, enzymes, bio-plastics, heavy metals, and construction materials from sewage sludge in wastewater treatment plants. One such application of resource recovery/recycling of waste from sewage sludge is biochar production. Hence, recognizing sludge as a 'resource' and not as waste has effectively utilized the sludge in resource and energy recovery. Biochar is a biomaterial abundant in carbon, formed by the thermal breakdown of feedstock. Since the biochar has broad surface area, well-developed porous structure, and active functional group, it can be utilized as an adsorbent to eliminate pollutants from wastewater. Therefore, the biochar production from sewage sludge, which is regarded as waste, can be inventive [3].

Hence, in this present study, the biochar is prepared using sewage sludge and coconut shells by the co-pyrolysis process. The developed adsorbent is applied to adsorb dye. Methylene blue dye is used as a model pollutant. Adsorption is a surface phenomenon where the adsorbate (methylene blue) gets deposited on the adsorbent's (biochar) surface. Adsorption is regarded as an efficient and feasible technology for water/wastewater decontamination owing to its low cost, ease of application, flexibility, and sensitivity to hazardous compounds. The methylene blue can cause vomiting, inhalation problems, eyes burn, and methemoglobinemia when exposed to living beings. Hence, removing methylene blue dye is the prime concern [4]. Since the sewage sludge has less carbon content, there may be improper development of surface area, and the porous structure, which affects the adsorption capacity of the biochar. Hence, this limitation of sewage sludge biochar can be addressed by producing biochar by co-pyrolysis technique utilizing lignocellulosic biomass, which has abundant carbon content. Hence, the coconut shell is used as lignocellulosic biomass since it is abundantly available in region of south canara. Therefore, the objective of the study is to effectively implement a Circular Economy in wastewater treatment by synthesis of sewage sludge biochar using the co-pyrolysis technique and its potential application as an adsorbent to eliminate the model pollutant methylene blue.

2. Materials and methods

2.1. Biochar preparation

The sewage sludge was collected from the NITK, Surathkal campus wastewater treatment plant, while the coconut shells were brought from the Surathkal local market. The sewage sludge and coconut shell were cleaned with deionized water, sun-dried, and then dried in a hot-air oven for 1 hour at 100 °C. The dried raw materials were crushed to a 100 mesh size for further use. Two different biochars were prepared to understand the effect of co-pyrolysis using coconut shells; 1.Pure sewage sludge biochar 2. sewage sludge and coconut shell (1:1) biochar. The pyrolysis process was used to synthesize biochar at a temperature of 600°C. In the pyrolysis experiment, a certain quantity of sample is placed in a closed container and heated in a muffle furnace for 2 hours. Nitrogen was pumped into the muffle furnace to provide an inert environment. After the stated period, the biochar was allowed to cool naturally before being collected. [5].

2.2. The yield of biochar

The biochar yield was computed as the percentage of the mass of produced biochar to the mass of pyrolysis feedstocks, as shown below.

$$y(\%) = \frac{m_1}{m_2} \times 100 \qquad (1)$$

Where y is the yield of biochar; $m_1$is mass of biochar produced; m_2 is the mass of raw materials mix

2.3. Characterisation of biochar

The characterisation of the biochar was done using various techniques. Fourier transform infrared

(FTIR) was used to assess the functional groups present on the biochar surface, and the spectrum was acquired in the wavelength range of $400cm^{-1}$ to $4000cm^{-1}$. The biochar's morphological structure is evaluated using scanning electron microscopy (SEM) with an energy dispersive X-ray spectrometer (EDS).

2.4. Adsorption experiment
The batch adsorption studies were performed using methylene blue dye to evaluate biochar's adsorption ability. All batch experiments were conducted under pH 9 with 50mL solution, and the samples were kept in an incubator orbital shaker at a speed of 120 rpm.

In conical flasks, 0.2g of biochar was introduced to a 50 mL methylene blue solution at different concentrations (10mg/L, 25mg/L, 50mg/L) for the adsorption kinetic study. The samples were taken at varied intervals and filtered by using a vacuum filter through the use of a 0.45 µm membrane. The concentration of methylene blue in the filtrate was determined using a spectrophotometer with a wavelength of 665nm. The influence of pH on methylene blue adsorption was tested by changing the pH from 2 to 13 using 0.1M NaOH and 0.1M HCl solution. The biochar dose was varied from 2g/L to 12g/L to determine the influence of biochar dose on adsorption capacity by keeping a fixed adsorbate concentration of 50mg/L. The adsorption capacity and removal efficiency are both determined using the equations (2-4) as follows:

Adsorption capacity (mg/g), $$q_t = \frac{C_o - C_t}{m} \times v \quad (2)$$

$$q_e = \frac{C_o - C_e}{m} \times v \quad (3)$$

Removal efficiency (%), $$R = \frac{C_o - C_e}{C_o} \times 100 \quad (4)$$

here, the q_t (mg/g) and q_e (mg/g) adsorption capacity of biochar at a contact time t and equilibrium respectively; C_o, C_t, C_e(mg/L) represents the initial concentration of adsorbate, adsorbate concentration at the contact time t and equilibrium respectively; V (L) represents the volume of solution; m(g) is the mass of biochar added to solution; R(%) is the percentage removal of adsorbent.

3. Results and discussion

3.1. The yield of biochar.
Figure 1 depicts the biochar yield produced from a mixture of coconut shell and sewage sludge (1:1) and pure sewage sludge.. The decrease in biochar yield is found when mixed with coconut shells. This could be because coconut shells contain more organic matter and lower ash content than sewage sludge [6]. But from the experiment results shown in figure 2, the biochar with coconut shell showed a good adsorption capacity when compared to biochar made with sewage sludge. Therefore the biochar prepared with coconut shell and sewage sludge was used for further studies.

3.2.Charcterisation of biochar
The functional groups present in biochar prior to methylene blue adsorption and later were investigated using FTIR, is depicted in figure 3. The biochar produced by co-pyrolyzed biochar showed an abundance of functional groups. The strong, broad FTIR spectra at wavelength 3435 cm^{-1} are primarily responsible for the stretching vibration of the hydroxyl group (OH^{-1}) with intermolecular bonding. The absorbance peak between 1600 to 1500 cm^{-1} shows the existence of an aromatic C=C ring stretching. So, the peak at 1589 cm^{-1} is recognized as stretching of the aromatic C=C ring. The sharp peak with medium intensity at 1255 cm^{-1} illustrates the C-O stretching of alcohol, ether, and amides. It is seen in figure 5b. the peaks of hydroxyl groups and carbonyl groups shifted to 3328 cm^{-1}

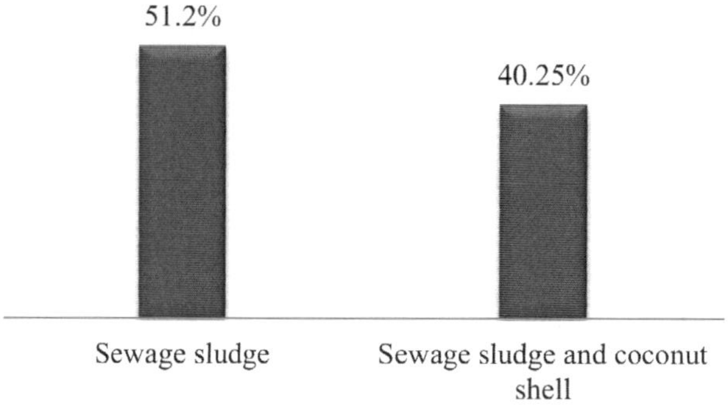

Figure 1. Yield of biochar

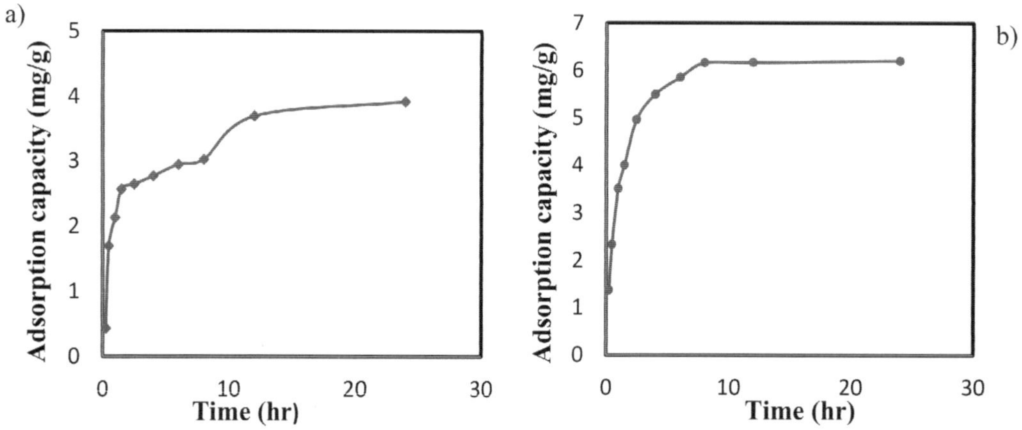

Figure 2. The adsorption capacity of biochar a) sewage sludge b) sewage sludge and coconut shell

and some of the peaks disappeared. This suggested that the biochar functional groups could have a role in the methylene blue adsorption process [7].

The SEM-EDS analysis was done on biochar before and after adsorption to determine the morphology, as shown in figure 4. The micrographs of SEM analysis on biochar structure show the development of pores on the surface. The biochar prepared from sewage sludge and coconut shell shows an uneven surface with cracks, pores, and voids on the surface. The pore formation can be explained due to the breaking of lignin and cellulose and the decomposition of volatiles, which have developed active sites, responsible for adsorption [8]. The biochar surface after methylene blue adsorption becomes comparatively smoother and compact. This might have happened due to the clogging of pores by methylene blue molecules. The SEM-EDS analysis revealed the composition of biochar, and it is reported in Table 1. The biochar is composed of carbon (72.8%), oxygen (16.6%), and nitrogen (3.7%). It also contained small traces of Mg, Al, Fe, Ca, P, and Cu since the sewage sludge is a complex substance that includes organic compounds, micropollutants, trace elements, micro-organisms, etc. The high amount of carbon in the sample confirms the development of the pores of the biochar. Similar results were reported with biochar prepared using seaweed [8] and co-pyrolysis of waste tires and sewage sludge [9].

Table 1. EDS analysis Composition of biochar

Element	C	N	O	Mg	Al	Si	P	Ca	Fe	Cu
Weight %	72.9	3.7	16.6	0.5	0.6	1.4	1.1	0.6	0.8	01.6

a)

b)

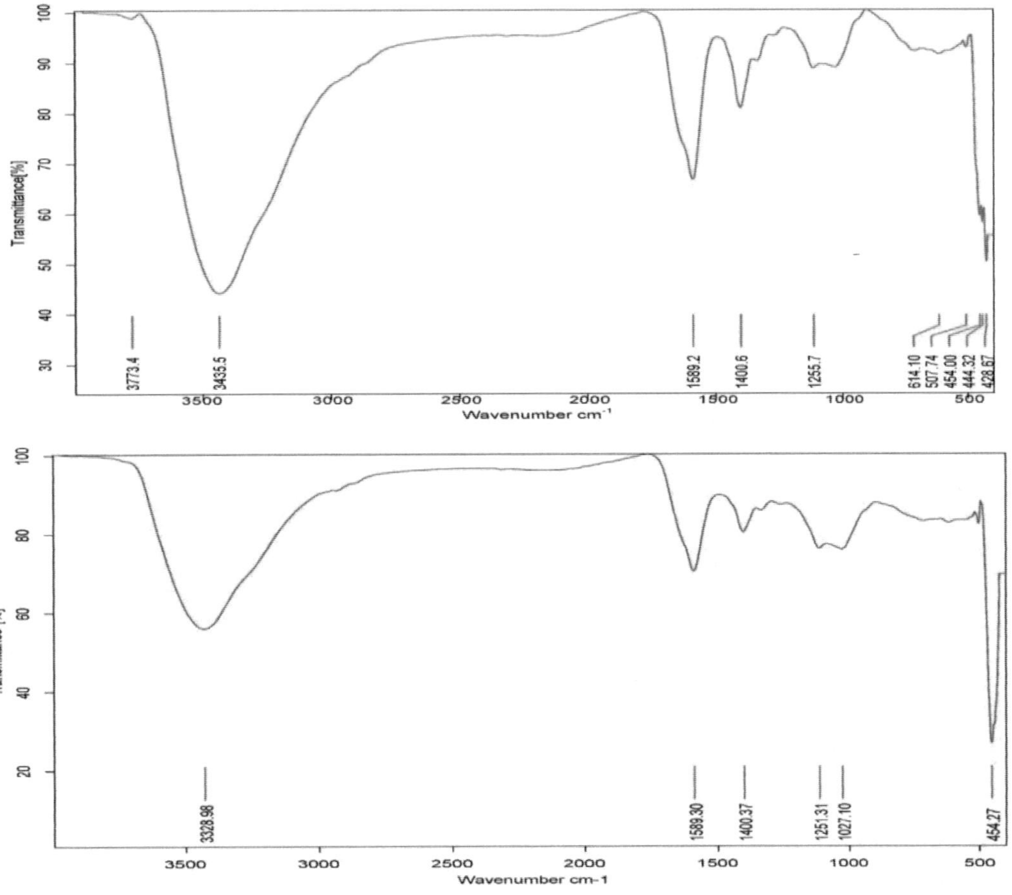

Figure 3. FTIR Spectrum a) before b) after Methylene blue adsorption

a) b)

c)

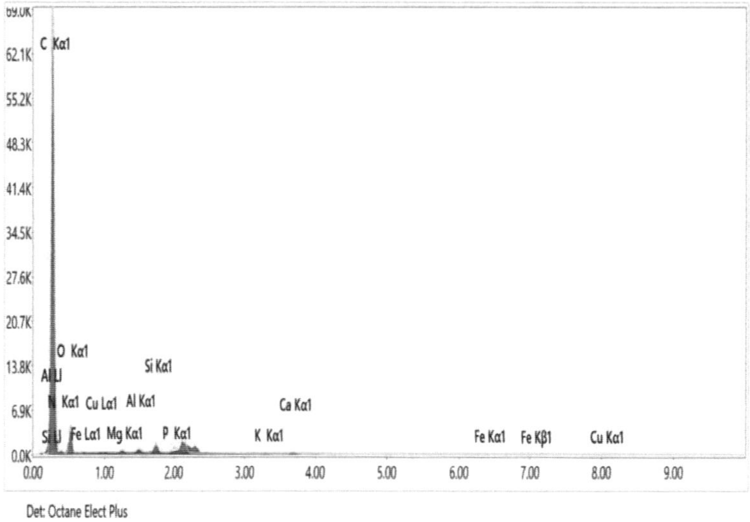

Figure 4. SEM images of biochar (a) before adsorption (b) after adsorption (c) EDS spectra of biochar composition

3.3. Adsorption experiment

*3.3.1.Effect of contact time***.** The impact of contact time on methylene blue percentage removal and adsorption capacity is depicted in Figure 5. The contact time is a requisite feature in determining the extent of the adsorption process. The adsorption occurred rapidly until 2.5hrs, and the percentage removal reached 97.28%. The adsorption capacity was also 11.978 mg/g. However, extending the contact time furthur did not result in a considerable increase in adsorption capacity, which is discussed as below.

Initially, the rapid increment in adsorption capacity was noticed as contact time increased owed to the existence of more active sites for adsorption to take place on the adsorbent surface. However, no notable change in adsorption was observed as contact time was further increased beyond 2.5hrs. This might be attributed to a deficit of active sites on the adsorbent's surface, and the adsorbent might have reached equilibrium. A significant change in adsorption was not seen when contact time increased beyond 2.5hrs. Hence it can be concluded that the equilibrium was reached, and the optimum time was judged to be 2.5hrs for the adsorption process [10].

3.3.2.Effect of biochar dose. The essential factor, the biochar dosage, influences methylene blue adsorption on the biochar. The influence of biochar dose on methylene blue percentage removal and adsorption capacity is depicted in Figure 6. When biochar dose varied from 2g/L to 12g/L, the percentage removal of the methylene blue increased from 87.1% to 98.72%, but the adsorption capacity of biochar reduced to 4.11 mg/g. The increment in the biochar dosage provides additional specific active sites for adsorption to take place, but at the higher concentration, the adsorbent might undergo agglomeration, and due to this, the overall active sites might have decreased thereby decreasing the adsorption capacity [11]. When the biochar dose is 4g/L, the efficiency of biochar reaches above 90%, and the adsorption capacity was also 11.57mg/g. Hence the suitable biochar dosage is considered to be 4g/L.

3.3.3.Effect of solution pH. Another essential factor influencing adsorption onto biochar is pH. Figure 7 depicts biochar's percentage removal and adsorption capacity at pH levels from 2 to 13. From the experimental values, the increase in methylene blue percentage removal from 26.92 % to 99.58 % was reported as the pH rose from 2 to 11. The increase in the percentage removal illustrates that the alkaline condition promotes methylene blue adsorption on biochar. The pH-induced effect of methylene blue adsorption on biochar can be summed up as follows. At lower pH, the biochar surface was occupied by H^+ ions making the biochar surface positively charged. Also, methylene blue is a cationic dye and is positively charged. Since the adsorbent surface and adsorbate are positively charged, a repulsive force existed between them, which might have restrained the adsorption of methylene blue. But as the pH was raised, the H^+ ions present in the solution were lessened, resulting in the increased adsorption capacity of biochar [12]. Since the removal efficiency reached 91.06% at pH 9, all the experiments were conducted by adjusting pH at 9.

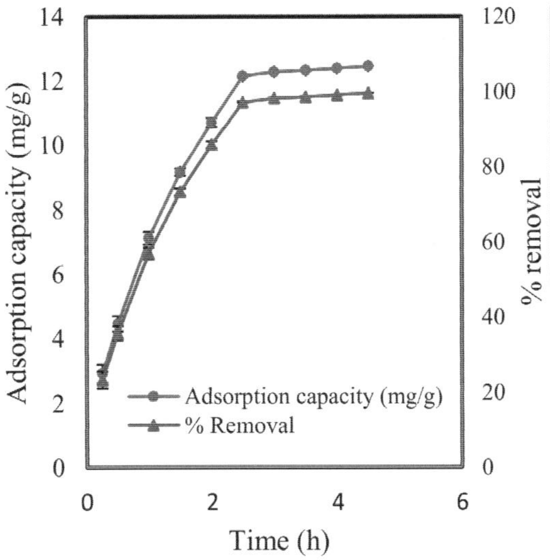

Figure 5. Effect of contact time

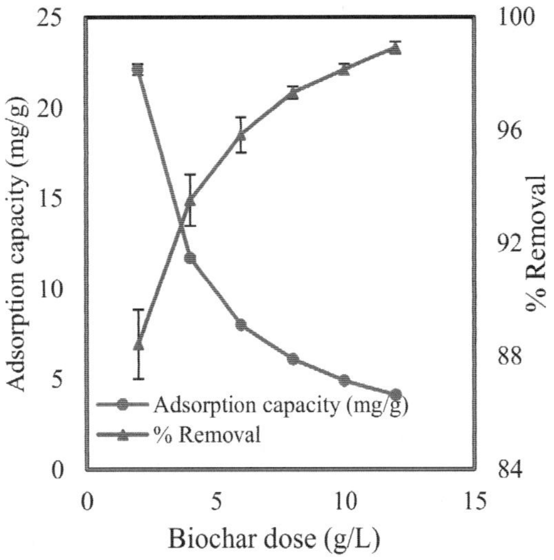

Figure 6. Effect of biochar dosage

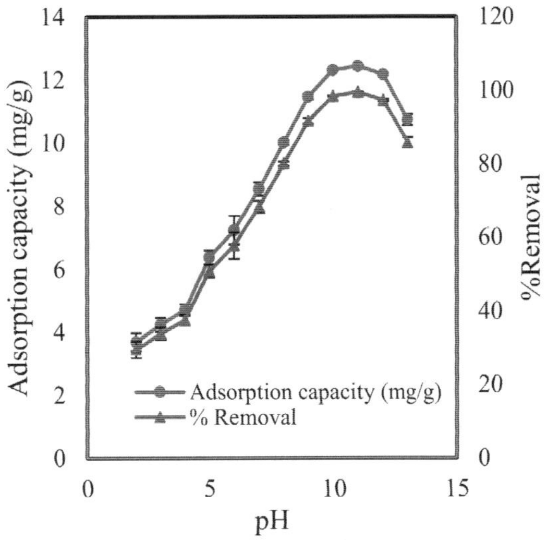

Figure 7. Effect of pH

3.3.4. Kinetic studies. The adsorption kinetic studies provide details of the reaction rate or diffusion, which shows how faster the adsorption process occurs. It also gives information on the mechanism of adsorption involving mass transfer, diffusion, and interaction of solute particles on surface sites. The

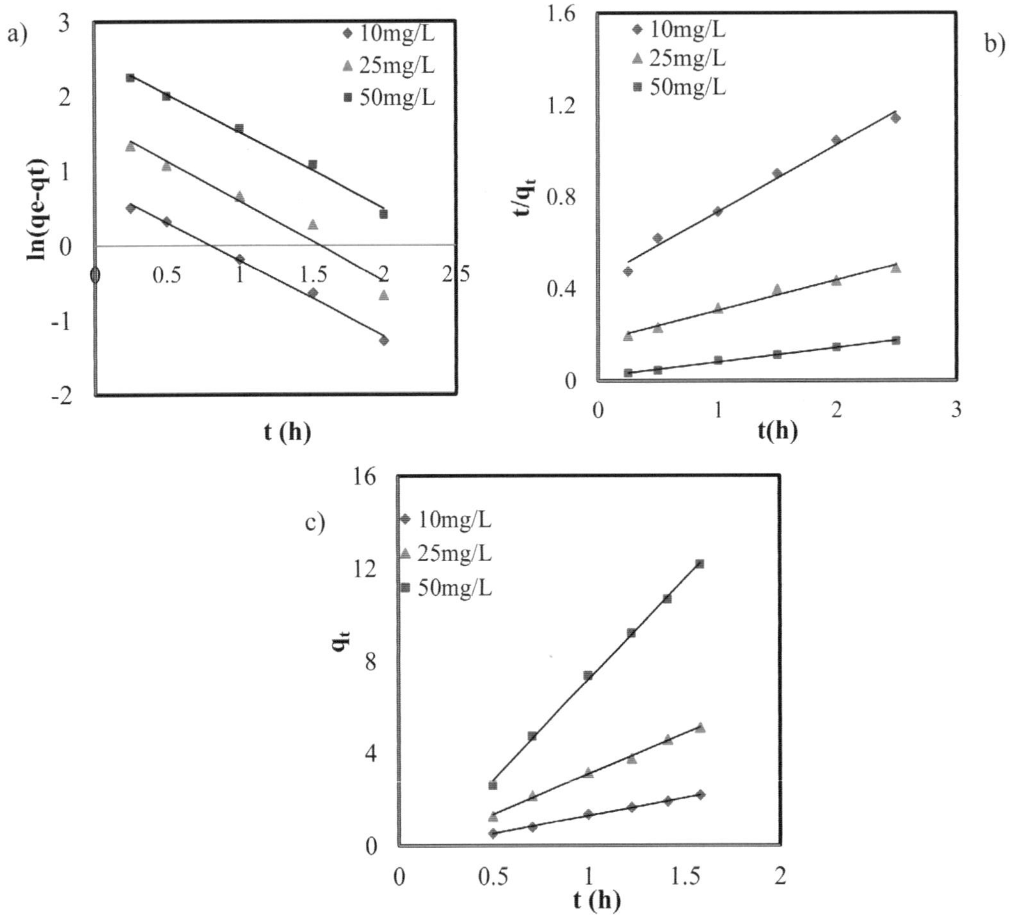

Figure 8. Kinetic data for methylene blue adsorption on biochar a) pseudo-first-order model b) pseudo-second-order model c) particle diffusion model

adsorption of ions or molecules on the adsorbate is usually explained as a mass-transfer system followed by surface interaction between adsorbate and adsorbent. The adsorption mass transfer system comprises three steps: 1. diffusion of molecules or ions in solution 2. diffusion in water film 3. diffusion in liquid contained pores also called intraparticle diffusion [13]. Because bulk liquid and film diffusion contribute less to diffusion rate, intraparticle diffusion, — in other words, diffusion through the surface and pores, is regarded the rate-limiting step. It is commonly represented as the intraparticle diffusion kinetic model. The reaction-controlled kinetics are represented by the pseudo-first-order and pseudo-second-order kinetic models, which illustrate that adsorption itself is a rate-limiting step [13].

In the present study, kinetic models such as pseudo-first-order, pseudo-second-order, and intraparticle diffusion models were used to understand the adsorption kinetics and mechanism of methylene blue adsorption. The equations (5-7) are as follows:

Pseudo-first-order: $$log(q_e - q_t) = log\, q_e - \frac{K_1}{2.303} \times t \quad (5)$$

Pseudo-second-order $$\frac{t}{q_t} = \frac{1}{K_2 q_e^2} + \frac{t}{q_e} \quad (6)$$

Intra-particle diffusion : $q_t = K_d t^{1/2}$ +C (7)

Where, q_t (mg/g) and, q_e(mg/g) are adsorption capacity of biochar at a contact time t and equilibrium time. K_1, K_2 and K_d are the rate constants of pseudo-first-order, pseudo-second-order, and intra-particle models, respectively.

The graphs of the kinetic study are depicted in figure 8 and the kinetic parameters are reported in table 2. From the data, the pseudo-first-order model has less regression coefficient ($R^2 < 0.97$) and the experimental q_e is not in good agreement with calculated q_e values, which shows that this model is not appropriate for methylene blue adsorption on biochar. But the regression co-efficient obtained by pseudo-second-order is greater than 0.98 for all methylene blue concentrations and also the q_e value is closer to experimental values. This suggested that the pseudo-second-order model presented the best correlation with experimental data and is a better-suited kinetic model to illustrate methylene blue adsorption on biochar. According to the assumptions of this model, it is possible to deduce that chemisorption, which incorporates valence forces in the creation of chemical bonds between adsorbate and adsorbent, governs overall adsorption. Therefore adsorption might be result of chemical reaction between methylene blue and the biochar. Similar findings were seen for methylene blue adsorption biochar prepared from sewage sludge [14].

Table 2. Adsorption kinetic parameters for methylene blue adsorption on biochar

concentration (mg/L)	pseudo-first-order			pseudo-second-order			Intra particle diffusion		
	q_e (mg/g)	K_1(hr^{-1})	R^2	q_e(mg/g)	K_2(hr^{-1})	R^2	K_d	C	R^2
10	3.038	-0.1819	0.9715	3.439	0.1901	0.9863	0.3417	0.2139	0.9873
25	7.276	-0.1925	0.9376	7.541	0.1018	0.9831	0.7727	0.6449	0.9905
50	17.199	-0.1838	0.9737	15.949	0.216	0.9949	1.9213	1.0624	0.9828

3.3.5 Isotherm studies. Adsorption isotherms, which are a typical feature of a particular adsorbent-adsorbate interaction at a specific environmental condition, are commonly used to investigate the process of adsorption. These isotherms are often used to characterize the interaction between the adsorbed and dissolved concentrations at equilibrium. In this study, three isotherm models were used to describe adsorbate-adsorbent interaction; Langmuir, Freundlich, and Temkin.

The Langmuir isotherm model is centred on monolayer adsorption, homogeneous sites, and uniform structure of adsorbent. The Freundlich isotherm model is an empirical method for heterogeneous systems that describes multilayer coverage on the adsorption sites, whereas the Temkin isotherm is formed on the idea that molecule adsorption on the surface reduces linearly with surface coverage due to adsorbate-adsorbent interaction [15]. The equations (8-10) of the isotherms are as follows:

Langmuir isotherm: $$\frac{C_e}{q_e} = \frac{C_e}{q_m} + \frac{1}{q_m K_L} \quad (8)$$

Freundlich isotherm: $$log q_e = log K_F + \frac{1}{n} log C_e \quad (9)$$

Temkin isotherm: $$q_e = \frac{RT}{b_T} ln K_T + \frac{RT}{b_T} ln C_e \quad (10)$$

The results of various isotherms are plotted and shown in figure 9. The Langmuir isotherm showed a more significant regression coefficient ($R^2 > 0.99$) than the other two isotherm models.The

IOP Conf. Series: Earth and Environmental Science 1084 (2022) 012063 doi:10.1088/1755-1315/1084/1/012063

experimental data also showed better fitting with the Langmuir isotherm, which indicates that the adsorbent surface is uniform; only monolayer adsorption takes place, which involves both physical and chemical adsorption. Also, there is no interaction between adjacent molecules adsorbed molecules when a molecule occupies an active site.

The constant separation factor (R_L) equation (11) is used to determine whether the process is favorable or unfavorable.

$$R_L = \frac{1}{1+K_L C_O} \tag{11}$$

C_O is the largest initial methylene blue concentration (mg/L)
When $R_L = 0$ shows irreversible adsorption;
$0 < R_L < 1$ shows favourable adsorption;
$R_L > 1$ shows unfavourable adsorption;
$R_L = 1$ linear adsorption.
In the present study, the $R_L < 1$ shows the adsorption is favourable.

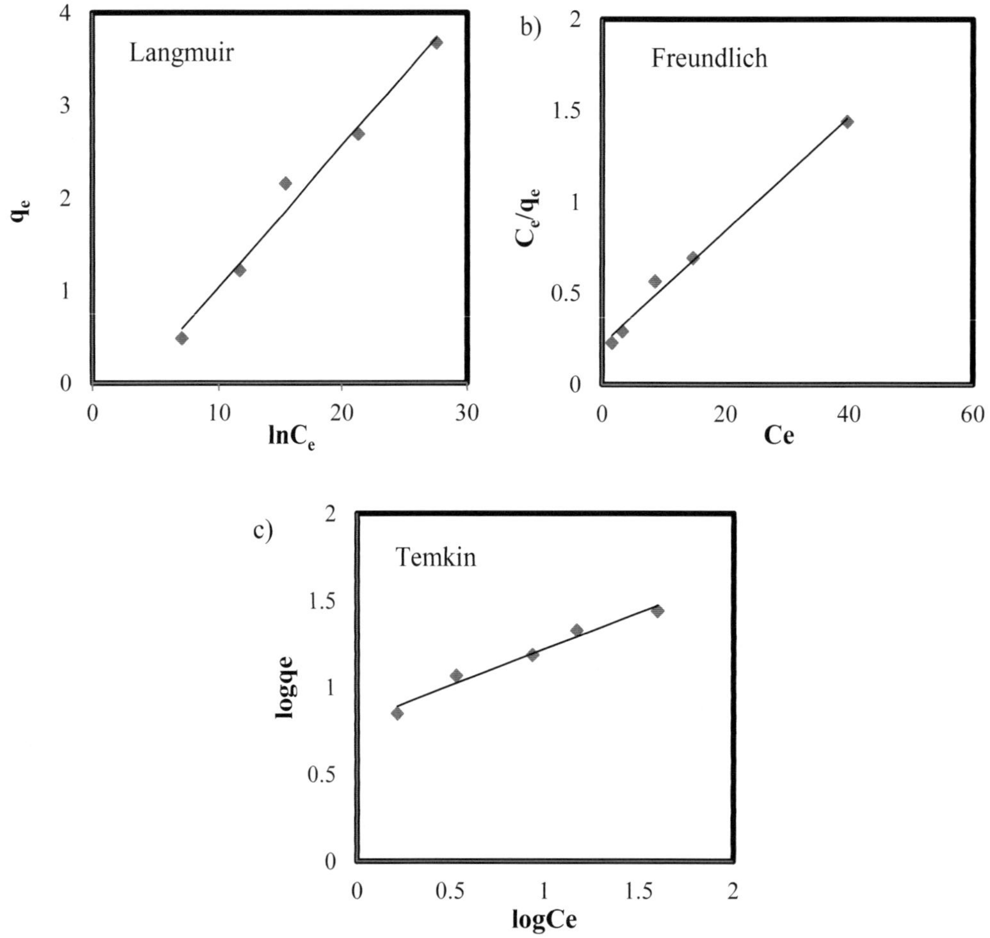

Figure 9. Isotherm graphs for methylene blue adsorption on biochar.

Table 3. Isotherm parameters for methylene blue adsorption on biochar

Concentration (mg/L)	Langmuir isotherm			Freundlich isotherm			Temkin isotherm		
	q_{max} (mg/g)	K_L	R^2	n	K_F	R^2	B^T	K_T	R^2
25	31.64	0.157	0.9904	0.418	2.234	0.9729	0.1534	0.039	0.9811

4. Conclusion

The conversion of sewage sludge into biochar by adding lignocellulosic biomass (coconut shell) to enhance the adsorption capacity was a good possibility to utilize the waste effectively. The removal efficiency of 97.28% was achieved with a contact time of 2.5h with a 4g/L biochar dose. The Langmuir isotherm was determined to be a good fit by isotherm investigations, with the maximum monolayer adsorption of 31.64mg/g. The kinetic analysis suggested that the pseudo-second-order model best matches the experimental findings, and chemical interactions between adsorbate and adsorbent might be possibly reason for adsorption. From FTIR analysis, the reduction in the peak intensity and shifts in the peak revealed that adsorption might be due to pie-pie stacking interaction, and also the O-H and C-H functional groups might have played an important role in adsorption.

References

[1] CPCB. (2021). National Inventory of Sewage Treatment Plants March 2021 *March*, **183**.

[2] Smol, M., Adam, C., & Preisner, M. 2020 Circular economy model framework in the European water and wastewater sector. *Journal of Material Cycles and Waste Management*, **22(3),** 682–697. https://doi.org/10.1007/s10163-019-00960-z

[3] Singh, S., Kumar, V., Dhanjal, D. S., Datta, S., Bhatia, D., Dhiman, J., Samuel, J., Prasad, R., & Singh, J. 2020 A sustainable paradigm of sewage sludge biochar: Valorization, opportunities, challenges and future prospects. In *Journal of Cleaner Production* **269** Elsevier Ltd. https://doi.org/10.1016/j.jclepro.2020.122259

[4] Rafatullah, M., Sulaiman, O., Hashim, R., & Ahmad, A. 2010 Adsorption of methylene blue on low-cost adsorbents: A review. *Journal of Hazardous Materials* **177(1–3)** 70–80. https://doi.org/10.1016/j.jhazmat.2009.12.047

[5] Yin, Q., Liu, M., & Ren, H. 2019 Biochar produced from the co-pyrolysis of sewage sludge and walnut shell for ammonium and phosphate adsorption from water. *Journal of Environmental Management* **249** 109410. https://doi.org/10.1016/j.jenvman.2019.109410

[6] Huang, H. jun, Yang, T., Lai, F. ying, & Wu, G. qiang 2017 Co-pyrolysis of sewage sludge and sawdust/rice straw for the production of biochar *Journal of Analytical and Applied Pyrolysis*, **125** 61–68. https://doi.org/10.1016/j.jaap.2017.04.018

[7] Aljeboree, A. M., Alshirifi, A. N., & Alkaim, A. F. 2017 Kinetics and equilibrium study for the adsorption of textile dyes on coconut shell activated carbon. *Arabian Journal of Chemistry*, **10** S3381–S3393. https://doi.org/10.1016/j.arabjc.2014.01.020

[8] Ahmed, M. J., Okoye, P. U., Hummadi, E. H., & Hameed, B. H. 2019 High-performance porous biochar from the pyrolysis of natural and renewable seaweed (Gelidiella acerosa) and its application for the adsorption of methylene blue. *Bioresource Technology*, **278** 159–164. https://doi.org/10.1016/j.biortech.2019.01.054

[9] Fan, X., Zhang, J., Xie, Y., Xu, D., Liu, Y., Liu, J., & Hou, J. 2021 Biochar produced from the co-pyrolysis of sewage sludge and waste tires for cadmium and tetracycline adsorption from water. *Water Science and Technology* **83(6)** 1429–1445. https://doi.org/10.2166/wst.2021.058

[10] Sha, Q., Xie, H., Liu, W., Yang, D., He, Y., Yang, C., Wang, N., & Ge, C. 2021 Removal of fluoride using platanus acerifoli leaves biochar–an efficient and low-cost application in wastewater treatment. *Environmental Technology (United Kingdom)*. https://doi.org/10.1080/09593330.2021.1964002

[11] Lei, T., Li, S. J., Jiang, F., Ren, Z. X., Wang, L. L., Yang, X. J., Tang, L. H., & Wang, S. X. 2019 Adsorption of Cadmium Ions from an Aqueous Solution on a Highly Stable Dopamine-Modified Magnetic Nano-Adsorbent. *Nanoscale Research Letters* **14(1)** https://doi.org/10.1186/s11671-019-3154-0

[12] Fan, S., Tang, J., Wang, Y., Li, H., Zhang, H., Tang, J., Wang, Z., & Li, X. 2016 Biochar prepared from co-pyrolysis of municipal sewage sludge and tea waste for the adsorption of methylene blue from aqueous solutions: Kinetics, isotherm, thermodynamic and mechanism. *Journal of Molecular Liquids* **220** 432–441. https://doi.org/10.1016/j.molliq.2016.04.107

[13] Wang, J., & Guo, X. 2020 Adsorption kinetic models: Physical meanings, applications, and solving methods. *Journal of Hazardous Materials* **390** 122156. https://doi.org/10.1016/j.jhazmat.2020.122156

[14] Fan, S., Wang, Y., Wang, Z., Tang, J., Tang, J., & Li, X. 2017 Removal of methylene blue from aqueous solution by sewage sludge-derived biochar: Adsorption kinetics, equilibrium, thermodynamics and mechanism. *Journal of Environmental Chemical Engineering* **5(1)** 601–611. https://doi.org/10.1016/j.jece.2016.12.019

[15] Ghorbani-Khosrowshahi, S., & Behnajady, M. A. 2016 Chromium(VI) adsorption from aqueous solution by prepared biochar from Onopordom Heteracanthom. *International Journal of Environmental Science and Technology* **13(7)** 1803–1814 https://doi.org/10.1007/s13762-016-0978-3

Investigation into the Compressive Stress on Lumbar Spine among Tractor Drivers during different Ride Conditions

Chander Prakash[1], Lakwinder Pal Singh[2] and Ajay Gupta[2]
[1] Research Scholar, Department of Industrial and Production Engineering, Dr BR Ambedkar National Institute of Technology Jalandhar, Punjab, India
[2] Associate Professor, Department of Industrial and Production Engineering, Dr BR Ambedkar National Institute of Technology Jalandhar, Punjab, India

Email: chanderpra95@gmail.com

Abstract. The aim of this investigation is to figure out the stress (compressive) on the lumbar spine of tractor driver posture among different ride conditions i.e., RPM (Revolutions per minute), Postures, and Road Conditions). The spinal stress under various riding conditions was assessed using the daily equivalent static compression dose (Sed). Design Expert, a statistical programme that combines Taguchi's L9 Orthogonal Array Method, is used to determine the experimental runs with different parameters such as varying the tractor speed in RPM (700, 1000 & 1300), driver posture (Lean forward without backrest-P1, vertical erect without backrest-P2, vertical erect with backrest-P3), selected road conditions (village road, brick road and farm road). Analysis of the overall influence of riding conditions among all the experiments was done by calculating signal-to-noise (S/N) ratios. A regression model has been developed to analyze the Sed response and the desirability technique was used to minimize the Sed response. The daily dose of static compression on lumber spine was found between 0.35 and 0.67 MPa, which comes within the category of "moderate adverse health effect" as defined by ISO2631-5: 2004. Statistically, the RPM and Road type conditions were found highly significant with the percentage contribution is 77% and 21% respectively. As comfort is crucial for the tractor driver. The Sed values can be used for tractor seat design for more comfort. The approach is novel as per the Indian context for the tractor driver's comfort. The methods of work-study and sampling in a larger context are well established.

Keywords: Agriculture, Tractor, Whole body vibration, Design Expert.

1. Introduction

Agriculture farmers engage maximum duration of time per day almost throughout the year, the uses of tractor during the busiest day up to 12-14 h per day [1]. They exposed to high vibrations which may lead to some adverse health effect or back related problems [2] [3]. Low motor vehicle (tractor) is extensively used for daily life in transportation system such as sowing, ploughing and toeing heavy implements in the agriculture practices. Domestic tractor sales increased to 61,249 units in December 2020, up from 42,806 units in December 2019, according to statistics from the Tractor & Mechanization Association (TMA) [4]. Tractor vibration exposure is high with comparison to on-road vehicles [5]. In tractor, the low-frequency vibration is transmitted to the driver's body mostly via the driver's seat as a result of the terrain-tyre interaction. [6]. Existing the natural frequencies among drivers there are many direct effects laid to the ride comfort due to the low-frequencies vibration [7]. Most critical ranges of natural frequencies to the human body are (0.5-10 Hz), which may cause some injury due to resonance [8]. Usually, driver exposed prolonged exposure under the exposure of vibration during different farm operations [9]. Tractor drivers may also suffer musculoskeletal disorders [10], motion sickness [11], cardiovascular problem [7], and neurological problems [12] as a result of these work exposures. Therefore, the seat must be well-designed to avoid vibrations in this frequency band. Consequently, vibration reductions of much greater magnitude are clearly needed, and several methods for achieving these huge advantages have been explored. Many research has studied exposure to vibrations as a possible cause of health problems [13] [3]. In which, most of the studies have been performed on the control environment or simulators to study the various input parameters such as., vibration magnitude [14], backrest studies [15], postures [16], tyre pressure [17], seat design [18], excitation type [19], position of foot [20] on ride comfort. Despite this, several studies have been conducted in real world environments [6] taking into account factors such as vehicle speed and terrain, as well as the variability of driver postures.

From the literature analysis, it has been analysed that the posture variability with the different road terrain and RPM is yet unexplored as far as occupational studies concern. So, this paper is focused on the tractor for analysing the seat vibrations, with varying the driver posture, RPM and road type. The objective of this study, to determine the static equivalent compression dose among the driver during tractor driving.

2. Methodology

2.1. Tractor and Seat

The experiments were carried out using a 50 horsepower 2015 model tractor 'T'. The tyre inflation pressure was maintained in accordance with the tractor catalog's recommended level. The study was conducted in north region Ludhiana, Punjab (India). Selected tractor "T" specification has been mentioned in table 1. The tractor seat suspension system is consisting two tension spring with the hydraulic damper. The tractor seat is originally manufactured by the tractor industries, no changes has been made on the seat from the last 7 years. During the towing agriculture operations, average working time of tractor is 8-11 hours daily. And the working time is varying with respect to different agriculture task.

Table 1. Tractor "T" Specifications

No. of Cylinder	3
Horse Power	50 HP
Engine Capacity	3147 cc
Engine Rated RPM	2200 RPM
PTO HP	42.5
Gear Box	8 Forward+ 2 Reverse
Forward and Backword Speed	31.51-12.67 kmph
Fuel Tank	60 lt.
Wheel Base	2090 mm
Lifting Capacity	1400

2.2 Experiment Design

In this study, three input parameters were taking for the experimentation i.e., RPM, different terrains and driving postures. Three levels of the RPM (700, 1000 & 1300) were considered to study the effect of varying RPM on vibration transmission [21]. These variations of RPM's were taking to drive easily on the three different terrains such as, Tar Road (1); Brick Road (2) & Farm Road (1). These speed levels were chosen based on a series of random pilot tests. Furthermore, the data gathered during the pilot trial is not included in this research. Pilot experiments were done to ensure the tractor drive easily on the different terrains with the selected RPM levels. Furthermore, three postures were considered during the whole experimentation. Different selected postures are vertical erect with backrest

(P1-1); slightly forward without backrest (P2-2); lean forward without backrest (P3-3) as mentioned in figure 1 (B).

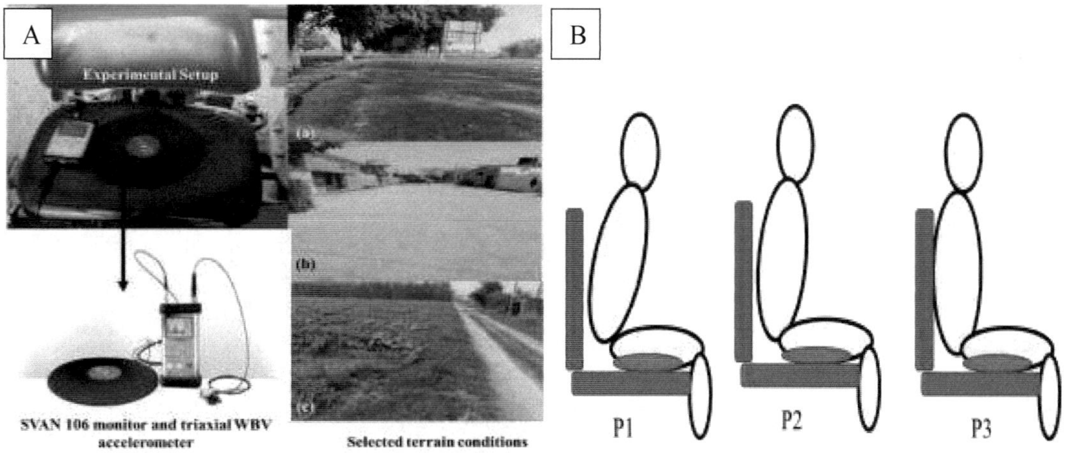

Figure 1. (A) Representation of experimental setup and selected terrain conditions (a, b & c); (B) selected postures (P1, P2 & P3) for the study

2.2.1 Subject

A 25-year-old man with weight of 79 kg, a height of 1.531 m, and a BMI of 33.09 kg/m^2 was used as the test subject. The subject was indeed for driving tractor is seven (7) years, with an average of five (5) hours of tractor driving each day. The subject has signed a consent form in order to receive ethical clearance for the tests. Before the experiment, the individual was informed of the goal of the research.

2.2.2 Instrument

A SVAN 106 human vibration sensor was installed on the seat of tractor as shown in figure 1 (a). Weighted acceleration among the translation axes were measured on the driver seat. The whole measurement and calculation were carried out as per ISO 2631-1 (1997).

2.2.3 Daily equivalent static compression dose calculation

The S_{ed} was calculated from the vector sum of D_k exposures and the unit is Megapascal (MPa). As a general rule, all of the values were based on an 8-hour daily value called Dk (8), and the unit is m/s^2. In the end, all of the S_{ed} values were equivalent to an 8-hour day (S_{ed} (8)).

The Dk, Dk (8) and Sed (8) are calculated as per ISO 2631-5, 2004. The formula of these parameters is mentioned in equations (1) and (2).

$$D_k = \left[\sum_i A_{ik}^6\right]^{\frac{1}{6}} \quad , \quad D_k(8) = D_k \left[\frac{480\ minutes}{measurement}\right]^{\frac{1}{6}} \tag{1}$$

$$S_{ed}(8) = ((m_k D_k(8))^6)^{\frac{1}{6}} \quad \text{for k = x,y,z} \tag{2}$$

A_{ik} is the peak of response acceleration $a_{ik}(t)$. The values for m_k are 0.015, 0.035 and 0.032 MPa/(m/s^2) for the x, y and z axes respectively.

Measured vibration data transfer to the SVANPC++ software for the data analysis. It provides the weighted root mean square values od vibration and maximum acceleration peaks towards the respective axes. Further, In Minitab 16 Taguchi's L9 orthogonal array method was executed to obtain the S/N ratio, and ANOVA analysis.

2.3 DoE (Design of Experiment)

Using Taguchi's L9 orthogonal array, the primary experiments were planned in a methodical fashion as mentioned in table 2. There were a minimum number of trials required to understand how each design parameter affected Sed response by using this orthogonal array. For each experiment, the orthogonal array method was also determined in terms of S/N [22]. The S/N ratios were calculated in order to get the best riding condition and minimize Sed response. A high S/N ratio means that the signal is much stronger than the deviation effects of noise parameters [23]. The S/N ratio is divided into three categories: better is smaller, better is bigger, and better is nominal. The goal of the research was to reduce output responses in regard to design factors, hence the 'smaller is better' approach was addressed.

Moreover, by adopting the "smaller-the-better" option and the S/N ratios calculated via mathematical equation (3) as mentioned below:

$$(S/N) = -10\log\left[1/R(Y1^2+Y2^2+\cdots+Yn^2)\right] \quad (3)$$

To establish the mean S_{ed} reaction, each experiment was replicated with three times trails.

Table 2. Taguchi's L9 Orthogonal Array

	No. of Experiment								
Input Parameter	1	2	3	4	5	6	7	8	9
RPM (level)	1	1	1	2	2	2	3	3	3
Posture (level)	1	2	3	1	2	3	1	2	3
Terrain (type)	1	2	3	2	3	1	3	1	2

This study has followed the Declaration of Helsinki (WMA, 2013), and the protocol was approved by the Research advisory of Dr. BR Ambedkar National Institute of Technology Jalandhar, Punjab (India) [vide letter no. NITJ/IPE/RA/128]

3. Results & Discussion

The objective of present study is the influence of input parameters over the output parameter. Moreover, determine the optimum ride conditions to enhance the minimization of output response. In the Taguchi method, for output response characteristics are mainly 2 types i.e., noise parameter, and another is signal or control [24]. The signal or control parameters are flexible for the operator. But the noise parameter is uncertain due to the uncontrolled turbulence existing during the working environment. Consequently, control parameters plays and important role towards the output response. In the present study, three control parameters are RPM, postures and road conditions.

3.1 S/N Ratio (Signal-to-Noise Ratio)

Employing Taguchi's L9 orthogonal array method, to analysed SNRs for each combination. Each experiment was carried out three times (T1-T3) with the identical settings, and the average Sed was calculated. As shown in figure 3, the mean response of the three trails has been taken for the further analysis as mentioned in Table 3. The maximum value of S/N indicates the optimum level of each parameter.

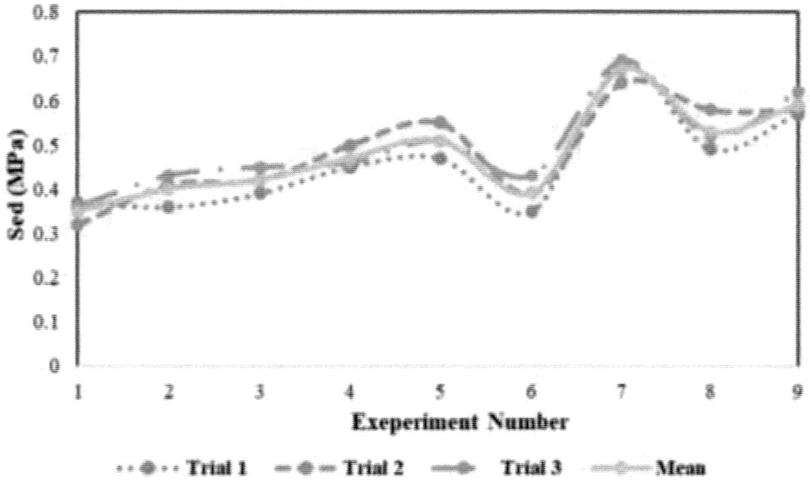

Figure 2. S_{ed} response with Taguchi's L9 Combination

Table 3. Experiment design and it's respective results

Experiment No.	RPM	Posture	Road Type	T1	T2	T3	S_{ed} (mean)	S/N Ratio
1	700	1	1	0.37	0.32	0.36	**0.35**	**9.13232**
2	700	2	2	0.36	0.41	0.43	0.40	7.95055
3	700	3	3	0.39	0.42	0.45	0.42	7.52958
4	1000	1	2	0.45	0.5	0.46	0.47	6.55261
5	1000	2	3	0.47	0.55	0.51	0.51	5.86228
6	1000	3	1	0.35	0.39	0.43	0.39	8.17045
7	1300	1	3	0.68	0.64	0.69	0.67	3.47025
8	1300	2	1	0.49	0.58	0.52	0.53	5.50905
9	1300	3	2	0.57	0.58	0.62	0.59	4.59664

The mean of output response S_{ed} ranges was 0.35-0.67 m/s^2 and the S/N ratios were found 3.4703-9.1323 (dB). As illustrated in Figure 3, S/N ratios were examined to shown the major effect plot with respect to RPM, postures, and road conditions. The maximum S/N ratio value with minimum S_{ed} was obtained in experiment combination is 1. And minimum S/N ratio value represents the conditions at maximum S_{ed} was found in experiment number 7.

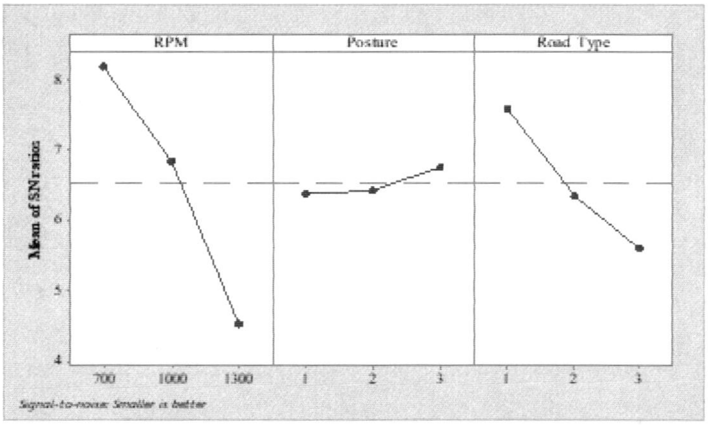

Figure. 3. Main effect plot for S/N ratios

There was a significant relationship between bus and road type [25]. This indicates that when RPM and different road condition varying, the daily equivalent static compression dosage (S_{ed}) increased. An increase in vibration magnitude may lead to strong acceleration peaks as the forward speed increases [24]. Constantly being subjected to excessive acceleration for long periods of time is unhealthy for tractor drivers' health and performance [26]. Figure demonstrates the S_{ed} response has been raised with increasing the RPM levels. Furthermore, at the posture 3 the S_{ed} response is found minimum. While in the road type, the S_{ed} response is drastically increasing from low to high with varying the selected terrains.

3.2 ANOVA Analysis

Analysis of variance (ANOVA) was calculated to represent the statistical significance of each design parameter the daily equivalent static compression dosage. Sequential sums of squares, F-values, adjusted mean squares, P-values, delta values, % contributions (P), and rankings for all of the variables were computed and briefly summarised in Table 4. The statistical significance was determined and compared to the tabulated value at a 95% confidence level. It can be seen that the RPM and road type had a big impact on the S_{ed} response. However, the posture of the driver was found to have no effect on the S_{ed} response. According to Equation (4), the proportion of each parameter's impact on the S_{ed} was calculated.

$$P\,\% = \frac{\text{Seq SS}_\text{I}}{\text{Seq SS}_\text{T}} \times 100 \tag{4}$$

Where Seq SS_I indicates the Individual sequential sum of squares and SS_T indicated the sequential sum of squares.

In the ANOVA analysis, Table 4 shows the Sed response was highly affected with the RPM (77.01%), road type (21.10%), and posture (1.56%). Each parameter's delta value was computed as the difference between its maximum and minimum S/N ratio values.

Table 4. ANOVA Analysis

Source	DF	Seq SS	Adj SS	Adj MS	F-value	P-value	P (%)
RPM	2	0.066756	0.066756	0.033378	231.08	0.004	77.01
Posture	2	0.001356	0.001356	0.000678	4.69	0.176	1.56
Road Type	2	0.018289	0.018289	0.009144	63.31	0.016	21.10
Residual Error	2	0.000289	0.000289	0.000144			0.33
Total	8	0.086689					100

Table 5. Response Table for Means

Levels	RPM	Posture	Road Conditions
1	0.3900	0.4967	0.4233
2	0.4567	0.4800	0.4867
3	0.5967	0.4667	0.5333
Delta	0.2067	0.0300	0.1100
Rank	1	3	2

RPM, posture, and road type each had a delta value of 0.207, 0.030, and 0.110, which supports the findings of the computed contribution %age; the delta value shows each parameter's ranking in terms of its influence on S_{ed} reaction, with rank 1 being the one with the highest delta value shown in Table 5.

3.3 Regression Modelling

To analyze the design characteristic, the regression model was developed to find out the best fit and minimum S_{ed} response among the selected experiments. Equation (5) depicted the regression model as below:

$$S_{ed} = 0.0567 + 0.000344\ \text{RPM} - 0.0150\ \text{Posture} + 0.0550\ \text{Road Type} \tag{5}$$

Table 6. Experimental and Predicted response with error contribution

RPM	Posture	Road Type	Experimental S_{ed}	Predicted S_{ed}	Error %
700	1	1	0.35	0.338	1.2
700	2	2	0.40	0.378	2.2
700	3	3	0.42	0.418	0.2
1000	1	2	0.47	0.496	-2.6
1000	2	3	0.51	0.536	-2.6
1000	3	1	0.39	0.411	-2.1
1300	1	3	0.67	0.654	1.6
1300	2	1	0.53	0.529	0.1
1300	3	2	0.59	0.569	2.1
		Average error percentage			**1.11**

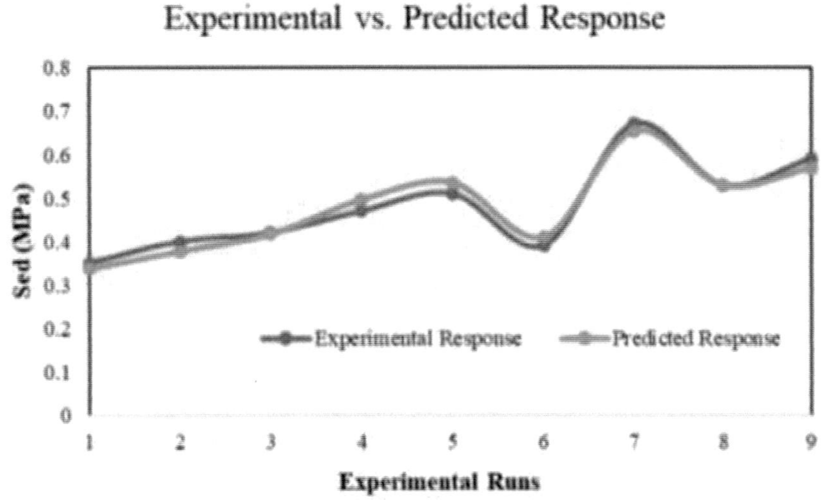

Figure. 4. S_{ed} trend with a comparison of actual and predicted response

The mean of the S_{ed} response has been mentioned in Table 6, It was compared the experimental and prediction response with each experimental error. The prediction response was calculated by using equation 5. It is discovered a strong correlation between experimental and predicted values with a mean error of 1.11%, the R^2 value is 99.7% and R^2 (adj.) is 98.7%. The trend of experimental and predicted response has been shown in figure 4. In the result, the optimum level of each input is RPM, posture, and road type as 700, Lean forward without backrest (P1) and the road type is village road (1).

4. Conclusion and Recommendations

From the above given results, there are some conclusions have been made and structured below:

- The DoE design, Taguchi defines the best and optimum experiment runs for optimization purposes. It saves effort, time, and money.
- RPM and road types were shown to be statistically significant parameters($p \leq 0.05$) in determining the S_{ed} response, with a percentage contribution of 77.01% and 21.1% respectively, in influencing the S_{ed} response.
- The proposed regression model exhibited a high degree of correlation between the actual and projected S_{ed} response, with a mean error of 1.11 between the actual and projected S_{ed} response.
- The manufacturing of the tractor seat can be done by taking the optimized values form this study.

Despite from this, there is still need to do the more research with varying the different cushion, seat suspension system, AI tools for the data analysis etc., for the tractor drivers. That can improve the comfort on the tractor seat.

References

[1] E. Sorainen, J. Penttinen, M. Kallio, E. Rytkönen, and K. Taattola, "Whole-Body Vibration of Tractor Drivers During Harrowing," *Am. Ind. Hyg. Assoc. J.*, vol. 59, no. 9, pp. 642–644, Sep. 1998, doi: 10.1080/15428119891010820.

[2] M. Bovenzi and A. Betta, "Low-back disorders in agricultural tractor drivers exposed to whole-body vibration and postural stress," *Appl. Ergon.*, vol. 25, no. 4, pp. 231–241, 1994, doi: https://doi.org/10.1016/0003-6870(94)90004-3.

[3] A. J. Scarlett, J. S. Price, and R. M. Stayner, "Whole-body vibration: Evaluation of emission and exposure levels arising from agricultural tractors," *J. Terramechanics*, vol. 44, no. 1, pp. 65–73, 2007, doi: 10.1016/j.jterra.2006.01.006.

[4] "The Future Of Agriculture Mechanization Will Be Data Driven.pdf."

[5] C. Mehta and V. K. Tewari, "Seating discomfort for tractor operators - A critical review," *Int. J. Ind. Ergon.*, vol. 25, pp. 661–674, Jul. 2000, doi: 10.1016/S0169-8141(99)00054-2.

[6] T. E. FAIRLEY†, "Predicting the discomfort caused by tractor vibration," *Ergonomics*, vol. 38, no. 10, pp. 2091–2106, Oct. 1995, doi: 10.1080/00140139508925253.

[7] M. Griffin, "Handbook of Human Vibration Academic Press London 1990," 1990.

[8] A. Kumar, P. Mahajan, D. Mohan, and M. Varghese, "IT—Information Technology and the Human Interface: Tractor Vibration Severity and Driver Health: a Study from Rural India," *J. Agric. Eng. Res.*, vol. 80, no. 4, pp. 313–328, 2001, doi: https://doi.org/10.1006/jaer.2001.0755.

[9] R. Hildebrand, E. Keskinen, and J. A. R. Navarrete, "Vehicle vibrating on a soft compacting soil half-space: Ground vibrations, terrain damage, and vehicle vibrations," *J. Terramechanics*, vol. 45, no. 4, pp. 121–136, 2008, doi: 10.1016/j.jterra.2008.09.003.

[10] M. J. Griffin, "Discomfort from feeling vehicle vibration," *Veh. Syst. Dyn.*, vol. 45, no. 7–8, pp. 679–698, Jul. 2007, doi: 10.1080/00423110701422426.

[11] S. Kwaku Essien, C. Trask, M. Khan, C. Boden, and B. Bath, "Association Between Whole-Body Vibration and Low-Back Disorders in Farmers: A Scoping Review," *J. Agromedicine*, vol. 23, no. 1, pp. 105–120, Jan. 2018, doi: 10.1080/1059924X.2017.1383333.

[12] L. Moggio, A. de Sire, N. Marotta, A. Demeco, and A. Ammendolia, "Vibration therapy role in neurological diseases rehabilitation: an umbrella review of systematic reviews," *Disabil. Rehabil.*, pp. 1–9, Jul. 2021, doi: 10.1080/09638288.2021.1946175.

[13] M. Cutini, C. Costa, and C. Bisaglia, "Development of a simplified method for evaluating agricultural tractor's operator whole body vibration," *J. Terramechanics*, vol. 63, pp. 23–32, Feb. 2016, doi: 10.1016/j.jterra.2015.11.001.

[14] H. Çiloğlu, M. Alziadeh, A. Mohany, and H. A. Kishawy, "Assessment of the whole body vibration exposure and the dynamic seat comfort in passenger aircraft," *Int. J. Ind. Ergon.*, vol. 45, pp. 116–123, 2015.

[15] N. Nawayseh and M. J. Griffin, "A model of the vertical apparent mass and the fore-and-aft cross-axis apparent mass of the human body during vertical whole-body vibration," *J. Sound Vib.*, vol. 319, no. 1–2, pp. 719–730, 2009, doi: 10.1016/j.jsv.2008.05.030.

[16] S. Rakheja, R. G. Dong, S. Patra, P. É. Boileau, P. Marcotte, and C. Warren, "Biodynamics of the human body under whole-body vibration: Synthesis of the reported data," *Int. J. Ind. Ergon.*, vol.

40, no. 6, pp. 710–732, 2010, doi: 10.1016/j.ergon.2010.06.005.
[17] L. M. Sherwin, P. M. O. Owende, C. L. Kanali, J. Lyons, and S. M. Ward, "Influence of tyre inflation pressure on whole-body vibrations transmitted to the operator in a cut-to-length timber harvester," *Appl. Ergon.*, vol. 35, no. 3, pp. 253–261, 2004, doi: https://doi.org/10.1016/j.apergo.2004.02.002.
[18] N. Nawayseh and M. J. Griffin, "Tri-axial forces at the seat and backrest during whole-body vertical vibration," *J. Sound Vib.*, vol. 277, no. 1, pp. 309–326, 2004, doi: https://doi.org/10.1016/j.jsv.2003.09.048.
[19] A. Anand and S. K. C. P., "A Review of Whole-Body Vibration of Seated Occupants in Moving Vehicles," *SSRN Electron. J.*, vol. 1, no. 1, pp. 347–352, 2021, doi: 10.2139/ssrn.3794558.
[20] S. RAKHEJA, I. HARU, and P.-É. BOILEAU, "SEATED OCCUPANT APPARENT MASS CHARACTERISTICS UNDER AUTOMOTIVE POSTURES AND VERTICAL VIBRATION," *J. Sound Vib.*, vol. 253, no. 1, pp. 57–75, 2002, doi: https://doi.org/10.1006/jsvi.2001.4249.
[21] A. Taghizadeh Alisaraei, "Analysis of annoying shocks transferred from tractor seat using vibration signals and statistical methods," *Comput. Electron. Agric.*, vol. 141, pp. 160–170, Sep. 2017, doi: 10.1016/j.compag.2017.07.020.
[22] M.-L. Huang, Y.-H. Hung, and Z.-S. Yang, "Validation of a method using Taguchi, response surface, neural network, and genetic algorithm," *Measurement*, vol. 94, no. C, pp. 284–294, 2016, doi: 10.1016/j.measurement.2016.08.006.
[23] M. Tjantelé, "Parameter design using the Taguchi methodology," *Microelectron. Eng.*, vol. 10, no. 3, pp. 277–286, 1991, doi: https://doi.org/10.1016/0167-9317(91)90028-C.
[24] E. Bagci and Ş. Aykut, "A study of Taguchi optimization method for identifying optimum surface roughness in CNC face milling of cobalt-based alloy (stellite 6)," *Int. J. Adv. Manuf. Technol.*, vol. 29, no. 9, p. 940, 2005, doi: 10.1007/s00170-005-2616-y.
[25] O. Thamsuwan, R. Blood, R. Ching, L. Boyle, and P. Johnson, "Whole body vibration exposures in bus drivers: A comparison between a high-floor coach and a low-floor city bus," *Int. J. Ind. Ergon.*, vol. 43, pp. 9–17, Jan. 2013, doi: 10.1016/j.ergon.2012.10.003.
[26] B. Sam and K. Kathirvel, "Vibration Characteristics of Walking and Riding Type Power Tillers," *Biosyst. Eng.*, vol. 95, pp. 517–528, Dec. 2006, doi: 10.1016/j.biosystemseng.2006.08.013.

Transport Infrastructure Investment: A Comparative Analysis between China, Russia and India

T Bangaraju [1], P Aneja[2] and B Mishra[3]

[1] Professor, School of Business, University of Petroleum and Energy Studies, Dehradun
[2]Assistant Professor, School of Business, University of Petroleum and Energy Studies, Dehradun
[3] Student, School of Business, University of Petroleum and Energy Studies, Dehradun

E-mail: bangarraju@gmail.com; paneja@ddn.upes.ac.in; biswajeet.mishra824@gmail.com

Abstract: China, Russia and India, three major members of BRICS, share a strong trade relationship in over a duration of time. These countries have shown remarkable growth in varioussectors overtaking EU and North American countries. Infrastructure investment is a crucial factor to give a momentum to progress of any country. This study attempts to understand the long run and short run effect of total inland infrastructure investment on freight growth in the said countries. In this study, Autoregressive Distributed Lag [ARDL] bounds testing approach was used to examine the impact of total inland transport infrastructure investment on total inland freight transport in the emerging economies, viz. China, Russia and India. Results indicate thatChina has higher positive affect of inland infrastructure investment on freight growth rate in theshort run while India has higher positive affect in long run. Russia, owing to lower population, trade and industrialization growth showed to have lower freight growth in both short run and long run.

Keywords: Transport, infrastructure, freight, Investment

1. **Introduction**

Transport infrastructure investment [TTI] is an important component of infrastructure investment which covers development of new and old construction, upgradation activity related to roads, railways, inland waterways, seaports, and airports. Effective financial framework for the investment in these sectors gives a positive progressive economic growth. The investment has given a positive angle towards regional development and growth. Development of regional freight transportation hubs, electric vehicle recharge stations, effective maintenance of the infrastructure, capacities to move the cargo are the need of the hour for any developing and developed nation [1]. As discussed, there is a need to estimate the scope for infrastructure investment in transportation domain for development of any region or a country. The transport infrastructure investment is a critical component for any the country's economy which plays a major factor towards GDP growth. There are many current instances of delays and losses due to infrastructure failures for freight and passenger traffic [2]. Thus, there is also need for understanding critical infrastructure investment and areas of investment to avoid traffic disruptions. The transport investment is crucial for GDP and employment which in turn effect the economy of any nation [3]. The study suggests a strong positive relationship where transport infrastructure could be a vital factor to bring positive economic ecosystem. Thereby, this research attempts to explore the relationship between Transport Infrastructure investment, GDP and total inland freight transport. The countries chosen are

India, China, and Russia which are high emerging economies and part of BRICS trade block. Brazil and South Africa were not considered due to lack of data availability. The literature review is discussed in the second section followed by data and methods. The analysis and discussion are briefed in the fourth section flowed by conclusion in the fifth section.

2. Literature review

FHFYThe literature review is vast in the domain of infrastructure investments; however, the section mainly discusses about various transport infrastructure investment and related issues. Analytical hierarchy process and cognitive maps were used to evaluate various alternatives of projects for transport investments in Turkey [4]. Since major investments are under political control, the author has attemptedto make decision support model to decide on investment decision rather than traditional cost benefit analysis. Stef and Saskia attempted to analyse transport infrastructure investment decisions when there is demand uncertainty [5]. Merits and demerits of overinvestments are brought out in couple of models of demand uncertainty. Built operate and transfer policy for transport infrastructure investment has beensuggested in [6] based on study in OECD countries. The relationship between debt-to-GDP ratio and transport was explored by developing linear equation by taking into options of budget financing, privatesector financing under PFI method and private sector financing under BOT scheme. Regional competitiveness and its relationship with transport investments was discussed [7]. Need for accessibility indicator was brought forward to measure the competitiveness. For a better sustainable transportation future, various preferences for public transport investments under budget constraints were empirically investigated in [8]. Mixed Logit models' analysis gave light rail transit preference over bus rapid transit in capital cities of Australia. Again, in Australia, larger economic benefits due to transport investments was investigated for Sydney North West Rail Link Project in [9].
The study resulted in comprehending that accident reduction, travel cost savings, road decongestion were major benefits otherthan GDP due to infrastructure investments. A couple of studies were done to comprehend about transport investments in China. Cross-region analysis was done to investigate relation between economicdevelopment and transport investment [10]. The comparative cross-regional study reveals a positive trend in economic returns due to extra transport investment. Similarly, in [11] empirical findings significant spatial variations in productivity effects due to infrastructure investments in the respective Chinese regions. The rise of central China was accounted due to transport investments for better facilities. Spatial computable general equilibrium model was deployed to explore wider economic impacts due to planned transport infrastructure projects in Norway [12]. Certain project specific characteristics were major factors towards wider economic impacts of transport infrastructure projects.Study from Latvia, suggests critical role of transport infrastructure investment as significant towards employment and transport system [13]. The article focuses on need for methodology towards measuringtransport infrastructure development performance which could be applicable systematically and useful for transport investment decisions makers. For Sustainable development, the need for road, rail and transport infrastructure was envisaged for Slovakia in [14]. The GDP growth led to burden on road network which in turn suggest for greater investment in transport infrastructure. Similar study in Belgium, the GDP growth is found to be positively impacted by transport infrastructure investments, Length of motorways and rail networks apart from traditional trade openness and technological change [15]. Data envelopment analysis was done to rank New Zealand, Japan, Central European and eastern European countries to examine the efficiency of total transport investment [16]. Good governance played an important factor in transport investment decisions which in turn leads to better efficiency in investments. Potential accessibility indicator role towards infrastructure investment is investigated in Poland to verify the regional growth [17].
Presence of positive correlation exist between accessibility and regional employment due to transport infrastructure investments. The need for large transport infrastructure investments in China and Europe

is envisaged for improving performances of supply chains in [18]. Results signify that intermodal inland road/rail routes could prove better alternative thanthe maritime routes for better supply chain efficiency between Chin and Europe through efficient Transport infrastructure investments. Multiple discrete extreme value model was deployed to evaluate rail infrastructure investment impact on society in Argentina [19]. Environmental benefit is a key outputdesired in transport infrastructure investment as discussed in [20]. The need for alignment of Industries productivity along with transport infrastructure investment is desired for consistent profits. In recent study, impact TTI on regional convergence and GDP growth is brought forward through literature review which shows causal relationship between them and presence spatial spillovers analysis further demonstrate positive relationship in [21]. Strong evidence of bidirectional relationship between tourism and highway, unidirectional relationship between railway and tourism using granger causality isenvisaged in China from the study [22]. Study from Seoul region, explores relationship between transport infrastructure investment and its impact on firm productivity through level agglomeration in [23]. Strong positive relationship is evidenced from the study which implies economic advantage for thecountry and region. The need for investment policies to address freight externalities and their impact on economy competitiveness was brought forward.

3. Data and Methods

The yearly data has been obtained Organization for Economic Development website for the year 2004 to 2018 [24]. The Autoregressive Distributed Lag [ARDL] bounds testing approach was applied to examine the impact of total inland transport infrastructure investment on total inland freight transport in the selectedeconomies, viz. China, Russia and India. The ARDL modelling approach was proposed by Pesaran [25]. This approach is based on the impact of short-run and long-run independent variables on dependentvariables and short-run impact of independent variables on dependent variable, regardless whichever variables are I[0], I[1] and are fractionally integrated. Moreover, this technique is more appropriate for the sample samples.

As primary step, the ARDL estimation takes a following form:

$$\Delta \mathrm{InTIFT}_t = \beta_0 + \sum_{i=1}^{p} \delta_1 \Delta \mathrm{InTIFT}_{t-i} + \sum_{i=1}^{p} \delta_2 \Delta \mathrm{InTIFT}_{t-i} + \beta_1 \mathrm{InTIFT}_{t-i} + \varepsilon_t \quad (1)$$

Where, TIFT and TITII represents total inland freight transport and total inland transport infrastructure investment in the selected countries. Ln and Δ denotes natural log and first difference operator, respectively. Under the ARDL procedure, the above equation [1] is estimated to relationship of long-run between the variables using F-test among the lagged level variables for establishing joint significance of coefficients i.e., H0: $\beta 1 = \beta 2 = 0$ against the alternative H1: β 1 □ β2 □ 0. The null hypothesis of the no cointegrations is rejected, provided the F-statistics obtained is greater than the lower-bound and higher-bound critical values. This suggests, presence of cointegration association between the variables.

The conditional ARDL long-run model after establishing cointegration is estimated as:

$$\Delta \mathrm{InTIFT}_t = \beta_0 + \sum_{i=1}^{p} \beta_1 \mathrm{InTIFT}_{t-i} + \sum_{i=1}^{p} \beta_2 \mathrm{InTIFII}_{t-i} + \varepsilon_t \quad (2)$$

The optimal lag length for the ARDL estimation is selected using Akaike Information Criterion [AIC]. Finally, the short-run dynamic parameters are estimated using an error correction term [ECT]. This is specified as follows:

$$\text{In}\Delta\text{TIFT}_t = \beta_0 + \sum_{i=1}^{p} \delta_1 \Delta\text{InTITII}_{t-i} + \phi\text{ECT}_{t-1} + \varepsilon_t. \quad (3)$$

Where, δ_1 is the short-run coefficient. The swiftness of adjustment of a parameter is displayed by the ECT term. This demonstrates how fast the series can return to its equilibrium. The coefficient should besignificant and have negative sign. Ramsay reset tests, serial correlation and heteroscedasticity tests performed were to check the admissibility of the model. In order to estimate the stability of long run andshort-run coefficients of the ARDL model, the CUSUM and CUSUMQ tests were deployed.

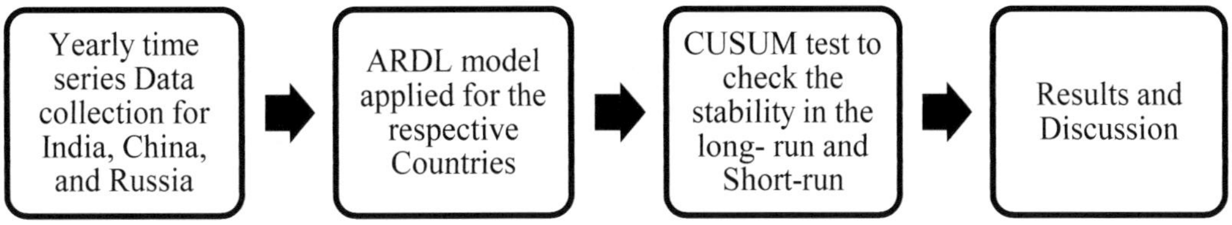

Figure 1. Flow chart of analysis

4. Analysis and results

4.1 Stationarity test

There could be misleading inferences due to use of non-stationary variables in any historical series model [26]. The order of integration is critical requisite for the cointegration test [27]. Therefore, the study employs the Augmented Dicker Fuller [ADF] test to examine stationarity condition of the variables and the order of integration The ADF test results are shown in the Table 1. The results from the analysis confirm that variables, TIFT and TITII are not stationary at level in the case of each economyviz. China, Russia and India. The null hypothesis of non-stationary cannot be rejected at level, whereasthe null hypothesis of non-stationary has been rejected in the first differences of the time-series, implyingthe variables are found to be stationary at its first differences and are integrated of order one, I[1] for allthe countries under study.

Table 1. Results of augmented dickey fuller test

Variables	Levels	First Difference	Decision
China			
TIFT	-2.014458	-2.731484***	I[1]
TITII	-2.116412	-2.899394***	I[1]
Russia			
TIFT	-0.645500	-3.195295**	I[1]
TITII	-1.981131	-2.837578***	I[1]
India			
TIFT	-1.754336	-2.925385***	I[1]
TITII	-2.213102	-3.234385**	I[1]

Note: ** &*** indicates significance at the five and ten per cent level, respectively. Optimal lag length is determined by the Akaike

Information Criterion [AIC].

4.2 ARDL Cointegration Bound Test

The bound test approach is deployed to test the presence of cointegration in the time series. The Akaike Information Criteria [AIC] was used to select the suitable lag length. The calculated F-statistics are found to be greater than F-critical values at 1% and 5% level for China, Russia and India respectively as shown in Table 2. The empirical result supported the rejection of the null hypothesis of no cointegration and confirmed the existence of a long-run relationship between the total inland freight transport and total inland infrastructure investment in the selected countries.

Table 2. ARDL co-integration bounds test

China		
F-Statistic: 6.391171**	Critical Value	
	Lower Bound	Upper Bound
Sig. level 10%	3.303	3.797
Sig. level 5%	4.09	4.663
Sig. level 1%	6.027	6.76
Russia		
F-Statistic: 6.028313**	Critical Value	
	Lower Bound	Upper Bound
Sig. level 10%	3.303	3.797
Sig. level 5%	4.09	4.663
Sig. level 1%	6.027	6.76
India		
F-Statistic: 14.15162*	Critical Value	
	Lower Bound	Upper Bound
Sig. level 10%	3.303	3.797
Sig. level 5%	4.09	4.663
Sig. level 1%	6.027	6.76

Note: * and ** indicates that values from computed statistic falls above the upper bonds value at 1% and 5% percent significance level, respectively.

4.3 ARDL Long-run Estimates

After established the cointegration among the variables, the selected ARDL long-run process based on AIC is estimated and the results are shown in the Table 3.

Table 3. Estimated Long-Run ARDL Model

China			
Selected ARDL [1 1] Model			
Regressor	**Coefficient**	**t-statistic**	**Prob. value**
TITII	0.478566	2.631291**	0.0273
Constant	3.488737	0.695729	0.5042
Russia			
Selected ARDL [1 3] Model			
Regressor	**Coefficient**	**t-statistic**	**Prob. value**
TITII	0.288786	2.320407**	0.0316

Constant	8.394612	0.397701	0.7073
India			
Selected ARDL [1 2] Model			
Regressor	**Coefficient**	**t-statistic**	**Prob. value**
TITII	0.556311	10.18994*	0.0000
Constant	1.632760	1.237489	0.2558
Note: * and ** – indicates significance at one and five per cent level, respectively. Optimal lag length is determined by the Akaike InformationCriterion [AIC].			

Table 3 display the results from the ARDL long-run approach. Total inland freight transport is used as dependent variable and Total Inland Infrastructure Investment is used as independent variable in this study. Results of ARDL model indicate that Total Inland Infrastructure Investment has positive and significant effect on Total inland freight transport in the long run. This implies that one unit increase inTotal Inland Infrastructure Investment will increase the Total inland freight transport by 0.47 units in China.

For Russia, the ARDL results indicate that Total Inland Infrastructure Investment has positive and significant effect on Total inland freight transport in the long run. This implies that one unit increase inTotal Inland Infrastructure Investment will increase the Total inland freight transport by 0.28 units in Russia.

For India, the ARDL results indicate that Total Inland Infrastructure Investment has positive and significant effect on Total inland freight transport in the long run. This implies that one unit increase inTotal Inland Infrastructure Investment will increase the Total inland freight transport by 0.55 units in India.

4.4 ARDL Short-run Estimates

Table 4. Estimated Short-Run ARDL Model

China			
Regressor	**Coefficient**	**t-statistic**	**Prob. value**
Δln[TITII] t	0.247946	2.450978**	0.0153
ECTt-1	-0.379808	-2.258531**	0.0403
Serial Correlation LM Test [F-statistics]	= 0.101323	[0.9049]	
Heteroscedasticity Test [F-statistics]	= 0.259292	[0.6217]	
Functional Form Test [F-statistics]	= 2.334930	[0.1672]	
Russia			
Regressor	**Coefficient**	**t-statistic**	**Prob. value**
Δln[TITII] t	0.065258	1.956443***	0.0978
Δln[TITII] t-1	0.063467	1.858560	0.1222
Δln[TITII] t-2	0.055607	1.945258***	0.0982

ECTt-1	-0.122630	-2.078200***	0.0923
Serial Correlation LM Test [F-statistics] = 1.428387 [0.2980]			
Heteroscedasticity Test [F-statistics] = 0.841915 [0.3857]			
Functional Form Test [F-statistics] = 1.747311 [0.2567]			
India			
Regressor	**Coefficient**	**t-statistic**	**Prob. value**
Δln[TITII] t-1	0.064134	1.281796	0.2407
Δln[TITII] t-2	0.080907	2.314209**	0.0402
ECTt-1	-0.375590	-7.388154*	0.0002
Diagnostic Tests			
Serial Correlation LM Test [F-statistics] = 0.158646 [0.7042]			
Heteroscedasticity Test [F-statistics] = 2.741786 [0.1321]			
Functional Form Test [F-statistics] = 1.096138 [0.3354]			

Note: *, ** and *** – indicates significance at one, five and ten per cent level, respectively. Figures in brackets indicate p-value. Optimal lag length is determined by the Akaike Information Criterion [AIC].

The short-run coefficients with the error correction terms are reported in Table 4. The error correction terms [ECT] are found to be negative and all the estimates were found to be statistically significant. This implying that the total inland freight transport converges on its long-run equilibrium with the speedadjustment via the channel of total inland infrastructure investment.

For China, the coefficient of TITII is found to be significant and positive at 5% percent level. This implying that 1% increase in total inland infrastructure investment will increase the total inland freight transport by 0.24% in the short run.

For Russia, the coefficient of TITII is found to be significant and positive at 10% percent level, implyingthat 1% increase in total inland infrastructure investment will increase the total inland freight transport by 0.05% in the short run.

For India, the coefficient of TITII is found to be significant and positive at 5% percent level. This implying that 1% increase in total inland infrastructure investment will increase the total inland freight transport by 0.08% in the short run.

4.5 Diagnostics test

Further, Table 4 shows the result of Breusch-Godfrey test for serial correlation, ARCH-LM test for Heteroskedasticity and Ramsey test for specification. The Breusch Godfrey test result indicates that there is no problem of serial correlation as probability value is higher than 0.05 in all the ARDL estimations. The ARCH-LM test results indicate that there is no problem of Heteroskedasticity in all theARDL estimations as probability value is higher than 0.05 in all the ARDL estimations. In addition, theRamsey RESET test indicates that the ARDL models are correctly specified. Therefore, the sequence ofdiagnostic tests employed indicates that ARDL models reliable and properly specified.

Finally, the study applied the CUSUM [Cumulative Sum of Recursive Residuals] and CUSUMSQ

[Cumulative Sum of Squares of Recursive Residuals] plots to examine the stability of the ARDL models and shown in fig 2, 3 and 4. If the plots of the CUSUM and CUSUMSQ statistics stay within the critical bounds of five per cent level of significance, then the ARDL models are stable. Figure 2, 3 and 4 show that CUSUM and CUSUMSQ statistics for China, Russia and India respectively. The values obtained are within the 5% critical bounds, which suggests that long-run and short-run coefficients are stable based on ARDL calculations.

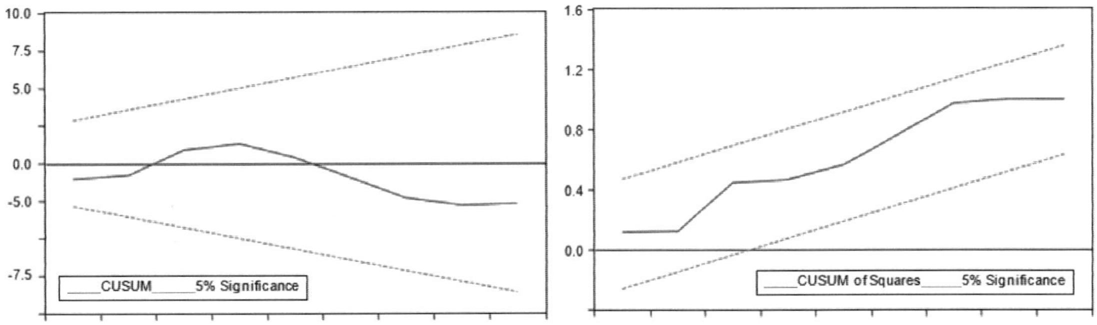

Figure 2. Cusum and Cusum of squares for China

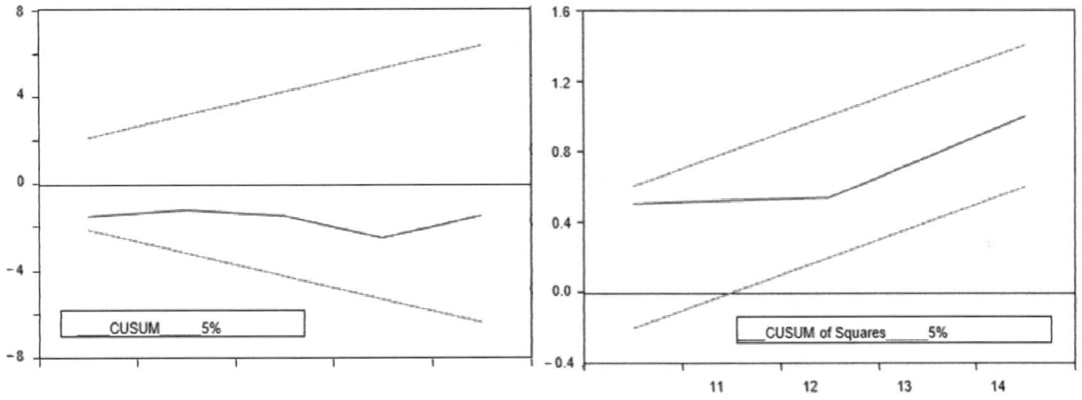

Figure 3. Cusum and Cusum of squares for Russia

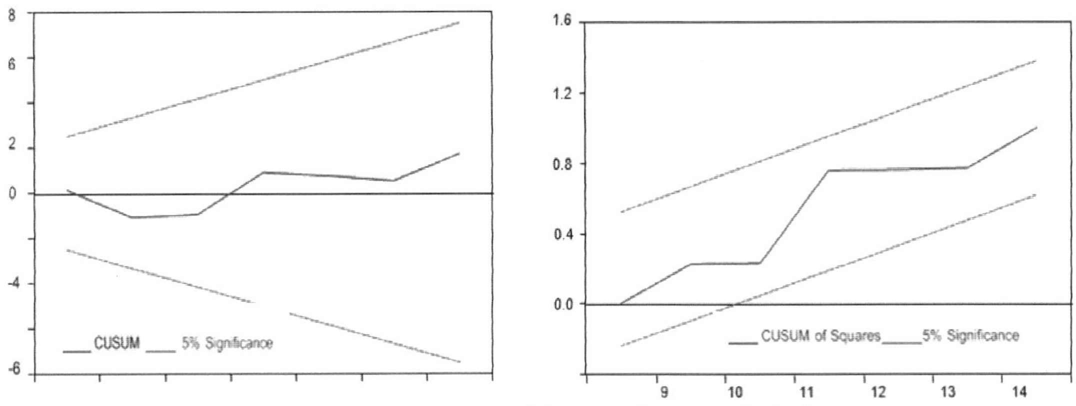

Figure 4. Cusum and Cusum of squares for India

5. Conclusions

The study clearly brings forth that Inland infrastructure investment plays a critical and positive role towards growth freight in the three BRICS member nations. On the long run, among the three countries,India had the highest growth, followed by China and Russia. Thereby, could be also a major driver of GDP for all the three countries. On the short run, China demonstrates massive positive affect of growthin freight market on account of transport infrastructure investment, followed by India and Russia respectively. This could also indicate that infrastructure investments in China bring out faster growth infreight markets due to high GDP, Industrialization, and higher investment efficiency. For India, in short-run the growth in freight markets appear slow but long run benefits seem to be higher which could be due to lower infrastructure investment efficiency and trade. Russia having lower population, trade and lower industrialisation growth was found to have lower freight growth in both short run and long run. There is a need to evaluate the TTI investment efficiency towards sustainability and profitability in the three countries apart the GDP growth. Further, the efficiency in roadways, railways, airways would givebetter decision taking opportunity for the respective countries.

References

[1] Belenky AS, Fedin GG, Kornhauser AL. Estimating the Needed Volume of Investment in a Public– Private Partnership to Develop a Regional Energy/Freight Transportation Infrastructure. International Journal of Public Administration 2019; 42[15-16]:1275–310.

[2] Marieke S., van der Tuin, Pel AJ. The disruption transport model: computing user delays resulting from infrastructure failures for multi-modal passenger & freight traffic. Eur. Transp. Res. Rev. 2020; 12[1].

[3] Venezia E. Transport investments: are they crucial for the economic system? IJEME 2011; 2[4]:351.

[4] Caliskan N. A decision support approach for the evaluation of transport investment alternatives. European Journal of Operational Research 2006; 175[3]:1696–704.

[5] Proost S, van der Loo S. Transport Infrastructure Investment and Demand Uncertainty. Journal of Intelligent Transportation Systems 2010; 14[3]:129–39.

[6] Lavee D, Beniad G, Solomon C. The Effect of Investment in Transportation Infrastructure on the Debt-to-GDP Ratio. Transport Reviews 2011; 31[6]:769–89.

[7] Kiel J, Smith R, Ubbels B. The Impact of Transport Investments on Competitiveness. TransportationResearch Procedia 2014; 1[1]:77–88.

[8] Hensher DA, Ho C, Mulley C. Identifying preferences for public transport investments under a constrained budget. Transportation Research Part A: Policy and Practice 2015; 72:27–46.

[9] Legaspi J, Hensher D, Wang B. Estimating the wider economic benefits of transport investments: Thecase of the Sydney North West Rail Link project. Case Studies on Transport Policy 2015;3[2]:182–95.

[10] Shi W, Bang H-S, Li KX. A cross-region analysis of the output elasticity of transport investment inChina. Maritime Policy & Management 2016; 43[2]:222–41.

[11] Yu N, Jong M de, Storm S, Mi J. The growth impact of transport infrastructure investment: A regional analysis for China [1978–2008]. Policy and Society 2012; 31[1]:25–38.

[12] Hansen W, Johansen BG. Regional repercussions of new transport infrastructure investments: An SCGE model analysis of wider economic impacts. Research in Transportation Economics 2017; 63:38–49.

[13] Skorobogatova O, Kuzmina-Merlino I. Transport Infrastructure Development Performance. Procedia Engineering 2017; 178:319–29.

[14] Varjan P, Rovňaníková D, Gnap J. Examining Changes in GDP on the Demand for Road

Freight Transport. Procedia Engineering 2017; 192:911–6.
[15] Meersman H, Nazemzadeh M. The contribution of transport infrastructure to economic activity: Thecase of Belgium. Case Studies on Transport Policy 2017; 5[2]:316–24.
[16] Kyriacou AP, Muinelo-Gallo L, Roca-Sagalés O. The efficiency of transport infrastructure investment and the role of government quality: An empirical analysis. Transport Policy 2019; 74:93–102.
[17] Rokicki B, Stępniak M. Major transport infrastructure investment and regional economic development – An accessibility-based approach. Journal of Transport Geography 2018; 72:36–49.
[18] Wiegmans B, Janic M. Analysis, modeling, and assessing performances of supply chains served bylong-distance freight transport corridors. International Journal of Sustainable Transportation 2019; 13[4]:278–93.
[19] Tapia RJ, Jong G de, Larranaga AM, Bettella Cybis HB. Application of MDCEV to infrastructure planning in regional freight transport. Transportation Research Part A: Policy and Practice 2020; 133:255–71.
[20] Cho S, Choi K. Transport accessibility and economic growth: Implications for sustainable transport infrastructure investments. International Journal of Sustainable Transportation 2021; 15[8]:641–52.
[21] Fageda X, Olivieri C. Infrastructure Transport Investments, Economic Growth and Regional Convergence. In: International Encyclopedia of Transportation. Elsevier; 2021. p. 2–5.
[22] Zhang J, Zhang Y. Tourism, transport infrastructure and income inequality: A panel data analysis ofChina. Current Issues in Tourism 2022; 25[10]:1607–26.
[23] Lee JK. Transport infrastructure investment, accessibility change and firm productivity: Evidence from the Seoul region. Journal of Transport Geography 2021; 96:103182.
[24] Organisation for Economic Co-operation and Development. ITF Transport Outlook 2019. 2019th ed. [Place of publication not identified]: OECD Publishing; 2019.
[25] Pesaran MH, Shin Y, Smith RJ. Bounds testing approaches to the analysis of level relationships. J. Appl. Econ. 2001; 16[3]:289–326.
[26] Libanio GA. Unit roots in macroeconomic time series: theory, implications, and evidence. nova Economia_Belo Horizonte 2005; 15[3]:145–76.
[27] John Glynn, Nelson Perera, Reetu Verma. Unit Root Tests and Structural Breaks: A Survey with Applications. Journal of Quantitative Methods for Economics and Business Administration 2007; 3[1]:67–79.

An Investigation of Sick Building Syndrome (SBS) affecting the health of occupants in Western Maharashtra, India.

Amol Pawar[1] Anil Agarwal[2] Babalu Rajput[3] Sudarshan Kore[4]

1, 3 Associate Professor, School of Construction NICMAR University Pune, India
4 Assistant Professor, School of Construction NICMAR University Pune, India
2 Professor, School of Construction NICMAR University Pune, India,

apawar@nicmar.ac.in. 09850237189

Abstract. This paper highlights the problems caused due to lack of planning in designing the building envelope for a building. In recent years, as construction technology advances, designers are focusing on decreasing the weight of buildings for high-rise constructions. Also, the trends in the market are to plan buildings with extensive use of mechanical ventilation systems. So, in this regard, a study was done to find out the symptoms that are found in such types of buildings by restoring to more mechanical ventilation. It is a structured questionnaire-based study to discover the common symptoms of sick building syndrome (SBS). The locations which we chose for the survey included mixed environments like Dry climate, humid climate, hot climate, etc. Regions such as Karad, Kolhapur, and Vita had dry climates, while areas like Lonavala, Mumbai, and Airoli, had humid climates. Commonly found symptoms affected by the occupants were headaches, difficulty breathing, and concentration. The most significant factors are dust, glare, and light. Also, complaints of hay fever were given by the occupants. Headache, Difficulties in concentration, fatigue and cough are the prominent factors observed in the three months by the occupants. So it can be concluded that the mechanical ventilation should be aided with good recovery of the natural air inflow. Also, location-wise climatic conditions should be considered when planning the ventilation systems.

Keywords: Air conditioning, Air quality, Sick Building Syndrome (SBS), symptoms, causes.

1. Introduction

Indicators of Sick Building Syndrome (SBS). Sick building syndrome (also known as environmental disease, building-related disease [BRI], or multiple chemical sensitivity [MCS]) is a disease that affects some people after they are exposed to chemical, biological, or physical agents that are thought to be found in buildings, according to some clinicians. The term "sick building syndrome" refers to a set of circumstances (SBS). When building occupants experience acute health and comfort effects that appear to be related to the amount of time they spend in the building, but no specific disease or cause can be identified. The concerns could be separated into a single room or area or widespread throughout the building. Acute discomfort symptoms reported by building occupants include headache, eye, nose, or throat irritation, dry cough, dry or itchy skin, dizziness and nausea, concentration difficulty, fatigue, and odor sensitivity.

Sick Building Syndrome Causes

- A lack of ventilation.
- Chemical contaminants originating indoors
- Chemical contaminants from the environment:
- Biological pollutants

2. Review of Literature

(Joshi, 2008) reviewed the various aspects of critical interrelationships about the sick Building Syndrome (SBS). The author found that the presence of this symptom could be confirmed in people who lived in the buildings and via the IAQ (indoor air quality). Additionally, it was stated that this health hazard is rising, and managing this illness is a common practice.

(Baldry, Bain & Taylor, 1997) In a study focusing on new workplaces, poor planning is usually the underlying cause of sickness. Negative externalities may significantly impact the productivity of an organization. It was discovered that office exteriors, interiors, and ventilation are increasing the issue.

(Jaakkola, Heinonen & Seppänen, 1989) Studied the complaints, symptoms and various indoor air quality factors for a modern eight-floor building. There was a considerable variation in the temperature, and it was not under the occupants' control. They found that room temperature was the most factor in sick building syndrome (SBS). Linear correlation between these parameters existed, and they concluded that individual temperature control would improve thermal comfort.

(Lahtinen, Huuhtanen & Reijula, 1998) In their review of SBS studied the SBS from the occupational stress perspective. They found that psychological factors are pretty strongly associated with SBS. Also, they found that some research concluded that psychological factors alone are not responsible for SBS. The other factors include chemical, physical or biological. All these factors interact with each other.

(Mendell & Smith, 1990) Several earlier studies had looked at the correlation between ventilation system usage and work-based symptoms, with mixed results. It was noted that these facts were at odds with each other. So five additional studies were conducted. Air-conditioned buildings were frequently regarded as among those who suffered from headaches. Fouling of the air conditioning system was not taken into consideration.

(Zweers, Preller, Brunekreef & Boleij, 1992) Health and medical research showed that the quality of one's work-life varies with the quality of the environment. It was obtained by administering a survey on personal health and air quality complaints. More than 7,000 questionnaires were collected from all of 60 companies that were located throughout the country of the Netherlands. This research found that the symptoms were worse in air-conditioned structures. These were the primary factors: allergies, as well as respiratory symptoms, and control of workplace temperature.

(Gupta, Khare & Goyal, 2007) It was determined that an airtight envelope is essential in designing office buildings for decreasing thermal comfort. A survey was done on a multistoried air-conditioned building in Delhi to understand more about this. Researchers looked into the connection between SBS scores and CO2 levels. Quantitative analysis was used to check for SO2, NO2, and SPM (particulate matter), and CO. Third-floor levels are consistently higher than other levels. A correlation was discovered between CO2 and SBS, as higher levels of CO2 led to more SBS.

(Sundell et al., 2011) conducted a review of how ventilation rate affects health. The authors conducted a study in 2005 and discovered that symptoms of health issues related to SBS are dependent on ventilation rates. They also discovered that inflammation, respiratory infections, and asthma worsen as ventilation rates decrease. They go on to say that more research should be done to determine the relationship between ventilation rates and health under various climatic conditions. If measures like source control or air cleaning combat indoor pollution levels, ventilation rates must be increased.

(Menzies et al., 1993) carried out a study within three consecutive two-week blocks to study the effect of ventilation on SBS. This study was conducted on four buildings. The participants of the study were unaware of the change in ventilation rates. The symptoms asked were headaches, fatigue, difficulty concentrating, cough and irritation of the eyes, nose or throat. More than half the participants reported at least one symptom each week. They also found that any one symptom was closely associated with the relative humidity but not with temperature, air velocity or carbon dioxide concentration.

(Fanger, Lauridsen, Bluyssen & Clausen, 1988) Their study tried to quantify the pollution sources by new olf units in 20 randomly selected offices and halls in Copenhagen. Fifty-four judges visited these establishments to assess the conditions in different scenarios.

Djukanovic (2002) did a cost-benefit analysis of improved air quality in an air-conditioned office building. It was tested in terms of both temperature and cold climate. The annual energy cost and heating system cost were calculated, and it was discovered that the annual benefit was ten times greater than the increase in annual energy and maintenance costs. Pack back time to recover the initial investment of enhancing air quality was also observed to be less.

(Norhidayah, Chia-Kuang, Azhar & Nurulwahida, 2013) The study aimed to determine the relationship between indoor air quality (IAQ) parameters and SBS symptoms in three Malaysian buildings.

The buildings were chosen based on their history and occupancy.

A questionnaire-based survey was carried out, and it was discovered that the same factors were responsible for the SBS in all three buildings.

The three buildings' indoor air quality (IAQ) varied according to velocity, CO2 concentration, temperature, and relative humidity parameters. According to the findings, ventilation and the accumulation of potentially hazardous contaminants within the indoor environment are essential predictors of sick building syndromes.

(Burt, 1999) Low-frequency noise, centered on a frequency of about 7 Hz, was found in several office rooms investigated. Symptoms of infrasound exposure may include fatigue, headache, nausea, difficulty concentrating, disorientation, seasickness, digestive disorders, cough, vision problems, and dizziness, all of which are symptoms of the sick building syndrome. Many of the occupants displayed symptoms of this nature. It was discovered that the low-frequency component of ventilation noise is frequently amplified in tightly sealed rooms.

Individuals repeatedly or long-term exposure to such magnified infrared sound may develop an allergic reaction.

(Gou & Siu Yu Lau, 2012) sought to detect SBS in open-plan offices by incorporating design factors (indoor installations, workstation partitions, and operating windows) that predict SBS for architects and interior designers, as well as indoor environmental characteristics (thermal comfort, air quality, noise, and lighting) that contribute to SBS symptoms. This study employed a standard Building Use Studies (BUS) questionnaire, which included symptoms of sick building syndrome, environmental satisfaction and perception, and background information on respondents and their office space. The survey was conducted in 30 offices of office-related professionals in Hong Kong, with 469 Chinese workers participating.

3. METHODOLOGY

3.1 *Questionnaire*

This study involved a combination of different data collection methods, such as literature review, questionnaire etc. The first stage was a literature review followed by a questionnaire survey. This set of questionnaires was specially designed for a targeted group of respondents comprising people from different locations to have uniformity in results as far as locations are concerned. The questionnaire distribution to the respondents included physical distribution, distribution via the internet through google forms, and part was distributed through social media. The format of the questionnaire survey was designed in such a way that it gave deep insight into the various factors which to build up ill-health issues.

3.2 *Data Collection*

For collecting the responses, the questionnaire format was circulated partially through hard copies, partially through social media, and also with the help of the internet via google forms. During the distribution of the

questionnaire, care was taken about the demographic classification of the population, viz—age, gender, occupation, location etc.
The locations which we chose for the survey included mixed environments like Dry climate, humid climate, hot climate etc. Regions such as Karad, Kolhapur, and Vita have dry climates, while regions like Lonavala, Mumbai, etc., have humid climates.

3.3 Analysis of Data

RII is computed for each factor shown here in the equation. Kometa et al. (1994) and Sambasivan and Soon (2007) used the RII method to consider the causes' relative importance of sick building syndrome. This same methodology in this study was used.

$$\text{RII} = \frac{\sum W}{A*N}$$

Where W is the weight respondents gave to each attribute and ranges from 1 to 4 using the Likert type scale method, and N is the total number of the people who responded. A is the highest valve for an attribute of 4 in this case.

4. Results and Discussion

About one seventy-five (175) people from different categories were sent responses, out of which one hundred and twenty people responded. The respondents were considered from various age categories, different professions, gender and also from localities. The response rate is thus 68%, which is considered a good response in this type of survey. The interview questionnaire, which was thus developed, was circulated among various people, as mentioned above, to assess the perception regarding sick building syndrome amongst people and find out the most significant factors causing the phenomenon called sick building syndrome. The questionnaire was circulated through hard copy and social media like LinkedIn, email, WhatsApp messenger etc. Even these replies were accepted since they were done to preserve the quality of the opinions gathered in the survey. The number of respondents belonging to each locality is shown in following Table 1.

Table 1. Categorization of Respondents

Age Region	20-25	26-30	31-35	36-40	41-45	46-50	51& above
Pune	28	7	10	0	0	2	0
Kolhapur	7	4	5	3	1	0	2
Karad	11	5	8	2	1	5	3
Mumbai	5	0	0	0	0	1	0
Lonavala	9	1	0	0	0	0	0
Total	60	16	13	5	2	8	5

4.1 Research Findings

Respondent Experience on Part A

This questionnaire aims to obtain information about occupant background, the profession in which they are working, and various parameters which are the most significant factors affecting the human comfort and leading to the phenomenon called sick building syndrome. The following table gives the most significant factors causing human discomfort.

The table below shows the responses received from various respondents in the last three months for a residential environment. The priority index was calculated, and subsequently, RII was derived. The following parameters are significant in determining the factors that play a prominent role in sick building syndrome.

Table 2. Different Types of Factors Causing SBS

Symptoms	1	2	3	4	5	Priority	RII
Temperature variations	32	33	23	55	55	3.00	0806
Stuffy bad air	50	38	22	33	33	5.00	0675
Unpleasant smell	74	48	8	13	13	7.00	0513
Noise	45	20	45	33	33	4.00	0.733
Dust and Dirt	13	22	71	37	37	1.00	0.871
Dampness	49	48	22	24	24	6.00	0.640
Less light/glare	26	33	38	46	46	2.00	0.813

From the above table, we can determine the most significant factors causing the sick building syndrome by comparing these factors with the respective RII and their rank. Thus the most significant factors are enlisted below in following Table 3.

Table 3. Prominent Factors

Sr.No	inent Factors	RII	Priority
1	Dust and Dirt	0.871	1
2	Dim light/ Glare	0.813	2
3	Varying room temperature	0.806	3
4	Noise	0.733	4

Questionnaire Survey Part B:

The symptoms related to the disease were surveyed in the form of yes/no questions. Subsequently, the chi-square test was being conducted using SPSS since the questions were objective, without any weightage attached.

The procedure adopted for this test is as follows:

Step 1: hypothesis

H0: symptoms are dependent on the disease

H1: symptoms are independent of the disease

Step 2: data variables entered in SPSS

Step 3: output generated in SPSS

Step 4: analysis of output.

The output drawn was significant, meaning that the symptoms were independent of disease in most cases except in the case of hay fever.

Questionnaire Survey Part C

This questionnaire aims to obtain information about various effects occupants suffer from the above-listed factors, which were earlier described in part A. The following table gives the most significant effects due that lead to discomfort. The table below shows the responses received from various respondents in the last three months for a residential environment. The priority index was calculated, and subsequently, RII was derived. The following parameters are significant in establishing the prominent effects of sick building syndrome.

Table 4. Effects of Sick Building Syndrome

Symptoms	1	2	3	4	Rank	RII
Fatigue	25	42	34	42	3	0.79
Headache	12	13	77	43	1	0.91
Nausea/dizziness	76	41	24	3	11	0.50
Difficulty in concertation	15	18	78	32	2	0.80
Itching/burning	82	34	23	3	12	0.48
Eyes irritation	82	61	8	23	7	0.66
Running Nose	66	47	19	9	9	0.52
Dry throat	31	43	45	22	5	0.70
Cough	23	22	83	14	4	0.77
Dry skin	81	23	37	2	10	0.51
Scaling ears	30	46	66	3	6	0.69
Dry hands	60	34	13	37	8	0.65

The above table shows that the frequently causing symptoms due to various factors, as mentioned in table 4, are listed separately as per their highest rank.

Table 5. Prominent Effects

Sr No	Description of Symptoms	Priority	RII
1	Headache	1	0.919
2	Difficulties in concentrating	2	0.860
3	Fatigue	3	0.790
4	Cough	4	0.775

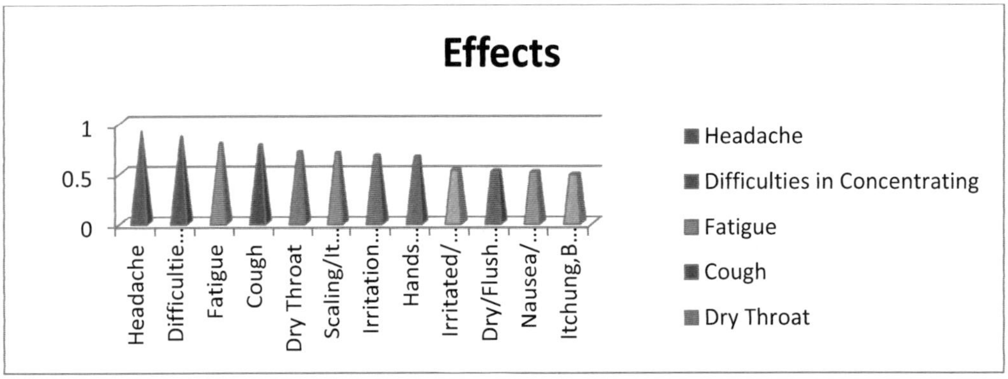

Figure 1. Effects of Sick Building Syndrome

5. Conclusion and Recommendations:

This paper aimed to identify the factors that cause the phenomenon known as sick building syndrome, as these factors have severe consequences for human health. When the causes of sick building syndrome are identified, the effects can be minimized or avoided. A total of seven factors were identified through a detailed literature review and a questionnaire distributed to various people, and they are as follows: varying room temperature, stuffy bad air, unpleasant smell, noise, dust and dirt, dampness, and dim light/glare. The study then quantified the relative importance of these factors and displayed their ranking, which was then grouped based on their rank. The relative importance index was used as a simple method of analysis. As previously discussed, this method addressed the most significant factors causing sick building syndrome and the potential effects on human health.

The critical step in addressing sick building syndrome symptoms is to eliminate their causes.
• Ensuring adequate heating in the property would then enhance the internal temperature of the house's surfaces and reduce the likelihood of condensation.
Also, ensure that the building is energy efficient by installing insulated walls and double-glazed windows to prevent heat from escaping from the property.
• Double glazing, loft insulation, and draft proofing will help reduce heat loss in a home.
Installing insulation will help maintain a higher temperature on the property's surfaces.
• Adequate ventilation is required to allow moisture to escape a property before it condenses.
Installing an energy-efficient extractor fan in the kitchen and bathroom can help to increase humidity and prevent condensation.

References

[1]. Baldry, C., Bain, P., & Taylor, P. (1997). Sick and tired? - —Working in the modern Office. *Work, Employment &and Society*, *11*(3), 519–-539. DOI: 10.1177/0950017097011003007.
[2]. Burt, T. (1996). Sick Building Syndrome: Acoustic aspects. *Indoor and Built Environment*, *5*(1), 44–59. https://doi.org/10.1177/1420326X9600500107doi:https://doi.org/10.1177/1420326X9600500107.
[3].Djukanovic, R., Wargaocki, P., & Fanger, P. O. (2002). Cost-benefit analysis of improved air quality in an office building. in *Proceedings of the Indoor Air, 2002*.
[4]. Fanger, P. O., Lauridsen, J., Bluyssen, P., & Clausen, G. (1988). Air pollution sources in offices and assembly halls, quantified by the olf unit. *Energy and Buildings*, *12*(1), 7–-19. DOI: 10.1016/0378-7788(88)90052-7.
[5].Gou, Z., & Siu Yu Siu-Yu Lau, Siu. Yu., S. (2012). Sick building syndrome in open-plan offices. *Journal of Facilities Management*, *10*(4), 256–-265. DOI: 10.1108/14725961211265729.
[6].Gupta, S., Khare, M., & Goyal, R. (2007). Sick building syndrome—A case study in a multistore centrally air-conditioned building in the Delhi City. *Building and Environment*, *42*(8), 2797–-2809. DOI: 10.1016/j.buildenv.2006.10.013
[7].Jaakkola, J. J. K., Heinonen, O. P., & Seppänen, O. (1989). Sick building syndrome, the sensation of dryness and thermal comfort about room temperature in an office building: Need for individual control of temperature. *Environment International*, *15*(1–6)(1-6), 163–-168. DOI: 10.1016/0160-4120(89)90022-6
[8].Joshi, S. M. (2008). The sick building syndrome. *Indian Journal Oof Occupational Aand Environmental Medicine*, *12*(2), 61–64. DOI: 10.4103/0019-5278.43262
[9]. Lahtinen, M., Huuhtanen, P., & Reijula, K. (1998). Sick Bbuilding Ssyndrome and Ppsychosocial Ffactors - —a Literature Review. *Indoor Air*, *8*(S4)(Suppl. 4), 71–-80. doi: 10.1111/j.1600-0668.1998.tb00012.x
[10]. Mendell, M. J., & Smith, A. H. (1990). A consistent pattern of elevated symptoms in air-conditioned office buildings: aA reanalysis of epidemiologic studies. *American Journal of Public Health*, *80*(10), 1193–-1199. DOI: 10.2105/ajph.80.10.1193
[11]. Menzies, R., Tamblyn, R., Farant, J. P., Hanley, J., Nunes, F., & Tamblyn, R. (1993). The Effect of Varying Levels of Outdoor-Air Supply on the Symptoms of Sick Building Syndrome. *New England Journal Of Medicine*, *328*(12), 821–-827. DOI: 10.1056/nejmNEJM199303253281201
[12]. Norhidayah, A., Chia-Kuang, L., Azhar, M. K., & Nurulwahida, S. (2013). Indoor Air Quality and Sick Building Syndrome in Three Selected Buildings. *Procedia Engineering*, *53*, 93–-98. DOI: 10.1016/j.proeng.2013.02.014
[13]. Sundell, J., Levin, H., Nazaroff, W. W., Cain, W. S., Fisk, W. J., & Grimsrud, D. T. et al., . . . Weschler, C. J. (2011). Ventilation rates and health: multidisciplinary review of the scientific literature. *Indoor Air*, *21*(3), 191–-204. DOI: 10.1111/j.1600-0668.2010.00703.x
[14]. United States Air and Rradiation. (2019), *(6609J)6609Jj* Research and Development Environmental Protection (MD-56) Agency February 1991 Indoor Air Facts No. 4 (revised) Sick Building Syndrome. (2019).

[15]. Zweers, T., Preller, L., Brunekreef, B., & Boleij, J. S. M. (1992). Health and Indoor Climate Complaints of 7043 office Workers in 61 Buildings in the Netherlands. *Indoor Air*, *2*(3), 127-–136. doi: 10.1111/j.1600-0668.1992.00001.x.

Analysis of Blended Concrete Cubes under Impact loading using ANSYS

Janhavi Singh[1] and **Shilpa Pal**[2]

Delhi Technological University, Delhi, India

Email: janhavisingh_2k20ste09@dtu.ac.in

Abstract. Blended concretes, with the partial replacement of the cement with secondary cementitious material like fly ash, rice husk ash, GGBS (Ground Granulated Blast Slag), silica fume etc. are gaining a wide range of applications in civil engineering nowadays. The behaviour of concrete structures has been studied by many researchers under different types of loads like impact load, wind load, earthquake load etc. Impact loading analysis has applications like ship impact resistance design for marine structures, impact-resistant structural design against military and terrorist attacks etc. This study aims to perform a FEM analysis of blended concrete cubes under impact load. Deformation and stress response has been obtained from the velocity impact simulations in ANSYS Explicit dynamics module. Parametric analysis has been done by changing the height of the impactor, shape of the impactor, boundary conditions and partial replacement of cement in the blended concrete with different supplementary cementitious material. It is observed that the strength of the blended concrete cube decreases with an increase in height of the impact.

1. Introduction

There is a growth in Portland cement demand in developing countries [1]. It is expected that the annual world cement production will shoot from 2540 million tonnes (2006) to around 3680-4380 million tonnes (2050) [2]. Industries' waste materials have the potential to be used in concrete as a replacement for aggregates or cement [1]. Some studies' results report that the cement can be replaced by various types of mineral admixtures [3]. Mineral admixtures such as metakaolin, GGBS (Ground Granulated Blast Slag) or Blast slag, Fly Ash, Rice Husk Ash, Palm Oil Fuel Ash and Silica Fume improve the quality of the concrete. They impart a significant influence on the hardened properties [4]. Partial replacement of the cement with such Supplementary Cementitious Materials (SCMs) brings down the consumption of cement and thus minimizes the impact on the environment [5]. Further use of such Supplementary Cementitious Materials helps to achieve a considerable amount of saving in cost and energy and promotes pollution reduction.[6]. So, it becomes clear that the blended concrete, in which the cement in the concrete is partially replaced by supplementary cementitious materials is gaining importance and the usage of such concretes should be extensively promoted in future. Thus there is a need to study such concretes under different types of loads. For our study, Fly ash and

Blast slag are used as supplementary cementitious materials, for partially replacing cement in the blended concrete. The cubes thus modelled in ANSYS are analysed for the impact loading. During the functional life of a structure or a structural element, there is a high probabilityof it being subjected to impact forces or repeated impacts. These forces are of various types and can occur due to a variety of reasons. For example- Moving vehicles may cause an impact accidentally on columns and walls [7]. Dropping of an object from height [8], water causing an impact on hydraulic structures, offshore structures or bridgepiers [9, 10], can be some incidences when the impact force comes into play.

2. Specimen and configuration

Finite Element simulations of the 3 blended concrete cubes of size 150 mm x 150 mm x 150 mm have been performed using commercially accessible ANSYS 17.2 software. 2 Binary Blends and 1 Ternary Blend of concrete have been considered in which the cement is partially replaced with Fly ash and Blast slag. Table 1 shows the percentage with which the cement is replaced with the Fly ash and Blast slag in the 3 blends of concrete. The CRM (Cement Replacement Material)/Binder ratio as shown in Table 1 has been taken from the work of Jeong-Eun Kim et al. [11].

Table 1. CRM/Binder ratio in the 3 Blends of Concrete

Name of the Blend	Fly ash / Binder	Blast Slag / Binder
BB1(Binary Blend 1)	20%	0%
BB2(Binary Blend 2)	25%	0%
TB1(Tertiary Blend 1)	25%	25%

Table 2. Properties of Blended Concrete

Properties	BB1	BB2	TB1
Density (kg/cum)	2344	2349	2344
Young's Modulus (GPa)	26.3	25.5	26.8
Poisson's ratio	0.205	0.184	0.199
Bulk Modulus (GPa)	14.859	13.449	14.839
Shear Modulus (GPa)	10.913	10.076	11.176
Compressive Strength (MPa)	44.6	41.2	42.9

A spherical ball of diameter 130 mm and a steel box of base dimensions 100 mm X 100 mm have been used as impactors. Both the impactors have the same volume and are made with structural steel. Properties of the structural steel are given in Table 3. For this study Density (kg/cum), Young's Modulus(GPa), Poisson's ratio and Compressive Yield Strength (MPa) of the 3 blended concrete are obtained from the experimental work of Jeong-Eun Kim et al. [11]. These properties are specified in Table 3. The FE analysis has been performed under 3 different heights of impact (3m, 3.5m and 4m) for 3 different concrete cubes, with 2 boundary conditions - base fixed and all 5 sides of the concrete cube fixed.

IOP Conf. Series: Earth and Environmental Science 1084 (2022) 012067 doi:10.1088/1755-1315/1084/1/012067

Table 3. General material properties of Structural Steel

Properties	Values
Density (kg/cum)	7850
Young's Modulus(GPa)	200
Poisson's ratio	0.3
Bulk Modulus (GPa)	166.67
Shear Modulus (GPa)	76.92
Tensile Yield Strength (MPa)	250
Compressive Yield Strength (MPa)	250
Ultimate Tensile Strength (MPa)	460

3. Finite Element Modelling of Specimen

ANSYS 17.2 has been used for FE Analysis performed on the 3 blended concrete cubes. Generally, FE analysis consists of 3 steps [12]

Step 1 Modelling the Cube and then meshing
Step 2 Assignment of the boundary conditions required
Step 3 Performing the Analysis of the model and then extracting the solution (Analysis types, loading, and solving)

In this study, for each blended concrete, 2 models have been made (for spherical impactor and box impactor). For each of the 2 models, the analysis has been done for 2 different boundary conditions. In one of the conditions, the cube is made fixed at the base. In the other condition, the cube is fixed at the base and 4 sides. In total, 12 models have been made and they have been analyzed for 3 different heights of impact (3m, 3.5m and 4m). The impactor is considered to be falling freely under gravity from these 3 heights. When a ball falls freely from a height under gravity, the velocity with which it touches the ground is the square root of the 2 times the product of the acceleration due to gravity and the height of the free fall. For our study, we have taken the value of acceleration due to gravity as 9.8m/sq sec. 7.688 m/s, 8.282m/s and 8.854 m/s are the velocities of impactor obtained for the heights of impact as 3m, 3.5m and 4m respectively. The Analysis of these models has been performed and maximum equivalent von misses stress and maximum total deformation have been computed under a variety of impact load conditions.

3.1 Meshing of Specimen and Body Interaction

For this study, explicit meshing property and 20 mm mesh size have been used for concrete and steel impactor. An explicit dynamics model has been used for the impact load analysis in ANSYS. In explicit dynamics, the main parameter chosen were end time duration and velocity inputs. The meshing of the cube model is shown in Figure 1a and Figure 1b.

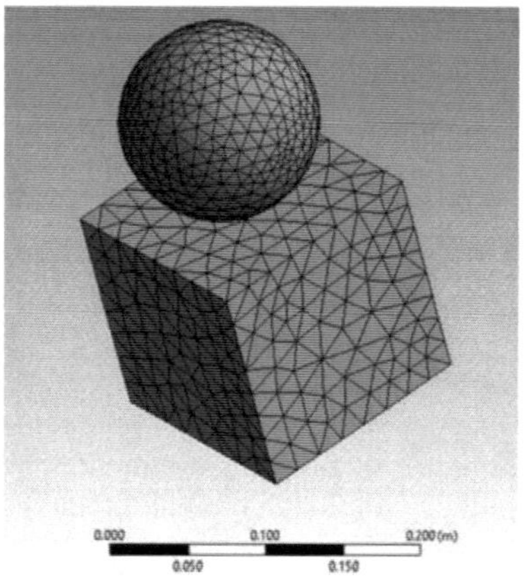

Figure 1a. FE Mesh- cube with sphere impact **Figure 1b.** FE Mesh-cube with box impact

3.2 Loading Conditions

Impact falling from varying heights was applied at the mid-point of the top surface of the cube. For the purpose of this study, the value of end time for loading is taken as 0.30 ms. The 3 velocity variations considered are 7.688 m/s, 8.282 m/s and 8.854 m/s.

4. Result and Discussion

4.1 Deflection and Stresses in the Blended Concrete under the Sphere shaped impactor

The maximum total deflection of cubes under impact was obtained aftersolving for the 3 blends.

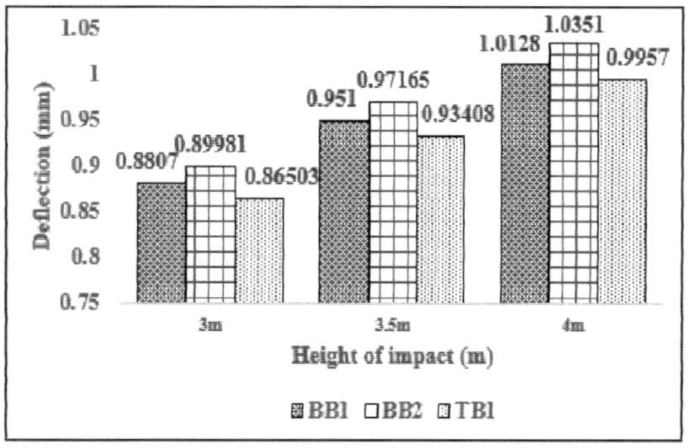

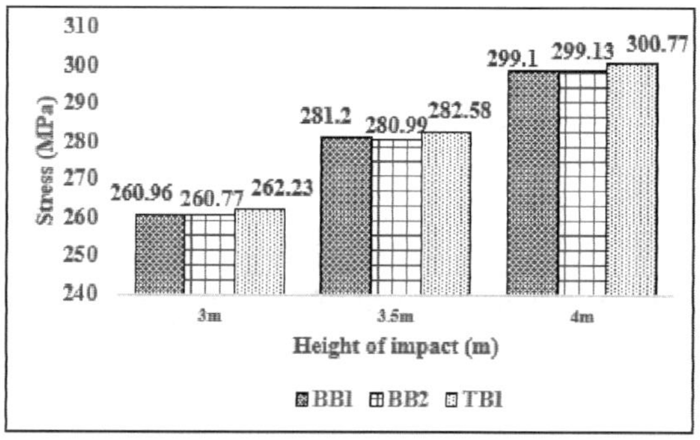

Figure 2a. Deflection- spherical impact- Base fixed **Figure 2b.** Stress - spherical impact-Base fixed

Figure 2a and Figure 2b show the comparative results of deflection and stress, respectively, obtained on varying the height of impact for the 3 blended concrete cubes with the base fixed, when the shape of the impactor is a sphere. Deflection in the cubes fixed at the base, in BB2 is 2.07% to 2.2 % more than that in BB1 and that in TB1 is 1.688% to 1.77% higher than that in BB1. As it can be inferred from the graph, stresses developed in BB1 and BB2 are almost the same with a slight percentage difference of 0.07% to 0.01%. Stress developed in TB1 is 0.49% to 0.56% more than that in BB1. However, on fixing the 5 sides, the deflection in the cubes increases by 99.78% to 103.705% and the stress decreases by 5.57% to 9.928%. Figure 3 and Figure 4 show the comparison between the deflection and stress respectively in the 3 Blended concretes, with the varying height of impact, under the sphere-shaped impactor when the base and the 5 sides are fixed respectively.

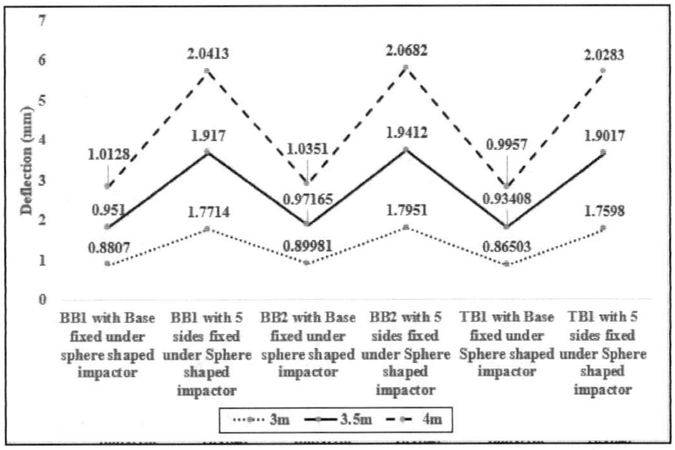

Figure 3. Comparison between the deflection under sphere-shaped impactor when the base and the 5 sides are fixed respectively

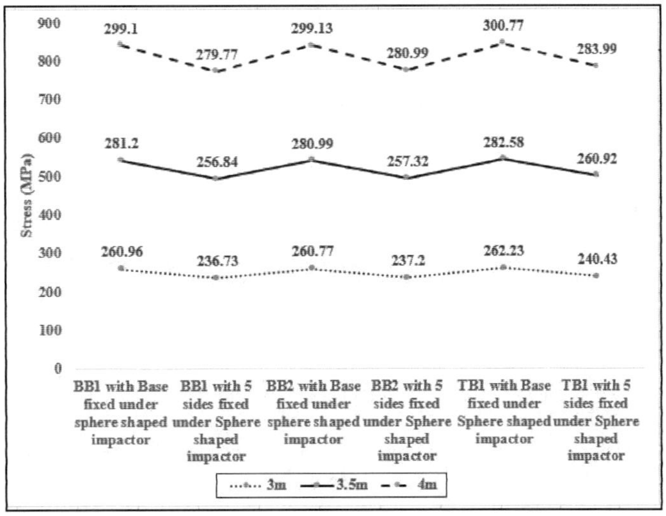

Figure 4. Comparison between the stress under sphere-shaped impactor when the base and the 5 sides are fixed respectively

4.2 Deflection and Stresses in the Blended Concrete under the Box shaped impactor

Under the Box-shaped impactor, the deflection in the cube, fixed at the base under different heights of impact in BB2 is about 2.5 % more than that in BB1 and that in TB1 is 1.67% lower than that in BB1. Stress developed in BB2 under different heights of impact is less than that in BB1, with a percentage difference of about 5.96%. Stress developed in TB1 under different heights of impact is about 3.31% more than that in BB1. However on fixing the 5 sides the deflection in the cubes decreases by 66.61% to 66.62% for BB1, 69.62% to 69.63% for BB2 and 62.63% to 62.65% for TB1. The decrease in the stress is 77.05% to 77.07% for BB1, 82.60% to 81.63% for BB2 and 74.72% to 74.75% for TB1.

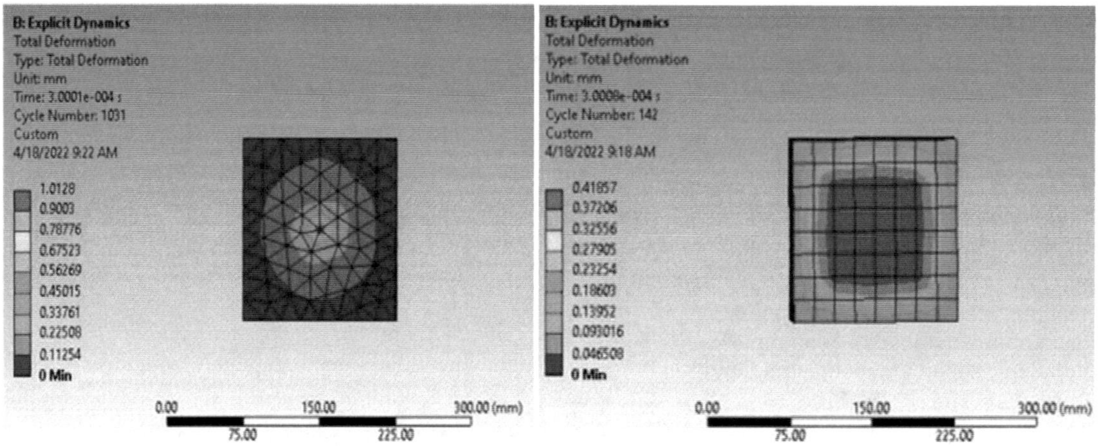

Figure 5. Deflection contours of BB1 cube under the height of impact 4m, under sphere-shaped impactor and box-shaped impactor

Deflection and stress contours of BB1 cube under the height of impact 4m, under sphere-shaped impactor and box-shaped impactor, are shown in Figure 5 and Figure 6 respectively. The maximum deformation and the maximum stress are obtained at the centre of the top surface of the cube.

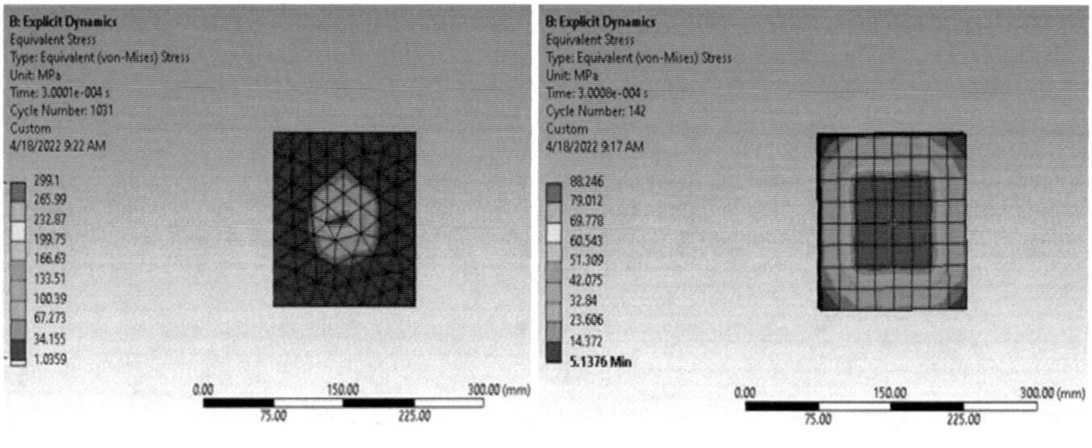

Figure 6. Stress contours of BB1 cube under the height of impact 4m, under sphere-shaped impactor and box-shaped impactor

Figure 7 and Figure 8 show the comparative analysis of the deflection and stress, respectively in the 3 blended concrete in various conditions, under Box-shaped impactor and Sphere-shaped impactor.

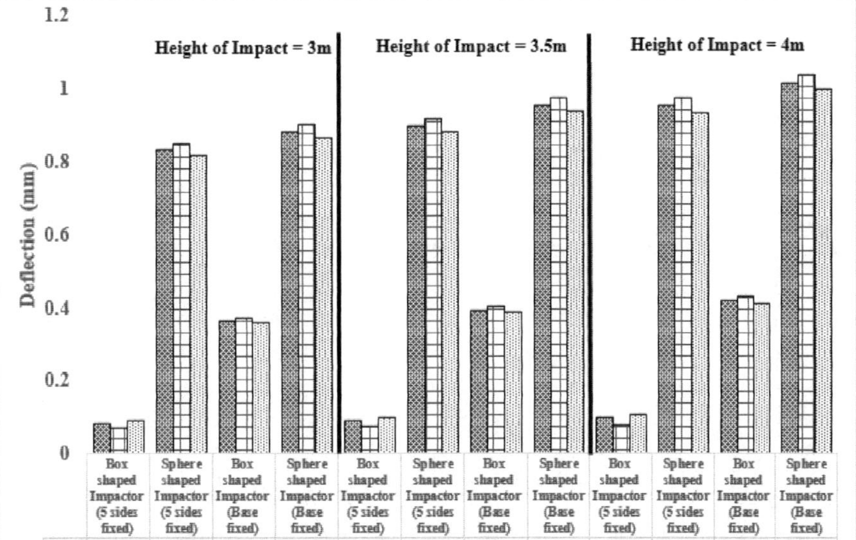

	Box shaped Impactor (5 sides fixed)	Sphere shaped Impactor (5 sides fixed)	Box shaped Impactor (Base fixed)	Sphere shaped Impactor (Base fixed)	Box shaped Impactor (5 sides fixed)	Sphere shaped Impactor (5 sides fixed)	Box shaped Impactor (Base fixed)	Sphere shaped Impactor (Base fixed)	Box shaped Impactor (5 sides fixed)	Sphere shaped Impactor (5 sides fixed)	Box shaped Impactor (Base fixed)	Sphere shaped Impactor (Base fixed)
BB1	0.083171	0.82985	0.36247	0.8807	0.089795	0.89607	0.39151	0.951	0.095962	0.94985	0.41857	1.0128
BB2	0.068362	0.84832	0.37154	0.89981	0.0737	0.91601	0.40131	0.97165	0.078858	0.97044	0.42905	1.0351
TB1	0.090089	0.81469	0.35642	0.86503	0.097264	0.87969	0.38498	0.93408	0.10394	0.93314	0.41158	0.9957

Figure 7. Comparative analysis of the deflection, in BB1, BB2 and TB1, for different heights of impact, in 2 different boundary conditions, under Box shaped impactor and Sphere shaped impactor

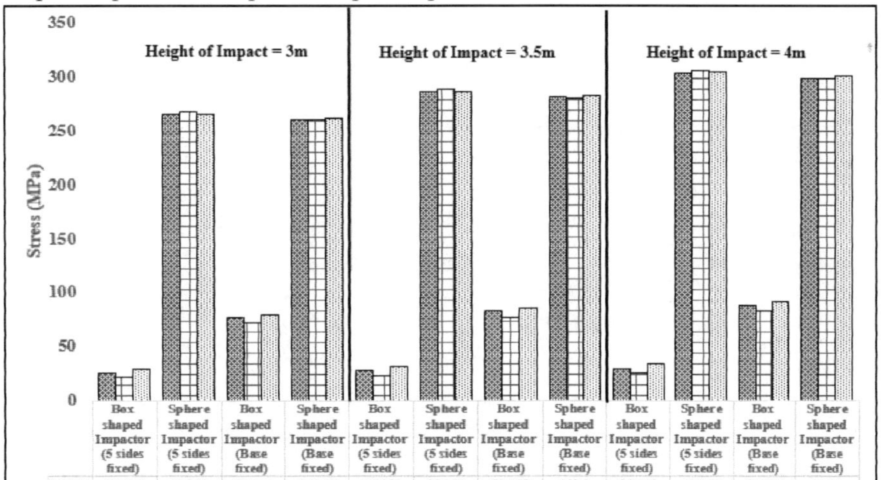

	Box shaped Impactor (5 sides fixed)	Sphere shaped Impactor (5 sides fixed)	Box shaped Impactor (Base fixed)	Sphere shaped Impactor (Base fixed)	Box shaped Impactor (5 sides fixed)	Sphere shaped Impactor (5 sides fixed)	Box shaped Impactor (Base fixed)	Sphere shaped Impactor (Base fixed)	Box shaped Impactor (5 sides fixed)	Sphere shaped Impactor (5 sides fixed)	Box shaped Impactor (Base fixed)	Sphere shaped Impactor (Base fixed)
BB1	25.524	266.01	76.438	260.96	27.558	286.61	82.552	281.2	29.451	303.75	88.246	299.1
BB2	21.836	268.06	71.881	260.77	23.576	288.82	77.627	280.99	25.197	305.83	82.977	299.13
TB1	29.513	266.09	78.966	262.23	31.862	286.71	85.283	282.58	34.048	304.61	91.167	300.77

Figure 8. Comparative analysis of the stress, in BB1, BB2 and TB1, for different heights of impact, in 2 different boundary conditions, under Box shaped impactor and Sphere shaped impactor

5. Result and Discussion

Using the finite element programme ANSYS, the impact behavior of the Blended concrete cubes with varying boundary conditions and shape of the impactor, under different impact velocities were analyzed. As a result of the findings, the following conclusions can be drawn:

- The deflection and the stress in the Blended concrete cubes increase with the increasing height of impact because as the height of impact increases, the impact energy increases.
- Cubes fixed on 5 sides show higher deflection and lower stresses. This is because less energy is dissipated when 5 sides of the cubes are fixed than when only the base of the cube is fixed.
- Less deflection and less stresses are developed when cubes are under impact of the Box shaped impactor than that of the spherical shaped impactor. This is because the area of application of the same impact increases in the former case.
- The Ternary Blended concrete comprising Blast slag and fly ash having developed higher stress and low deflection, shows better performance under impact load than the Binary blended concrete comprising only fly ash.

References

[1] Singh G and Bansal RS Effect of partial replacement of cement by silica fume and sand by quarry dust on strength and durability of concrete. 2017 Int. J. Eng. Res **4(9)** 414–418.

[2] Liu H, Bu Y, Nazari A, Sanjayan, JG and Shen, Z Low elastic modulus and expansive well cement system: the application of gypsum microsphere 2016 Constr Build Mater **106**, 27–34

[3] Bhargava GM and Kumar CJD State of art of usage of alternative materials in concrete 2015 Int. j. eng. **2(12)**,33–37

[4] Naveena SVD and Jagarapu DCK Experimental investigation on fibre reinforced beam column joint by partial replacement of cement with GGBS 2019 Int. j. recent technol **7(6)**, 215–219

[5] Dubey S, Chandak R and Yadav RK Experimental study of concrete with metakaolin as partial replacement of OPC 2015 Int. j. adv. eng. **2(6)**, 38–40

[6] Ahamed VS, Siddiraju S Study of strength of concrete with palm oil fuel ash as cement replacement 2016 Int. J. Civ. Eng. Technol **7(3)**, 337– 341

[7] Nili M and Afroughsabet V Combined effect of silica fume and steel fibers on the impact resistance and mechanical properties of concrete 2010 Int. J. Impact Eng. **37** 879–886

[8] Murali G and Ramprasad K A feasibility of enhancing the impact strength of novel layered two stage fibrous concrete slabs 2018 Eng. Struct. **175** 41–49

[9] Abid SR, Hilo A and Daek YH Experimental tests on the underwater abrasion of engineered cementitious composites 2018 Constr Build Mater **171** 779–792

[10] Abid SR, Shamkhi MS, Mahdi NS and Daek YH Hydro-abrasive resistance of engineered cementitious composites with PP and PVA fibers 2018 Constr Build Mater **187** 168–177

[11] Kim JE, Park WS, Jang Y, Kim SW, Sun-Woong Kim,Yi- Hyun Nam, Kim DG and Rokugo K Mechanical properties of energy efficient concretes made with binary, ternary, and quaternary cementitious blends of fly Ash, blast furnace slag, and silica fume 2016 Int. J. Concr. Struct. Mater. **10(3),** 97-108

[12] Armstrong CG Computer-Aided Design Butterworth-Heinemann Ltd 1994 (Department of Mechanical & Manufacturing Engineering, Queen's University of Belfast, Ashby Building, Stranmillis Road, Belfast BT9 5AH, UK) Volume 26

Compressive Strength of Waste PET Bottle Based Fibre

Devendra Kumar Somwanshi[1], Vijay Prakash Sharma[2], Sikha Sharma[3], Kalpit Jain[4]

[1,4]Assistant Professor, Poornima College of Engineering, Jaipur, India
[2] Assistant Professor, Manipal University Jaipur, Jaipur, India
[3]Assistant Professor, Poornima University, Jaipur, India

E-mail: [1]imdev.som@gmail.com, [2]vijayprakashsharma@gmail.com

Abstract. PET (polyethylene terephthalate) bottles have progressively turned into a crucial and fundamental piece of an everyday person's life. Thus the need emerges to course these waste plastic containers to their ideal use. To add to it, incorporation of waste PET bottles in concrete composites is a promising technique for developing sustainable composite materials which may be applied to the construction industry. Hence concrete with waste PET bottles proves not only as an effective plastic waste management technique, but also an approach towards economic, environment friendly and sustainable development of building materials in the future. In the research work, an experimental study has been carried out to analyze the possibility of using waste PET bottle fibres with 3 different aspect ratios of 10, 20, and 30, in concrete at different replacement dosages of 0.5%, 1%, and 1.5% of sand by weight, and compared against control mix with no replacements of sand.

Keyword: PET, Bottles, Fibres, Composite Materials

1. Introduction

The changed lifestyle and ceaselessly extending people has achieved a gigantic development in how much plastic waste creation. The yearly usage of plastic materials by and large has extended from 5 million tons during the 1950s to around 100 million tons in the new times, which has lead to plastic waste age in tremendous aggregates. It was evaluated in 2016 that 56 million tons of PET things are made consistently [1]. Anyway it is ensured that most thermoplastics would be capable, on a key level, be reused, regardless, Recycling of PET compartments is more sensible than a few other plastic applications because of the extraordinary worth of the gum and the essentially particular use of PET for water and carbonated soft drink bundling on a wide scale all around the planet [2].

This waste is either scorched or dumped in soil as landfill, the two of which are damaging for the environment. The consuming of plastic results in appearance of horrendous compound gases which directly cause air tainting and in the end lead to occasion of respiratory sicknesses on receptiveness with time. Of course, on being dumped in ground as landfill, plastic wastes achieves loss of soil readiness, decline in water saturation, contamination of ground water, awful waste, etc [3]. Along these lines, among different waste segments, plastic waste benefits excellent thought by temperance of its non-biodegradable property in view of which it causes defilement in the environment. In India around 40 million tons of solid waste is conveyed consistently. This is extending at a speed of 1.5 to 2% reliably. Waste plastics contain 12.3% of outright waste made most of which is from discarded water bottles. The waste PET containers can't be sorted out off by either dumping or consuming, as they produce uncontrolled fire on consuming and contaminate the soil and vegetation on dumping [4]. Polyethylene Terephthalate (PET) is the most routinely used thermoplastic polyester. PET is used fantastically as containers for taking care of carbonated and non-carbonated drinks. A direct polymer has incredible mechanical properties and extraordinary layered reliability under factor load. Semi clear thermoplastic polyester, intense, low gas permeability, falsely and thermally consistent, successfully dealt with and dealt with, mileage safe and non biodegradable is the ordinary ascribes of PET [5]. Considering its adaptability, it is also used in materials, films, utility item, athletic attire, etc. Food taking care of endeavors slant toward PET for all intents and purposes, significant solid areas for sterile, weight. This shows that it has a wide extent of purposes. PET has a spot with the thermoplastics with phenomenal genuine properties [6].

2. Literature Review

Betioli AM and Silva DA, 2005 examined their examination work in which they assessed the debasement and strength of PET filaments, acquired from the reused plastic jugs presented to solutions for a considerable length of five month, through Fourier-Transform Infrared spectroscopy (FTIR) and checking electron microscopy (SEM) [7]. Albano C, et-al, 2009 introduced the investigation of mechanical way of behaving of concrete with reused Polyethylene Therephtalate (PET), differing the water/concrete proportion (0.50 and 0.60), PET substance (10 and 20 vol. %) and the molecule size. The mixes were presented to various temperatures (200, 400, 600 degree C), and thus the impact of the warm debasement of PET in the substantial was examined [8]. Kim SB, et-al, 2010 introduced an exploratory examination to concentrate on the strength, malleability and disappointment method of reused PET fiber built up concrete that was completed. The researchers checked the presentation limit of reused PET fiber supported concrete, and contrasted it and that of polypropylene (PP) fiber built up concrete for fiber volume parts of 0.5%, 0.75%, and 1.0% [9]. Foti D, 2011 announced the aftereffects of certain tests performed on substantial examples supported with filaments produced using waste polyethylene terephthalate (PET) bottles which were gotten by basically cutting the bottles. Materials utilized for making of substantial blend included roundabout Ready blend concrete PET filaments 1% by volume of concrete, shifting in thickness acquired from single layer, twofold layer, and four layers of covering bottles [10].

3. Experimental Procedure

Conclusively 4 types of mixes were designed for each aspect ratio and 180 Specimens were cast in total consisting of cubes, cylinders and beams, each 60 in number.
Concrete mixes prepared are as follows:
Conventional mix without addition of PET fibers.

- Addition of PET fibre @ 0.5% partial replacement of sand with above aspect ratio
- Addition of PET fibre @ 1.0% partial replacement of sand with above aspect ratio
- Addition of PET fibre @ 1.5% partial replacement of sand with above aspect ratio

Table 1. Specimen Details

% PET fibre addition	Days of Curing	Compressive (Cubes)			Split Tensile (Cylinders)			Flexure (Beams)			Total
		Aspect Ratio-10			Aspect Ratio-20			Aspect Ratio-30			
		10	20	30	10	20	30	10	20	30	
0%	3		2			2			2		6
	7		2			2			2		6
	28		2			2			2		6
0.50%	3	2	2	2	2	2	2	2	2	2	18
	7	2	2	2	2	2	2	2	2	2	18
	28	2	2	2	2	2	2	2	2	2	18

1%	3	2	2	2	2	2	2	2	2	2	18
	7	2	2	2	2	2	2	2	2	2	18
	28	2	2	2	2	2	2	2	2	2	18
1.50%	3	2	2	2	2	2	2	2	2	2	18
	7	2	2	2	2	2	2	2	2	2	18
	28	2	2	2	2	2	2	2	2	2	18
			60			60			60		180

3.1. Workability Concrete

Workability of concrete is a significant property to decide prior to setting of concrete, A substantial is supposed to be serviceable in the event that it is effectively moved, set, compacted and completed with no isolation. Concrete with high compaction factor is supposed to be more useful. Table 2 shows values of compaction factor for the different values of PET fibre content in concrete.

Table 2. Compaction Factor of Concrete w.r.t. PET Fiber Percentage

PET Fiber %	Compaction Factor (AR-10)	Compaction Factor (AR-20)	Compaction Factor (AR-30)
0%	0.960	0.96	0.96
0.50%	0.868	0.857	0.843
1%	0.847	0.836	0.82
1.50%	0.823	0.799	0.771

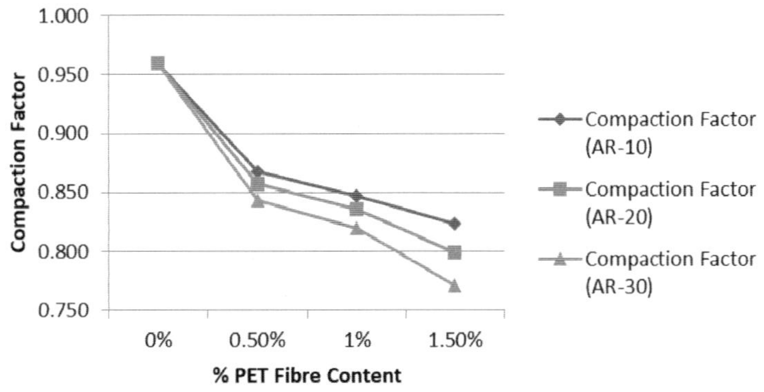

Figure 1. Compaction Factor of Concrete for different aspect ratios w.r.t. PET Fiber Percentage

Figure 1 shows the comparison of Compaction factor for various fibre content percentages. It is seen that as the steel fiber content in concrete increments compaction component of concrete declines in like manner consequently the usefulness diminishes.

3.2. Compressive Strength of Concrete

By and large augmentation in strength of concrete with age is noticed. Table 3 shows the suggested strength of concrete at various ages in examination with the strength at 28 days in the wake of projecting.

Table 3. Recommended Strength gain Percentage % with successive days

Age	Strength percentage %
1 day	16%
3 days	40%

7 days	65%
14 days	90%
28 days	99%

Table 4 shows the recommended strength for different grades of concrete in N/mm2 after 14 and 28 days of casting.

Table 4. Grades of Concrete and its Recommended Strength gain of 7 days & 28 days in N/mm^2

Grade of Concrete	Minimum compressive strength (N/mm^2) at 7 days	Specified characteristic compressive strength (N/mm^2) at 28 days
M15	10	15
M20	13.5	20
M25	17	25
M30	20	30
M35	23.5	35
M40	27	40
M45	30	45

Compressive strength of concrete is utmost property of concrete. Cubes of dimensions with 150×150×150 mm were cast and testes foe compressive strength on compression testing machine.

Table 5. Compressive Strength of Cubes (AR - 10)

% PET fibre addition	3 Days Compressive Strength			7 Days Compressive Strength			28 Days Compressive Strength		
	No. of Cubes	Compressive Strength	% Increased	No. of Cubes	Compressive Strength	% Increased	No. of Cubes	Compressive Strength	% Increased
0	2	20.00		2	32.00		2	39.56	
0.50	2	25.33	26.67	2	31.11	-2.78	2	40.00	1.12
1	2	28.89	44.44	2	34.22	6.94	2	42.22	6.74
1.50	2	23.11	15.56	2	32.89	2.78	2	40.00	1.12

Table 5 shows the upsides of compressive strength of concrete shapes and rate increase in strength of cubes after expansion of explicit substance of PET fiber of aspect ratio - 10, for curing of 3, 7 and 28 days. For 28 days restoring it was seen that compressive strength of concrete increase with expansion of PET fiber (0% to 1%) however decrease with additional option of fiber (1.5%).

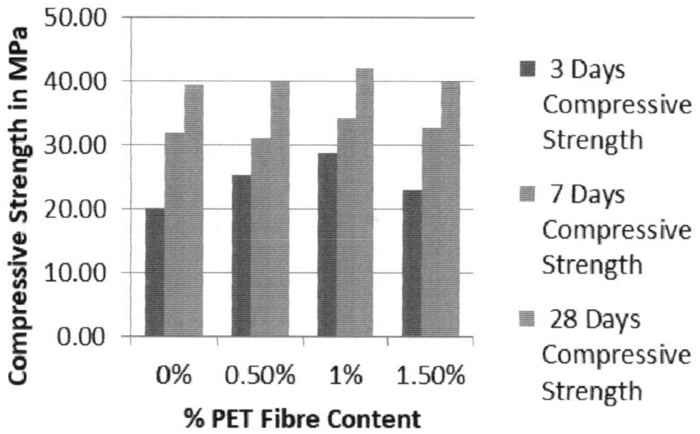

Figure 2. Compressive Strength of Cubes (Aspect Ratio - 10)

Table 6. Compressive Strength of Cubes (AR - 20)

% PET fibre addition	3 Days Compressive Strength			7 Days Compressive Strength			28 Days Compressive Strength		
	No. of Cubes	Compressive Strength	% Increased	No. of Cubes	Compressive Strength	% Increased	No. of Cubes	Compressive Strength	% Increased
0	2	20.00		2	32.00		2	39.56	
0.50	2	26.22	31.11	2	32.00	0.00	2	40.89	3.37
1	2	29.33	46.67	2	35.11	9.72	2	42.22	6.74
1.50	2	24.44	22.22	2	32.44	1.39	2	40.00	1.12

Table 6 shows the values of compressive strength of concrete cubes and percentage increment in strength of cubes after addition for specific content of PET fibre of aspect ratio - 20, for water curing of 3, 7 and 28 days. For 28 days curing it was observed that compressive strength of concrete increases with addition of PET fiber (0% to 1%) but decreases with further addition of fiber (1.5%).

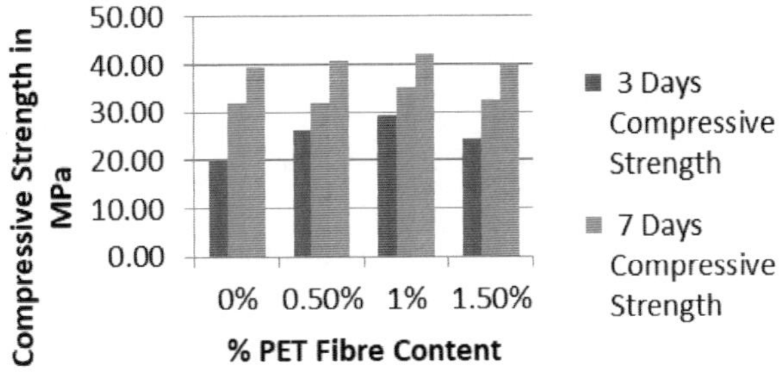

Figure 3. Compressive Strength of Cubes (Aspect Ratio - 20)

Table 7. Compressive Strength of Cubes (AR - 30)

% PET fibre addition	3 Days Compressive Strength			7 Days Compressive Strength			28 Days Compressive Strength		
	No. of Cubes	Compressive Strength	% Increased	No. of Cubes	Compressive Strength	% Increased	No. of Cubes	Compressive Strength	% Increased
0	2	20.00		2	32.00		2	39.56	
0.50	2	25.78	28.89	2	36.44	13.89	2	40.44	2.25
1	2	26.22	31.11	2	35.56	11.11	2	41.33	4.49
1.50	2	22.22	11.11	2	33.33	4.17	2	40.44	2.25

Table 7 shows the values of compressive strength of concrete cubes and percentage increment in strength of cubes after addition for specific content of PET fibre of aspect ratio - 30, for water curing of 3, 7 and 28 days. It was observed that compressive strength of concrete increases with addition of PET fibre up to 1%, but reduces on further addition of fibres (1.5%).

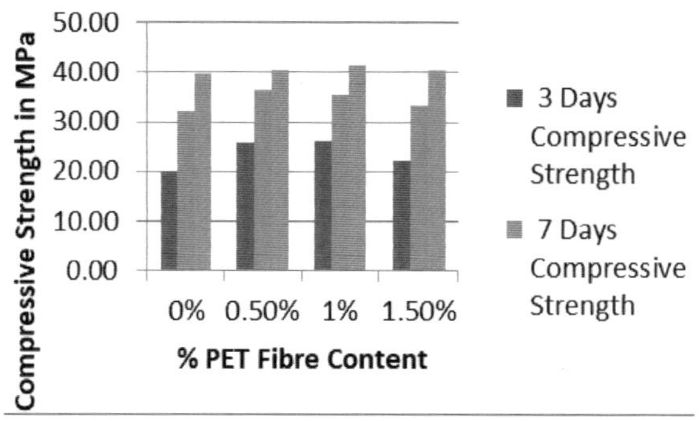

Figure 4. Compressive Strength of Cubes (Aspect Ratio-30)

Table 8. Compressive Strength of Cubes (28 days)

% PET fibre addition	Aspect Ratio - 10			Aspect Ratio - 20			Aspect Ratio - 30		
	No. of Cubes	Compressive Strength	% Increased	No. of Cubes	Compressive Strength	% Increased	No. of Cubes	Compressive Strength	% Increased
0	2	39.56		2	39.56		2	39.56	
0.50	2	40.00	1.12	2	40.89	3.37	2	40.44	2.25
1	2	42.22	6.74	2	42.22	6.74	2	41.33	4.49
1.50	2	40.00	1.12	2	40.00	1.12	2	40.44	2.25

Table 8 shows the values of compressive strength of concrete cubes and percentage increment in strength of cubes after addition for specific content of PET fibre of aspect ratio -10, 20 and 30, for water curing of 28 days. Though the change in strength was very small with change in aspect ratio of fibres, however, the strength of concrete was almost same for specimens with PET fibres with aspect ratio 10 and 20 and was little less for fibres with aspect ratio of 30.

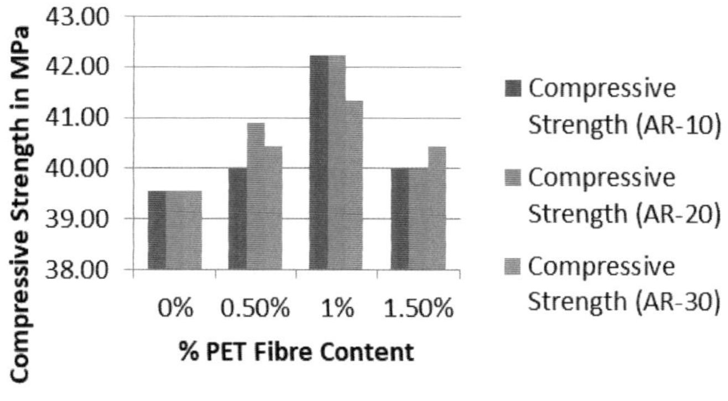

Figure 5. Compressive Strength of Cubes (28 Days)

4. Conclusion

In this research work, Mix-Design of M-30 grade concrete; reference IS 10262: 2009, having water-concrete proportion 0.45 is used for the experiment. Keeping any remaining constituent materials consistent waste PET fibres are penetrated at rates of (0% to 1.5%) while supplanting fine sand. Actually 180 specimens of PET Fiber Reinforced Concrete (PFRC) were projected with incredible accuracy and were restored for 3 days, 7 days and 28 days. Every one of the specimens were restored in water. During concreting/projecting of shapes, compaction factor test on new concrete was led for check of functionality with above rate (%) expansion of PET fiber for example (0% to 1.5%). After culmination of development time of concrete, compressive strength test was directed on every one of the specimens with individual date of casting. From the review it was seen that compressive strength expanded with rate (%) of PET fiber (0% to 1.0%) 3 days, 7 days and 28 days and from that point diminished on expansion of 1.5% PET fiber for all ages. It was additionally seen that ideal rate increase in compressive strength of concrete was 46.67 % for 3 days restoring, 9.72 % for 7 days relieving, and 6.74 % following 28 days restoring on supplanting sand by 1% PET strands (perspective proportion - 20).

References

[1] Albertsson-Ann, C.; Huang-Samuel, J. Degradable Polymers, Recycling and Plastics Waste Management; Royal Institute of Technology: Stockholm, Sweden; University of Connecticut: Storrs, CT, USA; Marcel Dekker: New York, NY, USA, 1995.

[2] Thompson, R.C.; Moore, C.J.; Saal, F.S.V.; Swan, S.H. Plastics, the environment and human health: Current consensus and future trends. Philos. Trans. R. Soc. B Biol. Sci. 2009, 364, 2153–2166.

[3] Dutta, S.; Nadaf, M.B.; Mandal, J.N. An Overview on the Use of Waste Plastic Bottles and Fly Ash in Civil Engineering Applications. Procedia Environ. Sci. 2016, 35, 681–691.

[4] Pardo-Bosch, F.; Aguado, A. Decision-making through Sustainability. In 6th IAHR International Symposium on Hydraulic Structures: Hydraulic Structures and Water System Management; Crookston, B., Tullis, B., Eds.; Utah State University: Logan, UT, USA, 2016; pp. 577–586. ISBN 978-1-884575-75-4.

[5] Borhan, T.M.; Dhaheer, M.S.A.; Mahdi, Z.A. Characteristics of Sustainable Self-compacting Concrete Reinforced by Fibres from Waste Materials. Arab. J. Sci. Eng. 2020, 45, 4359–4367.

[6] Bayramov, F.; Ta¸sdemir, C.; Ta¸sdemir, M. Optimisation of steel fibre reinforced concretes by means of statistical response surface method. Cem. Concr. Compos. 2004, 26, 665–675.

[7] Silva DA, Betioli AM, Gleize PJP, Roman HR, Gomes LA and Ribeiro JLD. Degradation of recycled PET fibers in Portland cement-based materials. Cement and Concrete Research. 2005; 35:1741-1746.

[8] Albano, C., Camacho, N., Hernández, M., Matheus, A. and Gutierrez, A.J.W.M., 2009. Influence of content and particle size of waste pet bottles on concrete behavior at different w/c ratios. Waste Management, 29(10), pp.2707-2716.

[9] Kim, S.B.; Yi, N.H.; Kim, H.Y.; Kim, J.H.J.; Song, Y.C. Material and structural performance evaluation of recycled PET fiber reinforced concrete. Cem. Concr. Compos. 2010, 32, 232–240.

[10] Foti D. (2011). Preliminary analysis of concrete reinforced with waste bottles PET fibers. Construction and Building Materials. 25(2011). pp. 1906-1915.

A Literature Review on Mechanical Properties of Waste PET Bottle Based Fibre

Devendra Kumar Somwanshi[1], Vijay Prakash Sharma[2], Sikha Sharma[3], Kalpit Jain[4]
[1,4]Assistant Professor, Poornima College of Engineering, Jaipur, India
[2] Assistant Professor, Manipal University Jaipur, Jaipur, India
[3]Assistant Professor, Poornima University, Jaipur, India

E-mail: [1]imdev.som@gmail.com, [2]vijayprakashsharma@gmail.com

Abstract. PET which stands for polyethylene terephthalate. It is a form of polyester just like the clothing fabric. PET bottles is a clear, strong, and lightweight plastic that is widely used for packaging foods and beverages, especially convenience-sized soft drinks, juices and water. Have increasingly become an indispensable and essential part of a common man's life. The joining of waste PET containers in substantial composites is a promising method for creating manageable composite materials which might be applied to the development business. Consequently, concrete with squander PET jugs demonstrates not just as a viable plastic waste administration procedure, yet in addition a methodology towards financial, climate well-disposed and reasonable advancement of building materials later on. In this review paper, properties of reinforced concrete made by the incorporation of waste PET bottle fibres have been analysed and discussed.

Keyword: Renewable Energy, Wind Energy, Hydropower, Biomass, Geothermal

1. Introduction

The changed way of life and perpetually expanding populace has brought about a huge expansion in the amount of plastic waste creation. Due to massive use of PET in plastic materials, utilization of such materials has been expanded to more than 100 million tons in the 20th century which has led to plastic waste age in gigantic sums. It was assessed in 2016 that 56 million tons of PET items are created every year [1]. However it is guaranteed that most thermoplastics would be able, on a fundamental level, be reused, in any case, Recycling of PET containers is more reasonable than some other plastic applications as a result of the great worth of the gum and the practically selective utilization of PET for water and carbonated soda pop packaging on a wide scale all over the planet [2].

This waste is either singed or unloaded in soil as landfill, the two of which are destructive for the climate. The consuming of plastic outcomes in arrival of destructive compound gases which straightforwardly cause air contamination and in the end lead to event of respiratory illnesses on openness with time. Then again, on being unloaded in ground as landfill, plastic squanders brings about loss of soil ripeness, decrease in water permeation, pollution of ground water, unfortunate waste and so on [3]. In this way, among various waste portions, plastic waste merits exceptional consideration by virtue of its non-biodegradable property because of which it causes contamination in the climate. Approximately 2% consistent growth is registered in generation of such waste in India only and now almost 50 million tons of such waste is generated every year in the country. Squander plastics comprise 12.3% of absolute waste created the greater part of which is from disposed of water bottles. The waste PET jugs can't be arranged off by one or the other unloading or consuming, as they produce uncontrolled fire on consuming and sully the dirt and vegetation on unloading [4].

The most regularly utilized thermoplastic polyester is Polyethylene Terephthalate one reason is that it is utilized transcendently as jugs for putting away carbonated and non-carbonated drinks. A straightforward polymer has great mechanical properties under factor load and great layered dependability. Semi translucent thermoplastic polyester, tough, low gas porousness, artificially and thermally steady, effectively handled and took care of, mileage safe and non-biodegradable are the normal attributes of PET [5]. In view of its flexibility, it is additionally utilized in materials, films, utility product, athletic apparel and so forth. Food handling enterprises incline toward PET as it is, major areas of strength for sterile, weight. This shows that it has a wide scope of uses. PET has a place with the thermoplastics with fantastic actual properties [6].

2. Literature Review

A detailed review of research papers, on Mechanical Performance Analysis of Waste PET bottles-based Fibre Reinforced Concrete published within the period from 2005 to 2020 has been undertaken. Authors studied the effects of using plastic on mechanical properties of concrete by varying percentages of plastic as fibre to improve the properties of concrete. The various percentages of fibre used for experimental work were 0%, 0.2%, 0.4%, 0.6%, 0.8% and 1.0%. It is evident from the above studies that strength of PET fibre induced concretes was less than that of conventional concrete. But, the authors did not discuss any literature review which may have led to poor knowledge for strength enhancing techniques while using PET fibres in concrete, and consequently no improvements were recorded for compressive strengths for concrete [1].

The inference derived from the research paper was that PET fibres are degraded in alkaline environment. However, the process of degradation of these fibres is a long process and hence takes time for significant effects. Therefore, the authors suggested to carry out experimental analysis for extended periods such as up to 365 days and further, so that effects of fibres over longer periods may be observed and inferences be made accordingly [2].

Paper reported the results of some tests performed on concrete specimens reinforced with fibres made from waste polyethylene terephthalate (PET) bottles which were obtained by simply cutting the bottles. Materials used for making of concrete mix included circular Ready mix concrete PET fibres 1% by volume of concrete, varying in thickness obtained from single layer, double layer, and four layers of overlapping bottles [3].

Authors discussed the experimental investigation that was carried out to study the mechanical behaviour of hybrid fibrous concrete reinforced with steel fibres and Recycled Polyethylene Terephthalate fibres. M 40 grade of concrete was used for this study in which the steel fibres were mixed at a percentage level of 0.5 and RPET fibres were mixed from 0 to 0.5%, at equal intervals of 0.1% [4].

Authors presented the experimental analysis carried out to investigate the compressive strength of concrete by replacing natural aggregates with recycled aggregates in various percentages of 25%, 50%, 75% and 100% of, accompanied by incorporation of PET fibres at 1% and 2% by volume in M-20 grade concrete. From the past literatures the authors learnt that with the use of binders the strengths of PET fibre reinforced concrete were improved as compared to conventional concrete mix [5].

Authors discussed the investigation which aimed at partial replacement of sand with 5 to 20% of waste plastic material in M25 grade of concrete to make an attempt to increase the ductile strength of concrete as it is generally found weak in tension. From past literatures the authors learnt that higher flexural strength for RCC beams could be achieved by adding 2% waste plastic fibres [6].

Authors investigated the effects of substitution of sand with PET processed particles. In this process cubic and cylindrical specimens with varying w/c ratios were prepared and tested for physical properties of fresh concrete. For testing the mechanical properties, the specimens were cured in specific standard conditions which gave interesting results [7].

Authors discussed the use of PET fibres as an alternative material in M-25 concrete by replacement of 0.5%, 1%, 2%, 4% & 6% of fine aggregate by waste PET fibres. Some other ideas such as PET panels and mattresses and direct use of PET bottles for construction of non-load bearing walls with appropriate fillers have also been discussed. The authors, from past literatures learnt about various benefits of using waste PET fibres in concrete, as they improve acoustic properties in concrete, reduce shrinkage by acting as crack arresters and reduce dead load of the structure by 2-6% approximately [8].

Authors investigated the experimentally evaluated effects of steel and PET (polyethylene terephthalate) fibres separately on the properties of both fresh and hardened Self-compacting concrete (SCC). From review of previous literatures the authors learnt that due to high cementitious content and fillers in SCC more early-age cracking of SCC (due to plastic shrinkage, as well as thermal stress) and more drying shrinkage, compared to normal concrete are witnessed as common characteristics. Therefore use of fibres in SCC was a good idea to promote ductility and to reduce early-age cracking [10].

The results showed a greater compressive strength for unreinforced concrete though, however PET fibre reinforced concrete showed satisfactory results for and was observed at a value of 40 MPa as compared to 42.4 for un-reinforced concrete. For four point bending test a maximum load bearing capacity of 15.18 KN was reported by the authors for PET fibres reinforced concrete as compared to 13.91 KN for unreinforced concrete specimens and consequently the strengths for the two types of specimens varied from a maximum value of 4.07 MPa to 3.77 MPa respectively. For first crack strength improved values were observed at a mean value of 4.78 MPa for PET fibre reinforced concrete against a value of 3.39 MPa for unreinforced concrete. For energy absorption capacities a mean value of 0.46 for ductility indices was reported for PET fibre reinforced concrete resulting in post crack energy of 1753 Nmm2 [11].

Authors discussed the experimental analysis carried out to study the compressive strength of M-20 grade concrete using plastic and a combination of plastic and steel fibres in the ratio of 2:1 in concrete. A controlled specimen with 0% fibre was also prepared for comparing the result values with that of fibre reinforced concrete. The objective of the paper was to produce fibre reinforced concrete using waste plastic and to optimize their use and resulting benefits in concrete. The results reported by the authors showed a decrease in 28 day strength for plastic fibre reinforced concrete when compared to 0% fibre mix. The 28 day strength was reduced from 630 KN for 0% fibre mix to 621, 596, 587, 577, and 570 KN for 0.2, 0.4, 0.6, 0.8 and 1% fibre induced mixes, respectively [12].

Materials used for experimental analysis included fine aggregate, coarse aggregate, and plastic aggregate. For Coarse Aggregate, tests conducted to study the properties were Abrasion Test (<25%) and Flakiness Index (<25%). For Fine Aggregate, tests for angularity (>45%), and water absorption test (<2%) were conducted. Plastic aggregates were formed from collecting, cleaning, shredding, heating and melting of waste polypropylene and PET plastics. It was suggested by the authors that 10% Fibre reinforced concrete may be used for non-structural elements and less important works such as walkways. In this paper testing of specimens was carried out while varying the % replacement by plastic aggregates to which the negative results were obtained [13].

Author presented an experimental study conducted to determine the compressive strength and adsorption of normal and PET fibre reinforced concrete (PFRC) for grades of M-20 and M-30 along with varying geometry and aspect ratio of fibres incorporated in PFRC. The dosage of these fibres was also varied from 0 to 3%. During the review of past literatures the authors were educated about some bitter properties such as low tensile and impact strength, brittleness, low resistance to cracks and high density. From review of past literatures the authors learnt that for replacement of fine aggregates, up to 20 %, by ground waste plastic, resulted in reduction of slump value by approximately 25% [14].

Authors discussed the experimental analysis carried out to study the compressive strength of M-20 grade concrete using plastic and a combination of plastic and steel fibres in the ratio of 2:1 in concrete. A controlled specimen with 0% fibre was also prepared for comparing the result values with that of fibre reinforced concrete. The objective of the paper was to produce fibre reinforced concrete using waste plastic and to optimize their use and resulting benefits in concrete. The results reported by the authors showed a decrease in 28 day strength for plastic fibre reinforced concrete when compared to 0% fibre mix. The 28 day strength was reduced from 630 KN for 0% fibre mix to 621, 596, 587, 577, and 570 KN for 0.2, 0.4, 0.6, 0.8 and 1% fibre induced mixes, respectively [16].

IOP Conf. Series: Earth and Environmental Science 1084 (2022) 012069 doi:10.1088/1755-1315/1084/1/012069

Table 1. Compressive Strength of Different Location

Sr. No.	Year & Location	Days of Testing	% PET added or Replaced	% Change in Compressive Strength	% Change in Split-Tensile Strength	% Change in Flexural Strength
1	2009, Venezuela	7, 14, 28	0, 10, 20	12.73(↓),45.45(↓)	10.71(↓), 19.64(↓)	8.24(↑), 1.18(↓)
2	2010, South Korea		0, 0.5, 0.75, 1	1.89(↓), 5.66(↓), 9.43(↓)	-	7.5(↑), 7.3(↑), 6.31(↑)
3	2011, Tamil Nadu, India	7,28	0,0.5	5.12 (↑)	1.75(↑)	-
4	2011, Portugal	7,28,63	0,0.5,1 & 1.5	1.93(↑),13.53(↓), 4.39(↓)		19.42(↑),33.98(↑)
5	2012, Maharashtra	7,14,128	0, 0.2, 0.4,0.6 & 0.8, 1	10.5(↓), 14.74(↓), 21.84(↓), 23.42(↓) & 22.07(↓)	(6.31(↑), 19.42(↑), 25.24(↑), 35.19(↑) & 19.53(↑)	-
6	2012, Brazil	28,150	0, 0.05 & 0.18, 0.3	0.65(↑), 3.01(↓),7.50(↓)	-	-
7	2013, Kerela, India	28	0,20	+21.51(↑)	22.04(↓)	3.641.18(↓)
8	2013, Tamil Nadu, India	28	0,0.1, 0.2, 0.3, 0.4,0.5	0.47(↑), 0.87 (↑), 0.19(↓),0.8(↓) ,3.85(↓)	3.84(↑), 24.5(↑), 30.18(↑),22.45(↑),13.2(↓)	1.71(↑), 2.39(↓), 3.24(↓), 4.44(↓), 11.09(↓)
9	2013, Chennai, India	28	0, 5, 10, 15, 20	2.47(↑), 4.13,7.42(↑), 4.67(↓)	0.99(↑), 3.94(↑), 13.79(↑),25.62(↓)	
10	2013, Iran	28	0, 5,10,15	+2.00(↑), 3.22(↓),10.73(↓)	2.5(↓), 7.50(↓), 15(↓)	4.32(↑), 5.60(↓), 13.12(↓)
11	2012, Tamil Nadu, India	7,14,28	0, 0.5, 1, 2, 4,6	(6.45(↑), 29.03(↑),29.03(↑), 22.58(↑), 19.35(↑)	5.85(↑), 9.04(↑), 13.30(↑), 10.64(↑), 9.04(↑)	40.63(↑), 62.5(↑), 78.13(↑), 84.38(↑),84.32(↑)
12	2013, Switzerland		0, 0.5, 1, 1.5	+ 9.09(↑), 0.45(↓),4.86(↓)	9.1(↑), 15.5(↑), 23.6(↑)	-
13	2014, Jabalpur, India	7,28	0,0.2 & 0.4, 0.6, 0.8, 1	4.17(↓), 7.92(↓), 12.5(↓),14.58(↓),16.67(↓)	-	-
14	2014, Uttar Pradesh, India	7,28,56	0,0.5, 0.75, 1	1.40(↑),4.20(↑), 3.80(↓)	-	-
15	2014, Maharashtra, India	28	0, 0.05, 1,1.5, 2,2.5,3	2.89(↑), 5.39(↑),2.52(↑), 3.59(↓), 14.03(↓),21.58(↓)	-	-
16	2014, Ghana	7,14,28	0, 10, 20, 30, 40, 50, 60	-	8.87(↓), 22.17(↓), 27.63(↓), 39.72(↓), 45.97(↓),54.03(↓)	7.02(↓), 14.73(↓), 21.57(↓), 33.05(↓), 39.55(↓),47.95(↓)
17	2015, Maharashtra, India	7,28	0, 10, 20, 30,40,50	23(↓), 29.8(↓), 40.75(↓), 44.3(↓),49.5(↓)	-	-
18	2015, Tamil Nadu, India	7,28	0,0.25,.05,1	3.02(↑), 1.5(↓),3.91(↓)	+3.35, ,1.67(↓),4.07(↓)	-
19	2015, Maharashtra, India	7,28	0, 10, 20, 30,40,50	32.7(↓), 42.5(↓), 53.2(↓),54.35(↓), 54.45(↓)	-	-
20	2015, Arunachal Pradesh, India	28	0,10,20	22.22(↓),33.59(↓)	-(18.8,33.76)	-

21	2015, Theissur	28	0,15	3.75(↓)	-8.33	14.43(↓)
22	2015, Australia		0,1,1.5	1.14(↑), 12.29(↓)	+(45.45, 45.45)	-
23	2015, Rajkot	28	0,0.5,1,2,4,6	, 27.98(↓), 27.98(↓), 28.62(↓), 8.56(↓), 18.14(↓), 3.56(↓)	-	-
24	2015, Sri Lanka	28	0,1,2,3	0.83(↑),0.33(↑), 10.45(↓)	+(4.35, 7.83, 29.13)	-
25	2016, Maharashtra, India	7,28	0,10,15	9.14(↑),16.85(↑)	-	-
26	2016, Algeria	7,28	0, 10, 20,30,40	22.50(↑),27.50(↑), 10(↑), 4(↑)	-	-
27	2016, Tamil Nadu, India	7,28	0,0.5,1, 1.5, 2	8.43(↑), 11.09(↑), 6.44(↑),4.46(↑)	+(21.38,33.33)	-

Authors concluded that strengths of specimens up to 10% replacements of natural aggregates with plastic aggregates were within permissible limi ts as per IS code whereas for 20% replacements it was found satisfactory for M20 grade concrete. With the increase in waste plastic content the density of the mix reduced proportionally [17].
The authors suggested that such low strength concrete mix may be used for mass concreting structures such as pavements, where high strengths are not required. In this paper the authors laid emphasis on preparation of green concrete with increasing concerns over environmental pollution [18].
Author discussed the study conducted to determine the compressive strength of concrete masonry units with plastic bottle cores and PET fibres in concrete. Centre voids were created with the use plastic bottles in concrete masonry units. Partial replacement of fine aggregate was done using PET fibres in varying quantities of 0.5 %, 1%, 2%, 4% and 6% of fine aggregate by weight [22].
Authors discussed about the suitability of PET fibres in concrete pipes that are otherwise reinforced with steel bars and used for carrying water. Thus, in this study authors carried out an experimental investigation to check for the feasibility of replacing steel bars with standard sized PET fibres, at different dosage levels increasing from 0 to 3%, at fixed intervals of 1%, for reinforcing concrete pipes at w/c ratios of 0.3 and 0.45 [23].
Authors discussed about the study that was carried out to investigate various physical and mechanical properties of concrete consisting of Polyethylene terephthalate (PET) and Low density Poly ethylene (LDPE) as fine aggregates, as a partial replacement for sand in various percentages of 0, 10, 20, 30 and 40%. Also these waste plastics were incorporated in concrete in fibre form at various percentages of 0, 0.5, 1, 1.5 and 2% by volume of sand to further study their effects in fibre form [25].
Table 1 shows the comparative analysis of compressive strength based on research on different locations. The positive in the percentages shows by the help of this sign (↑) and negative in the percentages shows by the help of (↓) this sign.

3. Strengths and Weakness

After, the review of research papers under the solid area, Workability and waste materials utilized, and part of steel fiber utilized protection from various synthetic assaults. We can ready to find, the qualities and shortcomings of different arrangement approaches used to address the issues.

3.1. Strengths

- Fibers treated with NaOH solution prior to use in concrete mix gave better results in terms of bonding strength with concrete and consequently higher compressive and split-tensile strengths were achieved [6].
- Replacement of Natural River Sand by ground plastic resulted in higher strengths as well as helped conservation of natural resources [8].
- With the increase in replacement of sand by grounded PET in concrete mix, the bulk density of concrete reduced proportionally [10].
- Addition of Glenium 51 improved the workability of concrete mix and thus higher values for slump and toughness and bending strengths were achieved in conjunction with increased deflection [9], [12].
- Use of waste PET in concrete mix reduced the water absorption of concrete by 57% approximately [19], [11].
- PET fibre addition results in reduction in shrinkage cracks and delaying of shrinkage time in concrete by 78.57% [7], [15].
- In PET fibre reinforced concrete the size of shrinkage cracks such as length and width reduced to 20% when compared to unreinforced concrete [24].
- Flexural toughness and impact strength of specimens increased by 15-20% on addition of PET fibres into the concrete mix [22].

- Addition of PET to concrete results in restraining segregation of fresh concrete mix [25].
- PET has good thermal and sound insulation properties and resistance to freeze-thaw cycle making [6], [21].
- Replacing coarse aggregates with plastic aggregates reduced the dry density and consequently the overall dead weight of the structural component thus making it light and easy to carry [18], [11], [9]
- Reduced internal pores and air cavities were observed on replacement of sand with PET fibres in concrete [23].
- Width of cracks occurring from shrinkage in concrete reduced significantly with the addition of Recycled PET fibres in concrete reducing ingress of water and other harmful chemicals, reducing its deterioration and hence making it durable [18].

3.2. Weakness

- Modulus of elasticity reduced proportionally with the increase in dosage of PET fibres [14].
- With the increase in PET fibre addition in concrete the workability of the mix reduced proportionally and significantly [7], [14].
- Fibres added to concrete mix undergo degradation over a period of time due to the presence of highly alkaline environment and therefore, comparatively reduced strengths were achieved at later ages for the mix [16].
- PET fibres added to concrete resulted in formation of non-homogeneous mixtures due to reduced workability which in turn prohibited proper mixing of the constituent materials added to the mix [19].

4. Conclusion

The review of research papers has been done in the space of mechanical execution investigation of waste PET container-based fiber supported cement to research and figure out flow difficulties and extent of work nearby. After the review, one issue was found in the literature, which depended on exploratory methodology. Study of literature was done inside and out of normal discoveries of research works, qualities and shortcomings and gaps to construct issue proclamation and goal.

Reference

[1] Pravin B. Waghmare, Atul M. Raghatate & Niraj D. Baraiya (2012) "Comparative Performance of Elevated Isolated Liquid Storage Tanks (With Shaft Staging)" International Journal of Advance technology in civil engineering, ISSN:2231-5721, Volume 1.

[2] Pelisser, F., Montedo, O. R. K., Gleize, P. J. P., & Roman, H. R. (2012). Mechanical properties of recycled PET fibers in concrete. Materials research, 15(4), 679-686.

[3] Foti, D. (2013). Use of recycled waste pet bottles fibers for the reinforcement of concrete. Composite Structures, 96, 396-404.

[4] Maruthachalam, D., & Muthukumar, J. (2013). Mechanical Performance of Recycled PET Fiber Reinforced Concrete with Low Volume Fraction. International Journal of Structural and Civil Engineering Research, 2(2), 101-108.

[5] Ramana, N. V., Harathi, R., Babu, S. N., & Babu, S. V. (2013). Regression Models to Evaluate Compressive Strength of Polyethylene Terephthalate (PET) Fibre Reinforced Recycle Aggregate Concrete. Int. J. Eng. Res. Dev, 8, 11-16.

[6] J.N.S.Suryanarayana Raju, M.Senthil Pandian,Mechanical Study on Concrete with Waste Plastic, International Journal of Research in Civil Engineering, Architecture & Design, Vol.1, Issue 1, pp.62-67,September, 2013.

[7] Rahmani, E., Dehestani, M., Beygi, M. H. A., Allahyari, H., & Nikbin, I. M. (2013). On the

mechanical properties of concrete containing waste PET particles. Construction and Building Materials, 47, 1302-1308.

[8] Chowdhury, S., Maniar, A. T., & Suganya, O. (2013). Polyethylene terephthalate (PET) waste as building solution. Int. J. Chem. Environ. Biol. Sci, 1(5).

[9] Irwan, J. M., Asyraf, R. M., Othman, N., Koh, K. H., Annas, M. M. K., & Faisal, S. K. (2013). The mechanical properties of PET fiber reinforced concrete from recycled bottle wastes. In Advanced Materials Research (Vol. 795, pp. 347-351). Trans Tech Publications Ltd.

[10] Madandoust, R., Ranjbar, M. M., & Moshiri, A. A. (2014). The effects of steel and pet fibers on the properties of fresh and hardened self-compacting concrete. Asian Journal of Civil Engineering (BHRC), 15(5), 671-682.

[11] Fraternali, F., Spadea, S., & Berardi, V. P. (2014). Effects of recycled PET fibres on the mechanical properties and seawater curing of Portland cement-based concretes. Construction and Building Materials, 61, 293-302.

[12] Khilesh, S. (2014). Study of Strength Property of Concrete Using Waste Plastics and Steel Fiber. International Journal of Engineering And Science, 3(5), 9-11.

[13] Patil, P. S., Mali, J. R., Tapkire, G. V., & Kumavat, H. R. (2014). Innovative techniques of waste plastic used in concrete mixture. International Journal of Research in Engineering and Technology, 3(9), 29-32.

[14] Khilesh sarwe, "Study of Strength Property of Concrete Using Waste Plastics and Steel Fiber",The international journal of engineering and science (ijes), volume 3, issue 5, pages 09-11, 2014.

[15] Ohemeng, E. A., Yalley, P. P., Dadzie, J., & Djokoto, S. D. (2014). Utilization of waste low density polyethylene in high strengths concrete pavement blocks production. Civil Environ. Res, 6(5), 126-135.

[16] Ariyamuthu A. and Varadarajan R. (2015),"Experimental study on effects of shredded waste plastics in rigid pavements." International Journal of Emerging Technology and Advanced Engineering (IJETAE) Volume 5,Issue 2, February 2015, ISSN 2250-2459

[17] Kasib R. Malak "Use of waste plastic in concrete mixture as aggregate replacement" International journal of engineering, education and technology, april 2015

[18] Ahmed, M., Muzafar, Dr., Raju, S., Siddi, 2013. Properties of Concrete by the Addition of Plastic Solid Waste, International Journal of Science and Research, pp1170 -1173

[19] Nidhi, M., Bini, Prof. Kumaran, M., 2015. Evaluation of Strength and Durability of Waste Plastic MixConcrete, International Journal of Research in Advent Technology,pp.34-40. Journal of Innovative Research in Science, Engineering and Technology, pp.3409-3413.

[20] Margareth da Silva Magalhães and Matheus Soares Viana Fernandes Bending Behaviour of Recycled PET Fiber Reinforced cement based composite. IACSIT International Journal of Engineering and Technology, Vol. 7, No. 4, August 2015

[21] Yin, S., Tuladhar, R., Shi, F., Combe, M., Collister, T., & Sivakugan, N. (2015). Use of macro plastic fibres in concrete: A review. Construction and Building Materials, 93, 180-188.

[22] Fataniya, R., Maaze, R., Kapadiya, K., & Pipalia, P. V. F. (2015). Experimental investigation of concrete masonry units with plastic bottle cores and PET fibers. International Journal for Scientific Research & Development (IJSRD), 3(04), 2321-0613.

[23] Rathnayaka, R. M. M. P., Malshan, W. V. D., & De Silva, S. (2015). Incorporating recycled PET fibres for concrete Cylindrical Culverts.

[24] Corinaldesi, V., & Nardinocchi, A. (2016). Mechanical characterization of Engineered Cement-based Composites prepared with hybrid fibres and expansive agent. Composites Part B: Engineering, 98, 389-396.

[25] Guendouz, M., Debieb, F., Boukendakdji, O., Kadri, E. H., Bentchikou, M., & Soualhi, H. (2016). Use of plastic waste in sand concrete. J. Mater. Environ. Sci, 7(2), 382-389.

Quality Assessment of Composts from different Organic sources

Parveen Parihar[1], Rakesh Choudhary[1]

[1]Bhartiya Skill Development University, Jaipur, India
E-mail: pparihar253@gmail.com

Abstract. Currently, environmental pollution is a global issue that is directly or indirectly related to human activities. Uses of chemical fertilizer also cause long-term environmental pollution. Concerning this, the composting technique for the management of organic waste is a good deal. This method or technique is better than chemical fertilizers as it is eco-friendly and environmentally friendly. The use of compost depends on its different quality parameters. The presented study shows the comparative analysis of three types of compost which were prepared by windrow type of composting. Composts were differentiated based on their organic matter source. Analysis of compost quality was done based on their macronutrient – Nitrogen, Phosphorus, and Potassium and the physical composition of compost. The result shows that all three types of composts match the quality standard of the Indian Fertilizer Control Order-1985. Compost from vegetable waste contains more nutrients as compared to others. All three were very less different in their nutrient composition. By comparing the composting time duration fruit waste is converted into compost earlier than others. This shows that vegetable waste gives high-quality compost that can be used as organic fertilizer. Fruits and farm waste are also useful but vegetable wastes give better quality compost concerning macronutrients.

1. Introduction

Biodegradable waste is the unavoidable part of total waste generated in India. It is worth reusing the waste for its proper maintenance. The use of technology for this should be in such a manner that it is environment-friendly, cost-effective, and less consuming of equipment[1].Composting is a naturally occurring biological degradation of organic compounds that is complete in the presence of the aerobic condition. This technique consists of complex metabolic reactions which are performed by different microorganisms in aerobic conditions and the presence of nitrogen and carbon. By the end of the composting process, a stable product is obtained called compost. The microorganisms in composting accelerate the conversion of complex organic compounds into simple humic substances. The process completes in two phases. Inthefirst phase, the microorganisms degrade the organic matter and in the second phase, the organic matter converts into a stable humic substance. This stable humic substance is useful as organic fertilizer or soil amendment [2]. For composting different types of biodegradable waste could be used. Compost from different types of organic waste is different in terms of its quality as there is a difference in the organic composition of waste. For better management of waste and application of compost in agriculture, it is necessary to identify which type of organic waste produces compost containing more nutrient contents for the plants and also give a better quality of compost [3]. To measure the quality of compost, stability and maturity plays an important role [4]. Several factors describe the stability, maturity, and quality of the compost. The presented study is about the factors which show the quality of compost. The quality of compost is very difficult to define. Compost quality covers both stability and maturity.Stability means the conversion of unstable complex organic matter into a simple stable composition. The degree of stability is estimated as the biodegradability of organic substances[5]. The term maturity is identified as the effect of compost on plants. Maturity degree is related to the absence of phytotoxicityincompost. If immature compost is applied tothesoil then it would give a negative impact on the soil as well as on plant growth. In our study, both these parameters were analyzed for the quality assessment of compost. Compost quality is also related to the macronutrient content present in compost which is Nitrogen, Phosphorus, and Potassium [6]. The benefit of using compost in agriculture is based on the availability of these nutrients forplant. The study is based on different types of organic waste that is generated in household. Fruits waste, vegetable waste, and farm waste was used for the assessment of quality. The standard for the compost quality is matched with the Indian Fertilizer Control Order-1985. Compost use also depends on how much time is needed to produce mature and stable compost. Our study also focuses on the time duration of completing the process of composting for different types of organic waste.

2. Methodology

For quality assessment of compost four types of organic waste were used for composting. Simple windrow type of composting was performed. Windrow composting is a simple method as it requires no specific equipment for conducting composting. For study food waste is used, which are fruits waste and vegetable waste. As food waste is a commonly generated waste that can create nuisance and pollution if not managed properly. Farm waste is also used for the study. It is also biodegradable so useful for composting [7]. For the quality analysis of compost three piles were prepared. Fruits waste, vegetable waste, and farm waste were put for composting separately respectively for their comparative analysis. Composting piles were prepared in the following manner. The composition of composting pile is shown in Table 1 and the chemical composition of organic waste used for composting is shown in Table 2.

Table 1. Composition of compost piles

Pile	Composition
Pile A	Fruits waste, soil, dry leaves, sawdust
Pile B	Vegetable waste, soil, dry leaves, sawdust
Pile C	Farm waste, soil, dry leaves, sawdust

Table 2. Chemical composition of organic waste

Type of Waste	Carbon Composition	Nitrogen Composition	% Moisture	Reference
Fruits waste	55.2	1.9	80.3	Tedesco *et.al*[8]
Veggies waste	42.8	1.84	40.5	Butnariu*et.al*[9]
Farm waste	34-41.5	2.8	4.2	Yadav *et.al*[10]

2.1 Parameters for starting Composting

To start composting optimal conditions are required. This contains proper C/N ratio, moisture, temperature, and aeration.

2.1.1 Carbon to Nitrogen ratio

In process of composting a Carbon Nitrogen ratio is the main factor which accelerates organic compounds for degradation. For an instant start of composting the required Carbon to Nitrogen, the ratio is 30:1 [11]. Carbon is needed as the energy source for the microorganism and nitrogen for the amino acid, nucleic acid, and protein synthesis. In our study, we maintain the C/N ratio of 30:1 in all four piles so that the process gets a quick acceleration.

2.1.2 Moisture

In composting, moisture is essential for the microorganism to transport the nutrients within cell walls. To balance the moisture in windrow type of composting is difficult. This can be done by adding adequate water while monitoring the piles. Average 40-60% moisture is needed in an active composting pile [12]. Following this in our study regular monitoring of moisture was done and water was poured to maintain the moisture in all piles.

2.1.3 Temperature

Temperature is a critical parameter in composting. The proper temperature of composting pile keeps the process active and balances the microorganism's growth. For composting average required temperature range is 40-60°C [13]. To maintain the average temperature regular turning of the piles was done in our experimental study.

2.1.4 Aeration

Oxygen in composting controls the other parameters in the process and also influence the growth of aerobic microorganisms. Oxygen significantly affects moisture content and loss of nitrogen in form of ammonia. Oxygen content in piles depends on the porosity in organic matter which means the size of the organic matter is directly linked with oxygen availability in composting piles. To maintain aeration in our study the organic matter was shredded into small pieces for the porosity and regular turning of organic matter was done throughout the process.

2.2 Quality parameters of compost

The product obtained from composting process is compost. This is useful in agriculture or as a soil treatment. Its use depends on its quality. It is also necessary to analyze the parameters for stability and maturity of the product obtained by composting before applying it to the field [2]. The main objective of compost quality is related to the nutrients present in compost that plants need for their growth. These nutrients are generally nitrogen, carbon, phosphorus, and potassium [14]. In our study quality of all the composts prepared was checked to make sure if the compost is useful for plants and soil health and if it contains the entire necessary nutrient in sufficient quantity. On this basis, we can analyze the quality of compost.

3. Result and Discussion

Three types of compost were prepared from three windrow piles. Composts were differentiated based on their organic source. Compost from fruit wastes was matured earliest out of the three wastes. The reason is the presence of much sugar in fruits which is easier for the microorganism to degrade [15]. During composting, fruit waste required less water to add up to the pile because the waste of fruits already contains almost 80% of water. Result also shows that waste from farms required maximum time to convert into compost. And also farm waste required more water to maintain moisture. Time taken by each composting pile to convert into mature compost is represented in Table 3.

Table 3. Time duration for composting of different waste.

Organic Waste	Duration required for composting
Fruits Waste	10 weeks
Vegetable Waste	12 weeks
Farm Waste	15 weeks

In our study, the quality measurement of composts was taken based on macronutrients like nitrogen, phosphorus, and potassium which are major constituents of fertilizers. Comparing the composts show that vegetable waste contains maximum nitrogen content. The physical composition of compost is also an important parameter for quality. The physical composition of all the compost is shown in the table **4.** The quantity of nutrients in all composts is represented inTable5.

Table 4. The physical composition of compost.

Organic Waste	Organic matter	C/N ratio	PH
Fruits waste	6.18%	11:1	8.41
Veggieswaste	6.56%	11:1	8.17
Farm waste	5.98%	12:1	7.2

Table 5. Nutrient composition of different compost

Organic Waste	Nitrogen	P_2O_5	K_2O
Fruitswaste	.32%	.14%	.26%
Veggieswaste	.34%	.18%	.32%
Farm waste	.28%	.12%	.24%

4. Conclusion

Composting is a simple natural process that produces natural organic fertilizer, compost. Usage of compost depends on its nutrient quality which is macronutrients Nitrogen, phosphorus, and potassium (NPK). This nutrient quality varies in compost prepared from different organic sources. Compost use also depends on its physical properties which are organic matter content, C/N ratio, and PH. From the above study of three types of composts, the quality analysis shows that all the composts are suitable for use on land. Their physical and chemical composition matches the standards of the Indian Fertilizer Control Order (FCO) 1985. Comparing the three composts after analysis it can be concluded that there is very less difference in the nutrient composition of all composts and the compost from vegetables contain more nutrients than the others. There is also a difference in the duration of the composting process fora different types of waste. In our study waste from fruits required minimum time in comparison to other wastes.

References

[1] A. Mandpe, S. Kumari, and S. Kumar, "Composting: A Sustainable Route for Processing of Biodegradable Waste in India," in *Organic Waste Composting through Nexus Thinking*, H. Hettiarachchi, S. Caucci, and K. Schwärzel, Eds. Cham: Springer International Publishing, 2020, pp. 39–60. doi: 10.1007/978-3-030-36283-6_3.

[2] P. Sisouvanh *et al.*, "Can Organic Amendments Improve Soil Physical Characteristics andIncrease Maize Performances in Contrasting Soil Water Regimes?," *Agriculture*, vol. 11, no. 2, p. 132, Feb. 2021, doi: 10.3390/agriculture11020132.

[3] M.Ajmalet al., "Optimization of pilot-scale in-vessel composting process for various agricultural wastes on elevated temperature by using Taguchi technique and compost quality assessment," Process Saf. Environ. Prot., vol. 140, pp. 34–45, Aug. 2020, doi: 10.1016/j.psep.2020.05.001.

[4] M. A. Rashwan, F. Naser Alkoaik, H. Abdel-Razzak Saleh, R. Blanqueza Fulleros, and M. Nagy Ibrahim, "Maturity and stability assessment of composted tomato residues and chicken manure using a rotary drum bioreactor," *J. Air Waste Manag. Assoc.*, vol. 71, no. 5, pp. 529–539, May 2021, doi: 10.1080/10962247.2020.1859416.

[5] S. Mahapatra, Md. H. Ali, and K. Samal, "Assessment of compost maturity-stability indices and recent development of composting bin," *Energy Nexus*, vol. 6, p. 100062, Jun. 2022, doi: 10.1016/j.nexus.2022.100062.

[6] E. D. Roy *et al.*, "Compost Quality and Markets Are Pivotal for Sustainability in Circular Food-Nutrient Systems: A Case Study of Sri Lanka," *Front. Sustain. Food Syst.*, vol. 5, p. 748391, Nov. 2021, doi: 10.3389/fsufs.2021.748391.

[7] U. De Corato, "Agricultural waste recycling in horticultural intensive farming systems by on-farm composting and compost-based tea application improves soil quality and plant health: A review under the perspective of a circular economy," *Sci. Total Environ.*, vol. 738, p. 139840, Oct. 2020, doi: 10.1016/j.scitotenv.2020.139840.

[8] D. E. A. Tedesco, S. Scarioni, A. Tava, S. Panseri, and A. Zuorro, "Fruit and Vegetable Wholesale Market Waste: Safety and Nutritional Characterisation for Their Potential Re-Use in Livestock Nutrition," *Sustainability*, vol. 13, no. 16, p. 9478, Aug. 2021, doi: 10.3390/su13169478.

[9] M. Butnariu and A. Butu, "Chemical Composition of Vegetables and their Products," in *Handbook of Food Chemistry*, P. C. K. Cheung, Ed. Berlin, Heidelberg: Springer Berlin Heidelberg, 2014, pp. 1–49. doi: 10.1007/978-3-642-41609-5_17-1.

[10] D. Yadav, L. Barbora, L. Rangan, and P. Mahanta, "Tea waste and food waste as a potential feedstock for biogas production," *Environ. Prog. Sustain. Energy*, vol. 35, no. 5, pp. 1247–1253, Sep. 2016, doi: 10.1002/ep.12337.

[11] M. A. Macias-Corral, J. A. Cueto-Wong, J. Morán-Martínez, and L. Reynoso-Cuevas, "Effect of different initial C/N ratio of cow manure and straw on microbial quality of compost," *Int. J. Recycl. Org. Waste Agric.*, vol. 8, no. S1, pp. 357–365, Dec. 2019, doi: 10.1007/s40093-019-00308-5.

[12] E. Kim, D.-H. Lee, S. Won, and H. Ahn, "Evaluation of Optimum Moisture Content for Composting of Beef Manure and Bedding Material Mixtures Using Oxygen Uptake Measurement," *Asian-Australas. J. Anim. Sci.*, vol. 29, no. 5, pp. 753–758, Nov. 2015, doi: 10.5713/ajas.15.0875.

[13] E. C. Chinakwe *et al.*, "Effect of Temperature Changes on the Bacterial and Fungal Succession Patterns during Composting of Some Organic Wastes in Greenhouse," *J. Adv. Microbiol.*, pp. 1–10, Feb. 2019, doi: 10.9734/jamb/2019/v15i130075.

[14] L. Lin, F. Xu, X. Ge, and Y. Li, "Biological treatment of organic materials for energy and nutrients production—Anaerobic digestion and composting," in *Advances in Bioenergy*, vol. 4, Elsevier, 2019, pp. 121–181. doi: 10.1016/bs.aibe.2019.04.002.
[15] F. Nenciu*et al.*, "Decentralized Processing Performance of Fruit and Vegetable Waste Discarded from Retail, Using an Automated Thermophilic Composting Technology," *Sustainability*, vol. 14, no. 5, p. 2835, Feb. 2022, doi: 10.3390/su14052835.

Tainted Wastewater Remediation through Inverted Multimedia Biofilter

Amit M Kharwade[1], Shalini Yadav[2], Ranjit Patil[3], Sunil K Sharma[4]

[1]Research Scholar, Department of Civil Engineering, Rabindranath Tagore University, Bhopal, India
[2]Professor, Department of Civil Engineering, Rabindranath Tagore University, Bhopal India
[3]Assistant Professor, Department of Civil Engineering, Priyadarshini Bhagwati College of Engineering, Nagpur, India
[4]Associate Professor, Department of Civil Engineering, Rabindranath Tagore University, Bhopal India.
Email: akharwade325@gmail.com

Abstract. The growth of industry leads to the contamination of surface as well as groundwater. Improper industrial wastewater disposal causes a serious problem of metal accumulation in the soil and freshwater. To tackle this issue there is a need of recycling and reuse wastewater using low-cost sustainable treatment. The main objective of this research is to check the feasibility of multimedia vermifilter for the efficiency removal of physicochemical parameter and fluoride metal ions of fluoride-tented used water (wastewater). For the treatment, we used fluoride contaminated used water (wastewater) from the fluoride contaminated site. In this study multimedia vermifilter integrated with moringa seed extract, activated charcoal powder, and earthworm Eisenia fetida were used as bio-media to treat contaminated used wastewater, In this experimental analysis, we checked the removal efficiency on different HLR in m3/day/m2. The result showed that percentage removal efficiency for BOD, COD, TDS and Fluoride using inverted multimedia filter was observed for the optimized hydraulic loading rate at 0.933m^3/day/m^2 was found to be BOD (81.64%), COD (51.47%), TDS (82.17%), Fluoride (35.81%) respectively. The earthworm, moringa seed extract, and activated charcoal play a vital role in the removal of Bioconversion of organic matter. Furthermore Treated effluent waste-water is then used for gardening and irrigation purposes.
Keyword: Fluoride, Moringa olifera ,Flyash, Charcoal, Esina Fetida

1. Introduction

The ecological system is affected due to insufficient management of waste handling, Wastewater production and processing it's a big problem in developing countries.[1] in preventing the health issue and To protect from hazards effect of waste on society the treatment of wastewater is important[2].The rise in activities of various industrial and urban sector, agricultural, metal exploration, power sector lead to the generation heavy metal in the air, soil surface, surface water, and underground water bodies which will lead to the detrimental effect on the environment.

Detailed investigation and monitoring of water bodies state that pollution in the water bodies is due to improper discharge activity[3].the people which are living in the urban and village community face the problem of water shortage due to mismanagement of wastewater, we have to develop cost-effective, environmentally friendly, efficient treatment technologies. [4]. Also, the heavy metal pollution in the surface and ground water bodies is the major issue that will cause a serious issue for the community as well as the ecology.
Fluorine is occurring naturally on the earth surface which contributes 6% to 9% of earth strata. Fluoride metal is generally occurring in the earth strata in the form of various mineral composition like in the form of fluorite, phosphate rocks, cryolites, apatites, mica, out of this fluorite, appetites and mica has general occurrence. [5] as per the recommendation given by WHO, the most appropriate limit of fluoride in water is 15 ppm. However, the Bureau of Indian Standards (BIS) has set a limit of 0.0005 and 0.0010g/L.[6] The possible remedial measure is an important concern. Treatment of such contaminated wastewater through convention treatment such as ultra-filtration, reverse osmosis, ion exchange, solvent extraction, sedimentation, and chemical precipitation, has some limitations besides high costs.[7]. For the requirement of cost effective, feasible and natural treatment of water or wastewater, moringa olifera seed and leaf powder is beneficial and used as strong coagulant. [6] Adsorption processes using natural adsorbents and agricultural waste are becoming new alternatives for removing fluoride from aqueous solutions. The acid-treated Moringa oleifera leaf powder showed better adsorption capacity than the alkali-treated Moringa oleifera leaf powder. [8] Moringa tree seeds can be used as a water-soluble extract when crushed into powder, providing an effective natural clarifying agent for highly turbid, untreated, pathogenic surface water.[9]
The principal of Vermifiltration work on the natural-based and cost-effective treatment compared to other conventional treatments.[10].Experimental work conducted on small scale to large scale, researcher studies shows it has high-efficiency removal rate on domestic as well as industrial based effluent for the parameter viz on COD, BOD and suspended solids , [11] in this principal earthworm play an important will dealing with organic pollutant, earthworm work through the general mechanism It is taken into the body by "uptake" and biodegradation of organic waste, heavy metals and solids from sewage, and "absorption" from the body wall. [12] Earthworms can accumulate heavy metals when exposed to heavy metal-contaminated soil[13] Enzymes can remove certain awkward pollutants by acting on them, precipitating them or converting them into their products.[14] so, for effective treatments for metal removal ,advanced Vermifiltration processes are the better options than the conventional treatment methods commonly adopted in wastewater treatment plants. [7].
Worldwide Several earthworm species are used as bio rectors in the Vermifiltration principal .out of these some give better results in treatments Eisenia Fetida (under the subclass Oligochaeta) is also known to be a surface resident and an efficient consumer of organic matter. E. Fetida is also a natural consumer of waste, including external solid waste, with good growth and fertility. Eisenia Fetida is the best selection among other category, being an epigeic, it will helps to convert bioorganic waste by converting into stable and rich component as plant part as main source of supplement [13]. A present Lab-scale analysis focused on the treatment of fluoride contaminated used wastewater in the Vidarbha region, Nagpur, Maharashtra, India. In this study multimedia vermifilter integrated with moringa seed carbon powder, fly ash, and earthworm E. Fetida as natural media to treat contaminated used wastewater. Heavy Essence present in water are dangerous and toxic and Must be removed from the water, by using natural bio-absorbent material we can stabilize it. Many studied reported junking of some heavy essence from water by Moringa oleifera.[7]

2. Material and Methods

2.1Formation of inverted Multimedia filter

In experimental studies, we used an inverted multimedia filter for the treatment of fluoride-tented used water(domestic), which was collected from the contaminated site. In transparent plastic bottle contain various media. At the bottom-most layered the soil with earthworm, Eisenia fetida were placed at depth of 4cm, above that 8cm depth of river sand is placed. The next layer is the grounded dry powder of Maringa seed extract (drumsticks) was provided at depth of 2.0 cm then the next layer above the moringa seed extracts is the 3cm depth layered of activated charcoal is placed. Above the charcoal powder, a layer aggregate of size10mm aggregate was placed at depth of 5cm and the next and last layered will be stone of size between(20mm-40mm) is provided. This multi-layered filter is then placed in the collection container where all the treated wastewater is been collected at the bottom of the chamber and treated water is then received through means of a manual tap. for various hydraulic loading rates, discharges were adjusted using manual tap. The collected effluent wastewater is then analyzed for demining the concentration of BOD, COD, TDS, and Fluorides .and checked for its suitability.

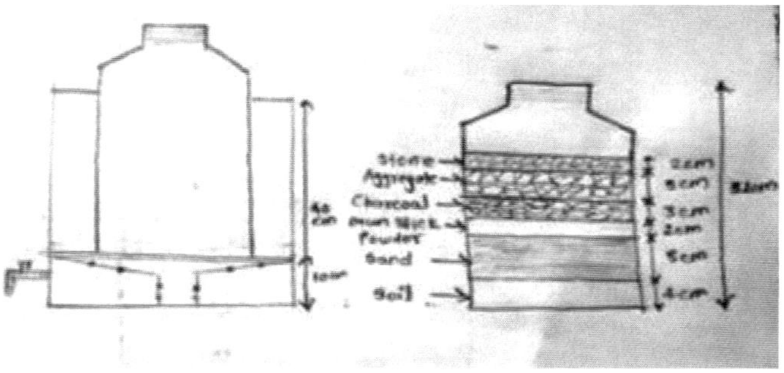

Figure 1. Experimental setup filter **Figure 2.**Detailing of Inverted multimedia

2.2 Media Preparation

Coconut shell is collected first then broken into small pieces. Followed by drying in sunlight and then grinding into a powdered form called coconut husk. This powdered form is then heated in the oven at110°C temperature. After cooling the dried materials were kept in the muffle furnace at 550°C for 3hrfor removal of other volatile impurities. This leads to the formation of fixed carbon (charcoal).for the moringa seed dry powder the drum stick were collected from kalmeshwar village at Nagpur, then moringa seed were extracted and crushed and grind to fine powder and then directly applied on in inverted multimedia filter. Table 1 and Table 2 shows the chemical composition of dry moringa seed and Activated carbon powder were examined using XRF spectrometer.

2.3 Earthworms collection and placement

The earthworm species E.Fetida is placed on the bottom along with cow dung, layered of multimedia filter which is collected from vermiculture plant at Wardha. Garden Soil and dry cow dung are rich in nutrient; this media help the earthworms to enhance their growth during the acclimatization period of the experiment. 10-15 days of acclimatization is provided to the earthworm. During this period care should be taken of moisture content 50-60% will be maintained.

2.4 Working of Inverted multimedia filter

Before starting the inverted media filter after the setup is ready. Collected tented Wastewater is sprinkled over the filter for 10 -15 days of acclimatization system.4 lit of wastewater is collected from the metal contaminated area and kept in a storage tank having a capacity 8 lit. Then flow is regulated by using manually tap for different hydraulic loading rate of 0.933 $m^3/day/m^2$,1.4 $m^3/day/m^2$ 1.75 $m^3/day/m^2$ the treated wastewater is then collected at different retention times of 15min,10min, and 8min through the filter. Flittered water is collected in the collection chamber. The Influent and effluent wastewater was collected in a separate plastic bottle and analyzed in the laboratory for various parameters. The parameter like BOD, COD, Total dissolve solid and fluoride.

Table 1. Element composition in dry Moringa seed powder

Element	% Mass in dry Moringa seed powder
Na	11.8
Mg	4.98
Al	0.589
Si	1.35
P	9.52
S	67
Cl	2.41
K	18.6
Ca	24
Mn	0.056
Fe	1.22

Table 2. Element composition in Activated carbon powder

Element	%Mass in Activated carbon powder (Coconut shell)
C	89.7%
H	2.21%
N	0.4%
O	6.11%
Si	0.28%

3. Result and Discussion

Table 3. Evaluation of Inverted multimedia vermifilter in various hydraulic loading rates $m^3/day/m^2$

Wastewater Parameter	Influent in mg/lit	Effluent in mg/lit forHLR =0.933 $m^3/day/m^2$	Effluent in mg/lit forHLR =1.4 $m^3/day/m^2$	Effluent in mg/litforHLR =1.75 $m^3/day/m^2$
BOD	74.5	13.675	16.5	23.2
COD	237	115	126.75	139.25
TDS	674.25	119.75	205.26	229.95
Fluoride	2.15	1.38	1.43	1.56

Mean concentration in mg/lit (no of sample=04)

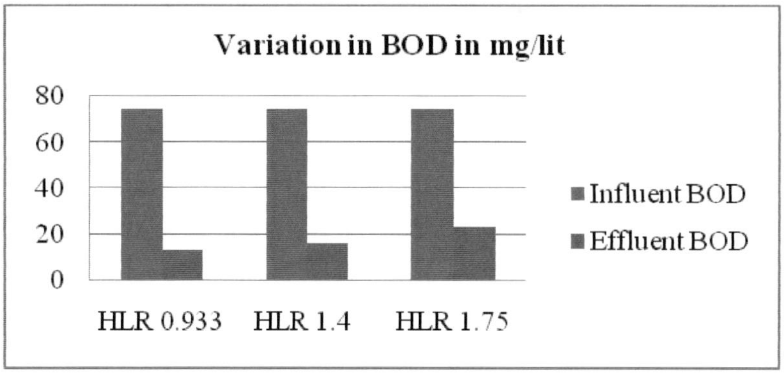

Figure 3. Show average variation in BOD for Different hydraulic loading rates in $m^3/day/m^2$

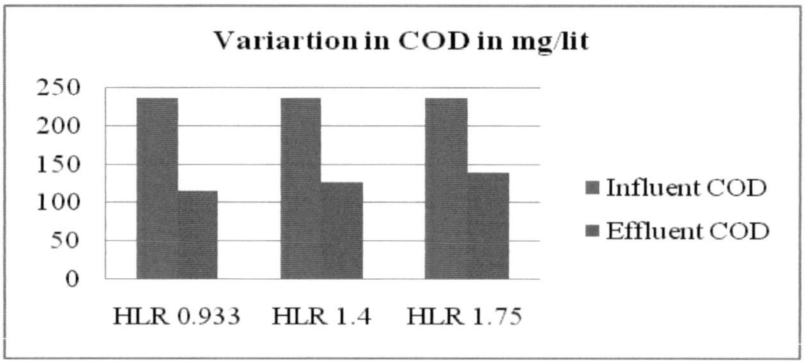

Figure 4. Show average variation in COD for Different hydraulic loading rates in $m^3/day/m^2$

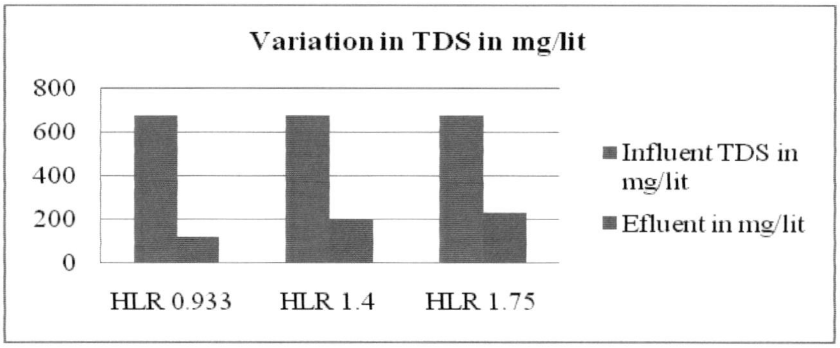

Figure 5. Show average variation in TDS for Different hydraulic loading rates in $m^3/day/m^2$

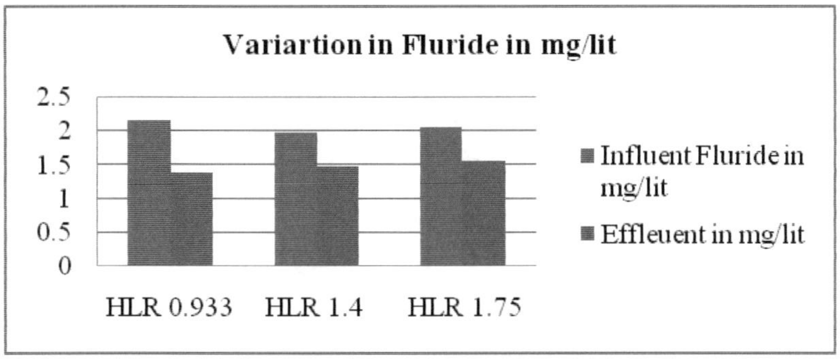

Figure 6. Show average variation in Fluoride for Different hydraulic loading rates in $m^3/day/m^2$

On the detailed analysis of inverted multimedia vermifilter for the various hydraulic loading rate (HRT) 0.933.1.4, 1.75 $m^3/day/m^2$. From table 3 it has been seen that inverted multimedia show the progressive result on removal efficiency on BOD, COD, TDS, and Fluoride. Soil Media, moringa seed extract, activated charcoal, and earthworm helps to reduce the chemical and biological parameter of effluent wastewater. From the figure 3, shows the percentage reduction on the BOD removal ranges from 68.85-81.64% The earthworm uses 'enzymatic action' (which works as a biological to increase the speed and quickness of biochemical reactions) to breakdown the wastewater organics, which is why BOD is removed in vermifilter.[18].Figure 4 shows a percentage reduction in the COD removal ranging from 41.24-51.27% Detention period has a significant impact on COD reduction. The longer the detention time, the more enzyme Earthworms secrete, which aids in the decomposition of several other compounds that microorganisms cannot digest. This could be related to an increase in the production of decomposer microbial biofilms in the geological system (gravels)[7]. Figure 5 shows a percentage reduction in the TDS removal ranging from 66-82.17% in worm filter, these solids are constantly eaten up by the earthworms. Figure 6 shows that fluoride metal reduction was found between 35.81% - 27.44% moringa olifera work as good absorbent media to remove heavy metal ions [7] Integrated moringa seed extract with charcoal give a significant reduction in physic-chemical characteristics of tented wastewater [9].

4. Conclusion

Improper industrial wastewater disposal causes a serious problem of metal accumulation in the soil and freshwater. To tackle with this issue there is a need of recycling and reuse of wastewater using low-cost sustainable treatment. According to a recent research, an inverted multimedia vermifilter shows a better result in the removal of a physical and chemical parameter. But very less removal was observed on fluoride removal. The study shows that at HLR0.933 $m^3/day/m^2$ gives good results in the BOD, COD, TDS, and Fluoride elimination. Compared to other HLR. The percentage removal rate for BOD (81.64%), COD (51.47%), TDS (82.17%), and Fluoride (35.84%) was obtained at HLR0.933$m^3/day/m^2$.values are in the permissible range so we recycled for irrigation and gardening purposes. As it has been stated, that the advanced multimedia filter works better than the conventional one. On the observation, a normal vertical filter gives a better result than an inverted one. Esina fetida earthworm facing challenges during activity in the inverted.

5. References

[1] Kumar.T, Bhargava .R Prasad H,Pruthi.V 2015,"Evaluation of Vermifiltration process using natural ingredients for effective wastewater treatment", *Journal of Ecological Engineering* 75 ,370–377.

[2] Samal K, Dash R R,Bhunia P 2017, "Treatment of wastewater by Vermifiltration integrated with macro phytofilter :A review", *Journal of Environmental Chemical Engineering* ,http://dx.doi.org/10.1016/j.jece.2017.04.026.

[3] P.D.Nemade, A.M.Kadam and H.S.Shankar 2009,"Wastewater renovation using constructed soil filter (CSF): A novel approach", *Journal of Hazardous Materials,* 170, 657–665.
[4] Sukhdeep Kaur Cheema P P2019, "Vermifiltration Using Garden Waste as Padding Media for Treatment of Dairy Wastewater",*ICSWMD,*21 ,51–58.
[5] John. Bentiland William Kwame Buah2016 ,"Removal of Fluoride in Water Using Activated Carbons Prepared From Selected Agricultural Waste Materials", *Journal of Environment andEarth Science* ,6.
[6] Kumar Rajeev andChawla Jyoti 2020 ," Carbon-Based Materials for De-Fluoridation of Water: Current Status and Challenges", DOI:http://dx.doi.org/10.5772/intechopen.90879.
[7] Eman noori ali 2020,"Removal of heavy metal from water and wastewater by using moringa oleifera",DOI:10.5772/intechopen.89769.
[8] Shabnam DanAmit Chattree 2018," Sorption of fluoride using chemically modifed Moringa oleifera leaves",*Applied Water Science* 76.
[9] Chepkwony G, Mwamburi E and Makumba B. 2019 ,"Effect of Charcoal Filters Integrate with Moringa oleifera Seed Extracts on Microbial Population in Water from Unprotected Sources of Kapseret Division, Kenya", *journal of advanced in microbiology,* 1-8 DOI:10.9734/JAMB/2019/44978.
[10] Adugnaa.A, Andrianisab.H, Konateb.YandMaigab.A2019,"Fate of filter materials and microbial communities during Vermifiltration Process", *Journal of Environmental Management* 242 98–105.
[11] Sinha K Bharambe G and Chaudhari U 2008,"Sewage treatment by Vermifiltration with synchronous treatment of sludge by earthworms: a low-cost sustainable technology over conventional systems with potential for decentralization", *Springer Science Business Media*, 28 409–420.
[12] Bhise H. S.1 , Anaokar G. S," Design and Suitability of Modular Vermifilter for Domestic Sewage Treatment", *International Journal of Emerging Engineering Research and Technology,* 344-51.
[13] Majumdar A, Barla A,Upadhyay M,Ghosh D, Chaudhuri.P, Srivastava S and Bose S 2018,"Vermiremediation of metal(loid)s via Eichhorniacrassipes phytomass extraction: A sustainable technique for plant amelioration", *Journal of Environmental Management*,220,118–125.
[14] Patnaik Aliva and Bidkar Dnyanraj Baban 2017," Bioremediation of Wastewater from Drain in Burla Town, Odisha, By Vermifiltration", *Asian Journal of Science and Technology 8,*5315-5323.
[15] Ntibrey R A K, Kuranchie F A, Gyasi S F 2020, "Antimicrobial and coagulation potential of Moringa oleifera seed powder coupled with sand filtration for treatment of bath wastewater from public senior high schools in Ghana", *Heliyon* 6.
[16] Ravikumar K and Prof.Sheeja A K 2013," Fluoride Removal from Water using *Moringa oleifera* Seed Coagulation and Double Filtration", *International Journal of Scientific &Engineering Research* 4.
[17] Winston Lik Khai Lim, Edwin Chin Yau Chung, Chien Hwa Chong, Nicholas Tze Kai Ong,Wee Seong Hew, Nurasatifah binti Kahar and Zhen Jie Goh2018" Removal of fluoride and aluminum using plant-based coagulants wrapped with fibrous thin film", Process Safety and Environmental Protection117 ,704–710.
[18] Sinha R.K. Bharambe G. Bapat P 2007, "Removal of High BOD And COD Loading ofPrimary Liquid Waste Product From Dairy Industry By Vermifiltration Technology Using Earthworm", *Indian J.EnvironmentalProtection*,27, 6.

Corrosion Analysis of RCC Beam using Simplified FEM Model

Yamuna Bhagwat, Gopinatha Nayak*, Poornachandra Pandit, Harithakrishnan, Aishwarya Lakshmi

Department of Civil Engineering, Manipal Institute of Technology, Manipal Academy of Higher Education, Manipal, Karnataka 576104, India.

E-mail:
{yamuna.bhagwat1@gmail.com,nayak.gopinath@manipal.edu*,pc.pandit@manipal.edu, harithakrishnan414@gmail.com, aishwarya.aishu437@gmail.com}

Abstract. The durability of the reinforced concrete structures highly influenced by many important factors including steel corrosion. Corrosion starts off with the availability of the water and oxygen, products formed due to this get deposited in the interface between the concrete and steel. As the depositing corrosion volume increases, an expansive pressure at the interface exerts on the surrounding concrete and eventually causes structural degradation, which reduces concrete and steel bond. This work focuses on a parametric study on rust expansion pressure to understand the effect of grade of concrete, diameter of the rebar, position of the rebar and c/d ratio using ABAQUS software. After the analysis, it is found that rust expansion pressure increases as the grade of concrete and c/d ratio increases, which means the pressure required to cause surface cracks due to corrosion is more in higher grades concrete and high c/d ratio. Position of the bar did not affect the corrosion pressure requirement to cause surface cracks. As the corrosion level increased, deflection of the beam also increased. But position of the corroded reinforcing bars in the beam slightly affects the deflection of the beam.

Keywords: Corrosion, ABAQUS, RCC Beam, Durability.

1. Introduction

Effect of corrosion in steel embedded concrete greatly affects the structure's serviceability. Passivation of steel in concrete is because of the alkalinity of concrete, but with aging of the concrete alkalinity reduces. Steel passivation is destroyed because of the ingress of external agents like sulphate, chloride, and carbon dioxide present in the water, aggregates, and other raw materials of the concrete and the environment where the structures come in direct or indirect contact. Concrete loses its pH value by local acidification, oxygen depletion and carbonation and in turn loses passivation of the steel[1]. The corrosion products formed due to these reasons get deposited on the steel surface and increase the volume of the steel bar by 2-6 times of original volume. This volume increase causes the radial pressure on the concrete and results in the microcracks formation adjacent to the steel. Extension of later cracks causes concrete spalling by breaking the bond between steel and concrete[2]. Concrete spalling exposes the steel in concrete; thus, deterioration increases and the structure loses its load bearing capacity and durability[3]. The concrete cover cracking time is calculated as a function of corrosion product formation and resistance occurred to form corrosion products[4]. As a result of corrosion, occurrence of larger

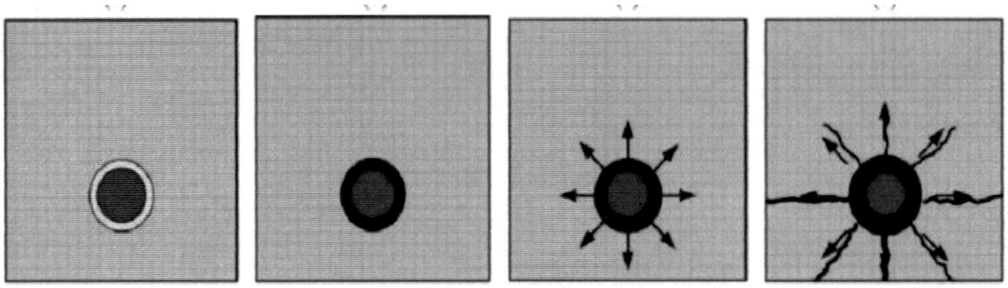

Figure 1: Uniformly distributed rebar corrosion and concrete cover cracking pattern [9]

voids in steel-concrete interface delays the cover cracking in concrete[5]. Corrosion initiation and concrete cover cracking pattern are represented in Figure 1.

Intensive efforts have been put forth to understand the corrosion effect on the reinforced structures considering load carrying capacity, bond strength, and cracking pattern as study parameters[6]. The brittle failures of the structures have been observed as an increase in the corrosion rate of the steel[7]. Hakan Yalciner et al. studied the corroded steel bar bond strength in concrete by pullout test. High strength concrete showed the brittle failure with an increase in strength of the concrete[8].

This paper concentrates on the studies conducted on the corrosion of rebars by analytical analysis and simulation. According to the study conducted, the distribution pattern of the rust on rebar plays a vital role in concrete cover cracking[10], [11]. Different methods have been adopted for corrosion modeling, such as thick cylinder method, modeling the increase in volume of corrosion product by using half elliptic radial displacements and uniform radial displacements on outer and inner semi circumferences[10]. Assuming uniformly distributed corrosion along and around the reinforcement, stresses of corrosion expansive pressure are directly applied. To define the non-uniform expansion pressure distribution of the rust products, a half ellipse shape with non-uniform radial displacement is used[9]. Amleh and Gosh[12] have modeled the reduction of friction and contact pressure at the concrete and steel interface due to corrosion using ABAQUS software and found that loss in bond strength is almost linear to the mass loss of reinforcements by corrosion. Emilie et al. [13] used the thermal analogy to define expanding nature of the corrosion product considering the non-uniform precipitation of corrosive products into surrounding concrete. Assuming constant coefficient of thermal expansion, temperature increment was used to simulate the expansive pressure of corrosion products. Du et al. [14] have done a finite element study to assess the effect caused by corrosion to the surrounding concrete, mainly cracking of the concrete cover due to corrosion. Effect of corrosion was modeled as internal pressure as well as radial expansion. The results obtained for internal pressure were validated by comparing it with the model proposed by Williamson and Clark[15] in which the uniform pressure was applied on the internal surface of the cylindrical holes, which were lined with PVC tube and by comparing with experimental results of accelerated corrosion tests in case of radial expansion. Under plane strain assumption, two-dimensional models were used to represent the specimens. The cracking of concrete was found to occur in four stages in case of radial expansion and depended only on the bar diameter and cover. The results were similar to that of internal pressure modeling. In this work, analysis is done by assuming corrosion of the reinforcing bar is uniform and the expansive pressure developed around the bar due to corrosion products is uniform. Concrete is considered as a homogeneous mass. The expansive pressure due to corrosion products, cover thickness, diameter of the rebar, location of the rebar, and concrete strength are the major parameters that affect the concrete cover cracking. So here, the effect of these parameters is studied considering the corrosion of the rebar.

Table 1: Concrete property parameters considered for simulation

Dilation Angle (in degrees)	**Eccentricity**	$\mathbf{f_{bo}/f_{co}}$	**K**	**Viscosity Parameter**
35	0.1	1.16	0.667	0.007985

f_{bo}=compression strength in the biaxial state, f_{co}=strength in the uniaxial state
K=shape factor of yielding surface

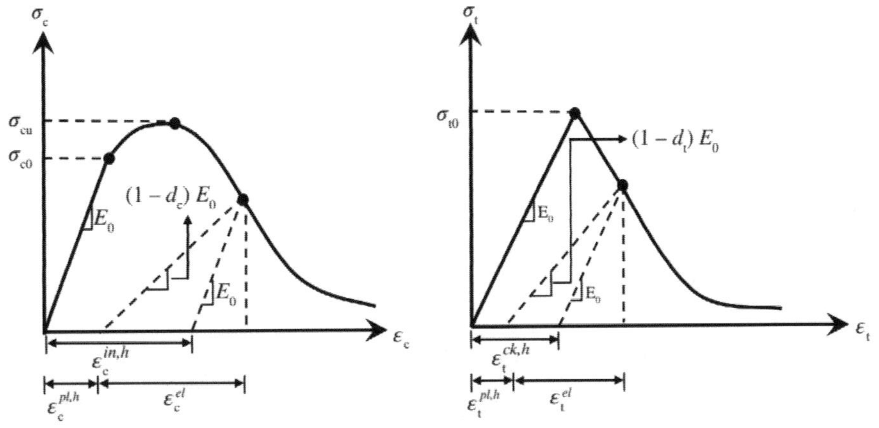

(a) Compression (b) Tension

Figure 2: Stress-strain curves for compression and tension in concrete as per CDP model

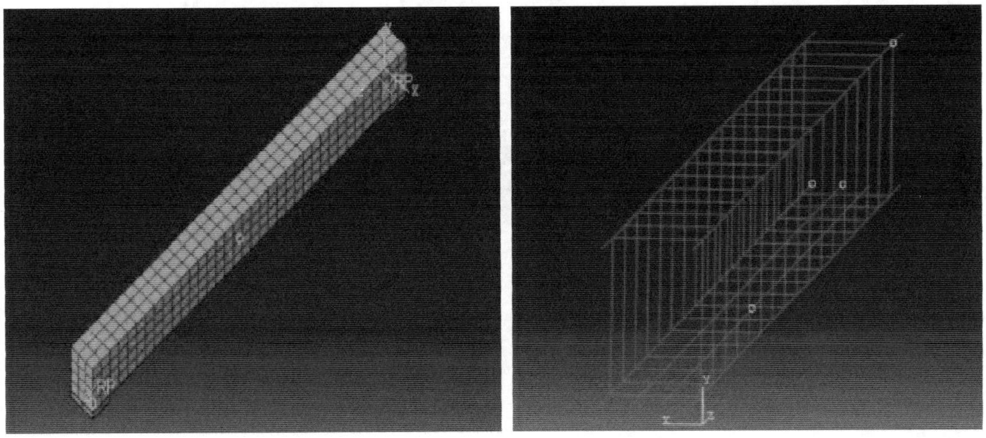

(a) Beam (b) Reinforement

Figure 3: Beam and reinforcement model

2. Methodology

Non-linear finite element analysis is conducted using ABAQUS software, which is widely used for finite element study. The type of analysis used was "Dynamic Explicit". The element used for modeling the concrete is C3D8R. This linear brick element has an 8-node, reduced integration and hour-glass control. For non-linear behaviour modeling of the concrete, elastic definition is used till the material reaches cracking stress and beyond that for non-linear behaviour concrete damage plasticity (CDP) model is employed. The stress and strain values are defined both for

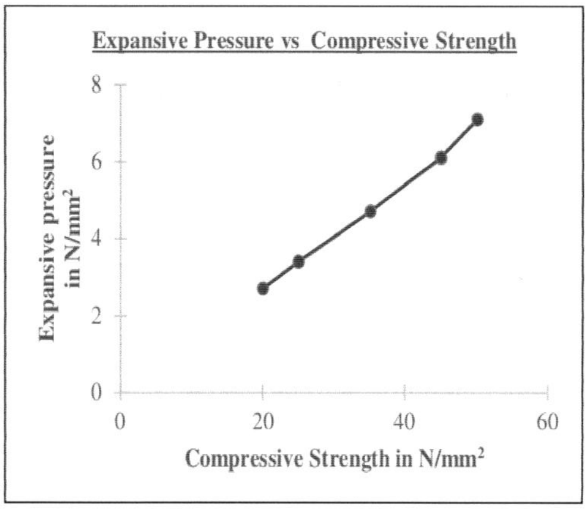

(a) expansive pressure v/s compressive strength

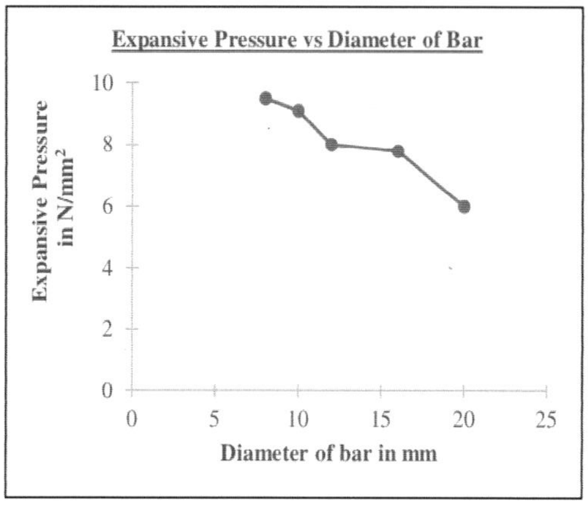

(b) expansive pressure v/s diameter of bar

Figure 4: Variation in corrosion expansive pressure with compressive strength and diameter of bar

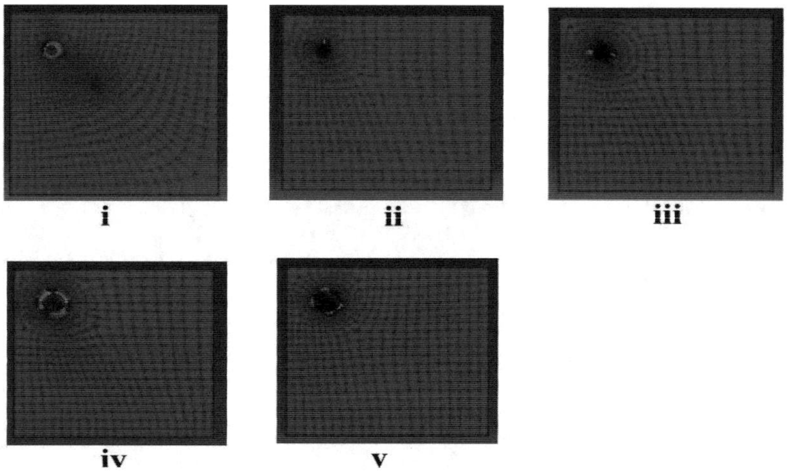

Figure 5: Damage model of rebar at corner for different diameters. i) 8 mm dia. bar ii) 10 mm dia. bar iii) 12 mm dia. bar iv) 16 mm dia. bar v) 20 mm dia. bar

tension and compression behaviour. The stress-strain plot of tensile and compressive behaviour of concrete subjected to uniaxial loading as per CDP model is shown in Figure 2. Element used to model the reinforcement is T3D2 which is a truss element. The grade of concrete considered is 40 MPa and Poisson's ratio of steel and concrete considered are 0.3 and 0.18 respectively. Modulus of elasticity is taken as $5000\sqrt{f_{ck}}$, where f_{ck} is the characteristic strength of concrete. A rust layer is modeled as a hollow cylinder with diameter of the rebar and assembled using tie constraint with the concrete. The uniform pressure is applied to simulate the expansive pressure developed because of increased corrosion product volume. Pressure exerted by the corrosion products is applied towards the cylindrical wall radially as internal pressure. Simply supported beam of dimension 250×375×3600 mm is considered for study with 12 mm dia. bars at all four corners and two numbers of 16 mm dia. bars at bottom middle portions, 8 mm dia. stirrups at a spacing of 200 mm c/c and clear cover of 30 mm is used. The beam and the reinforcement model are shown in Figure 3. To make the reinforcement and concrete function

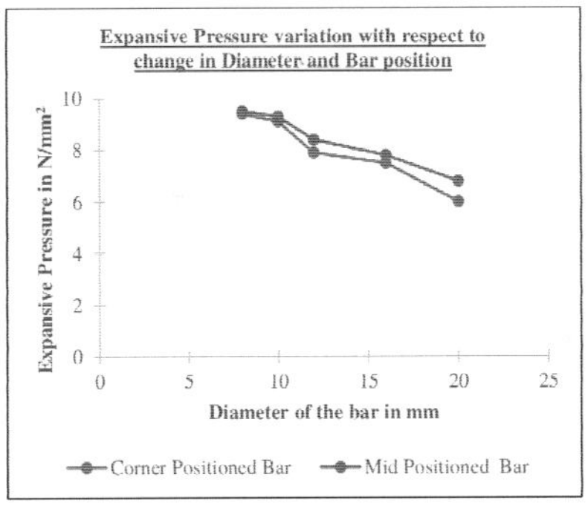

(a) expansive pressure v/s diameter of bar

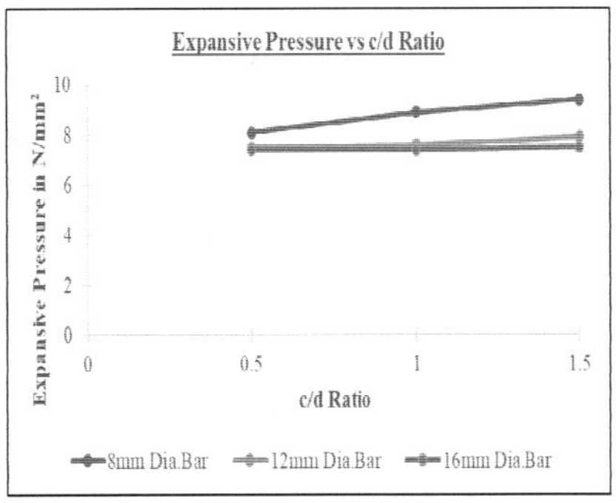

(b) expansive pressure v/s c/d ratio

Figure 6: Variation between corrosion expansive pressure and change in position of rebar and Variation between corrosion expansive pressure and c/d ratio

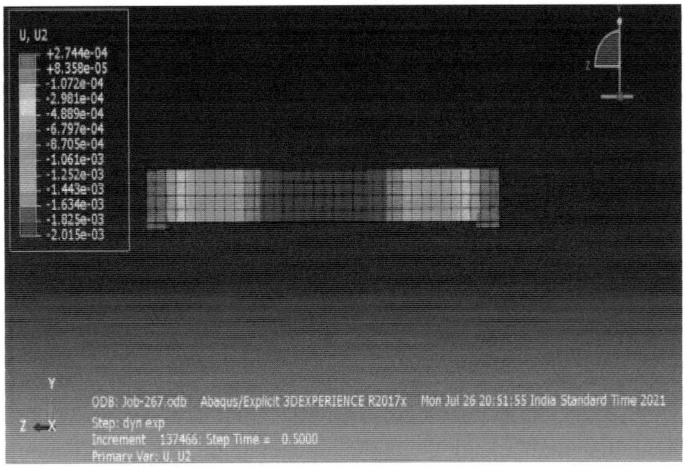

Figure 7: Deflection of the beam at 10% corrosion level (in meter)

together, embedded constraint is used. Other parameters considered for modeling are listed in Table 1.

3. Results and Discussion

Parametric study on rust expansion pressure is conducted to understand the effect on grade of concrete, diameter of the rebar, location of the rebar and cover to depth ratio.

3.1. Effect of grade of concrete

Grade of the concrete was varied and the expansion pressure was applied till the concrete cracks. The grades of the concrete used are 20, 25, 35, 45 and 50 MPa. The variation of expansion pressure with grade of concrete is shown in Figure 4(a). Increase in the grade of concrete increases strength of concrete and resisting capacity of the concrete against the pressure developed by the corrosion of reinforcement. So as the compressive strength increased, the pressure required to cause damage increased.

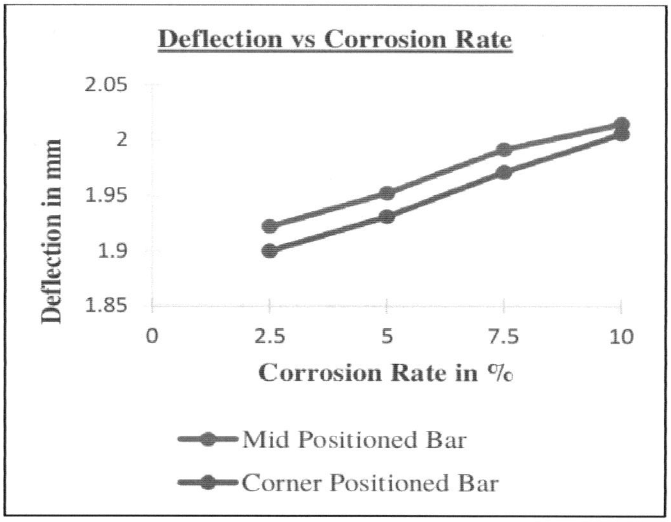

Figure 8: Comparison chart of deflection vs corrosion rate of the RCC beam

3.2. Effect of rebar diameter

The diameter of rebars considered for study is 8, 10, 12, 16 and 20 mm. To simulate the expansion pressure due to rust, hollow cylindrical element of diameter equal to the rebar diameter is used and pressure is applied on internal wall of the cylinder till concrete cracks. In this study, the expansion pressure is found to decrease with increase in diameter of rebar. The DAMAGET plot for 8, 10, 12, 16 and 20 mm diameter bar is shown in Figure 4(b). It is noticed that with increase in diameter of bar, the expansion pressure is found to decrease. As the diameter of bar increases, corrosion products produced is also more, thus requiring lower expansive pressure to initiate the damage.

3.3. Effect of rebar position

Hollow cylinders are placed at corners and also at the rebar's mid position, replacing reinforcement by hollow cylinder and applying the pressure on the internal wall of the cylinder. The diameter is varied, keeping all the parameters constant at corner and mid position separately, damage model of rebar at corner position shown in Figure 5. The corrosion pressure is found to be slightly on the higher side for mid-section bar compared to corner bar shown in Figure 6(a). The result shows that corner-bar corrosion is considered to be more critical since surface cracks occurred with less internal pressure than the internal pressure created at the bars located in middle portion.

3.4. Effect of c/d ratio

Variation of corrosion expansion pressure w.r.t c/d ratio is shown in Figure 6(b). Failure pressure is found to increase with increase in c/d ratio. As the thickness of cover increases, pressure required to develop cracks and to reach the surface of the beam is found to be high. This means having a large cover to depth ratio protects the reinforcement from corrosion failures. Use of high-grade concrete and cover thickness increases the durability of the structure with respect to corrosion.

3.5. Deflection of beam
The deflection of the beam was studied by considering corner located bars corrosion and middle located bars corrosion, to explore corroded bar location effect on beam deflection. First, only the bottom mid two 16 mm dia. bars are modeled as corroded and then 12 mm dia. corner bars. The deflection obtained is compared. As observed from the plot given in Figure 8, there is no much difference in deflection of beam when either corner or middle located bars are corroded. The deflection contour of the considered beam with corrosion rate of 10% is shown in Figure 7. The validation of the work is done by modeling and analysing in ABAQUS software, considering the RCC beam was used by Jung et al.[16] for their experimental work. The mid span deflection value obtained by the ABAQUS modeling and experiment conducted by Jung et al.[16] are found to be similar.

4. Conclusion
Following are the conclusions drawn from the aforementioned findings and discussion:

- Rust expansion pressure increased as the grade of concrete and c/d ratio increased.
- The pressure required to cause surface cracks due to corrosion is more in higher grade concrete and higher c/d ratio.
- Rust expansion pressure showed a decreasing trend when reinforcing bar diameter is increased.
- Position of a bar does not affect the corrosion severity since changing the location of the reinforcing bars didn't affect the magnitude of the expansion pressure much.

References
[1] Y. Zhou, B. Gencturk, K. Willam, and A. Attar (2015) "Carbonation-Induced and Chloride-Induced Corrosion in Reinforced Concrete Structures", J. Mater. Civ. Eng., vol. 27, no. 9, p. 04014245, doi: 10.1061/(asce)mt.1943-5533.0001209.
[2] Y. S. Choi, S. T. Yi, M. Y. Kim, W. Y. Jung, and E. I. Yang (2014) "Effect of corrosion method of the reinforcing bar on bond characteristics in reinforced concrete specimens", Constr. Build. Mater., vol. 54, pp. 180–189, doi: 10.1016/j.conbuildmat.2013.12.065.
[3] X. Zhang, Y. Zhang, B. Liu, B. Liu, W. Wu, and C. Yang (2021) "Corrosion-induced spalling of concrete cover and its effects on shear strength of RC beams", Eng. Fail. Anal., vol. 127, no. June, p. 105538, doi: 10.1016/j.engfailanal.2021.105538.
[4] I. Balafas and C. J. Burgoyne (2010) "Environmental effects on cover cracking due to corrosion", Cem. Concr. Res., vol. 40, no. 9, pp. 1429–1440, doi: 10.1016/j.cemconres.2010.05.003.
[5] I. Lau, C. Q. Li, and F. Chen (2020) "Analytical and Experimental Investigation on Corrosion-Induced Concrete Cracking", Int. J. Civ. Eng., vol. 18, no. 1, pp. 99–112, doi: 10.1007/s40999-019-00469-4.
[6] B. Sanz, J. Planas, and J. M. Sancho (2018) "Study of the loss of bond in reinforced concrete specimens with accelerated corrosion by means of push-out tests", Constr. Build. Mater., vol. 160, pp. 598–609, doi: 10.1016/j.conbuildmat.2017.11.093.
[7] A. A. Almusallam (2001) "Effect of degree of corrosion on the properties of reinforcing steel bars", Constr. Build. Mater., vol. 15, no. 8, pp. 361–368, doi: 10.1016/S0950-0618(01)00009-5.
[8] H. Yalciner, O. Eren, and S. Sensoy (2012) "An experimental study on the bond strength between reinforcement bars and concrete as a function of concrete cover, strength andcorrosion level", Cem. Concr. Res., vol. 42, no. 5, pp. 643–655, doi: 10.1016/j.cemconres.2012.01.003.
[9] X. Du, L. Jin, and R. Zhang (2014) "Modeling the cracking of cover concrete due to non-uniform corrosion of reinforcement", Corros. Sci., vol. 89, no. C, pp. 189202, doi:10.1016/j.corsci.2014.08.025.
[10] J. Zhang, X. Ling, and Z. Guan (2017) "Finite element modeling of concrete cover crack propagation due to non-uniform corrosion of reinforcement", Constr. Build. Mater., vol. 132, pp. 487–499, doi: 10.1016/j.conbuildmat.2016.12.019.
[11] K. Zhang, J. Xiao, Y. Zhao, and Q. Zhang (2019) "Analytical model for critical corrosion level of reinforcements to cause the cracking of concrete cover", Constr. Build. Mater., vol. 223, pp. 185–197, doi: 10.1016/j.conbuildmat.2019.06.210.
[12] L. Amleh and A. Ghosh (2006) "Modeling the effect of corrosion on bond strength at the steel-concrete interface with finite-element analysis", Can. J. Civ. Eng., vol. 33, no. 6, pp. 673–682, doi: 10.1139/L06-052.

[13] A. E. A. Thybo, A. Michel, and H. Stang (2017) "Smeared crack modelling approach for corrosion-induced concrete damage", Mater. Struct. Constr., vol. 50, no. 2, pp. 1–14, doi: 10.1617/s11527-017-0999-5.
[14] Y. G. Du, A. H. C. Chan, and L. A. Clark (2006) "Finite element analysis of the effects of radial expansion of corroded reinforcement", Comput. Struct., vol. 84, no. 13–14, pp. 917–929, 2006, doi: 10.1016/j.compstruc.2006.02.012.
[15] S. J. Williamson and L. A. Clark (2000) "Pressure required to cause cover cracking of concrete due to reinforcement corrosion", Mag. Concr. Res., vol. 52, no. 6, pp. 455–467, doi: 10.1680/macr.2000.52.6.455.
[16] J. S. Jung, B. Y. Lee, and K. S. Lee (2019) "Experimental Study on the Structural Performance Degradation of Corrosion-Damaged Reinforced Concrete Beams", Adv. Civ. Eng., vol.2019, no.2, doi: 10.1155/2019/9562574.

Review of open space rules and regulations and identification of specificities for plot-level open spaces to facilitate sustainable development: An Indian case

Sameer Gujar[1], **Amit Deshmukh**[2] **and Reshu Gupta**[3]

[1] Research Scholar, Department of Architecture and Planning, Visvesvaraya National Institute of Technology, South Ambazari Road, Nagpur, Maharashtra, India
[2] Department of Architecture and Planning, Visvesvaraya National Institute of Technology, South Ambazari Road, Nagpur, Maharashtra, India
[3] School of Architecture, CHRIST (Deemed to be) University, Kengeri campus, Kanmanike, Kumbalgodu, Mysore Road, Bengaluru, Karnataka, India
Email- sameergujar.vnit@gmail.com

Abstract
Rapid urbanization and an increase in the alteration of natural resources have led to climate crises, driving the need to promote sustainable development. Urban open space management plays a vital role in such scenarios. Research on urban open spaces has been mainly conducted at regional, municipal, and neighborhood scales. Rarely has the focus been on the plot-level potentials and management of open spaces. Therefore, the study looks into the Indian development control rules and regulations and identifies that although these stipulate the percentage of open space for development on each plot, specificities for open spaces are unclear. Further, the study analyses quantitative and qualitative aspects of open spaces for selected group housing schemes in Pune city. The inquiry shows that per capita open space in Pune is comparatively lower than national standards. The quantitative aspects include FSI, building ground coverage, built-up area, number of floors, and number of dwelling units, and each relates to open spaces in one way or another. The qualitative interpretations disclose that a plot-level open space can significantly impact the regional-level open space network. Hence, the research advocates a bottom-up approach wherein plot-level open space can become the focus in formulating new norms and policies for sustainable development.

1. Introduction

The world faces a real challenge of climatic crises. Developing green has become the central theme for designing the built environment. Nature-based systems are new emerging concepts for providing sustainable smart cities for the future. In balancing the built and the other, it has been well established that careful planning of a network of urban open spaces can strategically handle the urban ecology [1–4].

World Health Organization (WHO) states urban green space as 'all urban land covered by vegetation of any kind,' which may be a subset of open spaces. State University of New York (2010) has defined green space as any piece of land covered with vegetation. It refers to parks, golf courses, sports fields, and other open lands within the built-up area, which may or may not be publicly accessible. The Planning Institute of Australia (2009) has described open spaces as 'land reserved for formal or informal sports, recreation, preservation of natural environments, providing green space and managing urban stormwater.' Researchers and authorities interchangeably use urban open spaces, urban landscape, urban green, nature space, urban forest, amenity space, and the urban wilderness, encompassing various aspects of open spaces.

The landscape of open spaces can range from playing fields to highly maintained environments to relatively natural landscapes. These include green belts around the city region, green corridors, boulevards, gardens, paths, spaces between buildings, open space reservations, district parks or local parks, community spaces, cemeteries, recreational areas, amenity spaces, farmer's markets, backyards, and courtyards. It also contains pedestrian malls, plazas, schoolyards, streets, transit malls, town trails, vacant plots, and waterfronts. Open spaces also comprise agricultural lands, forests, undeveloped shorelines, scenic lands, public parks and preserves, water bodies, wetlands, streams, and floodplains [5]. It is essential to carefully classify open spaces, their functions, characteristics, and cultural understandings. Therefore, this study follows Indian rules and regulations, which define open spaces as open-to-sky spaces, excluding roads, used by the residents for recreation.

1.1. Background

During the colonization period, while developed nations explored new planning concepts of efficient land management and provided a better standard of living, providing ample open spaces, cities in developing nations faced densification with deteriorated living standards [6–8]. By the late 19th century, cities in developing countries independently framed a formal discipline for development. The focus on open spaces mostly involved planning large parks and public open spaces (POS) at the city level. Plot sizes were usually smaller during the initial planning stages, having controlled density. Open spaces at the plot level were provided only for light and ventilation. Besides, the development rules focused on the health aspects and ignored the recreational needs.

Over time, new concepts such as plot amalgamation and land pooling emerged, leading to larger plot sizes. New rules drafted for amalgamated plots addressed open space management, considering both facets of light, ventilation, and recreation [9,10]. With the increasing urbanization, planning notions like Floor Area Ratio (FAR) or Floor Space Index (FSI) cater to the increased urban population. At present, these allow high population densities on individual plots. These densities have directly manifested into vertical growth in the form of multistoried apartments. With ever-increasing densities and the culture of vertical towers in a plot, assessment for per capita open space has become an urgent need.

Internationally, the United Nations (UN) and the European Union (EU) have set the limits as 30 sq. m. and 26 sq. m. of per capita green spaces, respectively. On the other hand, the Leadership in Energy and Environment Design Neighborhood Development (LEED-ND) recommends that the green area per capita be more than 20 sq. m. Also, LEED-ND recommends maintaining a minimum of 1.25 ha of open space per 1000 residents, i.e., around 12.5 sq. m. per person, and specifies that access to open space is within 250 m of residential areas in the form of smaller parks. Nevertheless, green space includes all vegetation of all land types, including recreational POS, vacant lands, forests, and protected areas. Though the lack of norms internationally, few countries have set standards for their respective cities (Table 1).

Table 1. Standards for POS sizes in cities of Developed Countries [11].

Cities	POS Size	Target Population	m^2/person
1.Greater London	4 hectares	1000 residents	40
2.Edinburgh	2.9 hectares	1000 residents	29
3.Cambridge	4.6 hectares	1000 residents	46
4.Washington	3.8 hectares	1000 residents	38
5.Minneapolis	2 hectares	1000 residents	20
6.Los Angeles	4.85 hectares	1000 residents	48.5
7.Kansas City	3.64 hectares	1000 residents	36.4
8.Bristol	1.0 hectares	1000 residents	10

Table 1 shows that the POS size for various cities is different. However, researchers have argued that such cities have failed to achieve the set standards. Few studies point out that the measures fail to give quality of open spaces [12,13]. Research studies have shown that the accessibility standards to such public open spaces are inadequate and are needed to be scientifically verified. Also, these do not consider the people's changing demographic patterns, social behaviors, and physical capabilities [14,15]. Thus, this indicates the need to address the specificities of per capita open space at the plot level.

1.2. The Indian Scenario

As per the UN Department of Economic and Social Affairs 2018 report, it is projected that India will have added 416 million urban dwellers by 2050 [16]. Therefore, for such a large number of city dwellers, a sustainable urban environment shall become very important in the daily life of the people. One of the critical aspects is the open spaces. Such spaces impact a place's social, economic, and environmental settings. Play areas, active and passive recreational areas, community cultural parks, and centers provide social opportunities and health benefits. These also influence property values, probable tourism, and employment prospects and bring extra-economic profits. Nature reserves and botanical gardens are open spaces providing educational openings and ameliorating urban environments. In addition, open spaces are habitats for wildlife too.

The growing urbanization has led to a significant decrease in per capita green space in many Indian cities [17]. The Indian Government has introduced acts, policies, and programs, such as the Jawaharlal Nehru National Urban and Rural Mission (JnNURM) in 2005 and the Atal Mission for Rejuvenation and Urban Transformation (AMRUT) in 2015, for the development of infrastructure, green spaces, and parks in the cities. Furthermore, in 2015 the Government of India implemented the Smart Cities Mission in the selected cities with targeted sustainability goals, including preserving and developing city-level open spaces [18]. However, plot-level open spaces catering to the everyday life of people for respective densities are yet to be recognized. High-density areas do not negate the possibilities of reduced POS, but studies show that users in such dense areas frequently use such spaces [19,20]. Furthermore, a steady increase in mass housing development has led to an apartment-based housing culture. This increase has reduced private recreational spaces and increased the reliability on POS for recreation. Apart from these, today's cities have diverse functional needs from open spaces [21,22]. Therefore, a "one size fit all" would be prone to failure in such high-density areas. Thus, there is a need to review the current status of the Indian acts and rules concerning per capita open space for prospects in plot-level management.

2. Methodology

The study intends to understand the open space dynamics at the plot level. It conducts a detailed literature review of planning implementation tools, including the City Development Plan (CDP), which incorporates the Master Plan and the Development Control Rules (DCR). Firstly, this study involves a detailed study of the tiered legislative framework, its process, and its hierarchy. Later, the study reviews how each guideline addresses open spaces concerning their definitions, types, and respective norms. Based on this initial study, recreational or layout open spaces are studied further. Though local authorities follow a typical guideline for formulating the DCR, few of them have city-specific modifications. Cities that come under this list include state capitals such as Tawang, Bhopal, Chandigarh, Patna, Lucknow, Bangalore, Mumbai, Chennai, Agartala, Hyderabad, Kolkata, and Bhubaneshwar. Other cities on the list are Thiruvananthapuram, Nashik, Pune, Shimla, and Nagpur; Union territories such as New Delhi, Pondicherry, Daman, Diu, and Nagar haveli are also on the list. The researcher conducted a detailed analysis of the DCRs of these Indian cities and union territories for recreational open spaces.

Moreover, there is a need to understand the impact of the rules and regulations on the ground to identify specificities related to plot-level open spaces. Therefore, the study undertook a quantitative and qualitative assessment of open spaces at plot level for seven group housing schemes in Pune city, developed with the latest rules and regulations. The Indian Census 2011 has stated an average family size of 4.3 members irrespective of the size of their dwelling unit. Hence, the population density on each plot differs based on the number of dwelling units provided by the developer. To study this further, case examples with varied densities were selected.

Under the quantitative analysis, identified examples were examined for density parameters such as the total number of dwelling units, plot areas, built-up areas, building ground coverage, utilized FSI, and recreational open space to derive the resultant per capita open space. For all the schemes, all parameters fitted well within the city's DCR and building bye-laws.

IOP Conf. Series: Earth and Environmental Science 1084 (2022) 012073 doi:10.1088/1755-1315/1084/1/012073

The qualitative analysis looked at the aspects of open spaces that contribute to the overall sustainability of the development. There are three pillars of sustainability: social, economic, and environmental. The significance of open space at the plot level is more inclined to the residents for social sustainability. However, the absence of accessible recreational open spaces in immediate surroundings determines the discrete social values of plot-level open spaces. For economic sustainability, open spaces distinctly impact property values. However, it highly depends on the size of the plot and the corresponding size of the open spaces. The plot-level open spaces need to be large enough to impact the economy significantly. Natural assets like hills, forests, dense vegetation, rivers, lakes, other water bodies, and open spaces perform best in and around the site for environmental sustainability. The plot-level open spaces have a high potential to respond to these assets. Therefore, to understand the qualitative aspects of open spaces at the plot level, the researchers conducted field surveys and studied the city's development plan. A limited area of a 500m radius around the plot, corresponding to a person's average comfortable walking distance, is studied.

3. Study of Open Space Rules and Regulations in India

3.1 Rules and Regulations

There is a tiered legislative framework for India's planning and implementation of rules and regulations for urban planning; at the top is the Central or Union Government, followed by the State government and the local administrative body. The State Act allows the local executive body to formulate Regional Plans (RP) and CDPs. The first step taken by the State is to identify respective Regions for the formulation of RP. Then, the local body formulates the CDP once the State Government approves the RP. If the city does not have a designated Regional Plan, it can develop its CDP directly. The Union Ministry of Housing and Urban Affairs (MoHUA) has promulgated the Urban and Regional Development Plans Formulation & Implementation (URDPFI) guidelines for formulating the RP and CDP. Later, the local body frames appropriate DCR or Building Bye-laws to facilitate the implementation of the CDP. Like URDPFI guidelines, the Bureau of Indian Standards (BIS) formulates the National Building Code (NBC) to enable the local bodies to draft the DCR. There is constant review and amendments of the State Acts and the DCRs from time to time. For this, the local body updates the CDP after every ten years.

As per the respective State Acts, every individual in any local government authority area must follow the CDP and DCR to construct any structure and obtain prior permissions within the city limits from the local administration. Though all the local bodies and development authorities within the State follow the same State Act, they have the freedom to make a few alterations or additions to these rules as per the city's geographic, social and climatic requirements.

3.1.1 City Development Plan

The CDP, in unity with the URDPFI guidelines, indicates state-specific details regarding land management at the city level. It categorizes different Planning Units within the city ranging from sub-city to cluster or group level. It dictates various norms to implement activities such as Land Zoning and permissible Built-up Area. The CDP also states the maximum FAR or FSI in various zones depending on the transit typology. In addition, the CDP needs to consider the local demographics, climatic social structure, and economic aspects of the city. Thus, the CDP may adopt and add norms that help maintain local identity and care for local people's wellbeing.

3.1.2 Development Control Rules

The DCR has to be formulated based on the provisions and norms mentioned in the CDP. It also takes into consideration NBC. The DCR provides detailed guidelines and rules for built-up development regarding land utilization. These include open spaces, built-up areas, roads, amenities, and other services for various occupancies. It also mentions the detailed utilization of FAR or FSI within the city for those mentioned in the CDP.

3.1.3 Open Spaces definition

The State Acts mention open spaces as a part of amenity, defined as any useful facility for the city's residents. DCR defines open spaces as an area forming an integral part of the plot left open to the sky. Such open spaces do not include roads, which are open to the sky. The DCR further mentions that the primary purpose of these open spaces is to provide breathable spaces in the city that can provide enough light and ventilation into the adjoining built-up spaces.

3.1.4 Rules for Open Spaces

The CDP categorizes the open spaces as,
1) Recreational Space
2) Organized Green Space (OG)
3) Other Open Spaces (such as vacant lands / open spaces including flood plains, forest cover, etc. in plain areas)

The guidelines mention that a CDP should provide a desirable 10-12 sq. m. per person open space, which includes all the above categories for the whole city. However, protected and ecologically sensitive areas in hilly regions are over and above these considerations.

While there is no specific definition for recreation space, the guidelines explain OG further. The URDPFI guidelines list parks, playfields, specified parks, amusement parks, green gyms, maidan, multipurpose open spaces, botanical gardens, zoos, and traffic parks as OG. Table 2 indicates norms for OG in the different zone.

Table 2. Hierarchy and Norms for Organized Greens in Plain Areas [23].

Sr. No.	Planning Unit	OG space		
		Number of OG	Population Served per OG	Min. Area of each OG
1	Housing Cluster	3-4	5000	0.5 ha
2	Neighborhood	3-4	10500	1.0 ha
3	Community	2-3	1 Lakh	5.0 ha
4	District/ Zone	1	5 Lakh	25.0 ha
5	Sub-city center	1	10 Lakh	100.0 ha

The DCR categorizes open spaces as,
1) Mandatory Open Spaces
2) Interior Open Spaces (Chowks)
3) Amenity Open Spaces
4) Layout / Recreational Open Spaces (LOS)

Mandatory Open Spaces are the setbacks from the plot boundary meant to provide adequate light and ventilation to the structure. As per DCR, there should be a provision of an Interior Open Space (courtyard, inner chowk) when the whole of one side of one or more habituated rooms, including the kitchen. These habitable spaces are such that they are not abutting on either the front, rear, or side(s) open spaces. Amenity Open Spaces are open to sky parking spaces and other amenities. At the same time, LOS is a larger POS concerned with space for recreational purposes such as public parks. While the NBC guides provisions for Mandatory Open Spaces, Interior Open Spaces, and Amenity Open Spaces, the LOS rules shall be congruent with the norms mentioned in the CDP. However, the NBC suggests a minimum of 3 sq. m. per person open space in built-up areas. DCR state LOS norms in a certain percentage of the total plot area. These rules state the minimum sizes of these open spaces for various land use activities to avoid misuse of the net area. However, LOS is only applicable to residential areas, and hence, the study will focus on LOS norms in detail.

LOS rules apply for plot areas equal to or more than 0.2 ha (hectares). It assumes the mandatory open spaces are enough for recreational use in smaller plots. However, as larger plot areas would cater to more people, there is a need to provide adequate open space for recreation. Thus, LOS norms are laid specifically for Plot sub-division and Group Housing. The plot sub-division is the division of the reconstruction of existing plots, which may be an amalgamation of smaller plots or a division of larger plots. Group Housing deals with mass housing for more than 25 dwelling units and has a minimum plot size of 3000 sq. m. The norms for LOS are proposed based on plot areas. The percentage of LOS ranges from 10% to 25% of the total plot area. Few DCR also mentions the minimum area of the LOS. They also mention the minimum width to length ratio for the LOS to be optimal for the user and not remain a mere formality on paper.

3.2 Current Scenario of LOS Norms across India

Although DCRs must satisfy the open space norms in CDP, URDPFI, and NBC, many DCRs do not conform. The study of DCRs of 20 cities in India shows that about 85% have rules for LOS concerning Plot Division, whereas only 20% of the DCRs have mentioned LOS rules in Group Housing. Further, only 8 of these DCRs mention detailed norms for LOS in the Plot sub-division, while only 1 DCR mentions detailed norms for LOS in Group Housing. With time, respective authorities have made multiple amendments in the individual State Acts and DCRs but rarely have addressed the norms for open spaces. Few Indian cities tried innovating by providing more personal open spaces in residential areas. For example, Pune and Pimpri Chinchwad DCR had provisions for open-to-sky alternate terraces in apartment buildings. However, this provision was taken back in recent amendments, citing privacy issues and increased housing costs, leading to revenue loss to the builders.

The norms state that designers or builders must calculate the LOS based on the plot area. These LOS help the designer or builder get necessary clearances and sell properties at premium prices. Also, DCR makes provisions for the developer to buy additional FSI to increase the occupancy density with the plot, as permissible in the CDP. Today, owning an individual piece of land or house is becoming scarce due to increased land prices and changing demography of Indian families [24–26]. In many Indian cities, residential townships, a form of group housing, have emerged as the new trend in the housing sector to counter these issues. Nevertheless, the absence or unclear rules related to LOS have led to developers' pooling land for developing such townships [27,28]. This situation highlights a lack of coordination between the RP, CDP, and DCR and their implementation on the ground [29,30].

Although, the CDP and DCR dictate the maximum ground coverage for built-up for a developable plot which ranges from 40% to 75%, with large plots amounting to less ground coverage. However, the rules for LOS remain the same. Also, the restriction of built-up by controlling the ground coverage ratio only restricts the horizontal growth. The developer can still pursue vertical growth by buying additional FSI. Thus, with the additional built-up area, recreational open spaces remain the same, ignoring the aspect of density residing in such developments. Hence, this reduces the per capita open space, and some users may lack access to personal recreational open space. Furthermore, many researchers have stressed the importance of open spaces for disaster management [31,32]. Thus, such a provision may lead to haphazard development and may fail to accommodate many people during emergencies creating panic situations.

The CDPs and DCRs reflect a region's culture and climate; however, the role of open spaces is limited to natural light and ventilation, seldom acknowledging their role in recreation. Cities such as Chandigarh and Gandhinagar, planned post-Independence, have consciously planned POS. Chandigarh is one of the greenest cities in India. However, research indicates that the neighborhood walkability has been reduced over time [33]. There is a need for a more accessible POS in such a case. Also, in cities like Bangalore and Delhi, which perform better in per capita open space than other Indian cities, open spaces declined in the last few years [34–37]. Although the URDPFI guidelines set a target of 10 to 12 sq.m. of open space per person for Indian cities, the open space norms in the DCRs are inadequate to comply with them. Hence, there is a need for robust plot-level management guidelines in Indian cities to achieve these targets.

4. Quantitative and Qualitative Analysis of Plot-level Open Spaces

4.1 Quantitative Analysis

The study compared seven group housing case studies for their density parameters to understand the ground reality. Since the literature review revealed that Indian cities have similar DCRs, selecting case studies from a single city seemed appropriate. Therefore, the study analyses seven group housing schemes in Pune city. As seen in Figure 2, dwelling units differ on each plot irrespective of their plot sizes. For example, Case 2 has a smaller plot size and higher FSI than Case 3 or Case 4, but dwelling units are still lesser than the latter. The population density is comparatively higher on plots with more dwelling units. Therefore, higher FSI may not necessarily imply high density.

The primary analysis shows that no housing schemes achieve the per capita open spaces standards. The per capita recreation open spaces varied from 0.8 sq.m. per person to 3 sq.m. per person, whereas examples with higher population density had lower per capita recreational open spaces (Figure 1). The only exception is Case 1 because it provided the recreational open space on the podium, reducing the space utilization for amenities and increasing the same for the recreational open spaces. Similarly, in Case 2, recreational spaces are provided extensively on the terrace, resulting in higher per capita recreational open spaces. However, the study also highlights that larger recreation open spaces with plot sizes do not necessarily mean high per capita recreational open space. For example, as seen in Case 7, 16% of recreational open space, more than the stated 10% (as per Pune city DCR), is provided. However, the per capita recreational open space amounted to 0.8 sq.m. per person.

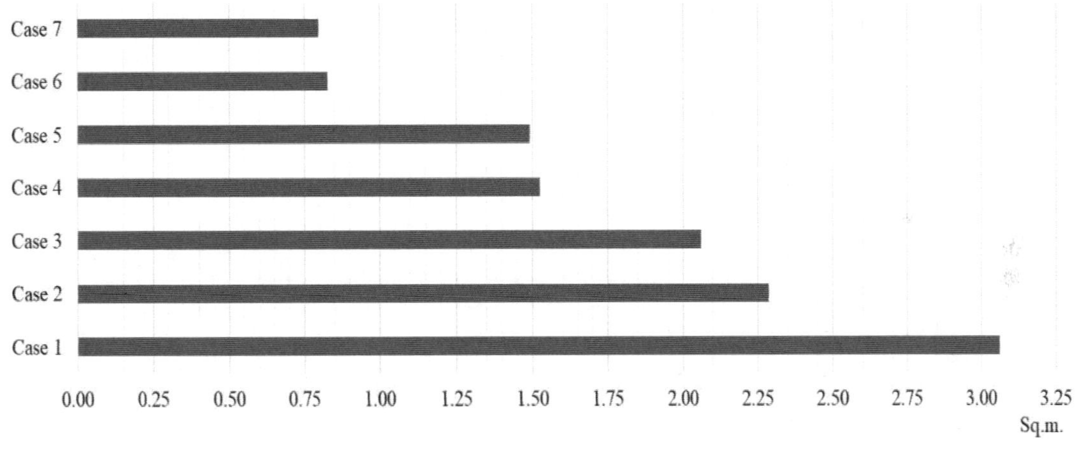

Figure 1. Per capita Recreation Open Space in Studied Cases.

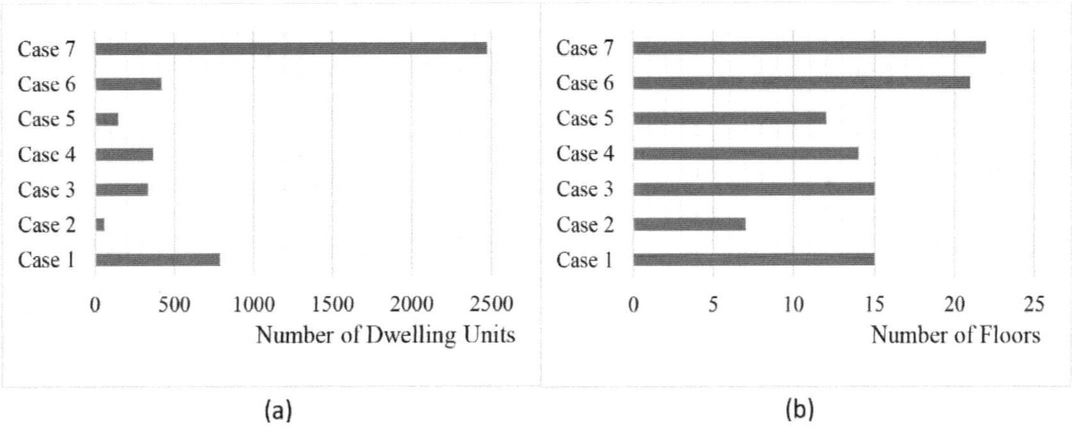

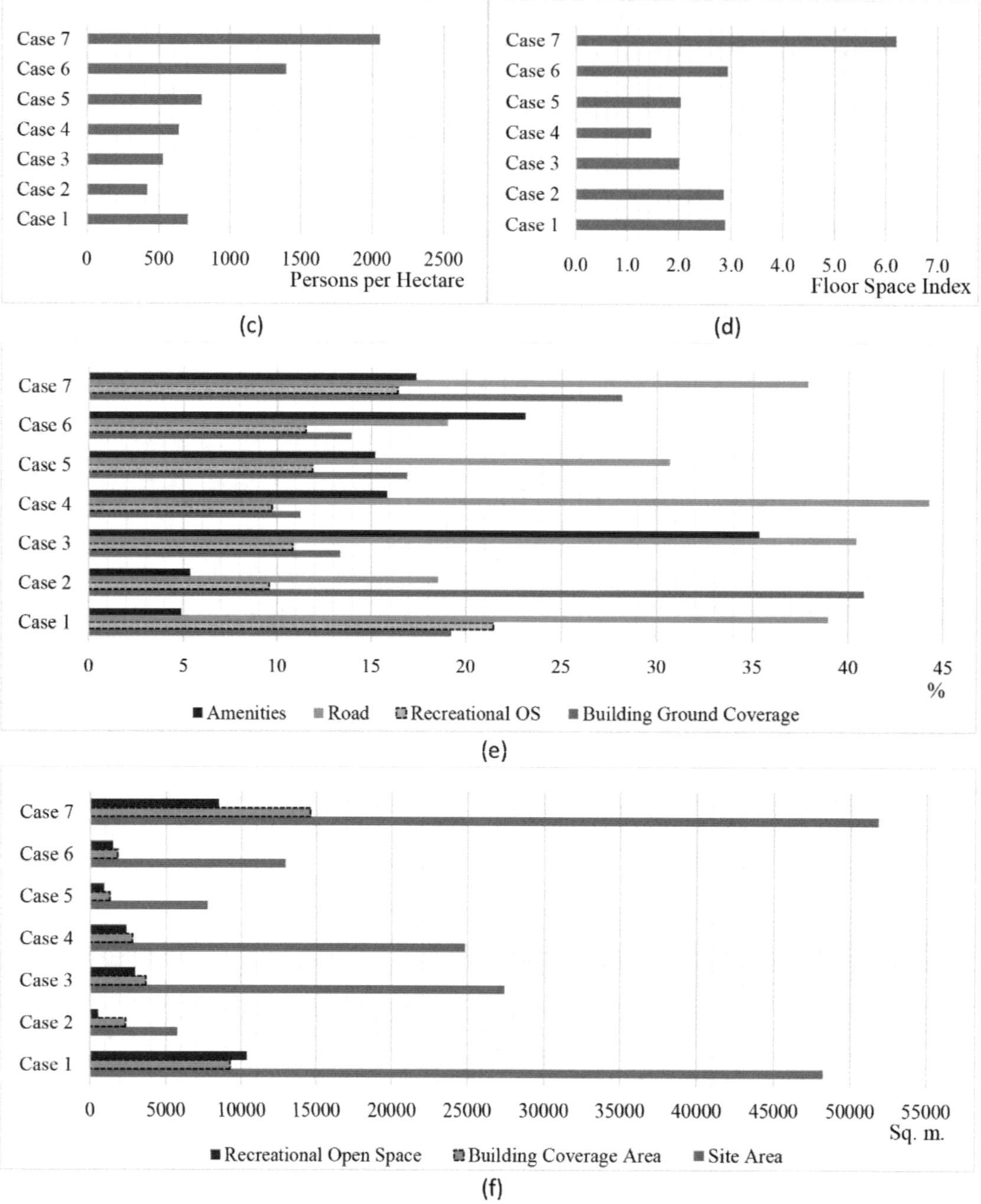

Figure 2. Quantitative Analysis of Identified Group Housing Projects.

Furthermore, the analysis states that amenities and roads do not necessarily affect a housing plot's per capita performance (Figure 2). Even though few housing schemes, such as Case 3 and Case 4, may benefit by reducing the area under roads, it would not change the per capita open space considerably. Therefore, it is imperative to plan the housing layouts efficiently to increase their per capita recreational open spaces.

4.2 Qualitative Analysis

An analysis of the immediate surrounding of each housing scheme shows that the plot level open spaces have some or other potential to contribute to the social, economic, and environmental sustainability of the city (Figure 3 and Figure 4). The housing schemes have a medium to high potential concerning social sustainability. The main reason for high potential in Case 3, Case 4, Case 5, and Case 7 is the unavailability of accessible recreational open spaces in the vicinity. Whereas, for Case 2, though the area has many recreational open spaces, almost all of them are private societies and inaccessible to other people; thus, the potential of plot-level open space increases. In Case 1, there are accessible recreational spaces; however, the hillock near the site, though it plays a role as a recreational open space, is inaccessible to people of all ages.

The housing schemes have a low potential for economic potential, except for Case 3 and Case 6. These exceptions have a large undeveloped residential zone in the vicinity, thus increasing the potential of plot-level open spaces in increasing the area's economic value. In other cases, there is high-density residential development due to which the potential of their plot-level open space to contribute to the economy of the area reduces considerably.

Around each studied case are environmental assets such as hillocks, *nullahs* (natural stormwater drains), dense vegetation areas, and national highways, which have regional impacts. The plot-level open spaces of the housing schemes have a high to very high potential to contribute to the existing network of these assets. For Case 1 and Case 2, which have natural assets in their respective vicinities, the potential of plot-level open spaces is high. Whereas for Case 4, Case 5, and Case 7, which have high-density built environments and natural assets around each site, the potential of plot-level open spaces is very high.

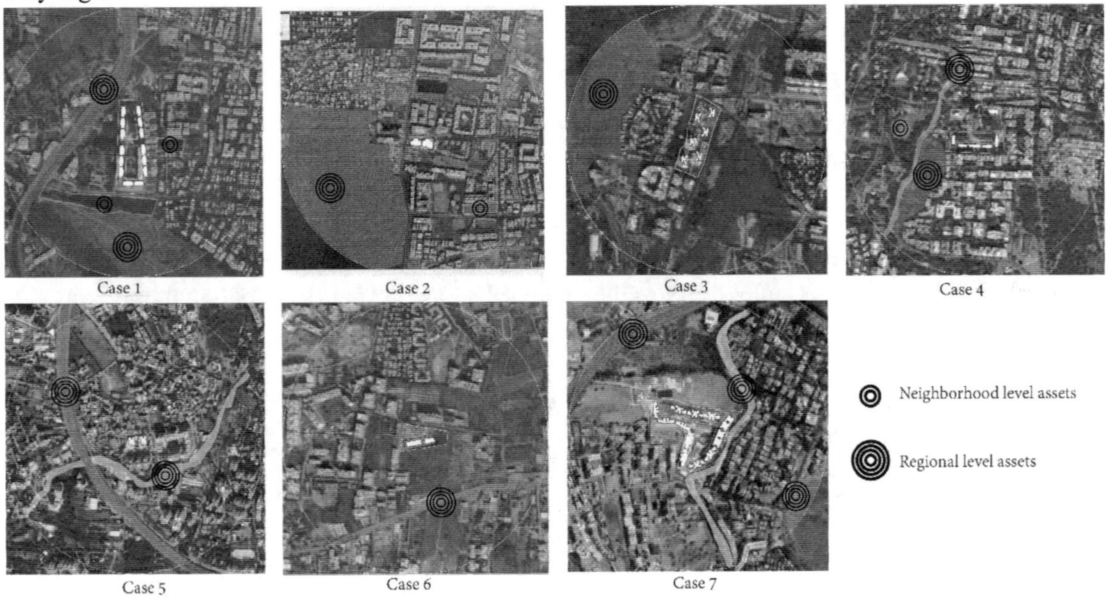

Figure 3. Qualitative Analysis of Identified Group Housing Projects in Pune [38].

Though these housing schemes have a good potential to contribute to the city's sustainability, none of the plot-level open spaces in these housing schemes respond to their respective sensitivities found in their immediate surroundings. A brief plot and open space shape analysis show the importance of building profiles in planning plot-level open spaces. For example, the open spaces provided in Case 5, Case 6, and Case 3 are highly affected by the site shape and building profiles, impacting the social and economic vertices. Thus, it also shows how designers and developers prefer built-form shapes over quality open spaces.

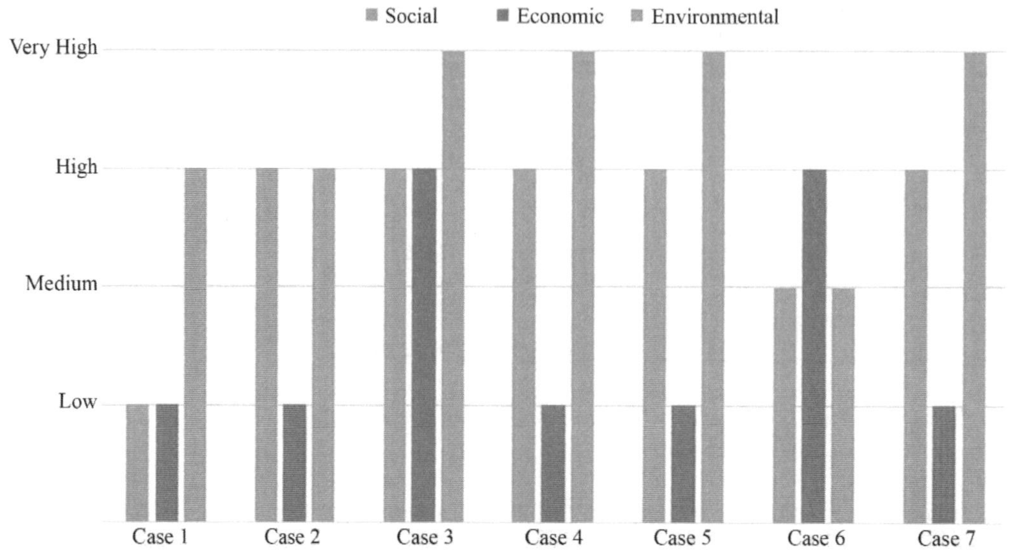

Figure 4. Impact of Plot-level open spaces on Social, Economic, and Environmental vertices of the city.

4.3 Discussion

The CDPs state the FSI, while the DCRs state the building ground coverage for specific plot sizes. Earlier research shows that more FSI leads to higher density scenarios, reducing per capita open space and compromising quality [39–43]. While planning at the site level, the study shows that density parameters significantly affect the per capita open space. For any built environment, density parameters include building ground coverage, FSI, number of floors, and number of dwelling units. As each of these parameters is interrelated and contributes to quantitative and qualitative aspects of recreational open spaces, the research advocates building design to consider them warily. Few cases in the current study demonstrate that more dwelling units decrease per capita open space values. Though FSI does impact the built-up area, the sizes of individual dwelling units are at the developer's discretion, resulting in a varied number of houses on similar plot sizes.

The DCR of Pune specifies 10% recreational open space for plot sizes of 2000 sq.m. and above. In most cases, the designers or developers prefer to centrally place this 10 % open space, thus restricting its potential. The DCR also sets building ground coverages for different plot sizes. Lower ground coverage leads to a better scope of open space development. The building's ground coverage impacts the building profiles, affecting the interior and corresponding open spaces. Apart from building ground coverage, the shape of the plot highly influences the building profile. Irregular plot profiles shall lead the designer to prioritize building profiles and layouts over the open spaces, which accommodate the plot's residual spaces, as seen in a couple of scenarios in this study. Therefore, it is at the designer's or the developer's disposal to design the building and define the relationship between its ground coverage and its profile.

Decision-makers must understand that plot-level alterations always impact the overall environment. Similarly, they need to consider the dense natural vegetation, *nullahs*, streams, and lakes within or around the site while designing plot-level open spaces. In addition, while managing open spaces at the plot level, decision-makers must consider the accessibility to existing open spaces and the visual built density around them, as these have social and economic implications. Furthermore, since plot-level open spaces can compensate for the environmental losses due to the new development, planning authorities need to look at hills, forests, and farmlands as a larger extensive network of the plot-level open spaces in proposed norms or strategies. The study reiterates the earlier research findings that open spaces at the plot level need to respond to the various man-made and natural assets in the immediate vicinity of the plot [2,19,44].

5. Conclusions, Future Research Scope, and Limitations

For Indian cities, as far as the open spaces are concerned, the study shows a stark disconnect between rules and regulations formulated by various statutory bodies. The provisions made in the DCRs are inadequate to achieve the targets mentioned in CDP, NBC, and URDPFI guidelines. Therefore, we need to look into plot-level dynamics for open spaces management. Each community will need to design a system of open spaces at the plot level to incorporate its various requirements. However, as this system never works in isolation, it has to structure from plot level to the neighborhood to the regional scales. Hence, there is a need to formulate guidelines for open spaces that remain interconnected between all scales and a bottom-up approach to open space management in Indian city planning. As per the study of the quantitative aspects, the per capita open spaces hint at deriving each plot's per capita built-up area. Also, it initiates the discussion of the importance of the contribution of dwelling unit sizes to the plot's population density—also the significance of the number of dwelling units over density parameters. The qualitative aspects of plot-level open spaces define plot-level open spaces' role in immediate surroundings. The study substantiates plot-level open spaces' importance in contributing to the city's social, economic, and environmental sustainability.

The study does not consider the surrounding spatial structure, built unbuilt mass, the scheme's location, and orientation in its larger context. It may directly guide the open space qualities at the plot level and thus need further study. Many other parameters may be listed similarly to find a relation between open spaces at plot level as against its context. Besides these qualitative aspects, the study acknowledges an interrelationship between quantitative aspects. Amidst new methods and technologies of incorporating recreational spaces in the plot, there is a need to understand the response of ground-level and podium-level open spaces to the overall sustainability. Also, can terraces and balconies contribute to this response in vertical development? Further research can help establish a matrix for sustainable development goals by identifying these links between various features of quantitative and qualitative understandings of plot-level open spaces to ensure better practices. Since the study identifies only a few specificities of plot-level open spaces, researchers may develop a more extensive and exhaustive data set for many cities to help formulate a robust open space management plan. Such a database will facilitate a bottom-up approach to achieving the city's sustainability goals.

References

[1] Andersson E, Borgström S and Mcphearson T 2017 Nature-Based Solutions to Climate Change Adaptation in Urban Areas 51–64

[2] Norton B A, Coutts A M, Livesley S J, Harris R J, Hunter A M and Williams N S G 2015 Planning for cooler cities: A framework to prioritise green infrastructure to mitigate high temperatures in urban landscapes *Landsc. Urban Plan.* **134** 127–38

[3] Coutts A M, Tapper N J, Beringer J, Loughnan M and Demuzere M 2013 Watering our cities: The capacity for Water Sensitive Urban Design to support urban cooling and improve human thermal comfort in the Australian context *Prog. Phys. Geogr.* **37** 2–28

[4] Shashua-Bar L, Pearlmutter D and Erell E 2011 The influence of trees and grass on outdoor thermal comfort in a hot-arid environment *Int. J. Climatol.* **31** 1498–506

[5] Rakhshandehroo M and Mohd Yusof M J 2014 Establishing new urban green spaces classification for Malaysian cities *Res. Gate* 1–14

[6] Stanley B, Stark B, Johnston K and Smith M 2012 Urban open spaces in historical perspective: A transdisciplinary typology and analysis *Urban Geogr.* **33** 1089–117

[7] Vyse H 2015 Ideal Homes in Ancient Egypt : the Archaeology of Social Aspiration *Oper. Carried*

Pyramids Gizeh 1837 237–41
[8] Vahia M N and Yadav N 2011 Reconstructing the History of Biotas *Soc. Evol. Hist.* **10** 421–34
[9] Marcus C C and Sarkissian W 1988 *Housing as if People Mattered: Guidelines for Medium-Density Family Housing*
[10] Macfadyen D 1970 *Sir Ebenezer Howard and the Town Planning Movement* (Manchester University Press)
[11] Maryanti M R, Khadijah H, Uzair A M and Ghazali M A R M M 2016 The urban green space provision using the standards approach: issues and challenges of its implementation in Malaysia *Sustain. Dev. Plan. VIII* **1** 369–79
[12] Hindley J 2007 A park for the 21st century: Observations on the transformation of mile end park *Capital. Nature, Social.* **18** 104–24
[13] Wolch J, Wilson J P and Fehrenbach J 2005 Parks and park funding in los angeles: An equity-mapping analysis *Urban Geogr.* **26** 4–35
[14] Byrne J and Sipe N 2010 *Green and open space planning for urban consolidation--A review of the literature and best practice* vol 11
[15] Næss P 2005 Residential location affects travel behavior - But how and why? The case of Copenhagen metropolitan area *Prog. Plann.* **63** 167–257
[16] Roy J, Tschakert P, Waisman H, Abdul Halim S, Antwi-Agyei P, Dasgupta P, Hayward B, Kanninen M, Liverman D, Okereke C, Pinho P, Riahi K and Suarez A 2018 Sustainable Development, Poverty Eradication and Reducing Inequalities *Global Warming of 1.5 °C an IPCC SPECIAL REPORT on the impacts of global warming of 1.5 °C above pre-industrial levels and related global greenhouse gas emission pathways, in the context of strengthening the global response to the threat of climate change* pp 445–538
[17] Govindarajulu D 2014 Urban green space planning for climate adaptation in Indian cities *Urban Clim.* **10** 35–41
[18] Ministry of Urban Development 2015 *Smart Cities Mission Statement and Guidelines* (India)
[19] Gulati R 2019 Neighborhood spaces in residential environments: Lessons for contemporary Indian context *Front. Archit. Res.* **9** 20–33
[20] Faragallah R N 2018 The impact of productive open spaces on urban sustainability: The case of El Mansheya Square – Alexandria *Alexandria Eng. J.* **57** 3969–76
[21] Yang J, Zhang F and Shi B 2019 Analysis of Open Space Types in Urban Centers Based on Functional Features *E3S Web Conf.* **79**
[22] Ibrahim F I, Omar D and Nik Mohamad N H 2019 Human Interaction In Urban Open Spaces *Environ. Proc. J.* **4** 188
[23] Town and Country Planning Organisation 2015 *Urban and Regional Development Plans Formulation and Implementation (URDPFI) Guidelines - Volume 1* (Ministry of Urban Development)
[24] Mahalik M K and Mallick H 2016 Are house prices guided by fundamentals or speculative factors? An empirical inquiry for India *Int. J. Econ. Policy Emerg. Econ.* **9** 47–64
[25] Nayak D K and Behera R N 2014 Changing household size in India: An inter-state comparison *Trans. Inst. Indian Geogr.* **36** 1–18
[26] Bandyopadhyay A and Saha A 2011 Distinctive demand and risk characteristics of residential housing loan market in India *J. Econ. Stud.* **38** 703–24
[27] Ram P and Needham B 2016 The provision of affordable housing in India: Are commercial developers interested? *Habitat Int.* **55** 100–8
[28] Ahluwalia I and Mohanty P K 2014 Unlocking Land Value for Financing Urban Development in India *Indian Counc. Res. Int. Econ. Relations* 1–13

[29] Kaushik I 2013 Challenges and Issues in Mega City Planning in India *Int. J. Sci. Eng. Res.* **1** 128–32
[30] Turok I 2014 *The Evolution of National Urban Policies - A Global Review* (UN-Habitat)
[31] Greeshma P and Kumar K M 2016 Disaster Resilience in Vulnerable cities through Neighbourhood Development: A case of Chennai *Procedia Technol.* **24** 1827–34
[32] Jayakody R R J C, Amarathunga D and Haigh R 2018 Integration of disaster management strategies with planning and designing public open spaces *Procedia Eng.* **212** 954–61
[33] Jigyasu N 2014 Re-Planning ‘Planned’ Public Spaces: The Neigbourhood Market of Sector 15 in Chandigarh *Creat. Sp.* **2** 105–25
[34] Ramachandra T V, Bharath H A, Gouri K and Vinay S 2017 Green Spaces in Bengaluru: Quantification through Geospatial Techniques *Indian For.* **143** 307–20
[35] Shahfahad, Kumari B, Tayyab M, Hang H T, Khan M F and Rahman A 2019 Assessment of public open spaces (POS) and landscape quality based on per capita POS index in Delhi, India *SN Appl. Sci.* **1** 1–13
[36] Ramaiah M and Avtar R 2019 Urban Green Spaces and Their Need in Cities of Rapidly Urbanizing India: A Review *Urban Sci.* **3** 94
[37] Choubey A N 2017 Urban green in Delhi : A temporal analysis (1995-2016) *Int. J. Acad. Res. Dev.* **2** 427–31
[38] Google Earth Pro (7.3.4.8248) 2021 Pune City Region, India, 18°31′13″N 73°51′24″E
[39] Dutta S, Bardhan S, Bhaduri S and Koduru S 2020 Understanding the relationship between density and neighbourhood environmental quality - A framework for assessing Indian cities *Int. J. Sustain. Dev. Plan.* **15** 1067–79
[40] Dutta S and Bardhan S 2017 Density and Neighbourhood Environmental Quality – A Comparative Study in the context of Indian Cities *Int. J. Emerg. Technol.* **8** 315–23
[41] Russo A and Cirella G T 2018 Modern compact cities: How much greenery do we need? *Int. J. Environ. Res. Public Health* **15**
[42] Haarhoff E, Beattie L and Dupuis A 2016 Does higher density housing enhance liveability? Case studies of housing intensification in Auckland *Cogent Soc. Sci.* **2** 1–16
[43] Dave S 2011 Neighbourhood density and social sustainability in cities of developing countries *Sustain. Dev.* **19** 189–205
[44] Aldous D E 2011 Planning Green Open Spaces for South East Asian Capital Cities *Citygreen* **01** 10

Nanotechnology and building construction: Towards effective stakeholder engagement

Swetha Priya Darshini Thammadi and Sateesh Kumar Pisini

Department of Civil Engineering, Fiji National University

Email: swetha.thammadi@fnu.ac.fj

Abstract. Innovative construction materials and techniques are imminent to survive the climate change associated extreme weather impacts. Nanotechnology in construction sector improves the comfortability and quality of life by enhancing the mechanical properties of cement and concrete, stronger and lighter structural composites, low maintenance coating, reducing the thermal transfer rate and/or insulation, and construction-related nano-sensors. Various types of nanoparticles used in construction sector are Nano-titania, Carbon nanotubes, Nano-silica, Copper and Clay nanoparticles, Zycosoil, and nanostructured metals such as Nano-ferric oxide, and Nano-aluminium oxide. Advantages of nanomaterials application in building construction are displayed in materials such as concrete, steel, timber, glass, insulations, coatings, energy and nanosensors. The objective of this study is to analyze the nanomaterials application in construction industry besides environmental, health and social impacts. The novelty of this study includes stakeholder engagement matrix for nanomaterials in construction sector. Although nanotechnology is in its infancy, there is a need to develop a framework for nanotechnology regulation especially in construction sector due to its impact on climate change and vice versa due to the significant contribution of construction sector to greenhouse gas emissions. Therefore, a nanotechnology regulation framework has been proposed for identification and effective stakeholder engagement. Implications of nanotechnology on construction materials, narrowing the nano-divide, scope for sustainable development and concepts of implementation have been discussed in detail.

1. Introduction

Increasing need for sustainable development in engineering endeavors over the previous decades has seen advancement in construction materials from both the chemical and physical point of view. Application of modern technology in construction industry has hypothetically stemmed by the increased demand for new, high quality and efficient materials along with a need to reduce the consumption of raw material and energy. One of the greatest modern technologies of the current century is "nanotechnology", which can be useful in addressing the needs of construction industry. Nanotechnology deals with organization and control of Nano-scale matter (typically with size ranging between 0.1 and 100 nm), which therefore may involve engineered manipulation at the atomic level [1]. Such an endeavor requires understanding materials at atomic scale [2]. Typically, there are two types of nanoparticles (NPs): (1) naturally occurring NPs (e.g., produced naturally in forest fires, volcanoes or as combustion by-products) and (2) engineered nanoparticles (ENPs) generally developed for specific application e.g., titanium dioxide – TiO_2, silica – SiO_2, iron oxide – Fe_2O_3, carbon nanotubes – CNTs, etc. The scope of nanotechnology in construction applications till date is to enhance the mechanical properties of cement and concrete, stronger and lighter structural composites, low maintenance coating, reducing the thermal transfer rate and/or insulation, and construction-related Nano-sensors [3]. The basic construction materials such as cement, concrete and steel will benefit from nanotechnology. The assimilation of nanoparticles will lead to stronger, more robust, self-healing, air purifying, fire resistant, easy to clean and quick compacting concrete. Nanotechnology has led to the creation of fire-resistant materials such as steel, long-lasting concrete, and other useful materials. The first bridge in Germany, the Gärtner Platzbrücke Berdige, was completed in 2007, and the Jubilee Church in Rome, which was completed in 2003, are both excellent examples of the use of nanotechnology and nanomaterials in engineering applications. It should be noted, however, that investment on nanotechnology research is vital, particularly for construction materials, in making new buildings more sophisticated and modern [4, 5].

Nanomaterials used in construction include Nano-titania (Nano-TiO_2), CNTs, Nano-silica (Nano-SiO_2), nanostructured metals such Nano-ferric oxide (Nano-Fe_2O_3), and carbon nanofibers (CNFs). Nanotechnology can elevate the potential of building materials by making it durable, strong, light weight, secure, energy efficient, ease of maintenance, cheaper and more sustainable. Limited use of nanomaterials is due to the lack of engineering norms and guidelines for the design and manufacturing of these materials [6,7]. This study aims to discuss in detail, the frequently used nanomaterials in construction industry and their application (Table 1) in first section. Potential health, environment and social impacts of nanomaterials in construction industry are discussed in the second section. Third section ascertains relevant regulations/guidelines for nanomaterials worldwide. The fourth section presents a comprehensive framework (Figure 1) considering the knowledge gaps in research and lack of Nano-product database in construction sector. Further, qualitative analysis performed using stakeholder engagement grid (Figure 2) demonstrates the interest and corresponding power of stakeholder to enact a regulation. Finally, potential for nanotechnology in developing, low developed and Small Island Developing States (SIDS), climate change impacts and Sustainable Development Goals (SDGs) are detailed in the last section.

Table 1. Application of nanomaterials in construction industry

Nanoparticle	Construction material	Applications
CNTs	Concrete	Structural health monitoring (MEMS/NEMS), prevents cracks, increased durability
	Ceramics	Improves mechanical and thermal properties
TiO_2	Concrete	Rapid hydration, self-cleaning and increased durability
	Glass	Super hydrophilicity, fouling-resistance and anti-fogging
	Coating	Self-cleaning, hydrophobic properties
SiO_2	Concrete	Consolidation in mechanical strength
	Ceramics	Coolant, fire resistance and light transmission
	Glass	Flame-proofing, self-cleansing and anti-reflection
Fe_2O_3	Concrete	Increased compressive strength and abrasion-resistance in concrete
Cu	Steel	Decreases surface unevenness, improves weldability, formability and corrosion resistance
Al_2O_3	Asphalt concrete timber	Increase in serviceability
	Concrete	Elasticity modulus increases [8]
	Bricks mortar	Enhancing compressive strength and resistance to abrasion
Clay nanoparticles	Concrete	Increases resistance to chloride penetration, improvement in self-compacting properties and mechanical performance, decrease in permeability and shrinkage [9, 10, 11]
Zycosoil	Asphalt concrete	Higher compaction and increase in fatigue life

1.1 Nanotechnology and Concrete

Concrete is one of the fundamental materials in construction industry. Inclusion of Nano-SiO_2 in concrete leads to densification at the Nano- and micro- scale, thus resulting in improved mechanical properties. Nano-SiO_2 in dispersion/slurry form improves self-compacting concrete's segregation resistance [12]. Improved strength of pre-existing concrete structural elements is observed with fiber wrapping [13]. Cracking, which is the most common problem experienced by the concrete structural elements, is typically treated with a fiber sheet (matrix) comprising of nano-SiO_2 particles and hardeners which breach and close surface cracks and strengthen the concrete thereby. Nano-modified concrete decreases the construction duration and labor, as well as relatively more economical – lower maintenance and repair cost.

Inclusion of CNT (1% by weight) has demonstrated increase in the compressive and flexural strength of concrete [14]. Also, oxidized Multi-Walled Nanotubes (MWNT's) reinforced concrete demonstrates enhanced compressive and flexural strengths relative to the unreinforced control samples [13].

1.2 Nanotechnology and Steel

Steel, one of the prime constituents in reinforced concrete and a key construction material, is likely to benefit from the advancements in nanotechnology. Research demonstrates that the assimilation of copper nanoparticles reduces the surface unevenness of steel, thus limiting the number of stress risers and hence cracking [15]. Moreover, research into the improvement of cementite phase of steel to nano-size has generated stronger cables. High strength steel cables are often applied in bridge construction and in pre-cast concrete tensioning. This improved cable material is likely to decrease the construction costs and duration, specifically in suspension bridges as the cables are run between span ends. Nanoparticles are dropping the effects of hydrogen embrittlement and enhancing the steel microstructure through the reduction of the effects of inter-granular cementite phase [16].

CNT composites are thought to be six times stronger than carbon fiber composites currently in the market [17]. Nanotubes are also far more flexible than carbon fiber, which breaks rapidly under pressure. Nanotubes can be compressed without breaking. When compared to traditional steel reinforced constructions, the reinforced CNT composite structure offers an estimated 50 - 150 fold improvement in tensile strength [13].

1.3 Nanotechnology and Timber

Timber, is another conventionally used sustainable building material, which is renewable, requires less energy to manufacture and stores CO_2, and has many advantages with commendable performance. It is durable, has appropriate mechanical and thermal properties, fire resistant and economically viable. Due to its vulnerability to biological attacks, surface coating and other protective treatments are required to improve the durability, surface characteristics and protection of the inner material. A timber coating alternative developed using Alumina and nano-SiO_2 amalgamated with hydrophobic polymers is applied on the surface of timber. This coating forms and invisible layer on the surface and also repels dirt, algae, oil, dust and water [18]. The surface quality is retained as the coating provides effective protection from UV light [19].

1.4 Nanotechnology and glass

Another application of nanotechnology is in the fire protective glass technology. An intumescent layer between nano-SiO_2 glass panels becomes and hard when heated [14]. On the other hand, nano-TiO_2 covers glazing due to its sterilizing and antifouling properties along with its capabilities to destroy organic pollutants, volatile organic compounds and bacterial membranes. NanoTiO_2 coating is applied to the exterior of a building to prevent the adhesion of contaminants, thereby reducing maintenance costs [20].

Nanotechnology has been evaluated to enhance the properties of glazing, particularly the heat gain and loss, by thin film coatings, thermo-chromic, photo-chromic, and electro-chromic technologies. Thin film coatings are spectrally sensitive surface applications for window glass which filter out infrared light to lessen heat gain in buildings [14]. Thermo-chromic glass, for example, adapts transparency in response to temperature heating applications, while photo-chromic glass varies its transparency in response to light intensity. Finally, the most promising adaptable glass technology is the electro-chromic glass. It is made to respond to variations in applied voltage with the use of a tungsten oxide layer, becoming more transparent at the touch of a button. All these applications are envisioned to cut energy use in cooling buildings and possibly assist in decreasing energy consumption in buildings [14].

1.5 Nanotechnology and insulations

The efficient use of energy is a significant aspect of sustainable development in buildings and construction industry. This can be accomplished by inert means through efficient insulation methods and materials. As nanotechnology progresses, insulation materials and applications will become more effective and less reliant on non-renewable resources. Production companies have evaluated that nanotechnology derived insulating materials are approximately 30% more efficient than conventional materials. Analysis demonstrates the possible application of nanoparticles to insulation materials: To develop a layer bound to the base material; chemical vapor deposition, meniscus, dip, plasma coating and spray can all be used to apply nanomaterials or nanoparticles to substrates. Because these materials can store still air, their surface-to-volume ratio improves. These nanoscale insulating materials can be sprayed as coatings or employed as thin films inserted between stiff panels [21].

A unique insulation material created after the advent of nanotechnology is Aerogel. It is an extremely low density solid, essentially a gel in which the liquid constituent has been substituted with gas. Dubbed “frozen smoke”, aerogel has a composition of just 5% solid and 95% air and is considered the lightest solid in the world and can carry more than 2000 times its own weight. Low heat conductivity, low sound velocity, and excellent transparency are among the physical attributes of aerogel. Because they are moisture sensitive, they are usually credited as being excellent for sandwiching between moisture-resistant wall panels. Skylights, windows, and translucent wall panels are some of the architectural applications of aerogel [22].

1.6 Nanotechnology and coatings

Coatings with nanoparticles or nano-layers have been devised for a variety of applications, including self-cleaning, thermal control, energy saving, anti-reflection coatings for glass/windows; easy-to-clean, antibacterial coatings for work surfaces, protective or anti-corrosion coatings for components, and more hard-wearing paints and anti-graffiti coating for buildings and structures. Research demonstrates that bimetal Nanoparticles such as Fe/Ag, Fe/Pd or Zn/Pd, may be useful as effective reducing agents and catalysts for a variety of environmental pollutants [23]. Nanostructured coatings can be utilized to reflect and transmit specific wavelength bands of light [24].

1.7 Nanotechnology and energy

The upsides of nanotechnology can be directed towards the reduction of energy consumption – because an estimated 50% of the total energy generated worldwide is consumed by buildings [25]. An efficient solution for this is insulation although the renovation of existing buildings for this is a tricky issue due to the limited space for manipulation and installation. However, an innovative energy concept offers a novel ultra-thin wall insulation comprising of a hydrophobic aerogel structure. Since it repels water, it is essentially mold free. Another solution is the use of nanoparticles in glazing technology, which opens the possibility of super-insulating windows [26]. Thus, buildings energy loss through walls and windows can be substantially minimized and the performance of energy storage and solar gains can be amplified, resulting in a reduction of energy consumption.

1.8 Nano sensors

Sensors / devices supported by nanotechnology has enormous potential for building intelligent materials and structures with the ability to "self-sensing" and "self-actuating." Sensors for nano and microelectromechanical systems (NEMS and MEMS) have been developed and used in designs to examine the environmental conditions (temperature, humidity, smoke, noise, etc.) and material / structural performance (stress, corrosion, cracks, strain, vibration, etc.) during the life span of the structure. Nanosensors ranging from 10^{-9}m to 10^{-5}m can be embedded in structures during construction [24].

2. Impacts of Nanomaterials

2.1 Health impacts

Nanotechnology from a general perspective has likely amplified the risk of exposure to nanoparticles through inhalation, ingestion and dermal using ENPs. The advantages of nanotechnology are no doubt enormous. However, an understanding of the toxic effects of nanomaterials in construction sector on human health helps decision making bodies, regulatory agencies and end-users.

TiO_2 nanoparticles commonly used in construction industry has toxic effects such as growth inhibition, photosynthetic activity is suppressed, acute lethality, lung inflammation, reactive oxygen species (ROS) can cause oxidative damage and it is bactericidal for gram-positive bacteria [27-30]. SiO_2 nanoparticles frequently used in construction applications has ROS toxicity, exposure to SiO_2 in crystalline form (amorphous SiO2 is less toxic) in workplaces – may result in silicosis in workers and lead to tuberculosis and lung cancer [31]. It can also cause DNA interaction, damage and changes in gene expression [32], membrane damage due to cell death [33] and also induce oxidative stress [34-36]. CNTs with inhalable size particles can cause pulmonary toxicity [37, 38], and exposure to CNTs during pregnancy may induce developmental toxicity in fetus [39]. CNTs display antibacterial properties and cell membrane damage [40, 41]. Cu and CuO nanoparticles cause DNA damage [28, 42].

2.2 Environmental impacts

Exposure to nanomaterials in construction industry is associated with activities such as drilling, cutting, machining and sanding resulting in risk of inhalation [43]. Monitoring air quality is essential as breathing nanoparticles poses serious health concerns generated during molding, production and coating nanomaterials in factory.

Nanoparticles are released either deliberately or involuntarily to the environment, when emissions are released to atmosphere, through the waste generated, or through polluted land or water remediation. Because nanoparticle supplemented goods contribute significantly to the nanoparticles deposited in the environment, it is possible to hypothesize that as the use of nanomaterials increases, the same can be implied for environmental nanoparticles [44]. In this scenario, the ENPs are the primary focus of risk assessments, but secondary nanoparticles must not be overlooked.

2.3 Social impacts

Nanotechnology offers novel technological solutions in various aspects of construction sector, however, it could possibly eliminate various trades due to mechanized technology in manufacturing and service sector jobs. Thus, loss of employment in primary and secondary sectors results in social inequalities.

3. Regulations/Guidelines on Nanomaterials

USEPA defines nanomaterials as a subset of ultrafine (1 and 100 nanometers) particles but with structural constituents lesser than 100 nanometers [45]. Nanoscale materials production and use regulated by Toxic Substances Control Act (TSCA) ensures that they are not harmful to human health and the environment [46]. On the other hand, Safe Work Australia, the Australian authority on workplace safety, has administrated ENPs under the same laws that regulate technologies, materials, substances and chemicals. This mandates that nanomaterials related risks are to be tackled by elimination, minimization, or communication. However, this means that managers have the responsibility of ensuring worker safety around nanomaterials, who lack knowledge of the possible risks of nanomaterials with rising trend in nanomaterial applications. IRSST (Institut de recherche Robert-Sauvé en santé et en sécurité du travail), Canada; Cordis, Europe; ABDi (Brazilian Agency for Industrial Development), Brazil; AIST (National Institute of Advanced Industrial Science and Technology), Japan;

ANSES (Agence nationale de sécurité sanitaire de l'alimentation), France are few examples of various organizations in major economies providing guidelines/standards for safe handling of nanomaterials and the risks associated with nanoparticles. In China, no guidelines on regulation of nanotechnology are present; but there are government funded organizations that perform research on the influence of nanotechnology on environmental, health, and safety. For instance, Nanomaterials toxic effects on cells are investigated by Chinese Academy of Sciences (CAS) and the National Centre for Nanoscience and Technology (NCNST) [47]. The toxicology studies on medicinal applications of proposed nanomaterials are led by the CAS, Beijing University, and other institutions [48].

The Working Party for Manufactured Nanomaterials (WPMN) was established by the Organization for Economic Cooperation and Development (OECD) in 2006 to evaluate the progression of nanomaterials in commercial applications. Its goal is to ensure that hazard, exposure, and risk assessments for manufactured nanomaterials are of the highest possible quality, scientifically sound, and consistent around the world [49]. WPMN, an OECD subsidiary, is an important regulatory agency for nanotechnology. The actions of the OECD National Coordinators of the Test Guidelines Programme complement WPMN efforts [50].

As scientists, citizens, and lawmakers contribute towards development of nanotechnology regulation, it would be beneficial to have a comprehensive framework based on strict policy implementation and governance. Further, an inclusive understanding of stakeholders using a stakeholder engagement grid has been developed which demonstrates a diverse range of organizations with varying power and interest in nanotechnology for case studies of USA (West), India (Asia) and Australia (Pacific) respectively. The current study aggregates and analyzes this information for an enhanced understanding and regulatory decision-making pertaining to nanotechnology development in developing, low developed and SIDS countries.

4. Framework for Stakeholder Engagement

As discussed in Table 1, application of nanomaterials in construction industry enhances the performance of heavy construction materials (cement, concrete, steel, timber) as well as building materials (insulation, glass, paints, coatings). As per CPWR (Centre for Construction Research and Training) in USA, 600 construction products have been alleged to contain nanomaterials. The specific nanoparticles present in each product are difficult to determine due to inadequate labeling of construction materials. Furthermore, the most prevalent marketing technique currently in use is the use of the term "nano" in the product name without any apparent qualities that promote Nano's enabling [43]. Hence, it can be perceived as not easy to obtain relevant information of nanomaterials in construction products due to lack of transparency in published data as expressed by various researchers. Demolition is an apparent example for difficulty in gathering material data regarding the ingredients of building elements [43, 51]. Therefore, a feasible framework has been developed considering the existing regulations in developed countries, discrepancies, identification of gaps in research and lack of nano-product database in construction sector.

The proposed framework (Figure 1) is based on strict policy implementation and governance. Governments need to identify the stakeholders and their respective responsibilities along with strict adherence to regulations. There should be no access to markets for products that are not enrolled in the database. The emissions from production of nanomaterials should be highly controlled and in conformance to all OHS safety standards. Green labelling should be mandatory for usage by public to ensure transparency and accountability. Information provided in the database should be accessible and public participation should be made mandatory in decision making. The novelty in nano-technology related industries requires the personnel involved in R&D or field applications, to have interdisciplinary background. Policies related to nanotechnology requires the collaboration amongst various stakeholders such as government, businesses, R&D agencies and citizens.

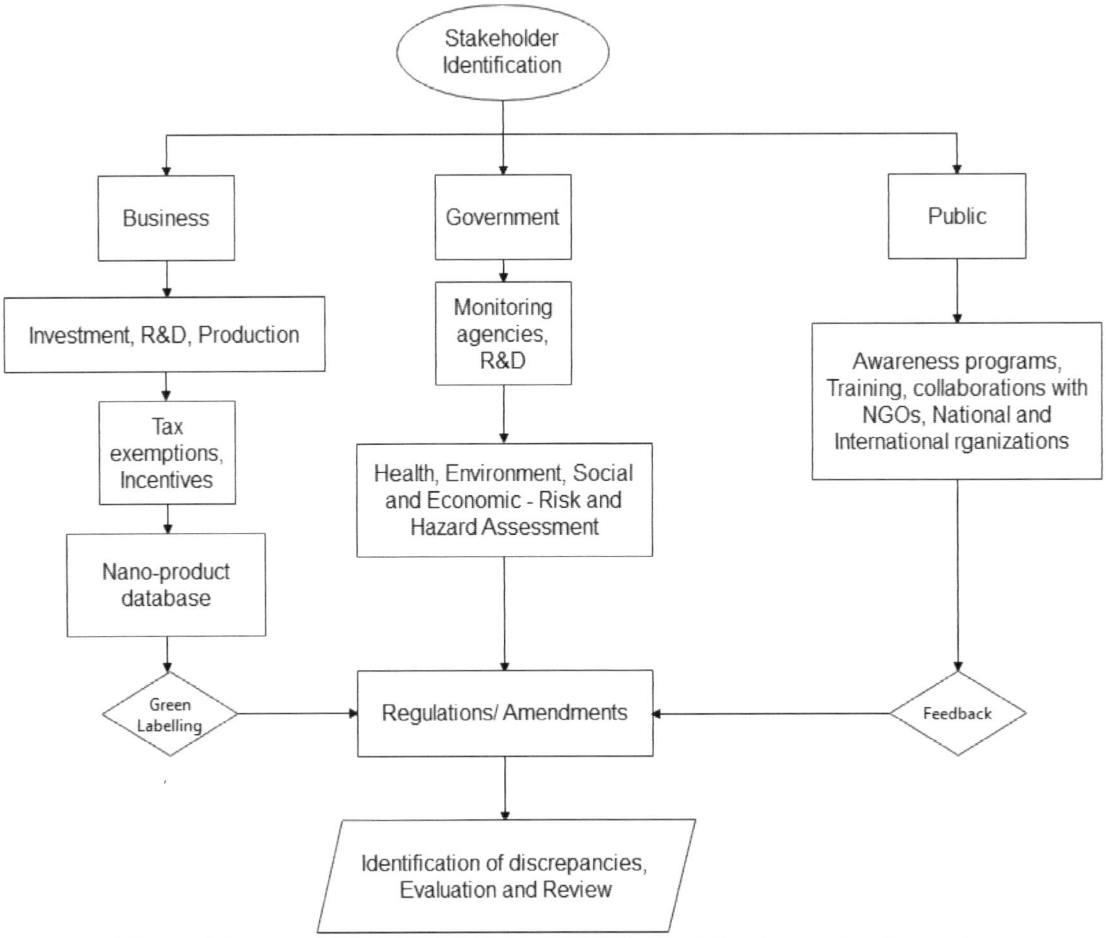

Figure 1. Proposed framework for nanomaterials regulation in construction sector

According to Mendelow's power-interest grid, the stakeholder matrix consists of a broad category of stakeholders (groups or agencies) grouped together based on attributes with interest in drafting nanotechnology regulation and authority to act. The stakeholder matrix is a qualitative analysis to demonstrate the interest and corresponding power to enact a regulation only and doesn't intend to enforce the level of interest or power for nanotechnology regulation.

The first quadrant of high power - high interest consists of those stakeholders with (high) power to formulate, enact and enforce regulations as well demonstrated (high) interest in nanotechnology regulation. The second quadrant of high power - low interest consists of those stakeholders with (high) power to formulate, enact and enforce regulations but lack demonstrated (low) interest in nanotechnology regulation or deferring the regulatory role to another agency. The third quadrant of high interest - low power consists of those stakeholders with (high) interest in nanotechnology regulation but lack (low) power to formulate, enact and enforce regulations. The fourth quadrant of low power - low interest consists of those stakeholders who lack the (low) power formulate, enact and enforce nanotechnology regulations and interest in nanotechnology to help agencies accomplish their missions but (low) interest in nanotechnology regulation. Table 2 demonstrates the Stakeholder engagement grid for case studies of USA (West), India (Asia) and Australia (Pacific) respectively.

Table 2. Stakeholder engagement grid

HIGH POWER, LOW INTEREST	**HIGH POWER, HIGH INTEREST**
USA: Office of S&T policy, Dept of Labour - OHS, US House of Representatives INDIA: MoEF, Centre for OHS, DAE, BIS Committee AUSTRALIA: Dept of Industry, Innovation, Climate change, Science, Research and Education, FSANZ, APVMA, Nanotechnology OHS	USA: US consumer product safety commission, EPA, USFDA, National Institute for OHS INDIA: DST, DSIR, DIT, DBT, MCIT, ICAR AUSTRALIA: Australian Research Council, NICNAS, CSIRO
LOW POWER, LOW INTEREST	**LOW POWER, HIGH INTEREST**
USA: Dept of Defence, Dept of Homeland Security, Dept of Education, Dept of Science, Dept of Justice, Dept of State, Dept of Transportation, Dept of Interior - USGS, Dept of Treasury, US Dept of Agriculture, US International Trade Commission, Nuclear Regulatory commission INDIA: Dept of Agriculture, Dept of Defense, NCL, Dept of Education AUSTRALIA: Dept of Defense, ETP, PACE	USA: National Nanotechnology Initiative, Dept of Energy, US Dept of Commerce, National Science Foundation, National Aeronautics and Space Administration \ INDIA: CTRI, NSAG, CSIR, NATAG, NISTADS, TERI, IITR, SERC, NRDC, NIOT, Gene Campaign, India Nano AUSTRALIA: Dept of Environment, Water, Heritage and Arts, FOE - Australia, NAG

Occupational Health and Safety (OHS), Ministry of Environment and Forests (MoEF), Department of Atomic Energy (DAE), Bureau of Indian Standards (BIS), Food Standards Australian New Zealand (FSANZ), Australian Pesticide and Veterinary Medicines Agency (APVMA), Environmental Protection Agency (EPA), Food and Drug Administration (FDA), Dept of Science and Technology (DST), Department of Scientific and Industrial Research (DSIR), Department of Information Technology (DIT), Department of Biotechnology (DBT), Ministry of Communication and Information Technology (MCIT), Indian Council of Agricultural Research (ICAR), National Industrial Chemical Notification and Assessment Scheme (NICNAS), Commonwealth Scientific and Industrial Research Organization (CSIRO), U.S. Geological Survey (USGS), National Chemical Laboratory (NCL), Enabling Technologies Policy (ETP), Public Awareness and Community Engagement (PACE), Clinical Trial Registry-India (CTRI), Nano Science Advisory Group (NSAG), Council for Scientific and Industrial Research (CSIR), Nano Applications and Technology Advisory Group (NATAG), National Institute of Science, Technology And Development Studies (NISTADS), The Energy and Resources Institute (TERI), Indian Institute of Toxicological Research (IITR), Structural Engineering Research Centre (SERC), National Research Development Corporation (NRDC), National Institute of Ocean Technology (NIOT), Friends of the Earth Australia (FOE), Nanotechnology Advisory Group (NAG)

5. Discussion and Concluding Remarks

Nanotechnology offers a plethora of new materials for different industries and allied sectors. Nanotechnology research in the building industry is still in its infancy. Its presence is significant because civil engineering is primarily associated with building materials such as steel, concrete glass, and timber. Nanotechnology has aided in the development of highly efficient and effective materials with superior technical qualities. The use of nanotechnology in the creation of self-cleaning, self-repairing concretes has substantially lowered the cost of structural maintenance. Nanotechnology has also aided in the improvement of material durability by offering fire resistance and anti-corrosive paints. Despite the fact that nanotechnology is being applied to every aspect of civil engineering, more research and understanding is needed to reap the benefits. Limited studies have been conducted to ascertain the scope of nanomaterials in rigid pavements, marine infrastructure, sustainable architecture, enhancing shear strength characteristics of soft soil and occupational health and safety. Smog-blasting application for pavement application has been developed which is not yet used widely [52]. Undoubtedly, Nanotechnology will facilitate the creation of materials with novel characteristics and high functionality. Nanotechnology-based materials also enhance the life of structures. With the application of nanotechnology, a paradigm shift in the construction sector has already begun, and the technology will completely transform the industry. If nanotechnology is used effectively, it will have a huge impact on technology and industry, as the use of nanomaterials in civil engineering or building construction industries is profound and promising and presents numerous opportunities and challenges.

In countries with advanced technology, such as USA, Japan and EU, the investments in nanotechnology from both private and public sources are growing year by year. Many countries around the world, including some developing countries, have since launched nanotechnology programs with varying degrees of success. Other developing nations at initial stages of nanotechnology regulation can embrace the challenges associated with perseverance and fervor by developing a robust nanoscience/nanotechnology program in their state through appropriate curriculum planning and design, quick and efficient regulation, financial investment, and collaborative partnerships with the private sector and donor agencies/institutions. Long-term economic benefits will undoubtedly boost the country's sustainability and global competitiveness.

There appears to be a need for nanotechnology awareness among construction professionals. To address this issue, comprehensive measures for targeted R&D, technology monitoring, and knowledge transfer in the construction industry are required. Low-income countries and SIDS should evaluate their tertiary education programs expeditiously to give the much-needed attention to nanomaterials testing, fabrication, and characterization using cutting-edge technology; otherwise, they risk promoting the 'nano divide,' from which they will suffer the most as consumers. Climate-vulnerable countries can emphasize public-private partnership models and collaborate for development and implementation of regulatory frameworks by the government agencies, institutions and stakeholders. If our dream of global integration is to become a reality, we must address the challenges of developing countries as soon as possible.

Innovative construction materials and techniques are imminent to survive the climate change associated extreme weather impacts such as coastal erosion, subsidence, flooding and drainage systems. Economically, insurance sector is looking forward for consideration of climate change impacts in premiums. Weather conditions have a compelling impact on construction activities and systems need to be set up to include climate-change associated risk mitigation. Nanotechnology has greatly impacted our lives by providing a possible alternative to conventional materials and thereby reducing greenhouse gases further to mitigate climate change and enhance climate change adaptation as well as resilience.
Nanomaterials technology in building and construction sector is a boon to mankind as it improves the comfortability and quality of life. Nanotechnology has the potential to combat the world's issues in the twenty-first century by offering alternative energy as a reasonable alternative for fossil fuels, hence reducing greenhouse gas emissions. Engineers face a challenge and an opportunity in the 2030 Agenda and UN Sustainable Development Goals (SDG). Sustainable cities and communities (SDG 11), clean energy (SDG 7) and climate action (SDG 13) are all attempts to better manage, design, and develop our structures and infrastructure. Investment in nanotechnology will stimulate economic growth that would otherwise be stifled by the recession. As a result, auxiliary businesses such as marketing for new products, waste recycling and disposal, and intellectual property disputes will be supported.

References

[1] Drexler K 1981 *Proceedings of the National Academy of Sciences* **78** 5275–5278
[2] Goddard W A Brenner D W Lyshevski S E Iafrate G J 2007 *CRC Press*
[3] Vigneshkumar 2014 *Discovery* **23** 75 8-12
[4] Bhuvaneshwari B Sasmal S Iyer N R 2011 Proc *WSEAS Int Conf EMESEG'11* pp. 230-235
[5] Naganathan S Singh C S J Shen Y W Kiat P E Thiruchelvam 2014 S *Adv. Mater. Res.***935**151–154
[6] Mamalis A G 2007 *J Mater Process Technol* **181** 52–58
[7] Rosenholm J M Sahlgren C Linden M *Current Drug Targets* **12** 8 pp. 1166–1186
[8] Li Z Wang H He S Lu S Wang M 2006b *Materials Letters* **60** 356-359
[9] Chang T P Shih J Y Yang K Hsiao T C 2007 *Journal of Materials Science* **42** 7478-7487
[10] He X Shi X 2008 *Journal of the Transportation Research Board* **2070** 13-21
[11] Morsy M S Aglan H A Abd-el-Razek M M 2009 *Construction and Building Materials* **23** 515-521
[12] Lau Kin T David H 2002 *Composites* **33** 4 263-277.
[13] Ganesh V Karthik 2009 *engineeringcivil.com*
[14] Mann S 2006 *Nanoforum Report* Accessed www.nanoforum.org May 30, 2008

[15] Ge Z Gao Z 2008 *Proc. First Int. Conf. on construction in developing countries* Karachi 4-5.
[16] Hemeida O 2010 Green nanoarchitecture
[17] Baughman R H Zakhidov A A de Heer W *Science* **297** 5582 787–792
[18] Mohamed 2015 A S Y *Proc. of the Int Conf on Env and Civil Eng, Pattaya, Thailand.*
[19] Mohajerani A Vajna J Ellcock R 2018 *Journal of Clean Production* **179** 292–307
[20] Beatty C 2006 *NSF Workshop on Nanotechnology, Material Science and Engineering, National Science Foundation*
[21] Sev A Ezel M 2014 *International Journal of Civil, Architectural, Structural and Construction Engineering* **8** 8
[22] Elvin G 2007 *Green technology forum, Indianapolis, Indiana*
[23] BASF http://www.basf.de
[24] Goddard III W A Brenner D W Lyshevski S E Iafrate G J 2004 *Cement and Concrete Research* **34** 1043-1049
[25] Geiker M R Andersen M M 2009 *Sustainability of construction materials*
[26] Hristijan Babuka 2016 *Sustainable Architecture*
[27] Aruoja V Dubourguier H C Kasemets K Kahru A 2009 *Science of the Total Env* **407** 1461-1468
[28] Blaise C Gagné F Férard J F Eullaffroy P 2008 *Environmental Toxicology* **23** 591-598.
[29] Rincón A G Pulgarin C 2004 *Applied Catalysis B: Environmental* **51** 283-302
[30] Zhao J Li N Wang S Zhao X Wang J Yan J Ruan J Wang H Hong F 2010 *Journal of Experimental Nanoscience* **5** 447-462
[31] Leung C C Yu I T Chen W 2012 *The Lancet* **379** 2008-2018
[32] Donaldson K Poland C A Schins R P F 2010 *Nanotoxicology* **4** 414-420.
[33] Lin W Huang Y W Zhou X D Ma Y 2006 *International Journal of Toxicology* **25** 451-457
[34] Wang F Gao F Lan M Yuan H Huang Y Liu J 2009 *Toxicology in Vitro* **23** 808-815
[35] Ye Y Liu J Chen M Sun L Lan M 2010a *Environmental Toxicology and Pharmacology* **29** 131-137
[36] Ye Y Liu J Xu J Sun L Chen M Lan M 2010b *Toxicology in Vitro* **24** 751-758
[37] Kobayashi N Izumi H Morimoto Y 2017 *Journal of Occupational Health* **59** 394 - 407
[38] Wei W Sethuraman A Jin C Monteiro-Riviere N A Narayan R J 2007 *Journal of Nanoscience and Nanotechnology* **7** 1284-1297
[39] Ema M Hougaard K S Kishimoto A Honda K 2016 *Nanotoxicology* **10** 391 - 412
[40] Kang S Pinault M Pfefferle L D Elimelech M 2007 *Langmuir* **23** 8670 - 8673
[41] Kang S Mauter M S Elimelech M 2009 *Env Sci & Tech* **43** 2648-2653
[42] Lee W M An Y J Yoon H Kweon H S 2008 *Environmental Toxicology and Chemistry* **27** 1915-1921
[43] Jones W Gibb A Goodier C Bust P 2017 *Construction Mngt and Economics* **35** 122-136
[44] Bhatt I Tripathi B N 2011 *Chemosphere* **82** 308–317
[45] Hegazy M Badawi A El Rehim S A Kamel W 2013 *Egyptian Journal of Petroleum* **22** 549–556.
[46] Filon F L Mauro M Adami G Bovenzi M Crosera M 2015 *Regul. Toxicol. Pharmacol* **72** 310–322
[47] Liu D 2008 *Nanotechnol. Law Bus.* **5** 465
[48] Bai C 2005 *Science* **309** 61–63
[49] Paranavithana S Mohajerani A 2006 *Resour. Conserv. Recycl.* **48** 1–12
[50] Rasmussen K González M Kearns P Sintes J R Rossi F Sayre P *Regul. Toxicol. Pharmacol.* **74** 147–160

[51] West G H Lippy B E Cooper M R Marsick D Burrelli L G Griffin K N Segrave A M 2016 *Journal of Nanoparticle Research* **18** 1-27
[52] Shen S Burton M Jobson B Haselbach L 2012 *Construction and Building Materials* **35** 874-883

Design and development of the divergent wind turbine

Himanshu Rajput[1], **Anil Gupta**[1], **Harihar Sah**[1], **Dr. Manoj Gattani**[1] **and Raj kumar Satankar**[2]
[1] Department of Civil Engineering, Poornima College of Engineering, Jaipur, Rajasthan, 302022, India
[2] Department of Mechanical Engineering, Poornima College of Engineering, Jaipur, Rajasthan, 302022, India
Guptaanil9812@gmail.com

Abstract. Wind energy is a prime source of renewable energy nowadays. Wind energy is converted to electrical energy with the help of wind turbines. There is various kind of wind turbines depending upon their axis and shape. The wind turbine which we have designed is a vertical axis helical wind turbine that is circular. Going from top to bottom, the diameter of the circular blades increases. The diameter at the top is the lowest and at the bottom it is maximum. Such a design is proposed to utilize the maximum wind pressure created by vehicles on road. Positive results have been received by testing the wind turbines on CFD simulation. Three different kinds of wind turbines have been tested under the same conditions on different parameters. Wind turbines having 4 blades have been compared with curved blade wind turbines with the respective amount of blades, and results are drawn.
Keywords: Savonius, Derrius, Divergent, turbine, Aerofoil

1. Introduction

All are aware of the fact that burning fossil fuels and using them continuously in different forms is leading to climate change and depletion of fossil fuels, hence many countries are adopting some renewable energy technology using renewable energy sources which are abundantly available on the planet. According to sources, almost 67% of the total world electricity production comes from fossil fuels, this is surely an indicator that we are losing fossil fuels year by year. So keeping all the facts and figures in mind, majorly countries are now switching to renewable energy sources for power generation, and one of them is wind sources. Today, the wind turbine has come so far that most researchers are studying the efficient wind turbine as using these turbines, kinetic energy can be converted into electrical energy. Since there are many types of wind turbines available depending on the axis of rotation and aerodynamic forces, so our prime focus of this research paper was to develop an efficient wind turbine with a great efficiency than traditional ones.

The most widespread type of wind turbine in use today is known as a horizontal axis wind turbine (HAT). In HATs, blades, fitted to the rotor might be in upwind or downwind directions and works as lift base turbines. When using HATs, the wind will go around both the upper and lower parts of a blade. Because of the curve of the blade, air moves over the top more quickly than it does over the bottom, which results in a region of lower pressure. The pressure differential that exists between the top and bottom of the blade results in a topward push (Hyams, 2012). HATs are utilized extensively in big wind farm applications in distant and offshore places where there is access to the wind that is both clean and unaffected by human activity. In contrast, HATs are not regarded as an efficient design for use in urban contexts because of the high value of cut-in wind speed, the turbulent character of the wind, and the public perception of these mechanical devices (Liu & Ho, 2016). Moreover, to turn the blades so that they face the direction of the wind, these wind turbines must have a separate yaw control system.

In Vertical axis wind turbines (VAT), the axis of rotation of a VAT is at a right angle to the directions that wind and ground are moving. Since these are very light in weight and smaller structures than HATs, these are easily able to be installed in urban areas where large and heavy structures cannot be installed (Eriksson et al., 2008).

Savonius wind turbines were invented by Savonius in the 1930s, that's why these were called savonius types on the name of the inventor. These are simple VAT that consists of two half-cylinders with different shapes i.e. convex and concave. These half-cylindrical sides face wind at the same time, though their orientation is opposite. The ease of construction is the Savonius rotor's most significant key differentiator. Since it does not require any complex procedures or specialized technical abilities to be fabricated, it is even capable of being produced in modest workshops found locally. Because of this, it is a viable option for applications in nations that are less developed. Savonius rotors are characterized by their high level of solidity and their modest rotational speeds. Because they can provide a high beginning torque, rotors with high solidity are well-suited for use in applications such as water pumping (Mathew & Philip, 2012). But these rotors have more disadvantages than advantages. These have very low efficiency since the power coefficient value is too low in these types of VAT and these are not compatible with aerodynamic forces.

Derrius wind turbines were invented by Derrius in the 1940s, that's why these were called Derrius types on the name of the inventor. These are much compatible with aerodynamic forces and these have simpler designs than savonius types. These types of turbines automatically reduce bending stress and there will only be tension force. But these turbines have some limitations as these types of turbines have very low starting torque, so they need external power from an external source.

2. Literature Review

In a research done by Magda Moh. M. Saad, a Norzelawati Asmuin, 2014, it was found that HATs are more efficient than VAT. HATs have an efficiency of 60% which is comparatively higher than the VAT. But due to the large size of HATs over VAT, VAT are more feasible and plantable than HATs. But the main advantage that VAT serve is their applicability in regions where there is low wind pressure. Based on the size, VATs are designed in such a way that they can be planted even on rooftops in comparison to large-sized HATs that require a lot of space. HATs are generally 100 - 120 m high and their blades range between 60 - 80 ms.

Further, a study was conducted by Rakesh Kumar, Kaamran Raahemifar, and Alan S. Fung on the urban use of VATs where they compared different kinds of VATs at different wind speeds, and the number of rotations was noted at the respective speeds. There, four kinds of wind turbines tested were Curved, Straight, Aerofoil, and Twisted and by the research, it was found that the most number of rotations were captured in curved in comparison to other turbines. In the second place there came the twisted one which was followed by the aerofoil and the least rotations were seen in the straight, bladed wind turbine.

Benini et al. compared two different VATs based on annual energy output (AEO) in their study which was conducted in Italy. The comparison of turbines was based on aerodynamic forces, one is drag-based turbines and the other one is lift-based turbines. Since the design of lift-based turbines is complex and collects more energy so it was expected to have a greater AEO. On the other hand, drag-based turbines gave more power outputs at low wind speeds, moreover, they had a capacity for self-starting.

As stated in the introduction part, there are two types of VATs i.e., Darrieus and Savonius but both have their design limitations. So to counter this limitation, the design of both types of wind turbines have been converted into a single hybrid design having all the advantage of individual design.

In continuation to this, Sharma et al. firstly constructed a three blades hybrid turbine having a Darrieus rotor on the top part of the Savonius rotor, and then, they analyzed its performance. They studied that, with the increase of the overlap ratio, torque and power coefficients continuously increase, but a point came were increasing in the overlap ratio showed decreasing in the power and torque coefficients, resulting in an optimum overlap ratio. Experimental investigation showed that the maximum power coefficient for the optimum overlap ratio was 0.53.

Islam et al. analyzed mainly three types of VAT models based on aerodynamic forces on a single design of a straight-blade Darrieus turbine. The momentum model was the first one that was used for analyzing the performance which was again categorized into three sub-parts for experimental work i.e., Single-stream tube model, multiple stream tube model, and double multiple stream tube models. After that, the vortex and cascade model was studied in which the high-speed tip ratio was the crucial factor for analyzing the performance. It was found that the cascade model was suitable for high-speed tip ratio, while it was not so much accurate. The vortex model was found to be the most accurate one among others.

Brusca et al. used a highly precise calculation code on different tube stream models which were used by Islam. They tried to connect the aspect ratio of a single-bladed VAT and the coefficient of power. Since, variations in aspect ratio lead to the variation of the Reynolds number, resulting in power coefficient variation and affecting the performance of the single-bladed VAT. An experimental study showed that the higher Reynolds number and lower aspect ratio increased the efficiency of the single VAT.

Since most of the researchers used the Navier-Stokes equation to study the CFD theory with different models, Iida et al. used a mechanism called large eddy simulation to improve the performance of single-bladed VAT. As there are so many variations in the angles of attack of VAT, so it was much needed to use the simulation with very good accuracy and high precision. They found that at a tip speed ratio of 3, the maximum power coefficient occurred while the low tip speed ratio was not good enough to improve the torque coefficient subsequently improving the performance.

3. Design

In the previous chapters, we have seen what the types of wind turbines used are in the current scenario. How the VATs are different from the HATs, on which parameters we can bifurcate the vertical and the HATs, and can draw a comparison to know which one is better. It was evident from the previous chapters that at comparatively low heights, VATs have an edge over horizontal ones. We have also seen the pros and cons of HATs when used at lower heights and know-how a VAT can replace the HAT by focusing on the areas which can be worked upon. Now the effectiveness of the vertical axis wind depends upon the shape of the blades. There are different kinds of shapes of the wind turbine such as straight bladed; circular bladed; helical shaped etc. But how much is each one of them effective, which is the most effective design for the maximum amount of rotations and torque. Hence, we have designed a vertical axis divergent wind turbine with different design features. When talking about the VATs, one most important factors that come into mind is the number of blades that the turbines have. In previous studies, it has been found that two-bladed wind turbines are more effective. Thus, further comparisons will be drawn based on the number of blades.

The present invention provides a wind turbine capable of generating power. The said wind turbine device includes a base. The turbine further includes a turbine shaft (S) mounted on the base (B) along a vertical axis of the shaft (S). The shaft (S) is configured for rotation in response to rotation by harvested wind energy in the said turbine. The turbine further includes divergent blades (B) configured around the shaft (S), including a lower portion (LB) and an upper portion (UB). The blades are adapted to harvest wind for driving the shaft, thereby allowing the shaft to generate mechanical energy. The width upper portion of the plurality of divergent blades is more than the width of the lower portion of the plurality of divergent blades, thereby ensuring a divergent shape of the said plurality of the blades. The turbine further includes a generator (G) operably connected to the shaft. The generator (G) is installed to convert the mechanical energy into electrical energy. Also the wind turbine has a helical shift during the upward movement of the blade (H). The blades are at some angle from top to bottom. Wherein a ratio of the width of the upper portion (UB) and lower portion (LB) is in the range of 1: 1.5 to 1:4.

IOP Conf. Series: Earth and Environmental Science 1084 (2022) 012075 doi:10.1088/1755-1315/1084/1/012075

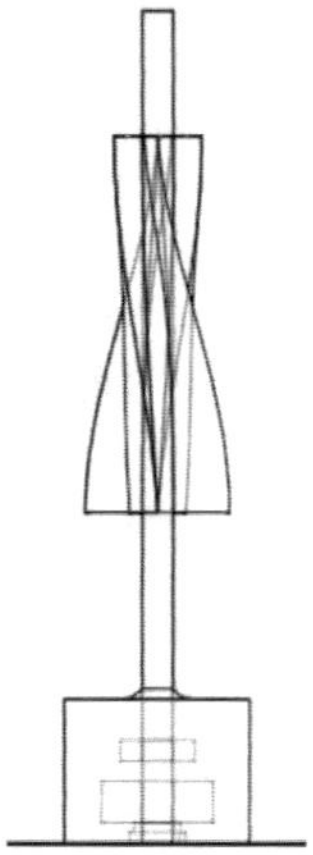

3D view

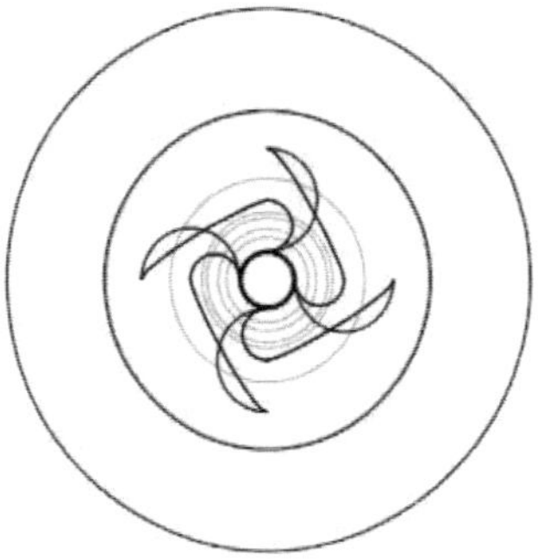

Top View

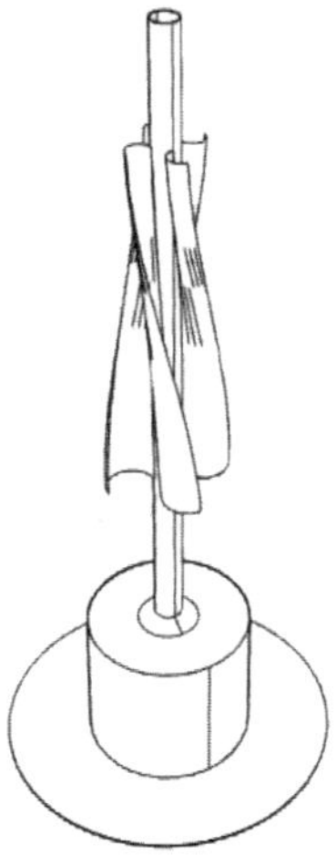

Front View

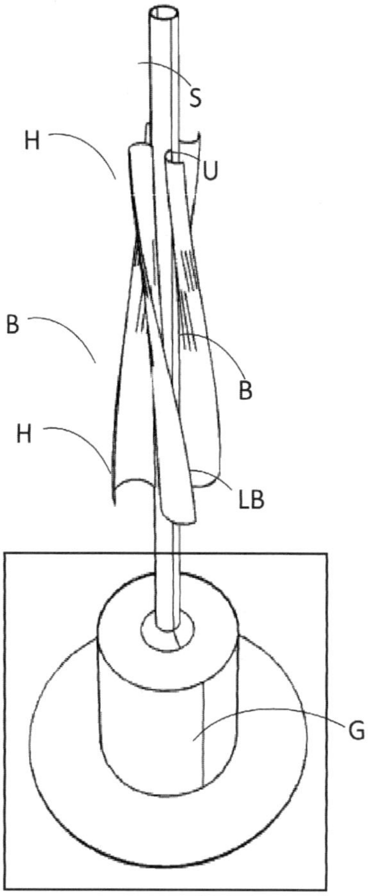

IOP Conf. Series: Earth and Environmental Science 1084 (2022) 012075 doi:10.1088/1755-1315/1084/1/012075

Table 1. Specifications of divergent blade

Diameter of the upper part of the blade	2.5cm
Diameter of the lower part of the blade	5cm
Height of the blades	13cm

4. Methodology

The proposed wind turbine includes a base. The wind turbine further includes a turbine shaft that is mounted on the base along a vertical axis of the turbine shaft. The turbine shaft is configured for rotation in response to rotation by harvested wind energy in the turbine. The proposed wind turbine further includes a divergent design of blades that are configured around the shaft. More specifically, one side of the divergent blades is attached to the shaft throughout the length of the shaft via a mechanical fixture. Each of the blades includes a lower portion and an upper portion. The blades are forming hollow semicircular-like portions at the lower portion and the upper portion. In the proposed wind turbine, the width of the upper portion is more than the width of the lower portion, thereby ensuring a divergent shape of the blades. The divergent blades are helical in shape. the design of the divergent blades being configured to harvest wind for driving the turbine shaft, thereby allowing the shaft to generate mechanical energy. The design of the divergent blades allows the air to enter from the lower portion of the blades, and the entered air leaves the blades from the upper portion of the blades, thereby allowing the blades (to create momentum for rotating the shaft at a faster rate. the proposed wind turbine includes a generator that is installed at the base. Specifically, the generator is operably connected to the turbine shaft. The generator is being installed to convert the mechanical energy into electrical energy. The proposed wind turbine further includes one or more heat concentrators which are installed at a bottom of the blades. In the turbine, one or more heat concentrators are installed to heat the nearby air present at the bottom of the blades. More specifically, heating increases the flow rate of the air at the bottom of the blades, thereby allowing the heated air to flow through the divergent blades from the lower portion to the upper portion at a faster rate. This together with the other aspects of the proposed wind turbine along with the various features of uniqueness that characterized the turbine are pointed out. For a better understanding of the proposed wind turbine, its operating advantages, and the specified objective attained by its uses, reference should be made to the accompanying descriptive matter in which there are illustrated.

The comparison of the proposed wind turbine with the circular axis wind turbine is going to be done on the basis of the wind velocity. Hence, the position of the wind blower is fixed and the wind velocity is also made constant for both the wind turbine. It was also seen that when the wind velocity increases the temperature at which wind will act on the turbine also increases gradually. It is one of the input data which is to be kept constant in order to get optimum results. Identification of the variables have been done which are stated as follows-

- Wind velocity ranging from 9 m/s to 26.8 m/s
- Temperature
- Shape of the blades of the wind turbine

The expected outcome is that the number of rotations achieved at different wind speed in the proposed divergent wind turbine will be higher than the traditional ones. And also the torque generated in the process will also be higher. Also the 2 bladed wind turbine is going to be more efficient than the 4 bladed turbine.

4.1 Development of the turbine

Table top models of wind turbines for laboratory testing were made of mentioned specifications. Both the traditional as well as the proposed wind turbine was 3D printed with a PLA material of same density and volume so that to compare both at standard parameters.

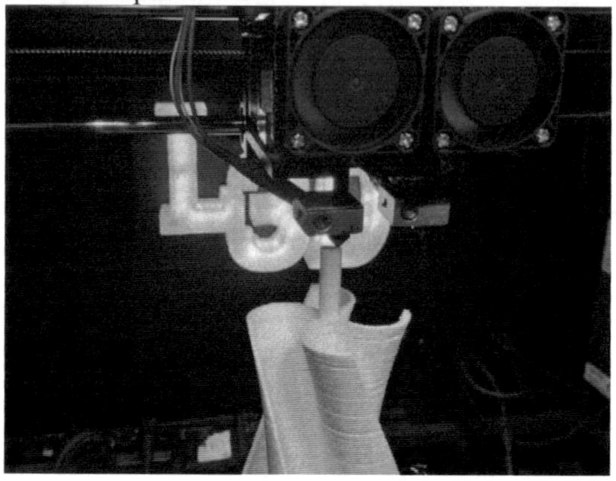

5. Experimental Result and Simulation Analysis

Tests and experimental analysis have been drawn on a comparatively small wind turbine but with having the same ratio as the original one. In the study that was conducted by Rakesh Kumar et al. where they compared different kinds of VATs at different wind speeds and a number of rotations were noted at respective speeds. There, four kinds of wind turbines tested were Curved, Straight, Aerofoil, and Twisted and by the research, it was found that the most number of rotations were captured in curved in comparison to other turbines. So the results have been carried out by comparing the proposed 4-bladed wind turbine with the curved one.

Comparisons drawn were two folded first on the Ansys software CFD simulation was run by taking wind as the fluid and secondly by creating an atmosphere for testing the turbines under real-time conditions.

5.1. Simulation Analysis

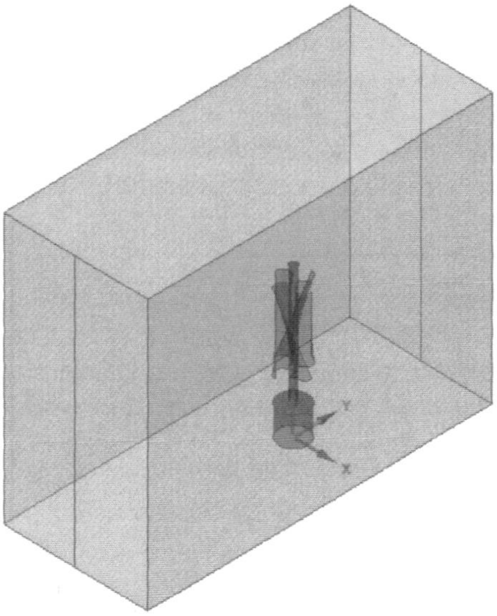

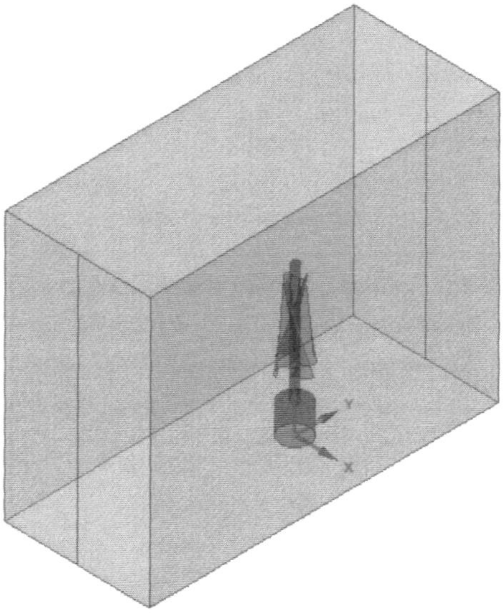

Simulation analysis was done on Ansys Software under following conditions.

Table 2. Conditions for Simulation on Ansys

Mesh Size @ blade, shaft	1.25 mm
Inflation layer @ blade	5 layer,
Inflation layer, first layer thickness	1 mm
Element Count	229165
Node Count	1349255
Maximum Skewness	0.88
Analysis Type	Steady
Precision	Double
Turbulence model	Standard k-epsilon model
Near Wall Treatment	Enhanced Wall function
Fluid	Air
Inlet	16.67 m/s, Normal to boundary
Outlet	atm pressure outlet
Blade	Movable wall @120 rpm

5.2. Simulation Results

```
Moments - Moment Center (0 0 0)
                        Moments (n-m)
Coefficients
Zone                    Pressure                                    Viscous                                     Total
Pressure                                    Viscous                                     Total
blade                   (90.534994 -51.288256 2.4734316)            (0.16829822 -0.034045766 0.032954512)
(90.703293 -51.322301 2.5063861)            (147.81224 -83.735927 4.0382557)            (0.2747726 -0.055584925
0.053803285)        (148.08701 -83.791512 4.092059)
------------------------ ------------------------------------------- -------------------------------------------
------------------------------------------- -------------------------------------------
------------------------------------------- -------------------------------------------
Net                     (90.534994 -51.288256 2.4734316)            (0.16829822 -0.034045766 0.032954512)
(90.703293 -51.322301 2.5063861)            (147.81224 -83.735927 4.0382557)            (0.2747726 -0.055584925
0.053803285)        (148.08701 -83.791512 4.092059)

Moments - Moment Center (0 0 0) Moment Axis (0 0 1)
                        Moments (n-m)                               Coefficients
Zone                    Pressure        Viscous         Total           Pressure        Viscous         Total
blade                   2.4734316       0.032954512     2.5063861       4.0382557       0.053803285     4.092059
------------------------ --------------- --------------- --------------- --------------- --------------- ---------------
Net                     2.4734316       0.032954512     2.5063861       4.0382557       0.053803285     4.092059
```

IOP Conf. Series: Earth and Environmental Science 1084 (2022) 012075 doi:10.1088/1755-1315/1084/1/012075

```
Moments - Moment Center (0 0 0)
                          Moments (n-m)
Coefficients
Zone                      Pressure                                             Viscous                                              Total
Pressure                                     Viscous                                             Total
blade                     (73.804649 20.705818 -3.0950048)                     (-0.0082544032 -0.37954956 0.052076917)
(73.796395 20.326269 -3.0429279)             (120.49739 33.805417 -5.0530691)                    (-0.013476577 -0.61967276
0.085023538)        (120.48391 33.185745 -4.9680455)
------------------------- --------------------------------------------- ---------------------------------------------
--------------------------------------------- ---------------------------------------------
--------------------------------------------- ---------------------------------------------
Net                       (73.804649 20.705818 -3.0950048)                     (-0.0082544032 -0.37954956 0.052076917)
(73.796395 20.326269 -3.0429279)             (120.49739 33.805417 -5.0530691)                    (-0.013476577 -0.61967276
0.085023538)        (120.48391 33.185745 -4.9680455)

Moments - Moment Center (0 0 0) Moment Axis (0 0 1)
                          Moments (n-m)                                 Coefficients
Zone                      Pressure        Viscous         Total           Pressure        Viscous         Total
blade                     -3.0950048      0.052076917     -3.0429279      -5.0530691      0.085023538     -4.9680455
------------------------- --------------- --------------- --------------- --------------- --------------- ---------------
Net                       -3.0950048      0.052076917     -3.0429279      -5.0530691      0.085023538     -4.9680455
```

Table3. Results after Simulation

Parameters	Normal blade wind turbine	Wind turbine with divergent blades of the present invention
Speed	120 rpm	120 rpm
Moment	**4.09 N-m**	**4.96 N-m**

Here a when CFD simulation was run on ansys software we clearly saw an increment in the moment generated. Also it was not a slight increment or just a fluke, we saw an increase of about 21% in the moment generated.

Real-Time Testing Results

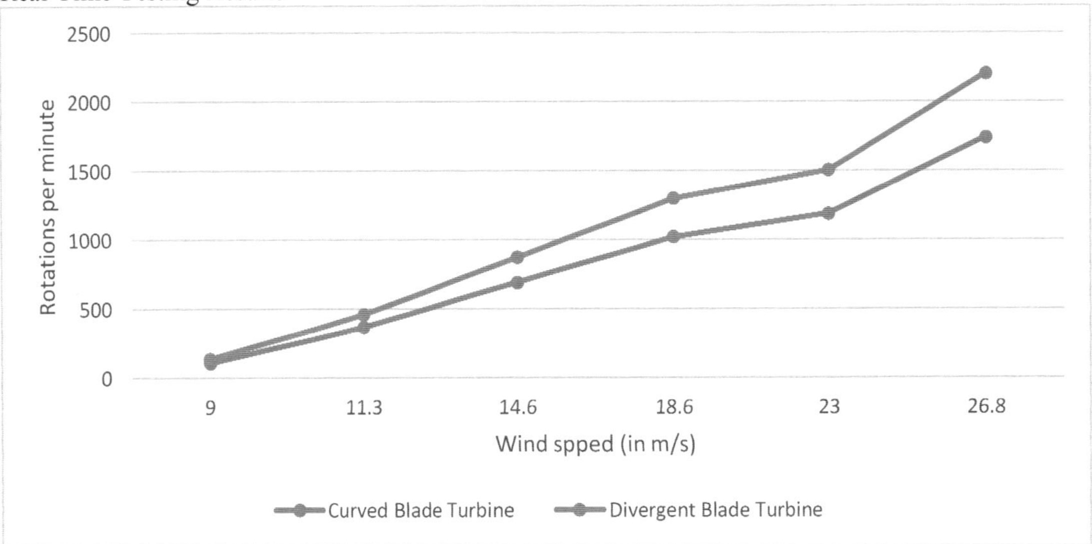

Figure 1. Variation of rotation per minute of curved and divergent blades to wind speed

Table 4. Comparison of traditional and divergent wind turbine

Wind Speed(m/s)	Rotations Per Minute		% Change
	Traditional Wind Turbine	Proposed Divergent Wind Turbine	
9.0	107	136	21.32% increase
11.3	368	460	20.00% increase
14.6	693	874	20.70% increase
18.6	1021	1301	21.52% increase
23.0	1189	1504	20.94% increase
26.8	1743	2204	20.90% increase

6. Conclusions

Wind energy is a prime source of renewable energy nowadays. Wind energy is converted to electrical energy with the help of wind turbines. There is various kind of wind turbines depending upon their axis and shape, and this shape justifies and shows how much wind energy it can convert into a mechanical one. Hence we have identified a design and have tested it under software conditions and real atmosphere. Its performance is recorded and compared to the conventionally superior considered curved bladed turbine. The following the results obtained

- We have tested 4 bladed wind turbine and compared it with a traditional wind turbine.
- Under the simulation results an increase of 21% in moment generated was seen.
- It was founded that 2 bladed wind turbine came out to be more efficient than the three and four bladed turbine.
- In our results we have found that the wind turbine which we are proposing is having a hike of 20 - 22% in number of rotations achieved.

References

[1] Alaimo, A., C., & Tumino, D. (2015). 3D CFD analysis of a vertical axis wind turbine. *Energies*, *8*(4), 3013–3033.

[2] Cheng, Q., Liu, & Yang, B. (2017). Aerodynamic analysis of a helical Vertical axis wind turbine. *Energies*, *10*(4).

[3] Delafin, P. L., Wang, L., & Kolios, A. (2016). Effect of the number of blades and solidity on the performance of a vertical axis wind turbine. *Journal of Physics: Conference Series*, *753*(2).

[4] Eriksson, S. & Leijon, M. (2008). Evaluation of different turbine concepts for wind power. *Renewable and Sustainable Energy Reviews*, *12*(5), 1419–1434.

[5] Hyams, M. A. (2012). Wind energy in the built environment. *Metropolitan Sustainability: Understanding and Improving the Urban Environment*, 457–499.

[6] Kumar, R., Raahemifar, & Fung, A. S. (2018). A critical review of vertical axis wind turbines for urban applications. *Renewable and Sustainable Energy Reviews*, *89*(March), 281–291.

[7] Liu, S. Y., & Ho, Y. F. (2016). Wind energy applications for Taiwan buildings: What are the challenges and strategies for small wind energy systems exploitation? *Renewable and Sustainable Energy Reviews*, *59*, 39–55.

[8] M. Saad, M. M. (2014). Comparison of Horizontal Axis Wind Turbines and Vertical Axis Wind Turbines. *IOSR Journal of Engineering*, *4*(8), 27–30.

[9] Mathew, S., & Philip, G. S. (2012). Wind turbines. Evolution, basic principles, and classifications. In *Comprehensive Renewable Energy* (Vol. 2). Elsevier Ltd.

[10] Scheurich, F., T. M., & Brown, R. E. (2010). The influence of blade curvature and helical blade twist on the performance of a vertical-axis wind turbine. *48th AIAA Aerospace Sciences Meeting Including the New Horizons Forum and Aerospace Exposition*, *January*, 1–16.
[11] Tjiu, W., Marnoto, K. (2015). Darrieus vertical axis wind turbine for power generation I: Assessment of Darrieus VAWT configurations. *Renewable Energy*, *75*, 50–67.
[12] Wang, Z., & Zhuang, M. (2018). Improvement of the aerodynamic performance of vertical axis wind turbines with leading-edge serrations and helical blades using CFD and Taguchi method. *Energy Conversion and Management*, *177*(May), 107–121.

Energy Efficient Buildings in India: Key Area and Challenges

Preeti Tarkar

Assistant Professor, Institute of Business Management, GLA University, Mathura, India
Email: preeti.tarkar@gmail.com

Abstract. A common dilemma for developing countries is to balance economic growth and environmental protection. Since the major issue in designing the building is energy conservation. The costing of building project is another issue in applying the principle of energy. A building uses a huge amount of energy during its lifetime. It is greater than the energy consumed in construction. There are numerous causes why building will use more energy during its life. Taking into account the building materials used in the built environment as well as the overall energy use should be taken into account when looking for ways to minimize energy consumption. In light of climate change, by 2030, India has to reduce the greenhouse gas emissions to 35%. Thus, the aim of this research is to is to find out the measures that can reduce the energy consumption of building not only at the time of construction but also lifetime. The study suggests various energy efficient measures for construction of building and for further functioning. In this paper the concept of green building was discussed to make a building energy efficient. Green building focuses on the reduction of carbon emissions and conservation of energy using a progressive difference-in-differences method. The study suggests the various energy saving concepts. Further, there are many green building rating systems to rate the building on various dimension to measure whether a building is environmental-friendly or not. In the end of the paper various issues involved in energy efficient building's construction are discussed.

Keywords. Energy efficiency; green buildings; green building rating system; challenges

1. Introduction

Climate change is a major concern for all in the world. It is taking place because of the greenhouse gas emissions from natural activities and human [1]. In recent years, improved living standards, rapid urbanization and continued industrialization have boosted energy consumption. Energy is required to perform all human activity and it is a driving force. The major source of energy is fossil fuels. Energy production relies primarily on traditional energy sources, which are the main causes of environmental pollution. For improving the building's environmental performance, building should be energy efficient. Thus, the consideration of all measures which can make energy efficient building are important. Buildings are large energy consumers throughout their life cycle. Civilized society is based on the buildings and it is an integral part. Buildings account for one-sixth of the world's freshwater extraction, one-quarter of logged timber, and two-fifths of energy flow and material [2]. Building construction is an energy-intensive process that consumes energy at every stage of the entire life cycle, from site clearance to operation and maintenance. This article is the political and social management of the construction of energy efficient buildings. Buildings account for about 40% of land energy consumption. In India, over the last three decades, the construction sector has consumed more energy than any other sector [3]. In the OECD region, the energy consumption in the building sector is very high at 25% to 40 % which is more than other sectors and the consumption of total energy is increasing [4].

India aims to reduce GDP emission strength by 33-35% by 2030 [5]. To achieve this goal, the building's energy efficiency needs to be improved, as all sectors, especially the building sector, consume more than 30% of India's total electricity [6]. Improving the energy efficiency of a building will reduce energy demand, conserve limited natural resources and control the emission of toxic gases like: CO2 emissions. This will improve the overall environmental performance of the building. Building construction involves a variety of activities. Planning, planning, execution, operation, and maintenance. Every stage of building construction consumes energy in some way. Energy sources used in building development include coal. It is used to produce building materials, oil and fuel for transportation and operational equipment, and power to power equipment. Many initiatives have taken by India to deal with the issue of climate change. The National Action Plan on Climate Change in India established eight national missions to protect the environment. These are the National Mission for Strategic Knowledge of Climate Change, National Mission for Sustainable Agriculture, National Mission for Ecosystem Conservation, the National Water Mission, and the Himalayan, the National Mission for Improving Energy Efficiency, Green India, National Solar Mission. These missions are operated under the regulatory and voluntary mechanisms and covers a wide range of problems, including forest management, agricultural environmental sustainability, knowledge generation, and demand-side energy management for sustainability.

2. Objectives and scope of the study

The main aim of the paper is to find out the measures that can reduce the energy consumption of building not only at the time of construction but also lifetime. This paper discusses green building concepts, green building rating systems, energy conservation measures and implementation and issues in construction of energy efficient buildings.

3. The energy consumption in residential buildings

The total energy consumed by residential building sector is less than 25 percent of the total electricity. In terms electricity consumption, building sector is the third largest consumer. 26 percent consumption has increased in between 2014 to 2017 [7]. Energy consumption has increased due to the population growth, urbanization and it created a large gap between the high demand and limited supply of the electricity. The demand of electricity is increasing mainly due to the change in living standards and high reliance on the affordability of the thermal comfort. The ultimate population of India is increasing and it is projected as by 2060, the population of India will be 1.72 billion. Urban expansion and urbanization are increasing rapidly. It is expected that two third population of India will live in urban area by 2050. The population of urban areas will rise 33% of the total population to 50% by 2050. In next twenty years, more than 70 million new urban units of houses will be added. It is estimated that two thirds of the built-up area will be constructed in next two decades [8].

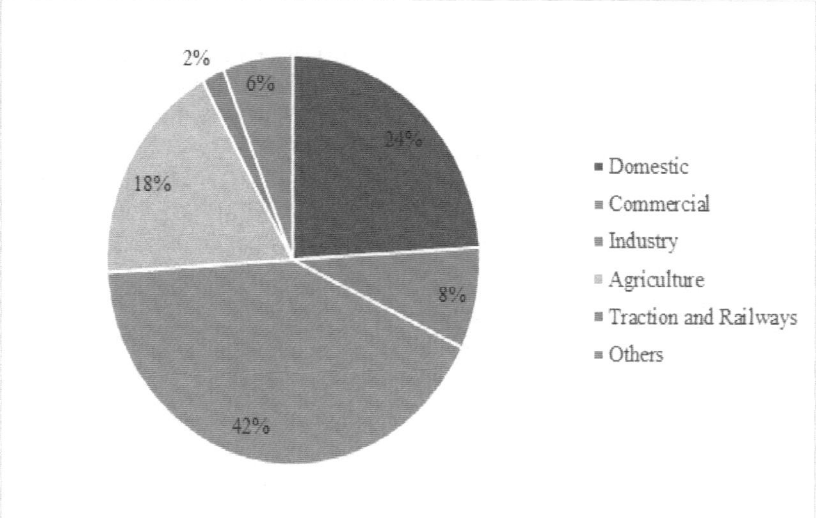

Figure.1 Electricity Consumption by sectors during 2018-19 [9]

4. Energy Efficient Buildings

In order to make buildings more energy efficient building many technological solutions are available. The concept of green building comprises many elements like: occupant`s health and safety, Design efficiency, Energy efficiency, sustainable site sustainable selection, Materials efficiency and Water efficiency. A building that assures the healthiest environment during its lifetime is known as green building or an energy efficient building. These buildings represent optimally utilize the water land and energy resources [10].

Green building design utilise the natural resource and conditions of the predeveloped site. Buildings may even become energy producers and can become energy efficient when it uses a properly sized and energy efficient heating/cooling system in a building, use minimum glass on east and west exposures, use solar energy for lighting and for water heating [11], use minimum light colour for roofing and wall finishing materials. Maximize light colours for wall finishing materials and roofing, maximum use of natural lighting and air etc. [10]

There are various benefits from the good environmental quality building that it can reduce the risk of asthmatic diseases and respiratory diseases. Good building design reduces the toxic gases emissions and improve the indoor air quality because the emission of these gases can have negative impact on health and on indoor air quality.

5. Various Energy Saving Concepts

Orientation: it is very important to decide the orientation of building in right direction because right orientation can increase the rate of day lighting and solar gain. South-facing windows in the Northern Hemisphere are the most exposed to the sunlight. On the other side, windows should be carefully designed in the west [12].

Site Selection: decision of selecting the site is very important and depends on the price. A bad decision can reduce various sustainable features. The main difference between the traditional and high-performance building are in terms of their locations [13].

Roof and Wall- the construction of wall and roof can decrease conducive warmth.

Energy Efficient Appliances- according to the BEE star rating. Extra performance is required for star rating. Use of energy efficient appliances enhances the energy conservation.

Ventilation, Heating and Air-Conditioning- Reducing the warmth load of the shape permits for the setup of a smaller heating and cooling gadget. The significance of amazing air flow structures is frequently disregarded at some point of the layout phase, however is a essential attention in inexperienced constructing.

Waste reduction- reduction of waste or proper treatment of waste creates a healthier environment. Proper disposal of cloth can reduce the quantity of cloths going into landfills. Degradable waste should be treated properly and it can be used to make compost.

6. Green Building Rating System

In order to sell the idea of strength green constructing idea, inexperienced constructing score gadget changed into advanced for construction sector. There are various rating systems are governed to rate the green buildings [14]. Green Rating for Integrated Habitat Assessment (GRIHA), TERI (The Energy and Resources Institute). This is mainly used for rating the building on various dimensions like water conservation, energy efficiency, use of renewable resources etc. It rates the building on the basis of building's overall environmental performance on 0-104 scale. SVAGRIHA (Small Versatile Affordable GRIHA) is used for small buildings construction. For Indian construction sector, GRIHA and SWAGRIHA are new constructing score gadget. On the premise of range of factors scored, a constructing may be rated among 1 to 5 stars. GRIHA changed into advanced through TERI and has now been followed through the Ministry of New and Renewable Energy (MNRE) because the National Rating System for inexperienced homes in India and to sell inexperienced homes in India and to supervise the diverse sports related to it, MNRE and TERI at the same time hooked up an independently Registered society referred to as ADARSH (Association for Development and Research of Sustainable

Habitats). ADARSH features as a platform for interplay among diverse stakeholders in addition to promotes GRIHA, SVAGRIHA and different comparable inexperienced constructing score structures in India while SVAGRIHA is a newly developed program specially designed for projects which are not more than 2500sq.mt i.e., homes with constructed up vicinity not more than 2500sq.mt [15].

7. Issues in creation of Energy Efficient Buildings

The primary issue with the construction of energy efficient buildings is non-implementation of the government policies. As an attempt of governance, it ought to be acted upon. There should be a proper implementation of the government policies. In sum, three sets of features are identified that inhibit such translation.

The demand of residential buildings is always high but it is not high in case of energy efficient buildings. The construction of energy efficient buildings is criticized by the cost concerns and short-termism. There is a problem of the balance in opinions of builders and users. Builders want reduce the cost of construction and user wants environmental efficient buildings.

In fact, this has led to concerns about energy costs and a focus on energy investment savings rather than energy. In addition, the cautious government has opted for general and indirect public regulatory activities. Rather than engaging in effective direct regulation, such as passing strict building codes and using the strong position of public institutions as building owners to drive improvement and act as a role model. It is clear that they favour ineffective tools such as financial incentives.

8. Conclusions

On the basis of literature review, the following important points have been drawn. Across the world numerous studies are conducted on the policies, rating systems and the ways to reduce the energy consumption in the buildings. Two groups are created on the basis of different mitigations around the world. First, regulatory viz. legislation supporting environmental protection, taxes, bans, controls on emissions, Subsidies and standards related to the equipment effectiveness, and second, voluntary. The second mitigation is the more effective and important.

Several rating systems are available worldwide. Mostly the focus of all rating systems are on the energy performance of a building, but none have a clear purpose for the environmental performance of a building. Since, the environmental performance of building and energy efficiency of building are interrelated, carbon emissions should be carefully considered when assessing the buildings.

Simulation and lifecycle analysis have proven to be very effective in making decisions among the various options available. The important thing which should be considered when designing an Energy efficient building is passive solar architecture. There should be more renewable sources of energy. Thus, the design of building should be to minimize the consumption of electricity and reduce the emission of toxic gases like carbon. This is the main objective of energy efficient buildings.

Environmental performance of a building can be improved by improving the energy efficiency of a building. There are two forms in which energy in Building is present: Operational energy and Embodied energy. By improving the efficiency and technology of the device, it is possible to reduce large operating energy. Low-energy materials made from solid waste are more advantageous because they save scarce natural resources and helps in reducing environmental pollution. For better results, we need to focus on cement, steel and brick, which have a high mass fraction of building materials.

A system approach is a possible measure to improve energy efficiency which ultimately improves the environmental performance. Therefore, building development's overall parameters need to be interrelated, and the building should be designed to minimize its environmental impact with optimal energy consumption and minimal cost.

References

[1] Chitre, S.P., 2016. Climate change & India: Paris & Beyond. *Available at SSRN 2799605*.

[2] Emmanuel, R., 2004. Estimating the environmental suitability of wall materials: preliminary results from Sri Lanka. *Building and Environment*, *39*(10), pp.1253-1261.

[3] Majumdar, M. ed., 2001. *Energy-efficient buildings in India*. The Energy and Resources Institute (TERI).
[4] OECD, 2003. Environmentally Sustainable Buildings. Challenges and Policies. Organisation for Economic Co-Operation and Development, Paris.
[5] India's Intended Nationally Determined Contribution. http:// www4.unfccc.int/submissions/INDC (15 May 2022, date last accessed).
[6]EnergyStatistics2020.http://mospi.nic.in/sites/default/files/publication_reports/Energy%20Statistics %202
[7] Dalal, R., Bansal, K. and Thapar, S., 2021. Bridging the energy gap of India's residential buildings by using rooftop solar PV systems for higher energy stars. *Clean Energy*, *5*(3), pp.423-432.
[8] McKinsey Global Institute. India's urban awakening: building inclusive cities, sustaining economic growth, www.mckinsey.com (2010, accessed 10 April 2022).
[9] CEA Annual Report 2019. https://cea.nic.in/annual-report (30 April 2022, date last accessed).
[10] Chandel, S.S., Sharma, A. and Marwaha, B.M., 2016. Review of energy efficiency initiatives and regulations for residential buildings in India. *Renewable and Sustainable Energy Reviews*, *54*, pp.1443-1458.
[11] Gupta, J. and Chakraborty, M., 2021. Energy efficiency in buildings. In *Sustainable Fuel Technologies Handbook* (pp. 457-480). Academic Press.
[12] Bano, F. and Kamal, M.A., 2016. Examining the role of building envelope for energy efficiency in office buildings in India. *Architecture Research*, *6*(5), pp.107-115.
[13] Chel, A. and Kaushik, G., 2018. Renewable energy technologies for sustainable development of energy efficient building. *Alexandria Engineering Journal*, *57*(2), pp.655-669.
[14] Tarkar, P., 2022. Role of green hospitals in sustainable construction: Benefits, rating systems and constraints. *Materials Today: Proceedings*, *60*, pp.247-252.
[15] Thesis Title "Commercial green building: the value added gap perspective" by Amaravadee Vinyangkoon

Numerical Analysis of RCC Beam Using ABAQUS

Dr. A. Hemamathi[1,*], Dr. Binu Sukumar[2], and R.B. Hambrish Guru Chantrakant[3]

[1]Associate Professor, Department of Civil Engineering, R.M.K Engineering College, Tamil Nadu, India.
[2] Professor, R.M.K Engineering College, Tamil Nadu, India
[3]UG Student, Department of Civil Engineering, R.M.K Engineering College, Tamil Nadu, India.
Corresponding author ahm.ce@rmkec.ac.in

Abstract. Experimental investigation of RCC beam provide an important understanding on its flexural properties. The stages involved from the design to casting, curing, testing and interpretation of results is time consuming and involves economy and labour. To reduce the above said challenges, numerical investigation seems to be an alternate without involvement of money and labour. Preliminary design stages involves design equation which are provided by the basic experiments conducted and they also give the basic information to develop the finite element model. This paper presents the numerical investigation of RCC beam to study the flexural properties using an effective finite element analysis tool ABAQUS. RCC beam prototypes were casted and their flexural properties were experimentally investigated. The results were validated through the finite element model by providing suitable properties for the material that are actually used for the prototype with the desired loading and support conditions. Beam of size 3000 x 230 x 300 mm of M20 grade concrete was casted and tested in the loading frame for flexure and deflection. The study focussed for under reinforced, balanced and over reinforced sections. Numerical analysis was performed using ABAQUS for model geometry and the results were compared with the experimental results. The results agreed with both the experimental and numerical values. Numerical results helps in developing reliable models which can reduce the number of required test specimens for conducting experiments and the model can be used for various study. Also from the finite element results the propagation of cracks can be well understood for the actual specimen.

Keywords: RCC beam, Numerical investigation, ABAQUS,

1. Introduction

In general, structures are subjected to different types of loading condition. The primary structural element of RC structure is a beam and it is subjected to transverse loading condition. It is a flexural member subjected to bending action under the action of transverse load [1]. A RC beam is considered to be statically determinate when the analytical calculations are performed with known equilibrium equations and it may be considered as a homogeneous section with simple geometry until the concrete in the beam remains un cracked [2]. The individual beam element can be tested experimentally to study the effects of various loads and the behaviour of material. The experimental testing method of ordinary concrete beam is costly method to understand the behaviour under transverse loading. Also, it is time

consuming process as the beam has to be cured for a period of time. This can be overcome by the use of computer software wherein the beam can be modelled at a swift pace making it a cost-effective process. Once the model is validated with the experimental results, it will be useful in refining analytical models, making good prediction on complex nonlinear behaviour of RC beams. [3]
ABAQUS is a powerful finite element tool to provide solution to engineering simulation programs. It can provide solution to simple linear analyses problems to the most challenging nonlinear simulations. ABAQUS can model any geometry virtually from its available extensive library. Also, it contains number of material models to simulate any engineering materials such as metals, rubber, polymers, composites, reinforced concrete, crushable and resilient foams, and geotechnical materials such as soils and rock. ABAQUS tool not only provide solution to structural (stress/displacement) problems, but also it can simulate diversified problems such as heat transfer, mass diffusion, thermal management of electrical components (coupled thermal-electrical analyses), acoustics, soil mechanics (coupled pore fluid-stress analyses), piezoelectric analysis, electromagnetic analysis, and fluid dynamics. Also, ABAQUS can be used to model multiple components associating each geometry and defining each component with the appropriate material models and specifying component interactions. In nonlinear analysis ABAQUS is capable of choosing load increments and convergence tolerances and continually adjusts them during the analysis to ensure that an accurate solution is obtained efficiently [15].
The present study aims at comparing the experimental results with the ABAQUS results. The study started with experimental testing of beam of size 3000 x 230 x 300 mm of M20 grade concrete for under, balanced, over reinforced sections. The numerical study was performed for the model considered in the experimental study using Finite Element Analysis (FEA) tool, ABAQUS. The numerical results obtained from ABAQUS tool was compared with the experimental results which showed good agreement between the results.

1.1 Significance of study

The study focuses on the experimental study and numerical study of RC beam. The deflection parameter was investigated and checked with IS code specification. Experimental investigation on beam involves time, material and facility for testing. This paper provides a basic study on the RC beam where a numerical investigation was validated by experiment. Further the numerical procedure can be utilized further for RC beam with any grade of concrete, grade of steel, diameter and number of rebars and type of beam.

2. Experimental Investigation

Mix design for M20 concrete was carried out as per IS 456: 2000 and IS 10262: 2019. Six cubes were casted and tested for 28 days. M20 grade of concrete was used for casting. After curing for 7 days, 3 cubes were tested for checking the initial strength of concrete which found to be 65% of the design value. The same procedure was done for the other 3 cubes after curing for 28 days. After the curing period of 28 days the beams were experimentally subjected to two-point loading by using hydraulic actuator as given in Figure 2A. Simply supported condition was used with a hinge on one side and roller on other side to attain an effective span of 3000 mm. The test specimen of dimension 230 x 300 mm was placed on the supports and mounted with the hydraulic actuator in the centre of test specimen. An intermediate beam was placed for two-point load action between the load jack and the test specimen with two roller supports on the specimen top surface.

2.1 Beam casting and testing

Reinforced concrete beam was designed as per limit state method using IS 456: 2000 for mild exposure condition. The reinforcement provided in the beam as given below and detailing is shown in Figure 1.

Tension reinforcement: 2 nos. of 12 mm diameter Fe 415 steel
Transverse reinforcement: 2 legged 8 mm diameter stirrups at 150 mm c/c
Cover to concrete: 40 mm

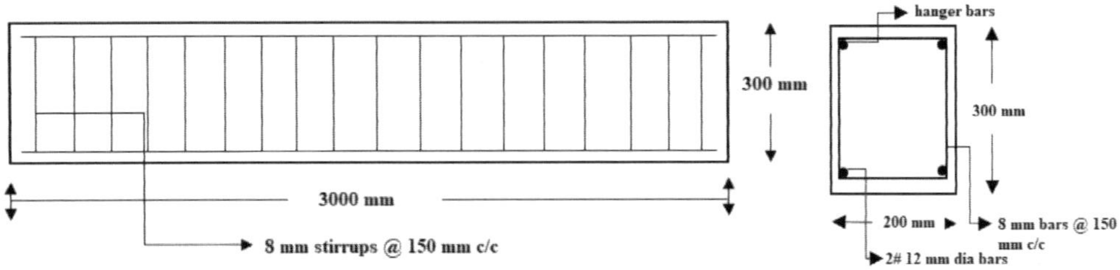

Figure 1. Detailing of the beam specimen

After curing the beam for 28 days the specimen was subjected to two-point loading using hydraulic jacks. Simply supported condition was used with a hinge on one side and roller on other side for an effective span of 3000 mm. The test specimen having dimensions of 230 x 300 mm was placed on the supports and the hydraulic jacks was mounted in the centre of test specimen.

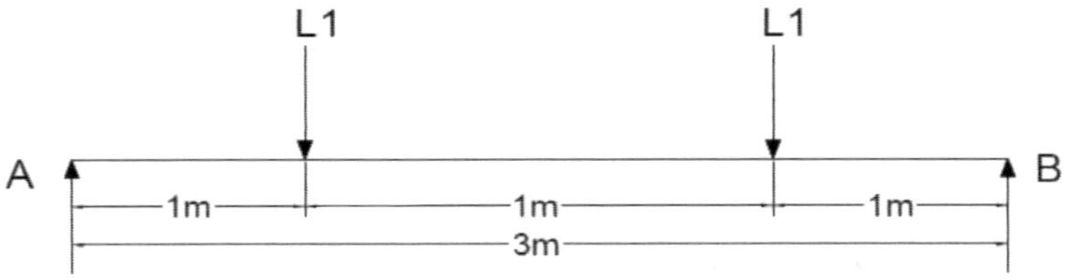

Figure 2A. Two point loading

Figure 2B. Testing of beam

2.1.1 Calculation of flexural strength from Lab Test

The beam is calculated for its flexural strength as per Indian standards. The Flexural Strength or Modulus of Rupture (f_b) is given in equation 1

$$f_b = \frac{Pl}{bd^2} \quad (1)$$

Where, b = width of specimen d = effective depth

l= supported length P = Maximum Load taken by the specimen.

2.1.2 Tensile Strength of Concrete

The flexural and splitting tensile strengths of the specimen was calculated as per the standard specification described in IS 516 and IS 5816 respectively. IS 456:2000 provides an estimate of the tensile strength from the compressive strength using the following formula given in equation 2

$$f_{cr} = 0.7\sqrt{f_{ck}} \quad (2)$$

Where f_{ck} is the characteristic cube compressive strength of concrete in N/mm^2.

f_{ck}= 20 N/mm^2

Flexural strength, $f_{cr} = 0.7\sqrt{20} = 3.3\ N/mm^2$

The test results of the beam are tabulated in Table 1

Table: 1 Beam test results

Beam ID	Beam 1	Beam 2	Beam 3
Beam Size (mm)	230*300*3000	230*300*3000	230*300*3000
Effective Depth D (mm)	280	280	280
Age At Testing (Days)	31	31	31
Area (Sq.mm)	64,400	64,400	64,400
Flexure Load (N)	18500	17800	13500
Flexural Strength ($F_b = Pl/Bd^2$) (MPa)	3.08	2.96	2.25
As Per IS 456	3.13	3.13	3.13
Nature Of Failure	FLEXURE	FLEXURE	FLEXURE

2.2 Theoretical calculation

The moment of resistance M_u is calculated as per IS 456: 2000 and is given in equation 3, where x_u is the neutral axis depth, b - width of specimen, d- effective depth and f_{ck} is the characteristic compressive strength of concrete

$$M_u = 0.36\frac{x_u}{d}\left(1 - 0.42\frac{x_u}{d}\right)bd^2 f_{ck} \quad (3)$$

= 0.36 x 0.46(1 – 0.42 x 0.46) 230 x 280 x 280 x 20

M_u = 48.18 kN.m

$$P_u = \frac{M_u}{lever\ arm} = \frac{48.18 \times 10^6}{1000} = 4.8\ t$$

2.3 Deflection

The deflection of the beam is calculated as per structural analysis calculation for the beam shown in Figure 3 and is given in equation 4.

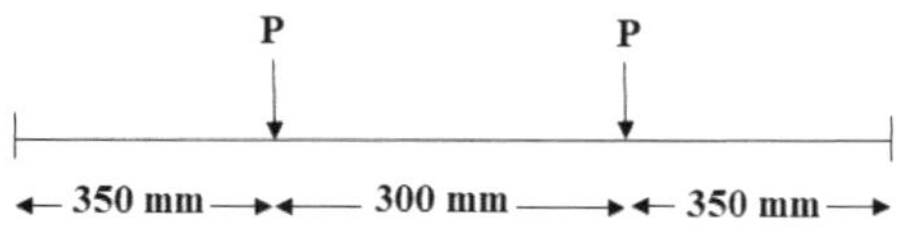

Figure 3. Loading on the beam

$$\delta_{max} = \frac{Pa}{24\,EI}(3L^2 - 4a^2) \tag{4}$$

$= 4.14$ mm

The deflection of beam may be generally limited to the following:

The final deflection due to all loads including the effects of temperature, creep and shrinkage and measured from the base level of the supports of Floors, roofs and all other horizontal members, should not normally exceed span / 250.
Span = 3000mm

Permissible deflection = 3000/250 = 12 mm

3. ABAQUS Modelling

ABAQUS is a finite element analysis software suite, and was introduced in 1978. It is a commercial software package for performing linear and nonlinear analysis. Based on the main objectives of this study, finite element model of reinforced concrete beam was developed three dimensionally and the process of ABAQUS is explained in the Figure 4 given below:

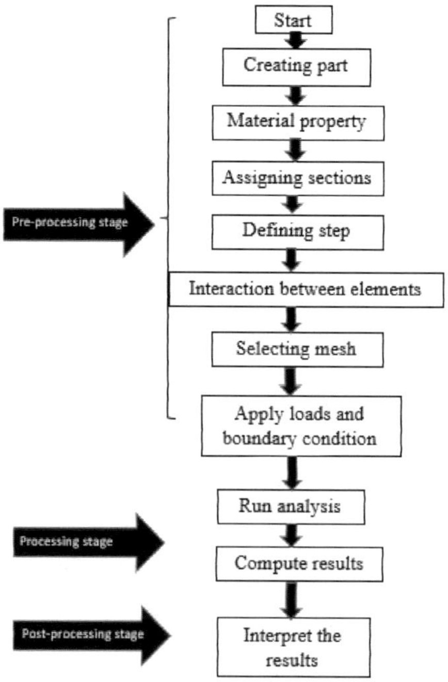

Figure 4. Flowchart of steps adopted in ABAQUS modelling

3.1 Creating part

The beam is modelled as a deformable solid body by creating part in two-dimensional profile. Then it was converted to three dimensions by extrusion. Three new part is created in the model BEAM_CONCRETE, REBAR_STEEL, STIRUPPS_STEEL. The extruded part is shown in figure 5. In this study concrete is modelled as eight noded element, steel is modelled as two noded beam element.

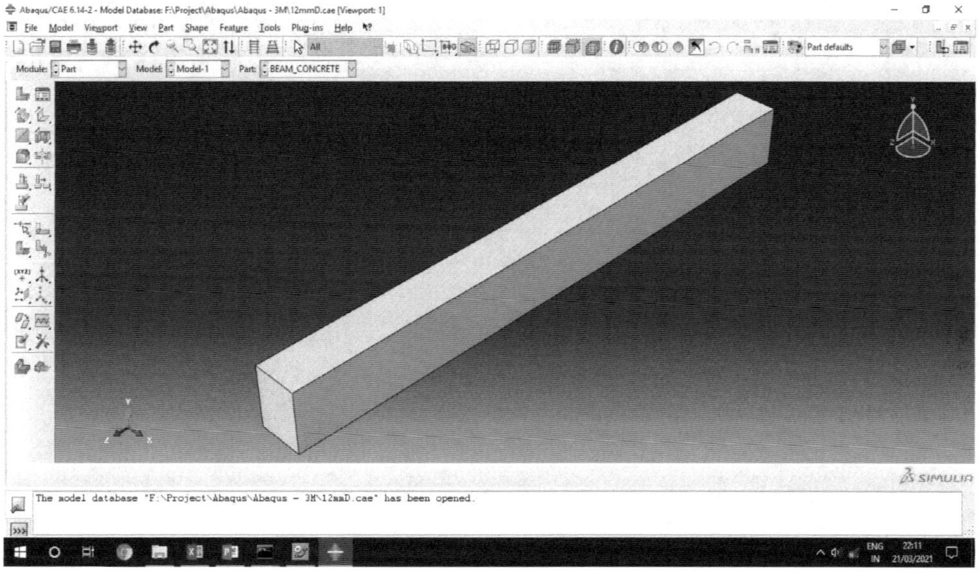

Figure 5. Extruded part

Beam of length 3 m and width 230 mm and depth 300 mm is created. Plane stress elements are commonly used to model beams in 2D as the beam is relatively long compared to its width. The reinforcement can be modelled with 2D and 1D elements as embedded reinforcement. The rebars were modelled as two –node beam elements connected to the nodes of adjacent solid elements.

3.2 Material property

The property of material used are defined as below in ABAQUS

1. Young's modulus of 22360 MPa for concrete and Young's modulus of 215000 MPa for steel.
2. Poisson's ratio of 0.3 for steel and 0.2 for concrete
3. Mass Density of 2400 kg/m^3 for concrete and Mass Density of 7460 kg/m^3 for steel.

3.3 Assigning section

A homogeneous solid section is created and assigned to the beam. The solid section created in the previous step was assigned as concrete and the two nodded beam element created for reinforcements were assigned to Steel. The assigned section is shown figure 6.

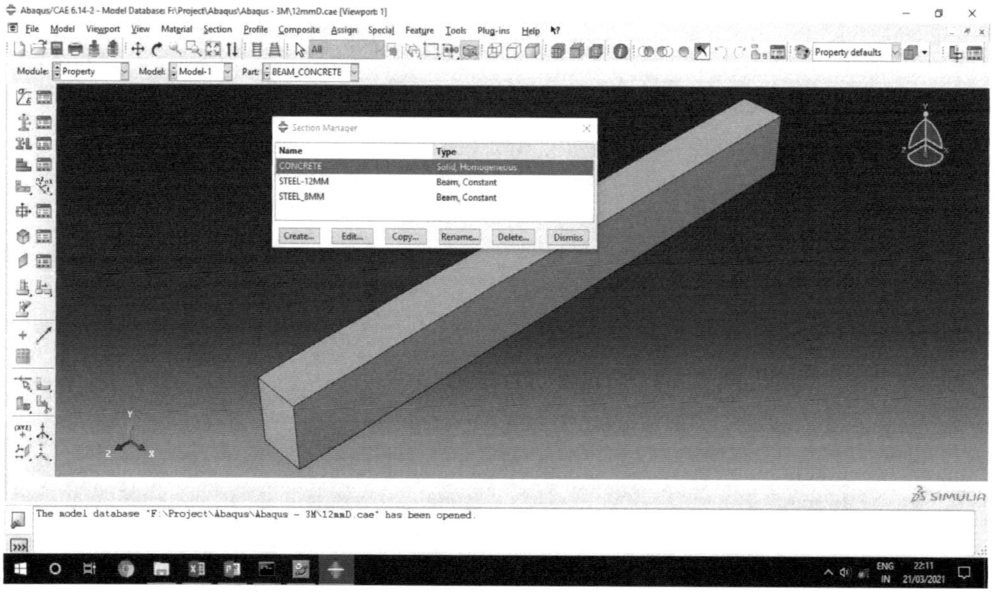

Figure 6. Material property assigned section

3.4 Step creation and assembling the model

In this simulation the static response of the simply supported beam is subjected to pressure load applied over the beam top. Consequently, this model will consist of two steps and the model is assembled as shown figure 7.

- An initial step, in which the boundary condition are applied that constraints two and of the simply supported beam.
- A general, static analysis step is applied with pressure load on top face of the beam

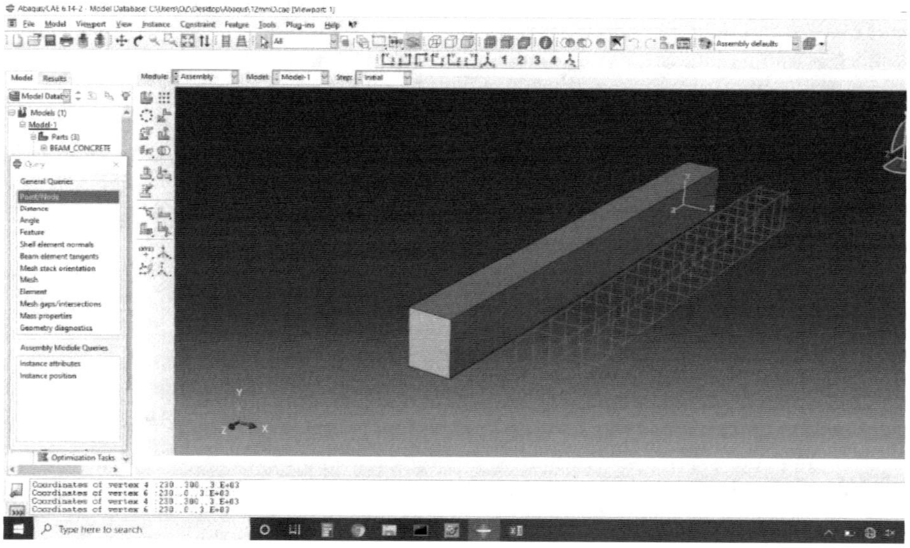

Figure 7. Assembling the model

3.5 Applying boundary condition and loading on the model

A new load is created in the model tree and load dialog box appears. In the dialog box, load is selected as POINT LOAD. Load will be selected a step 1 and load will be applied. In the selected Step list, Pressure (as the point load in this case is distributed throughout the surface area of the beam in contact) is applied on the beam. The magnitude of the load is selected as -50000 (CF2) and the negative sign indicates that load is applied along Y- axis. The beam with boundary condition is shown in figure 8.

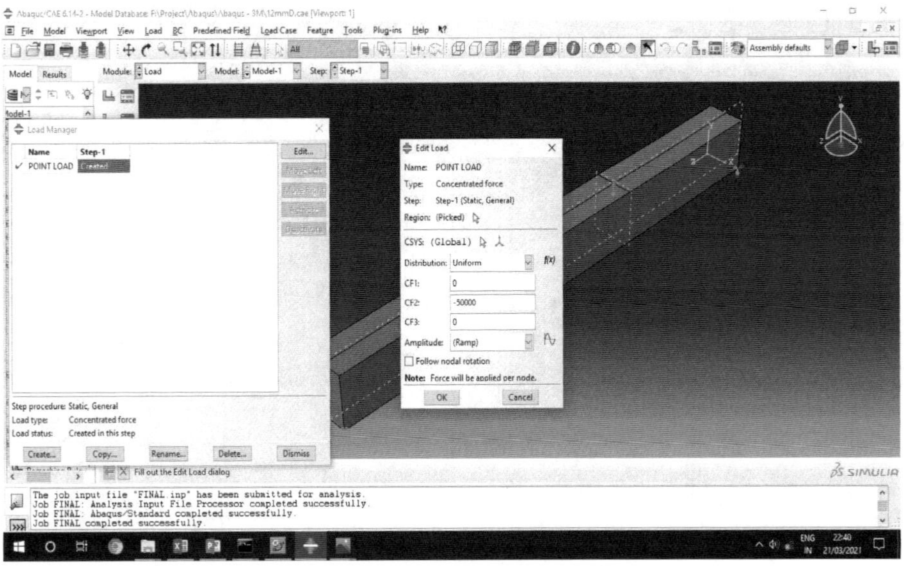

Figure 8. Applying boundary condition

3.6 Meshing the model

The process of generating nodes and elements is called Meshing. Elements are defined by generating nodes and connecting them to form a mesh. The finite element meshing is used to generate the mesh module. The meshing technique can be selected from ABAQUS that is used to create the mesh, the element shape, and the element type. Number of meshing techniques are available in ABAQUS to arrive at most converging solution. The default meshing technique assigned to the model is indicated by the colour of the model. If the model displays orange colour, it indicates that it cannot be meshed without assistance from the user. The meshed model is shown in figure 9.

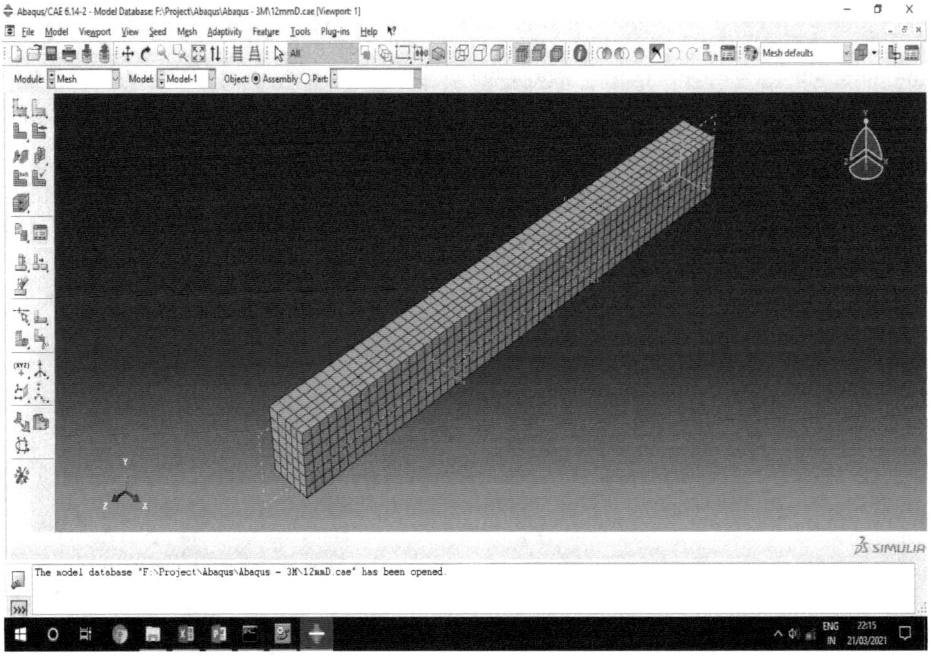

Figure 9. Meshed model

3.7 Creating and submitting an analysis

The model is created with all properties defined, assigned and meshed. The model is then analysed by running the analysis. Figure 10 shows the modelled specimen.

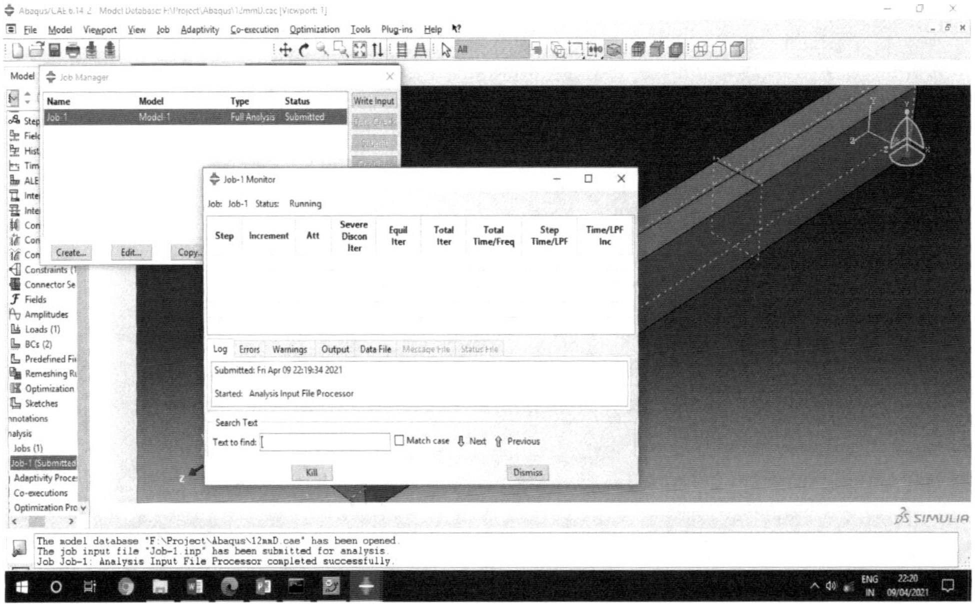

Figure 10. Running analysis

4. Evaluation of results

The beam was analysed by assigning the properties to the model. The analysis without aborting indicates the error free model. The results of the analysis are viewed in the Visualization module. The deflected shape is shown in figure 11. The results of deflection was then compared with the analytical and experimental results.

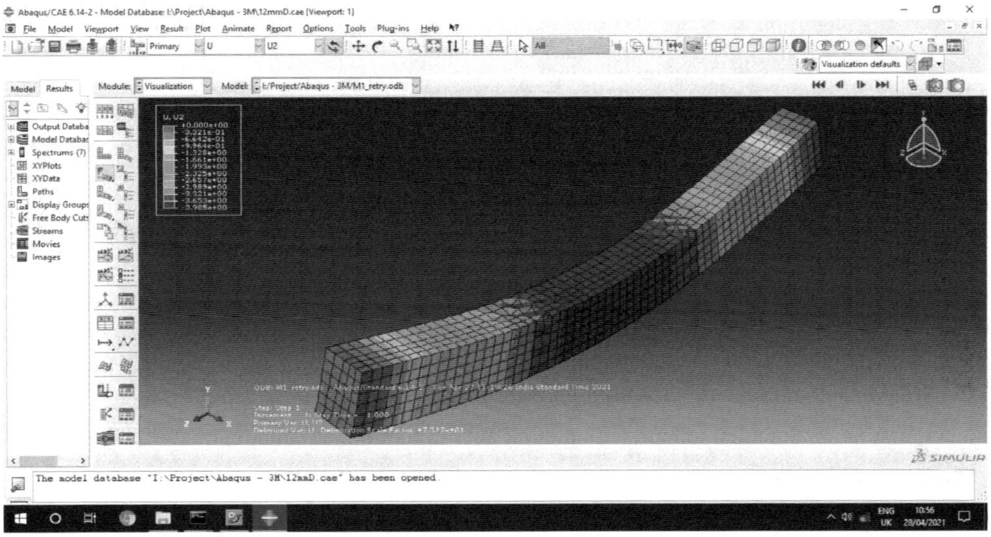

Figure 11. Deflected beam

5. Comparison of results

The analysed beam from ABAQUS was compared with the experimental and theoretical values as described in the work. It was found that the results have proved to be in par with the theoretical calculation and the properties assigned for both concrete and steel well behaved in the model and the results are tabulated in the table 2. The model was then re-assigned with two different tension reinforcement; viz, 2 # of 16 mm diameter bars and 2 # of 20 mm diameter bars. In both the cases the beam has shown the expected results.

Table: 2 Comparison of results

Beam ID	Beam 1	Beam 2	Beam 3
Beam Size (mm)	230*300*3000	230*300*3000	230*300*3000
Diameter of rod	12	16	20
Ast of rod mm^2	226.1	402.1	628.3
Theoretical displacement (mm)	4.14	4.14	4.14
Numerical displacement (MPa)	3.925	3.743	3.457
Experimental strength (mm)	2.76	-	-
Permissible displacement (mm)	12	12	12

6. Conclusions & Discussions

A RC beam of length 3 m, 230 mm wide and 300 mm deep was casted with 2 # 12 mm diameter bars as tension reinforcement and 2-legged 8 mm stirrups at 150 mm c/c was provided as transverse reinforcement on the beam. The beam was analysed theoretically as per specifications given in IS 456:2000 and numerically using finite element tool ABAQUS. The following conclusions have been arrived from the experimental testing and numerical evaluation:

- The beam was applied with two-point loading and the flexural strength of the beam was found experimentally which was in par with the specification given in IS 456: 2000
- The Experimental displacement and Numerical displacement for the beam reinforced with 12 mm diameter bar was studied and compared and they found to be similar.
- From the results of the 12 mm diameter bars used as tension reinforcement, numerical displacement for 16mm and 20mm reinforcement bars was done numerically using ABAQUS Modelling.
- The failure mechanism of a reinforced concrete beam is modelled quite well using FEA and the failure load predicted is very close to the failure the load measured during experimental testing.

References

[1] Sayed Mohamad Soleimani and Sajjad Sayyar Roudsari, "Analytical Study of Reinforced Concrete Beams Tested under Quasi-Static and Impact Loadings" *Journal of Applied Sciences,* Volume 9, 2019

[2] Dimosthenis Floros Olafur Agust Ingason, "Modelling and simulation of reinforced concrete beams" *Department of Applied Mechanics Division of Solid Mechanics Chalmers University Of Technology*, 2013

[3] T. Tejaswini, Dr.M.V.Rama Raju, "Analysis of RCC Beams using ABAQUS", International Journal of Innovations in Engineering and Technology (IJIET), Volume 5 Issue 3, June 2015.

[4] A.Hemamathi, Sanjith Murugavel , Binu Sukumar , M.Usha Rani, "Numerical Investigation of Precast Grouted Sleeve Connection under Cyclic Loading Using ABAQUS", *Journal of Physics: Conference Series,* 2070 (2021) 012178

[5] Sayed M S and Sajjad S R, "Analytical Study of Reinforced Concrete Beams Tested under Quasi-Static and Impact Loadings", *Applied Science, 2019.*

[6] S.V.Chaudhari, M.A.Chakrabarti, "Modeling of concrete for nonlinear analysis Using Finite Element Code ABAQUS", *International Journal of Computer Applications (0975 – 8887), Volume 44– No.7, April 2012.*

[7] T.S.Vishnu Kumar, N.Rushwanth Chowdary, "FINITE ELEMENT MODELING OF RCC BEAM BY USING ABAQUS", *International Journal of Creative Research Thoughts (IJCRT), Volume 8, Issue 7 July 2020.*

[8] Hamid S, Mahdi S, Amir H A, Mohammad A and Ali S, "Evaluation of reinforced concrete beam behaviour using finite element analysis by ABAQUS", *Scientific Research and Essays Vol. 7(21), pp. 2002-2009, 7 June 2012.*

[9] I. Saifullah, M. Nasir-uz-zaman, S.M.K. Uddin, M.A. Hossain and M.H. Rashid, "Experimental and Analytical Investigation of Flexural Behavior of Reinforced Concrete Beam", *International Journal of Engineering & Technology IJET-IJENS Vol: 11 No: 01, February 2011.*

[10] Swoo H L, Ali A, Kyung J, Hee D, "ABAQUS modeling for post-tensioned reinforced concrete beams", *Journal of Building Engineering Volume 30, July 2020.*

[11] Workshop 1, "Linear Static Analysis of a Cantilever Beam, Introduction to Abaqus", *D's Simulia*

[12] Indian Standard Plain and Reinforced Concrete - Code of Practice, Is. 456: 2000.

[13] Indian Standard Concrete Mix Proportioning – Guidelines (Second revision) IS 10262: 2019.

[14] Indian Standard Methods of Tests for Strength of Concrete IS 516: 2004.

[15] Dassault Systems "Getting Started with Abaqus/CAE Abaqus 2016".

Manufacturing of Brick from Medical Waste (MW)

Debasish Das[1], Debargha Banerjee[2], Swarup Rudra Paul[3], Shaswatika Pal[3], Jagriti Das[3], Archanava Chandra[3], Princu Kumar Singh[3] and Sujoy Adhikari[3]

[1] Assistant Professor and Head, Civil Engineering Department, JIS College of Engineering, Kalyani, WB
[2] Assistant Professor, Civil Engineering Department, JIS College of Engineering, Kalyani, WB
[3] Graduate Students, Civil Engineering Department, JIS College of Engineering, Kalyani, WB

Email.debasish.das@jiscollege.ac.in

Abstract. The environmental concern of Medical Waste (MW) generation has escalated to an alarming level due to the versatility and high demand in various applications due to COVID 19. Reusing MW to make construction material looks to be an environmentally acceptable solution to finding an effective way to use them. This is also because traditional building materials take a lot of energy to manufacture, which has numerous negative environmental consequences. This paper summarises prior investigations on repurposing various MW as a construction raw material and aggregate, as well as their quality, with a focus on bricks and paving blocks. This research begins by demonstrating the characteristics of plastics and the environmental effects of MW. After that, there will be a talk of MW reuse and how it influences total construction material quality. According to this study, only a few investigations on the use of MW in the production of paving blocks have been conducted. Furthermore, the majority of studies used compressive strength and water absorption as the principal criteria for evaluating brick and paving block quality. The compressive strength is found to be 12-13 N/mm^2 and water absorption is found to be less than 1% for the prepared sample.

1. Introduction

Masks and protective gear were worn once and then discarded in large quantities, eventually ending up in landfills or bodies of water. The exact amount of garbage generated by the epidemic is unknown, but experts estimate it to be enormous. Based on a projection for Italy, one study suggested that the globe may use up to 129 billion face masks every month [1]. India, with a population of over 1.3 billion people, has mandated the use of masks in public and created tens of millions of protective suits since the outbreak [2]. Between June and September, India's environmental regulator believes that 20,000 tonnes of coronavirus-related biomedical waste were created [3]. COVID related Bio-Medical waste generation statistics during 2020 is shown in Fig 1.

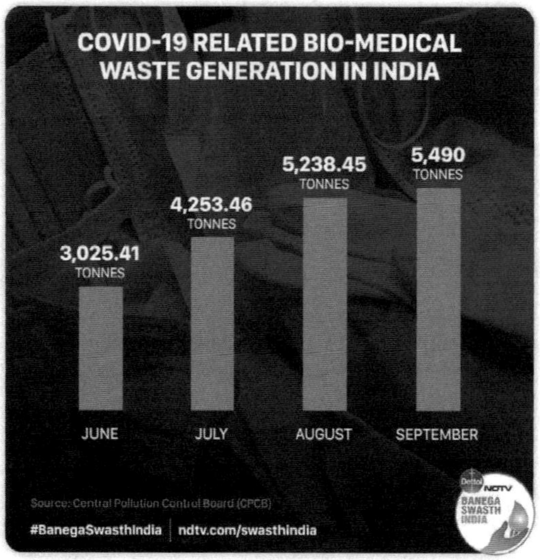

Figure 1. COVID-19 Related Bio-Medical Waste Generation in India

Source: Central Pollution Control Board (CPCB)

According to a recent UN Environment Program assessment [4,5], developing nations generate around 2.7 Kg MW per hospital per bed each day. Daily medical waste grew sixfold [6,7] in Wuhan, China, the pandemic's original epicentre, during the outbreak. From February to July of this year, health authorities in the United Kingdom delivered about 2.3 billion personal protective equipment (PPE) units [8,9]. According to experts, the sudden rise in single-use PPE usage has thwarted years of efforts to address the problem of non-degradable solid waste.

2. Methodology

As per the previous study from various researchers the following methodology is proposed for the current study.

i. Each square foot of these bricks will require 7 kg of MW. The current study [10] classifies surgical masks and PPE kits as MW.
ii. In order to begin full-scale brick production, 'Eco Bins' will be placed in various locations such as hospitals, police stations, and bus stops.
iii. "These eco-bins will gather personal protective equipment and face masks." It also features a side indication that indicates if the bin is full or not.
iv. Once the bin is filled, they will isolate the material for 72 hours before transporting it to their location, as per the Central Pollution Control Board (CPCB) requirements.
v. For manufacturing of such bricks around 20% of MW, 25% of paper waste, 5% of binder materials and clay will be mixed and various tests like compression test, water absorption test etc will be performed in the lab.

Methodology flow char is shown in Figure 2 below.

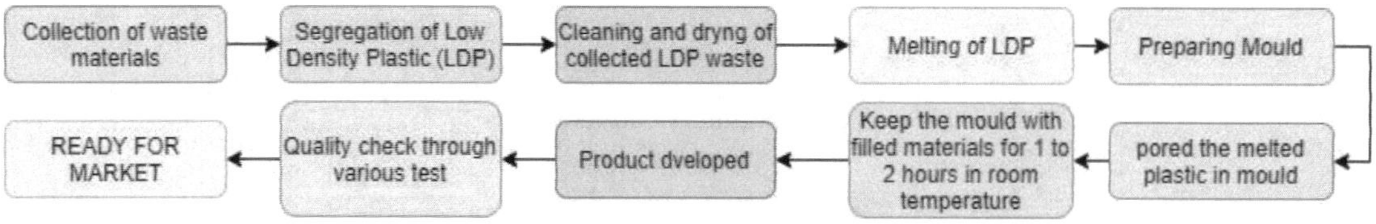

Figure 2. Methodology flow chart

3. Sample preparation

Number of samples are prepared considering the methodology as mentioned in Section 2. Disinfected mask and PPE kits are cuts into small pieces. Some paper waste also cut into small pieces and properly mixed with clay and binder materials. Sand is considered as the binder material. The cutting process is shown below in Figure 3.

Figure 3. Cutting process of MW

After the cutting process the chopped MW, binder materials and clay mixed to prepare the brick. Prepared sample brick is shown in Figure 4 below.

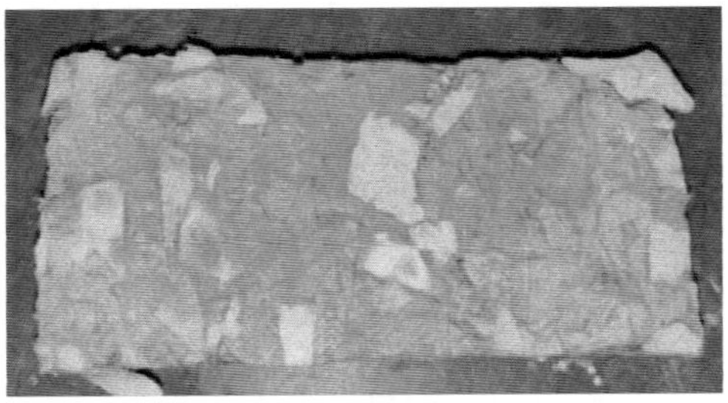

Figure 4. Prepared Sample

4. Laboratory testing

Compressive strength test and Water absorption tests are performed to check the quality of the prepared sample.

4.1. *Compressive strength test*

Bricks are commonly used to build load-bearing masonry walls, columns, and footings. Compressive loads are mostly experienced by these load-bearing masonry constructions. As a result, knowing the compressive strength of bricks is critical for determining their suitability for construction. Ball seats in the shape of a segment of a sphere, with the sphere's centre aligned with the plate's centre, are used in compression testing machines. The procedure for Compressive Strength Test is discussed below.[11]

- The sample is placed between the testing machine's plates with the flat face horizontal with mortar-filled face facing upwards.
- Apply axial load at a constant rate of 14 N/mm^2 per minute until failure occurs, then record the highest load at failure.
- The highest load at which the specimen fails to increase the indicator reading on the testing equipment any more is the load at failure.

4.2. *Water absorption test*
Water absorption tests on bricks are used to determine the bricks' durability, as well as their degree of burning, quality, and weathering behaviour. More resistant to freezing damage is a brick that absorbs less than 7% of its weight in water. Water absorption tests can be used to determine the degree of compactness since water is absorbed through pores in bricks. The ability of a brick to absorb water increases as the number of pores in the brick increases. Vitrified bricks are ones that absorb less than 3% of their weight in water as a result. This test determines the proportion of water absorption by bricks. Procedure of Water Absorption Test on Bricks [11].

- Dry the specimen at 105°C to 115°C in a vented oven until it reaches a fairly consistent bulk.
- Bring the specimen to room temperature and weigh it (M1) A specimen that is too hot to touch should not be used.
- Immerse the thoroughly dry specimen in clean water for 24 hours at 27+2°C.
- Remove the specimen from the water and wipe away any signs of moisture with a damp cloth before weighing it (M2). Eq. 1 shows how to calculate water absorption.

$$\textit{Water absorption value (W)} = \frac{M_2 - M_1}{M_1} X\ 100 \tag{1}$$

5. Result and Discussion
As per the tests mentioned in Section 4, Compressive Strength and Water Absorption value have been calculated. The compressive strength is found to be 12-13 N/mm^2 and Water Absorption value less than 1%. Further the obtained results for the prepared brick from MW are compared with the traditional bricks and a comparative study result is shown below in Table 1.

Table 1. Comparative study of various obtained results

Parameter	Traditional Brick	Brick from MW
Strength	7.5 to 10 N/mm^2	12-13 N/mm^2
Weight	3.5 to 4 Kg	2.2 to 2.5 Kg
Water absorption	Should not be more than 20% of its dry weight	less than 1%.
Cost	1st class brick cost around Rs. 10 per piece	Cost around Rs. 8 to 8.5 per piece
Transportation cost	Can carry 1000 number of brick in Pick up van	Can carry 1400-1500 number of bricks in pick up van

It is found from Table 1 that, bricks from MW are better than the traditional bricks in all the selected parameter.

6. Conclusion
Objective of the current study is to prepare construction materials from MW. Bricks are prepared from the MW and few tests have been performed as mentioned in Section 4. It is found that both the primary parameters i.e. Compressive Strength and Water Absorption value are better than the traditional brick. Therefore, the obtained comparative result show us a new direction in the real state segment for effective and economical construction works and also help to reduce the medical waste in post COVID scenario.

7. Reference
[1] G LaGrega, M. D., Buckingham, P. L. and J. C. Evans 2001 Hazardous Waste Management 2nd Edition, Mc-GrawHill.

[2] US Congress, Office of Technology Assessment 1988 Issuesin Medical Waste Management—Background Paper USGovernment Printing Office, Washington DC.

[3] M. M. Hassan, S. A. Ahmed K. A., Rahman and T. K.Biswas 2008 Pattern of Medical Waste Management: Exist-ing Scenario in Dhaka City, Bangladesh BMC PublicHealth, Vol. 8, pp. 36. doi:10.1186/1471-2458-8-36.

[4] Y.-C. Jang, C. Lee, O.-S. Yo on and H. Kim 2006 Medical Waste Management in Korea Journal of Environmental Management, Vol. 80, No. 2, pp. 107-115.doi:/10.1016/j.jenvman.2005.08.018.
[5] UAE Yellow Pages 2010 Clinics and Hospitals http://www.yellowpages.ae/category/clinics-/http://www.scribd.com /doc/4100881/.
[6] Middle East Health Care Intelligence 2010 http://biomedme.com/uae/dubaisclemenceau-medical-center-amongst-top-ten-best-hospitals-for-medical-tourism_6555.html.
[7] Medical Waste Management Act (MWMA) 2010 El Dorado County http://www.edcgov.us/Government/EMD/Hazardous Ma-terials/ Medcal Waste Management Act.aspx.
[8] US Environmental Protection Agency, US Environmental Protection Agency 2010 Medical Waste Tracking Act of1988 http://www.epa.gov/wastes/ nonhaz/industrial/ medical/track-ing.htm.
[9] Waste Med, Types of Medical Waste 2010 http://www.wastemed.com/types.htm *(18) (PDF) Medical Waste Management and Control*. Available from: https://www.researchgate.net/publication/258332243_Medical_Waste_Management_and_Control.
[10] Binish Desai 2020 India's 'Recycle Man' makes bricks from discarded face masks Link: https://www.thehindu.com/life-and-style/homes-and-gardens/binish-desai-has-designed-bricks-made-from-recycled-face-masks-during-covid-19/article32295016.ece.
[11] IS 1077 1992: Common Burnt Clay Building Bricks -Specification [CED 30: Clay and Stabilized Soil Products for Construction].

Acknowledgments
Authors wishing to acknowledge assistance or encouragement from colleagues, special work by technical staff, parent institute and nearby Medical College & Hospital for their support and cooperation for the completion of the current study.

Relationship of shading devices and its effects on daylight in Commercial buildings in Pune

Neha Joshi[1], and Pradnya Patki[2]

[1] MKSSS's Dr. Bhanuben Nanavati College of Architecture, Karve Nagar, Pune 52, India. Savitribai Phule Pune University.
[2] Bharatiya Kala Prasarini Sabha's College of Architecture, Tilak Road, Pune 30, India. Savitribai Phule Pune University.

Corrosponding author email-nehaknis@gmail.com

Abstract. Day lighting is a critical part of visual environment in workspaces. It creates a sense of well-being and contributes to the efficiency of the occupants. Research has established that solar radiation admitted through the windows have serious effects on visual comfort of the occupants in the indoor spaces. Use of solar shading devices can address glare problems. This signifies the need for design of optimal shading devices that increase daylight levels with desirable control on excessive sunlight with an aim to reduce glare and discomfort for the occupants. In this research various configurations of conventional shading devices used for commercial buildings located in Pune, India was examined. A typical commercial building is modelled using the simulation program Integrated Environmental Solutions (IES) where existing as well as different identified configuration of shading devices was analysed for daylight performance and visual comfort with respect to amount and uniformity of illumination with Radiance rendering tool. The analysis indicated that use of appropriate shading device result in improved visual comfort for occupants. The performance of different shading device presented aimed to help architects to select suitable shading device in order to contribute to sustainability aspects of office buildings in warm and humid climatic regions.

1. Introduction

Buildings are expected to provide shelter and ultimately maintain a high standard of occupant comfort. [31]. The energy consumption in the commercial sector in India is about 9% of the total electrical consumption which has seen a growth of 13.5% over the last four years. This increase is attributed to addition of energy intensive commercial units, of which office building constructed to cater for the information technology [IT] sector has major share. [26]. Buildings' energy efficiency improvement has become an international big deal for designers and researchers given the high potential for building modifications to reduce Green House Gas emissions. Energy consumption due to artificial lighting occupies a huge percentage of building energy consumption and is preferred as it creates good architectural light environment [21].

The building façade being one of the most important aspects of a building which play critical role in daylight utilization of the building. It is observed that "The building façade in has evolved over time with complexity accommodating a wide range of functions and performance requirements that include day-lighting, glare control, water and air infiltration, acoustical and thermal insulation, visual connectivity to outdoor environment, security and aesthetical appeal" [29]. Windows are recognized as one of the most critical components of building facades which act as a link between building with natural environment. They play an important role in meeting changing needs of occupants in light of work patterns maximizing overall building performance [11]. With the new technologies coming into play, the definition of window as an aperture in an opaque envelope does not completely stand true with the use of fully glazed skeleton structure and double-glazed facades [33]. With daylight windows are also essential for determining the occupant's level of satisfaction with the building. The daylight directly affects the human body and at the same time affects the mood, behaviour and hormonal balance of the occupants [25].

The different aspects of window design are size, configuration location and optical properties, which affect visual comfort and day lighting conditions [22]. In addition to that design parameters like window-to-wall ratio (WWR), thermal transmittance (U-value), shading, visual transmittance (VT) and daylight control are also critical [20]. Currently, commercial buildings are characterized by the presence of larger windows with big glazed surfaces resulting in energy losses due to overheating of indoor spaces due to direct solar penetration [37]. Increase in WWR does mean increase illuminance but if the building orientation and position are not considered it can lead to increased glare [1]. Shading devices can successfully reduce this glare and provide visual comfort [10]. It has been stated that geometry, control and properties of shading devices dictate daylight and thermal performance of a building [36]. Window systems designed to harvest natural light where high intensity of solar radiation in tropical countries like India often result in uncomfortable working conditions that include physical symptoms like headaches, itchy watering eyes, stress, difficulty in doing visual task, annoyance and glare [23]. Glare represents luminance induced sensation within the visual field which causes discomfort, annoyance loss in visibility and visual performance as it is significantly more than the level of luminance to which human eyes are adapted [5]. It is found that plane illuminance (WPI) which is more than 500 lux is too bright to work on computer however in tropical sky global illuminance is higher than 100 lux which cause glare and uncomfortable work environment indoors [20]. In tropical regions the challenge is not about achieving the required quantity of daylight but to control the quality of daylight with avoiding direct exposure to the sky component which could be achieved by using adequately designed shading devices [20].

2. Role of shading devices

It is a popular conception that large opening let ample daylight inside the workspace, but excessive solar radiation and direct view of the sky hamper the visibility and comfort of the occupants for which solar shades are needed [20]. Many times, use of solar shading is more acceptable to the architects than reducing the area of glazed surface, as it might lead to change in the architectural character of the building. At the same time, glare index values are independent of the window size for a particular set of sky luminance, room dimension and interior surface reflectance. Hence to meet a specific glare index, disconnecting the luminance by use of solar shading is helpful [12]. By choosing a shading device options that let in optimum daylight without causing glare, plays an important part in balancing daylight, thermal and visual comfort and energy use [28]. There are broadly two types of solar shading that are in use, the external and the internal solar shading. The external solar shades have proved to be more effective in reducing heat gains, glare and eventually energy demands for variety of climates [8].

At present, there are varied kinds of shading devices in the market and the choice is governed by many parameters [38]. Shading devices can be fixed or movable. The movable gives the occupant opportunity of adjust the shading devices as per requirement [9]. This study is limited to fixed solar shading device. There are basic three types of shading devices the horizontal shading device, vertical shading device and egg crate shading devices [15]. "The decision for the type of shading device depends on the amount of solar penetration required for the place and month. The solar potential must be carefully analyzed and designed for. To understand in a detailed manner is helpful to study the hourly values of irradiance" [35].

2.1 Influence of Orientation

Building orientation plays a critical role in choice of shading device. Looking at the solar geometry, it is evident that the exposure of each façade is different and is dictated by orientation of the façade. Hence each façade should be independently designed [18]. "As a rule of thumb in the northern hemisphere, horizontal louvers are an effective shading type for southern exposure, while vertical louvers are for east- and west-facing sun control" [27]. Even if northern façade receives limited solar radiation, with respect to tropics the northern façade should not be ignored as the solar exposure is considerable higher [2]. The northern façade receives the sun in the morning and late evening they are easily considered efficient for day lighting as mentioned in [7]. The south elevation in the northern hemisphere is comparatively easier to shade as a simple horizontal projection can easily shade the façade. The choice of shading device is a result of a combination of height of the window and solar angle [3]. The most difficult to shade orientations are east and west as the solar angle is low and horizontal projects are ineffective [13]. An added problem of glare is encountered for the west façade in the afternoon in summer especially. Hence it becomes critical that these two facades are well designed [17]. With respect to louvers, horizontal can be pre-dominant on all orientations while vertical louvers are restricted to the east and west orientations [33].

3. Study Context

Scenario of Information Technology (IT) parks in Pune: 36 Public IT parks are established by Maharashtra Industrial Development Corporation (MIDC) and The City and Industrial Development Corporation of Maharashtra (CIDCO). For getting private participation in creating world-class infrastructure for IT industry, 425 Private IT parks have been approved, out of which 88 IT parks have started functioning. Pune, a city in the state of Maharashtra, is the upcoming hub for growth in the commercial sector. 38 of such IT parks have been approved in Pune [24]. These IT parks are the energy guzzlers of the city. One of the important reasons for high energy consumption is the air-conditioning and artificial lighting requirement of the building. They consume 80% of the energy requirement of the building [30]. In US it was observed that application of daylight systems by architects and engineers was restricted and aspects like client's desires, building codes and simulation were considered as motivating factors for their use [4]. To understand the status of use of shading devices used in Pune, a visual study was done to assess the relationship between orientation and shading devices. The sample size taken was 40 Commercial buildings in the city of Pune at Magarpatta and Viman Nagar were visually examined to understand if orientation is considered while designing shading devices for a building façade.

The following types of shading devices were observed in the buildings

1. Overhang: Horizontal Projections over the windows
2. Recessed: Window projecting inwards
3. Fixed louvers: Horizontal slats fixed on the external face of the window
4. Green: Landscape features like trees acting as shading devices
5. Pergola: Battens parallel to the glazing area

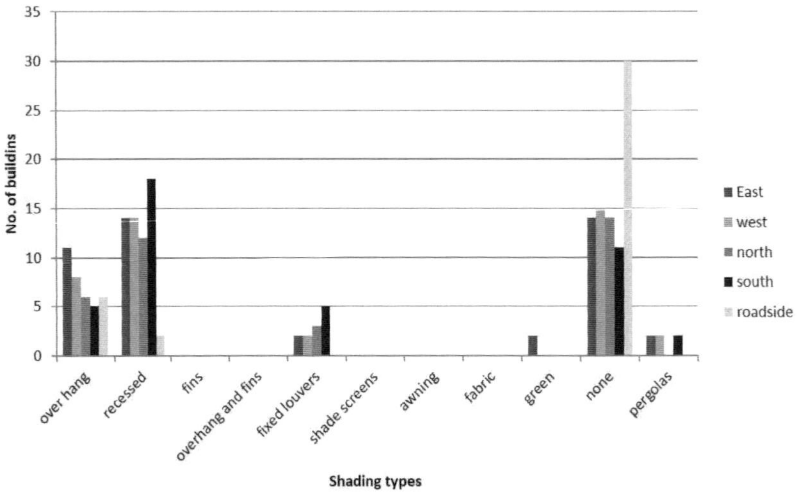

Figure 1. Comparative representation of shading devices used in various commercial buildings in Pune and the building orientation

Figure 2. Recessed windows.

Figure 3. Horizontal Projections.

Figure 4. Fixed Louvers.

Figure 5. Pergolas.

Limited variety of shading devices were found that included overhangs recessed and fixed louvers. Majority of the shading devices were designed considering the visual connectivity to the roads and building frontage design rather than orientation, azimuth and altitude angles. Many buildings were found with large glazed windows without shading device on the building front irrespective of orientation. Building envelop on sides other than front had recessed windows commonly used. The priority is being given to get the advantage of the road frontage as the fully glazed façade was largely found facing the road with no shading devices. None of the buildings used fins, overhangs, screens, awnings or fabric as a shading device. The sites facing the north took full advantage of orientations, but the rest three orientations were overlooked. Other than recessed, small projections, no new or innovative kinds of shading devices were used.

4. Materials and methods

To analyse role of shading devices a model was created of a representative office building located in Pune, and a series of daylight simulations on a software named Integrated Environmental Solutions (IES VE) is performed. An existing software company's office building is taken as the base case which is analysed for different configurations of shading devices with keeping other building parameters constant such as floor area, occupancy, envelope characteristics, building materials, working hours and window area. Efficiency of shading devices is examined in terms of illumination levels. IES uses RADIANCE as a daylighting simulation tool. *Radiance is internationally recognised as one of the leading lighting simulation tools available. IES have been using Radiance over a wide range of commercial projects for many years and to facilitate this work IES have integrated Radiance into the <virtual environment>* [14]. Software IES - Radiance has been selected as it is capable of analysing day lighting and thermal performance together and provides appropriate results to understand building performance [36]. The simulation is one of the most widespread Daylight simulation tool [34]. *RADIANCE is based on the backward ray tracing algorithm. This means that light rays are traced in the opposite direction to that which they naturally follow. The process starts from the eye (the viewpoint) and then traces the rays up to the light sources taking into account all physical interactions (reflection, refraction) with the surfaces of the objects composing the scene. Polarisation of light rays is not taken into account* [6]. The limitation of the software is that it is more accurate for diffused daylighting than for direct component [20].

In this study, the visual comfort examined with respect to amount and uniformity of illumination. Thus, the quantity and quality of daylight was studied. The effect of different shading device on illumination levels and daylight distribution of the building obtained and compared to the base case. As per the National Building code of India 2016, the optimum illumination range for computer work stations in office buildings is 300 – 500 lux. Beyond 750 lux is considered as high illumination levels and could lead to visual discomfort. With respect to glare the Illuminating Engineering Society defines glare as 1. Over Illumination and 2. Excessive contrast that is high range of illumination levels within the visual field. All the three types of visual discomfort will be studies in the paper. Also View through the façade after use of the shading device is also examined.

The output of the simulation study is in the form of Radiance generated plans used for analysis which shows various contours representing illumination levels in lux. The cool colours blue and green represent lower lux levels and the warmer colours represent higher level of illuminations. Optimum illumination level is marked in yellow within the black outline of the floor plate while excessive illumination and less illumination both are considered to cause visual discomfort [39]. The Simulation results represent results for the entire year and do not represent a particular month or weather condition.

4.1 Base Case

A 6 storied building having 2419.2 sqm floor area located in Pune is considered as the base case which represent a typical software company's office building having glass façade without provision of shading devices. The building is facing west with a fully glazed western façade while eastern façade having recessed windows 1.5 m X 1.7 m in size placed at an interval of 1.5 m (Figure 6). The characteristics of building material used are as shown in table

Table 1. Base Case description of Commercial building used in simulation.

Parameter	Description
Walls	Brick walls 0.3432 m thick, plastered both sides
Windows	Clear glazing, 6 mm thick with Visible light normal transmittance: 1
Occupancy	375 persons per floor (@ 9 sqm per person)
Lighting	Fluorescent lighting
Equipment	Equipment included laptops, UPS where only process energy is considered for analysis.
Air changes	6 (NBC)
Type of HVAC system	Split system with mechanical ventilation with ventilation mechanism that of air conditioning
Working hours	Monday to Saturday, 9.30 am to 7.30 pm.

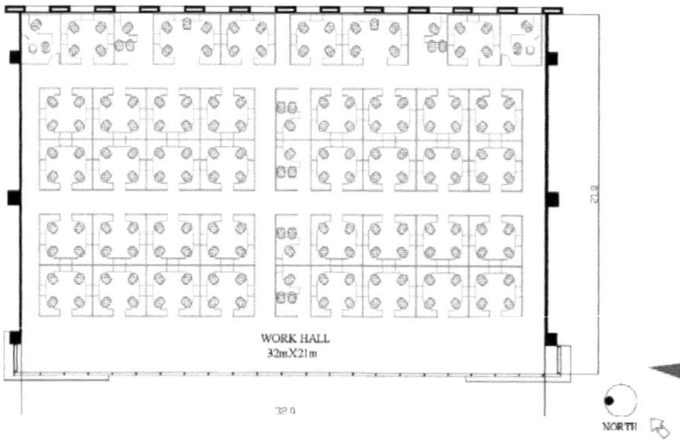

Figure 6. The open floor plan of the software company.

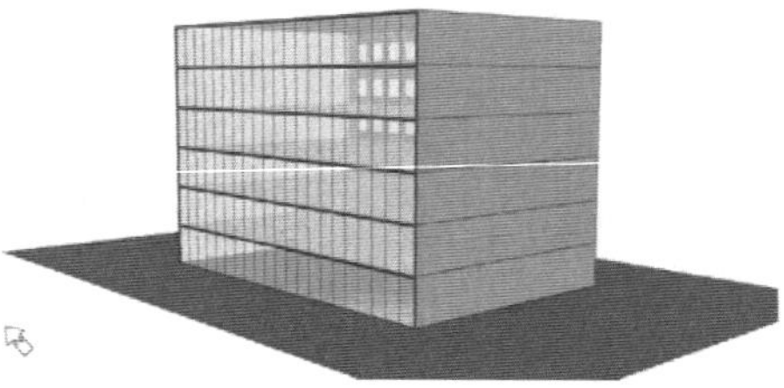

Figure 7. Model of the base case.

WWR for different facades on each orientation is obtained. The WWR on the west direction was found 100% as it is fully glazed while WWR is 30% on the east façade which has recessed windows of 1.5 m X 1.7 m placed at a distance of 1.5 m from each other.

5. Results

Base case: It is seen that optimum illumination are restricted to specific areas towards the central part. Periphery is over illuminated while it reduces drastically at the central part which can create visual discomfort. (Figure 8) As the building is completely glazed, the view is clear.

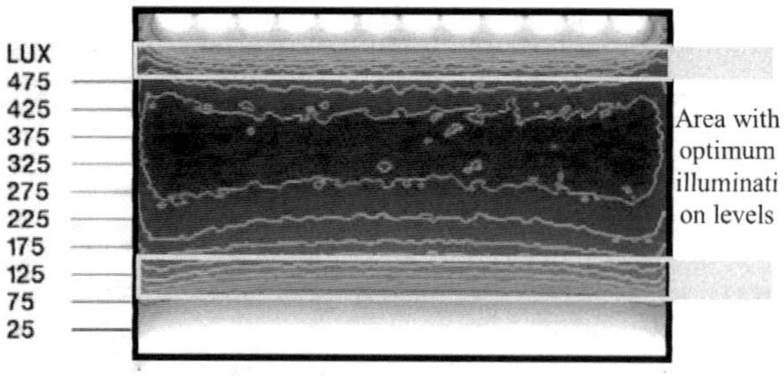

Figure 8. Lux levels in plan of Base Case Commercial building.

The performance of the building is further analysed with addition of different typologies of common shading devices used in Pune to examine the effect on illumination pattern. The result is shown in table 2.

Table 2. Results of simulation of proposed cases.

Shading Device	**Radiance Plan & observations**
Option 1 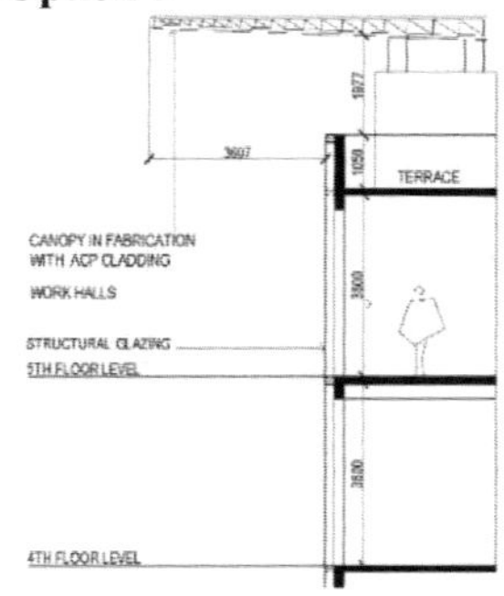 Section option 1 A large horizontal projection is taken at the terrace level of the building.	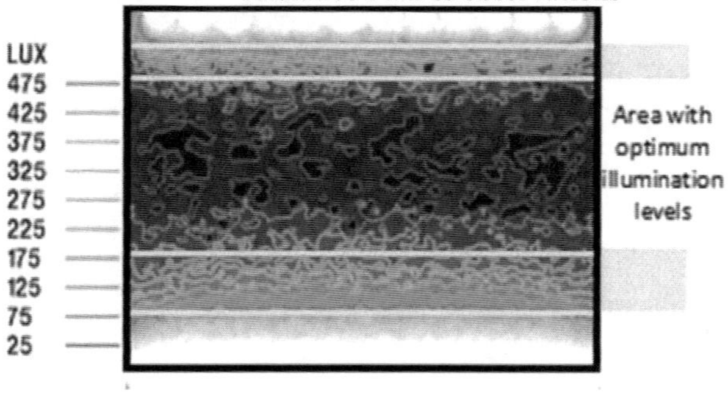 Lux levels in plan option 1 The visual discomfort persists at the periphery due to excess illumination. Daylight penetration has increased in the central part leading to patches of low illumination levels. View through the façade is clear.
Option 2 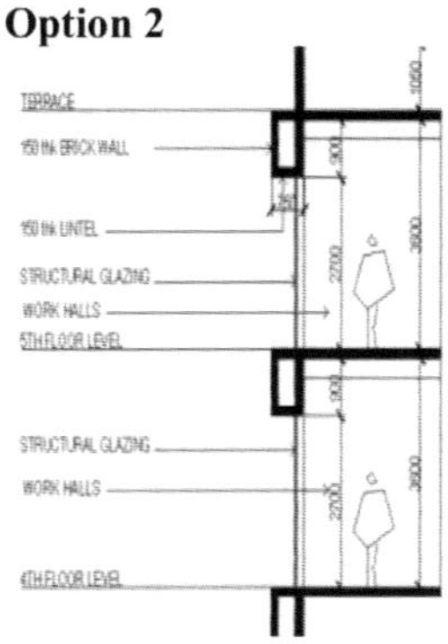 Section option 2 Small horizontal projection is taken.	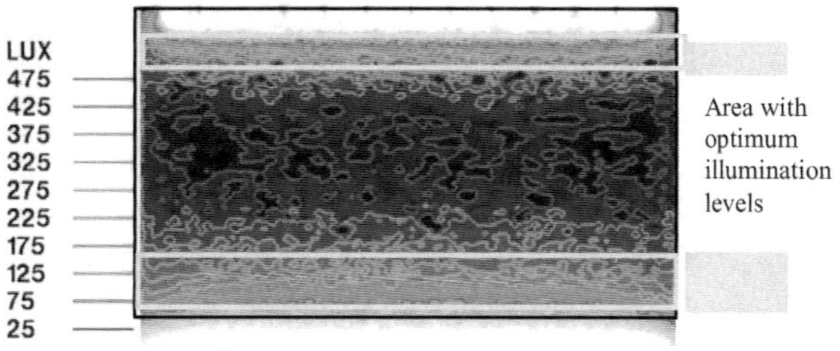 Lux levels in plan option 2 Central part has better illumination levels. The over illumination at periphery persists. View through the façade is clear.
Option 3 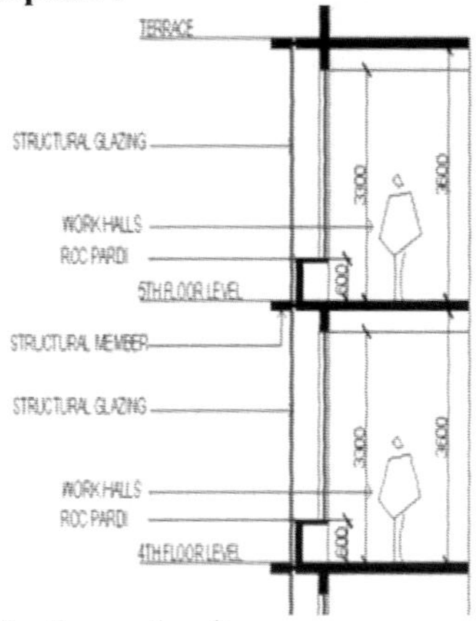 Section option 3 Recessed glazing	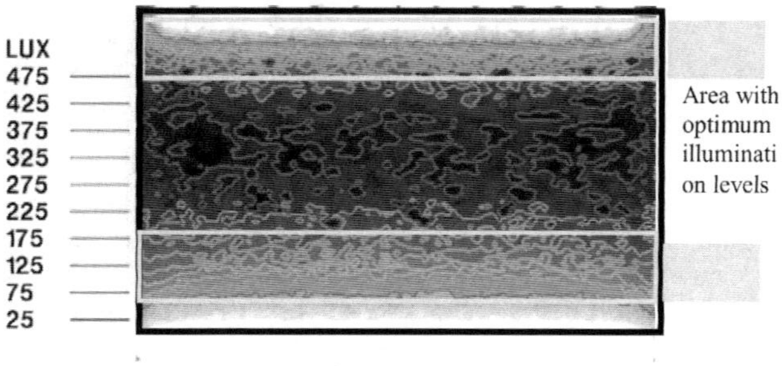 Lux levels in plan option 3 The area of optimum illumination has increased with reduction of over illumination at the periphery. The reduction of illumination levels from the periphery has become gradual. Contours of 150 – 250 have increased. View through the façade is clear.

Option 4

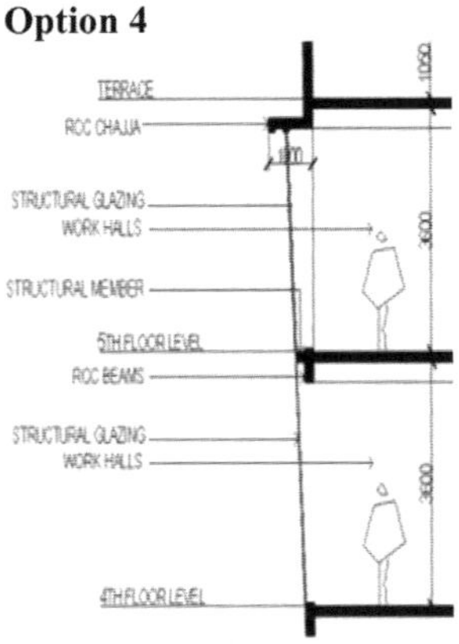

Section option 4
The façade is tilted towards the ground.

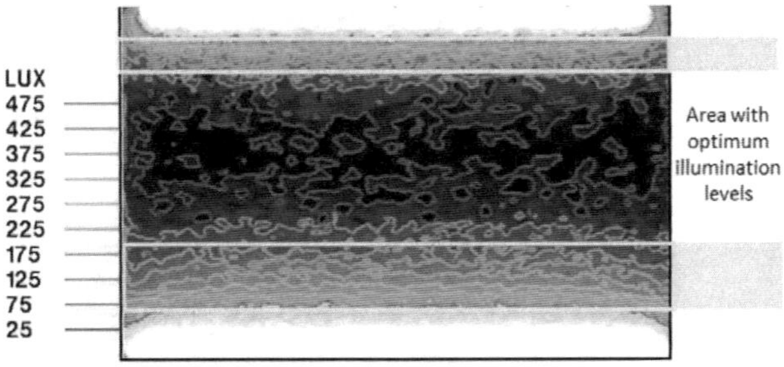

Lux levels in plan option 4.
Central band of low illumination levels have increased. No change in illumination levels Illumination contours. View through the façade is clear.

Option 5

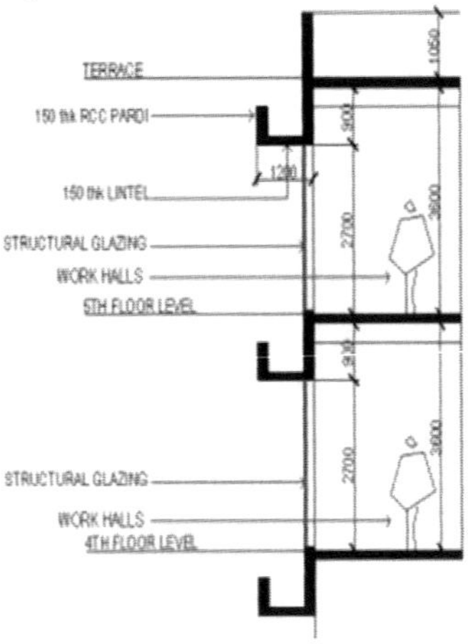

Section option 5

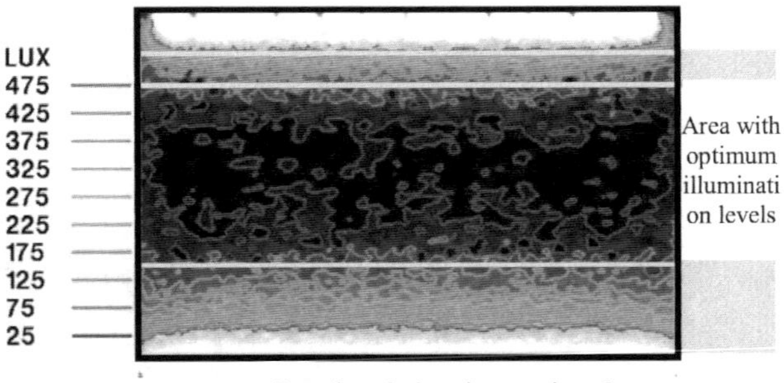

Lux levels in plan option 5
Better illumination levels with better daylight penetration. The patches of low illumination have decreased. The width of high illumination near the periphery has reduced.

Option 6

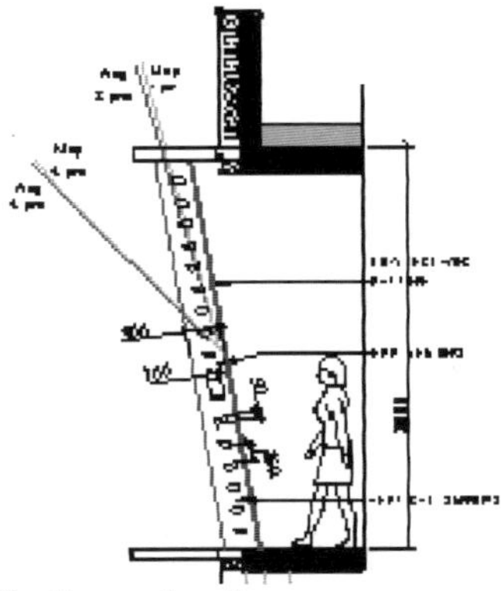

Section option 6
Horizontal slats are used to cover the glazing.

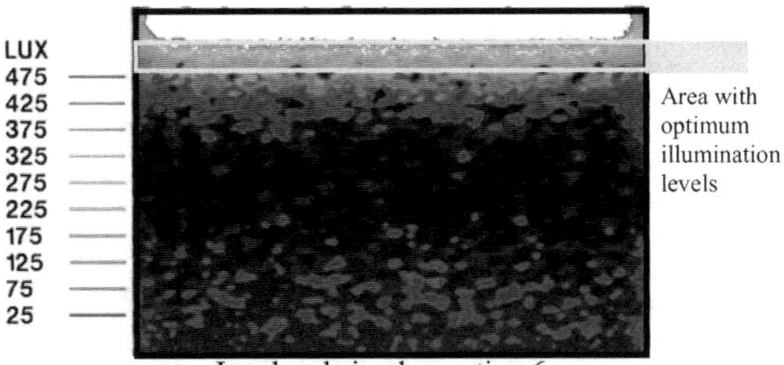

Lux levels in plan option 6
The illumination levels have reduced excessively, which increases the requirement of artificial lighting. View through the façade is obstructed.

IOP Conf. Series: Earth and Environmental Science 1084 (2022) 012079 doi:10.1088/1755-1315/1084/1/012079

Option 7

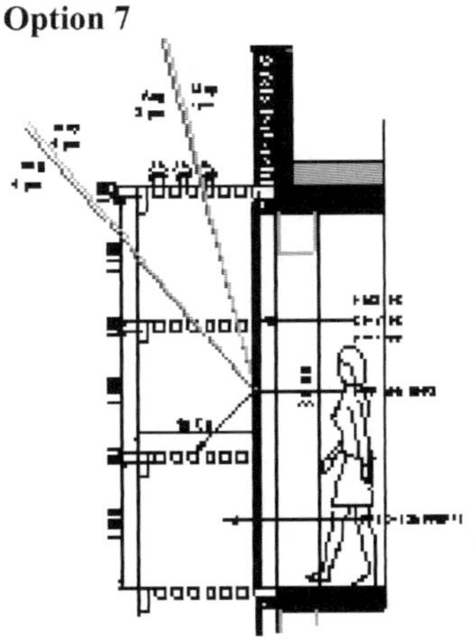

Section option 7
Horizontal Pergolas are used at different heights.

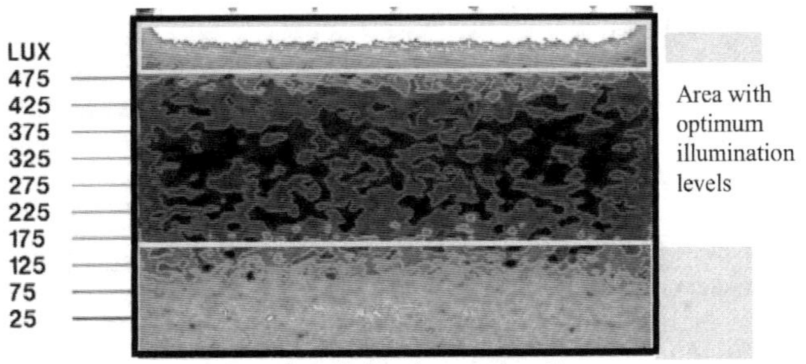

Lux levels in plan option 7

Better illumination levels on the floor plate with reduced peripheral glare and better day light penetration. The decrease of illumination levels from the periphery is gradual too. Minimal obstruction for the view through the facade

Option 8

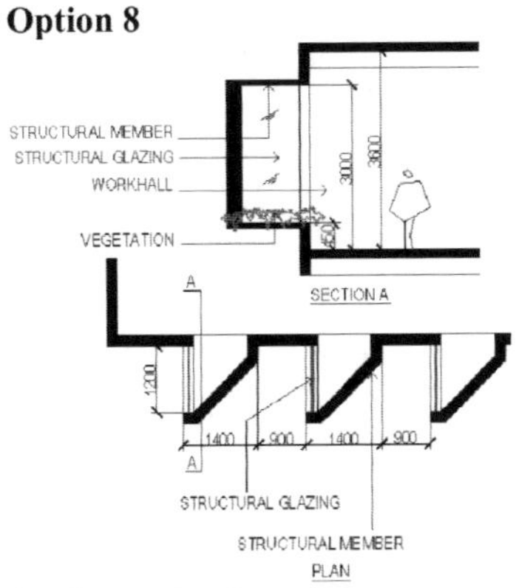

Section option 8
The façade face is modified to change the glazing orientation from west to south.

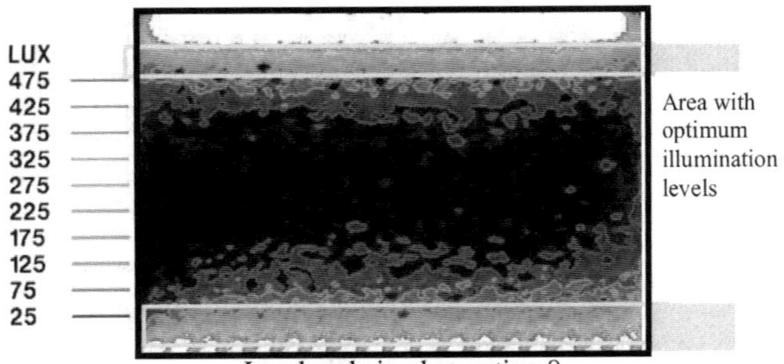

Lux levels in plan option 8

The peripheral high illumination is reduced but in patches. Daylight penetration is restricted. Central part is under illuminated. View through the façade is completely obstructed.

Option 9

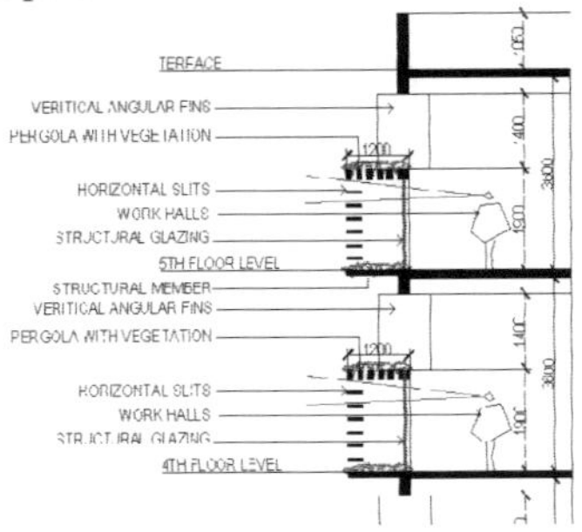

Section option 9
Combination of vertical angular fins with pergola and horizontal slits are used.

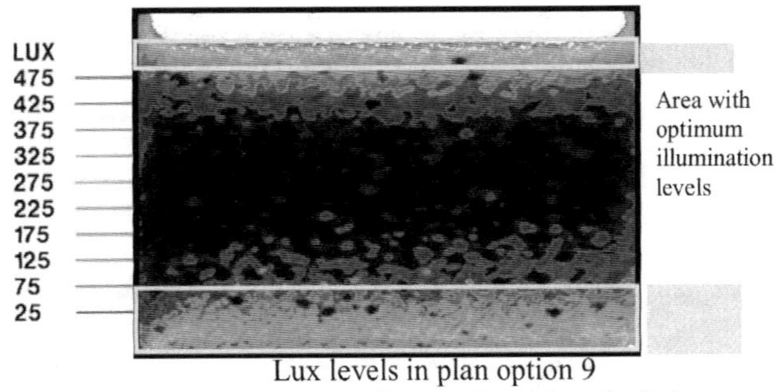

Lux levels in plan option 9

Peripheral illumination levels have reduced but daylight penetration in the interior of the floor plate is below the acceptable range.

6. Result & Discussion

The result of the research show the manner in which daylight results vary with change in shading devices. The Shading device options 1, 2, 3 and 4 are barely able to shade the façade, and perform the worst with daylight. In these options the workspace has maximum exposure to direct harsh sunlight which creates discomfort and glare. As the shading device design starts getting responsive to the climate, the illumination levels on the workspace start modulating. In option 7, 8 and 9, it is observed that the illumination levels become more stable and uniform and are well distributed for a major part of the floor. The high illumination glare is reduced considerably and daylight penetration has enhanced. This shows that the shading devices play a very crucial role on the daylight performance of a building. They affect the quantity and quality of daylight which is very crucial in daylight design.

Each façade is different and should be catered to in a unique way. The critical months and times for different places are different. They should be analyzed and then the building facade to be designed. In the practical world, the building orientation is rarely in the hands of the designer due to site constraints but catering the building façade accordingly is possible. Custom-made shading devices should be incorporated in the design stage itself so that their participation in the Daylight usability is more and the design is more responsible.

7. Conclusion

After doing the above research it was observed that first and foremost Shading devices are not a priority when façade design is considered for the city of Pune. The growing number of fully glazed IT buildings with no shading devices give us some idea to the energy consumption of the building. There are no trends in the design of shading devices and are a mere aesthetic input to the envelope which has no or very less effects on the building energy consumption.

It was assessed that use of shading devices does make prominent changes to the daylight penetration and has effective role to play in the daylight distribution. The option number 7 where the pergolas are spaced vertically at different heights, is able to give satisfactory results for illumination levels, daylight distribution is uniform without compromising the view as well. A building which can effectively use the daylight can certainly reduce the artificial lighting loads and create healthy spaces to result in higher productivity.

8. Acknowledgements

The authors would like to thank Dr. Vasudha Gokhale for her guidance while preparing this manuscript. The author would like to thank Dr. Anurag Kashyap Principal and the faculty of MKSSS's Dr. Bhanuben Nanavati College of Architecture for giving us the permission to use the IES software for simulation study.

References

[1] Alhagla K, Mansour A and Elbassuoni, R 2019 Optimizing windows for enhancing daylighting performance and energy saving *Alexandria Engineering Journal* 283 - 90

[2] Alibaba, H 2018 Solar control and shading strategies for double skin facades in hot climate 4 *Imperial Journal of Interdisciplinary Research (IJIR).*

[3] Architects T 2012 *The carbon neutral design project.* Retrieved from https://www.tboake.com/carbon-aia/strategies1b.html

[4] Azizkhani M and Haberl J 2021 Assessment and discussion of the level of the application of passive/natural systems and daylighting systems by practitioners in the US. *Science and Technology for the Built Environment* 109-128

[5] Carlucci S, Causone F, Rosa F and Pagliano L 2015 A review of indices for assessing visual comfort with a view to their use in optimization processes to support building integrated design *Renewable and Sustainable Energy Reviews*

[6] Compagnon D 2022 *RADIANCE: a simulation tool for daylighting systems* Retrieved from https://floyd.lbl.gov/radiance/refer/rc97tut.pdf

[7] CHPS BEST PRACTICES MANUAL 2002 in *Daylighting* chps Inc

[8] Dakheel J, and Aoul K 2017 Building Applications, Opportunities and Challenges of Active Shading Systems: A State-Of-The-Art Review *Energies*

[9] Datta G 2001 Effect of fixed horizontal louver shading devices on thermal perfomance of building by TRNSYS simulation *Renewable Energy 23*

[10] Denan Z, Hanita N and Majid A 2015 An Assessment on Glare from Daylight through Various Designs of Shading Devices in Hot Humid Climate, Case Study in Kuala Lumpur, Malaysia. *IACSIT International Journal of Engineering and Technology*

[11] Dodo Y, Kandar M, Ossen D, Dan'azimi J, Bornoma A and Abubakar A 2013 Importance of a View Window in Rating Green Office Buildings *Advanced Materials Research* 180 - 83
[12] Dubois C 2001 Impact of Shading Devices on Daylight Quality in Offices *Energy Build des*
[13] Emmanuel R and Baker K 2012 *Carbon management in built environment*
[14] Integrated Environmental Solutions Limited 2014 *Radiance IES User Guide.*
[15] Faisal G and Aldy P 2016 Typology of building shading elements on Jalan *Materials Science and Engineering* 128
[16] Grygierek J and Baranowski A 2011 Energy demand in the office buildings for various internal heat gain *Architecture Civil Engineering Environment*
[17] Florida solar energy center 2014 Retrieved from http://www.fsec.ucf.edu/en/consumer/buildings/homes/windows/shading.htm.
[18] Hamdani M, Bekkouche S, Benouaz T, Mohamed K and Benamrane N 2014 Orientation of buildings: predictive control based on the calculation of temperature and solar direct contribution *International Letters of Chemistry, Physics and Astronomy* 94 - 101
[19] Soo L and Kim G 2010 Predicted Performance of shading devices for healthy visual environment *Indoor and Built environment* 486-96
[20] Merghani, A., & Bahloul, S. 2017 Comparison Between Radiance Daylight Simulation Software Results and Measured on-Site Data. *Energy Simulation in High-rise Buildings*
[21] Hou Y 2018 Reasonable use of artificial lighting in building energy saving *AIP Conference Proceedings 1971*
[22] Hui S and Athanasios T 2010 A parametric analysis for the impact of facade design options on the daylighting performance of office spaces *International High Performance Buildings Conference Purdue*
[23] Iacomussi P, Radis M, Rossi G and Rossi L 2015 Visual Comfort with LED Lighting *6th International Building Physics Conference, IBPC 2015*
[24] MSME Development Institute 2016 *State Profile of Maharashtra*
[25] Jensen A 2021 *How Office Lighting Affects Productivity* Retrieved from https://www.andrewjensen.net/how-office-lighting-affects-productivity/
[26] Kumar S, Kapoor R, Rawal R, Seth S and Walia A 2010 Developing an Energy Conservation Building Code Implementation strategy in India *Summer Study on Energy Efficiency in Buildings 209-24*
[27] Lee K, Han K and Lee J 2017 The impact of shading type and azimuth orientation on the daylighting in a classroom–focusing on effectiveness of façade shading,comparing the results of DA and UDI *Energies 2017*
[28] Maria A, Gasparella A, Cappelletti F and Tzempelikos A 2018 Comfort and energy performance analysis of different glazing systems coupled with three shading control strategies *Science and Technology for the Built Environment* 545-58
[29] Arias A, Patterson M, Carlson A and Noble D 2015 Fundamentals in façade retrofit practice *Procedia Engineering* 934 – 41
[30] Nagaraju K 2015 An Overview of Energy Sector in India *International Journal of Science and Research* 1589 - 94
[31] O'Brien W 2013 Occupant-proof buildings: Can we design buildings that are robust against occupant behaviour? *Proceedings of BS2013* 1746-54
[32] Page J 1979 Systematic techniques of design for solar houses in high latitudes *Solar energy conversion* 629 - 713
[33] Ruck N 2000 Daylight in buildings - a sourcebook on daylighting systems and components *International Energy Agency Energy Conservation in Buildings and Community Systems Programme*
[34] Reinhart C 2004 Key findings from a online survey on the use of daylight simulation programs *Proceedings of eSim 2004* 1-8
[35] Sousa J 2012 Energy Simulation Software for Buildings: Review and comparison *Proc. of the First Int. Workshop on Information Technology for Energy Applications*
[36] Tzempelikos A and Athienitis A 2007 The impact of shading design and control on building cooling and lighting demand *Solar energy 81*
[37] Ulpiani G, Benedettelli M, Perna C and Berardo N 2017 Overheating phenomena induced by fully-glazed facades: Investigationof a sick building in Italy and assessment of the benefits achieved viamodel predictive control of the AC system *Solar Energy* 830-52
[38] Wienold J, Frontini F, Herkel S and Mende S 2011 Climate based simulation of different shading

IOP Conf. Series: Earth and Environmental Science 1084 (2022) 012079 doi:10.1088/1755-1315/1084/1/012079

device systems for comfort and energy demand *Building Simulation 2011: 12th Conference of International Building Performance Simulation Association* Sydney 2680 - 87
[39] Wilkins A and Boyce P 2018 Visual discomfort indoors. *Lighting Research and Technology* 980-0114

Response Spectrum Analysis of a RC Building Based on Complete Quadratic Combination (CQC) Method

Debargha Banerjee[1], Debasish Das[1], Uddipta Ghosh[1], Dina Debnath[2], Santanu Debnath[2], Snehasish Malakar[2], Ratnadwipa Das[2], Sanjoy Datta[2] and Syed Ealham Hossain[2]

[1] Assistant Professor, Civil Engineering Department, JIS College of Engineering, Kalyani, WB
[2]Graduate Students, Civil Engineering Department, JIS College of Engineering, Kalyani, WB

Email. debasish.das@jiscollege.ac.in

Abstract. In the realm of structural design, Seismic Analysis is crucial. Many structures have collapsed because of earthquakes; as a result of numerous studies, different methods for analysing and determining vibration modes of a structure have been created. Seismic analysis can be done in two ways: static analysis and dynamic analysis. The basic mode of vibration dominates the response in the static technique of seismic analysis, and the mass and stiffness of the structures are evenly distributed, resulting in a regular shape, however this assumption fails for irregular buildings. As a result, dynamic methods are used to solve this issue. Response spectrum is one such dynamic method that is relatively simple to utilize compared to others. The benefit of employing the seismic response spectrum method is that it aids in the prediction of displacement and member forces in structural systems. As a result, this method aids in determining peak structural responses within a linear range, which can then be used to calculate lateral forces created in structures due to earthquakes, allowing for more earthquake-resistant building design. Although this method is approximate, it is a cost-effective and beneficial method for preliminary design research. The response spectrum method is used on STAAD.PRO V8i software to design a G+4 residential structure in our study. The building's measurements are 16 meters long, 12 meters wide and 15 meters tall. The foundation was built to a depth of 1.5 meters. The structure was built to withstand earthquakes in seismic zone III on medium-density soil. Dead load (IS 875 Part I), Live load (IS 875 Part II), Seismic Load (IS 1893 Part I), and Wind Load are some of the load instances that were presented based on Indian Standard code (IS 875 Part III). To acquire the best response, the response spectrum modal combination Complete Quadratic Combination (CQC) is used. Various parameters like wind intensity, seismic parameters, dynamic weight, modal participation factors, base shear and peak storey shear are calculated. Base shear and peak storey shear are found to maximum in Mode 2 and between 5th and 6th respectively.

1. Introduction

Civil engineers' principal concern has become the static and dynamic analysis of a building. Seismic Analysis is a branch of structural analysis that involves calculating a building's response to earthquakes. In earthquake-prone areas, it is an element of the structural design, structural evaluation, and retrofit process. Response Spectrum Analytic (RSA) is a linear dynamic statistical analysis approach for determining the likely maximum seismic response of a basically elastic structure by measuring the contribution from each natural mode of vibration. By measuring pseudo-spectral acceleration, velocity, or displacement as a function of structural period of a given time history and damping level, RSA provides insight into dynamic behavior. Therefore, keeping the facts in mind, the current study on Linear Dynamic Seismic Analysis to calculate Storey peak shear, base shear is conducted by CQC method.

2. Literature review

It has been discovered that multistory buildings are particularly vulnerable to earthquake excitation, with a modal participation rate of 75% [1]. Another study found that when the structure is made up of Special Moment Resisting Frames (SMRF), the maximum storey displacement using Construction Spectrum Analysis (CSA) is 56% higher than the displacement with RSA [2]. In a conventional set up, base shear approval is more in Zone V and in the delicate soil. From the first to the last storey, the maximum displacement increases. Shear reverence is more prevalent in Zone V. Quakes have a big impact on unpredictable shapes, especially in high seismic zones. Base shear values will be greatly raised at upper stories due to RSA and Static Analysis (SA) [3]. When compared to dynamic analysis, static analysis yields superior results. Because the values of displacements are higher in static analysis than in dynamic analysis, the outcome is almost uneconomical. Considering these considerations, the current work used dynamic analysis [4]. The analysis included the storey deflection, storey drift, and base shear. A shear wall's presence, as well as its shape, is crucial. [5]. The bare frame has the lowest base shear, whereas the infill frame has the highest. In order to improve the value of bay dimension, time period grows by 1 to 30%, base shear increases by 10 to 60%, maximum displacement increases by 2 to 25%, and maximum drift increases by 1 to 28% in various types of multistory buildings [6]. It was also suggested that RSA be used in the case of tall, uneven, and essential constructions [7]. It is determined through static analysis that the structure's base shear increases as seismic zones increase. The displacement and storey drift of the building models grow as the seismic zone increases, according to static analysis. Storey shear, on the other hand, decreases as building heights rise. [8]. RSA gives a useful approximation of the structural safety and reliability of structures that have been subjected to earthquakes. RSA enables a clear knowledge of the contributions of various vibration modes to the overall seismic response of structures [9]. The RSA method is more precise since it addresses the dynamic character of the seismic load in comparison to the Seismic Coefficient Method (SCM). Base shear at the bottom storey is higher in RSA than in SCM [10]. It was discovered that buildings on medium soil face more base shear than those in hard soil. When a building is built on hard soil, it has less displacement and drifts than a building built on medium soil [11, 18]. The current study was conducted in the area with medium soil to obtain better base shear. IS 875 (Part 1), IS 875 (Part 2), IS 875 (Part 3) and IS 1893 (Part 1) codes are used to design the structure considering Dead Load (DL), Live Load (LL), Wind Load (WL) and Seismic Load (SL) respectively [12, 13, 14]. The analysis included the storey deflection, storey drift, and base shear. A shear wall's presence, as well as its shape, is crucial. They help to prevent high displacement caused by seismic forces [15]. Another study found that when a building's height increases, the axial compressive force decreases. Based on frequency data acquired from various software, buildings were discovered to be substantially stiffer [14]. The modal frequency is higher in higher modes, according to the findings. At greater seismic zones, base shear, lateral pressures, maximum storey displacement, storey shear, and overturning moment all increased. Because the equivalent static lateral force approach produces greater values of forces and moments, making construction uneconomical, the RSA method should be considered [16]. Another study found that the internal columns in all floor levels are the most affected by the compression forces caused by all load combinations [18]. For high-rise structures with vertical regularities, a dynamic analysis is required. The considered structure is stiff for earthquake excitation because the modal participation factor is greater than 75% in the upper mode [19].

3. Methodology

A G+4 RC structure with a height of 17.4 metres is considered for analysis. Analysis and design is followed accordingly with reference to IS Code 1893 (Part I), 2002. The current study is done to follow the steps given below:

Step 1: Preparing the building frame with shear wall using STAAD Pro Software and assign the properties of beam, column, shear wall and support. The building frame is shown below in Figure 1.

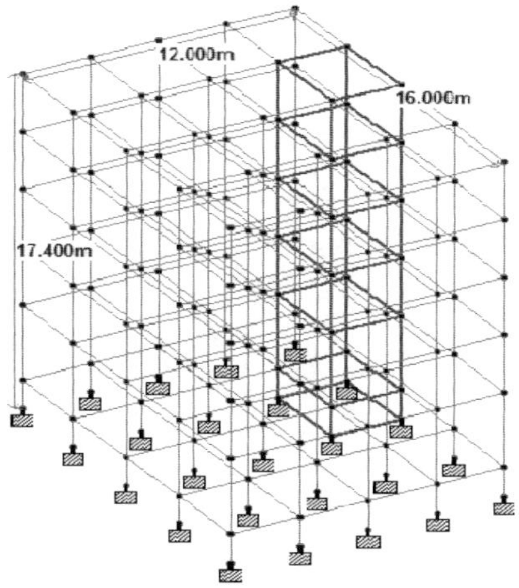

Figure 1. Isometric view of building frame with shear wall

Building frame consisting of width 12 m, length 16 m and height of 17.4 m from base level is design using Staad Pro software. The building specifications are given below in Table 1.

Table 1. Building section properties

S. No.	Section	Material	Member
1	0.35mx0.3m	Concrete	BEAM
2	0.60mx0.45m	Concrete	COLUMN
3	0.40mx0.35m	Concrete	TIEBEAM
4	0.45mx0.35m	Concrete	BOTTOMOF LIFTLEVEL
5	Surfacethickness0.2m	Concrete	SHEARWALL

The 3D view of the building frame is highlighted below in Figure 2.

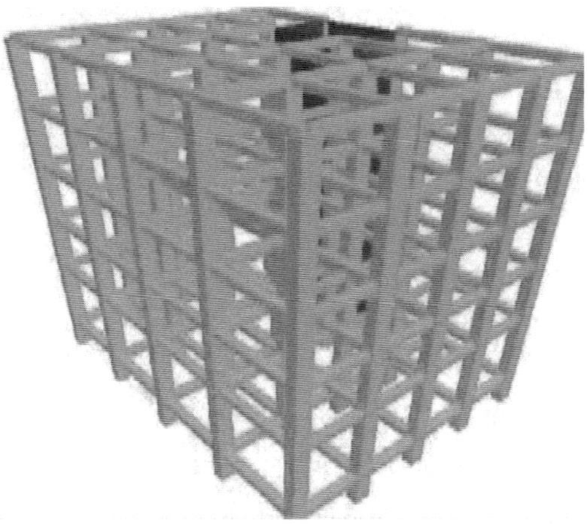

Figure2. 3D view of building frame along with shear wall

The 3D view is created using Staad Pro software and the shear wall is highlighted as the black portion in Figure 2.

Step 2: Assign the different types of loads i.e DL, LL, WL and SL as per IS 875 (Part 1), IS 875 (Part 2), IS 875 (Part 3) and IS 1893 (Part 1) codes respectively.
Member load and floor load under DL are shown in Figure 3 (member load UDL 15.5 N/mm) and Figure 4.

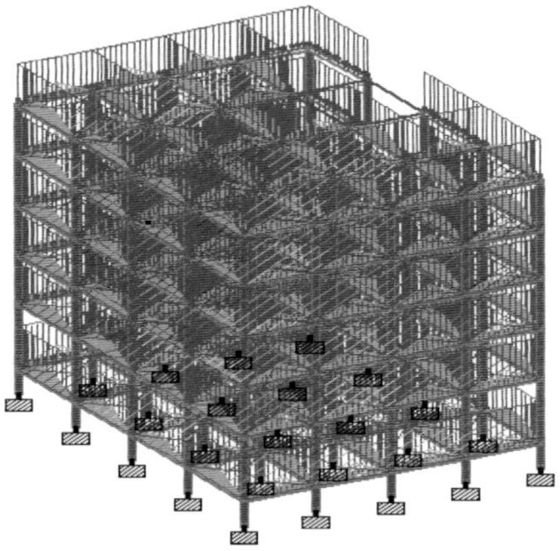

Figure 3. Assignment of Member Load

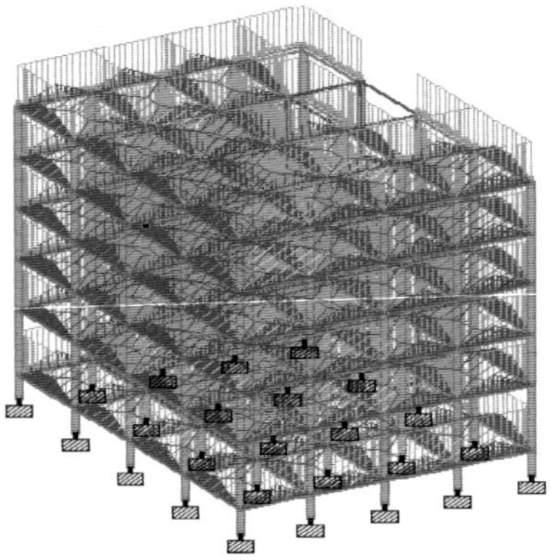

Figure 4. Assignment of Floor Load

LL including and excluding roof level are shown in Figure 5 and Figure 6 respectively. Figure 5 is obtained after applying a floor load of 2.2 kN/m at the building height from 1.75 m to 14.5m range in Y direction excluding the roof.

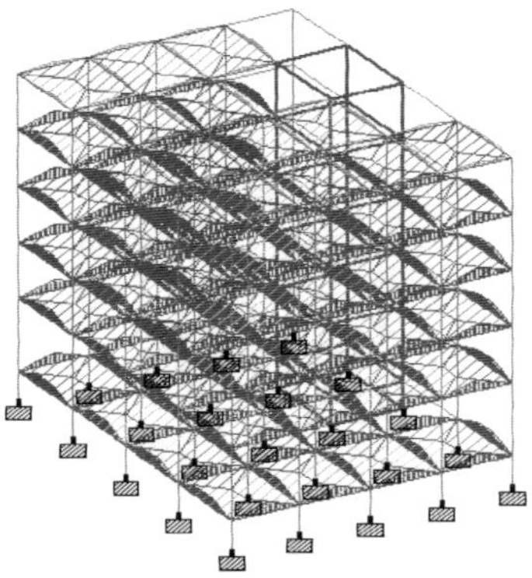

Figure 5. Assignment of LL excluding roof area

Figure 6 is obtained after applying a floor load of 0.85 kN/m at the roof level in Y direction.

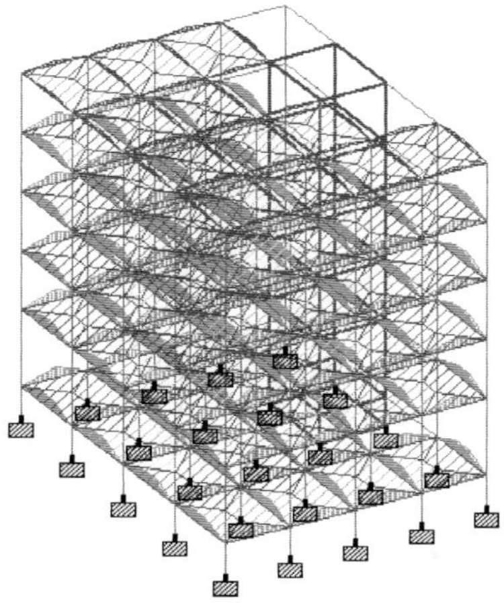

Figure 6. Assignment of LL in roof area

Lateral movement due to SL in building frame is shown from Figure 7 to Figure 10. SL in X+ and X– direction is shown in Figure 7 & Figure 8 and SL in Z+ and Z- direction is displayed in Figure 9 and Figure 10.

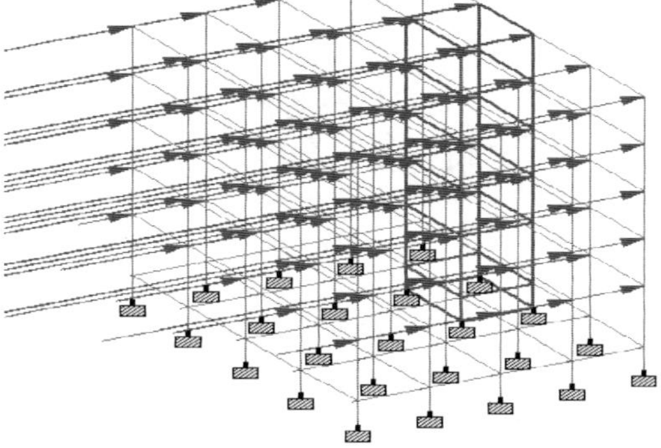

Figure 7. SL in X+ direction

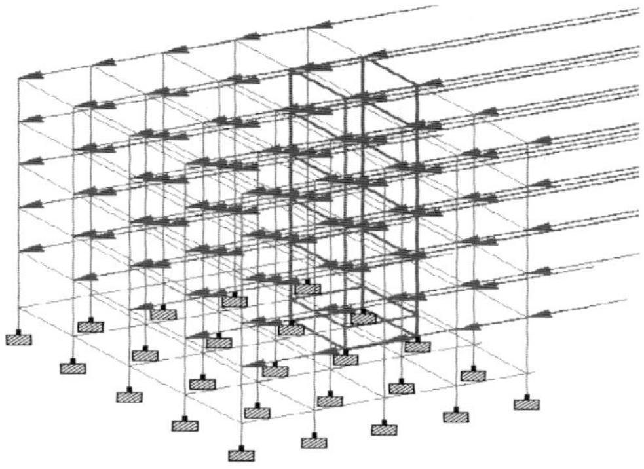

Figure 8. SL in X- direction

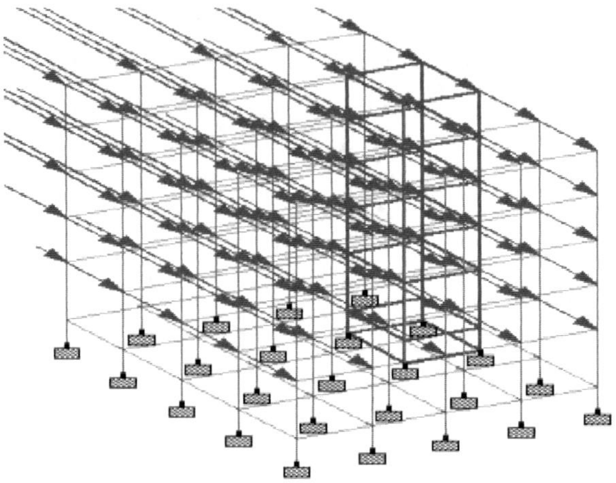

Figure 9. SL in Z+ direction

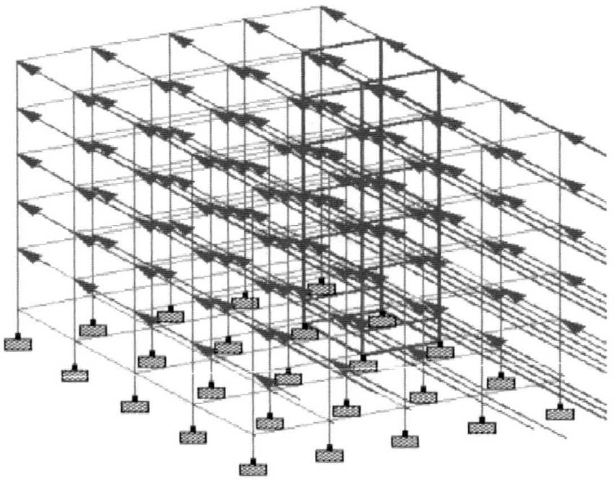

Figure 10. SL in Z- direction

Lateral movement due to WL in building frame is shown from Figure 11 to Figure 14. WL in X+ and X– direction is shown in Figure 11 & Figure 12 and WL in Z+ and Z- direction is displayed in Figure 13 and Figure 14.

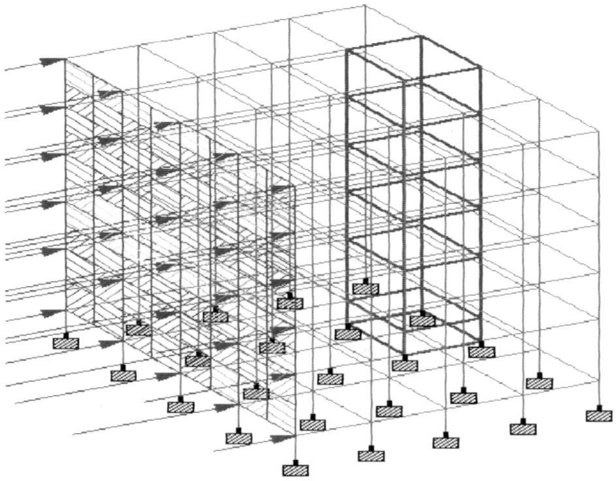

Figure 11. WL in X+ direction

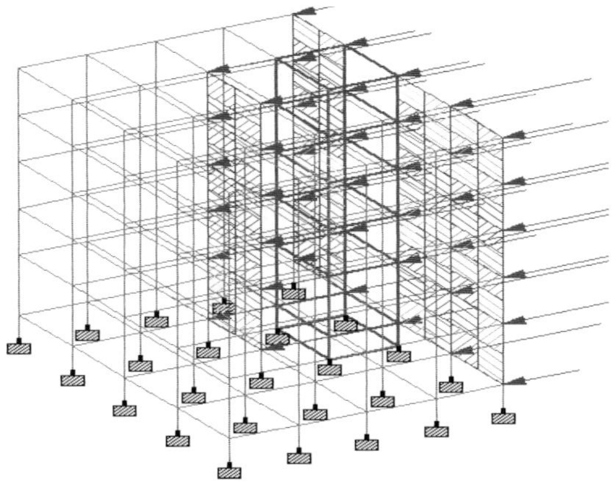

Figure 12. WL in X- direction

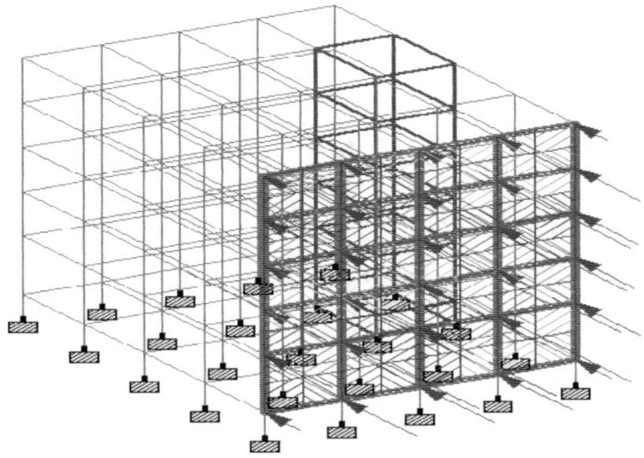

Figure 13. WL in Z+ direction

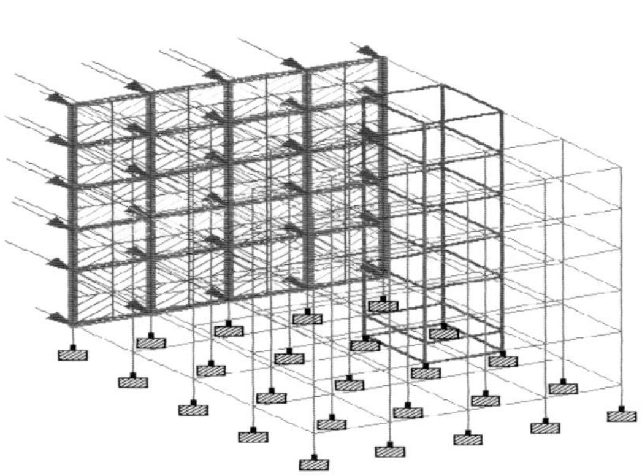

Figure 14. WL in Z- direction

Step 3: Obtained the value of base shear, peak storey shear, modal participation factor by analyzing the frame under seismic wave by RSA.

4. Result and discussion

Base on the methodology mentioned in section 3 (in Methodology section), the following analysis are made. Based on the referred methodology wind intensity, seismic parameters, dynamic weight, modal participation factors, base shear and peak storey shear are calculated, and subsequent results are shown in the following tables and figures highlighted below. Seismic parameters are assigned to calculate the peak storey shear, modal participation factor and base shear. The assigned seismic parameters are shown in Table 3 below.

4.1 Wind intensity
Wind intensity or wind pressure at location depends on, basic wind speed, type of structure, damping factor, risk factor, topography factor etc. Based on IS 875 (Part III) [20] wind intensities at different storey in the selected building structure are calculated using Staad Pro software. The obtained result for wind intensity is shown below in Table 2.

Table 2. Wind intensity value obtained from study

S. No.	Intensity	Height (in mm)
1	0.0000861	0
2	0.0000861	4572.000027
3	0.0000883	5558.77018
4	0.0000903	6545.539856
5	0.0000921	7532.310009
6	0.0000938	8519.080162
7	0.0000953	9505.849838
8	0.0000967	10492.60044
9	0.0000980	11479.39968
10	0.0000992	12466.19987
11	0.00010034	13452.89993
12	0.000101438	14439.70013
13	0.000102484	15426.50032
14	0.000103484	16413.20038
15	0.000104441	17399.99962

4.2 Seismic parameters
Standard values for the parameters like zone factor, response reduction factor for special masonry reinforced frame, soil type factor etc. have been selected based on IS 1893 (Part I). The assigned standard values for seismic parameters are shown below in Table 3.

Table 3. Analysis of seismic parameters (IS 1893-2002/2005)

Parameters	Value	Unit
Zone factor	0.16	
Response reduction factor (RF)	5	
Importance Factor (I)	1	
Rock and soil site factor (SS)	2	
Type of structure (ST)	1	
Damping ratio (DM)	0	
Period in X Direction (PX)		Seconds
Period in Z Direction(PZ)		Seconds
Depth of foundation(DT)	1500	mm
Ground level(GL)		mm
Spectral Acceleration(SA)	0	
Multiplying Factor for SA(DF)	0	

4.3 Dynamic weight calculation
Dynamic weight refers to the weight i.e. placed on an object and put it in motion. Dynamic weight is calculated based on CQC method and the obtained results are used to calculate different seismic parameters like base shear, peak storey shear etc. the calculated dynamic weight are shown below in Table 4.

Table 4. Calculation of dynamic weight

MODE	X	Y	Z	Weight
1	2.30520E+03	1.760062E-03	5.574606E+03	5.588115E+03
2	1.115090E+04	7.983828E-02	3.332371E+03	7.411074E+03
3	3.540734E+03	1.236973E-01	2.011241E+03	3.017814E+03
4	7.229958E+02	2.947209E-01	5.755830E+03	6.691264E+03
5	1.075260E+02	8.910577E-02	1.287810E+03	4.771087E+03
6	3.346359E+00	3.315472E-03	1.454272E+03	3.571485E+03

To evaluate the structure's overall response, all conceivable modes' responses must be combined. However, due to practical concerns such as computer run-time and memory, it is not always possible to run a structure for all potential modes. As a result, mode truncation is used. However, as can be seen, modal truncation produces mistakes in evaluating the structure's whole answer. The approach used to estimate the approximate response of the reduced modes is called "Missing Weight". The obtained (using CQC Method) missing weight and effective modal weight values are shown below in Table 5.

Table 5. Result of missing weight and effective modal weight

Various weights (in KN)	Direction		
	X	Y	Z
Dynamic Weight	2.544601E+04	1.920226E+04	1.920226E+04
Missing Weight	-7.615301E+03	-1.920167E+04	-5.484398E+03
Effective Modal Weight	1.783071E+04	5.924378E-01	1.371786E+04

4.4 Estimation of frequency and modal participation factor

When a structure is subjected to a force or displacement excitation in a certain direction, the modal participation factor or mass participation factor is a measure of how strongly a given mode contributes to the structure's reaction. Further, mass participation factors are depended on the frequency generated at different storey level based on the given data shown in Table 6. Modal participation factors are used to calculate the base shear. Base shear is the maximum expected lateral force that will occur due to seismic ground motion at the base of a structure. It depends upon the soil condition at the site [11, 17].

Table 6. Estimated frequency at different storey level

Mode	Frequency (in Hz)	Period seconds	Participation X (in %)	Participation Y (in %)	Participation Z (in %)	Type
1	1.103	0.906	9.059	0.000	29.031	Elastic
2	1.371	0.729	43.822	0.000	17.354	Elastic
3	1.799	0.556	13.915	0.001	10.474	Elastic
4	2.726	0.367	2.841	0.002	0.300	Elastic
5	3.045	0.328	0.423	0.000	6.707	Elastic
6	3.181	0.314	0.013	0.000	7.573	Elastic

The obtained participation factor based on frequency value at different time interval (in Table 6) are used to calculate the base shear as shown in Table 7 below.

4.5 Base shear calculation

The maximum expected lateral stress on the base of the structure owing to seismic activity is called base shear. It is estimated using the seismic zone, soil composition, the lateral force specified in the construction code and modal participation factor. The estimated base shear value is shown in Table 7 which is shown below:

Table 7. Obtained base shear value

	Mass Participation Factors			Percent			Base shear calculation		
Mode	X	Y	Z	Sum-X	Sum-Y	Sum-Z	X	Y	Z
1	9.06	0.00	29.03	9.059	0.00	29.031	55.34	0.00	0.00
2	43.82	0.00	17.35	52.881	0.00	46.385	332.59	0.00	0.00
3	13.91	0.00	10.47	66.796	0.001	56.859	138.59	0.00	0.00
4	2.84	0.00	0.30	69.637	0.003	57.159	28.92	0.00	0.00
5	0.42	0.00	6.71	70.060	0.003	63.865	4.30	0.00	0.00
6	0.01	0.00	7.57	70.073	0.003	71.439	0.13	0.00	0.000

The given input data (in X+, X-, Z+ and Z- direction) for CQC method and for calculating the base shear, is shown in Appendix I.

4.6 Peak Storey Shear

Peak storey shear is defined as the maximum shear developed at a particular storey level. Peak storey shear is used to calculate the base shear of any structure. Estimated peak storey shear value is shown below in Table 8.

Table 8. Estimated peak storey shear

Storey	Level in meter (m)	Peak storey shear in KN	
		X Direction	Y Direction
6th	17.40	0.00	58.46
5th	14.40	0.00	111.02
4th	11.40	0.00	156.56
3rd	8.40	0.00	197.55
2nd	5.40	0.00	227.45
1st	1.50	0.00	230.96
Base	0.00	0.00	230.96

It is obtained from Table 8 that; peak storey shear is developed between 5th and 6^{th}storey. The obtained peak storey shear thus utilizes to calculate the base shear in the subsequent sections.

5. Conclusion

Objectives of the current study is to calculate the wind intensity, seismic parameters, dynamic weight, modal participation factors, base shear and peak storey shear are calculated in the current study (mentioned in Section 4). Wind pressure in different floor heights are calculated based on basic wind speed, type of structure, damping factor, risk factor, topography factor and shown in Table 2. The obtained result for the selected objectives are shown from Table 4 to Table 8. It is observed that, base shear is found to maximum in Mode 2 (Table 7) and peak storey shear is developed between 5th and 6th storey (Table 8). Similar kind of analysis can also be performed for higher storey structures using linear dynamic analysis.

Reference

[1] Kangle S. R., Yerudkar D. S., 2020 Response Spectrum Analysis for Regular Multistory Structure in Seismic Zone III, International Journal of Engineering Research & Technology (IJERT), DOI:

[2] Thant N. N., Kyaw T. Y., 2019 Study on the Effect of Response Spectrum Analysis and Construction Sequence Analysis on Setback Steel Structure, International Journal of Trend in Scientific Research and Development, pp.1349-1355.

[3] Naresh M. V., Chari K. J. B., 2019 Study on Static and Dynamic Analysis of Multi-storied Building in Seismic Zones, International Journal of Recent Technology and Engineering (IJRTE), ISSN: 2277-3878.
[4] Yashaswini B. S., Dadapeer A. B. S., 2017 Comparative Study on Static and Dynamic Analysis of Multistoried Building Using ETABS, International Journal of Scientific Research in Science and Technology, ISSN: 2395-6011.
[5] Kumar N. V., 2017 Seismic performance of a reinforced, concrete multi-storey building, having circular shear wall and, square shear wall at core of the building, International Journal of Civil Engineering and Technology.
[6] Patil R.S., Pawar V. D., 2020 Response Spectrum Analysis and Comparison of Seismic Parameters of Low-rise, High-rise and Asymmetrical RC Structure with and without Infill for Different Bay Dimensions, International Research Journal of Engineering and Technology.
[7] Bello M., Adedeji A.A, Rahmon R.O., Kamal M.A., 2018 Dynamic Analysis of Multi-Storey Building under Seismic Excitation by Response Spectrum Method using ETABS, USEP: Journal of Research Information in Civil Engineering.
[8] Rajeswari P., Neelakantam A. Koti, 2019 Seismic analysis and design of multistorey building in different seismic zones by using ETABS, International Research Journal of Engineering and Technology (IRJET).
[9] Kadhum A. K., Abdul-Razzaq K. S., 2018 Effect of seismic load on reinforced concrete multistory building from economical point of view, International Journal of Civil Engineering and Technology (IJCIET).
[10] Choudhary S. N., Bokare P.S, 2018 Dynamic Analysis of Multistory Building Using Response Spectrum Method and Seismic Coefficient Method -A Comparison, IJISET - International Journal of Innovative Science, Engineering & Technology.
[11] Pradeep D., Chethan V. R., Ashwini B. T., 2018 Response spectrum analysis of multistorey building with floating columns, International Journal for Research Trends and Innovation.
[12] IS 875 (Part-I) Bureau of Indian Standards 1987 Code of Practice for Design Loads (Other than Earthquake) for Buildings and Structures: Dead Loads-Unit Weights of Building Materials and Stored Materials (Second Revision). UDC 624.042: 006.76.
[13] IS 875 (Part-II) Bureau of Indian Standards 1987 Code of Practice for Design Loads for Buildings and Structures: Imposed Loads (Second Revision). UDC 624.042.3: 006.76.
[14] IS 1893 (Part I) Bureau of Indian Standards 2002, "Criteria for Earthquake Resistant Design of Structures", ICS 91.120.25.
[15] Kumar N. V., Prakash D. A., 2017 Seismic Performance of a reinforced concrete multi storey building having circular shear wall and square shear wall at core of the building, IJCIET, e-ISSN:- 0976 – 6308 , p – ISSN : 0976 – 6316.
[16] Tondon B. K., Needhidasan S., 2018 Seismic Analysis of Multi Storied Building in Different Zones, International Journal of Trend in Scientific Research and Development (IJTSRD).
[17] Joshi A., Safdari M., 2018 A review: Sesimic analysis of multistorey building having floating columns, International Journal of Research Science & Management.
[18] Hassaballa A. E., Adam F. M., Ismaeil M. A., 2013 Seismic Analysis of a Reinforced Concrete building by Response Spectrum Method, IOSR Journal of Engineering.
[19] Firoj M., Singh S. K., 2018 Response Spectrum Analysis for Irregular Multi-Storey Structure in Seismic Zone V, 16th Symposium on Earthquake Engineering.
[20] IS 875 (Part-III), 2015, Bureau of Indian Standards, "Design Loads (other than Earthquake) for Building and Structure-code of practice", ICS: 91.100.10.

Acknowledgments
Authors wishing to acknowledge assistance or encouragement from colleagues, special work by technical staff, parent institute and nearby Medical College & Hospital for their support and cooperation for the completion of the current study.

AUTHOR INDEX